# Interactions in the Marine Benthos

## Global Patterns and Processes

This book synthesises the marine component of the Aquatic Biodiversity and Ecosystems Conference (ABEC) 2015 which was held to assess scientific progress over the past 25 years. It provides a comprehensive and global review of work since the 1992 Systematics Association publication of *Plant–Animal Interactions in the Marine Benthos*. Taking a regional and, where appropriate, habitat perspective, it considers patterns and processes shaping marine biodiversity and ecosystem functioning around the world. The volume analyses abiotic and biotic interactions and the factors determining distribution patterns, community structure and ecosystem functioning of marine systems. It explores how phylogeography and biogeographic processes influence assemblage composition and, hence, community structure and how this is shaped by environmental factors and biological interactions. This book is ideal for researchers, advanced undergraduate and postgraduate students and academics studying marine ecosystems. It would be of interest for conservation practitioners managing marine habitats and ecosystems.

STEPHEN J. HAWKINS is Emeritus Professor of Natural Sciences at the University of Southampton and Lankester Research Fellow at the Marine Biological Association of the United Kingdom. He has worked on coastal biodiversity and ecosystems for over 40 years.

KATRIN BOHN is a marine ecology specialist at Natural England, specialising in condition assessments of marine protected areas, rocky shore monitoring and the ecology of invasive non-native species. Previously, she held research posts at the universities of Portsmouth and Southampton.

LOUISE B. FIRTH is a Lecturer in marine ecology at the University of Plymouth, whose research interests include community dynamics and global climate change in both natural and artificial environments.

GRAY A. WILLIAMS is the Director of The Swire Institute of Marine Science at The University of Hong Kong, and his research is focussed on the ecology of shores in South-East Asia.

# The Systematics Association Special Volume Series

SERIES EDITOR

## DAVID J. GOWER

*Department of Life Sciences, The Natural History Museum, London, UK*

The Systematics Association promotes all aspects of systematic biology by organising conferences and workshops on key themes in systematics, running annual lecture series, publishing books and a newsletter, and awarding grants in support of systematics research. Membership of the association is open globally to professionals and amateurs with an interest in any branch of biology, including palaeobiology. Members are entitled to attend conferences at discounted rates, to apply for grants and to receive the newsletter and mailed information; they also receive a generous discount on the purchase of all volumes produced by the association.

The first of the Systematics Association's publications, *The New Systematics* (1940), was a classic work edited by its then-president Sir Julian Huxley. Since then, more than 70 volumes have been published, often in rapidly expanding areas of science where a modern synthesis is required.

The association encourages researchers to organise symposia that result in multi-authored volumes. In 1997 the association organised the first of its international biennial conferences. This and subsequent biennial conferences, which are designed to provide for systematists of all kinds, included themed symposia that resulted in further publications. The association also publishes volumes that are not specifically linked to meetings, and encourages new publications (including textbooks) in a broad range of systematics topics.

More information about the Systematics Association and its publications can be found at our website: www.systass.org

Previous Systematics Association publications are listed after the index for this volume.

Systematics Association Special Volumes published by Cambridge University Press:

78. *Climate Change, Ecology and Systematics* (2011)
    Trevor R. Hodkinson, Michael B. Jones, Stephen Waldren and John A. N. Parnell
79. *Biogeography of Microscopic Organisms: Is Everything Small Everywhere?* (2011)
    Diego Fontaneto
80. *Flowers on the Tree of Life* (2011)
    Livia Wanntorp and Louis Ronse De Craene
81. *Evolution of Plant–Pollinator Relationships* (2011)
    Sébastien Patiny
82. *Biotic Evolution and Environmental Change in Southeast Asia* (2012)
    David J. Gower, Kenneth G. Johnson, James E. Richardson, Brian R. Rosen, Lukas Rüber and Suzanne T. Williams
83. *Early Events in Monocot Evolution* (2013)
    Paul Wilkin and Simon J. Mayo
84. *Descriptive Taxonomy: The Foundation of Biodiversity Research* (2014)
    Mark F. Watson, Chris Lyal and Colin Pendry
85. *Next Generation Systematics* (2016)
    Peter D. Olson, Joseph Hughes and James A. Cotton
86. *The Future of Phylogenetic Systematics: The Legacy of Willi Hennig* (2016)
    David Williams, Michael Schmitt and Quentin Wheeler
87. *Interactions in the Marine Benthos: Global Patterns and Processes* (2019)
    Stephen J. Hawkins, Katrin Bohn, Louise B. Firth and Gray A. Williams

SYSTEMATICS ASSOCIATION SPECIAL VOLUME 87

# Interactions in the Marine Benthos

## Global Patterns and Processes


EDITED BY

STEPHEN J. HAWKINS
*The Marine Biological Association and the University of Southampton*

KATRIN BOHN
*Natural England*

LOUISE B. FIRTH
*University of Plymouth*

GRAY A. WILLIAMS
*The University of Hong Kong*



THE Systematics ASSOCIATION

CAMBRIDGE
UNIVERSITY PRESS

**CAMBRIDGE**
**UNIVERSITY PRESS**

University Printing House, Cambridge CB2 8BS, United Kingdom

One Liberty Plaza, 20th Floor, New York, NY 10006, USA

477 Williamstown Road, Port Melbourne, VIC 3207, Australia

314–321, 3rd Floor, Plot 3, Splendor Forum, Jasola District Centre, New Delhi – 110025, India

79 Anson Road, #06-04/06, Singapore 079906

Cambridge University Press is part of the University of Cambridge.

It furthers the University's mission by disseminating knowledge in the pursuit of education, learning, and research at the highest international levels of excellence.

www.cambridge.org
Information on this title: www.cambridge.org/9781108416085
DOI: 10.1017/9781108235792



First published 2019

Printed in the United Kingdom by TJ International Ltd. Padstow Cornwall

*A catalogue record for this publication is available from the British Library.*

*Library of Congress Cataloging-in-Publication Data*
Names: Hawkins, S. J. (Stephen J.), editor.
Title: Interactions in the marine benthos : global patterns and processes / edited by Stephen J. Hawkins, Katrin Bohn, Louise B. Firth, Gray A. Williams.
Description: Cambridge, United Kingdom ; New York, NY : Cambridge University Press, 2019. | Includes bibliographical references and index.
Identifiers: LCCN 2019009293 | ISBN 9781108416085 (hardback : alk. paper) | ISBN 9781108402620 (pbk. : alk. paper)
Subjects: LCSH: Benthos. | Phylogeography. | Coastal ecology. | Intertidal ecology.
Classification: LCC QH91.8.B4 I67 2019 | DDC 576.8/809–dc23
LC record available at https://lccn.loc.gov/2019009293

ISBN 978-1-108-41608-5 Hardback

# Contents

# Preface

This volume stems from a workshop at the Aquatic Biodiversity and Ecosystems Conference (ABEC) held in Liverpool in August/September 2015. The meeting was a 25-year anniversary follow-up to the seminal Systematics Association 'Plant–Animal Interactions in the Marine Benthos' conference hosted by the University of Liverpool in 1990 (largely organised by Stephen J. Hawkins and Gray A. Williams; published in the Systematics Association series edited by David M. John, Stephen J. Hawkins and James H. Price). The 1990 conference was a one-off meeting that focussed on biotic interactions between benthic marine organisms, whereas the ABEC meeting broadened this to include freshwater systems.

The aims of ABEC 2015 were to revisit some of the themes of the Plant–Animal Interactions meeting, broaden to other environmental and trophic interactions and review scientific advances in the field after 25 years. The 2015 meeting was therefore designed for aquatic scientists (both marine and freshwater) with a focus on evolution, interactions and long-term change – particularly how these shape patterns of biodiversity and the relationships between biodiversity and ecosystem functioning. The scope was deliberately broad, encompassing natural and highly modified habitats in both marine and freshwater aquatic systems. Ecology and evolutionary biology were the key focus, but the meeting also catered for an interdisciplinary approach to studying and managing aquatic ecosystems, especially with the aim of unravelling local- and regional-scale impacts from global climate-driven change. A workshop was convened at the end of the conference to provide the main platform for this volume, taking a regional and habitat-based view of interactions – both abiotic and biotic – and their role in shaping patterns in intertidal and shallow water systems around the world (see Chapter 1 for details), as well as selected habitats and special systems.

The ABEC conference was once again hosted by the University of Liverpool between 30 August and 4 September 2015 and was jointly supported by the universities of Plymouth and Southampton. The organising committee reflected the breadth of the meeting, being led by Louise B. Firth (University of Plymouth) and including Stephen J. Hawkins, Hanna Schuster, Katrin Bohn (University of Southampton), Antony M. Knights (University of Plymouth), Leonie Robinson (University of Liverpool), Nessa O'Connor (Queen's University Belfast), Ian Donohue (Trinity College Dublin), Martin Genner (University of Bristol), Gray A. Williams (The University of Hong Kong), Iwan Jones (Queen Mary University of London) and Chris Frid (Griffith University). Over 300 delegates attended the conference, coming from over 40 countries and representing all continents. The meeting also marked 40 years of Steve Hawkins commencing work on limpet–algal interactions at the University of Liverpool as an undergraduate in 1975 and so provided an opportune time to reconvene in Liverpool.

The conference was supported by the Systematics Association (who also supported the production of this volume), the British Ecological Society, the British Phycological Society, Elsevier, the Fisheries Society of the British Isles, the Freshwater Biological Association, Garland Science, the Irish Environmental Protection Agency, the Marine Biological Association of the United Kingdom, the MARS Network, the National Oceanographic Centre, Oxford University Press, PDMS Isle of Man, the Royal Society Publishing, the Scottish Association of Marine Science, Taylor & Francis and the universities of Liverpool, Plymouth and Southampton. Assembly of this volume was much helped by Anaëlle Lemasson and in the latter stages of indexing and proofs by Leoni Adams. We alos thank CUP for superb copy editing (Abigail Neale) and support throughout the editorial process – and much patience, especially Annie Toynbee.

**Louise B. Firth, Katrin Bohn,**
**Gray A. Williams and Stephen J. Hawkins**

# Contributors

**Per Åberg**
Department of Marine Sciences, University of Gothenburg

**Leoni C. Adams**
Ocean and Earth Science, University of Southampton and the Marine Biological Association of the United Kingdom

**Moisés A. Aguilera**
Department of Marine Biology, Catholic University of the North and Marine Ecology, Centre for Advanced Studies in Arid Zones

**Laura Airoldi**
Department of Experimental Evolutionary Biology and Inter-Departmental Research Centre for Environmental Science, University of Bologna

**A. Louise Allcock**
Ryan Institute, National University of Ireland Galway

**Francisco Arenas**
Interdisciplinary Centre of Marine and Environmental Research (CIIMAR)

**Maria Bagur**
Research and Education Group on Environmental Issues, Las Brusquitas Biological Station and the Argentine Museum of Natural Sciences Bernardino Rivadavia, National Scientific and Technical Research Council (CONICET)

**Trine Bekkby**
Norwegian Institute for Water Research

**Susan S. Bell**
Department of Integrative Biology, University of South Florida

**Lisandro Benedetti-Cecchi**
Department of Biology, University of Pisa

**Uta Berger**
Institute of Forest Growth and Computer Sciences, Dresden University of Technology

**Laura K. Blamey**
Department of Biological Sciences, University of Cape Town, and Department of Environment, University of Seychelles

**Diana M. Boaventura**
João de Deus School of Education and Marine and Environmental Sciences Centre, University of Lisbon

**Katrin Bohn**
Natural England and Ocean and Earth Science, University of Southampton

**C. Debora G. Borges**
Interdisciplinary Centre of Marine and Environmental Research of the University of Porto (CIIMAR-UP)

**Bernardo R. Broitman**
Department of Marine Ecology, Centre for Advanced Studies in Arid Zones

**Fabio Bulleri**
Department of Biology, University of Pisa

**Michael T. Burrows**
Department of Ecology, Scottish Association for Marine Science

**Sofia Calla**
The Argentine Museum of Natural Sciences Bernardino Rivadavia, CONICET

**Patricio A. Camus**
Department of Ecology, Catholoci University of the Holy Conception

**João J. Castro**
Department of Biology and Laboratory of Marine Sciences, School of Science and Technology, University of Évora and Marine and Environmental Sciences Centre

**Benny K. K. Chan**
Biodiversity Research Center, Academia Sinica, Taiwan

**M. Gee Chapman**
Life and Environmental Sciences, University of Sydney

**Hartvig Christie**
Norwegian Institute for Water Research

**Melinda A. Coleman**
Department of Primary Industries, NSW Fisheries and Marine Sciences, Southern Cross University

**Ross A. Coleman**
School of Biological Sciences, University of Southampton and School of Live and Environmental Sciences, University of Sydney

**Sean D. Connell**
School of Biological Sciences and Environment Institute, University of Adelaide

**Tasman P. Crowe**
Earth Institute and School of Biology and Environmental Science, University College Dublin

**Teresa Cruz**
Department of Biology and Laboratory of Marine Sciences, School of Science and Technology, University of Évora and Marine and Environmental Sciences Centre

**Maria Cecilia Dalton**
The Argentine Museum of Natural Sciences Bernardino Rivadavia, CONICET

**Mark S. Davies**
Applied Sciences, University of Sunderland

**Yun-Wei Dong**
State Key Laboratory of Marine Environmental Science, Xiamen University

**Steven R. Dudgeon**
Department of Biology, California State University, Northridge

**Graham Epstein**
Ocean and Earth Science, University of Southampton and the Marine Biological Association of the United Kingdom

**Ally J. Evans**
Institute of Biological, Environmental and Rural Sciences, Aberystwyth University, Ocean and Earth Science, University of Southampton, and the Marine Biological Association of the United Kingdom

**João Faria**
Centre for Ecology, Evolution and Environmental Changes and Department of Biology, University of the Azores

**Phillip B. Fenberg**
Ocean and Earth Science, University of Southampton

**João G. Ferreira**
Department of Applied Environmental Science, Stockholm University, and School of Ocean Sciences, Bangor University

**Louise B. Firth**
School of Biological and Marine Sciences, University of Plymouth

**Simonetta Fraschetti**
Department of Biological and Environmental Sciences and Technologies, University of Salento

**Natalie J. Frost**
ABP Marine Environmental Research, Southampton

**Jon Grant**
Department of Oceanography, Dalhousie University

**John N. Griffin**
Department of Biosciences, Swansea University

**Tamar Guy-Haim**
GEOMAR Helmholtz Centre of Ocean Research and Marine Biology, National Institute of Oceanography, Israel Oceanographic and Limnological Research

**ME Hanley**
School of Biological and Marine Sciences, University of Plymouth

**Richard G. Hartnoll**
Port Erin Marine Laboratory and School of Environmental Sciences, University of Liverpool

**Stephen J. Hawkins**
Ocean and Earth Science and School of Biological Sciences, University of Southampton, Port Erin Marine Laboratory, University of Liverpool, School of Ocean Sciences, Bangor University, and the Marine Biological Association of the United Kingdom

**Eliza C. Heery**
Department of Biological Sciences, National University of Singapore, and Department of Biology and Friday Harbor Laboratories, University of Washington

**Roger J. H. Herbert**
Department of Life and Environmental Sciences, Bournemouth University

**Keith Hiscock**
The Marine Biological Association of the United Kingdom

**Mark Huxham**
School of Applied Sciences, Edinburgh Napier University

**Kieran Hyder**
Centre for Environment, Fisheries and Aquaculture Sciences and School of Environmental Sciences, University of East Anglia

**Stuart R. Jenkins**
Port Erin Marine Laboratory, University of Liverpool, and School of Ocean Sciences, Bangor University

**Ladd E. Johnson**
Department of Biology and Québec-Océan, Université Laval

**Mark P. Johnson**
Ryan Institute, National University of Ireland Galway

**Anaëlle J. Lemasson**
School of Biological and Marine Sciences, University of Plymouth, and the Marine Biological Association of the United Kingdom

**Fernando P. Lima**
Research Centre in Biodiversity and Genetic Resources, University of Porto

**Kathleen A. MacGregor**
Department of Biology and Québec-Océan, Université Laval

**Gustavo M. Martins**
Centre for Ecology, Evolution and Environmental Changes and Department of Biology, University of the Azores

**Patricia Masterson-Algar**
School of Healthcare Sciences, Bangor University

**Christopher D. McQuaid**
Department of Zoology and Entomology, Rhodes University

**Bruce A. Menge**
Department of Integrative Biology, Oregon State University

**Nova Mieszkowska**
School of Environmental Sciences, University of Liverpool, and the Marine Biological Association of the United Kingdom

**Pippa J. Moore**
Institute of Biological, Environmental and Rural Sciences, Aberystwyth University

**Paula S. Moschella**
The Mediterranean Science Commission (CIESM), Monaco

**Richard J. Murphy**
Australian Centre for Field Robotics (J18), The University of Sydney

**Ivan Nagelkerken**
School of Biological Sciences and Environment Institute, University of Adelaide

**Carla A. Narvaez**
Department of Biology, Villanova University

**Gillian M. Notman**
Department of Science, Natural Resources and Outdoor Studies, University of Cumbria

**Kathryn E. Pack**
Ocean and Earth Science, University of Southampton, and the Marine Biological Association of the United Kingdom

**Maria Gabriela Palomo**
Research and Education Group on Environmental Issues, Las Brusquitas Biological Station and the Argentine Museum of Natural Sciences Bernardino Rivadavia, CONICET

**Federica G. Pannacciulli**
ENEA, Marine Environment Research Centre

**Ohad Peleg**
Marine Biology, National Institute of Oceanography, Israel Oceanographic and Limnological Research and Marine Biology, Charney School of Marine Science, University of Haifa

**Peter S. Petraitis**
Department of Biology, University of Pennsylvania

**Pedro A. Ribeiro**
Marine and Environmental Sciences Centre, University of the Azores

**Gil Rilov**
Marine Biology, National Institute of Oceanography, Israel Oceanographic and Limnological Research and Marine Biology, Charney School of Marine Science, University of Haifa

**Bayden D. Russell**
The Swire Institute of Marine Science and the School of Biological Sciences, The University of Hong Kong

**Antonio M. Santos**
Research Centre in Biodiversity and Genetic Resources, University of Porto

**David R. Schiel**
School of Biological Sciences, University of Canterbury, New Zealand

**Hanna Schuster**
Cambridge Environmental Assessments, RSK ADAS and Ocean and Earth Science, University of Southampton

**Kenneth P. Sebens**
Department of Biology, Friday Harbor Laboratories and School of Aquatic and Fisheries Science, University of Washington

**Nick Shears**
School of Biological Sciences, University of Auckland

**Ana C. F. Silva**
Civil Engineering Research and Innovation for Sustainability, University of Lisbon, School of Biological and Marine Sciences, University of Plymouth, and the Marine Biological Association of the United Kingdom

**Martin W. Skov**
School of Ocean Sciences, Bangor University

**Paul V. R. Snelgrove**
Ocean Sciences and Biology, Memorial University of Newfoundland

**Sabrina Andrea Soria**
Research and Education Group on Environmental Issues, Las Brusquitas Biological Station and the Argentine Museum of Natural Sciences Bernardino Rivadavia, CONICET

**Wayne P. Sousa**
Department of Integrative Biology, University of California, Berkeley

**Heather Sugden**
School of Natural and Environmental Sciences, The Dove Marine Laboratory, Newcastle University

**Thew S. Suskiewicz**
Department of Biology and Québec-Océan, Université Laval

**Antonio Terlizzi**
Department of Biological and Environmental Sciences and Technologies, University of Salento, and Department of Life Sciences, University of Trieste and Stazione Zoologica Anton Dohrn

**Richard C. Thompson**
Port Erin Marine Laboratory, University of Liverpool, and School of Biological and Marine Sciences, University of Plymouth

**A. J. Underwood**
School of Life and Environmental Sciences, University of Sydney

**Maria Vale**
Centre for Ecology, Evolution and Environmental Changes and Department of Biology, University of the Azores and Ocean and Earth Science, University of Southampton

**Julio A. Vásquez**
Department of Marine Biology, Catholic University of the North, Marine Ecology, Center for Advanced Studies in Arid Zones and Faculty of Philosophy and Sciences, Universidad Peruana Cayetano Heredia

**Adriana Vergés**
Centre for Marine Science & Innovation School of Biological, Earth and Environmental Sciences, UNSW Sydney

**Kringpaka Wangkulangkul**
School of Ocean Sciences, Bangor University

**Thomas Wernberg**
UWA Oceans Institute and School of Biological Sciences, University of Western Australia

**David S. Wethey**
Department of Biological Sciences, University of South Carolina

**Sarah A. Woodin**
Department of Biological Sciences, University of South Carolina

**Gray A. Williams**
The Swire Institute of Marine Science and School of Biological Sciences, The University of Hong Kong

**Edward J. G. Wort**
Ocean and Earth Science, University of Southampton

Chapter 1

# Introduction

Louise B. Firth, Katrin Bohn, Gray A. Williams and Stephen J. Hawkins

At the end of the 2015 Aquatic Biodiversity and Ecosystems Conference, a day was set aside for a workshop following up on the 1990 Plant–Animal Interactions meeting and its associated Systematics Association book – *Plant–Animal Interactions in the Marine Benthos* (John et al., 1992). Talks given throughout the 2015 conference also informed the present volume and its chapters. The 2015 workshop took a comparative approach with a series of informal presentations and discussion sessions from selected participants from around the world. The general aim was to take a regionally based view of the role of interactions in setting distribution patterns, community structure and functioning of shallow-water marine ecosystems. The coverage was predominantly coastal, down to the limit of light penetration. Most contributions were from those working on rocky intertidal and subtidal habitats, reflecting the size (and willingness to contribute) of the research community coupled with the greater tradition of experimental approaches to examine interactions on more tractable hard substrata. In addition, mangroves, biofilms and the deep sea were also considered as special systems that are ubiquitous across several oceans where significant advances have been made and, therefore, warranted inclusion. Recent advances in remotely operated vehicles, for example, have increased the scope for observation and experiment in the deep sea (Johnson et al., 2013); whereas mangroves are important ecosystem engineers which provide important ecosystem services, but are declining globally (Polidoro et al., 2010; Chee et al., 2017). Biofilms were also included as a subject given their global distribution and importance as the site of first settlement of macrobenthic organisms and as a food source for grazers (Abreu et al., 2007). While this volume does not feature any chapters specifically on artificial structures, ocean sprawl or eco-engineering, a large number of talks and posters at the conference dealt with these emerging issues, reflecting their global importance (see Firth et al., 2016; Bishop et al., 2017 and Strain et al., 2018 for reviews). A notable omission is coral reefs, which were not covered because they already have a well-established community of research workers and deserve a volume in their own right. Inevitably, there are gaps in coverage reflecting difficulties in soliciting and delivering input, especially on soft shores as well as certain geographic locations. Coverage in 1992 and 2018 is shown on the maps in Figure 1.1.

The workshop itself also provided an opportunity to reinforce the desired approach for the contribution to the invited regional reviews. Authors were asked to consider a variety of key aspects including: phylogeographic and biogeographic processes establishing the species pool in a particular region, the patterns of distribution in response to geographic and local environmental gradients (including the role of abiotic and biotic interactions in determining those patterns), the

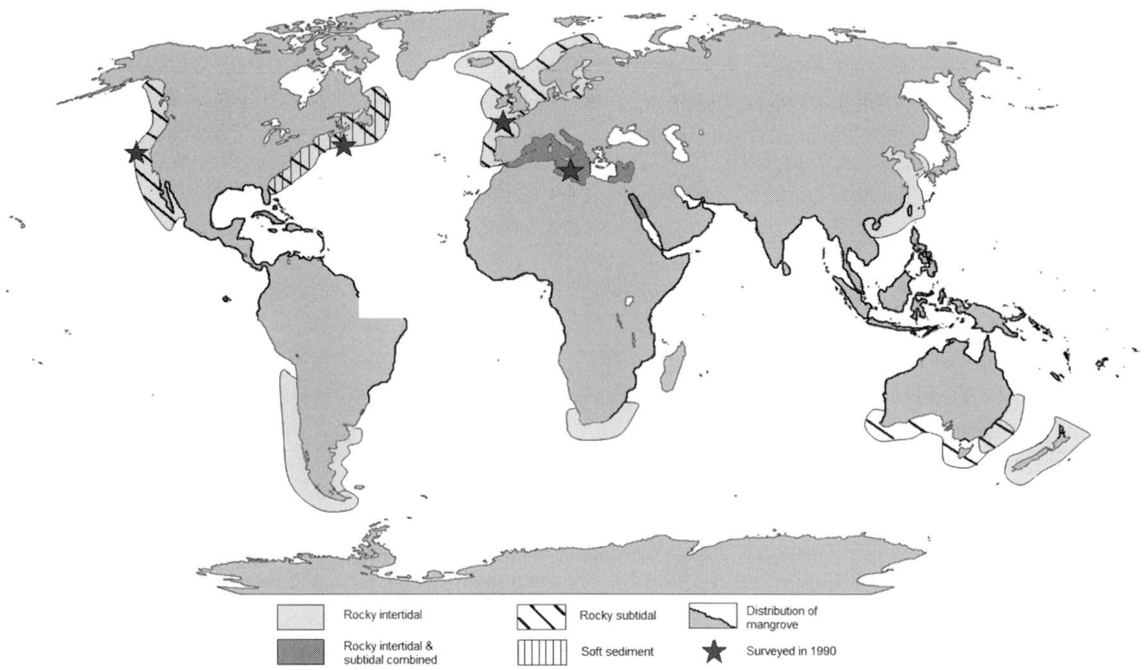


**Fig. 1.1** Coverage of areas reviewed in this volume compared to John et al. (1992; indicated by a star).

role of mesoscale processes such as upwelling and coastal hydrography and morphology in setting distribution patterns and abundance, the influence of non-native invasive species, and the consequences of patterns of community structure for emergent ecosystem functioning. They were also asked to identify knowledge gaps. Accordingly, Chapter 2 led by three of the editors of the volume followed this format. The rest of the volume is a tribute to the originality and diversity of approaches adopted by the authors or, perhaps more likely, the inability of the editors to herd cats.

Chapter 2 (by Hawkins et al.) is based on the opening plenary given by Hawkins on the intertidal zone of the north-east Atlantic and provides an update to Hawkins et al. (1992). This chapter is larger than the others as it deliberately digs back into the extensive history of study of the north-east Atlantic region. The next cluster of chapters looks in turn at the subtidal zone in the north-east atlantic (Chapter 3, by Hiscock et al.), as well as the intertidal (focussing on abiotic and biotic processes and anthropogenic effects, Chapter 4, by Dudgeon and Petraitis) and the subtidal (Chapter 5 by Johnson et al.) rocky habitats of the north-west Atlantic. There were no specific chapters on subtidal habitats in *Plant–Animal Interactions in the Marine Benthos* (John et al., 1992), generally being subsumed in chapters focussing more on the intertidal zone; therefore, the subtidal chapters in this volume serve as a useful reference for this system. Depositing shores of the north-west Atlantic are covered in Chapter 6 (by Woodin et al.), led by a pioneer of the experimental approach on soft shores. The less-studied south-west Atlantic coast of South America is covered next (Chapter 7, by Palomo et al.), focussing on the coast of Argentina and highlighting some of the unique patterns of biodiversity on hard shores, which show higher diversity at higher latitudes in contrast to generally expected patterns. The Mediterranean Sea is then the focus of Chapters 8 and 9. In Chapter 8, Benedetti-Cecchi et al. discuss the explosion of experimental work since the late 1980s that has been conducted to untangle the important processes shaping coastal systems – a very different approach from the classical phytosociology of the 1960s and 1970s (Ewald, 2003). Attendance at the Liverpool meeting in 1990 provided a major stimulus to the lead author (Benedetti-Cecchi) and his subsequent wider team to undertake field experiments. Chapter 9 (by Rilov et al.) then details the interactions underway in the Levant as a consequence of Lessepsian migration via the Suez canal, coupled with rapid climate warming (Rilov et al., 2018). The Levantine basin in the southeastern Mediterranean is the world's most invaded marine ecosystem (Edelist et al., 2013), and as such provides a unique opportunity not only to experimentally consider the negative impacts of non-native species (Reise et al., 2006; Zwerschke et al., 2016), but also their potential positive impacts on ecosystem functioning and emergent services (Borsje et al., 2011; Schlaepfer et al., 2011; Epstein 2017; Lemasson et al., 2017) in a highly degraded ecosystem.

Attention turns to the Pacific with chapters on the well-studied north-east Pacific coasts of North America: the intertidal zone is covered in Chapter 10 (by Fenberg and Menge), focussing on factors affecting biogeographic distributions and alteration by human impacts, while the subtidal zone is dealt within Chapter 11 (by Heery and Sebens), highlighting the interactions of biotic factors with environmental context. The south-east Pacific coast is then considered, focussing on the role of top-down and bottom-up forcing on ecological interactions along rocky shores in Chapter 12 (by Aguilera et al.). These coasts are strongly influenced by upwelling systems, which provides an overarching backdrop to the processes operating in these areas.

Focus then turns to the transition between the Atlantic and Indo-Pacific (Chapter 13, by McQuaid and Blamey) along the coasts of Southern Africa. There is a clear gap in coverage between this area and the north-west Pacific coastlines of greater China which are considered next (focussing on biogeography, phylogeography, biological processes and human impacts, Chapter 14, by Williams et al.). In addition to the reasonably well-studied shores of Hong Kong, more recent work on mainland China and Taiwan are included, giving much greater coverage, but still highlighting a lack of knowledge in this region. The shores of Australia and New Zealand are then considered, with Schiel et al. (Chapter 15) focussing on the rocky intertidal zone, while Connell et al. (Chapter 16) review work on the kelp forests of the temperate subtidal zone, attempting to predict the future sustainability of these key ecosystems.

Specific habitats are then considered in turn: mangroves (focussing on the importance of positive interactions, Chapter 17, by Huxham et al.), biofilms (Chapter 18, by Schuster et al.) and the deep sea (Chapter 19, by Allcock and Johnson). Finally, Chapter 20 is an attempt at synthesis in part addressed by a structured questionnaire to lead authors.

We hope that the reader will see much progress since *Plant–Animal Interactions in the Marine Benthos* (John et al., 1992). The diversity of approaches by the different authors reflects differences in scientific outlook, philosophy and approach. Work in North America, which has also strongly influenced approaches in Chile, tends to progress from a theoretical perspective. This contrasts with the British school stemming from Orton, Southward and Crisp, Kitching, and Ebling (strongly influenced by the French workers Fischer-Piette and Hatton, see Hawkins et al., 2016 for review) that can be best considered as experimental natural history. Underwood has made a huge contribution in advocating logical design and experimental rigour in hard reef ecology (e.g., Underwood, 1991, 1994, 1997, 2000); and this approach has infiltrated to most corners of the world (e.g., Williams and Morritt, 1995; Benedetti-Cecchi et al., 1996; Crowe and Underwood, 1999; Airoldi, 2000; Bulleri et al., 2002; Knights et al., 2006, 2012; O'Connor et al., 2006; Firth and Crowe, 2008, 2010; Iveša et al., 2010; Marzinelli et al., 2011; Knights, 2012; Bohn et al., 2013; Jackson et al., 2013; McManus et al., 2017) from its roots in Australia. Perhaps importantly, this book also reveals regions which are poorly known, or understudied. There is, for example, very little contribution from the world's hotspot of marine biodiversity in South East Asia; nor is there the representation one would expect, given their wide distribution and large extent, for soft shore habitats. Above all, we hope that the chapters provide a thought-provoking overview of particular regions and habitat, and point the way forwards for subsequent efforts to increase coverage in less well-studied regions.

In addition to the present book, there were a range of other outputs from this conference that were edited by members of the organising committee. A special issue of *Journal of Experimental Marine Biology and Ecology* was edited by Antony Knights, Louise Firth and Bayden Russell, and featured reviews including the effect of ocean sprawl on ecological connectivity (Bishop et al., 2017) and soft sediment habitats (Heery et al., 2017), ocean acidification impacts on ecosystem service provision in oysters (Lemasson et al., 2017), nutrient flux across the land–sea interface (Moss, 2017) and a global analysis of the role of kelp forests as biogenic habitat formers (Teagle et al., 2017). Research articles included environmental factors affecting host–parasite interactions (Firth et al., 2017), habitat complexity of artificial structures affecting biodiversity (Lavender et al., 2017, Loke et al., 2017), impacts of climate change on intertidal ectotherm behaviour (Ng et al., 2017) and larval metamorphosis in response to biofilm cues (de Brito Simith et al., 2017).

A special issue of *Oceanography and Marine Biology: An Annual Review* also stemming from the meeting was edited by Steve Hawkins, Ally Evans, Andrew Dale, Louise Firth, David Hughes and Philip Smith and featured reviews on herbivorous starfish (sea star) (Martinez et al., 2017), intertidal boulder fields (Chapman, 2017), ecological dominance on rocky shores (Rius et al., 2017), the distribution, current threats and conservation status of giant clams (Neo et al., 2017), the role of anthropogenic activities in affecting the establishment of non-indigenous species post arrival (Johnston et al., 2017) and herbivore effects on seaweed invasions (Enge et al., 2017).

We hope that this book will be of interest to marine ecologists, conservationists and managers alike, and will inspire further research where knowledge gaps have been identified.

## REFERENCES


Abreu, P. C., Ballester, E. L., Odebrecht, C. et al. (2007). Importance of biofilm as food source for shrimp (*Farfantepenaeus paulensis*) evaluated by stable isotopes ($\delta^{13}C$ and $\delta^{15}N$). *Journal of Experimental Marine Biology and Ecology*, **347**, 88–96.

Airoldi, L. (2000). Effects of disturbance, life histories, and overgrowth on coexistence of algal crusts and turfs. *Ecology*, **81**, 798–814.

Benedetti-Cecchi, L., Airoldi, L., Abbiati, M. and Cinelli, F. (1996). Estimating the abundance of benthic invertebrates: a comparison of procedures and variability between observers. *Marine Ecology Progress Series*, **138**, 93–101.

Bishop, M. J., Mayer-Pinto, M., Airoldi, L. et al. (2017). Effects of ocean sprawl on ecological connectivity: impacts and solutions. *Journal of Experimental Marine Biology and Ecology*, **492**, 7–30.

Bohn, K., Richardson, C. A. and Jenkins, S. R. (2013). Larval microhabitat associations of the non-native gastropod *Crepidula fornicata* and effects on recruitment success in the intertidal zone. *Journal of Experimental Marine Biology and Ecology*, **448**, 289–97.

Borsje, B. W., van Wesenbeeck, B. K., Dekker, F. et al. (2011). How ecological engineering can serve in coastal protection. *Ecological Engineering*, **37**, 113–22.

de Brito Simith, D. D. J., Abrunhosa, F. A. and Diele, K. (2017). Metamorphosis of the edible mangrove crab *Ucides cordatus* (Ucididae) in response to benthic microbial biofilms. *Journal of Experimental Marine Biology and Ecology*, **492**, 132–40.

Bulleri, F., Benedetti-Cecchi, L., Acunto, S., Cinelli, F. and Hawkins, S. J. (2002). The influence of canopy algae on vertical patterns of distribution of low-shore assemblages on rocky coasts in the northwest Mediterranean. *Journal of Experimental Marine Biology and Ecology*, **267**, 89–106.

Chapman, M. G. (2017). Intertidal boulder-fields: A much neglected, but ecologically important, intertidal habitat. *Oceanography and Marine Biology: An Annual Review*, **55**: 35–54.

Chee, S. Y., Othman, A. G., Sim, Y. K., Adam, A. N. M. and Firth, L. B. (2017). Land reclamation and artificial islands: Walking the tightrope between development and conservation. *Global Ecology and Conservation*, **12**, 80–95.

Crowe, T. P. and Underwood, A. J. (1999). Differences in dispersal of an intertidal gastropod in two habitats: the need for and design of repeated experimental transplantation. *Journal of Experimental Marine Biology and Ecology*, **237**, 31–60.

Edelist, D., Rilov, G., Golani, D., Carlton, J. T. and Spanier, E. (2013). Restructuring the sea: profound shifts in the world's most invaded marine ecosystem. *Diversity and Distributions*, **19**, 69–77.

Enge, S., Sagerman, J., Wikstrom, S. A. and Pavia, H. (2017). A review of herbivore effects on seaweed invasions. *Oceanography and Marine Biology: An Annual Review*, **55**, 421–40.

Epstein, G. (2017). Invasive alien species management: a personal impasse. *Frontiers in Environmental Science*, **5**, 68.

Ewald, J. (2003). A critique for phytosociology. *Journal of Vegetation Science* **14**, 291–6.

Firth, L. B., Grant, L. M., Crowe, T. P. et al. (2017). Factors affecting the prevalence of the trematode parasite *Echinostephilla patellae* (Lebour, 1911) in the limpet *Patella vulgata* (L.). *Journal of Experimental Marine Biology and Ecology,* **492**, 99–104.

Firth, L. B. and Crowe, T. P. (2008). Large-scale coexistence and small-scale segregation of key species on rocky shores. *Hydrobiologia*, **614**, 233–41.

Firth, L. B. and Crowe, T. P. (2010). Competition and habitat suitability: small-scale segregation underpins large-scale coexistence of key species on temperate rocky shores. *Oecologia*, **162**, 163–74.

Firth, L. B., Knights, A. M., Bridger, D. et al. (2016). Ocean sprawl: challenges and opportunities for biodiversity management in a changing world. *Oceanography and Marine Biology*, **54**, 201–78.

Hawkins, S. J., Hartnoll, R. G., Kain, J. M. and Norton, T. A. (1992). Plant–Animal Interactions on Hard Substrata in the Northeast Atlantic. In D. M. John, S. J. Hawkins and J. H. Price, eds. *Plant–Animal Interactions in the Marine Benthos*, Oxford University Press, Oxford, pp. 1–32.

Hawkins, S. J., Mieszkowska, N., Firth, L. B. et al. (2016). Looking backwards to look forwards: the role of natural history in temperate reef ecology. *Marine and Freshwater Research*, **67**, 1–13.

Heery, E. C., Bishop, M. J., Critchley, L. P. et al. (2017). Identifying the consequences of ocean sprawl for sedimentary habitats. *Journal of Experimental Marine Biology and Ecology*, **492**, 31–48.

Ivesa, L., Chapman, M. G., Underwood, A. J. and Murphy, R. J. (2010). Differential patterns of distribution of limpets on intertidal seawalls: experimental investigation of the roles of recruitment, survival and competition. *Marine Ecology Progress Series*, **407**, 55–69.

Jackson, A. C., Murphy, R. J. and Underwood, A. J. (2013). Biofilms on rocky shores: Influences of rockpools, local moisture and temperature. *Journal of Experimental Marine Biology and Ecology*, **443**, 46–55.

John, D. M., Hawkins, S. J. and Price, J. H., eds. (1992). *Plant–Animal Interactions in the Marine Benthos*. Oxford University Press, Oxford.

Johnson, M. P., White, M., Wilson, A. et al. (2013). A vertical wall dominated by *Acesta excavata* and *Neopycnodonte zibrowii*, part of an undersampled group of deep-sea habitats. *PLoS ONE*, **8**, e79917.

Johnston, E. L., Dafforn, K. A., Clark, G. F., Rius, M. and Floerl, O. (2017). How anthropogenic activities affect the establishment and spread of non-indigenous species post-arrival. *Oceanography and Marine Biology: An Annual Review*, **55**, 389–420.

Knights, A. M. (2012). Spatial variation in body size and reproductive condition of subtidal mussels: considerations for sustainable management. *Fisheries Research*, **113**, 45–54.

Knights, A. M., Crowe, T. P. and Burnell, G. (2006). Mechanisms of larval transport: vertical distribution of bivalve larvae varies with tidal conditions. *Marine Ecology Progress Series*, **326**, 167–74.

Knights, A. M., Firth, L. B. and Walters, K. (2012). Interactions between multiple recruitment drivers: post-settlement predation mortality and flow-mediated recruitment. *PLoS ONE*, **7**, e35096.

Lavender, J. T., Dafforn, K. A., Bishop, M. J. and Johnston, E. L. (2017). Small-scale habitat complexity of artificial turf influences the development of associated invertebrate assemblages. *Journal of Experimental Marine Biology and Ecology*, **492**, 105–12.

Lemasson, A. J., Fletcher, S., Hall-Spencer, J. M. and Knights, A. M. (2017). Linking the biological impacts of ocean acidification on oysters to changes in ecosystem services: a review. *Journal of Experimental Marine Biology and Ecology*, **492**, 49–62.

Loke, L. H., Bouma, T. J. and Todd, P. A. (2017). The effects of manipulating microhabitat size and variability on tropical seawall biodiversity: field and flume experiments. *Journal of Experimental Marine Biology and Ecology*, **492**, 113–20.

Martinez, A. S., Byrne, M. and Coleman, R. A. (2017). Filling in the grazing puzzle: a synthesis of herbivory in starfish. *Oceanography and Marine Biology: An Annual Review*, **55**, 1–34.

Marzinelli, E. M., Underwood, A. J. and Coleman, R. A. (2011). Modified habitats influence kelp epibiota via direct and indirect effects. *PLoS ONE*, **6**, e21936.

McManus, R. S., Archibald, N., Comber, S. et al. (2017). Partial replacement of cement for waste aggregates in concrete coastal and marine infrastructure: a foundation for ecological enhancement? *Ecological Engineering*, **120**, 655–67.

Moss, B. (2017). Marine reptiles, birds and mammals and nutrient transfers among the seas and the land: an appraisal of current knowledge. *Journal of Experimental Marine Biology and Ecology*, **492**, 63–80.

Neo, M. L., Wabnitz, C. C. C., Braley, R. D. et al. (2017). Giant clams (Bivalvia: Cardiidae: Tridacninae): a comprehensive update of species and their distribution, current threats and conservation status. *Oceanography and Marine Biology: An Annual Review*, **55**, 87–388.

Ng, T. P., Lau, S. L., Seuront, L. et al. (2017). Linking behaviour and climate change in intertidal ectotherms: insights from littorinid snails. *Journal of Experimental Marine Biology and Ecology*, **492**, 121–31.

O'Connor, N. E., Crowe, T. P. and McGrath, D. (2006). Effects of epibiotic algae on the survival, biomass and recruitment of mussels, *Mytilus* L. (Bivalvia: Mollusca). *Journal of Experimental Marine Biology and Ecology*, **328**, 265–76.

Polidoro, B. A., Carpenter, K. E., Collins, L. et al. (2010). The loss of species: mangrove extinction risk and geographic areas of global concern. *PLoS ONE*, **5**(4), e10095.

Reise, K., Olenin, S. and Thieltges, D. W. (2006). Are aliens threatening European aquatic coastal ecosystems? *Helgoland Marine Research*, **60**(2), 77.

Rilov, G., Peleg, O., Yeruham, E. et al. (2018). Alien turf: overfishing, overgrazing and invader domination in south-eastern Levant reef ecosystems. *Aquatic Conservation: Marine and Freshwater Ecosystems*, **28**, 351–69.

Rius, M., Teske, P. R., Manriquez, P. H. et al. (2017). Ecological dominance along rocky shores, with a focus on intertidal ascidians. *Oceanography and Marine Biology: An Annual Review*, **55**, 55–86.

Schlaepfer, M. A., Sax, D. F. and Olden, J. D. (2011). The potential conservation value of non-native species. *Conservation Biology*, **25**, 428–37.

Strain, E. M., Olabarria, C., Mayer-Pinto, M. et al. (2018). Eco-engineering urban infrastructure for marine and coastal biodiversity: Which interventions have the greatest ecological benefit? *Journal of Applied Ecology*, **55**, 426–41.

Teagle, H., Hawkins, S. J., Moore, P. J. and Smale, D. A. (2017). The role of kelp species as biogenic habitat formers in coastal marine ecosystems. *Journal of Experimental Marine Biology and Ecology*, **492**, 81–98.

Underwood, A. J. (1991). The logic of ecological experiments: a case history from studies of the distribution of macro-algae on rocky intertidal shores. *Journal of the Marine Biological Association of the United Kingdom*, **71**, 841–66.

Underwood, A. J. (1994). On beyond BACI: sampling designs that might reliably detect environmental disturbances. *Ecological Applications*, **4**, 3–15.

Underwood, A. J. (1997). *Experiments in Ecology: Their Logical Design and Interpretation Using Analysis of Variance*. Cambridge University Press, Cambridge.

Underwood, A. J. (2000). Experimental ecology of rocky intertidal habitats: what are we learning?. *Journal of Experimental Marine Biology and Ecology*, **250**, 51–76.

Williams, G. A. and Morritt, D. (1995). Habitat partitioning and thermal tolerance in a tropical limpet, *Cellana grata*. *Marine Ecology Progress Series*, **124**, 89–103.

Zwerschke, N., Emmerson, M. C., Roberts, D. and O'Connor, N. E. (2016). Benthic assemblages associated with native and non-native oysters are similar. *Marine Pollution Bulletin*, **111**, 305–10.

Chapter 2

# The Intertidal Zone of the North-East Atlantic Region

*Pattern and Process*

Stephen J. Hawkins, Kathryn E. Pack, Louise B. Firth, Nova Mieszkowska, Ally J. Evans, Gustavo M. Martins, Per Åberg, Leoni C. Adams, Francisco Arenas, Diana M. Boaventura, Katrin Bohn, C. Debora G. Borges, João J. Castro, Ross A. Coleman, Tasman P. Crowe, Teresa Cruz, Mark S. Davies, Graham Epstein, João Faria, João G. Ferreira, Natalie J. Frost, John N. Griffin, ME Hanley, Roger J. H. Herbert, Kieran Hyder, Mark P. Johnson, Fernando P. Lima, Patricia Masterson-Algar, Pippa J. Moore, Paula S. Moschella, Gillian M. Notman, Federica G. Pannacciulli, Pedro A. Ribeiro, Antonio M. Santos, Ana C. F. Silva, Martin W. Skov, Heather Sugden, Maria Vale, Kringpaka Wangkulangkul, Edward J. G. Wort, Richard C. Thompson, Richard G. Hartnoll, Michael T. Burrows and Stuart R. Jenkins

## 2.1 Introduction

In this chapter we consider the rocky intertidal zone of the region from Morocco and the Strait of Gibraltar to the Arctic coasts of Norway and Iceland. The offshore islands of the Azores are included, but not the southern Macaronesian islands of Madeira and the Canaries (but see Lawson and Norton, 1971; Hawkins et al., 2000). This chapter follows from Hawkins et al. (1992); but we broaden the scope beyond plant–animal interactions, while restricting focus to the intertidal zone. Much new work has been done since the 1990s in the region, particularly in extending observational and experimental studies to the Iberian Peninsula (e.g., Boaventura et al., 2002a, 2002b, 2002c, 2003; Lima et al., 2006, 2007; Schmidt et al., 2008; Vinagre et al., 2015).

We start with an overview of the history of study of the region and then outline the environmental setting and phylogeographic processes that have shaped the current biogeographic patterns. We briefly summarise distribution patterns in response to local environmental gradients at the community level, before outlining work on the underlying processes causing these patterns. We then consider mesoscale processes (10–100s of km), such as upwelling in Iberia and coastal configuration and nearshore hydrography around Britain and Ireland that can determine pattern through bottom-up nutrient forcing and modification of larval distribution and hence recruitment regimes. The impacts of non-native species in the region are briefly considered. Finally, the role biodiversity plays in shaping ecological processes and hence ecosystem functioning is discussed. We conclude by making suggestions for further work, highlighting the special features of the north-east Atlantic and giving a prognosis of future changes.

## 2.2 History of Study

The intertidal zone of the north-east Atlantic has a long history of ecological and biogeographic study. This started with formal descriptions of distribution patterns, particularly the phenomenon of vertical zonation and how it is modified by exposure to wave action (Audouin and Edwards, 1833). There followed a further century of descriptive, largely qualitative, studies with various classificatory schemes (early work is reviewed by Southward, 1958), perhaps reaching their zenith in the work of the Stephensons (Stephenson and Stephenson, 1949, 1972) and the seminal book by Lewis on the British Isles (Lewis, 1964). Similar approaches were adopted elsewhere in Europe: in the Iberian Peninsula (Ardré, 1969, 1971; Saldanha, 1974), France (Fischer-Piette, 1929) and Scandinavia (Børgesen, 1908; Børgesen and Jónsson, 1908).

With time, a more quantitative approach was adopted (e.g., Southward and Orton, 1954), including early application of multivariate statistical approaches to aid classification of geographic and local distribution patterns (e.g., Van den Hoek and Donze, 1967; Russell et al., 1971; Van den Hoek, 1984; Breeman, 1988; Lüning, 1990). Pioneering studies of geographic patterns of distribution of individual species were also made in the north-east Atlantic (e.g., Forbes, 1858; Orton, 1920; Nobre, 1940; Moore and Kitching, 1939). Fischer-Piette (1936) systematically studied both sides of the English Channel (La Manche) and extended observations throughout the Iberian Peninsula (Fischer-Piette, 1957, 1958, 1963) to what was then French and Spanish North Africa (Fischer-Piette and Prenant, 1957). Crisp and Southward took a similar approach in the British Isles and Ireland (Southward and Crisp, 1954; Crisp and Southward, 1958) including work with Fischer-Piette (Crisp and Fischer-Piette, 1959). Fischer-Piette and Hatton also undertook some of the earliest studies describing the dynamics of shore communities, realising early on the importance of recruitment fluctuations (Hatton and Fischer-Piette, 1932; Hatton, 1938). Southward (1967, 1980, 1991) started various time series, charting the influence of climate fluctuations on intertidal species, building on Southward and Crisp (1954, 1956). Lewis (1976) and co-workers also initiated work on long-term change in key species (Lewis, 1976; Bowman and Lewis, 1977; Kendall et al., 1985).

Beyond descriptive studies, the north-east Atlantic was the site of some of the earliest manipulative field experiments on rocky shores (e.g., Hatton, 1938; Fischer-Piette, 1948; Lodge, 1948), demonstrating the processes setting local distribution patterns in response to the gradients of stress associated vertically with tidal elevation and horizontally with wave exposure. These experiments investigated the role of biological interactions such as competition (Hatton, 1938, Connell, 1961a), facilitation (Hatton, 1938), grazing (Jones, 1946, 1948; Conway, 1946; Lodge, 1948; Burrows and Lodge, 1951; Southward, 1964), predation (Connell, 1961b) and biological disturbance by sweeping of algal canopies (Hatton, 1938). A variety of pioneering expeditionary experimental studies were made in both the intertidal and subtidal of Lough Hyne in West Cork, Ireland from the 1940s onwards (Kitching et al., 1959, 1966, 1976; Kitching and Ebling, 1961, 1967; Ebling et al., 1962; Goss-Custard et al., 1979; Kitching, 1987a, 1987b) exploring the role of physical and biological factors including competition, predation and behaviour in setting distribution patterns. The importance of settlement behaviour was also demonstrated in barnacles (gregarious settlement; Crisp and Knight-Jones, 1953; Knight-Jones, 1953) and other invertebrates (Wilson, 1968). Laboratory experimentation investigated the causes of both local and geographic distributions (e.g., Southward, 1958), including how organisms responded to environmental stressors associated with periods of emersion, such as desiccation and temperature (Foster, 1971).

## 2.3 Environmental Context, Phylogeography and Biogeography

Rocky shores predominate on the coastlines of much of the north-east Atlantic region (Emery and Kuhn, 1982), which have a strongly erosive regime due to strong wave action arising from oceanic swell driven by south-westerly and westerly wind, and the large fetch on the Icelandic, Iberian, French, west Irish, south-west coasts of the United Kingdom and on the outer islands of Britain and Norway. Extensive areas of exposed sandy beach occur in south-west Spain and Portugal, central Portugal and the Bay of Biscay. The more enclosed shores of the Irish and North seas, eastern English Channel and the south-east coast of Iceland are largely made up of areas of mobile shingle and gravel which support little macrobiota (Ingólfsson, 2006). Sandy and muddy shores predominate in more sheltered areas. The southern North Sea and eastern English Channel plus much of the northern Irish Sea mainly comprise soft eroding coastline, but with considerable areas of artificial hard substrata due to port systems and sea defences (Firth et al., 2016a). Much of the region is macrotidal (4–10 m) with very low amplitude tides (<0.5 m) occurring in limited areas (southern Norway and west Sweden), especially around amphidromic points (e.g., south-west Scotland and north-west Ireland, central English Channel). Deep water abuts the Moroccan and Iberian coasts, leading to extensive wind-driven upwelling of cold, nutrient-rich water as far north as Galicia (Fraga, 1981) in late spring and summer (Pires et al., 2016). For much of the rest of the European coastline an extensive shallow continental shelf means that little deep upwelling occurs (Southward et al., 1995). The Azores, as recent volcanic islands adjacent to the Mid-Atlantic Ridge, do not have a continental shelf and are surrounded by deep, clear, oligotrophic oceanic waters with a small tidal range (2–4 m), but the littoral zone is extended by vigorous wave action (Santos et al., 1995; Hawkins et al., 2000).

The flora and fauna of the north-east Atlantic have been shaped by successive periods of glacial expansion and retraction during warmer interglacials for the last 3 million years (Maggs et al., 2008). In the 1850s, Forbes (1858 – reproduced in Hiscock et al., 2004) realised that the British Isles was an area of overlap between warm-water "southern types" on the south and west coasts, and cold-water species to the north and east. Many clades invaded the north-east Atlantic from the North Pacific in a trans-Arctic interchange approximately 3.8 million years ago when the Arctic Ocean was relatively ice-free (Wares and Cunningham, 2001). These include many genera of fucoids and laminarians among the seaweeds. Similarly, littorinids, thaid whelks and balanoid barnacles all have North Pacific origins. Some species such as *Semibalanus balanoides* still occur on the shores of the North Pacific in both America and Asia, exhibiting pan-boreal distributions (Crickenberger and Wethey, 2017). As a consequence, the shores of both Atlantic seaboards have many species in common; although diversity is higher in the north-east than the north-west Atlantic (Jenkins et al., 2008a) because recolonisation following ice ages occurred mainly from east to west, probably via Iceland and Greenland to North America (Wares and Cunningham, 2001). There is also a strong element of the flora and fauna whose origins are primarily from the warm waters in more southerly latitudes of the Atlantic. These include seaweeds of the genera *Sargassum* and *Cystoseira* plus *Bifurcaria bifurcata*. Many red algae also probably have south Atlantic origins (Lüning, 1990), as well as major animal clades including patellid limpets, trochids (*Phorcus,* previously called *Osilinus,* and *Steromphala*, previously called *Gibbula*) and chthamalid barnacles. Thus, the north-east Atlantic is an area where clades from different oceans overlap and interact (Hawkins et al., 2009; Mieszkowska and Sugden, 2016).

Brown algae are more prevalent at higher latitudes (Hawkins et al., 1992; Jenkins et al., 2008a), probably as an outcome of these combined phylogeographic processes and the cooler thermal environment in boreal oceanic waters. Another process shaping allopatric speciation has been the opening and closing of the Mediterranean, coupled with the complex hydrography of the Strait of Gibraltar, with the influence of the Atlantic ceasing in modern times around the Alboran Front (Pannacciulli et al., 1997). Successive waves of colonisation of the Atlantic Islands (Macaronesia) have also occurred, exemplified by *Phorcus* spp. (Donald et al., 2012).

Continuing southwards from south-west Britain, *B. bifurcata* and *Cystoseira* spp. become more common, occurring in rock pools and forming conspicuous zones on open rock. From Brittany southward the stalked barnacle *Pollicipes pollicipes* dominates the lower parts of exposed shores, providing a valuable fishery (Barnes, 1996; Van Syoc et al., 2010). *Sabellaria alveolata* also forms extensive reefs in the Bay of Mont Saint Michel and further south throughout France, Spain and Portugal (Anadon, 1981; Gruet, 1986; Dubois et al., 2002, 2006; Ayata et al., 2009) and into the Mediterranean (Bertocci et al., 2017).

Further south in Europe the diversity of grazing patellid limpets increases (Ballantine, 1961), with *Siphonaria* being present from its poleward range limit in northern Portugal. There are also several species of trochids, grapsid crabs and herbivorous fish such as the salema (*Sarpa salpa*) on southern European coasts. Thus, grazing pressure is much greater in southern Europe than in the north (Franco et al., 2015).

From the north of the British Isles polewards, species diversity declines as many Lusitanian invertebrates have failed to colonise that far north following the end of the last ice age. For example, the poleward range edge of the polychaete worm *S. alveolata* occurs in south-west Scotland (Cunningham et al., 1984; Firth et al., 2015), and *Patella ulyssiponensis* reaches Shetland and southern Norway. The poleward range edge of *Chthamalus stellatus* is in Shetland, and for *Chthamalus montagui* occurs in Orkney; but neither species reaches Norway. The northern range edge of *Steromphala umbilicalis* also occurs in Orkney (Southward et al., 1995; Hawkins et al., 2009).

At higher latitudes dominance by fucoid algae can extend out onto exposed shores (Ballantine, 1961). *Fucus distichus* thrives in a highshore band on the most exposed shores on the north of mainland Britain and throughout Norway. Other *Fucus* species, considered part of the *F. distichus* complex, occur in more sheltered locations in Norway (e.g., *F. evanescens*). Fucoid algae retreat into more sheltered conditions at lower latitudes in Europe. In the south and west of Britain, patches of fucoids occur on moderately exposed shores, with dense cover only occurring on the most sheltered shores. The non-bladdered dwarf form of *Fucus vesiculosus* (var. *evesiculosus*) occurs on more exposed shores, forming a habitat mosaic with *Mytilus* spp. (Crowe et al., 2011). Further south, successive species of *Fucus* become scarcer and disappear from the shore, leaving only *Fucus spiralis* and the recently described *Fucus guiryi* occurring in patches between high- and mid-shore (Zardi et al., 2011). *Pelvetia canaliculata*, *F. vesiculosus*, *Ascophyllum nodosum*, *F. serratus* and *Himanthalia elongata* are absent on the open coast around the Basque country. They then reappear further south, along with kelps, along the colder, upwelling-influenced Galician and northern Portuguese coasts where precipitation is high (Southward et al., 1995). Large range extensions and retractions of fucoids have been reported along the north coast of Spain and Portugal in response to climate fluctuations and more recent climate change (Arrontes, 1993; Lima et al., 2009; Duarte et al., 2013; Nicastro et al., 2013).

The most southerly location to show a classic (*sensu* Stephenson) sheltered shore zonation pattern of *P. canaliculata* – *F. spiralis* – *F. vesiculosus* – *A. nodosum* – *F. serratus* – laminarians (*Laminaria digitata*, *Saccorhiza polyschides*) is Viana de Castelo in northern Portugal (Boaventura et al., 2002b). In southern Portugal *F. vesiculosus* is only found in estuarine refuges; it has disappeared from such refuges in Morocco in recent years (Nicastro et al., 2013). *Carcinus maenas* and *Littorina littorea* also occur in estuarine refuges towards their southern limits (Reid, 1996; Queiroga et al., 2006).

The rocky shores of the Azores bear some similarities to those of southern Portugal (Hawkins et al., 1990), but certain key taxa have failed to colonise the islands (Hawkins et al., 2000). Species that are absent include *C. montagui*, *Patella piperata*, *Siphonaria pectinata* and trochids of the genera *Phorcus* and *Steromphala* (although *Phorcus sauciatus* has recently colonised Santa Maria (Ávila et al., 2015) and São Miguel (Vale, personal communication). Many of these species occur in both Madeira and the Canaries (i.e., trochids, *P. piperata*) or the Canaries alone (*C. montagui*). Historically, the Lusitanian limpet *Patella rustica* was found in southern and central Portugal; it was absent from Galicia and Asturias but reappeared in the warm waters of the inner Bay of Biscay on the Basque coast (Nobre, 1940; Fischer-Piette, 1955; Southward et al., 1995). In recent years, this distributional gap has been bridged, linked to relaxation of upwelling (Lima et al., 2006). Despite this break in distribution being known to occur for at least 70 years, no evidence of genetic differentiation could be found between Biscayan populations and those in Portugal (Ribeiro et al., 2010). There are also Macaronesian endemics such as *Patella aspera*, the *Patella candei* complex (Côrte-Real et al., 1996; Weber and Hawkins, 2002, 2005; Faria et al., 2017, 2018; González-Lorenzo et al., 2015), *Megabalanus azoricus* and *Tectarius striatus*. The only canopy-forming species in the intertidal is *F. spiralis* (probably the newly described *F. guiryi*). *Cystoseira* canopies can form very low on the shore, in midshore rock pools and in the shallow subtidal zone of most exposed coasts.

There have been considerable fluctuations in climate in the north-east Atlantic region during the period of formal scientific study from the middle of the nineteenth century onwards, with alternations of colder (either side of the First World War [1900s–1920s], 1960s to mid-1980s) and warmer climatic periods (1930s–1950s), upon which more recent rapid warming due to anthropogenic climate change have been superimposed (Southward et al., 2004; Philippart et al., 2011; Birchenough et al., 2015). These have driven fluctuations in relative abundance and geographic distribution of species (Southward et al., 1995; Mieszkowska et al., 2006; Hawkins et al., 2008, 2009). There have been both retreats of the poleward or leading range edge of Lusitanian species during colder periods and advances during warmer periods, especially during the more recent period of rapid warming (Figure 2.1). Conversely, boreal species have thrived during colder periods, with some retraction of the trailing range limits in recent years. One of the most well-documented examples is the fluctuation in abundance of warm- (*Chthamalus* spp.) and cold-water (*S. balanoides*) barnacles in the south-west of England from the 1950s to date (Southward and Crisp, 1954, 1956; Southward, 1967, 1991;

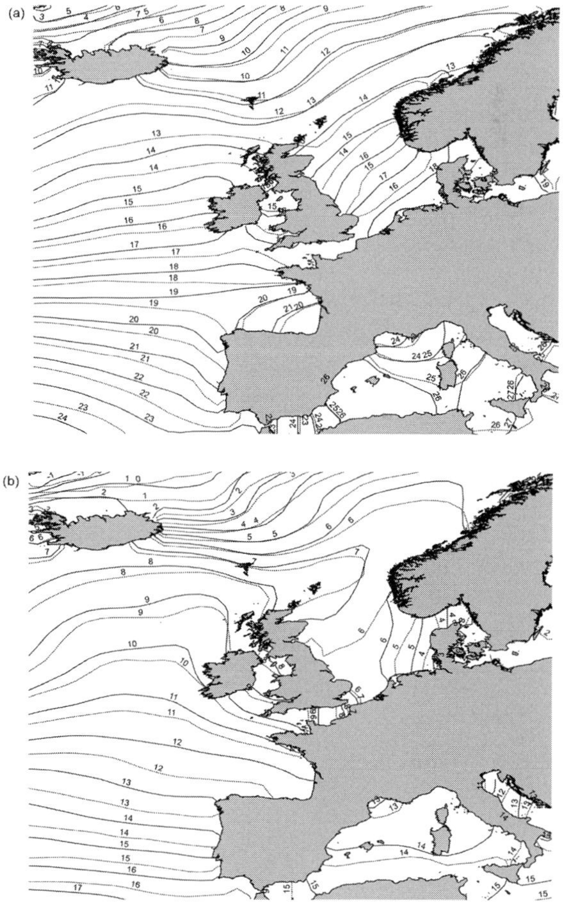


**Fig. 2.1** Mean summer maxima (a) and winter minima (b) of sea surface temperature isotherms between 1960 and 1985 (dotted) and 2000 and 2015 (solid). Data from Hadley Centre Sea Ice and Sea Surface temperature data set (HADISST), Rayner et al. (2003).

Hawkins et al., 2008, 2009; Philippart et al., 2011; Mieszkowska et al., 2014a). During warmer years *Chthamalus* appears to be released from competition by early mortality of juvenile *S. balanoides* (Poloczanska et al., 2008), while *S. balanoides* does well in colder years (Wethey et al., 2011). *S. balanoides* is prone to recruitment failure towards its southern range edge (Jenkins et al., 2000; Svensson et al., 2005; Rognstad et al., 2014); this is because reproduction is inhibited in warmer winters (Rognstad et al., 2014), perhaps coupled with mismatches of larval release with the spring phytoplankton bloom (e.g., Hawkins and Hartnoll, 1982a) being more likely after warmer, windier North Atlantic oscillation-positive index winters (Broitman et al., 2008). Both *Chthamalus* species have more broods and an extended breeding season further south in the range (Burrows et al., 1992; O'Riordan et al., 2004) and hence are likely to have more broods during warmer years towards the poleward leading range edge. These changes in abundance and reproductive output have led to range extensions in the Irish Sea (*C. stellatus*) and North Sea (*C. montagui*) (Hawkins et al., 2009). They have not been observed in the English Channel (Herbert et al., 2007) where hydrographic barriers coupled with Allee effects, the lack of suitable substrata and the polishing action of sand on rocks prevent settlement, probably preclude further spread eastwards (Herbert and Hawkins, 2006; Keith et al., 2011). *S. balanoides* has shown a major contraction in range (Wethey et al., 2011) at the trailing southern range limit in France and north-west Spain; in north-west Spain it has contracted from widespread populations in the 1950s and 1960s to a single reproductive population in the Ría de Arousa in 2006 (Wethey and Woodin, 2008). The Lusitanian barnacle *Perforatus perforatus* (formerly *Balanus perforatus*) has shown range extensions at its north-eastern leading range limit in the English Channel (Herbert et al., 2003).

Around the British Isles and Ireland extensions in the leading range edges of a suite of Lusitanian species have been idiosyncratic and aphasic, reflecting the life history characteristics of individual species, including habitat requirements and dispersal capability (Hawkins et al., 2008, 2009; Keith et al., 2011). Trochids with a short planktonic larval phase have shown the capability for the greatest expansion at the northern and north-eastern leading range edges (Mieszkowska et al., 2005, 2007; Hawkins et al., 2009, 2016a). Consolidation of range extensions probably occurs by successful colonisation of shores that are a short distance beyond the existing range limit as soon as local climatic conditions become suitable for survival. *Phorcus lineatus* has recolonised the northern sector of its distributional range following the trim-back that occurred in the extreme winter of 1962/1963 and the subsequent cold winters of the 1960s, 1970s and early 1980s; extensions of the leading range edges have now occurred beyond its previous limits in North Wales and the English Channel (Mieszkowska et al., 2007). *Steromphala umbilcalis* has been even more successful, the leading range edge extending much further east into the eastern English Channel and along the north coast of Scotland (Mieszkowska et al., 2006). There have been fewer retractions in the trailing range limits of boreal species compared to expansions of southern species (Hawkins et al., 2009; Burrows et al., 2011). The boreal kelp *Alaria esculenta* is now much less common than it once was (e.g., in Ireland, Simkanin et al., 2005), with a retraction at its southern range limit. This species disappeared from much of the western English Channel in the 1950s and, interestingly, did not reappear when it got colder in the 1960s (Hiscock et al., 2004).

After decades of gradual warming, the cold winter of 2009–10 saw enhanced recruitment of the northern barnacle *S. balanoides*, subsequently leading to re-expansion of its range in France and Iberia, from where it had previously retracted (Wethey et al., 2011). The southern barnacles *C. montagui* and *C. stellatus* did not suffer adult mortality; however, recruitment failed at their northern limit. The southern limpet *P. rustica*, the soft-shore polychaete *Diopatra* and the mussel *Mytilus galloprovincialis* showed no change in distribution and no adult mortality at their northern limits (Wethey et al., 2011). *Testudinalis testudinala* has disappeared from the Isle of Man near its former southern range limit (Forbes, 1858; Hawkins et al., 2009) with a current trailing range edge currently in northern Scotland and Northern Ireland. Filling gaps in the distribution of *S. alveolata* has occurred in recent years following the cold winter of 1962/1963, aided by the presence of artificial habitat (Firth et al., 2015). Such artificial habitat may have also enabled range extensions in the English Channel and along the Belgian coastline (Johannesson and Warmoes, 1990; Moschella et al., 2005; Hawkins et al., 2008; Keith et al., 2011; Firth et al., 2013a).

There have been similar shifts in distribution patterns in Portugal and Spain, with advances of Lusitanian species and retreat of boreal species (Lima et al., 2006, 2007). Responses to climate may be more complex in upwelling areas such as the Iberian Peninsula. Upwelling may intensify in a warmer, windier world (Bakun, 1990), leading to persistence and even increases of boreal, subtidal and lowshore species (e.g., Lima et al., 2009). In contrast, higher shore species may suffer from warmer air temperatures. There is currently debate about whether upwelling has and will increase (Miranda et al., 2013; Ribeiro et al., 2016) or decrease (Lemos and Pires, 2004; Pardo et al., 2011) along the Iberian Peninsula (Santos et al., 2011; Varela et al., 2015). Recent observations suggest that several kelp species (*S. polyschides*, *Laminaria ochroleuca* and *Laminaria hyperborea*) have undergone range contractions and/or declines in abundance in recent decades in response to seawater warming along the Iberian Peninsula (see review by Smale et al., 2013). It is very likely that kelp forest biomass and productivity will diminish under warmer, stormier conditions; unfortunately, direct measurements of kelp forest structure, biodiversity, productivity, detritus production and export, and resistance and resilience to perturbation along a regional-scale temperature gradient along the north-east Atlantic coastline are lacking (Smale et al., 2013).

Latitudinal gradients of sea and air temperatures (Figure 2.1) are highly modified at the local scale by the topography of the shore, providing microclimatic refuges from high aerial temperatures in areas of shade or water retention (Figure 2.2) (Seabra et al., 2011; Lima et al., 2016) with consequences for the biota (Firth et al., 2016b). Emersion stress can also be modified by species interactions, particularly shading by canopy algae that might enable persistence of cold-temperate species (e.g., *Patella vulgata* under *F. vesiculosus*; Moore et al., 2007; Marzinelli et al., 2012). Loss or thinning of macroalgal canopies will have severe consequences for understorey species (Jenkins et al., 1999a, 1999b, 2004, 2005; Pocklington et al., 2017). Many lowshore species are found higher up the shore under canopies; loss of canopy leads to death of these species, especially juveniles (Marzinelli et al., 2012); particularly on sheltered shores where the long-lived canopy species *A. nodosum* often dominates, this can lead to long-term changes in community structure (Jenkins et al., 2004; Cervin et al., 2005; Ingólfsson and Hawkins, 2008).

## 2.4 Distribution Patterns in Response to Local Gradients of Tidal Elevation and Wave Action

The vertical distribution patterns of rocky intertidal species have been well described for the British Isles plus the Atlantic and Channel coasts of France. In sheltered conditions there is strong vertical zonation of algal species, all of which can form dense canopies of up to 100 per cent cover. *P. canaliculata* and *F. spiralis* dominate the highshore, followed by a band of *F. vesiculosus* giving way to a midshore dominated by the large, long-lived canopy-forming species *A. nodosum*, being succeeded in turn by *F. serratus* and then kelps (usually *L. digitata* but on extremely sheltered or boulder shores *Saccharina latissima,* formerly known as *Laminaria saccharina*) from the lowshore into the shallow subtidal zone. There is some local variation in these patterns (see Lewis, 1964 for details), but *A. nodosum* dominance in the midshore region is ubiquitous in sheltered areas of high salinity.

On moderately exposed shores, fucoid cover diminishes, with *A. nodosum* becoming increasingly stunted before disappearing and being replaced by a patchy mosaic of *F. vesiculosus*, barnacles and limpets. Highshore *P. canaliculata*

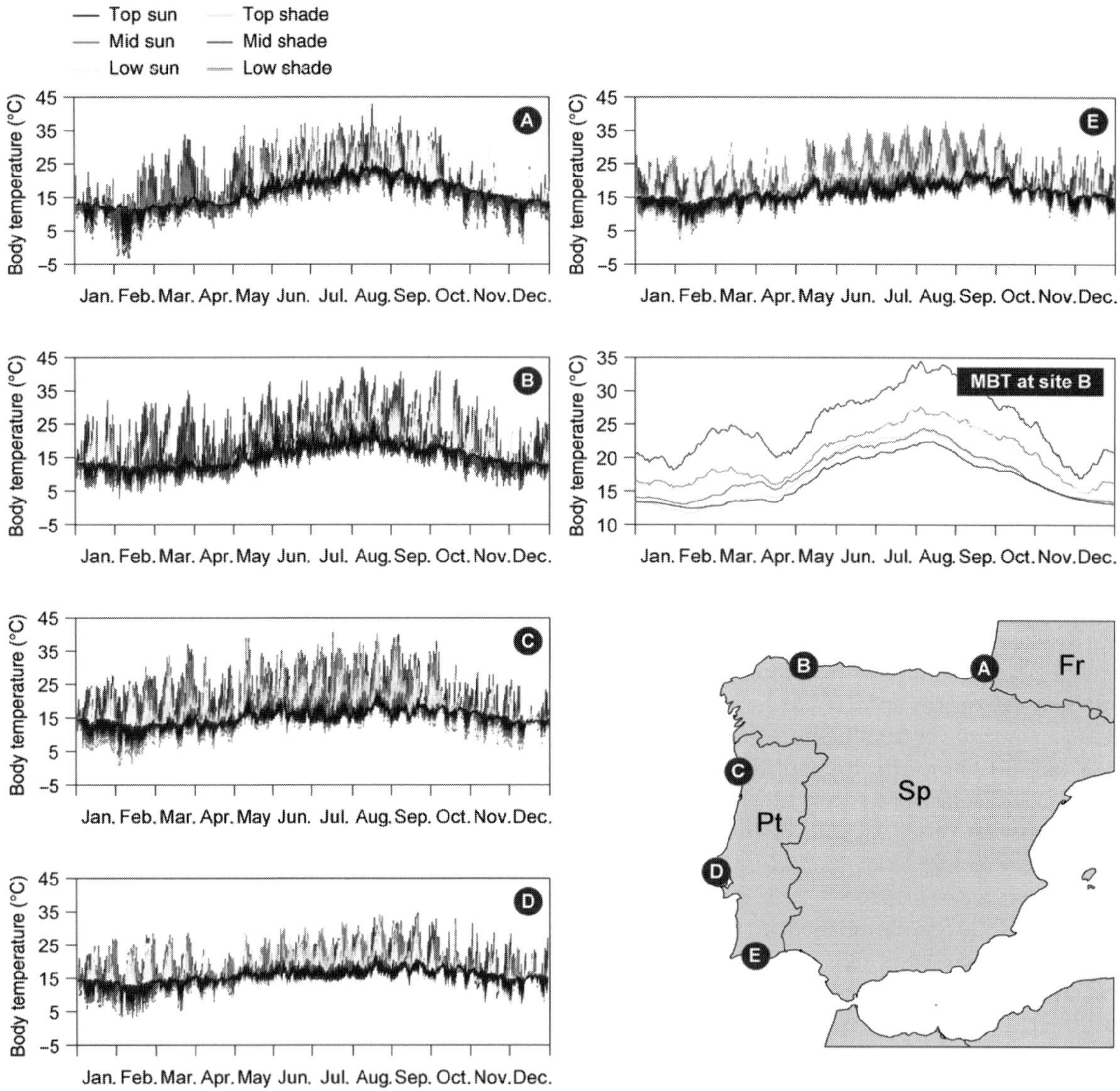


**Fig. 2.2** Body temperature profiles from 'robolimpets' deployed in different microhabitats (depicted by different line colours) recorded at five sites along the Atlantic coast of the Iberian Peninsula (A–E); 30-day average of daily maxima at La Caridad (mean body temperature at site B). *A black and white version of this figure will appear in some formats. For the colour version, please refer to the plate section.*

and *F. spiralis* become rarer with increasing exposure, as does *F. serratus* lower on the shore. In wave-exposed conditions *F. serratus* gives way to *H. elongata*, with only turf-forming algae occurring in extremely exposed conditions. In northern and western Britain (except the English Channel), a band of *A. esculenta* can occur in the most exposed conditions around the low-water spring-tide mark. On many exposed shores *Mytilus* spp. can form extensive beds; but they do not always occur because of mesoscale processes, especially larval supply and retention, that can influence distribution patterns (Wangkulankul, 2016; Wangkulankul et al., 2016). On the most exposed shores and those that are steeply sloping, fucoids become scarce. The exposed shore form of *F. distichus* occurs from Orkney northwards (with some relict populations on extremely exposed shores in Ireland and the Hebrides) (Powell, 1957, 1963).

At the northern end of the range considered here, on the shores of Norway the tidal range is small, approximately 0.5 m, resulting in tight, monospecific vertical zones on many rocky shores. *L. digitata* and *L. hyperborea* are present in high abundances along the exposed coastlines, with *S. latissima* being the most dominant kelp along the more sheltered shores of southern Norway, where *A. nodosum* is also abundant. In west Sweden, with a small tidal range ($<0.3$ m) and often larger irregular changes in the water level, the zonation pattern is similar to that described for macrotidal areas (see earlier), although *P. canaliculata* is missing. At a first glance these extremely narrow zonation patterns look very different to those on the macrotidal shores. However, for *A. nodosum*, observations show that the morphology of individuals as well as the demography and life history traits of the populations in west Sweden are remarkably similar to those on the British Isles (Åberg and Pavia, 1997; Svensson et al., 2009). *A. esculenta* is found in high densities in the Lofoten Islands. The fucoids *P. canaliculata, F. spiralis, F. vesiculosus* and *F. serratus* are also present in high abundances along the Norwegian coastline. The boreal *S. balanoides* is the only barnacle found on rocky shores, with *P. vulgata*, *T. testudinalis* and the kelp-dwelling *Patella pellucida* being the species of limpet present. *L. littorea, Littorina saxatilis* and *Melarhaphe neritoides* decrease in abundance as latitude increases along the Norwegian coastline (Narayanaswamy et al., 2010).

These patterns can be modified by local topography, with differences between shaded north-facing and sunnier south-facing rocks (Firth et al., 2016b). North-facing surfaces are refuges from temperature-related stresses, with greater species richness.

### 2.4.1 Causes of Vertical Patterns of Distribution

Classic work on the roles of physical factors in setting upper limits and competition in setting lower limits of intertidal species, using barnacles as a model species, was done by Connell (1961a, 1961b) at Millport in Scotland. Schonbeck and Norton, also working in Scotland, showed that upper limits of intertidal fucoids could be set by physical factors, while lower limits were generally set by biological interactions (Schonbeck and Norton, 1980). This led to an emerging paradigm in the 1970s and early 1980s that upper limits of sessile intertidal species were generally set by physical factors and lower limits by biological interactions such as competition, grazing and predation. The importance of larval behaviour in sessile invertebrates (Knight-Jones, 1953; Crisp, 1955; Knights et al., 2006) and adult behaviour in mobile animals in directly setting distribution patterns has also been long recognised (Newell, 1979 for review).

Observations of proliferating cover of algae after mass mortality of limpets from toxic dispersants (Southward and Southward, 1978) or experimental removal (Jones, 1948; Lodge, 1948; Burrows and Lodge, 1951; Hawkins, 1981a, Hawkins et al., 1992) gave indications that the upper limits of lowshore seaweeds could also be set by biological factors such as grazing. There is evidence that zonation patterns of patchily occurring fucoids on more exposed shores was set by the rate of early growth and the probability of escape from limpet grazing (Hawkins, 1981a; Hawkins and Hartnoll, 1983a). Mixed zonation patterns could emerge following large-scale limpet removal experiments (Lodge, 1948; Burrows and Lodge, 1950).

Other experiments on fucoid-dominated shores showed that competition could also set the upper limits of zonation of species (Hawkins and Hartnoll, 1985). *A. nodosum* removal experiments on the Isle of Man, where there is a narrow band of *F. vesiculosus* above *A. nodosum* with a dense band of *F. serratus* below it, showed that, in the absence of *A. nodosum* canopy, both species could coexist on the midshore (Hawkins and Hartnoll, 1985; Jenkins et al., 1999c, 2004). Similar findings were observed in Iceland (Ingólfsson and Hawkins, 2008). Interestingly, Lewis (1964) suggested this was the case, but did not make any experimental tests of this hypothesis. Small-scale removals of *A. nodosum* in the Isle of Man, simulating natural disturbance events on sheltered shores demonstrated the mechanism by which patches of the lowshore *F. serratus* could persist among dense beds of the dominant *A. nodosum* (Cervin et al., 2005).

Broadscale experiments in Portugal and the British Isles showed that the upper limits of algal turfs, *F. serratus*, and *H. elongata* could be set by limpets grazing (Boaventura et al., 2002a). The rapid growth of the algal turf was shown to exclude limpets and prevent them from extending down the shore: low on the shore the rate of algal growth exceeded the ability of limpets to control it (see Hawkins and Hartnoll, 1983b for explanation of the causes of the Stephenson and Stephenson three-zone system). In the Azores, much of the midshore is dominated by extensive patches of algal turfs as a result of chronic exploitation of limpets that substantially reduce their numbers, particularly the larger *P. aspera* in the intertidal (Martins et al., 2008, 2010).

In the north-east Atlantic, it is clear that, while upper limits are mostly set directly by physical factors, this is not always the case. Highshore species (*P. canaliculata*, *F. spiralis*) plus broad-ranging species such as *P. vulgata* (Orton, 1920) and *S. balanoides* that extend higher up the shore (Connell, 1961a, 1961b) have their upper vertical limits set by physical factors (Hawkins and Hartnoll, 1985). This is often the case for essentially subtidal species just extending into the intertidal, such as *L. digitata*; although removal of *F. serratus* can lead to a modest upshore extension (Hawkins and Hartnoll, 1985) which can be subsequently trimmed back. Thus, the upper limits of *L. digitata* can be set by physical factors and/or by biological interactions, depending on the location and season (Araújo et al., 2016). Thus, there is mounting evidence that for many low- and mid-shore species upper vertical limits are set by biological interactions such as competition, where canopy cover is dense, or by grazing – especially for algal turfs on more exposed shores (Boaventura et al., 2002a).

It is instructive to review the relative role of physical factors and biological interactions by taking a slice down moderately exposed or exposed shores of the south and central British Isles (revisiting Hawkins and Hartnoll, 1983b). In the highshore region above the barnacle zone (the littoral fringe [Lewis, 1964]), diatoms and ephemeral algae (*Ulothrix*, *Prasiola*, *Blidingia* and *Porphyra* species) proliferate in the winter months (Hawkins and Hartnoll, 1983a), but die-off in the spring owing to warmer weather; only tough cyanobacteria (Thompson et al., 2005; Skov et al., 2010) and lichens remain over the summer. Ephemeral algae can, however, persist in cool summers and further north in the British Isles and in Norway and Iceland, or in areas of wet run-off. Littorinid snails (*M. neritoides*, the *L. saxatilis* complex) are numerous but restricted to pits and crevices by harsh conditions (Hawkins and Hartnoll, 1983b), resulting in grazing haloes (Stafford and Davies, 2005) that can expand or contract seasonally. Physical factors are paramount here, with grazing restricted locally to around refuges from harsh physical factors (but see Stafford and Davies, 2004). Competitive interactions between ephemeral seaweeds have rarely been explored experimentally (but see Viejo et al., 2008), but the succession that occurs seasonally suggests that some species can dominate once released from harsh physical conditions over the winter. Further south in Europe, such proliferation of ephemeral algae is less commonly observed, other than on the most wave exposed shores.

The midshore region (the eulittoral zone [Lewis, 1964]) is dominated by barnacles and, in some places, mussels (Ballantine, 1961). In extreme exposure, the small, bladderless morph of *F. vesiculosus* can occur, especially among mussel mosaics. On these shores, grazing by patellid limpets is the predominant interaction. Occasionally, patches of *F. vesiculosus* do escape from limpet grazing. Predation by dog whelks may also be important, although there has been little work published on their community role in the north-east Atlantic (but see Hawkins et al., 1994). Their foraging is often constrained by the physical environment, with excursions occurring from refuges such as crevices (Burrows and Hughes, 1989; Johnson et al., 1998). Competition for space can occur between barnacle species (Connell, 1961b). Surprisingly little work has been done on mussel-dominated shores in Europe (but see O'Connor et al., 2006, O'Connor and Crowe, 2007; Crowe et al., 2011; Wangkulangkul et al., 2016).

Lower on the shore, *F. serratus* (moderate exposure), *H. elongata* (exposed) and *Cystoseira abies-marina* (very exposed) become more prevalent among red algal turfs (e.g., *Chondrus crispus*,

*Mastocarpus stellatus* and *Corallina* spp.), which are the only lowshore algae present in the most exposed conditions. Here, the ability of algae to dominate space is the most important interaction, with grazing by *P. vulgata* or *P. ulyssiponensis* leading to small clearings among the turf. Around the low-water spring-tide level, the turf gives way to kelp (*L. digitata*, except in the most exposed locations where *A. esculenta* occurs in the British Isles). Removal of *L. digitata* on moderately exposed shores can lead to colonisation of opportunist kelps: *S. latissima* primarily from more sheltered conditions and *A. esculenta* from more exposed shores (Hawkins and Harkin, 1985). Removal of the *L. digitata* canopy also leads to the proliferation of a dense understorey of opportunistic species (*Ectocarpus spp.*, ephemeral green algae, *Palmaria palmata*) that swamps the few limpets (*P. ulyssiponensis*) and sponges (*Hymeniacidon perlevis, Halichondria panacea*) that are facilitated by the sweeping effects of the kelp lamina. Algal canopy effects are the predominant interaction either side of lowest astronomical tides (Hawkins and Hartnoll, 1983a).

### 2.4.2 Causes of Horizontal Patterns of Distribution

The interplay between the underlying physical gradient from sheltered bays to exposed headlands with biological interactions to set distribution patterns has long been a subject of study in the north-east Atlantic. It is important to emphasise that this environmental gradient – unlike the vertical gradient from low-water to high-water – cannot be considered a unidirectional stress gradient. Some species thrive in wave-exposed conditions; others are more successful in sheltered waters; while many have optima in moderate wave action (see Raffaelli and Hawkins, 1996 for further discussion).

Early explanations emphasised the importance of the direct effects of the physical environment in determining distributions, with fucoid algae being excluded from wave-exposed conditions by the direct effects of wave action (see Lewis, 1964 for review). While the exposure to wave action clearly has a role to play in determining fucoid survival (Jonsson et al., 2006), classic limpet removal experiments on the Isle of Man (Jones, 1946, 1948; Lodge, 1948; Burrows and Lodge, 1950; Southward, 1956, 1964) showed that grazing prevented establishment of fucoid seaweeds on barnacle-dominated shores. Observations following kills of limpets by excessive application of toxic dispersants (Southward and Southward, 1978) confirmed that limpet grazing prevented establishment of algae even on the most exposed British shores, such as Sennen Cove in west Cornwall (see also Hawkins et al., 1983, Hawkins and Southward, 1992). Jonsson et al. (2006) by a combination of field experiments, measurements of wave action and modelling showed that establishment of fucoids was prevented by limpet grazing, but persistence was limited by wave action.

The importance of limpet grazing (Hawkins, 1981a, 1981b) in generating small-scale patchiness on moderately exposed British shores (Hartnoll and Hawkins, 1980, 1985; Hawkins and Hartnoll, 1983b) has been the focus of much small-scale experimentation (described in Hawkins et al., 1992). This has shown the complex positive and negative interactions between limpets, barnacles and fucoids on moderately exposed shores (see Hawkins and Hartnoll, 1983a; Hartnoll and Hawkins, 1985; Hawkins et al., 1992), especially the role of barnacles in facilitating fucoid escapes from grazing (Hawkins, 1981a, 1981b) by providing refuges for germlings and impeding foraging movements of *Patella* spp. (Hawkins and Hartnoll, 1982b). Clumps of fucoids not only provide habitat for *Nucella lapillus* and the anemone *Actinia equina,* they also act as nursery areas for juvenile limpets (Hawkins and Hartnoll, 1983a).

Detailed spatial statistical analysis described the grain of patchiness and how small-scale surface topography could override biological provision of habitat (Johnson et al., 1997). Individual-based spatially explicit models (Johnson et al., 1998) coupled with the use of cellular automata (Burrows and Hawkins, 1998), showed the importance of limpet homing and aggregation behaviour in the maintenance of patchiness on moderately exposed shores. Exploitation of limpets in southern Europe and the Azores may also drive low- and mid-shore communities to diverge towards those dominated by algal turfs

(Martins et al., 2008, 2010). Ultimately, however, the patch dynamics on these shores are driven by recruitment fluctuations, especially of barnacles (Hawkins and Hartnoll, 1982a; Jenkins et al., 2000), but also of limpets and fucoids (Little et al., 2017). Thus, there is an external stochastic element leading to patch genesis and an internal deterministic element leading to patch demise (Hartnoll and Hawkins, 1985; Martins et al., 2018). This work has been criticised as lacking generality (Chapman, 1995) because it was primarily conducted on Isle of Man limestone ledges with limited replication. Subsequent work as part of the EUROROCK programme employed nested experimental designs to explore recruitment fluctuations (Jenkins et al., 2000; O'Riordan et al., 2004; Cruz et al., 2005) and grazing interactions (Jenkins et al., 2005; Coleman et al., 2006). This has shown the generality of process at a European scale, with midshore limpet grazing being important in controlling algae at all latitudes. EUROROCK work showed considerable interannual and locational variation in barnacle recruitment in both *Chthamalus* species (O'Riordan et al., 2004; Cruz et al., 2005) and *S. balanoides* (Jenkins et al., 2000). This broadscale work was turned into stage-structured models of population processes on a European scale (Hyder et al., 2001; Svensson et al., 2005, 2006).

Competition among grazing limpets was also analysed *in situ* through experimental work (Boaventura et al., 2002c, 2003; Firth et al., 2009; Firth and Crowe, 2010). Competitive interactions within and between size-classes of the limpet *Patella depressa* were investigated in central Portugal. Although both size-classes could negatively affect each other, the effect of large limpets on small was greater than the reverse. Large limpets were shown to be superior competitors that may modulate the abundance of small limpets on the shore. It is unlikely, however, that they will totally exclude small limpets due to intensity of competition within the large size-class. Niche differentiation and high recruitment at lowshore levels are other possible factors (Boaventura et al., 2003) that can contribute to reduce competition between the size-classes. Competitive interactions between *P. depressa* and *P. vulgata* were also examined close to the biogeographical limit of distribution of *P. vulgata*, in the north of Portugal. However, at the end of the experiment no significant differences were found in the effects of *P. depressa* on *P. vulgata* and vice versa (Boaventura et al., 2002c).

As with many limpets which are exploited worldwide, *P. vulgata* is a protandric sequential hermaphrodite that changes sex from male to female during its lifespan (Le Quesne and Hawkins, 2006; Guallart et al., 2013). Since exploitation of limpets is mainly size-selective, protandric limpet populations have been depleted of female cohorts which aggravates the harvesting pressure upon the stocks (OSPAR Commission 2010). Moreover, some limpet species have a fixed size at sex change, such as the limpet *Cymbula oculus*, while other species have plastic responses (Munday et al., 2006). For instance, experimental size-selective harvesting of the protandric limpet *P. vulgata* in the British Isles led to a decrease in shell size at sex change, suggesting an earlier switch of males to females to compensate for the harvested females (Borges at al., 2016): Size-dependent sex change was indicated by $L_{50}$ (the size at which there is a 50:50 sex ratio), occurring at smaller sizes in treatments than controls, suggesting an earlier switch to females. These results were consistent with those shown by the protandric *Patella ferruginea* where individuals in populations with a low density of larger individuals switch to female at smaller sizes (Rivera-Ingraham et al., 2011). In parallel, Borges et al. (2015), when investigating the relationships between *P. vulgata* density and sex ratios of range edge and central range populations in the British Isles and Portugal, found that lower densities at range edge populations were correlated to higher proportions of females. It was suggested that relaxation of resource limitation resulting from reduced intraspecific competition at lower density (see Boaventura et al., 2003) would allow earlier promotion from males to females, resulting in more females at lower densities (i.e., sex change is density dependent). In fact, Boaventura et al. (2003) suggested that the higher percentages of males found in increased *P. depressa* density treatments could possibly be explained by a suppression of protandry due to competition, although this species is not widely considered as protandrous (Borges et al., 2015).

While the ecological processes determining the structure and dynamics of exposed rocky shores have received considerable attention, sheltered shores, dominated by fucoid canopy algae, have seen less work. Here the balance between the ability of macroalgae to recruit and grow and the ability of grazers to prevent them is shifted in favour of algae (Jenkins et al., 2008a). Removal experiments have demonstrated the important structuring role of large canopy-forming fucoids on understorey communities relative to a much-reduced role in patellid limpets (Jenkins et al., 1999a, 1999b, 1999c, 2004; Cervin et al., 2005). Long-term experiments in the midshore *A. nodosum* zone revealed interesting indirect effects of the dominant canopy; removal led to short-term loss of red turf-forming algae, a switch to a *Fucus*-dominated canopy and, in the long term, a six-fold increase in available bare space and patellid limpet populations (Jenkins et al., 1999a, 2004). Work on sheltered shores has also explored the interaction between canopy algae and settling barnacles. While at exposed sites the sweeping action of algal fronds reduces barnacle settlement thereby contributing to small-scale patchiness (Hawkins, 1983), on sheltered shores fucoids can limit barnacle populations to very low levels. Interactions are complex, with negative effects of sweeping on new arrivals, especially by the non-bladdered *F. serratus*, but also positive effects on post-settlement survival, presumably through amelioration of emersed conditions, especially higher on the shore among *F. vesiculosus* and *F. spiralis* (Hawkins, 1983; Jenkins et al., 1999d; Jenkins and Hawkins, 2003). Experimental manipulations have also shown that *F. serratus* (but not *A. nodosum*) can limit settlement through blocking access of cyprids to the substratum (Jenkins et al., 1999d; Jenkins and Hawkins, 2003).

## 2.5 Differential Patterns and Processes in Rock Pool Habitats Compared to Emergent Rock

Rocky shores typically comprise a mosaic of habitats with varying degrees of heterogeneity. Such complexity and spatial heterogeneity are important factors in the structure and functioning of rocky shore communities, contributing to community diversity and species coexistence in marine benthic systems (Johnson et al., 2003; Kostylev et al., 2005). Rock pools are ubiquitous features of rocky shores across the north-east Atlantic and worldwide. In comparison to emergent rock substrata, relatively little is known about the processes determining the structure and functioning of biotic communities in pools (but see O'Connor and Crowe, 2005; Martins et al., 2007; Noël et al., 2009, 2010; Griffin et al., 2010; Vye et al., 2014). Rock pools provide important nursery grounds (Orton, 1929; Lewis and Bowman, 1975; Thompson, 1980; Delany et al., 1998; Dias et al., 2016) and refugia from both abiotic stress and biological interactions (Schonbeck and Norton, 1978) for a wide range of organisms. Rock pools can extend the upper vertical limits of many organisms that are susceptible to desiccation, and, while some species tend to aggregate in pools, other species avoid them (e.g., Goss-Custard et al., 1979; Araújo et al., 2006).

Patterns in the community structure of rock pools differ greatly from those on emergent rock (Underwood, 1973; Hawkins and Hartnoll, 1983b; Noël et al., 2009; Firth et al., 2013b), leading to variation in the processes, shaping biological communities among habitat types, such as parasitism (Crewe, 1951; Kollien, 1996; Firth et al., 2017), grazing (Noël et al., 2009) and competition (Firth et al., 2009). For instance, patellid limpets have disjunct distributions in relation to rock pools and emergent rock at different stages in their life history. *P. vulgata*, *P. ulyssiponensis* and *P. depressa* are known to recruit into rock pools (Bowman and Lewis, 1977; Delany et al., 2002). *P. ulyssiponensis* is predominantly found in pools as well as on lowshore areas of rock covered by crustose coralline algae throughout its life (Lewis, 1964; Fretter and Graham, 1976; Delany et al., 2002). In contrast, *P. vulgata* and *P. depressa* are thought to migrate out of rock pools and are more abundant on emergent rock than in pools (Davies, 1969; Firth and Crowe, 2008). Furthermore, on emergent rock, *P. vulgata* tends to aggregate underneath *Fucus* spp. clumps and suffers mortality if they are removed, while *P. depressa* does not (Moore et al., 2007). In an experiment investigating habitat preference (emergent rock versus rock pools) and competition (intraspecific and interspecific) between *P. vulgata* and *P. ulyssiponensis*, Firth and Crowe (2010) found that *P. ulyssiponensis* exhibited increased growth in higher intraspecific densities on emergent rock and reduced growth in higher intraspecific densities in rock pools, indicating some degree of intraspecific facilitation on open rock and intraspecific competition in pools. There was no evidence of interspecific competition in either habitat.

Despite rock pools offering a potential refuge from the harsh abiotic conditions encountered on emergent rock (e.g., temperature and desiccation stress), they may also become stressful environments, with large fluctuations in temperature, salinity, pH and dissolved oxygen (Pyefinch, 1943; Naylor and Slinn, 1958; Goss-Custard et al., 1979; Morris and Taylor, 1983), especially higher on the shore. Rock pools in the upper shore are exposed to longer periods of emersion and exhibit greater variability in environmental conditions than pools located lower on the shore (Pyefinch, 1943). Small and shallow rock pools can exhibit significantly different temperature profiles than larger deeper ones (Martins et al., 2007), potentially making them thermally stressful environments. This is likely to be exacerbated in more extreme climates (e.g., Chan, 2000; Firth and Williams, 2009). Despite exhibiting dramatic fluctuations in physico-chemical conditions, natural rock pools often support greater diversity, abundance and/or biomass of organisms than emergent rock (Goss-Custard et al., 1979). While this is a very common assertion in the literature, there is surprisingly little quantitative evidence in the literature to support it (but see Firth et al., 2013b, 2014a).

Rock pools are complex habitats that can vary greatly in their area, depth, topography, degree of shading and inclination of surrounding rocks. Of the limited number of studies carried out in the north-east Atlantic on the physical properties underpinning rock pool communities, surface area is considered to have little influence on the biological structure of pools (Martins et al., 2007), while shading (Pyefinch, 1943) and slope (Firth et al., 2014a) have been shown to influence diversity and abundance of species. Far more is known about the influence of rock pool depth on species performance, diversity and community composition (Goss-Custard et al., 1979; Moschella et al., 2005; Bussell et al., 2007; Martins et al., 2007; Firth et al., 2014a). The relationship between diversity and function varies with depth depending on the response variable measured. Some studies have found that species richness was positively correlated with increasing pool depth (Moschella et al., 2005; Martins et al., 2007), while others have found the opposite pattern (Bussell et al., 2007; Firth et al., 2014a).

The growing body of literature emerging on the processes shaping the patterns of distribution of organisms among habitat types on rocky shores is playing an important role in informing the engineering design of coastal and marine artificial structures (e.g., sea walls, breakwaters, docks, offshore renewables, oil and gas platforms) through ecological engineering. Ecological engineering is the design of sustainable ecosystems for the mutual benefit of both humans and nature (Mitsch, 2012). Traditionally, artificial marine structures were built with a single function in mind (defence); however, changes in attitudes due to the proliferation of such structures (Morris et al., 2016; Evans et al., 2017; Strain et al., 2019) are leading to a shift in the way artificial environments (often referred to as 'grey' spaces) are perceived and designed (to become 'blue' spaces) (Sutton-Grier et al., 2015; Firth et al., 2016a; Mayer-Pinto et al., 2017; Strain et al., 2017). This can be achieved through the design of multifunctional structures providing habitat for marine life, while simultaneously serving their primary engineering function. Rock pools can be created easily and cheaply on artificial structures through a range of different techniques, ranging from creating water-retaining lips in sea walls (Chapman and Blockley, 2009), affixing precast units to vertical sea walls (Browne and Chapman, 2011, 2014; Morris et al., 2017), drill-coring directly into the substratum (Martins et al., 2010, 2016; Firth et al., 2014b; Evans et al., 2016) to pouring concrete among the boulders of rock armouring (Firth et al., 2016b). Other (more expensive) options include the deployment of large-scale precast habitat-enhancement units such as the BIOBLOCK (Firth et al., 2014b) and ECOncrete's tide pools (Perkol-Finkel and Sella, 2015; www.econcretetech.com), which can replace boulders on rock armouring.

## 2.6 Latitudinal Patterns

The EUROROCK team used a simple approach based on limpets leaving grazing marks on wax discs (Thompson et al., 1997) to show that grazing pressure was less seasonal and more intense in southern Europe compared to further north (Jenkins et al., 2001). Such general patterns can, however, be modified by local topography, influencing foraging behaviour (Johnson et al., 2008). Using exclusion cages, limpets were shown to control algal vegetation on midshores in southern Portugal through to the Isle of Man (Jonsson et al., 2006). In general, this process was most deterministic in the Isle of Man (Figure 2.3) (Jenkins et al., 2005; Coleman et al., 2006) where propagule supply was high, with some escapes also occurring in grazed controls. Even in the British Isles, there were marked differences, with fucoid recruitment being more likely and more intense on the Isle of Man and much more patchy and stochastic in south-west England (Arrontes et al., 2004; Jenkins et al., 2005).

In southern and central Portugal, canopy-forming fucoids are restricted to *F. spiralis* and *F. guiryi* (Zardi et al., 2011). Other species of fucoid drop out in northern Portugal on the open coast, although *F. vesiculosus* can occur in estuarine refuges (Serrão et al., 1999; Pearson et al., 2000). The limited species pool, lack of propagule supply and physical stress probably all contribute to the absence of canopy-forming fucoids in southern Europe. Recent comparative survey and experimental work between the British Isles and Portugal have shown that reproductive output is less in Portugal (Figure 2.4) (Ferreira et al., 2015a) and there is greater physiological stress (Ferreira et al., 2014), which in turn interacts with propagule supply to affect the probability of fucoid

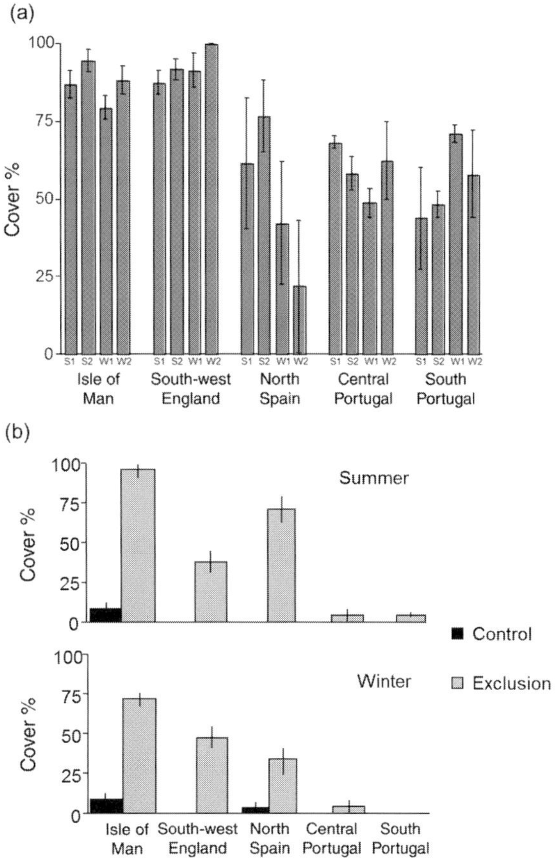


**Fig. 2.3** A continental-scale limpet exclusion experiment (see also Coleman et al., 2006). (a) Fucoid cover after 12 months shown as the difference between treatments and controls. Each bar represents a shore in each region with summer (S) and winter (W) starts shown. (b) Total algae cover after 12 months. In Portugal, this was primarily ephemeral algae and crusts. Winter and summer starts are shown.

survival and early growth to escape limpet grazing (Ferreira et al., 2015b). As referred to earlier, besides limpets, there are also several species of trochids, grapsid crabs and herbivorous fish such as the salema (*Sarpa salpa*) in southern European coasts, making grazing pressure much greater (Franco et al., 2015). Laboratory and in situ experiments with *F. serratus* suggested that three physical factors, solar irradiation, ocean and air temperatures, acting additively shape the distribution of this species in the Iberian Peninsula (Martínez et al., 2012). The future structure of large brown seaweeds on these

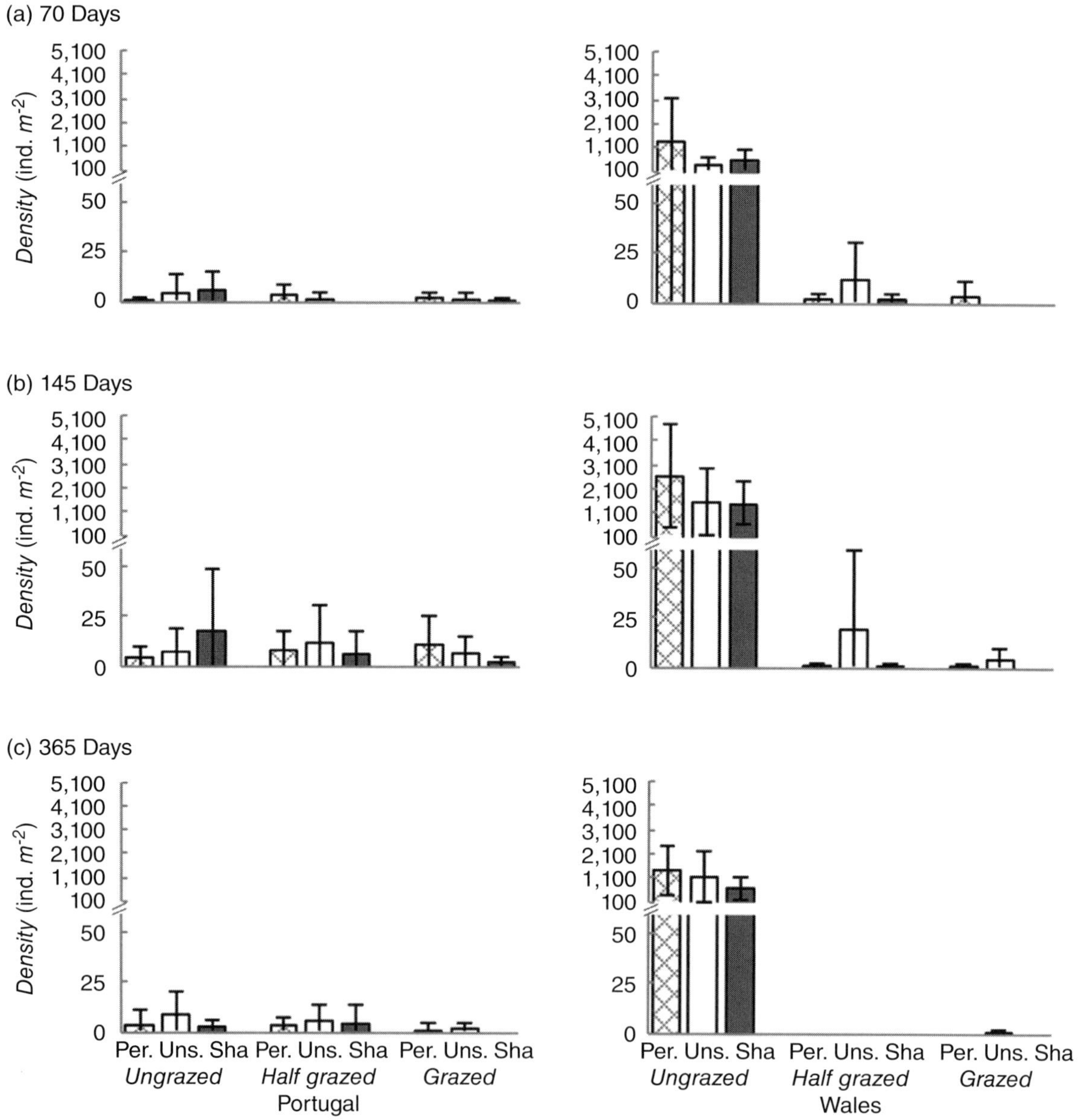


**Fig. 2.4** Densities of *Fucus* spp. with different grazing treatments after (a) 70, (b) 145 and (c) 365 days under different light intensities (Per.: perspex, procedure control; Uns.: unshaded, subjected to natural solar radiance levels; Sha.: shaded, under reduced solar radiance levels). Results demonstrate that reduced physical and biological pressures do not compensate for the lower supply of fucoid propagules at range edge populations in southern Europe (Portugal) compared to range centre populations in northern Europe (Wales). Data from Ferreira et al. (2015a).

Atlantic shores is predicted to change in the next few decades. The cold-temperate foundational seaweed *H. elongata* is predicted to become extinct at its current southern limit on the Iberian Peninsula in response to global warming, whereas the occurrence of southern Lusitanian *B. bifurcata* is expected to increase (Martínez et al., 2015).

## 2.7 Mesoscale Processes

Various processes operating at scales of 10s–100s of kilometres can influence species composition, community structure and functioning of ecosystems. These include upwelling and the topography of the coastline, both ultimately determined by the interaction of geomorphology and nearshore oceanographic processes. These in turn determine productivity of nearshore waters and the dispersal or retention of larvae.

In the north-east Atlantic there are extensive areas of upwelling off North Africa and on the coasts of Portugal and Spain (Southward et al., 1995; Jenkins et al., 2008a; Alvarez et al., 2011). In these regions, the bathymetry is steep with deep water close inshore, and wind-driven upwelling of nutrient-rich cold water occurs. On a geographic scale, this can lead to cold-water species persisting on the Asturian, Galician and Portuguese coasts that are absent in the warmer waters of the inner Bay of Biscay. Kelps, fucoid algae, littorinids and cold-water limpets such as *P. vulgata* are all able to survive in these areas (see Southward et al., 1995 for summary and review of early work). These shores are also more productive, supporting extensive raft culture of mussels in the Spanish Rias (Blanton et al., 1987; Figueiras et al., 2002). As upwelling waxes and wanes, advances and retreats of fucoid algae have been reported along the coast of the Basque Country and Asturias (Arrontes, 1993).

Conversely, relaxation of upwelling has been suggested as the likely cause for the advance of the Lusitanian species of limpet *P. rustica* into Galicia in the last two decades (Lima et al., 2006; Sousa et al., 2012). This species was previously only found in southern and mid-Portugal, being absent from northern Portugal, Galicia and Asturias before reappearing in high densities in the Basque Country. In recent years, this gap has been bridged (Lima et al., 2006) as the species has advanced. Interestingly, there is no evidence for genetic differences between *P. rustica* from the Basque region and those found in southern Portugal (Ribeiro et al., 2010), suggesting this gap – known to occur for at least eighty years – was either relatively new or occasionally breached.

Upwelling also influences recruitment regimes. Although most work on this topic has been done along the north Pacific (e.g., Roughgarden et al., 1988; Connolly and Roughgarden, 1998; Connolly et al., 2001), some work has been done in the north-east Atlantic. In Portugal, recruitment of barnacle (Cruz et al., 2005) and crab (Queiroga et al., 2007) larvae has been shown to be strongly affected by intensification and relaxation of upwelling. This leads to intensive recruitment of barnacles but with differential effects between species: *C. montagui* which has a reasonably short larval life (Burrows et al., 1999) is favoured in upwelling regions; whereas *C. stellatus* with a longer larval duration does less well, presumably dispersed offshore by upwelling. A complexity of factors interacts to create recruitment peaks, including onshore transport during relaxation of upwelling and timing of tide influencing settlement behaviour (Cruz, 1999; Cruz et al., 2005; Jacinto and Cruz, 2008; Trindade et al., 2016).

Along more complex coastlines, such as around the British Isles, the topography of the coastline (bays, headlands, concave and convex coastlines) interact with nearshore processes such as tidal mixing and fronts, with stratified water influencing larval retention and dispersal (Davies and Johnson, 2006; Jessopp et al., 2007). Jenkins et al. (2008b) used nearshore plankton trawls and daily surveys of settlement at an exposed headland and sheltered bay to explore the scale of dispersal of chthamalid barnacle larvae in two species differentially distributed with respect to exposure. He showed that patterns in larval distribution set when larvae were released (namely domination of *C. montagui* in shelter and *C. stellatus* in exposure, respectively) were lost when larvae were ready to settle, demonstrating a lack of local larval retention close to the release site. Enclosed waters with strong nutrient inputs from freshwater run-off are also much more productive. The Firth of Clyde is an example of an area with strong larval retention and highly productive nutrient-rich water. Very high barnacle settlement has been reported here (Connell, 1961a; Hills and Thomason, 2003) compared to other areas (Hawkins and Hartnoll, 1982b; Burrows et al., 2010). This strong recruitment in part explains the very clear-cut effect of competition setting the distribution patterns of barnacles, with *C. stellatus* (now known to be *C. montagui*) being outcompeted by densely settling *S. balanoides* (Connell, 1961b). The degree of coastal openness and regional variation in productivity (which at least in part is related to coastal topography) also influences barnacle larval nutritional reserves, or 'larval quality' (Giménez et al., 2017) with potential knock-on consequences for population- and community-level processes (Giménez and Jenkins, 2013; Torres et al., 2016).

Although studied much less, there is evidence that *S. alveolata* does well in large bays and inlets (Solway Firth, Liverpool Bay, Cardigan Bay, Bristol Channel, Baie de Mont St Michel, Rade de Brest) where both sediment supply is high and larvae are retained (Dubois, 2002, 2006; Firth et al., 2015).

Conversely, recruitment of species with planktonic larvae can often be low on isolated islands such as Rockall (Crisp, 1956; Johannesson, 1988), Lundy (Hawkins and Hiscock, 1983), the Scilly Isles (Crisp and Southward, 1958) and the Isle of Man (Hawkins and Hartnoll, 1982a; Jenkins et al., 2000). On Anglesey (Bennell, 1981; Hyder et al., 1998; Jenkins et al., 2008b) there are very sharp differences in recruitment on different sides of the island, with recruitment being particularly high in the Menai Strait and the south-west of the island. On the Isle of Man, recruitment of barnacles is higher on the west coast than the east coast due to wind-driven processes (Hawkins and Hartnoll, 1982a). There is also an area of exceedingly dense recruitment of *L. littorea* on the west coast opposite a stratified patch of water compared to elsewhere on the island (Norton et al., 1990).

Headlands can be important boundary areas influencing connectivity, often being the proximate cause of biogeographic range limits (Crisp and Southward, 1958; Herbert et al., 2009). In recent years, the poleward spread of the leading range edge of Lusitanian species has been shown to be strongly modulated by the interaction of length of larval life and behaviour with nearshore hydrography (e.g., Keith et al., 2011 in the English Channel) with differential capability for spread being exhibited by different species (Mieszkowska et al., 2006; Hawkins et al., 2008, 2009). The Llŷn Peninsula in North Wales has been a barrier for recolonisation by *P. lineatus* (Mieszkowska et al., 2007) and *P. depressa* following contraction during the cold spell of the 1960s to 1980s (Hawkins et al., 2008, 2009).

## 2.8 Non-native Invasive Species

The main non-native species found in high abundance in the rocky intertidal in the region are the seaweeds *Sargassum muticum* and *Undaria pinnatifida* (also found in the subtidal), the Australasian barnacle *Austrominius modestus* (formerly *Elminius*) and the Pacific oyster *Magallana gigas* (formerly known as *Crassostrea gigas*).

The Australasian barnacle, *A. modestus*, has been in Europe since the 1940s, first being recorded in the British Isles and subsequently spreading both northwards (Crisp, 1958) and southwards (Barnes and Barnes, 1963) in Europe. It is found primarily in estuaries, rias, harbours and sheltered bays (King et al., 1997; Lawson et al., 2004), settling on both natural and on temporary and artificial habitats, becoming less common on more exposed shores (Crisp, 1958). At a mesoscale it is particularly common on the coasts of the eastern English Channel, southern North Sea, the Irish Sea (Herbert et al., 2007) and the Bay of Saint Malo (O'Riordan and Ramsay, 1999). It is a multi-brooding species (O'Riordan and Murphy, 2000); thus, it is able to colonise free space made available by artificial structures (Bracewell et al., 2012), disturbance (Bracewell et al., 2013), predation by *N. lapillus* on the native species such as *S. balanoides* (Barnett, 1979) or intense intraspecific competition leading to hummocking and sloughing off *S. balanoides* (Barnes and Powell, 1950). Surprisingly few controlled experiments have been undertaken on interactions with other species. Observations in the North Sea suggest that recent expansion of *A. modestus* (potentially as a consequence of warming seas) has come at the expense of the native barnacles *S. balanoides* and *Balanus crenatus* (Witte et al., 2010), although experimental manipulations by Gallagher et al. (2016) in Ireland suggest *A. modestus* coexists without displacement of other intertidal species.

*S. muticum* invaded Europe in the 1970s (Critchley et al., 1983), spreading southwards and northwards from its point of introduction in France (Rueness, 1989). It can occur in the shallow subtidal zone as well as occupying rock pools (often as epibionts on the shells of limpets) on sheltered to moderately exposed shores. It is a highly seasonal species, overwintering in Europe as short shoots from which it proliferates to form dense canopies up to 100 per cent cover in rock pools during spring and summer, before dying back in the autumn (Engelen et al., 2015). Experimental manipulations in northern Spain have shown limited impacts on the native communities (Viejo, 1997), perhaps because of its autumn–winter dieback. In two out of three sites where *S. muticum* was experimentally removed, large leathery species (such as canopy-forming *B. bifurcata*) and total cover algae (excluding crusts and *S. muticum*) became more common (Olabarria et al., 2009). Functional impacts of this species have rarely been assessed (but see Vaz-Pinto et al., 2014; Vye et al., 2014). A recent experimental study in rock pools found that, even in the absence of large structural impacts associated with the invasion of *S. muticum*, the assemblages initially seeded with *S. muticum* presented higher rates of productivity and higher photosynthetic efficiency compared to non-invaded assemblages (Rodil et al., 2015a, 2015b; Vieira et al., 2017).

The invasive brown alga *U. pinnatifida* was first introduced in 1971 via accidental import with shellfish into French Mediterranean coastlines (Perez et al., 1981; Floc'h et al., 1991), followed by intentional introductions for mariculture into Brittany in 1981 (Perez et al., 1981). Across the north-east Atlantic, secondary spread is thought to have been facilitated by fouling on recreational and commercial vessels and transport to nearby ports and marinas (Fletcher and Farrell, 1999; Veiga et al., 2014). *U. pinnatifida* is now established across much of the north-east Atlantic (Epstein and Smale, 2017), where it is found most commonly on artificial substrates (Fletcher and Farrell, 1999; Farrell, 2003; Cremades et al., 2006; Farrell and Fletcher, 2006; Heiser et al., 2014; Veiga et al., 2014). Once established in these habitats, *U. pinnatifida* can spread onto natural substrates, and is predominantly found invading sheltered rocky reefs from the low intertidal to shallow subtidal zone (Floc'h et al., 1991; Castric-Fey et al., 1999; Cremades et al., 2006; Heiser et al., 2014; Minchin and Nunn, 2014; De Leij et al., 2017; Epstein and Smale, 2017).

Surveys of the rocky intertidal have identified that *U. pinnatifida* is more likely to occur, or is in higher abundance, where the native canopy is limited, which may be due to abiotic factors (Floc'h et al., 1996; Farrell, 2003; Cremades et al., 2006; Heiser et al., 2014; De Leij et al., 2017; Epstein and Smale, 2017). The recruitment of *U. pinnatifida* into mixed seaweed assemblages is often promoted by canopy disturbance; experimental clearances of native kelps within the intertidal have caused an increase in *U. pinnatifida* abundance and biomass (Floc'h et al., 1996; De Leij et al., 2017). A similar opportunistic life history has been recorded in subtidal habitats (Castric-Fey et al., 1993; Floc'h et al., 1996; Cremades et al., 2006; Heiser et al., 2014; De Leij et al., 2017), however, on artificial substrates, *U. pinnatifida* may be able to outcompete its native homologues (Farrell and Fletcher, 2006; Heiser et al., 2014). The impacts of these invasive species in adjacent subsidised ecosystems (by detritus) require further examination (Rodil et al., 2015a, 2015b).

The Pacific oyster, *M. gigas*, has now escaped from cultivation throughout Europe, being particularly dense in more sheltered and embayed locations (Lejart and Hily, 2005, 2011) where it can form a dense zone on the midshore of more sheltered shores and harbour walls, especially adjacent to oyster culture. There is no such native analogue in Europe, as the European oyster is primarily a subtidal species, occasionally being found in the low-intertidal zone. *M. gigas* can form dense reefs on top of sedimentary habitats in estuaries and bays (Lejart and Hily, 2005, 2011; Smaal et al., 2008; Herbert et al., 2016; Holm et al., 2016) displacing *Mytilus* spp. beds (Diederich, 2005; Fey et al., 2010; Holm et al., 2016). Again, surprisingly little work has been done on experimentally removing these oysters to test effects, but warming seas and dense source populations suggest expansion from frontier populations will continue at a rapid pace (Robins et al., 2017).

## 2.9 Biodiversity and Ecosystem Functioning

Considerable progress has been made in the study of biofilms (see Hill and Hawkins, 1991; Thompson et al., 1996 for methodological advances) since the late 1980s, including spatial and temporal patterns and ingestion by grazers (Hill and Hawkins, 1990), interaction with grazers (Hill and Hawkins, 1991; Jenkins and Hartnoll, 2001) and canopy-forming algae (Thompson et al., 2005), and providing cues for settling barnacles (Thompson et al., 1998). Their role in food provision for grazing gastropods (Thompson et al., 2000) and competition between grazers for microbial food has also been investigated (Boaventura et al., 2002b, 2003). Davies et al. (1992) showed that limpet mucus might also enhance available food as sedimenting phytoplankton would stick to mucus. This helped explain the greater diversity of diatoms in limpet guts than on rock chips viewed by scanning electron microscopy (Hill and Hawkins, 1991). Biofilms are energy rich (Davies et al., 1990; Davies and Blackwell, 2007), representing an important resource for grazers, and are likely to be formed at least in part through the movement of mucus trail-laying molluscs (Ng et al., 2013). The interactions between mucus-dominated biofilms and biotic processes such as settlement are poorly understood in the north-east Atlantic, although research elsewhere suggests significant activity (e.g., Johnson and Strathmann, 1989).

In an intensive experiment, Thompson et al. (2004) teased out the respective roles of bottom-up forcing by light and nutrients, and aphotic top–down control by limpet grazing in determining seasonal patterns of abundance of the photosynthetic component of biofilms. They emphasised the importance of 'lateral modification' by stressors in affecting total biomass and the relative importance of diatoms (dominating in late winter/early spring) and more tolerant cyanobacteria (dominating in the summer). This also showed the importance of grazing in controlling recruitment of fucoid germlings. A combination of grazing and desiccation stress may account for seasonal patterns of biofilm biomass throughout Europe (Thompson et al., 2000; Jenkins et al., 2001; Skov et al., 2010).

Rocky shore species and assemblages in the north-east Atlantic have been used as test systems to understand the relationship between biodiversity and ecosystem functioning in both field experiments (O'Connor and Crowe, 2005; Griffin et al., 2010) and laboratory mesocosms (Griffin et al., 2008, 2009; Russell et al., 2013). O'Connor and Crowe (2005) investigated the influence of diversity in an assemblage of three species of grazing gastropods (*P. ulyssiponensis*, *L. littorea* and *S. umbilicalis*) on productivity in shallow rock pools using algal biomass as a proxy. They showed strong idiosyncratic effects on algal biomass and productivity from removal of *P. ulyssiponensis*, known to be an important grazer in the region from previous experimental work (e.g., Boaventura et al., 2002a). Thus, identity of species was considered to be a primary factor influencing ecosystem functioning. Griffin et al. (2010) extended this work in rock pools denuded of all biota using the same three species of grazer to explore the influence of diversity during early succession. *P. ulyssiponensis* had the strongest effect in reducing algal growth, with little evidence of the influence of grazer diversity on algal biomass, diversity (both richness and evenness) and gross primary production. Griffin et al. (2009), however, showed in laboratory mesocosms an interaction between micro-habitat heterogeneity and grazer diversity on consumption of primary producers using three different species of grazer (*P. vulgata*, *L. littorea* and *S. umbilicalis*). On homogeneous substrata species identity was important, with individual grazers performing best on different preferred habitat types where they are more abundant. In heterogeneous habitats optimal conditions were present for all three species enabling spatial complementarity of feeding and the highest algal consumption rates by mixed assemblages.

At higher trophic levels, Griffin et al. (2008) explored the interaction between density and diversity of predatory crabs on consumption of lower trophic levels (grazers and space-occupying suspension feeders) in laboratory mesocosms. The effects of diversity were only apparent at high densities, with prey consumption by mixed assemblages of predators being double that of single species, reflecting functional niche differentiation between species. Arenas et al. (2009) examined the relationships between biomass, species richness, spatial aggregation and evenness and the productivity of natural macroalgal assemblages. They found the expected positive relationship between biomass and species richness. Additionally, results showed relationships between both spatial aggregation and evenness and some of the productivity-related variables analysed: assemblages with a higher degree of spatial aggregation had reduced light capture and photosynthetic efficiency, while increasing evenness increased maximum net primary productivity. Similar results were found by Rodriguez et al. (2016) with synthetic macroalgal communities. Prestes et al. (2017) highlighted the importance of species differentiation among algal species in their responses to environmental heterogeneity, thereby ensuring stability of standing stock of primary producers.

## 2.10 Progress and Knowledge Gaps

Since the 1990s (Hawkins et al., 1992) much progress has been made, especially in terms of geographical coverage. A field experimental approach has now spread throughout Europe, including work in Portugal, Spain, France, Ireland, the UK and Scandinavian countries, giving an increased understanding of interactions between species. Due to missionary work on experimental design by Underwood and Chapman through successive training courses and workshops, most work is now well designed and amenable to statistical analysis. Planned research over large scales testing the same hypotheses, using nested experimental or survey designs is still comparatively rare and difficult to fund (but see outputs from EUROROCK). It is hoped that this will not become even more difficult for UK scientists post-Brexit (Hawkins, 2017). The recent fashion for large-scale macro-ecological analyses of publicly available data or meta-analysis of published work, while contributing to high-profile papers and giving much insight (e.g., Burrows et al., 2011; Sunday et al., 2011, 2012; Poloczanska et al., 2013) has not encouraged small-scale in situ experimentation which is now viewed by some as rather old-fashioned. The way forward is to combine approaches and tools giving a broadscale perspective on patterns (such as remote sensing and macro-ecology) with regionally nested smaller-scale experiments on processes. The importance of the temporal as well as spatial scale of such experiments has also recently been highlighted (Jenkins and Uyà, 2016). Review of the literature has shown no change in the median duration of observation following experimental manipulation over the past three to four decades despite clear benefits from such an approach. Experimental work also needs to be set in a phylogeographic context to understand the origins and direction of movement of species, following deglaciation and more recent anthropogenic-driven climate change. Long-term and broadscale ecological data are also crucial to disentangle local and regional stressors from global change (Hawkins et al., 2008, 2009 2013, 2017; Mieszkowska et al., 2014b) and to understand how interactions may be altered in a rapidly changing part of the ocean (Burrows et al., 2011; Mieskowska and Sugden, 2016). Thus, there should still be a place for local experimentation, hopefully given elegance by insights from natural history (Hawkins et al., 2016b).

Techniques such as stable isotope analysis have been increasingly applied to rocky shores in the region (Dubois et al., 2007a, 2007b, 2007c; Lefebvre et al., 2009; Notman et al., 2016), giving insights into trophic interactions, especially the role of detritus. The consequences of changes in community structure for ecosystem functioning also need to be explored by measuring processes such as primary and secondary production, remineralisation, the fate of detritus and the role of import of pelagic productivity. Is any carbon produced in rocky coastal ecosystems sequestered in the long-term or does it just get stored in the short to medium term?

Recent work on micro-habitat modification of physical factors and consequences of microclimate for organisms at the individual level (Seabra et al., 2011; Lima et al., 2016) have major implications for understanding refugia from climate change-induced stress, as well as modification of outcomes of interactions and emerging community structure and functioning (Firth et al., 2016b). Some of these interactions will be modulated by the topographic template of the shore (Firth et al., 2016a); others will be modified by habitat-forming species (Moore et al., 2007; Marzinelli et al., 2012; Pocklington et al., 2017). In contrast, proliferating artificial structures, which often tend to have steeper gradients and less topographic complexity, will provide scant refuge from climate change (Chapman and Underwood, 2011; Firth et al., 2014a, 2014b, 2016a, 2016c), although porous structures do provide refuges (Sherrard et al., 2016).

The role of mobile predators, whether from on land (mainly birds) (Coleman et al., 1999, 2003; Kendall et al., 2004) or from the sea (fish, crabs, octopuses), must also not be ignored even though these are less tractable than sedentary invertebrates to experimentally manipulate. Recent work (Silva et al., 2008, 2010, 2014) has shown the important role crabs play in intertidal systems. With warming climate, herbivorous fish are likely to become much more important, especially in the south of the region (North Africa and Iberia) (Vergés et al., 2014; Franco et al., 2015; Gianni et al., 2017).

Despite the long history of both observational and experimental research in the region, host–parasite dynamics have been largely understudied. Some work has been done on species of commercial interest that are also found on rocky shores (e.g., the native oyster *Ostrea edulis*, Culloty et al., 1999, 2001, 2004 and the non-native oyster *M. gigas*, Holmes and Minchin, 1995), with comparatively little done on other species (but see James, 1968; Carrol et al., 1990; Elner and Raffaelli, 1980; Davies and Knowles, 2001; Powell and Rowley, 2008). Perhaps one of the best-studied systems is that of the trematode parasite *Echinostephilla patellae* (Lebour, 1911), which infects limpets and mussels as primary and secondary hosts and oystercatchers (*Haematopus ostralegus*) as definitive hosts (Kollien, 1996; Prinz et al., 2010a). Environmental factors affect the distribution and abundance of *E. patellae*, with a direct relationship between infection rates and increasing temperatures (Prinz et al., 2010b, 2011). Furthermore, prevalence in limpets appears to be higher in rock pools than emergent rock (Crewe, 1951; Thomas, 1965; Copeland et al., 1987, but see Kollien, 1996; Firth et al., 2017 for the opposite pattern) and is related to bird density and proximity to harbours (Copeland et al., 1987). More experimental work is required on host–parasite dynamics on rocky shores in the north-east Atlantic.

## 2.11 Overview and Synthesis

The north-east Atlantic region is an area where clades originating in the north Pacific (fucoids, balanoids, littorinids, thaids, laminarians) collide with clades from further south in the Atlantic (e.g., patellids, trochids, chthamalids). At high latitudes in the north, seaweeds dominate the midshore zone of all but the most exposed shores. Further south, midshore space-occupying invertebrates (mussels and barnacles) win, facilitated by grazing by patellid limpets that controls algal recruitment; propagule pressure is much less as fucoids become rarer, and juvenile growth is slower due to environmental stress, thereby reducing the probability of escapes from grazing (Figure 2.4) (Ferreira et al., 2014, 2015a, 2015b). Low on the shore seaweeds dominate space by forming algal turfs or kelp or fucoid canopies. These algae outpace the ability of grazing limpets to control them in the low-intertidal zone. *L. digitata* canopies can lead to rock covered by encrusting algae and sponges, facilitating limpets. If canopy is removed, then colonising ephemeral algae and turf-forming algae swamp the limpets. There is usually too much water movement immediately either side of low water for effective foraging by sea urchins. *Psammechinus miliaris* and *Echinus esculentus* only appear in the subtidal, and *Paracentrotus lividus* is confined to refuges in burrows relying mainly on the drift of food (Benedetti-Cecchi and Cinelli, 1995;

Boudouresque and Verlaque, 2007; Jacinto and Cruz, 2012).

High on the shore, physical factors dominate. At high latitudes in the north of the Atlantic, ephemeral algae are present all year round. Further south they are only present in the winter, dying-off in the summer. Grazing has limited effects, only occurring around refuges that littorinids maintain (Stafford and Davies, 2005; Skov et al., 2010, 2011).

Patterns are also strongly modified by mesoscale processes driven by upwelling that influences nutrient and larval supply (North Africa, Iberia) and coastal configuration, where embayed versus headlands also strongly influence larval supply (France northwards). In high-recruitment areas, interactions can be intense between space-occupying species, also driving predator abundance (e.g., dog whelks). Connell's (1961a) classic paper on competition was possible on the Isle of Cumbrae because space was almost saturated; elsewhere lower larval supply would have created less intense interactions, as shown by Gordon and Knights (2017) in Plymouth.

The north-east Atlantic has faster rates of warming than any other ocean, although the region south of Greenland and Iceland is undergoing cooling due to a climate-driven slowdown in the Atlantic meridional overturning circulation, causing a weakening in the Gulf Stream (Rahmstorf et al., 2015).

Species are responding to rapid alterations in the marine climate by adapting or exhibiting range shifts, or by becoming locally extinct. There is a high degree of spatial and temporal heterogeneity in the resultant impacts on marine communities due to the idiosyncratic responses of individual species. Warming seas have resulted in biogeographic range shifts of intertidal and subtidal species in coastal waters of the north-east Atlantic. The leading range edges of Lusitanian species are expanding, while the trailing edges of boreal species are retracting to higher latitudes, but with some cold-water species showing surprising resilience (Southward et al., 1995; Mieszkowska et al., 2006, 2014b; Lima et al., 2007; Hawkins et al., 2008, 2009; Wethey and Woodin, 2008; Mieszkowska and Sugden, 2016). In addition to changes in the distribution of species, community structure is also altering as species dominance and interactions change (Poloczanska et al., 2008; Hawkins et al., 2008, 2009; Mieszkowska et al., 2014b).

In a warming world the midshore of France and the British Isles are likely to show much less cover by large canopy-forming fucoids as harsher warmer, drier and stormier conditions coupled with increased grazing pressure from more grazing species reduces the probability of fucoids recruiting to form adult populations. Lowshore kelp forests will likely change with less *L. digitata* and *A. esculenta* and more *S. polyschides*. The late autumn to early spring window of dense ephemeral algal growth high on the shore (Hawkins and Hartnoll, 1983a) will also constrict, except in the north and in extreme exposure. These changes will have consequences for biodiversity (Thompson et al., 1996; Smale et al., 2013; Teagle et al., 2017) and productivity (Hawkins et al., 1992) – particularly the decrease in export of algal detritus (Notman et al., 2016). More shores will become dominated by suspension-feeding barnacles and mussels. Thus, there will be switches on many mid-latitude shores as many become net importers rather than exporters of energy (Hawkins et al., 2008, 2009).

## Dedication

This chapter is dedicated to pioneering giants of rocky shore research in the north-east: Edouard Fischer-Piette, Harry Hatton, Norman Jones, Dennis Crisp, Alan Southward and Jack Lewis. Most of the authors derive directly or indirectly from the J. H. Orton clade originating at the Marine Biological Association, Plymouth, and the Marine Biological Station, Port Erin.

### REFERENCES

Åberg, P. (1992). Size based demography of the seaweed *Ascophyllum nodosum* in stochastic environments. *Ecology*, **73**, 1488–501.

Åberg, P. and Pavia, H. (1997). Temporal and multiple scale spatial variation in juvenile and adult abundance of the brown alga *Ascophyllum nodosum*. *Marine Ecology Progress Series*, **158**, 111–19.

Alvarez, I., Gomez-Gesteira, M., deCastro, M., Lorenzo, M. N., Crespo, A. J. C. and Dias, J. M. (2011). Comparative analysis of upwelling influence between the western and northern coast of the Iberian Peninsula. *Continental Shelf Research*, **31**, 388–99.

Anadon, N. (1981). Contribution to the knowledge of the benthic fauna bentonica de Ria de Vigo III. Study of the reefs of *Sabellaria alveolata* (L.) (*Polchaeta, Sedentaria*). *Scientia Marina*, **45**, 105–22.

Araújo, R., Sousa-Pinto, I., Bárbara, I. and Quintino, V. (2006). Macroalgal communities of intertidal rockpools in the northwest coast of Portugal. *Acta Oecologica*, **30**, 192–202.

Araújo, R. M., Assis, J., Aguillar, R. et al. (2016). Status, trends and drivers of kelp forests in Europe: an expert assessment. *Biodiversity and Conservation*, **25**, 1319–48.

Ardré, F. (1969). *Contribution a l'étude des algues marines du Portugal: I. La flore*. Instituto Botânico de Faculdade de Ciências, Lissabon.

Ardré, F. (1971). *Contribution a l'étude des algues marines du Portugal: II. Ecologie et chorologie*. Centre d'Etudes et de Recherches Scientifiques, Biarritz.

Arenas, F., Rey, F. and Pinto, I. S. (2009). Diversity effects beyond species richness: evidence from intertidal macroalgal assemblages. *Marine Ecology Progress Series*, **381**, 99–108.

Arrontes, J. (1993). Nature of the distributional boundary of *Fucus serratus* on the north shore of Spain. *Marine Ecology Progress Series*, **93**, 183–93.

Arrontes, J., Arenas, F., Fernandez, C. et al. (2004). Effect of grazing by limpets on mid-shore species assemblages in northern Spain. *Marine Ecology Progress Series*, **277**, 117–33.

Audouin, J. V. and Edwards, H. M. (1833). Classification des Annélides, et Description de celles qui habitent les côtes de la France. *Annales des sciences naturelles: comprenant La physiologie animale et végétale, l'anatomie comparée des deux règnes, la zoologie, la botanique, la minéralogie et la géologie*, **28**, 187–247.

Ávila, S. P., Madeira, P., Rebelo, A. C. et al. (2015). *Phorcus sauciatus* (Koch, 1845) (*Gastropoda*: *Trochidae*) in Santa Maria, Azores archipelago: the onset of a biological invasion. *Journal of Molluscan Studies*, **81**, 516–21.

Ayata, S.-D., Ellien, C., Dumas, F., Dubois, S. and Thiébaut, E. (2009). Modelling larval dispersal and settlement of the reef-building polychaete *Sabellaria alveolata*: role of hydroclimatic processes on the sustainability of biogenic reefs. *Continental Shelf Research*, **29**, 1605–23.

Bakun, A. (1990). Global climate change and intensification of coastal ocean upwelling. *Science*, **247**, 198–201.

Ballantine, W. J. (1961). A biologically defined exposure scale for the comparative description of rocky shores. *Field Studies*, **1**, 1–19.

Barnes, H. and Barnes, M. (1963). *Elminius modestus* Darwin: further European records. *Progress in Oceanography*, **3**, 23–30.

Barnes, H. and Powell, H. T. (1950). The development, general morphology and subsequent elimination of barnacle populations, *Balanus crenatus* and *B. balanoides*, after a heavy initial settlement. *Journal of Animal Ecology*, **19**, 175–9.

Barnes, M. (1996). Pedunculate cirripedes of the genus *Pollicipes*. *Oceanography and Marine Biology: An Annual Review*, **34**, 303–94.

Barnett, B. E. (1979). A laboratory study of predation by the dog-whelk *Nucella lapillus* on the barnacles *Elminius modestus* and *Balanus balanoides*. *Journal of the Marine Biological Association of the United Kingdom*, **59**, 299–306.

Benedetti-Cecchi, L. and Cinelli, F. (1995). Habitat heterogeneity, sea urchin grazing and the distribution of algae in littoral rockpools on the west coast of Italy (western Mediterranean). *Marine Ecology Progress Series*, **126**, 203–12.

Bennell, S. J. (1981). Some observations on the littoral barnacle populations of North Wales. *Marine Environmental Research*, **5**, 227–40.

Bertocci, I., Badalamenti, F., Lo Brutto, S. et al. (2017). Reducing the data-deficiency of threatened European habitats: spatial variation of sabellariid worm reefs and associated fauna in the Sicily Channel, Mediterranean Sea. *Marine Environmental Research*, **130**, 325–37.

Birchenough, S. N. R., Reiss, H., Degraer, S. et al. (2015). Climate change and marine benthos: a review of existing research and future directions in the North Atlantic. *Wiley Interdisciplinary Reviews: Climate Change*, **6**, 203–23.

Blanton, J. O., Tenore, K. R., Castillejo, F., Atkinson, L. P., Schwing, F. B. and Lavin, A. (1987). The relationship of upwelling to mussel production in the rias on the western coast of Spain. *Journal of Marine Research*, **45**, 497–511.

Boaventura, D., Alexander, M., Della Santina, P. et al. (2002a). The effects of grazing on the distribution and composition of low-shore algal communities on the central coast of Portugal and on the southern coast of Britain. *Journal of Experimental Marine Biology and Ecology*, **267**, 185–206.

Boaventura, D., Da Fonseca, L. C. and Hawkins, S. J. (2003). Size matters: competition within populations of the limpet *Patella depressa*. *Journal of Animal Ecology*, **72**, 435–46.

Boaventura, D., Ré, P., Cancela da Fonseca, L. and Hawkins, S. J. (2002b). Intertidal rocky shore communities of the continental Portuguese coast: analysis of distribution patterns. *Marine Ecology*, **23**, 69–90.

Boaventura, D. M., Cancela da Fonseca, L. and Hawkins, S. J. (2002c). Analysis of competetive interactions between the limpets *Patella depressa* Pennant and *Patella vulgata* L. in the northern coast of Portugal. *Journal of Experimental Marine Biology and Ecology*, **271**, 171–88.

Borges, C. D. G., Doncaster, C. P., MacLean, M. A. and Hawkins, S. J. (2015). Broad-scale patterns of sex ratios in *Patella* spp.: a comparison of range edge and central range populations in the British Isles and Portugal. *Journal of the Marine Biological Association of the United Kingdom*, **95**, 1141–53.

Borges, C. D. G., Hawkins, S. J., Crowe, T. P. and Doncaster, C. P. (2016). The influence of simulated exploitation on *Patella vulgata* populations: protandric sex change is size-dependent. *Ecology and Evolution*, **2**, 514–31.

Børgesen, F. (1908). The Algae-Vegetation of the Faeröese Coasts. In *Botany of the Faeroes*, vol. 3. Nordisk Forlag, Copenhagen, pp. 339–532.

Børgesen, F. S. and Jónsson, H. (1908). The Distribution of the Marine Algae of the Arctic Sea, and of the Northernmost Part of the Atlantic. In E. Warming, ed. *Botany of the Faeroes*, vol. 3. Nordisk Forlag, Copenhagen, pp. 1–28.

Boudouresque, C. F. and Verlaque, M. (2007). Ecology of *Paracentrotus lividus*. In J. M. Lawrence, ed. *Edible Sea Urchins: Biology and Ecology*, Elsevier, Amsterdam, pp. 243–85

Bowman, R. S. and Lewis, J. R. (1977). Annual fluctuations in the recruitment of *Patella vulgata* L. *Journal of the Marine Biological Association of the United Kingdom*, **57**, 793–815.

Bracewell, S. A., Robinson, L. A., Firth, L. B. and Knights, A. M. (2013). Predicting free-space occupancy on novel artificial structures by an invasive intertidal barnacle using a removal experiment. *PLoS ONE*, **8**, e74457.

Bracewell, S. A., Spencer, M., Marrs, R. H., Iles, M. and Robinson, L. A. (2012). Cleft, crevice, or the inner thigh: 'another place' for the establishment of the invasive barnacle *Austrominius modestus* (Darwin, 1854). *PLoS ONE*, **7**, e48863.

Breeman, A. M. (1988). Relative importance of temperature and other factors in determining geographic boundaries of seaweeds: experimental and phenological evidence. *Helgoländer Meeresuntersuchungen*, **42**, 199–241.

Broitman, B. R., Mieszkowska, N., Helmuth, B. and Blanchette, C. A. (2008). Climate and recruitment of rocky shore intertidal invertebrates in the Eastern North Atlantic. *Ecology*, **89**, S81–S90.

Browne, M. A. and Chapman, M. G. (2011). Ecologically informed engineering reduces loss of intertidal biodiversity on artificial shorelines. *Environmental Science & Technology*, **45**, 8204–207.

Browne, M. A. and Chapman, M. G. (2014). Mitigating against the loss of species by adding artificial intertidal pools to existing seawalls. *Marine Ecology Progress Series*, **497**, 119–29.

Burrows, E. M. and Lodge, S. (1951). Autecology and the species problem in *Fucus*. *Journal of the Marine Biological Association of the United Kingdom*, **30**, 161–76.

Burrows, M. T. and Hawkins, S. J. (1998). Modelling patch dynamics on rocky shores using deterministic cellular automata. *Marine Ecology Progress Series*, **167**, 1–13.

Burrows, M. T., Hawkins, S. J. and Southward, A. J. (1992). A comparison of reproduction in co-occurring chthamalid barnacles, *Chthamalus stellatus* (Poli) and *Chthamalus montagui* Southward. *Journal of Experimental Marine Biology and Ecology*, **160**, 229–49.

Burrows, M. T., Hawkins, S. J. and Southward, A. J. (1999). Larval development of the intertidal barnacles *Chthamalus stellatus* and *Chthamalus montagui*. *Journal of the Marine Biological Association of the United Kingdom*, **79**, 93–101.

Burrows, M. T. and Hughes, R. N. (1989). Natural foraging of the dogwhelk, *Nucella lapillus* (Linnaeus); the weather and whether to feed. *Journal of Molluscan Studies*, **55**, 286–95.

Burrows, M. T., Jenkins, S. R., Robb, L. and Harvey, R. (2010). Spatial variation in size and density of adult and post-settlement *Semibalanus balanoides*: effects of oceanographic and local conditions. *Marine Ecology Progress Series*, **398**, 207–19.

Burrows, E. and Lodge, S. (1950). A note on the inter-relationships of *Patella*, *Balanus* and *Fucus* on a semi-exposed coast. *Reports of the Port Erin Marine Biological Station*, **62**, 30–4.

Burrows, M. T., Schoeman, D. S., Buckley, L. B. et al. (2011). The pace of shifting climate in marine and terrestrial ecosystems. *Science*, **334**, 652–5.

Bussell, J. A., Lucas, I. A. N. and Seed, R. (2007). Patterns in the invertebrate assemblage associated with *Corallina officinalis* in tide pools. *Journal of the Marine Biological Association of the United Kingdom*, **87**, 383–8.

Carrol, H., Montgomery, W. I. and Hanna, R. E. B. (1990). Dispersion and abundance of *Maritrema arenaria* in *Semibalanus balanoides* in North-East Ireland. *Journal of Helminthology*, **64**, 151.

Castric-Fey, A., Beaupoil, C., Bouchain, J., Pradier, E. and L'Hardy-Halos, M. T. (1999). The introduced alga *Undaria pinnatifida* (Laminariales, Alariaceae) in the rocky shore ecosystem of the St Malo area: growth rate and longevity of the sporophyte. *Botanica Marina* **42**, 71–82.

Castric-Fey, A., Girard, A. and L'Hardy-Halos, M. T. (1993). The distribution of *Undaria pinnatifida* (Phaeophyceae, Laminariales) on the Coast of St. Malo (Brittany, France). *Botanica Marina* **36**, 351–8.

Cervin, G., Åberg, P. and Jenkins, S. R. (2005). Small-scale disturbance in a stable canopy dominated community: implications for macroalgal recruitment and growth. *Marine Ecology Progress Series*, **305**, 31–40.

Chan, B. K. K. (2000). Diurnal physico-chemical variations in Hong Kong rockpools. *Asian Marine Biology*, **17**, 43–54.

Chapman, A. R. O. (1995). Functional ecology of fucoid algae: twenty-three years of progress. *Phycologia*, **34** (1), 1–32.

Chapman, M. G. and Blockley, D. J. (2009). Engineering novel habitats on urban infrastructure to increase intertidal biodiversity. *Oecologia*, **161**, 625–35.

Chapman, M. G. and Underwood, A. J. (2011). Evaluation of ecological engineering of "armoured" shorelines to improve their value as habitat. *Journal of Experimental Marine Biology and Ecology*, **400**(1–2), 302–13.

Coleman, R. A., Goss-Custard, J. D., Durell, S. and Hawkins, S. J. (1999). Limpet *Patella* spp. consumption by oystercatchers *Haematopus ostralegus*: a preference for solitary prey items. *Marine Ecology Progress Series*, **183**, 253–61.

Coleman, R. A., Salmon, N. and Hawkins, S. (2003). Sub-dispersive human disturbance of foraging oystercatchers *Haematopus ostralegus*. *Ardea*, **91**, 263–8.

Coleman, R. A., Underwood, A. J., Benedetti-Cecchi, L. et al. (2006). A continental scale evaluation of the role of limpet grazing on rocky shores. *Oecologia*, **147**, 556–64.

Connell, J. H. (1961a). Effects of competition, predation by *Thais lapillus*, and other factors on natural populations of the barnacle *Balanus balanoides*. *Ecological Monographs*, **31**, 61–104.

Connell, J. H. (1961b). The influence of interspecific competition and other factors on the distribution of the barnacle *Chthamalus stellatus*. *Ecology*, **42**, 710–23.

Connolly, S. R., Menge, B. A. and Roughgarden, J. (2001). A latitudinal gradient in recruitment of intertidal invertebrates in the northeast Pacific Ocean. *Ecology*, **82**, 1799–813.

Connolly, S. R. and Roughgarden, J. (1998). A latitudinal gradient in Northeast Pacific intertidal community structure: evidence for an oceanographically based synthesis of marine community theory. *The American Naturalist*, **151**, 311–26.

Conway, E. (1946). Browsing of *Patella*. *Nature*, **158**, 752–2.

Copeland, M. R., Montgomery, W. I. and Hanna, R. E. B. (1987). Ecology of a digenean infection, *Cercaria patellae* in *Patella vulgata* near Portavogie Harbour, Northern Ireland. *Journal of Helminthology*, **61**, 315.

Côrte-Real, H. B. S. M., Hawkins, S. J. and Thorpe, J. P. (1996). Population differentiation and taxonomic status of the exploited limpet *Patella candei* in the Macaronesian islands (Azores, Madeira, Canaries). *Marine Biology*, **125**, 141–52.

Cremades, J., Freire, O. and Peteiro, C. (2006). Biología, distribución e integración del alga alóctona *Undaria pinnatifida* (Laminariales, Phaeophyta) en las comunidades bentónicas de las costas de Galicia (NW de la Península Ibérica). *Anales del Jardín Botánico de Madrid*, **63**.

Crewe, W. (1951). The occurrence of *Cercaria patellae* lebour (Trematoda) and its effects on the host; with notes on some other helminth parasites of british limpets. *Parasitology*, **41**, 15.

Crickenberger, S. and Wethey, D. S. (2017). Reproductive physiology, temperature and biogeography: the role of fertilization in determining the distribution of the barnacle *Semibalanus balanoides*. *Journal of the Marine Biological Association of the United Kingdom*, **98**, 1411–24.

Crisp, D. J. (1955). The behaviour of barnacle cyprids in relation to water movement over a surface. *The Journal of Experimental Biology*, **32**, 569–90.

Crisp, D. J. (1956). The Intertidal Zoology of Rockall. In E. Fischer-Piette, ed. *Rockall*. Geoffrey Bles, London, pp. 177–9.

Crisp, D. J. (1958). The spread of *Elminius Modestus* Darwin in North-West Europe. *Journal of the Marine Biological Association of the United Kingdom*, **37**, 483.

Crisp, D. J. and Fischer-Piette, E. (1959). Repartition des principales especes interastidales de la Cote Atlantique Francaise en 1954–1955. *Annales de l'Institut Océanographique, Monaco*, **36**, 276–381.

Crisp, D. J. and Knight-Jones, E. W. (1953). The mechanism of aggregation in barnacle populations. *Journal of Animal Ecology*, **22**, 360–2.

Crisp, D. J. and Southward, A. J. (1958). The distribution of intertidal organisms along the coasts of the English Channel. *Journal of the Marine Biological Association of the United Kingdom*, **37**, 157–203.

Critchley, A. T., Farnham, W. F. and Morrell, S. L. (1983). A chronology of new European sites of attachment for the invasive brown alga, *Sargassum muticum*, 1973–1981. *Journal of the Marine Biological Association of the United Kingdom*, **63**, 799.

Crowe, T., Frost, N. and Hawkins, S. (2011). Interactive effects of losing key grazers and ecosystem engineers vary with environmental context. *Marine Ecology Progress Series*, **430**, 223–34.

Cruz, T. (1999). Settlement patterns of *Chthamalus* spp. at Praia da Oliveirinha (SW Portugal). *Acta Oecologica*, **20**, 285–7.

Cruz, T., Castro, J. J., Delany, J. et al. (2005). Tidal rates of settlement of the intertidal barnacles *Chthamalus stellatus* and *Chthamalus montagui* in Western Europe: the influence of the night/day cycle. *Journal of Experimental Marine Biology and Ecology*, **318**, 51–60.

Culloty, S. C., Cronin, M. A. and Mulcahy, M. F. (2001). An investigation into the relative resistance of Irish flat oysters *Ostrea edulis* L. to the parasite *Bonamia ostreae*. *Aquaculture*, **199**, 229–44.

Culloty, S. C., Cronin, M. A. and Mulcahy, M. F. (2004). Potential resistance of a number of populations of the oyster *Ostrea edulis* to the parasite *Bonamia ostreae*. *Aquaculture*, **237**, 41–58.

Culloty, S. C., Novoa, B., Pernas, M. et al. (1999). Susceptibility of a number of bivalve species to the protozoan parasite *Bonamia ostreae* and their ability to act as vectors for this parasite. *Diseases of Aquatic Organisms*, **37**, 73–80.

Cunningham, P. N., Hawkins, S. J., Jones, H. D. and Burrows, M. T. (1984). *The Geographical Distribution of* Sabellaria alveolata *(L.) in England, Wales and Scotland, with investigations into the community structure of, and the effects of trampling on* Sabellaria alveolata *colonies*, Nature Conservancy Council, Peterborough.

Davies, A. J. and Johnson, M. P. (2006). Coastline configuration disrupts the effects of large-scale climatic forcing, leading to divergent temporal trends in wave exposure. *Estuarine, Coastal and Shelf Science*, **69**, 643-8.

Davies, M. S. and Blackwell, J. (2007). Energy saving through trail following in a marine snail. *Proceedings of the Royal Society Series B: Biological Sciences*, 274, 1233–6.

Davies, M. S., Hawkins, S. J. and Jones, H. D. (1990). Mucus production and physiological energetics in *Patella vulgata* L. *Journal of Molluscan Studies*, **56**, 499–503.

Davies, M. S., Hawkins, S. J. and Jones, H. D. (1992). Pedal mucus and its influence on the microbial food supply of two intertidal gastropods, *Patella vulgata* L. and *Littorina littorea* L. *Journal of Experimental Marine Biology and Ecology*, **161**, 57–77.

Davies, M. S. and Knowles, A. J. (2001). Effects of trematode parasitism on the behaviour and ecology of a common marine snail (*Littorina littorea* (L.)). *Journal of Experimental Marine Biology and Ecology*, **260**, 155–67.

Davies, P. S. (1969). Physiological ecology of *Patella*. III. Desiccation effects. *Journal of the Marine Biological Association of the United Kingdom*, **49**, 291.

De Leij, R., Epstein, G., Brown, M. P. and Smale, D. A. (2017). The influence of native macroalgal canopies on the distribution and abundance of the non-native kelp *Undaria pinnatifida* in natural reef habitats. *Marine Biology*, **164**, 156.

Delany, J., Myers, A. and McGrath, D. (2002). *A Comparison of the Interactions of the Limpets* Patella vulgata *and* Patella ulyssiponensis *with Crustose Coralline Algae*, Royal Irish Academy, Dublin.

Delany, J., Myers, A. A. and McGrath, D. (1998). Recruitment, immigration and population structure of two coexisting limpet species in mid-shore tidepools, on the West Coast of Ireland. *Journal of Experimental Marine Biology and Ecology*, **221**, 221–30.

Dias, M., Roma, J., Fonseca, C. et al. (2016). Intertidal pools as alternative nursery habitats for coastal fishes. *Marine Biology Research*, **12**, 331–44.

Diederich, S. (2005). Differential recruitment of introduced Pacific oysters and native mussels at the North Sea coast: coexistence possible? *Journal of Sea Research*, **53**, 269–81.

Donald, K. M., Preston, J., Williams, S. T. et al. (2012). Phylogenetic relationships elucidate colonization patterns in the intertidal grazers *Osilinus philippi*, 1847 and *Phorcus risso*, 1826 (Gastropoda: Trochidae) in the northeastern Atlantic Ocean and Mediterranean Sea. *Molecular Phylogenetics and Evolution*, **62**, 35–45.

Duarte, L., Viejo, R. M., Martínez, B., deCastro, M., Gómez-Gesteira, M. and Gallardo, T. (2013). Recent and historical range shifts of two canopy-forming seaweeds in North Spain and the link with trends in sea surface temperature. *Acta Oecologica*, **51**, 1–10.

Dubois, S., Commito, J. A., Olivier, F. and Retière, C. (2006). Effects of epibionts on *Sabellaria alveolata* (L.) biogenic reefs and their associated fauna in the Bay of Mont Saint-Michel. *Estuarine, Coastal and Shelf Science*, **68**, 635–46.

Dubois, S., Jean-Louis, B., Bertrand, B. and Lefebvre, S. (2007a). Isotope trophic-step fractionation of suspension-feeding species: implications for food partitioning in coastal ecosystems. *Journal of Experimental Marine Biology and Ecology*, **351**, 121–8.

Dubois, S., Marin-Léal, J. C., Ropert, M. and Lefebvre, S. (2007b). Effects of oyster farming on macrofaunal assemblages associated with *Lanice conchilega* tubeworm populations: a trophic analysis using natural stable isotopes. *Aquaculture*, **271**, 336–49.

Dubois, S., Orvain, F., Marin-Léal, J. C., Ropert, M. and Lefebvre, S. (2007c). Small-scale spatial variability of food partitioning between cultivated oysters and associated suspension-feeding species, as revealed by stable isotopes. *Marine Ecology Progress Series*, **336**, 151–60.

Dubois, S., Retière, C. and Olivier, F. (2002). Biodiversity associated with *Sabellaria alveolata* (Polychaeta: Sabellariidae) reefs: effects of human disturbances. *Journal of the Marine Biological Association of the United Kingdom*, **82**, 817–26.

Ebling, F. J., Sloane, J. F., Kitching, J. A. and Davies, H. M. (1962). The ecology of Lough Ine: XII. The distribution and characteristics of *Patella* species. *Journal of Animal Ecology*, **31**, 457–70.

Elner, R. W. and Raffaelli, D. G. (1980). Interactions between two marine snails, *Littorina rudis* Maton and *Littorina nigrolineata* Gray, a predator, *Carcinus maenas* (L.), and a parasite, *Microphallus similis* Jägerskiold. *Journal of Experimental Marine Biology and Ecology*, **43**, 151–60.

Emery, K. O. and Kuhn, G. G. (1982). Sea cliffs: their processes, profiles, and classification. *Geological Society of America Bulletin*, **93**, 644.

Engelen, A. H., Serebryakova, A., Ang, P. et al. (2015). Circumglobal invasion by the brown seaweed *Sargassum muticum*. *Oceanography and Marine Biology: An Annual Review*, **53**, 81–126.

Epstein, G. and Smale, D. A. (2017). *Undaria pinnatifida*: a case study to highlight challenges in marine invasion ecology and management. *Ecology and Evolution*, **7**, 8624–42.

Evans, A. J., Firth, L. B., Hawkins, S. J., Morris, E. S., Goudge, H. and Moore, P. J. (2016). Drill-cored rockpools: an effective method of ecological enhancement on artificial structures. *Marine and Freshwater Research*, **67**, 123.

Evans, A. J., Garrod, B., Firth, L. B. et al. (2017). Stakeholder priorities for multi-functional coastal defence developments and steps to effective implementation. *Marine Policy*, **75**, 143–55.

Faria, J., Martins, G. M., Pita, A. et al. (2017). Disentangling the genetic and morphological structure of *Patella candei* complex in Macaronesia (NE Atlantic). *Ecology and Evolution*, **7**, 6125–40.

Faria, J., Pita, A., Martins, G. M. et al. (2018). Inbreeding in the exploited limpet *Patella aspera* across the Macaronesia archipelagos (NE Atlantic): implications for conservation. *Fisheries Research*, **198**, 180–8.

Farrell, P. (2003). *A Study of the Recently Introduced Macroalga* Undaria pinnatifida *(Harvey) Suringar (Phaeophyceae Laminariales) in the British Isles*, University of Portsmouth, Portsmouth.

Farrell, P. and Fletcher, R. L. (2006). An investigation of dispersal of the introduced brown alga *Undaria pinnatifida* (Harvey) Suringar and its competition with some species on the man-made structures of Torquay Marina (Devon, UK). *Journal of Experimental Marine Biology and Ecology*, **334**, 236–43.

Ferreira, J. G., Arenas, F., Martínez, B., Hawkins, S. J. and Jenkins, S. R. (2014). Physiological response of fucoid algae to environmental stress: comparing range centre and southern populations. *New Phytologist*, **202**, 1157–72.

Ferreira, J. G., Hawkins, S. and Jenkins, S. R. (2015a). Physical and biological control of fucoid recruitment in range edge and range centre populations. *Marine Ecology Progress Series*, **518**, 85–94.

Ferreira, J. G., Hawkins, S. J. and Jenkins, S. R. (2015b). Patterns of reproductive traits of fucoid species in core and marginal populations. *European Journal of Phycology*, **50**, 457–68.

Fey, F., Dankers, N., Steenbergen, J. and Goudswaard, K. (2010). Development and distribution of the non-indigenous Pacific oyster (*Crassostrea gigas*) in the Dutch Wadden Sea. *Aquaculture International*, **18**, 45–59.

Figueiras, F. G., Labarta, U. and Fernández Reiriz, M. J. (2002). Coastal upwelling, primary production and mussel growth in the Rías Baixas of Galicia. *Hydrobiologia*, **484**, 121–31.

Firth, L. B. and Crowe, T. P. (2008). Large-scale coexistence and small-scale segregation of key species on rocky shores. *Hydrobiologia*, **614**, 233–41.

Firth, L. B. and Crowe, T. P. (2010). Competition and habitat suitability: small-scale segregation underpins large-scale coexistence of key species on temperate rocky shores. *Oecologia*, **162**, 163–74.

Firth, L. B., Crowe, T. P., Moore, P., Thompson, R. C. and Hawkins, S. J. (2009). Predicting impacts of climate-induced range expansion: an experimental framework and a test involving key grazers on temperate rocky shores. *Global Change Biology*, **15**, 1413–22.

Firth, L. B., Grant, L. M., Crowe, T. P. et al. (2017). Factors affecting the prevalence of the trematode parasite *Echinostephilla patellae* (Lebour, 1911) in the limpet *Patella vulgata* (L.). *Journal of Experimental Marine Biology and Ecology*, **492**, 99–104.

Firth, L. B., Mieszkowska, N., Grant, L. M. et al. (2015). Historical comparisons reveal multiple drivers of decadal change of an ecosystem engineer at the range edge. *Ecology and Evolution*, **5**, 3210–22.

Firth, L. B., Mieszkowska, N., Thompson, R. C. and Hawkins, S. J. (2013a). Climate change and adaptational impacts in coastal systems: the case of sea defences. *Environmental Science: Processes & Impacts*, **15**, 1665.

Firth, L. B., Thompson, R. C., White, R. et al. (2013b). Promoting biodiversity on artificial structures: can natural habitats be replicated? *Diversity and Distributions* **19**, 1275–83.

Firth, L. B., Schofield, M., White, F. J., Skov, M. W. and Hawkins, S. J. (2014a). Biodiversity in intertidal rockpools: informing engineering criteria for artificial habitat enhancement in the built environment. *Marine Environmental Research*, **102**, 122–30.

Firth, L. B., Thompson, R. C., Bohn, K. et al. (2014b). Between a rock and a hard place: environmental and engineering considerations when designing coastal defence structures. *Coastal Engineering*, **87**, 122–35.

Firth, L. B., Knights, A. M., Bridger, D. et al. (2016a). Ocean Sprawl: Challenges and Opportunities for Biodiversity Management in a Changing World. In *Oceanography and Marine Biology*. CRC Press, Boca Raton, FL, **54**, pp. 201–78.

Firth, L. B., White, F. J., Schofield, M. et al. (2016b). Facing the future: the importance of substratum features for ecological engineering of artificial habitats in the rocky intertidal. *Marine and Freshwater Research*, **67**, 131.

Firth, L. B., Browne, K. A., Knights, A. M., Hawkins, S. J. and Nash, R. (2016c). Eco-engineered rockpools: a concrete solution to biodiversity loss and urban sprawl in the marine environment. *Environmental Research Letters*, **11**, 094015.

Firth, L. B. and Williams, G. A. (2009). The influence of multiple environmental stressors on the limpet *Cellana toreuma* during the summer monsoon season in Hong Kong. *Journal of Experimental Marine Biology and Ecology*, **375**, 70–5.

Fischer-Piette, E. (1929). Recherches de bionomie et d'océanographie littorales sur la Rance et le littoral de la Manche. *Annales de l'Institut Océanographique, Monaco*, **5**, 201–429.

Fischer-Piette, E. (1936). Études sur la biogéographie intercotidale des deux rives de la Manche. *Journal of the Linnean Society of London, Zoology*, **40**, 181–272.

Fischer-Piette, E. (1948). Sur les éléments de prospérité des Patelles et sur leur spécificité. *Journal de Conchyliologie*, **88**, 45–96.

Fischer-Piette, E. (1955). Repartition, le long des cotes septentrionales de l'Espagne, des principales especes peuplant les rochers intercotidaux. *Annual Insitutional Ocenaographique Monaco*, **31**, 37–124.

Fischer-Piette, E. (1957). Sur des déplacements de frontièrs biogéographies, observés au long des côtes ibériques dans le domaine intercotidal. *Publicaciones del Instituto de Biologia Aplicada*, **XXVI**, 35–40.

Fischer-Piette, E. (1958). Sur l'écologie intercotidale oust-ibérique. *Comptes Rendus de l'Académie des Sciences*, **246**, 1301–303.

Fischer-Piette, E. (1963). La distribution des principaux organismes intercotidaux Nord- Ibériques en 1954–55. *Annales de l'Intitut Océangraphique, Mónaco*, **XL**, 165–312.

Fischer-Piette, E. and Prenant, M. (1957). Quelques données ecologiques sur les cirripedes intercotidaux du Portugal, de l'Espagne du sud et du nord du Maroc. *Bulletin du Centre d'Etudes et Recherches Scientifiques, Biarritz*, **1**, 361–8.

Fletcher, R. L. and Farrell, P. (1999) Introduced brown algae in the North East Atlantic, with particular respect to *Undaria pinnatifida* (Harvey) Suringar. *Helgolander Meeresuntersuchungen*, **52**, 259–75.

Floc'h, J.-Y., Pajot, R. and Mouret, V. (1996). *Undaria pinnatifida* (Laminariales, Phaeophyta) 12 years after its introduction into the Atlantic Ocean. *Hydrobiologia*, **326**, 217–22.

Floc'h, J. Y., Pajot, R. and Wallentinus, I. (1991). The Japanese brown alga *Undaria pinnatifida* on the coast of France and its possible establishment in European waters. *ICES Journal of Marine Science*, **47**, 379–90.

Forbes, E. (1858). The distribution of marine life, illustrated chiefly by fishes and molluscs and radiata. In A. K. Johnston, ed. *Physical Atlas*. William Blackwood & Sons, Edinburgh, pp. 99–101.

Foster, B. A. (1971). Desiccation as a factor in the intertidal zonation of barnacles. *Marine Biology*, **8**, 12–29.

Fraga, F. (1981). Upwelling off the Galician coast, Northwest Spain. In F. A. Richards, ed. *Coastal Upwelling*, American Geophysical Union, Washington, DC, pp. 176–82.

Franco, J. N., Wernberg, T., Bertocci, I. et al. (2015). Herbivory drives kelp recruits into 'hiding' in a warm ocean climate. *Marine Ecology Progress Series*, **536**, 1-9.

Fretter, V. and Graham, A. (1976). The prosobranch molluscs of Britain and Denmark I: Pleurotamariacea, Fissurellacea and Patellacea. *Journal of Molluscan Studies*, (Suppl. 1), 1–37.

Gallagher, M. C., Culloty, S., McAllen, R. and O'Riordan, R. (2016). Room for one more? Coexistence of native and non-indigenous barnacle species. *Biological Invasions*, **18**, 3033–46.

Gianni, F., Bartolini, F., Pey, A. et al. (2017). Threats to large brown algal forests in temperate seas: the overlooked role of native herbivorous fish. *Scientific Reports*, **7**, 6012.

Giménez, L. and Jenkins, S. R. (2013). Combining traits and density to model recruitment of sessile organisms. *PLoS ONE*, **8**, e57849.

Giménez, L., Torres, G., Pettersen, A., Burrows, M., Estevez, A. and Jenkins, S. R. (2017). Scale-dependent natural variation in larval nutritional reserves in a marine invertebrate: implications for recruitment and cross-ecosystem coupling. *Marine Ecology Progress Series*, **570**, 141–55.

González-Lorenzo, G., Mesa Hernández, E., Pérez-Dionis, G., Brito Hernández, A., Galván Santos, B. and Barquín Diez, J. (2015). Ineffective conservation threatens *Patella candei*, an endangered limpet endemic to the Macaronesian islands. *Biological Conservation*, **192**, 428–35.

Gordon, J. M. and Knights, A. M. (2017). Revisiting Connell: competition but not as we know it. *Journal of the Marine Biological Association of the United Kingdom*, **98**, 1–9.

Goss-Custard, S., Jones, J., Kitching, J. A. and Norton, T. A. (1979). Tide pools of Carrigathorna and Barloge Creek. *Philosophical Transactions of the Royal Society B: Biological Sciences*, **287**, 1–44.

Griffin, J., Noël, L., Crowe, T. et al. (2010). Consumer effects on ecosystem functioning in rockpools: roles of species richness and composition. *Marine Ecology Progress Series*, **420**, 45–56.

Griffin, J. N., de la Haye, K. L., Hawkins, S. J., Thompson, R. C. and Jenkins, S. R. (2008). Predator diversity and ecosystem functioning: density modifies the effect of resource partitioning. *Ecology*, **89**, 298–305.

Griffin, J. N., Jenkins, S. R., Gamfeldt, L., Jones, D., Hawkins, S. J. and Thompson, R. C. (2009). Spatial heterogeneity increases the importance of species richness for an ecosystem process. *Oikos*, **118**, 1335–42.

Gruet, Y. (1986). Spatio-temporal changes of sabellarian reefs built by the sedentary polychaete *Sabellaria alveolata* (Linné). *Marine Ecology*, **7**, 303–19.

Guallart, J., Calvo, M., Acevedo, I. and Templado, J. (2013). Two-way sex change in the endangered limpet *Patella ferruginea* (Mollusca, Gastropoda). *Invertebrate Reproduction and Development*, **57**, 247–53.

Hartnoll, R. G. and Hawkins, S. J. (1980). Monitoring rocky-shore communities: a critical look at spatial and temporal variation. *Helgoländer Meeresuntersuchungen*, **33**, 484–94.

Hartnoll, R. G. and Hawkins, S. J. (1985). Patchiness and fluctuations on moderately exposed rocky shores. *Ophelia*, **24**, 53–63.

Hatton, H. (1938). Essais de bionomie explicative sur quelques espkces intercotidales d'algues et d'animaux. *Annales de l'Institut Océanographique, Monaco*, **17**, 241–348.

Hatton, H. and Fischer-Piette, E. (1932). Observations et experiences sur le peuplement des cotes rocheuses par les Cirripedes. *Bulletin de l'Institut océanographique, Monaco*, **592**, 15.

Hawkins, S. J. (1981a). The influence of *Patella* grazing on the fucoid-barnacle mosaic on moderately exposed rocky shores. *Kieler Meeresforsch*, **33**, 537–43.

Hawkins, S. J. (1981b). The influence of season and barnacles on the algal colonization of *Patella vulgata* exclusion areas. *Journal of the Marine Biological Association of the United Kingdom*, **61**, 1.

Hawkins, S. J. (1983). Interactions of *Patella* and macroalgae with settling *Semibalanus balanoides* (L.). *Journal of Experimental Marine Biology and Ecology*, **71**, 55–72.

Hawkins, S. J. (2017) Editorial: ecological processes are not bound by borders: Implications for marine conservation in a post-Brexit world. *Aquatic Conservation*, **27**, 904–08.

Hawkins, S. J., Burnay, L. P., Neto, A. I., Tristao da Cunha, R. and Frias Martins, A. M. (1990). A description of the zonation patterns of molluscs and other important biota on the south coast of Sao Miguel, Azores. *Acoreana*, 1990 supplement, 21–38.

Hawkins, S. J., Corte-Real, H. B. S. M., Pannacciulli, F. G., Weber, L. C. and Bishop, J. D. D. (2000). Thoughts on the ecology and evolution of the intertidal biota of the Azores and other Atlantic islands. *Hydrobiologia*, **440**, 3–17.

Hawkins, S. J., Evans, A., Firth, L. B. et al. (2016a). Impacts and Effects of Ocean Warming on Intertidal Rocky Habitats. In D. Laffoley and J. M. Baxter, eds. *Explaining Ocean Warming: Causes, Scale, Effects and Consequences, Full report*, ICUN, Gland, CH, pp. 147–76.

Hawkins, S. J., Evans, A. J., Mieszkowska, N. et al. (2017). Distinguishing globally-driven changes from regional- and local-scale impacts: the case for long-term and broad-scale studies of recovery from pollution. *Marine Pollution Bulletin*, 124, 573–86.

Hawkins, S. J., Firth, L. B., McHugh, M. et al. (2013). Data rescue and re-use: recycling old information to address new policy concerns. *Marine Policy*, **42**, 91–8.

Hawkins, S. J. and Harkin, E. (1985). Preliminary canopy removal in algal dominated communities low on the shore and in the shallow subtidal. *Botanica Marina*, **28**, 223–30.

Hawkins, S. J. and Hartnoll, R. (1985). Factors determining the upper limits of intertidal canopy-forming algae. *Marine Ecology Progress Series*, **20**, 265–71.

Hawkins, S. J. and Hartnoll, R. G. (1982a). The influence of barnacle cover on the numbers, growth and behaviour of *Patella vulgata* on a vertical pier. *Journal of the Marine Biological Association of the United Kingdom*, **62**, 855.

Hawkins, S. J. and Hartnoll, R. G. (1982b). Settlement patterns of *Semibalanus balanoides* (L.) in the Isle of Man (1977–1981). *Journal of Experimental Marine Biology and Ecology*, **62**, 271–83.

Hawkins, S. J. and Hartnoll, R. G. (1983a). Changes in a rocky shore community: an evaluation of monitoring. *Marine Environmental Research*, **9**, 131–81.

Hawkins, S. J. and Hartnoll, R. G. (1983b). Grazing of intertidal algae by marine invertebrates. *Oceanography and Marine Biology: An Annual Review*, **21**, 195–282.

Hawkins, S. J., Hartnoll, R. G., Kain, J. M. and Norton, T. A. (1992). Plant–Animal Interactions on Hard Substrata in the Northeast Atlantic. In D. M. John, S. J. Hawkins and J. H. Price, eds. *Plant–Animal Interactions in the Marine Benthos*, Oxford University Press, Oxford, pp. 1–32.

Hawkins, S. J. and Hiscock, K. (1983). Anomalies in the abundance of common eulittoral gastropods with planktonic larvae on Lundy Island, Bristol Channel. *Journal of Molluscan Studies*, **49**, 86–8.

Hawkins, S. J., Mieszkowska, N., Firth, L. B. et al. (2016b). Looking backwards to look forwards: the role of natural history in temperate reef ecology. *Marine and Freshwater Research*, **67**, 1–13.

Hawkins, S. J., Moore, P. J., Burrows, M. T. et al. (2008). Complex interactions in a rapidly changing world: responses of rocky shore communities to recent climate change. *Climate Research*, **37**, 123–33.

Hawkins, S. J., Proud, S. V., Spence, S. K. and Southward, A. J. (1994). From the Individual to the Community and Beyond: Water Quality, Stress Indicators and Key Species in Coastal Ecosystems. In D. W. Sutcliffe, ed. *Water Quality and Stress Indicators in Marine and Freshwater Ecosystems: Linking Levels of Organisation Individuals, Populations, Communities*, Freshwater Biological Association, Ambleside, pp. 35–62.

Hawkins, S. J. and Southward, A. J. (1992). The Torrey Canyon oil spill: recovery of rocky shore communities. In *Restoring the Nation's Marine Environment. Maryland: Proceedings of the Symposium on Marine Habitat Restoration*, Sea Grant Publication, National Oceanic and Atmospheric Administration, Maryland Sea Grant College, 584–631.

Hawkins, S. J., Southward, A. J. and Barrett, R. L. (1983). Population structure of *Patella vulgata* during succession on rocky shores in South-west England. *Oceanologica Acta*, Special, 103–07.

Hawkins, S. J., Sugden, H. E., Mieszkowska, N. et al. (2009). Consequences of climate-driven biodiversity changes for ecosystem functioning of North European rocky shores. *Marine Ecology Progress Series*, **396**, 245–59.

Heiser, S., Hall-Spencer, J. M. and Hiscock, K. (2014). Assessing the extent of establishment of *Undaria pinnatifida* in Plymouth Sound Special Area of Conservation, UK. *Marine Biodiversity Records*, 7, e93.

Herbert, R. J. H. and Hawkins, S. J. (2006). Effect of rock type on the recruitment and early mortality of the barnacle *Chthamalus montagui*. *Journal of Experimental Marine Biology and Ecology*, **334**, 96–108.

Herbert, R. J. H., Hawkins, S. J., Sheader, M. and Southward, A. J. (2003). Range extension and reproduction of the barnacle *Balanus perforatus* in the eastern English Channel. *Journal of the Marine Biological Association of the United Kingdom*, **83**, 73–82.

Herbert, R. J. H., Humphreys, J., Davies, C. J., Roberts, C., Fletcher, S. and Crowe, T. P. (2016). Ecological impacts of non-native Pacific oysters (*Crassostrea gigas*) and management measures for protected areas in Europe. *Biodiversity and Conservation*, **25**, 2835–65.

Herbert, R. J. H., Southward, A. J., Clarke, R. T. Sheader, M. and Hawkins, S. J. (2009). Persistent border, an analysis of the geographic boundary of an intertidal species. *Marine Ecology Progress Series*, **379**, 135–50.

Herbert, R. J. H., Southward, A. J., Sheader, M. and Hawkins, S. J. (2007). Influence of recruitment and temperature on distribution of intertidal barnacles in the English Channel. *Journal of the Marine Biological Association of the United Kingdom*, **87**, 487.

Hill, A. S. and Hawkins, S. J. (1990). An investigation of methods for sampling microbial films on rocky shores. *Journal of the Marine Biological Association of the United Kingdom*, **70**, 77–88.

Hill, A. S. and Hawkins, S. J. (1991). Seasonal and spatial variation of epilithic microalgae distribution and abundance and its ingestion by *Patella vulgata* on a moderately exposed rocky shore. *Journal of the Marine Biological Association of the United Kingdom*, **71**, 403–23.

Hills, J. M. and Thomason, J. C. (2003). The 'ghost of settlement past' determines mortality and fecundity in the barnacle, *Semibalanus balanoides*. *Oikos*, **101**, 529–38.

Hiscock, K., Southward, A., Tittley, I. and Hawkins, S. (2004). Effects of changing temperature on benthic marine life in Britain and Ireland. *Aquatic Conservation: Marine and Freshwater Ecosystems*, **14**, 333–62.

Holm, M. W., Davids, J. K., Dolmer, P., Holmes, E. and Nielsen, T. (2016). Coexistence of Pacific oyster *Crassostrea gigas* (Thunberg, 1793) and blue mussels *Mytilus edulis* Linnaeus, 1758 on a sheltered intertidal bivalve bed? *Aquatic Invasions*, **11**, 155–65.

Holmes, J. M. C. and Minchin, D. (1995). Two exotic copepods imported into Ireland with the Pacific oyster *Crassostrea gigas* (Thunberg). *The Irish Naturalists' Journal*, **25**, 17–20.

Hyder, K., Åberg, P., Johnson, M. P. and Hawkins, S. J. (2001). Models of open populations with space-limited recruitment: extension of theory and application to the barnacle *Chthamalus montagui*: Modelling barnacle populations. *Journal of Animal Ecology*, **70**, 853–63.

Hyder, K., Johnson, M. P., Hawkins, S. J. and Gurney, W. S. (1998). Barnacle demography: evidence for an existing model and spatial scales of variation. *Marine Ecology Progress Series*, **174**, 89–99.

Ingólfsson, A. (2006). The intertidal seashore of Iceland and its animal communities. *Zoology of Iceland I*, **7**, 1–85.

Ingólfsson, A. and Hawkins, S. J. (2008). Slow recovery from disturbance: a 20 year study of *Ascophyllum* canopy clearances. *Journal of the Marine Biological Association of the United Kingdom*, **88**, 689–91.

Jacinto, D. and Cruz, T. (2008). Tidal settlement of the intertidal barnacles *Chthamalus* spp. in SW portugal: interaction between diel and semi-lunar cycles. *Marine Ecology Progress Series*, **366**, 129–35.

Jacinto, D. and Cruz, T. (2012). *Paracentrotus lividus* (Echinodermata: Echinoidea) attachment force and burrowing behavior in rocky shores of SW Portugal. *Zoosymposia*, **7**, 231–40.

James, B. L. (1968). The occurrence of larval Digenea in ten species of intertidal prosobranch molluscs in Cardigan Bay. *Journal of Natural History*, **2**, 329–43.

Jenkins, S. R., Åberg, P., Cervin, G. et al. (2000). Spatial and temporal variation in settlement and recruitment of the intertidal barnacle *Semibalanus balanoides* (L.) (Crustacea: Cirripedia) over a European scale. *Journal of Experimental Marine Biology and Ecology*, **243**, 209–25.

Jenkins, S. R., Arenas, F., Arrontes, J. et al. (2001). European-scale analysis of seasonal variability in limpet grazing activity and microalgal abundance. *Marine Ecology Progress Series*, **211**, 193–203.

Jenkins, S. R., Coleman, R. A., Santina, P. D., Hawkins, S. J., Burrows, M. T. and Hartnoll, R. G. (2005). Regional scale differences in the determinism of grazing effects in the rocky intertidal. *Marine Ecology Progress Series*, **287**, 77–86.

Jenkins, S. R. and Hartnoll, R. G. (2001). Food supply, grazing activity and growth rate in the limpet *Patella vulgata* L.: a comparison between exposed and sheltered shores. *Journal of Experimental Marine Biology and Ecology*, **258**(1), 123–39.

Jenkins, S. R. and Hawkins, S. J. (2003). Barnacle larval supply to sheltered rocky shores: a limiting factor? *Hydrobiologia*, **503**(1–3), 143–51.

Jenkins, S. R., Hawkins, S. J. and Norton, T. A. (1999a). Direct and indirect effects of a macroalgal canopy and limpet grazing in structuring a sheltered intertidal community. *Marine Ecology Progress Series*, **188**, 81–92.

Jenkins, S. R., Hawkins, S. J. and Norton, T. A. (1999b). Interaction between a fucoid canopy and limpet grazing in structuring a low shore intertidal community. *Journal of Experimental Marine Biology and Ecology*, **233**, 41–63.

Jenkins, S. R., Moore, P., Burrows, M. T. et al. (2008a). Comparative ecology of north Atlantic shores: do differences in players matter for process? *Ecology*, **89**, S3–S23.

Jenkins, S. R., Murua, J. and Burrows, M. T. (2008b). Temporal changes in the strength of density-dependent mortality and growth in intertidal barnacles. *Journal of Animal Ecology*, **77**, 573–84.

Jenkins, S. R., Norton, T. A. and Hawkins, S. J. (1999c). Interactions between canopy forming algae in the eulittoral zone of sheltered rocky shores on the Isle of Man. *Journal of the Marine Biological Association of the United Kingdom*, **79**, 341–9.

Jenkins, S. R., Norton, T. A. and Hawkins, S. J. (1999d). Settlement and post-settlement interactions between *Semibalanus balanoides* (L.) (Crustacea: Cirripedia) and three species of fucoid canopy algae. *Journal of Experimental Marine Biology and Ecology*, **236**, 49–67.

Jenkins, S. R., Norton, T. A. and Hawkins, S. J. (2004). Long term effects of *Ascophyllum nodosum* canopy removal on mid shore community structure. *Journal of the Marine Biological Association of the United Kingdom*, **84**, 327–9.

Jenkins, S. R. and Uyà, M. (2016). Temporal scale of field experiments in benthic ecology. *Marine Ecology Progress Series*, **547**, 273–86.

Jessopp, M., Mulholland, O., McAllen, R., Johnson, M., Crowe, T. and Allcock, A. (2007). Coastline configuration as a determinant of structure in larval assemblages. *Marine Ecology Progress Series*, **352**, 67–75.

Johannesson, K. (1988). The paradox of Rockall: why is a brooding gastropod (*Littorina saxatilis*) more widespread than one having a planktonic larval dispersal stage (*L. littorea*)? *Marine Biology*, **99**, 507–13.

Johannesson, K. and Warmoes, T. (1990). Rapid colonization of Belgian breakwaters by the direct developer, *Littorina saxatilis* (Olivi) (Prosobranchia, Mollusca). *Hydrobiologia*, **193**, 99–108.

Johnson, L. E. and Strathmann, R. R. (1989). Settling barnacle larvae avoid substrata previously occupied by a mobile predator. *Journal of Experimental Marine Biology and Ecology*, **128**, 87–103.

Johnson, M., Burrows, M., Hartnoll, R. and Hawkins, S. (1997). Spatial structure on moderately exposed rocky shores:patch scales and the interactions between limpets and algae. *Marine Ecology Progress Series*, **160**, 209–15.

Johnson, M., Burrows, M. and Hawkins, S. (1998). Individual based simulations of the direct and indirect effects of limpets on a rocky shore *Fucus* mosaic. *Marine Ecology Progress Series*, **169**, 179–88.

Johnson, M. P., Frost, N. J., Mosley, M. W. J., Roberts, M. F. and Hawkins, S. J. (2003). The area-independent effects of habitat complexity on biodiversity vary between regions. *Ecology Letters*, **6**, 126–32.

Johnson, M. P., Hanley, M. E., Frost, N. J., Mosley, M. W. J. and Hawkins, S. J. (2008). The persistent spatial patchiness of limpet grazing. *Journal of Experimental Marine Biology and Ecology*, **365**, 136–41.

Jones, N. (1948). Observations and experiments on the biology of Patella vulgata at Port St. Mary, Isle of Man. In *Proceedings and Transactions of the Liverpool Biological Society*, pp. 60–77.

Jones, N. S. (1946). Browsing of *Patella*. *Nature*, **158**, 557–8.

Jonsson, P. R., Granhag, L., Moschella, P. S., Åberg, P., Hawkins, S. J. and Thompson, R. C. (2006). Interactions between wave action and grazing control the distribution of intertidal macroalgae. *Ecology*, **87**, 1169–78.

Keith, S. A., Herbert, R. J. H., Norton, P. A., Hawkins, S. J. and Newton, A. C. (2011). Individualistic species limitations of climate-induced range expansions generated by meso-scale dispersal barriers: dispersal barriers limit range expansions. *Diversity and Distributions*, **17**, 275–86.

Kendall, M. A., Bowman, R. S., Williamson, P. and Lewis, J. R. (1985). Annual variation in the recruitment of *Semibalanus Balanoides* on the north Yorkshire coast 1969–1981. *Journal of the Marine Biological Association of the United Kingdom*, **65**, 1009.

Kendall, M. A., Burrows, M. T., Southward, A. J. and Hawkins, S. J. (2004). Predicting the effects of marine climate change on the invertebrate prey of the birds of rocky shores. *IBIS*, **146**, 40–7.

King, P. A., Keogh, E. and McGrath, D. (1997). The current status of the exotic barnacle *Elminius modestus* Darwin in Galway Bay, Ireland. *The Irish Naturalists' Journal*, **17**, 365–9.

Kitching, J. A. (1987a). Ecological studies at Lough Hyne. *Advances in Ecological Research*, **17**, 115–86.

Kitching, J. A. (1987b). The flora and fauna associated with *Himanthalia elongata* (L.) S. F. Gray in relation to water current and wave action in the Lough Hyne marine nature reserve. *Estuarine, Coastal and Shelf Science*, **25**, 663–76.

Kitching, J. A. and Ebling, F. J. (1961). The Ecology of Lough Ine. *Journal of Animal Ecology*, **30**, 373–83.

Kitching, J. A. and Ebling, F. J. (1967). Ecological studies at Lough Ine. *Advances in Ecological Research*, **4**, 197–291.

Kitching, J. A., Ebling, F. J., Gamble, J. C., Hoare, R., McLeod, A. and Norton, T. A. (1976). The ecology of Lough Ine. XIX. Seasonal changes in the Western Trough. *Journal of Animal Ecology*, **45**, 731–58.

Kitching, J. A., Muntz, L. and Ebling, F. J. (1966). The Ecology of Lough Ine. XV. The ecological significance of shell and body forms in Nucella. *Journal of Animal Ecology*, **35**, 113–26.

Kitching, J. A., Sloane, J. F. and Ebling, F. J. (1959). The Ecology of Lough Ine: VIII. Mussels and Their Predators. *Journal of Animal Ecology*, **28**, 331–41.

Knight-Jones, E. W. (1953). Laboratory experiments on gregariousness during setting in *Balanus balanoides* and other barnacles. *Journal of Experimental Biology*, **30**, 584–98.

Knights, A., Crowe, T. and Burnell, G. (2006). Mechanisms of larval transport: vertical distribution of bivalve larvae varies with tidal conditions. *Marine Ecology Progress Series*, **326**, 167–74.

Kollien, A. H. (1996). *Cercaria patellae* Lebour, 1911 developing in *Patella vulgata* is the cercaria of *Echinostephilla patellae* (Lebour, 1911) n. comb. (Digenea, Philophthalmidae). *Systematic Parasitology*, **34**, 11–25.

Kostylev, V. E., Erlandsson, J., Ming, M. Y. and Williams, G. A. (2005). The relative importance of habitat complexity and surface area in assessing biodiversity: Fractal application on rocky shores. *Ecological Complexity*, **2**, 272–86.

Lawson, G. W. and Norton, T. A. (1971). Some observations on littoral and sublittoral zonation at Teneriffe (Canary Isles). *Botanica Marina*, **14**, 116–20.

Lawson, J., Davenport, J. and Whitaker, A. (2004). Barnacle distribution in Lough Hyne Marine Nature Reserve: a new baseline and an account of invasion by the introduced Australasian species *Elminius modestus* Darwin. *Estuarine, Coastal and Shelf Science*, **60**, 729–35.

Le Quesne, W. J. F. and Hawkins, S. J. (2006). Direct observations of protandrous sex change in the patellid limpet *Patella vulgata*. *Journal of the Marine Biological Association of the United Kingdom*, **86**, 161–2.

Lebour, M. V. (1911). A review of the British marine *Cercariae*. *Parasitology*, **4**, 416.

Lefebvre, S., Marín Leal, J. C., Dubois, S. et al. (2009). Seasonal dynamics of trophic relationships among co-occurring suspension-feeders in two shellfish culture dominated ecosystems. *Estuarine, Coastal and Shelf Science*, **82**, 415–25.

Lejart, M. and Hily, C. (2005). Proliferation of *Crassostrea gigas* (Thunberg) in the Bay of Brest: First estimations of the stock and its impact on the global functioning of the ecosystem. In *8th International Conference on Shellfish Restoration, Brest*.

Lejart, M. and Hily, C. (2011). Differential response of benthic macrofauna to the formation of novel oyster reefs (*Crassostrea gigas*, Thunberg) on soft and rocky substrate in the intertidal of the Bay of Brest, France. *Journal of Sea Research*, **65**, 84–93.

Lemos, R. T. and Pires, H. O. (2004). The upwelling regime off the West Portuguese Coast, 1941–2000. *International Journal of Climatology*, **24**, 511–24.

Lewis, J. R. (1964). *The Ecology of the Rocky Shores*, English Universities Press, London.

Lewis, J. R. (1976). Long-term ecological surveillance: practical realities in the rocky littoral. *Oceanography and Marine Biology: An Annual Review*, **14**, 371–90.

Lewis, J. R. and Bowman, R. S. (1975). Local habitat-induced variations in the population dynamics of *Patella vulgata* L. *Journal of Experimental Marine Biology and Ecology*, **17**, 165–203.

Lima, F. P., Gomes, F., Seabra, R. et al. (2016). Loss of thermal refugia near equatorial range limits. *Global Change Biology*, **22**, 254–63.

Lima, F. P., Queiroz, N., Ribeiro, P. A., Hawkins, S. J. and Santos, A. M. (2006). Recent changes in the distribution of a marine gastropod, *Patella rustica* Linnaeus, 1758, and their relationship to unusual climatic events. *Journal of Biogeography*, **33**, 812–22.

Lima, F. P., Queiroz, N., Ribeiro, P. A., Xavier, R., Hawkins, S. J. and Santos, A. M. (2009). First record of *Halidrys siliquosa* on the Portuguese coast: counter-intuitive range expansion? *Marine Biodiversity Records*, **2**, e1.

Lima, F. P., Ribeiro, P. A., Queiroz, N., Hawkins, S. J. and Santos, A. M. (2007). Do distributional shifts of northern and southern species of algae match the warming pattern? *Global Change Biology*, **13**, 2592–604.

Little, C., Trowbridge, C. D., Pilling, G. M., Stirling, P., Morritt, D. and Williams, G. A. (2017). Long-term fluctuations in intertidal communities in an Irish sea-lough: Limpet-fucoid cycles. *Estuarine, Coastal and Shelf Science*, **196**, 70–82.

Lodge, S. M. (1948). Algal growth in the absence of *Patella* on an experimental strip of foreshore, Port St Mary, Isle of Man. In *Proceedings and Transactions of the Liverpool Biological Society*, pp. 78–85.

Lüning, K. (1990). *Seaweeds: Their Environment, Biogeography, and Ecophysiology*, John Wiley & Sons, Hoboken, NJ.

Maggs, C. A., Castilho, R., Foltz, D. et al. (2008). Evaluating signatures of glacial refugia for North Atlantic benthic marine taxa. *Ecology*, **89**, S108–22.

Martínez, B., Arenas, F., Rubal, M. et al. (2012). Physical factors driving intertidal macroalgae distribution: physiological stress of a dominant fucoid at its southern limit. *Oecologia*, **170**, 341–53.

Martínez, B., Arenas, F., Trilla, A., Viejo, R. M. and Carreño, F. (2015). Combining physiological threshold knowledge to species distribution models is key to improving forecasts of the future niche for macroalgae. *Global Change Biology*, **21**, 1422–33.

Martins, G. M., Arenas, F., Tuya, F., Ramírez, R., Neto, A. I. and Jenkins, S. R. (2018). Successional convergence in experimentally disturbed intertidal communities. *Oecologia*, **186**, 507–16.

Martins, G. M., Hawkins, S. J., Thompson, R. C. and Jenkins, S. R. (2007). Community structure and functioning in intertidal rock pools: effects of pool size and shore height at different successional stages. *Marine Ecology Progress Series*, **329**, 43–55.

Martins, G. M., Jenkins, S. R., Hawkins, S. J., Neto, A. I. and Thompson, R. C. (2008). Exploitation of rocky intertidal grazers: population status and potential impacts on community structure and functioning. *Aquatic Biology*, **3**, 1–10.

Martins, G. M., Thompson, R. C., Neto, A. I., Hawkins, S. J. and Jenkins, S. R. (2010). Enhancing stocks of the exploited limpet *Patella candei* d'Orbigny via modifications in coastal engineering. *Biological Conservation*, **143**, 203–11.

Martins, G. M., Jenkins, S. R., Neto, A. I., Hawkins, S. J. and Thompson, R. C. (2016). Long-term modifications of coastal defenses enhance marine biodiversity. *Environmental Conservation*, **43**, 109–16.

Marzinelli, E. M., Burrows, M. T., Jackson, A. C. and Mayer-Pinto, M. (2012). Positive and negative effects of habitat-forming algae on survival, growth and intra-specific competition of limpets. *PLoS ONE*, **7**, e51601.

Mayer-Pinto, M., Johnston, E. L., Bugnot, A. B. et al. (2017). Building 'blue': an eco-engineering framework for foreshore developments. *Journal of Environmental Management*, **189**, 109–14.

Mieszkowska, N., Burrows, M. T., Pannacciulli, F. G. and Hawkins, S. J. (2014a). Multidecadal signals within co-occurring intertidal barnacles *Semibalanus balanoides* and *Chthamalus* spp. linked to the Atlantic multidecadal oscillation. *Journal of Marine Systems*, **133**, 70–6.

Mieszkowska, N., Hawkins, S. J., Burrows, M. T. and Kendall, M. A. (2007). Long-term changes in the geographic distribution and population structures of *Osilinus lineatus* (Gastropoda: Trochidae) in Britain and Ireland. *Journal of the Marine Biological Association of the United Kingdom*, **87**, 537.

Mieszkowska, N., Kendall, M. A., Hawkins, S. J. et al. (2006). Changes in the range of some common rocky shore species In Britain – a response to climate change? *Hydrobiologia*, **555**, 241–51.

Mieszkowska, N., Leaper, R., Moore, P. et al. (2005). *Marine Biodiversity and Climate Change: Assessing and Predicting the Influence of Climatic Change Using Intertidal Rocky Shore Biota*, Marine Biological Association of the United Kingdom 20. Occasional Publications, Plymouth.

Mieszkowska, N., Sugden, H., Firth, L. B. and Hawkins, S. J. (2014b). The role of sustained observations in tracking impacts of environmental change on marine biodiversity and ecosystems. *Philosophical Transactions of the Royal Society A: Mathematical, Physical and Engineering Sciences*, 372.

Mieszkowska, N. and Sugden, H. E. (2016). Climate-driven range shifts within benthic habitats across a marine biogeographic transition zone. *Advances in Ecological Research*, 55, 325–69.

Minchin, D. and Nunn, J. (2014). The invasive brown alga *Undaria pinnatifida* (Harvey) Suringar, 1873 (Laminariales: Alariaceae), spreads northwards in Europe. *BioInvasions Records*, **3**, 57–63.

Miranda, P., Alves, J. and Serra, N. (2013). Climate change and upwelling: response of Iberian upwelling to atmospheric forcing in a regional climate scenario. *Climate Dynamics*, **40**, 2813–24.

Mitsch, W. J. (2012). What is ecological engineering? *Ecological Engineering*, **45**, 5–12.

Moore, P., Hawkins, S. J. and Thompson, R. C. (2007). Role of biological habitat amelioration in altering the relative responses of congeneric species to climate change. *Marine Ecology Progress Series*, **334**, 11–19.

Moore, H. B. and Kitching, J. A. (1939). The biology of *Chthamalus stellatus* (Poli). *Journal of the Marine Biological Association of the United Kingdom*, **23**, 521–41.

Morris, R. L., Chapman, M. G., Firth, L. B. and Coleman, R. A. (2017). Increasing habitat complexity on seawalls: investigating large- and small-scale effects on fish assemblages. *Ecology and Evolution*, **7**(22), 9567–79.

Morris, R. L., Deavin, G., Hemelryk Donald, S. and Coleman, R. A. (2016). Eco-engineering in urbanised coastal systems: consideration of social values. *Ecological Management & Restoration*, **17**, 33–9.

Morris, S. and Taylor, A. C. (1983). Diurnal and seasonal variation in physico-chemical conditions within intertidal rockpools. *Estuarine, Coastal and Shelf Science*, **17**, 339–55.

Moschella, P. S., Abbiati, M., Åberg, P. et al. (2005). Low-crested coastal defence structures as artificial habitats for marine life: using ecological criteria in design. *Coastal Engineering*, **52**, 1053–71.

Munday, P. L., Buston, P. M. and Warner, R. R. (2006). Diversity and flexibility of sex-change strategies in animals. *Trends in Ecology & Evolution*, **21**, 89–95.

Narayanaswamy, B. E., Renaud, P. E., Duineveld, G. C. et al. (2010). Biodiversity trends along the western European margin. *PLoS ONE*, **5**, e14295.

Naylor, E. and Slinn, D. J. (1958). Observations on the ecology of some brackish water organisms in pools at Scarlett Point, Isle of Man. *The Journal of Animal Ecology*, **27**, 15.

Newell, R. C. (1979). *Biology of Intertidal Organisms*, Marine Ecological Surveys, Ltd, Sittingbourne.

Ng, T. P. T., Saltin, S. H., Davies, M. S., Johannesson, K., Stafford, R. and Williams, G. A. (2013). Snails and their trails: the multiple functions of trail-following in gastropods. *Biological Reviews*, **88**, 683–700.

Nicastro, K. R., Zardi, G. I., Teixeira, S., Neiva, J., Serrão, E. A. and Pearson, G. A. (2013). Shift happens: trailing edge contraction associated with recent warming trends threatens a distinct genetic lineage in the marine macroalga *Fucus vesiculosus*. *BMC Biology*, **11**, 6.

Nobre, A. (1940). *Fauna malacológica de Portugal, 1: Moluscos marinhos e de águas salobras*. A. Nobre, Porto.

Noël, L. M. L. J., Griffin, J. N., Thompson, R. C. et al. (2010). Assessment of a field incubation method estimating primary productivity in rockpool communities. *Estuarine, Coastal and Shelf Science*, **88**, 153–9.

Noël, L. M. L. J., Hawkins, S. J., Jenkins, S. R. and Thompson, R. C. (2009). Grazing dynamics in intertidal rockpools: connectivity of microhabitats. *Journal of Experimental Marine Biology and Ecology*, **370**, 9–17.

Norton, T. A., Hawkins, S. J., Manley, N. L., Williams, G. A. and Watson, D. C. (1990). Scraping a living: a review of littorinid grazing. *Hydrobiologia*, **193**, 117–38.

Notman, G., McGill, R., Hawkins, S. and Burrows, M. (2016). Macroalgae contribute to the diet of *Patella vulgata* from contrasting conditions of latitude and wave exposure in the UK. *Marine Ecology Progress Series*, **549**, 113–23.

O'Connor, N. E. and Crowe, T. P. (2005). Biodiversity loss and ecosystem functioning: distinguishing between number and identity of species. *Ecology*, **86**, 1783–96.

O'Connor, N. E. and Crowe, T. P. (2007). Biodiversity among mussels: separating the influence of sizes of mussels from the ages of patches. *Journal of the Marine Biological Association of the United Kingdom*, **87**, 551.

O'Connor, N. E., Crowe, T. P. and McGrath, D. (2006). Effects of epibiotic algae on the survival, biomass and recruitment of mussels, *Mytilus* L. (Bivalvia: Mollusca). *Journal of Experimental Marine Biology and Ecology*, **328**, 265–76.

O'Riordan, R. M., Arenas, F., Arrontes, J. et al. (2004). Spatial variation in the recruitment of the intertidal barnacles *Chthamalus montagui* Southward and *Chthamalus stellatus* (Poli) (Crustacea: Cirripedia) over an European scale. *Journal of Experimental Marine Biology and Ecology*, **304**, 243–64.

O'Riordan, R. M. and Murphy, O. (2000). Variation in the reproductive cycle of *Elminius modestus* in southern Ireland. *Journal of the Marine Biological Association of the United Kingdom*, **80**, 607–16.

O'Riordan, R. M. and Ramsay, N. F. (1999). The current distribution and abundance of the Australasian barnacle *Elminius modestus* in Portugal. *Journal of the Marine Biological Association of the United Kingdom*, **79**, 937–9.

Olabarria, C., Rodil, I. F., Incera, M. and Troncoso, J. S. (2009). Limited impact of *Sargassum muticum* on native algal assemblages from rocky intertidal shores. *Marine Environmental Research*, **67**, 153–8.

Orton, J. H. (1920). Sea-temperature, breeding and distribution in marine animals. *Journal of the Marine Biological Association of the United Kingdom*, **12**, 339.

Orton, J. H. (1929). Observations on *Patella vulgata*. Part III. Habitat and Habits. *Journal of the Marine Biological Association of the United Kingdom*, **16**, 277.

OSPAR Commission (2010). *Background Document for Azorean Limpet* Patella aspera, Biodiversity Series, OSPAR, Paris.

Pannacciulli, F. G., Bishop, J. D. D. and Hawkins, S. J. (1997). Genetic structure of populations of two species of *Chthamalus* (Crustacea: Cirripedia) in the north-east Atlantic and Mediterranean. *Marine Biology*, **128**, 73–82.

Pardo, P. C., Padín, X. A., Gilcoto, M., Farina-Busto, L. and Pérez, F. F. (2011). Evolution of upwelling systems coupled to the long-term variability in sea surface temperature and Ekman transport. *Climate Research*, **48**, 231–46.

Pearson, G., Kautsky, L. and Serrão, E. (2000). Recent evolution in Baltic *Fucus vesiculosus:* reduced tolerance to emersion stresses compared to intertidal (North Sea) populations. *Marine Ecology Progress Series*, **202**, 67–79.

Perez, R., Lee, J. Y. and Juge, C. (1981). Observations sur la biologie de l'algue japonaise *Undaria pinnatifida* (Harvey) Suringar introduite accidentellement dans l'Etang de Thau. *Science et Peche*, **325**, 1–12.

Perkol-Finkel, S. and Sella, I. (2015). Harnessing urban coastal infrastructure for ecological enhancement. *Proceedings of the Institution of Civil Engineers – Maritime Engineering*, **168**, 102–10.

Philippart, C. J. M., Anadón, R., Danovaro, R. et al. (2011). Impacts of climate change on European marine ecosystems: Observations, expectations and indicators. *Journal of Experimental Marine Biology and Ecology*, **400**, 52–69.

Pires, A. C., Nolasco, R., Rocha, A., Ramos, A. M. and Dubert, J. (2016). Climate change in the Iberian Upwelling System: a numerical study using GCM downscaling. *Climate Dynamics*, **47**, 451–64.

Pocklington, J. B., Jenkins, S. R., Bellgrove, A. et al. (2017). Disturbance alters ecosystem engineering by a canopy-forming alga. *Journal of the Marine Biological Association of the United Kingdom*, **98**, 687–98.

Poloczanska, E. S., Brown, C. J., Sydeman, W. J. et al. (2013). Global imprint of climate change on marine life. *Nature Climate Change*, **3**, 919–25.

Poloczanska, E. S., Hawkins, S. J., Southward, A. J. and Burrows, M. T. (2008). Modeling the response of populations of competing species to climate change. *Ecology*, **89**, 3138–49.

Powell, A. and Rowley, A. F. (2008). Tissue changes in the shore crab *Carcinus maenas* as a result of infection by the parasitic barnacle *Sacculina carcini*. *Diseases of Aquatic Organisms*, **80**, 75–9.

Powell, H. T. (1957). Studies in the genus *Fucus* L. II: distribution and ecology of forms of *Fucus distichus* L. Emend Powell in Britain and Ireland. *Journal of the Marine Biological Association of the United Kingdom*, **36**, 663–93.

Powell, H. T. (1963). New records of *Fucus distichus* subspecies for the Shetland and Orkney islands. *British Phycological Bulletin*, **2**, 247–54.

Prestes, A. C., Cacabelos, E., Neto, A. I. and Martins, G. M. (2017). Temporal stability in macroalgal assemblage standing stock despite high species turnover. *Marine Ecology Progress Series*, **567**, 249–56.

Prinz, K., Kelly, T., O'Riordan, R. and Culloty, S. (2010b). Temporal variation in prevalence and cercarial development of *Echinostephilla patellae* (Digenea, Philophthalmidae) in the intertidal gastropod *Patella vulgata*. *Acta Parasitologica*, **55**, 39–44.

Prinz, K., Kelly, T. C., O'Riordan, R. M. and Culloty, S. C. (2010a). Occurrence of macroparasites in four common intertidal molluscs on the south coast of Ireland. *Marine Biodiversity Records*, **3**, e89.

Prinz, K., Kelly, T. C., O'Riordan, R. M. and Culloty, S. C. (2011). Factors influencing cercarial emergence and settlement in the digenean trematode *Parorchis acanthus* (Philophthalmidae). *Journal of the Marine Biological Association of the United Kingdom*, **91**, 1673–9.

Pyefinch, K. A. (1943). The intertidal ecology of Bardsey Island, North Wales, with special reference to the recolonization of rock surfaces, and the rock-pool environment. *The Journal of Animal Ecology*, **12**, 82.

Queiroga, H., Almeida, M. J., Alpuim, T. et al. (2006). Tide and wind control of megalopal supply to estuarine crab populations on the Portuguese west coast. *Marine Ecology Progress Series*, **307**, 21–36.

Queiroga, H., Cruz, T., dos Santos, A. et al. (2007). Oceanographic and behavioural processes affecting invertebrate larval dispersal and supply in the western Iberia upwelling ecosystem. *Progress in Oceanography*, **74**, 174–91.

Raffaelli, D. and Hawkins, S. (1996). *Intertidal Ecology*, Springer Netherlands, Dordrecht.

Rahmstorf, S., Box, J. E., Feulner, G. et al. (2015). Exceptional twentieth-century slowdown in Atlantic Ocean overturning circulation. *Nature Climate Change*, **5**, 475–80.

Rayner, N. A. A., Parker, D. E., Horton, E. B. et al. (2003). Global analyses of sea surface temperature, sea ice, and night marine air temperature since the late nineteenth century. *Journal of Geophysical Research: Atmospheres*, **108**(D14).

Reid, D. G. (1996). *Systematics and evolution of Littorina*, The Ray Society Publications, London.

Ribeiro, P. A., Branco, M., Hawkins, S. J. and Santos, A. M. (2010). Recent changes in the distribution of a marine gastropod, *Patella rustica*, across the Iberian Atlantic coast did not result in diminished genetic diversity or increased connectivity: Population genetics of *Patella rustica*. *Journal of Biogeography*, **37**, 1782–96.

Ribeiro, S., Amorim, A., Abrantes, F. and Ellegaard, M. (2016). Environmental change in the Western Iberia upwelling ecosystem since the preindustrial period revealed by dinoflagellate cyst records. *The Holocene*, **26**, 874–89.

Rivera-Ingraham, G. A., Espinosa, F. and García-Gómez, J. C. (2011). Environmentally mediated sex change in the endangered limpet *Patella ferruginea* (Gastropoda: Patellidae). *Journal of Molluscan Studies*, **77**, 226–31.

Robins, P., Tita, A., King, J. and Jenkins, S. (2017). Predicting the dispersal of wild Pacific oysters *Crassostrea gigas* (Thunberg, 1793) from an existing frontier population – a numerical study. *Aquatic Invasions*, **12**, 117–31.

Rodil, I. F., Lucena-Moya, P., Olabarria, C. and Arenas, F. (2015a). Alteration of macroalgal subsidies by climate-associated stressors affects behavior of wrack-reliant beach consumers. *Ecosystems*, **18**, 428–40.

Rodil, I. F., Olabarria, C., Lastra, M. and Arenas, F. (2015b). Combined effects of wrack identity and solar radiation on associated beach macrofaunal assemblages. *Marine Ecology Progress Series*, **531**, 167–78.

Rodriguez, S., Martín, A. P., Sousa-Pinto, I. and Arenas, F. (2016). Biodiversity effects on macroalgal productivity: exploring the roles of richness, evenness and species traits. *Marine Ecology Progress Series*, **562**, 79–91.

Rognstad, R., Wethey, D. and Hilbish, T. (2014). Connectivity and population repatriation: limitations of climate and input into the larval pool. *Marine Ecology Progress Series*, **495**, 175–83.

Roughgarden, J., Gaines, S. and Possingham, H. (1988). Recruitment dynamics in complex life cycles. *Science*, **241**, 1460–6.

Rueness, J. (1989). *Sargassum muticum* and other introduced Japanese macroalgae: biological pollution of European coasts. *Marine Pollution Bulletin*, **20**, 173–6.

Russell, B. D., Connell, S. D., Findlay, H. S., Tait, K., Widdicombe, S. and Mieszkowska, N. (2013). Ocean acidification and rising temperatures may increase biofilm primary productivity but decrease grazer consumption. *Philosophical Transactions of the Royal Society B: Biological Sciences*, **368**, 20120438.

Russell, F. S., Southward, A. J., Boalch, G. T. and Butler, E. I. (1971). Changes in biological conditions in the English Channel off Plymouth during the last half century. *Nature*, **234**, 468–70.

Saldanha, L. (1974). Estudo do povoamento dos horizontes superiores da rocha litoral da costa da Arrábida. *Arquivos do Museu Bocage, Segunda Série*. **1**, 1–382.

Santos, F., Gómez-Gesteira, M., deCastro, M. and Álvarez, I. (2011). Upwelling along the western coast of the Iberian Peninsula: dependence of trends on fitting strategy. *Climate Research*, **48**, 213–8.

Santos, R. S., Hawkins, S., Monteiro, L. R., Alves, M. and Isidro, E. J. (1995). Marine research, resources and conservation in the Azores. *Aquatic Conservation: Marine and Freshwater Ecosystems*, **5**(4), 311–54.

Schmidt, P. S., Serrão, E. A., Pearson, G. A. et al. (2008). Ecological genetics in the north Atlantic: environmental gradients and adaptation at specific loci. *Ecology*, **89**, S91–S107.

Schonbeck, M. and Norton, T. A. (1978). Factors controlling the upper limits of fucoid algae on the shore. *Journal of Experimental Marine Biology and Ecology*, **31**, 303–13.

Schonbeck, M. W. and Norton, T. A. (1980). Factors controlling the lower limits of fucoid algae on the shore. *Journal of Experimental Marine Biology and Ecology*, **43**, 131–50.

Seabra, R., Wethey, D. S., Santos, A. M. and Lima, F. P. (2011). Side matters: microhabitat influence on intertidal heat stress over a large geographical scale. *Journal of Experimental Marine Biology and Ecology*, **400**, 200–08.

Serrão, E. A., Brawley, S. H., Hedman, J., Kautsky, L. and Samuelsson, G. (1999). Reproductive success of *Fucus vesiculosus* (Phaeophyceae) in the Baltic Sea. *Journal of Phycology*, **35**, 254–69.

Sherrard, T. R. W., Hawkins, S. J., Barfield, P., Kitou, M., Bray, S. and Osborne, P. E. (2016). Hidden biodiversity in cryptic habitats provided by porous coastal defence structures. *Coastal Engineering*, **118**, 12–20.

Silva, A., Hawkins, S., Clarke, K., Boaventura, D. and Thompson, R. (2010). Preferential feeding by the crab *Necora puber* on differing sizes of the intertidal limpet *Patella vulgata*. *Marine Ecology Progress Series*, **416**, 179–88.

Silva, A. C. F., Boaventura, D. M., Thompson, R. C. and Hawkins, S. J. (2014). Spatial and temporal patterns of subtidal and intertidal crabs excursions. *Journal of Sea Research*, **85**, 343–8.

Silva, A. C. F., Hawkins, S. J., Boaventura, D. M. and Thompson, R. C. (2008). Predation by small mobile aquatic predators regulates populations of the intertidal limpet *Patella vulgata* (L.). *Journal of Experimental Marine Biology and Ecology*, **367**, 259–65.

Simkanin, C., Power, A. M., Myers, A. et al. (2005). Using historical data to detect temporal changes in the abundances of intertidal species on Irish shores. *Journal of the Marine Biological Association of the United Kingdom*, **85**, 1329.

Skov, M., Volkelt-Igoe, M., Hawkins, S., Jesus, B., Thompson, R. and Doncaster, C. (2010). Past and present grazing boosts the photo-autotrophic biomass of biofilms. *Marine Ecology Progress Series*, **401**, 101–11.

Skov, M. W., Hawkins, S. J., Volkelt-Igoe, M., Pike, J., Thompson, R. C. and Doncaster, C. P. (2011). Patchiness in resource distribution mitigates habitat loss: insights from high-shore grazers. *Ecosphere*, **2**, art60.

Smaal, A. C., Kater, B. J. and Wijsman, J. (2008). Introduction, establishment and expansion of the Pacific oyster *Crassostrea gigas* in the Oosterschelde (SW Netherlands). *Helgoland Marine Research*, **63**, 75.

Smale, D. A., Burrows, M. T., Moore, P., O'Connor, N. and Hawkins, S. J. (2013). Threats and knowledge gaps for ecosystem services provided by kelp forests: a northeast Atlantic perspective. *Ecology and Evolution*, **3**, 4016–38.

Sousa, L. L., Seabra, R., Wethey, D. S. et al. (2012). Fate of a climate-driven colonisation: demography of newly established populations of the limpet *Patella rustica* Linnaeus, 1758, in northern Portugal. *Journal of Experimental Marine Biology and Ecology*, **438**, 68–75.

Southward, A. J. (1956). The population balance between limpets and seaweeds on wave-beaten rocky shores. *Annual Report Marine Biological Station, Port Erin*, **68**, 20–9.

Southward, A. J. (1958). Note on the temperature tolerances of some intertidal animals in relation to environmental temperatures and geographical distribution. *Journal of the Marine Biological Association of the United Kingdom*, **37**, 49.

Southward, A. J. (1964). Distribution and Ecology of the Hermit Crab *Clibanarius erythropus* in the Western Channel. In D. J. Crisp, ed. *Grazing in Terrestrial and Marine Environments*, Blackwell Science, Oxford, pp. 265-73.

Southward, A. J. (1967). Recent changes in abundance of intertidal barnacles in south-west England: a possible effect of climatic deterioration. *Journal of the Marine Biological Association of the United Kingdom*, **47**, 81.

Southward, A. J. (1980). The Western English Channel – an inconstant ecosystem? *Nature*, **285**, 361–6.

Southward, A. J. (1991). Forty years of changes in species composition and population density of barnacles on a rocky shore near Plymouth. *Journal of the Marine Biological Association of the United Kingdom*, **71**, 495.

Southward, A. J. and Crisp, D. J. (1954). Recent changes in the distribution of the intertidal barnacles *Chthamalus stellatus* poli and *Balanus balanoides* L. in the British Isles. *The Journal of Animal Ecology*, **23**, 163.

Southward, A. J. and Crisp, D. J. (1956). Fluctuations in the distribution and abundance of intertidal barnacles. *Journal of the Marine Biological Association of the United Kingdom*, **35**, 211.

Southward, A. J., Hawkins, S. J. and Burrows, M. T. (1995). Seventy years' observations of changes in distribution and abundance of zooplankton and intertidal organisms in the western English Channel in relation to rising sea temperature. *Journal of Thermal Biology*, **20**, 127–55.

Southward, A. J., Langmead, O., Hardman-Mountford, N. J. et al. (2004). Long-term oceanographic and ecological research in the Western English Channel. *Advances in Marine Biology*, 47, 1–105.

Southward, A. J. and Orton, J. H. (1954). The effects of wave-action on the distribution and numbers of the commoner plants and animals living on the Plymouth breakwater. *Journal of the Marine Biological Association of the United Kingdom*, **33**, 1.

Southward, A. J. and Southward, E. C. (1978). Recolonization of rocky shores in Cornwall after use of toxic dispersants to clean up the *Torrey Canyon* spill. *Journal of the Fisheries Research Board of Canada*, **35**, 682–706.

Stafford, R. and Davies, M. S. (2004). Temperature and desiccation do not affect aggregation behaviour in high shore littorinids in north-east England. *Journal of Negative Results: Ecology and Evolutionary Biology*, **1**, 16–20.

Stafford, R. and Davies, M. S. (2005). Spatial patchiness of epilithic biofilm caused by refuge-inhabiting high shore gastropods. *Hydrobiologia*, **545**, 279–87.

Stephenson, T. A. and Stephenson, A. (1949). The universal features of zonation between tide-marks on rocky coasts. *The Journal of Ecology*, **37**, 289–305.

Stephenson, T. A. and Stephenson, A. (1972). *Life between Tidemarks on Rocky Shores*, W. H. Freeman & Co Ltd, New York.

Strain, E. M. A., Alexander, K. A., Kienker, S. et al. (2019). Urban blue: a global analysis of the factors shaping people's perceptions of the marine environment and ecological engineering in harbours. *Science of the Total Environment*, **658**, 1293–305.

Strain, E. M. A., Olabarria, C., Mayer-Pinto, M. et al. (2017). Eco-engineering urban infrastructure for marine and coastal biodiversity: which interventions have the greatest ecological benefit? *Journal of Applied Ecology*, **55**, 426–41.

Sunday, J. M., Bates, A. E. and Dulvy, N. K. (2011). Global analysis of thermal tolerance and latitude in ectotherms. *Proceedings of the Royal Society of London B: Biological Sciences*, **278**, 1823–30.

Sunday, J. M., Bates, A. E. and Dulvy, N. K. (2012). Thermal tolerance and the global redistribution of animals. *Nature Climate Change*, **2**, 686–90.

Sutton-Grier, A. E., Wowk, K. and Bamford, H. (2015). Future of our coasts: the potential for natural and hybrid infrastructure to enhance the resilience of our coastal communities, economies and ecosystems. *Environmental Science & Policy*, **51**, 137–48.

Svensson, C. J., Jenkins, S. R., Hawkins, S. J. and Åberg, P. (2005). Population resistance to climate change: modelling the effects of low recruitment in open populations. *Oecologia*, **142**, 117–26.

Svensson, C. J., Johansson, E. and Åberg, P. (2006). Competing species in a changing climate: effects of recruitment disturbances on two interacting barnacle species. *Journal of Animal Ecology*, **75**, 765–76.

Svensson, C. J., Pavia, H. and Åberg P. (2009). Robustness in life history of the brown seaweed *Ascophyllum nodosum* (Fucales, Phaeophyceae) across large scales: effects of spatially and temporally induced variability on population growth. *Marine Biology*, **156**, 1139–48.

Teagle, H., Hawkins, S. J., Moore, P. J. and Smale, D. A. (2017). The role of kelp species as biogenic habitat formers in coastal marine ecosystems. *Journal of Experimental Marine Biology and Ecology*, **492**, 81–98.

Thomas, M. L. H. (1965). Observations on the occurrence of *Cercaria patellae* Lebour in *Patella vulgata* L. on the Inner Farne. *Transactions of the Natural History Society of Northumberland, Durham and Newcastle-upon-Tyne*, **15**, 140–6.

Thompson, G. B. (1980). Distribution and population dynamics of the limpet *Patella vulgata* L. in Bantry Bay. *Journal of Experimental Marine Biology and Ecology*, **45**, 173–217.

Thompson, R. C., Johnson, L. E. and Hawkins, S. J. (1997). A method for spatial and temporal assessment of gastropod grazing intensity in the field: the use of radula scrapes on wax surfaces. *Journal of Experimental Marine Biology and Ecology*, **218**, 63–76.

Thompson, R. C., Moschella, P. S., Jenkins, S. R., Norton, T. A. and Hawkins, S. J. (2005). Differences in photosynthetic marine biofilms between sheltered and moderately exposed rocky shores. *Marine Ecology Progress Series*, **296**, 53–63.

Thompson, R. C., Norton, T. A. and Hawkins, S. J. (1998). The influence of epilithic microbial films on the settlement of *Semibalanus balanoides* cyprids – a comparison between laboratory and field experiments. *Hydrobiologia*, **375**, 203–16.

Thompson, R. C., Norton, T. A. and Hawkins, S. J. (2004). Physical stress andbiological regulation control pattern and process in benthic biofilms. *Ecology*, **85**, 1372–82.

Thompson, R. C., Norton, T. A., Roberts, M. F. and Hawkins, S. J. (2000). Feast or famine for intertidal-grazing molluscs: a mis-match between seasonal variations in grazing intensityand the abundance of microbial resources. *Hydrobiologia*, **440**, 357–66.

Thompson, R. C., Wilson, B. J., Tobin, M. L., Hill, A. S. and Hawkins, S. J. (1996). Biologically generated habitat provision and diversity of rocky shore organisms at a hierarchy of spatial scales. *Journal of Experimental Marine Biology and Ecology*, **202**, 73–84.

Torres, G., Giménez, L., Pettersen, A., Bue, M., Burrows, M. and Jenkins, Sr. (2016). Persistent and context-dependent effects of the larval feeding environment on post-metamorphic performance through the adult stage. *Marine Ecology Progress Series*, **545**, 147–60.

Trindade, A., Peliz, J., Diaz, J., Lamas, L., Oliveira, P. and Cruz, T. (2016). Cross-shore transport in a daily varying upwelling regime: a case study of barnacle larvae on the southwestern Iberian coast. *Continental Shelf Research*, **127**, 12–27.

Underwood, A. J. (1973). Studies on the zonation of intertidal prosobranchs (Gastropoda: Prosobranchia) in the region of Heybrook Bay, Plymouth. *Journal of Animal Ecology*, **42**, 353–72.

Van den Hoek, C. (1984). World-wide latitudinal and longitudinal seaweed distribution patterns and their possible causes, as illustrated by the distribution of Rhodophytan genera. *Helgoländer Meeresuntersuchungen*, **38**, 227–57.

Van den Hoek, C. and Donze, M. (1967). Algal phytogeography of the European Atlantic coasts. *Blumea*, **15**, 63–85.

Van Syoc, R. J., Fernandes, J. N., Carrison, D. A. and Grosberg, R. K. (2010). Molecular phylogenetics and biogeography of *Pollicipes* (Crustacea: Cirripedia), a Tethyan relict. *Journal of Experimental Marine Biology and Ecology*, **392**, 193–9.

Varela, R., Álvarez, I., Santos, F. and Gómez-Gesteira, M. (2015). Has upwelling strengthened along worldwide coasts over 1982-2010? *Scientific Reports*, **5**, 10016.

Vaz-Pinto, F., Olabarria, C. and Arenas, F. (2014). Ecosystem functioning impacts of the invasive seaweed *Sargassum muticum* (Fucales, Phaeophyceae). *Journal of Phycology*, **50**, 108–16.

Veiga, P., Torres, A. C., Rubal, M., Troncoso, J. and Sousa-Pinto, I. (2014). The invasive kelp *Undaria pinnatifida* (Laminariales, Ochrophyta) along the north coast of Portugal: distribution model versus field observations. *Marine Pollution Bulletin*, **84**, 363–5.

Vergés, A., Steinberg, P. D., Hay, M. E. et al. (2014). The tropicalization of temperate marine ecosystems: climate-mediated changes in herbivory and community phase shifts. *Proceedings of the Royal Society B: Biological Sciences*, **281**, 20140846.

Vieira, R., Pinto, I. S. and Arenas, F. (2017). The role of nutrient enrichment in the invasion process in intertidal rockpools. *Hydrobiologia*, **797**, 183–98.

Viejo, R. M. (1997). The effects of colonization by *Sargassum muticum* on tidepool macroalgal assemblages. *Journal of the Marine Biological Association of the United Kingdom*, **77**, 325.

Viejo, R. M., Arenas, F., Fernández, C. and Gómez, M. (2008). Mechanisms of succession along the emersion gradient in intertidal rocky shore assemblages. *Oikos*, **117**, 376–89.

Vinagre, C., Mendonça, V., Narciso, L. and Madeira, C. (2015). Food web of the intertidal rocky shore of the west Portuguese coast – determined by stable isotope analysis. *Marine Environmental Research*, **110**, 53–60.

Vye, S. R., Emmerson, M. C., Arenas, F., Dick, J. T. A. and O'Connor, N. E. (2014). Stressor intensity determines antagonistic interactions between species invasion and multiple stressor effects on ecosystem functioning. *Oikos*, **124**, 1005–12.

Wangkulankul, K. (2016). Community level effects of variable recruitment of a key species *Mytilus edulis* L. in the rocky intertidal. PhD, Bangor University.

Wangkulankul, K., Hawkins, S. J. and Jenkins, S. R. (2016). The influence of mussel-modified habitat on *Fucus serratus* L. a rocky intertidal canopy-forming macroalga. *Journal of Experimental Marine Biology and Ecology*, **481**, 63-70.

Wares, J. P. and Cunningham, C. W. (2001). Phylogeography and historical ecology of the north Atlantic intertidal. *Evolution*, **55**, 2455–69.

Weber, L. I. and Hawkins, S. J. (2002). Evolution of the limpet *Patella candei* d'Orbigny (Mollusca, Patellidae) in Atlantic archipelagos: human intervention and natural processes. *Biological Journal of the Linnean Society*, **77**, 341–53.

Weber, L. I. and Hawkins, S. J. (2005). *Patella aspera* and *P. ulyssiponensis*: genetic evidence of speciation in the North-east Atlantic. *Marine Biology*, **147**, 153–62.

Wethey, D. S. and Woodin, S. A. (2008). Ecological Hindcasting of Biogeographic Responses to Climate Change in the European Intertidal Zone. In J. Davenport, G. M. Burnell, T. Cross, et al., eds. *Challenges to Marine Ecosystems: Proceedings of the 41st European Marine Biology Symposium*, Springer Netherlands, Dordrecht, pp. 139–51.

Wethey, D. S., Woodin, S. A., Hilbish, T. J., Jones, S. J., Lima, F. P. and Brannock, P. M. (2011). Response of intertidal populations to climate: Effects of extreme events versus long term change. *Journal of Experimental Marine Biology and Ecology*, **400**, 132–44.

Wilson, D. P. (1968). The settlement behaviour of the larvae of *Sabellaria alveolata* (L.). *Journal of the Marine Biological Association of the United Kingdom*, **48**, 387.

Witte, S., Buschbaum, C., van Beusekom, J. E. E. and Reise, K. (2010). Does climatic warming explain why an introduced barnacle finally takes over after a lag of more than 50 years? *Biological Invasions*, **12**, 3579–89.

Zardi, G. I., Nicastro, K. R., Canovas, F. et al. (2011). Adaptive traits are maintained on steep selective gradients despite gene flow and hybridization in the intertidal zone. *PLoS ONE*, **6**, e19402.

Chapter 3

# The Ecology of Rocky Subtidal Habitats of the North-East Atlantic

Keith Hiscock, Hartvig Christie and Trine Bekkby

## 3.1 Introduction

Ecological studies of subtidal hard substratum in temperate regions essentially started with the work of Kitching (1941) on the effects of light and wave action on kelp forest communities on the west coast of Scotland, carried out between 1932 and 1934. Much work has subsequently been done to better understand what lives where and how distinctive assemblages of species are distributed in relation to environmental conditions in the temperate regions of the world.

Rocky subtidal communities are among the most species-rich and productive ecosystems on the planet (e.g., Christie et al., 2009; Teagle et al., 2017). Assemblages of species that rely on the presence of biological structures, such as those associated with kelp plants, have been described (Moore, 1973; Schultze et al., 1990; Christie et al., 2007; Norderhaug et al., 2014). Impacts of grazing (scraping of seabed biota especially by sea urchins) have been observed, documented and experimented on (reviewed in Kraufvelin, 2017). Less clear are what are natural long-term fluctuations or change that override local impacts. This review emphasises work in the subtidal rocky habitats of the north-east Atlantic, with a focus on kelp forests (see Section 3.4) and the threats they face. We deal mostly with natural substrata, but also mention what can be learned by studying succession on artificial surfaces.

## 3.2 Environmental Interactions: Abiotic Factors Structuring Rocky Subtidal Communities

On a geographical scale, the distribution of species is determined by the major oceans they are confined within, and by more local conditions such as the temperatures and food availability they are attuned to thrive or survive in. Additionally, there may be barriers to spread, such as extensive areas of sediment between rocky seabeds preventing larvae from reaching a suitable substratum to settle on in a different region or subregion during their lifespan. In the north-east Atlantic, three major biogeographical regions are identified: a warm temperate 'Lusitanian' region that influences Spain and Portugal including the Azores; a colder 'Boreal' region in the Irish Sea, off the northern seaboard of Britain and in the North Sea; and an 'Arctic' region that includes northern Norway and northern Iceland. In between are Lusitanean–Boreal, Boreal–Lusitanean and Arctic–Boreal regions. These regions are described and mapped in Hiscock (1998, 2018).

Seabed type, illumination (the amount and type of light reaching the seabed depending on depth, water turbidity and shading), strength of wave action and tidal currents, and water salinity are among the local conditions that create a

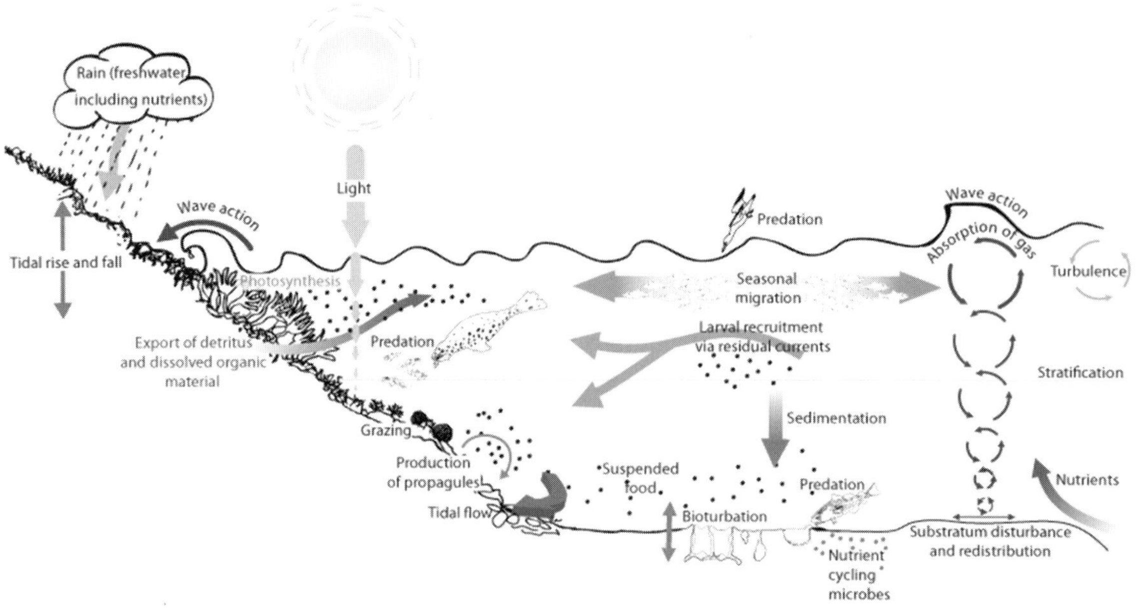


**Fig. 3.1** A summary of the main physical, chemical and biological processes that influence shallow seabed communities on the open coast of the north-east Atlantic. From Hiscock (2018).

particular underwater habitat, and therefore determine the community of species that lives there (Figure 3.1; see Hiscock, 1983, 1985). In addition to grazing, species interactions, habitat size and complexity are important factors for species richness and abundance (Teagle et al., 2017), and can dictate the type of biotic interactions that occur there, such as predation, competition and succession.

### 3.2.1 Light and Depth

The vertical zonation of species on rocky subtidal habitats is ultimately determined mainly by the attenuation of light with depth (Hiscock, 1985). At a depth where about 1 per cent of surface illumination persists, the kelp forest becomes a park and foliose algae become dominant to a depth where about 0.1 per cent of surface light is present, below which only sessile animals dominate the rocks. What those depths are in metres below sea level depends on the turbidity of the water, on grazing that may reduce downward extent of algae, and on competition with suspension feeding animals that may be abundant in strong water movement.

The zone dominated by dense kelp forest is mainly in the shallow parts of the kelp zone (e.g., 0–15 m depth), while, at the deepest part, the kelps are more scattered and smaller in individual size. This is important both for production and for associated organisms. The large shallow kelps with luxuriant associations of epiphytic growth on the stipes host a high number of animals (Christie et al., 2003; Teagle et al., 2017), while the small kelps at the deeper zones have fewer associated organisms. The rocky surface between the kelps may also be substrata for smaller seaweeds and sessile animals. Animals predominate below the zone dominated by algae. This is the circalittoral zone and examples of species present there in different biotopes can be found in Hiscock (1985, 2018) and in the descriptions of separate circalittoral biotopes located via web-based resources (e.g., MarLIN www.marlin.ac.uk; European Union Nature Information System [EUNIS]: https://eunis.eea.europa.eu/).

### 3.2.2 Wave Action and Topographical Features

Within the vertical zones established mainly by light, the composition of plants and animals living on the rocky seabed will be largely determined by the strength of wave action and tidal streams. Wave action is a potentially destructive force: to survive in very wave-exposed conditions, species need to be able to resist mechanical disturbance or seek shelter during storms. With increasing shelter from wave action, rich communities may develop, but where wave action and tidal currents are slight, siltation may occur and reduce species attached to the rocky surface. While wave action may be the dominant water movement factor in very exposed conditions, the strength of tidal currents is often a determining factor for the development of particular communities (see Hiscock, 1983). The degree of exposure to wave action also has subtle effects on kelp forest growth, distribution and density (Kain, 1971; Bekkby et al., 2009, 2014). Water flow has a great impact on macroalgae, directly (breakage and dislodgement) or indirectly, affecting e.g., photosynthesis, nutrient uptake, transport across boundary layers, settlement and recruitment, and growth (Hepburn et al., 2007). For instance, wave action moves the algal fronds, thereby maximising the area available to trap light and ensuring a high nutrient flux (Lobban and Harrison, 1994). Macroalgae also show variation in characteristics and morphology with different levels of wave exposure (Pedersen et al., 2012; Bekkby et al., 2014).

Local shading by topographical features (caves, overhangs) can influence species distribution on subtidal rocky habitats. Such features reduce the ability of algae to grow, and promote the settlement of species such as the corals *Alcyonium hibernicum*, *Hoplangia durotrix* and *Leptopsammia pruvoti*, which favour overhangs and caves. Wave surge (for instance, at the ends of seaward-facing gullies) is also an important modifying factor and increases the force of wave action greatly. Where wave surge occurs, thriving communities of highly tolerant species that benefit from the large amount of food brought to them can be found. There, scour can be important, especially where there are cobbles and pebbles that are moved by wave action. The community that develops in such gullies is the 'surge gully community' described in Hiscock and Mitchell (1980). The walls of gullies and caves may show a distinct zonation from bare rock at the base, through sparse fast-growing species such as barnacles and tube worms, to more long-lived species such as colonial sea squirts and bryozoans above the level of scour. Such zonation was described from studies in a submarine gulley by Kitching et al. (1934). At the other end of the spectrum of wave and tide exposure, very sheltered sites and sites where turbidity is high are likely to suffer siltation on upward-facing surfaces, with consequent smothering of all species but those that can protrude above the silt. One such location is Abereiddy Quarry in west Wales (Hiscock and Hoare, 1975).

## 3.3 Species Interactions: Biotic Factors Structuring Rocky Subtidal Communities

In rich coastal ecosystems, ongoing processes such as competition for space, grazing and predation are important for the structure and functioning of the system (Figure 3.1). Provision of habitat by specific species is therefore crucial.

### 3.3.1 Habitat Provision

There are many species that, by their presence, facilitate the occurrence of associated species, often creating rich assemblages or providing a habitat for species that would not otherwise occur. Some are structurally important such as kelp plants and horse mussels (*Modiolus modiolus*). Some may provide an obligatory substratum for commensal species.

Kelp plants have been greatly studied as providing a habitat for a large number of associated species (reviewed by Teagle et al., 2017). Fronds in particular are an important host: for instance, in Norway, on average 130 species and 8,000 individuals are recorded on individual *Laminaria hyperborea* plants (Christie et al., 2003). Kelp holdfasts, which serve as the attachment point of the plant to resist wave action, represent structurally complex habitats with nooks and crannies for other sessile species to settle on and for mobile species – such as amphipod and worms – to shelter in and avoid predation. Moore (1973) recorded 387 macrofauna and meiofauna species associated with the holdfasts of *L. hyperborea* from north-east England. Different macrophytes support different faunal occurrences, and, as such, coastal community diversity and abundance is dependent the provision of habitat by a variety of macrophyte species (Christie et al., 2007, 2009). The large three-dimensional kelp beds provide both food and shelter for a number of fish species (Norderhaug et al., 2005), while some juvenile fish are more dependent on seeking shelter when mainly feeding on pelagic food items.

While most of the species living attached to or within the kelp are not, some may be harmful to the host, and many of these relationships are specific between host and tenant, such as the attachment of the hydroid *Obelia geniculata* and the bryozoan *Membranipora membranacea* specifically to fronds of *L. hyperborea*, *Saccorhiza polyschides* and *Saccharina latissima*. Such 'fouling' by epibiota can be an impediment to photosynthesis, and while some annuals algae may die-off with their associated epibiota, perennials get rid of their fouling communities by shedding their fronds once a year in spring or autumn, regrowing 'clean' fronds, such as *L. hyperborea* which sheds and regrows its fronds in spring each year.

Among the animal species that are structurally important and provide space for a rich associated community on rocky and sediment habitats, horse mussel beds (*M. modiolus*) are particularly widespread in northern waters of the north-east Atlantic (see Fariñas-Franco et al., 2013). They provide a complex habitat with shelter from predation for some species and a substratum for the attachment of others. However, it is the turf of erect branching bryozoans and hydrozoans that are most widespread on circalittoral rocks and that shelter a diverse community of polychaetes, crustaceans and molluscs. Some species require the presence of a specific host. For instance, the barnacle *Adna anglica* grows only on stony corals especially *Caryophyllia smithii*. The barnacle most likely benefits by being raised above the substratum into water currents but kept clear of silt, although there seems no benefit to the coral. The pink sea fan *Eunicella verrucosa* provides a substratum for species that may benefit by being raised up into water currents to feed or that feed on the sea fan. Animals that live specifically on the sea fans include the sea fan anemone *Amphianthus dohrnii* and the barnacle *Hesperibalanus fallax* (also on plastic and hydroids) while those that feed specifically on the tissue but seem to cause no conspicuous damage are the sea slug *Tritonia nilsodhneri* and the false cowrie *Simnia hiscocki*. Other species take advantage of the branches by wrapping the tendrils of their egg cases around them: the catsharks *Scyliorhinus canicula* and *Scyliorhinus stellaris*.

### 3.3.2 Trophic Interactions

Feed or be fed upon is, perhaps, the most expected interaction with the environment and between animal species. One of the most dramatic changes that occur in shallow subtidal rock communities is the spring outburst of hydroids – often rapidly followed by the outburst of sea slugs that eat them (Hiscock, 2018). The variety of sea slugs is very large, and their feeding specificity to a particular species (usually a hydroid but also other cnidarians) is remarkable. Sporadic settlements of mussels *Mytilus edulis* may become a dominant feature of subtidal rocks, but these will most likely be consumed by dense bands of the starfish *Asterias rubens*, leaving settlement space for whatever sessile species have larvae in the plankton at that time. Starfish also feed on clumps of horse mussels *M. modiolus*, depriving species that require shelter between the mussel shells. Other feeding activities are continuous, with crustacean and worm species living among foliose algae and in the bryozoan/hydrozoan turf being subject to grazing by wrasse and larger decapod crustaceans in particular. Many species, such as squat lobsters *Galathea* sp(p). and juvenile lobsters *Homarus gammarus* and *Palinurus elephas,* hide away under boulders or in fissures during the day to avoid being eaten by fish or larger lobsters. The extent to which mobile predatory species that interact with rocky subtidal benthos, especially wrasse, are prey to larger predatory fish (like cod) and seals (*Halichoerus grypus* and *Phoca vitulina*) is difficult to know: the study by Brown et al. (2012) did not find the otoliths of any wrasse species in faecal samples from seals, but the study was mainly from samples in northern

Britain where wrasse are sparse. However, the remains of other benthic species – short-spined and long-spined sea scorpions (*Scorpaena* spp.) – were found. Seals may anyway take only parts of the flesh and organs, consuming no or few bones.

Species living attached to rocky substrata in the coastal zone rely greatly on organisms (plankton) and organic matter suspended in the water column for food. That food is either brought to them by water currents (passive suspension feeding), or collected by some pumping action (active suspension feeding), or by a combination of both. Some of that organic material is detrital, derived from primary production by seaweeds.

A number of feeding strategies by mesograzers and macrograzers occur, revealed by laboratory experiments and field studies based on stable isotope measurements (see Fredriksen, 2003; Christie and Norderhaug, 2017). While many ephemeral algae are grazed directly, the perennial species are less palatable and mainly become a food source after being detached and broken down to smaller particulate organic carbon enriched by microorganisms. Particularly for kelp forests (grazing of kelp is further discussed in Section 3.4), the primary production is very high, higher than the consumption by the herbivores, leading to a high export of kelp material providing food for animals living outside the kelp forests and also outside rocky coastal habitats.

### 3.3.3 Symbiosis (Mutualism and Commensalism)

Many of the reasons for species living together and presumably interacting require more natural history observation than experimentation. Conger eels, *Conger conger*, for example, live in holes in the reef and are often accompanied by prawns *Palaemon serratus*, spider crabs *Inachus* sp (p). and squat lobsters *Galathea* sp(p). Those species may 'know' they are not favoured food for the conger, and so gain protection by living in its proximity. Some, perhaps the prawns, may also perform a cleaning function for the conger. While cleaning is a much studied and photographed activity in tropical reef species, it is less observed and studied in temperate regions. Nevertheless, cleaning of parasites, such as copepods, and more generally of mucus and detritus is known to occur in the rocky subtidal waters of the north-east Atlantic. For instance, the observant scuba diver will see benthic fish being cleaned by other fish – usually ballan wrasse *Labrus bergylta* by rock cook wrasse *Centrolabrus exoletus*. The extent of this cleaning behaviour has led to a fishery and captive breeding programmes for wrasse, then introduced into salmon farm cages to clean fish lice from the salmon. Cleaner shrimps occur in the north-east Atlantic, with *Periclimenes sagittifer* associated with snakelocks anemones *Anemonia viridis*. There is a remarkable image by Paul Naylor of a spider crab *Inachus* sp. that has hitched a ride on a John Dory *Zeus faber* and is cleaning the parasites from its flank. Sharp-eyed scuba divers with cameras will, no doubt, observe and photograph more of such interactions. Further, habitats that provide a variety of niches will provide microhabitats for a number of species living close together. Seaweeds of different structure or other differences in characteristics will create habitats for a diverse community over small spatial scales. Species living together where some are facilitated by another species was described earlier.

### 3.3.4 Competition for Space

Occupying space may be considered a mainly passive interaction between sessile species: get there first and you win your place, but get there later and you may have to overgrow whatever is already settled in order to succeed. Such succession may produce a 'layered' community, with the initial fast-growing colonising species (for instance, tube worms or barnacles) being overgrown by erect branching bryozoans and hydroids. Some species actively compete for space. For instance, sea anemones that reproduce by longitudinal fission or by budding may be aggressive and produce specialised 'catch' tentacles to at least ward-off rivals, resulting in a clear demarcation between separate clones with a 'no-man's-land' in between (K. Hiscock, own observations). If competition for space is moderated by some grazing, free surfaces are continuously being made available for settlement and there is potential for a wide range of species to settle and create a high biodiversity. Sebens

(1985) observed: 'If the space-clearing factor acts preferentially on the competitively dominant species (e.g., selective predators), then lower total rates of disturbance might produce the maximum diversity' (p. 361).

### 3.3.5 Succession and Community Change

Change continuously happens in subtidal rock communities. Whether it is seasonal change, change resulting from events such as destructive storms, or change that follows mass settlement of, for instance, mussels, there will be colonisation of free spaces and there will be interaction between species competing for that space. The broad 'picture' of succession and competition on hard substrata has been described my many authors and is summarised in Sebens (1985). Sebens identified that a true climax community will not develop if there is grazing by sea urchins which frees-up space and invites competition from colonising species (see Section 3.3.6 on recruitment and Section 3.4 on sea urchin–kelp interaction). Those who need to better understand the character of 'change' (for instance, those who are charged with interpreting the results of monitoring to assess favourable conservation status) should read papers that address interactions and variability by Sebens and his co-workers.

A great deal was learned about colonisation and succession on hard substrata when a disused frigate, ex-*HMS Scylla* was sunk to the seabed in south-west England in March 2004 (Hiscock et al., 2010). In the first two years, unexpected settlements of the sea urchin *Psammechinus miliaris* (which is normally only seen under boulders in south-west England, during the day at least) and of the queen scallop *Aequipecten opercularis* occurred. Other species settled more-or-less as expected, with, for example, plumose anemones *Metridium senile* settling in September of the first year, growing rapidly and reproducing by basal fission, while dead men's fingers *Alcyonium digitatum* did not settle until 2005; both growing to become dominant on parts of the reef. Elegant anemones *Sagartia elegans* settled in the first year to compete between separate clones for space. Notably, pink sea fans, *E. verrucosa* – a protected species in Britain – settled in August 2007; the

**Fig. 3.2** The bridge of ex-*HMS Scylla* after five-and-a-half years on the seabed. Colonisation of this deliberately placed frigate revealed a great deal about colonisation processes (see Hiscock et al., 2010). Image from Keith Hiscock.

nearest colonies were about 50 m from the sunken vessel attached to natural rock. Sea urchins *Echinus esculentus* were slow to settle and show themselves as large individuals, but never became space-makers. Over the winter of 2004/2005, the sea urchins (*P. miliaris*) had 'free rein' and, one year on from the vessel being placed on the seabed, it had been largely grazed clean. Wrasse species were slow to colonise the reef, but, following their arrival, they, along with other fish species, feasted on the urchins, exerting an important population control. Under absent/reduced grazing, colonisation took-off. Some species were ephemeral, colonising the reef and being present for a few months or years before disappearing or becoming less abundant. Four to five years after *HMS Scylla* was placed on the seabed, almost all of the species that were to eventually characterise the reef had settled (Figure 3.2).

Artificial structures can offer alternative hard substrata in the subtidal zone, and be the scene of colonisation and succession. In a fjord site, for instance, artificial structures were initially dominated by the growth of the tunicate *Ciona intestinalis*, and later by a more diverse assemblage of sessile animals. These structures were then gradually invaded by fish, mainly gobiids and wrasses (unpublished report from a southern Norwegian fjord). In more open coastal sites further north, *C. intestinalis*, *Pomatoceros* sp. and some filamentous algae represented the initial growth, while kelps and codfish invaded at a later stage.

### 3.3.6 Recruitment

Rocky coasts may differ from other habitats because of their physical longevity. However, many of the seaweed habitats found along rocky coasts are of limited and often seasonal duration. Associated mobile animals have to be able to colonise whenever suitable substrata (especially seasonal growths of seaweeds) emerge, and must colonise another habitat when the existing habitat suddenly disappears. Different experiments with artificial seaweeds have found the kelp/seaweed fauna to be highly mobile, regularly swimming or drifting between their preferred habitats (Norderhaug et al., 2002; Jørgensen and Christie, 2003; Waage-Nielsen et al., 2003). The mobility of the animals living on the seasonal epiphytic algae is relatively higher than those that prefer more persistent habitats. This strategy of high mobility ensures a high diversity of fauna can rapidly exploit habitats of limited duration; however, it also leads to the risk of being transported out of the seaweed beds (Christie and Kraufvelin, 2004).

What is greatly forgotten in studies of colonisation (especially of recovery following damage) and in making predictions about effects of climate change on species distribution is that not all low-mobility species (sessile, sedentary or slow moving) have larvae or propagules that are capable of travelling significant (>1 km) distances to colonise new areas or extend distributional ranges. Add to that the natural barriers to spread (extensive areas of unsuitable substrata, physiographic features that deflect current flow offshore, etc.) and the likelihood of rapid spread or recovery after loss of many species becomes questionable. The differences in dispersal capability between species have been known for many years (see Kinlan and Gaines, 2003). Algal spores and zygotes especially seem to travel only short distances (although other methods of dispersal such as rafting occur). As more studies are published (for instance, Nickols et al., 2015), the high likelihood that recruits will stay close to their natal location is becoming clearer and scientifically naive concepts of connectivity and networks of Marine Protected Areas look untenable, when considering many of those low-mobility species at least.

## 3.4 The Case of Kelp Forests

### 3.4.1 Sea Urchins Densities on Rocky Subtidal Seabeds

Different species of sea urchins are naturally found on north-east Atlantic rocky subtidal seabeds in low densities (Lawrence, 1975; Skadsheim et al., 1995; Sjøtun et al., 2006). But, sometimes, the sea urchins become abundant and reach high densities (in places, more than 100 individuals per square metre) that graze down macroalgae (e.g., Sivertsen, 1997) and other attached species. For instance, in heavily grazed areas in Lough Hyne (Ine) in south-west Ireland, Kitching and Thain (1983) recorded as many as sixty-one *Paracentrotus lividus* per square metre, leaving rocks dominated by calcareous algae. Densities do not need to be very high to result in urchin barrens. Comely and Ansell (1988) report up to four *E. esculentus* per square metre at Millport in the Firth of Clyde, although Bishop and Earll (1984) report densities of only 0.1–1.3 urchins per square metre at St Abbs in south-east Scotland. Such densities of *E. esculentus* appear the norm on open or semi-enclosed rocky coasts in most parts of Scotland, and result in urchin barrens on rock and bare kelp stipes. In other parts of Britain, the density of *E. esculentus* seems much lower and, although local grazed patches are seen, there are not urchin barrens. For instance, 0.06–0.1 per square metre seems the likely density at Skomer in west Wales (Burton et al., 2015) and, although this density is half that recorded there in 1981 (Bishop and Earll, 1984), it seems similar to other parts of south-west Britain (K. Hiscock, own observations). Norderhaug and Christie (2009) have reviewed sea urchin grazing in the north-east Atlantic from the Canary Islands to Spitsbergen, where grazing may be overwhelmingly important in some regions and changes the distribution, abundance, species distribution and richness of rocky subtidal communities. The species that most conspicuously graze algae, barnacles and erect bryozoans in the shallow subtidal are the sea urchins *Echinus acutus* (Scandinavia only), *Strongylocentrotus droebachiensis* (from Orkney northwards to Spitsbergen), *E. esculentus* (all coasts), *P. miliaris* (mainly in temperate regions) and *P. lividus* (parts of western Ireland and further south to the Mediterranean). There are other species that may be significant grazers in subtidal rocky areas including chitons, small limpets and other gastropods (see Norderhaug and Christie, 2009; Teagle et al., 2017).

### 3.4.2 Sea Urchins: Key Grazers of Kelp

Jones and Kain (1967) first demonstrated the importance of urchin grazing in determining the downward extent of kelp (*L. hyperborea*) in the north-east Atlantic. At Port Erin on the Isle of Man, high numbers of *E. esculentus* were removed from a strip about 10 m wide extending to 11 m below chart datum level. In the lower 3 m, at the start of the experiment, there were no foliose algae. During the three-year course of the experiment, nearly 3,000 sea urchins were removed. *L. hyperborea* plants then developed to the bottom of the strip and remained for the duration of the experiment: their absence had been due to grazing. In a similar study, after removing sea urchins, Leinaas and Christie (1996) found an initial growth of filamentous algae, rapidly outcompeted by the kelp *S. latissima*. After about four years, succession to the dominant kelp *L. hyperborea* took place. A similar macroalgal succession took place along the Norwegian coast by deploying large artificial structures (unpublished results). By removing *L. hyperborea* beds (kelp harvest, see Christie et al., 1998), the new generation of kelps rapidly overgrew early successional seaweeds and recovered in size and biomass, but the succession of associated organisms and a total community recovery was a process of longer duration.

Among a number of grazing events reported (see Norderhaug and Christie, 2009), the most extensive one is the green sea urchin (*S. droebachiensis*) – first reported in the early 1980s (see Sivertsen, 1997) – grazing many million tonnes of kelp (*L. hyperborea* and *S. latissima*) along the northern Norwegian coast (Figure 3.3). This event is still maintaining large barren coastal areas (see later). The effects of grazing by high abundances of urchins are stark, replacing highly productive kelp forests with desert-like barren grounds. For instance, *L. hyperborea* plants have stipes bare of the usual attached algae, entire kelp forest can be removed, and rock below the kelp canopy and deeper is dominated by pink encrusting algae without sponges and without the rich turf of erect bryozoans, hydrozoans and other erect species that occur away from urchin grazing. The rock looks bare and species richness and abundance is greatly reduced – especially that which is normally hidden within the turf of algae or sessile fauna.

**Fig. 3.3** Forest of *L. hyperborea* with the rocks below heavily grazed by the sea urchin *S. droebachiensis*. Photographed on the Norwegian coast. Image from Hartvig Christie.

### 3.4.3 Phase Shift: From Kelp Forest to Barrens, and Back

Kelp forest and sea urchin-dominated barrens are two alternative stable states of an ecosystem (Steneck et al., 2013; Ling et al., 2015). There are several studies on why urchin 'infestations' occur and how they are reduced. For instance, grazing of kelp has been caused by reduced predation on sea urchins due to over-exploitation of top predators (Leleu et al., 2012; Steneck et al., 2013). In the north-east Atlantic, crabs have been found to be important predators on sea urchins (Clemente et al., 2007; Fagerli et al., 2014). Environmental factors may also influence the cold-water species *S. droebachiensis,* which may have had a higher survival during the relatively cold period of the

1960s and 1970s when destructive grazing first started in Norway. Since then, the sea urchins occurred in high densities and resulted in a large-scale and persistent grazing of the kelp forests in sheltered and moderately wave-exposed areas north of 63°N and all the way into Russian waters (Sivertsen, 1997; Norderhaug and Christie, 2009). Skadsheim et al. (1995) have reported the first findings of population reductions of sea urchins in parts of mid Norway, with a possible explanation related to an infestation by the endoparasitic nematode *Echinomermella matsi* or other waterborne agents. More recently, a gradual northward decline in sea urchin populations followed by kelp forest recovery has been documented. Field studies and field experiments have linked this to increased ocean temperatures (Fagerli et al., 2013; Rinde et al., 2014) coupled with increased predation pressure on the sea urchins by crabs (Norderhaug and Christie, 2009; Fagerli et al., 2014). Whatever the reason for the decline in abundance of urchins, kelp forest and associated macroalgae, macrofauna and fish expanded their distribution into former barren grounds. Additionally, experimental removals of sea urchins (Leinaas and Christie, 1996, but also more recent unpublished experiments with quicklime and artificial substrata) have confirmed a rapid recovery of seaweeds and succession back to former ecosystem conditions. However, recently, heavy grazing by high densities of *E. acutus* has been observed as an increasing problem in western Norwegian fjords below the thermocline/halocline (unpublished observations). In Ireland, at Lough Hyne (Ine), *P. lividus* were once abundant (until at least the early 1970s), controlling the growth of shallow subtidal algae (Kitching and Ebling, 1961). By the late 1990s, the structure and size of the population had changed substantially, with a much lower abundance of urchins and higher cover of foliose and filamentous algae (Barnes and Crook, 2001). Such changes were also observed in other populations in western Ireland, and were considered the result of large-scale mortality of adult urchins combined with recruitment failure: perhaps due to this species being at the northern limits of its geographical distribution.

In recent years, interdisciplinary collaborations have focussed on marine ecosystem services. Important services are associated with kelp forests, and, as such, they have been attributed economic values (Smale et al., 2013). Kelp forests contribute to bioremediation, carbon storage and sequestration, production to different levels in the food chain, fishery of commercial species and to tourism/recreation (see, for instance, Birkett et al., 1998).

## 3.5 Human Impacts

Humans are a species that interact with seabed marine life mainly through coastal constructions, offshore structures, fishing, harvesting, pollution and the introduction of non-native species.

Seaweed communities are impacted directly through harvesting (e.g., *Laminaria digitata* harvesting in France and *L. hyperborea* in Norway for alginate production (Christie et al., 1998; Davoult et al., 2011). Despite being regulated, kelp harvest affects parts of the kelp bed, but the areas recover over a few years. The effects of natural seaweed exploitation are dependent on how large a part of the standing stock is harvested and how often. In addition to the intended removal of kelps, harvesting can have a longer-term effect on associated fauna. The effects on rocky coast ecosystems may be of a permanent character if the same areas are harvested at regular intervals after recovery.

Harbours, breakwaters and other constructions may create unnatural habitats that introduce hard substratum into areas (such as estuaries and bays) where it did not previously exist (Foster et al., 2016). Offshore structures attract species with good dispersal capabilities and are colonised rapidly – often acting as stepping stones for non-native species which can then spread to new environments. Non-native species may fit in to natural communities but a few may become dominant and compete with native species: for instance, the slipper limpet *Crepidula fornicata* and wakame (kelp) *Undaria pinnatifida* in the UK (Epstein and Smale, 2017a, 2017b).

Rocky subtidal habitats are often protected by their broken nature from the destructive effects of fishing, including overfishing and the use of heavy mobile fishing gear. Where reefs are reasonably flat, gear such as scallop dredges may impinge on the reef and damage it by displacing and breaking attached species. Fishing activity may damage reef communities, including species with life history traits that suggest they will not recover: for instance, some branching sponges and anthozoans (although information on larval dispersal distances, growth rates and longevity is often very sparse and conclusions have to be reached on observations of persistence). Not knowing what the interactions between species are, and what the knock-on effects of those interactions might be, makes management for biodiversity conservation and sustainable fisheries problematic. One such dilemma relates to a novel fishery for wrasse (used as cleaner fish in salmon farms, see Section 3.3.3). Wrasse are known to be territorial and to occupy small spatial areas (see, for instance, Villegas-Rios et al., 2013), meaning that local depletions are highly likely. Because the larval dispersal capabilities of the wrasse are not known, its population recovery potential following depletion is difficult to estimate (although dispersal distance may be limited, and therefore genetic isolation may occur, see Skiftesvik et al., 2015). Because wrasse are important space clearers, feeding vigorously on crustaceans, worms and gastropods in the algal and animal turf attached to rocks (and also urchins, see Section 3.3.5), they allow a higher diversity of species to colonise. If populations are depleted, the extent of small turf species may become larger and other organisms feeding on wrasse, such as some seabird species (shags, cormorants) and perhaps seals, may have less food available to them. However, these knock-on effects for biodiversity conservation seem only minor. Fishing may further impact subtidal ecosystems by removing key predators. Indeed, overfishing of larger predatory fish may have crucial effects by allowing their prey organisms to thrive. As mentioned earlier, increases in sea urchin may occur where fish (including shellfish) large enough to feed on them have been reduced in abundance or extirpated.

Eutrophication is a worldwide problem (see Filbee-Dexter and Wernberg, 2018) harming rocky subtidal species, and especially perennial algal communities such as kelp beds. In eutrophic conditions, kelp beds are overgrown and outcompeted by turf algae that form a new stable, but less complex and diverse, state.

## 3.6 Large-Scale Latitudinal Gradients and Ocean Warming

The geographical distributions of species that characterise subtidal rocky area are well documented in the north-east Atlantic but may change as seas get warmer (for Great Britain see Hiscock et al., 2004, for Norway see Rinde et al. 2014, for changes in kelp species distribution in the southern part of the Atlantic European coast, see Araújo et al., 2016). Some likely changes might be seen as distributional boundaries shift (warmer water species might extend northwards and/or become more abundant where they occur; colder water species might become less abundant at their southern limits and/or their southern boundary will retreat northwards). In particular, important consequences for the ecosystem as a whole might occur, especially involving species interactions. The habitat provision by the warm-water kelp *Laminaria ochroleuca*, with its smaller holdfast and smoother stipe, is likely to be much less than that provided by *L. hyperborea* (Arnold et al., 2016; Teagle and Smale, 2018). Cold-water early successional species such as *Alaria esculenta* may no longer occur (Hawkins and Harkin, 1985), changing successional pathways. Those impacts might be very substantial.

For a stark example of significant change, the increase in the rate of flow of the East Australian Current which carries warm water south along the eastern coast of Australia has led to the east coast of Tasmania becoming both warmer and saltier, with mean trends of 2.28°C/century and 0.34 of salinity/century over the 1944–2002 period. That warmer water has resulted in changes to seabed ecosystems that are especially related to the increased southerly extent and abundance of sea urchins (*Centrostephanus rodgersii*), resulting in catastrophic overgrazing of productive Tasmanian kelp beds, leading to loss of biodiversity and important rocky reef ecosystem services. The impact of increased abundance of urchins were compounded by overfishing of spiny lobsters that eat urchins (see Ling et al., 2009). By comparison, changes in the north-east Atlantic, for the moment, have been fewer and less dramatic (see also Chapter 16 by Connell et al.).

Along the Norwegian coast, the cold-water sea urchins *S. droebachiensis* are retreating northwards as seawater temperature increases (Rinde et al., 2014), allowing degraded kelp forests to recover in the southern part of the large barren grounds. The retreat of sea urchins and the recovery of kelp is also enhanced by the temperate predatory crabs *Carcinus maenas* and *Cancer pagurus* extending their distribution northwards (Fagerli et al., 2014) and eating the sea urchins efficiently. Although the recovery of kelp forests in the former barrens in the southern part of Norway may be considered positive, as structural complexity and the provision of ecosystem services improve, the implications of the range extension of crabs leading to regime shift along large coastal areas may be cause for concern.

A scenario to think about in Britain is that *P. miliaris* and small *E. esculentus* are currently important grazing species in some sea lochs – perhaps because wrasse occur in low abundance in colder waters. If increasing seawater temperature leads to more urchin-eating wrasse, will urchins become less abundant and urchin barrens become recolonised?

## 3.7 Conclusions

There are many studies that describe the character and composition of the plant and animal communities that populate rocky subtidal habitats. Those studies have culminated in the identification of distinctive biotopes (habitats and their associated species) catalogued in EUNIS (https://eunis.eea.europa.eu/). How those biotopes 'work' is important to understand for marine environment management and protection.

We have given here an introduction to how the organisms respond to different environmental conditions occurring in the subtidal zone, how it in turns influences the assemblages of organisms, and how key organisms found in this level thrive there and interact either through facilitation, competition or via interactions up and down the food chain. More recent field experiments and small-scale laboratory studies have revealed a further insight into how these systems are organised.

This basic knowledge of interactions that we have outlined gives us a context to help understand how human activities create the drivers of disturbances that may affect systems. But, with a number of stressors working in the same area, a particular challenge is to understand how these factors work together and lead to either synergistic or antagonistic effects. Investigating or observing interactions between single species fails to see a bigger picture, where several factors come together to cause change. Similarly, predicting likely change in biology as a result of change in one environmental factor may be confounded because other factors have also changed. Synergism can cause 'ecological surprises', where unexpected regime shifts occur quickly because a tipping point is exceeded (Crain et al., 2008). Expect such surprises and do not necessarily expect to be able to explain them. For instance, the change from kelp to turf beds has, in many cases, been connected to a synergy between eutrophication, temperature increase and overfishing of species that consume important grazers. Learning the primary and secondary responses of communities of species to a set of stressors may make mitigation possible, including managing human activities to minimise damage to marine life and habitats. To do that means better understanding the tolerance and recovery potential of species and biotopes to human activities, including climate change.

## REFERENCES

Araújo, R. M., Assis, J., Aguillar, R. et al. (2016). Status, trends and drivers of kelp forests in Europe: an expert assessment. *Biodiversity Conservation*, **25** (7), 1319–48.

Arnold, M., Teagle, H., Brown, M. P. and Smale, D. A. (2016). The structure of biogenic habitat and epibiotic assemblages associated with the global invasive kelp *Undaria pinnatifida* in comparison to native macroalgae. *Biological Invasions,* **18**.

Barnes, D. K. A. and Crook, A. C. (2001). Implications of temporal and spatial variability on *Paracentrotus lividus* populations to the associated commercial coastal fishery. *Hydrobiologia,* **465** (1), 95–101.

Bekkby, T., Rinde, E., Erikstad, L. and Bakkestuen, V. (2009). Spatial predictive distribution modelling of the kelp species *Laminaria hyperborea*. *ICES Journal of Marine Science,* **66**, 2106–15.

Bekkby, T., Rinde, E., Gundersen, H., Norderhaug, K. M., Gitmark, J. and Christie H. (2014). Length, strength and water flow – the relative importance of wave and current exposure on kelp *Laminaria hyperborea* morphology. *Marine Ecology Progress Series,* **506**, 61–70.

Birkett, D. A., Maggs, C. A., Dring, M. J., Boaden, P. J. S. and Seed, R. (1998). *Infralittoral Reef Biotopes with Kelp Species (Volume VII): An Overview of Dynamic and Sensitivity Characteristics for Conservation Management of Marine SACs*. Scottish Association of Marine Science (UK Marine SACs Project), Oban, p. 174.

Bishop, G. M. and Earll, R. (1984). Studies on the populations of *Echinus esculentus* at the St. Abbs and Skomer Voluntary Nature Reserves. *Progress in Underwater Science*, **9**, 53–66.

Brown, S. L., Bearhop, S., Harrdod, C. and McDonald, R. A. (2012). A review of spatial and temporal variation in grey and common seal diet in the United Kingdom and Ireland. *Journal of the Marine Biological Association of the United Kingdom,* 92, 1711–22.

Burton, M., Lock, K., Jones, J. and Newman, P. (2015). Skomer Marine Conservation Zone. Distribution and abundance of *Echinus esculentus* and selected starfish species. Natural Resources Wales Evidence Report No. 158.

Christie, H., Fredriksen, S. and Rinde, E. (1998). Regrowth of kelp and colonization of epiphyte and fauna community after kelp trawling at the coast of Norway. *Hydrobiologia,* **375**/376, 49–58.

Christie, H., Jorgensen, N. M., Norderhaug, K. M. and Waage-Nielsen, E. (2003). Species distribution and habitat exploitation of fauna associated with kelp (*Laminaria hyperborea*) along the Norwegian coast. *Journal of the Marine Biological Association of the United Kingdom,* **83**, 687–99.

Christie H., Jørgensen N. M. and Norderhaug, K. M. (2007). Bushy or smooth, high or low; importance of habitat architecture and vertical level for distribution of fauna on kelp. *Journal of Sea Research,* **58**, 198–208.

Christie, H. and Kraufvelin, P. (2004). Mechanisms regulating amphipod population density within macroalgal communities with restricted predator impact. *Scientia Marina,* **68**, 189–98.

Christie, H. and Norderhaug, K. M. (2017). Secondary Production. In E. Olafsson, ed. *Marine Macrophytes as Foundation Species*. Science Publishers, CRC Press, Taylor & Francis Group, Boca Raton, FL, pp. 161–79.

Christie, H., Norderhaug, K. M. and Fredriksen, S. (2009). Macrophytes as habitat for fauna. *Marine Ecology Progress Series,* **396**, 221–33.

Clemente, S., Hernandez, J. C., Toledo, K. and Brito, A. (2007). Predation upon *Diadema* aff. *antillarum* in barren grounds in the Canary Islands. *Scientia Marina,* **71** (4), 745–54.

Comely, C. A. and Ansell, A. D. (1988). Population density and growth of *Echinus esculentus* L. on the Scottish west coast. *Estuarine Coastal and Shelf Science,* **27** (3), 311–34.

Crain, C. M., Kroeker, K. and Halpern, B. S. (2008). Interactive and cumulative effects of multiple human stressors in marine systems. *Ecology Letters,* **11**, 1304–15.

Davoult, D., Engel, C. R., Arzel, P., Knoch, D. and Laurans, M. (2011). Environmental factors and commercial harvesting: exploring possible links behind the decline of the kelp *Laminaria digitata* in Brittany, France. *CBM-Cahiers de Biologie Marine*, **52** (4), 429.

Epstein, G. and Smale, D. A. (2017a). Environmental and ecological factors influencing the spillover of the non-native kelp, *Undaria pinnatifida*, from marinas into natural rocky reef communities. *Biological Invasions,* 20 (4), 1049–72.

Epstein, G. and Smale D. A. (2017b). *Undaria pinnatifida*: a case study to highlight challenges in marine invasion ecology and management. *Ecology and Evolution,* **7** (20), 8624–42.

Fagerli, C. W., Norderhaug, K. M. and Christie, H. (2013). Lack of sea urchin settlement may explain kelp forest recovery in overgrazed areas in Norway. *Marine Ecology Progress Series*, **488**, 119–32.

Fagerli, C. W., Norderhaug, K. M., Christie, H., Pedersen, M. F. and Fredriksen, S. (2014). Predators of the destructive sea urchin grazer *Strongylocentrotus droebachiensis* on the Norwegian coast. *Marine Ecology Progress Series*, **502**, 207–18.

Fariñas-Franco, J. M., Allcock, L., Smyth, D. and Roberts, D. (2013). Community convergence and recruitment of keystone species as performance indicators of artificial reefs. *Journal of Sea Research*, **78**, 59–74.

Filbee-Dexter, K. and Wernberg, T. (2018). Rise of turfs: a new battlefront for globally declining kelp forests. *Bioscience*, **68** (2), 64–76.

Foster, V., Giesler, R. J., Wilson, A. M. W., Nall, C. R. and Cook, E. J. (2016). Identifying the physical features of marina infrastructure associated with the presence of non-native species in the UK. *Marine biology*, **163** (8), 173.

Fredriksen, S. (2003). Food web studies in a Norwegian kelp forest based on stable isotope (delta 13C and delta 15N) analysis. *Marine Ecology Progress Series*, **260**, 71–81.

Hawkins, S. J. and Harkin, E. (1985). Preliminary canopy removal experiments in algal dominated communities low on the shore and in the shallow subtidal on the Isle of Man. *Botanica Marina* **28** (6), 223–30.

Hepburn, C. D., Holborow, J. D., Wing, S. R., Frew, R. D. and Hurd, C. L. (2007). Exposure to waves enhances the growth rate and nitrogen status of the giant kelp *Macrocystis pyrifera*. *Marine Ecology Progress Series*, **339**, 99–108.

Hiscock, K. (1983). Water Movement. In R. Earll and D. G. Erwin, eds. *Sublittoral Ecology*. Oxford University Press, Oxford, pp. 58–96.

Hiscock, K. (1985). Aspects of the Ecology of Rocky Sublittoral Areas. In P. G. Moore and R. Seed, eds. *The Ecology of Rocky Coasts*. Hodder & Stoughton, London, pp. 290–328.

Hiscock, K., ed. (1998). *Marine Nature Conservation Review: Benthic Marine Ecosystems: A Review of Current Knowledge for Great Britain and the North-East Atlantic*. Joint Nature Conservation Committee, Peterborough.

Hiscock, K. (2018). *Exploring Britain's Hidden World: A Natural History of Seabed Habitats*. Wild Nature Press, Plymouth.

Hiscock, K. and Hoare, R. (1975). The ecology of sublittoral communities at Abereiddy Quarry, Pembrokeshire. *Journal of the Marine Biological Association of the United Kingdom*, **55**, 833–64.

Hiscock, K. and Mitchell, R. (1980). The Description and Classification of Sublittoral Epibenthic Ecosystems. In W. F. Farnham, D. E. G. Irvine and J. H. Price, eds. *The Shore Environment: Ecosystems*, vol. 2. Academic Press, London, pp. 323–70.

Hiscock, K., Sharrock, S., Highfield, J. and Snelling, D. (2010). Colonisation of an artificial reef in south-west England – ex-HMS Scylla. *Journal of the Marine Biological Association of the United Kingdom*, **90**, 69–94.

Hiscock, K., Southward, A. J., Tittley, I. and Hawkins, S. J. (2004). Effect of changing temperature on benthic marine life in Britain and Ireland. *Aquatic Conservation*, **14**, 333–62.

Jones, N. S. and Kain, J. M. (1967). Subtidal algal colonization following the removal of *Echinus*. *Helgoland Marine Research*, **15** (1), 460–6.

Jørgensen, N. M. and Christie, H. (2003). Diurnal, horizontal and vertical dispersal of kelp-associated fauna. *Hydrobiologia*, **503**, 69–76.

Kain, J. M. (1971). The biology of *Laminaria hyperborea*. VI. Some Norwegian populations. *Journal of the Marine Biological Association of the United Kingdom*, **51**, 387–408.

Kinlan, B. P. and Gaines, S. D. (2003) Propagule dispersal in marine and terrestrial environments: a community perspective. *Ecology*, **84**, 2007–20.

Kitching, J. A. (1941). Studies in sublittoral ecology. III *Laminaria* forest on the west coast of Scotland; a study of zonation in relation to wave action and illumination. *Biological Bulletin of the Marine Biology Laboratory Woods Hole*, **80**, 324–37.

Kitching, J. A. and Ebling, F. J. (1961). The ecology of Lough Ine. X1. The control of algae by *Paracentrotus lividus* (Echinoidea). *Journal of Animal Ecology*, **30**, 373–83.

Kitching, J. A., Macan, T. T. and Gilson, H. C. (1934). Studies in sublittoral ecology. I. A submarine gully in Wembury Bay, south Devon. *Journal of the Marine Biological Association of the United Kingdom*, **19**, 677–705.

Kitching, J. A. and Thain, V. M. (1983). The ecological impact of the sea urchin *Paracentrotus lividus* (Lamarck) in Lough Ine, Ireland. *Philosophical Transactions of the Royal Society of London B*, **300**, 513–52.

Kraufvelin, P. (2017) Macroalgal grazing by the green sea urchin: born to consume resources. *Marine Biology*, **164**, 132.

Lawrence, J. M. (1975). On the relationship between marine plants and sea urchins. *Oceanography and Marine Biology Annual Review*, **13**, 213–86.

Leinaas, H. P. and Christie, H. (1996). Effects of removing sea urchins (*Strongylocentrotus droebachiensis*): stability of the barren state and succession of kelp forest recovery in the east Atlantic. *Oecologia*, **105**, 524–36.

Leleu, K., Remy-Zephir, B., Grace, R. and Costello, M. J. (2012). Mapping habitats in a marine reserve showed how a 30-year trophic cascade altered ecosystem structure. *Biological Conservation*, **155**, 193–201.

Ling, S. D., Scheibling, R. E., Rassweiler, A. et al. (2015). Global regime shift dynamics of catastrophic sea urchin overgrazing. *Philosophical Transactions of the Royal Society of London B*, **370**.

Ling, S. D., Johnson, C. R., Frusher, S. D. and Ridgeway, K. R. (2009). Overfishing reduces resilience of kelp beds to climate-driven catastrophic phase shift, *Proceedings of the National Academy of Sciences USA*, **106**, 22341–5.

Lobban, C. S. and Harrison, P. J. (1994). *Seaweed Ecology and Physiology: 1*. Cambridge University Press, Cambridge.

Moore, P. G. (1973). The kelp fauna of northeast Britain. I. Introduction and the physical environment. *Journal of Experimental Marine Biology and Ecology*, **13**, 97–125.

Nickols, K. J., White, J. W., Largier, J. L. and Gaylord, B. (2015). Marine population connectivity: reconciling large-scale dispersal and high self-retention. *American Naturalist*, **185** (2), 196–211.

Norderhaug, K. M., Christie, H. and Rinde, E. (2002). Colonisation of kelp imitations by epiphyte and holdfast fauna; a study of mobility patterns. *Marine Biology*, **141**, 965–73.

Norderhaug, K. M., Christie, H., Fosså, J. H. and Fredriksen, S. (2005). Fish–macrofauna interactions in a kelp (*Laminaria hyperborea*) forest. *Journal of the Marine Biological Association of the United Kingdom*, **85**, 1279–86.

Norderhaug, K. M. and Christie, H. C. (2009). Sea urchin grazing and kelp re-vegetation in the NE Atlantic. *Marine Biology Research*, **5**, 515–28.

Norderhaug, K. M., Christie, H., Rinde, E., Gundersen, H. and Bekkby, T. (2014). Importance of wave and current exposure to fauna communities in *Laminaria hyperborea* kelp forests. *Marine Ecology Progress Series*, **502**, 295–301.

Pedersen, M. F., Nejrup, L. B., Fredriksen, S, Christie, H. and Norderhaug, K. M. (2012). Effects of wave exposure on population structure, demography, biomass and productivity in kelp *Laminaria hyperborea*. *Marine Ecology Progress Series*, **451**, 45–60.

Rinde, E., Christie, H., Fagerli, C. W. et al. (2014). The influence of physical factors on kelp and sea urchin distribution in previously and still grazed areas in the NE Atlantic. *PLoS ONE*, 9, e0100222, http://dx.doi.org/10.1371/journal.pone.0100222.

Schultze, K., Janke, K., Krüß, A. and Weidemann, W. (1990). The macrofauna and macroflora associated with *Laminaria digitata* and *L. hyperborea* at the island of Helgoland (German Bight, North Sea). *Helgolander Meeresunters*, **44**, 39–51.

Sebens, K. P. (1985). Community Ecology of Vertical Rock Walls in the Gulf of Maine, U.S.A.: Small-scale Processes and Alternative Community States. In P. G. Moore and R. Seed, eds. *The Ecology of Rocky Coasts*. Hodder & Stoughton, London, pp. 346–71.

Sivertsen, K. (1997). Geographical and environmental factors affecting the distribution of kelp beds and barren grounds and changes in biota associated with kelp reduction at sites along the Norwegian coast. *Canadian Journal of Fisheries and Aquatic Sciences*, **54**, 2872–87.

Sjøtun, K., Christie, H. and Fossa, J. H. (2006). The combined effect of canopy shading and sea urchin grazing on recruitment in kelp forest (*Laminaria hyperborea*). *Marine Biology Research* **2**, 24–32.

Skadsheim, A., Christie, H. and Leinaas, H. P. (1995). Population reductions of *Strongylocentrotus droebachiensis* (Echinodermata) in Norway and the distribution of its endoparasite *Echinomermella matsi* (Nematoda). *Marine Ecology Progress Series*, **119**, 199–209.

Skiftesvik, A. B., Durif, C. M. F., Bjelland, R. M. and Browman, H. I. (2015). Distribution and habitat preferences of five species of wrasse (Family Labridae) in a Norwegian fjord. *ICES Journal of Marine Science*, **72**, 890–9.

Smale, D. A., Burrows, M. T., Moore, P., O'Connor, N. and Hawkins, S. J. (2013). Threats and knowledge gaps for ecosystem services provided by kelp forests: a northeast Atlantic perspective. *Ecology & Evolution*, **3**, 4016–38.

Steneck, R. S., Leland, A., McNaught, D. C. and Vavrinec, J. (2013) Ecosystem flips, locks, and feedbacks: the lasting effects of fisheries on Maine's kelp forest ecosystem. *Bulletin of Marine Science*, **89**, 31–55.

Teagle, H., Hawkins, S. J., Moore, P. J. and Smale, D. A. (2017). The role of kelp species as biogenic habitat formers in coastal marine ecosystems. *Journal of Experimental Marine Biology and Ecology*, **492**, 81–98.

Teagle, H. and Smale D. A. (2018). Climate-driven substitution of habitat-forming species leads to reduced biodiversity within a temperate marine community. *Diversity & Distribution*, **24**, 1367–80.

Villegas-Rios, D., Alos, J., March, D., Palmer, M., Mucientes, G. and Saborido-Rey, F. (2013). Home range and diel behaviour of the ballan wrasse, *Labrus bergylta*, determined by acoustic telemetry. *Journal of Sea Research*, **80**, 61–71.

Waage-Nielsen, E., Christie, H. and Rinde, E. (2003). Short term dispersal of kelp fauna to cleared (kelp harvested) areas. *Hydrobiologia*, **503**, 77–91.

Chapter 4

# Rocky Intertidal Shores of the North-West Atlantic Ocean

Steven R. Dudgeon and Peter S. Petraitis

## 4.1 Introduction

Intertidal communities on hard substrates of the north-west Atlantic Ocean have long been a model system for study of the ecological processes that generate and maintain patterns in natural ecosystems (e.g., Menge, 1976; Lubchenco and Menge, 1978). A large part of the success of this model system is because the rocky intertidal zone of north-west Atlantic shores is a low diversity ecosystem (Witman et al., 2004) and a relatively simple system, which makes it a relatively easy and tractable system for experimentation. For example, the earliest published food web for the rocky shore community of the north-west Atlantic Ocean was published in 1916 (Colton, 1916; Fisher, 2005). The system also contains a number of sharp and well-defined gradients. Unlike some other well-studied systems, such as the south-east coast of Australia and the west coast of the USA, tides in the north-west Atlantic are semi-diurnal (except for portions of the Gulf of Saint Lawrence and the north-east coast of Newfoundland where mixed tides prevail; see Figure 6.3) with the lowest tides of the month usually occurring early in the morning or late in the afternoon. As a result, organisms tend to be buffered against the worst effects of summer but exposed to the harshest winter conditions. There are also strong environmental gradients over relatively short distances at multiple scales along the east coast of North America over which changing rates of ecological processes can be estimated. In addition to the gradient in tidal elevation, which all temperate rocky shores share, there are clear gradients latitudinally and within regions with respect to ice scour, substrate type and availability, hydrodynamic exposure, temperature, salinity and nutrients. These gradients have afforded ecologists the opportunity to study, using a single regional species pool, changes in community dynamics associated with changing biotic processes in different abiotic environments.

As sharp as these environmental gradients in space are, the principal signature shaping the ecology and evolution of the north-west Atlantic Ocean, arguably, has been the magnitude of biotic and abiotic change through time. In an earlier review of north-west Atlantic rocky shores, Vadas and Elner (1992) opened with a quote from Johnson and Skutch's (1928) paper describing algal communities of the littoral zone: 'The components of any given vegetational belt may, of course, change markedly from century to century' (p. 34). Apparently, this idea has captured the minds of marine ecologists of the region for the better part of a century. Perhaps more than any other coastal region, the north-west Atlantic Ocean is the poster child for environmental change on post-Miocene evolutionary, historical and recent, ecological timescales (Carlton, 1982; Vermeij, 2001; Carlton and Cohen,

2003; Mills et al., 2013; Pershing et al., 2015). Beginning with species invasions from the trans-Arctic interchange in the late Pliocene (Vermeij, 1991) to extreme climate change throughout the Pleistocene, culminating in the most recent glacial maximum 18,000 years BP, the western North Atlantic has been impacted more than any other marine biome on the planet (CLIMAP Project Members, 1976). Historical processes resulted in the Gulf of Maine (GOM) being among the least species-rich marine ecosystems on the planet (Vermeij, 2001; Witman et al., 2004).

Dramatic change in the ecological environment of north-west Atlantic shores has continued in recent history to the present. Two of the most important consumers on rocky shores of the north-west Atlantic Ocean, the predatory green crab, *Carcinus maenas*, and the herbivorous gastropod, *Littorina littorea*, are not native to North America, having arrived 200 and 500 years ago, respectively, consistent with human-mediated transport in ships (Carlton, 1982; Carlton and Cohen, 2003; Blakeslee et al., 2008). As expected in low-diversity systems (e.g., Stachowicz et al., 1999), these invaders have had large impacts, especially in New England, which suggests that rocky intertidal shores looked very different in colonial times (Lotze and Milewski, 2004). In just the past few decades, several more algal and invertebrate species have invaded north-west Atlantic shores (e.g., Harris and Jones, 2005; Dijkstra et al., 2007; Osman and Whitlach, 2007; Mathieson et al., 2008; Sephton et al., 2011; Johnson et al., 2012; O'Connor, 2014). With respect to recent abiotic environmental changes, poorly buffered water masses in the GOM are particularly vulnerable to ocean acidification (OA; Wang et al., 2013), and they have warmed more rapidly than 99.9 per cent of the global oceans since 2004 (Pershing et al., 2015). Warming has demonstrable effects on species distributions and community dynamics (Wethey, 2002; Burrows et al., 2011; Kordas et al., 2015; Wernberg et al., 2016). Climate change poses an emerging threat to the stability and production of coastal ecosystems in the region if key species lack genetic variation for plasticity or adaptation (but see Wagner, 2017).

In this context of continual environmental change, we review the ecology of hard substrate intertidal ecosystems of the north-west Atlantic Ocean. First, we describe the phylogeographic histories of representative key species that have shaped this ecosystem and contemporary biogeographic patterns of distribution and abundance. Then, we review the models that explain these patterns updated by what has been learned the past twenty-five years about ecological processes influencing north-west Atlantic rocky shores.

Our focus emphasises three emerging themes. We think these themes have come to the forefront because of the low biodiversity of the system, the seasonal and spatial gradients of environmental harshness and the high variability of recruitment. Thus, the first theme, not surprisingly, is the role of variability in ecological processes, such as predation, competition, facilitation and recruitment. Marine ecologists since the early work of Connell and Paine, have focused on the average effect, and this bias is evident in the rise of meta-analyses, which are based on the assumption that the standardised average effect is the appropriate measure by which to evaluate the role of various ecological processes (Petraitis, 1998; Figure 4.1). Clearly, in the north-west Atlantic, with its low biodiversity and large variability in both abiotic and biotic factors, there is growing recognition that the noise in these processes has important consequences separately from overall average effects. We examine spatial and temporal variability in these processes and the scales at which they occur. In regard to variation in time, specific questions of interest include whether (1) variability is caused by ecological processes tracking the environment and (2) community states we observe are contingent upon rare, large-scale events such as ice scour, heat waves or other extreme events that may leave signatures on present-day ecosystems (Figure 4.1). The underlying premise is that the meaningful scale on which ecological processes on north-west Atlantic shores may be understood is likely much larger or longer than a typical experiment that spans two to three years and 10 m at a single site or a few sites spread over several kilometres.

The second theme involves studies that explored intra- and inter-specific interactions in

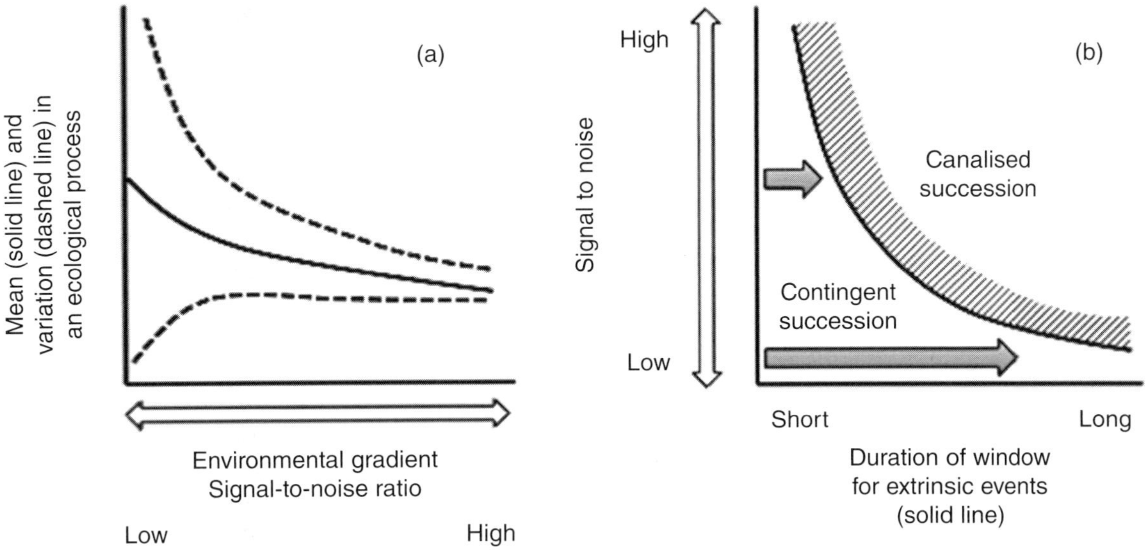


**Fig. 4.1** The effect of variation in ecological processes on the switch from contingent to canalised succession. (a) The ratio of signal to noise in an ecological process may change across an environmental gradient. The signal is the unstandardised average effect and the noise is a measure of variability (e.g., the variance, range or standard deviation). The average effect size is a measure of the signal-to-noise ratio. The environmental gradient could be the change in conditions, such as wave surge from sheltered to exposed shores. (b) The signal-to-noise ratio affects the opportunity for extrinsic factors to influence the course of succession. The solid line shows the point at which there is a switch from canalised to contingent succession. However, there is no reason that this switch could not be gradual and the line would then be a band. Extrinsic factors include such events as winter ice scour, seasonal delivery of larvae and unseasonable weather. Note that short versus long duration is relative to the average lifespans of the organisms in the system and not absolute.

greater depth. These studies include examination of indirect effects, both density- and trait-mediated, and how these effects contribute to the generation and maintenance of community patterns. These approaches do not deny the large amount of variability in ecological processes, but rather emphasise that an in-depth understanding of short-term and small-scale interactions may provide an explanation for some of the larger-scale variation.

The third theme is, of course, the role of anthropogenic activities, both directly through top-down harvesting, bottom-up eutrophication and as vectors for species introductions, and indirectly through climate change. Human activities often increase supplies of nutrients from the bottom-up and/or cause changes in physical and chemical variables associated with climate change that modify species interactions to influence outcomes of community structure.

We conclude by identifying major gaps in our knowledge of rocky shore ecosystems of the north-west Atlantic Ocean. Some of the gaps arise because the three emerging themes have not yet been fully explored. However, we will also explore an area that we think will become increasingly important. That is the contribution of the detrital ('brown') web (*sensu* Rooney and McCann, 2012) to ecosystem production, function and community structure of rocky shores through integration with the photosynthetic energy channel ('green'; *sensu* Rooney and McCann, 2012) of food webs. Herbivores are considered a primary agent of consumer control on New England rocky shores (Lubchenco, 1978; Bertness et al., 2002; Silliman et al., 2013); however, most algal production (~60 per cent) enters into detrital, rather than trophic, food webs (Vadas and Elner, 1992; Vadas et al., 2004) and its contribution to fuelling secondary production through the detrital pathway is unknown.

## 4.2 North-West Atlantic Biogeography and Phylogeographic Patterns

Biogeographic regions in which species assemblages evolve are most fruitfully defined by four primary components: (1) the thermal environment, (2) the extent of a type of contiguous coastal area within a given thermal environment, (3) the geographic isolation of coastal areas and (4) the geological time necessary for biotas to evolve distinctness (Adey and Steneck, 2001). The physical land- and sea-scapes of the north-west Atlantic Ocean with respect to coastal geomorphology, tidal current patterns, oceanic currents and corresponding sea surface temperatures (SSTs) and frequency of ice cover that define the components of a biogeographic region are summarised in detail from the southern tip of Florida to Newfoundland, Canada in Chapter 6. Coastal habitats with hard substrate are rare south of Long Island Sound, and so these habitats occupy only a fraction of the northern part of the coastline from Florida to Newfoundland. Accordingly, our contribution focusses on this region of hard substrate north of Long Island Sound in the north-west Atlantic.

The two principal physical factors affecting rocky shores from north to south are ice scour and availability of hard substrate (Vadas and Elner, 1992). Rocky shores in the Canadian Maritimes are regularly scoured by pack ice that forms each year (Bergeron and Bourget, 1984; Bourget et al., 1985), whereas ice scour occurs infrequently in southern New England (Wethey, 1985) and is intermediate in severity and frequency on southern shores of Nova Scotia, the Bay of Fundy and the northern GOM (Markham, 1980; Bedford Institute of Oceanography, 1987, McCook, 1992; Petraitis et al., 2009a). Along the same north–south axis, rocky shorelines that characterise the Canadian Maritimes and the GOM increasingly give way to mixtures of boulder and sandflats, mudflats and salt marshes south of Cape Cod. These two factors set the boundaries for hard substrate coastal ecosystems occupied by persistent, perennial species assemblages that are the subject of this chapter. This region corresponds to the sub-Arctic thermogeographic region defined by Adey and Steneck (2001) as persistent habitable regions with summer temperatures ranging between 7 and 15 °C and winter temperatures ranging between −1.8 and 0 °C. The core of Briggs' (1974) coastal biogeographic region for the western North Atlantic Ocean corresponds with the subarctic thermogeographic region of Adey and Steneck (2001).

Groups of taxa known to have originated in the temperate North Atlantic during the Oligocene or Miocene epochs were probably derived from tropical ancestors by means of cold adaptation (Vermeij, 2001). However, many of the species that originated in the North Atlantic went extinct by the mid-Pliocene, especially on the east coast of North America. The hard substrate of the north-west Atlantic was recolonised by species invading from the Pacific Ocean via the Trans-Arctic Interchange beginning about 3.5 Mya (Vermeij, 2001). For example, approximately 83 per cent of rocky intertidal molluscs in New England and the Canadian Maritimes arrived from the Pacific, or are descendants from invaders from the Pacific Ocean (Vermeij, 1991).

Glaciation during the Pleistocene was responsible for much of the patterns of extinction and recolonisation of north-west Atlantic shores (Maggs et al., 2008). Differences in the hypothesised extents of ice sheets suggest locations of potential glacial refugia along the coasts and those areas occurred mostly in the eastern North Atlantic. In the western Atlantic, the region stretching from the Carolinas into Florida is suggested as the main southern refugium, although hard substrate may have been limiting and a second refugium has been suggested on the Grand Banks at the eastern tip of Newfoundland (Maggs et al., 2008). In a study of six cold-temperate amphi-Atlantic invertebrate species (the mussel *Mytilus edulis*, the whelk *Nucella lapillus*, the winkle *Littorina obtusata*, the barnacle *Semibalanus balanoides*, the sea star *Asterias rubens* and the isopod *Idotea balthica*), Wares and Cunningham (2001) found evidence that European populations of all six species survived the last glacial maximum (LGM). Moreover, five (all but

*M. edulis*) of the six species showed evidence of recolonisation from Europe. *M. edulis* and one of two clades of *S. balanoides* are suggested to have persisted on the American side prior to the LGM (Wares and Cunningham, 2001). The hermit crab, *Pagurus longicarpus*, is suggested to have had both a southern refugium in the Gulf of Mexico and a periglacial refugium in the Canadian Maritimes (Young et al., 2002).

Two of the most interesting phylogeographic histories are the invasions of north-west Atlantic shores by the green crab, *C. maenas*, and the periwinkle snail, *L. littorea* (Carlton, 1982; Carlton and Cohen, 2003; Blakeslee et al., 2008). Native to Europe, green crabs were first reported in New York in 1817 (Say, 1817). They rapidly spread northward to southern Massachusetts, but thereafter range expansion slowed, entering the southern portion of the GOM in the early twentieth century. It took approximately fifty years for *C. maenas* to spread through the northern GOM, reaching New Brunswick, Canada by 1951 and the Atlantic coast of Nova Scotia by 1956 (Roman, 2006; Breen and Metaxas, 2009). The strong genetic cline in Nova Scotia and higher than expected genetic diversity at the northern range edge suggests the range expansion to Cape Breton and Prince Edward Island in the 1990s was the result of cryptic invasions by shipping from northern Europe (Roman, 2006).

*L. littorea* was first reported in Pictou, Nova Scotia, Canada in 1840 (Carlton, 1982; Reid, 1996). It spread to Halifax by 1857, the Bay of Fundy by the early 1860s, Penobscot Bay in the mid-coast of Maine by 1870 and quickly spread through the southern GOM to Cape Cod by 1872 before expanding into the mid-Atlantic region in the twentieth century (Steneck and Carlton, 2001). Mitochondrial sequences of both *L. littorea* and its trematode parasite, *Cryptocotyle lingua* from both Europe and North America indicate that it was introduced to North America from Europe within the past 500 years (Blakeslee et al., 2008; Brawley et al., 2009).

Thus, two of the most important consumers on north-west Atlantic shores are very recent arrivals, and they continually provide novel challenges to native species and the existing food web structure as their range has expanded into new habitats.

## 4.3 Spatial and Temporal Variability in Pattern and Process on North-West Atlantic Rocky Shores

Notwithstanding the recent invasions by exotic species (discussed later), the patterns of species distributions described by Vadas and Elner (1992, their figures 2.1 and 2.2) correspond well with those prevailing on north-west Atlantic rocky shores. On both exposed and sheltered shores, they distinguished two habitats, perennial stands of macroalgae and persistent patches of free space. The patches of free space cycle annually through stages of ephemeral algae, barnacles and free space due to the sequential foraging activities of the principally herbivorous *L. littorea* and the carnivorous *N. lapillus* (Vadas, 1992).

Perhaps better known is the pattern across wave exposure where more persistent communities differ between exposed and sheltered shores. Sheltered shores are occupied by a small number of invertebrate and algal species, of which the fucoid rockweed, *Ascophyllum nodosum*, (hereafter, *Ascophyllum*) is the most common foundation species of the community on the mid-shore (Lewis, 1964; Stephenson and Stephenson, 1972; Bertness et al., 1999). Historically, barnacles (*S. balanoides*) and mussels (*Mytilus* spp.) occurred within and between *Ascophyllum* stands (Lewis, 1964), although mussels have declined throughout the GOM over the last several decades (Petraitis and Dudgeon, 2015; Sorte et al., 2017). Mussels continue to persist in muddy habitats in the low-intertidal zone of sheltered shores (S. Dudgeon and P. Petraitis, own observations).

Sheltered shores are characterised by low diversity, both in species richness and evenness. *Ascophyllum* is the most abundant species (in biomass) on north-west Atlantic rocky shores with standing crops commonly exceeding 20 kg of fresh weight·m$^{-2}$ on sheltered shores (Petraitis and Dudgeon, 1999; Vadas et al., 2004). The encrusting red alga, *Hildenbrandia rosea*, covers much of the substrate underneath *Ascophyllum*, but other macroalgae (e.g., *Fucus vesiculosus* and *Chondrus crispus*) are uncommon. *L. littorea*, *L.*

*obtusata* and, increasingly, *C. maenas* (Mutti et al., unpublished data) are abundant mobile consumers, whereas *N. lapillus* and the limpet *Tectura testudinalis* are present in low numbers (see Petraitis and Vidargas, 2006; Petraitis et al., 2008, 2009b).

Moderately exposed and exposed rocky shores harbour a richer assemblage of species. Mussels and barnacles are predicted to occupy the most exposed shores where water motion hampers activities of their consumers and they outcompete macroalgae (Lubchenco and Menge, 1978; Bertness et al., 2004a), but perennial macroalgae can be very abundant in the low and mid-intertidal zones, even at very exposed sites like Pemaquid Point, when mussels were abundant in the early 1970s (Vadas and Manzer, 1971). In general, the low-intertidal zone consists of a carpet of red macroalgae, including *C. crispus*, *Mastocarpus stellatus* and, at some sites, *Palmaria palmata* and with kelps occupying the sublittoral fringe (Dudgeon et al., 1999). *Mytilus* occurs in the low and mid-intertidal zones when it is present. The two most conspicuous differences from sheltered shores are the sizes of *Ascophyllum* and *Mytilus*. On exposed shores, individuals of *Ascophyllum*, while infrequent, are quite short and individual mussels are small. On sheltered shores, the pattern is reversed; both *Ascophyllum* and mussels are large (S. Dudgeon and P. Petraitis, own obervations).

The mid-intertidal zone of exposed shores is a mix of *Fucus* species. *Fucus distichus* subsp. *evanescens* inhabits the lowest portion of the mid intertidal, followed further up the shore by *F. vesiculosus* and *F. spiralis*. A fifth fucoid species, *F. distichus*, occurs only in rock pools in Maine and Nova Scotia (Chapman and Johnson, 1990; Pearson and Brawley, 1998), but occurs on emergent rock in Massachusetts (G. Trussell, personal communication). *Fucus serratus* is not native to North America, but invaded the Canadian Maritimes 150 years ago, and has migrated southwards along the Atlantic coast of Nova Scotia where the current southern limit resides near Yarmouth (Johnson et al., 2012). It has not yet entered the Bay of Fundy.

The high-intertidal zone is dominated by *S. balanoides*, interspersed with patches of ephemeral algae, and the supralittoral is covered by cyanobacteria. As on sheltered shores, periwinkles (*L. littorea*) are the most abundant mobile consumer. Other herbivorous gastropods (e.g., the snails *L. obtusata* and *Lacuna vincta*, which are seasonally abundant, and the limpet *T. testudinalis*) and two predators, the dog whelk (*N. lapillus*) and the green crab (*C. maenas*) are also common intertidally. Sea stars and sea urchins are rare in the intertidal zone.

Qualitative descriptions imply that patterns of distribution and abundance across spatial scales of wave exposure, or oceanographic regions, differ, on average, because species abundances vary relatively little through time. However, our long-term sampling data collected from unmanipulated plots around Swans Island, Maine are highly variable, suggesting inferences about ecological processes are very much coloured by when and where experiments are done. Sampling data show highly skewed patterns of abundance with as much of the variation due to differences among years as among small-scale spatial patterns among sites (see Table 4.1). Over nearly a twenty-year time span (1997–2016), densities of snails (*L. littorea*, *L. obtusata*, *N. lapillus* and *T. testudinalis*), barnacles (*S. balanoides*) and fucoids (*Ascophyllum* and *F. vesiculosus*) varied over 1–2 orders of magnitude and, on average, in 36 per cent of the sampled quadrats, one or more species was not present (quadrat size: 0.25 $m^2$; range across species: 0.4–74.2 per cent). Densities of mussels and recruitment of mussels, barnacles and fucoids (*M. edulis*, *S. balanoides* and *Ascophyllum*, respectively) varied from 2 to almost 4 orders of magnitude. The proportion of mussel cohorts consumed per twelve weeks by green crabs and dog whelks ranges from 0.0 to $>$0.50 among years in the same unmanipulated plot at nine of twelve sites and overshadows average rates of predation in unmanipulated plots among sites (Figure 4.2). Nevertheless, the striking differences in ranges for densities of consumer species and rates of consumption of mussels versus recruitment of mussels, barnacles and fucoids suggest recruitment has the potential to overwhelm top-down control. These observations echo earlier conclusions about supply-side effects in marine systems (Fairweather, 1988; Roughgarden et al., 1988; Underwood and Fairweather, 1989; Connolly and Roughgarden, 1998).

**Table 4.1** Estimates of distributions and of per cent variance components for common species in the GOM. Data are from the twelve control plots that have been sampled by Dudgeon and Petraitis since 1997. Columns labelled Min., Max., N, and Not present give the minimum value, the maximum value, the total number of sampled quadrats and the percentage of quadrats in which a species was not present, respectively. Columns labelled Year, Bay, Year x bay, Site(bay) and Residual give the variance components as percentages among years between 1997 and 2015, among four bays, and three sites nested within each bay; components were estimated using REML ANOVA models. Location of plots, layout of the experimental design, details of the sampling procedures and some of the data are available online (Dudgeon and Petraitis, 2001; Petraitis and Vidargas, 2006; Petraitis et al., 2008, 2009b).

| Species[a] | Average | Median | Min. | Max. | N | Not present (%) | Year (%) | Bay (%) | Year x bay (%) | Site (bay) (%) | Residual (%) |
|---|---|---|---|---|---|---|---|---|---|---|---|
| *L. littorea* | 26.6 | 21.7 | 0.0 | 103.3 | 252 | 0.40 | 16.91 | 0.00 | 0.00 | 40.43 | 42.66 |
| *L. obtusata* | 6.5 | 4.0 | 0.0 | 47.0 | 252 | 1.98 | 31.68 | 0.35 | 5.79 | 21.33 | 40.84 |
| *N. lapillus* | 1.5 | 0.7 | 0.0 | 27.7 | 252 | 21.83 | 13.06 | 6.63 | 12.31 | 22.98 | 45.02 |
| *T. testudinalis* | 0.4 | 0.0 | 0.0 | 10.4 | 252 | 52.78 | 18.37 | 0.00 | 4.96 | 34.46 | 42.22 |
| *Ascophyllum* (>2 cm in length) | 8.0 | 6.7 | 0.0 | 30.0 | 120 | 9.17 | 0.00 | 3.59 | 7.64 | 33.96 | 54.80 |
| *Ascophyllum* (< 2 cm in length) | 5.9 | 0.0 | 0.0 | 128.3 | 120 | 74.17 | 38.20 | 0.00 | 11.20 | 15.90 | 34.70 |
| *F. vesiculosus* (>2 cm in length) | 2.9 | 0.0 | 0.0 | 41.7 | 120 | 55.00 | 8.14 | 10.92 | 0.00 | 23.80 | 57.14 |
| *F. vesiculosus* (<2 cm in length) | 5.9 | 0.0 | 0.0 | 131.7 | 120 | 73.33 | 9.12 | 0.00 | 4.82 | 15.05 | 71.01 |
| *M. edulis* (>1 cm in length) | 14.7 | 0.0 | 0.0 | 350.0 | 120 | 70.83 | 35.07 | 9.97 | 2.61 | 15.50 | 36.85 |
| *M. edulis* (<1 cm in length) | 50.8 | 0.0 | 0.0 | 2,108.3 | 120 | 90.83 | 52.17 | 0.00 | 0.00 | 1.13 | 46.70 |
| *S. balanoides* (adult)[b] | 1.0 | 0.5 | 0.0 | 9.7 | 252 | 36.51 | 27.60 | 3.99 | 1.22 | 12.50 | 54.70 |
| *S. balanoides* (young of the year)[b] | 3.6 | 1.4 | 0.0 | 33.8 | 252 | 25.00 | 11.01 | 3.30 | 4.62 | 49.81 | 31.27 |
| Fucoid recruitment (mostly *Ascophyllum*)[c] | 78.7 | 32.0 | 0.0 | 974.0 | 153 | 11.11 | 17.22 | 0.00 | 17.26 | 9.39 | 56.12 |
| *S. balanoides* recruitment[d] | 160.9 | 71.0 | 0.0 | 1,062.0 | 153 | 18.95 | 23.56 | 0.13 | 5.48 | 20.53 | 50.30 |
| *M. edulis* recruitment[e] | 168.8 | 33.0 | 0.0 | 6,603.0 | 139 | 0.72 | 52.79 | 3.30 | 1.05 | 17.39 | 25.47 |
| **Average** | | | | | | 36.17 | 23.66 | 2.81 | 5.26 | 22.28 | 45.99 |
| **Median** | | | | | | | 18.37 | 0.35 | 4.82 | 20.53 | 45.02 |
| **Min.** | | | | | | | 0.00 | 0.00 | 0.00 | 1.13 | 25.47 |
| **Max.** | | | | | | | 52.79 | 10.92 | 17.26 | 49.81 | 71.01 |

[a] A single datum is the average of three to five 0.25 $m^2$ quadrats per plot unless otherwise noted.
[b] Datum is the average of six 4 $cm^2$ quadrats per plot.
[c] Datum is the number of zygotes and germlings per $cm^2$ per eight weeks.
[d] Datum is the number of recruits per 100 $cm^2$ per eight weeks.
[e] Datum is the number of recruits per 40 $cm^2$ per twelve weeks.

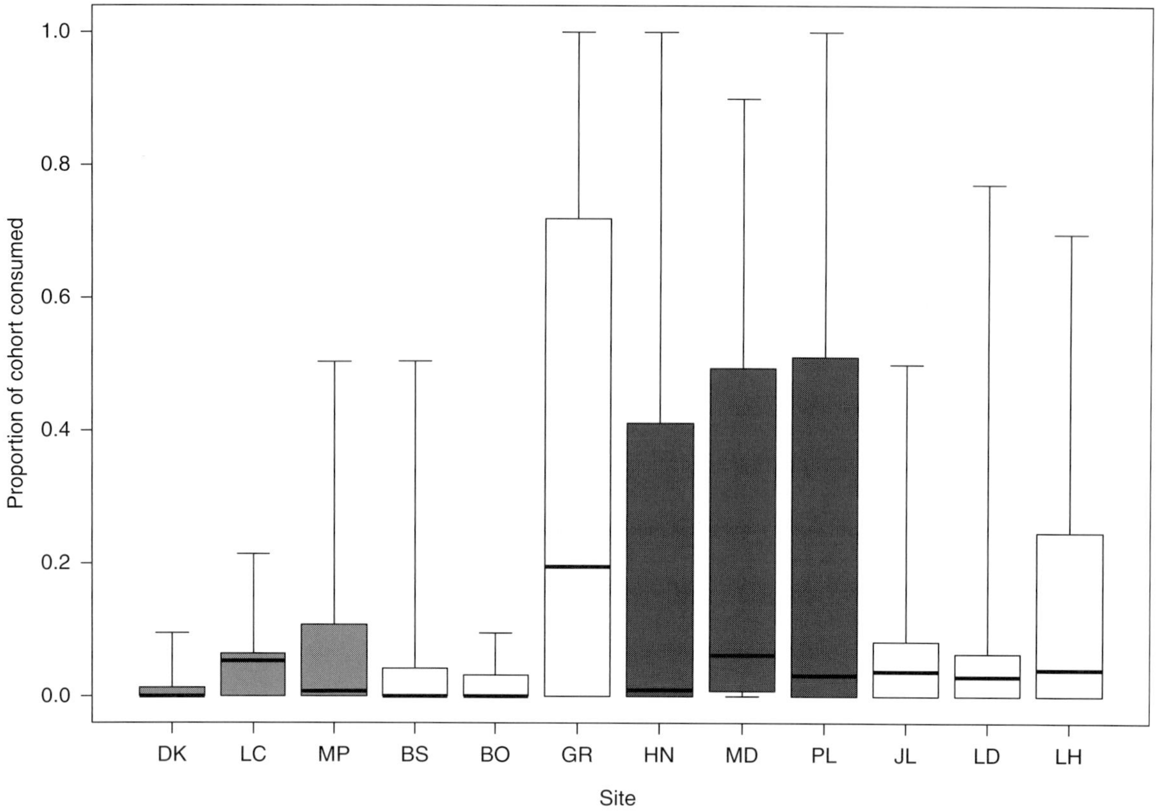


**Fig. 4.2** The proportion of annual cohorts of fifteen mussels (*M. edulis*) consumed either by dog whelks (*N. lapillus*) or green crabs (*C. maenas*) in unmanipulated control plots over twelve years of identical experiments at twelve sites on Swans Island, ME. Boxes represent the interquartile range, black lines the median and whiskers the range. Box shadings represent the bay that each site is located in: grey (Burnt Coat Harbour), light grey (Mackerel Cove), dark grey (Seal Cove) and white (Toothacher Cove). Two letter site labels correspond to the following sites on Swans Island: DK = Dick's, LC = Long Cove, MP = Mill Pond, BS = Basil, BO = Bob's, GR = Grace, HN = Hen, MD = MacDuffie's, PL = Pole 24, JL = July 4, LD = Ledges and LH = Little House Cove. Site locations are available elsewhere (Petraitis and Dudgeon, 2015, supplement 1).

Variance components analyses reveal the major sources are variation among years (average: 24 per cent, range: 0–53 per cent), variation among sites nested within bays (average: 22 per cent, range: 1–50 per cent) and residual variation (average: 46 per cent; range: 25–71 per cent). Sites nested within bays are about 500–2,000 m apart and the bays themselves are >1,000 m apart. The spatial and residual variation suggests experiments should be spread over at least 100–1,000 m and well replicated to truly capture the inherent spatial variation.

Many studies provide evidence that ecological processes vary at multiple spatial scales from metres to hundreds of kilometres and create variation in outcomes of interactions affecting community structure. One class of hypotheses for these patterns of variation at different spatial scales is the environmental stress models, which posit differences in the average effect of ecological processes operating across different scales are the important drivers (Menge and Sutherland, 1976, 1987). Perhaps the best known is the model of predator-mediated coexistence in which the activities of predators and other consumers are modulated by hydrodynamic forces of waves on rocky shores. This 'consumer stress' model hypothesises that mussels competitively dominate wave-exposed sites where water motion hampers the predatory activities of crabs, snails and seastars (Menge, 1976, 1983; Lubchenco and Menge, 1978). On exposed shores with intermediate levels of wave surge, the average effect of consumers is greater and thus leads to a more diverse assemblage of macroalgae and invertebrates. On sheltered shores with very little wave surge, predators control abundances of mussels and barnacles and macroalgae dominate. Similarly, the average intensity of grazing by herbivores, most notably the abundant periwinkle gastropod, *L. littorea*, varies with water motion and thus the strongest effects occur on sheltered shores (Lubchenco and Menge, 1978; Silliman et al., 2013; Dudgeon and Petraitis, unpublished data). Grazing on sheltered shores reduces ephemeral algal species richness and abundance (Lubchenco, 1978; Worm et al., 2002), resulting in a low-diversity assemblage of perennial macroalgae of large biomass (Lubchenco, 1983; Vadas and Elner, 1992). In areas with intense grazing pressure, even erect macroalgae can be eliminated from sheltered shores (Bertness et al., 1983; Petraitis, 1987).

The consumer stress model applied to northwest Atlantic rocky shores has been modified to include interaction with bottom-up effects and regional oceanographic variation (Leonard et al., 1998; Worm et al., 2002; Bryson et al., 2014). Modification of the model was needed to explain the co-occurrence of fucoids (*Ascophyllum* in particular) and mussels in sheltered bays and estuaries. This pattern has been well known for many years (e.g., Lewis, 1964), although not part of Menge and Lubchenco's original model because they did not examine very sheltered sites. At the larger scale, the geographic spread among sites of different exposure used by Menge and Lubchenco partially confounded the effects of wave exposure with regional differences – in particular, the well-known break in ecological processes at Penobscot Bay (Stephenson et al., 2009; Bryson et al., 2014). These processes, too, differ in their average effect at the spatial scales at which they operate.

Recognition that bottom-up effects interact with top-down effects to influence the dynamics of rocky shore communities afforded models accommodating a greater variety of the patterns observed. Bottom-up effects, broadly defined, include fluxes of propagules and larvae driven by current flows, as well as supplies of resources (e.g., inorganic nutrients), and here we provide three examples of the interaction between bottom-up and top-down effects in the north-western Atlantic Ocean. First, it is well known that swift tidal current flows (~1 m/s) over sheltered sites can provide a large flux of larvae and reduce foraging by consumers (Leonard et al., 1998). Enhanced recruitment and less predation of mussels in swift tidal currents enable mussels to occupy estuarine sites sheltered from wave stress. This pattern is common in narrow bays of coastal Maine. In contrast, limited recruitment of mussels and barnacles at sheltered sites with slow-flowing currents cannot keep pace with rates of predation and, consequently, fucoid macroalgae occupy these environments.

A second example involves how top-down and bottom-up effects interact with respect to the well-known hump-shaped curve of diversity (Paine and Vadas, 1969; Connell, 1978; Lubchenco, 1978; Petraitis et al., 1989). Local-scale diversity is highest at intermediate levels of either consumer activity or resource supply (Tilman, 1982; Kassen et al., 2000). For instance, rock pools on rocky shores in New England were shown to have hump-shaped diversity curves with respect to densities of the herbivorous gastropod, *L. littorea* (Lubchenco, 1978). In combination, consumer and resource effects act synergistically on North Atlantic rocky shores (Worm et al., 2002); herbivores reduce macroalgal diversity when resource supply and productivity are low, but increase diversity when resource supply and productivity are high. Consumers and nutrients in combination have bigger effects on algal communities than either does alone.

Finally, regional oceanographic patterns can alter the delivery of propagules and thus affect top-down control. For example, the outcome of similar experiments of consumer effects on community development differs in the southern and northern GOM (Petraitis and Dudgeon, 1999, 2004, 2005; Dudgeon and Petraitis, 2001; Bertness et al., 2002, 2004a, 2004b, Petraitis et al., 2009a). This suggests regional-scale factors have a role in local-scale community dynamics even when different regions share the same species

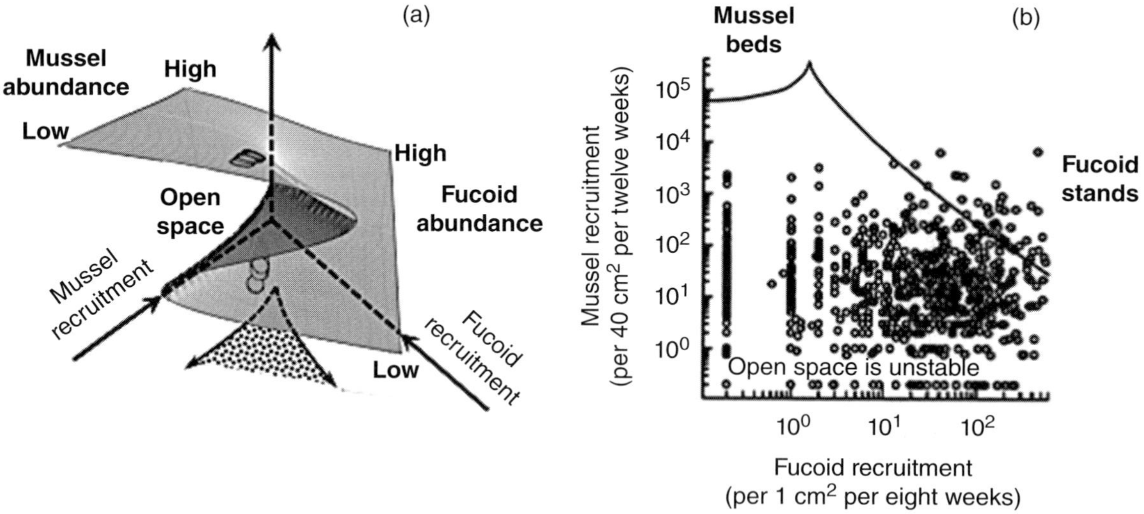


**Fig. 4.3** A cusp catastrophe with mussel and fucoid recruitment as drivers. (a) A three-dimensional plot of a cusp catastrophe. The surface shows the position of the equilibrium states of mussel beds and fucoid stands; the area inside the fold defines the unstable equilibria, which are open space. The shaded projection onto the XY plane gives the combinations of mussel and fucoid recruitment that supports two stable states. Note that low recruitment rates are towards the origin. Only one stable state is possible outside of the shaded area on the XY plane. (b) The estimated lines defining the cusp separate single and two stable state outcomes with respect to mussel and fucoid recruitment. Each open circle is the recruitment by mussels and fucoids to one of sixty experimental plots in each year from 1997 to 2015. For the cases in which recruitment was zero, data points were set to 0.2 in order to be plotted on the log–log scale. Estimated lines for the cusp are based on cover by barnacles, mussels and fucoids and are taken from figure 4A in Petraitis and Dudgeon (2015).

assemblage. Bryson et al. (2014) reported differences in the strength of consumer effects and abundance of barnacle and mussel recruits in disturbed plots between southern and northern GOM sites that led to different successional trajectories during recovery. Whereas community dynamics at sites in the southern GOM (south and west of Penobscot Bay) were consistent with predictions of the consumer-stress model, those in the northern GOM (east of Great Wass Island) were not. The differences were attributed largely to regional oceanographic differences that nullify the effect of wave exposure at sites in the north and limit recruitment of invertebrate larvae across exposures (Bryson et al., 2014).

An entirely different class of hypotheses emphasises the variability in ecological processes as the drivers of community dynamics and structure. For example, Petraitis and colleagues suggested variation in not only bottom-up and top-down processes (e.g., recruitment and predation), but also abiotic stresses (e.g., ice scour) drives the co-occurrence of mussel beds and fucoid stands in sheltered bays (Petraitis and Latham, 1999; Petraitis, 2013; Petraitis and Dudgeon, 2015). The possibility that mussel beds and fucoid stands are different communities capable of inhabiting the same environment is based on the theory of alternative stable states (Petraitis and Latham, 1999, Figure 4.3). Petraitis and colleagues hypothesised that ice scour opened areas for recolonisation by fucoids and mussels and whichever species was able to gain a foothold would persist as a stable state because of positive feedbacks. Mussel beds would persist because established beds can resist predator attacks (e.g., Petraitis, 1987) and enhance mussel recruitment. Fucoid stands would persist because established stands provide habitat for mussel predators and enhance self-recruitment (Petraitis and Dudgeon, 1999; Dudgeon and Petraitis, 2001). A patch of bare space, newly opened by ice scour, is an unstable state that exists in between the stable states of mussel beds and fucoid stands. The initial trajectory towards one or the other stable state is put into motion by the vagaries of recruitment and predation. The underlying model of alternative stable states, however, is deterministic.

The success of mussels versus fucoids is likely dependent on the spatial extent of ice scour. Mussels can only establish if an ice scour event is large enough so there are areas that are beyond the reach of mussel predators, which avoid open areas and tend to occur in fucoid stands. In contrast, the success of fucoids in open areas depends on the proximity of established stands from which refuge is provided for mussel predators and a source of fucoid zygotes. In other environments (i.e., different parameters), it is possible only a single stable state – that is, mussel beds or fucoid stands – will persist at a particular site (Bertness et al., 2002, 2004a).

There are common features but also conceptual differences between models and hypotheses based on average effects and those based on variation in effects. Traditional models of consumer stress and environmental stress (e.g., Menge and Sutherland, 1976, 1987) assume that effects 'average out' over space and time. For rocky intertidal communities in particular, models of average effects emphasise the cause and effect linkages between low wave exposure and high predator activity, and between coastal oceanography and recruitment. In contrast, models based on the idea that variation can override average effects leads to the possibility of, not only contingent succession, which is stochastic, but also models of alternative or multiple stable states, which are deterministic.

The conceptual differences between models of average effects and variation effects arise from preconceptions about the consistency of biological processes relative to environmental factors. This is, in part, a matter of perception about the relative importance of the signal to the noise. In average effect models, environmental drivers of species interactions are assumed to be consistent in time and space, resulting in predictable patterns among sites in different environments and across various gradients of stress. This is likely to be true for moderately to very exposed shores where average effects of wave surge occur on a larger spatial and temporal scale than the foraging behaviour of predators. Wave surge has a strong signal relative to the weak and idiosyncratic signal of foraging effects, which can be modified by indirect interactions (see Section 4.4 on the complexities of predator–prey interactions). In addition, ecological processes such as succession tend to be 'faster' and, thus, close the window of opportunity for the vagaries of environmental factors and ecological processes to have an effect (Figure 4.1). For example, *Fucus* recovery occurs within one to two years on exposed shores (Bertness et al., 2004a; Dudgeon and Petraitis, unpublished data) while it can take over twelve years in sheltered bays (Jenkins et al., 2004; Dudgeon and Petraitis, unpublished data).

In contrast, models of variation effects depend on environmental and ecological processes occurring on similar scales. Mussel beds and fucoid stands have the potential for being alternative community states in sheltered bays because ice scour, which sets the system to the unstable state of open patches, has a spatial extent and temporal duration that are on the same scales as mortality and recruitment patterns of fucoids and mussels. Ice accumulation is episodic and pulses of scouring remove organisms, creating patches typically tens of square metres in size (P. Petraitis, unpublished data). Severe pulse events may clear organisms from hundreds of square metres of rock along a shoreline in single or contiguous patches (Markham, 1980; Bedford Institute of Oceanography, 1987; McCook, 1992; Petraitis et al., 2009a). Variation in settlement of propagules and the activities of consumers on timescales of weeks likewise occurs over tens of square metres of shoreline (Mutti et al., unpublished data). Variation in recruitment and mortality among patches of different sizes can leave strong signatures, causing mosaics of different communities within the same environment at these spatial scales.

The idea that variation can be an important determinant of community dynamics is not a new idea, although earlier studies on rocky shores emphasised variation on a very restricted spatial scale. For example, Bertness (1989) showed that density dependence of survival of the barnacle *S. balanoides* is either positive or negative depending on environmental characteristics. In physically benign habitats, high recruitment of barnacles leads to massive density-dependent mortality. In harsh, stressful habitats, high recruitment thermally buffers individuals from physical stress and enhances survivorship. Along similar lines, Dudgeon et al. (1999) showed the strength of competition between the red macroalgae, *C. crispus* and *M. stellatus*, on moderately exposed rocky shores varies with tidal elevation, which is related to variation in environmental harshness (Dudgeon et al., 1999). In physically benign habitats of the low-intertidal zone favourable for algal growth, *C. crispus* competitively excludes the upright frond stage of *M. stellatus*. As little as 0.5 m higher on the shore, growth of *C. crispus* is limited in the more stressful habitat and the two species coexist.

Some of the best evidence for how variation in processes can drive community dynamics comes from studies done over larger spatial scales. For example, regional differences in thermal stress and predation rates between regions north and south of Cape Cod switch interactions between adult barnacles (*S. balanoides*) and *Ascophyllum* (Leonard, 2000). Fucoid canopies buffer against thermal stress in the south, increasing survivorship of barnacles, whereas facilitation does not occur in the north. Leonard's (2000) study suggests that the interspecific interaction between barnacles and fucoids tracks environmental change. Facilitation south of Cape Cod occurred in 1995, an extremely warm year, but not in 1996, a cooler year in which only negative interspecific interactions were observed across latitudes north and south of Cape Cod.

Interspecific interactions between juvenile fucoids and newly settled barnacles in the GOM also vary regionally because of differences in the timing of recruitment. *Ascophyllum* reproduces in spring, and zygotes settle several weeks earlier in southern populations than in northern populations (Kordas and Dudgeon, 2009, 2011). Likewise, barnacles settle earlier in the southern areas but on a slightly different schedule. In southern areas, *Ascophyllum* settles first and the barnacles that settle afterwards kill nearby *Ascophyllum* germlings by growing over them (Kordas and Dudgeon, 2011). In more northern areas, barnacles settle before *Ascophyllum* and grow more slowly, and thus tend to facilitate germling survival.

Taken together, the work of Leonard on adults and of Kordas and Dudgeon on juveniles shows how annual and regional variation plays an important role in co-occurrence of fucoids and barnacles.

More importantly, these studies and others that emphasise the role of variation underline the dangers of thinking only in terms of average effects. The assumption that environmental drivers are strong, act consistently and average out over large spatial and temporal scales leads, not only to the notion that any place on the shore is as good as another to set-up an experiment, but also to the idea that there is little benefit gained from employing more than the bare minimum of replication that is needed to get a significant result (for further discussion of this problem, see Bourget and Fortin, 1995). Low replication at a few sites over a short time interval is likely to overestimate the strength of average effects relative to the amount of natural variation because of spatial and temporal autocorrelation. Replication across multiple scales of space and time are required if we wish to obtain realistic estimates of both the average effects and of the variability of environmental factors.

## 4.4 Complexity of Predator–Prey Interactions and Indirect Effects

It is well established that consumers play a central role in the dynamics and structure of north-west Atlantic rocky shores. An extreme and compelling example is how predators can control development of benthic fouling communities solely by the consumption of recruiting juveniles, leading to different assemblages in different locations (Osman and Whitlach, 2004). The direct effects of consumer–prey interactions limit abundances of competitively dominant species (mussels on emergent substrata, *Ulva* spp. in rock pools) on less-exposed open coast, river sites or in rock pools

(Menge, 1976; Lubchenco, 1978; Lubchenco and Menge, 1978; Bertness et al., 2002, 2004a). Reduced abundance of competitively dominant mussels, in turn, enables several species of macroalgae to benefit indirectly and occupy rocky intertidal shores. Trophic cascades on New England rocky shores appear to extend from avian predators (seagulls, *Larus* spp.) to crabs, mussels, snails and algae (Ellis et al., 2005, 2007).

Earlier studies of trophic cascades emphasised the density-mediated effect of predator–prey interactions as well as their average effects. Studies of the past two decades have dug deeply into mechanisms underlying interactions between predator and prey and have uncovered an astonishing complexity of responses and behaviours that adds to the sources of variation that contribute to community dynamics. These responses enhance the variability of consumer effects on rocky shore communities (Berlow, 1999; Petraitis et al., 2009a).

Inducible defences provide a good example of this. Inducible defences appear widespread among invertebrate species that serve as prey to predatory crabs (Palmer, 1990; Trussell, 1996, 2000; Leonard et al., 1999; Trussell and Smith, 2000). Gastropod prey species (e.g., *Littorina* spp., *N. lapillus*) of crabs produce thicker shells in the presence of crab effluent that functions as a cue of risk (Palmer, 1990; Trussell, 2000). Along the same lines, the mussel (*M. edulis*) produces thicker shells, more shell mass and more byssal threads to firmly attach to the substrate at sites with higher predation risk from green crabs (Leonard et al., 1999). Whereas *M. edulis* throughout New England respond to risk cues from green crabs, only *M. edulis* from southern New England populations that have been in contact with *Hemigrapsus sanguineus*, thicken shells in response to *Hemigrapsus* risk cues (Freeman and Byers, 2006). The invasive Asian shore crab was first reported in New Jersey in 1988, implying that the evolution of inducible shell thickening in *M. edulis* occurred within fifteen years of the introduction of *H. sanguineus* (Freeman and Byers, 2006). We expect that differential responses will lead to variable interactions between mussels and crabs.

Indicators of risk of predation, like abundances of crab species and historical contact with invasive crabs (*C. maenas* and *H. sanguineus*) are greater in the southern GOM (west and south of the Penobscot River; Seeley, 1986; Trussell and Smith, 2000; Bourdeau and O'Connor, 2003; Freeman and Byers, 2006; Kraemer et al., 2007; Griffen and Byers, 2009; Epifanio, 2013). Snails show geographic variation in predator defences and inducibility of those defences that is consistent with the highest risk of predation in the southern GOM, a pattern exemplified by *L. obtusata*. Snails from southern GOM populations produce thicker, stronger, heavier shells and a smaller body size compared with snails from northern GOM populations (Trussell, 2000). Reciprocal transplant experiments showed that the geographic pattern could be explained by phenotypic plasticity. Snails exposed to the effluent of the green crab, *C. maenas*, rapidly increase shell thickness in magnitude comparable to both historical transitions in thickness attributed to selection (Seeley, 1986) and clinal variation predicted from seawater temperatures (Trussell and Smith, 2000). A trade-off between reduced growth and increased shell thickness appears similar among snails from southern and northern regions of the GOM. Reduced vulnerability to predation outweighs the costs of reduced body size and fecundity (Trussell and Nicklin, 2002).

Similar responses of slowed growth and thickened shells in response to green crab effluent have been observed in developing juvenile *L. littorea* and *N. lapillus* (Palmer, 1990; Trussell et al., 2003; Large and Smee, 2013) and attributed to predation by *C. maenas* (Vermeij, 1982). Developmental plasticity induced in juveniles appears to be an important mechanism to deter predators. However, inferences about changing shell architecture in response to shell-crushing crab predators are complicated by comparison of historical (i.e., prior to green crab introduction) with recent populations of *N. lapillus* from the very same sites and by chromosomal differences (Vazquez, 2015). Fisher et al. (2009) observed that shell lengths of *N. lapillus* consistently increased from 1915–1922 to 2007 at nineteen sites in the GOM. The increase in shell size ranged from 7.7 per cent at wave-exposed sites to 27.2 per cent at wave-sheltered sites. Several plausible alternative hypotheses could explain the observed size increases.

Importantly, aperture size and increases in lip thickness do not show disproportionate changes after correcting for increases in shell length over the past century following the introduction of *C. maenas*, as predicted by a hypothesis of selection for shell defensive traits. More interestingly, the response of dog whelks appears to be linked to differences in chromosomal races in the GOM, with some races being more responsive than others (Vazquez, 2015, K. Vazquez and P. Petraitis, unpublished manuscript).

The effect that inducible defences have at the community level, assuming they have been triggered broadly (but see later discussion), is to dampen the density-mediated direct and indirect effects on the community. Variation in risk, patchiness in the ephemeral presence of cues, the ability of snails to detect those cues and induce defences place-to-place within regions of the GOM generate variability in both the effectiveness of induced defences and density-mediated consumer effects. For instance, an indirect effect of spionid polychaetes living within shells of *N. lapillus* is to compromise shell defences against crab predators (Fisher, 2010).

Indirect effects can also be mediated by behavioural traits of prey in the presence of, but not consumed by, predators. Behavioural traits of prey responding to risk cues from crab predators are termed trait-mediated indirect interactions (TMIIs; Abrams et al., 1996). TMIIs may compensate for the dampening effect that inducible defences have on density-mediated indirect effects (Trussell et al., 2003), suggesting that they may be an equivalently important cause of indirect effects in a trophic cascade (Trussell et al., 2004).

On New England rocky shores, the presence of green crab risk cues reduces foraging activity by both carnivorous (*N. lapillus*) and herbivorous (*Littorina* spp.) gastropods (Trussell et al., 2003). Reduced consumption rates of both barnacles and fucoids by their respective gastropod consumers leads to greater abundances of both taxa through the indirect effect of green crab presence. Nevertheless, the foraging preference of dog whelks for barnacles remains in the presence of crabs (Dernbach and Freeman, 2015). Snails that forage less grow more slowly. Predation risk, therefore, reduces the quantity and efficiency of trophic energy transfer in the rocky shore food web, with implications for food web length and structure (Trussell et al., 2006).

Prey that reduce foraging in the presence of a predator must eventually weigh the trade-off between metabolic demand and safety. Well-fed dog whelks forage very little in the presence of predators, whereas starved ones forage a lot (Matassa and Trussell, 2014). Thus, the physiological state of the prey influences choice of behaviours, causing temporal variability in foraging even under consistent risk of predation. Matassa and Trussell (2014) showed that foraging by *N. lapillus* is a dynamic activity that fluctuates with perceived predation risk as frequently as day by day. Fed and starved dog whelks converge in their foraging behaviours under constant low or high risks of predation, but behave differently when risk is high and variable in time.

Field and laboratory experiments have shown that strong and consistent changes in snail behaviour can occur from the mere presence of crab predators (Trussell et al., 2003, 2004, Matassa and Trussell, 2014, 2015). Non-consumptive effects and TMIIs in the context of a typical rocky shore merits discussion. Experimental set-ups confining experimental subjects and concentrating risk cues in small rock pools, boxes on emergent substrata or laboratory mesocosms were clearly needed to demonstrate cause and effect. However, in nature, crabs are patchily distributed and extremely mobile. It is likely that crab effluent cued on by snails is ephemeral and highly variable in space and time on emergent substrata. The dynamic response by gastropod prey to these cues may generate variability in both their induced defences and foraging activity in space and time. How these responses average out or increase variability depends on the foraging range of crabs and the arena of sensitivity by prey to cues. This could vary on a very small scale. We suggest that TMIIs are an important driver of the variability component of predation in structuring north-west Atlantic rocky shore communities. Because TMIIs can be as strong as density effects (Trussell et al., 2004) and prey behaviours can fluctuate in time, in contrast to consumed prey, TMIIs should contribute more than density-mediated effects to variability in the strength of predation.

## 4.5 Direct and Indirect Effects of Humans on North-West Atlantic Rocky Shores

Perhaps the most striking realisation of the past thirty years comes from evidence documenting how strongly humans interact, both directly and indirectly, with ecosystem processes on rocky shores. Humans interact directly with coastal ecosystems through harvesting, eutrophication and as vectors for the introduction of non-native species. The indirect effects, principally warming and OA, manifest from anthropogenically driven climate change.

Humans have had a long history of consistently interacting with species and processes of north-west Atlantic marine ecosystems. That history, summarised for the Quoddy region in the outer Bay of Fundy, Canada in an elegant study by Lotze and Milewski (2004), can be distinguished by two periods: the indigenous hunter-gatherer period from 3,000 to 1,000 yr BP and the period of European colonisation of North America, especially the last 350 years. The early indigenous period can be characterised as consistent, low-impact use of marine resources across all trophic levels. Faunal remains from archaeological sites include some vertebrates, but mostly marine invertebrates, especially soft shell clams (*Mya arenaria*), yet there is no evidence of extinctions, changes in size or composition over 3,000 years (Bourque, 1995; Steneck, 1997; Spiess and Lewis, 2001; Lotze and Milewski, 2004). The diversity of intertidal animal species used includes (in addition to soft shell clams), blue mussels, whelks, sea urchins and barnacles. Indigenous peoples certainly made use of the marine littoral zone, but whether their actions shaped species abundances and community composition is uncertain.

With respect to use of resources, the period of European colonisation to the present can be characterised as increasingly intense and efficient harvesting, beginning at the top of food webs and serially exploiting lower trophic levels (Lotze and Milewski, 2004). Overfishing is among the clearest examples of humans transforming coastal oceanic ecosystems (Steneck, 1997; Jackson et al., 2001; Steneck and Carlton, 2001; Myers and Worm, 2005). The effects of serial depletion of apex vertebrate predators, especially Atlantic cod (*Gadus morhua*), but including other fishes, shorebirds and marine mammals, prior to and during the twentieth century eventually triggered a trophic cascade. Changes included increased abundances of lobsters, crabs and urchins and loss of kelp and other non-calcareous perennial algae associated with shifts to communities dominated by crustose coralline algae (Jackson, 2001; Steneck and Carlton, 2001). Fishing down the food web to include the harvest of sea urchins recovered kelp beds that provided habitat for lobsters and crabs, further increasing their abundances in the absence of large, predatory finfishes.

The trophic cascade from overfishing dramatically altered subtidal habitats, but data that speak directly to how it may have affected rocky intertidal habitats are lacking. Extremely mobile invertebrates, such as crabs, are known to follow tidal currents while foraging (Cohen et al., 1995) and may seek shelter on wave-protected shores among the large canopies of the fucoid *Ascophyllum*, especially in areas of kelp loss in the subtidal. Increased numbers of crabs in the intertidal zone have been observed in recent years (Mutti et al., unpublished data) and may be linked to observed declines in populations of *M. edulis* in the GOM (Sorte et al., 2017), especially among recruit and juvenile size classes (Petraitis and Dudgeon, 2015).

In addition to the direct effect that fishing has on ecosystems from the top-down, fishing also has an effect from the bottom-up. Overfishing at higher trophic levels has left in its wake a single commercial fishery for lobster on which coastal economies in eastern North America now depend for 80 per cent of their total fishery income (Boudreau and Worm, 2010; Steneck et al., 2011). The bait and trap method of fishing for lobster has been viewed as effectively 'farming lobsters' (Grabowski et al., 2010), and, as with many commercial farming practices, creates the potential for eutrophication of coastal ecosystems. The input of bait to the benthos by the lobster fishery (>75,000 metric tons per year of herring alone; Grabowski et al., 2010) is massive and, in 2016, a shortage of herring resulted in a temporary closure of the herring fishery. The mass of herring used as bait is at least two-fold the harvested mass of lobsters (Harnish and Willison, 2009) and much of the input to the coast comes from fishing offshore, providing a mechanism of increasing energy and nutrient subsidies transferred into the coastal ecosystems of the GOM (Fogarty et al., 2012). Much of this excess bait probably subsidises populations of lobster, crab and other species (Grabowski et al., 2010). Reduced competition and predation on crabs by the harvest of lobsters coupled with subsidising crab populations with lobster bait provide additional mechanisms, facilitating increased populations of crabs inhabiting littoral habitats.

Eutrophic conditions have developed from a multitude of other activities as human populations increased along north-west Atlantic shores, including municipal waste water treatment, runoff from artificial agricultural fertilisers and wood debris from logging, among other point sources emitting compounds rich in nitrogen and phosphorus (Cooper and Brush, 1993; Valiela et al., 1997; Lotze and Milewski, 2004). Although eutrophication is most pronounced in estuaries near urban centres (see Chapter 6) and declines with distance from these epicentres, its effects have been characterised on rocky shores. Compared to oligotrophic (control treatment) sites, eutrophic (or experimental treatment) sites on north-west Atlantic rocky shores have reduced canopies of perennial macroalgae (e.g., fucoids), potentially great abundances of ephemeral algae, especially bloom-forming ulvoids, and increased abundances of suspension-feeding benthic invertebrates, but greatly reduced diversity with communities increasingly dominated by single species of algae or invertebrate (Worm et al., 1999, 2001; Lotze and Worm, 2000; Lotze and Milewski, 2004; Worm and Lotze, 2006). The loss of perennial fucoid macroalgae under eutrophic conditions is particularly problematic. Fucoid canopies provide habitat for a wide variety of sessile and mobile animals and several macroalgae. The positive effects fucoids, like *Ascophyllum*, have increases diversity (Bertness et al., 1999) of an otherwise species-impoverished ecosystem (Witman et al., 2004).

The summed effects of human-mediated top-down and bottom-up effects is transformation of north-west Atlantic rocky shores to systems largely driven by bottom-up processes that generate a low-diversity assemblage of small, ephemeral, yet highly productive species. Harvesting now occurs across all trophic levels of north-west Atlantic rocky shores, including primary producers. For instance, the foundation species of sheltered rocky shores, *Ascophyllum*, is an increasingly important economic resource with harvests as high as 37,000 tonnes in 2007 in Atlantic Canada alone (Ugarte et al., 2007). With the top trophic levels already depleted, harvest of lower trophic levels and perennial primary producer species will reduce standing stock in the community and likely reinforce ecosystem production being fuelled by small, ephemeral algae, invertebrates and microbes.

Humans are as effective as vectors of introduction to add species to coastal ecosystems as they are at removing them. In so doing, existing interactions between species and community dynamics are destabilised for generations to follow. The last few decades have consisted of frequent invasions to north-west Atlantic rocky shores with a phyletic distribution of exotic species comparable in diversity to that of the taxa native to the region. As of 2010 in the GOM alone, there were at least sixty-four confirmed invasions by marine species (not including cryptogenic species for which native or introduced status cannot be determined) spread across seventeen phyla (Pappal, 2010). Among the well-known examples besides *C. maenas* and *L. littorea* are thirteen crustaceans including the brachyuran crab, *H. sanguineus*, no fewer than seven ascidian species, *Membranipora membranacea* (Bryozoa), five brown seaweeds including *F. serratus* and *Colpomenia peregrina*, the green seaweed, *Codium fragile*, and fourteen species of red seaweeds, including *Grateloupia turuturu* (the largest red seaweed; Mathieson et al., 2008), *Gracilaria vermiculophylla* and *Dasysiphonia* (=*Heterosiphonia*) *japonica*. Many of these species are abundant, have large ecological impacts and are displacing native and/or other introduced species (e.g., Johnson et al., 2012; O'Connor, 2014).

Three factors stand out as explanations for the propensity for invasion to north-west Atlantic rocky shores. The first is, obviously, the importance of the east coast of North America in global trade via shipping. The frequent arrivals of ships provide countless opportunities for exotic species introductions. Data showing that the majority of exotic species in the north-west Atlantic come from Europe and ~25 per cent from Asia is consistent with this hypothesis (Pappal, 2010).

Second is the extremely low diversity of north-west Atlantic rocky shores, providing ample opportunity for exotic species to become established. Stachowicz et al. (1999) showed that an exotic colonial ascidian, *Botrylloides diegensis*, survived better in low biodiversity, compared to high biodiversity habitats because in richer habitats space (the limiting resource) is used more completely and efficiently by the additional species.

Third is climate change and potential habitat modifications that may disadvantage native species and provide advantage for exotic species. Exotic species may benefit from changing environmental conditions in the new habitat that coincide with their physiological tolerances. This is thought to explain the range expansion and greater abundance of *C. maenas* in the northern GOM in both the 1950s (Glude, 1955) and again in the past decade (Mutti et al., unpublished data). Exotic species also benefit from extended periods and new routes for shipping as the climate warms (Pappal, 2010). This may have played a role in the repeated introductions and associated range expansion of *C. maenas* in the Canadian Maritimes discussed earlier (Roman, 2006).

What are the community-level consequences of introducing non-native species? By virtue of humans' role as vectors of repeated species introductions, we are likely a strong force, increasing the magnitude of variability among interspecific interactions, which is reflected in the dynamics of rocky shore communities.

Human activities are increasingly affecting rocky shore ecosystems of the north-west Atlantic indirectly from anthropogenically driven climate change. Two parameters on the rise that threaten the integrity of rocky shores in the region are warming and OA. Ocean warming, driven by climate change, has increased SST 0.23°C/year since 2004, a faster rate of change than 99.9 per cent of the global ocean (Mills et al., 2013; Pershing et al., 2015). The GOM is especially vulnerable to OA because of the low buffering capacity of shelf waters, inputs from the associated watershed and a more acidic Labrador Coastal Current (Wang et al., 2013). As observed on temperate rocky shores of the north-east Pacific Ocean (Wootton et al., 2008), continued pressing of the ecosystem by increasing SST and dissolved $pCO_2$ provides another mechanism to shift species composition to further favour greater occupation by macroalgae at the expense of calcifying invertebrate taxa, which comprise the majority of common species on north-west Atlantic rocky shores.

Non-calcifying macroalgae stand to benefit from climate change in several ways. Increasing $pCO_2$ in seawater increases the supply of dissolved inorganic carbon resources taken up for photosynthesis and growth (Kübler et al., 1999; Kübler and Dudgeon, 2015), whereas the consequent reduction in seawater pH adversely affects their calcified consumers (Harley et al., 2012; Kroeker et al., 2013). Thus, a simple prediction is increased standing stock biomass of macroalgae by means of enhanced growth and reduced losses to herbivory under continued OA.

Even with the rapid rate of warming in the GOM the past decade, macroalgae living there and in the Canadian Maritimes will probably benefit from rising SST, although some variance attributed to idiosyncratic responses among species should be expected. Increased metabolic rates at warmer temperatures will enhance macroalgal uptake of the increased supply of inorganic nutrients provided by OA and eutrophication. Studies of several common and ecologically important red seaweeds, kelps and fucoids indicate capacities for either ecotypic differentiation (Gerard, 1988) or sustained thermal acclimation to the warmer temperatures expected in the coming decades (Davison, 1987; Davison et al., 1991; Kuebler et al., 1991; Kübler and Davison, 1993, 1995; Dudgeon et al., 1995; Keser et al., 2005).

Only in southern New England (south and west of Cape Cod) are perennial macroalgae on rocky shores likely to be adversely affected by rising SST in the coming decades. At the southern limit in Long Island Sound, growth of *Ascophyllum* over twenty-four years was positively associated with a 1.6°C rise in temperature to the present summer ambient temperature of 23°C (Keser et al., 2005). However, nearby populations in the thermal plume of a power plant rapidly decreased growth and increased mortality at temperatures above 25°C. The distribution limits of perennial seaweeds at their respective southern range edges will likely shift slightly northward in the western North Atlantic, but much less so than northward edge range shifts predicted on European shores (e.g., Jueterbock et al., 2013).

Nevertheless, the macroalgae that are likely to continue to dominate rocky shores of southern New England are ephemeral, bloom-forming seaweeds, like *Ulva* spp. These species are likely to persist and replace long-lived perennial species (Thornber et al., 2013). Indeed, short-lived, ephemeral species of macroalgae are likely to increase on all shores of the north-west Atlantic under conditions of warming, OA and eutrophication. In the past, these species were seasonal, but now, in many places in the GOM Maine, these species now persist throughout the year (J. Kübler, personal communication).

In contrast to many macroalgae, the combined effects of warming and OA from climate change have severe negative effects on many marine invertebrates, especially among calcifying species (Kroeker et al., 2013). For example, we have seen declines in the abundance of gastropods whose larvae are planktonic and begin calcification while in the water column (e.g., *L. littorea* and *T. testudinalis*) but not in barnacles, which do not calcify while planktonic (Petraitis and Dudgeon, unpublished manuscript).

Blue mussels of the genus *Mytilus* are probably the best-studied group of the effects of OA and warming in the north-west Atlantic, and here we focus on how mussels have responded to climate change. Mussels are not only a good representative of calcifying invertebrates on north-west Atlantic rocky shores, but also an ecologically important group. *M. edulis* shows adaptive phenotypic plasticity and variation in plasticity to thermal stress from warming (Lesser, 2016). However, mussels exposed to both future expected OA conditions and thermal stress showed evidence of metabolic depression, suggesting dampening of the acclimation response. The metabolic depression affects phenotypes of traits other than calcification. *Mytilus* under OA also produce fewer, weaker byssal threads weakening attachment strength to the shore (O'Donnell et al., 2013; Zhao et al., 2017). Interestingly, however, the ability of *M. edulis* to tolerate OA is greatly enhanced when mussels have a rich food supply (Thomsen et al., 2013). Jones et al. (2010) observed that warming above 32°C on consecutive exposures increased expression of heat shock protein 70, and resulted in high mortality rates that presumably underlie the range contraction of *M. edulis* in recent decades at its southern range edge.

## 4.6 Looking Forward: A Synthesis of Community and Ecosystem Ecology across Scales

Rocky shores of the north-west Atlantic Ocean have served as a natural laboratory for study of the ecological processes that generate and maintain patterns of species distribution and abundance for over fifty years. These shores are likely to continue service as a useful laboratory for field ecology in the coming decades because its past and present may provide insight into the future of coastal marine ecosystems in the Anthropocene. Humans have played a major and important role in shaping the communities inhabiting the hard substrate littoral systems in the north-west Atlantic for centuries (Jackson et al., 2001; Steneck and Carlton, 2001; Lotze and Milewski, 2004) and provide lessons to be learned for managing ecosystems moving forwards. Likewise, the present rate of climate change, especially in the GOM, is rapid (Pershing et al., 2015) and appropriately designed experiments offer a glimpse of how future climate change may modify ecological processes and the relationships among them to reshape patterns and production of coastal ecosystems with implications for strategies of ecosystem-based management of coastal resources.

History shows that the north-west Atlantic coast is a biome under continual change and the present rates of biotic and abiotic change give every indication that the rate of change will be accelerating. In this context, we identify the gaps in our knowledge about rocky shore ecosystems of the north-west Atlantic in an attempt to get ahead in our ability to understand and predict the effects of this accelerating change. First, we reiterate that future studies need to place greater emphasis on estimating magnitudes of variability in responses of variables measured. It is crucial to understand the role of variability in shaping ecosystem dynamics by direct comparison with average effects. Sadly, this is not a new message (Bourget and Fortin, 1995). This is especially needed as environmental conditions continue to change and estimates of average effects of ecological processes become less meaningful. The body of work on north-west Atlantic rocky shores has focussed primarily on how consumers, resources, non-trophic interactions, supplies of recruiting propagules and the interactions among them differ in their average effect across persistent environmental gradients. Studies designed to estimate both average and variability effects as drivers of community dynamics will enable direct comparison and the conditions under which each is important.

Second, data are needed that explicitly link trait-mediated indirect effects to community-level patterns in situ to determine whether TMIIs contribute to average effects or variability effects in natural ecosystems. Evidence for indirect community-level effects of TMIIs comes from mesocosms, rock pools or experimental chambers in the field. It is unknown if those effects would manifest with mobile predators and prey at larger spatial scales. The temporal dynamics of sensitive prey responses to risk cues of predators based on the physiological state of the prey (Matassa and Trussell, 2014) coupled with a clumped distribution and mobility of crab predators suggest highly variable prey foraging behaviour. Depending on the shape of the function that links prey foraging behaviour to indirect effects, the variability of outcomes of indirect effects at the community level may be an important driver of the system. Most importantly, it is not at all clear if indirect effects will dampen or enhance the effects of variability across larger scales. This remains an open question.

Third, the merging of community and ecosystem ecology perspectives is an important synthesis that has been slow to develop in the study of north-west Atlantic rocky shores. Paine's seminal works (1966, 1969) on Pacific rocky shores effectively decoupled community and ecosystem ecology perspectives in studies of hard substrate marine communities for many years. However, much stands to be gained in our understanding of north-west Atlantic rocky shores by a new marriage of these two approaches. This is particularly important on the rocky shores in the north-west Atlantic Ocean, given the twin demons of high production (Mann, 1973; Vadas et al., 2004) and low diversity (Witman et al., 2004) that likely drive the underlying resilience of the ecosystem.

One avenue to unite the two is via linkage of green and brown trophic webs. Trophic ecology is among the most studied aspects of north-west Atlantic rocky shores, and herbivory is often invoked as a process controlling seaweed distribution and abundance (Lubchenco, 1978; Bertness et al., 2002; Silliman et al., 2013). Indeed, our current understanding of rocky shore food web structure is based solely on transfer to top trophic levels from living macroalgae by means of herbivory (i.e., the 'green food web', *sensu* Rooney and McCann, 2012). Yet, rocky shores in the north-west Atlantic Ocean are typically occupied by macroalgae with a large standing stock and high productivity. The lowest reaches of the intertidal and shallow subtidal zones are dominated by kelps with fresh weight biomass as high as 30–32 kg·m$^{-2}$ and annual productivity exceeding 1,000 g of carbon·m$^{-2}$·year$^{-1}$ (Mann, 1973). Mid-intertidal zones dominated by the fucoid alga, *Ascophyllum*, commonly have fresh biomass standing stocks of 25–28 kg·m$^{-2}$ with net annual primary production of 14–15 kg·m$^{-2}$·year$^{-1}$, or 894 g of carbon·m$^{-2}$·year$^{-1}$ (Vadas et al. 2004; P. Petraitis, S. Dudgeon, unpublished data). Vadas et al. (2004) reported that ~60 per cent of standing biomass annually enters detrital pools. This is likely an underestimate of detrital contribution by *Ascophyllum* since at least 10 per cent of vegetative frond biomass annually enters detrital pools solely from monthly shedding of the epidermis that reduces epiphyte loads (Halat et al., 2015; Garbary et al., 2017). Subsidies contributed to detrital pools by the fucoid, *F. vesiculosus*, along a 14–15 km peninsula at Schoodic Point, Maine during the fall reproductive period (September–December) were estimated to be 610 kg of carbon and 45 kg of nitrogen solely from egg release (Muhlin et al., 2011).

Despite this massive subsidy of detritus to the nearshore ecosystem, its role in providing a source of energy at the basal trophic level (the so-called brown food web, *sensu* Rooney and McCann, 2012) to fuel secondary production in food webs is poorly understood. Variation in algal detritus has been shown to cause spatiotemporal variation in surface deposit-feeding annelid assemblages in soft sediments (Kelaher and Levinton, 2003). Detrital and photosynthetic energy channels may be integrated in multiple ways to fuel secondary production. For instance, as shown by Kelaher and Levinton (2003), consumption of particulate organic matter (POM) provides a means for direct utilisation of detritus by higher trophic levels. Alternatively, microbial decomposition of detrital POM provides a mechanism to regenerate nutrients subsequently taken up by photoautotrophs.

The merging of community and ecosystem approaches to understand rocky shore ecosystems will enable testing of fundamental, yet unresolved, questions about the north-west Atlantic coast. What are the relative magnitudes of macroalgal biomass transferred to consumers versus the detrital pool? Which pathway of trophic transfer of algae, via herbivory or detrital POM, fuels greater primary and secondary production and drives patterns of distribution and abundance on North Atlantic rocky shores?

Integration and feedback between 'green' and 'brown' basal trophic levels may affect aggregate community properties such as the magnitude of total production, its variability and enhanced resilience. These properties may depend upon species at higher trophic levels changing the energy source upon which they rely as the biomasses of detritus and photoautotrophs vary in time, or the spatial coupling among rocky shore and other types of habitat that results from the transfer of organic matter between them. For example, coastal currents transport drifting, bloom-forming algae from source populations to coastal habitats that typically have little algae (Lyons et al., 2009). Seabirds act as a vector to transfer nutrients from sea to land to affect plant species composition (Ellis et al., 2006). Temporal shifts between utilisation of 'green' and 'brown' energy sources provides yet another source of variability of ecological processes on rocky shores. However, effects on community-level patterns from such shifts cannot be predicted without a deeper understanding of the functional relationships between them. Community patterns may not change with shifting use of detrital and photosynthetic energy channels if they compensate one another. Alternatively, community properties may be dampened or enhanced by such shifts. While the role of subsidies into the rocky intertidal zone from not only subtidal systems, but also other systems such as soft-bottom communities is certainly an area of promising research, many of the ideas of marine subsidies harken back to the earlier ground-breaking work of Gary Polis (Polis et al., 1997).

## 4.7 Concluding Remarks

The flora and fauna of the rocky intertidal shores of the north-west Atlantic have been a rich source for marine ecology since the experimental work of the early pioneers (e.g., Bourget, Chapman, Harris, Lubchenco, Menge and Vadas). These communities have continued to be used as a model ecosystem not only for undertaking new and novel experimental approaches, but also for the development of new ideas and concepts. The simplicity of the ecosystem certainly played a part in the successes of early researchers and in the combined productivity of recent investigators. It is ironic that such an impoverished flora and fauna has been such a rich source of ecological insights about how species interact (e.g., direct, indirect and trait-mediated effects, and inducible defences) and community-level effects (e.g., trophic cascades, bottom-up effects and development of alternative states).

Many of these insights have been cast in terms of average effects. The belief that things 'average out' is based on the naive assumption that systems like the rocky intertidal ecosystem of the north-west Atlantic are ergodic. Ergodic systems have three properties that have important consequences for how ecologists view the world, which are (1) averaging over space gives the same answer as averaging over time, (2) the long-term state of the system does not depend on initial conditions and (3) the short-term (or small-scale) average at one time (or place) will be the same over any other time interval (or spatial scale). Nearly all of the experimental research that has been done on rocky intertidal shores in the north-west Atlantic has been done on the scale of tens of metres and over three to five years. Some studies have replicated the same design at several sites spread over tens to hundreds of kilometres. Thus, the broader generalisations about the importance of various ecological processes are based on the assumption that the average is a good estimator and things do, indeed, average out over time and space (also see Gaines and Denny's (1993) cautionary comments about reliance on average values when extreme values may be more important). We are also well aware that we live in a glass house and that all of these shortcomings apply to our own work, most of which has been done on a single island slightly more than 200 $km^2$ in area. Nevertheless, our experiments spanning two decades replicated at the scale of hundreds of metres has shown us that variability is as, or more, important than the average effect, but likely goes undetected in short-term experiments and when the spatial scale of replication is much less than the spatial scale of experimental treatments.

The role of spatial and temporal variability will continue to be a major experimental and conceptual challenge for ecology in the next decade. The challenge will be especially large for rocky intertidal ecology because, as we have shown, recruitment can vary over five orders of magnitude and significant sources of variation occur at all spatial and temporal scales (Petraitis and Dudgeon, 2015). In addition, how variation in recruitment translates into in situ interactions cannot be easily predicted because variation at one level can either enhance or dampen 'down-stream' processes (Methratta and Petraitis, 2008). In general, the relationship between any two processes or variables (e.g., recruitment and predation rate, or SST and predation rate) can be either magnified or dampened from expectation if the functional relationship between them is non-linear; a consequence known as Jensen's inequality. For the relationship, $y = f(x)$, if $f(x)$ is convex, then $\bar{y} \geq f(\bar{x})$, and if $f(x)$ is concave, then $\bar{y} \leq f(\bar{x})$; where $\bar{y} = \sum p_i f(x_i)$, $f(\bar{x}) = f(\sum p_i x_i)$ and $p_i$ equals the proportion or probability of the occurrence of $x_i$. As a first approximation, size of the bias is proportional to variance, although skewness and kurtosis can have a big effect (Ruel and Ayres, 1999; Martin and Huey, 2008; Denny, 2017). If Jensen's inequality plays a role in the relationship between environmental variables and per capita effects, then environmental variability may profoundly alter the shape and development of rocky shore communities.

Scale is also a consideration for the role of anthropogenic activities and the linkage between brown and green webs, which are the two other challenges mentioned in this chapter. Anthropogenic activities and food web linkages are defined by specific biological and chemical processes but an ecological appreciation of their importance to the flora and fauna of north-western Atlantic shores depends on understanding the scale and rates at which OA, changes in SST and the transfer of organic material occur. Additionally, different anthropogenic activities differ with respect to the spatial and temporal scales at which they directly impact natural ecosystems and how they interact with each other and natural variability across these scales to indirectly influence the dynamics of ecosystems (Hawkins et al., 2013). Our continued reliance on production by Atlantic shores requires management that will be increasingly contingent upon long-term and broadscale studies that enable disentangling the interactions among anthropogenic and natural drivers to attribute relative strengths of their effects on natural ecosystems (Hawkins et al., 2017).

Global climate models predict increasing frequencies, intensities and durations of climatic extremes with global warming in the coming decades (Easterling et al., 2000; Meehl and Tebaldi, 2004; Rahmstorf and Coumou, 2011; Field et al., 2012). Increase in extreme events and overall environmental variability is likely to strengthen the role that variability of ecological processes plays in shaping the structure and dynamics of ecosystems, and we suggest that ecologists should increasingly pay attention to that variability (see Ruel and Ayres, 1999; Harley and Paine, 2009; Bozinovic et al., 2011; Thompson et al., 2013). For example, variation in temperature is expected to pose a greater risk to invertebrate species than the average effect of warming (Stachowicz et al., 2002; Vasseur et al., 2014). As humans exert increasing pressure on coastal ecosystems in the years ahead, both directly and indirectly, we should expect variability effects to usurp average effects as the principal drivers of rocky shore ecosystem dynamics of the north-west Atlantic Ocean.

## Acknowledgement

This material is based upon work supported by the US National Science Foundation under grant nos DEB-1020480 and DEB-1555641.

### REFERENCES

Abrams, P. A., Menge, B. A., Mittelbach, G. G., Spiller, D. and Yodzis, P. (1996). The Role of Indirect Effects in Food Webs. In G. Polis and K. Winemiller, eds. *Food Webs: Dynamics and Structure*. Chapman Hall, New York, pp. 371–95.

Adey, W. H. and Steneck, R. S. (2001). Thermogeography over time creates biogeographic regions: a temperature/space/time-integrated model and an abundance-weighted test for benthic marine algae. *Journal of Phycology*, **37**, 677–98.

Bedford Institute of Oceanography. (1987). *State of the Ocean: March. Weekly Briefing 24 April 1987*. Bedford Institute of Oceanography, Bedford, Nova Scotia.

Bergeron, P. and Bourget, E. (1984). The effect of cold temperatures and ice on intertidal populations in northern regions, particularly in the St. Lawrence estuary. *Oceanis*, **10**, 259–78.

Berlow, E. L. (1999). Strong effects of weak interactions in ecological communities. *Nature*, **398**, 330–4.

Bertness, M. D. (1989). Intraspecific competition and facilitation in a northern acorn barnacle population. *Ecology*, **70**, 257–68.

Bertness, M. D., Leonard, G. H., Levine, J. M., Schmidt, P. R. and Ingraham, A. O. (1999). Testing the relative contribution of positive and negative interactions in rocky intertidal communities. *Ecology*, **80**, 2711–26.

Bertness, M. D., Trussell, G. C., Ewanchuk, P. J. and Silliman, B. R. (2002). Do alternate stable community states exist in the Gulf of Maine rocky intertidal zone? *Ecology*, **83** (12), 3434–48.

Bertness, M. D., Trussell, G. C., Ewanchuk, P. J., Silliman, B. R. and Crain, C. M. (2004a). Consumer-controlled community states on Gulf of Maine rocky shores. *Ecology*, **85**, 1321–31.

Bertness, M. D., Trussell, G. C., Ewanchuk, P. J. and Silliman, B. R. (2004b). Do alternate stable community states exist in the Gulf of Maine rocky intertidal zone? *Ecology*, **85**, 1165–7.

Bertness, M. D., Yund, P. O. and Brown, A. F. (1983). Snail grazing and the abundance of algal crusts on a sheltered New England rocky beach. *Journal of Experimental Marine Biology and Ecology*, **71**, 147–64.

Blakeslee, A. M., Byers, J. E. and Lesser, M. P. (2008). Solving cryptogenic histories using host and parasite molecular genetics: the resolution of *Littorina littorea's* North American origin. *Molecular Ecology*, **17**, 3684–96.

Boudreau, S. A. and Worm, B. (2010). Top-down control of lobster in the Gulf of Maine: insights from local ecological knowledge and research surveys. *Marine Ecology Progress Series*, **403**, 181–91.

Bourdeau, P. E. and O'Connor, N. J. (2003). Predation by the nonindigenous Asian shore crab *Hemigrapsus sanguineus* on macroalgae and molluscs. *Northeastern Naturalist*, **10**, 319–34.

Bourget, E. and Fortin, M. J. (1995). A commentary on current approaches in the aquatic sciences. *Hydrobiologia*, **300**, 1–16.

Bourget, E., Archambault, D. and Bergeron, P. (1985). Effet des proprieties hivernales sur les peuplements epibenthiques intertidaux dans un milieu subarctique, l'estuaire du Saint-Laurent. *Le Naturaliste Canadiens, Revue d'Ecologie et de Systématique*, **112**, 131–42.

Bourque, B. J. (1995). *Diversity and Complexity in Prehistoric Maritime Societies: A Gulf of Maine Perspective.* Plenum Press, New York.

Bozinovic, F., Bastías, D. A., Boher, F., Clavijo-Baquet, S., Estay, S. A. and Angilletta, Jr. M. J. (2011). The mean and variance of environmental temperature interact to determine physiological tolerance and fitness. *Physiological and Biochemical Zoology*, **84**, 543–52.

Brawley, S. H., Coyer, J. A., Blakeslee, A. M. H. et al. (2009). Historical invasions of the intertidal zone of Atlantic North America associated with distinctive patterns of trade and emigration. *Proceedings of the National Academy of Sciences USA*, **106**, 8239–44.

Breen, E. and Metaxas, A. (2009). Effects of non-indigenous *Carcinus maenas* on the growth and condition of juvenile *Cancer irroratus*. *Journal of Experimental Marine Biology and Ecology*, **377**, 12–19.

Briggs, J. (1974). *Marine Zoogeography*. New York: McGraw-Hill.

Bryson, E. S., Trussell, G. C. and Ewanchuk, P. J. (2014). Broad-scale geographic variation in the organization of rocky intertidal communities in the Gulf of Maine. *Ecological Monographs*, **84**, 579–97.

Burrows, M., Schoeman, D. S., Buckley, L. B. et al. (2011). The pace of shifting climate in marine and terrestrial ecosystems. *Science*, **334**, 652–5.

Carlton, J. (1982). The historical biogeography of *Littorina littorea* on the Atlantic coast of North America, and implications for the interpretation of the structure of New England intertidal communities. *Malacology Review*, **15**, 146.

Carlton, J. T. and Cohen, A. N. (2003). Episodic global dispersal in shallow water marine organisms: the case history of the European shore crabs *Carcinus maenas* and *C. aestuarii*. *Journal of Biogeography*, **30**, 1809–20.

Chapman, A. R. O. and Johnson, C. R. (1990). Disturbance and Organization of Macroalgal Assemblages in the Northwest Atlantic. In A. R. O. Chapman and A. J. Underwood, eds. *Determinants of Structure in Intertidal and Subtidal Macroalgal Assemblages*. Kluwer, Dordrecht, pp. 77–121.

Cohen, A. N., Carlton, J. T. and Fountain, M. C. (1995). Introduction, dispersal and potential impacts of the green crab *Carcinus maenas* in San Francisco Bay, California. *Marine Biology*, **122**(2), 225–37.

Colton, H. S. (1916). On some varieties of *Thais lapillus* in the Mount Desert region, a study of individual ecology. *Proceedings of the Academy of Natural Sciences of Philadelphia*, **68**, 440–54.

Connell, J. H. (1978). Diversity in tropical rain forests and coral reefs. *Science*, **199**, 1302–10.

Connolly, S. R. and Roughgarden, J. (1998). A latitudinal gradient in northeast Pacific intertidal community structure: Evidence for an oceanographically based synthesis of marine community theory. *American Naturalist*, **151**, 311–26.

Cooper, S. R. and Brush, G. S. (1993). A 2,500-year history of anoxia and eutrophication in Chesapeake Bay. *Estuaries*, **16**, 617–26.

Dijkstra, J. Harris, L. G. and Westerman, E. (2007). Distribution and long-term temporal patterns of four invasive colonial ascidians in the Gulf of Maine. *Journal of Experimental Marine Biology and Ecology*, **342**, 61–8.

CLIMAP Project Members. (1976). The surface of the ice-age Earth. *Science*, **191**, 1131–7.

Davison, I. R. (1987). Adaptation of photosynthesis in *Laminaria saccharina* (Phaeophyta) to change in growth temperature. *Journal of Phycology*, **23** (2), 273–83.

Davison, I. R., Greene, R. M. and Podolak, E. J. (1991). Temperature acclimation of respiration and photosynthesis in the brown alga *Laminaria saccharina*. *Marine Biology*, **110**, 449–54.

Denny, M. W. (2017). The fallacy of the average: on the ubiquity, utility and continuing novelty of Jensen's inequality. *Journal of Expermintal Biology*, **220**, 139–46.

Dernbach, E. M. and Freeman, A. S. (2015). Foraging preference of whelks *Nucella lapillus* is robust to influences of wave exposure and predator cues. *Marine Ecology Progress Series*, **540**, 135–44.

Dudgeon, S. R. and Petraitis, P. S. (2001). Scale-dependent recruitment and divergence of intertidal communities. *Ecology*, **82**, 991–1006.

Dudgeon, S. R., Kübler, J. E., Vadas, R. L. and Davison, I. R. (1995). Physiological tolerances to environmental variation in intertidal red algae: does thallus morphology matter? *Marine Ecology Progress Series*, **117**, 193–206.

Dudgeon, S. R., Steneck, R. S., Davison, I. R. and Vadas, R. L. (1999). Coexistence of similar species in a space-limited intertidal zone. *Ecological Monographs*, **69**, 331–52.

Easterling, D. R., Meehl, G. A., Parmesan, C., Changnon, S. A., Karl, T. R. and Mearns, L. O. (2000). Climate extremes: observations, modelling, and impacts. *Science*, **289**, 2068–74.

Ellis, J. C., Chen, W., O'Keefe, B., Shulman, M. J. and Witman, J. D. (2005). Predation by gulls on crabs in the rocky intertidal and shallow subtidal zones of the Gulf of Maine. *Journal of Experimental Marine Biology and Ecology*, **324**, 31–43.

Ellis, J. C., Fariña, J. M. and Witman, J. D. (2006). Nutrient transfer from sea to land: the case of gulls and cormorants in the Gulf of Maine. *Journal of Animal Ecology*, **75**, 565–74.

Ellis, J. C., Shulman, M. J., Wood, M., Witman, J. D. and Lozyniak, S. (2007). Regulation of intertidal food webs by avian predators on New England rocky shores. *Ecology*, **88**(4), 853–63.

Epifanio, C. E. (2013). Invasion biology of the Asian shore crab *Hemigrapsus sanguineus*: a review. *Journal of Experimental Marine Biology and Ecology*, **441**, 33–49.

Fairweather, P. G. (1988). Consequences of supply-side ecology – manipulating the recruitment of intertidal barnacles affects the intensity of predation upon them. *Biological Bulletin*, **175**, 349–54.

Field, C. B., Barros, V., Stocker, T. F. et al. (2012). Managing the Risks of Extreme Events and Disasters to Advance Climate Change Adaptation. In *A Special Report of Working Groups I and II of the Intergovernmental Panel on Climate Change*. Cambridge University Press, Cambridge.

Fisher, J. A. D. (2005). Exploring ecology's attic: overlooked ideas on intertidal food webs. *Bulletin of the Ecological Society of America*, **86**, 145–51.

Fisher, J. A. D. (2010). Parasite-like associations in rocky intertidal assemblages: implications for escalated gastropod defences. *Marine Ecology Progress Series*, **399**, 199–209.

Fisher, J. A. D., Rhile, E. C., Liu, H. and Petraitis, P. S. (2009). An intertidal snail shows a dramatic size increase over the past century. *Proceedings of the National Academy of Sciences USA*, **106**, 5209–12.

Fogarty, M. J., Townsend, D. and Klein, E. (2012). Advances in understanding ecosystem structure and function in the Gulf of Maine. *American Fisheries Symposium*, **79**, 261–72.

Freeman, A. S. and Byers, J. E. (2006). Divergent induced responses to an invasive predator in marine mussel populations. *Science*, **313**, 831–3.

Gaines, S. D. and Denny, M. W. (1993). The largest, smallest, highest, lowest, longest, shortest – extremes in ecology. *Ecology*, **74** (6), 1677–92.

Garbary, D. J., Galway, M. E. and Halat, L. (2017). Response to Ugarte et al.: *Ascophyllum* (phaeophyceae) annually contributes over 100% of its vegetative biomass to detritus. *Phycologia*, **56** (1), 114–16.

Gerard, V. (1988). Ecotypic differentiation in light-related traits of the kelp *Laminaria saccharina*. *Marine Biology*, **97** (1), 25–36.

Glude, J. B. (1955). The effects of temperature and predators on the abundance of the soft-shell clam, *Mya arenaria* in New England. *Transactions of the American Fisheries Society*, **84**, 12–26.

Grabowski, J. H., Clesceri, E. J., Baukus, E. J., Gaudette, J., Weber, M. and Yund, P. O. (2010). Use of herring bait to farm lobsters in the Gulf of Maine. *PLoS ONE*, **5**, e10188.

Griffen, B. and Byers, J. E. (2009). Community impacts of two invasive crabs: the interactive roles of density, prey recruitment and indirect effects. *Biological Invasions*, **11** (4), 927–40.

Halat, L., Galway, M. E., Gitto, S. and Garbary D. J. (2015). Epidermal shedding in *Ascophyllum nodosum* (Phaeophyceae): seasonality, productivity and relationship to harvesting. *Phycologia*, **54**, 599–608.

Harley, C. D. G. and Paine, R. T. (2009). Contingencies and compounded rare perturbations dictate sudden distributional shifts during periods of gradual climate change. *Proceedings of the National Academy of Sciences USA*, **106**, 11172–6.

Harley, C. D. G., Anderson, K. M., Demes, K. W. et al. (2012). Effects of climate change on global seaweed communities. *Journal of Phycology*, **48** (5), 1064–78.

Harnish, L. and Willison, J. H. M. (2009). Efficiency of bait usage in the Nova Scotia lobster fisher: a first look. *Journal of Cleaner Production*, **17**, 345–7.

Harris, L. G. and Jones, A. C. (2005). Temperature, herbivory and epibiont acquisition as factors controlling the distribution and ecological role of an invasive seaweed. *Biological Invasions* **7**, 913–24.

Hawkins, S. J., Firth, L. B., McHugh, M. et al. (2013). Data rescue and reuse: recycling old information for new concerns. *Marine Policy*, **42**, 91–8.

Hawkins, S. J., Evans, A. J., Firth, L. B. et al. (2017). Distinguishing globally-driven changes from regional- and local-scale impacts: the case for long-term and broad-scale studies of recovery from pollution. *Marine Pollution Bulletin*, **124**, 573–86.

Jackson, J. B. C. (2001). What was natural in the coastal oceans? *Proceedings of the National Academy of Sciences USA*, **98** (10), 5411–18.

Jackson, J. B. C., Kirby, M. X. Berger, W. H. et al. (2001). Historical overfishing and the recent collapse of coastal ecosystems. *Science*, **293**, 629–38.

Jenkins, S. R., Norton, T. A. and Hawkins, S. J. (2004). Long term effects of *Ascophyllum nodosum* canopy removal on mid shore community structure. *Journal of the Marine Biological Association of the U.K.*, **84**, 327–9.

Johnson, L. E., Brawley, S. H. and Adey, W. H. (2012). Secondary spread of invasive species: historic patterns and underlying mechanisms of the continuing invasion of the European rockweed *Fucus serratus* in eastern North America. *Biological Invasions*, **14**, 79–97.

Johnson, D. S. and Skutch, A. F. (1928). Littoral vegetation on a headland of Mt. Desert Island, Maine. II. Tidepools and the environment and classification of submersible plant communities. *Ecology*, **9**, 307–38.

Jones, S. J., Lima, F. P. and Wethey, D. S. (2010). Rising environmental temperatures and biogeography: poleward range contraction of the blue mussel, *Mytilus edulis* L., in the western Atlantic. *Journal of Biogeography*, **37**, 2243–59.

Jueterbock, A., Tyberghein, L., Verbruggen, H., Coyer, J. A., Olsen, J. L. and Hoarau, G. (2013). Climate change impact on seaweed meadow distribution in the North Atlantic rocky intertidal. *Ecology and Evolution*, **3**, 1356–73.

Kassen, R., Buckling, A., Bell, G. and Rainey, P. B. (2000). Diversity peaks at intermediate productivity in a laboratory microcosm. *Nature*, **406**, 508–12.

Kelaher, B. P. and Levinton, J. S. (2003). Variation in detrital enrichment causes spatio-temporal variation in soft sediment assemblages. *Marine Ecology Progress Series*, **261**, 85–97.

Keser, M., Swenarton, J. T. and Foertch, J. F. (2005). Effects of thermal input and climate change on growth of *Ascophyllum nodosum* (Fucales, Phaeophyceae) in eastern Long Island Sound (USA). *Journal of Sea Research*, **54**, 211–20.

Kordas, R. L. and Dudgeon, S. R. (2009). Modelling variation in interaction strength between barnacles and fucoids. *Oecologia*, **158**, 717–31.

Kordas, R. L. and Dudgeon, S. R. (2011). Dynamics of species interaction strength in space, time and with developmental stage. *Proceedings of the Royal Society of London B*, **278**, 1804–13.

Kordas, R. L., Dudgeon, S. R., Storey, S. and Harley, C. D. G. (2015). Intertidal community responses to field-based experimental warming. *Oikos*, **124**, 888–98.

Kraemer, G. P., Sellberg, M., Gordon, A. and Main, J. (2007). Eight-year record of *Hemigrapsus sanguineus* (Asian shore crab) invasion in western Long Island Sound estuary. *Northeastern Naturalist*, **14**, 207–24.

Kroeker, K. J., Kordas, R. L., Crim, R. et al. (2013). Impacts of ocean acidification on marine organisms: quantifying sensitivities and interaction with warming. *Global Change Biology*, **19** (6), 1884–96.

Kübler, J. E. and Davison, I. R. (1993). High-temperature tolerance of photosynthesis in the red alga, *Chondrus crispus*. *Marine Biology*, **117**, 327–36.

Kübler, J. E. and Davison, I. R. (1995). Thermal acclimation of light-use characteristics of *Chondrus crispus* (Rhodophyta). *European Journal of Phycology*, **30**, 189–95.

Kübler, J. E. and Dudgeon, S. R. (2015). Predicting effects of ocean acidification and warming on algae lacking carbon-concentrating mechanisms. *PLoS ONE*, **10** (7), e0132806.

Kübler, J. E., Johnston, A. M. and Raven, J. A. (1999). The effects of reduced and elevated $CO_2$ and $O_2$ on the seaweed, *Lomentaria articulata*. *Plant, Cell & Environment*, **22**, 1303–10.

Kuebler, J. E., Davison, I. R. and Yarish, C. (1991). Photosynthetic temperature adaptation in the red algae, *Lomentaria baileyana* and *Lomentaria orcadensis*. *British Phycological Journal*, **26**, 9–19.

Large, S. I. and Smee, D. L. (2013). Biogeographic variation in behavioural and morphological responses to predation risk. *Oecologia*, **171**, 961–9.

Leonard, G. H. (2000). Latitudinal variation in species interactions: a test in the New England rocky intertidal zone. *Ecology*, **81**, 1015–30.

Leonard, G. H., Levine, J. M., Schmidt, P. R. and Bertness, M. D. (1998). Flow-driven variation in intertidal community structure in a Maine estuary. *Ecology*, **79**, 1395–411.

Leonard, G. H., Bertness, M. D. and Yund, P. O. (1999). Crab predation, waterborne cues, and inducible defences in the blue mussel, *Mytilus edulis*. *Ecology*, **80** (1), 1–14.

Lesser, M. P. (2016). Climate change stressors cause metabolic depression in the blue mussel, *Mytilus edulis*, from the Gulf of Maine. *Limnology and Oceanography*, **61**, 1705–17.

Lewis, J. R. (1964). *The Ecology of Rocky Shores*. English University Press, London.

Lotze, H. K. and Milewski, I. (2004). Two centuries of multiple human impacts and successive changes in a north Atlantic food web. *Ecological Applications*, **14** (5), 1428–47.

Lotze, H. K. and Worm, B. (2000). Variable and complementary effects of herbivores on different life stages of bloom-forming macroalgae. *Marine Ecology Progress Series*, **200**, 167–75.

Lubchenco, J. (1978). Plant species diversity in a marine intertidal community: importance of herbivore food preferences and algal competitive abilities. *American Naturalist*, **112**, 23–39.

Lubchenco, J. (1983). *Littorina* and *Fucus*: effects of herbivores, substratum heterogeneity, and plant escapes during succession. *Ecology*, **64**, 1116–23.

Lubchenco, J. and Menge, B. A. (1978). Community development and persistence in a low rocky intertidal zone. *Ecological Monographs*, **59**, 67–94.

Lyons, P. Thornber, C., Portnoy, J. and Gwilliam, E. (2009). Dynamics of macroalgal blooms along the Cape Cod National Seashore. *Northeastern Naturalist*, **16** (1), 53–63.

Maggs, C. A., Castihlo, R., Foltz, D. et al. (2008). Evaluating signatures of glacial refugia for North Atlantic benthic marine taxa. *Ecology*, **89** (11), S108–22.

Mann, K. H. (1973). Seaweeds: their productivity and strategy for growth. *Science*, **182**, 975–81.

Markham, W. E. (1980). *Ice Atlas: Eastern Canadian Seaboard*. Canadian Government Publishing Centre, Hull, Quebec.

Martin, T. L. and Huey, R. B. (2008). Why "suboptimal" is optimal: Jensen's inequality and ectotherm thermal preferences. *American Naturalist*, **171**, E102–18.

Matassa, C. M. and Trussell, G. C. (2014). Prey state shapes the effects of temporal variation in predation risk. *Proceedings of the Royal Society of London B*, **281**, 20141952.

Matassa, C. M. and Trussell, G. C. (2015). Effects of predation risk across a latitudinal temperature gradient. *Oecologia*, **177**, 775–84.

Mathieson, A. C., Dawes, C. J., Pederson, J. Gladych, R. A. and Carlton, J. T. (2008). The Asian red seaweed *Grateloupia turuturu* (Rhodophyta) invades the Gulf of Maine. *Biological Invasions*, **10**, 985–8.

McCook, L. J. (1992) Species interactions and community structure during succession following massive ice scour of a rocky intertidal seashore. PhD, Dalhousie University.

Meehl, G. A. and Tebaldi, C. (2004). More intense, more frequent, and longer lasting heat waves in the 21st century. *Science*, **305**, 994–7.

Menge, B. A. (1976). Organization of the New England rocky intertidal community: role of predation, competition, and environmental heterogeneity. *Ecological Monographs*, **46**, 355–93.

Menge, B. A. (1983). Components of predation intensity in the low zone of the New England rocky intertidal region. *Oecologia*, **58**, 141–55.

Menge, B. A. and Sutherland, J. P. (1976). Species diversity gradients: Synthesis of roles of predation, competition and temporal heterogeneity. *American Naturalist*, **110**, 351–69.

Menge, B. A. and Sutherland, J. P. (1987). Community regulation: variation in disturbance, competition, and predation in relation to environmental stress and recruitment. *American Naturalist*, **130**, 730–57.

Methratta, E. T. and Petraitis, P. S. (2008). Propagation of scale-dependent effects from recruits to adults in barnacles and seaweeds. *Ecology*, **89** (11), 3128–37.

Mills, K. E., Pershing, A. J., Brown, C. J. et al. (2013). Fisheries management in a changing climate: lessons from the 2012 ocean heat wave in the Northwest Atlantic. *Oceanography*, **26**, 191–5.

Muhlin, J. F., Coleman, M. A., Rees, T. A. V. and Brawley, S. H. (2011). Modelling of reproduction in the intertidal macrophyte *Fucus vesiculosus* and implications for spatial subsidies in the nearshore environment. *Marine Ecology Progress Series*, **440**, 79–94.

Myers, R. A. and Worm, B. (2005). Extinction, survival or recovery of large predatory fishes. *Philosophical Transactions of the Royal Society B*, **360**, 13–20.

O'Connor, N. J. (2014). Invasion dynamics on a temperate rocky shore: from early invasion to establishment of a marine invader. *Biological Invasions*, **16**, 73–87.

O'Donnell, M. J., George, M. N. and Carrington, E. (2013). Mussel byssus attachment weakened by ocean acidification. *Nature Climate Change*, **3** (6), 587–90.

Osman, R. W. and Whitlach, R. B. (2004). The control of the development of a marine benthic community by predation on recruits. *Journal of Experimental Marine Biology and Ecology*, **311**, 117–45.

Osman, R. W. and Whitlach, R. B. (2007). Variation in the ability of *Didemnum* sp. to invade established communities. *Journal of Experimental Marine Biology and Ecology* **342**, 40–53.

Paine, R. T. (1966). Food web complexity and species diversity. *American Naturalist*, **100**, 65–75.

Paine, R. T. (1969). A note on trophic complexity and community stability. *American Naturalist*, **103**, 91–3.

Paine, R. T. and Vadas, R. L. (1969). Effects of grazing by sea urchins, *Strongylocentrotus* spp., on benthic algal populations. *Limnology and Oceanography*, **14**, 710–19.

Palmer, A. R. (1990). Effect of crab effluent and scent of damaged conspecifics on feeding, growth, and shell morphology of the Atlantic dogwhelk *Nucella lapillus* (L.). *Hydrobiologia*, **193**, 155–82.

Pappal, A. (2010). *State of the Gulf of Maine Report: Marine Invasive Species*. Massachusetts Office of Coastal Zone Management and Gulf of Maine Council on the Marine Environment, Boston.

Pearson, G. A. and Brawley, S. H. (1998). Control of gamete release in fucoid algae: sensing hydrodynamic conditions via carbon acquisition. *Ecology* **79** (5), 1725–39.

Pershing, A. J., Alexander, M., Hernandez, C. M. et al. (2015). Slow adaptation in the face of rapid warming leads to collapse of the Gulf of Maine cod fishery. *Science*, **350**, 809–12.

Petraitis, P. S. (1987). Factors organizing rocky intertidal communities of New England: herbivory and predation in sheltered bays. *Journal of Experimental Marine Biology and Ecology*, **109**, 117–36.

Petraitis, P. S. (1998). How Can We Compare the Importance of Ecological Processes If We Never Ask, "Compared to What?" In W. Resetarits and J. Bernardo, eds. *Experimental Ecology, Issues and Perspectives*. Oxford University Press, Oxford, pp. 183–201.

Petraitis, P. S. (2013). *Multiple Stable States in Natural Ecosystems*. Oxford University Press, Oxford.

Petraitis, P. S. and Dudgeon, S. R. (1999). Experimental evidence for the origin of alternative communities on rocky intertidal shores. *Oikos*, **84**, 239–45.

Petraitis, P. S. and Dudgeon, S. R. (2004). Do alternate stable community states exist in the Gulf of Maine rocky intertidal zone? Comment. *Ecology*, **85**, 1160–5.

Petraitis, P. S. and Dudgeon, S. R. (2005). Divergent succession and implications for alternative states on rocky intertidal shores. *Journal of Experimental Marine Biology and Ecology*, **326**, 14–26.

Petraitis, P. S. and Dudgeon, S. R. (2015). Variation in recruitment and the establishment of alternative stable states. *Ecology*, **96**, 3186–96.

Petraitis, P. S. and Latham, R. E. (1999). The importance of scale in testing the origins of alternative community states. *Ecology*, **80**, 429–42.

Petraitis, P. S. and Vidargas, N. (2006). Marine intertidal organisms found in experimental clearings on sheltered shores in the Gulf of Maine, USA (Ecological Archives E087–047). *Ecology*, **87**, 795.

Petraitis, P. S., Latham, R. E. and Niesenbaum, R. A. (1989). The maintenance of species diversity by disturbance. *Quarterly Reviews in Biology*, **64**, 393–418.

Petraitis, P. S., Liu, H. and Rhile, E. C. (2008). Densities and cover data for intertidal organisms in the Gulf of Maine, USA, from 2003 to 2007. *Ecology*, **89**, 588.

Petraitis, P. S., Methratta, E. T., Rhile, E. C., Vidargas, N. A. and Dudgeon, S. R. (2009a). Experimental confirmation of multiple community states in a marine ecosystem. *Oecologia*, **161**, 139–48.

Petraitis, P. S., Liu, H. and Rhile, E. C. (2009b). Barnacle, fucoid, and mussel recruitment in the Gulf of Maine, USA, from 1997 to 2007. *Ecology*, **90**, 571.

Polis, G. A., Anderson, W. B. and Holt, R. D. (1997). Toward an integration of landscape and food web ecology: The dynamics of spatially subsidized food webs. *Annual Review of Ecology and Systematics*, **28**, 289–316.

Rahmstorf, S. and Coumou, D. (2011). Increase of extreme events in a warming world. *Proceedings of the National Academy of Sciences USA*, **108**, 17905–9.

Reid, D. G. (1996). *Systematics and Evolution of Littorina*. The Ray Society, Andover.

Roman, J. (2006). Diluting the founder effect: cryptic invasions expand a marine invader's range. *Proceedings of the Royal Society of London B*, **273**, 2453–9.

Rooney, N. and McCann, K. S. (2012). Integrating food web diversity, structure and stability. *Trends in Ecology and Evolution*, **27**, 40–6.

Roughgarden, J., Gaines, S. D. and Possingham, H. (1988). Recruitment dynamics in complex life cycles. *Science*, **241**, 1460–6.

Ruel, J. J. and Ayres, M. P. (1999). Jensen's inequality predicts effects of environmental variation. *Trends in Ecology and Evolution*, **14** (9), 361–6.

Say, T. (1817). An account of the Crustacea of the United States. *Journal of the Academy of Natural Sciences of Philadelphia*, **1**, 57–63.

Seeley, R. H. (1986). Intense natural selection caused a rapid morphological transition in a living marine snail. *Proceedings of the National Academy of Sciences USA*, **83**, 6897–901.

Sephton, D., Vercaemer, B., Nicolas, J. M. and Keays, J. (2011). Monitoring for invasive tunicates in Nova Scotia, Canada (2006–2009). *Aquatic Invasions*, **6** (4), 391–403.

Silliman, B., McCoy, M. W., Trussell, G. C. et al. (2013). Non-linear interactions between consumers and flow determine the probability of plant community dominance on Maine rocky shores. *PLoS ONE*, **8** (8), e67625.

Sorte, C. J. B., Davison, V. E., Franklin, M. C. et al. (2017). Long-term declines in an intertidal foundation species parallel shifts in community composition. *Global Change Biology*, **23** (1), 341–52.

Spiess, A. E. and Lewis, R. A. (2001). The Turner farm fauna: 5000 years of hunting and fishing in Penobscot Bay, Maine. *Occasional Publications in Maine Archaeology*, **11**, 1–177.

Stachowicz, J. J., Whitlach, R. B. and Osman, R. W. (1999). Species diversity and invasion resistance in a marine ecosystem. *Science*, **286**, 1577–9.

Stachowicz, J. J., Terwin, J. R., Whitlatch, R. B. and Osman, R. W. (2002). Linking climate change and biological invasions: ocean warming facilitates nonindigenous species invasions. *Proceedings of the National Academy of Sciences USA*, **99**, 15497–500.

Steneck, R. S. (1997). Fisheries-Induced Biological Changes to the Structure and Function of the Gulf of Maine Ecosystem. Plenary Paper. In G. T Wallace and E. F Braasch, eds. *Proceedings of the Gulf of Maine Ecosystem Dynamics Scientific Symposium and Workshop*, Report 91-1 – Regional Association for Research in the Gulf of Maine. RARGOM, Hanover, NH, pp. 151–65.

Steneck, R. S. and Carlton, J. T. (2001). Human Alterations of Marine Communities: Students Beware! In M. D. Bertness, S. D. Gaines and M. E. Hay, eds. *Marine Community Ecology*. Sinauer Publishers, Sunderland, MA, pp. 445–68.

Steneck, R. S., Hughes, T. P., Cinner, J. E. et al. (2011). Creation of a gilded trap by the high economic value of the Maine lobster fishery. *Conservation Biology*, **25** (5), 904–12.

Stephenson, T. A. and Stephenson, A. (1972). *Life between Tidemarks on Rocky Shores*. Freeman Publishers, San Francisco.

Stephenson, E. H., Steneck, R. S. and Seeley, R. H. (2009). Possible temperature limits to range expansion of non-native Asian shore crabs in Maine. *Journal of Experimental Marine Biology and Ecology*, **375**, 21–31.

Thompson, R. M., Beardall, J., Beringer, J., Grace, M. and Sardina, P. (2013) Means and extremes: building variability into community-level climate change experiments. *Ecology Letters*, **16**, 799–806.

Thomsen, J., Casties, I., Pansch, C., Körtzinger, A. and Melzner, F. (2013). Food availability outweighs ocean acidification effects in juvenile *Mytilus edulis*: laboratory and field experiments. *Global Change Biology*, **19**, 1017–27.

Thornber, C. S., Rinehart, S., Guidone, M. et al. (2013). Ecological dynamics of Ulva macroalgal blooms. *Phycologia*, **52** (4), S111.

Tilman, D. (1982). *Resource Competition and Community Structure*. Princeton University Press, Princeton, NJ.

Trussell, G. C. (1996). Phenotypic plasticity in an intertidal snail: the effects of a common crab predator. *Evolution*, **50**, 448–54.

Trussell, G. C. (2000). Predator-induced plasticity and morphological trade-offs in latitudinally separated populations of *Littorina obtusata*. *Evolutionary Ecology Research*, **2**, 803–22.

Trussell, G. C. and Nicklin, M. O. (2002). Cue sensitivity, inducible defence, and trade-offs: the influence of contrasting invasion histories between a crab predator and a marine snail. *Ecology*, **83**, 1635–47.

Trussell, G. C. and Smith, L. D. (2000). Induced defences in response to an invading crab predator: an explanation of historical and geographic phenotypic change. *Proceedings of the National Academy of Sciences USA*, **97**, 2123–7.

Trussell, G. C., Ewanchuk, P. J. and Bertness, M. D. (2003). Trait-mediated effects in rocky intertidal food chains: predator risk cues alter prey feeding rates. *Ecology*, **84** (3), 629–40.

Trussell, G. C., Ewanchuk, P. J., Bertness, M. D. and Silliman, B. R. (2004). Trophic cascades in rocky shore tidepools: distinguishing lethal and nonlethal effects. *Oecologia*, **139**, 427–32.

Trussell, G. C., Ewanchuk, P. J. and Matassa, C. M. (2006). The fear of being eaten reduces energy transfer in a simple food chain. *Ecology*, **87** (12), 2979–84.

Ugarte, R. A., Critchley, A., Serdynska, A. R. and Deveau, J. P. (2007) Changes in composition of rockweed (*Ascophyllum nodosum*) beds due to possible recent increase in sea temperature in Eastern Canada. *Journal of Applied Phycology*, **21**, 591–8.

Underwood, A. J. and Fairweather, P. G. (1989). Supply-side ecology and benthic marine assemblages. *Trends in Ecology and Evolution*, **4**, 16–20.

Vadas, R. L., Sr. (1992). *Littorinid Grazing and Algal Patch Dynamics*. In J. Grahame, P. J. Mill and D. G. Reid, eds. *Proceedings of the Third International Symposium on Littorinid Biology*. The Malacological Society of London, London.

Vadas, R. L., Sr. and Elner, R. W. (1992). Plant–Animal Interactions in the Northwest Atlantic. In D. M. John, S. J. Hawkins and J. H. Price, eds. *Plant–Animal Interactions in the Marine Benthos*. The Systematics Association Special Volume No. 46, Clarendon Press, Oxford, pp. 33–60.

Vadas, R. L. and Manzer, F. (1971). The Use of Aerial Colour Photography for Studies on Rocky Intertidal Benthic Marine Algae. In A. Anson, ed. *Proceedings of the Third Biennial Workshop on Aerial Colour Photography in the Plant Sciences and Related Fields*. American Society of Photogrammetry, Bethesda, MA, pp. 255–66.

Vadas, R. L., Sr., Wright, W. A. and Beal, B. F. (2004) Biomass and productivity of intertidal rockweeds (*Ascophyllum nodosum* LeJolis) in Cobscook Bay. *Northeastern Naturalist*, **11** (Special Issue), 123–42.

Valiela, I., McClellan, J., Hauxwell, J., Behr, P. J., Hersh, D. and Foreman, K. (1997). Macroalgal blooms in shallow estuaries: Controls and ecophysiological and ecosystem consequences. *Limnology and Oceanography*, **42**, 1105–18.

Vasseur, D. A., DeLong, J. P., Gilbert, B. et al. (2014). Increased temperature variation poses a greater risk to species than climate warming. *Proceedings of the Royal Society of London B*, **281**, 20132612.

Vazquez, K. E. (2015). Phenotypic variation in the dogwhelk, *Nucella lapillus*: an integration of ecology, karyotype, and phenotypic plasticity. PhD, University of Pennsylvania.

Vermeij, G. J. (1982). Phenotypic evolution in a poorly dispersing snail after arrival of a predator. *Nature*, **299**, 349–50.

Vermeij, G. J. (1991). Anatomy of an invasion: the trans-Arctic interchange. *Paleobiology*, **17**, 281–307.

Vermeij, G. J. (2001). Community Assembly in the Sea: Geologic History of the Living Shore Biota. In M. D. Bertness, S. D. Gaines and M. E. Hay, eds, *Marine Community Ecology*. Sinauer Publishers, Sunderland, MA, pp. 39–60.

Wagner, A. (2017). The White-Knight Hypothesis, or does the environment limit innovations? *Trends in Ecology and Evolution*, **32** (2), 131–40.

Wang, Z. A., Wanninkhof, R., Cai, W.-J. et al. (2013). The marine inorganic carbon system along the Gulf of Mexico and Atlantic coasts of the United States: Insights from a transregional coastal carbon study. *Limnology and Oceanography*, **58** (1), 325–42.

Wares, J. P. and Cunningham, C. W. (2001). Phylogeography and historical ecology of the North Atlantic intertidal. *Evolution*, **55** (12), 2455–69.

Wernberg, T., Bennett, S., Babcock, R. C. et al. (2016). Climate-driven regime shift of a temperate marine ecosystem. *Science*, **353**, 169–72.

Wethey, D. S. (1985). Catastrophe, extinction, and species diversity: a rocky intertidal example. *Ecology*, **66**, 445–56.

Wethey, D. S. (2002). Biogeography, competition, and microclimate: the barnacle *Chthamalus fragilis* in New England. *Integrative and Comparative Biology*, **42**, 872–80.

Witman, J. D., Etter, R. J. and Smith, F. (2004). The relationship between regional and local species diversity in marine benthic communities: a global perspective. *Proceedings of the Natural Academy of Sciences USA*, **101**, 15664–9.

Wootton, J. T., Pfister, C. A. and Forester, J. D. (2008). Dynamic patterns and ecological impacts of declining ocean pH in a high-resolution multi-year dataset. *Proceedings of the National Academy of Sciences USA*, **105**, 18848–53.

Worm, B. and Lotze, H. K. (2006). Effects of eutrophication, grazing, and algal blooms on rocky shores. *Limnology and Oceanography*, **51** (1, part 2), 569–79.

Worm, B., Lotze, H. K., Boström, C., Engkvist, R., Labanauskas, V. and Sommer, U. (1999). Marine diversity shift linked to interactions among grazers, nutrients and dormant propagules. *Marine Ecology Progress Series*, **185**, 309–14.

Worm, B., Lotze, H. K. and Sommer, U. (2001). Algal propagule banks modify competition, consumer and resource control on Baltic rocky shore. *Oecologia*, **128**, 281–93.

Worm, B., Lotze, H. K., Hillebrand, H. and Sommer, U. (2002). Consumer versus resource control of species diversity and ecosystem functioning. *Nature*, **417**, 848–51.

Young, A. M., Torres, C., Mack, J. E. and Cunningham, C. W. (2002). Morphological and genetic evidence for vicariance and refugium in Atlantic and Gulf of Mexico populations of the hermit crab, *Pagurus longicarpus*. *Marine Biology*, **140** (5), 1059–66.

Zhao, X. G., Guo, C., Han, Y. et al. (2017). Ocean acidification decreases mussel byssal attachment strength and induces molecular byssal responses. *Marine Ecology Progress Series*, 565, 67–77.

# Chapter 5

# Subtidal Rocky Shores of the North-West Atlantic Ocean

## *The Complex Ecology of a Simple Ecosystem*

Ladd E. Johnson, Kathleen A. MacGregor, Carla A. Narvaez and Thew S. Suskiewicz

## 5.1 Introduction

The north-west Atlantic Ocean includes diverse nearshore ecosystems ranging from the subtropical conditions of south-eastern Florida to the subarctic environments of Greenland and eastern Canada. Subtidal rocky habitats are, however, limited to the more northern latitudes above 40°N. At lower latitudes, bottom communities are dominated by sand with occasional rocky outcrops in deeper waters. These latter habitats are heavily influenced by the Gulf Stream, which brings both warmer waters and associated species. These assemblages appear to be dominated by subtropical macroalgae, which provide habitat for a number of fish species, again of subtropical origin (Reed, 2004; Reed and Ross 2005). Unfortunately, little is known regarding the ecology of these assemblages. Oysters can also provide hard substratum by creating biogenic reefs in shallower waters (Beck et al., 2011).

In contrast, extensive rocky shores exist north of New Jersey, extending from New York to the Arctic and including biogeographic provinces ranging from warm temperate to subarctic biomes (Figure 5.1). Nearshore communities found below Cape Cod (Massachusetts) are still, however, heavily influenced by the warm waters of the Gulf Stream. North of Cape Cod, the Gulf Stream has less influence, although periodic intrusions of Gulf Stream waters are the major driver of ecosystem dynamics on the shores of Atlantic Nova Scotia (see later).

Earlier reviews (Himmelman, 1991; Mathieson et al., 1991; Vadas and Elner, 1992) have provided excellent overviews of the epifauna and epiflora of subtidal communities of the north-west Atlantic Ocean as well as general information on the geology and abiotic environment of this region. Although described at times as 'diverse', these communities are in fact depauperate relative to other ocean basins, including the shores of the north-eastern Atlantic Ocean (Jenkins et al., 2008; Chapter 2). Indeed, only Arctic ecosystems appear less diverse (Tittensor et al., 2010). These reviews have also described the major distributional patterns of subtidal communities and the processes generating them. As might be expected given the difficulties of underwater research, much of the knowledge on which these generalisations were based was gained largely through the efforts of a limited number of scientists who developed long-term programmes specialising in underwater research or from areas near marine stations (e.g., Isle of

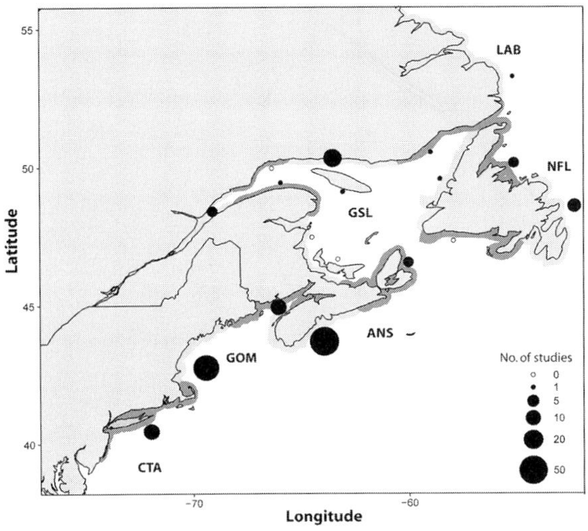


**Fig. 5.1** Map summarising the spatial distribution of studies of shallow subtidal rocky habitats across eighteen coastal sections of the north-west Atlantic Ocean. Shaded sections of coast (alternating light and dark grey) indicate the sections used to separate studies, except for Prince Edward Island and Anticosti Island (each comprising their own sections), which are unshaded. Bubble size represents the total number of studies found that refer to sites within that section of coastline. Hollow white bubbles represent sections with zero studies. The five larger regions discussed in the text are Labrador (LAB), including northern Labrador; Newfoundland (NFL); the Gulf of Saint Lawrence (GSL); Atlantic Nova Scotia (ANS); the Gulf of Maine (GOM) and cold temperate Atlantic (CTA). Warm temperate Atlantic is not shown as it has little shallow subtidal rocky habitat (see text for details). Studies were selected for this analysis based on an initial literature search in Web of Science followed by the addition of relevant cited articles (L. Johnson, unpublished data). Articles spanning a large spatial area with sites in multiple sections of coastline are counted multiple times.

Shoals, New Hampshire and Maine). Still, these pioneering efforts have provided a general portrait of this system. A major consensus of this pre-1990s work, based mostly in Atlantic Nova Scotia, would be that the system's principal characteristics are largely determined by the interactions between kelp and other large brown algae and the green sea urchin *Strongylocentrotus droebachiensis* (Figure 5.2a and 5.2b), the dominant herbivore in the system. High urchin abundance leads to areas dominated by crustose coralline algae (CCA), which can resist chronic grazing. In such areas, fleshy macroalgae are limited to species that are chemically or morphologically defended, e.g., the kelp *Agarum clathratum* (Himmelman, 1991; Gagnon et al., 2005; Adey and Hayek, 2011; Blain and Gagnon, 2014). Where urchins are rare, kelp beds readily form, providing both local primary production and habitat for invertebrates and fish (Bégin et al., 2004; Kelly et al., 2011). The determinants of these two contrasting states have been a preoccupation of researchers both in the north-west Atlantic Ocean and elsewhere across the globe, and there is still widespread concern over the extent and direction of such regime shifts in marine ecosystems (Filbee-Dexter and Scheibling, 2014a; Ling et al., 2015) due to the loss of ecosystem services provided by kelp beds and their associated fauna.

Here we build on these earlier syntheses by reviewing research in subtidal rocky communities conducted over the past twenty-five years. As before, our knowledge is still incomplete due to limitations in the number of studies and restrictions in geographic coverage (Figure 5.1). However, great progress has been made in our understanding of these communities, especially kelp–urchin dynamics. In this latter regard, we now know that the basic story is much more complex than once thought, with the dynamics varying markedly among different regions – the details of these differences are a major focus of this review. In addition, emerging issues have become apparent in the past decades, including the switch from finfish fisheries to invertebrate fisheries, the role of non-indigenous species (NIS) and biological invasions, and, finally and inevitably, the effects of climate change in altering the underlying dynamics of north-west Atlantic Ocean ecosystems. We begin, however, with a basic review of the biogeography of the system and a description of its unique feature, sea ice.

## 5.2 A Brief Biogeography of the North-West Atlantic Ocean

The rocky shores of the north-west Atlantic Ocean can be broadly characterised into four distinct biomes: warm temperate, cold temperate,

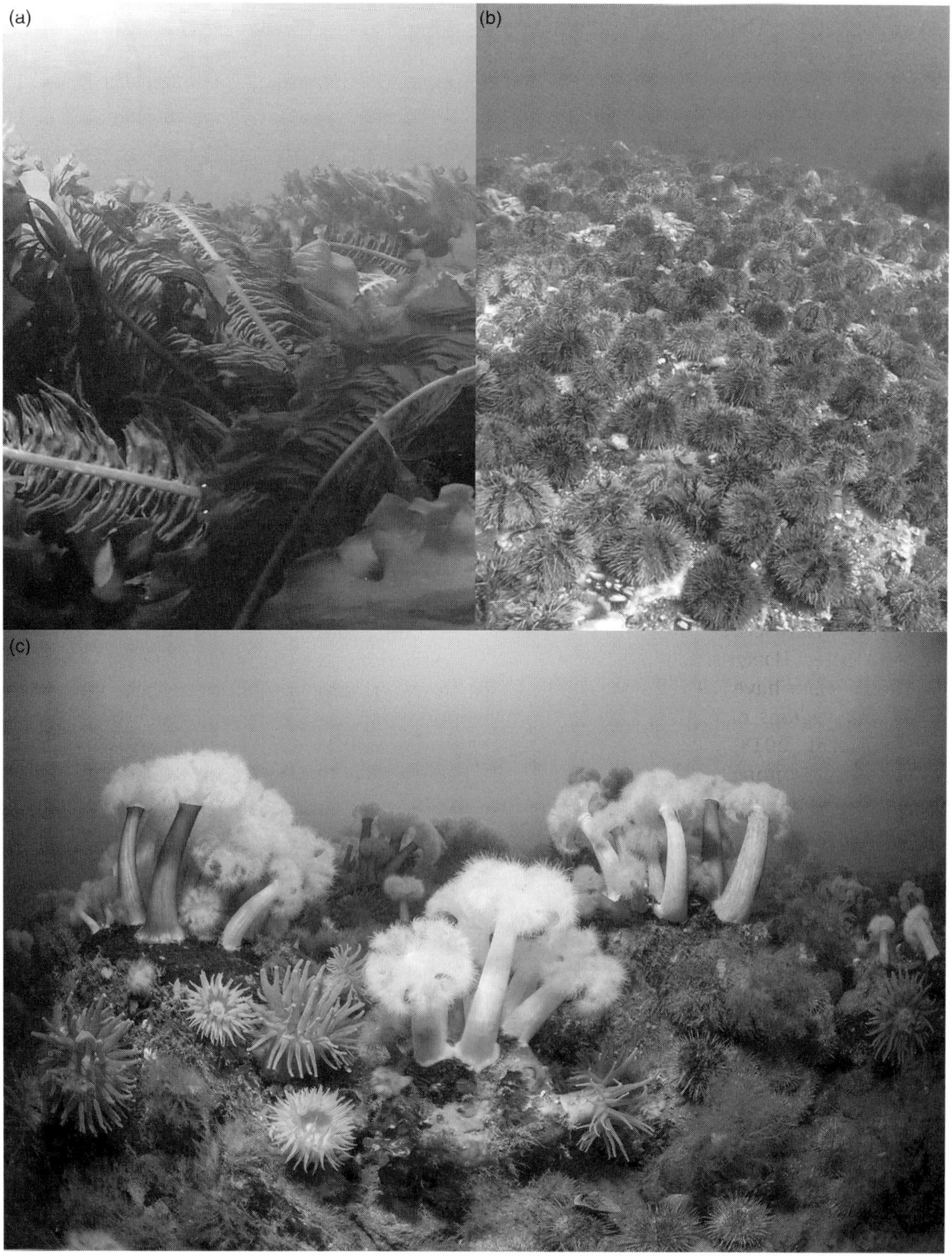


**Fig. 5.2** Distinct communities in the northern Gulf of Saint Lawrence: (a) kelp bed dominated by *Alaria esculenta* in the shallow subtidal fringe; (b) a classic shallow urchin barren ground on a bedrock bottom with high densities of urchins, high cover of coralline algae and isolated plants of chemically or physically defended macroalgae; (c) an invertebrate-dominated area near Les Escoumins (Quebec) characterised by a relatively steep slope of mixed cobble, boulder and bedrock. This complex three-dimensional bottom hosts an impressively rich and diverse invertebrate community. This well-known dive site is an urchin barren in the sense that it lacks any significant kelp and that it contains relatively high densities of urchins, but its aspect is anything but barren. Photo credits: (a) Kathleen MacGregor, (b) Carla Narvaez and (c) Robert LaSalle. *A black and white version of this figure will appear in some formats. For the colour version, please refer to the plate section.*

boreal and subarctic. These biomes are defined by sharp changes in temperature at their northern and southern limits created by major oceanic currents. Each of these biomes has its own distinct assemblage of flora and fauna, and, while a species strongly associated with one biome can often be found in adjacent biomes, its abundance relative to other guild members tends to be very different between biomes. Traditionally, the geography of these biomes is defined as follows: warm temperate – central Florida to Cape Hatteras (North Carolina); cold temperate – Cape Hatteras to Cape Cod (Massachusetts); boreal – Cape Cod to the central Gulf of Saint Lawrence (GSL; Quebec/southern Newfoundland); and subarctic – central GSL to Northern Labrador, including Newfoundland and the southern tip of Greenland (Adey and Hayek, 2011). The boundaries between these biomes may, however, have begun to shift in the past few decades. As evidence for this shift, northern distributional limits of classically boreal species such as the American lobster, *Homarus americanus*, and crabs in the genus *Cancer* have shifted into the historically subarctic regions of Newfoundland and the GSL (Pinsky et al., 2013). Similarly, the northern range of the cold temperate barnacle *Chthamalus fragilis* has expanded into the boreal waters of the southern Gulf of Maine (GOM; Govindarajan et al., 2015).

The warm temperate biome is principally composed of sandy and other soft-bottom habitats with hard substrata limited to shellfish reefs in shallow protected waters and rocky outcroppings in deeper waters. The abiotic environment is dominated by the Gulf Stream, which flows close to the coastline and provides warm surface waters throughout the year. The southern extent of the cold temperate biome is found near Cape Hatteras, where the Gulf Stream moves further offshore. Extensive terrestrial input comes from the Chesapeake Bay, the Delaware River and Hudson River/Long Island Sound estuaries, resulting in turbid, sediment-laden waters. Most of the shoreline (e.g., the Jersey Shore and the Atlantic coast of Long Island) consists of sandy barrier islands with little natural hard substrata. Extensive rocky outcroppings are only found along the more northern shores of this biome near Rhode Island. Cape Cod, the southernmost extent of the boreal biome, is the major coastal geographic feature that defines the southern limit of the GOM. This region is characterised by extensive rugged and rocky shores formed by past glacial processes. The GOM is dominated by two currents: the warmer western Maine boundary current and the much colder eastern Maine coastal current, which flows from the Bay of Fundy (a northward extension of the GOM into Canada). The boreal biome continues up the western Atlantic coast along the shore of Nova Scotia, but also includes the southern GSL, a large inland sea. Extreme thermal conditions characterise this latter area, as historically it is the southernmost limit of sea ice, but water temperatures in summer can be as high as those seen at the southern edge of the cold temperate biome (Virginia). Finally, the subarctic biome is influenced by the cold Labrador Current, which flows south from the Arctic along Labrador and the eastern edge of Newfoundland. This current also enters the GSL and creates subarctic conditions in the northern part of the gulf. It can transport icebergs and sea ice, which can be important disturbance agents structuring the communities of nearshore ecosystems (see Section 5.3).

## 5.3 Sea Ice, An Almost Unique Feature of the North-West Atlantic Ocean

Whereas sea ice plays a central role in shaping ecological assemblages of nearshore environments in polar zones (Gutt, 2001; Johnson, 2007; Smale et al., 2008), its importance in temperate latitudes is limited to the eastern margins of continents in the northern hemisphere, specifically the subarctic coasts of Canada, Greenland and Russia. Despite its obvious role as an agent of disturbance and its less obvious role as a moderator of environmental conditions (e.g., light and water motion), it has been a relatively understudied subject in temperate latitudes due to the remote access to such shores. Indeed, there appears to be little published information (at least in English) on the ecological impacts of sea

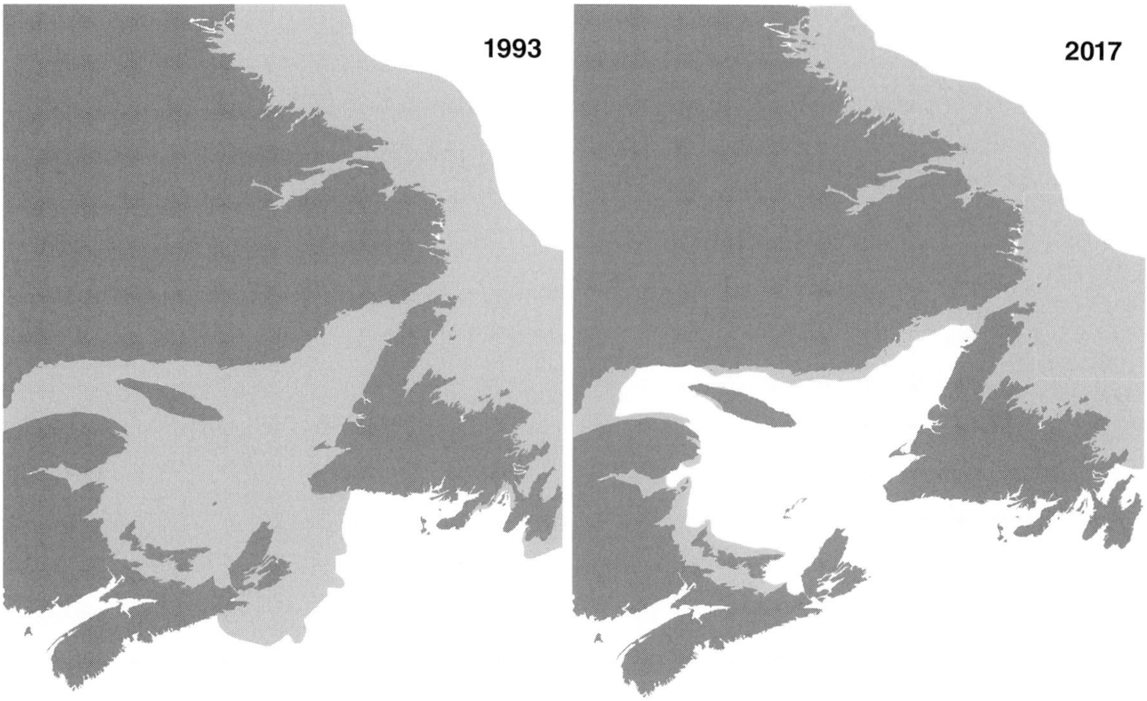


**Fig. 5.3** Sea ice cover (light grey) in eastern Canada during the first week of February, illustrating the extreme variation in ice cover between years. Data obtained from maps published by the Canadian Ice Service (2017) and its precursor, Ice Central (1993), from remote sensing systems and reconnaissance aircraft.

ice in the Okhotsk Sea or the Pacific coast of the Kamchatka Peninsula (Russia) (Zenkevitch, 1963). A handful of studies have, however, examined ice scour in the north-west Atlantic Ocean, both in the past (e.g., Keats et al., 1985; Bergeron and Bourget, 1986) and more recently (e.g., McCook and Chapman, 1997; Scrosati, 2013), although most studies have focussed on intertidal habitats, which are more influenced by sea ice (Chapter 4). The effects of sea ice are, of course, only prevalent in the northern biomes of the north-west Atlantic Ocean, but there is considerable inter-annual variation in its southern reach, which can extend as far as central Nova Scotia in exceptional years (Figure 5.3). In broader geographic terms, seasonal ice cover can even affect more sheltered areas in the Bay of Fundy and GOM, and, overall, there are approximately 30,000 km of coastline potentially affected by this factor in the north-west Atlantic Ocean. The duration and thickness of ice predictably increases along this latitudinal gradient, but considerable variation also exists among years at the same location (Figure 5.3) as well as among various topographical features (Figure 5.4). For example, protected embayments typically freeze early and remain frozen throughout the winter, while nearby wave-swept rocky points may alternate between being scoured by unattached sea ice and ice-free depending on oceanic conditions (Gutt, 2001; Figure 5.4).

Sea ice can impact nearshore assemblages both directly and indirectly. Ice can contact the bottom and crush or dislodge organisms living there, but, because ice floats, disturbance from ice scour is generally limited to shallower depths. Large icebergs are the exception to this rule, and their disturbance of benthic communities occasionally extends below the euphotic zone (i.e., >50 m; Conlan et al., 1998). Frozen in place by pack ice during the winter, they drift southward

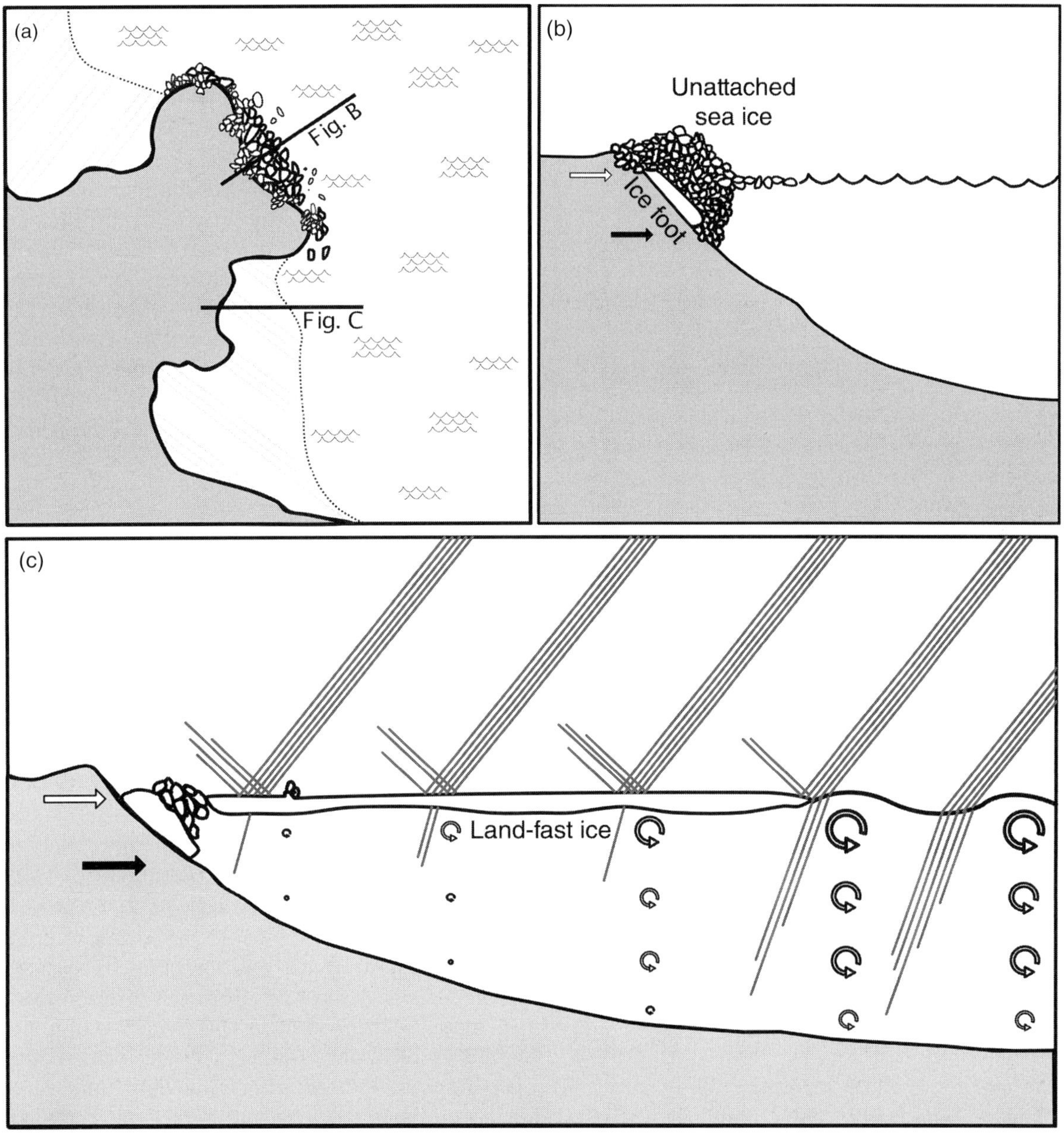


**Fig. 5.4** Types of ice affecting nearshore ecosystems. White and black arrows (in b and c) represent high and low tidemarks, respectively. (a) A bird's-eye view of different kinds of ice forming along a wave-swept point, with unattached sea ice along the exposed point and land-fast ice inside protected embayments. (b) Unattached sea ice piled onshore and below the low tide mark by waves and strong onshore winds. The ice foot, when present, protects the intertidal zone from sea ice scour. (c) Land-fast ice can extend large distances from shore and rapidly attenuates both light (diagonal lines) and orbital wave velocities (circular arrows).

each spring along the Labrador Current and, in exceptional years, can reach the coast of Nova Scotia. Iceberg disturbance is thus likely common throughout the subarctic rocky shores of Labrador and Newfoundland, and, depending on the orientation of the shoreline and the surrounding bathymetry, some subarctic shorelines may be scoured by icebergs annually. Locations in

**Fig. 5.5** The extent of sea ice in a representative nearshore environment (Pointe Mitis, Quebec). (a) The shore on a clear fall day; the shallow subtidal area around this point is mostly rocky, and the outcroppings and small islets that characterise this point are clearly visible. (b) The same area seen in winter. Sea ice cover is remarkably dynamic on this local scale, shifting on and off the shore with the tides and dominant winds. See Figure 5.3 for a larger scale. Photo credits: (a) Ladd Johnson and (b) John Price.

the lee of prevailing currents may, however, see little or no iceberg contact over decadal timescales (Barnes, 2016).

In the shallow subtidal zone, extensive ice scour can result from the accumulation of fragmented sea ice (Figure 5.5). The tides, currents, wind and waves are the drivers of this disturbance, and in arctic regions ice floes can pile up to a depth of tens of metres (Figure 5.4b; Conlan et al., 1998), shifting with fluctuations in water motion, furthering their scouring effects. The bottom can, however, be protected from the direct effects of ice scour if ice occurs as an 'ice foot', which forms as a layer of ice on the rock surface, as 'anchor ice', which extends from the surface all the way to the benthos, or finally as 'land-fast ice', which attaches to the shore and does not move with the winds or currents (Figure 5.4a).

Sea ice can also have several important indirect impacts on rocky subtidal communities. Ice attenuates both light and ocean waves (Figure 5.4c), sometimes reducing each by two orders of magnitude or more (Squire et al., 1995; Zhao et al., 2010). Reduced light will obviously reduce photosynthesis and thus growth of seaweeds, and reduced water motion can affect nutrient exchange and herbivore behaviour (see later). Finally, drifting ice can carry and redeposit non-conglomerated substrata, even boulder-sized material, into new areas.

Despite the overwhelming influences that sea ice can have on these areas, we know comparatively little about its role in structuring subtidal marine communities. Ice scour, for example, can remove macrofauna and flora, but studies have reached different conclusions about whether this increases or decreases biodiversity within the affected area. Keats et al. (1985) found that, in years where ice scour was low, a single species of kelp outcompeted all other macroalgae, leading to lower overall species richness. In contrast, another study showed both positive and negative effects (Peterson, 1977). Perhaps counter-intuitively, the rapid formation of land-fast ice can limit ice scour within an area by protecting rock surfaces from the direct impact of loose floating sea ice, thereby changing the effect of ice from a direct, physical disturbance to one with indirect effects on light and water motion. Currents, wind and shore topography all greatly influence the formation and persistence of land-fast ice, creating a situation where locations within the same latitudinal region may experience drastically different ice conditions (Squire, 2007). These differences will need to be incorporated into climate models, which predict drastic changes in ice formation and cover throughout the northern portion of the northwest Atlantic Ocean (Flato and Brown, 2007). We clearly need a more thorough understanding of ice and its impacts on subtidal rocky communities to understand how these changes will impact this large region.

## 5.4 Kelp in the North-West Atlantic Ocean – An Unrealised Potential

Kelps, large brown algae in the order Laminariales, are a highly productive component of subtidal communities (Mann, 1973; Krumhansl et al., 2016) that can potentially occur wherever sufficient light reaches a rocky bottom. As their high rates of primary productivity and structural complexity provide food and habitat to multiple species (Bégin et al., 2004; Teagle et al., 2017), they can play a key role in determining the diversity and productivity of coastal ecosystems. Despite having a potentially wide distribution in the north-west Atlantic Ocean, kelp are often severely constrained, and many areas that appear suitable for their growth are mostly devoid of these important primary producers. Within the north-west Atlantic Ocean, abiotic conditions appear to determine the geographic limits at the largest scale, whereas biotic interactions are more important at smaller scales.

High summer water temperatures set the southern limit for kelp, currently in Long Island Sound, New York (Gerard, 1997). At temperatures above 18°C, respiration exceeds primary productivity (Humm, 1969) and plants senesce. Whereas populations can survive brief warm periods, kelp cannot survive above these temperatures for more than a few weeks (Simonson et al., 2015). As water temperatures rise and summer warming events become more frequent and prolonged, the southern limit of kelp may shift northward (Merzouk and Johnson, 2011). Interestingly, kelp appear to have no physiological minimum temperature limit, as populations occur in the Arctic where temperatures never rise much above 0°C. Low temperature can, however, indirectly influence kelp when it leads to the formation of sea ice, which reduces levels of light and water motion (see earlier). The former is of obvious importance as light can be greatly attenuated by ice, often dropping to levels near zero (Chapman and Lindley, 1980) and thereby limiting primary productivity. Water motion is, however, also a driver of kelp productivity (Hurd, 2000) and can also reduce grazing activity (Gagnon et al., 2006; Lauzon-Guay and Scheibling, 2007; Frey and Gagnon, 2015). In the north-west Atlantic Ocean seasonal ice cover can occur for short periods (typically along shores protected from waves) into the southern GOM, but in more northern latitudes (e.g., off the coast of Labrador) ice cover may persist for more than half the year (see Figure 5.3). Kelp abundance and vertical distribution in western Greenland are positively correlated with the length of the ice-free period and are predicted to increase as temperatures rise (Krause-Jensen et al., 2012). In these more northern regions, a shift of several weeks towards earlier break-up and later formation of ice could result in a doubling of solar irradiance to the benthos, with the potential to dramatically alter these ecosystems.

Between the southern limit set by summertime temperatures and the extreme north where light attenuation by ice limits primary productivity, there exists a broad area where kelp distribution appears restricted mostly by biological interactions. For the latter half of the last century, abundant urchin populations created intense grazing pressure on all fleshy seaweeds, resulting in large urchin barrens in otherwise suitable kelp habitat (but see Gagnon et al., 2005; Adey and Hayek, 2011 and Blain and Gagnon, 2014 for descriptions and discussion of the distribution and ecological role of chemically defended kelp and other large brown algae in these areas, described as more 'savannah-like' than 'barren' [Adey and Hayek, 2011]). The phenomenon of regime shifts between urchin-dominated and kelp-dominated states (Ling et al., 2015) in this environment is treated in detail later, both generally for the north-west Atlantic Ocean and specifically for three relatively well-studied regions – the GOM, the GSL and the Atlantic shores of Nova Scotia (ANS) where the dynamics of kelp bed recovery vary considerably (see Section 5.6). Beyond the complexity of the kelp–urchin interaction, biological invasions by both a fouling invertebrate (the bryozoan *Membranipora membranacea*) and an opportunistic seaweed (the green alga *Codium fragile*) have also affected kelp beds (see Section 5.8).

In spite of the limited distribution of kelp within the north-west Atlantic Ocean, there is increasing evidence that kelp beds can provide an important subsidy to both adjacent and distant areas (Krumhansl and Scheibling, 2012; but see Nadon and Himmelman, 2006). Kelp detritus ranges in size from small fragments to entire plants and is generated by frond breakage or dislodgment from the substratum. Loss of kelp biomass is greatest during storms (Filbee-Dexter and Scheibling, 2012), especially in locations where there are biotic factors that reduce the strength of the blade (e.g., herbivory by snails; Krumhansl et al., 2011) or increase drag (e.g., encrustation by bivalves and bryozoans; Krumhansl et al., 2011). Several studies have documented dislodgment rates, and those from the north-west Atlantic Ocean (Nova Scotia: Krumhansl and Scheibling, 2011 and Chapman, 1984; Rhode Island: Brady-Campbell et al., 1984) estimate that 6.2–49 per cent of annual productivity is exported in this way (Krumhansl and Scheibling, 2012). A less studied, but important, pathway of kelp subsidy is the faecal material produced by urchins (Mamelona and Pelletier, 2005; Sauchyn and Scheibling, 2009a). When high densities of urchins accumulate in grazing fronts and graze through kelp beds, faecal pellet production can be a significant source of organic matter for benthic populations in rocky subtidal and sedimentary habitats (Sauchyn and Scheibling, 2009b; Sauchyn et al., 2011).

The fate of detrital kelp is less well known, but urchins in adjacent barrens can receive macroalgal subsidies from adjacent kelp beds (Britton-Simmons et al., 2009; Filbee-Dexter and Scheibling, 2012). In shallow waters these subsidies can prevent destructive grazing by urchins by supplying them with adequate food resources (Harrold and Reed, 1985; Vanderklift and Wernberg, 2008). However, accumulations of kelp detritus in deeper waters in Nova Scotia have been associated with the formation of destructive grazing fronts in adjacent shallower waters, suggesting that urchin populations in deeper waters play an important role in driving urchin–kelp dynamics (Filbee-Dexter and Scheibling, 2017). At a larger scale, detrital kelp is less abundant on exposed coastlines compared to bays and sheltered locations (Filbee-Dexter and Scheibling, 2016) and appears to decrease with distance from the source, as reflected by fatty acid analysis of urchin gonads and gut contents (Kelly et al., 2012; Filbee-Dexter and Scheibling, 2014b).

## 5.5 Green Sea Urchins – A Plague of 'Submarine Rodents'

High-densities of urchins are a defining feature of barren grounds, but the driving forces creating dense populations in the north-west Atlantic Ocean are not well understood. Settlement is variable from site to site and year to year but appears to be concentrated in early to late summer (June–August) (Balch and Scheibling, 2000; Lambert and Harris, 2000; Jennings and Hunt, 2010). Although few recruitment studies have been conducted in the GSL, recognizable cohorts of small individuals have been interpreted as an indication of widespread high recruitment in certain years (Himmelman et al., 1983a, 1983b). However, efforts to measure settlement in the GSL in 2011–2013 found extremely low numbers of settlers, with none being detected in two of the three years of sampling (K. LeGault, unpublished data). In Nova Scotia, high recruitment observed in 1960 was associated with high spring temperatures (Hart and Scheibling, 1988).

Post-recruitment mortality has a potentially large influence in controlling urchin density, most notably in the GOM (see Section 5.6). Work north of the GOM has identified several predators of small urchins, including rock crab, sea stars, several species of fish, lobster, carnivorous worms and large urchins (Himmelman, 1991; Scheibling and Hamm, 1991; Jennings and Hunt, 2010; LeGault and Hunt, 2016). In the GSL, salinity has been identified as a driving force behind the size structure of certain urchin populations, since small individuals are more sensitive to low salinity than larger urchins (Himmelman et al., 1984). This selective mortality appears to account for differences in the population structure along the Saint Lawrence maritime estuary, where small urchins are rare or absent at locations subject to periodic hypo-osmotic conditions (Himmelman et al., 1983b).

Finally, urchins can survive for long periods on little to no food, subsisting on calcified encrusting algae and the biofilm that develops on such surfaces (Lang and Mann, 1976). This ability allows infrequent pulses of high recruitment to accumulate, facilitating the development and maintenance of high urchin densities in areas denuded of fleshy macroalgae. Urchins are unique in this sense, being consumers that do not crash with the exhaustion of their food supply.

The importance of variation in urchin behaviour in influencing and structuring benthic communities is becoming more and more apparent. Green sea urchins have two distinct behavioural modes in their life history (Himmelman, 1986). Small (test diameter <10–15 mm) urchins are cryptic, remaining hidden and moving very little. Above this size, urchins switch behaviour and begin actively foraging (Himmelman, 1986; Scheibling and Hamm, 1991; Dumont et al., 2004b). This phenomenon is due to the presence of predators such as crab and fish (Scheibling and Hamm, 1991), but also larger conspecifics (Jennings and Hunt, 2014) which represent a significant threat for small urchins; only when individuals reach a size protected from predation do they begin to actively forage (Dumont et al., 2004b; Jennings and Hunt, 2014; LeGault and Hunt, 2016; Narvaez, 2019). This ontogenetic switch from a cryptic and sedentary strategy to active foraging affects growth rates (Himmelman et al., 1983b, Narvaez, 2019) and may contribute to the size distribution of populations of urchins in different subtidal habitats.

Aggregations of urchins are a key element of urchin–kelp dynamics, as they lead to the formation of destructive feeding fronts, which can completely graze kelp beds and shift subtidal rocky systems into a barrens state. Early studies proposed several drivers of urchin aggregation, including protection from predation and attraction to food (Breen and Mann, 1976a; Bernstein et al., 1983). Subsequent studies (e.g., Vadas et al., 1986) have demonstrated that urchin aggregations are driven principally by the presence of food, specifically kelp. Although avoidance of predator cues and aggregation in crevices or on tank walls in the presence of predators has been demonstrated, principally in the lab (Scheibling and Hamm, 1991), results in the field are equivocal and show more limited effects of predators (Harding and Scheibling, 2015). Moreover, it remains unknown whether the mechanism driving aggregation on food under field conditions is detection of chemical cues (known to be employed by other marine benthic invertebrate species; e.g., Wyeth et al., 2006, and inferred in lab studies of urchins, e.g., Bernstein et al., 1983) or whether some other mechanism is used. Evidence to date suggests that aggregations can largely be explained by random movements, with a cessation of movement once contact with food is made (K. MacGregor, unpublished data).

Models of the formation of feeding fronts have shown that incorporating simple behaviours such as making movement vary with food availability can lead to aggregation of individuals (Abraham, 2007). Further work points to the movement and foraging behaviour of green sea urchins as crucial factors determining whether a feeding front will form (Lauzon-Guay et al., 2009; Feehan et al., 2012a). Due to the difficulty of tagging or tracking individual urchins in situ, there are few estimates of movement parameters derived from direct measurements of individual movements (Dumont et al., 2006; Lauzon-Guay et al., 2006). These early results highlight, however, the important role of environmental modification of behaviour.

A behavioural avoidance of areas of high water motion by urchins has been inferred in the past to explain the restriction of kelp to shallow exposed areas, which may represent refuges from urchin grazing (Himmelman, 1984; Himmelman and Nédélec, 1990; Adey and Hayek, 2011). Observations of grazing fronts have indeed shown that they advance much more slowly as a function of increasing urchin density in the front, which in turn varies inversely with significant wave height (Lauzon-Guay and Scheibling, 2007). The decreased density of urchins in the front probably represents movement away from this high-risk area. Tidal currents also affect urchin movement, with faster current speeds reducing movement and influencing direction

(Morse and Hunt, 2013). Green sea urchins avoid areas swept by algal fronds, indicating that avoidance behaviour plays a significant role in creating algal refuges in areas of intense grazing pressure (Konar and Estes, 2003; Gagnon et al., 2003, 2006). Urchin avoidance of soft substrata such as sand is also often invoked as a mechanism for creation of refuges for algae in deeper waters (e.g., Adey and Hayek, 2011), and observations with other species indicate that this may be the case (Laur et al., 1986). However, current work in the GSL indicates that, although green urchins are found on lower densities on soft substratum, it does not prevent their movements, and thus such 'barriers' are unlikely to provide refuges at small spatial scales (MacGregor, 2019).

## 5.6 Kelp–Urchin Interactions – Three Systems, Three Dynamics

The striking differences between luxuriant kelp beds and the CCA that dominate urchin barrens has made temperate rocky subtidal communities a classic example of alternative states in ecological systems (Lawrence, 1975; Filbee-Dexter and Scheibling, 2014a; Ling et al., 2015). In the North Pacific, sea otters control the shift between these states – a textbook example of top-down control of community structure, in this case a trophic cascade mediated by a keystone predator (Estes and Duggins, 1995). In the north-west Atlantic Ocean, there are three systems where the dynamics between kelp beds and urchins are controlled by different factors. In ANS, the system operates in a cyclical manner driven by the interaction between a pathogen, its urchin host and environmental forcing. In the GOM, the system has passed through three different phases, where anthropogenic factors (specifically fishing) triggered the trophic cascade that 'flipped' each phase and appears to have now 'locked' it into a seaweed-dominated state. Finally, in the GSL and Newfoundland, the system, albeit much less well studied, is clearly locked in an urchin-barren phase.

### 5.6.1 Atlantic Nova Scotia – Cycling for Now

The appeal of the simplified scenario that explained the dynamics in the North Pacific appears to have led to its rapid application to ANS in the 1970s and 1980s when population outbreaks of the green sea urchin were thought to be caused by a release from predation by American lobsters due to commercial exploitation (see Breen and Mann, 1976b; Mann and Breen, 1972; Wharton and Mann, 1981; Mann, 1982). However, this hypothesis was not supported by subsequent laboratory or field studies (see Miller, 1985a; Elner and Vadas, 1990; Scheibling, 1996). Instead, the collapse of the urchin population in the mid-1980s due to an amoeboid disease (Miller, 1985b) clearly showed that other processes were important. Indeed, a substantial body of information has been amassed over the past twenty-five years, principally by Robert Scheibling and his collaborators, that has convincingly demonstrated that, in this system, episodic outbreaks of a disease (a paramoebiasis) replaces the functional role of predation as the major agent controlling urchin populations (Feehan and Scheibling, 2014b; Feehan et al., 2016; Figure 5.6a).

In the 1950s Nova Scotia was described as kelp-dominated (MacFarlane, 1952), but by the mid-1960s the description changed to a mosaic of urchin barrens and kelp beds (Breen and Mann, 1976a). By the early 1970s, dense aggregations of urchins were documented to have overgrazed *Laminaria digitata* and *Saccharina latissima* beds and, although there is no direct evidence, this urchin outbreak was attributed either to the decline of predatory fishes or mass urchin recruitment events (Hart and Scheibling, 1988; Steneck et al., 2004). Barrens dominated until the early 1980s, when hurricanes coupled with high water temperatures facilitated outbreaks of an amoeboid disease (*Pamamoeba invadens*) causing mass urchin mortality (Scheibling, 1984). The disappearance of urchins from shallow waters enabled kelp beds to quickly re-establish (Miller, 1985b; Scheibling, 1986). Urchin populations in deeper waters were, however, able to survive the mass mortality events due to the thermal refuge

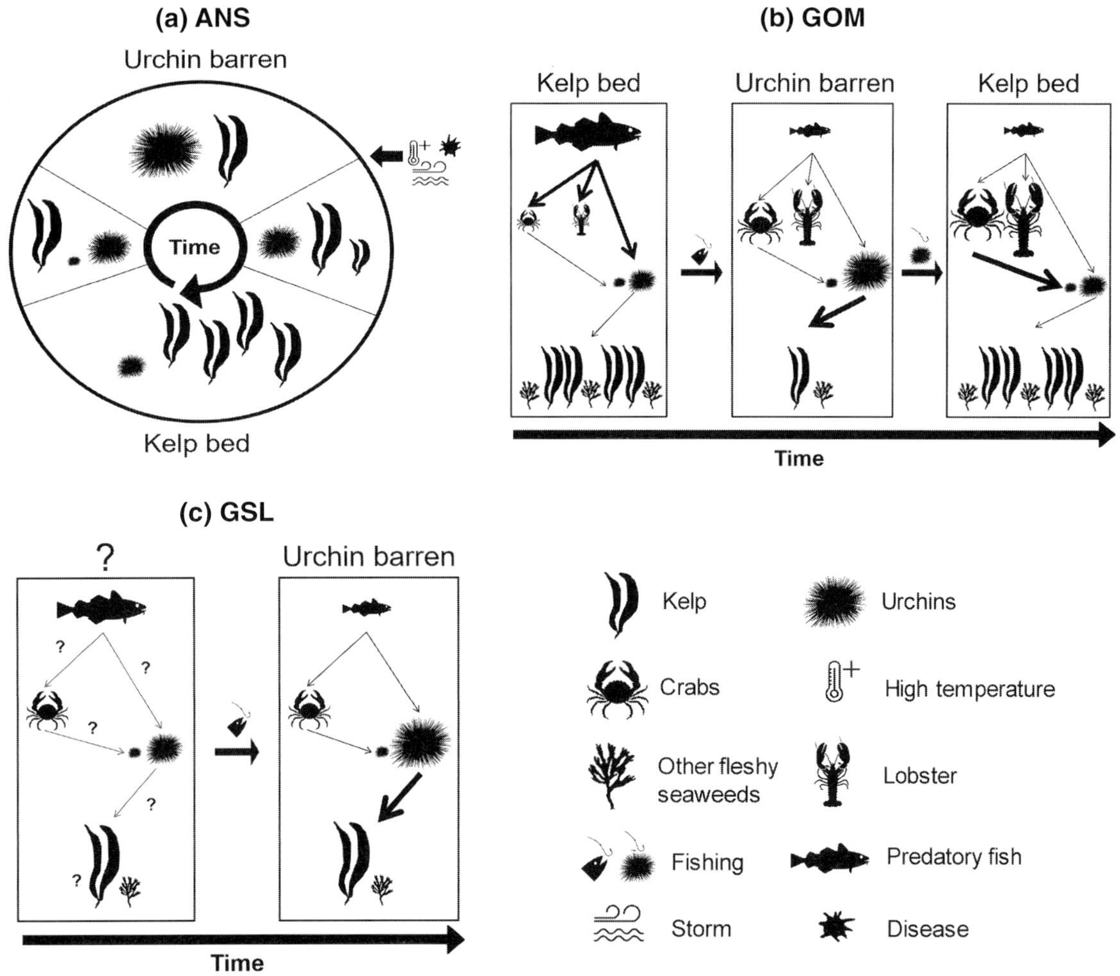


**Fig. 5.6** Urchin–kelp dynamics in three regions of the north-west Atlantic Ocean. (a) Cyclic changes in state in ANS, (b) progressive changes in state ('flips' and 'locks' *sensu* Steneck et al., 2013) in the GOM and (c) possible shifts in state over time in the GSL. Size of fauna pictograms reflects abundance except when two urchins are pictured side by side, in which case the smaller urchin represents recruits. For seaweed, the number of pictograms represents abundance except when two kelp are pictured side by side, in which case the smaller kelp represents recruits. Arrow thickness within individual states represents the intensity of the effect and question marks represent uncertainty or a lack of information.

provided by the low temperatures there (Scheibling and Stephenson, 1984; Scheibling and Hennigar, 1997). Thus, by the early 1990s, urchins migrating from deep waters (and/or recruitment events) created new accumulations within and along the margins of kelp beds that led to the formation of new barrens (Lauzon-Guay et al., 2009). This time, however, urchin barrens were rapidly decimated by the widespread recurrence of the parameboid disease in 1995 and 1999, killing most of the urchins down to a depth of 25 m (Scheibling et al., 1999; Filbee-Dexter and Scheibling, 2014a). In the following years, several disease outbreaks occurred, although they affected smaller areas (Feehan and Scheibling, 2014a). Consequently, kelp beds now dominate rocky subtidal environments, even though invasive species have negatively affected kelp recovery in many locations (see Section 5.8). Barrens with adult urchins remain, but they are sparsely distributed along the central and south-western shores of Nova Scotia (Feehan and Scheibling, 2014b).

Paramoebiasis outbreaks are more common now compared to previous decades, and, although the source populations of *P. invadens* are still unknown, negative growth in the laboratory at temperatures <2°C suggests that it is not indigenous to the Atlantic coast (Jellet and Scheibling, 1988; Buchwald et al., 2015). Mathematical models, later validated with field data, showed that disease outbreak is associated with both the intensity and proximity of tropical cyclones that convey the disease and the concomitant increase in water temperatures, which favours the propagation of the disease (Scheibling and Lauzon-Guay, 2010; Feehan et al., 2016). The increase in the frequency of urchin 'killer storms' since 1980 is a trend that might accelerate with future global warming conditions, so an increase in the frequency of disease outbreaks is expected in the future (Scheibling et al., 2013). However, even if the potential delivery of the pathogen is increasing, the spatial spread of the disease might be limited by the density of susceptible host populations (Scheibling and Lauzon-Guay, 2010; Scheibling et al., 2010; Feehan et al., 2012b). Moreover, another consequence of ocean warming conditions is the recent shift observed from luxurious kelp beds to rocky reefs dominated by opportunistic turf-forming and invasive algae (Filbee-Dexter et al., 2016).

The kelp–urchin system of ANS is likely now one of the most well-understood marine ecosystems in the world, rivalling even that of the kelp–urchin–sea otter system of the Aleutian Islands in the northern Pacific Ocean. However, although this explanation works well in ANS, it does not apply to adjacent regions of the north-west Atlantic Ocean, in particular the GOM and the GSL, where differences in coastal oceanography (water temperature, ocean currents) lead to a low occurrence or absence of paramoebiasis (Feehan et al., 2012b).

### 5.6.2 GOM – Locked in a Kelp State

The GOM has been well studied since the early 1980s by Robert Steneck and colleagues, and the dynamics have been well described recently (Steneck et al., 2013; Figure 5.6b). In summary, when predatory fish were abundant in costal zones, kelp beds and barren grounds occurred in a patchwork within rocky subtidal habitats (Steneck et al., 2004). However, after intense finfish fishing, the numbers and size of cod and other groundfish declined rapidly, leading to a functional extirpation of these predatory fishes by the mid-1990s (see Section 5.7.1). The lack of large urchin predators led to an increase in the grazing pressure by high densities of urchins, so an expansion of barrens occurred. The shift towards an urchin barren-dominated state left kelp beds restricted to shallow, wave-exposed sites, where urchin densities were low (Steneck and Dethier, 1994). Urchins dominated rocky subtidal habitats until the late 1980s, when urchin populations were overexploited by high levels of harvesting to satisfy the demand of Asian markets for the edible gonads, known as 'uni' by the Japanese. Landings began in 1987, peaked in 1992 and, even after strict restraints were imposed to reduce the rate of catch in 1994, had dropped by ~70 per cent by 1999. The Maine urchin fishery has thus become a classic example of a boom-and-bust fishery and is yet another step in the sequential overexploitation of urchins across the globe (Berkes et al., 2006).

The reduction of this dominant benthic consumer led to a massive re-establishment of kelp in the late 1990s. The expectation that urchin populations would then slowly return was, however, never met, and kelp populations continue to dominate, albeit slowly ceding ground to other species of seaweeds (Steneck et al., 2013). The resistance of this system to cycle back to urchin domination, even partially, appears to be due to the emergence of invertebrate predators, especially decapod crabs (see later), that are capable of directly controlling juvenile urchins and indirectly the kelp bed state itself, after the system was 'flipped' by overfishing (Steneck et al., 2004).

Corticated macroalgae (e.g., *Chondrus crispus*) and filamentous algae have been steadily increasing after this phase-shift in the 1990s (Steneck et al., 2013). High densities of newly settled crabs (*Cancer* sp.) are primarily found in these habitats, so this increase in corticated and filamentous algae is believed to have increased the carrying capacity for crabs by providing protection from small visual predators (Palma and Steneck, 2001). The role of Jonah (*Cancer borealis*) and rock (*Cancer irroratus*) crabs as important predators of newly settled urchins has been demonstrated in caging experiments (Scheibling and Hamm, 1991). Field surveys found significantly lower survivorship of newly settled urchins in understory algae of kelp beds than in adjacent barrens (Steneck et al., 2013), although holdfasts of *S. latissima* can, ironically, provide juvenile urchins a refuge from crab predation (Feehan et al., 2014). Further evidence of the importance of crab predation comes from efforts to repopulate areas where urchins had been previously overexploited, as urchins relocated within kelp beds were rapidly consumed by rock crabs (Steneck et al., 2013). The emerging role of crabs as apex predators that limit the recruitment and post-settlement survival of urchins has thus been proposed as an explanation for the failure of high levels of urchin recruitment to produce high densities of adult urchins in kelp beds, and the GOM now seems to be 'locked' in a seaweed-dominated state (Steneck et al., 2004, 2013).

### 5.6.3 GSL – Locked in an Urchin State

Considerably less work has been done in the third north-west Atlantic Ocean system, the GSL, and, once again, what is known is largely due to work by a small research group, in this case John Himmelman and collaborators. More importantly, the dynamics of this system appear to be quite different than the other two previously described, primarily in that there seems to have been no observed shifts in the rocky bottom ecosystems. Unlike other systems, the GSL appears to be locked in a stable, urchin-dominated state. It remains unclear whether the extirpation of cod triggered urchin population outbreaks in this region, as little information is available previous to cod overexploitation. An early account (Dawson, 1867) stated that urchins were 'common' and found in 'great abundance', enough so that he referred to them as 'submarine rodents'. Likewise, the more recent recollections of Walter Adey, who explored the GSL and Newfoundland in the mid-1960s, indicate that the system was already dominated by urchin barrens at that time (Adey and Hayek, 2011; Figure 5.6c). Surveys in the early 1980s recorded urchin barrens in most locations, with kelp beds generally restricted to depths of less than 3 m on wave-exposed shores (Himmelman et al., 1983b; Himmelman, 1986, 1991). More recent surveys in 2011 revealed little change over thirty years (Narvaez, 2019). These surveys included multiple sites across much of the northern GSL, but the majority of the work in the intervening years has been limited to the Mingan Archipelago, a group of islands that form a national park centred on the northern shore of the GSL.

One of the reasons that the GSL has resisted flipping towards a kelp-dominated state is that, nowadays, there seems to be no urchin population control, either by predators, disease or fisheries. Although urchin predators exist (i.e., decapod crustaceans, fishes and birds [Himmelman and Steele, 1971]), their impact in structuring benthic communities is reduced, as they do not appear to be major urchin predators (Himmelman, 1991). Given the cold waters of this subarctic environment, it is not surprising that disease processes are not important, and the only urchin die-off reported from this area was after an unusual deposition of ice-scoured kelp on top of urchins (Dumont et al., 2004a). Although a small amount of urchin harvesting does occur in the northern GSL, the potential for large-scale urchin harvesting is limited for several reasons. First, gonads develop in late winter when harvesting is difficult due to ice and poor weather. Second, they are often small and of poor quality due to the high urchin densities and low food availability. Finally, much of the coast is remote, and there is a lack of infrastructure for processing and transportation of gonads, which are primarily consumed in Asia and must be very fresh to have any value.

Kelp beds in this region are restricted to shallow fringes in places where they find refuge from urchin grazing (Himmelman and Lavergne, 1985; C. Narvaez, unpublished data). Refuge can be provided by physical factors such as wave action (see Section 5.5), an idea consistent with observations in the Mingan Archipelago where the size of shallow kelp beds increases with increasing exposure (Himmelman, 1991). Low salinity is another physical factor that reduces urchin density, and thus their grazing capacities, in shallow waters near freshwater discharges. In the Saint Lawrence Estuary, seasonal discharge of fresh water from rivers has been shown to reduce urchin survival, particularly of small sizes, as they are subject to a higher osmotic stress than larger urchins (Himmelman et al., 1983b, 1984). Therefore, at locations where salinity is reduced constantly or seasonally, kelp can find a refuge due to their higher osmotic stress resistance compared to urchins (Himmelman and Lavergne, 1985). Finally, ice abrasion has been suggested as a driver of low urchin densities in shallow waters (Himmelman and Lavergne, 1985). After ice disturbance, the faster recruitment and growth rates of algae compared to urchins could allow algae to recolonise the disturbed zone. Although no experimental studies have directly tested this idea, removal of all urchins from a large area facilitated kelp recruitment to a depth of 10 m (Himmelman et al., 1983a). The generality of this response to disturbance remains to be seen; however, as the recruitment of kelp and other large brown algae may depend on characteristics of the sites such as the distance to source populations of algae, the size and location of the disturbed patch and the density of urchins in the surrounding area.

The presence of grazing-resistant algae in the lower algal fringe has been suggested as another mechanism that provides refuge for kelp (Himmelman and Lavergne, 1985; Gagnon et al., 2004). This zone is characterised by a generally lower urchin density and the presence of large brown algae (*Agarum cribosum* and *Desmarestia viridis*) and sometimes smaller red algae. *A. cribosum* has been experimentally shown to reduce the shoreward movement of urchins, thus effectively reducing their ability to penetrate shallower kelp beds and form grazing fronts (Gagnon et al., 2004). The presence of this grazing-resistant algal fringe has been associated with the long-term stability of some kelp beds in the Mingan Archipelago, where no dramatic shifts in kelp distribution have been described – analysis of aerial photographs taken since the creation of the park in the early 1980s has shown almost no changes in the distribution of kelp beds (P. Gagnon, unpublished data). Like the GOM, this system also seems locked for the time being, but, in this case, in an urchin-dominated state.

## 5.7 Food Web Alterations – Emergence and Expansion of Invertebrate Fisheries

The collapse of the groundfish fishery resulted in major ecosystem changes in the north-west Atlantic Ocean due to subsequent trophic cascades (Worm and Myers, 2003; Frank et al., 2005). Following the loss of apex predators, ecosystems shifted towards alternate states dominated by planktivorous forage fishes and large invertebrates (Lotze and Milewski, 2004; Savenkoff et al., 2007a, 2007b; Bundy et al., 2009; Kenneth et al., 2011). Fisheries have compensated by expanding towards alternative species at lower trophic levels (Lotze and Milewski, 2004; Anderson et al., 2011), a phenomenon known as 'fishing down the food web' (*sensu* Pauly, 1998). Data from the late 2000s show that invertebrate fisheries have increased five-fold globally since the 1950s (Anderson et al., 2011), and the economic value of the invertebrate fisheries that emerged after the groundfish collapse, like the northern snow crab (*Chionoecetes opilio*) and northern shrimp (*Pandalus borealis*), now exceeds that of the groundfish fisheries. From an ecological perspective, however, the extraction of predatory finfish has important consequences on food webs, affecting biological and functional diversity (Jackson et al., 2001; Frank et al., 2005; McCain et al., 2016).

### 5.7.1 Cod

Overfishing of finfish stocks, in particular the Atlantic cod (*Gadus morhua*), in the north-west Atlantic Ocean is a textbook example of a trophic cascade, centred primarily around the total collapse of the cod fishery, which led eventually to a moratorium in Canada in the early 1990s. It has essentially remained in place as the stocks have not yet recovered – and may never do so. There are several explanations that have been considered, including elevated natural mortality of large fish and reduced recruitment rates (Shelton et al., 2006). During this time, grey seal (*Halichoerus grypus*) populations have increased, and there is a growing consensus that predation by seals might contribute to the lack of recovery of certain groundfish stocks (Bundy et al., 2009). A recent study showed that, in areas where cod and white hake aggregate, they represent an average 68 per cent of the grey seal diet, a much higher level than reported for other regions (Hammill et al., 2014). Another source of mortality is the nine-fold increase in forage fish biomass compared to pre-collapse years. Small, pelagic forage fish, like the Atlantic herring (*Clupea harengus*), that were once groundfish prey are now restricting recruitment of large fish species by feeding upon early life stages, a phenomenon termed predator–prey reversal (Fauchald, 2010).

The GOM is likely the region were consequences of the overexploitation of predatory fish in rocky habitats has been best documented. Archaeological evidence from middens in this region suggests that predatory finfish were very abundant and large in size (e.g., in some cases 80 per cent of the excavated bones came from cod, and specimens measuring 1–1.5 m in length were common; Bourque, 1995). However, introduction of more sophisticated fishing technology in the 1930s and targeting of spawning aggregations led to a sharp decrease in size and numbers of fish caught. In spite of this trend, finfish fisheries continued, resulting in the ecological extinction of the top predator, Atlantic cod, and several other benthic fish species, including the haddock (*Melanogrammus aeglefinus*), Atlantic halibut (*Hippoglossus hippoglossus*) and Atlantic wolfish (*Anarhichas lupus*) by the mid-1980s and early 1990s (Steneck et al., 2004, 2013; Frank et al., 2005). Reduction in average body size, as also observed in the north-east Atlantic (Genner et al., 2010), also diminished the ability of cod to consume urchins and control their populations.

The role of large predatory fishes in structuring ecosystems has long been recognised, and thus the weakening of the top-down control of lower trophic levels following their extirpation in coastal zones (Jackson et al., 2001; Worm and Myers, 2003; Frank et al., 2005; Estes et al., 2011) was not surprising. The most obvious change was the marked increase in the abundance of macroinvertebrates that were once common prey of benthic fish communities after the fisheries collapse (Jackson et al., 2001; Boudreau and Worm, 2010). The large increases in the abundance of lobsters and urchins in the coastal zone supported lucrative fisheries for both species, the former sustainable and the latter not (see Sections 5.7.2 and 5.6.3, respectively, for details). Striking evidence for the magnitude of these effects comes from studies of several remote, offshore pinnacles that never experienced the level of fishing pressure that the coastal environment received. Comparisons of these offshore sites with inshore sites show cod to be more than an order of magnitude more abundant on the pinnacles and kelp cover over three times higher. Correspondingly, both lobsters and urchins were rarely seen at these sites (Witman and Sebens, 1992; Vadas and Steneck, 1995).

### 5.7.2 Decapods: Lobster and Crab

American lobster landings have increased steadily over the last four decades (Steneck and Wahle, 2013), reaching record values in the USA and Canada in recent years (Figure 5.7). The high abundance of lobster in spite of heavy fishing was first an enigma to managers (Miller, 1994; Steneck, 2006a), but several studies have now convincingly argued that the extirpation of large predators is the primary explanation for the increase in lobster abundance (i.e., Boudreau

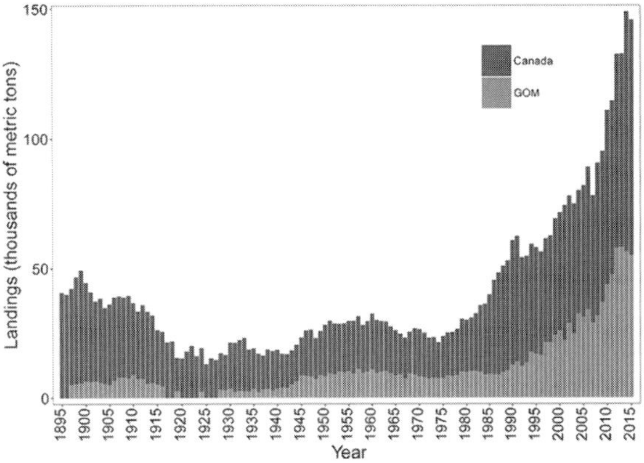


**Fig. 5.7** Lobster landings for the GOM and the east coast of Canada. Data were available from 1895 to 2015, but some years are missing for the GOM (Maine Department of Marine Resources, www.maine.gov/dmr/commercial-fishing/landings/index.html, and Fisheries and Oceans Canada, www.dfo-mpo.gc.ca/stats/commercial/land-debarq-eng.htm).

and Worm, 2010; Steneck and Wahle, 2013; Boudreau et al., 2015).

When predators were abundant, lobsters were vulnerable to predation even as adults, whereas, under the current environment, the risk of predation is confined to a brief period after settlement (Wahle and Steneck, 1992). The demographic bottleneck of lobsters is now restricted to the availability of suitable habitat for post-larval settlers, and this recruitment-driven demography (Palma et al., 1999) represents a shift in the ecological processes that govern lobster population dynamics, one where localised hotspots of recruitment create regions with very high adult lobster density (Wahle et al., 2013). A recent meta-analysis (Boudreau et al., 2015) found evidence for both top-down (predator-release) and bottom-up (large-scale climate) regulation of lobster abundance in the north-west Atlantic Ocean. Top-down effects, mostly on juveniles, dominated demographics at the cold and warm extremes of lobster distribution, whereas bottom-up effects influenced recruitment in warmer regions. At a local scale, however, the distribution and abundance of lobster is now likely a consequence of several processes in which recruitment, predation and shelter competition all play a role (Steneck and Wahle, 2013).

The American lobster has a complex life cycle and undergoes ecological niche shifts throughout its ontogeny. Pelagic post-larval stages settle in shallow waters (<20 m; Wahle et al., 2013) and actively select substrata that provide small shelters (i.e., cobbles) where they are safe from predators (Wahle and Steneck, 1992; Palma et al., 1998). Recruits exhibit unusually low settlement mortality (Wahle et al., 2004), so their densities can reach high values. Young-of-year can live at densities exceeding 5 $m^{-2}$ (Wahle et al., 2013), but, as they increase in size, intraspecific competition for shelter intensifies and strong agonistic behaviour, including cannibalism, occurs at high densities (Steneck, 2006b). Larger lobsters are less vulnerable to predation, and, as competition increases, their range of movement also increases. At harvestable sizes, their annual migration range is 5 km, and, at reproductive sizes, this range expands to 20–30 km (Campbell and Stasko, 1985). Thus, offshore regions are dominated by large, reproductive-phase lobsters, even if there is little apparent settlement in this habitat (Steneck, 2006a).

Sea temperature in the north-west Atlantic Ocean is expected to rise at a rate higher than the global average (Hoegh-Guldberg and Bruno, 2010), but its specific effect on the different life stages of lobsters is not yet fully understood (Quinn, 2017). Current average summer temperatures in coastal Maine are already close to the optimum of lobster larvae for this region (Annis, 2005), so, at higher temperatures, settlement might decrease. In general, the duration of lobster larval stages is reduced with increased water temperature. However, in terms of development time, larvae with a 'warm-origin' (i.e., southern GOM) have a different functional response to temperature change than larvae with a 'cold-origin' (i.e., northern GSL), suggesting that 'cold-origin' lobster larvae may have undergone a counter-gradient adaptation for development in colder waters (Quinn et al., 2013). Higher water temperatures have also increased the concern about potential increases in the incidence of lobster diseases (Shields, 2013), especially epizootic shell disease (Smolowitz et al., 2014), which is caused by bacteria that opportunistically exploit vulnerable individuals, resulting in abnormal moulting and death (Sindermann, 1990). In southern New England, this disease was initially thought to have caused a fishery collapse in the late 1990s (Cobb and Castro, 2006), but this event was later attributed to the combined effects of diminished larval supply and disease (Wahle et al., 2009). By 2003 the disease had spread to Cape Cod Bay and to the nearshore areas of the GOM (Glenn and Pugh, 2006) but did not penetrate coastal Maine or Canada. The role of increased water temperature in outbreaks of this disease remains speculative (Glenn and Pugh, 2006), and significant uncertainties remain about the causes of outbreaks and the mechanisms of transmission (Wahle et al., 2009). In contrast, the abnormally high temperatures that occurred in the GOM in 2012 (2°C above mean temperature from 1982 to 2011) triggered the inshore movement of lobsters three weeks earlier than in previous years and increased their moulting rate, thereby boosting numbers of legal-size lobster available for the fishery (Mills et al., 2013).

The recent increase in lobster populations has outpaced fishing capacity, evident in the steadily increasing catch per unit effort in the last two decades in the USA and Canada (3.6- and 4.8-fold increase since 1975, respectively). The American lobster fishery is likely the only long-exploited fishery in the world that has higher landings today than ever before (Steneck and Wahle, 2013; Figure 5.7). Part of the increase in lobster populations in the GOM is due to increased food availability due to the use of high-quality bait (i.e., Atlantic herring [*C. harengus*]) in lobster traps (Grabowski et al., 2010). Juvenile lobsters enter traps to consume the bait and then either escape or are released because they are under the legal size limit. Fisheries in the GOM are thus in some way a form of 'lobster ranching' where populations are subsidised by imported food resources that increase the productivity of the fishery (Grabowski et al., 2010).

The American lobster coexists with other decapods, like the rock crab (*C. irroratus*) and the Jonah crab (*C. borealis*), which are both ecologically and economically important large-bodied invertebrates. After the GOM 'flipped' towards kelp beds (see earlier), abundance of *C. borealis* increased. Based on bottom-trawl surveys, average *C. borealis* density for the period 1996–2001 was five times higher than in the 1973–1995 period (Steneck et al., 2013). These three species occupy similar habitats and have similar diets, but studies have shown that the lobster acts as both a dominant competitor and an intraguild predator, driving crabs to use alternative habitats to minimise competition and predation (Richards, 1992; Moody and Steneck, 1993; Sainte-Marie and Chabot, 2002). Rock crabs have higher fecundity and an earlier onset of maturity than lobsters, but, because settlers are less selective of substrata (i.e., settling in cobble and sand) and suffer higher post-settlement mortality compared to lobsters, adult population densities are comparable (Palma et al., 1998). Behavioural modification in crabs also helps to minimise predation. For example, both crabs exhibit diurnal activity, while lobster have mostly nocturnal peak activities (Novak, 2004), and, although both lobster and rock crabs prefer hard relative to soft substrata, the latter are displaced by lobsters when shelter availability is limited. Among the Jonah crab, the non-indigenous green crab (*Carcinus maenas*) and the rock crab, the latter is the preferred prey of lobsters (Jones and Shulman, 2008) and exhibits an escape response to avoid them (Novak, 2004). The increased habitat complexity provided by macroalgae in kelp beds provides microhabitats that might play an important role in the survival of subordinate crab species. Small- and medium-sized rock crabs can be found in the kelp canopy in the field, and laboratory experiments have shown that crabs tend to move up kelp when lobster are present, presumably reducing interspecific competition and avoiding predation by occupying three-dimensional space (Wells et al., 2010). However, lobsters are now considered to be a more benign predator in coastal GOM because much of their food now comes from the subsidies provided by lobster trap bait (Grabowski et al., 2010).

Interestingly, Jonah and rock crabs were historically considered a by-catch in nearshore lobster fisheries, and even a nuisance when they consumed bait. However, in the late 1980s as stocks of more popular crabs became depleted, fishers began to exploit these species, which could supplement the catch value when large enough. During winter and spring, when lobster catches are low, fishermen have even been known to target Jonah crabs. In Canada, Jonah and rock crabs are both managed fisheries, and both species are sold as seafood as well as used for lobster bait (Rondeau et al., 2014; Fisheries and Oceans Canada, 2016). However, little information exists for the Jonah crab in Canadian waters, and knowledge of its life history is limited to studies in New England and Chesapeake Bay (Fisheries and Oceans Canada, 2009). In the USA, there is no stock assessment or fishery management of Jonah crab, due in part to limited data on its population, growth rates and distribution (Swenton et al., 2014). Concern has also been raised regarding *Cancer* crab fisheries as they increase in the USA and Canada (Anderson et al., 2011), but studies on their potential remain limited (Robichaud and Frail, 2006; Swenton et al., 2014).

## 5.8 NIS and Biological Invasions – Unwelcome Additions

Since European exploration and subsequent colonisation, extensive vectors have existed for the introduction of NIS. Early sailing vessels frequently contained diverse sessile communities attached to their hulls, and the use of solid ballast (i.e., stones – collected along one shoreline and jettisoned upon arrival at another) was commonplace into the early twentieth century (Carlton, 1999; Johnson et al., 2012). Later on, as shipping became mechanised, ballast stones were replaced with untreated ballast water, and the larval stages of marine organisms could be transported thousands of kilometres in just a week or two (Seebens et al., 2013; Chan et al., 2015). The aquaculture and seafood industries have served as yet another vector. Shellfish and other seafood imported from other continents, particularly Asia, were frequently packed with non-native seaweeds, which could also contain other biological hitchhikers. Today we have identified many of these potential vectors and have taken regulatory steps to limit new introductions. Perhaps what is most surprising is that the north-west Atlantic Ocean has so few invasive marine species despite several centuries of potential exchange with other continents (Carlton, 1999).

Of the estimated 535 species of macroalgae in the north-west Atlantic Ocean, just thirty-two have been categorised as invasive or introduced (Mathieson, 2016). Of these, only a handful have had large negative ecological impacts (i.e., displacing native species or reducing localised productivity) (Davidson et al., 2015). The NIS with the greatest ecological impact has been *C. fragile*, which was introduced in New York in the 1950s, most likely as a hitchhiker on cultivated oysters from Japan. Since then, it has expanded slowly northward (Carlton and Scanlon, 1985), reaching the southern GSL in the mid-1990s (Lyons and Scheibling, 2009). It is largely opportunistic, colonising recently disturbed areas, often reaching extremely high densities in localised areas and displacing native seaweeds, particularly kelp (Scheibling and Gagnon, 2006). *Fucus serratus* – first detected in 1868 – has undergone several range expansions (Johnson et al., 2012). It is the dominant intertidal seaweed in many wave-exposed areas but can also be dominant in subtidal habitats (Adey and Hayek, 2011). *Grateloupia turuturu*, a more recent invasive seaweed, first detected in 1994 in Boston, is the largest red alga in the world and has a broad thermal tolerance (Mathieson et al., 2008). Another newcomer, *Dasysiphonia* (*Heterosiphonia*) *japonica,* arrived from Asia approximately ten years ago and has already dominated certain subtidal zones (median cover of 14 per cent and maximum 80 per cent) and contributes up to 65 per cent of beach wrack in some locations (Newton et al., 2013). Its potential distribution could range across the north-west Atlantic Ocean from Florida to Newfoundland.

Several abundant non-indigenous invertebrates in the north-west Atlantic Ocean have been present for over a century and have enormous impacts on benthic assemblages. The European green crab (*C. maenas*) was introduced in New York in the mid-1820s and expanded northward over subsequent decades. Although its spread appeared to stop due to colder temperatures and circulation patterns (Byers and Pringle, 2006), a more recent introduction of a more cold-tolerant genotype (Roman, 2006) has expanded its range to southern Newfoundland. The green crab is a voracious predator with a diverse diet (Carlton and Cohen, 2003; Breen and Metaxas, 2009), and when abundant, it can destroy epibenthic habitats like eelgrass and mussel beds (Garbary et al., 2014; Neckles, 2015). Historic increases in green crab populations in the GOM have been tightly correlated with warmer than average winters and reduced local ice cover (Glude, 1955; Welch, 1968), suggesting further increases in range and impacts with global warming. The herbivorous snail *Littorina littorea* is another non-indigenous invertebrate that has been present in this ecosystem for well over a century (Chapman et al., 2007). Originally from Europe, it frequently occurs in densities of hundreds per square metre and is the most ecologically important grazer in the intertidal and shallow subtidal environments from the central eastern United States up through Nova Scotia (Jenkins et al., 2008). More recently, the Asian shore crab *Hemigrapsus sanguineus* has also invaded the north-west Atlantic Ocean (Epifanio, 2013). Originally from eastern Asia, this crab persists from the low intertidal to the shallow subtidal environment, and can become very abundant in cobble habitats, outnumbering and outcompeting the indigenous rock crab *C. irroratus* (McDermott, 1998; Jensen et al., 2002).

Perhaps more frequently than any other group, non-indigenous bryozoans and ascidians have become established in the north-west Atlantic Ocean (Ramsey et al., 2008), and several have had substantial impacts on local ecosystems as well as aquaculture operations (McKindsey et al., 2009). *M. membranacea*, a bryozoan native to the north-western Pacific Ocean, was first observed in the GOM in 1987 (Berman et al., 1992) and has since spread throughout the boreal region, extending from the east coast of the USA to Newfoundland (Watanabe et al., 2009; Caines and Gagnon, 2012). It grows epiphytically on seaweeds, especially kelp, and colonies can completely overgrow blades, blocking light and nutrient exchange and causing blades to become brittle and break either as a direct result of natural wave action or in combination with grazing (O'Brien et al., 2013). This bryozoan is a major concern for kelp farming and harvesting operations throughout New England and Canada because blades encrusted with *M. membranacea* are unmarketable.

Both the growth rate and dispersal potential of *M. membranacea* are tightly correlated to water temperature; warmer winter water temperatures yield more propagules whereas warmer summer temperatures result in larger colonies (Saunders and Metaxas, 2009). Episodic outbreaks of *M. membranacea* aided by warmer water temperatures have resulted in the defoliation of kelp beds throughout ANS (Watanabe et al., 2009; Scheibling and Gagnon, 2009) and the GOM (Lambert et al., 1992; Harris and Tyrrell, 2001). When *M. membranacea* infestation is severe, massive loss of standing kelp biomass can potentially accelerate the destructive grazing behaviour of urchins (Scheibling et al., 1999), lowering local productivity and diversity, and indirectly facilitating invasion by *C. fragile* (Levin et al., 2002; Scheibling and Gagnon, 2009) and other invasive species such as the tunicate *Didemnum vexillum* (Bullard et al., 2007).

In addition to *M. membranacea*, numerous non-indigenous ascidians (both colonial and solitary) have become widespread throughout the north-west Atlantic Ocean (Dijkstra et al., 2007; Carman et al., 2009; Sephton et al., 2011). These species are frequently fouling organisms in harbours and on aquaculture operations (Ramsey et al., 2008), but can also compete for space on rocky bottoms (Ma et al., 2017).

## 5.9 Diversity Hotspots – Beyond the Kelp–Urchin Arena and Back

Although the majority of research on subtidal rocky systems in the north-west Atlantic Ocean has focussed on kelp beds and urchin barrens as the dominant subtidal communities (Figure 5.1a and 5.1b), there are a number of other communities that occur within the north-west Atlantic Ocean that add greatly to the biodiversity and functioning of these ecosystems. Many of these communities, although considered to be restricted in spatial extent (an assumption that has not been confirmed for most assemblages in most areas), include a multitude of species and can represent diversity hotspots. These predominately invertebrate assemblages exist in areas where macroalgal growth is limited (primarily by light and herbivory and to a lesser extent temperature) and where physical disturbance and predation are also limited. In such rocky subtidal areas, the limitation or complete exclusion of macroalgal growth allows a large diversity of mainly filter- and suspension-feeding invertebrates to colonise the substratum. In the north-west Atlantic Ocean, there are three representative habitats in this category: (1) vertical rock walls in shallow subtidal zones where macroalgae are limited by reduced light (Miller and Etter, 2008), (2) shallow bivalve beds where macroalgae are limited by grazing, e.g., horse mussel beds (Witman, 1987) or by temperature, light or lack of suitable substrata in highly turbid coastal areas, e.g., oyster reefs and (3) rocky outcrops below the lower limits of significant light penetration where communities of invertebrates such as deep-sea corals dominate.

### 5.9.1 Rock Walls – Neglected by Scientists?

Because of the dense invertebrate communities found on shallow vertical areas, rock walls are often favourite recreational dive sites, but they are only rarely the focus of scientific efforts. Studies of these communities in the GOM have confirmed that rock walls are indeed more diverse than adjacent horizontal surfaces, with four times as many species found on vertical surfaces than on horizontal surfaces and many species found only on the vertical wall (Miller and Etter, 2011). Two types of vertical surfaces have been identified in the GOM, those that are accessible to urchin grazing and those that are not (Sebens, 1985), with corresponding differences in invertebrate communities. The factors evoked to explain the development and persistence of these diverse communities are shading, grazing, sedimentation, disturbance and water flow, but only rarely has experimental work attempted to untangle the roles of these factors, an understandable situation given the difficulty of underwater research, especially in colder waters. In communities with low urchin densities (1–2 ind.m$^{-2}$), experimental manipulations have clearly shown that differences in invertebrate communities between adjacent vertical and horizontal surfaces are due to the reduced light levels on vertical surfaces that limit growth of macroalgae (Miller and Etter, 2008) and not to predation. The absence of an effect of grazing and disturbance, as compared to earlier studies in the GOM (Sebens, 1985), is perhaps related to the decrease in urchin density and the overall shift towards a more macroalgal-dominated state (see Section 5.6.2). The community structure on rock walls is also determined by differences in flow (Leichter and Witman, 1997), affecting both feeding and recruitment (Palardy and Witman, 2014), and by competition for space (Genovese and Witman, 1999). Rock walls likely represent a habitat protected by their vertical orientation from disturbance, particularly sedimentation and scouring (Miller and Etter, 2008). These conditions allow increased recruitment and survival of small early life stages of sessile invertebrates (Palardy and Witman, 2014). This process could be particularly true in more northern regions, where scouring and disturbance by ice is more important.

In the GSL, popular dive sites are cliff or wall habitats, dominated by filter and suspension feeders such as anemones, soft corals, sponges and tunicates (Figure 5.1c). Community patterns in rocky subtidal habitats in the vast maritime estuary of the Saint Lawrence River were described by Himmelman and Lavergne (1985) who noted the presence of a zone of filter-feeding invertebrates in deeper urchin-dominated areas. In these zones, urchin densities are high enough to limit algal abundance but lower than in urchin barrens. However, these zones have received even less attention than in the GOM, and almost no literature exists that examines the processes at work in structuring these diverse communities. Indeed, research on subtidal sessile invertebrates appears to be largely one of two types: either studies of the physiological, ecological or morphological aspects of a single species (with little attention to spatial patterns of distribution and the role of the species in the community) or a focus on invasive species and fouling communities on artificial structures (e.g., Mathieson et al., 1991; Dijkstra and Harris, 2009).

By providing shaded and/or vertical hard substrata for invertebrate communities, artificial structures have expanded the amount of habitat available for these filter- and suspension-feeding communities in certain areas while also perhaps facilitating the spread of invasive species. Indeed, the recognition that marine species were being transferred from their native range into new areas has set the stage for a growing body of research on species interactions and community dynamics of fouling communities. A long-term comparison in the GOM, for example, found that, although the foundation species in these fouling communities has shifted from mussels to the introduced ascidian, *Didemnum vexilum*, the overall diversity and associated abundance of the invertebrate community has in fact increased with this shift (Dijkstra and Harris, 2009). Unfortunately, the relative ease of conducting studies on fouling communities (e.g., the use of settling plates), especially on artificial structures (i.e., docks), may have diminished interest in studying invertebrate communities on natural substrata.

### 5.9.2 Bivalves as Foundation Species – Mussel Beds and Oyster Reefs

Both mussel beds and oyster reefs provide three-dimensional structure that creates habitat for many other invertebrate species. These small interstitial spaces and crevices provide critical protection from predators for many small species and juvenile life stages. This attribute gives bivalve beds the potential to be diversity hotspots.

In the GOM, urchin grazing in shallow subtidal areas can facilitate the persistence of extensive beds of horse mussels (*Modiolus modiolus*) by limiting kelp growth on mussel shells and thereby preventing their dislodgment due to drag forces experienced during storms (Witman, 1987). This results in a zonation in certain areas whereby kelp are restricted to shallow waters with their lower limit set by grazing. This process thus determines the upper limit for horse mussel beds, which then can extend down to depths of 18–20 m. In the Bay of Fundy, 1,500 mussel beds with a total area of 11,670,000 $m^2$ have been identified (Kenchington, 2014), illustrating well their potential contribution to ecosystem function. These reefs are of special conservation concern worldwide because of the diversity of their associated invertebrate communities and because of their susceptibility to damage from trawling activities (Cook et al., 2013). Outside of the GOM, very little is known of the extent and importance of subtidal mussel beds. An exception is the Mingan Archipelago in the northern GSL where sea stars are voracious predators of mussel, and intense foraging by *Leptasterias polaris* and *Asterias rubens* limits the distribution of mussel beds to only a few metres in depth (Gaymer et al., 2001a) and eliminates any beds that occasionally form in deeper waters. Behavioural studies in Newfoundland have demonstrated that wave action can, however, modulate the ability of *A. rubens* to explore and localise mussel prey (St-Pierre and Gagnon, 2015a).

Further south, oyster reefs can be found in areas where macroalgae are limited by light (shallow high-turbidity areas) and on unstable subtidal rocky substrata (e.g., cobble). Native oyster reefs provide many ecosystem services in the shallow coastal areas where they are found, including water filtration, creation of habitat for other species, erosion protection for shorelines and fisheries (Beck et al., 2011). In fact, there is historical evidence for the occurrence of vast reefs in shallow temperate areas around the globe, indicating that shellfish reefs may have been the temperate equivalent of coral reefs, critical for creating habitat and maintaining biodiversity (Beck et al., 2011). In the north-west Atlantic Ocean, oyster reefs have received considerable attention, and restoration efforts are active in many regions, particularly more southern biomes. In areas from Florida to Cape Cod, native reefs of the eastern oyster (*Crassostrea virginica*) have declined 90–99 per cent, and only small remnant or restored populations exist. In more northern areas, populations are smaller but have declined by a lesser extent (Beck et al., 2011). In eastern Canada, oysters grow in both intertidal and subtidal habitats and can be found on the shores of Nova Scotia, on Cape Breton Island and in the Bay of Fundy. Their extent appears, however, to be much diminished from historical times as the abundance of oyster shells in middens suggests that they were harvested extensively by indigenous people (Kenchington, 2014).

### 5.9.3 Deep Rocky Habitats – Islands in a Sea of Mud

Below the shallow subtidal zone where light is attenuated to levels below the compensation point for photosynthesis, rocky substrata of all orientations begin to be dominated by filter- and suspension-feeding invertebrates. In Nova Scotia, a recent survey of tunicate beds concluded that this switch from algal-dominated to invertebrate-dominated habitat structure was associated with a high diversity of other invertebrates (Francis et al., 2014). There are, unfortunately, very few other studies from the north-west Atlantic Ocean describing the rocky subtidal zones below the limits of non-technical scuba diving but above the areas commonly surveyed from oceanographic vessels. This area, the twilight zone between the shallower zone accessible by scuba and deeper zones reached by dredging and other traditional bottom sampling methods, is only just beginning to be explored (e.g., Filbee-Dexter and Scheibling, 2017). Diverse invertebrate assemblages such as that described by Francis et al. (2014) are certainly more common than represented in the literature.

In recent years, interest in and knowledge of the invertebrate communities of north-west Atlantic Ocean deep waters, especially those associated with deep-water corals, have grown. These areas of high diversity of deep-water species, first described during the Agassiz surveys in 1888, are primarily known from by-catch associated with dredging and trawling, although video surveys have added to our knowledge in recent years. Although not all deep-water anthozoans are associated with rocky substrata (e.g., sea pen meadows on muddy or sandy bottoms), many of the larger habitat-forming species (corals, gorgonians, sea fans, horny corals and sea feathers) are intimately tied to hard substrata. Mirroring patterns observed in shallow subtidal zones, the greatest diversity is found on rocky bottoms (Baker et al., 2012). Many of these species are long lived and slow growing and are therefore particularly vulnerable to damage from fishing gear (Kenchington, 2014). Deep corals have been found throughout the north-west Atlantic Ocean, from the Florida and Carolinian deep reefs (Reed, 2004; Reed and Ross, 2005) to Newfoundland, Labrador and the Arctic (Gilkinson and Edinger, 2009). These corals are substratum-limited due to the paucity of hard substrata on deep continental shelf edges and slopes (Gilkinson and Edinger, 2009); in fact, these hard-bottom-associated corals have been termed 'islands in a sea of mud' (Wahl, 2009). Concern over these fragile diversity hotspots in deep coastal waters has prompted global actions, including the drafting of a conservation strategy for eastern Canada (Fisheries and Oceans Canada, 2015).

### 5.9.4 Return to the Kelp–Urchin Arena – Are Barren Grounds that Barren?

The previous subsections have explored areas where macroalgal growth is limited and kelps, in particular, are excluded for largely abiotic reasons (e.g., light, substratum). Explicit in this discussion was a designation of these areas as being outside of urchin barrens. But what about within urchin barrens, where macroalgae are largely excluded through intense grazing by urchins? Are these areas devoid of life, as the name 'barren grounds' suggests? As discussed for rock walls earlier, horizontal surfaces in barren grounds support many fewer filter-feeding invertebrates than adjacent vertical walls, but the more relevant comparison of interest should be with an adjacent kelp bed or forest. Surprisingly few people have examined this question worldwide. In Chile, successional patterns were different between a kelp bed and adjacent barren ground, but this was due to differences in which species were present and not to a general lack of species in the barren ground (Uribe et al., 2015). In Alaska, the invertebrate communities inhabiting holdfasts of the same species of kelp did not differ depending on whether the holdfast came from a dense kelp forest or from a lone plant growing in an isolated location in a barren ground (Schuster and Konar, 2014). Very few studies in the north-west Atlantic Ocean have explicitly addressed the question of comparing invertebrate diversity between kelp beds and barren grounds. Bégin et al. (2004) found that invertebrate assemblages associated with kelp beds did indeed differ from barren ground invertebrate communities in the GSL; however, this difference was driven by species living directly on algal fronds, and no differences were detected between invertebrate assemblages in holdfasts and on the bottom in the kelp bed when compared to the barren ground community. Kelly et al. (2011) also identified differences in invertebrate assemblages associated with different algal assemblages; like Bégin et al. (2004), these overall differences were driven by differences in abundance and biomass of several key species, however, and not by unique or missing species, although lower richness and diversity (but not evenness) were observed in barren grounds.

These findings point to the role that habitat complexity and structural diversity play in determining the abundance and diversity of organisms found in a certain habitat. In marine systems, the role of macroalgae in providing structural complexity (e.g., refuges, protection from predation) is perhaps more important than their role of providing primary production. Diversity in different habitats may thus depend largely on the quality of the structure provided by foundation species (Teagle et al., 2017). If so, it is entirely possible that so-called barren grounds will, depending on their underlying physical structure, differ in the invertebrate assemblage that they support. Rhodolith beds, a habitat created by accumulation of unattached non-geniculate coralline algae that resemble small, bumpy pebbles, are a perfect example of this possibility. The high densities of urchins and lack of fleshy macroalgae associated with this habitat would make them appear at first glance to be just another type of barren ground. However, the highly complex structure that they create abounds with crevices and interstitial spaces (Foster, 2001), which can host an abundant invertebrate assemblage (Foster, 2001; Gagnon et al., 2012; Adey et al., 2015). Rhodolith beds are consequently known as highly diverse areas and are widely distributed in the world's oceans – indeed, Foster (2001) classes them as one of the 'big four' marine macrophyte communities in terms of area covered globally, along with kelp beds, seagrass meadows and the more traditional non-geniculate coralline reefs, i.e., classic barren grounds.

Even in more traditional nearshore urchin barrens, the role of bottom structure is apparent. In the Mingan Archipelago in the GSL, brittle stars are abundant. The swiss-cheese-like structure of the bottom allows brittle stars (Drolet et al., 2004) and other invertebrates (Gaymer et al., 2001b) to escape predation by sea stars or urchins by sheltering in crevices. This situation contrasts with urchin barrens where the substratum is a flat bedrock shelf with minimal relief for providing refuge from predators. Highly textured bottom substrata are common in the GSL and Labrador, while smooth subtidal rocky shelves are relatively rare, and it is possible that these differences in bottom composition drive patterns in 'barren' ground diversity with the highest contrast existing in areas where no habitat complexity aside from macroalgae exists. Rhodolith beds may modulate this situation by provided additional biogenic structure – indeed, several rhodolith beds have been identified in Newfoundland in nearshore rocky areas composed principally of bedrock shelves (Gagnon et al., 2012; Adey et al., 2015). Other important nearshore rocky habitats are areas of coralline-encrusted cobbles. These have been well studied in ANS, where the interstitial spaces created by the overlapping of cobbles support abundant populations of invertebrates, including molluscan mesograzers (Scheibling et al., 2009a). The development of macroalgal beds after urchin die-offs is strikingly different on cobble beds (Scheibling and Raymond, 1990) as large macroalgae are removed by hydrodynamic dislodgment (Scheibling et al., 2009b) and urchin grazing is limited by the presence of small predators, resulting in a community of sparse, small algae (Scheibling and Raymond, 1990). It remains to be seen, however, if these patterns are widely observed elsewhere in the north-west Atlantic Ocean.

## 5.10 The Future – Global Change and Research Needs in the Next Quarter Century

As the research covered in the preceding pages has amply shown, we have made great progress over the past twenty-five years in term of describing the diversity of the north-west Atlantic Ocean, not so much in terms of species diversity, but more in terms of habitat and ecosystem diversity. Not only do we know more about the occurrence of different kinds of habitats, but also how the dynamics of a given habitat can vary from one place to another and from one time to another. Granted, variation in space and time is not a novel observation in ecological studies, but in the north-west Atlantic Ocean, we now have a much better understanding of the different processes involved and how variation in the processes themselves can drive community dynamics over ecosystem scales. This progress is, however, dwarfed by what still needs to be done, both to expand the geographic extent of our knowledge and to incorporate global change, including climate change and homogenising of our coastal flora and fauna by the introduction of NIS (see Section 5.8).

### 5.10.1 Geographic Knowledge Gaps – Diving is Difficult

Although subtidal habitats appear tantalisingly within reach of the long-established technology of scuba diving, the reality is that accessing them from the shore or small boats is a laborious and expensive process, especially as diving activities are becoming more and more regulated, adding to costs (e.g., training, equipment acquisition and maintenance) and logistics (e.g., support personnel). It is not surprising then that most of our knowledge of the patterns and process of benthic communities on hard substrata still comes from studies of the intertidal zone (e.g., Jenkins et al., 2008; Hawkins et al., 2009; Chapter 4) in spite of its unique abiotic conditions (e.g., emersion during low tide, high exposure to wave action) and limited areal extent.

Nevertheless, a small number of researchers have maintained active diving programmes in the north-west Atlantic Ocean over the past two to four decades. Most, however, have retired or are nearing retirement, and it seems unlikely that they will be easily replaced given the logistic complexity and high expense of diving operations, not to mention the challenges of diving in very cold waters – summer temperatures in subarctic regions average 4–8°C and rarely go above 10°C. Drysuits are the fashion, and numb hands the norm.

These concerns are all multiplied in the more remote northern areas of the north-west Atlantic Ocean. As illustrated in Figure 5.1, research effort (as measured by the number of publications citing underwater research) falls off dramatically as one moves northward through the north-west Atlantic Ocean, and long stretches of coast – some over 400 km long – are represented by a single study or no study at all. Much of our knowledge comes from university-based research, and, for practical reasons, research is often done close to campus (an exception would be decades of research done by Himmelman and colleagues in the Mingan Archipelago, almost 1,000 km from their home campus, Laval University in Quebec City). The low number of such institutions in northern areas severely limits our knowledge of the generality of patterns seen at one or two sites. As an example, there is just one university, Memorial University, in all of Newfoundland, with just two campuses located on more than 17,000 km of coastline. Likewise, only one campus, Université du Quebec à Rimouski, is located on the entire GSL coast of Quebec. With multiple universities (e.g., University of Maine, University of New Hampshire, Northwestern University, University of Massachusetts and Boston University), university-owned marine stations (e.g., Isle of Shoals, Marine Science Center, Darling Marine Center and Coastal Studies Center) and private institutions (e.g., Woods Hole Oceanographic Institution, Gulf of Maine Research Institute and Bigelow Laboratories), it is not surprising that our knowledge of the GOM far exceeds that of the GSL.

A final problem, the lack of coordinated research across ecosystems of interest, is not unique to either underwater research or the north-west Atlantic Ocean. Collaborations to compare patterns and processes across these vast areas have been largely missing due both to the reality of limited funding and the complexity of logistics, but also to the nature of scientists who prefer to follow their individual research programmes (but see Balch et al., 1998). As laudable as such curiosity-based investigations are, subtidal researchers could take a lesson from oceanographers, who have learned to coordinate their activities around expensive infrastructure to undertake 'big science'. Likewise, the coordinated research programmes of European states (e.g., EUROROCK; Jenkins et al., 2005), privately (e.g., PISCO) or publicly funded (e.g., ZEN) networks offer different models for approaching this problem. The recent creation of KEEN (Kelp Ecosystem Ecology Network) is a promising possibility for coordinating nearshore research in temperate zones. The heroic efforts of Walter Adey to sample subtidal sites across vast stretches of unexplored coastlines in the north-west Atlantic Ocean, including the rarely visited Labrador coast, also offer an inspiring example of what a single individual can accomplish (Adey and Hayek, 2011).

One bright spot in nearshore underwater research is the advances in remotely operated vehicles or towed camera systems, which allow much greater areas to be surveyed. Moreover, these instruments can reach depths that are normally beyond the range of scuba divers. Although generally limited to descriptive work, this approach can allow us to extrapolate processes documented in detail in smaller areas to scales more amenable for estimating ecosystem processes. An excellent example is the study of the distribution and consumption of drift kelp in Nova Scotia (Filbee-Dexter and Scheibling, 2014b). Likewise, increased bottom mapping using side-scan sonar and lidar can also provide basic information for determining the extent of both dominant (e.g., horse mussel beds) and rare (e.g., deep-sea corals) habitats.

### 5.10.2 Global Change

No modern discussion of ecosystem research can avoid including the importance of global processes in influencing the structure and function of biological communities. Coastal marine ecosystems are no exception, and several reviews over the past decade have provided a broad picture of the alterations expected with climate change (Harley et al., 2006; Hawkins et al., 2008, 2009; Doney et al., 2012), biological invasions (Ruiz et al., 2000) and possible synergies of biological invasions occurring with or because of climate change (Occhipinti-Ambrogi, 2007; Walther et al., 2009). Unfortunately, ideas and concerns have run far ahead of field studies demonstrating clear effects of either of these aspects of global change, but there is wide consensus that changes have occurred, will continue to occur and will likely negatively affect many species, habitats and ecosystems.

A likely outcome of warmer water temperatures is a poleward shift of species distributions, a pattern observed on other coasts including in the north-east Atlantic Ocean (Lima et al., 2007; Pinsky et al., 2013). While some evidence for this trend has been observed in the north-west Atlantic Ocean for a barnacle and several decapods (see Section 5.2), no general shifts have been observed. A recent effort to assess the extent of this phenomenon in kelp (Merzouk and Johnson, 2011) found no evidence for such change, although this was attributed largely to a lack of adequate time series and large-scale sampling of species distributions (see Section 5.10.1), a problem similar to that in the north-eastern Atlantic Ocean (Smale et al., 2013). A recent global synthesis of changes in kelp abundance (Krumhansl et al., 2016) suggests that, in some locations within the GOM, a reduction in kelp cover has occurred, but is primarily because of biological interactions. Several studies have, however, highlighted how increased water temperatures may cause future problems or exacerbate existing ones. For example, increased water temperature will likely allow further range expansion of the invasive bryozoan *Membranipora membranecea* and increase its impact on kelp populations (Saunders and Metaxas, 2009).

Climate change appears to be altering the frequency and intensity of hurricanes in coastal ANS, altering the disease dynamic controlling kelp–urchin interactions there (Scheibling and Lauzon-Guay, 2010). In the last four to six decades, biomass of kelp beds in Nova Scotia have declined by 85–99 per cent, and recent studies have linked this massive kelp reduction with synergistic processes driven by global warming (Filbee-Dexter et al., 2016). Kelp beds naturally experience canopy loss when temperatures and storm intensities increase in late summer and winter, respectively. However, this process, when coupled with positive feedback mechanisms that reduce seasonal regeneration of kelp beds, leads to the establishment of an alternative invasive/turf-algae-dominated state. For example, rising temperatures indirectly affect kelp tissue loss and mortality by increasing growth rates of the invasive bryozoan *Membranipora membranecea* (Saunders et al., 2010; see Section 5.8) and decreasing kelp density concentrates the grazing of the snail *Lacuna vincta*, which accelerates kelp loss even more (O'Brien et al., 2015). Furthermore, mass mortalities of urchins caused by recurrent outbreaks of disease in recent years have eliminated the few urchins that may have limited turf and other understory algae (Sumi and Scheibling, 2005) and has rendered the kelp bed community more vulnerable to a shift towards a turf-dominated state (Filbee-Dexter et al., 2016).

High population densities of the American lobster together with ocean warming may increase the physiological stress and disease occurrence in the warmer zones of its range (Wahle et al., 2015; Quinn, 2017; Section 5.7.2). Further, there might be a potential increase in the diversity of predatory fishes due to warmer temperatures; for example, there has been an increase in abundance of the red hake (*Urophycis chuss*) in the GOM (Nye et al., 2009). Finally, warmer waters will undoubtedly reduce the duration and extent of sea ice cover in northern areas of the north-west Atlantic Ocean. While this change may increase the vertical distribution and productivity of kelp in deeper waters (Krause-Jensen et al., 2012), reductions in land-fast ice might increase disturbance by ice scour in shallower water due to increased abrasion of unattached sea ice driven by waves and tides (see Section 5.3).

Other changes may occur with increasing temperature, but some caution is required before assuming such changes will necessarily affect biological processes and communities. An example is the potential effect on consumption rates of urchins. This rate process is a fundamental variable controlling the abilities of urchins to destructively graze kelp beds and maintain urchin barrens (see Section 5.4) and would be expected to increase with increasing water temperatures. A recent review of literature (Suskiewicz and Johnson, 2017) found no evidence to support this prediction for the green sea urchin, at least within the range of reasonable temperature changes. Similarly, feeding rates of the ubiquitous predatory sea star *A. rubens* (Gaymer et al., 2001a) also seem not to be affected by water temperature (St-Pierre and Gagnon, 2015b). Unfortunately, all the studies examined were conducted in the laboratory and there is an obvious need to address this question under field conditions as well. Finally, there continues to be uncertainty with regards to the role climate change may play in the recovery of fisheries in the North Atlantic Ocean (Mieszkowska et al., 2009), a concern that certainly applies to depleted cod stocks in the north-west Atlantic Ocean.

## 5.11 Conclusion

The north-west Atlantic Ocean represents a vast part of the Atlantic Ocean with extensive rocky habitats across nearshore environments in much of its northern biomes. In spite of the relative simplicity of the plant and animal assemblages that form the benthic communities in the north-west Atlantic Ocean, our understanding of their dynamics remains limited due to natural variation driven by ordinary (e.g., wave action, temperature) and extraordinary (e.g., hurricanes, sea ice) factors. Perhaps more importantly, human-induced influences, including overfishing, biological invasions and climate change, are also causing variation in this system, sometimes to extreme levels. Further progress will require additional studies, both detailed research on environmental influences on species interactions and extrapolation of these processes to larger scales (e.g., connectivity of metapopulations; sources and sinks of productivity). Ultimately, we can then better compare patterns and processes in the north-west Atlantic Ocean with other regions of the world to address questions of both local and global importance.

## Acknowledgements

The authors wish to thank Steve Hawkins for the invitation to contribute this chapter and Katrin Bohn for her assistance and patience in the editorial process. Past discussions with John Himmelman, Bob Steneck, Bob Scheibling, Jean-Sébastian Lauzon-Guay, Anna Metaxas, Jarrett Byrnes, Patrick Gagnon, Bernard Sainte-Marie and Walter Adey have contributed to the refinement of our thoughts and ideas. Comments on an earlier draft by Bob Scheibling and John Himmelman added conceptual and factual improvements. The Canadian Natural Sciences and Engineering Research Council (NSERC) Strategic Project Grant (396880-10) and Discovery Grant (LEJ) provided funding, and Fisheries and Oceans Canada and Quebec-Océan provided logistic support for our reseach in the GSL – our thanks to all.

## REFERENCES

Abraham, E. R. (2007). Sea-urchin feeding fronts. *Ecological Complexity*, **4**, 161–8.

Adey, W., Halfar, J., Humphreys, A. et al. (2015). Subarctic rhodolith beds promote longevity of crustose coralline algal buildups and their climate archiving potential. *Palaios*, **30**, 281–93.

Adey, W. H. and Hayek, L.-A. C. (2011). Elucidating marine biogeography with macrophytes: Quantitative analysis of the North Atlantic supports the thermogeographic model and demonstrates a distinct subarctic region in the Northwestern Atlantic. *Northeastern Naturalist*, **18**, 1–128.

Anderson, S. C., Flemming, J. M., Watson, R. and Lotze, H. K. (2011). Rapid global expansion of invertebrate fisheries: Trends, drivers, and ecosystem effects. *PLoS ONE*, **6**, e14735.

Annis, E. R. (2005). Temperature effects on the vertical distribution of lobster postlarvae (*Homarus americanus*). *Limnology and Oceanography*, **50**, 1972–82.

Baker, K. D., Wareham, V. E., Snelgrove, P. V. R. et al. (2012). Distributional patterns of deep-sea coral assemblages in three submarine canyons off Newfoundland, Canada. *Marine Ecology Progress Series*, **445**, 235–49.

Balch, T. and Scheibling, R.E. (2000). Temporal and spatial variability in settlement and recruitment of echinoderms in kelp beds and barrens in Nova Scotia. *Marine Ecology Progress Series*, **205**, 139–54.

Balch, T., Scheibling, R. E., Harris, L. G., Chester, C. C. and Robinson, S. M. C. (1998). Variation in Settlement of *Strongylocentrotus droebachiensis* in the Northwest Atlantic: Effects of Spatial Scale and Sampling Method. In R. Mooi and M. Telford, eds. *Echinoderms: San Francisco*. Balkema, Rotterdam, pp. 555–60.

Barnes, D. K. (2016). Iceberg killing fields limit huge potential for benthic blue carbon in Antarctic shallows. *Global Change Biology*, **23**, 2649–59.

Beck, M. W., Brumbaugh, R. D., Airoldi, L. et al. (2011). Oyster reefs at risk and recommendations for conservation, restoration, and management. *BioScience*, **61**, 107–16.

Bégin, C., Johnson, L. E. and Himmelman, J. H. (2004). Macroalgal canopies: Distribution and diversity of associated invertebrates and effects on the recruitment and growth of mussels. *Marine Ecology Progress Series*, **271**, 121–32.

Bergeron, P. and Bourget E. (1986). Shore topography and spatial partitioning of crevice refuges by sessile epibenthos in an ice disturbed environment. *Marine Ecology Progress Series*, **28**, 129–45.

Berkes, F., Hughes, T. P., Steneck, R. S. et al. (2006). Globalization, roving bandits, and marine resources. *Science*, **311**, 1557–8.

Berman, J., Harris, L., Lambert, W., Buttrick, M. and Dufresne, M. (1992). Recent invasions of the Gulf of Maine: Three contrasting ecological histories. *Conservation Biology*, **6**, 435–41.

Bernstein, B. B., Schroeter, S. C. and Mann, K. H. (1983). Sea urchin (*Strongylocentrotus droebachiensis*) aggregating behavior investigated by a subtidal multifactorial experiment. *Canadian Journal of Fisheries and Aquatic Sciences*, **40**, 1975–86.

Blain, C. and Gagnon, P. (2014). Canopy-forming seaweeds in urchin-dominated systems in eastern Canada: structuring forces or simple prey for keystone grazers? *PLoS ONE*, **9**, e98204.

Boudreau, S. A., Anderson, S. C. and Worm, B. (2015). Top-down and bottom-up forces interact at thermal range extremes on American lobster. *Journal of Animal Ecology*, **84**, 840–50.

Boudreau, S. A. and Worm, B. (2010). Top-down control of lobster in the Gulf of Maine: Insights from local ecological knowledge and research surveys. *Marine Ecology Progress Series*, **403**, 181–91.

Bourque, B. J. (1995). *Diversity and Complexity in Prehistoric Maritime Societies: A Gulf of Maine Perspective*. Plenum Press, New York.

Brady-Campbell, M. M., Campbell, D. B. and Harlin, M. M. (1984). Productivity of kelp (*Laminaria* spp.) near the southern limit in the Northwestern Atlantic Ocean. *Marine Ecology Progress Series*, **18**, 79–88.

Breen, P. A. and Mann, K. H. (1976a). Destructive grazing of kelp by sea urchins in eastern Canada. *Journal of the Fisheries Resource Board of Canada*, **33**, 1278–83.

Breen, P. A. and Mann, K. H. (1976b). Changing lobster abundance and the destruction of kelp beds by sea urchins. *Marine Biology*, **34**, 137–42.

Breen, E. and Metaxas, A. (2009). Effects of juvenile non-indigenous *Carcinus maenas* on the growth and condition of juvenile *Cancer irroratus*. *Journal of Experimental Marine Biology and Ecology*, **377**, 12–19.

Britton-Simmons, K., Foley, G. and Okamoto, D. (2009). Spatial subsidy in the subtidal zone: utilization of drift algae by a deep subtidal sea urchin. *Aquatic Biology*, **5**, 233–43.

Buchwald, R. T., Feehan, C. J., Scheibling, R. E. and Simpson, A. G. B. (2015). Low temperature tolerance of a sea urchin pathogen: Implications for benthic community dynamics in a warming ocean. *Journal of Experimental Marine Biology and Ecology*, **469**, 1–9.

Bullard, S. G., Lambert, G., Carman, M. R. et al. (2007). The colonial ascidian *Didemnum* sp. A: current distribution, basic biology and potential threat to marine communities of the northeast and west coasts of North America. *Journal of Experimental Marine Biology and Ecology*, **342**, 99–108.

Bundy, A., Heymans, J. J., Morissette, L. and Savenkoff, C. (2009). Seals, cod and forage fish: a comparative exploration of variations in the theme of stock collapse and ecosystem change in four Northwest Atlantic ecosystems. *Progress in Oceanography*, **81**, 188–206.

Byers, J. E. and Pringle, J. M. (2006). Going against the flow: retention, range limits and invasions in advective environments. *Marine Ecology Progress Series*, **313**, 27–41.

Caines, S. and Gagnon, P. (2012). Population dynamics of the invasive bryozoan *Membranipora membranacea* along a 450-km latitudinal range in the subarctic northwestern Atlantic. *Marine Biology*, **159**, 1817–32.

Campbell, A. and Stasko, A. B. (1985). Movements of tagged American lobsters, *Homarus americanus*, off southwestern Nova Scotia. *Canadian Journal of Fisheries and Aquatic Sciences*, **42**, 229–38.

Carlton, J. T. (1999). Molluscan invasions in marine and estuarine communities. *Malacologia*, **41**, 439–54.

Carlton, J. T. and Cohen, A. N. (2003). Episodic global dispersal in shallow water marine organisms: the case history of the European shore crabs *Carcinus maenas* and *C. aestuarii*. *Journal of Biogeography*, **30**, 1809–20.

Carlton, J. T. and Scanlon, J. A. (1985). Progression and dispersal of an introduced alga: *Codium fragile* ssp. (Chlorophyta) on the Atlantic coast of North America. *Botanica Marina*, **28**, 155–66.

Carman, M. R., Hoagland, K. E., Greenbeach, E. and Grunden, D. W. (2009). Tunicate faunas of two North Atlantic-New England islands: Martha's Vineyard, Massachusetts, and Block Island, Rhode Island. *Aquatic Invasions*, **4**, 65–70.

Chan, F. T., MacIsaac, H. J. and Bailey, S. A. (2015). Relative importance of vessel hull fouling and ballast water as transport vectors of nonindigenous species to the Canadian Arctic. *Canadian Journal of Fisheries and Aquatic Sciences*, **72**, 1230–42.

Chapman, A. R. O. (1984). Reproduction, recruitment and mortality in two species of *Laminaria* in southwest Nova Scotia. *Journal of Experimental Marine Biology and Ecology*, **78**, 99–109.

Chapman, A. R. O. and Lindley, J. E. (1980). Seasonal growth of *Laminaria solidungula* in the Canadian high Arctic in relation to irradiance and dissolved nutrient concentrations. *Marine Biology*, **57**, 1–5.

Chapman, J. W., Carlton, J. T., Bellinger, M. R. and Blakeslee, A. M. H. (2007). Premature refutation of a human-mediated marine species introduction: the case history of the marine snail *Littorina littorea* in the Northwestern Atlantic. *Biological Invasions*, **9**, 995–1008.

Cobb, J. S. and Castro, K. M. (2006). *Shell Disease in Lobsters: A Synthesis*. New England Lobster Research Initiative, Narragansett, RI, and Rhode Island Sea Grant, Narragansett.

Conlan, K. E., Lenihan, H. S., Kvitek, R. G. and Oliver, J. S. (1998). Ice scour disturbance to benthic communities in the Canadian high Arctic. *Marine Ecology Progress Series*, **166**, 1–16.

Cook, R., Fariñas-Franco, J. M., Gell, F. R. et al. (2013). The substantial first impact of bottom fishing on rare biodiversity hotspots: A dilemma for evidence-based conservation. *PLoS ONE*, **8**, e69904.

Davidson, A. D., Campbell, M. L. and Hewitt, C. L. (2015). Assessing the impacts of nonindigenous marine macroalgae: an update of current knowledge. *Botanica Marina*, **52**, 55–79.

Dawson, J. W. (1867). The food of the common sea-urchin. *The American Naturalist*, **1**, 124–5.

Dijkstra, J., Harris, L. G. and Westerman, E. (2007). Distribution and long-term temporal patterns of four invasive colonial ascidians in the Gulf of Maine. *Journal of Experimental Marine Biology and Ecology*, **342**, 61–8.

Dijkstra, J. A. and Harris, L. G. (2009). Maintenance of diversity altered by a shift in dominant species: implications for species coexistence. *Marine Ecology Progress Series*, **387**, 71–80.

Doney, S. C., Ruckelshaus, M., Duffy, J. E. et al. (2012). Climate change impacts on marine ecosystems. *Annual Review of Marine Science*, **4**, 11–37.

Drolet, D., Himmelman, J. H. and Rochette, R. (2004). Use of refuges by the ophiuroid *Ophiopholis aculeata*: contrasting effects of substratum complexity on predation risk from two predators. *Marine Ecology Progress Series*, **284**, 173–83.

Dumont, C. P., Himmelman, J. H. and Russell, M. P. (2004a). Sea urchin mass mortality associated with algal debris from ice scour. In *Proceedings of the 11th International Echinoderm Conference*, 177–82.

Dumont, C. P., Himmelman, J. H. and Russell, M. P. (2004b). Size-specific movement of green sea urchins *Strongylocentrotus droebachiensis* on urchin barrens in eastern Canada. *Marine Ecology Progress Series*, **276**, 93–101.

Dumont, C. P., Himmelman, J. H. and Russell, M. P. (2006). Daily movement of the sea urchin *Strongylocentrotus droebachiensis* in different subtidal habitats in eastern Canada. *Marine Ecology Progress Series*, **317**, 87–99.

Elner, R. W. and Vadas Sr, R. L. (1990). Inference in ecology: the sea urchin phenomenon in the northwestern Atlantic. *The American Naturalist*, **136**, 108–25.

Epifanio, C. E. (2013). Invasion biology of the Asian shore crab *Hemigrapsus sanguineus*: a review. *Journal of Experimental Marine Biology and Ecology*, **441**, 33–49.

Estes, J. A. and Duggins, D. O. (1995). Sea otters and kelp forests in Alaska: generality and variation in a community ecological paradigm. *Ecological Monographs*, **65**, 75–100.

Estes, J. A., Terborgh, J., Brashares, J. S. et al. (2011). Trophic downgrading of planet Earth. *Science*, **333**, 301–6.

Fauchald, P. (2010). Predator-prey reversal: a possible mechanism for ecosystem hysteresis in the North Sea. *Ecology*, **91**, 2191–7.

Feehan, C. J., Francis, F. T. Y. and Scheibling, R. E. (2014). Harbouring the enemy: kelp holdfasts protect juvenile sea urchins from predatory crabs. *Marine Ecology Progress Series*, **514**, 149–61.

Feehan, C. J. and Scheibling, R. E. (2014a). Disease as a control of sea urchin populations in Nova Scotian kelp beds. *Marine Ecology Progress Series*, **500**, 149–58.

Feehan, C. J. and Scheibling, R. E. (2014b). Effects of sea urchin disease on coastal marine ecosystems. *Marine Biology*, **161**, 1467–85.

Feehan, C. J., Scheibling, R. E., Brown, M. S. and Thompson, K. R. (2016). Marine epizootics linked to storms: mechanisms of pathogen introduction and persistence inferred from coupled physical and biological time-series. *Limnology and Oceanography*, **61**, 316–29.

Feehan, C. J., Scheibling, R. E. and Lauzon-Guay, J.-S. (2012a). Aggregative feeding behavior in sea urchins leads to destructive grazing in a Nova Scotian kelp bed. *Marine Ecology Progress Series*, **444**, 69–83.

Feehan, C. J., Scheibling, R. E. and Lauzon-Guay, J.-S. (2012b). An outbreak of sea urchin disease associated with a recent hurricane: support for the "killer storm hypothesis" on a local scale. *Journal of Experimental Marine Biology and Ecology*, **413**, 159–68.

Filbee-Dexter, K., Feehan, C. J. and Scheibling, R. E. (2016). Large-scale degradation of a kelp ecosystem in an ocean warming hotspot. *Marine Ecology Progress Series*, **543**, 141–52.

Filbee-Dexter, K. and Scheibling, R. E. (2012). Hurricane-mediated defoliation of kelp beds and pulsed delivery of kelp detritus to offshore sedimentary habitats. *Marine Ecology Progress Series*, **455**, 51–64.

Filbee-Dexter, K. and Scheibling, R. E. (2014a). Sea urchin barrens as alternative stable states of collapsed kelp ecosystems. *Marine Ecology Progress Series*, **495**, 1–25.

Filbee-Dexter, K. and Scheibling, R. E. (2014b). Detrital kelp subsidy supports high reproductive condition of deep-living sea urchins in a sedimentary basin. *Aquatic Biology*, **23**, 71–86.

Filbee-Dexter, K. and Scheibling, R.E. (2016). Spatial patterns and predictors of drift algal subsidies in deep subtidal environments. *Estuaries and Coasts*, **39**, 1724–34.

Filbee-Dexter, K. and Scheibling, R. E. (2017). The present is the key to the past: linking regime shifts in kelp beds to the distribution of deep-living sea urchins. *Ecology*, **98**, 253–64.

Fisheries and Oceans Canada. (2009). *Assessment of Jonah Crab in Lobster Fishing Area 41 (4X + 5Zc)*. Canadian Science Advisory Secretariat, Ottawa, Science Advisory Report, 2009/034.

Fisheries and Oceans Canada. (2015). *Coral and Sponge Conservation Strategy for Eastern Canada 2015*. Canadian Science Advisory Secretariat, Ottawa.

Fisheries and Oceans Canada. (2016). *Update of the Fishery Indicators for Rock Crab (Cancer irroratus) in the Southern Gulf of St. Lawrence*. Canadian Science Advisory Secretariat, Ottawa, Secretariat Science Response, 2016/053.

Flato, G. M. and Brown, R. D. (2007) Variability and climate sensitivity of landfast Arctic sea ice. *Journal of Geophysical Research*, **101**, 25,767-25,777.

Foster, M. S. (2001). Rhodoliths: between rocks and soft places. *Journal of Phycology*, **37**, 659–67.

Francis, F. T., Filbee-Dexter, K. and Scheibling, R. E. (2014). Stalked tunicates *Boltenia ovifera* form biogenic habitat in the rocky subtidal zone of Nova Scotia. *Marine Biology*, **161**, 1375–83.

Frank, K. T., Petrie, B., Choi, J. S. and Leggett, W.C. (2005). Trophic cascades in a formerly cod-dominated ecosystem. *Science*, **308**, 1621–3.

Frey, D. L. and Gagnon, P. (2015). Thermal and hydrodynamic environments mediate individual and aggregative feeding of a functionally important omnivore in reef communities. *PLoS ONE*, **10**, e0118583.

Gagnon, P., Himmelman, J. H. and Johnson, L. E. (2003). Algal colonization in urchin barrens: defense by association during recruitment of the brown alga *Agarum cribrosum*. *Journal of Experimental Marine Biology and Ecology*, **290**, 179–96.

Gagnon, P., Himmelman, J. H. and Johnson, L. E. (2004). Temporal variation in community interfaces: kelp-bed boundary dynamics adjacent to persistent urchin barrens. *Marine Biology*, **144**, 1191–203.

Gagnon, P., Johnson, L. E. and Himmelman, J. H. (2005). Kelp patch dynamics in the face of intense herbivory: stability of *Agarum clathratum* (Phaeophyta) stands and associated flora on urchin barrens. *Journal of Phycology*, **41**, 498–505.

Gagnon, P., Matheson, K. and Stapleton, M. (2012). Variation in rhodolith morphology and biogenic potential of newly discovered rhodolith beds in

Newfoundland and Labrador (Canada). *Botanica Marina*, **55**, 85–99.

Gagnon, P., St-Hilaire-Gravel, L. V., Himmelman, J. H. and Johnson, L. E. (2006). Organismal defenses versus environmentally mediated protection from herbivores: unraveling the puzzling case of *Desmarestia viridis* (Phaeophyta). *Journal of Experimental Marine Biology and Ecology*, **334**, 10–19.

Garbary, D. J., Miller, A. G., Williams, J. and Seymour, N. R. (2014). Drastic decline of an extensive eelgrass bed in Nova Scotia due to the activity of the invasive green crab (*Carcinus maenas*). *Marine Biology*, **161**, 3–15.

Gaymer, C. F., Himmelman, J. H. and Johnson, L. E. (2001a). Distribution and feeding ecology of the seastars *Leptasterias polaris* and *Asterias vulgaris* in the northern Gulf of St Lawrence, Canada. *Journal of the Marine Biological Association UK*, **81**, 827–43.

Gaymer, C. F., Himmelman, J. H. and Johnson, L. E. (2001b). Use of prey resources by the seastars *Leptasterias polaris* and *Asterias vulgaris*: a comparison between field observations and laboratory experiments. *Journal of Experimental Marine Biology and Ecology*, **262**, 13–30.

Genner, M. J., Sims, D. W., Southward, A. J. et al. (2010). Body size-dependent responses of a marine fish assemblage to climate change and fishing over a century-long scale. *Global Change Biology*, **16**, 517–27.

Genovese, S. J. and Witman, J. D. (1999). Interactive effects of flow speed and particle concentration on growth rates of an active suspension feeder. *Limnology and Oceanography*, **44**, 1120–31.

Gerard, V. A. (1997) The role of nitrogen nutrition in high-temperature tolerance of the kelp, *Laminaria saccharina* (Chromophyta). *Journal of Phycology*. **33**, 800–10.

Gilkinson, K. and Edinger, E. (2009). The ecology of deep-sea corals of Newfoundland and Labrador waters: biogeography, life history, biogeochemistry and relation to fishes. *Canadian Technical Report of Fisheries and Aquatic Sciences*, no. 2830.

Glenn, R. P. and Pugh, T .L. (2006). Epizootic shell disease in American lobster (*Homarus americanus*) in Massachusetts coastal waters: interactions of temperature, maturity, and intermolt duration. *Journal of Crustacean Biology*, **26**, 639–45.

Glude, J. B. (1955). The effects of temperature and predators on the abundance of the soft-shell clam, *Mya arenaria*, in New England. *Transactions of the American Fisheries Society*, **84**, 13–26.

Govindarajan, A. F., Bukša, F., Bockrath, K., Wares, J. P. and Pineda, J. (2015). Phylogeographic structure and northward range expansion in the barnacle *Chthamalus fragilis*. *PeerJ*, **3**, 3926.

Grabowski, J. H., Clesceri, E. J., Baukus, A. J. et al. (2010). Use of herring bait to farm lobsters in the Gulf of Maine. *PLoS ONE*, **5**, e10188.

Gutt, J. (2001). On the direct impact of ice on marine benthic communities: a review. *Polar Biology*, **24**, 553–64.

Hammill, M. O., Stenson, G. B., Buren, A. D. and Koen-Alonso, M. (2014). Feeding by grey seals on endangered stocks of Atlantic cod and white hake. *ICES Journal of Marine Science*, **71**, 1332–41.

Harding, A. P. C. and Scheibling, R. E. (2015). Feed or flee: effect of a predation-risk cue on sea urchin foraging activity. *Journal of Experimental Marine Biology and Ecology*, **466**, 59–69.

Harley, C. D. G., Hughes, A. R., Hultgren, K. M. et al. (2006). The impacts of climate change in coastal marine systems. *Ecology Letters*, **9**, 228–41.

Harris, L. G. and Tyrrell, M. C. (2001). Changing community states in the Gulf of Maine: synergism between invaders, overfishing and climate change. *Biological Invasions*, **3**, 9–21.

Harrold, C. and Reed, D. C. (1985). Food availability, sea urchin grazing, and kelp forest community structure. *Ecology*, **66**, 1160–9.

Hart, M. W. and Scheibling, R. E. (1988). Heat waves, baby booms, and the destruction of kelp beds by sea urchins. *Marine Biology*, **99**, 167–76.

Hawkins, S. J., Moore, P. J., Burrows, M. T. et al. (2008). Complex interactions in a rapidly changing world: responses of rocky shore communities to recent climate change. *Climate Research*, **37**, 123–33.

Hawkins, S. J., Sugden, H. E., Mieszkowska, N. et al. (2009). Consequences of climate-driven biodiversity changes for ecosystem functioning of North European rocky shores. *Marine Ecology Progress Series*, **396**, 245–59.

Himmelman, J. H. (1984). Urchin feeding and macroalgal distribution in Newfoundland, eastern Canada. *Le Naturaliste canadien*, **111**, 337–48.

Himmelman, J. H. (1986). Population biology of green sea urchins on rocky barrens. *Marine Ecology Progress Series*, **33**, 295–306.

Himmelman, J. H. (1991). Diving observations of subtidal communities in the northern Gulf of St. Lawrence. *Canadian Special Publications of Fisheries and Aquatic Science*, **113**, 319–32.

Himmelman, J. H., Cardinal, A. and Bourget, E. (1983a). Community development following removal of urchins, *Strongylocentrotus droebachiensis*, from the rocky subtidal zone of the St. Lawrence Estuary, eastern Canada. *Oecologia*, **59**, 27–39.

Himmelman, J. H., Guderley, H., Vignault, G., Droliin, D. G. and Wells, P. G. (1984). Response of the sea urchin, *Strongylocentrotus droebachiensis*, to reduced salinities: importance of size, acclimation, and inter-population differences. *Canadian Journal of Zoology*, **62**, 1015–21.

Himmelman, J. H. and Lavergne, Y. (1985). Organization of rocky subtidal communities in the St. Lawrence Estuary. *Le Naturaliste Canadien*, **112**, 143–54.

Himmelman, J. H., Lavergne, Y., Axelsen, F., Cardinal, A. and Bourget, E. (1983b). Sea urchins in the Saint Lawrence Estuary: their abundance, size-structure, and suitability for commercial exploitation. *Canadian Journal of Fisheries and Aquatic Sciences*, **40**, 474–86.

Himmelman, J. H. and Nédélec, H. (1990). Urchin foraging and algal survival strategies in intensely grazed communities in eastern Canada. *Canadian Journal of Fisheries and Aquatic Sciences*, **47**, 1011–26.

Himmelman, J. H. and Steele, D. H. (1971). Food and predators of the green sea urchin *Strongylocentrotus droebachiensis* in Newfoundland waters. *Marine Biology*, **9**, 315–22.

Hoegh-Guldberg, O. and Bruno, J. F. (2010). The impact of climate change on the world's marine ecosystems. *Science*, **328**, 1523–8.

Humm, H. J. (1969). Distribution of marine algae along the Atlantic coast of North America. *Phycologia*, **7**, 43–53.

Hurd, C. L. (2000). Water motion, marine macroalgal physiology, and production. *Journal of Phycology*, **36**, 453–72.

Jackson, J. B. C., Kirby, M. X., Berger, W. H. et al. (2001). Historical overfishing and the recent collapse of coastal ecosystems. *Science*, **293**, 629–37.

Jellett, J. F. and Scheibling, R. E. (1988). Effect of temperature and prey availability on growth of *Paramoeba invadens* in monoxenic culture. *Applied and Environmental Microbiology*, **54**, 1848–54.

Jenkins, S. R., Coleman, R. A., Della Santina, P. et al. (2005). Regional scale differences in the determinism of grazing effects in the rocky intertidal. *Marine Ecology Progress Series*, **287**, 77–86.

Jenkins, S. R., Moore, P., Burrows, M. T. et al. (2008). Comparative ecology of north Atlantic shores: do differences in players matter for process? *Ecology*, **89**, S3–S23.

Jennings, L. B. and Hunt, H. L. (2010). Settlement, recruitment and potential predators and competitors of juvenile echinoderms in the rocky subtidal zone. *Marine Biology*, **157**, 307–16.

Jennings, L. B. and Hunt, H. L. (2014). Spatial patterns in early post-settlement processes of the green sea urchin *Strongylocentrotus droebachiensis*. *Marine Ecology Progress Series*, **502**, 219–28.

Jensen, G. C., McDonald, P. S. and Armstrong, D. A. (2002). East meets west: Competitive interactions between green crab *Carcinus maenas*, and native and introduced shore crab *Hemigrapsus* spp. *Marine Ecology Progress Series*, **225**, 251–62.

Johnson, L. E. (2007). Ice Scour. In M. W. Denny and S. D. Gaines, eds. *Encyclopedia of Tidepools and Rocky Shores*. University of California Press, Berkeley, pp. 289–91.

Johnson, L. E., Brawley, S. H. and Adey, W. H. (2012). Secondary spread of invasive species: historic patterns and underlying mechanisms of the continuing invasion of the European rockweed *Fucus serratus* in eastern North America. *Biological Invasions*, **14**, 79–97.

Jones, P. L. and Shulman, M. J. (2008). Subtidal-intertidal trophic links: American lobsters [*Homarus americanus* (Milne-Edwards)] forage in the intertidal zone on nocturnal high tides. *Journal of Experimental Marine Biology and Ecology*, **361**, 98–103.

Keats, D. W., South, G. R. and Steele, D. H. (1985). Algal biomass and diversity in the upper subtidal at a pack-ice disturbed site in eastern Newfoundland. *Marine Ecology Progress Series*, **25**, 151–8.

Kelly, J. R., Krumhansl, K. A. and Scheibling, R. E. (2012). Drift algal subsidies to sea urchins in low-productivity habitats. *Marine Ecology Progress Series*, **452**, 145–57.

Kelly, J. R., Scheibling, R. E. and Balch, T. (2011). Invasion-mediated shifts in the macrobenthic assemblage of a rocky subtidal ecosystem. *Marine Ecology Progress Series*, **437**, 69–78.

Kenchington, E. (2014). A general overview of benthic ecological or biological significant areas (EBSAs) in maritimes region. *Canadian Technical Report of Fisheries and Aquatic Sciences*, **3072**, iv, 45.

Kenneth, F., Petrie, B., Fisher, J. A. D. and Leggett, W. C. (2011). Transient dynamics of an altered large marine ecosystem. *Nature*, **477**, 86–9.

Konar, B. and Estes, J. A. (2003). The stability of boundary regions between kelp beds and deforested areas. *Ecology*, **84**, 174–85.

Krause-Jensen, D., Marba, N., Olesen, B. et al. (2012). Seasonal sea ice cover as principal driver of spatial and temporal variation in depth extension and annual production of kelp in Greenland. *Global Change Biology*, **18**, 2981–94.

Krumhansl, K. A., Lee, J. M. and Scheibling, R. E. (2011). Grazing damage and encrustation by an invasive bryozoan reduce the ability of kelps to withstand breakage by waves. *Journal of Experimental Marine Biology and Ecology*, **407**, 12–18.

Krumhansl, K. A., Okamoto, D. K., Rassweiler, A. et al. (2016). Global patterns of kelp forest change over the past half-century. *Proceedings of the National Academy of Science*, **113**, 13785–90.

Krumhansl, K. A. and Scheibling, R. E. (2011). Detrital production in Nova Scotian kelp beds: patterns and processes. *Marine Ecology Progress Series*, **421**, 67–82.

Krumhansl, K. A. and Scheibling, R. E. (2012). Production and fate of kelp detritus. *Marine Ecology Progress Series*, **467**, 281–302.

Lambert, D. M. and Harris, L. G. (2000). Larval settlement of the green sea urchin, *Strongylocentrotus droebachiensis*, in the southern Gulf of Maine. *Invertebrate Biology*, **119**, 403–9.

Lambert, W. J., Levin, P. S. and Berman, J. (1992). Changes in the structure of a New England (USA) kelp bed: the effects of an introduced species? *Marine Ecology Progress Series*, **88**, 303–7.

Lang, C. and Mann, K. H. (1976). Changes in sea urchin populations after the destruction of kelp beds. *Marine Biology*, **36**, 321–6.

Laur, D. R., Ebeling, A. W. and Reed, D. C. (1986). Experimental evaluations of substrate types as barriers to sea urchin (*Strongylocentrotus* spp.) movement. *Marine Biology*, **93**, 209–15.

Lauzon-Guay, J.-S. and Scheibling, R. E. (2007). Seasonal variation in movement, aggregation and destructive grazing of the green sea urchin (*Strongylocentrotus droebachiensis*) in relation to wave action and sea temperature. *Marine Biology*, **151**, 2109–18.

Lauzon-Guay, J.-S., Scheibling, R. E. and Barbeau, M. A. (2006). Movement patterns in the green sea urchin, *Strongylocentrotus droebachiensis*. *Journal of the Marine Biological Association of the UK*, **86**, 167–74.

Lauzon-Guay, J. S., Scheibling, R. E. and Barbeau, M. A. (2009). Modelling phase shifts in a rocky subtidal ecosystem. *Marine Ecology Progress Series*, **375**, 25–39.

Lawrence, J. M. (1975). On the relationships between marine plants and sea urchins. *Oceanography and Marine Biology: An Annual Review*, **13**, 213–86.

LeGault, K. N. and Hunt, H. L. (2016). Cannibalism among green sea urchins *Strongylocentrotus droebachiensis* in the laboratory and field. *Marine Ecology Progress Series*, **542**, 1–12.

Leichter, J. J. and Witman, J. D. (1997). Water flow over subtidal rock walls: relation to distributions and growth rates of sessile suspension feeders in the Gulf of Maine: water flow and growth rates. *Journal of Experimental Marine Biology and Ecology*, **209**, 293–307.

Levin, P. S., Coyer, J. A., Petrik, R. and Good, T. P. (2002). Community-wide effects of nonindigenous species on temperate rocky reefs. *Ecology*, **83**, 3182–93.

Lima, F. P., Ribeiro, P. A., Queiroz, N., Hawkins, S. J. and Santos, A. M. (2007). Do distributional shifts of northern and southern species of algae match the warming pattern? *Global Change Biology*, **13**, 2592–604.

Ling, S. D., Scheibling, R. E., Rassweiler, A. et al. (2015). Global regime shift dynamics of catastrophic sea urchin overgrazing. *Philosophical Transactions of the Royal Society B: Biological Sciences*, **370**, 20130269.

Lotze, H. and Milewski, I. (2004). Two centuries of multiple human impacts and successive changes in a North Atlantic food web. *Ecological Applications*, **14**, 1428–47.

Lyons, D. A. and Scheibling, R. E. (2009). Range expansion by invasive marine algae: rates and patterns of spread at a regional scale. *Diversity and Distributions*, **15**, 762–75.

Ma, K. C. K., Deibel, D., Law, K. K. M. et al. (2017). Richness and zoogeography of ascidians (Tunicata: Ascidiacea) in eastern Canada. *Canadian Journal of Zoology*, **95**, 51–9.

MacFarlane, C. (1952). A survey of certain seaweeds of commercial importance in southwest Nova Scotia. *Canadian Journal of Botany*, **30**, 78–97.

MacGregor, K. A. (2019). Individual- and population-level responses to the environment: Environmental modification of movement behaviour in the green sea urchin, *Strongylocentrotus droebachiensis*. PhD, Laval University.

Mamelona, J. and Pelletier, M. (2005). Green urchin as a significant source of fecal particulate organic matter within nearshore benthic ecosystems. *Journal of Experimental Marine Biology and Ecology*, **314**, 163–74.

Mann, K. H. (1973). Seaweeds: their productivity and strategy for growth. *Science*, **182**, 975–81.

Mann, K. H. (1982). Kelp, sea urchins and predators: a review of strong interactions in rocky subtidal systems of eastern Canada, 1970–1980. *Netherlands Journal of Sea Research*, **16**, 414–23.

Mann, K. H. and Breen, P. A. (1972). The relation between lobster abundance, sea urchins, and kelp beds. *Journal of the Fisheries Research Board of Canada*, **29**, 603–5.

Mathieson, A. C. (2016) Rapid assessment survey of fouling and introduced seaweeds from southern Maine to Rhode Island. *Rhodora*, **118**, 113–47.

Mathieson, A. C., Dawes, C. J., Pederson, J., Gladych, R. A. and Carlton, J. T. (2008). The Asian red seaweed *Grateloupia turuturu* (Rhodophyta) invades the Gulf of Maine. *Biological Invasions*, **10**, 985–8.

Mathieson, A. C., Penniman, C. A. and Harris, L. G. (1991). Northwest Atlantic Rocky Shore Ecology. In

A. C. Mathieson and P. H. Neinhuis, eds. *Intertidal and Littoral Systems: Ecosystems of the World 24*. Elsevier, Amsterdam, pp. 109–91.

McCain, J. S. P., Cull, D. J., Schneider, D. C. and Lotze, H. K. (2016). Long-term shift in coastal fish communities before and after the collapse of Atlantic cod (*Gadus morhua*). *ICES Journal of Marine Science*, **73**, 1415–26.

McCook, L. J. and Chapman, A. R. O. (1997). Patterns and variations in natural succession following massive ice-scour of a rocky intertidal seashore. *Journal of Experimental Marine Biology and Ecology*, **214**, 121–47.

McDermott, J. (1998). The western Pacific brachyuran (*Hemigrapsus sanguineus*: Grapsidae), in its new habitat along the Atlantic coast of the United States: geographic distribution and ecology. *ICES Journal of Marine Science*, **55**, 289–98.

McKindsey, C. W., Lecuona, M., Huot, M. and Weise, A. M. (2009). Biodeposit production and benthic loading by farmed mussels and associated tunicate epifauna in Prince Edward Island. *Aquaculture*, **295**, 44–51.

Merzouk, A. and Johnson, L. E. (2011). Kelp distribution in the northwest Atlantic Ocean under a changing climate. *Journal of Experimental Marine Biology and Ecology*, **400**, 90–8.

Mieszkowska, N., Genner, M. J., Hawkins, S. J. and Sims, D. W. (2009). Effects of climate change and commercial fishing on Atlantic cod *Gadus morhua*. *Advances in Marine Biology*, **56**, 213–73.

Miller, B. (1994). Why are there so many American lobsters? *Lobster Newsletter*, **7**, 14–15.

Miller, R. J. (1985a). Seaweeds, sea urchins, and lobsters: a reappraisal. *Canadian Journal of Fisheries and Aquatic Sciences*, **42**, 2061–72.

Miller, R. J. (1985b). Succession in sea urchin and seaweed abundance in Nova Scotia, Canada. *Marine Biology*, **84**, 275–86.

Miller, R. J. and Etter, R. J. (2008). Shading facilitates sessile invertebrate dominance in the rocky subtidal Gulf of Maine. *Ecology*, **89**, 452–62.

Miller, R. J. and Etter, R. J. (2011). Rock walls: small-scale diversity hotspots in the subtidal Gulf of Maine. *Marine Ecology Progress Series*, **425**, 153–65.

Mills, K. E., Pershing, A. and Brown, C. (2013). Fisheries management in a changing climate: lessons from the 2012 ocean heat wave in the Northwest Atlantic. *Oceanography*, **26**, 191–5.

Moody, K. E. and Steneck, R. S. (1993). Mechanisms of predation among large decapod crustaceans of the Gulf of Maine coast: functional vs. phylogenetic patterns. *Journal of Experimental Marine Biology and Ecology*, **168**, 111–24.

Morse, B. L. and Hunt, H. L. (2013). Effect of unidirectional water currents on displacement behaviour of the green sea urchin *Strongylocentrotus droebachiensis*. *Journal of the Marine Biological Association of the UK*, **93**, 1923–8.

Nadon, M.-O. and Himmelman, J. H. (2006). Stable isotopes in subtidal food webs: have enriched carbon ratios in benthic consumers been misinterpreted? *Limnology and Oceanography*, **51**, 2828–36.

Narvaez, C. A. (2019). Green urchin demography in a subartic ecosystem: patterns and processes. PhD, Laval University.

Neckles, H. A. (2015). Loss of eelgrass in Casco Bay, Maine, linked to green crab disturbance. *Northeastern Naturalist*, **22**, 478–500.

Newton, C., Bracken, M. E. S., McConville, M., Rodrigue, K. and Thornber, C. S. (2013). Invasion of the red seaweed *Heterosiphonia japonica* spans biogeographic provinces in the western north Atlantic Ocean. *PLoS ONE*, **8**, e62261.

Novak, M. (2004). Diurnal activity in a group of Gulf of Maine decapods. *Crustaceana*, **77**, 603–20.

Nye, J. A., Link, J. S., Hare, J. A. and Overholtz, W. J. (2009). Changing spatial distribution of fish stocks in relation to climate and population size on the Northeast United States continental shelf. *Marine Ecology Progress Series*, **393**, 111–29.

O'Brien, J. M., Krumhansl, K. A. and Scheibling, R. E. (2013). Invasive bryozoan alters interaction between a native grazer and its algal food. *Journal of the Marine Biological Association of the UK*, **93**, 1393–400.

O'Brien, J. M., Scheibling, R. E. and Krumhansl, K. A. (2015). Positive feedback between large-scale disturbance and density-dependent grazing decreases resilience of a kelp bed ecosystem. *Marine Ecology Progress Series*, **522**, 1–13.

Occhipinti-Ambrogi, A. (2007). Global change and marine communities: alien species and climate change. *Marine Pollution Bulletin*, **55**, 342–52.

Palardy, J. E. and Witman, J. D. (2014). Flow, recruitment limitation, and the maintenance of diversity in marine benthic communities. *Ecology*, **95**, 286–97.

Palma, A. T. and Steneck, R. S. (2001). Does variable coloration in juvenile marine crabs reduce risk of visual predation? *Ecology*, **82**, 2961–7.

Palma, A. T., Steneck, R. S. and Wilson, C. J. (1999). Settlement-driven, multiscale demographic patterns of large benthic decapods in the Gulf of Maine. *Journal of Experimental Marine Biology and Ecology*, **241**, 107–36.

Palma, A. T., Wahle, R. A. and Steneck, R. S. (1998). Different early post-settlement strategies between

American lobsters *Homarus americanus* and rock crabs *Cancer irroratus* in the Gulf of Maine. *Marine Ecology Progress Series*, **162**, 215–25.

Pauly, D. (1998). Fishing down marine food webs. *Science*, **279**, 860–3.

Peterson, G. H. (1977). Biological Effects of Sea-Ice and Icebergs in Greenland. In M. L. Dunbar, ed. *Polar Oceans*. Arctic Institute of North America, Calgary, pp. 319–20.

Pinsky, M. L., Worm, B., Fogarty, M. J., Sarmiento, J. L. and Levin, S. A. (2013). Marine taxa track local climate velocities. *Science*, **341**, 1239–42.

Quinn, B. K. (2017). Threshold temperatures for performance and survival of American lobster larvae: a review of current knowledge and implications to modeling impacts of climate change. *Fisheries Research*, **186**, 383–96.

Quinn, B. K., Rochette, R., Ouellet, P. and Sainte-Marie, B. (2013). Effect of temperature on development rate of larvae from cold-water American lobster (*Homarus americanus*). *Journal of Crustacean Biology*, **33**, 527–36.

Ramsey, A., Davidson, J., Landry, T. and Arsenault, G. (2008). Process of invasiveness among exotic tunicates in Prince Edward Island, Canada. *Biological Invasions*, **10**, 1311–16.

Reed, J. K. (2004). *Deep-Water Coral Reefs of Florida, Georgia and South Carolina: A Summary of the Distribution, Habitat, and Associated Fauna*. South Atlantic Fishery Management Council, Fort Pierce, FL.

Reed, J. K. and Ross, S. W. (2005). Deep-water reefs off the southeastern US: recent discoveries and research. *Journal of Marine Education*, **21**, 33–7.

Richards, A. R. (1992). Habitat selection and predator avoidance: ontogenetic shifts in habitat use by the Jonah crab *Cancer borealis* (Stimpson). *Journal of Experimental Marine Biology and Ecology*, **156**, 187–97.

Robichaud, D. A. and Frail, C. (2006). Development of Jonah crab, *Cancer borealis*, and rock crab, *Cancer irroratus*, fisheries in the Bay of Fundy (LFAs 35-38) and off southwest Nova Scotia (LFA 34): from exploratory to commercial status (1995–2004). *Canadian Manuscript Report Fisheries and Aquatic Science*, 2775.

Roman, J. (2006). Diluting the founder effect: Cryptic invasions expand a marine invader's range. *Proceedings of the Royal Society B*, **273**, 2453–9.

Rondeau, A., Hanson, J. M. and Comeau, M. (2014). *Rock Crab,* Cancer irroratus, *Fishery and Stock Status in the Southern Gulf of St. Lawrence: LFA 23, 24, 25, 26A and 26B*. Canadian Science Advisory Secretariat, Ottawa, Research Document 2014/032, vi, 52.

Ruiz, G. M., Fofonoff, P. W., Carlton, J. T., Wonham, M. J. and Hines, A. H. (2000). Invasion of coastal marine communities in North America: apparent patterns, processes, and biases. *Annual Review of Ecology and Systematics*, **31**, 481–531.

Sainte-Marie, B. and Chabot, D. (2002). Ontogenetic shifts in natural diet during benthic stages of American lobster (*Homarus americanus*), off the Magdalen Islands. *Fisheries Bulletin*, **100**, 106–16.

Sauchyn, L. K., Lauzon-Guay, J. S. and Scheibling, R. E. (2011). Sea urchin fecal production and accumulation in a rocky subtidal ecosystem. *Aquatic Biology*, **13**, 215–23.

Sauchyn, L. K. and Scheibling, R. E. (2009a). Degradation of sea urchin feces in a rocky subtidal ecosystem: implications for nutrient cycling and energy flow. *Aquatic Biology*, **6**, 99–108.

Sauchyn, L. K. and Scheibling, R. E. (2009b). Fecal production by sea urchins in native and invaded algal beds. *Marine Ecology Progress Series*, **396**, 35–48.

Saunders, M. I. and Metaxas, A. (2009). Effects of temperature, size, and food on the growth of *Membranipora membranacea* in laboratory and field studies. *Marine Biology*, **156**, 2267–76.

Saunders, M. I., Metaxas, A. and Filgueira, R. (2010). Implications of warming temperatures for population outbreaks of a nonindigenous species (*Membranipora membranacea*, Bryozoa) in rocky subtidal ecosystems. *Limnology and Oceanography*, **55**, 1627–42.

Savenkoff, C., Castonguay, M., Chabot, D. et al. (2007a). Changes in the northern Gulf of St. Lawrence ecosystem estimated by inverse modelling: Evidence of a fishery-induced regime shift? *Estuarine, Coastal and Shelf Science*, **73**, 711–24.

Savenkoff, C., Swain, D. P., Hanson, J. M. et al. (2007b). Effects of fishing and predation in a heavily exploited ecosystem: Comparing periods before and after the collapse of groundfish in the southern Gulf of St. Lawrence (Canada). *Ecological Modelling*, **204**, 115–28.

Scheibling, R. E. (1984). Echinoids, epizootics and ecological stability in the rocky subtidal off Nova Scotia, Canada. *Helgolander Meeresuntersuchungen*, **37**, 233–42.

Scheibling, R. E. (1986). Increased macroalgal abundance following mass mortalities of sea urchins (*Strongylocentrotus droebachiensis*) along the Atlantic coast of Nova Scotia. *Oecologia*, **68**, 186–98.

Scheibling, R. E. (1996). The role of predation in regulating sea urchin populations in eastern Canada. *Oceanologica Acta*, **19**, 421–30.

Scheibling, R. E., Feehan, C. and Lauzon-Guay, J.-S. (2010). Disease outbreaks associated with recent hurricanes cause mass mortality of sea urchins in Nova Scotia. *Marine Ecology Progress Series*, **408**, 109–16.

Scheibling, R. E., Feehan, C. J. and Lauzon-Guay, J.-S. (2013). Climate Change, Disease and the Dynamics of a Kelp-Bed Ecosystem in Nova Scotia. In J. Fernandez-Palacios, L. de Nascimento, J. C. Hernandez et al., eds. *Climate Change Perspectives from the Atlantic: Past, Present and Future*. Servicio de Publicaciones, Universidad de La Laguna, Santa Cruz de Tenerife.

Scheibling, R. E. and Gagnon, P. (2006). Competitive interactions between the invasive green alga *Codium fragile* ssp. *tomentosoides* and native canopy-forming seaweeds in Nova Scotia (Canada). *Marine Ecology Progress Series*, **325**, 1–14.

Scheibling, R. E. and Gagnon, P. (2009). Temperature-mediated outbreak dynamics of the invasive bryozoan *Membranipora membranacea* in Nova Scotian kelp beds. *Marine Ecology Progress Series*, **390**, 1–13.

Scheibling, R. E. and Hamm, J. (1991). Interactions between sea urchins (*Strongylocentrotus droebachiensis*) and their predators in field and laboratory experiments. *Marine Biology*, **110**, 105–16.

Scheibling, R. E. and Hennigar, A. (1997). Recurrent outbreaks of disease in sea urchins *Strongylocentrotus droebachiensis* in Nova Scotia: evidence for a link with large-scale meteorologic and oceanographic events. *Marine Ecology Progress Series*, **152**, 155–65.

Scheibling, R. E., Hennigar, A. W. and Balch, T. (1999). Destructive grazing, epiphytism, and disease: the dynamics of sea urchin-kelp interactions in Nova Scotia. *Canadian Journal of Fisheries and Aquatic Sciences*, **56**, 2300–14.

Scheibling, R. E., Kelly, N. E. and Raymond, B. G. (2009a). Herbivory and community composition on a subtidal cobble bed. *Marine Ecology Progress Series*, **382**, 113–28.

Scheibling, R. E., Kelly, N. E. and Raymond, B. G. (2009b). Physical disturbance and community organization on a subtidal cobble bed. *Journal of Experimental Marine Biology and Ecology*, **368**, 94–100.

Scheibling, R. E. and Lauzon-Guay, J.-S. (2010). Killer storms: North Atlantic hurricanes and disease outbreaks in sea urchins. *Limnology and Oceanography*, **55**, 2331–8.

Scheibling, R. E. and Raymond, B. G. (1990). Community dynamics on a subtidal cobble bed following mass mortalities of sea urchins. *Marine Ecology Progress Series*, **63**, 127–45.

Scheibling, R. E. and Stephenson, R. L. (1984). Mass mortality of *Strongylocentrotus droebachiensis* (Echinodermata: Echinoidea) off Nova Scotia, Canada. *Marine Biology*, **78**, 153–64.

Schuster, M. and Konar, B. (2014). Foliose algal assemblages and deforested barren areas: phlorotannin content, sea urchin grazing and holdfast community structure in the Aleutian dragon kelp, *Eualaria fistulosa*. *Marine Biology*, **161**, 2319–32.

Scrosati, R. A. (2013). Patchy mussel dominance on ice-scoured rocky shores in Atlantic Canada. *Oceanarium*, **43**, 251–2.

Sebens, K. P. (1985). The ecology of the rocky subtidal zone: the subtidal rock surfaces in New England support a diversity of encrusting species that compete for space and that recolonize patches cleared through predation. *American Scientist*, **73**, 548–57.

Seebens, H., Gastner, M. T. and Blasius, B. (2013). The risk of marine bioinvasion caused by global shipping. *Ecology Letters*, **16**, 782–90.

Sephton, D., Vercaemer, B., Nicolas, J. M. and Keays, J. (2011). Monitoring for invasive tunicates in Nova Scotia, Canada (2006–2009). *Aquatic Invasions*, **6**, 391–403.

Shelton, A. O., Witman, D., Woodby, D. A. and Hebert, K. (2006). Evaluating age determination and spatial patterns of growth in red sea urchins in southeast Alaska. *Transactions of the American Fisheries Society*, **135**, 1670–80.

Shields, J. D. (2013). Complex etiologies of emerging diseases in lobsters (*Homarus americanus*) from Long Island Sound. *Canadian Journal of Fisheries and Aquatic Sciences*, **70**, 1576–87.

Simonson, E. J., Scheibling, R. E. and Metaxas, A. (2015). Kelp in hot water: I. Warming seawater temperature induces weakening and loss of kelp tissue. *Marine Ecology Progress Series*, **537**, 89–104.

Sindermann, C. (1990). *Principal Diseases of Marine Fish and Shellfish*, 2nd edn. Academic Press, San Diego, CA.

Smale, D. A., Brown, K. M., Barnes, D. K. A., Fraser, K. P. P. and Clarke, A. (2008). Ice scour disturbance in Antarctic waters. *Science*, **321**, 371.

Smale, D. A., Burrows, M. T., Moore, P., O'Connor, N. and Hawkins, S. J. (2013). Threats and knowledge gaps for ecosystem services provided by kelp forests: A northeast Atlantic perspective. *Ecology and Evolution*, **3**, 4016–38.

Smolowitz, R., Quinn, R. A., Cawthorn, R. J., Summerfield, R. L. and Chistoserdov, A. Y. (2014). Pathology of two forms of shell disease of the American lobster *Homarus americanus* Milne Edwards in Atlantic Canada. *Journal of Fish Diseases*, **37**, 577–81.

Squire, V. A. (2007). Of ocean waves and sea-ice revisited. *Cold Regions Science and Technology*, **49**, 110–13.

Squire, V. A., Dugan, J. P., Wadhams, P., Rottier, P. J. and Liu, A. K. (1995). Of ocean waves and sea-ice revisited. *Annual Review of Fluid Mechanics*, **27**, 115–68.

St-Pierre, A. P. and Gagnon, P. (2015a). Wave action and starvation modulate intra-annual variation in displacement, microhabitat selection, and ability to contact prey in the common sea star, *Asterias rubens* Linnaeus. *Journal of Experimental Marine Biology and Ecology*, **467**, 95–107.

St-Pierre, A. P. and Gagnon, P. (2015b). Effects of temperature, body size, and starvation on feeding in a major echinoderm predator. *Marine Biology*, **162**, 1125–35.

Steneck, R. S. (2006a). Is the American lobster, *Homarus americanus*, overfished? A review of overfishing with an ecologically based perspective. *Bulletin of Marine Science*, **78**, 607–32.

Steneck, R. S. (2006b). Possible demographic consequences of intraspecific shelter competition among American lobsters. *Journal of Crustacean Biology*, **26**, 628–38.

Steneck, R. S. and Dethier, M. N. (1994). A functional group approach to the structure of algal-dominated communites. *Oikos*, **69**, 476–98.

Steneck, R. S., Leland, A., Mcnaught, D. C. and Vavrinec, J. (2013). Ecosystem flips, locks, and feedbacks: the lasting effects of fisheries on Maine kelp forest ecosystem. *Bulletin of Marine Science*, **89**, 31–55.

Steneck, R. S., Vavrinec, J. and Leland, A. V. (2004). Accelerating trophic-level dysfunction in kelp forest ecosystems of the western north Atlantic. *Ecosystems*, **7**, 323–32.

Steneck, R. S. and Wahle, R. A. (2013). American lobster dynamics in a brave new ocean. *Canadian Journal of Fisheries and Aquatic Sciences*, **70**, 1612–24.

Sumi, C. B. T. and Scheibling, R. E. (2005). Role of grazing by sea urchins *Strongylocentrotus droebachiensis* in regulating the invasive alga *Codium fragile* ssp. *tomentosoides* in Nova Scotia. *Marine Ecology Progress Series*, **292**, 203–12.

Suskiewicz, T. S. and Johnson, L. E. (2017). Consumption rates of a key marine herbivore: a review of the extrinsic and intrinsic control of feeding in the green sea urchin. *Marine Biology*, **164**, 131–43.

Swenton, R., Borden, D., Busby, J. et al. (2014). *Jonah Crab Fishery: A Briefing for the Atlantic States Marine Fisheries Commission*. Atlantic States Marine Fisheries Commission, Arlington, VA.

Teagle, H., Hawkins, S. J., Moore, P. J. and Smale, D. A. (2017). The role of kelp species as biogenic habitat formers in coastal marine ecosystems. *Journal of Experimental Marine Biology and Ecology*, **492**, 81–98.

Tittensor, D. P., Mora, C., Jetz, W. et al. (2010). Global patterns and predictors of marine biodiversity across taxa. *Nature*, **466**, 1098–103.

Uribe, R. A., Ortiz, M., Macaya, E. C. and Pacheco, A. S. (2015). Successional patterns of hard-bottom macrobenthic communities at kelp bed (*Lessonia trabeculata*) and barren ground sublittoral systems. *Journal of Experimental Marine Biology and Ecology*, **472**, 180–8.

Vadas Sr., R. L. and Elner, R. W. (1992). Plant–Animal Interactions in the North-West Atlantic. In D. M. John, S. J. Hawkins and J. H. Price, eds. *Plant–Animal Interactions in the Marine Benthos; Systematics Association Special Volume No. 46*. The Systematics Association Special Volume No. 46, Clarendon Press, Oxford, pp. 33–60.

Vadas, R. L., Elner, R. W., Garwood, P. E. and Babb, I. G. (1986). Experimental evaluation of aggregation behavior in the sea urchin *Strongylocentrotus droebachiensis*. *Marine Biology*, **90**, 433–48.

Vadas, R. L. and Steneck, R. S. (1995). Overfishing and Inferences in Kelp–Sea Urchin Interactions. In H. R. Skjoldal, C. Hopkins, K. E. Erikstad and H. P. Leinaas, eds. *Ecology of Fjords and Coastal Waters*. Elsevier Science B.V., Amsterdam, pp. 509–24.

Vanderklift, M. A. and Wernberg, T. (2008). Detached kelps from distant sources are a food subsidy for sea urchins. *Oecologia*, **157**, 327–35.

Wahl, M. (2009). *Marine Hard Bottom Communities: Patterns, Dynamics, Diversity, and Change*. Springer, New York.

Wahle, R. A., Bergeron, C., Tremblay, J. et al. (2013). The geography and bathymetry of American lobster benthic recruitment as measured by diver-based suction sampling and passive collectors. *Marine Biology Research*, **9**, 42–58.

Wahle, R. A., Gibson, M. and Fogarty, M. (2009). Distinguishing disease impacts from larval supply effects in a lobster fishery collapse. *Marine Ecology Progress Series*, **376**, 185–92.

Wahle, R. A., Incze, L. S. and Fogarty, M. J. (2004). First projections of American lobster fishery recuitment using a settlement index and variable growth. *Bulletin of Marine Science*, **74**, 101–14.

Wahle, R. A., Dellinger, L., Olszewski, S. and Jekielek, P. (2015). American lobster nurseries of southern New England receding in the face of climate change. *ICES Journal of Marine Science*, **72** (Suppl. 1), i69–i78.

Wahle, R. A. and Steneck, R. S. (1992). Habitat restrictions in early benthic life: experiments on habitat selection and in situ predation with the American lobster. *Journal of Experimental Marine Biology and Ecology*, **157**, 91–114.

Walther G. R., Roques A., Hulme P. E. et al. (2009). Alien species in a warmer world: risks and opportunities. *Trends in Ecology & Evolution*, **24**, 686–93.

Watanabe, S., Scheibling, R. E. and Metaxas, A. (2009). Contrasting patterns of spread in interacting invasive species: *Membranipora membranacea* and *Codium fragile* off Nova Scotia. *Biological Invasions*, **12**, 2329–42.

Welch, W. R. (1968). Changes in abundance of the green crab, *Carcinus maenas* (L.), in relation to recent temperature changes. *Fishery Bulletin*, **67**, 337–45.

Wells, R. J. D., Steneck, R. S. and Palma, A. T. (2010). Three-dimensional resource partitioning between American lobster (*Homarus americanus*) and rock crab (*Cancer irroratus*) in a subtidal kelp forest. *Journal of Experimental Marine Biology and Ecology*, **384**, 1–6.

Wharton, W. G. and Mann, K. H. (1981). Relationship between destructive grazing by the sea urchin, *Strongylocentrotus droebachiensis*, and the abundance of American lobster, *Homarus americanus*, on the Atlantic coast of Nova Scotis. *Canadian Journal of Fisheries and Aquatic Sciences*, **38**, 1339–49.

Witman, J. D. (1987). Coexistence: Storms, grazing, mutualism, and the zonation of kelps and mussels. *Ecological Monographs*, **57**, 167–87.

Witman, J. D. and Sebens, K.P. (1992). Regional variation in fish predation intensity: a historical perspective in the Gulf of Maine. *Oecologia*, **90**, 305–15.

Worm, B. and Myers, R. A. (2003). Meta-Analysis of cod-shrimp interactions reveals top-down control in oceanic food webs. *Ecology*, **84**, 162–73.

Wyeth, R. C., Woodward, O. M. and Willows, A. O. D. (2006). Orientation and navigation relative to water flow, prey, conspecifics, and predators by the nudibranch mollusc *Tritonia diomedea*. *Biological Bulletin*, **210**, 97–108.

Zenkevitch, L. (1963). The Sea of Okhotsk. In *Biology of the Seas of the U.S.S.R.* Translated by S. Botcharsyava. George Allen and Unwin Ltd, London, pp. 783–817.

Zhao, J.-P., Li, T., Barber, D. et al. (2010). Attenuation of lateral propagating light in sea ice measured with an artificial lamp in winter Arctic. *Cold Regions Science and Technology*, **61**, 6–12.

# Chapter 6

# Shallow Water Muddy Sands of the North-West Atlantic Ocean

## *Latitudinal Patterns in Interactions and Processes*

Sarah A. Woodin, Susan S. Bell, Jon Grant, Paul V. R. Snelgrove and David S. Wethey

This chapter summarises current knowledge of processes that determine patterns of distribution and abundance of macroinfauna from the southern tip of Florida in the USA to Newfoundland in Canada, ~25°N–52°N. We focus on intertidal and shallow subtidal (~5 m depth) muddy sands and sandy muds because the majority of experimental data for the northwestern Atlantic coast are from those habitats. The general theme is latitudinal change in processes, mechanisms and patterns and their biogeographic distribution. We focus especially on large-scale patterns across latitudes because these are most likely to alter as climate changes intensify. We describe the following major biogeographic patterns that appear to be supported in the literature: (1) large disturbance predators, so clearly important in the mid-Atlantic and south along the northwestern Atlantic coast, appear much less so north of Cape Cod, where thermal regimes drive their disappearance; (2) important, large digging predators from Delaware Bay (39.25°N) southwards dramatically reduce densities of infauna, often by an order of magnitude or more, generally reducing the importance of strong competitive interactions; (3) the presence of refugia from disturbance, such as rooted plants, drives spatial patterns in this region; (4) reduced water clarity from ~31°N to 33.8°N yields a section of coastline depauperate in rooted plants such as *Halodule* and *Zostera,* and particularly in the south and mid-Atlantic, rising seawater temperatures strongly affect the extent and diversity of rooted plants; (5) dramatic changes in average tidal regimes from south to north result in much greater exposure to aerial conditions in the north, magnifying the latitudinal change in surface seawater temperature (SST); (6) presence of ice cover in the north and (7) the Boston–Washington, DC megalopolis accentuates human signatures expressed as eutrophication between 36.5°N and 42.6°N. Finally, based upon predicted climate conditions for the twenty-first century, we discuss potential changes in these seven patterns across latitudes that may alter processes that determine macroinfaunal abundance and distribution.

## 6.1 The Physical Landscape

A wide diversity of coastal geomorphology, climate regimes, benthic substrates and magnitude of human influence characterise the northwestern Atlantic coastlines of Canada and the USA (see descriptions in Dame et al., 2000; Roman et al., 2000; Bricker et al., 2008). We summarise

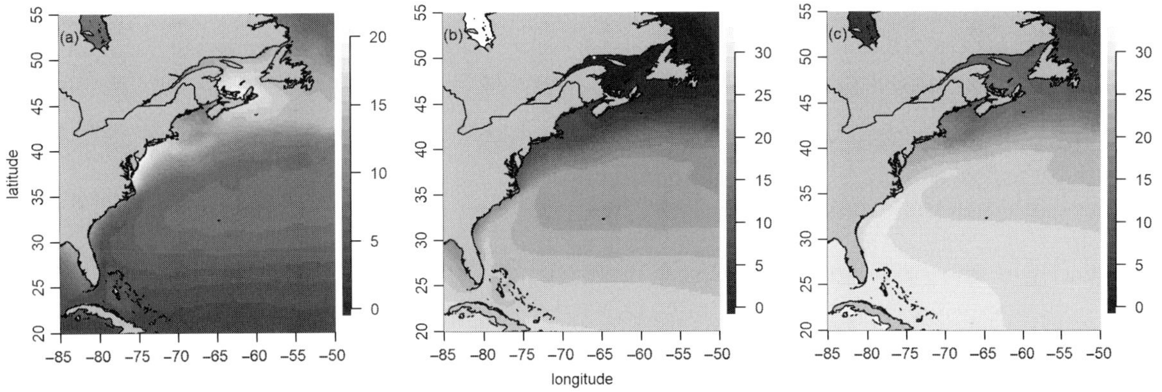


**Fig. 6.1** Mean sea surface temperature 1982–2011. (a) Yearly range; (b) winter mean; (c) summer mean. Data from Banzon et al., 2014. Source: ftp://eclipse.ncdc.noaa.gov/pub/OI-daily-v2/climatology/. *A black and white version of this figure will appear in some formats. For the colour version, please refer to the plate section.*

this information because these features define both the physical context and potential drivers of biotic patterns.

### 6.1.1 Thermal Patterns

The thermal gradients south to north are relatively linear (Shearman and Lentz, 2010), with dramatic seasonal changes, characteristic of western sides of ocean basins (Jenkins et al., 2008). The warm Gulf Stream, originating in the Gulf of Mexico, strongly influences the southern coast, hugging the coastline from 25°N to 35°N. Cooler temperatures occur northwards as the Gulf Stream moves further offshore north of Cape Hatteras (Figures 6.1 and 6.2). Offshore, the Gulf Stream meets with the Gulf of Saint Lawrence (GSL) outflow and Labrador Current (Figure 6.2). The cold Labrador Current, which originates between Labrador and Greenland, reinforces the latitudinal trend in temperatures as it moves southwards. The annual SST range for coastal habitats in the GSL exceeds 15°C, in contrast to the 10–15°C range in the Gulf of Maine (GOM; Figure 6.1a) and an annual range south of Cape Hatteras (35°N) of only 8–10°C. The sharp thermal discontinuity at Cape Hatteras (35°N) in winter reflects the influence of the Gulf Stream (Figure 6.1b), which shifts to near Cape Cod in summer (42°N) (Figure 6.1c). This discontinuity drives the large annual SST range (>15°C) of the coast of the Mid Atlantic Bight between the two capes.

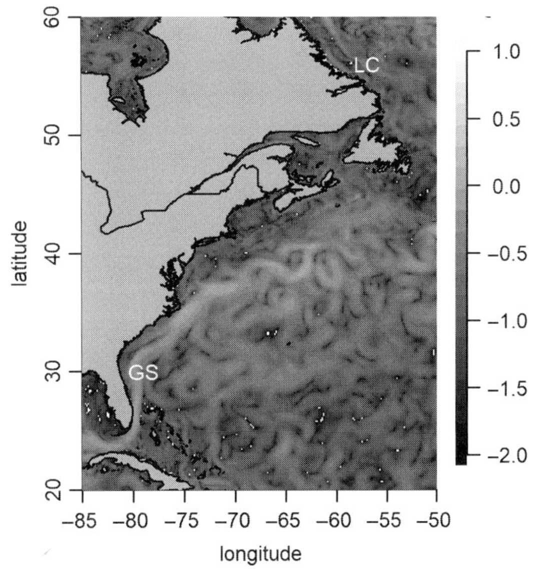


**Fig. 6.2** Surface currents in the north-west Atlantic. The cold-water Labrador Current (LC) flows south and the warm-water Gulf Stream (GS) flows north. Colours are $\log_{10}$ (velocities) (m $s^{-1}$) on January 1, 2012. Source: NCOM Global (Martin et al., 2009). *A black and white version of this figure will appear in some formats. For the colour version, please refer to the plate section.*

### 6.1.2 Currents

The pattern of currents, in conjunction with temperature, has significance for benthic organisms both in terms of discontinuities created by two opposing currents as well as in directionality. The northwards moving Gulf Stream and the southwards current of the Mid Atlantic Bight both deflect off Cape Hatteras, forming the Hatteras Front, a semi-permeable boundary to larvae of benthic organisms (Pappalardo et al., 2015). This front coincides with a change in thermal regime that apparently limits ranges of northern species more than those of southern species (Pappalardo et al., 2015). GOM currents are strongly influenced by the cold Labrador Current; they deflect offshore near Cape Cod, with limited flow around the tip of Cape Cod and south-westwards (Jennings et al., 2009). The much warmer waters of the Mid Atlantic Bight create the thermal discontinuity at Cape Cod. The currents in the GOM and the Mid Atlantic Bight generally flow southwards with deflections at the capes. Propagules of benthic species therefore generally move northwards from Cape Canaveral in Florida to the Hatteras Front where offshore deflection occurs. Noting the directions and deflections summarised earlier, some pockets of retention occur in the Bay of Fundy, Georges Bank and Nantucket Shoals as a result of glacial bank features. More northern range limits of macroinfauna occur near Cape Cod than southern limits, and northern limits occur more frequently at Cape Cod than at Cape Hatteras, perhaps reflecting the importance of the summer versus winter thermal discontinuity or the thermal range of the Mid Atlantic Bight (Hutchins, 1947; Pappalardo et al., 2015; Figure 6.1a).

### 6.1.3 Tides

Tidal amplitudes increase latitudinally, culminating in the largest tides in the world in the Bay of Fundy (Figure 6.3). Semi-diurnal tides occur on the outer northwestern Atlantic coast from southern Newfoundland to the tip of Florida, but mixed tides occur in the big estuaries (Pamlico Sound, Chesapeake Bay, Long Island Sound, GSL) and the north-east coast of Newfoundland (Figure 6.3a). Tidal amplitudes exceed 1 m only in the GOM, Bay of Fundy, the inner GSL and some tidal nodes such as near Savannah, Georgia (Figure 6.3b). Large regions of the coast experience similar timing of tides, with the largest geographic changes at Cape Cod and in the GSL (see timing contour lines on Figure 6.3b). Delays in tides also occur in the large estuaries along the coast. Wind

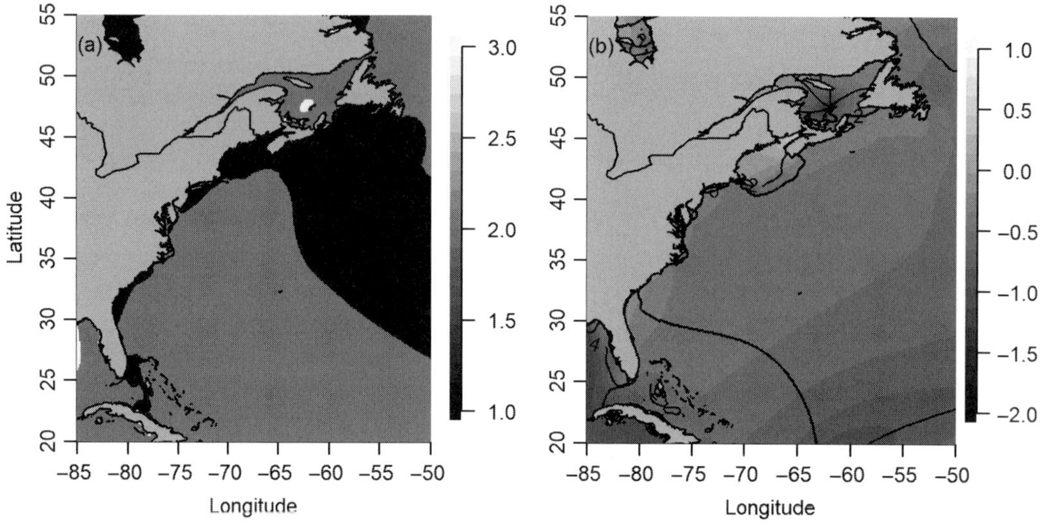


**Fig. 6.3** Tidal types and amplitudes. (a) Tide type (semi-diurnal, mixed or diurnal), calculated from tidal constituents via form ratio $F = (K1 + O1)/(M2 + S2)$. $F < 0.25$: semi-diurnal, $0.25 < F < 3$: mixed and $F > 3$: diurnal. In the figure scale: 1: semi-diurnal, 2: mixed and 3: diurnal. (b) Amplitude and phase of M2 tides on the North American Atlantic Coast. Colour scale is $\log_{10}$ (amplitude) and contour lines are M2 tide phases in hours. Source: TPXO7 Atlantic Ocean Atlas http://volkov.oce.orst.edu/tides/atlas.html (Egbert and Erofeeva, 2002). *A black and white version of this figure will appear in some formats. For the colour version, please refer to the plate section.*

generally determines water level along southern coasts because of their low tidal amplitudes.

### 6.1.4 Coastal Geomorphology

The predominance of rocky shores and gravel sediments from southern New England northwards define a dominant latitudinal feature of the north-western Atlantic coast, the result of exposed granitic bedrock, postglacial history and wave energy, although muddy bays commonly occur between rocky headlands. High wave energy, particularly north of Cape Cod, promotes rocky intertidal and coarse sedimentary environments, often dominated by sands and mixed with glacial till except in the most sheltered areas.

The high energy, predominantly rocky north-east coast of Newfoundland experiences the most days with waves over 3 m (Figure 6.4). Sandy sediments and rocky sandstone shorelines dominate the southern GSL, an area with very few days with waves >3 m (Figure 6.4). Sedimentary habitats predominate along the shallower and wider shelf south of Cape Cod, many sheltered by offshore islands or large shallow embayments with narrow openings such as Buzzards Bay and Long Island Sound.

### 6.1.5 Ice

Ice cover greatly disturbs the sediment and rock habitats of the more northern locales of the north-western Atlantic coasts (Gordon and Desplanque, 1983; Wethey, 1985). Icebergs from Greenland glaciers ground out along Newfoundland's east coast, often near the shore. In addition, pack ice flowing down along the Labrador coast to the north shore of Newfoundland, if pushed inshore by wind, can scrape and redistribute shallow sedimentary communities. Winter ice strongly influences the GSL (Saucier et al., 2003), as well as the east coast of Newfoundland (Johnston et al., 2005). Ice rafts in tidal channels and tidal flats transport sediment into marshes and can erode peat from creek banks (Pejrup and Andersen, 2000; Argow et al., 2011). Interannual variation in ice cover in Newfoundland correlates positively with the North Atlantic oscillation (Johnston et al., 2005), with heavy ice accumulation in North Atlantic oscillation-positive winters. For one to two months between December and February, ice cover can be extensive in the Bay of Fundy, GSL, and east and west coasts of Newfoundland (Figure 6.5). In the southern GSL, cold winter temperatures and even ice cover also

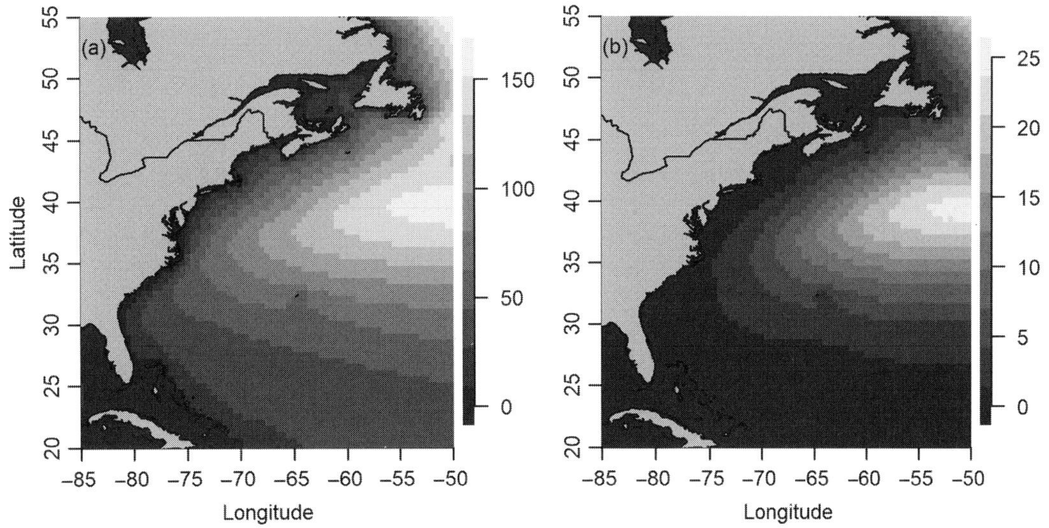


**Fig. 6.4** Mean number of days per year 1979–2009 with significant wave heights (a) >3 m and (b) >6 m. Note the difference in colour scales. Source: NOAA Wavewatch III hindcast with reanalysis winds ftp://polar.ncep.noaa.gov/pub/history/nopp-phase1/ (Chawla et al., 2011). *A black and white version of this figure will appear in some formats. For the colour version, please refer to the plate section.*

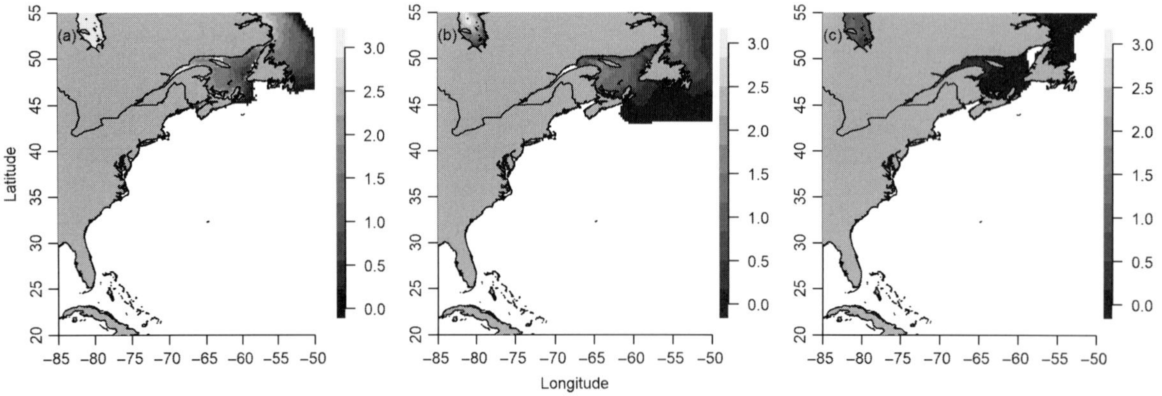


**Fig. 6.5** Winter ice cover in the north-west Atlantic (December–February). (a) 1980–2000, the average number of months of ice cover per year in satellite data from the National Snow and Ice Data Center (Sea Ice Index, dataset G02135); (b) 1980-2000, CMIP5 hindcast average number of months of winter ice cover in historical simulations; (c) 2080-2099, CMIP5 prediction average number of months of winter ice cover in the RCP 8.5 concentration pathway simulations. CMIP5 models were from the Institute Pierre Simon Laplace (CM5A), Tokyo University (MIROC ESM), Max Planck Institute for Meteorology (MPI ESM-MR), Japan Meteorological Agency (MRI CGCM3) and the Royal Netherlands Meteorological Institute (HadGEM2-AO). Colour scale on NSIDC data is the average number of months of winter ice cover per year. *A black and white version of this figure will appear in some formats. For the colour version, please refer to the plate section.*

occur, but summer temperatures exceed those of Atlantic coastal waters just to the south.

### 6.1.6 Primary Productivity

Primary productivity here is represented by chlorophyll *a* concentration. Primary production is not a physical driver because it represents the sum of several biologically driven processes, including nutrient supply and grazing rates, as well as local ocean climate. In shallow waters in particular, the activities of large active benthic species can drive important aspects of primary productivity, such as nutrient release from sediments, and benthic primary productivity may also contribute significantly (e.g., Middelburg et al., 1997; Welsh, 2003; Caffrey et al., 2014; Woodin et al., 2016). Primary productivity, however, also potentially correlates with physical parameters such as tidal regime and temperature, and thus might be expected to reflect a similar latitudinal pattern, which would complicate discussion of drivers. Satellite imagery, at least, does not reveal such a clear latitudinal pattern (Figure 6.6a). In general, annual average primary productivity increases closer to the coast. The south coast of the GSL and inner estuaries exhibit the highest annual primary productivity within our focal region, with greater within-year variability within estuaries than at nearby outer coast sites (Figure 6.6b: Chesapeake and Delaware Bays and Long Island Sound vs Mid Atlantic Bight, and St Lawrence River mouth vs Newfoundland and Atlantic coast of Nova Scotia). Atlantic coastal southern Florida appears to have lower annual mean values.

### 6.1.7 Changing Regimes

Climate change is altering these thermal regimes with the greatest linear rate of change in SST (2°C per century) in the GOM from 1870 to 2015 (Figure 6.7). Relative to adjacent coastlines, the GOM has warmed much faster (2.3°C per decade since 2004, Pershing et al., 2015). The coast south of Cape Hatteras is cooling slightly in winter (Figure 6.7a), perhaps because movement of the Gulf Stream farther offshore allows cold northern currents to penetrate southwards (Lima and Wethey, 2012). The summer warming rate in the north exceeds the <1°C per century of the south, adding approximately twenty extreme hot days per decade in the Canadian Maritimes and in the GOM since 1982 (Lima and Wethey, 2012). Simultaneously, seasonal warming now occurs approximately ten days earlier per decade along this coast (Lima and Wethey, 2012). These

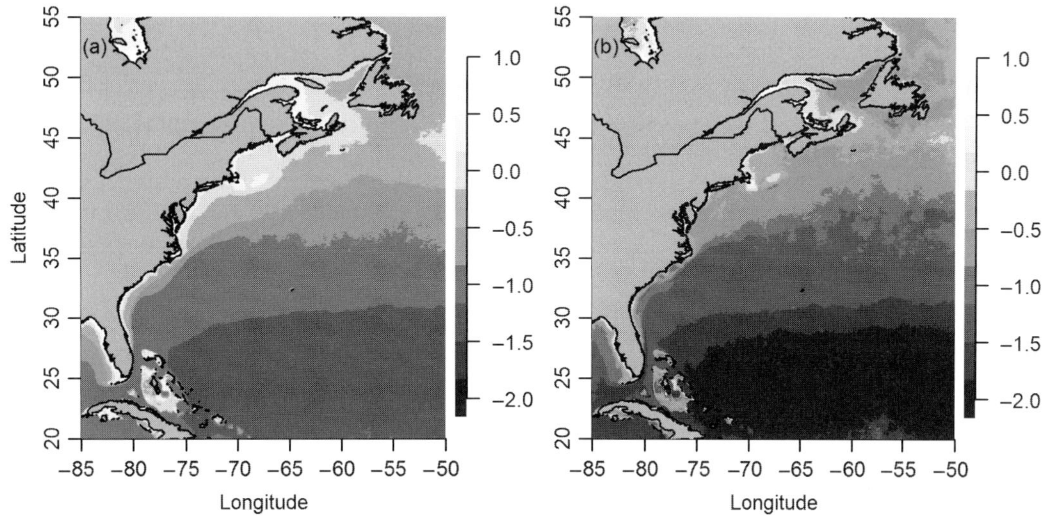


**Fig. 6.6** Chlorophyll *a* concentration from satellite imagery 2000–2015. (a) Climatological annual mean of chlorophyll *a* (mg/m$^3$). (b) Standard deviation of the monthly climatological averages of chlorophyll *a* (mg/m$^3$) from MODIS sensor on the Terra satellite (https://oceancolor.gsfc.nasa.gov). Data are expressed as log10. *A black and white version of this figure will appear in some formats. For the colour version, please refer to the plate section.*

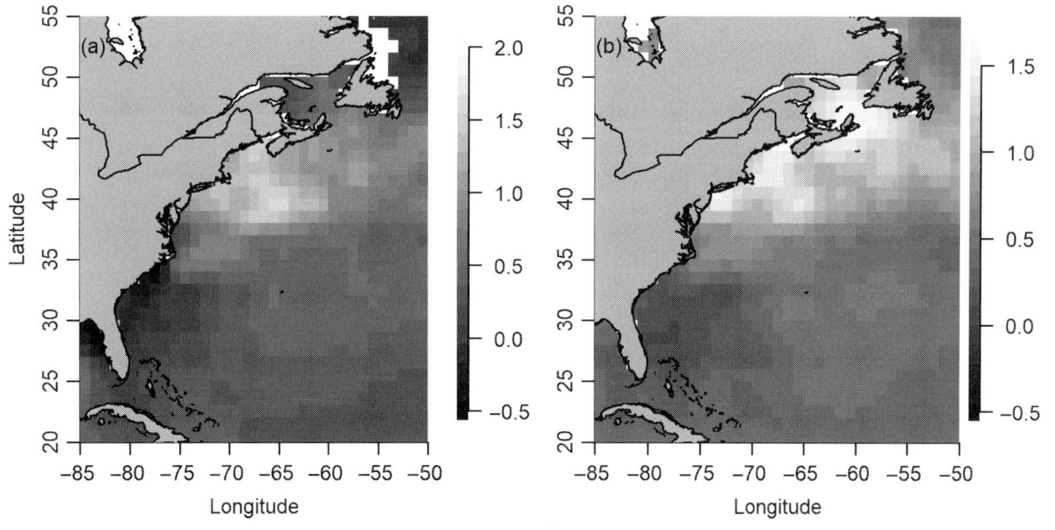


**Fig. 6.7** Linear rate of change of SST (°C per century) for the period 1870-2015. (a) Winter (December–February); (b) summer (June–August). Source: Hadley Centre HADISST (Rayner et al., 2003). www.metoffice.gov.uk/hadobs/hadisst/. *A black and white version of this figure will appear in some formats. For the colour version, please refer to the plate section.*

changes should magnify the existing difference between the south and the north in terms of annual ranges of SST (Figure 6.1).

The larger tidal amplitudes (Figure 6.3) and increased exposure to atmospheric conditions may further exaggerate the impacts of these changes in the north compared to the south, particularly given the loss of the insulating value of shore ice with warming (Partridge, 2000) and subsequent exposure of organisms to more extreme conditions.

The global climate models that include ice cover tend to underestimate winter ice cover by at least one month (Figure 6.5a and 6.5b) and predict very little ice cover for 2080–2100 – an average of less than one week per year in the GSL and north and east coasts of Newfoundland (Figure 6.5c). Models may underestimate ice cover, but they uniformly predict reduced ice at the end of the twenty-first century versus the present (Figure 6.5c).

In summary, the physical setting of shallow sandy mud and muddy sand habitats along the north-western Atlantic has strong latitudinal patterns in average SST, with important thermal breaks at Cape Hatteras and Cape Cod, and circulation reflecting this pattern. Tidal amplitudes and wave forces are larger, particularly north of Cape Cod and accelerate northwards, and ice cover only occurs in the north. In terms of changes to these parameters, both the northern Mid Atlantic Bight and the GOM have warmed significantly both in winter and summer over the past century; the GSL has had significant summer, but not winter, warming; and the southern part of the north-western Atlantic coast has more modest winter cooling. The implication is that, in terms of mechanisms, the range of physical forcing has been more extreme in the north than the south and, given aerial exposure during low tides, biotic exposure to this range is greater in the north than the south. An immediate question is whether processes that determine patterns of distribution and abundance of macroinfauna align latitudinally with these physical forces. In the next sections, we examine habitat types experienced by sedimentary fauna, and the biological forcing that interacts with the north–south physical regime.

## 6.2 Characteristic Ecosystems

The predominantly sandy sediments and sandstone rocky shorelines of the southern GSL experience very warm summers and thus a variety of more southern species, such as the eastern oyster, occur here, but are absent from the Bay of Fundy and Atlantic Nova Scotia. Extensive salt marshes dominate the coastline south of Long Island (42.08°N), often accompanied by barrier islands with open sand beaches. Estuaries occur throughout the coast, although most of the large river systems and coastal lagoons occur south of Long Island Sound (e.g., Chesapeake and Delaware bays). A prominent feature of the northern salt marshes is extensive ditching, begun in the seventeenth century to drain marshes for farming, but later expanded as a means of water and mosquito management. Salt marshes are more extensive in South Carolina and Georgia than in other parts of the coast, and their turbid waters exclude seagrasses (Roman et al., 2000). Tidal flats, the focus of our overview, are more extensive going north due to increased tidal range (Figure 6.3b).

While estuaries occur throughout the region, smaller river systems and coastal lagoons generally dominate north of New York. The shore of New Jersey marks a structural change, continuing southwards with wide lagoons and long barrier bodies. To the south, the presence of the large coastal plain more broadly facilitates wide, shallow lagoons fronted by long barrier islands. Other notable features of the mid-coast include Long Island Sound, a submerged basin, and Delaware and Chesapeake bays, both drowned river valleys. At the southern end of the survey area, the Saint John River at the Florida–Georgia border near Jacksonville represents a change from the extensive barrier islands of Georgia to the mainland beaches of northern Florida. Salt marsh and mangrove estuaries as well as barrier islands occur at Cape Canaveral and farther south in the Indian River Lagoon. A hardened intracoastal waterway and extensive urban and residential development characterises the large population centres of West Palm Beach to Miami.

As is detailed later, seagrasses are extremely important spatial determinants of density of macroinfauna because their physical attributes create refugia from digging predators. As noted earlier, turbid waters severely limit seagrass occurrence on the coasts of Georgia and South Carolina. *Zostera marina* is the only seagrass species present in euryhaline coastal waters from North Carolina northwards into Canada. *Halodule wrightii* is a co-dominant seagrass in the Indian River Lagoon in Florida, but also co-occurs with *Z. marina* in North Carolina. Within the Indian

River Lagoon in Florida, the rhizophytic alga, *Caulerpa prolifera*, and two additional seagrasses, *Syringodium filiforme* and *Thalassia testudinum* (Steward et al., 2005), also occur. Temperatures >30.0°C, which impair photosynthesis in *Z. marina*, set its southern limit and contribute to significant reductions in populations with rising temperatures, as was observed in North Carolina and Virginia in 2005 and 2010 (Moore et al., 2012). Populations of *Zostera* in Nova Scotia and Newfoundland are adapted to cold water temperatures and can complete their life cycle within the temperature range of 2–4°C. Thus, other abiotic factors may limit the northern distribution of *Zostera*, such as removal of above ground seagrass shoots by ice scouring (Robertson and Mann, 1984; Short and Wyllie-Echeverria, 1996). A combined response to temperature, nutrients and light availability limits geographic distributions of *H. wrightii* and *T. testudinum*; although low salinity may especially define the northern limit of *T. testudinum* (Moore, 1963).

## 6.3 Abundance Patterns and Mechanisms

Because physical forces, such as temperature and tidal regime, and habitat distributions change with latitude, an obvious prediction would be that changes in mechanisms that control abundance and distribution would also change with latitude. As is detailed in this section, this appears to be true.

### 6.3.1 Predator/Disturbance Agents and Latitude

Changes in the suite of predators that feed in sediments provide a major focus in understanding change in abundance and distribution of infauna with respect to latitude. Crudely, these predators can be divided into those that excavate sediments in search of prey and those that feed with minimal excavation. The minimal excavators include most shorebirds (Wilson, 1989; Daborn et al., 1993), flounder and other fishes (Burke, 1995; Kneib, 1988; Carlson et al., 1997; Packer et al., 1999) with the exception of cownose rays and some skates. The predators that excavate include both the benthic 'bulldozers' such as naticid snails (Wiltse, 1980; Commito, 1982) and horseshoe crabs (Botton, 1984), as well as the true excavators such as cownose rays (Sasko et al., 2006) and blue crabs (Virnstein, 1977), and perhaps the invasive green crab (Quijón and Snelgrove, 2005). These excavators can dig deeply into sediments to depths of 10 or 20 cm, consuming a variety of macroinfauna (Blundon and Kennedy, 1982a; Smith and Merriner, 1985). Importantly, the excavations displace infauna not eaten, both exposing them to the suite of smaller predators (fish, crabs, gastropods) that investigate such excavations (Risk and Craig, 1976; Oliver and Slattery, 1985; Hall, 1994), and potentially burying them in sometimes hypoxic spoil mounds created by the excavations.

Evidence produced over the last forty years indicates latitudinal trends in predator activity. Greater tidal amplitudes in the north suggest that epifaunal predators may have difficulty exploiting organisms living in the mid and upper intertidal, whereas exploitation by terrestrial predators such as shorebirds (Wilson, 1989) may be greater. In contrast, digging predators commonly occur during the warmer months of the year in sediments south of Cape Cod, particularly south of Long Island Sound, imposing a regime of frequent disturbance, resulting in a pockmarked landscape of pits and mounds (Cook, 1971; Woodin, 1978; Grant, 1983), and much reduced infaunal abundances. Low infaunal densities south of Cape Cod (often smaller by a factor of 5–10, Table 6.1) imply a reduced role for competitive interactions in controlling patterns of abundance and distribution than further north, except perhaps where refugia from predation support higher densities, presumably more competitive encounters, and potentially more limited resources. Such extensive and frequent excavation of sediments by predators, and the resulting landscapes, rarely occur north of Cape Cod (Ambrose, 1984).

One exception to the pattern of lesser disturbance by predators north of Cape Cod are sites with Atlantic sturgeon. These fish feed intensively in portions of the Bay of Fundy during spring spawning season and create obvious pits

**Table 6.1** Latitudinal patterns of infaunal abundance, species richness and species composition. Taxon of common species indicated by superscript: [a]amphipod, [b]bivalve, [o]oligochaete, [op]ophiuroid, [os]ostracod, [p]polychaete, [s]sipunculid, [c]cnidarian. 'R' is species richness. Infaunal densities are per square metre. 'na' not available. 'live' or 'fixed' refers to state of sample when sieved.

| Locale (reference) | Sieve size, season | Infaunal density $m^{-2}$ | R | Three most common species | Process affecting pattern according to author(s) |
|---|---|---|---|---|---|
| *North of Cape Cod* | | | | | |
| 47.72°N, 64.78°W (St Simon Bay, New Brunswick), Lu and Grant (2008) | 0.5 mm, live, June | 21,250 unvegetated | 37 | *Gemma gemma*[b], tubificids[o] | Tidal level (unvegetated area is at slightly higher tide level) |
| | | 16,333 seagrass bed (*Zostera*) | 43 | *Bittium alternatum*[g], *Odostomia* sp.[g] | |
| 47.72°N, 64.77°W (St Simon Bay, New Brunswick), Mallet et al. (2006) | 0.5 mm, live, September | 1,261–2,974 | 11–14 | *B. alternatum*[g], *Nereis diversicolor*[P], *G. gemma*[b] | Environmental stress including ice cover and low salinity pulse (22 psu) |
| 46.70°N, 64.85°W (mouth of Richibucto estuary, New Brunswick), Lu et al. (2008) | 0.5 mm, live, September—October | 19,934 | 35 | *Mediomastus ambiseta*[P], *Pygospio elegans*[P], *Exogone hebes*[P] | Densities lower with salinity flux, finer sediments and higher organic load |
| 45.15°N, 64.38°W (Avonport, Minas Basin, Bay of Fundy) Wilson (1989) | 0.25 mm, live, July | 55,325 | 6+ | *Corophium volutator*[a], *Aglaophamus neotenus*[P], *P. elegans*[P] | Fish, shorebirds, ice scour |
| 45.16°N, 64.32°W (Porter Point, Minas Basin, Bay of Fundy), Wilson (1991) | 0.25 mm, fixed, July | 49,263 | 5 | Ostracod (podocopid), *Heteromastus filiformis*[P], *Tharyx acutus*[P] | Fish, expected ice scour damage but not seen |
| 44.005°N, 69.66°W (Wiscasset, Maine), Ambrose (1984) | 0.5 mm, live, August | 31,000 | 12 | Oligochaetes, *C. volutator*[a], *Streblospio benedicti*[P] | Predatory infauna (*Nereis*[P], *Glycera*[P]) |

| | | | | | |
|---|---|---|---|---|---|
| 43.87°N, 64.81°W (Kejimkujik National Park Seaside, Nova Scotia), Wong and Dowd (2014) supplementary data | 0.5 mm, live, June–July | 33,392 intertidal sand<br>5,987 subtidal unvegetated<br>8,418 seagrass bed (*Z. marina* and *Ruppia maritima*) | | *Mya arenaria*[b], Nereidae[P], *Clymenella*[P] | Predation both on infauna and on predatory crabs, some suggestion of green crabs use of seagrass bed as refuge from predators such as gulls |
| 41.71°N, 70.33°W (Barnstable Harbor, Massachusetts), Whitlatch (1977) | 0.25 mm, live, summer | 44,672 (mean 16 stations) | 11.4 | *S. benedicti*[P], *Scolecolepides viridis*[P], *Eteone heteropoda*[P] | Temporal dominance function of zoogeographic origin |
| *Between Cape Cod and Cape Hatteras* | | | | | |
| 41.59°N, 70.64°W (Great Sippewissett Marsh, Falmouth, Massachusetts), Wiltse et al. (1984) | 0.5 mm, live, August | 9,066 | 8.8 | *S. benedicti*[P], *Paranais litoralis*[O], *Nematostella vectensis*[c] | Predation/disturbance by fish, shrimp and crabs |
| 40.96°N, 73.14°W (Flax Pond, Long Island Sound, New York), Cheng et al. (1993) | 0.5 mm, live, August (only abundant spp. counted) (mid intertidal, muddy sand) | 9,600–13,050 no cage, cage std dev. ~5,100 no difference, summer decline<br>24,400 June<br>13,050 August | | *S. benedicti*[P], *P. litoralis*[O], *G. gemma*[b] | Not predation, food quality decline over summer plus high temperatures? |
| 37.87°N, 75.36°W (Assateague Island, Virginia), Woodin (1978) | 1.0 mm, fixed, summer | 2,870 no *Diopatra*,<br>6,070 with *Diopatra* | 6.7<br>10 | *Spiochaetopterus oculatus*[P], *G. gemma*[b], *S. benedicti*[P] | Digging predators, partially countered by biogenic refugia |
| 38.12°N, 75.22°W (Chincoteague Bay, Maryland), Orth (1973) | 1.0 mm, live, July | 2,348 | 26 | *H. filiformis*[P], *Ampelisca vadorum*[a], *Lumbrineris tenuis*[P] | |
| *South of Cape Hatteras* | | | | | |
| 34.68°N, 76.60°W (Back Sound, North Carolina), Micheli et al. (2008) | 1 mm, live, May 1993<br>*Zostera* shallow subtidal | 4,936 *Zostera* bed<br>1,819 adjacent sand | na | Capitellidae[P], Spionidae[P], Arabellidae[P] | 'Habitat provision' |
| | *Halodule* bed intertidal | 2,889 *Halodule* bed<br>2,816 adjacent sand | | Spionidae[P], Arabellidae[P], Capitellidae[P] | |

(*cont*)

**Table 6.1** (cont.)

| Locale (reference) | Sieve size, season | Infaunal density $m^{-2}$ | R | Three most common species | Process affecting pattern according to author(s) |
|---|---|---|---|---|---|
| 33.34°N, 79.16°W (Debidue Creek, North Inlet, South Carolina), Fielman et al. (2001) | 0.25 mm, live, summer | 6,000 bare sand | 9 | *Tharyx maroni*$^{P}$, Spionidae$^{P}$, Capitellidae$^{P}$ | Digging predators and disturbance |
| 32.75°N, 79.89°W (reference saltmarsh creeks, Charleston Harbor, South Carolina), Lerberg et al. (2000) | 0.5 mm, live, summer | 6,716 | 21 | *S. benedicti*$^{P}$, *Monopylephorus rubroniveus*$^{O}$, *H. filiformis*$^{P}$ | Anthropogenic influences |
| 30.22°N, 81.30°W (shallow subtidal, <2 m depth, lower St John's River, Florida), Cooksey and Hyland (2007) | 0.5 mm, live, July 2000 and 2001 | 3,647–4,568 | 24 | *Mediomastus* sp.$^{P}$, *G. gemma*$^{b}$, *Phoronis* sp. | Dramatic decrease in density (357 $m^{-2}$) July 2002, apparent eutrophication event |
| 28.42°N, 80.61°W (Northern Indian River Lagoon, Florida), Grizzle (1981, 1984) | 0.5 mm, live, summer data 1977–1979 | 2,766–3,994 | 40 | *Phasolion cryptus*$^{S}$, *Clymenella mucosa*$^{P}$, *Capitella capitata*$^{P}$ | ???? |
| 27.35°N, 80.17°W (Indian River Lagocn, Florida), Nelson (1981) | 1.0 mm, live, May–July 1979 | 3,458–16,933 inside bed *H. wrightii* | na | Polychaetes, Amphipods, Tanaids | Fish, crustacean predators |

5–15 cm in diameter and up to 6 cm deep. Historically, much higher population levels of sturgeon along the eastern seaboard (Lotze, 2010) presumably produced greater impacts, but disturbance by sturgeon today is much less than that seen south of Cape Cod, comprising only 2–6 per cent of the sediment surface area (Pearson et al., 2007). In contrast, in Virginia, blue crabs, horseshoe crabs and skates and rays forage from May to August, and disturb over 25 per cent and up to 75 per cent of the sediment surfaces on average on any transect date (Woodin, 1978).

The geographic change in the pattern of disturbance reflects the northern limit of two of the major disturbance agents of the north-western Atlantic, the cownose ray *Rhinoptera bonasus* and the blue crab *Callinectes sapidus* (Schwartz, 1990; Johnson, 2015) (Table 6.2). Temperature appears to define the northern limits of both species (Table 6.2). *Rhinoptera* migrates southwards to Florida in the autumn, rarely occurring in water $<15^{\circ}$C (Poulakis, 2013). Temperature similarly limits *Callinectes,* which will not moult at $<10^{\circ}$C and larval failure occurs $<19^{\circ}$C (Brylawski and Miller, 2003). The geographic locations of their respective northern range limits will likely change in the future, but neither typically occurs north of Cape Cod at present and abundances are reduced between Delaware Bay and Cape Cod. However, grazing and bioturbation by the mud snail *Ilyanassa obsoleta* provides one source of disturbance that spans the full latitudinal range of our review. Although somewhat unusual in its form of disturbance and comparatively small size, *Ilyanassa* nonetheless impacts benthic microflora, larval recruitment and adult infauna (Hunt et al., 1987; Coffin et al., 2012). Importantly, however, it neither excavates the entire infaunal community nor facilitates predation by secondary predators as seen with *Callinectes* and *Rhinoptera.*

The correlation between latitudinal change in disturbance and increased infaunal abundances north of Cape Cod (Table 6.1) has been known for decades (Woodin, 1976; Ambrose, 1984) and is consistent with a reduced role for digging predators north of Cape Cod. It also supports our prediction of distinct latitudinal patterns of distribution and abundance along the north-western Atlantic coast, given the combination of compression of thermal zones and physical capes separating water masses from south to north, with possible major changes in dominant ecological mechanisms associated with changing physical conditions near Cape Hatteras and Cape Cod (e.g., Jennings et al., 2009). Given such clear patterns, biologists have responded predictably in looking for areas with higher abundances and presumed refugia, installing refugia mimics and, of course, excluding predators to observe pattern change (Table 6.3).

### 6.3.2 Refugia

A latitudinal gradient in availability of refugia influences biotic interactions along the north-western Atlantic coast. Living organisms create a variety of biotic structural additions to sediments, including dense root and rhizome mats of seagrasses and clumps of tubes of large tube builders such as those of the onuphid polychaete, *Diopatra cuprea*. These structures inhibit excavation by all but the largest predators and thus create refugia with reduced mortality (Blundon and Kennedy, 1982b; Peterson, 1982). As expected, higher densities of infauna and greater rates of survivorship characterise presumed refugia, at least south of Cape Cod (Table 6.3, part A). Interestingly, within the same water depth and site, macroinfaunal communities in seagrass beds and clumps of *Diopatra* tubes closely resemble those of adjacent bare sediments, differing largely in density only, suggesting that refugia differ from surrounding areas in infaunal mortality but not in other habitat characteristics. Some changes in density rankings do occur with higher densities, however, reflective of both competition among and predation by infauna (Luckenbach, 1987) and perhaps some inhibition of larger infauna by the roots and tubes (Brenchley, 1982).

Between southern Georgia and Cape Cod, densities of *Diopatra* are sufficiently high to create refugia with lowered rates of excavation. No equivalent dense large tube builders occur beyond the northern limit of *Diopatra* near Cape Cod (42.05°N, 70.65°W) (Pettibone, 1963). In Florida in the south, *Diopatra* rarely occurs at sufficiently high densities to serve as an effective refuge (Ban and Nelson, 1987; Berke, 2012) (Table 6.2). Seagrass taxa, however, provide a

**Table 6.2** Limitations of biotic drivers of patterns of infaunal abundance

| Species | Trophic mode | Range limitation | Range | Reference |
|---|---|---|---|---|
| *Disturbance* | | | | |
| *R. bonasus* | Predator (limited by mouth gape) | <15°C mortality during cold fronts of several days in Florida | Gulf of Mexico to southern New England, not very abundant in Indian River Lagoon (Snelson and Williams, 1981) | Poulakis (2013), Shaw et al. (2016) |
| *C. sapidus* | Predator (limited by claw force and sediment depth) | <19°C larval development failure<br><3°C crabs <30 mm high mortality<br><5°C females high mortality<br><10°C growth cessation<br>≥39°C lethal | Ephemeral to Nova Scotia, normal limit Cape Cod to northern Argentina | Johnson (2015), Brylawski and Miller (2003), Jensen et al. (2005), Mazzotti et al. (2006) |
| *Refuge* | | | | |
| *Z. marina* | Primary producer | >30°C photosynthesis is inhibited<br>Turbidity exclusion<br><br>Sea ice? | North Carolina to Nova Scotia,<br>absent South Carolina and Georgia<br><br>— | Moore and Jarvis (2008)<br>Roman et al. (2000),<br>Robertson and Mann (1984), Short and Wyllie-Echeverria (1996) |
| *H. wrightii* | Primary producer | Cold hardy, tolerates 0°C for short time periods; declining seagrass at 18.5°C, typically found in range of 20–30°C | North Carolina to Brazil (48.33°W, 25.33°S) | Zieman (1982), Hicks et al. (1998), Sordo et al. (2011) |
| *T. testudinum* | Primary producer | 20–30°C, tolerates 35°C for short time | Indian River Lagoon, Florida to Venezuela | Zieman (1982) |
| Rhizophytic algae | Primary producer | 14.5–33°C west coast of Florida;<br>20–31°C east coast of Florida to Keys | Jupiter Inlet and Indian River Lagoon, Florida, southern extent not known | Dawes et al. (1979), Demes et al. (2010) |
| *Diopatra cuprea* | Omnivore | Reproductive failure in cold summers? | Duxbury, Massachusetts to perhaps Brazil (undergoing systematic revision) | Pettibone (1963), Berke (2012) |

**Table 6.3** Effects of refugia (part A) and removal of epibenthic predators (part B). Infaunal densities are per square metre. 'na', not available.

Part A

| Location | Sieve size, other details | Without refuge | With refuge | Predator |
|---|---|---|---|---|
| *North of Cape Cod* | | | | |
| 47.72°N, 64.78°W (St Simon Bay, New Brunswick), Lu and Grant (2008) | 0.5 mm, live sieve, infaunal densities, June | 22,155 unvegetated | 24,850 seagrass bed (*Z. marina*) | na |
| 43.87°N, 64.81°W (Kejimkujik National Park Seaside, Nova Scotia), Wong and Dowd (2014) supplementary data | 0.5 mm, live sieve, infaunal densities, June–July, inner lagoon sites | 5,987 unvegetated | 8,418 seagrass bed (*Z. marina* and *R. maritima*) | Green crabs, gulls on crabs |
| *South of Cape Cod* | | | | |
| 37.26°N, 76.39°W (Sandy Point, York River, Virgina), Orth (1977) | 1.0 mm, fixed, infaunal densities, July 1972 | 1,039 unvegetated | 47,197 *Z. marina*, centre of bed | Large digging predators |
| 37.87°N, 75.36°W (Assateague Island, Virginia), Woodin (1978) | 1.0 mm, fixed, infaunal densities summer | 2,870 (0 *Diopatra*/0.01 $m^2$), 3,170 (1 *Diopatra*/0.01 $m^2$) | 6,070 (6 *Diopatra*/0.01 $m^2$) | Large digging predators |
| 34.71°N, 76.82°W (Pine Knoll Shores, Bogue Sound, North Carolina), Peterson (1982) | Densities of *Mercenaria mercenaria* and *Chione cancellata*, *Chione* burrows more shallowly | Sand only: *Mercenaria* 0.4 $m^{-2}$, *Chione* 0.4 $m^{-2}$ | Partial cover *H. wrightii*: *Mercenaria* 4.2 $m^{-2}$, *Chione* 5.2 $m^{-2}$<br>Thick cover *H. wrightii*: *Mercenaria* 11.3 $m^{-2}$, *Chione* 10.3 $m^{-2}$ | Whelks, *C. sapidus* |
| 34.71°N, 76.82°W (Pine Knoll Shores, Bogue Sound, North Carolina) Goshima and Peterson (2012) | *M. mercenaria* only | Sand only: 64% survivorship (2.5 months) | *H. wrightii* bed: 95% survivorship (2.5 months) | Whelks, primarily *Busycon carica* |
| 34.68°N, 76.60°W (Back Sound, North Carolina), Micheli et al. (2008) | 1 mm, live sieve, infaunal densities, May 1993, sand sites adjacent to seagrass beds | 1,819 unvegetated<br>2,816 unvegetated | 4,936 *Zostera* bed<br>2,889 *Halodule* bed | 'Habitat provision' |

Table 6.3 (cont.)

| Location | Sieve size, other details | Control | Manipulation control | Manipulation | Agent being manipulated |
|---|---|---|---|---|---|
| Part B | | | | | |
| *North of Cape Cod* | | | | | |
| 44.67°N, 63.57°W, Dartmouth Harbor Nova Scotia, Wong (2013) | Laboratory experiments, predator *Carcinus maenas*, manipulated rhizome and blade density, *Mya* as prey (18-25 mm length), little effect of blades | Ratio of inter-rhizome spaces to crab claw size is large, probability of capture/encounter <0.2 | Ratio of inter-rhizome spaces to crab claw size is small, probability of capture/encounter >0.8 | Density of below-ground rhizome structure | *C. maenas*, two carapace sizes (55–60, 70–80 mm) |
| 45.16°N, 64.32°W, (Porter Point, Minas Basin, Bay of Fundy), Wilson (1991) | June–July *T. acutus*<br>June–August *T. acutus*, 0.25 mm, fixed | 1,787<br>87.5 | 725 partial cage<br>975 partial cage<br>750 bird removal | 5,500 full cage<br>3,750 full cage | Fish<br>Fish and shorebirds |
| 44.005°N, 69.66°W (Wiscasset, Maine), Ambrose (1984) | 0.5mm, live, infaunal densities, June–October | 16,977 | 20,567 partial cage | 38,032 full cage | Fish and gulls |
| 44.97°N, 67.05°W, (Passamaquoddy and Cobscook Bays, Maine), Beal (2006) | May–November *M. arenaria* survival | | 31.9% open enclosures | 71.9% complete enclosures | Epibenthic predators including *C. maenas* |
| *Cape Cod to Cape Hatteras* | | | | | |
| 41.59°N, 70.64°W, (Great Sippewissett Marsh, Falmouth, Massachusetts), Wiltse et al. (1984) | 0.5 mm, live, infaunal densities, May–August | 1,440 | 1,084 partial cage | 70,151 full cage | Fish, shrimp, crabs |
| 37.30°N, 76.38°W, (Browns Bay, Virginia), Orth (1977) | 0.5 mm, fixed, infaunal densities, August–November 1974 | 32,200 control *Zostera* bed | 12,843 *Zostera* bed with clipped blades | 52,700 full cage within *Zostera* bed | Large digging predators, wave action |
| 37.87°N, 75.36°W (Assateague Island, Virginia), Woodin (1978) | 1.0 mm, fixed, infaunal densities, May–October 1974 | 2,870 (0 *Diopatra* per 0.01 $m^2$) | 6,070 (6 *Diopatra* per 0.01 $m^2$) | 7,400 (6 *Diopatra* mimics per 0.01 $m^2$) | Digging predators plus other predators |

| | | | | | |
|---|---|---|---|---|---|
| 37.87°N, 75.36°W (Assateague Island, Virginia), Woodin (1981) | 1.0 mm, fixed, infaunal densities, all 0 *Diopatra* per 0.01 $m^2$, April–July 1976 | 4,625 control | 7,600 *Limulus* and ray removal | 35,050 full cage | Digging predators plus other predators |
| *South of Cape Hatteras* | | | | | |
| 34.70°N, 76.82°W (Bogue Sound, North Carolina), Peterson (1982) | Survivorship of *M. mercenaria* and *C. cancellata*, July–November 1978 | 6% *Mercenaria*, 4% *Chione, Halodule* removed, control | 16% *Mercenaria*, 16% *Chione* partial cage, *Halodule* removal | 91% *Mercenaria*, 74% *Chione* full cage, *Halodule* removal | Whelks (primarily *B. carica*) |
| 34.70°N, 76.82°W (Pine Knoll Shores, Bogue Sound, North Carolina), Goshima and Peterson (2012) | Survivorship of *M. mercenaria* (40–80 mm), 2.5 months (July–October 1995) | 64% survivorship sand only | 92.5% survivorship trimmed *Halodule*, no blades | 95% survivorship *H. wrightii* bed | Whelks, primarily *B. carica* |
| 27.50°N, 80.29°W (Indian River Lagoon, Florida), Ban and Nelson (1987) | Laboratory experiment, *Diopatra* tube and *Halodule* additions vs predator addition (8 cm *Callinectes*), survivorship of 20 *Mulinia lateralis* (prey)(10–20 mm) | 6.5% 0 tubes | 17% 4 tubes per 0.01 $m^2$ | 43.5% 10 tubes per 0.01 $m^2$<br>71.5% *Halodule* mat<br>82% *Halodule* plus 4 tubes | *Callinectes* predation vs refuge |
| 27.535°N, 80.17°W (Indian River Lagoon, Florida), Nelson (1981) | 1.0mm, live sieve, infaunal densities, field experiments in *H. wrightii* bed, May | 16,933 unmanipulated | 14,236 cage with no predator | 3,658 cage with *Penaeus duorarum* | Shrimp predators have effect |
| 27.535°N, 80.17°W (Indian River Lagoon, Florida), Nelson (1981) | 1.0mm, live sieve, infaunal densities, field experiments in *H. wrightii* bed, June | 7,524 inside bed of *H. wrightii* | 4,058 cage with no predator | 3,302 cage with *Paleamonetes intermedius* | Shrimp predators have effect |

major source of both above- and below-ground structure in muddy sand sediments from the Indian River Lagoon in Florida north to Newfoundland. They appear to act as refugia from Cape Cod southwards (Table 6.3, part A). Unlike the eastern Pacific of North America and eastern Atlantic of European coasts, however, seagrasses along the western Atlantic coast of North America do not inhabit intertidal areas, although they may become exposed during extreme weather/tide conditions.

The importance of seagrasses as refugia depends on both the seagrass blades and the below-ground structure (Ban and Nelson, 1987), which vary substantially across species. Orth et al. (1984), however, reported that, even in the absence of dense mats, seagrass roots and rhizomes may prevent sediment disturbance within seagrass beds by ray foraging. Specifically, seagrass canopies and rhizomes may limit feeding on infauna by nektonic predators such as fish, crabs and rays, whereas rhizomes also impede digging predators such as whelks, rays and blue crabs (Brenchley, 1981; Peterson, 1982; Wong, 2013) (Table 6.3, part B). Fauna living below the depth of rhizome penetration may thus find refuge from predators.

Overall patterns of macroinfaunal abundance in seagrass beds relative to nearby sands align with observations of limited disturbance/predation by large digging predators north of Cape Cod with more intensive disturbance/predation in more southern regions. North of Cape Cod, where seagrass beds play a reduced role as refugia, infaunal densities are generally similar in bare sands and seagrass sediments (Table 6.3). South of Cape Cod, higher abundances of infauna have been reported in beds of both *Z. marina* and *H. wrightii* compared to those in nearby open sand flats (Table 6.3, part A). Likewise, most manipulative studies support the idea of seagrass as a refuge from large diggers (Table 6.3, part A), generally by assessing density levels and survivorship of infauna. In North Carolina, two seagrass taxa (i.e., *Z. marina* and *H. wrightii*), occupying different depths, appear to offer refuge to macroinfauna. Along the eastern coast of Florida, examination of sites containing any of three seagrass species (*H. wrightii, S. filiforme* and *T. testudinum*) often did not show expected patterns of greater infaunal densities in seagrass sediments versus bare sands. Instead, trophic studies revealed a complex suite of small decapod predators/disturbers associated with seagrass vegetation that feed upon macroinfauna in seagrass sediments in this region (e.g., Nelson, 1981). Similarly, in Nova Scotia, predation by the invasive green crab *C. maenas* apparently reduces infaunal densities inside seagrass beds relative to nearby intertidal sandflats because crabs in eelgrass have reduced susceptibility to gull predation relative to when feeding outside the seagrass bed (Wong and Dowd, 2014). Thus, localised foraging by decapods within seagrass habitats may mute any prey density increases within seagrass refugia from large disturbance agents, but this does not obscure the disturbance effect south of Cape Cod except in Florida.

### 6.3.3 Predator Manipulations

Manipulations of predator/disturbance agent abundance support the importance of such predators as mortality agents and the mechanism driving low infaunal abundances south of Cape Cod (Table 6.3, part B). Removal of predators typically results in increased densities of macroinfauna in areas without refugia relative to controls (Woodin, 1981; Wiltse et al., 1984; Ban and Nelson, 1987). Interestingly, in some cases, only slightly higher densities of macroinfauna or survivorship inside cages than those within seagrass beds offers further evidence of seagrass effectiveness as a refuge against large digging predators (Orth, 1977; Nelson, 1981; Peterson, 1982) (Table 6.3, part B). Additionally, experimental explorations of the refuge effect by insertion of artificial rhizome mat or dense large artificial tubes support the hypothesis that such structural elements reduce mortality (Woodin, 1978; Ban and Nelson, 1987). Behavioural observations add further support; shorebirds, for example, avoid foraging within dense tubes of *Diopatra* (Luckenbach, 1984).

As seen in the south, exclusion of predators north of Cape Cod also increases infaunal densities and survivorship (Table 6.3, part B), but densities in control areas are typically much greater than those in the south (Table 6.1). Unlike south of Cape Cod, excluding predators rarely alters the appearance of the habitat north of Cape Cod because the excluded predators are not excavators. The list of predators used in experimental manipulations north of Cape Cod includes shorebirds and infauna such as glycerid polychaetes, fish, gulls and the green crab *C. maenas* (Table 6.3, part B). South of Cape Cod, manipulated species include skates and rays, the blue crab *C. sapidus,* the horseshoe crab *Limulus polyphemus,* fish, shrimp and whelks *Busycon* spp. (Table 6.3, part B).

### 6.3.4 Predation on Juveniles

Predation on juveniles appears equally important both north and south of Cape Cod. Interesting, the predators on small prey such as infaunal juveniles enhance the effect of large digging predators or decimation by extreme events by feeding on individuals either exposed by the excavations or physical disturbance or emerging due to stress. If there is a size-related refuge from predation, such as the ability to withdraw deep into the sediment or produce a more robust shell, and if predator densities are low enough that some individuals escape (e.g., Beukema et al., 1998), high densities of animals should occupy the spatial refuge in the deeper sediment and provide evidence for competitive interactions among infauna. However, if (1) no size refuge is available, (2) only very large individuals escape or (3) predation is so intense on small sizes that almost none escape, then much of the sediment will appear underutilised with low densities and few large or deep organisms. The general geographic pattern on the north-western Atlantic coast suggests low density and mostly shallow-living infauna south of Cape Cod, consistent with the inability of most species to exploit a refuge of size or depth in sediment in the south.

North of Cape Cod, several abundant bivalve species, such as *M. mercenaria* and *M. arenaria,* are known to have a size escape from predation (Beal, 2006; Wong, 2013). Size refugia are, however, also known south of Cape Cod. For example, mouth gape size and crushing force limits predation by the cownose ray, *R. bonasus,* one of the abundant digging predators south of Cape Cod, to clams <31 mm in shell depth, and they feed primarily on individuals <26 mm shell depth (Fisher et al., 2011). Given size refugia both north and south of Cape Cod, but much lower bivalve densities south of Cape Cod, suggest that predation is so intense that few individuals survive the vulnerable size period except within refugia (Table 6.3).

### 6.3.5 Pulse Predation

Most of the predators discussed already display strong seasonal patterns of abundance. The highest impacts of the horseshoe crab and Atlantic sturgeon, for example, occur during spring spawning when their activities concentrate in the intertidal and shallow subtidal (Botton, 1984; Pearson et al., 2007). Neither cownose rays nor blue crabs feed in the mid-north-western Atlantic in late autumn to spring when water temperatures fall below 10–15°C (Jensen et al., 2005; Fisher et al., 2011). The extensive marshes and tidal creeks of the north-western Atlantic serve as nursery areas to a variety of fish, crabs, shrimp, etc. that feed as adults in the subtidal, but in very shallow water as juveniles (Ogburn et al., 1988). Migrating shorebirds exert perhaps the most dramatic and best-known example of pulse predation, particularly during autumn migration, when high densities of birds visit tidal flats (O'Reilly et al., 1995). In the upper Bay of Fundy, the migrating population of 800,000–1,400,000 semipalmated sandpipers, *Calidris pusilla,* feed on infauna, with peak abundances in early August during the autumn migration (Hicklin, 1987). Preferential feeding on the amphipod *C. volutator* dramatically reduces adult densities (Wilson, 1989; Cheverie et al., 2014). The Bay of Fundy also illustrates a trophic cascade attributable to shorebirds, where removal of *Corophium* by shorebirds releases amphipod grazing on diatom biofilms which thereby increases sediment stabilisation (Daborn et al., 1993, see also European studies by van Gils et al., 2012 and Cheverie et al., 2014).

### 6.3.6 Physical Disturbance

Very large-scale disturbances that cause massive mortality of infauna do not clearly vary with latitude, although some mortality sources, such as ice and iceberg scour, are uniquely northern

(Dashtgard et al., 2014), as are extreme cold events that constrain the northern edges of some subtropical species (e.g., Firth et al., 2011; Urian et al., 2011; Cavanaugh et al., 2014). *Z. marina* has now partially recovered from the disease that wiped out large swaths of seagrass in the 1930s across a broad range of latitudes spanning the mid to northern Atlantic coasts, thereby impacting a major type of refugia for infauna. Extreme events such as dramatic thermal or rainfall anomalies, storms with associated deposition or erosion or episodes of hypoxia result in large-scale defaunation (see Boesch et al., 1976; Simon and Dauer, 1977), but link to latitude only via those that correlate with human population densities. The events associated with large human population centres are more common south of Cape Cod where these populations occur (see human impacts on hypoxia, sedimentation, pollutants, etc.). Such events result in very large areas of destruction that create a very different landscape, at least in terms of scale, and open up a wide range of faunal responses (see Norkko et al., 2006 for a non-Atlantic example).

The mortalities reported for infauna associated with ice often appear less severe than expected, although Hicklin et al. (1980) reported dramatic reductions in spring *Corophium* abundances following a winter with ice cover (November: 9,011 $m^{-2}$, March: 385 $m^{-2}$). In many cases, no decline in infaunal abundances occurred, however, even with >1 m thick ice cover and movement of ice blocks, perhaps because frozen blocks of ice and sediment simply moved fauna from one area to another (Wilson, 1991). Gerwing et al. (2015), for example, reported a relatively modest effect of ice and other winter disturbances on intertidal mudflat infauna in the Bay of Fundy, and ice-related variables were poor predictors of mortality in abundant amphipods (Drolet et al., 2013). Drolet et al. (2013) attributed winter population reductions to depletion of energy stores rather than ice scour. Partridge (2000) found little evidence to link mortality to ice and her sediment temperature measurements suggested that ice might actually insulate sediments from atmospheric temperature extremes during low tide, as seen with the development ice cover on rocky shores (Pineda et al., 2005; Scrosati and Eckersley, 2007). Thus, the data available suggest that the mosaic of unimpacted areas, perhaps in combination with redeposition of surviving scraped fauna and protection by shore ice may be sufficient to support rapid recovery post-disturbance. This may explain why so few studies have shown clear effects of ice scour. Alternatively, as the data from European studies imply, it may be that a really severe winter has not been studied in the north-western Atlantic.

The more extensive studies of impacts of ice on European shores suggest that ice and/or low temperature result in declines in macrofauna only after particularly bad winters (Armonies et al., 2001). The effects of cold winters on adults and recruitment, however, differed considerably among species (Strasser et al., 2001a, 2001b). During the extreme winter of 1995–1996, direct ice scour killed Wadden Sea mussel beds and the emergent tube building polychaete *Lanice conchilega*; low temperatures killed cockles, but another common infaunal bivalve, *M. arenaria*, was unaffected (e.g., Strasser and Pieloth, 2001; Strasser et al., 2001b). This same severe winter initially reduced lugworms *Arenicola marina* by 50 per cent, but increased recruitment (Reise et al., 2001). Other severe winters greatly reduced cockle populations, but with weaker effects on mussels (Beukema et al., 2010). Death of epifauna and infauna can cascade through higher trophic levels such as oystercatchers, which suffered mass mortality, most likely from starvation after an extreme winter in 2012 (Schwemmer et al., 2014). Populations often recover quite slowly following such severe winters, especially if impacts span large portions of the geographic range. For example, following the winter of 1962–1963 some European populations of rocky shore species with limited dispersal, such as *Phorcus lineatus* and *Patella depressa*, were still expanding into their historical ranges forty years later (Mieszkowska et al., 2007; Hawkins et al., 2008). Presumably, some of these effects may also occur in North American systems strongly impacted by ice (e.g., GSL, east coast of Newfoundland), particularly given some overlap in species composition (e.g., *M. arenaria*), but further studies are needed to confirm this assumption.

### 6.3.7 Competitive Interactions

Given low densities of populations outside of refugia south of Cape Cod (Table 6.1), the small number of reports of competitive interactions, except among seagrasses, is hardly surprising. In North Carolina, *Halodule* and *Zostera* appear to interact along their zonation boundary (Micheli et al., 2008). Reports of dense populations of rhizophytic algae pre-empting space previously occupied by *H. wrightii* suggest competition for space (Stafford and Bell, 2006). In dense spionid polychaete populations in Barnstable Harbor, Massachusetts, north of Cape Cod, larger *Spio setosa* rip the feeding palps off neighbouring smaller *S. benedicti*, vacating feeding space for *Spio* (Whitlatch, 1976). Similarly, dense adult populations of *Corophium* in the Bay of Fundy aggressively exclude juveniles from predation refugia (Wilson, 1989). In the Wadden Sea in Europe, significantly higher mortality of *Corophium* occurs in the presence of either the lugworm *A. marina* or the cockle *Cerastoderma edule* or both (Flach and de Bruin, 1994). Both stimulate *Corophium* migration, apparently by disrupting and burying their burrows, increasing their susceptibility to epifaunal predators such as shrimp. We lack parallel information in north-western Atlantic interactions, despite similarity in players.

Apparent competition for food resources occurs among dense populations of suspension feeders, again based on data predominately from north of Cape Cod where high densities are common (Bradley and Cooke, 1959), as well as from dense experimental or culture conditions (Bricelj, 1992). Beal (2006), however, reported that predation on the clam, *M. arenaria*, influenced both survival and growth to a much larger extent than clam density per se in experimental plots on the Maine coast, despite a discernible competitive effect at elevated densities. Similarly, Möller and Rosenberg (1983) suggested that large populations of *Mya* in Sweden from the previous year depressed recruitment of both *Mya* and the cockle *Cerastoderma*, an adult–larval interaction supported experimentally by André and Rosenberg (1991).

### 6.3.8 Physics of Redistribution and Connectivity

Phylogenetic information provides additional insight into exchange of individuals along the north-western Atlantic coast and also hints at potential dispersal barriers. Physical transport can play a critical role in dispersing juveniles, particularly within bays (Emerson and Grant, 1991; Hunt and Mullineaux, 2002; Jennings and Hunt, 2009), sometimes facilitated by bulldozing (Dunn et al., 1999). Across regional spatial scales, the semipermeable barriers to dispersal at the convergences of water masses and thermal changes at both Cape Hatteras and Cape Cod clearly influence species distributions, as explored in detail by Pappalardo et al. (2015). Using genomics, Jennings et al. (2009) looked for phylogenetic breaks based on two mitochondrial genes of the maldanid polychaete, *Clymenella torquata*, particularly around Cape Cod. The short larval duration of this species, only a few days, considered in tandem with a life span of only a few years, suggests a species with potentially limited dispersal (Mangum, 1964). Jennings et al. (2009) found such a break south of Cape Cod, where diverging water currents from the GOM move eastwards away from the coast and south-westwards into the Mid Atlantic Bight. The strength of the haplotype divergence suggests little gene flow between populations on Long Island Sound southwards and those in Rhode Island and north into the GOM. Based on limited evidence, gene flow moves principally south, congruent with water currents (Jennings et al., 2009) and consistent with suggestions of differences in probability of survival of northern versus southern physiological races (Engle and Summers, 1999). Interestingly, *Z. marina* reportedly originated from Pacific populations and arrived in the Atlantic through Arctic passages and, according to the authors, may still be exchanging propagules between the north-eastern Pacific and the north-western Atlantic (Olsen et al., 2004). Its current southern limit is now North Carolina.

Range expansions of both native and invasive species inform the pattern of connectivity among populations along the north-west Atlantic coast.

Along the east coast of North America approximately 10 per cent of invasions are intra-coast i.e., from another location on the east coast of North America, while only 1 per cent are intra-coast along the west coast of North America (Ruiz et al., 2000). This difference in terms of intra-coastal invasions is presumably the result of dramatic differences in topography of the two coasts. Protected shallow-water sedimentary habitats are common along the east coast of North America, but often separated by several hundred kilometres along the west coast of North America (Emmett et al., 2000).

### 6.3.9 Colonisation and Glaciation

The proximity of sedimentary habitats to one another in the north-west Atlantic is very different from those of rocky shore habitats and appears to have had dramatic impacts on habitat availability during the last glacial maximum (LGM). Natural rock habitats are common in the northern coast but are absent south of Montauk Point, Long Island (41.07°N). As described earlier, shallow muddy sandy sedimentary habitats exist along the entire coast. As a consequence, during the LGM, approximately 20,000 years ago, when the ice sheet extended to just south of Long Island (40.5°N, 73.9°W), past the southern extent of natural rocky shore and eastwards several hundred kilometres offshore, the natural rock habitats were covered by ice (Dyke, 2004). There is some evidence that the Grand Banks of Newfoundland served as a glacial refuge (e.g., Maggs et al., 2008), and some offshore reefs and sea mounts may have been habitable to the east and south, but, given the colonisation history, such refugia appear to have been rare. In consequence, some rocky shore species may have survived, but many appear to be colonisers from Europe, probably from unglaciated rocky shores in Iberia (Ingólfsson, 1992; Wares and Cunningham, 2001). In contrast, during the last glacial maximum abundant sedimentary refugia existed south of the ice sheet along the coast. Repopulation by sedimentary organisms was most likely from those extant southern populations and constituted intra-coastal expansions (e.g., St-Onge et al., 2013). The invasive species literature is consistent with this perspective (Ruiz et al., 2000, 2015), but no systematic phylogenetic study of a range of species has been done for sedimentary infauna of the coasts of the North Atlantic.

The similarities and dissimilarities of both rocky shores and sedimentary habitats of the North Atlantic were extensively reviewed by Jenkins et al. (2008). The predominant taxa are very similar between the north-west and north-east Atlantic rocky shores, with one notable exception, the patellid limpets, which are absent from the north-west Atlantic (Jenkins et al., 2008). This similarity is as expected given post-glaciation recolonisation via stepping stones from European shores. In contrast, sedimentary landscapes of the north-west and north-east Atlantic appear strikingly different, especially those south of Cape Cod, where, as discussed earlier, large disturbance agents such as *C. sapidus, L. polyphemus,* skates and rays such as *R. bonasus,* are key players in determining abundance (Table 6.3). Such sediment disturbance agents exist in the north-east Atlantic but not in sufficient density or size to have impacts equivalent to those documented in the north-west Atlantic (Vermeij et al., 2008). In some cases, such as the skates and rays, this may be due to human exploitation, since they were common as fisheries catches along European coasts until the 1990s (Dulvy et al., 2000). In other cases such as *Limulus* they have never been present in Europe.

## 6.4 Humans as Experimental Disturbance Agents

### 6.4.1 Overview

Extensive and often extreme human impacts on coastal habitats include sediment elimination through hardened shorelines, destruction of submerged aquatic vegetation (SAV) and modifications of geomorphology with coastal protection. Large-scale human activities within sedimentary environments span the gamut from bait harvesting and clam digging to urban sewage, contaminant discharge and dredging. Consistent with other sections of this chapter, the emphasis is on latitudinal and biogeographic trends in human impacts. Within the plethora of human impacts, the focus here is on those most likely to alter dramatically processes controlling distributions and abundance.

Disturbance by humans encompasses the full range of largely negative effects from habitat removal to species removal. Some direct effects, such as elimination of eelgrass by dredging or other direct methods, contrast with indirect effects, such as eutrophication, resulting in SAV loss. Similar direct and indirect impacts occur for species and biodiversity. Overharvesting can directly eradicate species, but short-term low oxygen conditions (e.g., Garlo et al., 1979) can produce the same result. Both may impact simultaneously numerous species because harvesting of infauna typically involves excavation (e.g., hydraulic dredges) and thus collateral mortality and damage (Hall, 1994; Thrush and Dayton, 2002). In contrast to disturbance, some human interventions add habitat and species, although often with negative consequences. Substrate addition through jetties, wharves and dredge spoils can create new habitat directly, but also indirectly via nuisance algae (e.g., *Ulva*) that bloom under enhanced nutrient conditions. For better or for worse, invasive species literally add to local species, with innumerable examples in the marine environment as discussed later.

Most experimental marine ecology utilises controlled experiments involving changes in densities, habitats or species. The scale is typically small, although not always (e.g., Thrush et al., 2014), making extension to larger-scales problematic (see Thrush et al., 2000). The 'unintentional' manipulations linked to human impacts, in contrast, often span large spatial scales. The scale of the impact, as for other disturbance agents, is associated with the density, activity rate and size of the disturbance agents, in this case humans and their instruments. For example, intensive boating activities modify oyster reefs by dislodging oyster clumps from seaward margins, decreasing coverage of oyster reefs along major boating channels (Garvis et al., 2015). Declining oyster habitat can reduce the area utilised by some predators of infauna (e.g., blue crabs) as foraging or retreat sites. In some cases, human introductions of new taxa, such as green crabs occupying northern latitudes (Carlton and Cohen, 2003), may fill 'unoccupied' niches (soft sediment crab predator) with potential ripple effects to the whole community.

Human populations produce a huge latitudinal signature. The Boston–Washington megalopolis with more than 50 million people is in dramatic contrast to less densely populated coastlines to the south and north. A second population concentration occurs in south Florida where with modifications for water management, eutrophication is a pressing problem. The highest eutrophication levels occur in the Chesapeake Bay region (Bricker et al., 2008), with implications for the oxygen regime, status of benthic communities and health of seagrasses as detailed later.

### 6.4.2 Bait Digging and Harvesting

Human fishing impacts on intertidal and shallow subtidal systems typically occur either through bait digging, often for polychaetes (e.g., *Glycera*, *Nereis*), or harvesting (e.g., crabs, clams). Broadly speaking, these impacts occur through overharvesting, damage to the target species and collateral effects on non-target species. Given the documented impact of digging predators (Table 6.3), harvesting by excavation of large areas presumably lowers abundances and numbers of species above and beyond impacts on the target species. Simulated commercial bait collection in Maine supports this prediction (Brown and Wilson, 1997); similarly, clam digging reduces amphipod abundances (Logan, 2005) and causes significant collateral damage to populations of *Mya* (Emerson et al., 1990; Ambrose et al., 1998). These studies, along with numerous studies of bait digging and clamming in Europe (e.g., Cryer et al., 1987; Beukema, 1995; Sheehan et al., 2010), western North America (Peterson, 1977; Griffiths et al., 2006) and Australia (Skilleter et al., 2005) support the assertion of measurable, but modest, community-level impacts (e.g., decreased abundances of some species, reduced diversity, relatively rapid recovery) associated with these activities, a generalisation supported by a meta-analysis of impacts of intertidal rakes and dredges (Kaiser et al., 2006).

Removal of large epifaunal (e.g., horseshoe crabs, blue crabs, whelks) and mobile infaunal (e.g., sandworms, bloodworms) predators by harvesting, whether for medical use, food or bait, could have cascading effects on other ecosystem components with larger impacts (see Quijón and Snelgrove, 2005 for a subtidal example). Large predators such as horseshoe crabs have significant impacts on infauna, particularly small bivalves (Botton, 1984). Blue crabs play a similar role, preying on infauna living near the surface (Blundon and Kennedy, 1982b; Ban and Nelson, 1987) in shallow coastal sediments. Whelks are also important bivalve predators (Peterson, 1982). Mobile infaunal predators such as *Glycera* or *Nereis* (Ambrose, 1984) or nemerteans such as *Lineus* could potentially produce similar cascading effects, but we currently lack data.

Fishing for any of these taxa clearly alters species composition on local scales through habitat disturbance (Hall, 1994; Thrush and Dayton, 2002). Whether changes in predator abundance through fishing have significantly altered Atlantic shoreline ecosystems indirectly through food web effects and whether such alteration changes from north to south remains unclear; we expect, however, the impact of removing the larger disturbance agents of the south (horseshoe crabs, blue crabs, cownose rays) to be dramatic. The data to date suggest very high susceptibility of whelk and horseshoe crab populations to harvesting, with dramatic population declines resulting from harvest (Shuster and Botton, 1985; Widener and Barlow, 1999; Shalack et al., 2011). A proposed fishery for cownose rays, if instituted, will likely decimate the population (Fisher et al., 2013), as appears to be happening in Central and South America where it is listed as near threatened due to fishing pressure (IUCN Red List: www.iucnredlist.org/details/60128/0, accessed July 10, 2016).

### 6.4.3 Eutrophication

The highly visible and proximate nature of eutrophication and associated hypoxia near human populations has catalysed a rich body of benthic literature worldwide (e.g., Pearson and Rosenberg, 1978; Diaz and Rosenberg, 2008) and great public concern over 'dead zones'. Small-scale enrichment experiments (<1 m) as well as seasonal phytodetrital pulses (10s–100s of metres) offer insights on the relative effects of organic matter and reduced oxygen over broader spatial scales. As suggested for eutrophication, experimental small-scale phytodetritus addition often reduces diversity and opportunistic macro-infauna increase in abundance (Oviatt et al., 1986; Kelaher and Levinton, 2003; Quijón et al., 2008).

Over larger spatial scales (kilometres) benthic response to eutrophication follows a relatively predictable trajectory characterised by reduced diversity and elevated abundances of opportunists, unless anoxia eliminates metazoans altogether. In the absence of hypoxia, this added productivity can enhance secondary production and increase abundances of some taxa while altering food webs (Pearson and Rosenberg, 1978). The effects of severe hypoxia are relatively simple in wiping out most everything locally; however, intermediate eutrophication with less severe hypoxia may produce more complex outcomes involving interaction with predators (Levinton and Kelaher, 2004; Quijón and Snelgrove, 2008).

Geographically, the magnitude and severity of eutrophication along the North American coast correlates strongly with human coastal population size, as noted earlier. A recent review on eutrophication in US estuarine regions (Bricker et al., 2008) provides an excellent comparative analysis of the eastern seaboard from Florida to Maine. They note low levels of eutrophication in small, deep and well-flushed estuaries along high-energy coasts for the North Atlantic, and high levels of eutrophication in large and poorly flushed mid-Atlantic estuaries, reflecting somewhat lower energy and complex shorelines, and moderate to low eutrophication in medium sized but well-flushed estuaries with reduced energy along the southern seaboard.

This pattern raises the obvious question of whether coastal eutrophication, which tends to homogenise local communities by selecting for opportunists, also homogenises latitudinal gradients. Eutrophication likely leads to loss of both large tube-building species and seagrasses, the sources of biogenic refugia, and thus can drive a reduction in spatial complexity (Table 6.3). The added complications of hypoxia and fisheries removal of top predators in the same geographic locations all layer further complexity on the interspersed patchwork of eutrophic and non-eutrophic areas.

### 6.4.4 Human Impacts on Shorebirds

Shorebirds are important pulse predators along their migratory flyways, and populations in New England and the Maritimes appear to be in decline (Bart et al., 2007), perhaps reflecting vulnerability to human impacts. Pfister et al. (1992) demonstrate reduced use of feeding areas by migrating shorebirds in Massachusetts consistent with human recreational use, and potential feedback to declining shorebird populations. Similarly, various wading birds graze on horseshoe crab eggs in Delaware Bay, but human harvest has diminished egg availability and resulted in declines in species such as red knot (Niles et al., 2009), a species simultaneously under threat due to apparent temperature-induced changes in bill length which alters prey availability (van Gils et al., 2016). In other cases, eutrophication drives increased nuisance algae in the intertidal, hindering shorebird probing for prey (Green et al., 2015). Reduced shorebird predation reduces their potential influence on trophic dynamics. For example, reduced seabird grazing on the amphipod *C. volutator* in the Bay of Fundy potentially increases benthic microalgal biomass (Daborn et al., 1993), as noted earlier.

### 6.4.5 Fragmentation of Seagrass Habitat

Habitat fragmentation of seagrasses occurs when seagrass patchiness increases and cover decreases. A variety of human activities, including boating and fishing, have been linked to SAV fragmentation, whereby removal of seagrass above- and below-ground structure separates seagrass patches and increases the proportion of bare sediment (Bell et al., 2001). Seagrass loss can reduce sediment binding and increase sediment movement or resuspension, also altering light regime and water flow. Although natural forces, such as ray feeding and hydrodynamic forces (Fonseca and Bell, 1998; Townsend and Fonseca, 1998), can fragment habitat, increasing evidence from large-scale studies of seagrass spatial distribution points to human contributions to fragmentation. Surveys of benthic organisms in areas before and after seagrass loss, as well as studies that follow response of benthic organisms as vegetation cover increases (e.g., seagrass restoration), point to potential infaunal response to habitat fragmentation (Irlandi, 1994; Bell et al., 2006). In terms of their utility as refugia for infauna, small patches are ineffective in reducing mortality of adult *M. mercenaria* (Irlandi, 1997); therefore, as fragmentation increases, the availability of refugia from predators declines. Additionally, continuous beds may provide corridors for predatory blue crabs as they move among oyster reefs, suggesting important impacts of fragmentation of seagrass on trophic flow on oyster reefs (Micheli and Peterson, 1999).

### 6.4.6 Invasive Species

Human introduction is typically the source of taxa from other shores in coastal ecosystems (Carlton and Cohen, 2003). A recent compilation of species from the southern USA reported forty-six non-native invertebrate coastal species between Florida and South Carolina (Hymel, 2009). Many species may not cause major problems upon their arrival, in contrast to others that may significantly alter their new environments. In some cases, initial problems may be reduced or abate over time (e.g., Rossong et al., 2011a), and an invasive species can produce positive effects, such as the polychaete *Marenzelleria* (Norkko et al., 2012) which mitigates hypoxia in the Baltic. Or new arrivals may replace native species with relatively modest effects on associated taxa in one location and devastating effects in another (the common reed *Phragmites australis*: Posey et al., 2003 versus Lambert et al., 2010). Other species can be much more problematic. The well-studied green crab illustrates possible impacts ranging from competitive interactions to refuge destruction to changes in trophodynamics.

The green crab, *C. maenas,* appeared along selected areas of the US eastern seaboard from South Carolina to Newfoundland (Roman, 2006) through multiple invasions that began in the 1800s (Carlton and Cohen, 2003). Green crab habitat spans from protected low-energy systems to semi-exposed rocky shores (Grosholz and Ruiz, 1995) to depths of 6 m or more. Some populations occur primarily subtidally; others follow tidal movements as they feed (Cohen et al., 1995). A voracious predator, green crabs can alter food webs by outcompeting native species (Grosholz et al., 2000; McDonald et al., 2001), altering mollusc shell thickness (Grosholz and Ruiz, 1995) and reducing the abundance of a dominant prey item such as the soft shell clam (*M. arenaria*) (Floyd and Williams, 2004).

The broad geographic range invaded by green crabs on the east coast of North America has resulted in numerous species interactions. In the southern part of this range, adult blue crabs are thought to limit expansion of the green crab southwards. However, juvenile green crabs outcompeted juveniles of *C. sapidus* and *H. sanguineus* in laboratory experiments (MacDonald et al., 2007). Further north in their range, the presence of green crabs resulted in reduced growth in juvenile rock crabs *Cancer irroratus* (Breen and Metaxas, 2009). Laboratory studies also showed that green crab outcompeted juvenile (28–53 mm carapace length [CL]), and sub-adult (55–70 mm CL) lobster *Homarus americanus* for food resources (Rossong et al., 2006; Williams et al., 2006); juvenile lobsters also decreased foraging activity and increased shelter use in the presence of green crabs (Rossong et al., 2011b).

Benthic community analyses showed fewer species in mud, sand and eelgrass sites heavily populated by green crabs compared to sites without green crabs, although results depended on the taxa involved (Rossong, 2016). In some areas of eastern North America the decline of eelgrass (*Z. marina*) beds correlated with the introduction and establishment of green crabs (Garbary et al., 2004, 2014; Malyshev and Quijón, 2011). Green crabs often prey on infauna by uprooting eelgrass shoots, thereby decreasing available shelter refugia for organisms and increasing their vulnerability (Cohen et al., 1995; Davis et al., 1998). Green crabs also tear and cut the bundle sheath of eelgrass while foraging and burrowing (Davis et al., 1998). More recently, Matheson et al. (2016) linked changes in juvenile fish abundance to green crab's reducing eelgrass abundance, a critical habitat for fishes.

The relative success of green crabs towards the northern end of their distribution in comparison with the southern distributional range, where thermal stress increases (Cohen et al., 1995) along with higher densities of blue crabs, one of its predators, suggests non-uniform impacts throughout its range and that consequences may increase in severity in more northern areas. Whether this prediction holds will become clearer over time, but limited evidence suggests that some green crab impacts may relax over time (Rossong et al., 2011a). Wong and Dowd (2014) suggest, via a food web model, minimal trophic impact of removal of green crab from intertidal sand in Nova Scotia.

## 6.5 Future Conditions

The ocean will likely warm significantly by the end of the twenty-first century under the Intergovernmental Panel on Climate Change representative concentration pathway 8.5, which assumes a global energy imbalance of 8.5W $m^{-2}$ by 2100. The 10°C, 20°C and 25°C SST contours will likely move northwards as much as 5° latitude both in winter and summer (Figure 6.8). This large change in SST will presumably cause parallel, large-scale northwards shifts in the geographic distribution of species in which temperature affects survival or reproduction (e.g., Hutchins, 1947; Hall, 1964; Wethey et al., 2011). These shifts will reset long-recognised boundaries of biogeographic provinces. Cape Hatteras (35°N) and Cape Cod (42°N) were considered biogeographic boundaries in the nineteenth and twentieth centuries (e.g., Milne-Edwards, 1838; Dana, 1853; Stephenson and Stephenson, 1954; Hall, 1964) because they demarcated large temperature discontinuities (Figure 6.1). But these biogeographic boundaries may become a historical footnote. For example, the Virginian biogeographic province will no longer reside between Cape Hatteras and Cape Cod (Campbell and Valentine, 1977), assuming a 5° latitudinal shift of temperature contours.

### 6.5.1 Effects on Biogenic Refuges

With rising temperatures, the distribution of tube-dwellers such as *Diopatra* will likely expand northwards; however, the expected shift in seagrasses will be much more important,

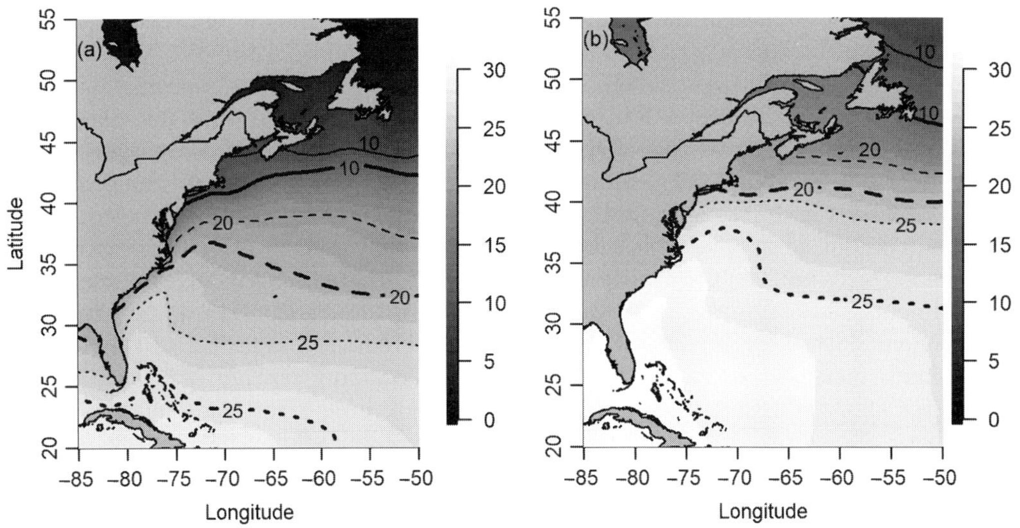


**Fig. 6.8** Ensemble projections of (a) mean winter (December–February) and (b) summer (June–August) SST in 2080–2100 under the Intergovernmental Panel on Climate Change representative concentration pathway 8.5 (8.5W $m^{-2}$ global energy imbalance by 2100). The colour scale is in °C. Blue contour lines are for the period 1980–2000, black contour lines are for the period 2080–2100, dotted contours are at 25°C, dashed contours at 20°C and solid contours at 10°C. Models were from the Beijing Climate Center (BCC CSM1–1), NASA Goddard Institute of Space Sciences (GISS E2R), Institute of Atmospheric Physics – Chinese Academy of Sciences (LASG FGOALS-g2), Tokyo University (MIROC ESM), Max Planck Institute for Meteorology (MPI ESM-MR), Japan Meteorological Agency (MRI CGCM3) and National Center for Atmospheric Research (CCSM4). *A black and white version of this figure will appear in some formats. For the colour version, please refer to the plate section.*

involving expansions of species in the south and in the north. In North Carolina, *H. wrightii* will expand its distribution to greater depths, assuming higher temperatures lead to loss of *Z. marina* (Moore and Jarvis, 2008) and water clarity remains similar to current conditions. Moore et al. (2012, 2014) reported that sustained summer temperatures in both 2005 and 2010 exceeding 30°C resulted in loss of *Z. marina* in Chesapeake Bay. Loss of *Z. marina* and expansion of *H. wrightii* will eventually alter both above- (decrease in canopy height) and below-ground (rhizome reduced) plant structure. Micheli et al. (2008) commented on differences in macroinfauna in sediments collected from *H. wrightii* and *Z. marina* seagrass beds, but attributing these differences specifically to seagrass type or water depth will require further research. In terms of the northern limit of *Zostera* if ice scour sets the northern limit, as suggested by Robertson and Mann (1984) and Short and Wyllie-Echeverria (1996), then projected ice reduction (Figure 6.5) may result in northwards expansion of *Zostera*.

The northern extent of *T. testudinum* will be greater in low-energy coastlines, although not in areas with substantial freshwater input and low salinities, and/or a turbid water column. Thus, the range of *T. testudinum* will probably not move incrementally northwards as temperatures rise, given the high-energy characteristics of the coastline along northern Florida and the freshwater inflow and turbid water conditions of coastal Georgia and South Carolina. If propagules arrive in North Carolina, likely via boating activities or uprooted individuals or seeds driven northwards during hurricanes (e.g., Bell et al., 2008), then *T. testudinum* may find suitable sites for establishment. Given the availability of appropriate sediments and water column characteristics, the range of subtropical rhizophytic algae may also expand northwards, leading to decreasing seagrass and marked changes in the vegetative canopy and below-ground structure. The calcareous skeleton of rhizophytic algae can contribute to sediment formation upon their death, thereby changing sediment physical features.

With accelerating sea-level rise (SLR), seagrasses will experience increased water depth and therefore a reduction in light reaching the bottom sediments along the deep edges of seagrass distribution. In some cases this effect will shift their distribution shorewards. Thus, the ranges of all seagrass species might shift shorewards without loss of any species. Migration of seagrasses shorewards will require availability of suitable sediments and assumes no intervention by anthropogenic activities or infrastructure, such as ongoing increased shore hardening in storm-impacted areas (Enwright et al., 2016).

### 6.5.2 Disturbance Agents

The ranges of both cownose rays and blue crabs will likely expand northwards into the GOM. Blue crabs already occur as vagrants north of Cape Cod, but larval thermal tolerances apparently preclude establishment of populations (Johnson, 2015). Given their vagility they should be able to forage successfully across the intertidal during high tide. If this foraging occurs, then it will reduce densities of infauna, including soft-shelled clams preferred by both cownose rays and blue crabs, presumably to the densities currently seen in the south, altering the latitudinal pattern (Table 6.1). Patterns of infauna north of Cape Cod may then more closely resemble those now seen in the south, with much higher densities and survivorships inside spatial refugia versus outside (Table 6.3, part A).

Some analyses conclude that increased winter precipitation will reduce salinity in large estuaries such as the Chesapeake (e.g., Najjar et al., 2010). Increased summer drought frequencies and SLR in the same region may also increase numbers of month-long saltwater intrusion events (Najjar et al., 2010). These projections point to increases in both high and low extremes of salinity over the next century. This fluctuation will likely dramatically reduce the success of species that use estuaries and tidal creeks as nursery grounds and, depending on the duration of salinity reductions, simultaneously reduce the diversity and abundance of infauna.

Ocean warming in winter will likely reduce the occurrence of sea ice in the GSL and Newfoundland over the next century, thereby dramatically reducing winter ice scour effects, and also ice-foot formation. With less thermally insulating ice cover, infaunal and epifaunal populations, such as green crabs, will likely experience greater low-temperature extremes than they encounter today. In contrast, removal of *Z. marina* by ice disturbance will likely be reduced.

In sum, with climate change and particularly with shifts in thermal regimes and consequent shifts in the distribution of disturbance agents, the area north of Cape Cod is likely to be exposed to much higher rates of predation on a broader range of prey sizes and depths, resulting in lowered infaunal densities and fewer escapes in size. More problematic are predictions for areas south of Cape Cod. *Zostera* is expected to be much more restricted southwards due to increases in water temperature. It is unclear whether the probable seagrass replacement, *Halodule,* is as effective a refuge provider as *Zostera*. Given the number of large estuaries combined with the Washington–Boston megalopolis and burgeoning coastal development everywhere, the stresses of increased salinity fluctuations, eutrophication, dredging, fishing pressures, etc. may reduce the predator populations which currently appear to drive the composition and abundance of the infaunal communities. The prey populations, however, may also be dramatically impacted, reducing coastal assemblages to those that can respond rapidly to short-lived favourable conditions, i.e., opportunists.

## Acknowledgements

The project was supported by grants from the National Aeronautics and Space Administration (NNX11AP77G to DSW and SAW), NSF (OCE1129401 to DSW and SAW) and the Gulf of Mexico Research Initiative (GOMR12017_VI_A159 to SSB), and by the Natural Sciences and Engineering Research Council of Canada (NSERC)-Cooke Industrial Research Chair in Sustainable Aquaculture (JG) and the NSERC Canadian Healthy Oceans Network (PVRS). This article is Contribution 93 in Ecological Forecasting from the University of South Carolina.

## REFERENCES


Ambrose, W. G. (1984). Influences of predatory polychaetes and epibenthic predators on the structure of a soft-bottom community in a Maine estuary. *Journal of Experimental Marine Biology and Ecology*, **81**, 115–45.

Ambrose, W. G., Dawson, M., Gailey, C. et al. (1998). Effects of baitworm digging on the soft-shelled clam, *Mya arenaria*, in Maine: shell damage and exposure on the sediment surface. *Journal of Shellfish Research*, **17**, 1043–9.

André, C. and Rosenberg, R. (1991). Adult-larval interactions in the suspension-feeding bivalves *Cerastoderma edule* and *Mya arenaria*. *Marine Ecology Progress Series*, **71**, 227–34.

Argow, B. A., Hughes, Z. J. and FitzGerald, D. M. (2011). Ice raft formation, sediment load, and theoretical potential for ice-rafted sediment influx on northern coastal wetlands. *Continental Shelf Research*, **31**, 1294–305.

Armonies, W., Herre, E. and Sturm, M. (2001). Effects of the severe winter 1995/96 on the benthic macrofauna of the Wadden Sea and the coastal North Sea near the island of Sylt. *Helgoland Marine Research*, **55**, 170–5.

Ban, S. and Nelson, W. G. (1987). Role of *Diopatra cuprea* Bosc (Polychaeta: Onuphidae) tubes in structuring a subtropical infaunal community. *Bulletin of Marine Science*, **40**, 11–21.

Banzon, V. F., Reynolds, R. W., Stokes, D. and Xue, Y. (2014). A 1/4° spatial resolution daily sea surface temperature climatology based on a blended satellite and in situ analysis. *Journal of Climate*, **27**, 8221–8.

Bart, J., Brown, S., Harrington, B. and Guy Morrison, R. I. (2007). Survey trends of North American shorebirds: population declines or shifting distributions? *Journal of Avian Biology*, **38**, 73–82.

Beal, B. F. (2006). Relative importance of predation and intraspecific competition in regulating growth and survival of juveniles of the soft-shell clam, *Mya arenaria* L., at several spatial scales. *Journal of Experimental Marine Biology and Ecology*, **336**, 1–17.

Bell, S. S., Brooks, R. A., Robbins, B. D., Fonseca, M. S. and Hall, M.O. (2001). Faunal response to fragmentation in seagrass habitats: implications for restoration efforts in a marine environment. *Biological Conservation*, **100**, 115–23.

Bell, S. S., Fonseca, M. S. and Stafford, N. B. (2006). Seagrass Ecology: New Contributions from a Landscape Perspective. In T. Larcum, R. J. Orth and C. M. Duarte, eds. *Seagrass Biology: A Treatise*. Kluwer, Netherlands, pp. 625–45.

Bell, S. S., Fonseca, M. S. and Kenworthy, W. J. (2008). Dynamics of a subtropical seagrass landscape: links between disturbance and mobile seed banks. *Landscape Ecology*, **23**, 67–74.

Berke, S. (2012). Biogeographic variability in ecosystem engineering: patterns in the abundance and behavior of the tube-building polychaete *Diopatra cuprea*. *Marine Ecology Progress Series*, **447**, 1–13.

Beukema, J. J. (1995). Long-term effects of mechanical harvesting of lugworms *Arenicola marina* on the zoobenthic community of a tidal flat in the Wadden Sea. *Netherlands Journal of Sea Research*, **33**, 219–27.

Beukema, J. J., Honkoop, P. J. C. and Dekker, R. (1998). Recruitment in *Macoma balthica* after mild and cold winters and its possible control by egg production and shrimp predation. *Hydrobiologica*, **375**/376, 23–34.

Beukema, J. J., Dekker, R. and Philippart, C. J. M. (2010). Long-term variability in bivalve recruitment, mortality, and growth and their contribution to fluctuations in food stocks of shellfish-eating birds. *Marine Ecology Progress Series*, **414**, 117–30.

Blundon, J. A. and Kennedy, V. S. (1982a). Mechanical and behavioral aspects of blue crab, *Callinectes sapidus* (Rathbun), predation on Chesapeake Bay bivalves. *Journal of Experimental Marine Biology and Ecology*, **65**, 47–65.

Blundon, J. A. and Kennedy, V. S. (1982b). Refuges for infaunal bivalves from blue crab, *Callinectes sapidus* (Rathbun), predation in Chesapeake Bay. *Journal of Experimental Marine Biology and Ecology*, **65**, 67–81.

Boesch, D. F., Diaz, R. J. and Virnstein, R. W. (1976). Effects of tropical storm Agnes on soft-bottom microbenthic communities of the James and York estuaries and the lower Chesapeake Bay. *Chesapeake Science*, **17**, 246–59.

Botton, M. L. (1984). The importance of predation by horseshoe crabs, *Limulus polyphemus*, to an intertidal sand flat community. *Journal of Marine Research*, **42**, 139–61.

Bradley, W. and Cooke, P. (1959). Living and ancient populations of the clam *Gemma gemma* in a marine coast tidal flat. *U.S. Fish and Wildlife Fish Bulletin no.*, **137**(58), 305–34.

Breen, E. and Metaxas, A. (2009). Effects of juvenile non-indigenous *Carcinus maenas* on the growth and condition of juvenile *Cancer irroratus*. *Journal of Experimental Marine Biology and Ecology*, **377**, 12–19.

Brenchley, G. A. (1981). Disturbance and community structure: and experimental study of bioturbation in marine soft-bottom environments. *Journal of Marine Research*, **39**, 767–90.

Brenchley, G. A. (1982). Mechanisms of spatial competition in marine soft-bottom communities. *Journal of Experimental Marine Biology and Ecology*, **60**, 17–33.

Bricelj, V. M. (1992). Aspects of the biology of the northern quahog, *Mercenaria mercenaria*, with emphasis on growth and survival during early life history. In M. A. Rice and D. Grossman-Garber, eds. *Proceedings of the Second Rhode Island Shellfish Industry Conference*. Rhode Island Sea Grant, Narragansett, pp. 29–48.

Bricker, S. B., Longstaff, B., Dennison, W. and Jones, A. (2008). Harmful Algae: effects of nutrient enrichment in the nation's estuaries: a decade of change. *Harmful Algae*, **8**, 21–32.

Brown, B. and Herbert Wilson Jr., W. (1997). The role of commercial digging of mudflats as an agent for change of infaunal intertidal populations. *Journal of Experimental Marine Biology and Ecology*, **218**, 49–61.

Brylawski, B. J. and Miller, T. J. (2003). Bioenergetic modeling of the blue crab (*Callinectes sapidus*) using the Fish Bioenergetics (3.0) computer program. *Bulletin of Marine Science*, **72**, 491–504.

Burke, J. S. (1995). Role of feeding and prey distribution of summer and southern flounder in selection of estuarine nursery habitats. *Journal of Fish Biology*, **47** (3), 355–66.

Caffrey, J. M., Murrell, M. C., Amacker, K. S., Harper, J. W., Phipps, S. and Woodrey, M. S. (2014). Seasonal and inter-annual patterns in primary production, respiration, and net ecosystem metabolism in three estuaries in the northeast Gulf of Mexico. *Estuaries and Coasts*, **37**(1), 222–41.

Campbell, C. A. and Valentine, J. W. (1977). Comparability of modern and ancient marine faunal provinces. *Paleobiology*, **3**, 49–57.

Carlson, J. K., Randall, T. A. and Mroczka, M. E. (1997). Feeding habits of winter flounder (*Pleuronectes americanus*) in a habitat exposed to anthropogenic disturbance. *Journal of Northwest Atlantic Fishery Science*, **21**, 65–73.

Carlton, J. T. and Cohen, A. N. (2003). Episodic global dispersal in shallow water marine organisms: the case history of the European shore crabs *Carcinus maenas* and *C. aestuarii*. *Journal of Biogeography*, **30**, 1809–20.

Cavanaugh, K. C., Kellner, J. R., Forde, A. J. et al. (2014). Poleward expansion of mangroves is a threshold response to decreased frequency of cold events. *Proceedings of the National Academy of Sciences*, **111**, 723–7.

Chawla, A., Spindler, D. and Tolman, H. L. (2011). A thirty year wave hindcast using the latest ncep climate forecast system reanalysis winds. In the *12th International Workshop on Wave Hindcasting and Forecasting & 3rd Coastal Hazards Symposium. Kohala Coast, Hawaii, National Centers for Environmental Prediction.*

Cheng, I.-J., Levinton, J. S., McCartney, M., Martinez, D. and Weissburg, M. J. (1993). A bioassay approach to seasonal variation in the nutritional value of sediment. *Marine Ecology Progress Series*, **94**, 275–85.

Cheverie, A. V., Hamilton, D. J., Coffin, M. R. S. and Barbeau, M. A. (2014). Effects of shorebird predation and snail abundance on an intertidal mudflat community. *Journal of Sea Research*, **92**, 102–14.

Coffin, M. R. S., Barbeau, M. A., Hamilton, D. J. and Drolet, D. (2012). Effect of the mud snail *Ilyanassa obsoleta* on vital rates of the intertidal amphipod *Corophium volutator*. *Journal of Experimental Marine Biology and Ecology*, **418-419**, 12–23.

Cohen, A. N., Carlton, J. T. and Fountain, M. C. (1995). Introduction, dispersal and potential impacts of the green crab *Carcinus maenas* in San Francisco Bay, California. *Marine Biology*, **122**, 225–37.

Commito, J. (1982). Effects of *Lunatia heros* predation on the population dynamics of *Mya arenaria* and *Macoma balthica* in Maine, USA. *Marine Biology*, **69**, 187–93.

Cook, D. O. (1971). Depressions in shallow marine sediment made by benthic fish. *Journal of Sedimentary Research*, **41**(2), 577–8.

Cooksey, C. and Hyland, J. (2007). Sediment quality of the Lower St. Johns River, Florida: an integrative assessment of benthic fauna, sediment-associated stressors, and general habitat characteristics. *Marine Pollution Bulletin*, **54**, 9–21.

Cryer, M., Whittle, G. N. and Williams, R. (1987). The impact of bait collection by anglers on marine intertidal invertebrates. *Biological Conservation*, **42**, 83–93.

Daborn, G. R., Amos, C. L. and Brylinsky, M. (1993). An ecological cascade effect: migratory birds affect stability of intertidal sediments. *Limnology and Oceanography*, **38**, 225–31.

Dame, R., Alber, M., Allen, D. et al. (2000). Estuaries of the South Atlantic coast of North America: their geographical signatures. *Estuaries*, **23**, 793–819.

Dana, J. D. (1853). On an isothermal oceanic chart, illustrating the geographical distribution of marine animals. *American Journal of Science*, **16**, 153–67.

Dashtgard, S. E., Pearson, N. J. and Gingras, M. K. (2014). Sedimentology, Ichnology, Ecology and Anthropogenic Modification of Muddy Tidal Flats in a Cold-Temperate Environment: Chignecto Bay, Canada. In I. P. Martini and H. R. Wanless, eds. *Sedimentary Coastal Zones from High to Low Latitudes: Similarities and Differences*. Geological Society, London.

Special Publications. London, Geological Society of London, 388, pp. 229-45.

Davis, R. C., Short, F. T. and Burdick, D. M. (1998). Quantifying the effects of green crab damage to eelgrass transplants. *Restoration Ecology*, **6**, 297–302.

Dawes, C. J., Bird, K., Durako, M., Goddard, R., Hoffman, W. and McIntosh, R. (1979). Chemical fluctuations due to seasonal and cropping effects on an algal-seagrass community. *Aquatic Botany*, **6**, 79–86.

Demes, K. W., Littler, M. M. and Littler, D. S. (2010). Comparative phosphate acquisition in giant-celled rhizophytic algae (Bryopsidales, Chlorophyta): Fleshy vs. calcified forms. *Aquatic Botany*, **92**, 157–60.

Diaz, R. J. and Rosenberg, R. (2008). Spreading dead zones and consequences for marine ecosystems. *Science*, **321**, 926–9.

Drolet, D., Kennedy, K. and Barbeau, M. A. (2013). Winter population dynamics and survival strategies of the intertidal mudflat amphipod *Corophium volutator* (Pallas). *Journal of Experimental Marine Biology and Ecology*, **441**, 126–37.

Dulvy, N., Metcalfe, J., Glanville, J., Pawson, M. and Reynolds, J. (2000). Fishery stability, local extinctions, and shifts in community structure in skates. *Conservation Biology*, **14**, 283–93.

Dunn, R. R., Mullineaux, L. S. and Mills, S. W. (1999). Resuspension of postlarval soft-shell clams *Mya arenaria* through disturbance by the mud snail *Ilyanassa obsoleta*. *Marine Ecology Progress Series*, **180**, 223–32.

Dyke, A. S. (2004). An outline of North American Deglaciation with emphasis on central and northern Canada. *Developments in Quaternary Science*, **2**, 371–406.

Egbert, G. D. and Erofeeva, S. Y. (2002). Efficient inverse modeling of barotropic ocean tides. *Journal of Atmospheric and Ocean Technology*, **19**, 183–204.

Emerson, C. W., Grant, J. and Rowell, T. W. (1990). Indirect effects of clam digging on the viability of soft-shelled clams *Mya arenaria*. *Netherlands Journal of Sea Research*, **27**, 109–18.

Emerson, C. W. and Grant, J. (1991). The control of soft-shell clam (*Mya arenaria*) recruitment on intertidal sandflats by bedload sediment transport. *Limnology and Oceanography*, **36**, 1288–300.

Emmett, R., Llanso, R., Newton, J. et al. (2000). Geographic signatures of North American West Coast estuaries. *Estuaries*, **23**(6), 765–92.

Engle, V. D. and Summers, J. K. (1999). Latitudinal gradients in benthic community composition in Western Atlantic estuaries. *Journal of Biogeography*, **26**, 1007–23.

Enwright, N. M., Griffith, K. T. and Osland, M. J. (2016). Barriers to and opportunities for landward migration of coastal wetlands with sea-level rise. *Frontiers in Ecology and the Environment*, **14**, 307–16.

Firth, L. B., Knights, A. M. and Bell, S. S. (2011). Air temperature and winter mortality: implications for persistence of the invasive mussel, *Perna viridis* in the intertidal zone of the south-eastern United States. *Journal of Experimental Marine Biology and Ecology*, **400**, 250–6.

Fielman, K. T., Woodin, S. A. and Lincoln, D. E. (2001). Polychaete indicator species as a source of natural halogenated organic compounds in marine sediments. *Environmental Toxicology and Chemistry*, **20**, 738–47.

Fisher, R. A., Call, G. C. and Grubbs, R. D. (2011). Cownose ray (*Rhinoptera bonasus*) predation relative to bivalve ontogeny. *Journal of Shellfish Research*, **30**, 187–96.

Fisher, R. A., Call, G. C. and Grubbs, R. D. (2013). Age, growth, and reproductive biology of cownose rays in Chesapeake Bay. *Marine and Coastal Fisheries: Dynamics, Management, and Ecosystem Science*, **5**, 224–35.

Flach, E. and de Bruin, W. (1994). Does the activity of cockles, *Cerastoderma edule* (L.) and lugworms, *Arenicola marina* L., make *Corophium volutator* Pallas more vulnerable to epibenthic predators: a case of interaction modification? *Journal of Experimental Marine Biology and Ecology*, **182**, 265–85.

Floyd T. and Williams, J. (2004). Impact of green crab (*Carcinus maenas* L.) predation on a population of soft-shell clams (*Mya arenaria* L.) in the southern gulf of Maine. *Journal of Shellfish Research*, **23**, 457–62.

Fonseca, M. S. and Bell, S. S. (1998). Influence of physical setting on seagrass landscapes near Beaufort, North Carolina, USA. *Marine Ecology Progress Series*, **171**, 109–21.

Garbary D. G., Miller, A. G. and Seymour, N. S. (2004). Destruction of Eelgrass Beds in Nova Scotia by the Invasive Green Crab. In A. R. Hanson, ed. Status and Conservation of Eelgrass (*Zostera marina*) in Eastern Canada. Canadian Wildlife Service Technical Report, Toronto, no. 412, pp. 13–14.

Garbary, D. J., Miller, A. G., Williams, J. and Seymour, N. R. (2014). Drastic decline of an extensive eelgrass bed in Nova Scotia due to the activity of the invasive green crab (*Carcinus maenas*). *Marine Biology*, **161**, 3–15.

Garlo, E. V., Milstein, C. B. and Jahn, A. E. (1979). Impact of hypoxic conditions in the vicinity of Little Egg Inlet, New Jersey in summer 1976. *Estuarine Coastal Marine Science*, **8**, 421–32.

Garvis, S. K., Sacks, P. E., and Walters, L. J. (2015). Formation, movement, and restoration of dead intertidal oyster reefs in Canaveral National Seashore and Mosquito Lagoon, Florida. *Journal of Shellfish Research*, **34**, 251–8.

Gerwing, T. G., Drolet, D. and Barbeau, M. A. (2015). Resilience of an intertidal infaunal community to winter stressors. *Journal of Sea Research*, **97**, 40–9.

Gordon, Jr, D. C. and Desplanque, C. (1983). Dynamics and environmental effects of ice in the Cumberland Basin of the Bay of Fundy. *Canadian Journal of Fisheries and Aquatic Sciences*, **40**, 1331–42.

Goshima, S. and Peterson, C. H. (2012). Both below- and aboveground shoalgrass structure influence whelk predation on hard clams. *Marine Ecology Progress Series*, **451**, 75–92.

Grant, J. (1983). The relative magnitude of biological and physical sediment reworking in an intertidal community. *Journal of Marine Research*, **41**, 673–89.

Green, L., Blumstein, D. T. and Fong, P. (2015). Macroalgal mats in a eutrophic estuary obscure visual foraging cues and increase variability in prey availability for some shorebirds. *Estuaries and Coasts*, **38**, 917–26.

Griffiths, J., Dethier, M. N., Newsom, A. et al. (2006). Invertebrate community responses to recreational clam digging. *Marine Biology*, **149**, 1489–97.

Grizzle, R. E. (1981). Opportunistic species of macrobenthos in a sewage-polluted lagoon, and an analysis of the indicator concept. Masters, University of Central Florida.

Grizzle, R. E. (1984). Pollution indicator species of macrobenthos in a coastal lagoon. *Marine Ecology Progress Series*, **18**, 191–200.

Grosholz, E. D. and Ruiz, G. M. (1995). Spread and potential impact of the recently introduced European green crab, *Carcinus maenas*, in central California. *Marine Biology*, **122**, 239–47.

Grosholz, E. D., Ruiz, G. M., Dean, C. A., Shirley, K. A., Maron, J. L. and Connors, P. G. (2000). The implication of a nonindigenous marine predator in a California bay. *Ecology*, **81**, 1206–24.

Hall, C. J. (1964). Shallow water marine climates and molluscan provinces. *Ecology*, **45**, 226–34.

Hall, S. J. (1994). Physical Disturbance and Marine Benthic Communities: Life in Unconsolidated Sediments. In A. D. Ansell, R. N. Gibson and M. Barnes, eds. *Oceanography and Marine Biology*. UCL Press, London, 32, 179–239.

Hawkins, S. J., Moore, P. J., Burrows, M. T. et al. (2008). Complex interactions in a rapidly changing world: responses of rocky shore communities to recent climate change. *Climate Research*, **37**, 123–33.

Hicklin, P. W. (1987). The migration of shorebirds in the Bay of Fundy. *Wilson Bulletin*, **99** (4), 540–70.

Hicklin, P. W., Linkletter, L. E. and Peer, D. L. (1980). Distribution and abundance of *Corophium volutator* (Pallas), *Macoma balthica* (L.) and *Heteromastus filiformis* (Claparède) in the intertidal zone of Cumberland Basin and Shepody Bay, Bay of Fundy. *Canadian Technical Report of Fisheries and Aquatic Sciences*. 965. Dartmouth, Nova Scotia : Fisheries and Oceans Canada

Hicks, D. W., Onuf, C. P. and Tunnell, J. W. (1998). Response of shoal grass, *Halodule wrightii*, to extreme winter conditions in the Lower Laguna Madre, Texas. *Aquatic Botany*, **62**, 107–14.

Hunt, H. L. and Mullineaux, L. S. (2002). The roles of predation and postlarval transport in recruitment of the soft shell clam (*Mya arenaria*). *Limnology and Oceanography*, **47**, 151–64.

Hunt, J. H., Ambrose, W. G. and Peterson, C. H. (1987). Effects of the gastropod, *Ilyanassa obsoleta* (Say) and the bivalve, *Mercenaria mercenaria* (L.), on larval settlement and juvenile recruitment of infauna. *Journal of Experimental Marine Biology and Ecology*, **108**, 229–40.

Hutchins, L. W. (1947). The bases for temperature zonation in geographic distribution. *Ecological Monographs*, **17**, 325–35.

Hymel, S. N. (2009). *Inventory of Marine and Estuarine Benthic Macroinvertebrates for Nine Southeast Coast Network Parks*, Natural Resource Report, NPS/SECN/NRR —2009/121, IUCN Red List. International Union for Conservation of Nature, Gland, www.iucnredlist.org/details/60128/0, accessed July 10, 2016.

Ingólfsson, A. (1992). The origin of the rocky shore fauna of Iceland and the Canadian Maritimes. *Journal of Biogeography*, **19**(6), 705–12.

Irlandi, E. A. (1994). Large- and small-scale effects of habitat structure on rates of predation: how percent coverage of seagrass affects rates of predation and siphon nipping on an infaunal bivalve. *Oecologia*, **98**(2), 176–83.

Irlandi, E. A. (1997). Seagrass patch size and survivorship of an infaunal bivalve. *Oikos*, **78**, 511–18.

Jenkins, S. R., Moore, P., Burrows, M. T. et al. (2008). Comparative ecology of North Atlantic shores: do differences in players matter for process? *Ecology*, **89** (11 Supplement), S3–S23.

Jennings, L. B. and Hunt, H. L. (2009). Distances of dispersal of juvenile bivalves (*Mya arenaria* (Linnaeus), *Mercenaria mercenaria* (Linnaeus), *Gemma gemma* (Totten)). *Journal of Experimental Marine Biology and Ecology*, **376**, 76–84.

Jennings, R. M., Shank, T. M., Mullineaux, L. S. and Halanych, K. M. (2009). Assessment of the Cape Cod phylogeographic break using the bamboo worm *Clymenella torquata* reveals the role of regional water masses in dispersal. *Journal of Heredity*, **100**, 86–96.

Jensen, O. P., Seppelt, R., Miller, T. J. and Bauer, L. J. (2005). Winter distribution of blue crab *Callinectes sapidus* in Chesapeake Bay: application and cross-validation of a two-stage generalized additive model. *Marine Ecology Progress Series,* **299**, 239–55.

Johnson, D. S. (2015). The savory swimmer swims north: a northern range extension of the blue crab? *Journal of Crustacean Biology,* **35**, 105–10.

Johnston, D. W., Friedlaender, A. S., Torres, L. G. and Lavigne, D. M. (2005). Variation in ice cover on the east coast of Canada from 1969 to 2002: climate variability and implications for harp and hooded seals. *Climate Research,* **29**, 209–22.

Kaiser, M. J., Clarke, K. R., Hinz, H., Austen, M. C. V., Somerfield, P. J. and Karakassis, I. (2006). Global analysis of response and recovery of benthic biota to fishing. *Marine Ecology Progress Series,* **311**, 1–14.

Kelaher, B. P. and Levinton, J. S. (2003). Variation in detrital enrichment causes spatiotemporal variation in soft-sediment assemblages. *Marine Ecology Progress Series,* **261**, 85–97.

Kneib, R. T. (1988). Testing for indirect effects of predation in an intertidal soft-bottom community. *Ecology,* **69**, 1795–805.

Lambert, A. M., Dudley, T. L. and Saltonstall, K. (2010). Ecology and impacts of the large-statured invasive grasses *Arundo donax* and *Phragmites australis* in North America. *Invasive Plant Science and Management,* **3**, 489–94.

Lerberg, S. B., Holland, A. F. and Sanger, D. M. (2000). Responses of tidal creek macrobenthic communities to the effects of watershed development. *Estuaries,* **23**, 838–53.

Levinton, J. and Kelaher, B. (2004). Opposing organizing forces of deposit-feeding marine communities. *Journal of Experimental Marine Biology and Ecology,* **300**, 65–82.

Lima, F. P. and Wethey, D. S. (2012). Three decades of high-resolution coastal sea surface temperatures reveal more than warming. *Nature Communications,* **3**, 704.

Logan, J. M. (2005). Effects of clam digging on benthic macroinvertebrate community structure in a Maine mudflat. *Northeastern Naturalist,* **12**, 315–24.

Lotze, H. K. (2010). Historical reconstruction of human-induced changes in U.S. estuaries. *Oceanography and Marine Biology: An Annual Review,* **48**, 265–336.

Lu, L. and Grant, J. (2008). Recolonization of intertidal infauna in relation to organic deposition at an oyster farm in Atlantic Canada – a field experiment. *Estuaries and Coasts,* **31**, 767–75.

Lu, L., Grant, J. and Barrell, J. (2008). Macrofaunal spatial patterns in relationship to environmental variables in the Richibucto Estuary, New Brunswick, Canada. *Estuaries and Coasts,* **31**, 994–1005.

Luckenbach, M. W. (1984). Biogenic structure and foraging by five species of shorebirds (Charadrii). *Estuarine, Coastal and Shelf Science,* **19**, 691–6.

Luckenbach, M. W. (1987). Effects of adult infauna on new recruits: implications for the role of biogenic refuges. *Journal of Experimental Marine Biology and Ecology,* **105**, 197–206.

MacDonald, J. A., Roudez, R., Glover, T. and Weis, J. S. (2007). The invasive green crab and Japanese shore crab: behavioral interactions with a native crab species, the blue crab. *Biological Invasions,* **9**, 837–48.

Maggs, C. A., Castilho, R., Foltz, D. et al. (2008). Evaluating signatures of glacial refugia for North Atlantic benthic marine taxa. *Ecology,* **89**, S108–22.

Mallet, A. L., Carver, C. E. and Landry, T. (2006). Impact of suspended and off-bottom Eastern oyster culture on the benthic environment in eastern Canada. *Aquaculture,* **255**, 362–73.

Malyshev, A. and Quijón, P. A. (2011). Disruption of essential habitat by a coastal invader: new evidence of the effects of green crabs on eelgrass beds. *ICES Journal of Marine Science,* **68**, 1852–6.

Mangum, C. P. (1964). Studies on speciation in maldanid polychaetes of the North American Atlantic coast II. Distribution and competitive interaction of five sympatric species. *Limnology and Oceanography,* **9**, 12–26.

Martin, P. J., Barron, C. N., Smedstad, L. F. et al. (2009). *User's Manual for the Navy Coastal Ocean Model (NCOM) version 4.0*, Naval Research Laboratory Report NRL/MR/7320–09-9151. United States Naval Research Laboratory, Washington, DC.

Matheson, K., McKenzie, C. H., Gregory, R. S. et al. (2016). Linking eelgrass decline and impacts on associated fish communities to European green crab *Carcinus maenas* invasion. *Marine Ecology Progress Series,* **548**, 31–45.

Mazzotti, F. J., Pearlstine, L. G., Barnes, T. et al. (2006). *Stressor Response Model for the Blue Crab,* Callinectes sapidus, *Technical Assistance for an Ecological Evaluation of the Southwest Florida Feasibility Study*. University of Florida, Florida Lauderdale Research and Education Center, Fort Lauderdale, FL.

McDonald, P. S., Jensen, G. and Armstrong, D. A. (2001). The competitive and predatory impacts of the nonindigenous crab *Carcinus maenas* (L.) on early benthic phase Dungeness crab *Cancer magister* Dana. *Journal of Experimental Marine Biology and Ecology,* **258**, 39–54.

Micheli, F. and Peterson, C. H. (1999). Estuarine vegetated habitats as corridors for predator movements. *Conservation Biology,* **13**, 869–81.

Micheli, F., Bishop, M. J., Peterson, C. H. and Rivera, J. (2008). Alteration of seagrass species composition and function over two decades. *Ecological Monographs*, **78**, 225–44.

Middelburg, J. J., Soetaert, K. and Herman, P. M. J. (1997). Empirical relationships for use in global diagenetic models. *Deep Sea Research Part I: Oceanographic Research Papers*, **44**, 327–44.

Mieszkowska, N., Hawkins, S. J., Burrows, M. T. and Kendall, M. A. (2007). Long-term changes in the geographic distribution and population structures of *Osilinus lineatus* (Gastropoda: Trochidae) in Britain and Ireland. *Journal of the Marine Biological Association of the United Kingdom*, **87**, 537–45.

Milne-Edwards, H. (1838). Mémoire sur la distribution géographique des Crustacés. *Annales des Sciences Naturelles Zoologie, Series 2*, **10**, 129–74.

Möller, P. and Rosenberg, R. (1983). Recruitment, abundance and production of *Mya arenaria* in marine shallow waters, western Sweden. *Ophelia*, **22**, 33–55.

Moore, D. R. (1963). Distribution of the sea grass, *Thalassia*, in the United States. *Bulletin of Marine Science of the Gulf and Caribbean*, **13**, 329–42.

Moore, K. A. and Jarvis, J. C. (2008). Environmental factors affecting recent summertime eelgrass diebacks in the lower Chesapeake Bay: implications for long-term persistence. *Journal of Coastal Research, Special Issue* **55**, 135–47.

Moore, K. A., Shields, E. C., Parrish, D. B. and Orth, R. J. (2012). Eelgrass survival in two contrasting systems: role of turbidity and summer water temperatures. *Marine Ecology Progress Series*, **448**, 247–58.

Moore, K. A., Shields, E. C. and Parrish, D. B. (2014). Impacts of varying estuarine temperature and light conditions on *Zostera marina* (eelgrass) and its interactions with *Ruppia maritima* (widgeongrass). *Estuaries and Coasts*, **37**, 20–30.

Najjar, R. G., Pyke, C. R., Adams, M. B. et al. (2010). Potential climate change impacts on the Chesapeake Bay. *Estuarine Coastal and Shelf Science*, **86**, 1–20.

Nelson, W. G. (1981). Experimental studies of decapod and fish predation on seagrass macrobenthos. *Marine Ecology Progress Series*, **5**, 141–49.

Niles, L. J., Bart, J., Sitters, H. P. et al. (2009). Effects of horseshoe crab harvest in Delaware Bay on Red Knots: Are harvest restrictions working? *Bioscience*, **59**, 153–64.

Norkko, A., Rosenberg, R., Thrush, S. F. and Whitlatch, R. B. (2006). Scale- and intensity-dependent disturbance determines the magnitude of opportunistic response. *Journal of Experimental Marine Biology and Ecology*, **330**, 195–207.

Norkko, J., Reed, D. C., Timmermann, K. et al. (2012). A welcome can of worms? Hypoxia mitigation by an invasive species. *Global Change Biology*, **18**, 422–34.

Ogburn, M. V., Allen, D. M. and Michener, W. K. (1988). *Fishes, Shrimps, and Crabs of the North Inlet Estuary, SC: A Four-Year Seine and Trawl Survey*, Baruch Institute Technical Report 88-1. Baruch Institute of Coastal Ecology, Georgetown, SC.

Oliver, J. S. and Slattery, P. N. (1985). Effects of crustacean predators on species composition and population structure of soft-bodied infauna from McMurdo Sound, Antarctica. *Ophelia*, **24**, 155–75.

O'Reilly, K. M. and Wingfield, J. C. (1995). Spring and autumn migration in arctic shorebirds: same distance, different strategies. *American Zoologist*, **35**, 222–33.

Olsen, J. L., Stam, W. T., Coyer, J. A. et al. (2004). North Atlantic phylogeography and large-scale population differentiation of the seagrass *Zostera marina* L. *Molecular Ecology*, **13**, 1923–41.

Orth, R. J. (1973). Benthic infauna of eelgrass, *Zostera marina*, beds. *Chesapeake Science*, **14**, 258–69.

Orth, R. J. (1977). The Importance of Sediment Stability in Seagrass Communities. In B. C. Coull, ed. *The Ecology of Marine Benthos*. Hobcaw Barony, Georgetown, SC, 6, 281–300.

Orth, R. J., Heck, Jr, K. L. and van Montfrans, J. (1984). Faunal communities in seagrass beds: a review of the influence of plant structure and prey characteristics on predator–prey relationships. *Estuaries*, **7**, 339–50.

Oviatt, C. A., Keller, A. A., Sampou, P. A. and Beatty, L. L. (1986). Patterns of productivity during eutrophication: a mesocosm experiment. *Marine Ecology Progress Series*, **28**, 69–80.

Packer, D. B., Griesbach, S. J., Berrien, P. L., Zetlin, C. A., Johnson, D. L. and Morse, W. W. (1999). *Summer Flounder, Paralichthys dentatus, Life History and Habitat Characteristics*, Northeast Region, National Marine Fisheries Service. National Oceanic and Atmospheric Administration, Woods Hole, MA.

Pappalardo, P., Pringle, J. M., Wares, J. P. and Byers, J. E. (2015). The location, strength, and mechanisms behind marine biogeographic boundaries of the east coast of North America. *Ecography*, **38**, 722–31.

Partridge, V. A. (2000). Aspects of the winter ecology and spring recolonization of the Windsor mudflat. Masters, Acadia University.

Pearson, N. J., Gingras, M. K., Armitage, I. A. and Pemberton, S. G. (2007). Significance of Atlantic sturgeon feeding excavations, Mary's Point, Bay of Fundy, New Brunswick, Canada. *Palaios*, **22**, 457–64.

Pearson, T. H. and Rosenberg, R. (1978). Macrobenthic succession in relation to organic enrichment and pollution of the marine environment. *Oceanography and Marine Biology Annual Review*, **16**, 229–311.

Pershing, A. J., Alexander, M. A., Hernandez, C. M. et al. (2015). Slow adaptation in the face of rapid warming leads to collapse of the Gulf of Maine cod fishery. *Science*, **350**, 809–12.

Pejrup, M. and Andersen, T. J. (2000). The influence of ice on sediment transport, deposition and reworking in a temperate mudflat area, the Danish Wadden Sea. *Continental Shelf Research*, **20**, 1621–34.

Peterson, C. H. (1977). Competitive organisation of the soft bottom macrobenthic communities of Southern California Lagoons. *Marine Biology*, **43**, 343–59.

Peterson, C. H. (1982). Clam predation by whelks (*Busycon* spp.): experimental tests of the importance of prey size, prey density, and seagrass cover. *Marine Biology*, **66**, 158–70.

Pettibone, M. H. (1963). Marine polychaete worms of the New England region. I. Aphroditidae through Trochochaetidae. *Bulletin of the United States National Museum*, **227**, 1–356.

Pfister, C., Harrington, B. A. and Lavine, M. (1992). The impact of human disturbance on shorebirds at a migration staging area. *Biological Conservation*, **60**, 115–26.

Pineda, J., DiBacco, C. and Starczak, V. (2005). Barnacle larvae in ice: survival, reproduction, and time to postsettlement metamorphosis. *Limnology and Oceanography*, **50**, 1520–8.

Posey, M. H., Alphin, T. D., Meyer, D. L. and Johnson, J. M. (2003). Benthic communities of common reed *Phragmites australis* and marsh cordgrass *Spartina alterniflora* marshes in Chesapeake Bay. *Marine Ecology Progress Series*, **261**, 51–61.

Poulakis, G. R. (2013). Reproductive biology of the Cownose Ray in the Charlotte Harbor Estuarine System, Florida. *Marine and Coastal Fisheries: Dynamics, Management, and Ecosystem Science*, **5**, 159–73.

Quijón, P. A., Kelly, M. C. and Snelgrove, P. V. R. (2008). The role of sinking phytodetritus in structuring shallow-water benthic communities. *Journal of Experimental Marine Biology and Ecology*, **366**, 134–45.

Quijón, P. A. and Snelgrove, P. V. R. (2005). Predation regulation of sedimentary faunal structure: potential effects of a fishery-induced switch in predators in a Newfoundland sub-Arctic fjord. *Oecologia*, **144**, 125–36.

Quijón, P. A. and Snelgrove, P. V. R. (2008). Trophic complexity in marine sediments. *Marine Ecology Progress Series*, **371**, 85–9.

Rayner, N. A., Parker, D. E., Horton, E. B. et al. (2003). Global analysis of sea surface temperature, sea ice, and night marine air temperature since the late nineteenth century. *Journal of Geophysical Research*, **108** (D14), 4407.

Reise, K., Simon, M. and Herre, E. (2001). Density-dependent recruitment after winter disturbance on tidal flats by the lugworm *Arenicola marina*. *Helgoland Marine Research*, **55**, 161–5.

Risk, M. J. and Craig, H. D. (1976). Flatfish feeding traces in the Minas Basin. *Journal of Sedimentary Research*, **46**, 411–13.

Robertson, A. I. and Mann, K. H. (1984). Disturbance by ice and life-history adaptations of the seagrass *Zostera marina*. *Marine Biology*, **80**, 131–41.

Roman, C. T., Jaworski, N., Short, F. T., Findlay, S. and Warren, R. S. (2000). Estuaries of the northeastern United States: habitat and land use signatures. *Estuaries*, **23**, 743–64.

Roman, J. (2006). Diluting the founder effect: cryptic invasions expand a marine invader's home range. *Proceedings of the Royal Society B: Biological Sciences*, 273, 1–7.

Rossong, M. A., Williams, P. J., Comeau, M., Mitchell, S. C. and Apaloo, J. (2006). Agonistic interactions between the invasive green crab, *Carcinus maenas* (Linnaeus) and juvenile American lobster, *Homarus americanus* (Milne Edwards). *Journal of Experimental Marine Biology and Ecology*, **329**, 281–8

Rossong, M. A. (2016). Impacts of newly established non-indigenous green crab (*Carinus maenas*) on native fauna in Placentia Bay, Newfoundland. PhD, Memorial University of Newfoundland.

Rossong, M. A., Barrett, T. J., Quijón, P. A. and Snelgrove, P. V. R. (2011a). Regional differences in foraging behaviour and morphology of invasive green crab (*Carcinus maenas*) populations in Atlantic Canada. *Biological Invasions*, **14**, 659–69.

Rossong, M. A., Quijón, P. A., Williams, P. J. and Snelgrove, P. V. R. (2011b). Foraging and shelter behaviour of juvenile American lobster (*Homarus americanus*): the influence of a non-indigenous crab. *Journal of Experimental Marine Biology and Ecology*, **403**, 75–80.

Ruiz, G. M., Fofonoff, P. W., Carlton, J. T., Wonham, M. J. and Hines, A. H. (2000). Invasion of coastal marine communities in North America: apparent patterns, processes, and biases. *Annual Review of Ecology and Systematics*, **31**, 481–531.

Ruiz, G. M., Fofonoff, P. W., Steves, B. P. and Carlton, J. T. (2015). Invasion history and vector dynamics in coastal marine ecosystems: a North American perspective. *Aquatic Ecosystem Health & Management*, **18**, 299–311.

St-Onge, P., Sévigny, J.-M., Strasser, C. and Tremblay, R. (2013). Strong population differentiation of softshell clams (*Mya arenaria*) sampled across seven biogeographic marine ecoregions: possible selection and isolation by distance. *Marine Biology*, **160**, 1065–81.

Sasko, D. E., Dean, M. N., Motta, P. J. and Hueter, R. E. (2006). Prey capture behavior and kinematics of the Atlantic cownose ray, *Rhinoptera bonasus*. *Zoology*, **109**, 171–81.

Saucier, F. J., Roy, F. and Gilbert, D. (2003). Modeling the formation and circulation processes of water masses and sea ice in the Gulf of St Lawrence, Canada. *Journal of Geophysical Research*, **108**(C8), 3269.

Schwartz, F. J. (1990). Mass migratory congregations and movements of several species of cownose rays, genus *Rhinoptera*: a world-wide review. *Journal of the Elisha Mitchell Scientific Society*, **106**, 10–13.

Schwemmer, P., Hälterlein, B., Geiter, O., Günther, K., Corman, V. M. and Garthe, S. (2014). Weather-related winter mortality of Eurasian oystercatchers (*Haematopus ostralegus*) in the northeastern Wadden Sea. *Waterbirds*, **37**, 319–30.

Scrosati, R. A. and Eckersley, L. K. (2007). Thermal insulation of the intertidal zone by the ice foot. *Journal of Sea Research*, **58**, 331–4.

Shalack, J. D., Power, A. J. and Walker, R. L. (2011). Hand harvesting quickly depletes intertidal whelk populations. *American Malacological Bulletin*, **29**, 37–50.

Shaw, A. L., Frazier, B. S., Kucklick, J. R. and Sancho, G. (2016). Trophic ecology of a predatory community in a shallow-water, high-salinity estuary assessed by stable isotope analysis. *Marine and Coastal Fisheries: Dynamics, Management, and Ecosystem Science*, **8**, 46–61.

Shearman, R. K. and Lentz, S. J. (2010). Long-term sea surface temperature variability along the U. S. east coast. *Journal of Physical Oceanography*, **40**, 1004–17.

Sheehan, E. V., Coleman, R. A., Thompson, R. C. and Attrill, M. J. (2010). Crab-tiling reduces the diversity of estuarine infauna. *Marine Ecology Progress Series*, **411**, 137–48.

Short, F. T. and Wyllie-Echeverria, S. (1996). Natural and human-induced disturbance of seagrasses. *Environmental Conservation*, **23**, 17–27.

Shuster, C. N. and Botton, M. L. (1985). A contribution to the population biology of horseshoe crabs, *Limulus polyphemus* (L.), in Delaware Bay. *Estuaries*, **8**, 363–72.

Simon, J. L. and Dauer, D. M. (1977). Re-establishment of a Benthic Community Following Natural Defaunation. In B. C. Coull, ed. *The Ecology of Marine Benthos*. University of South Carolina Press, Columbia, 6, 139–54.

Skilleter, G. A., Zharikov, Y., Cameron, B. and McPhee, D. P. (2005). Effects of harvesting callianassid (ghost) shrimps on subtropical benthic communities. *Journal of Experimental Marine Biology and Ecology*, **320**, 133–58.

Smith, J. W. and Merriner, J. V. (1985). Food habits and feeding behavior of the Cownose Ray, *Rhinoptera bonasus*, in Lower Chesapeake Bay. *Estuaries*, **8**(3), 305–10.

Snelson, Jr, F. F. and Williams, S. E. (1981). Notes on the occurrence, distribution, and biology of elasmobranch fishes in the Indian River Lagoon system, Florida. *Estuaries*, **4**, 110–20.

Sordo, L., Fournier, J., de Oliveira, V. M., Gern, F., de Castro Panizza, A. and da Cunha Lana, L. (2011). Temporal variations in morphology and biomass of vulnerable *Halodule wrightii* meadows at their southernmost distribution limit in the southwestern Atlantic. *Botanica Marina*, **54**, 13–21.

Stafford, N. B. and Bell, S. S. (2006). Space competition between seagrass and *Caulerpa prolifera* (Forsskaal) Lamouroux following simulated disturbances in Lassing Park, FL. *Journal of Experimental Marine Biology and Ecology*, **333**, 49–57.

Stephenson, T. A. and Stephenson, A. (1954). Life between tide-marks in North America. IIIB. Nova Scotia and Prince Edward Island: geographical features of the region. *Journal of Ecology*, **42**, 46–70.

Steward, J. S., Virnstein, R. W., Morris, L. J. and Lowe, E. F. (2005). Setting seagrass depth, coverage, and light targets for the Indian River Lagoon system, Florida. *Estuaries*, **28**, 923–35.

Strasser, M. T., Hertlein, A. and Reise, K. (2001a). Differential recruitment of bivalve species in the northern Wadden Sea after the severe winter of 1995/96 and of subsequent milder winters. *Helgoland Marine Research*, **55**, 182–9.

Strasser, M., Reinwald, T. and Reise, K. (2001b). Differential effects of the severe winter of 1995/96 on the intertidal bivalves *Mytilus edulis*, *Cerastoderm edule* and *Mya arenaria* in the northern Wadden Sea. *Helgoland Marine Research*, **55**, 190–7.

Strasser, M. and Pieloth, U. (2001). Recolonization pattern of the polychaete *Lanice conchilega* on an intertidal sand flat following the severe winter of 1995/96. *Helgoland Marine Research*, **55**, 176–81.

Thrush, S. F. and Dayton, P. K. (2002). Disturbance to marine benthic habitats by trawling and dredging: implications for marine biodiversity. *Annual Review of Ecology and Systematics*, **33**, 449–73.

Thrush, S. F., Hewitt, J. E., Cummings, V. J., Green, M. O., Funnell, G. A. and Wilkinson, M. R. (2000). The generality of field experiments: interactions between local and broad-scale processes. *Ecology*, **81**, 399–415.

Thrush, S. F., Hewitt, J. E., Parkes, S. et al. (2014). Experimenting with ecosystem interaction networks in search of threshold potentials in real world marine ecosystems. *Ecology*, **95**, 1451–7.

Townsend, E. C. and Fonseca, M. S. (1998). Bioturbation as a potential mechanism influencing spatial heterogeneity of North Carolina seagrass beds. *Marine Ecology Progress Series*, **169**, 123–32.

Urian, A. G., Hatle, J. D. and Gilg, M. R. (2011). Thermal constraints for range expansion of the invasive green mussel, *Perna viridis*, in the southeastern United States. *Journal of Experimental Zoology*, **315**, 12–21.

van Gils, J. A., van der Geest, M., Jansen, E. J., Govers, L. L., de Fouw, J. and Piersma, T. (2012). Trophic cascade induced by molluscivore predator alters porewater biogeochemistry via competitive release of prey. *Ecology*, **93**, 1143–52.

van Gils, J. A., Lisovski, S., Lok, T. et al. (2016). Body shrinkage due to Arctic warming reduces red knot fitness in tropical wintering range. *Science*, **352**, 819–21.

Vermeij, G. J., Dietl, G. P. and Reid, D. G. (2008). The trans-Atlantic history of diversity and body size in ecological guilds. *Ecology*, **89**(Suppl.), S39–52.

Virnstein, R. W. (1977). The importance of predation by crabs and fishes on benthic fauna in Chesapeake Bay. *Ecology*, **58**, 1199–217.

Wares, J. P. and Cunningham, C. W. (2001). Phylogeography and historical ecology of the North Atlantic intertidal. *Evolution*, **55**(12), 2455–69.

Welsh, D. T. (2003). It's a dirty job but someone has to do it: the role of marine benthic macrofauna in organic matter turnover and nutrient recycling to the water column. *Chemistry & Ecology*, **19**(5), 321–42.

Wethey, D. S. (1985). Catastrophe, extinction, and species diversity: a rocky intertidal example. *Ecology*, **66**, 445–6.

Wethey, D. S., Woodin, S. A., Hilbish, T. J., Jones, S. J., Lima, F. P. and Brannock, P. M. (2011). Response of intertidal populations to climate: effects of extreme events versus long term change. *Journal of Experimental Marine Biology and Ecology*, **400**, 132–44.

Whitlatch, R. B. (1976). Seasonality, species diversity and patterns of resource utilization in a marine deposit-feeding community. PhD, University of Chicago.

Whitlatch, R. B. (1977). Seasonal changes in the community structure of the macrobenthos inhabiting the intertidal sand and mud flats of Barnstable Harbor, Massachusetts. *Biological Bulletin*, **152**, 275–94.

Widener, J. W. and Barlow, R. B. (1999). Decline of a horseshoe crab population on Cape Cod. *Biological Bulletin*, **197**, 300–2.

Williams, P. J., Floyd, T. A. and Rossong, M. A. (2006). Agonistic interactions between invasive green crabs, *Carcinus maenas* (Linnaeus), and sub-adult American lobsters, *Homarus americanus* (Milne Edwards). *Journal of Experimental Marine Biology and Ecology*, **329**, 66–74.

Wilson, Jr, W. H. (1989). Predation and the mediation of intraspecific competition in an infaunal community in the Bay of Fundy. *Journal of Experimental Marine Biology and Ecology*, **132**, 221–45.

Wilson, Jr, W. H. (1991). The importance of epibenthic predation and ice disturbance in a Bay of Fundy mudflat. *Ophelia*, (Suppl. 5):507–14.

Wiltse, W. (1980). Effects of *Polinices duplicatus* (Gastropoda: Naticidae) on infaunal community structure at Barnstable Harbor, Massachusetts, USA. *Marine Biology*, **56**, 301–10.

Wiltse, W., Foreman, K., Teal, J. and Valiela, I. (1984). Effects of predators and food resources on the macrobenthos of salt marsh creeks. *Journal of Marine Research*, **42**, 923–42.

Wong, M. C. (2013). Green crab (*Carcinus maenas* [Linnaeus, 1758]) foraging on soft-shell clams (*Mya arenaria* Linnaeus, 1758) across seagrass complexity: Behavioural mechanisms and a new habitat complexity index. *Journal of Experimental Marine Biology and Ecology*, **446**, 139–50.

Wong, M. C. and Dowd, M. (2014). Role of invasive green crabs in the food web of an intertidal sand flat determined from field observations and a dynamic simulation model. *Estuaries and Coasts*, **37**, 1004–16.

Woodin, S. A. (1976). Adult-larval interactions in dense infaunal assemblages: patterns of abundance. *Journal of Marine Research*, **34**, 25–41.

Woodin, S. A. (1978). Refuges, disturbance, and community structure: a marine soft-bottom example. *Ecology*, **59**, 274–84.

Woodin, S. A. (1981). Disturbance and community structure in a shallow water sand flat. *Ecology*, **62**, 1052–66.

Woodin, S. A., Volkenborn, N., Pilditch, C. A. et al. (2016). Same pattern, different mechanism: locking onto the role of key species in seafloor ecosystem process. *Scientific Reports*, **6**, 26678.

Zieman, J. C. (1982). *The Ecology of the Seagrasses of South FLORIDA: A Community Profile*, FWS/OBS-82/25. U.S Fish Wildlife Service, Biological Service Program, Washington, DC.

Chapter 7

# Biodiversity and Interactions on the Intertidal Rocky Shores of Argentina (South-West Atlantic)

Maria Gabriela Palomo, Maria Bagur, Sofia Calla, Maria Cecilia Dalton, Sabrina Andrea Soria and Stephen J. Hawkins

## 7.1 Introduction

Here, we review the rocky intertidal zone and shallow subtidal of the Argentinian coast. We start by describing patterns of distribution of biodiversity before considering interactions among species that shape these patterns. Non-native and invasive species are then discussed and put into the context of global change and other local impacts. We conclude by considering the special features of Argentinean rocky shores which are shaped by their environmental setting and phylogeographic history leading to low diversity, missing functional groups for some taxa and a gradient of increasing diversity towards higher latitudes in the south.

## 7.2 Biodiversity and Patterns of Distribution

### 7.2.1 Generalities

The Argentinean coast extends 7,000 km from Río de la Plata (RP) (36°S, 56°W) to Tierra del Fuego (TF) (54°S, 64°W) (Martos and Piccolo, 1988, Figure 7.1). Intertidal rocky platforms increase in frequency and extent from north to south along the coast.

In Buenos Aires province the rocky coastal area is restricted (rarely exceeding 1 km in length), discontinuous and frequently fragmented by urban areas (Miloslavich et al., 2016) and artificial coastal defences (Marcomini and López, 1993). Most of the intertidal platforms in this province are composed of consolidated sediments (e.g., limestones, sandstones, calcretes) that support both epilithic and burrowing endolithic biota (Amor et al., 1991; Bagur et al., 2013, 2014); the only exceptions being a handful of hard, metamorphic rock (orthoquartzite) outcrops adjoining the city of Mar del Plata (Jaubet and Genzano, 2011; Gutiérrez et al., 2015) that only support epilithic biota. There, most ecological research has been adjacent to urban centres such as Mar del Plata, Quequén, Viedma and Ushuaia (see Figure 7.2). Research on the rocky shores of Buenos Aires province started in the 1960s with descriptive studies of zonation patterns, trophic relationships and studies focussed on commercial species (e.g., Olivier et al., 1966b, 1968, 1972; Olivier and Penchaszadeh, 1968). In subsequent decades, research focussed on mussel beds and their associated communities (e.g., Penchaszadeh, 1973; Nugent,

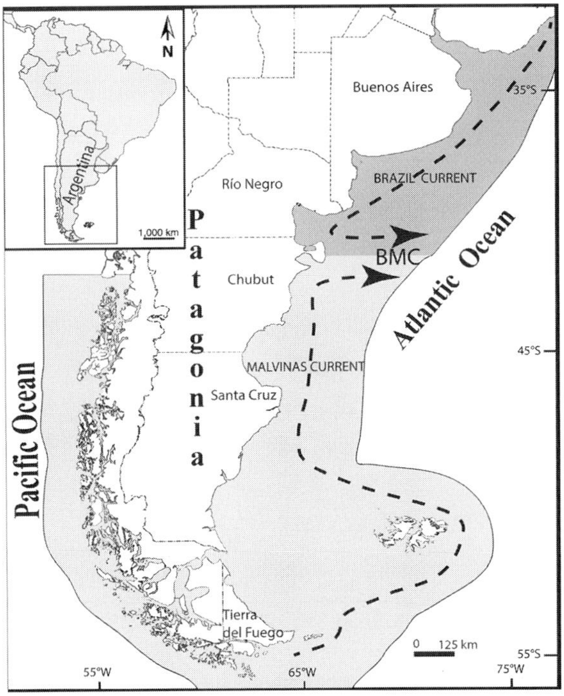


**Fig. 7.1** The Argentinian coast. Shading denotes biogeographic provinces. Dark grey: Argentine province (36–43°S); light grey: Magellanic province (43–56°S). Political provinces are labelled on the land. Arrows denote the warm-water Brazil Current and the cold-water Malvinas Current system. BMC denotes the Brazil–Malvinas confluence of these two systems. Modified from Liuzzi et al. (2011).

1986; Scelzo et al., 1996) and more recently on the relationships between environment, community structure and ecosystem functioning (e.g., López Gappa et al., 1990, 1993; Amor et al., 1991; Elías et al., 2001, 2003; Bagur et al., 2014; Arribas et al., 2015; Gutiérrez et al., 2015).

Similarly, in Patagonia, early studies described the intertidal communities and zonation patterns mainly in the gulfs of San Matías, Nuevo and San José (e.g., Ringuelet et al., 1962; Olivier et al., 1966a; Otaegui and Zaixso, 1974). In recent decades, species data have been updated to include new biogeographic information, mainly for magellanic fishes of commercial interest (e.g., Bastida et al., 1992; Roux and Bremec, 1996).

Oceanographic studies of the Argentinian shelf suggest that the principal circulation is characterised by the confluence of the cold, northwards-flowing, Malvinas Current and the warm, southwards-flowing, Brazil Current (Garzoli and Garraffo, 1989; Bastida et al., 1992; Brandini et al., 2000; Acha et al., 2004; Lucas et al., 2005). The Brazil and the Malvinas currents converge at the Brazil–Malvinas confluence (BMC) along the continental margin of South America within latitudes 35–45°S (Garzoli and Garraffo, 1989; García et al., 2004; Figure 7.1). At the BMC, the mixing of subtropical and sub-Antarctic waters determines the physico-chemical characteristics of the coastal waters, leading to high concentrations of nutrients and enhancing productivity in this region (Campagna et al., 2005).

## 7.2.2 Biogeographic Provinces

### 7.2.2.1 Argentine Province

Biodiversity studies of the continental shelf showed that the fauna in Argentina can be divided into two biogeographic provinces: the Argentine and Magellanic provinces (Piola and Rivas, 1997; Balech and Ehrlich, 2008, Figure 7.1). The Argentine province belongs to the eco-region of the warm temperate south-western Atlantic (Spalding et al., 2007), extending from 36 to 43°S, including the provinces of Buenos Aires and Río Negro and the north of Chubut province (Figure 7.1). The coasts in the Argentine province are gently sloping and characterised by semi-diurnal, low-amplitude tides (mean: 0.80 m). Water temperature ranges from 8 to 23°C with latitude and season (Boschi, 1979). The Magellanic province extends from 43 to 56°S along southern Chubut, as well as Santa Cruz and TF, including the Malvinas (the Falkland Islands) and southern Chile (Boschi, 1979; Spalding et al., 2007; Balech and Ehrlich, 2008; Figure 7.1). This province is characterised by macrotidal regimes, with tidal amplitudes of more than 8 m in Patagonia on shores characterised by extensive wave abraded platforms. Water temperature ranges from 4 to 16°C (Boschi, 1979). Shores in this province have gravelly seabeds, and experience strong, dry, offshore westerly winds and low salinity waters from the Malvinas Current (Campagna et al., 2005; Balech and Ehrlich, 2008; Piola et al., 2009).

**Fig. 7.2** Examples of intertidal rocky shores in Argentina: (a) Mar del Plata (38°01′S, 57°31′W) and (b) Quequén (38°34′S, 58°40′W) in Buenos Aires province; and (c) El Espigón (41°08′S, 63°02′W) in Río Negro province and (d) Ushuaia (54°49′S, 68°11′W) in Tierra del Fuego province. *A black and white version of this figure will appear in some formats. For the colour version, please refer to the plate section.*

Three characteristic zones (high, mid and low) can be observed on Argentinean intertidal rocky shores. Along with tidal height, wave exposure and desiccation are ultimate physical factors determining species distributions directly and indirectly by modulating biological interactions (Olivier et al., 1966b, 1970; Penchaszadeh, 1973, Zamponi and Genzano, 1992; Bertness et al., 2006).

### 7.2.2.2 Magellanic Province

The distribution of species show a similar zonation pattern throughout the province of Buenos Aires; the high intertidal is characterised by the limpet, *Siphonaria lessonii,* and barnacles (mainly the invasive *Balanus glandula*), together with a few scattered mussels (*Brachidontes rodriguezii*). The mid intertidal is dominated by extensive beds of the mussel, *B. rodriguezii,* with small associated invertebrates such as limpets, anemones, polychaetes and algae such as *Ulva* (formerly *Enteromorpha*) *intestinalis* and *Polysiphonia* sp. Finally, the low intertidal is characterised by the dominance of the coralline algae, *Corallina officinalis*, with the mussels *B. rodriguezii* and *Mytilus edulis*.[1] A higher diversity of invertebrates and algae is present in the low intertidal, including chitons, anemones, polychaetes, asteroids, limpets, crabs and other crustaceans (Olivier et al., 1966b; López Gappa et al., 1993; Bertness et al., 2006, Bazterrica et al., 2007; Carreto et al., 2007; Miloslavich et al., 2016; Soria et al., 2017.; see Table 7.1 for species information).

In central Patagonia few species inhabit the high-intertidal zone, which is generally covered by films of cyanobacteria. The mid intertidal is dominated by the mussel, *P. purpuratus* and the low intertidal by *C. officinalis*. Molluscs such as *S. lessoni*, *Trophon* sp. and *Lithophaga patagonica* are common at the mid and low intertidal zone. The low intertidal is occupied by mussels such as *M. edulis* and *Aulacomya atra* plus hydrozoans, crabs, bryozoans, polychaetes and sponges (Cuevas et al., 2006; Carreto et al., 2007; see Table 7.1). In southern Patagonia, rocky shores are exposed to adverse physical conditions caused by strong winds and

[1] This species is also often referred to as *Mytilus platensis* or *Mytilus edulis platensis*. Hereafter, we only refer to it as *Mytilus edulis*.

**Table 7.1** Main species found on Argentinian rocky shores: distribution and zonation in the intertidal. AP: Argentine province; MP: Magellanic province; RJ: Rio de Janeiro (22°S); SCA: Santa Catarina (27°S); RGS: Rio Grande do Sul (30°S); BA: Buenos Aires; SMG: San Matías Gulf (41°S); CH: Chubut, PV: Península Valdes (42°S); SC: Santa Cruz; SJG: San Jorge Gulf (46°S), PD: Puerto Deseado (47°S); RG: Río Grande (53°S), IM/FI: Islas Malvinas/Falkland Islands (51°S); IC: Isla Chiloé (42°S). Used here are the current accepted taxonomic names.

| Major taxonomic group | Species | Distribution in South America | Zonation on the rocky shore | Reference |
|---|---|---|---|---|
| Actinaria | *Anthothoe chilensis* (Lesson, 1830) | South Pacific Ocean and Chile and from South Brazil to TF | Mid and low intertidal and subtidal | Excoffon et al. (1997), Zamponi et al. (1998), Excoffon et al. (1999) |
| Asteroidea | *Anasterias antarctica* (Perrier, 1875) | Central and south Patagonia | Low intertidal and subtidal of Patagonia | Salvat (1985), Arribas et al. (2017) |
| Bivalvia | *B. rodriguezii* (d'Orbigny, 1846) | South Brazil to CH | Mid intertidal of BA and central Patagonia | Olivier et al. (1966b), Penchaszadeh (1973), Borthagaray and Carranza (2007), Adami (2008), Arribas et al. (2013), Trovant et al. (2015) |
| | *M. edulis platensis* (d'Orbigny, 1846) | South Brazil to TF | Mid intertidal in Patagonia | Pastorino (1995), Adami et al. (2013), Arribas et al. (2013), Trovant et al. (2015) |
| | *Perumytilus purpuratus* (Lamarck 1819) | South Chile to Ecuador and Golfo San Matias to TF | Mid and low intertidal in southern Patagonia | Marincovich (1973) |
| Cephalopoda | *Octopus tehuelchus* (d'Orbigny, 1834) | South Brazil to PV, CH | Low intertidal and subtidal of GSM, northern Patagonia | Haimovici and Perez (1991), Iribarne (1991b), Pastorino (1995) |
| Chlorophyta | *Cladophora* sp. (Kützing, 1843), *Ulva* sp. (Linneaus, 1753) | Atlantic coast, from north BA to TF | On mussel beds in mid- and low-intertidal | López Gappa et al. (1993), Boraso and Zaixso (2011), Liuzzi et al. (2011) |
| Gastropoda | *Fissurela radiosa tixierae* (Métivier, 1969) | IC, Chile to TF and GSM, CH | Mid intertidal in Patagonia | Pastorino (1995), Olivares Paz (2007) |
| | *Nacella deaurata* (Gmelin, 1791) | Southern Chile and coast of TF | Mid and low intertidal in Patagonia | Valdovinos and Rüth (2005), De Aranzamendi et al. (2009), Aldea and Rosenfeld (2011) |

*(cont)*

Table 7.1 (cont.)

| Major taxonomic group | Species | Distribution in South America | Zonation on the rocky shore | Reference |
|---|---|---|---|---|
| | *Nacella magellanica* (Gmelin, 1791) | Southern Chile to BA, IM/FI | Mid and low intertidal in Patagonia | Pastorino (1995), De Aranzamendi et al. (2009) |
| | *S. lessonii* (Blainville, 1827) | SCA, Brazil to TF and IM/FI | High and mid intertidal in mussel beds | Olivier and Penchaszadeh (1968), Marincovich (1973), Pastorino (1995), Bazterrica et al. (2007), Nuñez et al. (2015), Soria et al. (2017) |
| | *Tegula patagonica* (d'Orbigny, 1835) | RJ, Brazil to TF | Low intertidal and subtidal in Patagonia | Pastorino (1995) |
| | *Trophon geversianus* (Pallas, 1774) | South BA to TF and sub-Antarctic islands | Low intertidal and subtidal mussel beds in Patagonia | Pastorino (1995), Pastorino et al. (2014) |
| Malacostraca | *Cyrtograpsus altimanus* (Rathbun, 1914) | From RGS, Brazil, to GSJ, CH and SC | Low intertidal and subtidal | Boschi et al. (1992), Vinuesa (2005) |
| | *Cyrtograpsus angulatus* (Dana, 1851) | Discontinuously in South Pacific Ocean, from RJ, Brazil, to PD, SC | Low intertidal and subtidal | Vinuesa (2005) |
| Polychaeta | *Boccardia polybranchia* (Haswell, 1885) | Worldwide in warm and temperate seas | In sediments within mussel beds in ABP | Elías et al. (2003), Vallarino and Elías (2006), Orensanz et al. (2002) |
| | *Harmothoe magellanica* (McIntosh, 1885) | Antarctic, sub-Antarctic and MBP | Low intertidal and subtidal in Patagonia | Barnich et al. (2006, 2012), Orensanz et al. (2002) |
| | *Marphysa aenea* (Blanchard in Gay, 1849) | From Uruguay to IM/FI | Mid and low intertidal in Patagonia | Orensanz (1974) |
| | *Syllis gracilis* (Grube, 1840) | Worldwide in warm and temperate seas | In sediments within mussel beds in ABP | Vallarino et al. (2006), Orensanz et al. (2002) |
| | *Syllis prolixa* (Ehlers, 1901) | Worldwide in warm and temperate seas | In sediments within mussel beds in ABP | Elías et al. (2006, 2009) |
| Polyplacophora | *Chaetopleura isabellei* (d'Orbigny, 1841) | SCA, Brazil to CH | Low intertidal | Pastorino (1995), López Gappa and Tablado (1997) |
| Rhodophycophyta | *Ceramium* sp. (Roth, 1797), *C. officinalis* (Linnaeus, 1758), *Polysiphonia* sp. (Greville, 1823), *Pyropia leucosticta* (Thuret, 1863) | Atlantic coast, from north BA province to TF | Low intertidal of BA and Patagonia | López Gappa et al. (1993), Bertness et al. (2006), Boraso and Zaixso (2011), Liuzzi et al. (2011), Miloslavich et al. (2016) |

high desiccation rates (Bertness et. al., 2006). The high intertidal zone is covered by lichens and some chlorophytes. The mid intertidal level is completely dominated by *P. purpuratus* with some *S. lessonii* and *Ulva* spp. The low intertidal is characterised by *P. purpuratus, M. platensis, A. atra* and gastropods such as *N. magellanica* and *F. radiosa tixierae*; the algae mainly consist of *C. officinalis* and *Codium fragile* (Carreto et al., 2007; see Table 7.1).

In TF (southernmost Argentina, see Figure 7.1) the high intertidal zone is dominated by the barnacle *Notochthamalus scabrosus*, the mid-intertidal zone by the mussels *P. purpuratus*, *Mytilus chilensis* (which are co-dominants, Calcagno et al., 2012) and *A. atra*, while the lower intertidal zone is covered by the barnacle *Notobalanus flosculus* and by crustose coralline algae (Curelovich et al., 2018; Zaixso et al., 2017).

### 7.2.3 Biodiversity Patterns

The diversity patterns of some taxonomic groups along the Argentinean coast show an increase in richness at high latitudes. For example, the Argentine biogeographic province has a lower richness of polychaetes than the Magellanic biogeographic province (Orensanz, 1974; Elías et al., 2001; Bremec and Giberto, 2006; Bremec and Schejter, 2010). In the mussel beds of Buenos Aires province, syllids, cirratullids and spionids are abundant (Elías et al., 2006); whereas, in Patagonia, in addition to these groups, eunicids, onuphids, nereids and scale-worms such as polynoids occur (Orensanz, 1974; San Martín et al., 2005; Elías et al., 2017). A similar pattern is observed for the macroalgae. The marine flora of the coast of Argentina changes with latitude, with the low salinity plume of the RP having little influence (Boraso and Zaixso, 2008). In the Argentine Biogeographic province, the common seaweeds are Chlorophycophyta (*Ulva* formerly *Enteromorpha* sp., *Cladophora* sp.) and some Rhodophycophyta (*P. leucosticta, Ceramium* sp., *C. officinalis, Polysiphonia* sp.; López Gappa et al., 1993; see Table 7.1). In the Magellanic Biogeographic province, algal richness, mainly of green algae, increases at higher latitudes (Liuzzi et al., 2011).

The observed distribution patterns in the Argentinean rocky intertidal indicate an opposite trend to the general paradigm of higher diversity at lower latitudes (Gaston, 2000; Hillebrand, 2004): with lower diversity lower towards the equator. This contrasts with the Pacific coast of Chile where diversity increases equator-wards (e.g., in marine isopods and peracarid crustaceans; Benedetti-Cecchi et al., 2010; Liuzzi et al., 2011; Rivadeneira et al., 2011; Chapter 12; J. J. Cruz Motta, personal communication), and patterns elsewhere in the world (e.g., north-east Atlantic, Chapter 2; north-east and north-west Pacific, Chapters 10 and 14).

The rocky shores of Argentina, despite their geographic extent, are characterised by low biodiversity and low biomass when compared with other parts of the world (Wieters et al., 2012). Different studies of the Argentinean rocky coast show that, as in most other parts of the world, the structure of intertidal communities is strongly correlated to environmental conditions such as precipitation and strong southerly desiccating winds (45–140 km/h) (Arribas et al., 2013, 2015; Gutiérrez et al., 2015). The influence of ecosystem engineers, such as mussels (Gutiérrez et al., 2003; Sousa et al., 2009; Bagur et al., 2013), in turn affected by the recruitment of larvae (Arribas et al., 2015), helps modulate these harsh environmental conditions. These interacting factors have all been shown to play a role in structuring communities at the local scale (Bertness et al., 2006; Hidalgo et al., 2007). However, some processes operating at local scales such as organic and inorganic pollution, primary productivity and nutrient loading also can scale-up to determine distribution patterns at broader regional scales (Benedetti-Cecchi et al., 2010).

## 7.3 Factors Influencing Species Distribution Patterns on Argentinean Rocky Shores

### 7.3.1 Facilitation

The clearest examples of facilitation in Argentinean rocky shores are the numerous invertebrates and algae that use the extensive mussel beds as a place to live and as a refuge from predators and harsh environmental conditions (Adami et al., 2004; Borthagaray and Carranza, 2007;

Silliman et al., 2011; Arribas et al., 2013;, Bagur et al., 2016). The mid and high intertidal platforms in Argentinean rocky shores are dominated by bed-forming mussels, either *B. rodriguezii* (36–41°S), *P. purpuratus* (43–52°S) or both (41–43°S; see Arribas et al., 2013, Trovant et al., 2015). Mussels are known to physically alter their environments (via physical ecosystem engineering; *sensu* Jones et al., 1994) by modifying abiotic conditions and resources available to other organisms (Gutiérrez et al., 2003, 2011). Flow speeds, sediment transport, temperature and moisture fluctuations are lower in the interstices of mussel beds relative to adjacent bare rock (e.g., Carrington et al., 2008; Silliman et al., 2011; Bagur et al., 2016). These two species of mussels (*B. rodriguezii* and *P. purpuratus*) support different assemblages of invertebrates, even where they co-occur, suggesting they can affect abiotic conditions and/or modulate resources to other organisms in a different way (Arribas et al., 2013).

The importance of facilitation is most clearly seen on the rocky intertidal communities of Patagonia which are exposed to harsh physical conditions caused by dry, strong southern trade winds (mean speed 45 km/h, gusts up to 140 km/h, Paruelo et al., 1998) which result in intense desiccation of intertidal organisms (Bertness et al., 2006). The high desiccation rates make Patagonian rocky shores one of the most physically demanding rocky intertidal habitats in the world (Bertness et al., 2006; Hidalgo et al., 2007; Silliman et al., 2011). This highly stressful physical environment restricts the assemblage of mobile invertebrates to those living among the mussel matrix, outside of which they would die within 15 min of air exposure (Bertness et al., 2006; Silliman et al., 2011). In these extreme climates, the role of ecosystem engineers (*sensu* Jones et al., 1994) such as mussels is key to maintaining diversity by positive interactions that ameliorate physical stress (Figure 7.3; Silliman et al., 2011).

Similar facilitation is provided by dense mats of coralline algae in the low-intertidal zone (Bertness et al., 2006; Liuzzi and López Gappa, 2008) and by boring bivalves on soft rock intertidal shores (Bagur et al., 2016). Up to eleven endolithic macrofaunal species can be found on shores consisting of consolidated sediments (Bagur et al., 2014); their boreholes are frequently occupied by many different invertebrate species. In particular, the piddock, *L. patagonica* (a long-lived boring bivalve, abundant on consolidated-sediment shores in Argentina; Bagur et al., 2013, 2014) can create habitat for around thirty invertebrate species (Bagur et al., 2016). Interestingly, where both mussels and piddocks coexist, the species associated with piddocks are broadly similar to those found within mussel beds (*P. purpuratus*). Nevertheless, the species composition differs between the two ecosystem engineers. The distinctive habitat patches created by each engineer add exclusive subsets of species, which implies that the mussel and *L. patagonica* patches contribute in a complementary fashion to overall species richness in the intertidal landscape (Bagur et al., 2016).

A 'fouling cascade' – another case of facilitation by which epibionts facilitate secondary colonisation by other epibionts – has been recorded in Mar del Plata (Buenos Aires Province; Gutiérrez and Palomo, 2016). The presence of epibiotic barnacles (*B. glandula*) on mussel (*B. rodriguezii*) shells promoted the subsequent fouling by an ephemeral red algae (*Porphyra* sp.). This fouling cascade could explain the non-random aggregation of multiple epibiotic species onto a proportionally few individuals of the mussel host species (Gutiérrez and Palomo, 2016).

## 7.3.2 Competition between Species

Despite its possible importance in regulating species population densities on rocky shores worldwide (Connell, 1961; Dayton, 1971; Lubchenco, 1980), there are very few studies investigating competition between species on Argentinean rocky shores, with the exception of some pioneering studies in Argentinean Patagonia (Arribas et al., 2016). In this region, *B. rodriguezii* and *P. purpuratus,* the two super-abundant mussels that form extensive mussel beds along Argentinean coasts, coexist between 41 and 43°S (Trovant et al., 2013, 2015; Arribas et al., 2016; see also work by McQuaid and Blamey in Chapter 13). Within their coexistence zone (La Lobería) mussel densities were manipulated to test the effect of intra- and inter-specific competition on growth and mortality of both bivalve species (Arribas et al., 2016). Only the

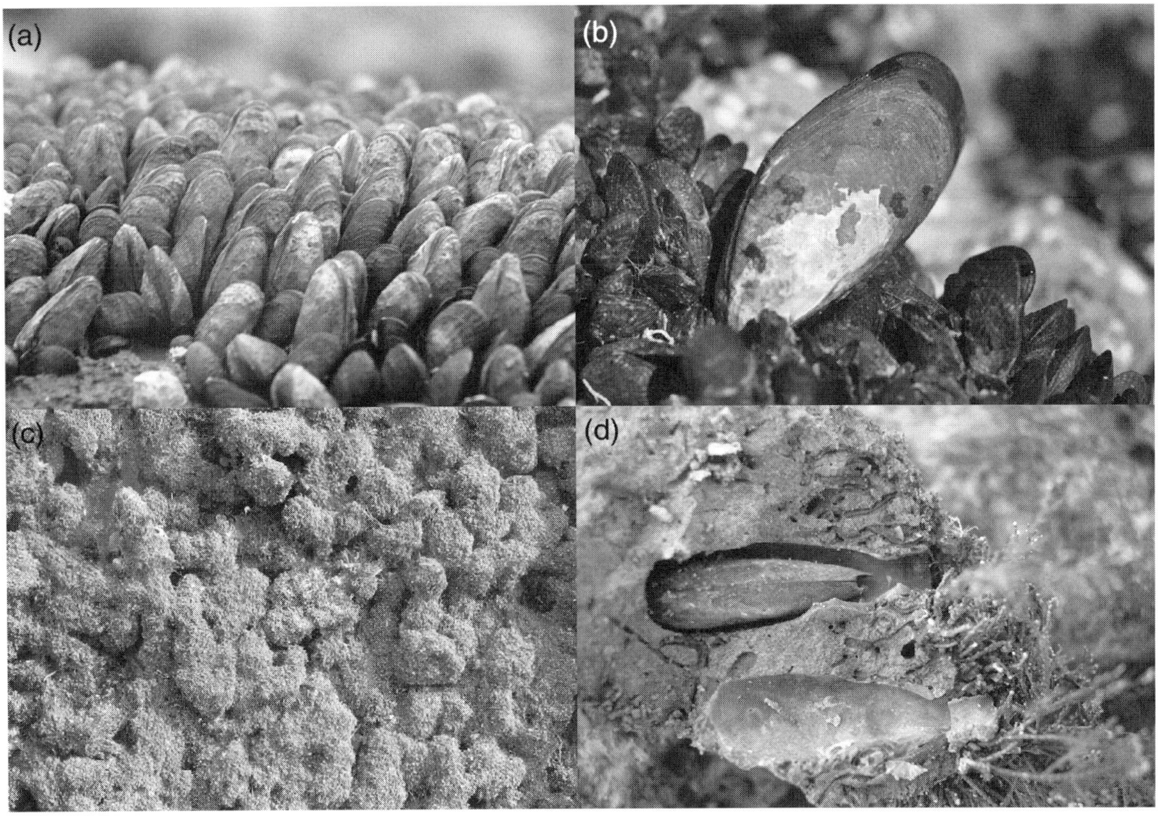


**Fig. 7.3** Common intertidal ecosystem engineers on Argentinian rocky shores. (a) The mussel *P. purpuratus*; (b) the mussel *M. edulis platensis* (emerging from a *P. purpuratus* mussel bed); (c) the alga *C. officinalis* (in a 0.5 m × 0.5 m quadrat); (d) the boring bivalve *L. patagonica*. Photo credits: (a and b) www.proyectosub.com.ar and (c and d) María Bagur. *A black and white version of this figure will appear in some formats. For the colour version, please refer to the plate section.*

width of *P. purpuratus* showed an increase in the presence of *B. rodriguezii*, and mortality remained the same in all treatments. Many replicates were lost in this experiment and thus further studies are needed. Nevertheless, this preliminary study suggests that limited supply of *P. purpuratus* larvae or local tidal conditions that differentially affect each species could be leading to weak competition between both mussel species, thus allowing their coexistence as space is undersaturated.

The non-indigenous marine macroalgae *Sciyzymenia dubyi* and *Ahnfeltiopsis* sp. are rapidly spreading in the rocky intertidal in Mar del Plata. Although their impact is currently unknown, they could potentially displace the native mussel *B. rodriguezii* (Palomo et al., 2016). These non-native algal species can also be potential competitors for space with native seasonal macroalgae, with possible negative consequences for the intertidal assemblages, although further studies are needed to test these hypotheses. Similarly, the non-native barnacle *B. glandula* – introduced to Argentina in the 1970s – has probably changed the community structure, with more pronounced effects in the upper-intertidal zone (where it is more abundant) by outcompeting native mussels that were previously occupying this zone (Schwindt et al., 2014).

### 7.3.3 Influence of Grazing

Similar to other rocky shores in the world, limpets (*S. lessonii, Nacella* spp., *F. radiosa tixierae*), snails (*T. patagonica*) and chitons (*C. isabellei, Plaxiphora aurata*) are the most common grazers in Argentina (Paine, 1980; Bazterrica et al. 2007; Adami, 2008) (Figure 7.4). Herbivorous amphipods and isopods are mesograzers also frequently present as

**Fig. 7.4** Common intertidal grazers on Argentinian rocky shores. (a) The pulmonate limpet *S. lessonii*; (b) the true-limpet *N. magellanica*; (c) the snail *T. patagonica*; (d) the keyhole limpet, *F. radiosa tixierae*. Photo credits: www.proyectosub.com.ar. *A black and white version of this figure will appear in some formats. For the colour version, please refer to the plate section.*

epiphytes of macroalgae. However, there are only a few studies documenting the role of grazers in controlling different algal populations (Bazterrica et al., 2007; Adami, 2008; Teso et al., 2009). Furthermore, the basic general biology and ecology of many species is still poorly understood. More studies investigating, for example, the diet of different species and their impacts on the intertidal community are needed.

The pulmonate limpet *S. lessonii* is present on all Argentinian rocky shores, frequently reaching high densities (e.g., 4,000 ind/m$^2$ at the mid-intertidal zone in Quequén, Adami, 2008; up to 3,676 ind/m$^2$ within mussels in Mar del Plata, Olivier and Penchaszadeh, 1968; 914 ind/m$^2$ on the mid-intertidal zone in Mar del Plata, Soria et al., 2017, 150 ind/m$^2$ on the mid-intertidal zone in Chubut, Bazterrica et al., 2007). *S. lessonii* feed on macro- and micro-algae that grow attached to mussels and on the rock surface (Olivier and Penchaszadeh, 1968; Bastida et al., 1971), almost exclusively during high tides (Olivier and Penchaszadeh, 1968; López Gappa et al., 1996). *Ulva rigida* and other thin bladed *Ulva* spp (formerly called *Enteromorpha*) are the main items in its diet (Bastida et al., 1971). *S. lessonii* also appears to have an effect on the recruitment of the macroalgae *P. leucosticta* and *Petalonia fascia* and microalgae *Navicula* sp., all growing on mussel shell surfaces (see Adami, 2008). *S. lessonii* can forage for considerable distances in the intertidal (Olivier and Penchaszadeh, 1968), but it shows differential behaviour, probably in response to the environmental conditions in which it lives. Homing behaviour (i.e., returning to the same crevice every day) of this limpet can be sensitive to biotic variables such as food supply, as well as to different climate contexts, such as desiccation (Nuñez et al., 2014). However, this sensitivity can also change at different time scales (i.e., days or seasons; Nuñez et al., 2014).

The ecology, diet and feeding behaviour of the two most common limpets in Patagonia, *N. magellanica* and *F. radiosa tixierae*, and the snail *T. patagonica* are still poorly understood. *N. magellanica* has been reported in Chile to be an omnivorous grazer, feeding on green microalgae, microbivalves and foraminiferans (Andrade Diaz and Brey, 2014). Although it is usually reported as a grazer (e.g., Wieters et al., 2012), we are unaware of any studies on the diet of *F. radiosa tixierae*. Many fissurellids are sponge-feeders or detritivores (Fretter and Graham, 1962), although some species seem to be exclusively herbivorous and others have been observed scavenging on animal material (Hickman, 1998). The topshell *T. patagonica* is a very common and abundant snail in Patagonia. A subtidal population of this snail in Golfo Nuevo has been observed consuming the biofouling adhered to the surface of the invasive kelp *Undaria pinnatifida* (Teso et al., 2009). High densities of this snail were detected consuming *Undaria* mainly during the alga's senescent period (December–April), when the algae are presumable more palatable. Despite their distribution and abundance, more studies are needed to unravel the ecological role and interactions of these three grazers on Patagonian rocky shores.

A few studies have examined the role of grazers in maintening bare space on rocky shores (see Bazterrica et al., 2007; Adami, 2008; Curelovich et al., 2018). The effect of grazers seems to vary in different regions. In Quequén (Buenos Aires province), where the climate is temperate and humid, Adami (2008) excluded *S. lessoni* and consequently algal biomass increased five-fold after five months, suggesting that the grazing activity of the pulmonate limpet regulates the biomass of epibenthic algae (growing on the mussel *B. rodriguezii*) in this region. However, in Chubut province, where the climate is much drier than in Buenos Aires and winds are strong, the effects of limpets grazing (*S. lessoni* in the high and mid intertidal zones and *N. magellanica* in the low-intertidal zone) are weak in comparison to the physical stresses that largely determine community structure (see Bazterrica et al., 2007). Although the densities of grazers are similar to other temperate rocky shorelines where limpets are important in controlling algal domination of free rock space (e.g., the British Isles; see also Chapter 2), in Patagonia limpet grazing appears to have weak effects on algae at all intertidal heights. Tethering experiments have revealed that physical stress sets the upper distribution limit of both limpets (*S. lessoni* and *N. magellanica*; Bazterrica et al., 2007). Finally, the combined effect of the grazer guild (different limpets, snails and chitons) has recently been studied in TF (see Curelovich et al., 2018). This preliminary exclusion experiment showed that grazers increase the probability of recruitment of the barnacle *N. scabrosus*, in the mid-intertidal zone.

### 7.3.4 Predation

Argentinean rocky shores show one peculiarity in comparison with many other rocky shores in the world (Paine, 1966; Menge and Sutherland, 1976): there are few predators and their sizes are frequently small. Unlike the larger predators found at the same latitude on Chilean rocky shores, invertebrate predators in this system are diminutive (generally <2 cm; Hidalgo et al., 2007; Silliman et al., 2011). As such, and in contrast with the Pacific coast where predators have significant effects on community structure and diversity (e.g., Castilla and Durán, 1985; Paine et al., 1985), native predators do not appear to play a large ecological role in regulating community structure in most parts of Argentina (but see the effects of *T. geversianus* in TF later). The super-abundant mussels *B. rodriguezii* and *P. purpuratus* dominate the mid-intertidal zone, monopolising the rock with extensive mussel beds without any key predators regulating their populations. The dominant cover of these monocultures of mussels suggests that predators do not have a strong impact in the mid intertidal, perhaps due to their small size or due to the phylogeographic history of the region (Hidalgo et al., 2007).

As a general phenomenon, physical factors appear to be the dominating structuring force in these communities and are likely to be evolutionarily and ecologically responsible for weakening the effects of consumers (Bertness et al., 2006). In Patagonia, where winds are very strong and desiccation is the main factor affecting intertidal species (Bertness et al., 2006), predators are always associated with sheltered habitats during low tide (i.e., within mussel beds and coralline algae or under rocks; Iribarne, 1990; Hidalgo et al., 2007; Arribas et al., 2017). Crabs (*C. altimanus, C. angulatus, Halicarcinus planatus*) and sea stars (*A. antarctica*) dominate the predator guild (Hidalgo et al., 2007), with occasional carnivorous or scavenging snails (e.g., *T. geversianus, Pareuthria plumbea*), octopus (*O. tehuelchus,* only in northern Patagonia) and fish (*Bovichtus argentinus, Patagonotothen cornucola*) (see Figure 7.5). Other small carnivore invertebrates such as sea

**Fig. 7.5** Common intertidal predators on Argentinian rocky shores. (a) The crab *C. angulatus*; (b) the sea star *A. antarctica*; (c) the muricid snail *T. geversianus*; (d) the octopus *O. tehuelchus*. Photo credits: (a–c) www.proyectosub.com.ar and (d) María Bagur. *A black and white version of this figure will appear in some formats. For the colour version, please refer to the plate section.*

anemones (*Parabunodactys imperfecta*, *A. chilensis*), polychaetes (Eunicidae, Nereididae) and nemerteans are also present within the mussel bed (Iribarne, 1991a; Hidalgo et al., 2007; Rechimont, 2013; Arribas et al., 2017). In order to feed, these predators emerge from their refuges at high tide.

Sea birds such as the kelp gull (*Larus dominicanus*) and oystercatchers (*Haematopus palliates*, *Haematopus ater* and *Haematopus leucopodus*) occasionally feed in the rocky intertidal zone. Kelp gulls along the Argentinean coast are generalist foragers that feed mostly on intertidal invertebrates and fish; also taking advantage of artificial food sources such as refuse tips and fisheries discards (Giaccardi et al., 1997; Yorio et al., 1998; Bertellotti and Yorio, 1999). Shells of the mussel *P. purpuratus* have been observed in the faeces of kelp gulls at Río Negro province (own observations). Oystercatchers eat mussels (*M. edulis platensis*) and limpets in the rocky intertidal (Ferrari et al., 2016), and seem to regulate the population of the limpet *N. magellanica* on one rocky shore in Chubut province, restricting the limpets to cracks and vertical surfaces (Bazterrica et al., 2007).

Most of the native predators feed primarily on small, soft-bodied prey and have broad diets. The three most common intertidal crab species (*C. altimanus*, *C. angulatus* and *H. planatus*) are omnivorous, feeding on polychaetes, small crustaceans and detritus (Olivier et al., 1972; Scelzo and de Bastida Lichtschein, 1978; Vinuesa et al., 2011). The sea star *A. antarctica* consumes a wide range of prey (nineteen prey items observed in Santa Cruz province, Gil and Zaixso, 2008; and twelve in Río Negro, Arribas et al., 2017) including molluscs and crustaceans, and can be regarded as a generalist or opportunistic predator (Gil and Zaixso, 2008). The superabundant mussel *P. purpuratus* is its most common prey (Gil and Zaixso, 2008; Arribas et al., 2017). The small 'pulpito' or 'tehuelche octopus', *O. tehuelchus*, feeds mostly on crabs (with a preference for *Neohelice granulata* in northern Patagonia intertidal zones) and attacks small mussels (*B. rodriguezii*) only when there is no other available prey (Iribarne, 1991a). Gut content analysis of the intertidal fish *P. cornucola* has revealed that polychaetes and small crustaceans (isopods and amphipods) were its most important prey (Hidalgo et al., 2007).

The only specialist predator in the Argentinean intertidal is the muricid gastropod *T. geversianus*. *T. geversianus* feeds on epibenthic bivalves (*Brachidontes*, *Perumytilus*, *Mytilus*, *Aulacomya*, *Hiatella* and *Tawera*) and has a marked preference for *M. edulis platensis* in the low-intertidal zone (Gordillo and Amuchástegui, 1998; Andrade and Ríos, 2007; Curelovich et al., 2018). As in all muricid gastropods, *T. geversianus* drills holes in the shell of its prey (Gordillo and Amuchástegui, 1998; Gordillo and Archuby, 2012) through a chemical/mechanical process (Pío, 2010). Its preference for *M. edulis* is probably related to the thin and smooth shell of mussels, which could reduce drilling times and consequently speed-up the ingestion process (Gordillo and Amuchástegui, 1998; Andrade and Ríos, 2007). *T. geversianus* does not, however, seem to have large impacts on intertidal mussel populations in the majority of Patagonia. In contrast, in TF (at the very south of Argentina) where *M. chilensis* is abundant, *T. geversianus* reach large sizes (up to 75 mm) and occupy mid- and low-intertidal zones (Calcagno et al., 2012; Curelovich et al., 2016, 2018) where it may regulate the lower distribution limit of the mussels. *T. geversianus* is also present with high densities in the subtidal zone (Andrade and Ríos, 2007; Márquez et al 2015). Recent work has shown that whelk drilling rate decreases with increasing time of aerial exposure, and that permanently submerged gastropods invest less time consuming the same amount of food than individuals exposed to different aerial exposure times (see Curelovich et al., 2018). Finally, Curelovich and co-workers found a positive correlation between the size of predator and size of prey consumed. These results suggest that the strong predation pressure of *T. geversianus* on *M. chilensis* at the Beagle Channel (TF) prevents this mussel from monopolising space in the low-intertidal zone by controlling its abundance, further limiting the vertical distribution of the mussel.

## 7.4 Impacts of Invasive Species

The impacts of invasive species in Argentina have been little studied (see Table 7.2 for summary of

**Table 7.2** Main invasive species on the rocky shores of Argentina. Used here are the current accepted taxonomic names.

| Species | Origin | Distribution | Functional role and impact | Literature |
|---|---|---|---|---|
| *Amphibalanus amphitrite* (Cirripedia, Balanidae; Darwin, 1854) | Indo-Pacific | From 38 to 40°S | Occupation of hard substrata | Bastida (1971a, 1971b), López Gappa et al. (1997), Calcagno et al. (1997, 1998), Orensanz et al. (2002), Spivak (2005), Mendez et al. (2014) |
| *B. glandula* (Cirripedia, Balanidae; Darwin, 1854) | Pacific coast of North America | From 36 to 47°S (53°S only a few individuals were observed) | Occupation of hard substrata; displacement of local mussel beds | Spivak and L'Hoste (1976), Bastida et al. (1980), Vallarino and Elías (1997), Rico et al. (2001), Orensanz et al. (2002), Cuevas et al. (2006), Schwindt (2007) |
| *Boccardia proboscidea* (Polychaeta, Spionidae; Hartman, 1940) | California, Pacific coast | Mar del Plata (38°S), Puerto Madryn (42°S) | Space monopolisation and reduction of total individuals, total taxa and diversity | Garaffo et al. (2012), Jaubet et al. (2011, 2015), Elías et al. (2015, 2017) |
| *Carcinus maenas* (Malacostraca, Carcinidae; Linnaeus, 1758) | North-east Atlantic | From 44 to 47°S | Predation on slow-moving and sessile animals | Hidalgo et al. (2005, 2007), Torres and Gonzalez-Pisani (2016) |
| *Magallana gigas* (Bivalvia, Ostreidae; Thumberg, 1793) | East Asian coast | From 39 to 41°S | Provides hard substrate for settlement and habitat structure and refuge for epifaunal organisms; may impact local biodiversity | Dos Santos and Borges (1995), Escapa et al. (2004), Castaños et al. (2009), Dos Santos and Fiori (2010), Roche et al. (2010), Padilla (2010), Croce and Parodi (2012), Giberto et al. (2012), Mendez et al. (2015) |
| *U. pinnatifida* (Phaeophyta, Alariaceae; Suringar, 1873) | East Asian coast | From 38 to 47°S | Produces shadow; disturbs the bottom and benthic communities, causing loss in species richness and diversity of native seaweeds; potential food source for local grazers | Piriz and Casas (1994), Orensanz et al. (2002), Casas et al. (2004), Torres et al. (2004), Martin and Cuevas (2006), Wallentinus (2007), Raffo et al. (2009), Teso et al. (2009), Meretta et al. (2012), Dellatorre et al. (2014) |

work to date on key invasive species, including their functional role and impact). Most marine introductions are related to maritime transportation (Ruiz et al., 1997); but only non-native species that have managed to adapt and reproduce outside their native range and are ecologically and/or economically harmful are considered invasive (Boudouresque and Verlaque, 2002; Bax et al., 2003; Penchaszadeh, 2005). Although thirty-one non-native marine species have become established on the Argentinian coast (Orensanz et al., 2002; Boltovskoy et al., 2011), just a few are considered invasive.

Several factors seem to buffer the Argentinean coastline from marine non-native species; among others, the scarcity of hard substrata on the northern shores (Boltovskoy et al., 2011) and volume of trade. Compared to other countries, Argentina has low maritime transit, focussed on ten major ports and a few marinas associated with recreational activities (Boltovskoy, 2008). In terms of shipping traffic, the ports of Patagonia are secondary behind the ports of Buenos Aires, which are the most important (Mar del Plata marine port and Rio de la Plata freshwater port). Nevertheless, the number of exotic species is increasing in Patagonian shores (Schwindt et al., 2014), probably because of a greater extent of rocky shores found along the southern coasts.

### 7.4.1 Algae

The most probable vectors of introduction of non-native algae are the ballast water and fouling of cargo ships and fishing vessels (Orensanz et al., 2002). Invasive algae spread on exposed or submerged natural and artificial substrata (e.g., *U. pinnatifida*; Piriz and Casas, 1994) in intertidal zones near port cities following their introduction (e.g., *Anpheltiopsis* sp.; Becherucci et al., 2014; *Schizymenia dubyi*; Ramirez et al., 2012; Palomo et al., 2016).

*U. pinnatifida* is native of eastern Asia (Saito, 1975), originally from the temperate coasts of Japan, Korea and parts of China (Akiyama and Kurogi, 1982). In Argentina, it was first recorded on the wharf piles of Puerto Madryn, Chubut province (42.75°S) in 1992 (Piriz and Casas, 1994). This species is mainly found in subtidal zones, but is also present in the low-intertidal zone in Patagonia. In its first eight years in Argentina, the distribution of *U. pinnatifida* expanded 20 km northwards and 18 km southwards of the introduction site, occupying both artificial and natural substrata (Orensanz et al., 2002). Its rate of expansion has been calculated as ~50 km/year, despite the efforts made to control its spread by the Ministry of Environment of the Chubut province (Dellatorre et al., 2014). Two other notable non-native species of algae have rapidly spread on Argentinian shores following their introductions: *Ahnfeltiopsis* sp. and *S. dubyi*. *Ahnfeltiopsis* sp. was first detected in 2007 in Mar del Plata city (3 per cent cover), and by 2011 it was identified as one of the most abundant algae on artificial structures (breakwaters) of the area (11 per cent cover) (Becherucci et al., 2014). *S. dubyi* was first detected in 2010 20 km south from Mar del Plata in the lower-intertidal level (2 per cent cover) (Ramirez et al., 2012), and a year later had spread to other intertidal levels (5 per cent cover) (Palomo et al., 2016).

Large or erect invasive algae, such as *U. pinnatifida*, can form a canopy above the smaller native algae, effectively shading them; in essence providing a new functional group on Argentinian shores. Dislodgement of large thalli of non-native species by the tides or storms can disturb the benthos (Becherucci et al., 2014; Dellatorre et al., 2014). Invasive algae are likely to be a serious local threat because of competition with native seaweeds (e.g., *U. pinnatifida*; Casas et al., 2004; Torres et al., 2004; Irigoyen et al., 2011). On the other hand, they can also constitute a new potential food source for local grazers (e.g., *U. pinnatifida*, Teso et al., 2009; *Ahnfeltiopsis* sp. and *S. dubyi*, Palomo et al., 2016) and provide new habitat for invertebrates (e.g., *Ahnfeltiopsis* sp., Palomo et al., 2016).

### 7.4.2 Barnacles

Three non-native barnacles (i.e., *B. glandula*, *Balanus trigonus* and *A. amphitrite*) were first found in Mar del Plata harbour in the 1970s (Spivak and L'Hoste, 1976; Schwindt, 2007). These barnacles are now present on the rocky shores of most of Argentina, probably because of further spread by maritime transportation (FUNIBER, 2004; Cuevas et al., 2006). Their distribution has subsequently expanded from these initial source populations via natural drift of larvae (Spivak, 2005) and ships ballast water (Boltovskoy et al., 2011).

*A. amphitrite* was introduced into the Atlantic by ships from the Pacific (Zullo, 1992). Reported in the port of Mar del Plata (38°S) in the 1960s (Bastida, 1971a), *A. amphitrite* became a dominant member in the intertidal zone, alternating with degraded patches of the native dominant species of the intertidal community, the mussel *B. rodriguezii* (Bastida, 1971a; Bastida et al., 1971; Orensanz et al., 2002), but was later replaced as the dominant species by *B. glandula* (see later in this section). *A. amphitrite* remains present on artificial structures (Calcagno et al., 1997, 1998; López Gappa et al., 1997; Spivak, 2005) in subtidal zones in Mar del Plata and in the intertidal zone of Bahia San Blas (40°S) (Spivak, 2005).

*B. glandula* was originally from the north-east Pacific (Pilsbry, 1916), occurring at the high-intertidal level from Baja California to Alaska (Foster et al., 1991). After 1970, *B. glandula* invaded Argentina (Olivier et al., 1966a, 1966b; Penchaszadeh, 1973, Spivak and L'Hoste, 1976; Geller et al., 2008), subsequently spreading from the port of Mar del Plata (38°S) (Bastida et al., 1980), where it is now the dominant species in the upper-intertidal zone, and reaching Rio Grande, TF (53°S), where a few individuals were present in the mid-2000s (Schwindt, 2007). The rate of expansion of *B. glandula* has been estimated between 40 and 244 km/year (see Spivak, 2005; Cuevas et al., 2006 and Schwindt, 2007 for further information).

These invaders compete with both native and established non-natives species. In Mar del Plata *B. glandula* has displaced other introduced barnacles (e.g., *B. trigonus* and *A. amphitrite*) from sheltered intertidal zones because of an earlier recruitment in the winter, after which it rapidly dominated the available natural and artificial free space (Bastida, 1971b; Spivak et al., 1975; Calcagno et al., 1997), as well as native mussels (e.g. *B. rodriguezii* and *M. edulis platensis*) from the higher-intertidal zone of natural exposed rocky shores (Vallarino and Elías, 1997; Spivak, 2005). *B. glandula* has also been reported to displace other previously dominant species as a result of a spatial competition process (e.g. former dominant limpet *S. lessonii*, Cuevas et al., 2006). The dominance of invasive barnacles in the higher-intertidal zone is related to the absence of predators, mussels and algal competition (Menge, 1991) and to their tolerance to desiccation when adults (Calcagno and Luquet, 1997).

### 7.4.3 Other Invasive Invertebrates

*M. gigas* (previously *Crassostrea gigas)* is known as the 'Pacific oyster', this commercially important species originates from Japan and was introduced to Argentina in 1982. This fast-growing oyster is tolerant of a wide range of environmental conditions (Shatkin, 1997) and has invaded various coastal environments (Orensanz et al., 2002; Castaños et al., 2009; Dos Santos and Fiori, 2010; Roche et al., 2010). Its introduction in Argentina was initiated without official consent into San Blas Bay (40°S) by a local fishery with the intention of starting an oyster culture operation, which was abandoned a few months later (Pascual and Castaños, 2000; Orensanz et al., 2002; Escapa et al., 2004; Borges, 2005; Castaños et al., 2009). From 1998 to 1999, its production was re-implemented, this time with appropriate authorisation and hatchery-produced spat (CRIAR Program, IBMyP Almirante Storni) which were then outplanted to different points along the coast (Castaños et al., 2009; Dos Santos and Fiori, 2010; Roche et al., 2010). This 'controlled' reintroduction allowed it to cover all available hard substrata and gradually expand along the coast, forming beds or attaching to mussel beds (Escapa et al., 2004). In Argentina, *M. gigas* can have positive effects on local biodiversity (Escapa et al., 2004; Mendez et al., 2015) by providing habitat structure and refuges, and enhancing deposition and sediment stability, thereby facilitating the establishment of other species (Orensanz et al., 2002; Mendez et al., 2015) . However, it can also have negative effects by promoting the spread of additional non-native species, such as *Polysiphonia morrowii* which uses the hard shell of *M. gigas* as substratum on which to settle (Croce and Parodi, 2012; Raffo et al., 2014).

*B. proboscidea* was originally described from the west coast of California (Hartman, 1940). This invasive polychaete was reported in Argentinian waters because of the reefs it constructed in

2008 in Mar del Plata (Jaubet et al., 2011; Garaffo et al., 2012). According to Jaubet et al. (2011), this spionid polychaete builds biogenic reefs in areas organically enriched by sewage discharges, and can cover almost the entire impacted site.In Mar del Plata, densities are higher than in other countries where this worm occurs but no reefs are formed (Dorsey, 1982; Everett, 1991). The presence of reefs of *B. proboscidea* has caused a significant reduction of total individuals, total taxa and diversity in sewage-impacted sites (Elías et al., 2015). The associated fauna, formerly rich and diverse in impacted sites, showed a tendency to disappear as the ecosystem engineer *B. rodriguezii* was replaced by *B. proboscidea* (Elías et al., 2015).

*C. maenas* is a north-east Atlantic native (Monod, 1956; Forest and Gantes, 1960; Christiansen, 1969), but has spread to many places worldwide (north-west Atlantic, Southern Africa, north-west Pacific, Australia, Bax et al., 2003). In Argentina, invasion by *C. maenas* is currently underway following introduction in the 2000s by discharge of ballast waters (Cohen et al., 1995); although alternative vectors such as the fouled pipes of ships and seaweeds used as packaging for shellfish have been reported (Ashley et al., 2003; Carlton and Cohen, 2003, Hidalgo et al., 2005). It has recently been observed in Nuevo Gulf, Chubut (Torres and Gonzalez-Pisani, 2016). Its ecological effects remain locally unknown but it is a highly active predator (McDonald et al., 2001), feeding on slow-moving and sessile animals, including the superabundant mussel *P. purpuratus*. Because native intertidal organisms are dependent on mussel beds and coralline algae for shelter from desiccation, successful invasion of *C. maenas* may lead to a significant decrease in native diversity by consuming foundation species (Hidalgo et al., 2007).

The next listed species have being reported by Boltovskoy et al. (2011) or listed as exotic and cryptogenic species in the south-west Atlantic (including Uruguayan coast) by Orensanz et al. (2002) and in Patagonian ports by Schwindt et al. (2014): Phaeophyceae, *Sporochnus pedunculatus* (Boraso de Zaixso and Negri, 1997); Florideophyceae, *S. dubyi* (Palomo et al., 2016; Ramirez et al., 2012), *Anotrichium furcellatum* (Boraso de Zaixso and Akselman, 2005) and *P. morrowii* (Croce and Parodi, 2012; Raffo et al., 2014); Ascidiacea, *Styela clava* (Goldstien et al., 2011), *Ascidiella aspersa* (Tatián et al., 2010) and *Lissoclinum fragile* (Rico et al., 2012); Crustacea, *Pyromaia tuberculata* (Schejter et al., 2002) and *Palaemon macrodactylus* (Schejter et al., 2002; Spivak et al., 2006); Polychaeta, *Sabellaria wilsoni* (Chiaradia et al., 2007); Bivalvia, *Semimytilus algosus* (Bigatti et al., 2014); and Gastropoda, *Pleurobranchaea* sp. (Farias et al., 2015).

## 7.5 Anthropogenic Impacts and Global Change

Anthropogenic impacts on intertidal rocky shores in Argentina have been thoroughly studied. Contamination from sewage outfalls is one of the main anthropogenic stressors on biodiversity associated with mussels (López Gappa et al., 1990, 1993; Díaz et al., 2002; Elías et al., 2006; Becherucci et al., 2014). Another intertidal species responding to pollution is the limpet *S. lessonii*, whose shells showed concentrations two times lower of calcium and whose soft tissues were lighter at heavily polluted sites (Nuñez et al., 2012). In polluted, enriched areas, the time of recovery of benthic assemblages after a disturbance was lower than in non-enriched sites based on the observed shift in community composition (Becherucci et al., 2016).

Puerto Madryn in Patagonia is a city with ca. 100,000 permanent residents and ca. 250,000 tourists each year. As such, the rocky intertidal is potentially at risk from human trampling and perturbations from vehicles. Mendez et al. (2017) showed that disturbances by vehicles were infrequent and the benthic communities could recover quickly. In contrast, the effects of human trampling were continuous during the recreational season and the disturbed community associated with mussels did not recover before the next season. This study highlights the importance of establishing effective management actions to achieve the desired conservation goals and mitigate the consequences of anthropogenic disturbances on rocky intertidal habitats.

Climate change is occurring due to natural processes and accelerated by persistent anthropogenic changes in all environments, including the coasts (Wong et al., 2014). Rising sea levels, increasing temperatures, changes in precipitation, larger storm surges and increasing ocean acidity are considered the more important effects to coastal systems globally (Wong et al., 2014). Coasts can be affected by erosive processes due to different causes, although sea-level rise maximises erosion. On Argentinean coasts, sea level has followed the global tendency with a rise between 15 and 20 cm in the last century (Barros and Camilloni, 2016).

In Argentina, between 1996 and 2005, there was an increase of 7 per cent in the average number of storm surges (Fiore et al., 2009; Dragani et al., 2013). The frequency and duration of storm surges increase the energy dissipated by waves and, as a consequence, erosion processes drastically increase along the coast (D'Onofrio et al., 2008). Increased wave impacts are expected to cause the dislodgement of mussels attached to intertidal rocky substrata (see Denny and Gaylord, 2002); such dislodgement could vary depending on whether a mussel is attached to another mussel or to the rocky substrates (Gutiérrez et al., 2015). In addition, multilayered mussel hummocks (characterised by a higher density of mussels per area unit) show a higher incidence of mussel dislodgment by waves than nearby single-layered areas of the mussel bed (Gutiérrez et al., 2015). Under these conditions, the mussel beds – which are such an important characteristic of many Argentinian shores – will mostly occur in single layers and may become more fragmented, with a cascading impact on the assemblages associated with the mussels.

The gap in baseline scientific knowledge in this region impairs forecast of the effects of climate change (Turra et al., 2013). Nevertheless, several research groups in Argentina are beginning to work on the possible effects of climate change in relation to intertidal rocky shores species. Studies on potential impacts of the increasing temperature and acidity on rocky shore biodiversity and the interactions between species are important; therefore, further studies on the impacts of climate change related to the rocky shores of Argentina are needed if we are to develop a programme measuring the potential negative effects and aiding adaptation.

## 7.6 Knowledge Gaps and Concluding Comments

Although in recent years studies on the rocky intertidal have increased, further studies to understand broadscale patterns of biodiversity along the coast are required, coupled with phylogeographic work to understand the processes determining these patterns. In particular, post-glacial recolonisation and interchange between the Pacific and Atlantic oceans needs further study; this is probably ultimately responsible for the reverse diversity gradient (fewer species towards the equator) seen latitudinally on Argentinian shores. Proximately, the reduced salinity and high sediment loads of the Plate Estuary coupled with low amplitude tides may also restrict diversity. These low-diversity Argentinian shores have been successfully invaded by several species that now occupy major ecological roles and provide new functional groups on shores (i.e., large fleshy algae, oysters and barnacles).

Rocky shores are important areas that provide coastal protection for cities and many other ecosystem services. Research on ecosystem-based management and ecosystem engineers as bioprotective agents has not been explored in Patagonia or Buenos Aires province, where flooding and erosion are more intense in the last decades due to the combination of climate change and anthropogenic impacts. Species interactions and food webs on rocky shores have only been patchily studied. The assemblages and interactions of these rocky intertidal shores depend on physical or biological factors that differ at very small scales of metres. In the absence of large predators, facilitative processes seem very important in the harsh environment of the arid and windy intertidal zone. It is thus likely that interactions in a particular area will be locally context dependent.

Climate change effects and the impact of anthropogenic activities on the rocky intertidal have been scarcely explored. Finally, the connection between rocky intertidal shores along the coast and with deeper waters and distant areas is also a gap of knowledge in the south-western Atlantic. Mapping biodiversity and understanding ecosystem functioning of rocky shores are both essential for the effective management, sustainable development and conservation of the coasts of Argentina, in order to maintain its important ecosystem services.

## Acknowledgements

The research presented here was partly funded by Consejo Nacional de Investigaciones Científicas y Técnicas (CONICET) grants (PIP 11220110100024 and PICT 2468) to MGP. MGP was supported by the workshop organisation. SJH was supported by a Partnership for Observation of the Global Ocean Fellowship. MB, SC, SAS and MCD are supported by CONICET fellowships. This is a contribution to the programme of the Research and Education group in Environmental Themes.

### REFERENCES

Acha, E. M., Mianzan, H. W., Guerrero, R. A., Favero, M. and Bava, J. (2004). Marine fronts at the continental shelves of austral South America: physical and ecological processes. *Journal of Marine systems*, **44** (1–2), 83–105.

Adami, M. (2008). Efectos de la herbivoría de la lapa *Siphonaria lessoni* Blainville, 1824 (Gastropoda) sobre la comunidad asociada a *Brachidontes rodriguezii* (d'Orbigny, 1846) (Bivalvia). *Revista del Museo Argentino de Ciencias Naturales nueva serie*, **10** (2), 309–17.

Adami, M. L., Pastorino, G. and Orensanz, J. M. (2013). Phenotypic differentiation of ecologically significant Brachidontes species co-occurring in intertidal mussel beds from the southwestern Atlantic. *Malacologia*, **56** (1–2), 59–67.

Adami, M. L., Tablado, A. and López Gappa, J. (2004). Spatial and temporal variability in intertidal assemblages dominated by the mussel *Brachidontes rodriguezii* (d'Orbigny, 1846). *Hydrobiologia*, **520** (1–3), 49.

Akiyama, K. and Kurogi, M. (1982). Cultivation of *Undaria pinnatifida* (Harvey) Suringar, the decrease in crops from natural plants following crop increase from cultivation. *Bulletin of the Freshwater Fisheries Research Laboratory*, **44**, 91–100.

Aldea, C. and Rosenfeld, S. (2011). Macromoluscos intermareales de sustratos rocosos de la playa Buque Quemado, Estrecho de Magallanes, sur de Chile. *Revista de biología marina y oceanografía*, **46** (2), 115–24.

Amor, A., Armengol, M. L., Rodriguez, A. I. and Traversa, L. P. (1991). Intertidal endolithic fauna and it's relationship to the mineralogical, physical and chemical characteristics of the substrate. *Marine Biology*, **111** (2), 271–80.

Andrade, C. and Ríos, C. (2007). Estudio Experimental de los Hábitos Tróficos de Trophon Geversianus (Pallas 1774) (Gastropoda: Murcidae): Selección y Manipulación de Presas. *Anales del Instituto de la Patagonia*, **35** (1), 45–54.

Andrade Diaz, C. and Brey, T. (2014).Trophic ecology of limpets among rocky intertidal in Bahía Laredo, Strait of Magellan (Chile). *Anales del Instituto de la Patagonia*, **42** (2), 65–70.

Arribas, L. P., Bagur, M., Gutiérrez, J. L. and Palomo, M. G. (2015). Matching spatial scales of variation in mussel recruitment and adult densities across southwestern Atlantic rocky shores. *Journal of Sea Research*, **95**, 16–21.

Arribas, L. P., Bagur, M., Klein, E., Penchaszadeh, P. E. and Palomo, M. G. (2013). Geographic distribution of two mussel species and associated assemblages along the northern Argentinean coast. *Aquatic Biology*, **18** (1), 91–103.

Arribas, L. P., Bagur, M., Palomo, M. G. and Bigatti, G. (2017). Population biology of the sea star *Anasterias minuta* (Forcipulatida: Asteriidae) threatened by anthropogenic activities in rocky intertidal shores of San Matías Gulf, Patagonia, Argentina. *Revista de Biología Tropical*, **65** (1), 73–84.

Arribas, L. P., Bagur, M., Soria, S. A., Gutiérrez, J. L. and Palomo, M. G. (2016). Competition between mussels at the rocky intertidal zone of La Lobería, Río Negro, Argentina. *Revista del Museo Argentino de Ciencias Naturales*, **18** (1), 1–7.

Ashley, D. M., Moore, K. M. and Hewitt, L. C. (2003) Ships' sea chests: an overlooked transfer mechanism for non-indigenous marine species? *Baseline, Marine Pollution Bulletin* **46** (46), 1504–15.

Bagur, M., Gutiérrez, J. L., Arribas, L. P. and Palomo, M. G. (2014). Endolithic invertebrate communities and bioerosion rates in southwestern Atlantic intertidal consolidated sediments. *Marine Biology*, **161** (10), 2279–92.

Bagur, M., Gutiérrez, J. L., Arribas, L. P. and Palomo, M. G. (2016). Complementary influences of co-occurring physical ecosystem engineers on species richness: insights from a Patagonian rocky shore. *Biodiversity and Conservation*, **25** (13), 2787–802.

Bagur, M., Richardson, C. A., Gutiérrez, J. L., Arribas, L. P., Doldan, M. S. and Palomo, M. G. (2013). Age, growth and mortality in four populations of the boring bivalve *Lithophaga patagonica* from Argentina. *Journal of Sea Research*, **81**, 49–56.

Balech, E. and Ehrlich, M. D. (2008). Esquema biogeográfico del mar argentino. *Revista de investigación de desarrollo pesquero*, **19**, 45–75.

Barnich, R., Fiege, D., Micaletto, G. and Gambi, M. C. (2006). Redescription of Harmothoe spinosa Kinberg, (Polychaeta: Polynoidae) and related species from Subantarctic and Antarctic waters, with the erection of a new genus. *Journal of Natural History*, **40** (1–2), 33–75.

Barnich, R., Orensanz, J. M. and Fiege, D. (2012). Remarks on some scale worms (Polychaeta, Polynoidae) from the Southwest Atlantic with notes on the genus *Eucranta* Malmgren, 1866, and description of a new *Harmothoe* species. *Marine Biodiversity*, **42** (3), 395–410.

Barros, V. and Camilloni, I. (2016). *La Argentina y el cambio climático: de la física a la política*. EUDEBA, Buenos Aires, 286.

Bastida, R. (1971a). Las incrustaciones biológicas en el puerto de Mar del Plata, Período 1966/1967. *Revista del Museo de Ciencias Naturales." Bernardino Rivadavia", Hidrobiología*, **3** (2), 203–85.

Bastida, R. (1971b). Las incrustaciones biológicas en las costas argentinas. La fijación mensual en el puerto de Mar del Plata durante tres años consecutivos. *Corrosión y Protección*, **2** (1).

Bastida, R., Capezzani, A. and Torti, M. R. (1971). Fouling organisms in the port of Mar del Plata (Argentina). I. *Siphonaria lessoni*: ecological and biometric aspects. *Marine Biology*, **10** (4), 297–307.

Bastida, R., Roux, A. and Martinez, D. E. (1992). Benthic communities of the Argentine continental-shelf. *Oceanologica Acta*, **15** (6), 687–98.

Bastida, R., Trivi de Mandri, M., Lichtschein de Bastida, V. and Stupak, M. (1980). Ecological aspects of marine fouling at the port of Mar del Plata (Argentina). V Congreso Internacional de Corrosión Marina e Incrustaciones (Barcelona, España). In *Sección Biología Marina*, 299–320.

Bax, N., Williamson, A., Aguero, M., Gonzalez, E. and Geeves, W. (2003). Marine invasive alien species: a threat to global biodiversity. *Marine policy*, **27** (4), 313–23.

Bazterrica, M. C., Silliman, B. R., Hidalgo, F. J., Crain, C. M. and Bertness, M. D. (2007). Limpet grazing on a physically stressful Patagonian rocky shore. *Journal of Experimental Marine Biology and Ecology*, **353** (1), 22–34.

Becherucci M. E., Benavides H. and Vallarino E. A. (2014) Effect of taxonomic aggregation in macroalgae assemblages in a rocky shore of Mar del Plata, Argentina, Southwest Atlantic Ocean. *Thalassas* **30**, 9–20.

Becherucci, M. E., Llanos, E. N., Garaffo, G. V. and Vallarino, E. A. (2016). Succession in an intertidal benthic community affected by untreated sewage effluent: a case of study in the SW Atlantic shore. *Marine Pollution Bulletin*, **109** (1), 95–103.

Benedetti-Cecchi, L., Iken, K., Konar, B. et al. (2010). Spatial relationships between polychaete assemblages and environmental variables over broad geographical scales. *PLoS ONE*, **5** (9), e12946.

Bertellotti, M. and Yorio, P. (1999). Spatial and temporal patterns in the diet of the Kelp Gull in Patagonia.*Condor*, 790–8.

Bertness, M. D., Crain, C. M., Silliman, B. R. et al. (2006). The community structure of western Atlantic Patagonian rocky shores. *Ecological Monographs*, **76** (3), 439–60.

Bigatti, G., Signorelli, J. H. and Schwindt, E. (2014). Potential invasion of the Atlantic coast of South America by *Semimytilus algosus* (Gould, 1850). *Journal of the Marine Biological Association of the United Kingdom, 2009*, **89** (8), 1571–80.

Boudouresque, C. F. and Verlaque, M. (2002). Biological pollution in the Mediterranean Sea: invasive versus introduced macrophytes. *Marine Pollution Bulletin*, **44** (1), 32–8.

Boltovskoy, D., ed. (2008). *Atlas de sensibilidad ambiental de la costa y el mar argentino, secretaría de ambiente y desarrollo sustentable, Argentina*. Naval Hydrography Service, Buenos Aires.

Boltovskoy, D., Almada, P. and Correa, N. (2011). Biological invasions: assessment of threat from ballast-water discharge in Patagonian (Argentina) ports. *Environmental Science and Policy*, **14** (5), 578–83.

Boraso, A. and Zaixso, J. M. (2011). Algas marinas bentónicas. In *Atlas de Sensibilidad Ambiental de la Costa y el Mar Argentino*. Naval Hydrography Service, Buenos Aires.

Boraso de Zaixso, A. L. and Akselman, R. (2005). *Anotrichium furcellatum* (Ceramiaceae, Rhodophyta) en Argentina: Una posible especie invasora. *Boletín de la Sociedad Argentina de Botánica*, **40** (3–4), 207–13.

Boraso de Zaixso, A. L. and Negri, R. M. (1997). Presencia de *Sporochnus pedunculatus* (Sporochnales Phaeophycophyta) en la costa Argentina. *Physis A*, **54** (126/127), 23–4.

Borges, M. E. (2005). *La ostra del Pacífico, Crassostrea gigas (Thumberg, 1793) en la Bahía Anegada (Provincia de Buenos Aires). Invasores: Invertebrados exóticos en el Río de la Plata y región marina aledaña.* Penchaszadeh, ed. EUDEBA, Buenos Aires, pp. 310–56.

Borthagaray, A. I. and Carranza, A. (2007). Mussels as ecosystem engineers: their contribution to species richness in a rocky littoral community. *Acta Ocecanologica*, **31** (3), 243–50.

Boschi, E. E. (1979). Geographic distribution of Argentinian marine decapod crustaceans. *Bulletin of the Biological Society of Washington,* **3** (1), 134–43.

Boschi, E. E., Fischbach, C. E. and Iorio, M. I. (1992). Catálogo ilustrado de los crustáceos estomatópodos y decápodos marinos de Argentina.

Brandini, F. P., Boltovskoy, D., Piola, A. et al. (2000). Multiannual trends in fronts and distribution of nutrients and chlorophyll in the southwestern Atlantic (30–62 S). *Deep Sea Research Part I: Oceanographic Research Papers,* **47** (6), 1015–33.

Bremec, C. and Giberto, D. (2006). Polychaete assemblages in the Argentinean Biogeographical Province, between 34° and 38° S. *Scientia Marina,* **70** (S3), 249–58.

Bremec, C. and Schejter, L. (2010). Benthic diversity in a submarine canyon in the Argentine Sea. *Revista Chilena de Historia Natural,* **83** (3).

Calcagno, J. A., Curelovich, J. N., Fernandez, V. M., Thatje, S. and Lovrich, G. A. (2012). Effects of physical disturbance on a sub-Antarctic middle intertidal bivalve assemblage. *Marine Biology Research,* **8** (10), 937–53.

Calcagno, J. A., López Gappa, J. and Tablado, A. (1997). Growth and production of the barnacle *Balanus amphitrite* in an intertidal area affected by sewage pollution. *Journal of Crustacean Biology*, **17** (3), 417–23.

Calcagno, J. A., López Gappa, J. and Tablado, A. (1998). Population dynamics of the barnacle *Balanus amphitrite* in an intertidal area affected by sewage pollution. *Journal of Crustacean Biology*, **18** (1), 128–37.

Calcagno, J. A. and Luquet C. M. (1997). Influence of desiccation tolerance on the ecology of *Balanus amphitrite* Darwin, 1854 (Crustacea: Cirripedia). *Nauplius, Rio Grande*, **5** (2), 9–15.

Campagna, C., Verona, C. and Falabella, V. (2005). Situación ambiental en la ecorregión del Mar Argentino. In A. Brown, U. Martínez Ortiz, M. Acerbi and J. Corcuera, eds. *La Situación Ambiental Argentina 2005.* Fundación Vida Silvestre Argentina, Buenos Aires, pp. 323–37.

Carlton, J. T. and Cohen, A. N. (2003). Episodic global dispersal in shallow water marine organisms: the case history of the European shore crabs *Carcinus maenas and C. aestuarii. Journal of Biogeography*, **30** (12), 1809–20.

Carreto, J. I., Bremec, C. and Boschi, E. E. (2007). *El Mar Argentino y sus recursos pesqueros. Tomo 5. El ecosistema marino (No. M11 INI 18068).* Instituto Nacional de Investigación y Desarrollo Pesquero (INIDEP), Mar del Plata.

Carrington, E., Moeser, G. M., Thompson, S. B., Coutts, L. C. and Craig, C. A. (2008). Mussel attachment on rocky shores: the effect of flow on byssus production. *Integrative and Comparative Biology,* **48** (6), 801–7.

Casas, G., Scrosati, R. and Piriz, M. L. (2004). The invasive kelp *Undaria pinnatifida* (Phaeophyceae, Laminariales) reduces native seaweed diversity in Nuevo Gulf (Patagonia, Argentina). *Biological Invasions*, **6** (4), 411–16.

Castaños, C., Pascual, M. and Camacho, A. P. (2009). Reproductive biology of the nonnative oyster, *Crassostrea gigas* (Thunberg, 1793), as a key factor for its successful spread along the rocky shores of northern Patagonia, Argentina. *Journal of Shellfish Research*, **28** (4), 837–47.

Castilla, J. C. and Duran, L. R. (1985). Human exclusion from the rocky intertidal zone of central Chile: the effects on *Concholepas concholepas* (Gastropoda). *Oikos*, **45** (3), 391–9.

Chiaradia, N., Marchesi, C., Azzone, D. et al. (2007). An unexpected reef-building worm in Mar del Plata, Argentina (SW Atlantic). In *9th International Polychaete Conference*, 12–17.

Christiansen M. E. (1969) Crustacea Decapoda Brachyura. In *Marine invertebrates of Scandinavia.* Universitetsforlage, Oslo, pp. 49–51.

Cohen, A. N., Carlton, J. T. and Fountain, M. C. (1995). Introduction, dispersal and potential impacts of the green crab *Carcinus maenas* in San Francisco Bay, California. *Marine Biology,* **122** (2), 225–37.

Connell, J. H. (1961). The influence of interspecific competition and other factors on the distribution of the barnacle *Cthamalus stellatus*. *Ecology* **42**, 710–23

Croce, M. E. and Parodi, E. R. (2012). Seasonal dynamic of macroalgae in intertidal pools formed by beds of *Crassostrea gigas* (Mollusca, Bivalvia) on the north Patagonian Atlantic coast. *Botanica Marina*, **55** (1), 49–58.

Cuevas, J. M., Martin, J. P. and Bastida, R. (2006). Benthic community changes in a Patagonian intertidal: a forty years later comparison. *Thalassas*, **22** (1), 31–9.

Curelovich, J., Lovrich, G. A. and Calcagno, J. A. (2016). The role of the predator *Trophon geversianus* in an intertidal population of *Mytilus chilensis* in a rocky shore of the Beagle Channel, Tierra del Fuego, Argentina. *Marine Biology Research*, **12** (10), 1053–63.

Curelovich, J. N., Lovrich, G. A., Cueto, G. R. and Calcagno, J. A. (2018). Recruitment and zonation in a sub-Antarctic rocky intertidal community. *Journal of the Marine Biological Association of the United Kingdom*, **98** (2), 411–22.

D'Onofrio, E. E., Fiore, M. M. E. and Pousa, J. L., 2008. Changes in the regime of storm surges at Buenos Aires, Argentina. *Journal of Coastal Research* **24** (1A), 260–5.

Dayton, P. K. (1971). Competition, disturbance and community organisation: the provision and subsequent utilisation of space in a rocky intertidal community. *Ecological Monographs*, **41**, 351–89.

De Aranzamendi, M. C., Gardenal, C. N., Martin, J. P. and Bastida, R. (2009). Limpets of the genus *Nacella* (Patellogastropoda) from the Southwestern Atlantic: species identification based on molecular data. *Journal of Molluscan studies*, **75** (3), 241–51.

Dellatorre, F. G., Amoroso, R., Saravia, J. and Orensanz, J. M. (2014). Rapid expansion and potential range of the invasive kelp *Undaria pinnatifida* in the Southwest Atlantic. *Aquatic Invasions*, **9** (4).

Denny, M. W. and Gaylord, B. (2002). The mechanics of wave-swept algae. *Journal of Experimental Biology*, **205**, 1355–62.

Díaz, P., López Gappa, J. J. and Piriz, M. L. (2002). Symptoms of eutrophication in intertidal macroalgae assemblage of Nuevo Gulf (Patagonia, Argentina). *Botanica Marine*, **45**, 267–73.

Dorsey, J. H. (1982). The Ecology of *Australonereis ehlersi (Augener, 1913) and Ceratonereis erythraeensis Fauvel*, 1919 (Polychaeta, Nereidae) living offshore from the Werribee sewage-treatment farm, Port Phillip Bay, Victoria, Australia. PhD, University of Melbourne, Melbourne.

Dos Santos, E. P. and Borges, M. E. (1995).Contribución al conocimiento de la Bahía Anegada, partido de Patagones, prov. Buenos Aires. In *IV Congreso Latinoamericano sobre ciencias del Mar, Mar del Plata*, 70.

Dos Santos, E. P. and Fiori, S. M. (2010). Primer registro sobre la presencia de *Crassostrea gigas* (Thunberg, 1793) (Bivalvia: Ostreidae) en el estuario de Bahía Blanca (Argentina). *Comunicaciones de la Sociedad Malacológica del Uruguay*, **9** (93).

Dragani, W. C., Martin, P. B., Alonso, G., Codignotto, J. O., Prario, B. E. and Bacino, G. (2013). Wind wave climate change: Impacts on the littoral processes at the Northern Buenos Aires Province Coast, Argentina. *Climatic Change*, **121** (4), 649–60.

Elías, R., Bremec, C. S. and Vallarino, E. A. (2001). Polychaetes from a southwestern shallow shelf Atlantic area (Argentina, 38 S) affected by sewage discharge. *Revista chilena de historia natural*, **74** (3), 523–31.

Elías, R., Jaubet, M. L., Ferrando, A. and Bottero, M. A. S. (2017). Historia y perspectivas de los estudios sobre poliquetos en argentina. *Boletín del Instituto Oceanográfico de Venezuela*, **1** (1).

Elías, R., Jaubet, M. L., Llanos, E. N. et al. (2015). Effect of the invader *Boccardia proboscidea* (Polychaeta: Spionidae) on richness, diversity and structure of SW Atlantic epilithic intertidal community. *Marine Pollution Bulletin*, **91** (2), 530–6.

Elías, R., Rivero, M. S., Palacios, J. R. and Vallarino, E. A. (2006). Sewage-induced disturbance on polychaetes inhabiting intertidal mussel beds of *Brachidontes rodriguezii* off Mar del Plata (SW Atlantic, Argentina). *Scientia Marina*, **70** (S3), 187–96.

Elías, R., Rivero, M. S., Sanchez, M. A., Jaubet, L. and Vallarino, E. A. (2009). Do treatments of sewage plants really work? The intertidal mussels' community of the southwestern Atlantic shore (38°S, 57°W) as a case study. *Revista de biología marina y oceanografía*, **44** (2).

Elías, R., Rivero, M. S. and Vallarino, E. A. (2003). Sewage impact on the composition and distribution of Polychaeta associated to intertidal mussel beds of the Mar del Plata rocky shore, Argentina. *Iheringia. Série Zoologia*, **93** (3), 309–18.

Escapa, M., Isaacch, J. P., Daleo, P. et al. (2004). The distribution and ecological effects of the introduced Pacific oyster *Crassostrea gigas (*Thunberg, 1793) in northern Patagonia. *Journal of Shellfish Research*, **23** (3), 765–73.

Everett, R. A. (1991). Intertidal distribution of infauna in a central California lagoon: the role of seasonal blooms of macroalgae. *Journal of Experimental Marine Biology and Ecology*, **150** (2), 223–47.

Excoffon, A. C., Belém, M. J. C., Zamponi, M. O. and Schlenz, E. (1997). The validity of *Anthothoe chilensis* (Actiniaria, Sagartiidae) and its distribution in Southern Hemisphere. Iheringia, Sér. *Zool*, **82**, 107–18.

Excoffon, A. C., Genzano, G. N. and Zamponi, M. O. (1999). Macrobentos asociado con una población de *Anthothoe chilensis* (Lesson, 1830) (Cnidaria, Actiniaria) en el puerto de Mar del Plata, Argentina. *Ciencias Marinas*, **25** (2).

Farias, N. E., Obenat, S. and Goya, A. B. (2015). Outbreak of a neurotoxic side-gilled sea slug (Pleurobranchaea sp.) in Argentinian coasts. *New Zealand Journal of Zoology*, **42** (1), 51–6.

Ferrari, S., Lizarralde, Z., Pittaluga, S. and Albrieu, C. (2016). Dieta y comportamiento de alimentación del ostrero austral (*Haematopus leucopodus*) durante el período pos-reproductivo en el estuario del Río Gallegos, Patagonia, Argentina. *Ornitología Neotropical*, **26** (1), 39–49.

Fiore, M. M., D'Onofrio, E. E., Pousa, J. L., Schnack, E. J. and Bertola, G. R. (2009). Storm surges and coastal impacts at Mar del Plata, Argentina. *Continental Shelf Research*, **29** (14), 1643–9.

Forest, J. and Gantes, H. (1960) Sur une collection de Crustaces Decapodes marcheu archeus du Maroc. *Bulletin of the Museum National d'Histoire Naturelle (Paris) Ser*, **2** (32), 346–58.

Foster, M. S., Harrold, C. and Hardin, D. D. (1991). Point vs. photo quadrat estimates of the cover of sessile marine organisms. *Journal of Experimental Marine Biology and Ecology*, **146** (2), 193–203.

Fretter, V. and Graham, A. (1962). *British Prosobranch Molluscs: Their Functional Anatomy and Ecology*. London Ray Society, London.

FUNIBER. (2004). Informe final del Proyecto de Desarrollo del Área Industrial de Puerto Deseado, Provincia de Santa Cruz. Fundación Universitaria Iberoamericana, 88.

Garaffo, G. V., Jaubet, M. L., Sánchez, M. D. L. Á., Rivero, M. S., Vallarino, E. A. and Elías, R. (2012). Sewage-induced polychaete reefs in a SW Atlantic shore: rapid response to small-scale disturbance. *Marine ecology*, **33** (3), 272–9.

García, C. A., Sarma, Y. V. B., Mata, M. M. and García, V. M. (2004). Chlorophyll variability and eddies in the Brazil–Malvinas Confluence region. *Deep Sea Research Part II: Topical Studies in Oceanography,* **51** (1–3), 159–72.

Garzoli, S. L. and Garraffo, Z. (1989). Transports, frontal motions and eddies at the Brazil-Malvinas Currents Confluence. *Deep Sea Research Part A. Oceanographic Research Papers,* **36** (5), 681–703.

Gaston, K. J. (2000). Global patterns in biodiversity. *Nature*, **405**, 220–7.

Geller, J., Sotka, E. E., Kado, R., Palumbi, S. R. and Schwindt, E. (2008). Sources of invasions of a northeastern Pacific acorn barnacle, *Balanus glandula*, in Japan and Argentina. *Marine Ecology Progress Series*, **358**, 211–18.

Giaccardi, M., Yorio, P. and Lizurume, M. E. (1997). Patrones estacionales de abundancia de la gaviota cocinera (*Larus dominicanus*) en un basural patagónico y sus relaciones con el manejo de residuos urbanos y pesqueros. *Ornitologia Neotropical*, **8**, 77–84.

Giberto, D. A., Bremec, C. S., Schejter, L. et al. (2012). La ostra del pacífico *Crassostrea gigas* (Thunberg, 1793) en la provincia de Buenos Aires: reclutamientos naturales en Bahía Samborombón. *Revista de Investigación y Desarrollo Pesquero*, **21**, 21–30.

Gil, D. G. and Zaixso, H. E. (2008). Feeding ecology of the subantarctic sea star *Anasterias minuta* within tide pools in Patagonia, Argentina. *Revista de Biología Tropical*, **56** (3).

Goldstien, S. J., Dupont, L., Viard, F. et al. (2011). Global phylogeography of the widely introduced North West Pacific ascidian *Styela clava*. *PLoS ONE*, **6** (2), e16755.

Gordillo, S. and Amuchástegui, S. N. (1998). Estrategias de depredación del gastrópodo perforador *Trophon geversianus* (Pallas) (Muricoidea: Trophonidae). *Malacologia philadelphia*, **39**, 83–92.

Gordillo, S. and Archuby, F. (2012). Predation by drilling gastropods and asteroids upon mussels in rocky shallow shores of southernmost South America: paleontological implications. *Acta Palaeontologica Polonica*, **57** (3), 633–46.

Gutiérrez, J. L., Jones, C. G., Byers, J. E. et al. (2011). Physical Ecosystem Engineers and the Functioning of Estuaries and Coasts. In C. H. R. Heip, C. J. M. Philippart and J. J. Middelburg, eds. *Functioning of Estuaries and Coastal Ecosystems: Treatise on Estuarine and Coastal Science*, vol. 7. Elsevier, Amsterdam, pp. 53–81.

Gutiérrez, J. L., Jones, C. G., Strayer, D. L. and Iribarne, O. O. (2003). Mollusks as ecosystem engineers: the role of shell production in aquatic habitats. *Oikos,* **101** (1), 79–90.

Gutiérrez, J. L. and Palomo, M. G. (2016). Increased algal fouling on mussels with barnacle epibionts: a fouling cascade. *Journal of Sea Research*, **112**, 49–54.

Gutiérrez, J. L., Palomo, M. G., Bagur, M., Arribas, L. P. and Soria, S. A. (2015). Wave action limits crowding in an intertidal mussel. *Marine Ecology Progress Series*, **518**, 153–63.

Haimovici, M. and Perez, J. A. (1991). Coastal cephalopod fauna of southern Brazil. *Bulletin of Marine Science,* **49** (1–2), 221–30.

Hartman, O. (1940). *Boccardia proboscidea*, a new species of spionid worm from California. *Journal of the Washington Academy of Sciences*, **30** (9), 382–7.

Hickman, C. S. (1998). Superfamily Fissurelloidea. In P. L. Beesley, G. J. B. Ross and A. Wells, eds. *Mollusca: The Southern Synthesis*, vol. 5, part B. CSIRO Publishing, Melbourne, pp. viii, 565–1234.

Hidalgo, F. J., Barón, P. J. and Orensanz, J. M. L. (2005). A prediction come true: the green crab invades the Patagonian coast. *Biological Invasions*, **7** (3), 547–52.

Hidalgo, F. J., Silliman, B. R., Bazterrica, M. C. and Bertness, M. D. (2007). Predation on the rocky shores of Patagonia, Argentina. *Estuaries and Coasts*, **30** (5), 886–94.

Hillebrand, H. (2004). Strength, slope and variability of marine latitudinal gradients. *Marine Ecology Progress Series*, **273**, 251–68.

Iribarne, O. O. (1990). Use of shelter by the small Patagonian octopus *Octopus tehuelchus*: availability, selection and effects on fecundity. *Marine Ecology Progress Series*, **66**, 251–58.

Iribarne, O. O. (1991a). Intertidal harvest of the Patagonian octopus, *Octopus tehuelchus* (d'Orbigny). *Fisheries Research*, **12** (4), 375–90.

Iribarne, O. O. (1991b). Life history and distribution of the small south-western Atlantic octopus, *Octopus tehuelchus*. *Journal of Zoology*, **223** (4), 549–65.

Irigoyen, A. J., Eyras, C. and Parma, A. M. (2011). Alien algae *Undaria pinnatifida* causes habitat loss for rocky reef fishes in north Patagonia. *Biological Invasions*, **13** (1), 17–24.

Jaubet, M. L., Garaffo, G. V., Vallarino, E. A. and Elías, R. (2015). Invasive polychaete *Boccardia proboscidea* Hartman, 1940 (Polychaeta: Spionidae) in sewage-impacted areas of the SW Atlantic coasts: morphological and reproductive patterns. *Marine Ecology*, **36** (3), 611–22.

Jaubet, M. L. and Genzano, G. N. (2011). Seasonality and reproductive periods of the hydroid *Clytia gracilis* in temperate littoral ecosystems. Is asexual reproduction the prime mechanism in maintaining populations? *Marine Biology Research*, **7** (8), 804–11.

Jaubet, M. L., Sánchez, M. A., Rivero, M. S., Garaffo, G. V., Vallarino, E. A. and Elías, R. (2011). Intertidal biogenic reefs built by the polychaete *Boccardia proboscidea* in sewage-impacted areas of Argentina, SW Atlantic. *Marine Ecology*, **32** (2), 188–97.

Jones, C. G., Lawton, J. H. and Shachak, M. (1994). Organisms as Ecosystem Engineers. In *Ecosystem management*. Springer, New York, pp. 130–47.

Liuzzi, M. G. and López Gappa, J. (2008). Macrofaunal assemblages associated with coralline turf: species turnover and changes in structure at different spatial scales. *Marine Ecology Progress Series*, **363**, 147–56.

Liuzzi, M. G., López Gappa, J. and Piriz, M. L. (2011). Latitudinal gradients in macroalgal biodiversity in the Southwest Atlantic between 36° and 55° S. *Hydrobiologia*, **673** (1), 205–14.

López Gappa, J., Calcagno, J. A. and Tablado, A. (1997). Spatial pattern in a low-density population of the barnacle *Balanus amphitrite* Darwin. *Hydrobiologia*, **357** (1–3), 129–37.

López Gappa, J., Tablado, A. and Magaldi, N. (1996). Observations on activity pattern and resting site fidelity in the pulmonate limpet *Siphonaria lessoni*. *Thalassas*, **12**, 27–36.

López Gappa, J. L., Tablado, A. and Magaldi, N. H. (1990). Influence of sewage pollution on a rocky intertidal community dominated by the mytilid Brachidontes rodriguezii. *Marine Ecology Progress Series*, **63**, 163–75.

López Gappa, J. L., Tablado, A. and Magaldi, N. H. (1993). Seasonal changes in an intertidal community affected by sewage pollution. *Environmental Pollution*, **82** (2), 157–65.

López Gappa, J. J. and Tablado, A. (1997). Growth and production of an intertidal population of the chiton *Plaxiphora aurata* (Spalowski, 1795). *The Veliger*, **40**, 263–70.

Lubchenco, J. (1980). Algal zonation in the New England rocky intertidal community: an experimental analysis. *Ecology*, **61** (2), 333–44.

Lucas, A. J., Guerrero, R. A., Mianzan, H. W., Acha, E. M. and Lasta, C. A. (2005). Coastal oceanographic regimes of the northern Argentine continental shelf (34–43 S). *Estuarine, Coastal and Shelf Science*, **65** (3), 405–20.

Marcomini, S. C. and López, R. A. (1993). Coastal protection effects at Buenos Aires, Argentina. In *Coastal Zone '93*, 2724–38.

Marincovich, L. (1973). *Intertidal mollusks of Iquique, Chile*, vol. 16. Natural History Museum, Los Angeles County.

Márquez, F., Vilela, R. A. N., Lozada, M. and Bigatti, G. (2015). Morphological and behavioral differences in the gastropod *Trophon geversianus* associated to distinct environmental conditions, as revealed by a multidisciplinary approach. *Journal of Sea Research*, **95**, 239–47.

Martin, J. P. and Cuevas, J. M. (2006). First record of *Undaria pinnatifida* (Laminariales, Phaeophyta) in Southern Patagonia, Argentina. *Biological Invasions*, **8** (6), 1399–402.

Martos, P. and Piccolo, M. C. (1988). Hydrography of the Argentine continental shelf between 38 and 42 S. *Continental Shelf Research*, **8** (9), 1043–56.

McDonald, P. S., Jensen, G. C. and Armstrong, D. A. (2001). The competitive and predatory impacts of the nonindigenous crab *Carcinus maenas* (L.) on early benthic phase Dungeness crab Cancermagister Dana. *Journal of Experimental Marine Biology and Ecology*, **258** (1), 39–54.

Mendez, M. M., Livore, J. P., Calcagno, J. A. and Bigatti, G. (2017). Effects of recreational activities on Patagonian rocky shores. *Marine Environmental Research*, **130**, 213–20.

Mendez, M. M., Schwindt, E., Bortolus, A., Roche, A., Maggioni, M. and Narvarte, M. (2015). Ecological impacts of the austral-most population of *Crassostrea gigas* in South America: a matter of time? *Ecological research*, **30** (6), 979–87.

Mendez, M. M., Sueiro, M. C., Schwindt, E. and Bortolus, A. (2014). Invasive barnacle fouling on an endemic burrowing crab mobile basibionts as vectors to invade a suboptimal habitat. Universidad de Vigo; *THALASSAS (Santiago de Compostela*, **30** (1), 39–46.

Menge, B. A. (1991). Generalizing from experiments: is predation strong or weak in the New England rocky intertidal? *Oecologia*, **88** (1), 1–8.

Menge, B. A. and Sutherland, J. P. (1976). Species diversity gradients: synthesis of the roles of predation, competition, and temporal heterogeneity. *The American Naturalist*, **110** (973), 351–69.

Meretta, P. E., Matula, C. V. and Casas, G. (2012). Occurrence of the alien kelp *Undaria pinnatifida* (Laminariales, Phaeophyceae) in *Mar del Plata*, Argentina. *Bioinvasions Records*, **1** (1), 59–63.

Miloslavich, P., Cruz-Motta, J. J., Hernández, A. et al. (2016). Benthic Assemblages in South American Intertidal Rocky Shores: Biodiversity, Services, and Threats. In R. Riosmena-Rodríguez, ed. *Marine Benthos*. Nova Publishers, New York.

Monod, T. (1956). Hippidea et Brachyura ouestafricaines – Mémoires de l'Institut Français d'Afrique Noire. *Ifan-Dakar*, **45**, 1–674.

Nugent, P. (1986). Ecología y biología de los mejillinares de *Brachidontes rodriguezi* (D' Orbigny., 1846) en el litoral bonaerense. Doctoral thesis, Universidad Nacional de la Plata.

Nuñez, J. D., Iriarte, P. F., Ocampo, E. H., Iudica, C. and Cledón, M. (2015). Deepphylogeographic divergence among populations of limpet *Siphonaria lessoni* on the east and west coasts of South America. *Marine Biology*, **162** (3), 595–605.

Nuñez, J. D., Laitano, M. V. and Cledón, M. (2012). An intertidal limpet species as a bioindicator: pollution effects reflected by shell characteristics. *Ecological Indicators*, **14** (1), 178–83.

Nuñez, J. D., Ocampo, E. H. and Cledón, M. (2014). A geographic comparison of the resting site fidelity behaviour in an intertidal limpet: Correlation with biological and physical factors. *Journal of Sea Research*, **89**, 23–9.

Olivares Paz, A. N. (2007). *Sistemática molecular del género Fissurella en el Pacífico Sudoriental*. Universidad Santiago de Compostela, A Coruña.

Olivier, S. R., Bastida, R. and Torti, M. R. (1970). Las comunidades bentónicas de los alrededores de Mar del Plata (Argentina). In *Actas del IV Congreso Latinoamericano de Zoología. Caracas, Venezuela, II*, 559–93.

Olivier, S. R., de Paternoster, I. K. and Bastida, R. (1966a). Estudios biocenoticos en las costas de Chubut (Argentina) I. Zonación biocenologica de Puerto Pardelas (Golfo Nuevo). *Boletín del Instituto de Biología Marina*, **10**, 1–74.

Olivier S. R., Escofet A., Orensanz J. M., Pezzani S. E., Turro A. M. and Turro M. E. (1966b) Contribución al conocimiento de las comunidades bentónicas de Mar del Plata. I. Las costas rocosas entre playa grande y playa chica. *Anales de la comisión de investigaciones científicas de la provincia de Buenos Aires (Argentina)*, **7**, 185–206.

Olivier, S. R., Escofet, A., Penchaszadeh, P. E. and Orensanz, J. M. (1972). Estudios ecológicos de la región estuarial de Mar Chiquita (Buenos Aires, Argentina). I. Las comunidades bentónicas. [Ecological studies of the estuary region of Mar Chiquita (Buenos Aires, Argentina). I. The benthic communities]. *Anales de la Sociedad Científica Argentina*, **193** (5/6), 239–62.

Olivier, S. R. and Penchaszadeh, P. E. (1968). Observaciones sobre la ecología y biología de Siphonaria (*Pachysiphonaria lessoni*) (Blainville, 1824) (Gastropoda, Siphonariidae) en el litoral rocoso de Mar del Plata (Bs. As.). *Cahiers de Biologie Marine*, **9**, 469–91.

Orensanz, J. M. (1974). Los anélidos poliquetos de la Provincia Biogeografica Magallánica. I. In *Catálogo de las especies citadas hasta*.

Orensanz, J. M. L., Schwindt, E., Pastorino, G. et al. (2002). No longer the pristine confines of the world ocean: a survey of exotic marine species in the southwestern Atlantic. *Biological Invasions*, **4** (1–2), 115–43.

Otaegui, A. V. and Zaixso, H. E. (1974). Distribución vertical de los moluscos marinos del litoral rocoso de la ría de Puerto Deseado (Santa Cruz, Argentina): Una guía para reconocer los diferentes pisos y horizontes litorales. *Physis A*, 33 (86), 321–34.

Padilla, D. K. (2010). Context-dependent impacts of a non-native ecosystem engineer, the Pacific oyster *Crassostrea gigas*. *Integrative and Comparative Biology*, **50** (2), 213–25.

Paine, R. T. (1966). Food web complexity and species diversity. *The American Naturalist*, **100** (910), 65–75.

Paine, R. T. (1980). Food webs: linkage, interaction strength and community infrastructure. *Journal of Animal Ecology*, **49** (3), 667–85.

Paine, R. T., Castillo, J. C. and Cancino, J. (1985). Perturbation and recovery patterns of starfish-dominated intertidal assemblages in Chile, New Zealand, and Washington State. *The American Naturalist*, **125** (5), 679–91.

Palomo, M. G., Bagur, M., Quiroga, M., Soria, S. and Bugnot, A. (2016). Ecological impacts of two non-indigenous macroalgae on an urban rocky intertidal shore. *Marine Biology*, **163** (8), 178.

Paruelo, J. M., Beltran, A., Jobbagy, E., Sala, O. E. and Golluscio, R. A. (1998). The climate of Patagonia: general patterns and controls on biotic processes. *Ecología Austral*, **8** (2), 85–101.

Pascual, M. and Castaños, C. (2000). *Cultivo de ostras cóncavas en Argentina: desde el criadero hasta la cosecha en el mar*. Secretarıa de Agricultura, Pesca y Alimentación, Ganadería.

Pastorino, G. (1995). Moluscos costeros recientes de Puerto Pirámide, Chubut, Argentina (No. 93). Academia nacional de ciencias, Cordoba. *Miscelanea*, **93**, 1–30.

Pastorino, G., Pio, M. J. and Gimenez, J. (2014). The egg capsules and embryos of the Patagonian gastropod *Trophon plicatus* (Lightfoot, 1786) (Caenogastropoda: Trophoninae) with remarks on the taxonomy of the southwestern Atlantic Trophoninae. *Journal of Molluscan Studies*, **80** (2), 213–18.

Penchaszadeh, P. (1973). Ecología de la comunidad del mejillín (*Brachydontes rodriguezi* d'Orbigny.) en el mediolitoral rocoso de Mar del Plata (Argentina): el proceso de recolonización. *Physis A*, **32** (84), 51–64.

Penchaszadeh, P. E. (2005) *Invasores: Invertebrados exóticos en el Río de la Plata y región marina aledaña*. Eudeba, Buenos Aires, pp. 21–37.

Pilsbry, H. A. (1916). The sessile barnacles (Cirripedia) contained in the collections of the US National Museum; including a monograph of the American species. *Bulletin of the United States National Museum*, **93**, 1–366.

Pío, M. J. (2010). Anatomía e histología del Órgano perforador accesorio (ABO) del gasterópodo Trophon geversianus (Mollusca: Muricidae). Degree thesis, Universidad CAECE.

Piola, A. R. and Falabella, V. (2009). El mar patagónico. In *Atlas del Mar Patagónico: especies y espacios*. Wildlife Conservation Society and Birdlife Internacional, Buenos Aires, pp. 54–75.

Piola, A. R. and Rivas, A. (1997). Corrientes en la plataforma continental. *El mar argentino y sus recursos pesqueros*, **1**, 119–32.

Piriz, M. L. and Casas, G. (1994). Occurrence of *Undaria pinnatifida* in Golfo Nuevo, Argentina. *Applied Phycology Forum*, **10** (4).

Raffo, M. P., Eyras, M. C. and Iribarne, O. O. (2009). The invasion of *Undaria pinnatifida* to a *Macrocystis pyrifera* kelp in Patagonia (Argentina, south-west Atlantic). *Journal of the Marine Biological Association of the United Kingdom*, **89** (8), 1571–80.

Raffo, M. P., Russo, V. L. and Schwindt, E. (2014). Introduced and native species on rocky shore macroalgal assemblages: zonation patterns, composition and diversity. *Aquatic Botany*, **112**, 57–65.

Ramirez, M. E., Nuñez, J. D., Ocampo, E. H. et al. (2012). *Schizymenia dubyi* (Rhodophyta, Schizymeniaceae), a new introduced species in Argentina. *New Zealand Journal of Botany*, **50** (1), 51–8.

Rechimont, M. E., Galvan, D. E., Sueiro, M. C. et al. (2013). Benthic diversity and assemblage structure of a north Patagonian rocky shore: a monitoring legacy of the NaGISA project. *Journal of the Marine Biological Association of the United Kingdom*, **93** (8), 2049–58.

Rico, A., Lanas, P. and López Gappa, J. (2001). Temporal and spatial patterns in the recruitment of *Balanus glandula* and *Balanus laevis* (Crustacea, Cirripedia) in Comodoro Rivadavia harbor (Chubut, Argentina). *Revista del Museo Argentino de Ciencias Naturales nueva serie*, **3** (2), 175–9.

Rico, A., Peralta, R. and López Gappa, J. (2012). Succession in subtidal macrofouling assemblages of a Patagonian harbour (Argentina, SW Atlantic). *Helgoland Marine Research*, **66** (4), 577.

Ringuelet, R. A., Amor, A., Magaldi, N. and Pallares, R. (1962). Estudio ecológico de la fauna intercotidal de Puerto Deseado en febrero de 1961 (Santa Cruz, Argentina). *Physis*, **23** (64), 35–53.

Rivadeneira, M. M., Thiel, M., González, E. R. and Haye, P. A. (2011). An inverse latitudinal gradient of diversity of peracarid crustaceans along the Pacific Coast of South America: out of the deep south. *Global Ecology and Biogeography*, **20** (3), 437–48.

Roche, M. A., Narvarte, M. A., Maggioni, M. and Cardón, R. (2010) Monitoreo de la invasión de la ostra cóncava Crassostrea gigas en la costa norte de Rio Negro: estudio preliminar. In *IV Reunión Binacional de Ecología, Buenos Aires*.

Roux, A. M. and Bremec, C. S. (1996). *Comunidades bentónicas relevadas en las transecciones realizadas frente al Río de la Plata (35° 15'S), Mar del Plata (38° 10'S) y Península Valdés (42° 35'S), Argentina (No. 639.2 INI-INF 11)*. Secretaría de Agricultura, Pesca y Alimentación, Instituto Nacional de Investigación y Desarrollo Pesquero-INIDEP, Buenos Aires.

Ruiz, G. M., Carlton, J. T., Grosholz, E. D. and Hines, A. H. (1997). Global invasions of marine and estuarine habitats by non-indigenous species: mechanisms, extent, and consequences. *American Zoologist*, **37** (6), 621–32.

Saito, Y. (1975). Practical Significance of Algae in Japan. In *Advance of Phycology in Japan*. Dr. W. Junk b. v., The Hague, 304–20.

Salvat, M. B. (1985). Biología de la reproducción de Anasterias Minuta Perrier (Echinodermata, Asteroidea): Especie incubadora de las costas patagónicas. Doctoral dissertation, Facultad de Ciencias Exactas y Naturales. Universidad de Buenos Aires.

San Martín, A. M., Gerdes, D. and Arntz, W. E. (2005). Distributional patterns of shallow-water polychaetes in the Magellan region: a zoogeographical and ecological synopsis. *Scientia Marina*, **69** (S2), 123–33.

Scelzo, M. A., Elías, R., Vallarino, E. A., Charrier, M., Lucero, N. and Alvarez, F. (1996). Variación estacional de la estructura comunitaria del bivalvo intermareal Brachydontes rodriguezi (D'Orbigny, 1846) en sustratos artificiales (Mar del Plata, Argentina). *Neritica*, **10**, 87–102.

Scelzo, M. A. and Lichtschein de Bastida, V. (1978). Desarrollo larval y metamorfosis del cangrejo *Cyrtograpsus altimanus* Rathbun, 1914 (Brachyura, Grapsidae) en laboratorio, con observaciones sobre la ecología de la especie. *Physis A*, **38** (94), 103–26.

Schejter, L., Spivak, E. D. and Luppi, T. A. (2002). Presence of *Pyromaia tuberculata* (Lockington, 1877) adults and larvae in the Argentine continental shelf (Crustacea: Decapoda: Majoidea). *Proceedings of the Biological Society of Washington*, **115** (3), 605–10.

Schwindt, E. (2007). The invasion of the acorn barnacle *Balanus glandula* in the south-western Atlantic 40 years later. *Journal of the Marine Biological Association of the United Kingdom*, **87** (5), 1219–25.

Schwindt, E., López Gappa, J., Raffo, M. P. et al. (2014). Marine fouling invasions in ports of Patagonia (Argentina) with implications for legislation and monitoring programs. *Marine Environmental Research*, **99**, 60–8.

Shatkin, G. (1997). Considerations regarding the possible introduction of the Pacific oyster (*Crassostrea gigas*) to the Gulf of Maine: a review of global experience. *Journal of Shellfish Research*, **16**, 463–78.

Silliman, B. R., Bertness, M. D., Altieri, A. H. et al. (2011). Whole-community facilitation regulates biodiversity on Patagonian rocky shores. *PLoS ONE*, **6** (10), e24502.

Soria, S. A., Teso, V., Gutiérrez, J. L., Arribas, L. P., Scarabino, F. and Palomo, M. G. (2017). Variation in density, size, and morphology of the pulmonate limpet *Siphonaria lessonii* along the Southwestern Atlantic. *Journal of Sea Research*, **129**, 29–35.

Sousa, R., Gutiérrez, J. L. and Aldridge, D. C. (2009). Non-indigenous invasive bivalves as ecosystem engineers. *Biological Invasions*, **11** (10), 2367–85.

Spalding, M. D., Fox, H. E., Allen, G. R. et al. (2007). Marine ecoregions of the world: a bioregionalization of coastal and shelf areas. *BioScience*, **57** (7), 573–83.

Spivak, E., Bastida, R., L'Hoste, S. and Adabbo, H. (1975). Los organismos incrustantes del puerto de Mar del Plata. II. Biología y ecología de *Balanus amphitrite* y *B. trigonus* (Crustacea-Cirripedia). *LEMIT Anales*, **3**, 41–124.

Spivak, E. D. (2005). *Los cirripedios litorales (Cirripedia, Thoracica, Balanomorpha) de la región del Rio de la Plata y las Costas marinas adyacentes. Invasores: Invertebrados exóticos en el Río de la Plata y región marina aledaña.* Eudeba, Buenos Aires, pp. 215–309.

Spivak, E. D., Boschi, E. E. and Martorelli, S. R. (2006). presence of *Palaemon macrodactylus* Rathbun 1902 (Crustacea: decapoda: Caridea: palaemonidae) in Mar del plata harbor, Argentina: first record from southwestern Atlantic waters. *Biological Invasions*, **8** (4), 673–6.

Spivak, E. D. and L'Hoste, S. G., (1976). Presencia de cuatro especies de Balanus en la costa de la provincia de Buenos Aires: distribución y aspectos ecológicos. In *Mar del Plata (Argentina)*.

Tatián, M., Schwindt, E., Lagger, C. and Varela, M. M. (2010).Colonization of Patagonian harbours (SW Atlantic) by an invasive sea squirt. *Spixiana*, **33**, 111–17.

Teso, S. V., Bigatti, G., Casas, G. N., Piriz, M. L. and Penchaszadeh, P. E. (2009). Do native grazers from Patagonia, Argentina, consume the invasive kelp *Undaria pinnatifida*. *Revista del Museo Argentino de Ciencias Naturales*, **11** (1), 7–14.

Torres, A. I., Gil, M. N. and Esteves, J. L. (2004). Nutrient uptake rates by the alien alga *Undaria pinnatifida* (Phaeophyta) (Nuevo Gulf, Patagonia, Argentina) when exposed to diluted sewage effluent. *Hydrobiologia*, **520** (1), 1–6.

Torres, P. J. and Gonzalez-Pisani, X. (2016). Primer registro del cangrejo verde, *Carcinus maenas* (Linnaeus, 1758), en Golfo Nuevo, Argentina: un nuevo límite norte de distribución en costas patagónicas. *Ecología Austral*, **26** (2), 134–7.

Trovant, B., Orensanz, J. L., Ruzzante, D. E., Stotz, W. and Basso, N. G. (2015). Scorched mussels (Bivalvia: Mytilidae: Brachidontinae) from the temperate coasts of South America: phylogenetic relationships, trans-Pacific connections and the footprints of Quaternary glaciations. *Molecular Phylogenetics and Evolution*, **82**, 60–74.

Trovant, B., Ruzzante, D. E., Basso, N. G. and Orensanz, J. L. (2013). Distinctness, phylogenetic relations and biogeography of intertidal mussels (Brachidontes,

Mytilidae) from the south-western Atlantic. *Journal of the Marine Biological Association of the United Kingdom*, **93** (7), 1843–55.

Turra A., Cróquer A. Carranza A. et al. (2013). Global environmental changes: setting priorities for Latin American coastal habitats. *Global Change Biology*, **19**, 1965–9.

Valdovinos, C. and Rüth, M. (2005). Nacellidae limpets of the southern end of South America: taxonomy and distribution. *Revista Chilena de Historia Natural*, **78** (3).

Vallarino, E. A. and Elías, R. (1997). The dynamics of an introduced: *Balanus glandula* population in the south-western Atlantic rocky shores. The consequences on the intertidal community. *Marine Ecology*, **18** (4), 319–35.

Vallarino, E. A. and Elías, R. (2006). A paradox in intertidal mussel beds of the SW Atlantic: increased diversity and reduced variability associated with sewage pollution. *Current Trends in Ecology*, **1**, 77–91.

Vinuesa, J. H. (2005). Distribution of decapod and stomatopod crustaceans from San Jorge Gulf, Argentina. *Revista de Biología Marina y Oceanografía*, **40** (1), 7.

Vinuesa, J. H., Varisco, M. and Escriche, F. (2011). Settlement and recruitment of the crab *Halicarcinus planatus* (Crustacea: Decapoda: Hymenosomatidae) in Golfo San Jorge, Argentina. *Journal of the Marine Biological Association of the United Kingdom*, **91** (03), 685–90.

Wallentinus, I. (2007). *Alien species alert: Undaria pinnatifida:(wakame or japanese kelp)*, https://wedocs.unep.org/handle/20.500.11822/2331.

Wieters, E. A., McQuaid, C., Palomo, G., Pappalardo, P. and Navarrete, S. A. (2012). Biogeographical boundaries, functional group structure and diversity of rocky shore communities along the Argentinean coast. *PLoS ONE*, **7** (11), e49725.

Wong, P. P., Losada, I. J., Gattuso, J.-P. et al. (2014). Coastal Systems and Low-Lying Areas. In C. B. Field, V. R. Barros, D. J. Dokken et al., eds. *Climate Change 2014: Impacts, Adaptation, and Vulnerability. Part A: Global and Sectoral Aspects. Contribution of Working Group II to the Fifth Assessment Report of the Intergovernmental Panel on Climate Change* . Cambridge University Press, Cambridge, pp. 361–409.

Yorio, P., Bertellotti, M., Gandini, P. and Frere, E. (1998). Kelp gulls *Larus dominicanus* breeding on the Argentine coast: population status and relationship with coastal management and conservation. *Marine Ornithology*, **26** (1), 11–18.

Zaixso, H. E., Sar, A. M., Lizarralde, Z. I. and Martin, J. P. (2017). Asociaciones macrobentónicas con presencia de mitílidos de la bahía San Julián (Patagonia austral, Argentina). *Revista de biología marina y oceanografía*, **52** (2), 311–23.

Zamponi, M. O. and Genzano, G. N. (1992). La fauna asociada a Tubularia crocea (Agassiz, 1862) (Anthomedusae; Tubulariidae) y la aplicación de un método de cartificación. *Hidrobiológica*, **2** (1–2), 35–42.

Zamponi, M. O., Genzano, G. N., Acuña, F. G. and Excoffon, A. C. (1998). Studies of Benthic Cnidarian taxocenes along a transect off Mar del Plata (Argentina). *Russian Journal of Marine Biology*, **24** (1), 7–13.

Zullo, V. A. (1992). *Balanus trigonus* Darwin (Cirripedia, Balaninae) in the Atlantic basin: an introduced species? *Bulletin of Marine Science*, **50** (1), 66–74.

# Chapter 8

# Species Interactions and Regime Shifts in Intertidal and Subtidal Rocky Reefs of the Mediterranean Sea

Lisandro Benedetti-Cecchi, Laura Airoldi, Fabio Bulleri, Simonetta Fraschetti and Antonio Terlizzi

## 8.1 Introduction

The Mediterranean Sea is an enclosed basin connected to the Atlantic through the Strait of Gibraltar in the west and to the Sea of Marmara and the Black Sea through the Dardanelles in the east. Its surface waters cover 2,969,000 km$^2$, which makes the Mediterranean the largest enclosed sea in the world (Bianchi and Morri, 2000). A recent analysis has estimated the occurrence of some 17,000 marine species in the Mediterranean, with taxa such as the Phaeophyta, Rhodophyta and Porifera including more than 10 per cent of the total number of species known globally (Coll et al., 2010). These figures are remarkable, considering that the Mediterranean Sea covers less than 1 per cent of the surface and volume of the world's oceans.

The origin of the highly diversified Mediterranean biota is relatively recent and is mainly derived from the Atlantic Ocean. The Mediterranean Sea has undergone various geological, climatic and hydrological transformations that have contributed to generate the hotspot of marine biodiversity that we see today (Bianchi and Morri, 2000; Coll et al., 2010; Lejeusne et al., 2010). In particular, isolation from the Atlantic during the Messinian crisis (6 Mya) resulted in strong evaporation with consequent dramatic changes in climate, sea level and salinity (Bianchi and Morri, 2000). The Messinian crisis decimated the biota of the ancient Mediterranean, which was largely dominated by Indo-Pacific species of warm-water affinities. When the connection with the Atlantic was re-established (5 Mya), the newly colonising species mixed with those surviving the Messinian crisis. Alternating glacial and interglacial periods during the Quaternary, favouring the colonisation of boreal and subtropical species, further contributed to the diversification of the biota in the Mediterranean Sea. The opening of the Suez Canal in 1869 has also impacted on the native biodiversity of the Mediterranean through the introduction of new species from the Red Sea (Galil et al., 2014), which will likely increase in the near future following the expansion of the Suez Canal in 2015 (Galil et al., 2015b).

There is a long tradition of descriptive studies in the Mediterranean Sea, with detailed accounts on taxonomic composition and regional patterns of distribution of marine organisms that exploded in the 1960s owing to intensifying oceanographic cruises and the advent of scuba diving (reviewed in Coll et al., 2010). Our understanding of region-wide patterns of marine biodiversity in the Mediterranean Sea has increased considerably in the last fifteen years as a result of large-scale field surveys (Sala et al., 2011), extensive reviews of the literature (Bouillon et al., 2004; Danovaro et al., 2010; Martin and Giannoulaki, 2014; Telesca et al., 2015), modelling (Sarà et al., 2013; Marras et al., 2015) and synthesis of expert opinions (Micheli et al., 2013). This integration has generated new insights into the present distribution of the Mediterranean biota, as well as focussing attention on the main threats to marine biodiversity and the need to implement better conservation practices at the regional scale (Airoldi and Beck, 2007; Claudet and Fraschetti, 2010; Coll et al., 2010, 2012; Mouillot et al., 2011).

In contrast to descriptive studies, experimental ecology has only been introduced recently in the Mediterranean, with an initial focus on biological interactions. The first experiment that incorporated the logical requirements of replication, randomisation and independence examined the effect of removing a canopy-forming alga in littoral rock pools (Benedetti-Cecchi and Cinelli, 1992a). A parallel study in the same system expanded the range of species interactions examined to include the effects of herbivory and competition between algal turfs (see Connell et al., 2014, for a clarification of the term turf) and canopy recruitment (Benedetti-Cecchi and Cinelli, 1992b). Experimental work proliferated in the following years, examining species interactions in the context of ecological succession (Benedetti-Cecchi, 2000a, 2000b) and extending this approach to subtidal environments (Airoldi et al., 1995; Airoldi, 2000a). As we will discuss, some of these studies established the foundation for the theory that canopy-forming algae and turfs represent alternative states in shallow temperate rocky coasts under different disturbance and stress regimes (Airoldi, 1998, 2000b; Benedetti-Cecchi et al., 2001b), and that changes in sediment loads are one of the main triggers of these shifts in subtidal habitats (Airoldi et al., 1996; Airoldi and Cinelli, 1997; Airoldi, 2003; Irving et al., 2009). More recently, manipulations on subtidal rocky reefs have shown how low levels of herbivory in combination with the local increase of nutrients may foster the recovery of macroalgal canopies and associated biota (Guarnieri et al., 2014).

Experiments in the Mediterranean have also contributed to focus attention on the variance of ecological interactions (Benedetti-Cecchi, 2000c, 2003). Novel experimental designs have been developed that facilitate the separation of the effects of changing the mean intensity from the variance of biological interactions or any other spatially or temporally variable ecological process (Bertocci et al., 2005; Benedetti-Cecchi et al., 2006). These experiments revealed, for example, how spatial variance and mean intensity of grazing may interactively maintain spatially heterogeneous patterns of algal cover, illustrating the great potential for grazing to generate alternative states in marine benthic habitats (Benedetti-Cecchi et al., 2005).

Motivated by the need to understand the effects of intensifying anthropogenic impacts and climate change on marine biodiversity, an increasing number of studies are now examining species interactions in relation to regional stressors and global threats such as ocean warming, acidification, extreme climate events and biological invasions (Bulleri et al., 2016). A better understanding of biotic interactions in the Anthropocene is also essential to guide habitat rehabilitation and restoration efforts. Here, we provide an overview of this type of research that will likely characterise future experimental research in the Mediterranean and elsewhere, using regime shifts as a conceptual framework to guide this approach. The concept of regime shifts, the abrupt transition between alternative states, unifies key aspects of species interactions and their responses to degrading environmental conditions, including resilience, early warning signals of collapse, extinction and hysteresis, all of which have direct bearing on environmental management. To achieve this, we start with a brief introduction to regime shifts and the underlying theory, followed by a discussion of ongoing regime shifts in the Mediterranean; such as the transition from macroalgal forests to turf-dominated assemblages and the widespread collapse of sessile organisms in response to heatwaves, species invasions, infectious diseases and pest metabolites. We then examine the implications of threshold-like biological responses and hysteresis that are typically associated with regime shifts for habitat restoration and rehabilitation. Finally, we conclude with an overview of the research that is needed to understand the interplay between species interactions and rapid environmental change, for which the Mediterranean is providing several dramatic examples.

## 8.2 Regime Shifts (in a Nutshell)

Regime shifts are increasingly observed in systems as diverse as neuronal cells, natural ecosystems and human society (Scheffer, 2009). The ubiquity of these phenomena suggests common underlying causes, such as reinforcing feedbacks, leading to alternative states and tipping points. Shifts between these contrasting states often have undesired effects, for example, the transformation of vegetated arid regions into deserts (Reynolds et al., 2007), the transition from clean to turbid water in lakes (Scheffer, 2009) and the collapse of coral reefs and fish stocks (Mumby et al., 2007; deYoung et al., 2008), which are all associated with significant losses in ecosystem services (Rocha et al., 2015).

A regime shifts occurs if a system can exist in more than one state, and that the transition from one state to another is a non-linear response to changing endogenous or exogenous conditions. Critical transitions are a special case in which alternative states can co-occur under the same set of environmental conditions. Critical transitions and alternative stable states have their roots in catastrophe theory (Gilmore, 1981) and can be illustrated by the prototypical S-shaped bifurcation curve (Figure 8.1). A regime shift occurs when the state variable (e.g., population abundance) moves along the upper branch of the curve in response to changes in a condition variable (e.g., environmental perturbation). When a critical threshold is crossed, the variable switches to the alternative state on the lower branch of the curve (Figure 8.1). The condition variable must improve well beyond the critical threshold · to enable the recovery of the original state, a

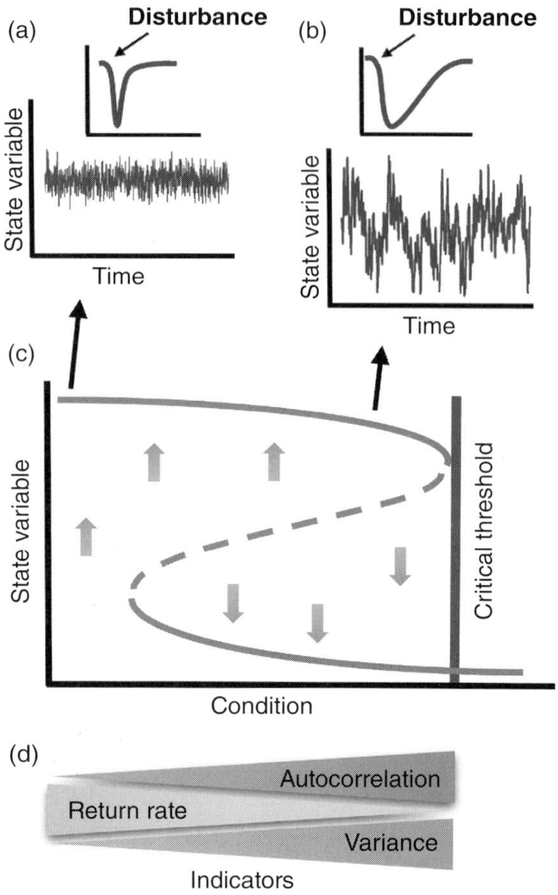


**Fig. 8.1** Bifurcation plot (S-shaped curve) showing alternative stable states and a critical threshold. (c) A regime shift occurs when the state variable (e.g., population abundance) moves along the upper branch of the curve in response to changes in a condition variable (e.g., environmental perturbation). When a critical threshold is crossed, the state variable jumps to the alternative state on the lower branch of the curve. The condition variable must improve well beyond the critical threshold point to enable the recovery of the original state, a phenomenon known as hysteresis. Solid lines show paths along stable equilibria, whereas the dashed line marks unstable equilibria separating the two alternative states. Arrows show the expected paths towards equilibrium for different starting points of the state variable. (a, b) illustrate how temporal fluctuations of the state variable become more variable and autocorrelated as the system approaches the threshold, (b) for conditions close to the threshold as opposed to (a) at a stable state. The time needed for the system to recover from a moderate disturbance (insets in a and b) is also expected to increase close to the threshold. (d) Expected changes of indicators. Modified from Benedetti-Cecchi et al. (2015).

phenomenon known as hysteresis. Reinforcing feedbacks contribute to the stability of the system in each condition (lines in Figure 8.1). The feedback mechanisms locking the system into the upper branch of the curve weaken as the system approaches the tipping point, so that a small perturbation is sufficient to flip the system to the alternative state, where other feedback mechanisms will operate. Between the alternative states, unstable conditions exist (the dashed line in Figure 8.1).

Not all ecosystems behave as predicted by catastrophe theory and some aspects of the theory are unrealistic. For example, we know that fixed equilibria are a rare case for species assemblages that instead oscillate considerably around shifting averages. Nevertheless, the theory retains its appeal because it provides testable predictions of what we should observe near a critical threshold – the so-called early warning signals of an impending regime shift. Specifically, generic indicators of incipient regime shifts have been derived from the prediction that recovery from small perturbations (e.g., a storm in a coastal system) should slow down when the system is already in a stressed state and close to a catastrophic transition. This effect can be measured directly as a decrease in return rate, or as an increase in its reciprocal, recovery time, which is the time required for the system to recover from small perturbations. Hence, recovery from small perturbations is expected to slow down in systems approaching a tipping point, a phenomenon known as critical slowing down (Figure 8.1). Indirect signatures of critical slowing down also emerge as characteristic fluctuations in time series or spatial data, such as enhanced autocorrelation, variance and skewness of the series (Carpenter and Brock, 2006; Scheffer et al., 2009; Drake and Griffen, 2010). Recent studies indicate that early warning signals also precede non-catastrophic transitions, extending the potential use of these indirect indicators of slowing down to a wide range of complex systems (Kéfi et al., 2013). Thus, critical thresholds are not a necessary condition for the presence of early warning signals, which may also show up under less complex, non-linear, dynamics.

Early warning signals are evaluated in analyses of time series and spatial data through simulations and via experiments (Scheffer et al., 2009; Drake and Griffen, 2010; Carpenter et al., 2011; Dai et al., 2012; Dakos et al., 2012; Veraart et al., 2012; Kéfi et al., 2014). Yet, field-based manipulations remain rare. To the best of our knowledge, only three studies have examined critical slowing down and early warning signals using field experiments. Carpenter and colleagues (2011) have shown how the manipulation of top predators can induce a shift in the food web structure of a lake and how the shift was forestalled by several early warning indicators. Sirota et al. (2013) observed increased temporal variance of dissolved oxygen in the aquatic environment that occurs within the leaves of the carnivorous plant, *Sarracenia purpurea*, as the system was experimentally forced towards a transition from aerobic to anaerobic conditions. Both these studies relied on natural, stochastic perturbations as the driving force generating fluctuations in state variables. Later, we describe a novel manipulative approach using multiple perturbations in a seven-year-long field experiment, where Benedetti-Cecchi et al. (2015) demonstrate how the transition from canopy-dominated to turf-forming algal assemblages can be anticipated by classical early warning signals under various regimes of background natural variability.

## 8.3 Loss of Macroalgal Canopies as a Regime Shift

Canopy algae such as kelps and fucoids are dominant habitat-forming species in rocky intertidal and shallow subtidal habitats along the world's temperate coasts (Chapman, 1995; Schiel and Foster, 2006; Bulleri et al., 2012; Crowe et al., 2013; Steneck and Johnson, 2014). Algal canopies provide food and habitat to diversified assemblages of understorey species and enhance coastal primary productivity (Dayton, 1985; Bianchelli et al., 2016). In many places, algal canopies are being replaced by turf-forming algae as a response to multiple stressors, including urbanisation, eutrophication and increasing sediment loads in coastal areas (Benedetti-Cecchi et al., 2001b; Airoldi and Beck, 2007; Strain et al., 2014). In the Mediterranean, the fucoid algae, *Cystoseira* spp., form dense canopies along many rocky intertidal and subtidal rocky coasts. These canopies maintain species-rich understorey assemblages of sessile invertebrates and smaller-sized algae by providing shade and reducing physical stress due to aerial exposure. These habitats are being lost at alarming rates (Airoldi et al., 2014) and manipulative experiments have demonstrated that these systems may switch towards the dominance of algal turfs if the canopy is removed or damaged (Benedetti-Cecchi et al., 2001a, 2012; Tamburello et al., 2013). These predictable responses to canopy manipulations have fostered the view that canopy-dominated rocky shores possess alternative states (Benedetti-Cecchi et al., 2015). In addition, alternative states can only occur if there are reinforcing mechanisms that stabilise the system in one condition or the other, and these mechanisms are present in both the canopy- and the turf-dominated states. *Cystoseira amentacea* has very limited dispersal due to large and negatively buoyant zygotes that settle within twelve hours of fertilisation (Susini et al., 2007). Limited dispersal, coupled with the possibility of juvenile plants fusing their holdfasts to reduce the risk of dislodgement result in reinforcing feedbacks that contribute to the persistence of the canopy-dominated state. Canopy loss promotes the invasion of algal turfs at the expense of understorey species. Once established, algal turfs may facilitate their own growth as they develop into an intricate mat by preventing canopy recruitment and ameliorating physical stress through water retention at low tide (Tamburello et al., 2013) and by limiting grazers and other competitors through the large amounts of sediments entrapped in their dense mats (Airoldi, 1998; Airoldi and Hawkins, 2007). Each state can, therefore, perpetuate itself in the absence of external perturbations or internal dynamics leading to population collapse (e.g., failure to recruit).

A simple model (Figure 8.2) has indicated the existence of a critical threshold in the *Cystoseira*–turf system, with a tipping point at about 75 per cent of canopy loss (Benedetti-Cecchi et al., 2015).

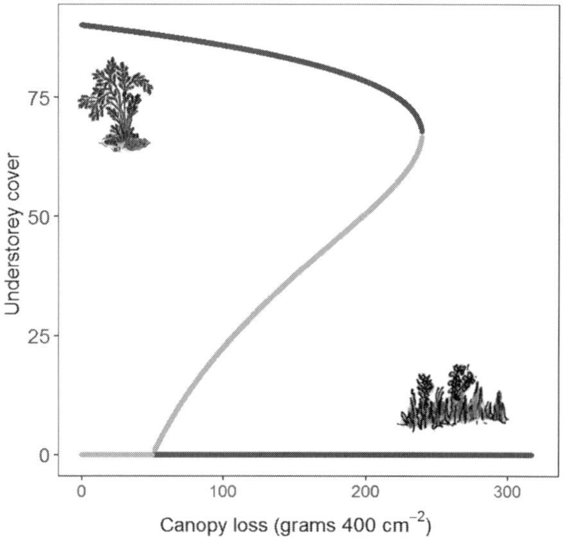


**Fig. 8.2** Bifurcation diagram of the *Cystoseira*–turf system. A transition from a canopy- (top left) to a turf-dominated (bottom right) assemblage with the collapse of understorey species is estimated to occur at 70–75 per cent of canopy loss. Modified from Benedetti-Cecchi et al. (2015).

A press perturbation experiment, in which canopy cover was clipped annually to varying degrees for seven years, was used to test this prediction and to probe early warning signals under natural conditions. In agreement with theoretical expectations, the experiment showed a rise in temporal autocorrelation and recovery time of understorey cover along the gradient of canopy perturbation signalling the approach of the tipping point (Figure 8.3a). Temporal variance, however, did not vary significantly in response to the perturbation (Figure 8.3a).

Extreme events and environmental anomalies are expected to become more common with climate change and how well early warning signals can perform in the presence of strong environmental fluctuations remains an open question (Dakos et al., 2013). Some systems can oscillate between alternative states over short time scales under strong environmental fluctuations, a phenomenon called flickering (Dakos et al., 2013). Benedetti-Cecchi et al. (2015) observed this phenomenon when small clearings designed to mimic the effects of strong storms were added

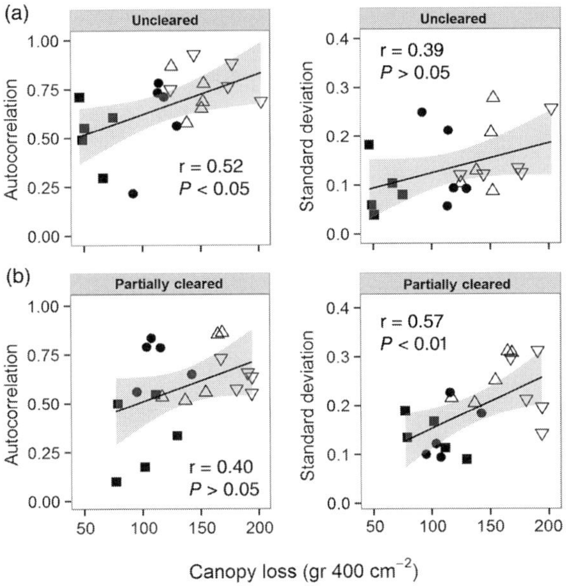


**Fig. 8.3** Performance of temporal autocorrelation and standard deviation as indicators of an impending regime shift following canopy loss either in (a) the presence or (b) the absence of additional perturbations that increase fluctuations in understorey cover. Bootstrap 95 per cent confidence intervals (grey bands) are shown around regression lines. Modified from Benedetti-Cecchi et al. (2015).

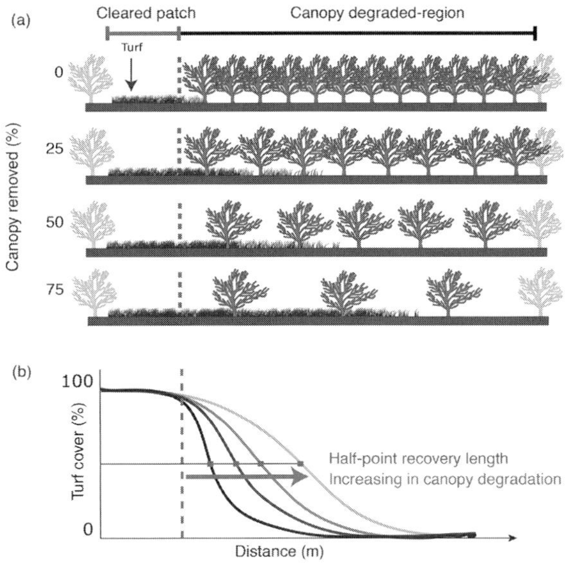


**Fig. 8.4** Experimental design to evaluate the recovery length as a spatial indicator of an impending regime shift. (a) Side view of experimental transects with increasing levels (0–75 per cent) of canopy thinning (light grey plants indicate that a full canopy extends beyond the transects). A cleared patch where all erect organisms were removed from the substratum was produced at the edge of each transect to facilitate the colonisation of algal turfs, which were expected to penetrate into the transects (the canopy degraded region) as a function of canopy removal (dark grey). (b) Qualitative outcomes of the experiment. The recovery length increased with increasing canopy degradation (from dark to light grey lines). Recovery length was measured as the distance from the edge of the cleared patch at which the percentage cover of algal turfs halved compared to the value observed in the clearing. This is also called the half-point recovery length (squares). Modified from Rindi et al. (2017).

to the canopy perturbation gradient. Understorey cover displayed strong temporal fluctuations and autocorrelation was unable to signal the approach of the tipping point in such experimental circumstances. In contrast, temporal variance in understorey cover was able to capture the fluctuating behaviour of the system, outperforming autocorrelation in terms of statistical power (Figure 8.3b; Benedetti-Cecchi et al., 2015). These results suggest that regime shifts may be anticipated under variable natural conditions using the appropriate indicator.

Classical indicators of regime shift can be derived from long-term observations, but extensive time series are usually difficult to obtain. Dai et al. (2013) have proposed recovery length, the spatial analogue of recovery time, as a novel indicator that does not require the need for lengthy time series. Recovery length is the distance necessary for connected populations to recover from a spatial perturbation, and Day and co-authors tested this indicator using microbial populations under controlled laboratory conditions. A field experiment with *C. amentacea* has provided a 'real world' test of this indicator (Rindi et al., 2017). The ability of algal turfs, allowed to colonise in cleared plots, to spread into adjacent areas occupied by *C. amentacea* was evaluated as a function of the amount of canopy removed (Figure 8.4a). The response variable (half-point recovery length) measured the distance into the area occupied by the algal canopy at which the percentage cover of algal turfs declined to half the value in the cleared plot. As expected, the ability of algal turfs to penetrate into algal canopies increased with the amount of *C. amentacea* removed (Figure 8.4b), i.e., the spatial scale at which the system recovered from the perturbation (the presence of invading algal turfs) increased with canopy degradation. Recovery length is applicable as an early warning signal whenever there are sharp boundaries separating distinct assemblages. Such boundaries are common in nature (e.g., transitions between algal turfs and forested areas occupied by canopy algae or seagrasses), so there is a great potential for using recovery length for rapid assessment of incipient regime shifts in the future.

## 8.4 Regime Shifts and Species Invasions

Biological invasions are globally acknowledged among the major threats to biodiversity (Mack et al., 2000). Nonetheless, the ecological impact of non-native species is often assumed rather than formally quantified (Mack et al., 2000; Pejchar and Mooney, 2009). The unambiguous assessment of the effects of non-native species on native species and communities can, in fact, only be achieved experimentally, an approach that has only been applied to a very small proportion of established non-native species worldwide. Given that natural systems are threatened by multiple human stressors, experiments manipulating the presence of the invader through its removal or addition are necessary to establish whether a non-native species should be regarded as a driver or a passenger of ecological change (Didham et al., 2005; Bulleri et al., 2010).

In the driver scenario, the non-native species is the ultimate cause of the alteration of native communities and, drawing from regime shift theory (and terminology), it could be regarded as the conditioning factor or perturbation that moves the system into the basin of attraction of an alternative state, generally characterised by lower diversity and productivity (Scheffer et al., 2009). In the passenger scenario, disruption of native communities by other human disturbances fosters the establishment and/or spread of the invader. Once established, the impact of such an opportunistic invader on native systems can vary from negligible (i.e., naturalisation) to the fundamental alteration of the structure and functioning of the system. In the first case, the invader plays a role neither in the shift to the alternative state nor in determining its degree of hysteresis. In the second case, the invader, although not being the key force moving the state variable towards the alternative attractor, could reduce the resistance and resilience of the system by altering existing feedbacks or introducing new, reinforcing, feedbacks once the shift has occurred. Using the 'ball-in-cup' analogy (Scheffer, 2009), the invader would promote the stability of the alternative state by enhancing the width and depth of the cup that represents the alternative state.

The role of non-native species as determinants of regime shifts has received some consideration, at least in terrestrial systems (Gaertner et al., 2014), while less attention has been devoted to assess how, and to what extent, non-native species can enhance the stability of alternative states. In the next paragraphs, focussing on Mediterranean rocky reef habitats, we provide some evidence that non-native species have the potential to cause the shift between macroalgal canopies and alternative states (i.e., encrusting coralline barrens), as well as to promote their stability (i.e., algal turfs).

The Mediterranean Sea is, indeed, a hotspot of invasion (Rilov and Galil, 2009; Zenetos et al., 2010; Galil et al., 2014). The most updated checklist of established non-native species includes 879 non-native species, belonging to a variety of phyla (Galil et al., 2014). Along with common vectors of introduction in marine environments, such as maritime traffic and aquaculture, high rates of introduction in the Mediterranean are largely the result of the opening of the Suez Canal that provided access for tropical species from the Red Sea (Galil et al., 2014). At the same time, the Mediterranean has been historically under pressure from a variety of other human activities that have caused major alterations to the structure and functioning of marine coastal systems (Airoldi and Beck, 2007; Micheli et al., 2013). This basin offers, therefore, a unique opportunity to explore the role of non-native species in regime shifts under a scenario of multiple stressors.

## 8.5 Non-native Species as Drivers of Regime Shift

The shift between subtidal macroalgal forests formed by brown seaweeds and encrusting coralline algal barrens is a classic example of regime shifts in marine environments (Steneck et al., 2002; Ling et al., 2015). This phenomenon has been documented along the temperate coasts of global oceans and seas and has been long recognised as the result of sea urchin overgrazing (Ling et al., 2015). In the last three decades, a plethora of studies have brought compelling evidence that grazing by two species of sea urchins, *Paracentrotus lividus* (Lamark, 1816) and *Arbacia lixula* (Linnaeus, 1758), is the main determinant of the formation and maintenance of barren areas on shallow Mediterranean rocky reefs (Verlaque et al., 2000; Hereu, 2006; Bulleri et al., 2009; Bulleri, 2013; Guarnieri et al., 2014).

More recently, however, extensive encrusting coralline barrens with no sea urchins have been reported on the Mediterranean coast of Turkey. Subsequent experimental work demonstrated that non-native rabbitfish were mostly responsible for the shift from macroalgal canopies formed by the fucoid, *Cystoseira* spp. to barrens (Sala et al., 2011). *Siganus luridus* (Rüppell, 1829) and *Siganus rivulatus* (Forsskål and Niebuhr, 1775) are Lessepsian migrants that were first recorded in the eastern Mediterranean in 1927 and 1956, respectively. Since then, these two fish have become dominant over the autochthonous parrotfish, *Sparisoma cretense* (Linnaeus, 1758), accounting for up to 90 per cent of the biomass of present-day herbivorous fish assemblages (Sala et al., 2011). Verges et al. (2014) have shown that the two species of Siganids are functionally distinct and feed complementarily on different life stages of erect macroalgae. Specifically, *S. luridus* feeds on adult macroalgae, while *S. rivulatus*, together with *S. cretense*, feeds on juvenile stages that are included in the epilithic algal matrix (Verges et al., 2014). Under these circumstances, *S. luridus* would represent the perturbation driving the shift from the forested to the barren state, while *S. rivulatus* and *S. cretense* lock the system into the degraded state by preventing the recovery of erect macroalgae.

The feeding complementarity among non-native species is strikingly similar to that documented for native assemblages of herbivores. Although somewhat controversial, the urchin, *P. lividus*, feeding mainly on erect macroalgae (Agnetta et al., 2015), has long been considered as the main species responsible for the formation of barrens. In contrast, the urchin, *A. lixula* and the limpet, *Patella caerulea*, which scrape the epilithic biofilm (and encrusting corallines in the case of the former species) off the substratum can efficiently halt the recovery of erect macroalgae, maintaining the stability of the barren state (Bulleri et al., 1999; Bonaviri et al., 2011; Piazzi et al., 2016). These two cases strongly suggest that complementarity in the feeding habit of herbivorous species is paramount for moving the system into the basin of attraction of the barrens state, and for promoting its stability. *P. lividus* can, for example, maintain high densities in barrens despite sparse food availability, likely due to its ability to switch from active feeding on attached macroalgae to passive feeding on drifting material, including algal fragments and particulate organic matter (Harrold and Reed, 1985; Bulleri et al., 1999) and ability to reallocate resources from reproductive to somatic tissues (McCarron et al., 2009). In the non-native scenario, rapid growth of the epilithic biofilm may indefinitely sustain *S. rivulatus* populations, guaranteeing the long-term persistence of barrens. The lack of suitable feeding resources may, however, enhance the migration of *S. luridus* towards areas colonised by erect macroalgae, ultimately expanding barren areas. This, coupled with the long-distance dispersal of the fishes larval phase (up to ~1,000 km), may result in the rapid expansion of the distribution of *S. luridus*. *S. luridus* has rapidly extended northwards in the last decade, likely favoured by seawater warming (Sala et al., 2011; Verges et al., 2014). A viable population of *S. luridus* has, for example, successfully established in Sicily since 2004 (Azzurro and Andaloro, 2004) and individuals have been more recently (2008) caught in trammel nets along the French Mediterranean coast, suggesting that the northward migration of this species is occurring at a very fast pace (Daniel et al., 2009). Under these circumstances, non-native herbivores might accelerate the ongoing decline of canopy-forming macroalgae in the northern Mediterranean.

## 8.6 Non-native Species Enhancing the Stability of the Alternate State

The green macroalga, *Caulerpa cylindracea* (Sonder), is one of the major pest species in the Mediterranean (Verlaque et al., 2000; Piazzi et al., 2005). This seaweed has become ubiquitous in coastal environments in both natural and artificial habitats by virtue of its rapid growth (up to 2 cm a day) and ability to reproduce both sexually and through fragmentation (Piazzi et al., 2005; Vaselli et al., 2008). Recent experimental work has shown that the direction and magnitude of the effects of this species on native rocky reef assemblages vary according to the background levels of cumulative human impacts (Bulleri et al., 2016). In particular, this species seems to exert negative effects on native communities at pristine sites, while its effects tend to become neutral or positive in degraded habitats. At the same time, *C. cylindracea* has been shown to be unable to establish when primary space is monopolised by lush canopies formed by either the seagrass, *Posidonia oceanica* (L.) Delile, or macroalgae belonging to the genera *Cystoseira* and *Halopythis* (Bulleri et al., 2010, 2016; Ceccherelli et al., 2014).

Experimental removal or natural loss of native macrophytes generally results in the prompt colonisation of space by opportunistic species, such as those that form algal turfs (Benedetti-Cecchi et al., 2001b; Bulleri et al., 2002; Mangialajo et al., 2008), and also by *C. cylindracea* (Bulleri et al., 2009, 2011, 2016; Ceccherelli et al., 2014). Once established, *C. cylindracea* has the potential to slow down or, indeed, halt the recovery of macroagal canopies, thus, locking the system into the alternative degraded state. These negative effects on the resilience of canopy stands do not require *C. cylindracea* to achieve dominance, as they have been recorded even when the seaweed only achieved relatively low cover (Bulleri et al., 2017). The main alteration in the physical characteristic of benthic environments invaded by *C. cylindracea* is the enhanced accumulation of sediments (Piazzi et al., 2007). The stolons of *C. cylindracea*, which form a complex web, can trap sediments and, thereby, limit gas exchange between the substrate and water column. Such a thick layer of sediments can reduce the recruitment of *Cystoseira*, preventing the recovery of the forested state (Balata et al., 2004). By promoting the dominance of algal turfs, *C. cylindracea*, therefore, actually fosters its own persistence. The complex three-dimensional (3D) matrix formed by the filamentous and articulated corallines that make up the turfs allows the efficient anchoring of the rhizoids of *C. cylindracea*, ultimately enhancing the spread of *Caulerpa* (Bulleri and Benedetti-Cecchi, 2008; Bulleri et al., 2009).

Other feedback mechanisms triggered by *C. cylindracea* must, however, operate to sustain the persistence of algal turfs, since loss of resilience of canopy stands has been documented even when the cover of *C. cylindracea* was too low to cause a biologically significant increase in sediment accumulation (Bulleri et al., 2017). Evidence from terrestrial environments suggest that invasive plants may trigger reinforcing feedbacks by altering the structure and functioning of microbial communities in the soil (Reinhart and Callaway, 2006; Gaertner et al., 2014). In contrast, the role of the microbial community in regulating the establishment and impacts of marine autotrophs has only recently received any attention. The microbial community associated with *C. cylindracea* has recently been shown to be very stable through time, suggesting that it could be involved in promoting the persistence of the invader itself (Rizzo et al., 2016) and, indirectly, the algal turfs (Bulleri et al., 2017). Assessing the role of the microbial community seems, therefore, a promising avenue of research to enhance our understanding of the forces regulating the establishment and impacts of non-native macroalgae, as well as their role in promoting the shift towards, or the maintenance of, alternative, degraded states.

## 8.7 Pest Metabolites Potentially Triggering Chemically Mediated Regime Shifts

Although the physical impact caused by the invasion of *C. cylindracea* on marine habitats and associated flora and fauna is widely studied, less explored are the subtle indirect impacts of molecules produced by invasive species. In this context, invasive species can represent new forms of chemical challenge, potentially able to determine shifts in trophic webs with consequent changes in the functioning of the invaded systems.

Caulerpenyne, caulerpine and caulerpicin, for example, are the most abundant and investigated secondary metabolites of *C. cylindracea*, and possess several biological properties, such as allelochemical, neurotoxic and cytotoxic activities (Raniello et al., 2007; Mollo et al., 2015), which may contribute to the invasion potential of the alga. In spite of the toxicity of these metabolites, and for reasons that still need to be fully elucidated, *C. cylindracea* has become an important food item in the diet of the sea bream, *Diplodus sargus,* which typically shows a varied and omnivorous diet (Terlizzi et al., 2011). More recent evidence indicates the presence of *C. cylindracea* in stomach contents of both native Mediterranean fish species (i.e., the sparids, *Spondyliosoma cantharus*, *Sarpa salpa* and *Diplodus vulgaris,* and the scarid, *S. cretense*) and the siganid Lessepsian migrant, *S. luridus*.

As a consequence of consuming *C. cylindracea*, the red pigment, caulerpin, accumulates in the tissues of *D. sargus*, representing a useful indicator of the exposure of this fish to the alien macroalga. By using caulerpin as a 'trophic marker', the ingestion of *C. cylindracea* has been related to the appearance of some cellular and physiological alterations in the fish, which include oxidative stress, the onset of genotoxic damage and the possible appearance of neurotoxic damage (Felline et al., 2012; Gorbi et al., 2014; Figure 8.5). Changes in the gonadosomatic index (GSI) of the fish have been also observed, indicating the alteration of gross gonad morphology, with possible implications on reproductive performance, which was further supported by the significant induction of the vitellogenin gene expression in fish with high hepatic content of caulerpin (Felline et al., 2012; Gorbi et al., 2014). All these observed changes in fish condition might lead to a detrimental health status and altered behaviour, potentially affecting the reproductive success of fish populations in the long term. Therefore, the invasion of *C. cylindracea*, through the entry of pest metabolites into food webs, may result in dramatic and abrupt changes in the dynamics of the top predators in the Mediterranean, including alteration of trophic webs and ecosystem functioning.

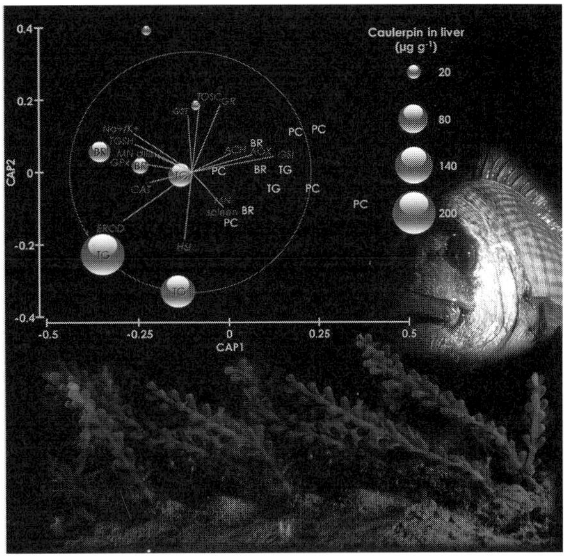


**Fig. 8.5** Canonical analysis of principal coordinates (CAP) representing differences in cellular and molecular responses across the different levels of caulerpin bioaccumulation in fish sampled in three different localities (indicated with BR, TG and PC). Grey circles indicating increasing caulerpin concentration in liver (μg·g$^{-1}$ dry weight) are superimposed on each individual. Grey lines show strength and direction of correlations of variables (biomarkers) with the two canonical axes. The variables included in the analysis are: EROD (7-ethoxyresorufin O-deethylase); AOX (Acyl CoA oxidase); GST (glutathione S-transferases); ACH (acetylcholinesterase); CAT (catalase); GR (glutathione reductase); TGSH (total glutathione); GPx (glutathione peroxidases); TOSC (total oxyradical scavenging); MN spleen and MN gills (micronuclei frequencies in spleen and gills); Na+/K+ (sodium-potassium-ATPase); HIS (hepatosomatic index). Modified from Felline et al. (2012).

## 8.8 Heat Waves and the Collapse of Sessile Invertebrates in Biogenic Habitats

In the Mediterranean Sea, thermal anomalies are one of the most dramatic examples of rapid environmental change causing ecological regime shifts (Garrabou et al., 2009; Rivetti et al., 2014, 2015). Changes in sea surface temperature (SST) have long been identified as one of the main physical drivers influencing both marine biodiversity and the services it provides. Ultimate direct or indirect consequences of thermal anomalies include alterations in the patterns of species distribution (Chevaldonne and Lejeusne, 2003), species' phenology (Puce and Bavestrello, 2009), spread in both pathogens and invasive species, and mass mortalities (Garrabou et al., 2009; Vendramin et al., 2013). Increases in SST are associated with modifications of the vertical thermal structure of the upper ocean, with possible ecological effects that are not limited to the first metres of the water column (Danovaro et al., 2001). In recent years, biogenic habitats such as coralligenous assemblages, which are hotspots of species diversity in the Mediterranean Sea, have been strongly affected by several mass mortality events (Martin and Giannoulaki, 2014; Rivetti et al., 2014; Figure 8.6). Gorgonians, scleractinians, zoantharians, sponges and bryozoans have been severely affected by thermal anomalies (Cerrano et al., 2000; Martin et al., 2002; Garrabou et al., 2009; Huete-Stauffer et al., 2011).

The two most dramatic thermal anomaly events in terms of geographic extension (1,000 km of coastline) and number of affected species (approximately thirty species including sponges, cnidarians, bivalves, ascidians and bryozoans) occurred during the summers of 1999 and 2003 along the north-western Mediterranean coasts (Figure 8.6). These two events coincided with temperatures of 3–4°C above the seasonal average (Cerrano et al., 2000). Almost all studies associated the mass mortalities observed to positive thermal anomalies in late summer and early autumn, when species are more exposed to energetic constraints due to the changing season (Kruzic and Popijac, 2015). However, the same regions affected by mass mortalities in different years exhibited differential responses among species, both in magnitude (ranging from low to high mortality), and in the affected depth range, indicating that factors other than mean temperature may be involved in the observed changes. The consequence is that, despite the cumulative evidence that thermal anomalies can determine abrupt shifts in sessile invertebrates, clear cause–effect relationships are still poorly understood and the transitions caused by heat waves towards less diversified and less productive assemblages (described earlier for macroalgal canopies) have been the focus of only a limited number of studies. The feedback mechanisms maintaining alternative states in these systems have never been clearly described so our current understanding of how thermal anomalies modify thresholds and erode resilience by impairing these mechanisms is poor. Experimental simulations of the effects of thermal anomalies on marine ecosystems can be complex (Rodolfo-Metalpa et al., 2006; Torrents et al., 2008; Ferrier-Pagés et al., 2009) and field-based manipulations have rarely been conducted.

Recently, Ponti et al. (2014) showed, with a manipulative study, that the presence of gorgonian forests mainly limits the growth of erect algae and enhances the abundance of encrusting algae and sessile invertebrates. Disappearances of gorgonians driven by thermal anomalies may, therefore, cause a shift from assemblages characterised by crustose coralline algae to filamentous algal assemblages, decreasing the complexity of biogenic (coralligenous) structures. Di Camillo and Cerrano (2015) documented two massive mortalities of gorgonians (in 2009 and 2011) and the pattern of recovery of the affected assemblages. Results show an impressive and fast shift from slow-growing and long-lived species to assemblages dominated by fast-growing and short-lived species such as stoloniferans, hydrozoans, mussels, algae, serpulids and bryozoans. As another example of dramatic shifts in species assemblages following ocean warming, Chevaldonne and Lejeusne (2003) described a transition from cold to warm affinity populations

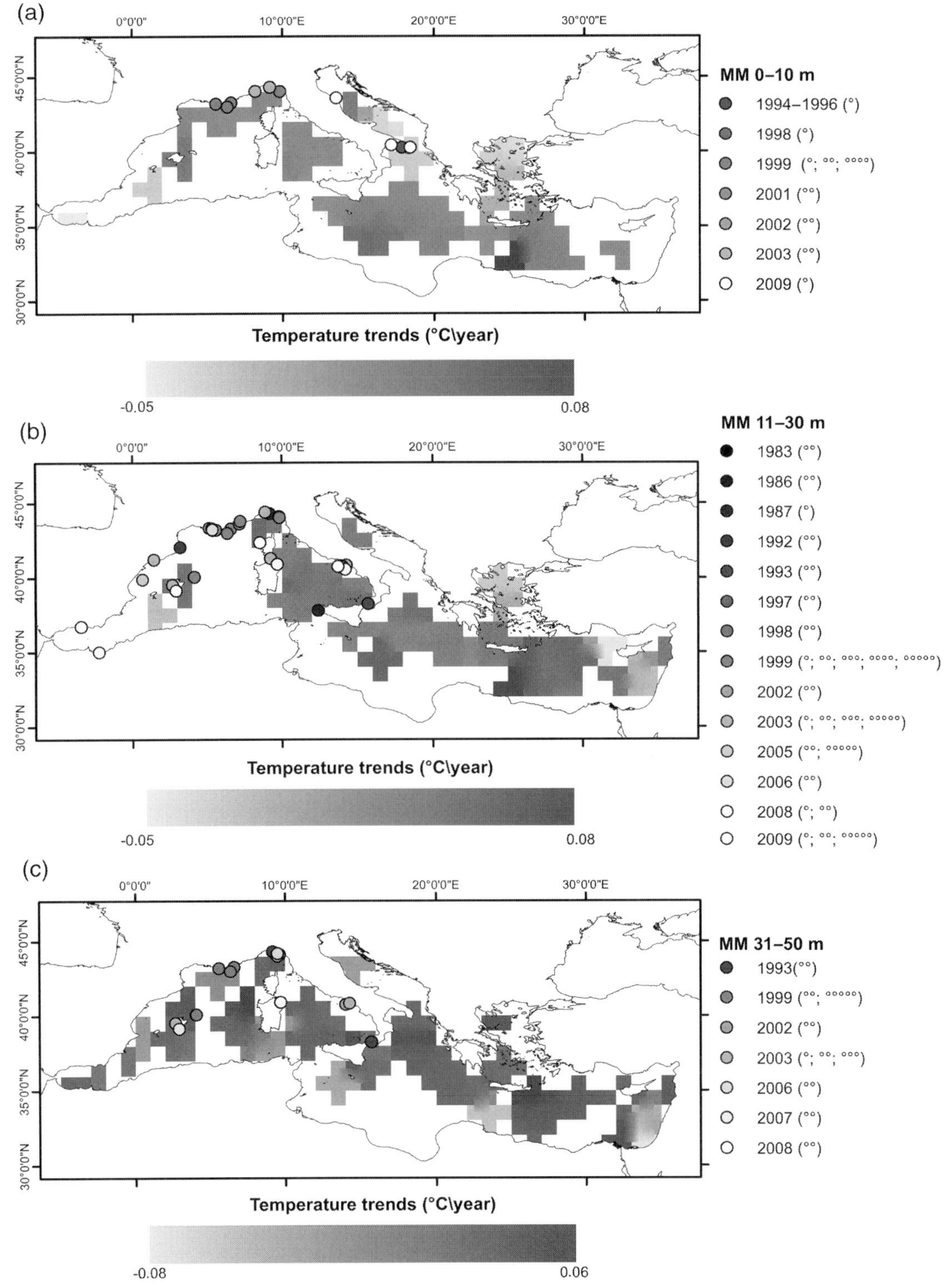


**Fig. 8.6** Temporal trends of SST and mass mortality events in the Mediterranean Sea. Temperature trends have been evaluated on grids of 1° latitude and 1° longitude through linear regression at (a) 0–10 m, (b) 11–30 m and (c) 31–50 m depth layers, using data from July to November for the years 1945–2011. Significant slopes are plotted in greyscale and nonsignificant ones in white. Symbols within and next to panels identify the location and year of occurrence of mass mortality events, respectively. Asterisks next to panels identify the taxa for which mass mortality was recorded: °sponges, °°cnidarians, °°° bryozoans, °°°°ascidians, °°°°°bivalves. Modified from Rivetti et al. (2014).

of endemic mysids in submarine caves over a period of five years. Pairaud et al. (2014) developed a methodology to assess the current risk of mass mortality (die-off events occurring at regional scales) associated with temperature increase along north-west Mediterranean continental coasts using a 3D regional ocean model to obtain the temperature conditions for the period 2001–2010, and validating the model outputs by comparing them with in situ observations in affected areas. When combined with information on species' thermal niches, the model successfully identified regions where mass mortality events had occurred in relation to thermal stress. This was the case of the red gorgonian *Paramuricea clavata* (Risso, 1826). These studies highlight that the effects of thermal anomalies can be detrimental for sessile invertebrates, even without considering the interaction with concurrent changes (e.g., acidification) that might occur in the future.

## 8.9 The Interplay of Warming and Infectious Diseases as Drivers of Population Collapse

Increased water temperature as a consequence of climate change can also be a major driver in determining the onset of disease outbreaks (Harvell et al., 1999). A wide range of infectious agents are characterised by optimal replication and increased pathogenicity under higher temperatures (Alker et al., 2001). The added effect of decreased immunocompetence of hosts due to thermal stress can result in the development of opportunistic infections. As a result, persistent abnormally high water temperatures, possibly in combination with other abiotic and biotic stressors, have been associated with several episodes of mass mortality among marine vertebrates and invertebrates (Perez et al., 2000), and have been hypothesised to contribute to the extensive declines of populations of *Cystoseira* spp. in the Mediterranean Sea by triggering changes in microbial communities (Strain et al., 2015; Mancuso et al., 2016). Global warming may also be responsible for changes in the distribution and abundance of metazoan (Klimpel and Palm, 2011) and protozoan (Ford and Chintala, 2006) parasites and such changes can have important consequences on human health and the seafood industry. Nematodes of the genus *Anisakis*, which parasitise marine fish as intermediate hosts and cetaceans as definitive hosts, are parasites of zoonotic interest, as they are responsible for human anisakiasis. Climate change may affect both the distribution and abundance of *Anisakis* spp. as a result of the migration of large cetaceans in more northern and southern habitats due to the melting of polar ice, leading to a higher abundance of *Anisakis* parasites in formerly unrepresented regions (Klimpel and Palm, 2011). Another example is the protozoan *Perkinsus marinus*, which affects growth and survival of the eastern oyster, *Crassostrea virginica*, leading to severe economic losses (Andrews, 1988; Burreson, 1996). Recent research indicates that this parasite has expanded its range northwards as a result of the warming trend in the north-eastern USA (Ford and Chintala, 2006).

During the last decades, reports of mass mortalities in Mediterranean fish (Marino and Azzurro, 2001; Kara et al., 2014) highlight the role of viral infections in threatening populations of ecologically and economically important species. Members of the genus *Epinephelus* are considered highly susceptible to a viral disease known as viral nervous necrosis (VNN). This disease is caused by a small ssRNA virus belonging to the genus *Betanodavirus* that is now widespread with different strains throughout the world's oceans (Munday et al., 2002), with each strain displaying different optimal replication temperatures (Iwamoto et al., 2000). The disease represents a major threat to the global marine aquaculture industry, and several episodes of mortality associated with VNN outbreaks have been reported in wild groupers from the Mediterranean and extra-Mediterranean waters (Marino and Azzurro, 2001; Gomez et al., 2009; Kara et al., 2014). A combination of spawning stress and abnormally high water temperatures have been suggested to trigger the infection (Kara et al., 2014).

A documented case of VNN in the Mediterranean reported a severe mass mortality among wild dusky grouper, *Epinephelus marginatus*, and

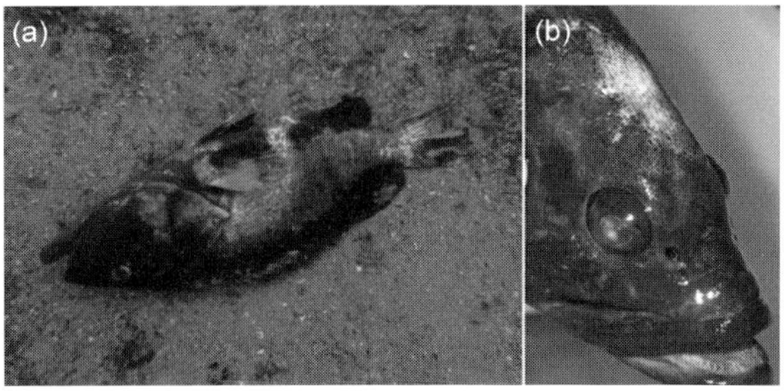


**Fig. 8.7** Adult specimens of dusky grouper *E. marginatus* infected by VNN found moribund on (a) the sea bottom and (b) displaying head skin erosion and corneal opacity. Modified from Vendramin et al. (2013).

golden grouper, *Epinephelus costae* (Vendramin et al., 2013; Figure 8.7). The outbreak was characterised by unusual and persistent high water temperatures (Vendramin et al., 2013). The *Betanodavirus* genotype isolated during the study displayed the highest optimal replication temperature (25–30°C) among *Betanodavirus* strains characterised to date (Hata et al., 2007) and could, therefore, represent an emerging threat in the face of current ocean warming.

A recent virological screening on different wild fish species in two marine protected areas (MPAs) showed that *Betanodavirus* is present with remarkable prevalence, indicating a disease threat for fish species that are susceptible to the infection (Tedesco, 2015). Although the application of protection in the MPAs should theoretically limit anthropogenic impacts, the presence of other stressors (i.e., increased water temperatures due to climate change) towards which MPAs offer no protection, may negatively affect the physiology of susceptible fish species and trigger a disease outbreak. In the case of *Betanodavirus*, its persistent circulation in wild fish populations, combined with abnormal climatic conditions, may impose limitations on the effective conservation of *Epinephelus* spp.

## 8.10 Conservation and Restoration Challenges

A recent review estimated that 10–50 per cent of damaged marine populations and ecosystems show some signs of recovery following adequate management and conservation measures, but the process can take up to many decades for long-lived species and complex ecosystems, and rarely do ecosystems recover to former levels of abundance (Lotze et al., 2011). Major identified drivers of recovery included the reduction of human impacts (especially exploitation, habitat loss and pollution), combined with favourable life-history strategies and environmental conditions, as well as awareness, legal protection and enforcement of management plans. The review also highlighted research gaps and challenges, including the need to identify thresholds and trade-offs in the management of the impacts of multiple stressors. A threshold can be set as the level of human-induced pressure (e.g., pollution) at which small changes produce substantial improvements in protecting an ecosystem's structural (e.g., diversity) and functional (e.g., resilience) attributes (Samhouri et al., 2010). This approach is based on the detection of non-linearities in relationships between ecosystem attributes and pressures. These relationships are, however, known in only a few cases, and often they focus on one direction only (increasing human pressures), without exploring the reverse pathways following management actions geared towards ecosystem recovery. Further, in natural systems, the cumulative impact from several stressors tends to result in synergistic and antagonistic effects rather that additive ones (Crain et al., 2008; Halpern et al., 2008), which has important management implications (Brown et al., 2013).

Where multiple stressors have additive effects, their management is relatively straightforward and consistent with approaches commonly focussing on one stressor at a time. Whenever stressors have synergistic or antagonistic interactions with other stressors, however, their management becomes more challenging. Of particular concern are co-occurring stressors, where multiple stressors reinforce the direction of change and reduce the likelihood that a system can retrace the same trajectory during restoration as during degradation. For example, fisheries' exploitation and increased nutrient loadings jointly affect food webs and production in estuaries, and each affects the sensitivity of species and ecosystems to the other stressor (Breitburg et al., 2009). As a result, there could be levels of fishery removals and nutrient loadings that lead to threshold responses and are resistant to remediation through fisheries and nutrient management (Scheffer et al., 2001; Breitburg and Reidel, 2013). Synergisms, in contrast, may respond quite favourably to removal of a single stressor, as long as the system has not passed a threshold into an alternative state. For example, enhanced nutrient loads and global climatic stressors synergistically drive the shift from algal canopies to turfs (Strain et al., 2014), but, if the system has not yet shifted, the reduction in the levels of local stressors (nutrient concentrations and sediment loads) can be used as an effective intervention to increase resilience of canopy algae to projected global climate stressors (high wave exposure and increasing SSTs; Strain et al., 2015). In the most severe cases, when ecosystems have passed a threshold such that their composition and/or function differs so much from the past that they can be defined as 'novel', recovery may become unlikely (Hobbs et al., 2009). For this reason, active restoration is becoming an integral part of conservation, natural resource management and sustainable development (Suding, 2011).

Restoration has been defined in multiple ways (Elliott et al., 2007). Here, we use the common definition 'the process of assisting the recovery of damaged, degraded, or destroyed ecosystems' (Hobbs et al., 2004), which views restoration as a broad term that encompasses multiple forms of interventions, from preventative management aimed at stress relief, to reparations involving habitat construction to enable the renewal of services in the form of a 'target-designed novel ecosystem' (Abelson et al., 2016). Restoration ecology is a relatively recent science in estuarine and marine systems, and wide gaps still exist among current implementation methods, approaches and standards, and the supporting science (Abelson et al., 2016). In the Mediterranean Sea, attempts at restoration started in the 1980s and mostly focussed on seagrass habitats (Giaccone and Calvo, 1980; Meinesz et al., 1991; Piazzi et al., 1998), while the restoration of other damaged ecosystems lags far behind.

In recent years, there has been growing interest in the restoration of *Cystoseira* forests on both natural and artificial substrata to counteract the local decline of these species (Falace et al., 2006; Susini et al., 2007; Perkol-Finkel et al., 2012; Gianni et al., 2013). This approach is being adopted because, although experiments consistently show that the reduction in the levels of local stressors (particularly sediment loads and nutrient concentrations) is an effective intervention to reverse local damages and increase resilience of canopy algae to projected global climate stressors (Sales et al., 2011; Strain et al., 2015), in many areas the damage is already so great that natural recovery is unlikely (Perkol-Finkel and Airoldi, 2010). *Cystoseira* spp. are, in fact, characterised by low dispersal ability, which reduces the potential for natural recovery of highly fragmented populations (Buonuomo et al., unpublished data). Gianni et al. (2013) reviewed the restoration potential for *Cystoseira* spp. in the Mediterranean Sea, and the technologies currently available or under development to promote restoration. Most studies have used adult plant transplantation on either natural or artificial substrata (Falace et al., 2006; Susini et al., 2007; Sales et al., 2011), or transplantation of juveniles either collected (Perkol-Finkel and Airoldi, 2010) or allowed to settle in the field on artificial panels (Perkol-Finkel et al., 2012), while tests of the potential to use gametes/zygotes are surprisingly limited. The outcomes of these pilot restoration tests have been promising but variable, with grazing being one of the most common causes of failure (Falace et al., 2006; Susini et al., 2007; Ferrario et al., 2016).

More recent work is focussing on the potential to garden canopy algae (as well as other potential target species, such as filter feeders) on artificial infrastructures, as part of an overall approach aiming to revitalise ecosystem functionality and services in marine urban areas (Perkol-Finkel et al., 2012). As currently designed, marine urbanisation (the development of artificial structures in marine environments) and other anthropogenic pressures on the coast are locally depleting many populations of canopy-forming algae (Benedetti-Cecchi et al., 2001b; Mangialajo et al., 2008; Perkol-Finkel and Airoldi, 2010;

Strain et al., 2015; Thibaut et al., 2015), while offering limited possibilities for populations to expand (Ferrario et al., 2016). Ongoing tests aim to identify novel designs for marine urbanisation, more compatible with the support/provisioning of vital marine ecosystems (Firth et al., 2014). These tests are complemented by analyses of the genetic structure and demographic connectivity of fragmented populations of *Cystoseira* to better understand their dispersal and recovery capabilities (Buonomo et al., 2017). Understanding population connectivity is fundamental to help select priority areas for conservation and restoration, as well as to guide a better design and planning of marine infrastructures to prevent deleterious ecological impacts such as, for example, the spread of non-indigenous species by serving as 'stepping stones' (Perkol-Finkel et al., 2012; Williams et al., 2016; Chapter 14).

## 8.11 Conclusions

Regime shifts are ubiquitous in marine ecosystems, ranging from the collapse of individual populations, such as commercial fish, to the disappearance of entire habitats, such as macroalgal forests and seagrasses (Rocha et al., 2015). Regime shifts have multiple co-occurring causes involving regional and global anthropogenic drivers (Mollmann et al., 2015). As extreme climate events become more frequent and human activities intensify both in coastal environments and in the open ocean, marine regime shifts will become more likely in the near future (Rocha et al., 2015). The Mediterranean Sea is particularly susceptible to such dramatic changes, since many of the processes that can drive regime shifts are magnified in this enclosed basin. Warming is occurring at a faster rate in the Mediterranean than elsewhere and heat waves are a major threat to shallow-water species assemblages (Vargas-Yanez et al., 2008; Wernberg et al., 2016). The rate of introduction of non-indigenous marine species documented in the last decades is unprecedented and the pace of marine invasions will likely accelerate with the doubling in size of the Suez Canal (Zenetos, 2010; Galil et al., 2015). These alterations are causing rapid changes to marine biodiversity, promoting 'no-analogue' assemblages (Williams and Jackson, 2007) and novel species interactions (Williams and Jackson, 2007; Alexander et al., 2016) with unknown effects on stability and resilience.

Novel species interactions will pose new challenges for anticipating regime shifts. We have shown how experiments can be designed to identify thresholds and to test for early warning signals in real-world conditions, and how long-term observations can help predict the conditions under which a regime shift can be expected to occur. Such thresholds are, however, unknown for many key habitats (e.g., seagrasses, biogenic formations, animal forests); the role of pest metabolites and infectious diseases in triggering regime shifts is still underappreciated and long-term observations remain rare. The lack of any clear understanding of the feedbacks involved in stabilising alternative states often limits the possibility of implementing effective restoration practices to counteract undesired shifts. Coordinated 'macroecological' experiments (Alexander et al., 2016), which also take into account the potential for novel interactions following species' range shifts, coupled with sustained large-scale biodiversity observations will be needed to overcome these limitations. Improved communication with managers and society at large will be essential to motivate support for these broadscale initiatives, which ultimately will have to become integrated into observatory networks to become globally relevant for the conservation of marine biodiversity.

## Acknowledgements

We sincerely thank Gray A. Williams for helpful comments and edits on the manuscript. We also thank Euromarine for support to the STRANGE (Status, trends and conservation of coastal marine biodiversity under global change scenarios) workshop, during which many of the ideas expressed in this paper have been clarified. Luca Rindi and Irene Rivetti helped with the preparation of the figures. Additional support during the preparation of this paper was provided by the research projects 'Coastal bioconstructions, structure, function and management' and TETRIS ('Observing, modelling and Testing synergies and TRade-offs for the adaptive management of multiple Impacts in coastal Systems') funded by the Italian Ministry of Education, University and Research.

## REFERENCES

Abelson, A., Halpern, B. S., Reed, D. C. et al. (2016). Upgrading marine ecosystem restoration using ecological-social concepts. *BioScience*, **66**, 156–16310.

Agnetta, D., Badalamenti, F., Ceccherelli, G., Di Trapani, F., Bonaviri, C. and Gianguzza, P. (2015). Role of two co-occurring Mediterranean sea urchins in the formation of barren from *Cystoseira* canopy. *Estuarine Coastal and Shelf Science*, **152**, 73–7.

Airoldi, L. (1998). Roles of disturbance, sediment stress, and substratum retention on spatial dominance in algal turf. *Ecology*, **79**, 2759–70.

Airoldi, L. (2000a). Effects of disturbance, life histories, and overgrowth on coexistence of algal crusts and turfs. *Ecology*, **81**, 798–814.

Airoldi, L. (2000b). Responses of algae with different life histories to temporal and spatial variability of disturbance in subtidal reefs. *Marine Ecology Progress Series*, **195**, 81–92

Airoldi, L. (2003). The effects of sedimentation on rocky coast assemblages. *Oceanography and Marine Biology*, **41**, 161–236

Airoldi, L., Ballesteros, E., Buonuomo, R. et al. (2014) Marine forests at risk: solutions to halt the loss and promote the recovery of Mediterranean canopy-forming seaweeds 2014. Regional Activity Centre for Specially Protected Areas (RAC/SPA). In *Proceedings of the 5th Mediterranean Symposium on Marine Vegetation*.

Airoldi, L. and Beck, M. W. (2007). Loss, status and trends for coastal marine habitats of Europe. *Oceanography and Marine Biology*, **45**, 345–405.

Airoldi, L. and Cinelli, F. (1997). Effect of sedimenentation on subtidal macroalgal assemblages: an experimental study from a mediterranean roky shore. *Journal of Experimental Marine Biology and Ecology*, **215**, 269–88.

Airoldi, L., Fabiano, M. and Cinelli, F. (1996). Sediment deposition and movement over a turf assemblage in a shallow rocky coastal area of the Ligurian Sea. *Marine Ecology Progress Series*, **133**, 241–51.

Airoldi, L. and Hawkins, S. J. (2007). Negative effects of sediment deposition on grazing activity and survival of the limpet Patella vulgata. *Marine Ecology Progress Series*, **332**, 235–40.

Airoldi, L., Rindi, F. and Cinelli, F. (1995). Structure, seasonal dynamics and reproductive phenology of a filamentous turf assemblage on a sediment influenced, rocky subtidal shore. *Botanica Marina*, **38**, 227–37.

Alexander, J. M., Diez, J. M., Hart, S. P. and Levine, J. M. (2016). When climate reshuffles competitors: a call for experimental macroecology. *Trends in Ecology and Evolution*, **31**, 831–41.

Alker, A. P., Smith, G. W. and Kim, K. (2001). Characterization of *Aspergillus sydowii* (Thom et Church), a fungal pathogen of Caribbean sea fan corals. *Hydrobiologia*, **460**, 105–11.

Andrews, J. D. (1988). Epizootiology of the disease caused by the oyster pathogen *Perkinsus marinus* and its effects on the oyster industry. *American Fisheries Society Special Publication*, **18**, 47–63.

Azzurro, E. and Andaloro, F. (2004). A new settled population of the lessepsian migrant *Siganus luridus* (Pisces: Siganidae) in Linosa Island – Sicily Strait. *Journal of the Marine Biological Association of the United Kingdom*, **84**, 819–21.

Balata, D., Piazzi, L. and Cinelli, F. (2004). A comparison among assemblages in areas invaded by *Caulerpa taxifolia* and *C. racemosa* on a subtidal Mediterranean rocky bottom. *Marine Ecology-Pubblicazioni Della Stazione Zoologica Di Napoli I*, **25**, 1–13.

Benedetti-Cecchi, L. (2000a). Predicting direct and indirect interactions during succession in a mid-littoral rocky shore assemblage. *Ecological Monographs*, **70**, 45–72.

Benedetti-Cecchi, L. (2000b). Priority effects, taxonomic resolution, and the prediction of variable patterns of colonisation of algae in littoral rock pools. *Oecologia*, **123**, 265–74.

Benedetti-Cecchi, L. (2000c). Variance in ecological consumer-resource interactions. *Nature*, **407**, 370–4.

Benedetti-Cecchi, L. (2003). The importance of the variance around the mean effect size of ecological processes. *Ecology*, **84**, 2335–46.

Benedetti-Cecchi, L., Bertocci, I., Vaselli, S. and Maggi, E. (2006). Temporal variance reverses the impact of high mean intensity of stress in climate change experiments. *Ecology*, **87**, 2489–99.

Benedetti-Cecchi, L. and Cinelli, F. (1992a). Canopy removal experiments in *Cystoseira*-dominated rockpools from the western coast of the Mediterranean (Ligurian Sea). *Journal of Experimental Marine Biology and Ecology*, **155**, 69–83.

Benedetti-Cecchi, L. and Cinelli, F. (1992b). Effects of canopy cover, herbivores and substratum type on patterns of *Cystoseira* spp. settlement and recruitment in littoral rockpools. *Marine Ecology Progress Series*, **90**, 183–91.

Benedetti-Cecchi, L., Pannacciulli, F., Bulleri, F. et al. (2001a). Predicting the consequences of anthropogenic disturbance: large-scale effects of loss of canopy algae on rocky shores. *Marine Ecology-Progress Series*, **214**, 137–50.

Benedetti-Cecchi, L., Pannacciulli, F., Bulleri, F. et al. (2001b). Predicting the consequences of anthropogenic disturbance: large-scale effects of loss of canopy algae on rocky shores. *Marine Ecology Progress Series*, **214**, 137–50.

Benedetti-Cecchi, L., Tamburello, L., Bulleri, F., Maggi, E., Gennusa, V. and Miller, M. (2012). Linking patterns and processes across scales: the application of scale-transition theory to algal dynamics on rocky shores. *Journal of Experimental Biology*, **215**, 977–85.

Benedetti-Cecchi, L., Tamburello, L., Maggi, E. and Bulleri, F. (2015). Experimental perturbations modify the performance of early warning indicators of regime shift. *Current Biology*, **25**, 1867–72.

Benedetti-Cecchi, L., Vaselli, S., Maggi, E. and Bertocci, I. (2005). Interactive effects of spatial variance and mean intensity of grazing on algal cover in rock pools. *Ecology*, **86**, 2212–22.

Bertocci, I., Maggi, E., Vaselli, S. and Benedetti-Cecchi, L. (2005). Contrasting effects of mean intensity and temporal variation of disturbance on a rocky seashore. *Ecology*, **86**, 2061–7.

Bianchelli, S., Buschi, E., Danovaro, R. and Pusceddu, A. (2016). Biodiversity loss and turnover in alternative states in the Mediterranean Sea: a case study on meiofauna. *Scientific Reports*, **6**, 34544, http://dx.doi.org/10.1038/srep34544.

Bianchi, C. N. and Morri, C. (2000). Marine biodiversity of the Mediterranean Sea: situation, problems and prospects for future research. *Marine Pollution Bulletin*, **40**, 367–76.

Bonaviri, C., Fernandez, T. V., Fanelli, G., Badalamenti, F. and Gianguzza, P. (2011). Leading role of the sea urchin *Arbacia lixula* in maintaining the barren state in southwestern Mediterranean. *Marine Biology*, **158**, 2505–13.

Bouillon, J., Medel, M. D., Pags, F., Gili, J. M., Boero, F. and Gravili, C. (2004). Fauna of the Mediterranean Hydrozoa. *Scientia Marina*, **68**, 39–206.

Breitburg, D. L., Craig, J. K., Fulford, R. S. et al. (2009). Nutrient enrichment and fisheries exploitation: interactive effects on estuarine living resources and their management. *Hydrobiologia*, **629**, 31–47.

Breitburg, D. L. and Reidel, G. F. (2013). *Multiple Stressors in Marine Systems*. E. Norse and L. Crowder, eds. Island Press, Washington, DC, 167–82.

Brown, C. J., Saunders, M. I., Possingham, H. P. and Richardson, A. J. (2013). Managing for interactions between local and global stressors of ecosystems. *PLoS ONE*, **8**, http://doi.org/10.1371/journal.pone.0065765.

Bulleri, F. (2013). Grazing by sea urchins at the margins of barren patches on Mediterranean rocky reefs. *Marine Biology*, **160**, 2493–501.

Bulleri, F., Alestra, T., Ceccherelli, G. et al. (2011). Determinants of *Caulerpa racemosa* distribution in the north-western Mediterranean. *Marine Ecology-Progress Series*, **431**, 55–67.

Bulleri, F., Badalamenti, F., Ivesa, L. et al. (2016). The effects of an invasive seaweed on native communities vary along a gradient of land-based human impacts. *PeerJ*, **4**, e1795.

Bulleri, F., Balata, D., Bertocci, I., Tamburello, L. and Benedetti-Cecchi, L. (2010). The seaweed *Caulerpa racemosa* on Mediterranean rocky reefs: from passenger to driver of ecological change. *Ecology*, **91**, 2205–12.

Bulleri, F. and Benedetti-Cecchi, L. (2008). Facilitation of the introduced green alga *Caulerpa racemosa* by resident algal turfs: experimental evaluation of underlying mechanisms. *Marine Ecology-Progress Series*, **364**, 77–86.

Bulleri, F., Benedetti-Cecchi, L., Acunto, S., Cinelli, F. and Hawkins, S. J. (2002). The influence of canopy algae on vertical patterns of distribution of low-shore assemblages on rocky coasts in the northwest Mediterranean. *Journal of Experimental Marine Biology and Ecology*, **267**, 89–106.

Bulleri, F., Benedetti-Cecchi, L., Ceccherelli, G. and Tamburello, L. (2017). A few is enough: a low cover of a non-native seaweed reduces the resilience of Mediterranean macroalgal stands to disturbances of varying extent. *Biological Invasions*, **19**, 2291–305.

Bulleri, F., Benedetti-Cecchi, L. and Cinelli, F. (1999). Grazing by the sea urchins *Arbacia lixula* L. and *Paracentrotus lividus* Lam. in the Northwest Mediterranean. *Journal of Experimental Marine Biology and Ecology*, **241**, 81–95.

Bulleri, F., Benedetti-Cecchi, L., Cusson, M. et al. (2012). Temporal stability of European rocky shore assemblages: variation across a latitudinal gradient and the role of habitat-formers. *Oikos*, **121**, 1801–9.

Bulleri, F., Tamburello, L. and Benedetti-Cecchi, L. (2009). Loss of consumers alters the effects of resident assemblages on the local spread of an introduced macroalga. *Oikos*, **118**, 269–79.

Buonomo, R., Assis, J., Fernandes, F., Engelen, A. H., Airoldi, L. and Serrao, E. A. (2017). Habitat continuity and stepping-stone oceanographic distances explain population genetic connectivity of the brown alga *Cystoseira amentacea*. *Molecular Ecology*, **26**, 766–80.

Burreson, E. M. A. (1996). Epizootiology of *Perkinsus marinus* disease of oysters in Chesapeake Bay, with emphasis on data since 1985. *Journal of Shellfish Research*, **15**, 17–34.

Carpenter, S. R. and Brock, W. A. (2006). Rising variance: a leading indicator of ecological transition. *Ecology Letters*, **9**, 308–15.

Carpenter, S. R., Cole, J. J., Pace, M. L. et al. (2011). Early warnings of regime shifts: a whole-ecosystem experiment. *Science*, **332**, 1079–82.

Ceccherelli, G., Pinna, S., Cusseddu, V. and Bulleri, F. (2014). The role of disturbance in promoting the spread of the invasive seaweed *Caulerpa racemosa* in seagrass meadows. *Biological Invasions*, **16**, 2737–45.

Cerrano, C., Bavestrello, G., Bianchi, C. N. et al. (2000). A catastrophic mass-mortality episode of gorgonians and other organisms in the Ligurian Sea (Northwestern Mediterranean), summer 1999. *Ecology Letters*, **3**, 284–93.

Chapman, A. R. O. (1995). Functional ecology of Fucoid algae – 23 years of progress. *Phycologia*, **34**, 1–32.

Chevaldonne, P. and Lejeusne, C. (2003). Regional warming-induced species shift in north-west Mediterranean marine caves. *Ecology Letters*, **6**, 371–9.

Claudet, J. and Fraschetti, S. (2010). Human-driven impacts on marine habitats: a regional meta-analysis in the Mediterranean Sea. *Biological Conservation*, **143**, 2195–206.

Coll, M., Piroddi, C. and Albouy, C. a. (2012). The Mediterranean Sea under siege: spatial overlap between marine biodiversity, cumulative threats and marine reserves. *Global Ecology and Biogeography*, **21**, 465–80.

Coll, M., Piroddi, C., Steenbeek, J. et al. (2010). The biodiversity of the Mediterranean Sea: estimates, patterns, and threats. *PLoS ONE*, **5**, e11842.

Connell, S. D., Foster, M. S. and Airoldi, L. (2014). What are algal turfs? Towards a better description of turfs. *Marine Ecology Progress Series*, **495**, 299–307.

Crain, C. M., Kroeker, K. and Halpern, B. S. (2008). Interactive and cumulative effects of multiple human stressors in marine systems. *Ecology Letters*, **11**, 1304–15.

Crowe, T. P., Cusson, M., Bulleri, F. et al. (2013). Large-scale variation in combined impacts of canopy loss and disturbance on community structure and ecosystem functioning. *PLoS ONE*, **8**, e66238.

Dai, L., Korolev, K. S. and Gore, J. (2013). Slower recovery in space before collapse of connected populations. *Nature*, **496**, 355–58.

Dai, L., Vorselen, D., Korolev, K. S. and Gore, J. (2012). Generic indicators for loss of resilience before a tipping point leading to population collapse. *Science*, **336**, 1175–7.

Dakos, V., Carpenter, S. R., Brock, W. A. et al. (2012). Methods for detecting early warnings of critical transitions in time series illustrated using simulated ecological data. *PLoS ONE*, **7**, http://dx.doi.org/10.1371/journal.pone.0041010.

Dakos, V., van Nes, E. H. and Scheffer, M. (2013). Flickering as an early warning signal. *Theoretical Ecology*, **6**, 309–17.

Daniel, B., Piro, S., Charbonnel, E., Francour, P. and Letourneur, Y. (2009). Lessepsian rabbitfish *Siganus luridus* reached the French Mediterranean coasts. *Cybium*, **33**, 163–4

Danovaro, R., Company, J. B., Corinaldesi, C. et al. (2010). Deep-sea biodiversity in the Mediterranean Sea: the known, the unknown, and the unknowable. *PLoS ONE*, **5**, http://dx.doi.org/10.1371/journal.pone.0011832.

Danovaro, R., Dell'Anno, A., Fabiano, M., Pusceddu, A. and Tselepides, A. (2001). Deep-sea ecosystem response to climate changes: the eastern Mediterranean case study. *Trends in Ecology and Evolution*, **16**, 505–10.

Dayton, P. K. (1985). Ecology of Kelp Communities. *Annual Review of Ecology and Systematics*, **16**, 215–45.

deYoung, B., Barange, M., Beaugrand, G. et al. (2008). Regime shifts in marine ecosystems: detection, prediction and management. *Trends in Ecology and Evolution*, **23**, 402–9.

Di Camillo, C. G. and Cerrano, C. (2015). Mass mortality events in the NW Adriatic Sea: phase shift from slow- to fast-growing organisms. *PLoS ONE*, **10**, e0126689.

Didham, R. K., Tylianakis, J. M., Hutchison, M. A., Ewers, R. M. and Gemmell, N. J. (2005). Are invasive species the drivers of ecological change? *Trends in Ecology and Evolution*, **20**, 470–4.

Drake, J. M. and Griffen, B. D. (2010). Early warning signals of extinction in deteriorating environments. *Nature*, **467**, 456–9.

Elliott, M., Burdon, D., Hemingway, K. L. and Apitz, S. E. (2007). Estuarine, coastal and marine ecosystem restoration: confusing management and science – a revision of concepts. *Estuarine, Coastal and Shelf Science*, **74**, 349–66.

Falace, A., Zanelli, E. and Bressan, G. (2006). Algal transplantation as a potential tool for artificial reef management and environmental mitigation. *Bulletin of Marine Science*, **78**, 161–6.

Felline, S., Caricato, R., Cutignano, A. et al. (2012). Subtle effects of biological invasions: Cellular and physiological responses of fish eating the exotic pest *Caulerpa racemosa*. *PLoS ONE*, **7**, http://dx.doi.org/10.1371/journal.pone.0038763.

Ferrario, F., Iveša, L., Jaklin, A., Perkol-Finkel, S. and Airoldi, L. (2016). The overlooked role of biotic factors in controlling the ecological performance of artificial marine habitats. *Journal of Applied Ecology*, **53**, 16–24.

Ferrier-Pagès, C., Tambutté, E., Zamoum, T. et al. (2009). Physiological response of the symbiotic gorgonian *Eunicella singularis* to a long-term temperature increase. *The Journal of experimental biology*, **212**, 3007–15.

Firth, L. B., Thompson, R. C., Bohn, K. et al. (2014). Between a rock and a hard place: Environmental and engineering considerations when designing coastal defence structures. *Coastal Engineering*, **87**, 122–35.

Ford, S. E. and Chintala, M. M. (2006). Northward expansion of a marine parasite: Testing the role of temperature adaptation. *Journal of Experimental Marine Biology and Ecology*, **339**, 226–35.

Gaertner, M., Biggs, R., Te Beest, M., Hui, C., Molofsky, J. and Richardson, D. M. (2014). Invasive plants as drivers of regime shifts: identifying high-priority invaders that alter feedback relationships. *Diversity and Distributions*, **20**, 733–44.

Galil, B. S., Boero, F., Campbell, M. L. et al. (2015). 'Double trouble': the expansion of the Suez Canal and marine bioinvasions in the Mediterranean Sea. *Biological Invasions*, **17**, 973–6.

Galil, B. S., Marchini, A., Occhipinti-Ambrogi, A. et al. (2014). International arrivals: widespread bioinvasions in European Seas. *Ethology Ecology and Evolution*, **26**, 152–71.

Garrabou, J., Coma, R., Bensoussan, N., Bally, M. and Chevaldonn (2009). Mass mortality in Northwestern Mediterranean rocky benthic communities: effects of the 2003 heat wave. *Global Change Biology*, **15**, 1090–103.

Giaccone, G. and Calvo, S. (1980). Restaurazione del manto vegetale mediante trapianto di *Posidonia oceanica* (Linneo) Delile. Risultati preliminari. *Memorie di Biologia Marina e Oceanografia*, Suppl., 207–11.

Gianni, F., Bartolini, F., Airoldi, L. et al. (2013). Conservation and restoration of marine forests in the Mediterranean Sea and the potential role of marine protected areas. *Advances in Oceanography and Limnology*, **4**, 83–101.

Gilmore, R. (1981). *Catastrophe Theory for Scientists and engineers*. John Wiley and Sons, New York.

Gomez, D. K., Matsuoka, S., Mori, K. I., Okinaka, Y., Park, S. C. and Nakai, T. (2009). Genetic analysis and pathogenicity of betanodavirus isolated from wild redspotted grouper *Epinephelus akaara* with clinical signs. *Archives of Virology*, **154**, 343–6.

Gorbi, S., Giuliani, M. E., Pittura, L. et al. (2014). Could molecular effects of *Caulerpa racemosa* metabolites modulate the impact on fish populations of *Diplodus sargus*? *Marine Environmental Research*, **96**, 2–11.

Guarnieri, G., Bevilacqua, S., Vignes, F. and Fraschetti, S. (2014). Grazer removal and nutrient enrichment as recovery enhancers for overexploited rocky subtidal habitats. *Oecologia*, **175**, 959–70.

Halpern, B. S., McLeod, K. L., Rosenberg, A. A. and Crowder, L. B. (2008). Managing for cumulative impacts in ecosystem-based management through ocean zoning. *Ocean and Coastal Management*, **51**, 203–11.

Harrold, C. and Reed, D. C. (1985). Food availability, sea-urchin grazing, and kelp forest community structure. *Ecology*, **66**, 1160–9.

Harvell, C. D., Kim, K., Burkholder, J. M. et al. (1999). Emerging marine diseases-climate links and anthropogenic factors. *Science*, **285**, 1505–10.

Hata, N., Okinaka, Y., Sakamoto, T., Iwamoto, T. and Nakai, T. (2007). Upper temperature limits for the multiplication of betanodaviruses. *Fish Pathology*, **42**, 225–8.

Hereu, B. (2006). Depletion of palatable algae by sea urchins and fishes in a Mediterranean subtidal community. *Marine Ecology-Progress Series*, **313**, 95–103.

Hobbs, R. J., Davis, M. A., Slobodkin, L. B. et al. (2004). Restoration ecology: the challenge of social values and expectations. *Frontiers in Ecology and the Environment*, **2**, 43.

Hobbs, R. J., Higgs, E. and Harris, J. A. (2009). Novel ecosystems: implications for conservation and restoration. *Trends in Ecology and Evolution*, **24**, 599–605.

Huete-Stauffer, C. A., Vielmini, I. B., Palma, M. A. et al. (2011). *Paramuricea clavata* (Anthozoa, Octocorallia) loss in the marine protected area of Tavolara (Sardinia, Italy) due to a mass mortality event. *Marine Ecology*, **32**, 107–16.

Irving, A. D., Balata, D., Colosio, F., Ferrando, G. A. and Airoldi, L. (2009). Light, sediment, temperature, and the early life-history of the habitat-forming alga *Cystoseira barbata*. *Marine Biology*, **156**, 1223–31.

Iwamoto, T., Nakai, T., Mori, K., Arimoto, M. and Furusawa, I. (2000). Cloning of the fish cell line SSN-1 for piscine nodaviruses. *Diseases of Aquatic Organisms*, **43**, 81–9.

Kara, H. M., Chaoui, L., Derbal, F. et al. (2014). Betanodavirus-associated mortalities of adult wild groupers *Epinephelus marginatus* (Lowe) and *Epinephelus costae* (Steindachner) in Algeria. *Journal of Fish Diseases*, **37**, 273–8.

Kéfi, S., Dakos, V., Scheffer, M., Van Nes, E. H. and Rietkerk, M. (2013). Early warning signals also precede non-catastrophic transitions. *Oikos*, **122**, 641–8.

Kéfi, S., Guttal, V., Brock, W. A. et al. (2014). Early warning signals of ecological transitions: methods for spatial patterns. *PLoS ONE*, **9**, http://dx.doi.org/10.1371/journal.pone.0092097.

Klimpel, S. and Palm, H. W. (2011). Anisakid nematode (Ascaridoidea) life cycles and distribution: increasing zoonotic potential in the time of climate change? *Progress in Parasitology*, **2**, 201–22.

Kruzic, P. and Popijac, A. (2015). Mass mortality events of the coral *Balanophyllia europaea* (Scleractinia, Dendrophylliidae) in the Mljet National Park (Eastern Adriatic Sea) caused by sea temperature anomalies. *Coral Reefs*, **34**, 109–18.

Lejeusne, C., Chevaldonné, P., Pergent-Martini, C., Boudouresque, C. F. and Pérez, T. (2010). Climate change effects on a miniature ocean: the highly diverse, highly impacted Mediterranean Sea. *Trends in Ecology and Evolution*, **25**, 250–60.

Ling, S. D., Scheibling, R. E., Rassweiler, A. et al. (2015). Global regime shift dynamics of catastrophic sea urchin overgrazing. *Philosophical Transactions of the Royal Society B: Biological Sciences*, **370**, http://dx.doi.org/10.1098/rstb.2013.0269.

Lotze, H. K., Coll, M., Magera, A. M., Ward-Paige, C. and Airoldi, L. (2011). Recovery of marine animal populations and ecosystems. *Trends in ecology and evolution*, **26**, 595–605.

Mack, R. N., Simberloff, D., Lonsdale, W. M., Evans, H., Clout, M. and Bazzaz, F. A. (2000). Biotic invasions: causes, epidemiology, global consequences, and control. *Ecological Applications*, **10**, 689–710.

Mancuso, F. P., D'Hondt, S., Willems, A., Airoldi, L. and De Clerck, O. (2016). Diversity and temporal dynamics of the epiphytic bacterial communities associated with the canopy-forming seaweed *Cystoseira compressa* (Esper) Gerloff and Nizamuddin. *Frontiers in Microbiology*, **7**, 476.

Mangialajo, L., Chiantore, M. and Cattaneo-Vietti, R. (2008). Loss of fucoid algae along a gradient of urbanisation, and structure of benthic assemblages. *Marine Ecology Progress Series*, **358**, 63–74.

Marino, G. and Azzurro, E. (2001). Nodavirus in dusky grouper (*Epinephelus marginatus* Lowe 1834) of the natural marine reserve of Ustica, south Thyrrenian Sea. *Biologia Marina Mediterranea*, **8**, 837–41.

Marras, S., Cucco, A., Antognarelli, F. et al. (2015). Predicting future thermal habitat suitability of competing native and invasive fish species: from metabolic scope to oceanographic modelling. *Conservation Physiology*, **3**, http://dx.doi.org/1410.1093/conphys/cou059.

Martin, C. S. and Giannoulaki, M. A. (2014). Coralligenous and maerl habitats: predictive modelling to identify their spatial distributions across the Mediterranean Sea. *Scientific Reports*, **4**, http://dx.doi.org/10.1038/srep05073.

Martin, Y., Bonnefont, J. L. and Chancerelle, L. (2002). Gorgonians mass mortality during the 1999 late summer in French Mediterranean coastal waters: The bacterial hypothesis. *Water Research*, **36**, 779–82.

McCarron, E., Burnell, G. and Mouzakitis, G. (2009). Growth assessment on three size classes of the purple sea urchin *Paracentrotus lividus* using continuous and intermittent feeding regimes. *Aquaculture*, **288**, 83–91.

Meinesz, A., Caye, G., Molenaar, H. and Loques, F. (1991). Growth and development in culture of orthotropic rhizomes of *Posidonia oceanica*. *Aquatic Botany*, **39**, 367–77.

Micheli, F., Halpern, B. S., Walbridge, S. et al. (2013). Cumulative human impacts on Mediterranean and Black Sea marine ecosystems: assessing current pressures and opportunities. *PLoS ONE*, **8**, e79889.

Mollmann, C., Folke, C., Edwards, M. and Conversi, A. (2015). Marine regime shifts around the globe: theory, drivers and impacts. *Philosophical Transactions of the Royal Society B-Biological Sciences*, **370**, http://dx.doi.org/10.1098/rstb.2013.0260.

Mollo, E., Cimino, G. and Ghiselin, M. T. (2015). Alien biomolecules: a new challenge for natural product chemists. *Biological Invasions*, **17**, 941–50.

Mouillot, D., Albouy, C. and Guilhaumon, F. A. (2011). Protected and threatened components of fish biodiversity in the Mediterranean Sea. *Current Biology*, **21**, 1044–50.

Mumby, P. J., Hastings, A. and Edwards, H. J. (2007). Thresholds and the resilience of Caribbean coral reefs. *Nature*, **450**, 98–101.

Munday, B. L., Kwang, J. and Moody, N. (2002). Betanodavirus infections of teleost fish: a review. *Journal of Fish Diseases*, **25**(3), 127–42.

Pairaud, I. L., Bensoussan, N., Garreau, P., Faure, V. and Garrabou, J. (2014). Impacts of climate change on coastal benthic ecosystems: assessing the current risk of mortality outbreaks associated with thermal stress in NW Mediterranean coastal areas Topical Collection on the 16th biennial workshop of the Joint Numerical Sea Modelling Group (JONSMOD) in Brest, France 21–23 May 2012. *Ocean Dynamics*, **64**, 103–15.

Pejchar, L. and Mooney, H. A. (2009). Invasive species, ecosystem services and human well-being. *Trends in Ecology and Evolution*, **24**, 497–504.

Perez, T., Garrabou, J., Sartoretto, S., Harmelin, J. G., Francour, P. and Vacelet, J. (2000). Mass mortality of marine invertebrates: an unprecedented event in the Northwestern Mediterranean. *Comptes Rendus de l'Academie des Sciences – Serie III*, **323**, 853–65.

Perkol-Finkel, S. and Airoldi, L. (2010). Loss and recovery potential of marine habitats: an experimental study of factors maintaining resilience in subtidal algal forests at the Adriatic sea. *PLoS ONE*, **5**, e1079110, http://doi.org/10.1371/journal.pone.0010791.

Perkol-Finkel, S., Ferrario, F., Nicotera, V. and Airoldi, L. (2012). Conservation challenges in urban seascapes: promoting the growth of threatened species on coastal infrastructures. *Journal of Applied Ecology*, **49**, 1457–66.

Piazzi, L., Balata, D., Foresi, L., Cristaudo, C. and Cinelli, F. (2007). Sediment as a constituent of Mediterranean benthic communities dominated by *Caulerpa racemosa* var. *cylindracea*. *Scientia Marina*, **71**, 129–35.

Piazzi, L., Balestri, E., Magri, M. and Cinelli, F. (1998). Experimental transplanting of *Posidonia oceanica* (L.) Delile into a disturbed habitat in the Mediterranean Sea. *Botanica Marina*, **41**, 593–602.

Piazzi, L., Bulleri, F. and Ceccherelli, G. (2016). Limpets compensate sea urchin decline and enhance the stability of rocky subtidal barrens. *Marine Environmental Research*, **115**, 49–55.

Piazzi, L., Meinesz, A., Verlaque, M. et al. (2005). Invasion of *Caulerpa racemosa* var. *cylindracea* (Caulerpales, Chlorophyta) in the Mediterranean Sea: an assessment of the spread. *Cryptogamie Algologie*, **26**, 189–202.

Ponti, M., Perlini, R. A., Ventra, V., Grech, D., Abbiati, M. and Cerrano, C. (2014). Ecological shifts in mediterranean coralligenous assemblages related to gorgonian forest loss. *PLoS ONE*, 9, http://dx.doi.org/10.1371/journal.pone.0102782.

Puce, S. and Bavestrello, G. A. (2009). Long-term changes in hydroid (Cnidaria, Hydrozoa) assemblages: effect of Mediterranean warming? *Marine Ecology*, **30**, 313–26.

Raniello, R., Mollo, E., Lorenti, M., Gavagnin, M. and Buia, M. C. (2007). Phytotoxic activity of caulerpenyne from the Mediterranean invasive variety of *Caulerpa racemosa*: a potential allelochemical. *Biological Invasions*, **9**, 361–8.

Reinhart, K. O. and Callaway, R. M. (2006). Soil biota and invasive plants. *New Phytologist*, **170**, 445–57.

Reynolds, J. F., Stafford Smith, D. M., Lambin, E. F. et al. (2007). Global desertification: building a science for dryland development. *Science*, **316**, 847–51.

Rilov, G. and Galil, B. S. (2009). Marine Bioinvasions in the Mediterranean Sea – History, Distribution and Ecology. In G. Rilov and J. A. Crooks, eds. *Biological Invasions in Marine Ecosystems*. Springer-Verlag, Berlin, pp. 549–75.

Rindi, L., Bello, M. D., Dai, L., Gore, J. and Benedetti-Cecchi, L. (2017). Direct observation of increasing recovery length before collapse of a marine benthic ecosystem. *Nature Ecology and Evolution*, 1, 153, http://dx.doi.org/10.1038/s41559-017-0153.

Rivetti, I., Boero, F., Fraschetti, S., Zambianchi, E. and Lionello, P. (2015). Anomalies of the upper water column in the Mediterranean Sea. *Global and Planetary Change*, **151**, 68–79.

Rivetti, I., Fraschetti, S., Lionello, P., Zambianchi, E. and Boero, F. (2014). Global warming and mass mortalities of benthic invertebrates in the Mediterranean Sea. *PLoS ONE*, **9**, 1–22.

Rizzo, L., Fraschetti, S., Alifano, P., Pizzolante, G. and Stabili, L. (2016). The alien species *Caulerpa cylindracea* and its associated bacteria in the Mediterranean Sea. *Marine Biology*, **163**, 1–12.

Rocha, J., Yletyinen, J., Biggs, R., Blenckner, T. and Peterson, G. (2015). Marine regime shifts: drivers and impacts on ecosystems services. *Philosophical Transactions of the Royal Society: Biological Sciences*, **370**, http://dx.doi.org/10.1098/rstb.2013.0273.

Rodolfo-Metalpa, R., Richard, C., Allemand, D., Bianchi, C. N., Morri, C. and Ferrier-Pags, C. (2006). Response of zooxanthellae in symbiosis with the Mediterranean corals *Cladocora caespitosa* and *Oculina patagonica* to elevated temperatures. *Marine Biology*, **150**, 45–55.

Sala, E., Kizilkaya, Z., Yildirim, D. and Ballesteros, E. (2011). Alien marine fishes deplete algal biomass in the Eastern Mediterranean. *PLoS ONE*, **6**, e17356.

Sales, M., Cebrian, E., Tomas, F. and Ballesteros, E. (2011). Pollution impacts and recovery potential in three species of the genus *Cystoseira* (Fucales,

Heterokontophyta). *Estuarine, Coastal and Shelf Science*, **92**, 347–57.

Samhouri, J. F., Levin, P. S. and Ainsworth, C. H. (2010). Identifying thresholds for ecosystem-based management. *PLoS ONE*, **5**, http://dx.doi.org/10.1371/journal.pone.0008907.

Sarà, G., Palmeri, V., Rinaldi, A., Montalto, V. and Helmuth, B. (2013). Predicting biological invasions in marine habitats through eco-physiological mechanistic models: a case study with the bivalve *Brachidontes pharaonis*. *Diversity and Distributions*, **19**, 1235–47.

Scheffer, M. (2009). *Critical Transitions in Nature and Society*. Princeton University Press, Princeton, NJ.

Scheffer, M., Bascompte, J., Brock, W. A. et al. (2009). Early-warning signals for critical transitions. *Nature*, **461**, 53–9.

Scheffer, M., Carpenter, S., Foley, J. A., Folke, C. and Walker, B. (2001). Catastrophic shifts in ecosystems. *Nature*, **413**, 591–6.

Schiel, D. R. and Foster, M. S. (2006). The population biology of large brown seaweeds: ecological consequences of multiphase life histories in dynamic coastal environments. *Annual Review of Ecology Evolution and Systematics*, **37**, 343–72.

Sirota, J., Baiser, B., Gotelli, N. J. and Ellison, A. M. (2013). Organic-matter loading determines regime shifts and alternative states in an aquatic ecosystem. *Proceedings of the National Academy of Science of the USA*, **110**, 7742–7.

Steneck, R. S., Graham, M. H., Bourque, B. J. et al. (2002). Kelp forest ecosystems: biodiversity, stability, resilience and future. *Environmental Conservation*, **29**, 436–59.

Steneck, R. S. and Johnson, C. R. (2014). Kelp Forests: Dynamic Patterns, Processes, and Feedbacks. In M. D. Bertness, J. F. Bruno, B. R. Silliman and J. J. Stachowicz, eds. *Marine Community Ecology and Conservation*. Sinauer Associates, Inc., Sunderland, MA, pp. 315–36.

Strain, E. M., Thomson, R. J., Micheli, F., Mancuso, F. P. and Airoldi, L. (2014). Identifying the interacting roles of stressors in driving the global loss of canopy-forming to mat-forming algae in marine ecosystems. *Global Change Biology*, **20**, 3300–12, http://dx.doi.org.10.1111/gcb.12619.

Strain, E. M. A., van Belzen, J., van Dalen, J., Bouma, T. J. and Airoldi, L. (2015). Management of local stressors can improve the resilience of marine canopy algae to global stressors. *PLoS ONE*, **10**, e0120837, http://dx.doi.org/10.1371/journal.pone.0120837.

Suding, K. N. (2011). Toward an era of restoration in ecology: successes, failures, and opportunities ahead. *Annual Review of Ecology, Evolution, and Systematics*, **42**, 465–87.

Susini, M. L., Mangialajo, L., Thibaut, T. and Meinesz, A. (2007). Development of a transplantation technique of *Cystoseira amentacea* var. stricta and *Cystoseira compressa*. *Hydrobiologia*, **580**, 241–4.

Tamburello, L., Bulleri, F., Bertocci, I., Maggi, E. and Benedetti-Cecchi, L. (2013). Reddened seascapes: experimentally induced shifts in 1/f spectra of spatial variability in rocky intertidal assemblages. *Ecology*, **94**, 1102–11.

Tedesco, P. (2015) Ecology of parasites and diseases of wild marine fauna. PhD, University of Salento.

Telesca, L., Belluscio, A., Criscoli, A. et al. (2015). Seagrass meadows (*Posidonia oceanica*) distribution and trajectories of change. *Scientific Reports*, **5**, http://dx.doi.org/10.1038/srep12505.

Terlizzi, A., Felline, S., Lionetto, M. G. et al. (2011). Detrimental physiological effects of the invasive alga *Caulerpa racemosa* on the Mediterranean white seabream *Diplodus sargus*. *Aquatic Biology*, **12**, 109–17.

Thibaut, T., Blanfune, A., Boudouresque, C. F. and Verlaque, M. (2015). Decline and local extinction of Fucales in the French Riviera: the harbinger of future extinctions? *Mediterranean Marine Science*, **16**, 206–24.

Torrents, O., Tambutté, E., Caminiti, N. and Garrabou, J. (2008). Upper thermal thresholds of shallow vs. deep populations of the precious Mediterranean red coral *Corallium rubrum* (L.): Assessing the potential effects of warming in the NW Mediterranean. *Journal of Experimental Marine Biology and Ecology*, **357**, 7–19.

Vargas-Yanez, M., Garcia, M. J., Salat, J., Garcia-Martinez, M. C., Pascual, J. and Moya, F. (2008). Warming trends and decadal variability in the Western Mediterranean shelf. *Global and Planetary Change*, **63**, 177–84.

Vaselli, S., Bulleri, F. and Benedetti-Cecchi, L. (2008). Hard coastal-defence structures as habitats for native and exotic rocky-bottom species. *Marine Environmental Research*, **66**, 395–403.

Vendramin, N., Patarnello, P., Toffan, A. et al. (2013). Viral encephalopathy and retinopathy in groupers (*Epinephelus* spp.) in southern Italy: a threat for wild endangered species? *BMC Veterinary Research*, **9**, 20, http://dx.doi.org/10.1186/1746-6148-9-20.

Veraart, A. J., Faassen, E. J., Dakos, V., van Nes, E. H., Lurling, M. and Scheffer, M. (2012). Recovery rates reflect distance to a tipping point in a living system. *Nature*, **481**, 357–9.

Verges, A., Tomas, F., Cebrian, E. et al. (2014). Tropical rabbitfish and the deforestation of a warming temperate sea. *Journal of Ecology*, **102**, 1518–27.

Verlaque, M., Boudouresque, C. F., Meinesz, A. and Gravez, V. (2000). The *Caulerpa racemosa* complex (Caulerpales, Ulvophyceae) in the Mediterranean Sea. *Botanica Marina*, **43**, 49–68.

Wernberg, T., Bennett, S., Babcock, R. C. et al. (2016). Climate-driven regime shift of a temperate marine ecosystem. *Science*, **353**, 169–72.

Williams, G. A., Helmuth, B., Russell, B. D., Dong, Y. W., Thiyagarajan, V. and Seuront, L. (2016). Meeting the climate change challenge: pressing issues in southern China and SE Asian coastal ecosystems. *Regional Studies in Marine Science*, **8**, 373–81.

Williams, J. W. and Jackson, S. T. (2007). Novel climates, no-analog communities, and ecological surprises. *Frontiers in Ecology and the Environment*, **5**, 475–82.

Zenetos, A. (2010). Trend in aliens species in the Mediterranean. An answer to Galil, 2009 Taking stock: inventory of alien species in the Mediterranean Sea. *Biological Invasions*, **12**, 3379–81.

Zenetos, A., Gofas, S., Verlaque, M. et al. (2010). Alien species in the Mediterranean Sea by 2010. A contribution to the application of European Union's Marine Strategy Framework Directive (MSFD). Part I. Spatial distribution. *Mediterranean Marine Science*, **11**, 381–493.

# Chapter 9

# The Restructuring of Levant Reefs by Aliens, Ocean Warming and Overfishing

## *Implications for Species Interactions and Ecosystem Functions*

Gil Rilov, Ohad Peleg and Tamar Guy-Haim

## 9.1 Introduction

In this chapter, we will examine the current knowledge on the community ecology of coastal reefs, both intertidal and subtidal, in the south-eastern Mediterranean, also known as the Levantine basin (Figure 9.1). This region is considered as a distinctive ecoregion within the Mediterranean (Spalding et al., 2007), and is the south-eastern (trailing) range edge of the distribution of most Mediterranean and Atlanto-Mediterranean species. The coastal water conditions in the basin are naturally extreme within this range, being the hottest, saltiest and least productive in the Mediterranean (Coll et al., 2010). Due to its relatively small size, the eastern Mediterranean waters also respond promptly to atmospheric variability (Castellari et al., 2000), and thus are prone to strong impacts of climate change. The eastern basin of the Mediterranean is generally poorer in native species than the western basin, and this has been previously explained by its great distance from the Atlantic (the native species' origin), its very low nutrient levels and its extreme physical conditions, as well as much less intensive sampling efforts (Por and Dimentman, 1989; Coll et al., 2010). By contrast, the shallow waters of the basin are being rapidly occupied in the last ~150 years by neobiota, mostly due to the introduction of Indo-Pacific species through the Suez Canal, in a process known as Lessepsian migration (Por, 1978). This invasion is mixing native species originating from the east Atlantic Ocean with tropical species, mostly of Indo-Pacific origin. This process, often referred to as 'tropicalisation' (Bianchi, 2007; Raitsos et al., 2010) reshuffles the Mediterranean food web as well as other interspecific interactions and results in mostly unknown consequences that are slowly but gradually being revealed.

Two rocky intertidal habitats are globally most abundant or most densely occurring in the Levantine Basin: vermetid reefs (Safriel, 1975; Chemello and Silenzi, 2011) and beachrocks (Vousdoukas et al., 2007). Vermetid reefs (also known as 'abrasion platforms') are biogenic reefs that form in warm temperate or tropical seas in areas where the coastal rocks are soft and erodible, and where sedentary, aggregating vermetid gastropods (*Dendropoma anguliferum* [Templado et al., 2016] *sensu lato* and *Vermetus triquctrus*) exist. In the Mediterranean, where the intertidal zone is usually narrow due to a microtidal regime, with a maximum tidal amplitude ca. 40 cm, vermetids are thought to be responsible for the formation of the wide and flat intertidal zone characterising the reefs (Safriel, 1974,

1975). The reefs have a biogenic crust (mostly *Dendropoma*) on top of the sedimentary bedrock that thickens towards the seaward edge, and is thought to compensate for the rapid erosion of the soft sedimentary rock below. Having no pelagic dispersal, *D. anguliferum* eggs hatch as crawling juveniles that settle on adults or in proximity to them (Calvo et al., 1998). This dispersal technique results in a formation of rims at the edge of the platforms that create shallow basins on the flat ('potholes'). These can hold water even during low tide and by that allow high biodiversity to exist on the reef (i.e., the vermetid functions as an ecosystem engineer, Figure 9.2; Safriel 1974, 1975). On Levant shores, the vermetid reef bedrock is made of sandstone (mainly the eolianite or 'kurkar' formation) or limestone. These reefs are most abundant on Israeli, Lebanese and Syrian shores, although they can also be found in Cyprus. They are not present in Turkey, and our knowledge of their presence on North African coasts is minimal (Chemello and Silenzi, 2011). Beachrock are much younger formations that are created by lithification of carbonate sediments at the sandy intertidal zone in warm seas with a small tidal range (Bricker, 1971) like the Mediterranean. Beachrocks are slightly tilted seaward and normally are not as wide as most vermetid reefs.

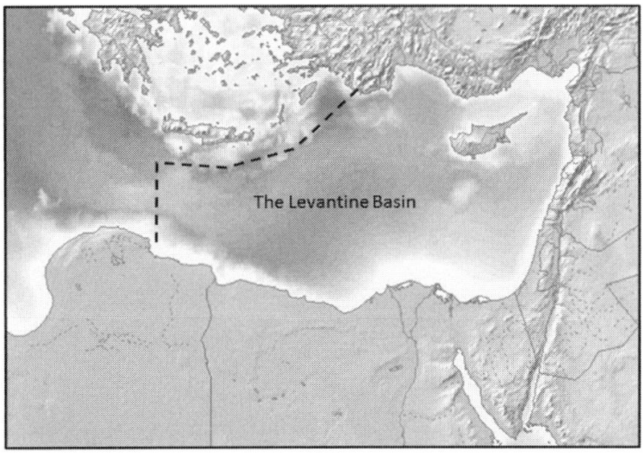


**Fig. 9.1** The Levantine Basin ecoregion

Historically (mostly during the crusaders time), rocks were excavated from vermetid reefs in Israel and have been used as a source of building material. Signs of excavations are visible in many places along the coast. The only biological resource that was collected for certain from this ecosystem was large predatory whelks such as *Stramonita haemastoma* (Linnaeus, 1767) and *Hexaplex trunculus* (Linnaeus, 1758). Extracts from their hypobranchial gland (known as the 'Tyrian purple') were used for dyeing robes of Levantine royalty during biblical and Roman times (until seventh century AD; Koren, 2005; Ziderman, 2008). Hundreds of individuals were needed to supply enough dye for a single robe; therefore, these whelks must have been extremely abundant historically. In some areas, artificial pools can be found on the reefs, and they are suspected

**Fig. 9.2** Photos of vermetid reefs in Achziv, north Israel. The platform can be flat with a rim at the edge or have several horizontal levels (terraces) that are riddled with small ridges created by *Dendropoma*.

to have been used to keep collected snails alive until processing.

Here, we will try to first illustrate how coastal reefs in the Levantine Basin might have looked like in the recent past, then describe their current ecological status, based on present-day surveys and monitoring projects, and finally attempt to predict what the near future might hold for them in the face of the increasing anthropogenic impacts that shape our oceans and coasts, mostly based on recent experimental work (Rilov et al., unpublished data). Unfortunately, little has been done experimentally in the past to test species interactions on the reefs in this region (Sala, 2004), but this is starting to change in studies conducted in the last few years. We will mostly present published and unpublished studies from the Israeli coast because this is the most studied area in the Levantine Basin and also because it serves as our primary ecological research playground. For intertidal habitats, the focus is on vermetid reefs because there is more information on them compared to other habitats such as beachrock.

## 9.2 Reefs of the Past

Unlike the case of the western Mediterranean, there is little information on the past (50–100 years back, i.e., pre-warming, and recent massive bioinvasions) composition, biotic interactions and ecological function of coastal reef communities (both intertidal and subtidal) in the Levant. Most knowledge on reef communities was descriptive until the 1970s, and until 2010 only a handful of quantitative studies exist. They either examined a few species or one taxonomic group at one or several sites, and usually for short time periods (e.g., on polychaetes; Ben Eliahu, 1975). There are, however, several taxonomic publications on several groups from the region, and there have been many publications describing new records of alien species (mostly fish, crustaceans, molluscs or macroalgae) that were introduced at high rates on the Levant shores, including on reefs. These studies sometimes included some reference to their biology (e.g., Galil et al., 1989; Golani and Diamant, 1991; Mienis, 2004; Gon and Golani, 2008; Hoffman et al., 2008b). However, there have been only very few studies on how these non-indigenous species (NIS) interact with the local communities, perhaps with the exception of the two invasive Red Sea rabbitfish, *Siganus rivulatus* and *Siganus luridus*, that were first documented on the Levant reefs in 1924 and 1955, respectively (Steinitz, 1927; Ben-Tuvia, 1964). As a consequence, the reef ecosystem in the Levant has been studied only relatively recently after several invaders have already become highly established in the system, i.e., only after the baseline has been shifted considerably, which poses a big challenge for conservation and management efforts (Pauly, 1995; Jackson et al., 2001; Knowlton and Jackson, 2008; Duarte et al., 2009). Therefore, little is known about how this environment has looked like or functioned without the influence of these NIS. The only ecological reference, perhaps, are more pristine reefs in the western Mediterranean or in the northern Aegean Sea.

### 9.2.1 Intertidal Reefs

In the early 1970s, Lipkin and Safriel (1971) provided the first detailed description of the ecological community living on intertidal reefs of the Levant coast, focussing on the zonation patterns measured in one site in central Israel, Mikhmoret, and adopting the universal zonation classification of Stephenson and Stephenson (1949) for their description. Zonation on this highly wave-exposed coast is shaped mostly by winds, wave action and barometric pressure rather than by the tide, which is very small in the Mediterranean (20–40 cm). In their paper, Lipkin and Safriel (1971), describe the vertical distribution of the characteristic biota of rocky intertidal zonation, with barnacles and periwinkles in the high- and mid-shore levels at the back of the platform (supralittoral and upper-midlittoral), and a cover of macroalgae and vermetids in the lowshore (mid- and lower-midlittoral), which is where most of the reef flat is located. There are also the typical mobile invertebrate consumers like limpets, topshells and crabs. Like the rest of the Mediterranean, there are no large predatory sea stars, but large whelks were probably important predators in the past. Studies in the 1960–1970s by Safriel et al. (1980b) describe that the whelk *S. haemastoma* was seen in large numbers in several rocky shore habitats on the Israeli coast, and it was suggested to affect the distribution of mussel beds of the indigenous species *Mytilaster minimus*.

Regarding dominant primary producers, Lipkin and Safriel (1971) described a lowshore community that is quite different from the one known from the western Mediterranean basin. In the western Mediterranean, brown algae of the genus *Cystoseira* are considered a dominant feature of the lower rocky intertidal shore where they form extensive canopy forests in pristine areas, and are therefore recognised as very important habitat formers (Benedetti-Cecchi et al., 2001; Bulleri et al., 2002). Today, these forests have declined considerably in many western Mediterranean regions and this reduction has been associated with habitat destruction, eutrophication and overgrazing by herbivores (see Blanfuné et al., 2016, and references therein, and Chapter 8). The only *Cystoseira* species mentioned in the Lipkin and Safriel (1971) zonation paper is *Cystoseira abrotanifolia* (this name is currently regarded as a taxonomic synonym of *Cystoseira foeniculacea* (Linnaeus) Greville on AlgaeBase) that ‘attains full size only in potholes about 1m deep, found on the platforms’. *Cystoseira* is not mentioned at all in a much later text listing the macroalgae found on abrasion platforms on the Israeli coast (Einav and Israel, 2007), suggesting that *Cystoseira* may not have been a prominent feature on the south-eastern Levant intertidal rocky shores in recent history. Instead, these reefs are dominated by many smaller algal species: mostly red algae calcifiers (e.g, *Jania rubens* and *Ellisolandia elongata*) or non-calcifiers (e.g., *Palisada perforata*, *Hypnea musciformis* and the tropical alien *Acanthophora nayadiformis*, mainly on the wave-exposed platform edge), the brown leafy calcareous species *Padina pavonica* on the platform flat, and in winter and spring the green *Ulva* spp. that can cover the entire platform. Low-laying turf (i.e., epilithic algal matrix [EAM], see Vergés et al., 2014b) is also a dominant feature.

In their seminal publication, Lipkin and Safriel (1971) hardly mention the presence or dominance of alien species, suggesting that they were very rare in this habitat and not an ecological issue in the rocky intertidal of the 1960s, when the survey that their paper is based on was conducted. Nonetheless, in the 1970s, Safriel and Sasson-Frostig (1988) experimentally tested the potential of the Red Sea mussel *Brachidontes pharaonis,* an old but, at the time, rare invader, to outcompete its native counterpart *M. minimus* and establish a large population on the Israeli shore. The alien mussel was already present in 1876 near the northern entrance of the Suez Canal (Pallary, 1912) but was kept at very low numbers for over a century. The authors found that it had undetectable recruitment rates, and, although it had shown strong negative effects on survival and growth of recruits of the native species, the authors concluded that *B. pharaonis* cannot become established on the Levantine shore due to its (untested) high density-independent mortality caused by exposure to either high wave action or sedimentation.

### 9.2.2 Subtidal Reefs

Levant subtidal reefs have even fewer published historical information on their ecological communities (than intertidal ones) that could teach us something about their pre-1980, more pristine past. This is true also for other areas of the eastern Mediterranean such as the Aegean Sea where quantitative data exists mostly from the last few decades (see Chintiroglou et al., 2005; Gerovasileiou et al., 2009; De Raedemaecker et al., 2010). A semi-quantitative study was conducted around Kos Island (at the south-eastern end of the Aegean Sea, just outside of the area defined as the Levantine Basin) in 1981 and repeated in 2013 (Bianchi et al., 2014). This study can be used to assess how communities in this region may have looked like before the recent increase in human pressure, the appearance of many alien species and the accelerated ocean warming. One important finding from that study is that, in the past, the Kos reefs were dominated by flourishing shallow brown algal (*Cystoseira*) forests. An even earlier study from Haifa Bay in north Israel undertook monthly benthic sampling (using dredge or a bottom grab) between April 1955 and October 1956. Sixty-two samples were taken from rocky areas at depth ranging 18–40m and both *Cystoseira* spp. and *Sargassum* spp. occurred in 25 per cent of these samples, indicating that they were indeed once abundant also on reefs in the very south-eastern edge of the Mediterranean (Edelstein, 1960). Although there is no past data on megafauna or apex predators (such as monk seals, sharks and sea turtles that are known to visit and feed on shallow reefs), we can assume that, similar to the western Mediterranean (Sala, 2004), they were once important components of the food web on Levant reefs. From fisheries reports of the 1930s on Palestine (Hornell, 1935) and Iskenderun Bay in Turkey (Gruvel, 1931) we know of an average annual landing of 2,000 turtles back then, meaning that they were very abundant in the region a century ago. Today, Levantine sea turtles still suffer from high mortality mostly caused by fishery activity (Levy et al., 2015).

## 9.3 Reefs of the Present

Today, we have more information on the ecological status and the ecological mechanisms of Levant intertidal and subtidal reefs. However, the spatial and temporal cover of the data is still very limited. This is probably partly due to the lingering geopolitical turmoil and tension in the region, which hinders science in the most affected countries and much needed cross-border collaboration. There is only a handful of published research on the current status and ecology of intertidal reefs in the Levantine Basin, but intensive and extensive monitoring and research in the last eight years along the Israeli coast can now shed light on the ecological status of vermetid reefs and the spatio-temporal dynamics of their communities. A summary of several of the findings stemming from this effort will be presented in this chapter. During the springs of 2007 and 2008, a collaboration of ecologists executed a basin-wide and detailed nearshore survey on the ecological state of the subtidal reefs of the Mediterranean that included surveys of fish, macroalgae and invertebrates (Sala et al., 2012). This effort was set to establish a baseline for Mediterranean reef health status and focussed on the effects of protection level on reef communities (Sala et al., 2012). The authors have identified three alternative community states: (1) reefs that are dominated by non-canopy algae and have large fish biomass, (2) reefs with abundant native algal canopies (mostly brown algae) and suspension feeders, and with lower fish biomass and (3) reefs with extensive barrens or turf algal cover and low fish biomass. Unfortunately, sampling in the Levant was minimal and did not include the south-eastern corner of the Mediterranean (i.e., eastern Turkey, Syria, Lebanon, Israel or Egypt) – although this is its fastest warming, most invaded region (Shaltout and Omstedt, 2014; Sisma-Ventura et al., 2014). The Israeli coast in the Levant is also the first stop for many Indo-Pacific invaders arriving directly from the Red Sea through the Suez Canal, many of which have not shown up yet anywhere else, making it the 'hottest hotspot' for marine alien species in the Mediterranean and probably globally (Edelist et al., 2013a; Katsanevakis et al., 2014). Nonetheless, in the last two decades, several published surveys and experiments in Turkey, Lebanon, Israel and Egypt provided a clearer picture of the status of the reefs (see later). Further, unpublished surveys and experiments conducted between 2010 and 2016 on the Israel coast provide a more holistic understanding on the dramatic changes that occurred and their ecological consequences. A summary of some of these more recent findings is presented in the next section.

### 9.3.1 Intertidal Reefs

Studies conducted on the ecology of rocky intertidal and shallow subtidal reefs along the Israeli coast in the 1990s have shown that the state of these reefs have already been transformed substantially since the 1960s. For example, contrary to the predictions made by Safriel and Sasson-Frostig (1988), extensive beds of the invasive Red Sea mussel, *B. pharaonis,* were found in the mid-1990s – about 120 years after the initial introduction – all along the Israeli coast (Rilov et al., 2004b). It has established large populations also as far north-west as Sicily (Sara et al., 2000). This mussel also occupied the vermetid reefs, a habitat where mussel beds were never seen on the Israeli coast, except in the southernmost vermetid reef in Palmachim (Safriel et al., 1980a). Rilov et al. (2004b) have shown that the populations of the native mussel *M. minimus* have declined over a period of five years, perhaps due to competition with the invasive mussel. At the same time, the important predator, the whelk *S. haemastoma* (Linnaeus, 1767), showed significant signs of rapid decline in the late 1990s, when populations were still abundant, but no recruitment of young was seen during those five years (Rilov et al., 2001). Furthermore, on many platforms along the coast, the rims formed at their edge by the vermetid ecosystem engineer, *D. anguliferum*, seemed eroded with no living individuals in many locations, suggesting that the populations of this reef builder might be under stress (Rilov et al., 2004b). Finally, in a survey conducted in 2000 in Mikhmoret (Hoffman et al., 2008a), the intertidal macroalga, *Halimeda tuna*, could not be found, although noted by Lipkin and Safriel (1971) in the 1960s and recorded in many other sites (~10 per cent frequency of occurrence in 1,262 sampled quadrats) between 1973 and 1995 (re analysis of the Lundberg [1996] report; Rilov unpublished data). Regular seasonal and annual monitoring surveys at many sites along the entire Israeli coast between 2009 and 2016 indicated that *S. haemastoma*, *D. anguliferum* and *H. tuna* were by then all but extinct in this region (Rilov, 2016). Similar situations were recently described for the Lebanese coast (Badreddine et al., 2019). None of these species are commercially harvested or otherwise exploited, suggesting that some other regional driver caused their collapse – perhaps ocean warming (Rilov, 2016). The predatory whelk *S. haemastoma* was not recorded during similar surveys in Cyprus and Crete in summer 2013, but was relatively abundant in two sites near Naples in south-western Italy where it was highly aggregated (Rilov, 2016). In the 1990s, when *S. haemastoma* was declining but still relatively abundant, a study showed that it and the other large whelk *H. trunculus* demonstrate habitat partitioning in intertidal and shallow subtidal rocky ecosystems on the Israeli coast, where the latter species occupies more sheltered or deeper habitats, most probably because of its lower surface adhesion tenacity and much longer food-handling time that precludes it from highly wave-exposed environments (Rilov et al., 2004a). At the same time, lab experiments showed that *S. haemastoma* favoured the invasive mussel, *B. pharaonis*, as prey over all tested native species (Rilov et al., 2002), which should have theoretically boosted its populations, so its collapse is a real puzzle. Surveys in the last few years show that *H. trunculus* is still present along the coast, although in low numbers (Rilov, 2016).

Interestingly, the presence of *Brachidontes* beds was found to be patchy in the 1990s along the Israeli coast (Rilov et al., 2004b), and was still patchy in the last decade, i.e., some reefs had high mussel bed cover and others had none, with no clear biogeographic pattern or obvious relation to specific shoreline features (Figure 9.3). Large variation in mussel bed cover along a single coast is usually related to coastal ocean features at the mesoscale like up-/down-welling regimes (Navarrete et al., 2005), small-scale variations in predation, competition and recruitment intensity (Rilov and Schiel, 2006, 2011; McQuaid et al.,

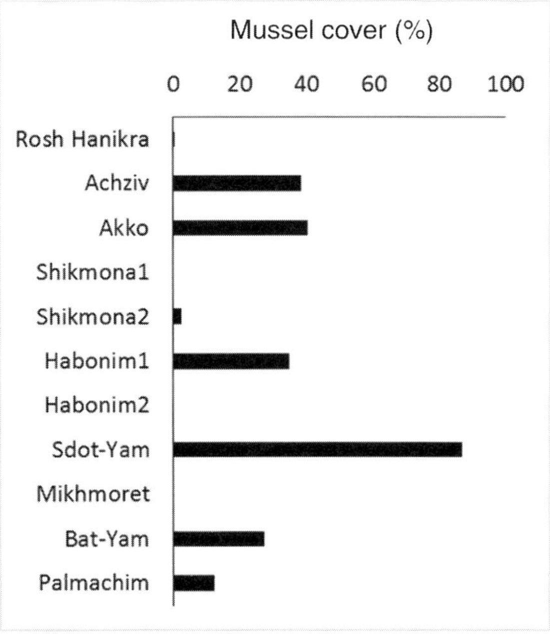


**Fig. 9.3** Percent cover of *B. pharaonis* on the platform centre of eleven study sites ordered from north to south (north at the top) in the autumn of 2009.

2015), or alternative stable states, e.g., with macroalgae (Petraitis and Dudgeon, 1999).

In order to test if the large differences in mussel cover were driven by variations in mussel recruitment or predation intensity, or the combination of both (as was shown for example in New Zealand, see Rilov and Schiel, 2011), mussel recruitment was followed for four years at four core sites located along the coast. At the same sites a short predation experiment was preformed using cages and fences similar to methods used in a New Zealand study (see Rilov and Schiel, 2006, 2011). The findings indicated that (1) recruitment level varied greatly among sites but was unrelated to adult mussel cover and (2) mortality of unprotected mussels was high at all sites and was probably driven by predation by fish and crabs, as whelks were absent (Figure 9.4). This suggests that there are no direct relationships between adult cover and either of the processes (recruitment or predation intensity).

The issue of zonation on the Israeli coast that was investigated by Lipkin and Safriel in the 1960s (1971) gained attention again in a recent study focussing on the mechanisms behind the distribution patterns of two chthamalid barnacles, *Chthamalus stellatus* and *Euraphia depressa*, with distinct habitat partitioning (Guy-Haim et al., 2015). *C. stellatus* extends from the high mid-tidal zone, above the algal belt that covers the platforms, to the supralittoral fringe, while *E. depressa* is restricted to the uppermost intertidal levels in wave-exposed areas and to cryptic habitats that are located within the belt of *C. stellatus* lower on the shore. Previous studies in the region have suggested that the reason for the fragmented distribution pattern of *E. depressa* is competitive displacement by the sympatric *C. stellatus* (Klepal and Barnes, 1975; Achituv and Safriel, 1980; Shemesh et al., 2009), fitting the common model of zonation suggested in the seminal work of Connell (1961, 1972). The model states that lower distribution limits are determined by biotic factors (competition and predation), while upper limits are set by physical factors. Using genetic markers to distinguish between the settled spats, recruitment measurements with settlement plates, and transplantation of settlers between zones, Guy-Haim et al. (2015) found two different settlement strategies: that of *E. depressa* was

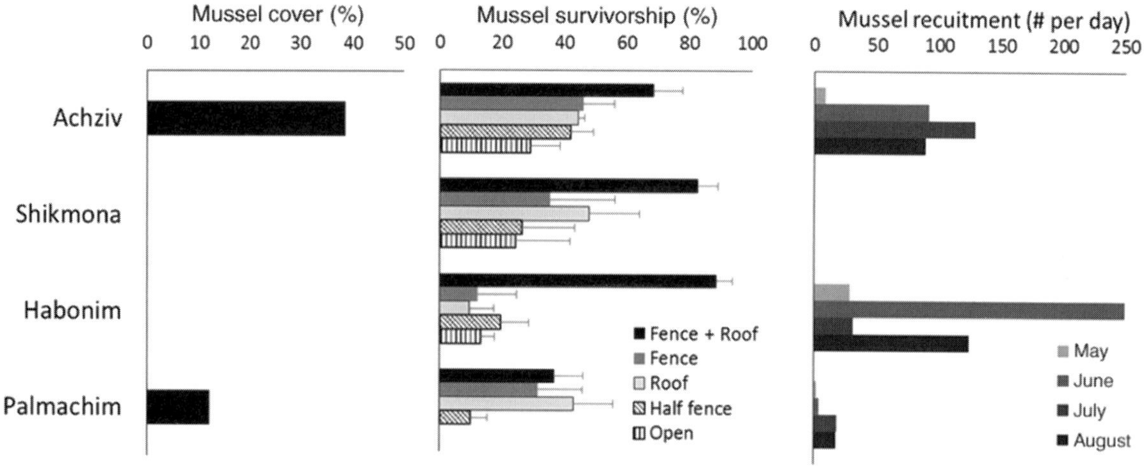


**Fig. 9.4** Mussel percent cover on the platform, survivorship rates under different treatments, and recruitment rates during spring and summer 2009. Survivorship rates were of mussels 1 cm long that were transplanted on 5 cm x 5 cm tiles and exposed to five treatments: fence + roof (protection from all macro-predators), fence (protection from crawling and walking predators such as crabs and snails), roof (protection from predators that feed from above, mainly fish), half fence (a procedural control open to all predators) and open (fully exposed to all predators), see Rilov and Schiel (2006) for details of the methods. Recruitment was measured into standard settlement collectors (Tuffies) that have been used in many regions worldwide and were deployed in monthly intervals (Connolly et al., 2001; Navarrete et al., 2005; Rilov et al., 2008).

habitat-specific, while settlement of *C. stellatus* was random. Shifting individuals of *C. stellatus* to either the high or cryptic zones resulted in high mortality in both habitats; however, exposing juveniles of *E. depressa* that had settled in artificially cryptic lowshore habitats (the upside-down sides of settlement plates) to *C. stellatus* presence had no effect on their survival. These findings did not agree with the Connell model that the inter-species boundary is determined by interspecific competition, implying that the competition model does not fit the Mediterranean intertidal barnacle zonation. Other models, dominated by physical forcing and pre-settlement recruitment-limiting factors may prevail in this ecosystem (Guy-Haim et al., 2015).

The community surveys conducted between 2009 and 2016 have revealed several interesting spatio-temporal community patterns. Although the shoreline distance between the furthest study sites is only about 100 km, multivariate analysis have shown significant differences between northern, central and southern sites, mostly in the communities at the edge of the platforms (Figure 9.5). This is mostly driven by the greater presence of fleshy red algae like *A. nayadiformis* and *Laurencia* spp., as well as the vermetid *V. triquetrus* in the northern sites compared to the central and southern sites.

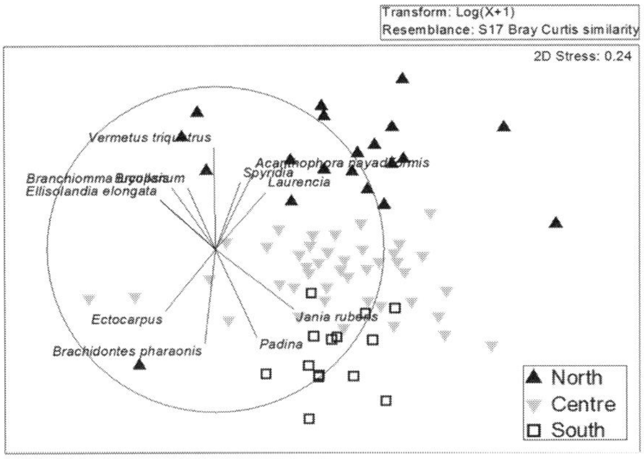


**Fig. 9.5** A multidimensional scaling ordination of site community similarity including all sites and years from three coastal regions on the Israeli coast (north – three sites, centre – six sites, south – two sites). Vectors of the dominant taxa driving the differences are projected on the ordination.

Clear seasonal patterns were also found, with higher taxa richness in the winter and spring compared to the summer and autumn months (Figure 9.6). The community structure is also considerably shifted from a winter–spring assemblage to a much poorer summer–autumn assemblage. Unfortunately, published work on seasonality of rocky intertidal assemblages are rare, especially in the Mediterranean, to allow a proper comparison of such temporal patterns, but, in cooler regions, summer is usually considered the more productive and richer season. It is possible that the very warm summer sea surface temperatures (SSTs) on the Levant coast are adversely affecting the more sensitive algal species, and the more resistant species prevail. Furthermore, between October and May, atmospheric conditions can generate unusually strong, dry, easterly winds that blow over the land and seaward (Saaroni et al., 1998). The high barometric pressure can lower water level so that even during high tide the platforms are exposed to air, leaving the platform flat exposed to dry (warm or cold) winds for long periods, sometimes days or even weeks. These prolonged desiccation events (PDEs) can have dire consequences for the local benthic community, causing massive algal bleaching and mortality of both algae and invertebrates. Reef structural complexity can create 'wet refugia' (tide pools and depressions); however, these can turn into 'traps' if the PDE lasts very long. The duration, frequency and strength of these extreme events (PDEs) are anticipated to increase under future climate change. Analysis of meteorological data have indeed shown that the frequency of synoptic systems that generate PDEs (mainly Red Sea trough systems) has more than doubled over the past forty years (Zamir et al., 2018), and PDEs are already a regularly occurring source of mass mortality on eastern Mediterranean shores. The characterisation and ecological impacts of these PDEs are presently the subject of intensive research on the Israeli coast.

## 9.3.2 Subtidal Reefs

Growing evidence over the past decade indicates that Levant reefs of today look very different from the reefs of the recent past largely due to

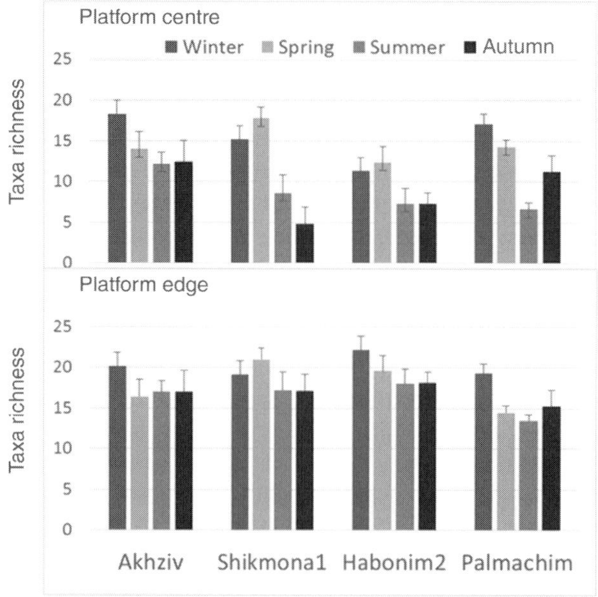


**Fig. 9.6** Average taxa richness in the four seasons over six years at the centre and edge of vermetid platforms at four sites on the Israeli coast. The error bars denote standard error.

overfishing, bioinvasions and climate change. The baseline of monitoring programmes established today are already far-shifted from any pristine state (Sala, 2004). Many taxonomic groups are becoming more and more dominated by alien species, and some of these species can have profound ecological impacts on the entire community. At the same time, many native species are becoming increasingly scarce. Unfortunately, published community surveys were absent from the Levant region until 2008, but several reef fish studies have shown some evidence of community transformation in the region before that. On the Israeli and Lebanese coasts, a few local surveys of fish assemblages have shown the domination of fish biomass by NIS – mostly the two highly abundant, herbivorous, Indo-Pacific rabbitfish species, *S. rivulatus* and *S. luridus* (Goren and Galil, 2001; Harmelin-Vivien et al., 2005). Interestingly, small, highly site-attached benthic species such as blennies, gobies and tripterygoids had no Indo-Pacific representatives in the assemblage, perhaps due to the lack of empty niches and high complementarity or their more restricted dispersal ability (Goren and Galil, 2001).

Extensive subtidal surveys of rocky reefs on the Israeli coast have confirmed the high presence of alien fish in the community. Predatory fish including both piscivores (such as groupers) and benthic feeders (such as sea breams) were, however, rare (Rilov, 2015). In a three-year study comparing areas inside and outside a small reserve (Achziv, north Israel), much higher densities of such predators were found inside the reserve compared to outside, indicating that overfishing has removed most of these predators from the Israeli reefs (Rilov et al., 2018). The low densities of native predators is not surprising given the increase in unregulated, poorly reported, nearshore artisanal and recreational fishing; sectors that mostly fish over reefs (Edelist et al., 2013b). The densities of the invasive siganids were, most of the time, also higher inside the reserve, suggesting that the presence of more predators in the reserve did not (perhaps surprisingly) adversely affect the abundance of the herbivorous invaders in the MPA, as was also found on coral reefs (Mumby et al., 2006; Hughes et al., 2007). Very recently, a new, potentially highly influential player, has started to frequently appear in the Levant reef ecological playground – the lionfish *Pterois miles*. This notorious Indo-Pacific predator showed up in increasing numbers in Cyprus, Lebanon and Turkey very recently in 2015–2016 (Bariche et al., 2013; Turan et al., 2014; Kletou et al., 2016), although it was already detected on the Israeli coast in the early 1990s (Golani and Sonin, 1992). This sudden population increase appears to contradict a 2014 model-based prediction that *P. miles* is unlikely to become established or invasive in the region (Johnston and Purkis, 2014). Lionfish (*Pterois volitans* and *P. miles*) invaded the Caribbean Sea in the 1980s, and since the 1990s have created an ecological havoc on the reefs there by greatly reducing the abundance and species richness of native fish species (Albins, 2015; Hixon et al., 2016). The impacts of a lionfish invasion on Levant benthic reef community are uncertain. However, it is not entirely implausible that this will be beneficial for the ecosystem, as it may reduce the massive populations of the highly invasive rabbitfish which causes serious ecological damage. If so, it would thereby expand the Indo-Pacific food web in the Mediterranean even further. This scenario deserves further study and a serious discussion among managers who for now encourage the culling of lionfish in the region. Similarly, it was recently suggested that, if the rabbitfish were to invade the Atlantic coral reefs, it could be ecologically beneficial as they could control excessive macroalgal growth that is detrimental to corals (Bellwood and Goatley, 2017).

Currently, the siganid invasions have profound impacts on shallow reef community structure in the Levant. These impacts may have begun half a century ago, but only recently they have been tested. One experimental study in south Turkey demonstrated that, when siganids are excluded (by cages) from small plots on reefs with barren seascape, erect brown algae quickly grow above the turf, indicating that, without the fish, canopy forests can prevail and these fish can graze these forests down to barrens (Sala et al., 2011). This effect was also demonstrated on the Israeli coast (Yeruham, 2013). Surveys in the north and south Aegean and the south-west corner of Turkey (which is considered part of the Levant) showed that rabbitfish are present only in the southern, warmer sites (Vergés et al., 2014b). Where rabbitfish are absent, the native herbivorous browser, *Sarpa salpa* is abundant and algae forests flourish. Where rabbitfish are present, however *S. salpa* was scarce, canopy algae were 65 per cent less abundant, and there was a 60 per cent lower overall benthic biomass and 40 per cent lower species richness (Vergés et al., 2014b). This community shift is part of a global phenomenon of tropicalisation of temperate reefs, where tropical herbivorous fish move into temperate regions that are warming due to climate change, causing regime shifts of benthic communities, from algal forests to barrens. The most extreme known case probably is the 'isoyake' in Japan, where coral reefs were found at sites that were dominated by kelp only two decades beforehand (Serisawa et al., 2004; Vergés et al., 2014a, 2016). Recent stable isotope analysis of native and invasive reef fish in Lebanon, including the rabbitfish, suggested a shift in the local food web dynamics (Fanelli et al., 2015). The high abundance of the rabbitfish in conjunction with the scarcity of piscivores may mean an overall contraction of the reef food web, which may be expressed in a faster turnover of algae biomass directly into detritus, with less energy reaching the higher trophic levels (as has been suggested by Goren and Galil, 2005). However, this needs to be tested by future research.

Another profound effect of the siganid invasion is the facilitation of other aliens by ichthyochory, live passage through fish digestive tracts, as was recently found by Guy-Haim et al. (2017). They found that the introduction and propagation of sixty-eight species of Indo-Pacific holobenthic foraminifera, which formerly could not be explained by known natural and anthropogenic vectors, can be linked to this newly identified bioinvasion vector. The herbivorous siganids incidentally ingest living epiphytic foraminifera and then defecate them unharmed, transporting them over long distances, across the Suez Canal and westward, in the Mediterranean Sea. In the eastern Mediterranean, some of these alien foraminifera are comprising up to 90 per cent of the foraminiferal assemblage in the rocky reefs (Hyams-Kaphzan et al., 2008). In the Aegean Sea, alien foraminifera predominates on the rocky reefs at formidable densities (Meriç et al., 2010). This overgrowth not only outcompetes the native foraminifera species, but it also dramatically changes the rocky reef habitats, with a consequent 'desertification' of the rocky reefs and a shift in the biodiversity from hard- to soft-bottom-dwelling species. Biodiversity shifts and potential losses are inevitable results of such a process.

Surprisingly, the most invasive seaweeds in the western Mediterranean, *Caulerpa taxifolia* and then *Caulerpa cylindracea* (formerly *Caulerpa racemosa* var. *cylindracea*) (Piazzi et al., 2001, 2016; Wright et al., 2007; Piazzi and Balata, 2008; Bulleri et al., 2010), are not considered abundant in the Levantine basin, and, although they can be found there (Cevik et al., 2012), they are possibly controlled by rabbitfish grazing (Lundberg et al., 1999). Instead, there are many other invasive (mostly Indo-Pacific) seaweeds in the region (Einav and Israel, 2008; Hoffman et al., 2008b; Israel et al., 2010), but little was known about their abundance or ecological role until recently. One of the most dominant Indo-Pacific invaders is the red calcareous alga, *Galaxaura rugosa,* which was reported to form large algae drifts on the shores of north Israel since 2003, creating large piles on the beach after storms (Hoffman et al., 2008b). Recent scuba diving and ROV surveys indicate that, in some areas in north Israel, large portions of the reefs are totally covered by this species (Rilov, unpublished data). Moreover, *G. rugosa* persists throughout the hot Levant summer while most native habitat-forming algal species shed their branches (and persist as dormant cauloids) or disappear between June and November/December. Gut content analysis suggests that the alien alga is rarely consumed by the rabbitfish (Garval, 2016; Rilov, unpublished data). This raises an important question: could this new algae invader compensate for some of the functions lost by the transformation of most Levant brown algal forests to turf barrens? Functions lost by the overgrazing of canopy algae like *Cystoseira* forests could include high primary production and standing stock (e.g., carbon uptake, biomass), the formation of a complex habitat (i.e., as ecosystem engineers) and an additional food source for other species. This question has recently been tested in several studies. One showed that the epibiotic assemblage on *G. rugosa* is as rich as that found on most native habitat-forming algae, but the epibiotic community structure is usually fundamentally different (Garval, 2016). For example, *G. rugosa* had much higher isopod populations than native hosts.

In situ incubation experiments comparing community structure and metabolic processes in plots dominated by either *Cystoseira*, turf or *Galaxaura* (Figure 9.7) in benthic metabolic chambers showed that, while biomass is partially regained (i.e., it is higher in *G. rugosa* plots than in grazed areas where turf prevails), community structure and metabolic functions in *G. rugosa* plots have taken a novel direction (Peleg, 2017; Peleg et al., in review; Figure 9.8). The *Cystoseira* plots were dominated by native species (even when excluding the habitat former from the analysis) and

**Fig. 9.7** Reef seascapes in the Haifa area dominated by (a) *Cystoseira* meadows in the late winter, (b) turf barrens and (c) *G. rugosa* shrubs. *A black and white version of this figure will appear in some formats. For the colour version, please refer to the plate section.*

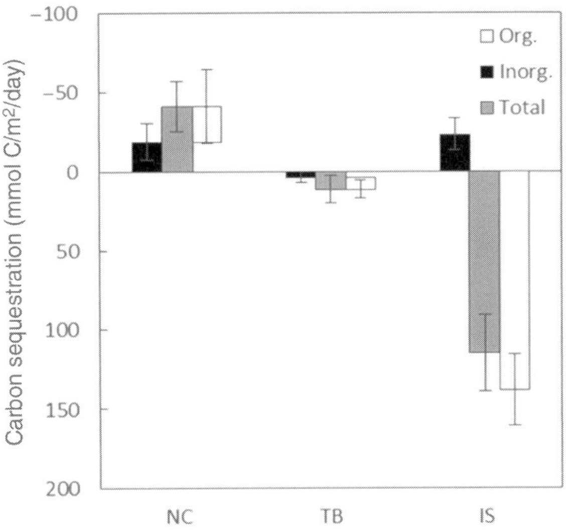


**Fig. 9.8** Diurnal carbon uptake balances (mmol C/m$^2$/day) calculated from in situ benthic incubations of the three benthic communities characterised by Peleg (2017) ($n$ = 6 per each community). NC: native canopy – algal *Cystoseira* forest; TB: turf barren – short often filamentous EAM; IS: invasive shrubs – *G. rugosa* and associated biota, mainly aliens. Total carbon sequestration (grey), and organic (org.) and inorganic/ sequestration (inorg.; i.e., photosynthesis/calcification, white and black bars, respectively). Negative values represent net dissolved inorganic carbon (DIC) carbon sequestration, while positive represent DIC production. Error bars represent the standard error.

gross production and respiration calculations showed that they are net autotrophic (Figure 9.8). Turf plots, not surprisingly, had low overall biomass and balanced metabolism (low and equal photosynthesis and respiration). *G. rugosa* plots were mostly dominated by alien species (many alien filter feeders and detritivores, e.g., *B. pharaonis* and *Cerithium scabridum*) and had higher biomass than turf plot. However, they were overall heterotrophic (more respiration than photosynthesis, Figure 9.8). This structural and metabolic transformation can have profound implications to the overall function of coastal areas, as macrophytes have a considerable role in overall marine metabolism and ocean carbon uptake (Duarte et al., 2013). Nonetheless, macroalgal-dominated habitats are not usually considered as carbon sinks, since they lack underground biomass (such as in seagrasses, mangroves and saltmarshes, see Hill et al., 2015) and, once seized, their carbon is being recycled back to the water column. Yet, if pre-impacted algal forests were perennial and had the ability to store carbon over many years, this shift from autotrophy to heterotrophy may indicate a shift from an ecosystem that is a carbon sink to a carbon source.

The near absence of the two snails, *D. anguliferum* and *S. haemastoma,* on the Israeli rocky intertidal coast, that was mentioned before, is part of a much larger phenomenon of recent population collapses of many native, non-harvested reef molluscs and two sea urchins (*Paracentrotus lividus* and *Arbacia lixula*; Rilov, 2016). Of fifty-nine molluscan species that were once described in the taxonomic literature as common on Levant reefs, thirty-eight were not found at all in recent surveys; by contrast, there was a total domination of NIS in the molluscan reef assemblage. The fast rise in SST in the Levant (1.5–3°C in the last three decades) was speculated to be a major driver that may have helped push the more temperature-sensitive invertebrates beyond their physiological tolerance limits, leading to these population collapses and possible extirpations (Rilov, 2016). Population collapses at the trailing (warm) edge of species distributions is considered as an advanced stage of climate-mediated range shifts in species distributions (Bates et al., 2014). Experimental work has indeed shown that at least the (almost complete) disappearance of *P. lividus* could be explained by ocean warming, as this urchin quickly died when temperatures crossed a 30.5°C threshold in both the field and lab (Yeruham et al., 2015). The contribution of resource competition with the invasive rabbitfish to its disappearance by grazing favourable algae to the degree that affects the urchins' fitness cannot be ruled out (Yeruham, 2013). However, in the central Aegean Sea, where water is cooler than along the Israeli coast, *S. luridus* and *P. lividus* still co-occur in high numbers even though turf barrens are present (Giakoumi, 2014), suggesting that competition with the fish is not the main cause of the urchin collapse, and temperature probably is. Interestingly, high densities of *P. lividus* (up to 63 m$^2$) were found in very recent urchin surveys (2013–2015) near Alexandria (Egypt), to the south-west of the present study region (Elmasry et al., 2015). In this area, the maximum summer temperatures are 29.5°C (Elmasry et al., 2015), still below the lethal temperature of 30.5°C (Yeruham et al., 2015). The footprint of climate change (mostly ocean warming) on Mediterranean Sea biota has been documented extensively in the past two decades, but primarily in the western basin and most notably during two severe and extensive heat waves in 1999 and 2003 that resulted in mass mortalities of a multitude of reef invertebrates (Rivetti et al., 2014; Marbà et al., 2015). The steady increase in SST, such as that occurring along the south-eastern Mediterranean coast, may result, not only in temporary declines in abundance through pulsed stress caused by heat waves, but also by a gradual elimination of species from regions altogether, similar to what is already happening on land (Walther et al., 2002; Chen et al., 2011). The species collapses documented by Rilov (2016) on the Levant coast may be the first evidence of that process; in this case, some species seem to be at the verge of a regional extinction and others may be close to a global extension (e.g., the Levantine clade of *D. anguliferum*, which genetic analysis indicate to be a distinct species Calvo et al., 2009).

All these findings suggest that bioinvasions, ocean warming and overfishing are acting together to drastically transform Levant reef ecosystems into a new, different ecosystem state. These reefs have gone through a considerable phase or even regime shift, in both physical conditions and community composition – from a warm temperate Atlantic biological realm to an almost tropical Indo-Pacific realm. Mediterranean reef communities have very rich and complex food webs of which we know very little about (Sala, 2004) and have been transforming from their pristine state for thousands of years (Jackson and Sala, 2001). We can therefore only suggest a very simplified version of the changes that the system has probably undergone in the past century or so: it probably was a system dominated by large predators (e.g., groupers) that controlled the densities of browsing herbivorous fish (mainly *S. salpa)* and omnivores (such as sparids) that may have foraged on grazing urchins, which would have left lush brown algal meadows intact. We now have far fewer predators, no urchins but very influential invasive Indo-Pacific grazers, very little canopy algae forests and a growing cover of mostly Indo-Pacific invasive algae (Figure 9.9a and 9.9b).

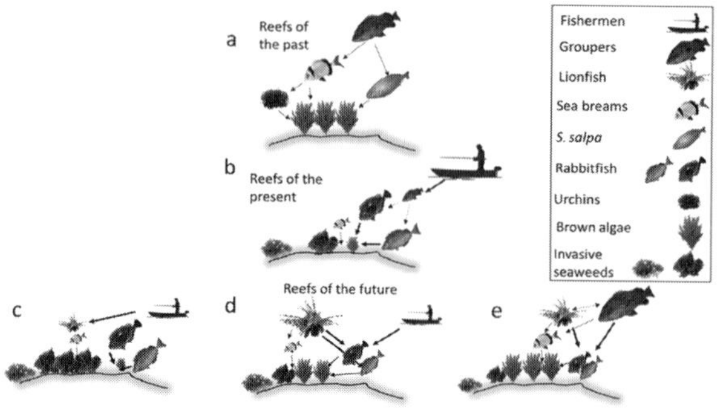


**Fig. 9.9** Simplified conceptual models of (a) past, (b) present and (c–e) possible futures of major components of Levant reef food web. (c) fished area (no targeting of rabbitfish), no extensive development of lionfish population or successful lionfish culling, resulting in invasive macroalgae domination; (d) fished area (also targeting rabbitfish), established lionfish population and no culling, resulting in reduced grazing and some brown algae forests recovery; (e) protected area, no lionfish culling (potential competition for prey between lionfish and groupers), resulting in reduced grazing and extensive recovery of brown algae forests.

## 9.4 Reefs of the Future

What does the future holds for the Levant reefs, and what can the present state of the Levant reefs teach us about the possible near future of reefs to the west? Considering the recent addition of a second pathway (canal) along part of the route of the Suez Canal, and thus the possible increase in invasion rates (Galil et al., 2015), and the likely ongoing anthropogenic carbon emissions into the atmosphere, resulting in further climate change impacts on the marine environment (Gattuso et al., 2015), the trends already presented are not expected to curb, but rather intensify. Many of the changes are unavoidable, for example, phenological shifts and, in extreme cases, like the urchins and possibly other species, the extirpation of temperature-sensitive species from the region altogether. Sea-level rise (SLR) and the increase in frequency of extreme events like storms, heat waves and, in the case of the eastern Mediterranean, the possible intensification of PDEs due to changes in the synoptic systems leading to strong, and prolonged easterly winds events, are further important climate change consequences.

### 9.4.1 Intertidal Reefs

Ocean warming is expected to continue and favour alien thermophilic species over sensitive native species. As air temperature is also predicted to continue to increase, augmented heat stress during low tide and calm seas is also expected. In experiments with intertidal gastropods living mostly at the back of Israeli vermetid reefs, limpets were exposed to a range of eleven different seawater temperatures between 15 and 35°C. Results have shown that the native intertidal limpet *Patella caerulea* had peak performance (respiration rate) at 29°C SST (Klein, 2015), compared to a peak at 25°C measured in its western Mediterranean counterpart (Bannister, 1974). This suggests that this species is capable of phenotypic or genotypic adaptation (selection) to local conditions, a hypothesis that needs to be tested using common garden experiments and genetic tools (e.g., Goldstien et al., 2006; Teske et al., 2011).

The Levant population peak performance of *P. caerulea* fits well with the maximum SST measured in the 1980s, prior to the fast warming of the past three decades. However *P. caerulea* is performing poorer under today's peak summer conditions (with some mortality occurring from 33°C upward), and thus its performance is expected to reduce further in the coming decades. By contrast, the alien Indo-Pacific limpet, *Cellana rota*, showed no decline in respiration rates and no mortality, even at the maximum tested SST temperatures of 35°C (Klein, 2015). This may suggest that overtime *C. rota* will be the dominate limpet on Levant rocky intertidal reefs. Should that materialise, it is unclear if this will result in one limpet species replacing another with identical ecological roles, with no major effects on food web and other functions, or in a modified mid-intertidal system, in case the species vary in their function.

With regard to extreme events, a recent analysis of weather in Israel over the past forty years showed that the occurrence of synoptic systems leading to PDEs have more than doubled in the winter time between the 1960s and the 2000s. This suggests that desiccation stress is probably increasing in the region (Zamir et al., 2018) in the season when biodiversity is highest. As mentioned before, under strong easterly winds and prolonged desiccation conditions, the vermetid reefs dry out and macroalgae die while mobile animals seek shelter first in shaded and later in wet microhabitats. Mobile animals can also be trapped inside small, shallow tidepools that can become extremely hot, salty and eventually hypoxic. Mortality in tidepools is often observed under such conditions (Rilov, personal observations). In the winter of 2014, a long PDE lasted for almost a month, which left the reef platforms exposed to air for weeks. This PDE exposed the reef biota to both extreme cold air conditions at night and dry warm conditions at daytime. The result was extensive algal bleaching and the eventual removal of most erect algal biomass in a season when these algae should peak in growth. If the trend of increased frequency of PDE-producing synoptic systems will continue, we can expect increased desiccation stress in this unique ecosystem with yet unknown ecological consequences that are now under extensive study.

SLR could counteract this kind of desiccation stress on the platform flat. In a sea with almost no tides where the extensive rocky intertidal platforms are located, mostly at the lowshore level, it is expected that even a mild increase of sea level will eventually drown the vermetid reefs, and no new reefs will be created in the new sea level because the reef building vermetid is now mostly extinct on the shore. Time series of sea level measurements between 1992 and 2012 indicated an average increase of about 6mm/year, totalling around 12 cm in twenty years, which is twice the global average (Rosen et al., 2013). In a region where the tidal range is 20–40 cm, this rise is already quite extensive, and may already have had some ecological impact. In a field experiment, further SLR was simulated to test its effects on the ecological community on the vermetid reef flat: intertidal communities were translocated to the shallow horizontal bedrock beyond the reef by drilling out 15 cm diameter rock cores with an intact typical lowshore intertidal assemblage on them and moving them into premade holes in the subtidal zone at 20 cm depth. Treatments included a control where cores were placed back to the intertidal, cores that were protected by cages from grazing by herbivorous fish (mainly the invasive rabbitfish), cores topped with half cages (procedural control) and cores that were fully exposed to grazing. After 100 days exposure to the experimental conditions, results indicated that, when intertidal community is fully submerged for long durations (i.e., following SLR) biodiversity of the platform assemblage reduces and the ecological community greatly transforms (Figure 9.10, Rilov and David unpublished data). This is due to a combination of intense grazing by the fish that remove all erect algae from the rocks, leaving only turf behind, and physiological sensitivity of intertidal algae to being fully submerged. A physiological analysis of *A. nayadiformis,* a typical Levant intertidal species that was claimed by Verlaque et al. (2015) to be an introduced species through an Egyptian 'intermediate canal' precursor to the Suez Canal, indicated that this species photosynthesises five times faster in air than in water, provided it is kept highly hydrated. This may be

**Fig. 9.10** Experimental intertidal cores in the SLR test. (a) Side and (b) top views of a core on day 0 of the experiment; (c) top view of a core protected by a cage on day 100; (d) top view of a core with no cage on day 100, with evidence of grazing down to turf. *A black and white version of this figure will appear in some formats. For the colour version, please refer to the plate section.*

the reason why it does not succeed in competing with subtidal algae (Einav and Beer, 1993). In the next few decades, it will be important to follow which climate change-driven stressor, SLR or PDEs, will be more influential on the ecological community of vermetid reefs – i.e., could PDEs still occur on the reef platform, and will they indeed intensify, fifty years from now, with a further rise of sea level?

### 9.4.2 Subtidal Reefs

Further warming and acidification of the oceans in the coming decades is expected to increase climate change impacts on the marine realm even under the most optimistic emission scenarios (Gattuso et al., 2015). The impacts of these processes are expected to be harshest in small enclosed basins like the Mediterranean, and specifically at the warm edge of species distributions where conditions are naturally more extreme, like the south-east corner of the Mediterranean – the Levant. A recent experiment testing the thermal tolerance of the abundant native red coralline macroalga *E. elongata* (formerly *Corallina elongata*), under a range of eleven temperatures between 15 and 35°C, showed that the growth season of this species may have already shrunk. Under future warming its performance may reduce further (Guy-Haim et al., 2016a). In Israel, this species naturally shifts today from a frondose to crustose-base form in the late spring and stay in this reduced form until the early winter (Guy-Haim et al., 2016a). However, in the northwestern Mediterranean, this species appears to stay in its frondose form throughout the summer (Guy-Haim and Rilov, personal observations), probably because it is colder there. *E. elongata* appears to have two physiological tipping points: (1) a metabolic breakdown at above 31°C, leading to thalli bleaching and necrosis and (2) a metabolic shift at 23°C (the average temperature in May and November), possibly promoting the seasonal algal shift from the frondose to the crustose form (Guy-Haim et al., 2016a). An evaluation of annual production rates in this study suggests a loss of approximately one third of the organic carbon and carbonate production contributed by corallines to the shallow Levantine coast in the upcoming decades under further warming

(Guy-Haim et al., 2016a). Guy-Haim et al. (2016a) predicted that, with continued warming, eastern Mediterranean corallines will experience a westward range contraction, initiated with phenological shifts, followed by performance declines and population decreases, and possibly ending with local extinctions. Similarly, when *Cystoseira rayssiae* performance was tested under past, present and future temperatures, its optimum temperature was found to be approximately 27°C, explaining why it switches to the dormant form (losing the branches) in early summer (Guy-Haim et al., 2016a). This, and recent mesocosm experiments (see later) suggested that the recent 2°C warming have already shortened the growth season of *Cystoseira* by one to two months (Guy-Haim et al., 2016b). As mentioned earlier, habitat-forming brown algae are already highly restricted in cover on Levant reefs due to rabbitfish overgrazing, and thus ocean warming may strike the final blow on this very important group. In contrast, the peak performance of the alien macroalga *G. rugosa* is around 33°C (Garval, 2016), explaining why it flourishes during the current much warmer summer conditions when most native species disappear. We can thus expect that *G. rugosa*, which seems to be rarely consumed by the rabbitfish, will increase its dominance on the reef with increased warming. Such tropicalisation driven by both increase temperature and the movement of aggressive herbivores is seen in other regions around the world (Vergés et al., 2014a, 2016).

Using a large, novel, flow-through benthic mesocosm system (benthocosm), following the conceptual guidelines of a recently developed benthocosm in Kiel, Germany (Wahl et al., 2015), long-term experiments at Israel Oceanographic and Limnological Research Institute tested the separate and combined effects of ocean warming and acidification on eastern Mediterranean shallow reef communities, focussing on a typical *Cystoseira* assemblage and allowing a natural recruitment of larvae from the open sea (Guy-Haim et al., 2016b). Temperature and pH treatments simulated the near-past, present and predicted future levels, and included also the natural daily fluctuations following ambient seawater conditions. While biodiversity (richness and Shannon diversity) did not change significantly with warming (+3°C) and acidification (–0.4 pH units; Figure 9.11a), community composition shifted from a native to a NIS dominance, and the abundance of calcifying invertebrate species increased (Figure 9.11a; Guy-Haim et al., 2016b). This increase in calcifying invertebrate biomass under increased $CO_2$ levels in the benthocosms is intriguing. It could be a consequence of the observed increase in growth of non-calcifying macroalgae epiphytes (Guy-Haim et al., 2016b) that can provide more habitat and food (see a recent similar finding sites near volcanic vents, Connell et al., 2017), and perhaps also be due to elevated pH levels at the microhabitat scale through increased photosynthetic activity (Guy-Haim, 2017; Guy-Haim et al. in review). Nonetheless, in summer, under combined warming and acidification, the overall balance in community functions measured by biogeochemical parameters presented a shift from an autotrophic to a heterotrophic system (Figure 9.11c), with a slight net dissolution balance (Figure 9.11d; Guy-Haim et al., 2016b).

The dramatic alterations in traits and functions, shown by both the in situ incubation experiments of native canopy, turf areas and invasive shrubs and the mesocosm experiments, mean that the Levant reefs are evidently going through a regime shift to a novel ecological state that will only intensify in the future. The evident collapses of multiple invertebrate species on the Israeli coast, described earlier, could be a warning sign for similar population collapses that may soon occur further west in the Mediterranean and affect the structure, function and services of reef ecosystems. Our experimental results indicate that whole community shifts can be expected soon in the Levant and then further west and north of this region as coastal waters are getting warmer. A possible loss of carbon sinks and 'gain' of sources should be considered by managers, in the light of the ever-growing atmospheric carbon concentration.

To conclude, because of centuries of human pressure on the coastal reefs, our understanding of Mediterranean food webs and ecosystem functions is 'based on a mix of unnatural, simplified communities dominated by small species, where

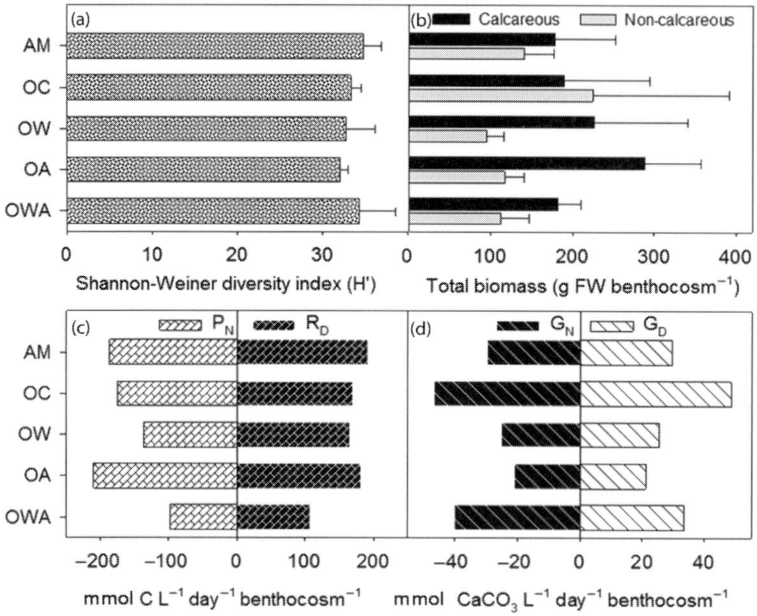


**Fig. 9.11** Taxa diversity, traits and functions in experimental benthic mesocosms, spring–summer 2015, under five conditions: AM (ambient), OC (cooling by 2°C), OW (warming by 3°C), OA (acidification by 0.4 pH units) and OWA (combined warming and acidification). (a) Taxa diversity; (b) calcareous and non-calcareous taxa biomass; (c) Diel daytime net photosynthesis ($P_N$) and night-time respiration ($R_D$) as a function of dissolved inorganic carbon; (d) Diel daytime net calcification ($G_D$) and night-time dissolution ($G_N$). *n* = 3. Error bars denote standard deviation.

megafauna has been virtually eliminated by overfishing' (Jackson and Sala, 2001). Now, as demonstrated here, many other native biotic components are also in decline while many are being replaced by alien invaders that may be similar or very different in their trophic and/or other functions. As both climate change and bioinvasions are for the most part beyond the control of most countries in the Levantine Basin in the near future, we can put our hope of maintaining some level of ecosystem resilience only on local management to somewhat buffer this multi-stressor scenario. This can be done by reducing local stressors like pollution and overfishing, and by that hopefully increasing ecosystems resilience to these regional and global stressors. Depending on management by local authorities in each country, several potential futures can be suggested for Levant reefs (Figure 9.9c and 9.9d). Under the mostly business-as-usual scenario that would be characterised by limited protection of predators, no developed fishery on rabbitfish and no extensive culling of lionfish should they become invasive, we can expect a simple food web and increased dominance of invasive species (Figure 8.9c). Under limited protection but intensive lionfish culling (should it become invasive and if it will indeed prey on rabbitfish) we may expect reduced rabbitfish populations and perhaps some recovery of brown algal forests (Figure 9.9d). Under extensive protection through an effective network of MPAs, irrespective of lionfish culling we can expect reduced rabbitfish populations and a major recovery of brown algal forests (Figure 9.9e). Recovery of brown algal (*Cystoseira*) forests will be possible only if ocean warming will not cause a total collapse of this group in the region. Finally, can we imagine a future where tropical coral reefs start developing on the Levant shores? In the past three decades, winter temperatures have increased from around 15 to 17°C (Rilov, unpublished data). Therefore, with the projected rapid ocean warming, when this minimum will cross above the presumed

**Fig. 9.12** A simplified conceptual model of a future Levant 'coral reef'.

18°C threshold (although this threshold may not be as rigid as presumed in the past; see, Coles and Fadlallah, 1991), some hardy Indo-Pacific reef-building corals with long larval duration that can cross the canal or last in the ballast waters of ships might start settling and establishing on Levant reefs (Figure 9.12). This is if current and near future maximum summer temperatures (32–34°C) will not become too warm even for these corals. Recent evidence suggests that some Red Sea corals may be preselected to withstand especially high temperatures and resist bleaching due to their evolutionary history (Fine et al., 2013); hence, they might be able to eventually conquer this new realm and bring the eastern Mediterranean to its past Tethys Ocean glory as a tropical sea.

## REFERENCES


Achituv, Y. and Safriel, U. (1980). *Euraphia depressa* (Poli) (Crustacea, Cirripedia), a recent Mediterranean colonizer of the Suez Canal. *Bulletin Marine Sciences*, **30**, 724–6.

Albins, M. A. (2015). Invasive Pacific lionfish *Pterois volitans* reduce abundance and species richness of native Bahamian coral-reef fishes. *Marine Ecology Progress Series*, **522**, 231–43.

Badreddine, A., Milazzo, M., Saab, M. A. A., Bitar, G. and Mangialajo, L. (2019). Threatened biogenic formations of the Mediterranean: current status and assessment of the vermetid reefs along the Lebanese coastline (Levant basin). *Ocean & Coastal Management*, **169**, 137–146.

Bannister, J. (1974). The respiration in air and in water of the limpets *Patella caerulea* (L.) and *Patella lusitanica* (Gmelin). *Comparative Biochemistry and Physiology Part A: Physiology*, **49**, 407–11.

Bariche, M., Torres, M. and Azzurro, E. (2013). The presence of the invasive Lionfish *Pterois miles* in the Mediterranean Sea. *Mediterranean Marine Science*, **14**, 292–4.

Bates, A. E., Pecl, G. T., Frusher, F. et al. (2014). Defining and observing stages of climate-mediated range shifts in marine systems. *Global Environmental Change*, **26**, 27–38.

Bellwood, D. R. and Goatley, C. H. R. (2017). Can biological invasions save Caribbean coral reefs? *Current Biology*, **27**, R13–14.

Ben Eliahu, M. N. (1975). Polychaete cryptofauna from rims of similar intertidal vermetid reefs on Mediterranean coast of Israel and in gulf of Elat – Nereidae (polychaeta-errantia). *Israel Journal of Zoology*, **24**, 177–91.

Ben-Tuvia, A. (1964). Two Siganid Fishes of Red Sea Origin in the Eastern Mediterranean. In *Bulletin*. Sea Fisheries Research Station (Haifa). Ministry of Agriculture, Department of Fisheries, Sea Fisheries Research Station, Haifa, pp. 1–9.

Benedetti-Cecchi, L., Pannacciulli, F., Bulleri, F. et al. (2001). Predicting the consequences of anthropogenic disturbance: large-scale effects of loss of canopy algae on rocky shores. *Marine Ecology-Progress Series*, **214**, 137–50.

Bianchi, C. N. (2007). Biodiversity issues for the forthcoming tropical Mediterranean Sea. *Hydrobiologia*, **580**, 7–21.

Bianchi, C. N., Corsini-Foka, M., Morri, C. and Zenetos, A. (2014). Thirty years after: dramatic change in the coastal marine ecosystems of Kos Island (Greece), 1981–2013. *Mediterranean Marine Science*, **15**, 482–97.

Blanfuné, A., Boudouresque, C.-F., Verlaque, M. and Thibaut, T. (2016). The fate of *Cystoseira crinita*, a forest-forming Fucale (Phaeophyceae, Stramenopiles), in France (North Western Mediterranean Sea). *Estuarine, Coastal and Shelf Science*, **181**, 196–208.

Bricker, O. P. (1971). Introduction: Beachrock and Intertidal Cement. In O. P. Bricker, ed. *Carbonate Cements*. John Hopkins Press, Baltimore, pp 1–3.

Bulleri, F., Balata, D., Bertocci, I., Tamburello, L. and Benedetti-Cecchi, L. (2010). The seaweed *Caulerpa racemosa* on Mediterranean rocky reefs: from passenger to driver of ecological change. *Ecology*, **91**, 2205–12.

Bulleri, F., Benedetti-Cecchi, L., Acunto, S., Cinelli, F. and Hawkins, S. J. (2002). The influence of canopy algae on vertical patterns of distribution of low-shore assemblages on rocky coasts in the northwest Mediterranean. *Journal of Experimental Marine Biology and Ecology*, **267**, 89–106.

Calvo, M., Templado, J., Oliverio, M. and Machordom, A. (2009). Hidden Mediterranean biodiversity:

molecular evidence for a cryptic species complex within the reef building vermetid gastropod *Dendropoma petraeum* (Mollusca: Caenogastropoda). *Biological Journal of the Linnean Society*, **96**, 898–912.

Calvo, M., Templado, J. and Penchaszadeh, P. E. 1998. Reproductive biology of the gregarious Mediterranean vermetid gastropod *Dendropoma petraeum*. *Journal of the Marine Biological Association of the United Kingdom*, **78**, 525–49.

Castellari, S., Pinardi, N. and Leaman, K. (2000). Simulation of water mass formation processes in the Mediterranean Sea: influence of the time frequency of the atmospheric forcing. *Journal of Geophysical Research: Oceans*, **105**, 24157–81.

Cevik, C., Cavas, L., Mavruk, S., Derici, O. B. and Cevik, F. (2012). Macrobenthic assemblages of newly introduced *Caulerpa taxifolia* from the Eastern Mediterranean coast of Turkey. *Biological Invasions*, **14**, 499–501.

Chemello, R. C. R. and Silenzi, S. (2011). Vermetid reefs in the Mediterranean Sea as archives of sea-level and surface temperature changes. *Chemistry and Ecology*, **27**, 121–7.

Chen, I. C., Hill, J. K., Ohlemuller, R., Roy, D. B. and Thomas, C. D. (2011). Rapid range shifts of species associated with high levels of climate warming. *Science*, **333**, 1024–6.

Chintiroglou, C., Antoniadou, C., Vafidis, D. and Koutsoubas, D. (2005). A review on the biodiversity of hard substrate invertebrate communities in the Aegean Sea. *Mediterranean Marine Science*, **6**, 51–62.

Coles, S. L. and Fadlallah, Y. H. (1991). Reef coral survival and mortality at low temperatures in the Arabian Gulf: new species-specific lower temperature limits. *Coral Reefs*, **9**, 231–7.

Coll, M., Piroddi, C., Steenbeek, J. et al. 2010. The biodiversity of the Mediterranean Sea: estimates, patterns, and threats. *PLoS ONE*, **5**, https://doi.org/10.1371/journal.pone.0011842.

Connell, J. H. (1961). The influence of interspecific competition and other factors on the distribution of the barnacle *Chthamalus stelattus*. *Ecology*, **42**, 710–23.

Connell, J. H. (1972). Community interactions on marine rocky intertidal shores. *Annual Review of Ecological Systems*, **3**, 169–92.

Connell, S. D., Doubleday, Z. A., Hamlyn, S. B. et al. (2017). How ocean acidification can benefit calcifiers. *Current Biology*, **27**, R95–6.

Connolly, S. R., Menge, B. A. and Roughgarden, J. (2001). A latitudinal gradient in recruitment of intertidal invertebrates in the northeast Pacific Ocean. *Ecology*, **82**, 1799–813.

De Raedemaecker, F., Miliou, A. and Perkins, R. (2010). Fish community structure on littoral rocky shores in the Eastern Aegean Sea: effects of exposure and substratum. *Estuarine Coastal and Shelf Science*, **90**, 35–44.

Duarte, C. M., Conley, D. J., Carstensen, J. and Sánchez-Camacho, M. (2009). Return to Neverland: shifting baselines affect eutrophication restoration targets. *Estuaries and Coasts*, **32**, 29–36.

Duarte, C. M., Losada, I. J., Hendriks, I. E., Mazarrasa, I. and Marbà, N. (2013). The role of coastal plant communities for climate change mitigation and adaptation. *Nature Climate Change*, **3**, 961–8.

Edelist, D., Rilov, G., Golani, D., Carleton, J. and Spanier, E. (2013a). Restructuring the Sea: profound shifts in the world's most invaded marine ecosystem. *Diversity and Distributions*, **19**, 69–77.

Edelist, D., Scheinin, A., Sonin, O. et al. (2013b). Israel: reconstructed estimates of total fisheries removals in the Mediterranean, 1950–2010. *Acta Adriatica*, **54**, 253–63.

Edelstein, T. (1960). *The Biology and Ecology of Deep Sea Algae of the Haifa Bay Area*. The Hebrew University of Jerusalem, Jerusalem.

Einav, R. and Beer, S. (1993). Photosynthesis in air and in water of *Acanthophora najadiformis* growing within a narrow zone of the intertidal. *Marine Biology*, **117**, 133–8.

Einav, R. and Israel, A. (2007). Seaweeds on the Abrasion Platforms of the Intertidal Zone of Eastern Mediterranean Shores. In *Algae and Cyanobacteria in Extreme Environments*. Springer, Berlin, pp. 193–207.

Einav, R. and Israel, A. (2008). Checklist of seaweeds from the Israeli Mediterranean: taxonomical and ecological approaches. *Israel Journal of Plant Sciences*, **56**, 127–84.

Elmasry, E., Abdel Razek, F. A., El-Sayed, A.-F. M., Omar, H. and Hamed, E. S. A. E. (2015). Abundance, size composition and benthic assemblages of two Mediterranean echinoids off the Egyptian coasts: *Paracentrotus lividus* and *Arbacia lixula*. *The Egyptian Journal of Aquatic Research*, **41**, 367–74.

Fanelli, E., Azzurro, E., Bariche, M., Cartes, J. E. and Maynou, F. (2015). Depicting the novel Eastern Mediterranean food web: a stable isotopes study following Lessepsian fish invasion. *Biological Invasions*, **17**, 2163–78.

Fine, M., Gildor, H. and Genin, A. (2013). A coral reef refuge in the Red Sea. *Global Change Biology*, **19**, 3640–7.

Galil, B., Pisanty, S., Spanier, E. and Tom, M. (1989). The indo-pacific lobster *Panulirus ornatus* (Crustacea, decapoda) – a new Lessepsian migrant to the eastern Mediterranean. *Israel Journal of Zoology*, **35**, 241–3.

Galil, B. S., Boero, F., Campbell, M. L. et al. (2015). 'Double trouble': the expansion of the Suez Canal and marine bioinvasions in the Mediterranean Sea. *Biological Invasions*, **17**, 973–6.

Garval, T. (2016). *Population Dynamics and Ecological Impacts of the Alien Macroalgae* Galaxaura rugosa *(J. Ellis & Solander) J.V.Lamouroux on the Israeli Shore*. University of Haifa, Haifa.

Gattuso, J.-P., Magnan, A., Billé, R. et al. (2015). Contrasting futures for ocean and society from different anthropogenic $CO_2$ emissions scenarios. *Science*, **349** (6243), http://dx.doi.org/10.1126/science.aac4722.

Gerovasileiou, V., Sini, M. I., Poursanidis, D. and Koutsoubas, D. (2009). Contribution to the Knowledge of Coralligenous Communities in the NE Aegean Sea. In *Proceedings of the 1st Mediterranean Symposium on the Conservation of the Coralligenous and other Calcareous Bio-concretions, Tabarka, Tunis*, pp. 205–7.

Giakoumi, S. (2014). Distribution patterns of the invasive herbivore *Siganus luridus* (Rüppell, 1829) and its relation to native benthic communities in the central Aegean Sea, Northeastern Mediterranean. *Marine Ecology*, **35**, 96–105.

Golani, D. and Diamant, A. (1991). Biology of the sweeper, *Pempheris vanicolensis* Cuvier & Valenciennes, a Lessepsian migrant in the eastern Mediterranean, with a comparison with the original Red Sea population. *Journal of Fish Biology*, **38**, 819–27.

Golani, D. and Sonin, O. (1992). New records of the Red Sea fishes, *Pterois miles* (Scorpaenidae) and *Pteragogus pelycus* (Labridae) from the eastern Mediterranean Sea. *Japanese Journal of Ichthyology*, **39**, 167–9.

Goldstien, S. J., Schiel, D. R. and Gemmell, N. J. (2006). Comparative phylogeography of coastal limpets across a marine disjunction in New Zealand. *Molecular Ecology*, **15**, 3259–68.

Gon, O. and Golani, D. (2008). Lessepsian migration of cardinalfishes (Telcostei, Apogonidae). *South African Journal of Botany*, **74**, 367.

Goren, M. and Galil, B. S. (2001). Fish biodiversity in the vermetid reef of Shiqmona (Israel). *Marine Ecology – Pubblicazioni Della Stazione Zoologica Di Napoli I*, **22**, 369–78.

Goren, M. and Galil, B. S. (2005). A review of changes in the fish assemblages of Levantine inland and marine ecosystems following the introduction of non-native fishes. *Journal of Applied Ichthyology*, **21**, 364–70.

Gruvel, A. (1931). *Les Etats de Syrie. Richesses marines et fluviales: Exploitation actuelle, Avenir*. Société des Editions Géographiques, Maritimes et Coloniales, Paris.

Guy-Haim, T. (2017) The impact of ocean warming and acidification on coastal benthic species and communities. PhD, University of Haifa.

Guy-Haim, T., Hyams-Kaphzan, J., Yeruham, E., Almogi-Labin, A. and Carlton, J. T. (2017). A novel marine bioinvasion vector: Ichthyochory, live passage through fish. *Limnology and Oceanography Letters*, **2**, 81–90.

Guy-Haim, T., Rilov, G. and Achituv, Y. (2015). Different settlement strategies explain intertidal zonation of barnacles in the Eastern Mediterranean. *Journal of Experimental Marine Biology and Ecology*, **463**, 125–34.

Guy-Haim, T., Silverman, J., Raddatz, S., Wahl, M., Israel, A. and Rilov, G. (2016a). The carbon turnover response to thermal stress of a dominant coralline alga on the fast warming Levant coast. *Limnology and Oceanography*, **61**, 1120–33.

Guy-Haim, T., Silverman, J., Raddatz, S., Wahl, M. and Rilov, G. (2016b). Shifted coastal communities and ecosystem functions under predicted warming and acidification. In *41th CIESM Congress, Kiel, Germany*.

Guy-Haim, T., Silverman, J., Wahl, M., Aguirre, J., Noisette, F. and Rilov, G. Epiphytes provide micro-scale refuge from ocean acidification: The dressed coralline hypothesis. *Journal of Ecology*, in review

Harmelin-Vivien, M. L., Bitar, G., Harmelin, J. G. and Monestiez, P. (2005). The littoral fish community of the Lebanese rocky coast (eastern Mediterranean Sea) with emphasis on Red Sea immigrants. *Biological Invasions*, **7**, 625–37.

Hill, R., Bellgrove, A., Macreadie, P. I. et al. (2015). Can macroalgae contribute to blue carbon? An Australian perspective. *Limnology and Oceanography*, **60**, 1689–706.

Hixon, M. A., Green, S. J., Albins, M. A., Akins, J. L. and Morris, Jr, J. A. (2016). Lionfish: a major marine invasion. *Marine Ecology Progress Series*, **558**, 161–5.

Hoffman, R., Dubinsky, Z., Israel, A. and Iluz, D. (2008a). The mysterious disappearance of *Halimeda tuna* from the intertidal zone along the Israeli Mediterranean. *Israel Journal of Ecology and Evolution*, **54**, 267–8.

Hoffman, R., Israel, A., Lipkin, Y., Dubinsky, Z. and Iluz, D. (2008b). First record of two seaweeds from the Israeli Mediterranean: *Galaxaura rugosa* (J. Ellis and Solander) JV Lamouroux (Rhodophyta) and Codium adhaerens C. Agardh (Chlorophyta). *Israel Journal of Plant Sciences*, **56**, 123–6.

Hornell, J. (1935). *Report on the Fisheries of Palestine*. Crown Agents of the Colonies, London, p. 65.

Hughes, T. P., Bellwood, D. R., Folke, C. S., McCook, L. J. and Pandolfi, J. M. (2007). No-take areas, herbivory and coral reef resilience. *Trends in Ecology & Evolution*, **22**, 1–3.

Hyams-Kaphzan, O., Almogi-Labin, A., Sivan, D. and Benjamini, C. (2008). Benthic foraminifera assemblage change along the southeastern Mediterranean inner shelf due to fall-off of Nile-derived siliciclastics. *Neues Jahrbuch für Geologie und Palaeontologie Abhandlungen*, **248**, 315–44.

Israel, A., Einav, R., Silva, P. C., Paz, G., Chacana, M. E. and Douek, J. (2010). First report of the seaweed *Codium parvulum* (Chlorophyta) in Mediterranean waters: recent blooms on the northern shores of Israel. *Phycologia*, **49**, 107–12.

Jackson, J. B. and Sala, E. (2001). Unnatural oceans. *Scientia Marina*, **65**, 273–81.

Jackson, J. B. C., Kirby, M. X., Berger, W. H. et al. (2001). Historical overfishing and the recent collapse of coastal ecosystems. *Science*, **293**, 629–38.

Johnston, M. and Purkis, S. (2014). Are lionfish set for a Mediterranean invasion? Modelling explains why this is unlikely to occur. *Marine Pollution Bulletin*, **88**, 138–47.

Katsanevakis, S., Coll, M., Piroddi, C. et al. (2014). Invading the Mediterranean Sea: biodiversity patterns shaped by human activities. *Frontiers in Marine Science*, **1**, https://doi.org/10.3389/fmars.2014.00032.

Klein, L. (2015). *The Effects of Temperature Changes on Grazer Gastropod Species along the Rocky Shores of the Eastern Mediterranean*. Bar-Ilan University, Ramat Gan.

Klepal, W. and Barnes, H. (1975). Further observations on the ecology of *Chthamalus depressus* (Poli). *Journal of Experimental Marine Biology and Ecology*, **17**, 269–96.

Kletou, D., Hall-Spencer, J. M. and Kleitou, P. (2016). A lionfish (*Pterois miles*) invasion has begun in the Mediterranean Sea. *Marine Biodiversity Records*, **9**, 1.

Knowlton, N. and Jackson, J. B. (2008). Shifting baselines, local impacts, and global change on coral reefs. *Plos Biology*, **6**, e54.

Koren, Z. C. (2005). The First Optimal All-Murex All-Natural Purple Dyeing in the Eastern Mediterranean in a Millennium and a Half. In *Dyes*. History and Archaeology Archetype Publications, London, pp. 136–49.

Levy, Y., Frid, O., Weinberger, A. et al. 2015. A small fishery with a high impact on sea turtle populations in the eastern Mediterranean. *Zoology in the Middle East*, **61**, 300–17.

Lipkin, Y. and Safriel, U. (1971). Intertidal zonation of the rocky shores at Mikhmoret (Mediterranean, Israel). *Journal of Ecology*, **59**, 1–30.

Lundberg, B. (1996). *Composition of the Seaweed Vegetation along the Mediterranean Coast of Israel*. Nature Reserves Authority, Jerusalem.

Lundberg, B., Payiatas, G. and Argyrou, M. (1999). Notes on the diet of the Lessepsian migrant herbivorous fishes, *Siganus luridus* and *S. rivulatus*, in Cyprus. *Israel Journal of Zoology*, **45**, 127–34.

Marbà, N., Jorda, G., Agusti, S., Girard, C. and Duarte, C. M. (2015). Footprints of climate change on Mediterranean Sea biota. *Frontiers in Marine Science*, **2**, https://doi.org/10.3389/fmars.2015.00056.

McQuaid, C. D., Porri, F., Nicastro, K. R. and Zardi, G. I. (2015). Simple, scale-dependent patterns emerge from very complex effects: an example from the intertidal mussels *Mytilus galloprovincialis* and *Perna perna*. *Oceanography and Marine Biology*, **53**, 127–56.

Meriç, E., Yoke⍰, M. B., Av⍰ar, N. and Bircan, C. (2010). An oasis for alien benthic Foraminifera in the Aegean Sea. *Aquatic Invasions*, **5**, 191–5.

Mienis, H. K. (2004). New data concerning the presence of Lessepsian and other Indo-Pacific migrants among the molluscs in the Mediterranean Sea with emphasis on the situation in Israel. *Turkish Journal of Aquatic Life*, **2**, 117–31.

Mumby, P. J., Dahlgren, C. P., Harborne, A. R. et al. 2006. Fishing, trophic cascades, and the process of grazing on coral reefs. *Science*, **311**, 98–101.

Navarrete, S. A., Wieters, E. A., Broitman, B. R. and Castilla, J. C. (2005). Scales of benthic-pelagic and the intensity of species interactions: from recruitment limitation to top-down control. *Proceedings of the National Academy of Sciences of the United States of America*, **102**, 18046–51.

Pallary, P. (1912). Catalogue des mollusques du littoral méditerranéen de l'Egypte. *Mémoires de l'Institut d'Egypte*, **7**, 69–207.

Pauly, D. (1995). Anecdotes and the shifting baseline syndrome of fisheries. *Trends in Ecology and Evolution*, **10**, 430.

Peleg, O. (2017). *Bioinvasions Drive Major Shifts in Levant Reef Community Structure and Ecosystem Functioning*. University of Haifa, Haifa.

Peleg, O., Guy-Haim, T., Yeruham, E., Silverman, J. and Rilov, G. Two-stage tropicalization inverts trophic state and carbon budget of shallow temperate rocky reefs. *Journal of Ecology*. **33**, 1000–13.

Petraitis, P. S. and Dudgeon, S. R. (1999). Experimental evidence for the origin of alternative communities on rocky intertidal shores. *Oikos*, **84**, 239–45.

Piazzi, L. and Balata, D. (2008). The spread of *Caulerpa racemosa* var. *cylindracea* in the Mediterranean Sea: An example of how biological invasions can influence beta diversity. *Marine Environmental Research*, **65**, 50–61.

Piazzi, L., Balata, D., Bulleri, F., Gennaro, P. and Ceccherelli, G. (2016). The invasion of *Caulerpa cylindracea* in the Mediterranean: the known, the unknown and the knowable. *Marine Biology*, **163**, 1–14.

Piazzi, L., Ceccherelli, G. and Cinelli, F. (2001). Threat to macroalgal diversity: effects of the introduced green alga *Caulerpa racemosa* in the Mediterranean. *Marine Ecology Progress Series*, **210**, 149–59.

Por, F. D. (1978). *Lessepsian Migration: The Influx of Red Sea Biota into the Mediterranean by Way of the Suez Canal*. Springer-Verlag, Berlin.

Por, F. D. and Dimentman, C. H. (1989). *The Legacy of Tethys: An Aquatic Biogeography of the Levant*. Kluwer Academic Publishers, Dordrecht.

Raitsos, D. E., Beaugrand, G., Georgopoulos, D. et al. 2010. Global climate change amplifies the entry of tropical species into the Eastern Mediterranean Sea. *Limnology and Oceanography*, **55**, 1478–84.

Rilov, G. (2015). *Test of Marine Reserves as a Management Tool for Marine Conservation on the Israeli Mediterranean Coast (Report Submitted to the Ministry of National Infrastructures, Energy and Water Resources)*. H34/2015, Israel Oceanographic and Limnological Research (IOLR), Haifa.

Rilov, G. (2016). Multi-species collapses at the warm edge of a warming sea. *Scientific Reports*, **6**, 36897.

Rilov, G., Benayahu, Y. and Gasith, A. (2001). Low abundance and skewed population structure of the whelk *Stramonita haemastoma* along the Israeli Mediterranean coast. *Marine Ecology Progress Series*, **218**, 189–202.

Rilov, G., Benayahu, Y. and Gasith, A. (2004a). Life on the edge: do biomechanical and behavioral adaptations to wave-exposure correlate with habitat partitioning in predatory whelks? *Marine Ecology Progress Series*, **282**, 193–204.

Rilov, G., Benayahu, Y. and Gasith, A. (2004b). Prolonged lag in population outbreak of an invasive mussel: a shifting-habitat model. *Biological Invasions*, **6**, 347–64.

Rilov, G., Dudas, S., Grantham, B., Menge, B. A., Lubchenco, J. and Schiel, R. D. (2008). The surf zone: a semi-permeable barrier to onshore recruitment of invertebrate larvae? *Journal of Experimental Marine Biology and Ecology*, **361**, 59–74.

Rilov, G., Gasith, A. and Benayahu, Y. (2002). Effect of an exotic prey on the feeding pattern of a predatory snail. *Marine Environmental Research*, **54**, 85–98.

Rilov, G., Peleg, O., Yeruham, E., Garval, T., Vichik, A. and Raveh, O. (2018). Alien turf: overfishing, overgrazing and invader domination in southeastern Levant reef ecosystems. *Aquatic Conservation: Marine and Freshwater Ecosystems*, **28**, 351–69. http://dx.doi.org/10.1002/aqc.2862.

Rilov, G. and Schiel, D. R. (2011). Community regulation: the relative importance of recruitment and predation intensity of an intertidal community dominant in a seascape context. *PLoS ONE*, **6**, https://doi.org/10.1371/journal.pone.0023958.

Rilov, G. and Schiel, R. D. (2006). Seascape-dependent subtidal-intertidal trophic linkages. *Ecology*, **87**, 731–44.

Rivetti, I., Fraschetti, S., Lionello, P., Zambianchi, E. and Boero, F. (2014). Global warming and mass mortalities of benthic invertebrates in the Mediterranean Sea. *PLoS ONE*, **9**, https://doi.org/10.1371/journal.pone.0115655.

Rosen, S. D., Raskin, L. and Galanti, B. (2013). Long-term characteristics of sea level, wave, wind and current at central Mediterranean coast of Israel from 20 years of data at GLOSS station 80 – Hadera. In *40th CIESM Congress, Marseille*.

Saaroni, H., Ziv, B., Bitan, A. and Alpert, P. (1998). Easterly wind storms over Israel. *Theoretical and Applied Climatology*, **59**, 61–77.

Safriel, U. N. (1974). Vermetid gastropods and intertidal reefs in Israel and Bermuda. *Science*, **186**, 1113–15.

Safriel, U. N. (1975). The role of vermetid gastropods in the formation of Mediterranean and Atlantic reefs. *Oecologia (Berlin)*, 20, 85–101.

Safriel, U. N., Felsenburg, T. and Gilboa, A. (1980a). The distribution of *Brachidontes variabilis* (Krauss) along the Red Sea coasts of Sinai. *Argamon Israel Jorournal of Malacology*, **7**, 31–43.

Safriel, U. N., Gilboa, A. and Felsenburg, T. (1980b). Distribution of rocky intertidal mussels in the Red Sea coasts of Sinai, the Suez Canal, and the Mediterranean coast of Israel, with special reference to recent colonizer. *Journal of Biogeography*, **7**, 39–62.

Safriel, U. N. and Sasson-Frostig, Z. (1988). Can colonizing mussel outcompete indigenous mussel? *Journal of Experimental Marine Biology and Ecology*, **117**, 211–26.

Sala, E. (2004). The past and present topology and structure of Mediterranean subtidal rocky-shore food webs. *Ecosystems*, **7**, 333–40.

Sala, E., Ballesteros, E., Dendrinos, P. et al. (2012). The structure of Mediterranean rocky reef ecosystems across environmental and human gradients, and conservation implications. *PLoS ONE*, **7**, https://doi.org/10.1371/journal.pone.0032742.

Sala, E., Kizilkaya, Z., Yildirim, D. and Ballesteros, E. (2011). Alien marine fishes deplete algal biomass in the eastern Mediterranean. *PLoS ONE*, **6**, https://doi.org/10.1371/journal.pone.0017356.

Sara, G., Romano, M. and Mazzola, A. (2000). The new Lessepsian entry *Brachidontes pharaonis* (Fischer P.,

1870) (Bivalvia, Mytilidae) in the western Mediterranean: a physiological analysis under varying natural conditions. *Journal of Shellfish Research*, **19**, 967–77.

Serisawa, Y., Imoto, Z., Ishikawa, T. and Ohno, M. (2004). Decline of the *Ecklonia cava* population associated with increased seawater temperatures in Tosa Bay, southern Japan. *Fisheries Science*, **70**, 189–91.

Shaltout, M. and Omstedt, A. (2014). Recent sea surface temperature trends and future scenarios for the Mediterranean Sea. *Oceanologia*, **56**, 411–43.

Shemesh, E., Huchon, D., Simon-Blecher, N. and Achituv, Y. (2009). The distribution and molecular diversity of the Eastern Atlantic and Mediterranean chthamalids (Crustacea, Cirripedia). *Zoologica Scripta*, **38**, 365–78.

Sisma-Ventura, G., Yam, R. and Shemesh, A. (2014). Recent unprecedented warming and oligotrophy of the eastern Mediterranean Sea within the last millennium. *Geophysical Research Letters*, **41**, 5158–66.

Spalding, M. D., Fox, H. E., Halpern, B. S. et al. 2007. Marine ecoregions of the world: A bioregionalization of coastal and shelf areas. *Bioscience*, **57**, 573–83.

Steinitz, W. (1927). Beiträge zur Kenntnis der Küstenfauna Palästinas. *Pubblicazioni Della Stazione Zoologica Di Napoli I*, 8, 331–53.

Stephenson, T. A. and Stephenson, A. (1949). The universal feature of zonation between the tide-marks on rocky coasts. *Journal of Ecology*, **36**, 289–305.

Templado, J., Richter, A. and Calvo, M. (2016). Reef building Mediterranean vermetid gastropods: disentangling the *Dendropoma petraeum* species complex. *Mediterranean Marine Science*, **17**(1), 13–31.

Teske, P. R., Papadopoulos, I., Mmonwa, K. L. et al. (2011). Climate-driven genetic divergence of limpets with different life histories across a southeast African marine biogeographic disjunction: different processes, same outcome. *Molecular Ecology*, **20**, 5025–41.

Turan, C., Ergüden, D., Gürlek, M., Yağlıoğlu, D., Uyan, A. and Uygur, N. (2014). First record of the Indo-Pacific lionfish *Pterois miles* (Bennett, 1828) (Osteichthyes: Scorpaenidae) for the Turkish marine waters. *Journal of Black Sea/Mediterranean Environment*, **20**(2).

Vergés, A., Doropoulos, C., Malcolm, H. A. et al. (2016). Long-term empirical evidence of ocean warming leading to tropicalization of fish communities, increased herbivory, and loss of kelp. *Proceedings of the National Academy of Sciences*, 201610725.

Vergés, A., Steinberg, P. D., Hay, M. E. et al. (2014a). The tropicalization of temperate marine ecosystems: climate-mediated changes in herbivory and community phase shifts. *Proceedings of the Royal Society B: Biological Sciences*, **281**, 20140846.

Vergés, A., Tomas, F., Cebrian, E. et al. (2014b). Tropical rabbitfish and the deforestation of a warming temperate sea. *Journal of Ecology*, **102**, 1518–27.

Verlaque, M., Ruitton, S., Mineur, F. and Boudouresque, C. F. (2015). *CIESM Atlas of Exotic Macrophytes in the Mediterranean Sea*. CIESM Publishers, Monaco.

Vousdoukas, M., Velegrakis, A. and Plomaritis, T. (2007). Beachrock occurrence, characteristics, formation mechanisms and impacts. *Earth-Science Reviews*, **85**, 23–46.

Wahl, M., Buchholz, B., Winde, V. et al. 2015. A mesocosm concept for the simulation of near-natural shallow underwater climates: The Kiel Outdoor Benthocosms (KOB). *Limnology and Oceanography: Methods*, **13**, 651–63.

Walther, G. R., Post, E., Convey, P. et al. 2002. Ecological responses to recent climate change. *Nature*, **416**, 389–95.

Wright, J. T., McKenzie, L. A. and Gribben, P. E. (2007). A decline in the abundance and condition of a native bivalve associated with *Caulerpa taxifolia* invasion. *Marine and Freshwater Research*, **58**, 263–72.

Yeruham, E. (2013). *Possible Explanations for Paracentrotus lividus (European purple sea urchin) Population Collapse in South-East Mediterranean*. Tel Aviv University, Tel Aviv.

Yeruham, E., Rilov, G., Shpigel, M. and Abelson, A. (2015). Collapse of the echinoid *Paracentrotus lividus* populations in the Eastern Mediterranean – result of climate change? *Scientific Reports*, **5**, 13479.

Zamir, R., Alpert, P. and Rilov, G. (2018). Increase in weather patterns generating extreme desiccation events: implications for Mediterranean rocky shore ecosystems. *Estuaries and Coasts*, http://dx.doi.org/10.1007/s12237-018-0408-5.

Ziderman, I. I. (2008). The biblical dye tekhelet and its use in Jewish textiles. *Dyes in History and Archaeology*, **21**, 36–44.

# Chapter 10

# North-East Pacific

## *Interactions on Intertidal Hard Substrata and Alteration by Human Impacts*

Phillip B. Fenberg and Bruce A. Menge

## 10.1 Introduction

The flora and fauna of the rocky intertidal zone are among the most biologically diverse on the planet – not in terms of species richness, but of the diversity and density of higher taxonomic categories. All of the major animal phyla can be found in the rocky intertidal, sometimes with representatives of each inhabiting a single rock or boulder. In addition to this phylogenetic diversity, the rocky intertidal may be one of the most ancient of habitats because it is a necessary and continual result of celestial mechanics set in motion prior to the diversification of life. When advising readers to '... look from the tide pool to the stars and then back to the tide pool again', Steinbeck and Ricketts (1951) seemed to underline this point while advocating for a holistic approach to ecological research. A few years earlier, Ricketts and Calvin set out to comprehensively document what was then known about the ecology of the north-eastern Pacific (NEP) rocky intertidal in their classic book, *Between Pacific Tides* (Ricketts et al., 1985).

Since the first edition of *Between Pacific Tides* (1939), the NEP has become widely recognised as an ideal natural laboratory for experimental ecologists and as a platform for more observationally focussed ecologists seeking to understand macroecological and biogeographical patterns. It is, of course, outside the scope of this chapter to attempt a comprehensive review of this extensive research. Rather, we focus on a couple of broad topics that are central to our current understanding of fundamental ecological, evolutionary and conservation topics that have benefitted from NEP rocky intertidal case studies. The first half of the chapter deals with recent work on the biotic and abiotic factors influencing patterns of range wide abundance and distribution of species, and how such patterns are being affected by human impacts. The second half reviews the latest research on the role of direct and indirect human impacts on top-down and bottom-up control of rocky intertidal community structure and functioning.

## 10.2 Biogeographic Context

The NEP coast spans from the Baja California peninsula (Pacific coast) to Alaska (Figure 10.1). The southern boundary of the NEP is well defined by a biogeographic break at Magdalena Bay (~25°N). The region from 25°N to –5°S (including the Gulf of California) constitutes the tropical eastern Pacific (TEP), while south of –5°S (Paita, Peru) to Tierra del Fuego is the south-eastern

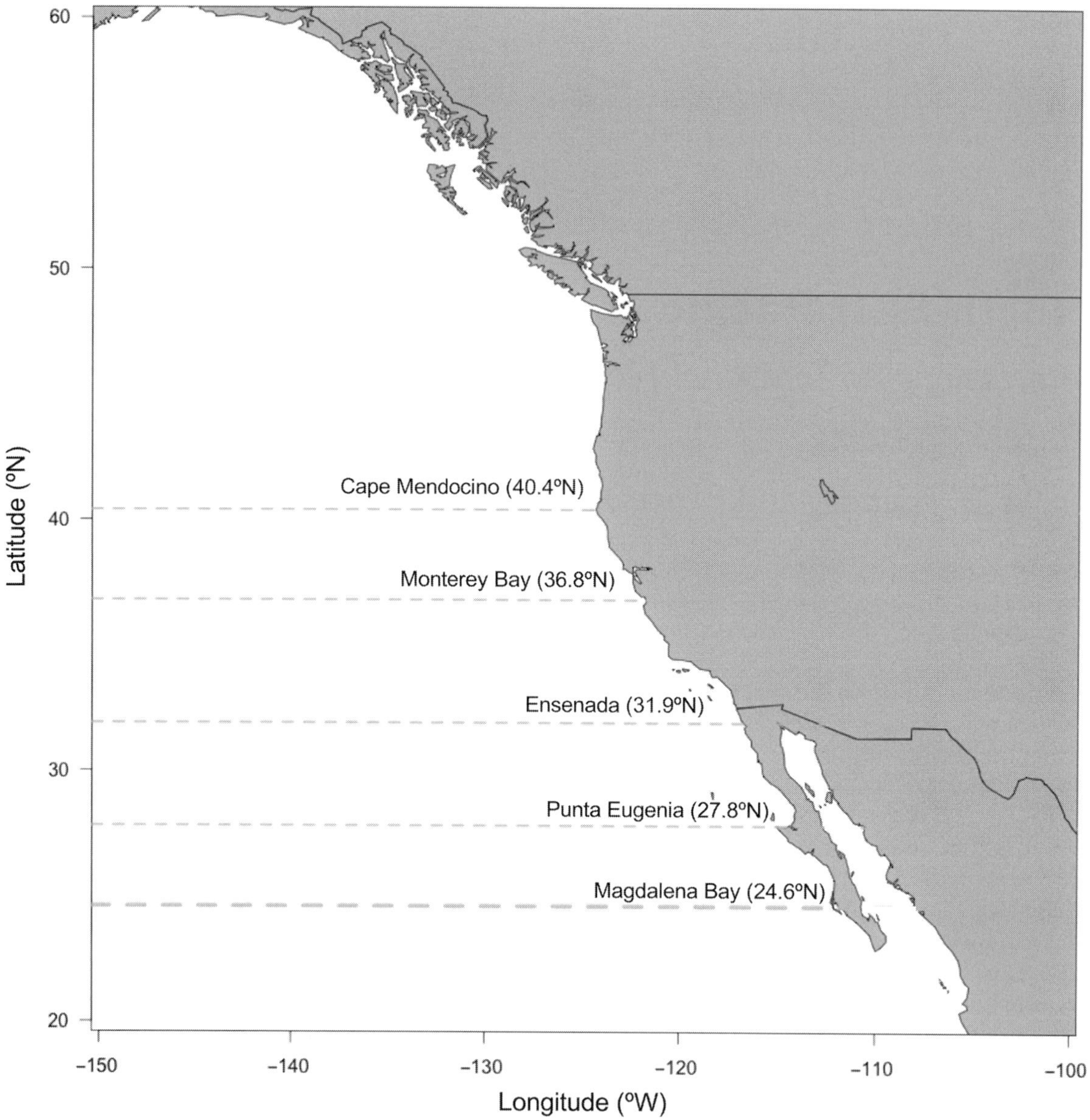


**Fig. 10.1** Map of the NEP with latitudes of locations mentioned in the text. The thick dashed line at Magdalena Bay marks the southern edge of the NEP.

Pacific (covered in Chapter 12). Very few rocky intertidal species commonly found in the NEP or the south-eastern Pacific have ranges that extend into the TEP (and vice versa), giving the eastern Pacific rocky intertidal a well-defined biogeographic structure at this broadscale (Fenberg and Rivadeneira, 2019).

## 10.3 Range-Wide Abundance Patterns

The largely north–south trending NEP coastline essentially means that species restricted to the intertidal zone have a linear, one-dimensional geographic range, with single geographic locations representing their northern and southern limits and range centres (assuming there are no major range discontinuities). As a result, ecologists have utilised NEP rocky intertidal species to ask basic biogeographic questions that are more difficult to assess for species with relatively complex geographic ranges (e.g., birds and mammals or pelagic and deep-sea species). For example, Sagarin and Gaines (2002a) used NEP rocky intertidal species to test the hypothesis that individuals of a species are more common in the centre of their geographic ranges, commonly referred to as the 'abundant centre hypothesis' (ACH; Brown, 1984). Of the twelve species analysed throughout their ranges, only two had their highest abundances in their range centre. Prior to this study, the ACH was a common assumption in the ecology and evolution literature. Subsequent tests of the ACH of NEP rocky intertidal species have been also shown to be equivocal (Sagarin and Gaines, 2002b; Fenberg et al., 2014), and, at least in one case, an observed abundant centre is thought to be partly due to human impacts. This case is the owl limpet, *Lottia gigantea*, which is an ecologically important space occupier in the mid-/high-intertidal zone (Stimson, 1970; Lindberg et al., 1998; Fenberg and Rivadeneira, 2011; Fenberg, 2013). It ranges along the Pacific coast of central Baja Sur (26.05°N) to Mendocino County in northern California (39.25°N). Individuals are highly abundant towards the centre of its geographic range (southern California/northern Baja), where *L. gigantea* populations have experienced a long history of size-selective harvesting pressure (Kido and Murray, 2003; Sagarin et al., 2007; Erlandson et al., 2011; Fenberg and Roy, 2012). Females are highly territorial, defending their algal 'garden' from intraspecific (usually smaller males) and interspecific competitors (Stimson, 1970; Wright, 1989). Territory size (i.e., home range size) is highly correlated with shell size (Fenberg, 2013). Thus, the number of individuals physically able to occupy a given area will be higher for harvested populations because small- to medium-sized individuals have smaller home ranges/territories and are not generally harvested. Therefore, harvested populations (primarily in the range centre) should have a higher density of individuals compared with populations that do not experience the effects of harvesting, contributing to the observed abundant centre pattern (Fenberg and Rivadeneira, 2011).

Regardless of whether species exhibit an abundant centre, a more meaningful question to ask is how range-wide abundance patterns are driven by biotic, abiotic and increasingly, anthropogenic factors (see earlier) over space and time (Sagarin et al., 2006). Understanding the causes of species' abundance patterns will give us valuable insights to basic yet poorly understood concepts in ecology and evolution, such as: (1) what causes species range limits?; (2) how does the shape of a species' range-wide abundance pattern influence probability of range expansion and/or contraction over time?; (3) do ecosystem processes (e.g., nutrient cycling, predator–prey interactions) vary across spatial gradients in concert with range-wide abundance patterns of keystone or foundation species? and (4), how do ecological and evolutionary traits such as body size, trophic level, dispersal potential and phylogenetic relationships interact with physical variables like sea surface temperature (SST) and habitat availability to shape range-wide abundance patterns?

Focussing such questions on species restricted to a single habitat type (e.g., rocky intertidal) across a large spatial gradient, such as the NEP, will provide ecologists with a useful model system to inform hypothesis testing across other habitat types and taxonomic groups. Indeed, much of the raw biological and physical data are already available from long-term monitoring and biodiversity survey projects, most notably from the Multi-Agency Rocky Intertidal Network (www.marine.gov) and the Partnership for Interdisciplinary Studies of Coastal Oceans (www.piscoweb.org). These consortia have surveyed (and resurveyed) over 100 rocky intertidal sites across the NEP, from central Baja to south-east Alaska. These ongoing surveys are used for monitoring species of conservation concern; (Miner et al., 2006; Sagarin et al., 2007; Crosson et al., 2014; Jurgens et al., 2015; Raimondi et al., 2015) and for providing the baseline information for large-scale biogeographic studies (Blanchette et al., 2008; Fenberg et al., 2015; Rivadeneira et al., 2015).

## 10.4 Northern Range Limits and Dynamics

Identifying the biological and physical factors that drive species range limits has been an ongoing focus of ecology and evolutionary biology (Sexton et al., 2009). This emphasis is partly because fundamental patterns of the biosphere, including global patterns of diversity and biogeographic structure are a function of the geographic distribution of species range limits. Thus, a general understanding of the factors that influence range limits are needed to explain these patterns and their macro-evolutionary origins. In a more contemporary sense, the causes of range limits are also needed to understand the susceptibility of species' ranges to alteration in response to modern climate change and other human impacts (e.g., habitat alteration). Given their relatively simple geographic ranges, NEP rocky intertidal species have become useful model organisms for such studies over the past decade, and some interesting patterns are beginning to emerge.

Broadly speaking, there are three causal mechanisms that may be used to explain range limits (Holt, 2003; Dawson et al., 2010): physical barriers to dispersal, genetic impoverishment and migration load. For NEP rocky intertidal species, physical barriers to dispersal are usually associated with habitat gaps (e.g., long stretches of sandy beach), areas of intense upwelling and/or abrupt SST changes (e.g., Point Conception and Cape Mendocino in California and Punta Eugenia in Baja). Some of these locations are well-known biogeographic and phylogeographic breaks, meaning that they often coincide with species range limits and population genetic divergence within species (Blanchette et al., 2008; Kelly and Palumbi, 2010; Fenberg et al., 2015). Whether these locations coincide with a species' border or phylogeographic break partly depends on its dispersal capability (Wares et al., 2001; Gaines et al., 2009; Pelc et al., 2009; Fenberg et al., 2015). Assessing the relative importance of which variables limit dispersal at the range edge (and, hence, set range limits) can be especially challenging for species for which their ranges end in locations without obvious physical barriers. Range edge recruitment success (particularly for species with pelagic larvae) may be subject to a complicated set of species-specific interactions between nearshore oceanic processes and factors such as the density of source populations, ability for self-recruitment, coastal geography, habitat availability, thermal tolerances, season and post-settlement survival. Gaines et al. (2009) provide a nice overview of dispersal and geographic range limits with good examples from the NEP rocky intertidal (focussing on Point Conception). Therefore, we will build on their review by largely focussing on a few case studies published from 2010 onwards, particularly for species that have experienced dynamic northern range limits where there are no obvious physical barriers to dispersal.

The northern range limit of the barnacle *Tetraclita rubescens* has expanded by at least 200 km since the 1970s, from San Francisco (37.8°N) to the Oregon/California border (42°N; Dawson et al., 2010). During this period of expansion, individuals also increased in abundance within the expanded portion of its range (Dawson et al., 2010). Overall, however, this species is substantially more abundant in the central and southern portions of its range (Sagarin and Gaines, 2002a; Blanchette et al., 2008). Using these abundance and distribution data in combination with a population genetics and theoretical framework, Dawson et al. (2010) have inferred the most likely causes of the range limit and recent expansion. They argue that a physical barrier to dispersal is not the underlying cause of its range limit considering the recent range expansion, the largely uninterrupted extent of rocky intertidal habitat in the expanded range, the long dispersal potential (18–26 days), lack of genetic population structure and high genetic diversity. They also reject genetic impoverishment as a potential cause. Genetic impoverishment may cause range limits if immigration to the range edge population(s) is low and must therefore be sustained by self-recruitment. Populations sustained by self-recruitment may have limited genetic variance needed to adapt to novel conditions (i.e., beyond the limit; Holt, 2003). In addition, range edges subjected to genetic impoverishment would be characterised by small peripheral populations with low genetic diversity and high genetic differentiation compared to more central populations. *T. rubescens* has high genetic diversity and lack of population structure found throughout its range, thus genetic impoverishment is thought to be an unlikely cause for the limit.

Rather, Dawson et al. (2010) suggest that migration load is the more likely candidate. If range edge populations are demographically dependent on immigration from range centre populations, then an influx of locally maladapted alleles may inhibit the opportunity for selection at the range edge, limiting the genetic scope for adaptation beyond the range edge. This genetic constraint is commonly referred to as 'migration load' (Dawson et al., 2010). Migration load would be observable if range edge populations are genetically diverse and weakly differentiated from the rest of the range. Range edge populations experiencing migration load would need to somehow overcome the flow of maladaptive alleles and attain the genetic capacity to adapt to the presumably harsh physical or biological environment beyond the range edge for expansion to occur. If, however, the environment beyond the range limit changes enough that adaptation is no longer needed, then range expansion is more likely.

The post 1970s range expansion of *T. rubescens* coincides with a SST warming of ~0.8°C along the California coast (Sagarin et al., 1999). Throughout its range, *T. rubescens* is most abundant in the relatively warm waters of southern California and Baja Mexico (Sagarin and Gaines, 2002a; Blanchette et al., 2008), suggesting that it is well adapted to warmer conditions. Thus, modern warming beyond the historical range edge would help facilitate a range expansion without the need to overcome any maladapted alleles that may have precluded populations to expand prior to the 1970s. If climate change relaxes the negative impacts of migration load (and migration load is a common cause of range limits), then, all else being equal, poleward range expansions are to be expected for NEP rocky intertidal species and generally across habitat types.

Although migration load is argued to be the primary cause of the northern range boundary for *T. rubescens*, a closer look at its range-wide abundance pattern reveals that there could also be at least one physical barrier to dispersal at play. A steep drop in the percentage cover of *T. rubescens* occurs at the northern end of Monterey Bay at Santa Cruz (36.95°N; Figure 10.2). From Santa Cruz to the modern limit, *T. rubescens* individuals average a cover of 0.0004 per cent, compared to 0.01 per cent in the central and southern portion of its range (data courtesy of www.pacificrockyintertidal.org). Assuming a predominately unidirectional flow of larvae from central to northern populations, then Monterey Bay could be a 'soft' barrier to dispersal – limiting, but not eliminating immigration towards the northern range edge. Migration load and the soft barrier may therefore act as a two-pronged cause for the range limit. Interestingly, *T. rubescens* (along with other 'southern' species) has increased in abundance at the southern end of Monterey Bay (at Hopkins Marine Station) in recent decades, presumably due to climate warming (Sagarin et al., 1999). This increase in adult abundance should also be mirrored by an increase in larval abundance around Monterey Bay; effectively increasing the chances for larvae to breach the soft barrier, providing more recruits for northern populations and increasing opportunities for range expansion.

Monterey Bay also appears to have an effect on the abundance and recruitment of northern populations of at least two other NEP rocky intertidal species. Like *T. rubescens*, the abundance (adults and juveniles) of the limpets *L. gigantea* and *Lottia scabra* drops sharply from Santa Cruz (at the northern end of Monterey Bay; Figure 10.2) to their respective northern limits (Gilman, 2006; Lehman, 2010; Fenberg and Rivadeneira, 2011; Shanks et al., 2014). Shanks et al. (2014) suggest that the drop in abundance and recruitment north of Monterey Bay is linked with a northern edge of where the kelp, *Macrocystis pyrifera*, form extensive beds. Sites associated with *M. pyrifera* beds (south of Santa Cruz) have higher juvenile and adult abundances compared to their northern ranges, a pattern mirrored by *L. gigantea* (Fenberg and Rivadeneira, 2011). Shanks et al. (2014)

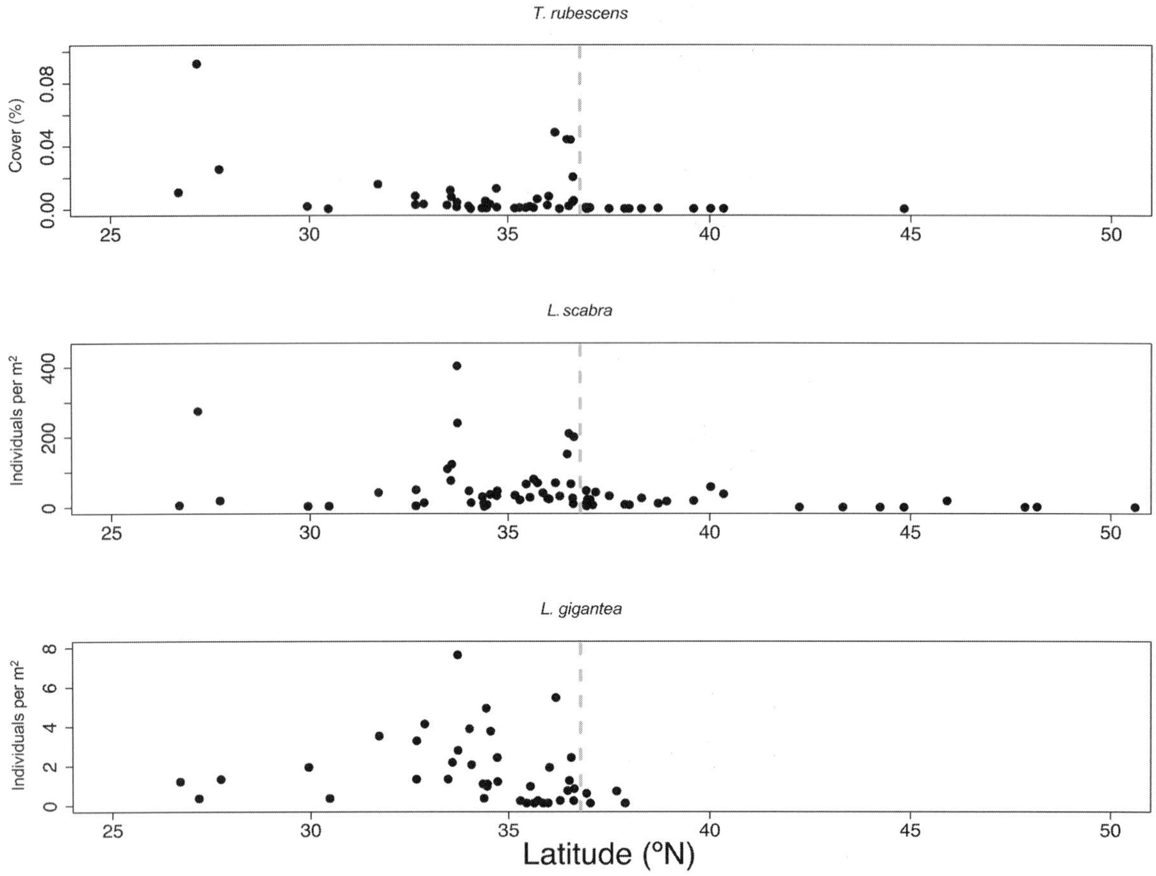


**Fig. 10.2** Range-wide abundance patterns of *T. rubescens, L. scabra* and *L. gigantea* (data courtesy of www.pacificrockyintertidal.org). The dashed lines represent Monterey Bay (~36.8°N) where abrupt declines in abundance coincide for each species leading up to their respective northern range limits.

hypothesise that *M. pyrifera* beds act to limit alongshore flow (Gaylord et al., 2007, 2012; Nickols et al., 2012) and, therefore, help increase the local retention of larvae. Without these beds, larvae may be transferred away from suitable rocky shore habitat, leading to high larval wastage. The kelp *Nereocystis luetkeana* is more commonly found north of Santa Cruz, but Shanks et al. (2014) suggest this alga is less capable of restricting offshore flow of larvae because (1) it has only one stipe (compared to multiple stipes of *M. pyrifera*), (2) they generally form smaller beds closer to shore and (3) they almost completely die back during the winter (which, at least for *L. gigantea,* happens to coincide with the spawning season; Kido and Murray, 2003). If further study supports this *M. pyrifera*/larval retention hypothesis, then it would suggest that the northern ranges of these NEP limpets (and possibly *T. rubescens* and other similarly ranging species) could be modulated, at least in part, by this biological interaction. Alternatively, the long sandy beach of Monterey Bay (~50 km) could act as a partial dispersal barrier for the limpets, which have relatively short pelagic larval durations (~6–14 days; Kay and Emlet, 2002; Dawson et al., 2014).

Much of what we know about the potential causes of range limits of NEP species come from research on the northern limits of a handful of species that primarily range in the central and northern portions of the NEP (California, Oregon, Washington), where, not coincidentally, many of the rocky shore ecologists tend to live. Far less is known about the drivers of northern limits of NEP species that primarily range into the southern part of the NEP along the Pacific coast of Baja California. For example, the historical range of the predatory gastropod *Mexacanthina lugubris* used to be entirely confined to Baja California, from Ensenada (31.9°N) to Magdalena Bay (24.6°N). Much like *T. rubescens* and other NEP species (e.g., *Kelletia kelletia*; Zacherl et al., 2003), the range limit of *M. lugubris* has expanded northwards by ~300 km into southern California (33.5°N) since the 1970s (Fenberg et al., 2014). While the range expansion could be a result of warming temperatures and individuals are simply following the edge of their thermal tolerance, other ecological, physical and anthropogenic factors also likely contribute to its recent northern range expansion.

Phylogeographically, the range of *M. lugubris* can be divided into well-defined northern and southern clades, separated by a break (estimated divergence time ~417 Kya) in the mid-peninsular region at Punta Eugenia (27.8°N). This well-defined genetic structure is thought to be due in part to its low dispersal potential (juveniles emerge directly from egg capsules attached to rock). Interestingly, the break is geographically concordant (although divergence times vary) with other phylogeographic breaks for a diverse set of marine and terrestrial species (see later, this chapter). For *M. lugubris*, the northern clade can be characterised by low haplotype and nucleotide diversity, especially in the expanded portion of its range (which is genetic evidence for the expansion) and high local abundances at the historical northern limit. Northern clade populations reside in a region of extensive rocky shore, especially in the vicinity around its historical northern limit, which coincides with a stretch of nearly 100 km of largely uninterrupted rocky shore. Moreover, its primary prey, the barnacle *Chthamalus fissus* is densely populated within the northern clade, again, particularly at the historical northern limit of *M. lugubris*. Fenberg et al. (2014) suggest that the wealth of habitat/food availability at the historic northern limit has helped to facilitate the high local abundance of *M. lugubris*, providing a strong source population that fuelled the northward expansion.

Much like *T. rubescens*, where high and increasing population densities towards its historical northern range edge may have facilitated the expansion, high abundances at the historic northern limit of *M. lugubris* in conjunction with a warming climate and bountiful habitat and food availability may have led to its range expansion. Importantly, however, warming temperatures alone will not necessarily cause a range edge expansion. Whether or not the range of a species expands will also depend upon its underlying meta-population dynamics, range-wide abundance patterns, and habitat and food availability, which can be highly species specific and geographically dependent (Fenberg et al., 2014). For example, intensive human size-selective harvesting of the highly competitive space occupier, *L. gigantea*, may have had an indirect positive effect on the abundance of *M. lugubris* at its historic northern limit. Size-selective harvest of *L. gigantea* could lead to competitive release, allowing other (non-harvested) space competitors like *C. fissus* to become more common. This increase in prey density might subsequently allow its predator, *M. lugubris*, to become more abundant. Thus, the anomalously high local abundances at its historic northern limit and even its subsequent range expansion may be an indirect and partial result of non-climate-related human impacts on species interactions.

The interaction between direct and indirect effects that result in the expansion of a species' range are likely to have ecological and evolutionary impacts for other species as well. These impacts may be particularly noticeable if the expanding species is abundant and a predator, like *M. lugubris*. For example, *C. fissus* alters the shape of its operculum openings when exposed to intense predation pressure by *M. lugubris* (Jarrett, 2009). Morphs with a narrow or bent operculum are more difficult for *M. lugubris* to prey upon. These defensive morphs are significantly more common at the historical northern range limit of *M. lugubris* where both species have high densities. This suggests that high densities of *M. lugubris* drives morphology of its prey. Accordingly, the defensive strategy is not common at a site in the expanded range, where *M. lugubris* abundance is not as high (Jarrett, 2009). But if *M. lugubris* increases in abundance within its expanded range in the coming years as expected, then it would logically follow that the defensive *C. fissus* morphs will become more common as well. If selection is intense and geographically widespread enough, then this predator–prey interaction may select for microevolutionary change in *C. fissus* morphology within the co-distributed portion of their range.

To put into perspective what we have learned about northern range expansions, it is useful to contrast it with examples of northern range contractions (or local extinctions towards the range edge). For example, both the pelagically dispersing limpets, *L. gigantea*, and *L. scabra*, have very low adult and juvenile abundances towards their respective northern range edges (see earlier). The size structure of their northern populations (north of Monterey Bay) tend to be heavily skewed towards medium- to large-sized individuals at low densities. Similarly, populations within the central range of *Tegula funebralis* (northern Baja California and southern California) consist of high densities of juveniles and adults, whereas more northerly populations (Oregon to Vancouver Island) have a patchy distribution, variable local abundances and recruitment appears to be more sporadic (Frank, 1975; Sagarin and Gaines, 2002a; Cooper and Shanks, 2011). Because individuals can live for twenty years or longer (Paine, 1969; Frank, 1975; Cooper and Shanks, 2011; Fenberg and Roy, 2012) and juveniles are rare, it is thought these northern populations are demographically unstable and prone to local extinctions. Studies of other NEP gastropods (e.g., *Haliotis* spp.) with similar reproductive life histories show that close proximity of reproductive individuals is key for successful fertilisation (Catton and Rogers-Bennett, 2013). Thus, low densities at the range edge should be related to low population fitness and increased probability of range contraction and localised extinctions due to Allee effects (Courchamp et al., 2008). Accordingly, the range of *L. gigantea* has contracted by a least two degrees of latitude in the last few decades and *L. scabra* has experienced a number of local extinctions at its northern range edge (Fenberg and Rivadeneira, 2011; Shanks et al., 2014).

## 10.5 Southern Range Limits and Dynamics

In stark contrast to what we know of northern range edge dynamics for some NEP species, there are very few case studies that have examined southern range edge dynamics. This is partly because climate change research generally predicts a northward range shift for NEP species, meaning that northern range edges have been more extensively monitored. In addition, many NEP species have their southern limits clustered within a relatively small and largely inaccessible region of the central/southern Baja coastline, between Punta Eugenia (27.8°N) and Magdalena Bay (24.6°N). For example, the ranges of *L. gigantea*, *L. scabra*, *T. rubescens*, *T. funebralis* and *M. lugubris* (already discussed), and many others, end around Magdalena Bay. A review of the geographic ranges of NEP rocky intertidal gastropods (not including meiofaunal species, those primarily living subtidally or in multiple habitat types) reveals no species that range into the TEP and about half the species in the data set have southern range limits in the region between Punta Eugenia and Magdalena Bay (Fenberg and Rivadeneira, 2019). Among molluscs, Magdalena Bay tends to be more commonly associated with the southern range limits of rocky intertidal gastropods whereas Punta Eugenia (or Isla Cedros) is more often cited as a southern limit for bivalves (which tend to live in more soft-sediment habitats; Bernard et al., 1991). Interestingly, Punta Eugenia is also a well-known phylogeographic break for a number of rocky intertidal species, including isopods, fish and molluscs (Riddle et al., 2000; Bernardi et al., 2003; Riginos, 2005; Hurtado et al., 2010; Fenberg et al., 2014; Dolby et al., 2015). The region between Punta Eugenia and Magdalena Bay may therefore be described as both a biogeographic and phylogeographic transition zone across a diverse set of taxa, acting as a barrier to (1) the movement of individuals between populations of some species, leading to genetically distinct lineages (i.e., a phylogeographic break) and (2) the movement of individuals further south into the TEP.

Despite the concordance of these biogeographic and phylogeographic patterns, relatively little research has been dedicated to their causes. Biologically and physically, the Pacific coast of the Baja peninsula is a story of two halves, with the region near Punta Eugenia marking the dividing line. At this latitude, there is a transition from mid-latitude to tropical conditions, offshore flow of the southward-flowing California Current, cyclonic eddies north and south of Punta Eugenia, and variation in upwelling regimes (Hewitt, 1981; Zaytsev et al., 2003; Herrera-Cervantes et al., 2014). On average, SST is 3.5°C warmer along the southern half of Baja California compared to the northern half. But variation in oceanic conditions is not the whole story. Only 25 per cent of the coastline south of Punta Eugenia to the tip of the peninsula is habitable for rocky intertidal species compared to 51 per cent for the northern half (data from Fenberg et al., 2014). The rather abrupt transition from a cool (with multiple areas of upwelling) and mostly rocky coastline in the north to one with warm waters dominated by sandy beaches (including a 115 km uninterrupted stretch of sandy beach just north of Magdalena Bay) makes the Pacific coast of southern Baja challenging in terms of both thermal stress and dispersal opportunities for NEP rocky intertidal species. The combined effect of these physical factors should correspond to abrupt declines in adult and juvenile abundance for NEP species with their southern range edge in this region, as observed for *L. gigantea* (Fenberg and Rivadeneira, 2011).

Given this, we should not expect southern range limit expansions of NEP species with limits between Punta Eugenia and Magdalena Bay as a potential result of climate change. Rather, we would expect these limits to either be temporally stable or to contract northwards (Fenberg et al., 2014). On the other hand, if warm-adapted TEP rocky intertidal species from the south (and in the Gulf of California) can overcome the habitat barriers, then we might hypothesise a gradual northward shift of the boundary between tropical and temperate fauna over the next century as warming progresses. Indeed, some of the warm protected bays along the Pacific coast of southern and central Baja California already act as northern refugia for some TEP species (Reid, 2002). This provides a source population from which to disperse when/if the outer coast becomes more thermally suitable (e.g., during El Niño years). But if TEP species are not able to overcome the habitat barriers, then we might expect a net loss of biodiversity at this northern edge of the tropics.

## 10.6 Range Dynamics over Paleontological Time Scales

This discussion has largely focussed on range dynamics over contemporary time scales. But we cannot make any general conclusions without taking into account processes occurring over much longer time scales. For example, classically it has been assumed that the poleward ranges of species from cool, temperate regions, such as those occurring in the northern part of the NEP, follow a basic 'expansion–contraction' (EC) model in response to respective interglacial and glacial periods. However, individualistic (i.e., species-specific) responses to environmental change may be more common than initially expected (Valentine and Jablonski, 1993; Stewart and Lister, 2001). Reliance on the fossil record alone to document NEP rocky intertidal range shifts over the Pleistocene can be problematic because (1) the relatively high-energy environment of the rocky shore is not ideal for preservation, (2) many Pleistocene records are biased towards relatively warm interglacial periods and (3) mixing of fauna from different time periods within individual fossil terraces may be more common than originally thought (Muhs et al., 2014). Rather, a combined approach that takes into consideration the fossil record, molecular phylogeography and other data sources is preferred. For example, Hellberg et al. (2001) use molecular, fossil and morphological evidence to support a late-Pleistocene northern range expansion in response to warming waters after the most recent glacial retreat for the direct-developing rocky intertidal gastropod, *Acanthinucella spirata*. While this supports the EC model (and climate change-mediated morphological evolution), subsequent tests of other NEP rocky intertidal species are equivocal. Although only based on mtDNA markers, Marko et al. (2010) found that only 5 of their 14 sampled NEP rocky intertidal species showed molecular evidence of a postglacial northern range expansion. Interestingly, the mussel *Mytilus californianus* exhibits evidence of persistence in British Columbia and south-east Alaska during the last glacial maximum (LGM; ~20,000 years ago), while its main predator, the keystone species, *Pisaster ochraceus*, was likely absent from the region during this period (Marko et al., 2010). These contrasting histories must have had cascading effects for northern communities during the LGM. For example, without predation pressure setting its lower limit (Paine, 1966), *M. californianus* may have been able to exist lower on the shore and even subtidally (Paine, 1976; Marko et al., 2010). The subtidal refuge in turn would have helped ameliorate the undoubtedly cold/freezing aerial temperatures during the LGM and helped sustain its presence in the northern region. This scenario of an interaction between the physical environment (e.g., temperature/glaciation) differentially affecting the northern range histories of *M. californianus* and *P. ochraceus* in conjunction with the predation release facilitating a potential subtidal refuge for *M. californianus* highlights the complicated factors that shape NEP rocky intertidal communities over long time scales – which nicely sets up the next half of this chapter.

## 10.7 Human Impacts on Top-Down Control of NEP Rocky Intertidal Community Structure

Robert Paine's classic predator-removal experiments in northern Washington State showed that ecological communities can be structured, in part, by top-down control (Paine, 1966). As the original 'keystone' species, Paine showed that *P. ochraceus* can significantly affect the diversity, abundance and distribution of species in the community. As a voracious predator, *P. ochraceus* prevents the space occupier, *M. californianus*, from dominating the rocky intertidal. If *P. ochraceus* is removed, then *M. californianus* expands the cover of its beds to lower-intertidal levels, edging out individuals that predominantly live on rock. However, many of these species can also be found living on or within the mussel beds, usually as juveniles or small adults, albeit at different abundances (Lohse, 1993; Lafferty and Suchanek, 2016).

This pioneering research not only highlighted the utility of the NEP rocky intertidal as a natural laboratory for ecological research, but Paine's ideas have since been confirmed for other keystone species in various marine and terrestrial systems (e.g., Mills et al., 1993; Power et al., 1996) – showing that single species (usually predators) can have an outsized effect on the overall structure and functioning of an ecosystem. This extensive body of research gives us a framework with which to make predictions on how direct or indirect human impacts on keystone species may have cascading ecological effects on the rest of the community. These days, the most worrisome indirect impacts on NEP rocky intertidal species are those associated with climate change and ocean acidification. Direct human impacts, on the other hand, usually take the form of size-selective harvesting. For the remainder of this section, we focus on how indirect and direct human impacts on species capable of top-down control have affected the NEP rocky intertidal.

## 10.8 Predator Responses to Altered Environments: Implications for Climate Change

The impacts of climate change on ecological communities are difficult to predict because indirect effects can generate ecological surprises resulting from complex webs of direct interactions (Gilman et al., 2010). However, ecological theory is available upon which to begin building a conceptual framework for expected change. Specifically, environmental stress theory makes predictions of how predators and prey should respond to increases in, for example, thermal stress (Menge and Sutherland, 1987; Menge and Olson, 1990; Menge et al., 2002; Bruno et al., 2003). The consumer stress model (CSM) assumes that predators are more sensitive to stress than are prey, while the alternative Prey Stress Model (PSM) assumes the opposite, that prey are more sensitive than are predators. Both versions assume that consumers feed primarily on sessile prey. Thus, the CSM predicts that, with increasing stress, predator effects on prey will weaken, while, in the PSM, increasing stress should strengthen the effects of consumers on prey.

A primary effect of climate change will be through altered temperature regimes (IPCC, 2014), which can impose physiological stress on organisms. Coastal temperatures along the US west coast have increased by up to 3°C across the continental shelf (Mote et al., 2010), while inner shelf temperatures have decreased (Menge, unpublished data) due to intensifying upwelling resulting from increases in temperature over land (Iles et al., 2012). Thus, intertidal organisms face stress at both ends of the temperature gradient (Sanford, 2002). In the next section, we provide examples indicating how such temperature stresses may affect top-down and bottom-up processes in rocky intertidal communities.

## 10.9 Temperature Stress

### 10.9.1 Effects on Whelks

Rocky intertidal species are exposed to both water and aerial temperatures on a daily basis, making them susceptible to the combined effects of oceanic and atmospheric climate change. But whether a particular species will respond differently to aerial versus water temperature may depend upon a number of factors, including vertical distribution on the shore, the timing of tidal cycles on a daily or seasonal basis, and behavioral and physiological response during periods of emersion/immersion. How water and aerial temperatures (and their subsequent influence on body temperature) affect species interactions, such as predation rates, will be a key step towards understanding how climate change affects top-down processes in the rocky intertidal. A laboratory study by Yamane and Gilman (2009) found that the growth and feeding rate of the predatory gastropod, *Nucella ostrina*, was low for individuals with high emersed body temperatures. This is in contrast to the observed increased feeding and growth rates for individuals in warm immersion temperatures. This observation suggests that the primary prey of *N. ostrina*, the barnacle *Balanus glandula*, may suffer higher mortality during periods of warm immersion, but a respite from predation during periods of warm emersion (Yamane and Gilman, 2009).

Despite the often clear results, such studies often leave us wondering whether they correlate with more realistic temporal environmental fluctuations and extreme events. In other words, most laboratory studies test for the effects of constant temperature treatments while ignoring patterns of temperature variability (Thompson et al., 2013). Taking such systems into the field can directly address this issue. For example, a field study of *N. ostrina* in Oregon revealed that whelk feeding and metabolic rates were both elevated by increased temperatures (Dahlhoff et al., 2001). Thermal stress (low-tide air temperature), whelk body temperatures and biochemical indices of whelk conditions varied with site, wave exposure and prey abundance, indicating that physiological stress interacts with prey abundance to modify foraging activity of whelks.

Although Dahlhoff et al. (2001) showed that whelk foraging and physiology were responsive to environmental stress (warm temperatures), the extent to which this altered impacts on prey populations was unclear. In another field experiment, Menge et al. (2002) tested the effects of thermal stress on a whelk–barnacle interaction by modifying the local thermal climate. The technique used cages containing rocks with barnacles on them that were fastened to concrete blocks. Treatments included one cage without and two with whelks, one of which had a small shelter fastened to the concrete block inside the cage. The experiment was run twice, once during a warm, sunny summer and once during a cooler, rainy summer. Besides the provision of the small shelter, the CSM was tested by placing cages on sunny and shady sides of blocks, and on high (more stressful) and low (less stressful) zones on the shore. Results showed that, as expected, whelk predation was reduced in the absence of shelter on the sunny side of the blocks. Surprisingly, height on the shore had no effect and predation rate was actually higher during the warmer summer than during the cooler summer, suggesting, as was observed by Dahlhoff et al. (2001), that, at least up to a point, whelk foraging activity is enhanced by warmer temperatures. In a simultaneous experiment, limpet grazing on algae showed identical, and even stronger effects of thermal stress on grazing rate (Menge et al., 2002).

### 10.9.2 Effects on Sea Stars

The ochre sea star *P. ochraceus* is a key top-down force in communities along the west coast of North America. Part of its dominant role is based on its relatively large size and its ability to adhere powerfully to rock surfaces and prey. Adults can reach large sizes, ranging up to several kilograms in wet mass, and, as a consequence of its apparent robustness, ecologists were slow to investigate its sensitivity to thermal stress. The first, and most surprising, study was that of Sanford (1999). Taking his cue from observations that *P. ochraceus* foraging activity and height on the shore appeared to cycle in concert with upwelling events, he carried out field and laboratory experiments to test the effects of water temperature on sea star feeding rates. Since upwelling brings cold water to the surface, he hypothesised that feeding intensity on transplanted clumps of mussels would decline during upwelling (cold) and increase during relaxation or downwelling (warm) events. Results supported his hypothesis. In both field and lab experiments, *P. ochraceus* fed slower in cold periods (or controlled cold sea water tables), faster in warmer periods (or warmer sea water tables) and, in the laboratory, shifted from fast to slow feeding rates as the water temperature was shifted from warmer to cooler. In a separate experiment, he showed an identical effect in whelks *Nucella canaliculata*. Based on these data, he proposed a conceptual model suggesting that feeding rate gains and metabolic rate losses with increasing temperature led to an optimal range of 8–15.5°C for obtaining energy for growth and reproduction.

In a later study, Petes et al. (2008) tested the effect of height on the shore, and, thus, presumably, of thermal stress on *P. ochraceus* and *M. californianus*. Transplanted mussels were caged with and without sea stars, and survival, reproduction (mussels only), growth and heat-shock proteins (HSPs) were sampled in summer. Results indicated that, as predicted by the CSM, sea stars were stressed more than their prey. High caged sea stars all died, while low ones survived. Low caged sea stars without food had elevated Hsp70 relative to fed sea stars. In contrast, all mussels in sea star cages survived. Those in high cages reproduced earlier in summer, and had elevated Hsp70 relative to low caged mussels. Hence, although exposure to more stressful conditions high on the shore clearly stressed both predator and prey, prey survived while predators died, indicating that prey endured stress better than predators.

More recently, Monaco et al. (2016) examined the *P. ochraceus* vs *M. californianus* interaction using a physiological and behavioural approach. Using biomimetic sea stars and mussels (i.e., field-deployed models of each with thermistors embedded within them that mimicked thermal changes of living animals through time), they modelled thermal performance curves for air and water. They found that thermal 'performance' of *P. ochraceus* was higher than that of their main prey, *M. californianus*. That is, the sea star seemed to tolerate thermal stresses better than did the mussel. They suggested that this result was more consistent with the PSM than with the CSM, and that the difference between their result and Petes et al. (2008) was that the cages used in the latter prevented behavioural responses (e.g., moving to crevices for low tides) by the sea stars. This interesting result, however, did not involve a test of the relative ecological performances of sea stars and mussels. In a CSM scenario, the predicted outcome is that prey survival is better in stressful than in benign conditions, while in a PSM scenario prey survival is worse in stressful than in benign conditions (Menge and Olson, 1990). Previous tests of predation rates along a wave-stress gradient in one of the same sites used by Monaco et al. (Strawberry Hill, Oregon; Menge et al., 1994) were consistent with the CSM model predictions. Thus, the Monaco et al. (2016) study suggests that the physiological mechanism proposed by the CSM (See Section 10.8) may be incorrect. Instead, the primary mechanism underlying the differences in CSM vs PSM may be that, unlike mobile consumers, sessile prey do not have a behavioural capacity (e.g., seeking low-stress sub-habitats) to avoid stress.

To what extent are predation rates (or other species interactions) affected by the level of temporal coincidence between aerial and water temperature values? That is, how are predation rates affected if physiologically stressful aerial and water temperatures occur during the same period (temporal coincidence) versus following in time but not overlapping (non-coincidence)? In a laboratory experiment, Pincebourde et al. (2012) found that per capita feeding rates of *P. ochraceus* were at their lowest if a period of warm aerial temperature was immediately followed by a period of warm water temperature (non-coincidence). They suggest that the negative physiological effect of warm aerial temperatures will last longer and recovery will be hampered if followed by a period of warm water. On the other hand, if a cold water period occurs after a period of aerial temperature stress, then individuals may be able to compensate or recover more quickly. Understanding the interaction of the timing of multiple stressors on individual species and their communities is especially important in the context of climate change.

In a field study, Szathmary et al. (2009) investigated the effect of aerial temperature exposure during low tide on foraging activity on succeeding days and found that the number of foraging *P. ochraceus* was inversely related to the average maximum temperature on the previous day. Thus, as suggested by Sanford's (2002) model, *P. ochraceus* foraging activity is inhibited by thermal extremes at both ends of the range, hot and cold. Consistent with this result, in laboratory experiments, Pincebourde et al. (2008) found that *P. ochraceus* growth rate also varied inversely with temperature. Clearly, this fearsome, robust predator is sensitive to the thermal climate. Since regional air temperatures are increasing and water temperatures are cooling due to increased upwelling (Iles et al., 2012), it seems inevitable that sea star activity patterns will change, with overall declines in predation impact.

While these remarks suggest that, in the short term, *P. ochraceus* (and likely whelk) predation will decrease, the implication of these studies is that such changes will be subtle. However, some conditions could engender catastrophic changes comparable to those induced by Paine (1966, 1974) and others in removal experiments. Between 2013 and 2014, massive die-offs were documented in *P. ochraceus* populations (and other sea stars) at many NEP rocky intertidal sites (Stokstad, 2014; Jurgens et al., 2015; Menge et al., 2016; www.pacificrockyintertidal.org). Over the course of a few days, individuals developed lesions, lost the structural integrity of their limbs, changed their behavior and ultimately died. Field observations showed that larger individuals ('adults') developed wasting symptoms before juveniles, but, as the former disappeared, rates of disease increased in juveniles as well (Menge et al., 2016). Similarly, laboratory studies of individuals from the San Juan Islands in Washington indicate that larger individuals developed signs of wasting before juveniles, but that juveniles died at a faster rate once they showed symptoms (Eisenlord et al., 2016). This 'sea star wasting syndrome' (SSWD) has been linked to a virus called sea star-associated densovirus (Hewson et al., 2014). Although the ultimate cause of the outbreak is still unclear, some evidence implicates warming as a factor (Blanchette et al., 2005; Stokstad, 2014; Eisenlord et al., 2016). For example, previous outbreaks recorded in southern California were associated with warming events (e.g., the 1997 El Niño event; Blanchette et al., 2005). Eisenlord et al. (2016) found that individuals from a warm temperature treatment had higher mortality rates than those in cooler treatments, suggesting that the disease progresses faster in warm water (also observed by Kohl et al., 2016). On the other hand, field sites from Oregon through central California had cooler water temperatures during the peak of SSWD in 2014, making it somewhat difficult to confirm if temperature is a driver of the die-offs (Menge et al., 2016). Currently, researchers are pursuing a multifactor explanation of SSWD, including temperature and possibly ocean acidification as possible factors interacting to produce this epidemic. Clearly, however, some factor altered the interaction between sea stars and what had previously been a benign microbe with disastrous consequences.

Regardless of how much temperature influences the prevalence of wasting, we can draw conclusions from Paine's research (and that of others: Menge et al., 1994; Robles et al., 2009) that the depletion of these voracious predators would cause significant cascading effects on local ecological communities. Without *P. ochraceus*, *M. californianus* beds would likely expand their vertical distribution towards the lower shore and potentially to the shallow subtidal in some cases. However, whether or not such a response will be universal, i.e., occur across the full biogeographic range of coexistence between *P. ochraceus* and *M. californianus*, is unclear. Some *P. ochraceus* removal experiments did not result in swift takeover of the lower intertidal by mussels (S. Gravem, personal communication;Menge, unpublished data). Such variable responses seem at least partly related to direct and indirect effects of ocean conditions, including upwelling strength and magnitude of prey recruitment (Menge et al., 2004). In southern California, where upwelling is weak, predation rates were low (Menge et al., 2004). North of Point Conception, predation rates varied among sites as a function of sea star abundance, density of mussel recruitment and the level of phytoplankton available as food for filter-feeding mussels and barnacles. Recruitment and phytoplankton blooms are strongly dependent on upwelling conditions, generally being greatest at sites of intermittent upwelling (Menge et al., 2011; Menge and Menge, 2013).

Studies of indirect human impacts on species capable of top-down control are often focussed on predators, but, of course, indirect human impacts on species at intermediate levels of the food chain may also affect community structure. A good example from NEP rocky shores is the community level impacts of withering syndrome (WS) on the black abalone, *Haliotis cracherodii*. WS is caused by a gastrointestinal prokaryote that hinders digestion, which is known to occur in all five southern California abalone species (Friedman et al., 2002; Crosson et al., 2014). The primary symptom of infected animals is the shrinking and weakening of the foot, which can cause dislodgement from the substratum and death (Crosson et al., 2014). Like wasting disease in *P. ochraceus*, WS is thought to be induced, in part, by elevated water temperatures (Raimondi et al., 2002). Ben-Horin et al. (2013) measured the body temperatures of *H. cracherodii* individuals in the field and in the laboratory. They found that high daily temperature variability increased the susceptibility of individuals to infection, but that expression of WS only occurred during the warmest temperatures. The rocky intertidal thermal regime is highly variable at local and regional scales (Helmuth et al., 2006), meaning that transmission of the prokaryote will likely be widespread. However, the disease will only become symptomatic when temperatures exceed thresholds for triggering WS (Ben-Horin et al., 2013). For example, WS related die-offs of *H. cracherodii* in southern and central California seem to co-occur during El Niño years, at a site where warm water is discharged from a power plant and more often in the southern portion of its geographic range (Raimondi et al., 2002, 2015). These observations suggest that future El Niño years, other incidences of warm water discharge and climate change will increase the spread of WS in *H. cracherodii* and other susceptible abalone species. The combined effects of WS and harvesting in *H. cracherodii* have had such a negative impact that it is now listed as critically endangered by the International Union for Conservation of Nature (IUCN; Smith et al., 2003).

Community-level responses may be quite drastic following mass mortalities or local extinctions of keystone predators or grazers (or other functional groups). *H. cracherodii*, for example, are important grazers of drift algae, kelp and crustose coralline algae. Miner et al. (2006) conducted surveys before and after mass mortalities of *H. cracherodii* and found that community composition shifted from that dominated by bare rock and crustose coralline algae (good abalone habitat) to one consisting primarily of sessile invertebrates and sea urchins (poor

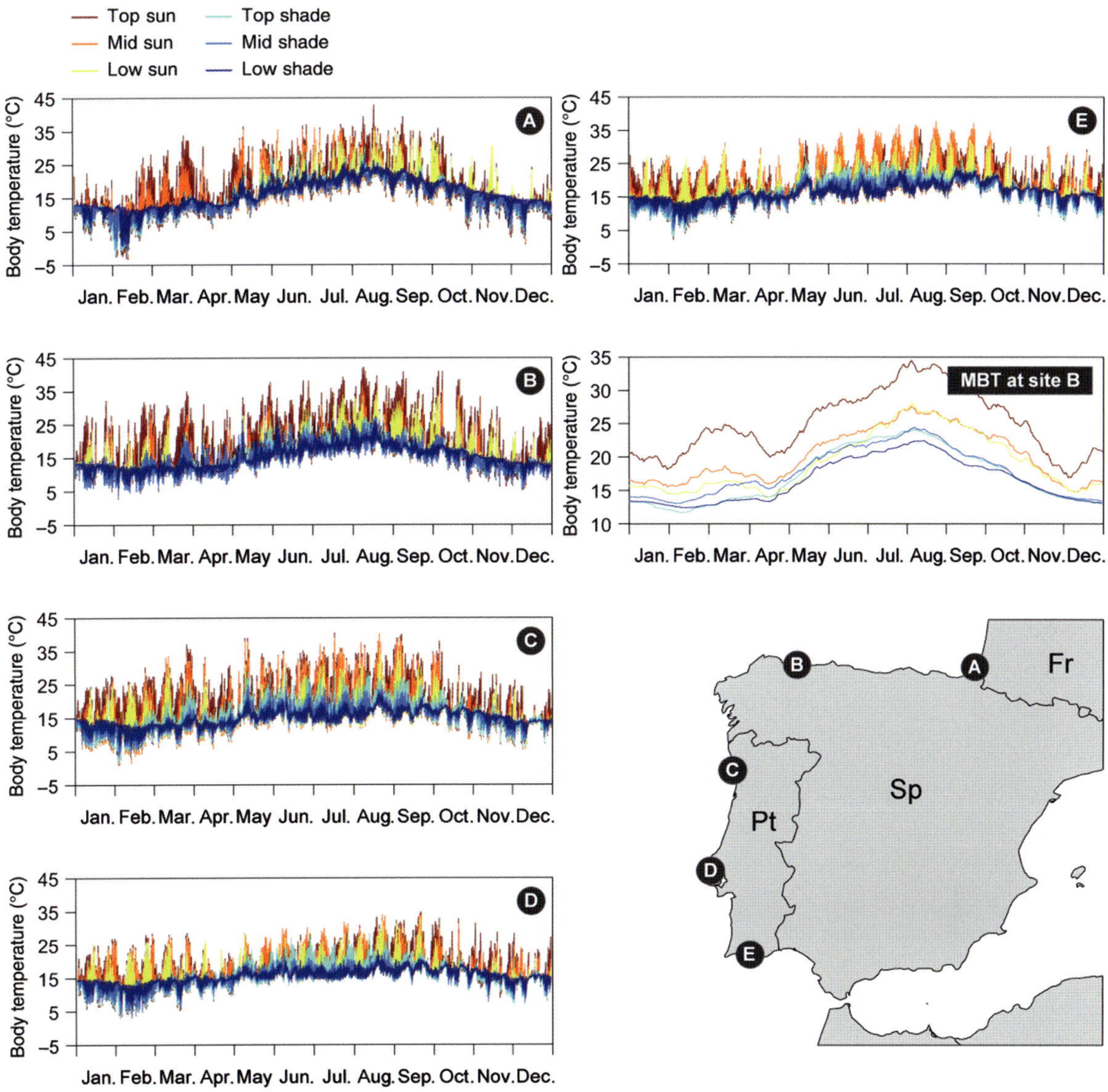


**Fig. 2.2** Body temperature profiles from 'robolimpets' deployed in different microhabitats (depicted by different line colours) recorded at five sites along the Atlantic coast of the Iberian Peninsula (A–E); 30-day average of daily maxima at La Caridad (mean body temperature at site B). *A black and white version of this figure will appear in some formats.*

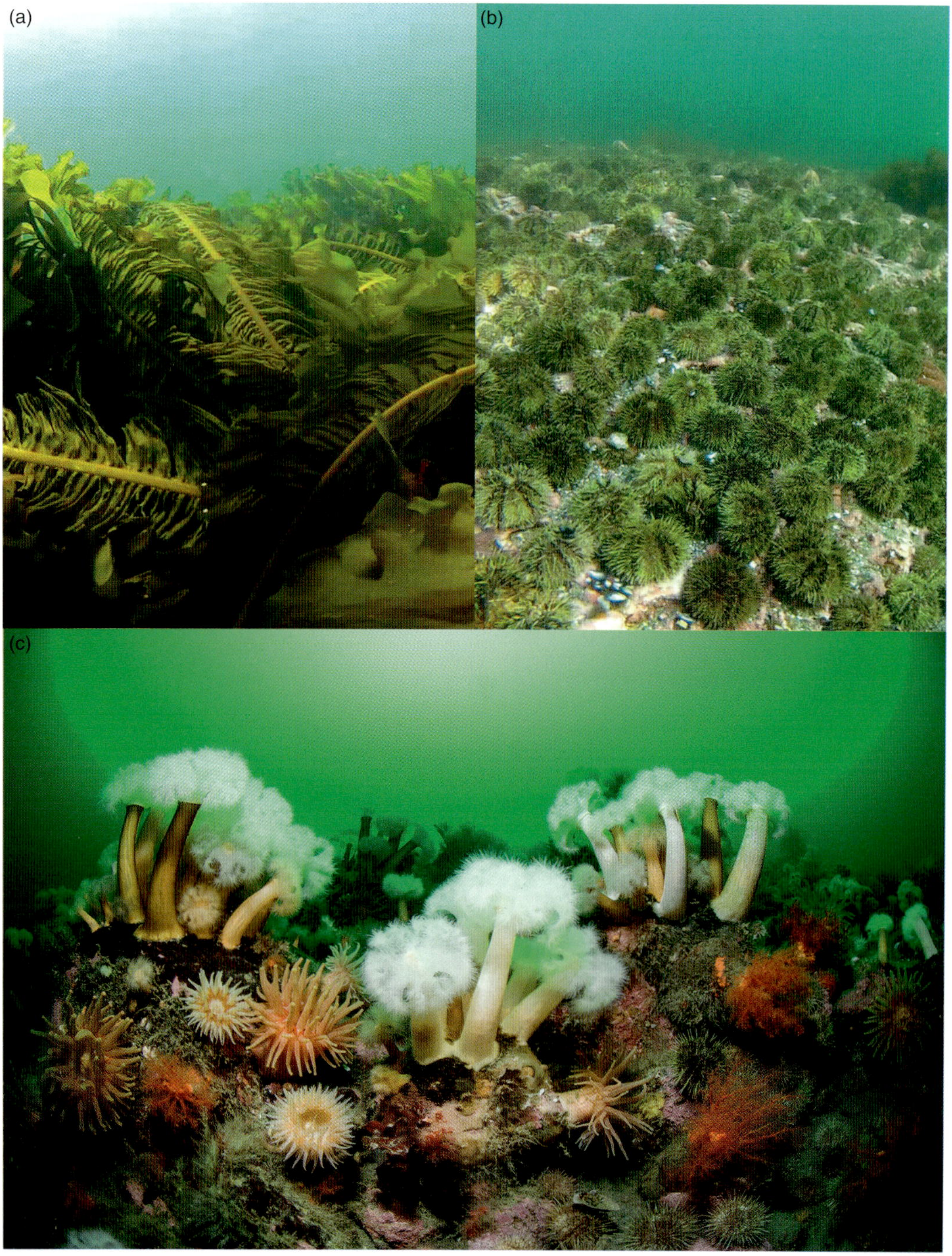


**Fig. 5.2** Distinct communities in the northern Gulf of Saint Lawrence: (a) kelp bed dominated by *Alaria esculenta* in the shallow subtidal fringe; (b) a classic shallow urchin barren ground on a bedrock bottom with high densities of urchins, high cover of coralline algae and isolated plants of chemically or physically defended macroalgae; (c) an invertebrate-dominated area near Les Escoumins (Quebec) characterised by a relatively steep slope of mixed cobble, boulder and bedrock. This complex three-dimensional bottom hosts an impressively rich and diverse invertebrate community. This well-known dive site is an urchin barren in the sense that it lacks any significant kelp and that it contains relatively high densities of urchins, but its aspect is anything but barren. Photo credits: (a) Kathleen MacGregor, (b) Carla Narvaez and (c) Robert LaSalle. *A black and white version of this figure will appear in some formats.*

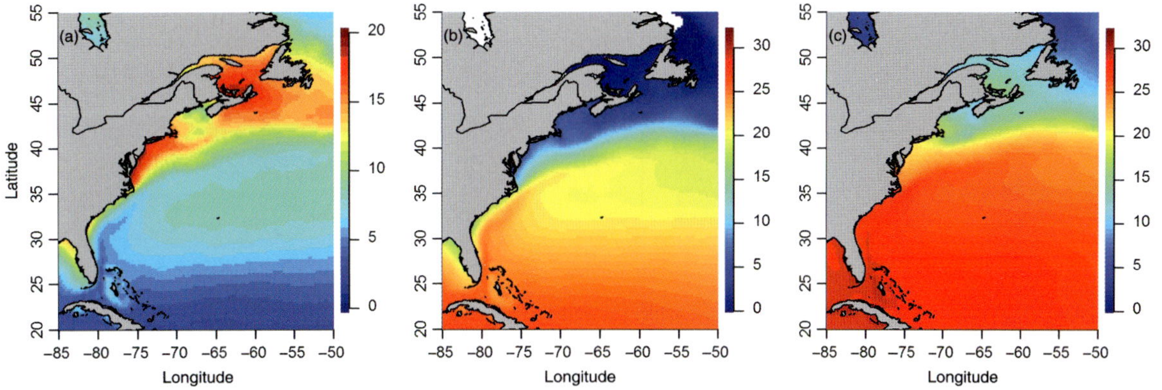


**Fig. 6.1** Mean sea surface temperature 1982–2011. (a) Yearly range; (b) winter mean; (c) summer mean. Data from Banzon et al., 2014. Source: ftp://eclipse.ncdc.noaa.gov/pub/OI-daily-v2/climatology/. *A black and white version of this figure will appear in some formats.*

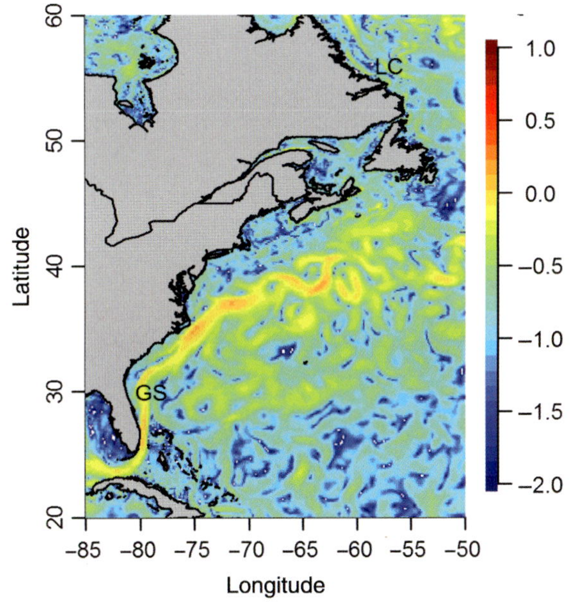


**Fig. 6.2** Surface currents in the north-west Atlantic. The cold-water Labrador Current (LC) flows south and the warm-water Gulf Stream (GS) flows north. Colours are $\log_{10}$ (velocities) (m $s^{-1}$) on January 1, 2012. Source NCOM Global (Martin et al., 2009). *A black and white version of this figure will appear in some formats.*

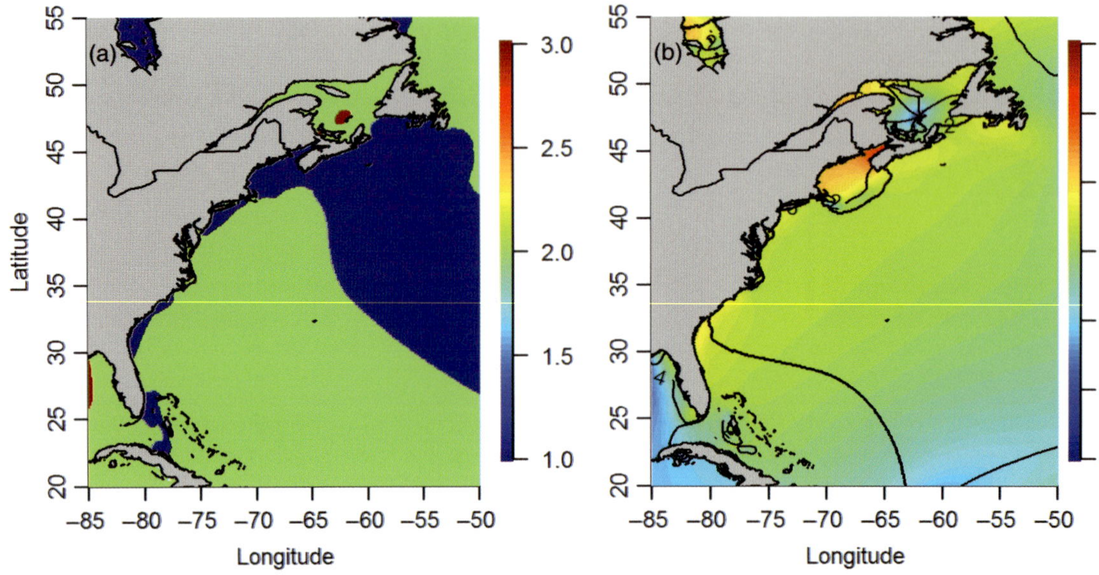


**Fig. 6.3** Tidal types and amplitudes. (a) Tide type (semi-diurnal, mixed or diurnal), calculated from tidal constituents via form ratio F = (K1 + O1)/(M2 + S2). $F < 0.25$: semi-diurnal, $0.25 < F < 3$: mixed and $F > 3$: diurnal. In the figure scale: 1: semi-diurnal, 2: mixed and 3: diurnal. (b) Amplitude and phase of M2 tides on the North American Atlantic Coast. Colour scale is $\log_{10}$ (amplitude) and contour lines are M2 tide phases in hours. Source: TPXO7 Atlantic Ocean Atlas http://volkov.oce.orst.edu/tides/atlas.html (Egbert and Erofeeva, 2002). *A black and white version of this figure will appear in some formats.*

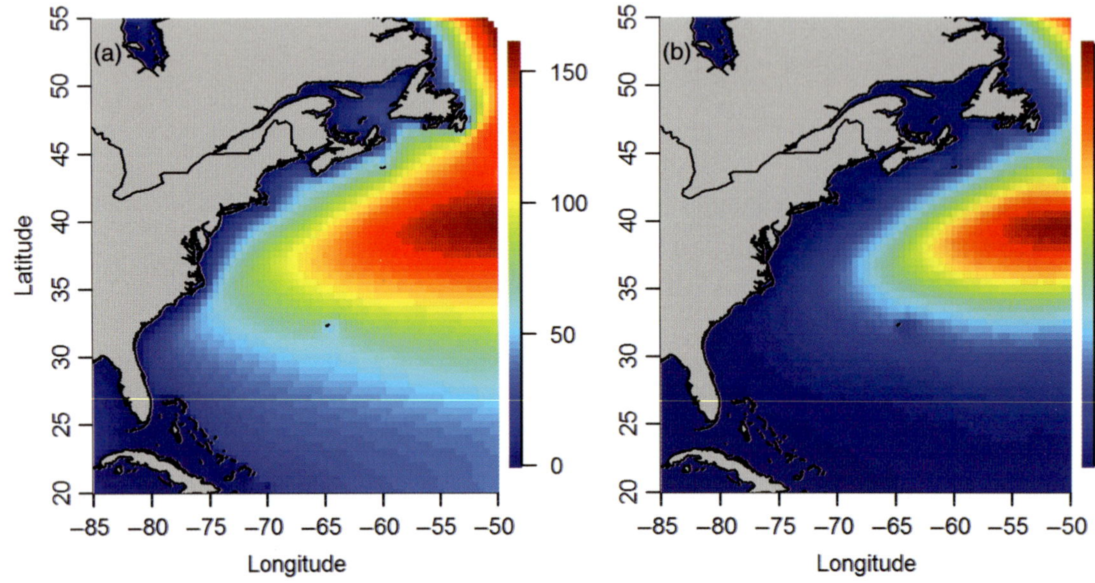


**Fig. 6.4** Mean number of days per year 1979–2009 with significant wave heights (a) >3 m and (b) >6 m. Note the difference in colour scales. Source: NOAA Wavewatch III hindcast with reanalysis winds ftp://polar.ncep.noaa.gov/pub/history/nopp-phase1/ (Chawla et al., 2011). *A black and white version of this figure will appear in some formats.*

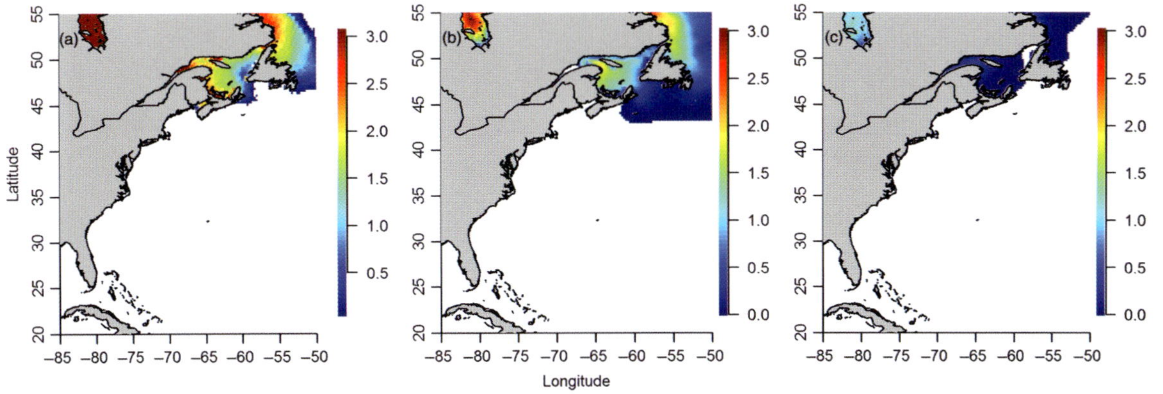


**Fig. 6.5** Winter ice cover in the north-west Atlantic (December–February). (a) 1980–2000, the average number of months of ice cover per year in satellite data from the National Snow and Ice Data Center (Sea Ice Index, dataset G02135); (b) 1980-2000, CMIP5 hindcast average number of months of winter ice cover in historical simulations; (c) 2080-2099, CMIP5 prediction average number of months of winter ice cover in the RCP 8.5 concentration pathway simulations. CMIP5 models were from the Institute Pierre Simon Laplace (CM5A), Tokyo University (MIROC ESM), Max Planck Institute for Meteorology (MPI ESM-MR), Japan Meteorological Agency (MRI CGCM3) and the Royal Netherlands Meteorological Institute (HadGEM2-AO). Colour scale on NSIDC data is the average number of months of winter ice cover per year. *A black and white version of this figure will appear in some formats.*

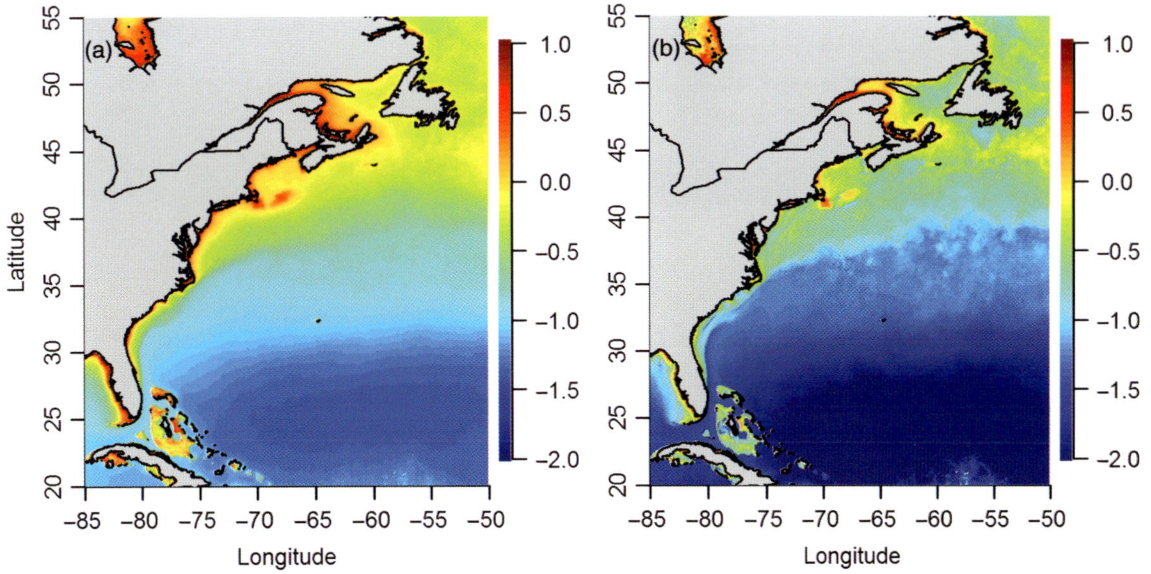


**Fig. 6.6** Chlorophyll *a* concentration from satellite imagery 2000–2015. (a) Climatological annual mean of chlorophyll *a* (mg/m$^3$). (b) Standard deviation of the monthly climatological averages of chlorophyll *a* (mg/m$^3$) from MODIS sensor on the Terra satellite (https://oceancolor.gsfc.nasa.gov). Data are expressed as log10. *A black and white version of this figure will appear in some formats.*

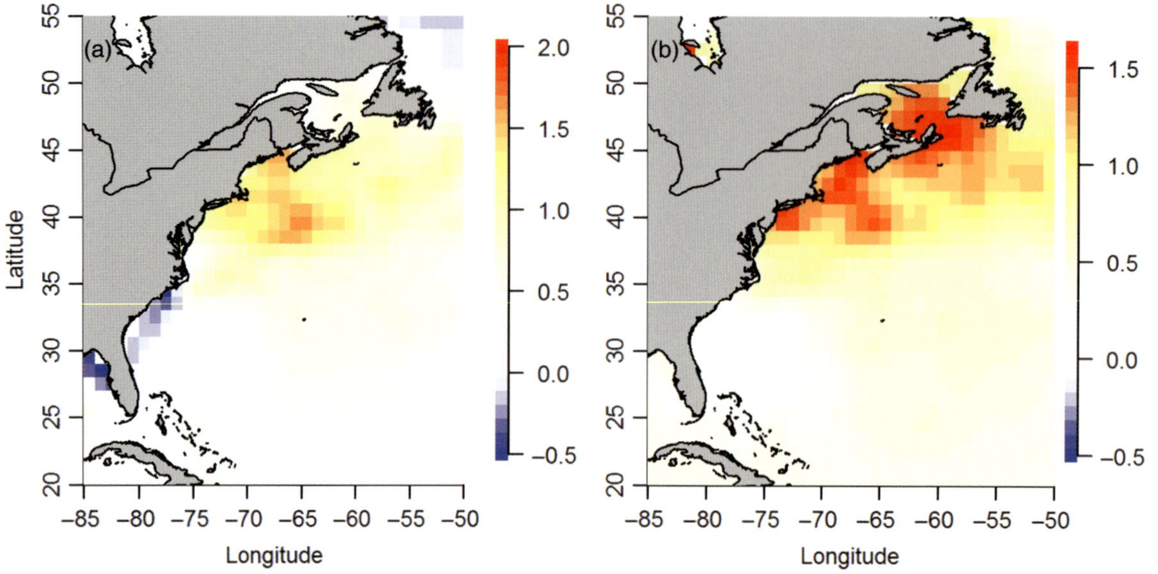


**Fig. 6.7** Linear rate of change of SST (°C per century) for the period 1870-2015. (a) Winter (December–February); (b) summer (June–August). Source: Hadley Centre HADISST (Rayner et al., 2003). www.metoffice.gov.uk/hadobs/hadisst/. *A black and white version of this figure will appear in some formats.*

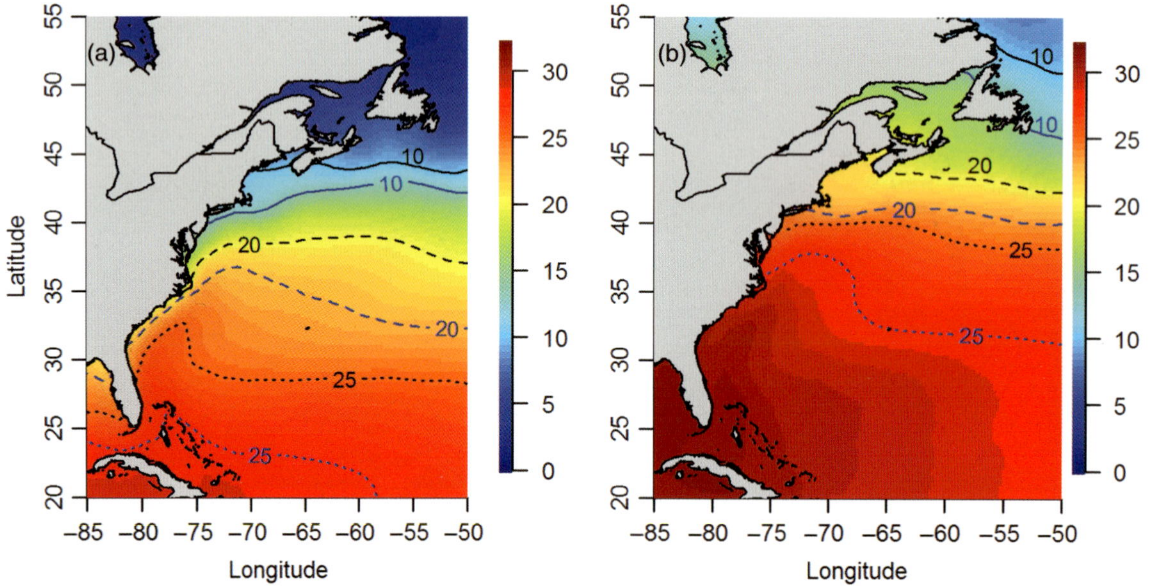


**Fig. 6.8** Ensemble projections of (a) mean winter (December–February) and (b) summer (June–August) SST in 2080–2100 under the Intergovernmental Panel on Climate Change representative concentration pathway 8.5 (8.5W $m^{-2}$ global energy imbalance by 2100). The colour scale is in °C. Blue contour lines are for the period 1980–2000, black contour lines are for the period 2080–2100, dotted contours are at 25°C, dashed contours at 20°C and solid contours at 10°C. Models were from the Beijing Climate Center (BCC CSM1–1), NASA Goddard Institute of Space Sciences (GISS E2R), Institute of Atmospheric Physics – Chinese Academy of Sciences (LASG FGOALS-g2), Tokyo University (MIROC ESM), Max Planck Institute for Meteorology (MPI ESM-MR), Japan Meteorological Agency (MRI CGCM3) and National Center for Atmospheric Research (CCSM4). *A black and white version of this figure will appear in some formats.*

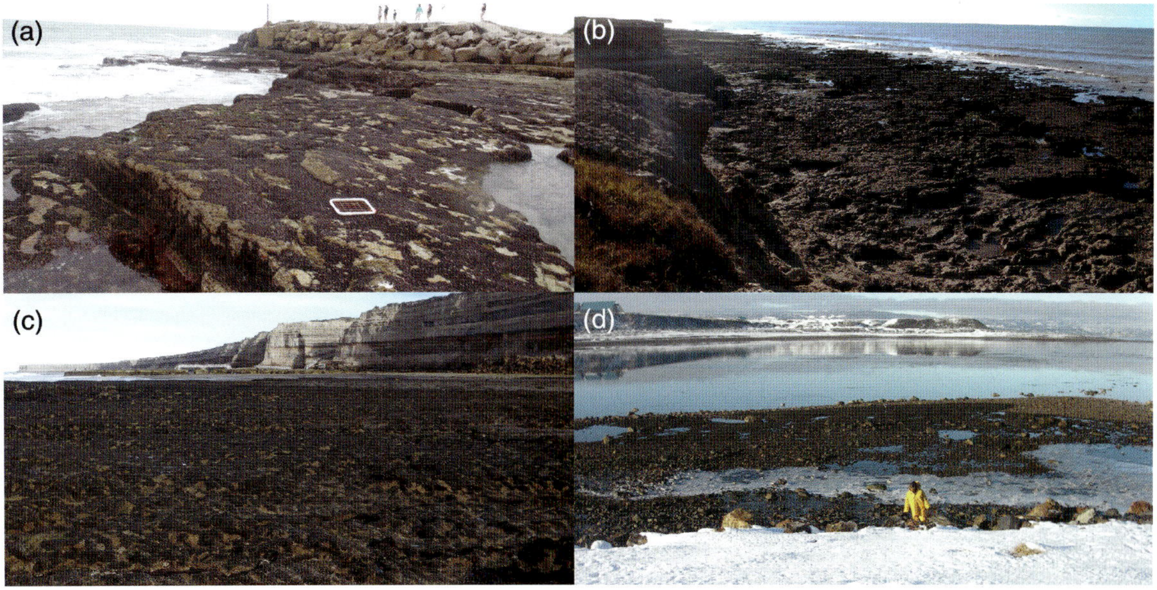


**Fig. 7.2** Examples of intertidal rocky shores in Argentina: (a) Mar del Plata (38°01´S, 57°31´W) and (b) Quequén (38°34'S, 58°40´W) in Buenos Aires province; and (c) El Espigón (41°08´S, 63°02´W) in Río Negro province and (d) Ushuaia (54°49´S, 68°11´W) in Tierra del Fuego province. *A black and white version of this figure will appear in some formats.*

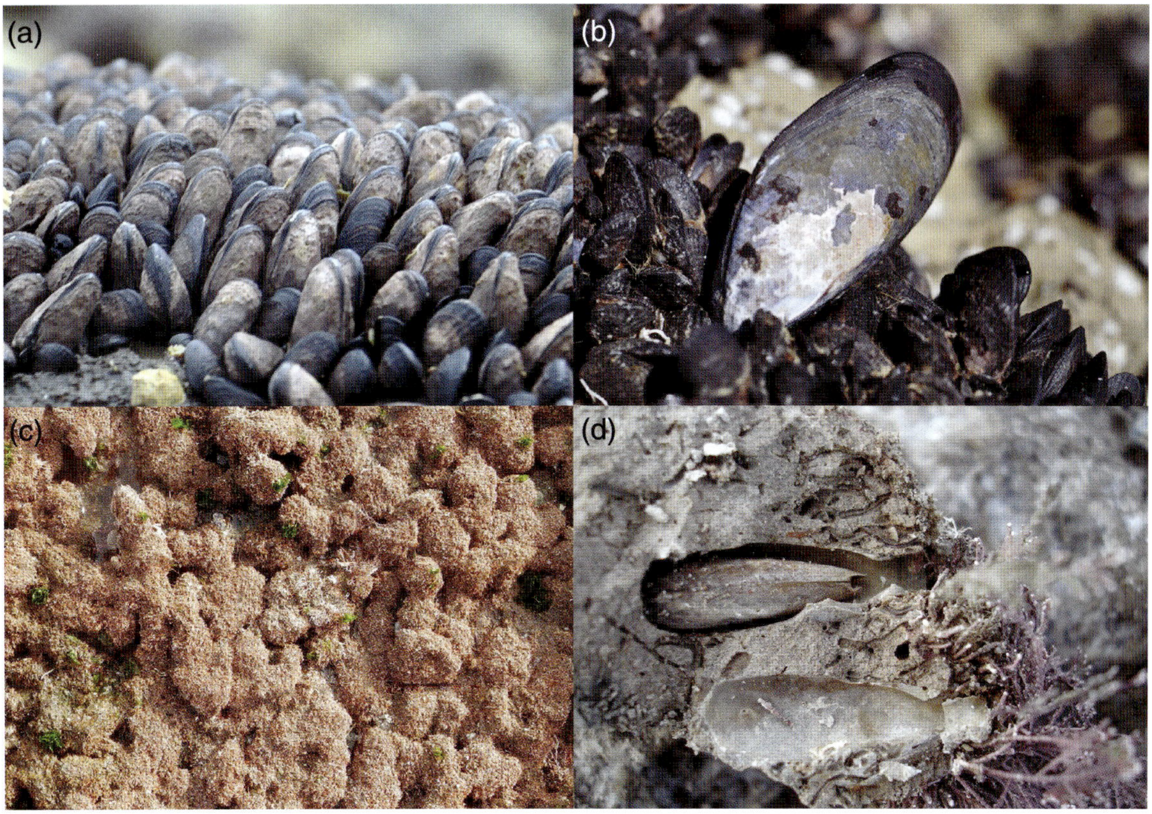


**Fig. 7.3** Common intertidal ecosystem engineers on Argentinian rocky shores. (a) The mussel *P. purpuratus*; (b) the mussel *M. edulis platensis* (emerging from a *P. purpuratus* mussel bed); (c) the alga *C. officinalis* (in a 0.5 x 0.5 m quadrat); (d) the boring bivalve *L. patagonica*. Photo credits: (a and b) www.proyectosub.com.ar and (c and d) María Bagur. *A black and white version of this figure will appear in some formats.*

**Fig. 7.4** Common intertidal grazers on Argentinian rocky shores. (a) The pulmonate limpet *S. lessonii*; (b) the true-limpet *N. magellanica*; (c) the snail *T. patagonica*; (d) the keyhole limpet, *F. radiosa tixierae*. Photo credits: www.proyectosub.com.ar. *A black and white version of this figure will appear in some formats.*

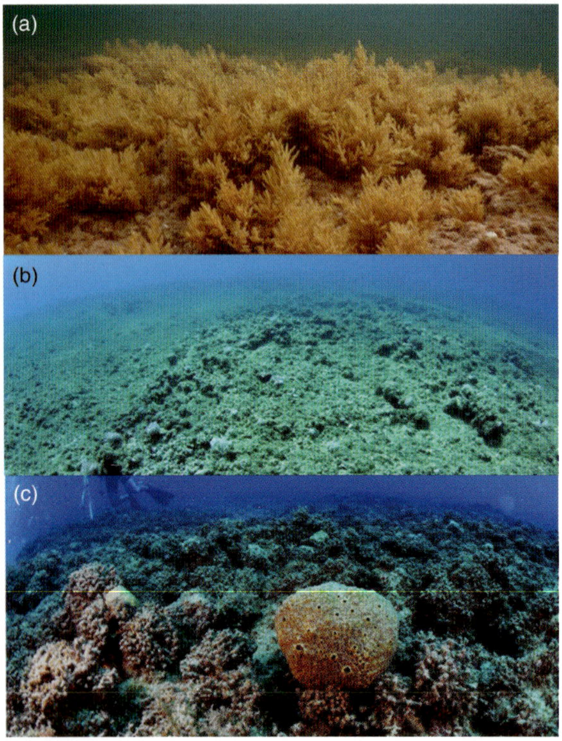


**Fig. 9.7** Reef seascapes in the Haifa area dominated by (a) *Cystoseira* meadows in the late winter, (b) turf barrens and (c) *G. rugosa* shrubs. *A black and white version of this figure will appear in some formats.*

**Fig. 7.5** Common intertidal predators on Argentinian rocky shores. (a) The crab *C. angulatus*; (b) the sea star *A. antarctica*; (c) the muricid snail *T. geversianus*; (d) the octopus *O. tehuelchus*. Photo credits: (a–c) www.proyectosub.com.ar and (d) María Bagur. *A black and white version of this figure will appear in some formats.*

**Fig. 9.10** Experimental intertidal cores in the SLR test. (a) Side and (b) top views of a core on day 0 of the experiment; (c) top view of a core protected by a cage on day 100; (d) top view of a core with no cage on day 100, with evidence of grazing down to turf. *A black and white version of this figure will appear in some formats.*

**Fig. 11.2** Pneumatocysts of two major canopy-forming kelp species in the north-east Pacific: (a) *Macrocystis*, which is most abundant in south of central California, and (b) *Nereocystis*, which primarily occurs north of central California. Photo from Rhoda Green. *A black and white version of this figure will appear in some formats.*

**Fig. 11.3** *P. helianthoides* in the Salish Sea at various stages of SSWD in 2013. Photo by Ed Gullekson. *A black and white version of this figure will appear in some formats.*

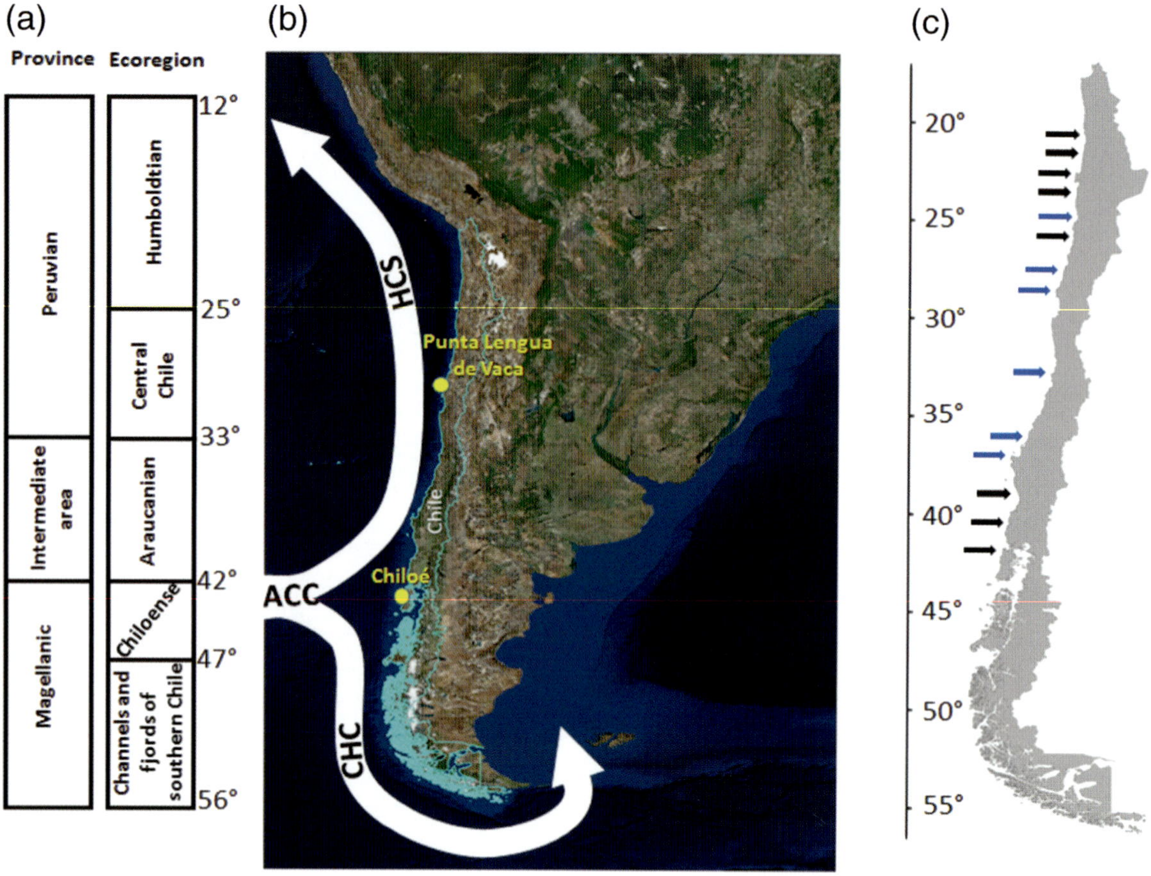


**Fig. 12.1** (a) Biogeographic provinces (after Camus, 2001; Thiel et al., 2007) and ecoregions (after Spalding et al., 2007) along the south-east Pacific coast, indicating their latitudinal boundaries. (b) Simplified view of the major cold flows (white arrows) off western South America, originating from the Antarctic Circumpolar Current (ACC) after reaching the Chilean coast: the poleward Cape Horn Current (CHC), encompassing the southern cone, and the equatorward (HCS), an eastern boundary current formed actually by coastal and oceanic branches. Yellow dots indicate the location of two places often referred to in the text because of their ecological or biogeographical importance. (c) Fine-grained latitudinal breaks in species diversity (after UACH, 2006) along the Chilean coast (excluding the area south of 45°S), detected in analyses combining multiple littoral taxa from rocky and sandy shores. Breaks (associated with the latitude axis on the left) correspond to significant spatial changes in beta diversity (black arrows) and in both alpha and beta diversity (blue arrows). *A black and white version of this figure will appear in some formats.*

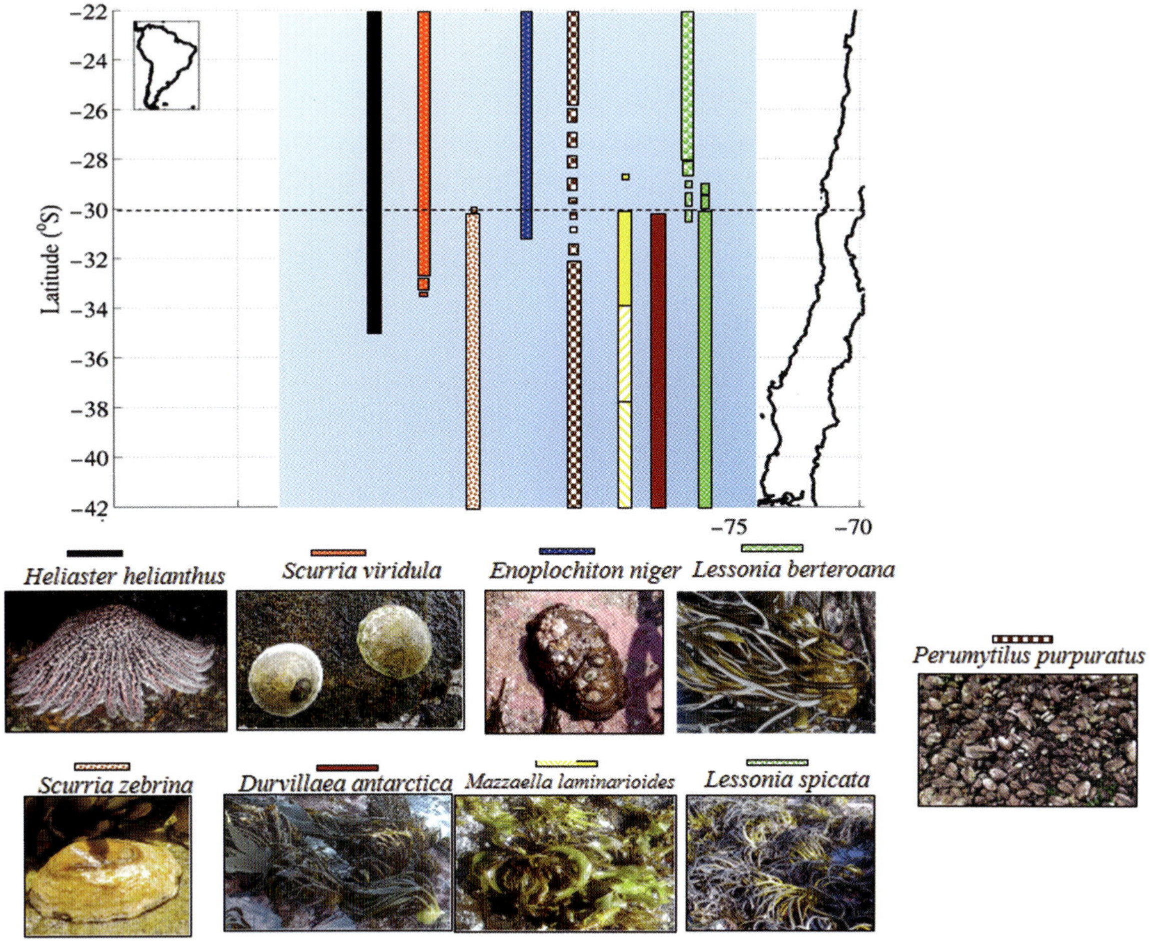


**Fig. 12.2** Scheme of the geographic distributional pattern of different intertidal species, (i.e., distributional species mosaic) present across north and central-south Chile. Most species find their polar or equatorial range edge between the biogeographic transitional zone at Punta Lengua de Vaca (PLV, 30°S), and near 36°S. The sunstar *Heliaster helianthus* and the grazer species; *Scurria viridula*, *Scurria zebrina* and *Enoplochiton niger* are the main consumers in intertidal levels across the range considered. The corticated alga *Mazzaella laminarioides* is dominant in mid- and high-intertidal levels south 30°S, where three clades are reported (see yellow bars, Montecinos et al., 2012). The fucoid *Durvillaea antarctica* and the kelp *Lessonia spicata* are also dominant south PLV in low-intertidal levels. North of this region, these algae are replaced by *Lessonia berteroana* (green bars, e.g., Tellier et al., 2011). Segmented bars correspond to small population densities, or small patches, of invertebrates and algae species present across the biographic transitional zone. For example, the mussel *Perumytilus purpuratus* (brown-grid bars) showed a discontinuous distribution across the transitional zone at 30°S, while a relatively continuous distribution is observed for this species both south and north of this range (Navarrete et al., 2005). *A black and white version of this figure will appear in some formats.*

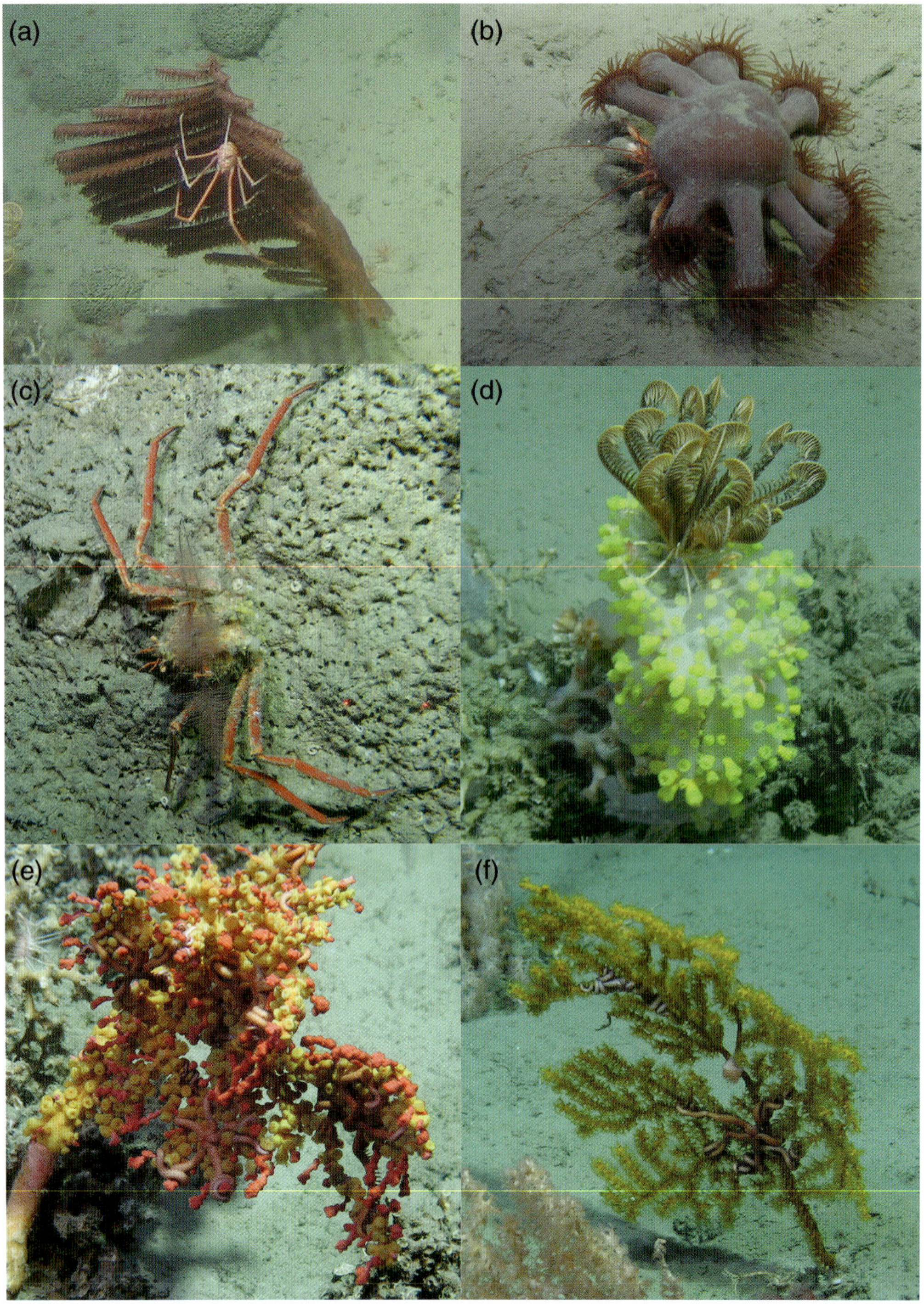


**Fig. 19.1** Animal–animal associations in the deep sea: (a) chirostylid on black coral *Bathypathes;* (b) the zoanthid *Epizoanthus paguriphilus* providing a shell for the hermit crab *Parapagurus pilosimanus*; (c) the carrier crab *Paramola cuvieri* carrying a black coral, probably *Parantipathes*; (d) yellow zoanthid on the sponge *Aphrocallistes beatrix*; (e) yellow zoanthid and brittle star *Asteroschema* on the bubblegum coral *Paragorgia*; (f) brittle star *Asteroschema* on the plexaurid coral *Paramuricea*. All photos taken by NUI Galway during cruise CE14009 aboard RV Celtic Explorer using ROV Holland I, copyright Marine Institute. *A black and white version of this figure will appear in some formats.*

abalone habitat). This community shift may also hinder population recovery following the mortality event because high cover of bare rock and crustose coralline algae are thought to be favourable for juvenile settlement and growth (Miner et al., 2006). Thus, WS and its cascading ecological impacts act as compounding factors limiting the potential recovery of *H. cracherodii*.

Besides humans (Butler et al., 2009), the primary predator of *H. cracherodii* and most observed source of mortality for individuals not affected by WS is the California sea otter (*Enhydra lutris nereis*), which is also listed as an endangered species by the IUCN (Doroff and Burdin, 2015). Such cases where both predator and prey are listed as threatened or endangered requires in-depth field studies quantifying how their interactions may affect their respective population dynamics. While these scenarios have been explored in a theoretical sense, the only empirical study to date has been conducted along the coast of California where *H. cracherodii* and *E. lutris nereis* are co-distributed. Raimondi et al. (2015) asked whether sea otter predation affects the demography and microhabitat distribution of abalone, potentially resulting in lower population density and/or greater usage of refuges where otters cannot easily reach (i.e., deep crevices in the rock). Indeed, they found that abalone did make greater usage of deep cracks and crevices at sites where sea otters were present, but, surprisingly, their population densities were also increased at such sites. Sea otter density may therefore have a positive effect on abalone populations, which is counter to a similar study of sea otter predation negatively impacting red abalone abundance (*H. rufescens*), which is co-distributed with *H. cracherodii* (Fanshawe et al., 2003). There are a few reasons why sea otters may have a positive effect on black abalone density, including the fact that both sea otters and abalone require rocky habitat, which are likely to include the microhabitats needed for abalone refuge. In addition, both sea otters and abalone require kelp, either as resting area and foraging habitat (otters) or as detritus for food (abalone). Moreover, we know from a classic trophic cascades case study (Estes and Duggins, 1995) that sea otter presence can increase kelp abundance (via predation control of kelp eating urchins). Hence, sea otters can indirectly increase the overall amount of food for black abalone. Management policy for black abalone must therefore take into account multiple species, including sea otters and kelp. Of course, this study was only possible within the portion of the *H. cracherodii* range without high incidence of WS. Whether the same positive relationship between abalone and sea otter density will also occur if WS becomes more widespread requires further study (Raimondi et al., 2015).

## 10.10 Harvesting

Thus far we have discussed how indirect human impacts (e.g., warmer or more extreme temperatures due to climate change) can affect species capable of top-down control of their community. But direct human impacts, like size-selective harvesting of such species has also been well studied. Limpets are ecologically important grazers and space occupiers of the NEP (Ricketts et al., 1985; Lindberg et al., 1998; Fenberg, 2013). *L. gigantea* is commonly size-selectively harvested within the central portion of its geographic range (~28–36°N) where it is one of the dominant space occupiers of the mid-/upper-intertidal (Stimson, 1970; Lindberg et al., 1998; Fenberg and Rivadeneira, 2011; Fenberg, 2013). As individuals grow larger, they proportionally increase the size of their territories/home ranges (Fenberg, 2013). The larger individuals in a population are usually territorial females and can occupy up to 0.20 $m^2$ (Figure 10.3a). Territory holders defend their algal 'garden' from intraspecific (usually smaller males) and interspecific competitors (Stimson, 1970; Wright, 1989) by 'bulldozing' individuals off their territory using the anterior portion of their shell. Other space occupiers such as mussels, barnacles and macroalgae can be seen on the edges of territories, as well as small- to medium-sized *L. gigantea* individuals and other limpet species. After a population has been size-selectively harvested, sessile species and smaller-sized *L. gigantea* individuals can invade previously occupied territories (Lindberg et al., 1998). If harvesting is consistent enough, then the space

**Fig. 10.3** (a) Home range scaling of *L. gigantea*. The black circles are individuals from sites where there is no size-selective harvesting. The grey circles are from sites where individuals are size-selectively harvested. (b) Box plots of the percentage of bare rock (suitable *L. gigantea* habitat) present in the mid-intertidal zone at harvested and protected sites. There is significantly more bare rock at protected sites ($P < 0.0001$).

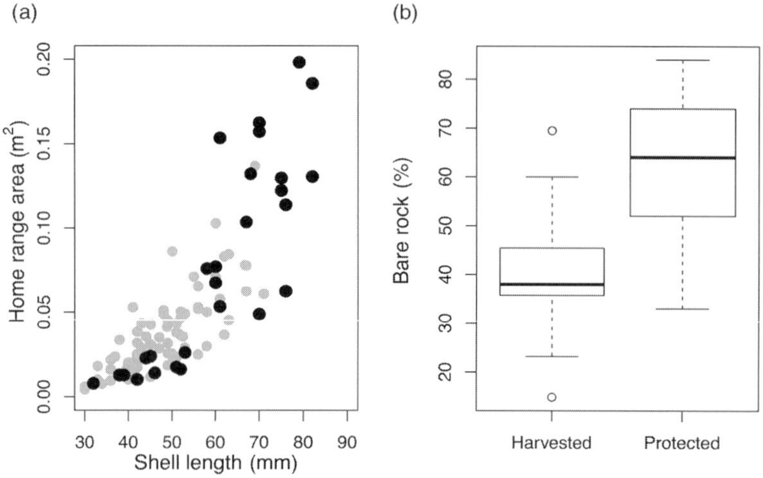


occupancy of a site can shift from one dominated by large grazers (largely consisting of *L. gigantea* home range area) to one dominated by other functional groups, like filter feeders (barnacles and mussels) and macroalgae. Accordingly, Fenberg (2008) showed that the mid-intertidal zone of sites in southern and central California where individuals of *L. gigantea* are commonly size-selectively harvested have significantly less area available as potential *L. gigantea* grazing space (measured as bare rock) compared to sites where harvesting is minimal (Figure 10.3b).

## 10.11 Human Impacts on Bottom-Up Control of NEP Rocky Intertidal Community Structure

The NEP is an eastern boundary upwelling system, defined by variable cross-shelf circulation due to upwelling/downwelling forcing (Checkley and Barth, 2009). Viewed in cross-section, coastal circulation patterns resemble a reversing conveyor belt. When winds blow equatorwards, surface waters flow equatorwards and, due to the Coriolis force, offshore. The pressure gradient created by these flows draws water from depth, bringing cold, high nutrient, low $O_2$ and high $CO_2$ water to the surface. When winds 'relax' (cease) or reverse, surface waters rebound coastwards and polewards, bringing warm, nutrient-depleted, high $O_2$ and low $CO_2$ water back to the coast. Between and within species patterns of geographic distribution and gene flow (i.e., biogeographic and phylogeographic breaks) often coincide with areas where upwelling is particularly strong, with predominantly equatorward currents (e.g., Cape Mendocino; Kelly and Palumbi, 2010; Fenberg et al., 2015). Recent research has demonstrated that, in addition to such large-scale patterns, seasonal and regional variation in upwelling/downwelling forcing are also strongly related to patterns of rocky intertidal community structure and functioning (Menge et al., 1997, 2003, 2004, 2015; Menge and Menge, 2013). Individual growth responses, species interactions and relative abundances of different trophic levels vary in response to the variation in upwelling cycles. Surprisingly, these life history and ecological responses are not necessarily at their highest in nearshore environments when upwelling is persistent.

Along a gradient from persistent downwelling to intermittent upwelling (i.e., alternating between upwelling and downwelling on ~7–10-day cycles) to persistent upwelling, Menge and Menge (2013) showed that ecological subsidies (phytoplankton

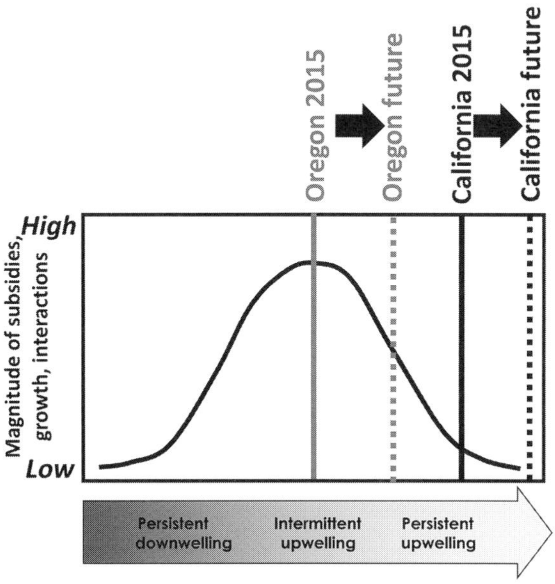


**Fig. 10.4** IUH showing predictions for changes in magnitude of ecological subsidies, growth rates of predators and prey and intensity of species interactions for Oregon and California coasts at the present and in the future.

abundance, prey recruitment rates), prey response (barnacle settlement and mussel growth) and ecological interactions (competition and predation rates) are at their highest with intermittent upwelling. The mechanisms proposed to underlie these patterns were summarised as the intermittent upwelling hypothesis, or IUH (Menge and Menge, 2013; Figure 10.4). Specifically, with persistent upwelling, nutrient-fuelled near-surface phytoplankton blooms and propagules are transported offshore away from coastal habitats. Persistent downwelling produces the opposite pattern, where surface waters flow shoreward but the comparative lack of nutrients prevents phytoplankton blooms from forming. The intermediate condition occurs when upwelling causes high nutrients to form phytoplankton blooms, which typically form within three to five days. Wind cessation or reversal leads to 'relaxation' of offshore flow or onshore flow, respectively, causing shoreward transport of near-surface larvae and phytoplankton back to the shore. Marine invertebrate larvae can have a pelagic larval duration ranging from days to weeks (Shanks, 2009). Thus, if larvae are developing during intermittent upwelling, then the frequent inshore and offshore movement could increase the odds that larvae are near rocky shore habitat when ready to settle. Indeed, sites associated with intermittent upwelling have increased phytoplankton/zooplankton abundance, recruitment and growth rates of invertebrates, and higher levels of competition/predation in rocky shore communities along the NEP, New Zealand and in other eastern boundary upwelling systems (Menge and Menge, 2013).

The IUH provides a well-supported mechanistic framework for how bottom-up processes can drive spatial and temporal variation in community structure in rocky intertidal systems, including along the NEP. Coupled with predictions of how climate change may affect upwelling/downwelling patterns, the IUH can also generate predictions about how ecosystem dynamics are likely to change in space. Future climate change projections predict that land temperatures will increase faster than coastal waters, creating a scenario favouring stronger upwelling-producing winds (Bakun et al., 2010; Di Lorenzo, 2015; Wang et al., 2015). In fact, there is evidence that upwelling has already intensified in most of the eastern boundary upwelling systems (Iles et al., 2012; Sydeman et al., 2014; Varela et al., 2015). Because upwelling brings cool water to the surface, increased upwelling may provide some refuge from SST warming in some locations. Generally, models predict that upwelling intensification will be particularly marked at higher latitudes, where coastal upwelling is generally less intense. This may cause homogenisation of upwelling-influenced habitats across latitudes and have cascading effects on the abundance and distribution of rocky intertidal flora and fauna (Wang et al., 2015). Specifically, the IUH predicts that under such poleward homogenisation of upwelling, the process rates that are high under intermittent upwelling will likely decrease as upwelling becomes more persistent and longer in duration. Along the west coast of North America, ecosystems that now typify much of the California coast would 'move northward', leading to reduced rates of recruitment, lower phytoplankton abundance, slower growth and weaker competition and predation (Menge and Menge, 2013; Figure 10.4).

An increase in upwelling also may exacerbate other emerging threats to coastal ecosystems, such as ocean acidification and anoxia because deeper waters have lower pH and oxygen levels than at the surface (Chan et al., 2008; Bakun et al., 2015). For example, historical *M. californianus* shells from midden sites (~1,000–2,420 years ago) in Washington State have significantly higher shell calcification rates and shell thicknesses compared to modern shells, which Pfister et al. (2016) attribute to declining pH in modern times. If their shells become thinner as a result of ocean acidification and are compounded by an increase in upwelling intensity, then, all else being equal, predation rates on mussels would likely increase. But as already noted, upwelling intensity and other interacting environmental drivers of NEP community structure do not often have clear latitudinal trends (Helmuth et al., 2006; Menge and Menge, 2013). For example, substrate temperature influences the internal body temperature of mussels, which can often be very different from measurements based on air or SST alone (Helmuth et al., 2016). Furthermore, potential drivers of community structure and individual species' performance may form mosaic patterns along the coast; meaning that abiotic or biotic factors vary in a spatially non-monotonic or patchy manner, as opposed to a uniform gradient. Areas of overlap of these biotic or abiotic factors can either create locations of susceptibility or resilience to environmental change (Kroeker et al., 2016). For example, Kroeker et al. (2016) showed that *M. californianus* growth rates are lowered and predation vulnerability is highest in locations with an overlap of high body temperatures, low pH and inconsistent food availability.

## 10.12 Conclusions

What is the relative influence of bottom-up/top-down control on the structure and functioning of ecological communities, and what physical and biological factors control the distribution and abundance of species over space and time? Our understanding of these two fundamental questions in ecology have benefitted greatly from examples from the NEP rocky intertidal. But if we were to somehow pursue this basic research along an NEP untouched by human influence, we probably still would not have complete answers. The NEP and all other coastlines of the globe are affected by direct (e.g., harvesting) and indirect human impacts (e.g., climate change, disease dynamics) that we know can have measureable ecological effects. Applied and basic ecologists must now add these extra layers of complexity to our models for understanding the dynamics of rocky intertidal systems.

## Acknowledgements

We would like to thank Stephen J. Hawkins for the invitation to contribute to this edited volume and Pete Raimondi and the team at www.pacificrockyintertidal.org for the data used in Figure 10.2.

### REFERENCES

Bakun, A., Black, B. A., Bograd, S. J. et al. (2015). Anticipated effects of climate change on coastal upwelling ecosystems. *Current Climate Change Reports*, **1**, 85–93.

Bakun, A., Field, D. B., Redondo-Rodriguez, A. and Weeks, S. J. (2010). Greenhouse gas, upwelling-favorable winds, and the future of coastal ocean upwelling ecosystems. *Global Change Biology*, **16**, 1213–28.

Ben-Horin, T., Lenihan, H. S. and Lafferty, K. D. (2013). Variable intertidal temperature explains why disease endangers black abalone. *Ecology*, **94**, 161–8.

Bernard, F. R., McKinnell, S. M. and Jamieson, G. S. (1991). *Distribution and Zoogeography of the Bivalvia of the Eastern Pacific Ocean*. Department of Fisheries and Oceans, Ottawa.

Bernardi, G., Findley, L., Rocha-Olivares, A. and Karl, S. (2003). Vicariance and dispersal across Baja California in disjunct marine fish populations. *Evolution*, **57**, 1599–609.

Blanchette, C. A., Melissa Miner, C., Raimondi, P. T., Lohse, D., Heady, K. E. and Broitman, B. R. (2008). Biogeographical patterns of rocky intertidal communities along the Pacific coast of North America. *Journal of Biogeography*, **35**, 1593–607.

Blanchette, C. A., Richards, D. V., Engle, J. M., Broitman, B. R. and Gaines, S. D. (2005) Regime shifts, community change and population booms of keystone predators at the Channel Islands. In *Proceedings of the California Channel Islands Symposium, Ventura*.

Brown, J. H. (1984). On the relationship between abundance and distribution of species. *The American Naturalist*, **124**, 255–79.

Bruno, J. F., Stachowicz, J. J. and Bertness, M. D. (2003). Inclusion of facilitation into ecological theory. *Trends in Ecology and Evolution*, **18**, 119–25.

Butler, J. A., DeVogelaere, R., Gustafson, C. et al. (2009) Status Review Report for Black Abalone (*Haliotis cracheroddii* Leach, 1814). In *N.M.F.S.* U.S. Department of Commerce and National Oceanic and Atmospheric Administration, Washington, DC.

Catton, C. A. and Rogers-Bennett, L. (2013). Assessing the recovery of pink abalone (*Haliotis corrugata*) by incorporating aggregation into a matrix model. *Journal of Shellfish Research*, **32**, 181–7.

Chan, F., Barth, J. A., Lubchenco, J. et al. (2008). Emergence of anoxia in the California current large marine ecosystem. *Science*, **319**, 920.

Checkley, D. M. and Barth, J. A. (2009). Patterns and processes in the California Current System. *Progress in Oceanography*, **83**, 49–64.

Cooper, E. E. and Shanks, A. L. (2011). Latitude and coastline shape correlate with age-structure of *Chlorostoma (Tegula) funebralis* populations. *Marine Ecology Progress Series*, **424**, 133–43.

Courchamp, F., Berec, L. and Gascoigne, J. (2008). *Allee Effects in Ecology and Conservation*. Oxford University Press, Oxford.

Crosson, L. M., Wight, N., VanBlaricom, G. R., Kiryu, I., Moore, J. D. and Friedman, C. S. (2014). Abalone withering syndrome: distribution, impacts, current diagnostic methods and new findings. *Diseases of Aquatic Organisms*, **108**, 261–70.

Dahlhoff, E. P., Buckley, B. A. and Menge, B. A. (2001). Physiology of the rocky intertidal predator *Nucella ostrina* along an environmental stress gradient. *Ecology*, **82**, 2816–29.

Dawson, M. N., Grosberg, R. K., Stuart, Y. E. and Sanford, E. (2010). Population genetic analysis of a recent range expansion: mechanisms regulating the poleward range limit in the volcano barnacle *Tetraclita rubescens*. *Molecular Ecology*, **19**, 1585–605.

Dawson, M. N., Hays, C. G., Grosberg, R. K. and Raimondi, P. T. (2014). Dispersal potential and population genetic structure in the marine intertidal of the eastern North Pacific. *Ecological Monographs*, **84**, 435–56.

Di Lorenzo, E. (2015). Climate science the future of coastal ocean upwelling. *Nature*, **518**.

Dolby, G. A., Bennett, S. E., Lira-Noriega, A., Wilder, B. T. and Munguía-Vega, A. (2015). Assessing the geological and climatic forcing of biodiversity and evolution surrounding the Gulf of California. *Journal of the Southwest*, **57**, 391–455.

Doroff, A. and Burdin, A. (2015) *Enhydra lutris*. The IUCN Red List of Threatened Species, e.T7750A21939518, www.iucnredlist.org/species/7750/21939518.

Eisenlord, M. E., Groner, M. L., Yoshioka, R. M. et al. (2016). Ochre star mortality during the 2014 wasting disease epizootic: role of population size structure and temperature. *Philosophical Transactions of the Royal Society B*, **371**, 20150212.

Erlandson, J. M., Braje, T. J., Rick, T. C. et al. (2011). 10,000 years of human predation and size changes in the owl limpet (*Lottia gigantea*) on San Miguel Island, California. *Journal of Archaeological Science*, **38**, 1127–34.

Estes, J. A. and Duggins, D. O. (1995). Sea otters and kelp forests in Alaska: generality and variation in a community ecological paradigm. *Ecological Monographs*, **65**, 75–100.

Fanshawe, S., Vanblaricom, G. R. and Shelly, A. A. (2003). Restored top carnivores as detriments to the performance of marine protected areas intended for fishery sustainability: a case study with red abalones and sea otters. *Conservation Biology*, **17**, 273–83.

Fenberg, P. B. (2008) *The Effects of Size-Selective Harvesting on the Population Biology and Ecology of a Sex-Changing Limpet Species,* Lottia gigantea. University of California, San Diego.

Fenberg, P. B. (2013). Intraspecific home range scaling: a case study from the owl limpet (*Lottia gigantea*). *Evolutionary Ecology Research*, **15**, 103–10.

Fenberg, P. B., Menge, B. A., Raimondi, P. T. and Rivadeneira, M. M. (2015). Biogeographic structure of the northeastern Pacific rocky intertidal: the role of upwelling and dispersal to drive patterns. *Ecography*, **38**, 83–95.

Fenberg, P. B., Posbic, K. and Hellberg, M. E. (2014). Historical and recent processes shaping the geographic range of a rocky intertidal gastropod: phylogeography, ecology, and habitat availability. *Ecology and Evolution*, **4**, 3244–55.

Fenberg, P. and Rivadeneira, M. M. (2019). On the importance of habitat continuity for delimiting biogeographic regions and shaping richness gradients. *Ecology Letters*, **22**, 664–73.

Fenberg, P. B. and Rivadeneira, M. M. (2011). Range limits and geographic patterns of abundance of the rocky intertidal owl limpet, *Lottia gigantea*. *Journal of Biogeography*, **38**, 2286–98.

Fenberg, P. B. and Roy, K. (2012). Anthropogenic harvesting pressure and changes in life history: insights from a rocky intertidal limpet. *The American Naturalist*, **180**, 200–10.

Frank, P. (1975). Latitudinal variation in the life history features of the black turban snail *Tegula funebralis* (Prosobranchia: Trochidae). *Marine Biology*, **31**, 181–92.

Friedman, C. S., Biggs, W., Shields, J. D. and Hedrick, R. P. (2002). Transmission of withering syndrome in black abalone, *Haliotis cracherodii* Leach. *Journal of Shellfish Research*, **21**, 817–24.

Gaines, S. D., Lester, S. E., Eckert, G., Kinlan, B. P., Sagarin, R. and Gaylord, B. (2009). Dispersal and Geographic Ranges in the Sea. In K. Roy and J. Witman, eds. *Marine Macroecology*. University of Chicago Press, Chicago, pp. 227–49.

Gaylord, B., Nickols, K. J. and Jurgens, L. (2012). Roles of transport and mixing processes in kelp forest ecology. *Journal of Experimental Biology*, **215**, 997–1007.

Gaylord, B., Rosman, J. H., Reed, D. C. et al. (2007). Spatial patterns of flow and their modification within and around a giant kelp forest. *Limnology and Oceanography*, **52**, 1838–52.

Gilman, S. E. (2006). The northern geographic range limit of the intertidal limpet *Collisella scabra*: a test of performance, recruitment, and temperature hypotheses. *Ecography*, **29**, 709–20.

Gilman, S. E., Urban, M. C., Tewksbury, J., Gilchrist, G. W. and Holt, R. D. (2010). A framework for community interactions under climate change. *Trends in Ecology and Evolution*, **25**, 325–31.

Hellberg, M. E., Balch, D. P. and Roy, K. (2001). Climate-driven range expansion and morphological evolution in a marine gastropod. *Science*, **292**, 1707–10.

Helmuth, B., Broitman, B. R., Blanchette, C. A. et al. (2006). Mosaic patterns of thermal stress in the rocky intertidal zone: implications for climate change. *Ecological Monographs*, **76**, 461–79.

Helmuth, B., Choi, F., Matzelle, A. et al. (2016). Long-term, high frequency in situ measurements of intertidal mussel bed temperatures using biomimetic sensors. *Scientific Data*, **3**.

Herrera-Cervantes, H., Lluch-Cota, S., Lluch-Cota, D. and Gutiérrez-de-Velasco, G. (2014). Interannual correlations between sea surface temperature and concentration of chlorophyll pigment off Punta Eugenia, Baja California, during different remote forcing conditions. *Ocean Science*, **10**, 345–55.

Hewitt, R. (1981). Eddies and speciation in the California current. *CalCOFI Report*, **22**, 96–8.

Hewson, I., Button, J. B., Gudenkauf, B. M. et al. (2014). Densovirus associated with sea-star wasting disease and mass mortality. *Proceedings of the National Academy of Sciences*, **111**, 17278–83.

Holt, R. D. (2003). On the evolutionary ecology of species' ranges. *Evolutionary Ecology Research*, **5**, 159–78.

Hurtado, L. A., Mateos, M. and Santamaria, C. A. (2010). Phylogeography of supralittoral rocky intertidal *Ligia* isopods in the Pacific region from central California to central Mexico. *PLoS ONE*, **5**, e11633.

Iles, A. C., Gouhier, T. C., Menge, B. A., Stewart, J. S., Haupt, A. J. and Lynch, M. C. (2012). Climate-driven trends and ecological implications of event-scale upwelling in the California current system. *Global Change Biology*, **18**, 783–96.

IPCC. (2014). *Climate Change 2014 – Impacts, Adaptation and Vulnerability: Regional Aspects*. Cambridge University Press, Cambridge.

Jarrett, J. N. (2009). Predator-induced defense in the barnacle *Chthamalus fissus*. *Journal of Crustacean Biology*, **29**, 329–33.

Jurgens, L. J., Rogers-Bennett, L., Raimondi, P. T. et al. (2015). Patterns of mass mortality among rocky shore invertebrates across 100 km of Northeastern Pacific coastline. *PLoS ONE*, **10**, e0126280.

Kay, M. C. and Emlet, R. B. (2002). Laboratory spawning, larval development, and metamorphosis of the limpets *Lottia digitalis* and *Lottia asmi* (Patellogastropoda, Lottiidae). *Invertebrate Biology*, **121**, 11–24.

Kelly, R. P. and Palumbi, S. R. (2010). Genetic structure among 50 species of the northeastern Pacific rocky intertidal community. *PLoS ONE*, **5**, e8594.

Kido, J. S. and Murray, S. N. (2003). Variation in owl limpet *Lottia gigantea* population structures, growth rates, and gonadal production on southern California rocky shores. *Marine Ecology Progress Series*, **257**, 111–24.

Kohl, W. T., McClure, T. I. and Miner, B. G. (2016). Decreased temperature facilitates short-term sea star wasting disease survival in the keystone intertidal sea star *Pisaster ochraceus*. *PLoS ONE*, **11**, e0153670.

Kroeker, K. J., Sanford, E., Rose, J. M. et al. (2016). Interacting environmental mosaics drive geographic variation in mussel performance and predation vulnerability. *Ecology Letters*, **19**, 771–9.

Lafferty, K. D. and Suchanek, T. H. (2016). Revisiting Paine's 1966 sea star removal experiment, the most-cited empirical article in The American Naturalist. *The American Naturalist*, **188**, 365–78.

Lehman, J. (2010). *Population Genetic Analysis of the Intertidal Limpet* Lottia scabra *and Inference of the Causes and Mechanisms of Range Limits*. University of California, Merced.

Lindberg, D. R., Estes, J. A. and Warheit, K. I. (1998). Human influences on trophic cascades along rocky shores. *Ecological Applications*, **8**, 880–90.

Lohse, D. P. (1993). The importance of secondary substratum in a rocky intertidal community. *Journal of Experimental Marine Biology and Ecology*, **166**, 1–17.

Marko, P. B., Hoffman, J. M., Emme, S. A. et al. (2010). The 'Expansion–Contraction'model of Pleistocene biogeography: rocky shores suffer a sea change? *Molecular Ecology*, **19**, 146–69.

Menge, B. A., Berlow, E. L., Blanchette, C. A., Navarrete, S. A. and Yamada, S. B. (1994). The keystone species concept: variation in interaction strength in a rocky intertidal habitat. *Ecological Monographs*, **64**, 249–86.

Menge, B. A., Blanchette, C., Raimondi, P. et al. (2004). Species interaction strength: testing model predictions along an upwelling gradient. *Ecological Monographs*, **74**, 663–84.

Menge, B. A., Cerny-Chipman, E. B., Johnson, A., Sullivan, J., Gravem, S. and Chan, F. (2016). Sea star wasting disease in the keystone predator *Pisaster ochraceus* in Oregon: Insights into differential population impacts, recovery, predation rate, and temperature effects from long-term research. *PLoS ONE*, **11**, e0153994.

Menge, B. A., Daley, B. A., Wheeler, P. A., Dahlhoff, E., Sanford, E. and Strub, P. T. (1997). Benthic–pelagic links and rocky intertidal communities: Bottom-up effects on top-down control? *Proceedings of the National Academy of Sciences*, **94**, 14530–5.

Menge, B. A., Gouhier, T. C., Freidenburg, T. and Lubchenco, J. (2011). Linking long-term, large-scale climatic and environmental variability to patterns of marine invertebrate recruitment: toward explaining 'unexplained' variation. *Journal of Experimental Marine Biology and Ecology*, **400**, 236–49.

Menge, B. A., Gouhier, T. C., Hacker, S. D., Chan, F. and Nielsen, K. J. (2015). Are meta-ecosystems organized hierarchically? A model and test in rocky intertidal habitats. *Ecological Monographs*, **85**, 213–33.

Menge, B. A., Lubchenco, J., Bracken, M. et al. (2003). Coastal oceanography sets the pace of rocky intertidal community dynamics. *Proceedings of the National Academy of Sciences*, **100**, 12229–34.

Menge, B. A. and Menge, D. N. (2013). Dynamics of coastal meta-ecosystems: the intermittent upwelling hypothesis and a test in rocky intertidal regions. *Ecological Monographs*, **83**, 283–310.

Menge, B. A. and Olson, A. M. (1990). Role of scale and environmental factors in regulation of community structure. *Trends in Ecology and Evolution*, **5**, 52–7.

Menge, B. A., Olson, A. M. and Dahlhoff, E. P. (2002). Environmental stress, bottom-up effects, and community dynamics: integrating molecular-physiological and ecological approaches. *Integrative and Comparative Biology*, **42**, 892–908.

Menge, B. A. and Sutherland, J. P. (1987). Community regulation: variation in disturbance, competition, and predation in relation to environmental stress and recruitment. *The American Naturalist*, **130**, 730–57.

Mills, L. S., Soulé, M. E. and Doak, D. F. (1993). The keystone-species concept in ecology and conservation. *BioScience*, **43**, 219–24.

Miner, C. M., Altstatt, J. M., Raimondi, P. T. and Minchinton, T. E. (2006). Recruitment failure and shifts in community structure following mass mortality limit recovery prospects of black abalone. *Marine Ecology Progress Series*, **327**, 107–17.

Monaco, C. J., Wethey, D. S. and Helmuth, B. (2016). Thermal sensitivity and the role of behavior in driving an intertidal predator–prey interaction. *Ecological Monographs*, **86**, 429–47.

Mote, P. W., Gavin, D., Huyer, A. et al. (2010) Climate Change in Oregon's Land and Marine Environments. In K. D. Dello and P. W. Mote, eds. *Oregon Climate Assessment Report*. Oregon State University, Corvallis.

Muhs, D. R., Groves, L. T. and Schumann, R. R. (2014). Interpreting the paleozoogeography and sea level history of thermally anomalous marine terrace faunas: a case study from the Last Interglacial Complex of San Clemente Island, California. *Monographs of the Western North American Naturalist*, **7**, 82–108.

Nickols, K., Gaylord, B. and Largier, J. (2012). The coastal boundary layer: predictable current structure decreases alongshore transport and alters scales of dispersal. *Marine Ecology Progress Series*, **464**, 17–35.

Paine, R. T. (1966). Food web complexity and species diversity. *The American Naturalist*, **100**, 65–75.

Paine, R. T. (1969). The *Pisaster-Tegula* interaction: prey patches, predator food preference, and intertidal community structure. *Ecology*, **50**, 950–61.

Paine, R. (1974). Intertidal community structure. *Oecologia*, **15**, 93–120.

Paine, R. (1976). Biological observations on a subtidal *Mytilus californianus* bed. *Veliger*, **19**, 125–30.

Pelc, R., Warner, R. and Gaines, S. (2009). Geographical patterns of genetic structure in marine species with contrasting life histories. *Journal of Biogeography*, **36**, 1881–90.

Petes, L. E., Mouchka, M. E., Milston-Clements, R. H., Momoda, T. S. and Menge, B. A. (2008). Effects of environmental stress on intertidal mussels and their sea star predators. *Oecologia*, **156**, 671–80.

Pfister, C. A., Roy, K., Wootton, J. T. et al. (2016) Historical baselines and the future of shell calcification for a foundation species in a changing ocean. *Proceedings of the Royal Society B*, 20160392.

Pincebourde, S., Sanford, E., Casas, J. and Helmuth, B. (2012). Temporal coincidence of environmental stress events modulates predation rates. *Ecology Letters*, **15**, 680–8.

Pincebourde, S., Sanford, E. and Helmuth, B. (2008). Body temperature during low tide alters the feeding performance of a top intertidal predator. *Limnology and Oceanography*, **53**, 1562–73.

Power, M. E., Tilman, D., Estes, J. A. et al. (1996). Challenges in the quest for keystones. *BioScience*, **46**, 609–20.

Raimondi, P., Jurgens, L. J. and Tinker, M. T. (2015). Evaluating potential conservation conflicts between two listed species: sea otters and black abalone. *Ecology*, **96**, 3102–8.

Raimondi, P. T., Wilson, M. C., Ambrose, R. F., Engle, J. M. and Minchinton, T. E. (2002). Continued declines of black abalone along the coast of California: are mass mortalities related to El Niño events? *Marine Ecology Progress Series* **242**, 143–52.

Reid, D. (2002). The genus *Nodilittorina* von Martens, 1897 (Gastropoda: Littorinidae) in the eastern Pacific Ocean, with a discussion of biogeographic provinces of the rocky-shore fauna. *Veliger*, **45**, 85–170.

Ricketts, E. F., Calvin, J., Hedgpeth, J. W. and Phillips, D. W. (1985). *Between Pacific Tides*, fifth edn. Stanford University Press, Palo Alto.

Riddle, B. R., Hafner, D. J., Alexander, L. F. and Jaeger, J. R. (2000). Cryptic vicariance in the historical assembly of a Baja California peninsular desert biota. *Proceedings of the National Academy of Sciences*, **97**, 14438–43.

Riginos, C. (2005). Cryptic vicariance in Gulf of California fishes parallels vicariant patterns found in Baja California mammals and reptiles. *Evolution*, **59**, 2678–90.

Rivadeneira, M. M., Alballay, A. H., Villafaña, J. A., Raimondi, P. T., Blanchette, C. A. and Fenberg, P. B. (2015). Geographic patterns of diversification and the latitudinal gradient of richness of rocky intertidal gastropods: the 'into the tropical museum'hypothesis. *Global Ecology and Biogeography*, **24**, 1149–58.

Robles, C. D., Desharnais, R. A., Garza, C., Donahue, M. J. and Martinez, C. A. (2009). Complex equilibria in the maintenance of boundaries: experiments with mussel beds. *Ecology*, **90**, 985–95.

Sagarin, R. D., Ambrose, R. F., Becker, B. J. et al. (2007). Ecological impacts on the limpet *Lottia gigantea* populations: human pressure over a broad scale on island and mainland intertidal zones. *Marine Biology*, **150**, 399–413.

Sagarin, R. D., Barry, J. P., Gilman, S. E. and Baxter, C. H. (1999). Climate-related change in an intertidal community over short and long time scales. *Ecological Monographs*, **69**, 465–90.

Sagarin, R. D. and Gaines, S. D. (2002a). Geographical abundance distributions of coastal invertebrates: Using one-dimensional ranges to test biogeographic hypotheses. *Journal of Biogeography*, **29**, 985–97.

Sagarin, R. D. and Gaines, S. D. (2002b). The 'abundant centre'distribution: to what extent is it a biogeographical rule? *Ecology Letters*, **5**, 137–47.

Sagarin, R. D., Gaines, S. D. and Gaylord, B. (2006). Moving beyond assumptions to understand abundance distributions across the ranges of species. *Trends in ecology and evolution*, **21**, 524–30.

Sanford, E. (1999). Regulation of keystone predation by small changes in ocean temperature. *Science*, **283**, 2095–7.

Sanford, E. (2002). Water temperature, predation, and the neglected role of physiological rate effects in rocky intertidal communities. *Integrative and Comparative Biology*, **42**, 881–91.

Sexton, J. P., McIntyre, P. J., Angert, A. L. and Rice, K. J. (2009). Evolution and ecology of species range limits. *Annual Review of Ecology, Evolution, and Systematics*, **40**, 415–36.

Shanks, A. L. (2009). Pelagic larval duration and dispersal distance revisited. *The Biological Bulletin*, **216**, 373–85.

Shanks, A. L., Walser, A. and Shanks, L. (2014). Population structure, northern range limit, and recruitment variation in the intertidal limpet *Lottia scabra*. *Marine Biology*, **161**, 1073–86.

Smith, G., Stamm, C. and Petrovic, F. (2003). *Haliotis cracherodii*. The IUCN Red List of Threatened Species, e.T41880A10566196, www.iucnredlist.org/species/41880/10566196.

Steinbeck, J. and Ricketts, E. F. (1951). *The Log from the Sea of Cortez: The Narrative Portion of the Book, Sea of Cortez*. Viking Press, New York.

Stewart, J. R. and Lister, A. M. (2001). Cryptic northern refugia and the origins of the modern biota. *Trends in Ecology and Evolution*, **16**, 608–13.

Stimson, J. (1970). Territorial behavior of the owl limpet, *Lottia gigantea*. *Ecology*, **51**, 113–18.

Stokstad, E. (2014). Death of the stars. *Science*, **344**, 464–7.

Sydeman, W. J., Garcia-Reyes, M., Schoeman, D. S. et al. (2014). Climate change and wind intensification in coastal upwelling ecosystems. *Science*, **345**, 77–80.

Szathmary, P. L., Helmuth, B. and Wethey, D. S. (2009). Climate change in the rocky intertidal zone: predicting and measuring the body temperature of a keystone predator. *Marine Ecology Progress Series*, **374**, 43–56.

Thompson, R. M., Beardall, J., Beringer, J., Grace, M. and Sardina, P. (2013). Means and extremes: building variability into community-level climate change experiments. *Ecology Letters*, **16**, 799–806.

Valentine, J. W. and Jablonski, D. (1993) Fossil Communities: Compositional Variation at Many Time Scales. In R. Ricklefs and D. Schluter, eds. *Species Diversity in Ecological Communities.* University of Chicago Press, Chicago, pp. 341–9.

Varela, R., Alvarez, I., Santos, F., deCastro, M. and Gomez-Gesteira, M. (2015). Has upwelling strengthened along worldwide coasts over 1982–2010? *Scientific Reports*, **5**.

Wang, D. W., Gouhier, T. C., Menge, B. A. and Ganguly, A. R. (2015). Intensification and spatial homogenization of coastal upwelling under climate change. *Nature*, **518**.

Wares, J. P., Gaines, S. and Cunningham, C. W. (2001). A comparative study of asymmetric migration events across a marine biogeographic boundary. *Evolution*, **55**, 295–306.

Wright, W. (1989). Intraspecific density mediates sex-change in the territorial patellacean limpet *Lottia gigantea*. *Marine Biology*, **100**, 353–64.

Yamane, L. and Gilman, S. E. (2009). Opposite responses by an intertidal predator to increasing aquatic and aerial temperatures. *Marine Ecology Progress Series*, **393**, 27–36.

Zacherl, D., Gaines, S. D. and Lonhart, S. I. (2003). The limits to biogeographical distributions: insights from the northward range extension of the marine snail, *Kelletia kelletii* (Forbes, 1852). *Journal of Biogeography*, **30**, 913–24.

Zaytsev, O., Cervantes-Duarte, R., Montante, O. and Gallegos-Garcia, A. (2003). Coastal upwelling activity on the Pacific shelf of the Baja California Peninsula. *Journal of Oceanography*, **59**, 489–502.

# Chapter 11

# The North-East Pacific

## *Interactions on Subtidal Hard Substrata*

Eliza C. Heery and Kenneth P. Sebens

## 11.1 Introduction

The north-east Pacific is a nutrient-rich and highly productive coastal region spanning from the Aleutian Islands of Alaska to the Baja Peninsula in Mexico (Figure 11.1; Foster, 1990). The region supports a variety of benthic habitat types (Lauerman et al., 1996; Tsurumi and Tunnicliffe, 2001; Love et al., 2006; Burd et al., 2008; McClain et al., 2009; Ruesink et al., 2010; Watson and Estes, 2011; Simkanin et al., 2012; Elahi et al., 2013; Piacenza et al., 2015), including extensive hard-substrate habitat in the shallow subtidal zone (Grantham et al., 2003; Britton-Simmons, 2006; Britton-Simmons et al., 2009; Konar et al., 2009; Watson and Estes, 2011; Elahi et al., 2013). Benthic assemblages in the north-east Pacific are unusually diverse compared with other temperate regions (Witman et al., 2004; Fautin et al., 2010), and comprise relatively large-bodied taxa (Estes et al., 2005; Vermeij, 2012), including a variety of large Laminariales (Foster and Schiel, 1985; Steneck et al., 2002; Witman et al., 2004). Several kelps of the north-east Pacific grow to form great canopies that tower tens of metres above the seafloor, providing three-dimensional structure for a vast array of flora and fauna (Gaylord et al., 2004; Siddon et al., 2008; Konar et al., 2009; Bertocci et al., 2015; Trebilco et al., 2015). These kelp forest ecosystems support complex trophic networks with a variety of higher-level consumers, including humans (Estes et al., 1998; Anthony et al., 2008; Szpak et al., 2013; Markel and Shurin, 2015), and may have facilitated human migration into North America in the late Pleistocene (Erlandson et al., 2007; Graham et al., 2009). High diversity and complex species interactions are also characteristic of the invertebrate-dominated vertical rock wall habitats that occur throughout the region, both within and separate from kelp forests (Witman et al., 2004; Elahi et al., 2013). Additionally, artificial structures, such as oil rigs, artificial reefs and offshore wave energy systems, increasingly act as habitat for hard-substrate assemblages in the north-east Pacific (Jessee et al., 1985; Love et al., 2006).

In this review, we present recent findings on the relationships among species and the factors that influence species interactions in shallow subtidal communities associated with hard substrate in the north-east Pacific. The geographic domain considered in this review includes nearshore areas from as far north as Prince William Sound, Alaska (60° 41'N latitude) to the tip of the Baja Peninsula, at Cabo San Lucas, Mexico (22° 53'N latitude). Literature on subtidal ecosystems in the Gulf of California (Thomson et al., 2000; Hernández-Carmona et al., 2001; Arreguın-Sánchez et al., 2002; Morales-Zárate et al., 2004; Riosmena-Rodríguez et al., 2005; Morzaria-Luna et al., 2013), the Bering Sea and the Chukchi

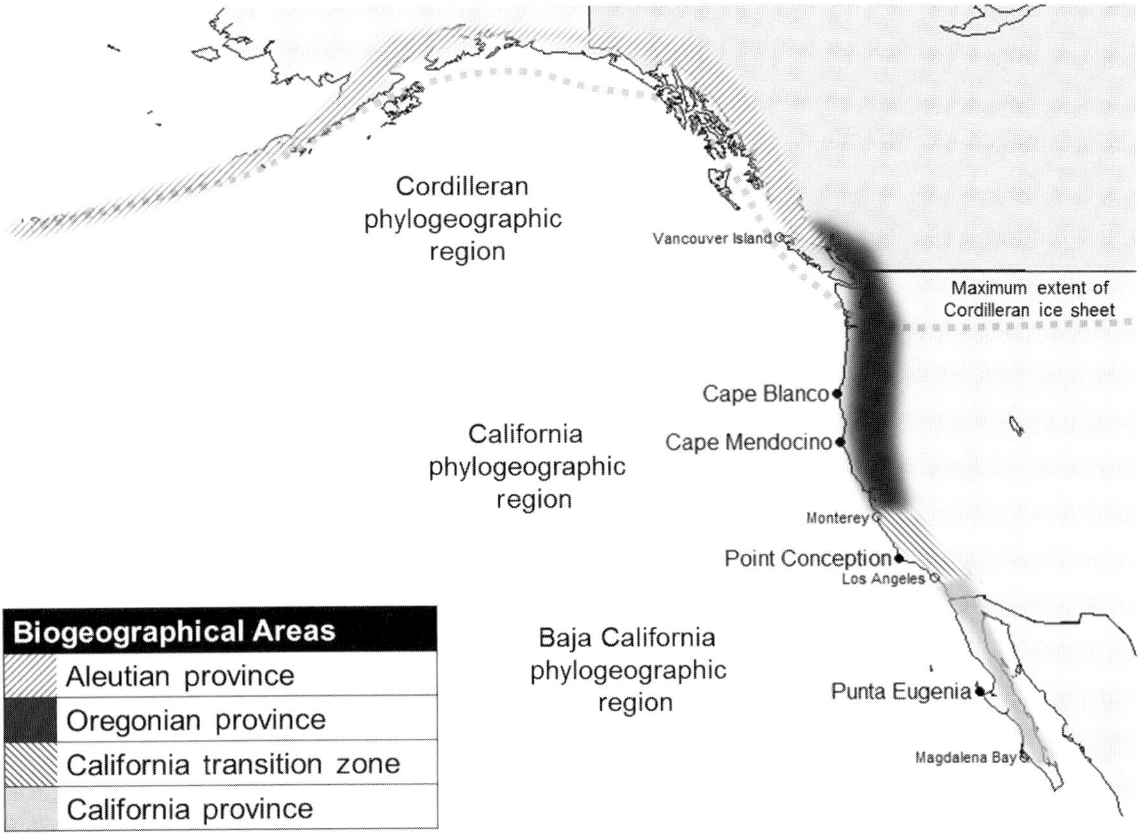


**Fig. 11.1** Map of the north-east Pacific, with the major biogeographical boundaries, as designated by Briggs and Bowen (2012) and the phylogeographic regions characterised by Dawson et al. (2006). The approximate maximum extent of the Cordilleran ice sheet is marked by the dashed line.

Sea, although extensive (Stoker, 1978; Carey, 1991; Gutt, 2001; Sirenko and Gagaev, 2007; McCormick-Ray et al., 2011; Blanchard, 2015; Datsky, 2015), is excluded from the present discussion. Furthermore, within the geographic domain considered here, we focus on research from the shallow subtidal zone, although deeper nearshore ecosystems (>30 m) in the region have also been explored in recent studies (Spalding et al., 2003; Dominik and Zimmerman, 2006; Lowe et al., 2015).

Rocky benthic assemblages in the north-east Pacific have been a focal point of subtidal research for many decades. The development of diving-based research methods in the 1950s and 1960s, and subsequent advances in diving technology (Lang et al., 2013), triggered a surge in published literature from the region (Neushul, 1967; North and Hubbs, 1968; North and Pearse, 1970; Dayton, 1975; Foreman, 1977; Calvin and Ellis, 1978; Estes et al., 1978; Lindstrom and Foreman, 1978). Findings from the 1960s through the 1990s have been summarised in numerous, excellent reviews by Dayton (1985), Schiel and Foster (1986), Tegner and Dayton (1987), Foster (1990, 1992) and Steneck et al. (2002). However, recent advances in computing, chemistry, genetics and other fields have since provided further insight into the interaction among species, and between species and the environment. North-east Pacific ecosystems have also undergone rapid and substantial abiotic and biotic shifts as a result of anthropogenic changes, particularly over the last decade (Gruber et al., 2012; Hewson et al., 2014;

Schultz et al., 2016). Our primary objective is to highlight recent advances in subtidal community ecology in the north-east Pacific and summarise major abiotic and biotic shifts currently influencing benthic assemblages in the region. We start by providing context with respect to the region's major oceanographic and climatic processes and important functional groups. This is followed by an overview of biogeographic and phylogeographic patterns, as well as assemblage patterns associated with temporal and spatial variation in meso-scale processes. We then discuss abiotic processes that are particularly important in shaping localised gradients in assemblage structure and close with an overview of the major anthropogenic changes affecting benthic assemblages in the north-east Pacific.

## 11.2 Oceanographic and Climatic Context

The north-east Pacific is characterised as warm- to cold-temperate and receives cold water from the North Pacific Current, which travels eastward between 42 and 52°N latitude (Sydeman et al., 2011). The North Pacific Current splits into two major coastal current systems as it approaches the North American coast (Strub et al., 2013). Approximately 60 per cent of the water delivered by the North Pacific Current diverges to the north along the coast of British Columbia and circulates in Gulf of Alaska as the Alaska Current (Royer, 1983; Freeland, 2006). Water diverging to the south forms the California Current (Lynn and Simpson, 1987), which travels as far as the Baja Peninsula (Checkley and Barth, 2009), with a major discontinuity in the vicinity south of Point Conception known as the southern California countercurrent (Figure 11.1).

Both the Alaska and California current systems are heavily influenced by seasonal upwelling. In spring–summer, cyclonic winds along the coast of Alaska and north to south winds along the remainder of the north-east Pacific coast push surface waters offshore through Ekman transport. The displaced surface water is then replaced with cold, nutrient-rich water from depth (Mann et al., 2006). Upwelling of deep-water layers fundamentally changes physical conditions in nearshore areas, increasing salinity, decreasing sea surface temperature (SST), reducing pH and increasing concentrations of nitrate and silicate nutrients that fuel planktonic productivity (Huyer, 1983). The strength of upwelling and extent of these changes is highly variable over multiple spatial scales. Upwelled water is diverted by coastal features such as headlands and peninsulas (Graham and Largier, 1997), bathymetric features (Rosenfeld et al., 1994) and fluvial input (Hickey and Banas, 2003), leading to localised differences in oceanographic conditions. Differences in upwelling strength between long stretches of coastline occur due to large-scale variation in surface winds and other climatic forces (Strub et al., 2013), and are important drivers of biogeographic and phylogeographic patterns (Dawson et al. 2006, discussed in Sections 11.4 and 11.5).

In addition to seasonal upwelling, oceanographic conditions in north-east Pacific nearshore environments are heavily influenced by inter-annual and decadal climatic processes, such as El Niño, the Pacific decadal oscillation (PDO) and the North Pacific gyre oscillation (NPGO). El Niño refers to periodic large-scale warming that occurs over the equatorial Pacific. It leads to a variety of changes in the north-east Pacific (Emery and Hamilton, 1985), including reduced upwelling intensity, higher SST, lower nutrient availability and stronger, more frequent storms (Tegner and Dayton, 1987). The PDO is an oceanographic index that represents a cyclical climatic pattern of alternating warm and cool phases in the North Pacific Ocean every twenty to thirty years (Mantua et al., 1997). The NPGO is an additional index that reflects low-frequency changes in upwelling strength, nutrient availability and phytoplankton abundance in surface water and is tied to variation in ocean gyre circulation (Di Lorenzo et al., 2008). Decadal and low-frequency variation in oceanographic conditions influence nearshore productivity (Nakamura et al., 1997; Francis et al., 1998; Di Lorenzo et al., 2010) and have considerable implications for benthic assemblages, which we discuss later.

In addition to these major meso-scale processes, recent work has identified a variety of novel oceanographic patterns and large-scale abiotic changes that likely affect benthic assemblages in the region. Some of these changes, such as rapid acidification of surface waters and greater fluvial influx from melting glaciers (Feely et al., 2008; Hood et al., 2009), have clear anthropogenic drivers. The mechanisms behind other changes are less certain. For instance, in late 2013, a large mass of unusually warm surface water formed over the north-east Pacific that had not been previously documented (Bond et al., 2015). The warm surface water anomaly, which became known as 'the Blob', persisted until late 2015, causing significant changes to nearshore SST. The Blob had considerable impacts on both pelagic and benthic ecosystems (Cavole et al., 2016; Basilio et al., 2017; Brodeur et al., 2018; Chauvet et al., 2018; Feehan et al., 2018; Pineda et al., 2018; Yang et al., 2018), although many effects and evaluations of potential drivers were still in the process of being documented in the literature at the time this edition went to press.

## 11.3 Major Functional Groups and Species Interactions

Subtidal hard-substrate biota in north-east Pacific ecosystems have commonly been characterised into several key functional groups (Steneck et al., 2002; Estes et al., 2004). Canopy-forming and understorey macroalgae play an important role in ecosystem dynamics, as they add structural complexity, serve as a source of primary production and a subsidy for adjacent ecosystems and have a variety of other functions (Bodkin, 1988; Duggins et al., 1989; Holbrook et al., 1990; Hamilton and Konar, 2007; Johnson et al., 2007; Miller and Gaylord, 2007; Arkema et al., 2009; Konar et al., 2009; Miller and Page, 2012; Rodriguez et al., 2013; Lowe et al., 2014; Trebilco et al., 2015; Clasen and Shurin, 2015). Kelp canopies significantly alter both the abiotic and biotic conditions for other organisms, reducing light availability and flow in benthic environments (Pearse and Hines, 1979). Reduced light from kelp canopies limits growth of understorey macrophytes and changes competitive interactions (Reed and Foster, 1984; Foster and Schiel, 1985; Stewart et al., 2009; Benes and Carpenter, 2015), while reduced flow increases the deposition of particulate matter and changes the conditions for macroalgal spore recruitment (Eckman et al., 1989; Gaylord et al., 2004).

Urchins and other consumers, such as chitons, are important drivers of macroalgal abundance and persistence (Estes and Palmisano, 1974; Dayton, 1975; Duggins, 1980; Watson and Estes, 2011), as well as invertebrate composition and assemblage structure (Levenbach, 2008; Elahi and Sebens, 2012, 2013). Extensive grazing by sea urchins can clear large patches on both horizontal and vertical subtidal rock surfaces (Elahi and Sebens, 2012). Urchin effects can be devastating, resulting in removal of most macroalgae and producing major changes in habitat complexity or 'architecture', which are often described as 'alternate states' of the subtidal community and can persist for decades (Simenstad et al., 1978). Removal of canopy-forming kelps also changes flow characteristics and causes benthic habitats to become less complicated hydrodynamically (shorter roughness elements), which may decrease flux of invertebrate larvae to hard surfaces (Graham and Sebens, 1996). Urchins are extreme generalists. Unlike most other benthic consumers that specialise on one or a few sessile species, urchins often remove all invertebrates except the largest cnidarians from rock surfaces. Species affected by urchin grazing include detritivores and suspension feeders, which may play an important role in food web dynamics, nutrient cycling and benthic–pelagic coupling (Levings et al., 1983; Duggins et al., 1989; Duggins and Eckman, 1997; Vetter, 1998). Urchins in turn serve as prey for a variety of secondary consumers, such as sea otters (Estes et al., 1998) and fish (Stephens et al., 2006; Davenport and Anderson, 2007), and can be relieved of predation pressure indirectly by top consumers, such as killer whales (Estes et al., 1998) and humans (Szpak et al., 2013).

Research in the north-east Pacific has emphasised the importance of top-down control in shaping benthic assemblages (Estes and Palmisano, 1974; Dayton, 1975; Duggins, 1980).

Theoretical development and empirical evidence supporting top-down models of kelp forest ecosystem dynamics originated in Alaska, where the re-establishment of sea otter populations had positive indirect effects on canopy-forming kelps via a trophic cascade (Duggins, 1980; Estes and Duggins, 1995). Trophic cascades involving sea otters, urchins and canopy-forming kelps became an iconic paradigm in the ecological literature and in public discourse (Estes et al., 1998, 2011). However, inconsistent findings from throughout the north-east Pacific, and limited evidence for top-down processes in California specifically, subsequently were fodder for lively debate (Foster and Schiel, 1988; Foster, 1990). Studies characterising the nature of kelp forest patch dynamics and top-down control have remained important topics in subtidal research in the region (Watson and Estes, 2011).

Rocky subtidal communities in the region have characteristics unlike those of adjacent rocky intertidal communities; zonation is broadened subtidally and monopolisation of space by single zone-forming sessile species is less evident (Connell and Keough, 1985). Community structure on subtidal rock surfaces is the result of intense competition for limited two-dimensional (2D) space (Sebens, 1985a, 1985b, 1986), and such space acquisition allows access to resources such as nutrients, prey or light. Researchers conducting subtidal research in the north-east Pacific often observe hundreds of species in just a few square metres of rock surface (Vance, 1988; Elahi and Sebens, 2012) and this diversity persists over long periods (Elahi et al., 2014). Coexistence on this relatively homogeneous resource (2D space) can result from a variety of mechanisms, including habitat specialisation, differential susceptibility to predation (Elahi and Sebens, 2012, 2013), interactions between adults and settling larvae (Tegner and Dayton, 1981), and combinations of inverse recruitment and competitive abilities under variable disturbance regimes (i.e., fugitive-dominant model: Armstrong, 1976; Connell and Slatyer, 1977; Hastings, 1988; Sebens, 1991; Tilman, 1994). Adaptive strategies for acquiring and maintaining space vary widely even within closely related taxa. For instance, algal crusts can restrict the settlement of kelp sporophytes, although coexistence is maintained by high densities of adult sporophytes relative to recruits (Okamoto et al., 2013).

Direct and indirect facilitative interactions have also been an important part of recent research on benthic community dynamics in the north-east Pacific. Crustose algae that grow on gastropod shells reduce the susceptibility of gastropods to predation by sea stars (Thornber, 2007). The inhibition of growth among understorey macroalgae by canopy-forming kelp may have a positive indirect effect on sessile invertebrates that compete with understorey macroalgae for space, and may be considerably more important for invertebrate assemblages than potential direct effects from kelp canopy cover (Arkema et al., 2009). Similarly, some canopy-forming kelps may facilitate red macroalgae by reducing competition from other understorey species (Benes and Carpenter, 2015). Levenbach (2008) found that urchins avoid contact with some benthic cnidarians, and this creates an associational refuge for filamentous and turf macroalgae in urchin barren habitats, even though the same Cnidarian species can have a negative effect on filamentous turf macroalgae in kelp forest habitats, where space is a limiting resource. The same mechanism can also support high densities of amphipods (Levenbach, 2008), which may additionally influence competitive interactions between understorey macroalgae (Levenbach, 2008).

## 11.4 Biogeographic Patterns

The region considered in this review includes three marine biogeographical provinces (Briggs, 1974) and one recently designated transition zone: the Aleutian province (north of approximately 50°N latitude), the Oregonian province (approximately 50–37°N latitude), the California transition zone (approximately 37–34°N latitude) and the Californian province (south of approximately 34°N latitude), which is also called the San Diegan province by some sources (Horn et al., 2006; Briggs and Bowen, 2012; Figure 11.1). The boundaries of these provinces are areas where the ranges of multiple marine species terminate (Briggs and Bowen, 2012). Smaller-scale (within-province) regions were also derived early-on based on the range limits of molluscs (Hall, 1964; Valentine, 1966), and appear to coincide relatively well with distribution patterns of a broader range of intertidal taxa (Blanchette et al., 2008). Although these finer biogeographical divisions have yet to be systematically tested using a broad array of subtidal taxa (although see review by Lindstrom, 2009), they coincide with oceanographic differences known to affect subtidal organisms (Botsford, 2001; Hickey and Banas, 2003) and may reflect biogeographic divisions in subtidal communities as well.

Numerous studies have documented latitudinal gradients in benthic assemblage structure and species interactions in the north-east Pacific (Scagel, 1963; Hall, 1964; Valentine, 1966; Addicott, 1970; Murray et al., 1980; Cowen, 1985; Blanchette et al., 2002; Valentine et al., 2002; Pelc et al., 2009; Cope and Haltuch, 2012; West et al., 2014; Zahn et al., 2016). Perhaps the most obvious differences in benthic assemblages with latitude are those associated with macroalgal composition (Druehl, 1970; Murray et al., 1980). *Macrocystis pyrifera* (Laminariaceae) (hereafter *Macrocystis*, Figure 11.2) is the predominant canopy-forming kelp species from Baja to Central California and is accompanied by a diverse array of understorey kelps, including *Eisenia arborea*, *Pterygophora californica*, *Dictyoneuropsis reticulate* and *Laminaria farlowii*, and various foliose red macroalgal species (Breda and Foster, 1985; Foster and Schiel, 1985; Graham, 2004). Further to the north, kelp beds are primarily formed by a different canopy-forming species, *Nereocystis luetkeana* (Laminariaceae) (hereafter *Nereocystis*, Figure 11.2; Abbott and Hollenberg, 1976; Foreman, 1984). *Nereocystis* beds are common on rocky shores from the San Francisco metropolitan area to Alaska (Druehl, 1970; Shaffer, 2000), although mixed stands of *Macrocystis* and *Nereocystis* also occur, particularly in northern California (Foster and Schiel, 1985; Bodkin, 1986; Stephens et al., 2006). An additional canopy-forming kelp, *Alaria fistulosa* (Alariaceae), is also found in Alaska and the Aleutian Islands (Dayton, 1975; Hamilton and Konar, 2007). *Nereocystis* and *A. fistulosa* kelp forests are also accompanied by a variety of understorey kelps, particularly from the genera *Saccharina*, *Laminaria*, *Costaria* and *Agarum* (Dean et al., 2000; Tegner and Dayton, 2000; Duggins et al., 2003; Hamilton and Konar, 2007), although *Pterogophora* is also common, particularly in Washington (Shaffer, 2000). While there are clear latitudinal gradients in the dominant canopy-forming and understorey species, many individual macrophyte species are broadly distributed (Druehl, 1970) and their geographic range can be patchy, leading to variation in macroalgal composition over time and space. For instance, *Macrocystis* has a discontinuous distribution along the north-east Pacific coast, forming expansive kelp forests in parts of British Columbia

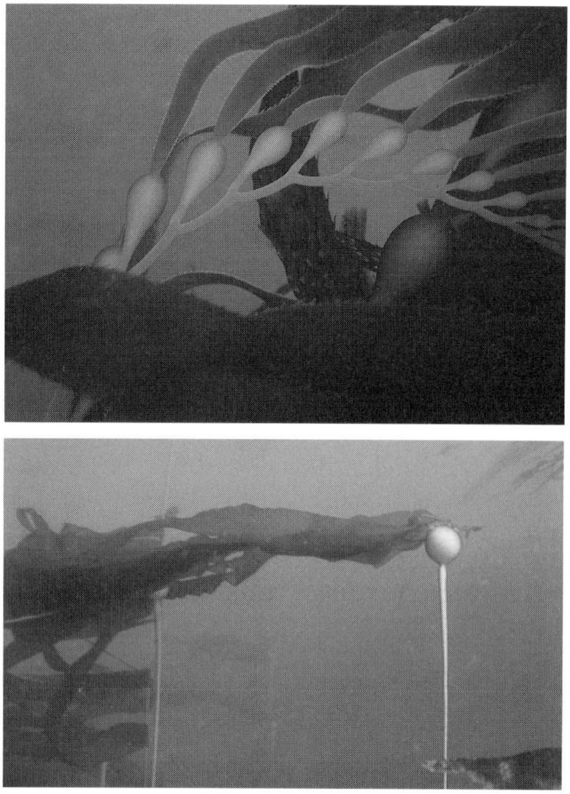

**Fig. 11.2** Pneumatocysts of two major canopy-forming kelp species in the north-east Pacific: (a) *Macrocystis*, which is most abundant in south of central California, and (b) *Nereocystis*, which primarily occurs north of central California. Photo from Rhoda Green. *A black and white version of this figure will appear in some formats. For the colour version, please refer to the plate section.*

(Watson and Estes, 2011; Clasen and Shurin, 2015), while being absent in much of Washington and Oregon. It should be noted that these distributions are already shifting as ocean temperature increases, as demonstrated by the northward movement of central California species (Sagarin et al., 1999).

North-east Pacific sessile invertebrate and demersal fish assemblages also differ considerably across a latitudinal gradient (Stephens et al., 2006; Briggs and Bowen, 2012), as do upper trophic levels. Several species of urchin (primarily Strongylocentrotidae) are found throughout much of the north-east Pacific, from Alaska to Baja, including the red urchin, *Strongylocentrotus franciscanus*, purple urchin, *Strongylocentrotus purpuratus*, and green urchin, *Strongylocentrotus droebachiensis* (Palumbi and Wilson, 1990; Debenham et al., 2000; Lester et al., 2007; Hammond and Hofmann, 2010). However, their enigmatic predator, the sea otter, *Enhydra lutris*, has a patchy distribution that extends only as far south as central California due to overharvest by the maritime fur trade (Estes and Palmisano, 1974; Estes et al., 1998). Urchin populations are also mediated by the predatory sea star, *Pycnopodia helianthoides* (Asteriidae) (Duggins, 1983), which is most abundant in northern waters (Shivji et al., 1983). In southern California and Baja, urchin populations are subject to top-down pressure from other predators, such as the California spiny lobster, *Panulirus interruptus* (Palinuridae) and California sheephead, *Semicossyphus pulcher* (Labridae) (Tegner and Dayton, 1981; Tegner and Levin, 1983).

Latitudinal differences in community composition in the north-east Pacific have historically been attributed to temperature (Vadas, 1972; Druehl, 1978; Lüning and Freshwater, 1988). This has been echoed in recent studies (Pondella et al., 2005; Matson and Edwards, 2007; Kordas et al., 2011; Zahn et al., 2016), although it is unclear whether temperature primarily influences benthic assemblages directly or through indirect mechanisms (Lubchenco et al., 1993). Warmer temperatures (and related nutrient reductions) limit growth of some kelps in the region (Vadas, 1972; Zimmerman and Kremer, 1986; Lüning and Freshwater, 1988), reduce reproductive success of limpets and other molluscs (Fritchman, 1961; Hall, 1964) and may set the northern range limit for many species. Temperature may also change consumer behaviour and abundance, leading to potential indirect temperature-driven controls of subtidal assemblage structure (Azad et al., 2011). Temperature is clearly among the drivers of latitudinal gradients in assemblage structure (Belanger et al., 2012) and of biogeographic patterns within subregions of the north-east Pacific, such as the Southern California Bight (Pondella et al., 2005; Zahn et al., 2016) and the Aleutian Islands (Miller and Estes, 1989). However, it is known to covary with a variety of other abiotic factors (reviewed by Wheeler and Neushul, 1981), including nutrients (Jackson, 1977) and light, which is reduced by phytoplankton blooms that occur in low temperature/high nutrient conditions. Discerning between the effects of these physical factors on macrophytes and other benthic species can be difficult (Dayton, 1985; Schiel and Foster, 2015).

Differences in the composition of canopy-forming and understorey kelps, primary consumers and higher-level consumers with latitude may be among reasons for observed discrepancies in the strength of top-down interactions throughout the region. Early work in Alaska demonstrated that otter-mediated trophic cascades enhanced the abundance of canopy-forming kelps by reducing densities of grazing urchins (Estes and Palmisano, 1974; Dayton, 1975; Duggins, 1980). Subsequent work has provided further support for the importance of top-down processes in Alaska (Estes and Duggins, 1995; Estes et al., 1998, 2004; Anthony et al., 2008), and the top-down paradigm from Alaska's kelp forests has since been extended well beyond its origins, to the world at large (Pace et al., 1999; Estes et al., 2011). There is now support for urchin-mediated trophic cascades from other parts of the north-east Pacific as well. In an exceptionally thorough study conducted over twenty-three years, Watson and Estes (2011) demonstrated trophic-mediated patterns in the distribution of kelp beds in two areas on the outer coast of Vancouver Island, British Columbia. The presence of sea otters limited the abundance of red urchins, which in turn led to a transition to an algal-dominated assemblage (Watson and Estes,

2011). In addition, rockfish inhabiting algal-dominated assemblages facilitated indirectly by otters had a higher mean trophic position and condition factor, and a smaller trophic niche space as they were able to specialise on fish and avoid lower trophic level prey items (Markel and Shurin, 2015). The dissolved organic matter produced by kelp patches also enhanced microbial activity, and the size of kelp forests was directly related to the abundance and growth rate of alginate-lyase-producing bacteria and to the grazing activity of zooplankton (Clasen and Shurin, 2015).

However, empirical support for keystone predators and trophic cascades has been less evident in California, raising questions over the generality of top-down control as an important driver across the region (Foster, 1990). While some studies found evidence of urchin barrens in California waters (North and Pearse, 1970; Ebeling et al., 1985), others found that urchin aggregations were relatively rare and did not lead to deforestation of *Macrocystis* beds (Foster and Schiel, 1988). Furthermore, given the small-scale variation in abiotic and biotic factors that influence kelp forest dynamics (Pearse and Hines, 1979; Harrold and Reed, 1985; Ebert and Russell, 1988; Laur et al., 1988), there were questions from researchers conducting field work in California about whether the top-down paradigm was in fact generalisable in other parts of the north-east Pacific, and in Alaska, where it was initially documented (Foster, 1990). In an important paper on this subject, Foster (1990) states, 'until unbiased site descriptions and appropriate experiments are available for the rest of the north-east Pacific, the question of generality for areas outside of California remains unanswered'.

There are numerous factors that may explain the discrepancy in observations regarding top-down control between southern and northern regions within the north-east Pacific. Human population density and exploitation of nearshore resources has differed with latitude. Southerly *Macrocystis* beds have a long history of human exploitation and a variety of industrial uses (Tegner and Dayton, 2000), while extraction of northern *Nereocystis* forests has been relatively limited (Springer et al., 2010), relieving it of a certain degree of anthropogenic top-down pressure. Exploitation of sea urchin populations has also varied spatially. In the Salish Sea, for instance, *S. franciscanus* has been harvested extensively, and populations are probably still at historically low levels. Perhaps as a result of low urchin population density, experimental studies to determine the potential effect of sea otters migrating into in the San Juan Archipelago showed that increased predation on urchins had minimal effects on community structure in the subtidal zone (Carter and VanBlaricom, 2002; Carter et al., 2007).

Latitudinal differences in the identify of natural predators of urchins is also likely to be important in determining the strength of top-down control. Dunn and Hovel (2019) found highly variable urchin mortality inflicted by California sheephead and a saturating functional response among predatory spiny lobsters relative to the density of urchin prey in southern California. Predator-induced density-dependent mortality was observed for purple urchins only, and occurred only below a certain threshold of urchin density (Dunn and Hovel, 2019). Such patterns among urchins and the prevailing urchin predators, as well as alternative trait-mediated trophic cascade pathways (Haggerty et al., 2018), could be major sources of variability in the strength of top-down control in the southern part of the north-east Pacific. By comparison, mass mortality of the predatory sea stars, *P. helianthoides*, has recently underscored the importance of top-down control and highlighted potential predator complementarity in northern parts of the north-east Pacific (Burt et al., 2018). The strength of top-down processes may change with changes in environmental conditions, however, as indicated by recent decoupling of predator and kelp population dynamics off the Washington coast (Shelton et al., 2018).

The relative importance of abiotic factors may also vary with latitude (Steneck et al., 2002; Low and Micheli, 2018; Okamoto et al., 2018), as could the relative susceptibility of northern versus southern species to abiotic disturbances. For instance, abiotic conditions in southern areas have a disproportionate effect on urchin survivorship and behaviour. Strongylocentrotidae species are more susceptible to storms (Ebeling et al., 1985) and disease (Lester et al., 2007) in the southern extent of their range. High-density aggregations of Strongylocentrotid urchins also disperse or decline rapidly following isolated deforestation events at southern sites (Watanabe and Harrold, 1991), but persist well after the disappearance of kelp beds in northern areas (Simenstad et al., 1978). Similarly, wave action during storms is a major driver of *Macrocystis* deforestation in California (Dayton et al., 1999). Storm-induced wave action is relatively intense in California compared with much of the north-east Pacific, but *Macrocystis* may also suffer high mortality from storm events because of its biomechanical properties. Compared with *Macrocystis*, *Nereocystis* thalli experience less drag from mid-water currents (Gaylord et al., 2004) and relatively limited mortality due to storms (Duggins et al., 2001).

It is possible that some physiological differences between *Nereocystis* and *Macrocystis* are, in turn, adaptive for differential levels of grazing intensity and overgrazing frequency. Studies of alternative life stages in *Nereocystis* suggest that their gametophyte stage may live endophytically in the tissue of filamentous red macroalgae (Garbary et al., 1999; Hubbard et al., 2004), although similar evidence for *Macrocystis* is lacking. Filamentous red macroalgae undergo lower grazing rates from *S. droebachiensis* than do other macrophytes (Heery et al., unpublished data) and could potentially serve as refugia during localised overgrazing events. Further, in *Nereocystis*, growth, spore release and flotation (via pneumatocysts) all rely on structures at or near the surface, well beyond the reach of urchins. Conversely, *Macrocystis* pneumatocysts are distributed throughout the plant, and both growth and spore release take place near the seafloor (Neushul and Haxo, 1963; Schmitz and Srivastava, 1976; Gaylord et al., 2004).

## 11.5 Phylogeographic Patterns

Localised patterns of diversity and community composition are heavily influenced by the regional species pool (Witman et al., 2004), which is the result of an ecosystem's evolutionary and paleogeographic history. The formation of present-day rocky subtidal assemblages in the north-east Pacific can be traced to the mid Eocene, when polar regions began to cool, leading to a southward shift of temperate zones (Briggs, 2003). Throughout subsequent periods of the Cenozoic, the formation and dissolution of the Bering land bridge, as well as extreme climatic gradients in polar regions, caused temperate biota in the Pacific and Atlantic to develop separately (Briggs, 2007). The opening of the Bering land bridge facilitated connectivity between the two regions relatively recently, starting between 5.3 and 3.5 million years ago (Gladenkov et al., 2002). However, gene flow during this time appears to primarily have occurred from the North Pacific to the North Atlantic, where species extinctions were prevalent (Jacobs et al., 2004). Although many Atlantic taxa are the result of this trans-Arctic interchange, few Pacific taxa are of Atlantic origin (Vermeij, 1991).

Several studies have sought to discern geographic patterns in the genetic structure of marine populations in the north-east Pacific (reviewed by Jacobs et al., 2004; Addison and Hart, 2005; Dawson et al., 2006). Gene flow across a species' range depends on dispersal potential and the extent of geographic isolation (Dawson et al. 2006). Phylogeographic patterns in the region are thus the result of a combination of species-specific life history traits, present-day meso-scale processes and past geologic, climatic and oceanographic factors (Jacobs et al., 2004; Dawson et al., 2006; Gottscho, 2016). Dawson et al. (2006) designate several relevant subregions within the north-east Pacific where certain processes have been particularly important in shaping genetic structure and that have thus been a focus in marine phylogeographic research: the Cordilleran, California and Pacific Baja California (Dawson et al., 2006).

The Cordilleran region encompasses the northern coastline from Alaska to northern Washington, where genetic structure of many species has been influenced by glaciation. Between 2.6 million years and 10,000 years ago, the Cordilleran ice sheet went through several cycles of expanding and contracting over the North American continent. At its maximum, the ice sheet reached as far south as northern Washington, creating the Puget Sound fjord system (Lyle et al., 2000). Glaciation caused SST to decrease (Dawson, 1992), leading to vicariance between northern and southern populations (Lindstrom et al., 1997; Wilson, 2006). The genetic structure expected to result from glaciation-induced divergence would have low genetic diversity and derived haplotypes in northern populations (Dawson et al., 2006). While this is evident in some species, there is considerable variation even among closely related taxa (Marko, 2004). The extent to which marine species were affected by glaciation may have depended on their specific fundamental niche. For instance, the genetic structure of several upper-intertidal species suggests that they recolonised from southern refuges following the last glacial maximum (LGM), when they may have been more affected by cold air temperatures than lower-intertidal invertebrates (Kelly and Palumbi, 2010). In addition to variable species responses to glaciation, the effects of multiple periods of glacial-induced divergence on genetic structure are complex. For instance, the Northern Clingfish (*Gobiesox maeandricus*) apparently found refugia from glaciation in the Strait of Georgia, as individuals there were found to have older haplotypes than the population that extends throughout the rest of the north-east Pacific (Hickerson and Ross, 2001). Alternatively, vicariance in northern and southern populations of Painted Greenling (*Oxylebius pictus*) was followed by secondary contact and high gene flow between isolated populations (Davis et al., 1981). Genetically distinct populations along a north–south axis can also occur as a result of current patterns within the Cordilleran region (Lindstrom et al., 1997; Rocha-Olivares and Vetter, 1999). The divergence of the North Pacific Current geographically overlaps with the southern extent of the LGM, further complicating the detection of phylogeographic patterns associated with glaciation (Dawson et al., 2006).

Phylogeographic patterns in the California and Pacific Baja California regions are the result of a combination of factors, but particularly tectonic activity and ocean currents (Alberto et al., 2011; Gottscho, 2016), although climatic conditions during glaciation were also important (Graham et al., 2009). Uplift and movement of the North American and Pacific plates have substantially altered coastal topography over the last few million years, leading to zoogeographic boundaries that are still evident in the genetic structure of marine populations (Gottscho, 2016). Several of these boundaries coincide with coastal features that act as upwelling jets, funnelling larvae along with upwelled water offshore (Sivasundar and Palumbi, 2010). Cape Mendocino, Point Conception and Punta Eugenia have all been highlighted as potential zoogeographic boundaries (Eberl et al., 2013), although Point Conception has been a point of some contention (Burton, 1998; Bernardi, 2000, 2005). Point Conception is considered to be a 'thermal barrier' for a variety of taxa, with warmer water to the south than to the north (Botsford and Lawrence, 2002). The area to the south of Point Conception has weaker winds during summer months and upwelling occurs at a different time entirely, in winter and early spring (Nidzieko and Largier, 2013). Genetic differences were originally detected north and south of Point Conception for rockfish (Love and Larson, 1978) and Painted Greenling (Davis et al., 1981). Subsequent studies on surfperch (Bernardi, 2000, 2005), gobies (Dawson, 2001) and macroalgae (Johansson et al., 2015) have indicated that a phylogeographic boundary is more evident further to the north, near Los Angeles.

Currents are generally highlighted in the literature as an important mechanism influencing larval dispersal of marine organisms in the north-east Pacific. Currents carry larvae of some organisms offshore and over relatively long distances (Gaines and Roughgarden, 1985; Gaylord and Gaines, 2000), while current variability and the behaviour and dispersal potential of larvae lead to small-scale variations in genetic structure (Palumbi, 2003; Kusumo et al., 2006; Mace and Morgan, 2006; Morgan et al., 2009) and even chaotic genetic patchiness (Selkoe et al., 2006). In addition to larvae, however, currents also transport other types of reproductive material that may link geographically separated populations. A recent effort to compile herbarium species records from throughout the region (Consortium for Pacific Northwest Herbarium, www.pnwherbaria.org) has highlighted disjunctive distributions for several macroalgal species between California and northern British Columbia. Using DNA-barcoding, Saunders (2014) identified thirty-three seaweed species that were found in both locations, despite their apparent absence along segments of coastline between these locations. This led to the development of a kelp conveyer hypothesis. Numerous invertebrates are associated with 'kelp rafting' (Hobday, 2000; Thiel and Gutow, 2005; Duggins et al., 2016) and some invertebrate species exhibit a similar disjunctive distribution pattern (Keever et al., 2009). This is an important area of future work.

## 11.6 Importance of Meso-scale Processes

North-east Pacific marine ecosystems are heavily influenced by a number of meso-scale processes. These processes cause strong temporal and spatial variation in abiotic factors, such as temperature and nutrients, that are important for benthic organisms. As marine assemblages respond to global change and a shifting climatic regime, elucidating the relationships between climatic forcing, oceanographic processes, population dynamics and community interactions in the north-east Pacific will be essential (King et al., 2011). Here we summarise current knowledge of the effects of meso-scale processes on benthic assemblages in the region.

### 11.6.1 Upwelling

Seasonal upwelling reduces SST and increases nutrient concentrations in the surface layers of nearshore habitats (Huyer, 1983). Nitrogen influx during upwelling is particularly important for macrophytes. *Macrocystis* receives the majority of its nitrates during upwelling periods (Fram et al., 2008), for instance, and nitrogen reserves acquired during this time may continue to fuel growth well after upwelling season has ended (Gerard, 1982). Nitrogen-limitation during years of low upwelling intensity may limit the formation of new beds and replenishment of existing stands of *Macrocystis* by reducing juvenile recruitment and survival (Hernández-Carmona et al., 2001). Additionally, upwelling events trigger peaks in phytoplankton abundance and primary productivity (Kokkinakis and Wheeler, 1987; Sackmann et al., 2004; Brzezinski and Washburn, 2011; Lucas et al., 2011), which could inhibit macroalgal growth by reducing light in benthos (Kavanaugh et al., 2009).

Upwelling-induced phytoplankton blooms have bottom-up effects in the rocky intertidal zone (Menge et al., 1997) and could provide an important food source for subtidal suspension feeders as well (Miller and Page, 2012). Recent studies using stable isotope analysis suggest that phytoplankton make-up the bulk of organic material consumed by suspension feeders in *Macrocystis* kelp forests of central and southern California (Page et al., 2008; Miller et al., 2013; Yorke et al., 2013; Koenigs et al., 2015). Direct evidence of this type of benthic–pelagic coupling is elusive, however, and benefits of upwelling-induced phytoplankton blooms for suspension feeders are not well documented (although see Sanford and Menge, 2001). Some suspension feeders may rely heavily on other sources of organic matter, such as kelp detritus (Duggins et al., 1989; Duggins and Eckman, 1997; Britton-Simmons et al., 2009; Galloway et al., 2013; Dethier et al., 2014), which are not directly tied to upwelling (Yorke et al., 2013). Work is needed, therefore, to understand possible bottom-up effects of upwelling in the north-east Pacific rocky subtidal.

Upwelling exhibits a high degree of spatial variation (Freeland and Denman, 1982) and differs in intensity between geographic sub-regions within the north-east Pacific. These sub-regions are separated by two coastal features in particular: Cape Blanco, a protruding headland in southern Oregon that acts as an upwelling jet, and Point Conception, the northern border of the Southern California Bight (Figure 11.1). North of Cape Blanco, mild surface winds result in relatively weak upwelling that is distinctly seasonal and gives way to a period of downwelling by early winter (Mann et al., 2006). Upwelling strength intensifies south of Cape Blanco and numerous coastal features, such as Cape Mendocino, Point Arena and Santa Cruz, form localised areas of high upwelling intensity (Connolly et al., 2001; Hickey and Banas, 2003; Sivasundar and Palumbi, 2010). The area to the south of Point Conception has considerably weaker winds and upwelling is less seasonal (Nidzieko and Largier, 2013). Oceanographic conditions in the Southern California Bight instead exhibit a strong east–west gradient as a result of competing influences from the California Current and southern California countercurrent (Dever, 2004). This pattern coincides with gradients in intertidal invertebrate communities (Blanchette et al., 2006) and fish assemblage structure (Hamilton et al., 2010).

The pelagic larval phase of many benthic invertebrates is transported offshore in areas of strong upwelling (Connolly and Roughgarden, 1998). For some species, potential negative effects of offshore transport are mediated by depth preferences among larvae (Miller and Morgan, 2013). In other species, recruitment is closely timed with periods of upwelling 'relaxation' (Botsford, 2001) and spatial patterns of recruitment success are inversely correlated with upwelling intensity (Menge et al., 1997). For instance, settlement of *Cancer* crabs coincides with relaxation of upwelling currents in areas where upwelling is strong, but is uncorrelated with relaxation and greater in areas where upwelled water is diverted and warmer waters remain (Wing et al., 1995, 1996).

Upwelling also varies temporally, which can affect the structure and functioning of benthic assemblages (Barth et al., 2007). Cyclical patterns in upwelling may be shifting as a result of climate change, with upwelling season starting later in the year and lasting for a shorter duration (Foreman et al., 2011). Oceanographers have begun to recognise the importance of upwelling-related phenology on pelagic ecosystems in the region (Bograd et al., 2009). Increased variability in the extent and timing of seasonal shifts facilitate distinct phytoplankton assemblages (Bograd et al., 2002), and delayed onset of upwelling appears to affect multiple trophic levels within pelagic food webs in the region (Brodeur et al., 2006; Mackas et al., 2006; Sydeman et al., 2006; Thomas and Brickley, 2006; Weise et al., 2006). Oceanographic time series data dating back to the 1960s suggest that upwelling season in the north-east Pacific is starting later in the year and lasting for a shorter duration (Bograd et al., 2009). Despite the known importance of upwelling for benthic organisms, phenological studies of benthic communities in the region are lacking. Such studies would require long-term data sets that simultaneously capture abundance and demographic parameters of resident species (Miller-Rushing et al., 2010), which should be a priority for future research. Phenology has the potential to affect benthic species interactions and community dynamics (Harley et al., 2006; Kordas et al., 2011), and will likely be an important factor for benthic communities in the north-east Pacific moving forwards.

### 11.6.2 Storm Events

Storms are another important seasonal processes influencing rocky subtidal assemblages (Tegner and Dayton, 1987). Autumn and winter storms are a major source of mortality for canopy-forming kelps the north-east Pacific (Rosenthal et al., 1974; Hawkes, 1981; Ebeling et al., 1985; Seymour et al., 1989; Duggins et al., 2003), and particularly in central California (Reed et al., 2011). While *Nereocystis* is relatively well adapted for mild to moderate storm-induced wave energy (Duggins et al., 2001), storm events commonly cause extensive deforestation of *Macrocystis* beds (Dayton et al., 1984). The negative effect of strong wave action during storms on net primary production of *Macrocystis* in central California is greater than that from grazing and nutrient limitation (Reed et al., 2011), which are known to be important in other parts of California and the north-east Pacific. Extreme wave action during storms can also cause high mortality of subtidal invertebrates, reducing the abundance of sea urchins, which decreases grazing pressure on macroalgae (Cowen et al., 1982; Ebeling et al., 1985).

The force imparted to benthic organisms by wave energy during storms depends on the cross-sectional surface area perpendicular to the direction of flow. Large flat objects and those that are perpendicular to flow are more likely to be dislodged than are smaller and more streamlined objects. Organisms extending higher off the bottom experience higher flow speeds, and thus higher drag forces. Benthic organisms also experience an upward force (lift) generated by the pressure differential set-up by moving water, which can rip them off the substratum (Denny, 1987), even when the organisms are arrayed in tight aggregations such as mussel beds (Carrington et al., 2009). Acceleration forces and impingement forces are also present (Gaylord et al., 2012), but impingement forces have only recently been measured accurately (Jensen and Denny, 2015, 2016) and were not found to be the largest forces imparted by waves. Rather, the speed of water within the broken wave produces greater drag forces and thus is more likely to cause dislodgement. High bidirectional flow under shoaling and breaking waves is likely to produce the highest forces on shallow subtidal organisms, especially during storms, and thus could be a major source of mortality.

Disturbance during storm events can remove a competitive dominant species, and allow recruitment and succession to generate a series of community stages before the dominant takes over again (Paine and Levin, 1981). Dayton (1971) recognised that the combination of wave forces and projectiles (floating wood) was an important determinant of community structure in north-east Pacific intertidal communities; in shallow subtidal communities rolling boulders may play a similar role (as in some intertidal sites; Sousa, 1984). Below about 10 m depth, wave-induced disturbance is minimal except for the removal of kelp holdfasts torn off by forces generated on near-surface blades. However, kelp beds experience radical community change following large storm events (Dayton et al., 1984), particularly in shallow areas (Dayton et al., 1999). Primary productivity by understorey macroalgae and phytoplankton increases following the removal of *Macrocystis* by winter storms (Miller et al., 2011). Extended periods without large waves can facilitate shifts in community structure in shallow areas, with greater dominance by low flow organisms such as sea cucumbers (Rassweiler et al., 2010).

The effects from storms vary depending on their magnitude and frequency (Seymour et al., 1989) and on the degree of exposure of benthic organisms (Tegner and Dayton, 1987). Storminess and storm-induced wave energy in the north-east Pacific have increased over recent decades (Bromirski et al., 2005). Increased storm frequency is expected to reduce the abundance of higher trophic groups that are associated with canopy-forming kelps, and decrease diversity and food web complexity (Byrnes et al., 2011). Increased storminess is closely tied with El Niño events (Dayton and Tegner, 1984), which also influence north-east Pacific benthic assemblages, and are discussed further.

### 11.6.3 El Niño

During an El Niño, upwelling intensity decreases, which causes higher SST and lower nutrient availability during typical upwelling months (Chavez et al., 1999). It also increases wave energy and the frequency and intensity of storms (Dayton et al., 1984), and is considered the greatest driver of coastal erosion in the north-east Pacific (Barnard et al., 2015). El Niño conditions cause deforestation of canopy-forming kelps and removal/mortality of benthic invertebrates (Tegner and Dayton, 1987; Glynn, 1988; Castilla and Camus, 1992; Tegner et al., 1997). For *Macrocystis*, El Niño conditions can cause complete deforestation across large areas, although the spatial scale of canopy loss is variable throughout the region and tends to be most extreme in southern parts of the north-east Pacific (Edwards and Estes, 2006).

Recolonisation of north-east Pacific canopy-forming kelps following an El Niño event is highly variable (Edwards and Estes, 2006). Frequently deforested kelp beds re-establish rapidly (Dayton, 1985; Ebeling et al., 1985; Tegner and Dayton, 1991), either by spore dispersal from other areas (Reed et al., 1988, 1997, 2011; Schroeter et al., 2015) or by seeding from dormant life stages (Dayton, 1985; Foster and Schiel, 1985; tom Dieck, 1993). Recolonisation may be heavily influenced by the remaining standing crop and recruitment, which are important determinants of net primary productivity in *Macrocystis* beds (Reed et al., 2008). Abiotic factors and combined anthropogenic impacts, such as limestone extraction, can also have negative effects on growth and recovery (Cavanaugh et al., 2011; Torres-Moye and Escofet, 2014). *Macrocystis* sporophytes may also survive El Niño warming in deep-water refugia (Ladah and Zertuche-González, 2004). Microscopic gametophytes may be able to persist during El Niño events, even when stands of adult sporophytes have been eliminated (Ladah and Zertuche-González, 1999). In some cases, however, kelp beds never re-establish at deforested sites (North et al., 1986). Very slow recovery is often evident near the southern range limit of *Macrocystis* (Hernández-Carmona et al., 2001; Edwards and Hernández-Carmona, 2005). This may be due at least in part to competition between macroalgal species. *E. arborea* stands that persist and recruit despite warm temperatures and intense storm events competitively exclude new recruits of *Macrocystis* in Baja California (Edwards and Hernández-Carmona, 2005). However, abiotic variables that influence recruitment and demographic connectivity are likely also important (Edwards and Estes, 2006). The probability of localised extinctions is lower among well-connected patches of *Macrocystis*, although extinction probability is also influenced by patch size (Castorani et al., 2015).

El Niño-related mortality events that are followed by rapid regrowth and recolonisation (Dayton et al., 1992) may be an important driver of temporal dynamics in north-east Pacific benthic environments. The impact of canopy loss on other macroalgal species varies. Kelp removal increases the availability of light to the seafloor and changes flow conditions (discussed later). Clark et al. (2004) found that complete removal of mixed stands of *Macrocystis* and *Pterygophora* led to the development of a new, smaller canopy formed by *Desmarestia lingulata*. Additionally, some red macroalgae respond quite slowly to the loss of canopy cover (Clark et al., 2004).

El Niño coincides with temporal variations in spawning, survivorship and even foraging behaviour for benthic fauna, particularly in southern parts of the north-east Pacific (Watson et al., 2010; Mazariegos-Villarreal et al., 2012; Funes-Rodríguez et al., 2015; De Anda-Montañez et al., 2016). For some invertebrates, such as the gastropod *Megastraea undosa*, El Niño actually facilitates recruitment events (Zacharias and Kushner, 2006). The mechanism behind this is unclear. Higher SSTs in early spring months could be a spawning cue (Fiedler et al., 1986). El Niño might also enhance the availability of food and post-settlement survival (Ladah and Zertuche-González, 1999) or support greater recruitment success via altered abiotic parameters. These peaks in recruitment do not enhance overall biomass of *M. undosa*, however, as larger adults also experience greater mortality during El Niño events (Zacharias and Kushner, 2006).

In northern parts of the north-east Pacific, El Niño can cause unusual current patterns that may alter connectivity and dispersal. Off British Columbia, the northward coastal current strengthens during El Niño years (Huyer et al., 2002). El Niño can also cause a poleward jet to form off Oregon and northern California (Kosro, 2002). This northward movement of surface water could support greater dispersal of southern species in northern areas within the north-east Pacific (See and Feist, 2010) and increase genetic connectivity among some benthic populations (Sunday et al., 2014). This may be part of why there are genetic similarities between some species in California and British Columbia (Behrens Yamada and Gillespie, 2008; Keever et al., 2009). It may also be among the drivers of the northward expansion of certain invasive species (Behrens Yamada and Gillespie, 2008; Yamada et al., 2015). For instance, Yamada et al. (2015) found that recruitment of *Carcina maenas* in Oregon and Washington occurred during warm conditions in El Niño years. In particular, the combination of El Niño and warm PDO phases may be particularly important for northward expansion of such species (Yamada et al., 2015), as well as for assemblage structure in more southern parts of the north-east Pacific (Parnell et al., 2010).

### 11.6.4 Decadal-Scale Oscillations

Fluctuations in ocean conditions occur over decadal scales in the North Pacific and are described by a variety of indices (Mantua et al., 1997; Forchhammer and Post, 2004; Di Lorenzo et al., 2008). The PDO index was derived in the late 1990s based on SST anomalies to represent a long-term cyclical process in the north-east Pacific in which ocean conditions alternate between warm and cool phases (Mantua et al.,

1997). The pattern reflected by PDO was clearly the result of climate-forcing, but the precise mechanism driving it was unclear (Francis et al., 1998; Mantua and Hare, 2002). After more than a decade of further work, it seems the pattern reflected by the PDO index is the result of three simultaneous processes: (1) variability in ocean and weather conditions caused by the semi-permanent low pressure centred over the Aleutian Islands, known as the Aleutian low, (2) re-emergence, a process by which restructured ocean layers maintain more constant SST than would be expected due to the thermal capacity of seawater alone and (3) long-term changes in the Kuroshio Current system on the western side of the North Pacific gyre (Newman et al., 2016). Newman et al. (2016) provide an excellent review of current knowledge of the processes that shape PDO cycles.

Regardless of the driving mechanisms, variability in nearshore conditions that have been attributed to PDO have significant implications for marine ecosystems (Hare and Mantua, 2000; Doney et al., 2012). Warm periods with a positive PDO index coincide with greater productivity in Alaska and reduced productivity in the California Current system (Francis et al., 1998). The opposite occurs during negative PDO periods. Temperature-induced changes in productivity may have considerable effects on pelagic food webs and on salmon populations (Hare and Mantua, 2000; Stenseth et al., 2002; Litzow et al., 2013). Warm PDO phases also correlate with higher fishery catches in Alaska and northern British Columbia (Okey et al., 2014), although effects on fish assemblages further to the south are less evident (Miller and McGowan, 2013). Blanchard (2015) found that the diversity and density of macrophytes Chukchi Sea, which is a subarctic region well outside the area considered in this review, was strongly correlated with the PDO index. Other studies in the north-east Pacific have noted that PDO may be among the variables that explain long-term changes in benthic assemblage structure and diversity (Smith et al., 2006; Parnell et al., 2010), but empirical support for this is lacking (Cavanaugh et al., 2011).

NPGO is another index of decadal oscillation based on sea surface height anomalies (Di Lorenzo et al., 2008). Positive NPGO values are associated with greater upwelling intensity, increased nutrient availability and higher productivity, as reflected by the concentration of chlorophyll *a* in surface layers (Di Lorenzo et al., 2010). NPGO reflects low-frequency fluctuations in ocean conditions in the north-east Pacific and is a good predictor of seasonal timing of phytoplankton blooms (Chenillat et al., 2012). Cavanaugh et al. (2011) found that NPGO correlated with decadal cycles in *Macrocystis* biomass in the Santa Barbara Channel Islands. Once El Niño-related mortality events are accounted for, NPGO may thus be a valuable index for characterising long-term dynamics of canopy-forming kelps (Cavanaugh et al., 2011; Bell et al., 2015). Evidence from intertidal and deep-sea environments in the north-east Pacific suggest that NPGO may also influence food availability and recruitment success of benthic suspension feeders. Menge et al. (2009) found greater recruitment of mussels on the Oregon coast during cooler, more productive NPGO phases. Kahn et al. (2012) found correlations between sponge density and NPGO in the Monterey Deep-Sea Fan, and suggested that recruitment success coincided with NPGO-related fluctuation of particulate organic carbon.

Information on the effects of decadal oscillations on subtidal hard-substrate assemblages is limited (Harley and Connell, 2009). Understanding the mechanisms behind large-scale climate indices, such as PDO and NPGO, is important for determining ecological responses to long-term environmental change (Hallett et al., 2004). Recent advances surrounding the mechanisms that drive decadal oscillations (Newman et al., 2016) may be a necessary step towards building knowledge of their effects on benthic ecosystems. One of the greatest challenges in characterising the effects of meso-scale processes is the relatively limited spatial and temporal scale that is inherent in most studies of benthic communities. Remote sensing of kelp beds is currently providing valuable data for studying these effects, however (Cavanaugh et al., 2010; Reed et al., 2016a). Such studies will be particularly important for improving our understanding of the complexity of responses among benthic assemblages to climate change, and the combination of interannual and decadal variation with long-term warming trends (Harley and Connell, 2009; Okey et al., 2014).

### 11.6.5 'The Blob'

The warm surface water anomaly dubbed 'the Blob' was first observed in late 2013 and caused unusually warm temperatures intermittently in the years that followed (Bond et al., 2015; Zaba and Rudnick, 2016). Early anecdotal accounts of southerly and even tropical organisms through the Gulf of Alaska, British Columbia and northern Washington (Peterson et al., 2015) suggest that the Blob has had considerable effects on marine ecosystems. Several recent studies have subsequently documented some of the effects on benthos. Most notably perhaps, warmer SST from the Blob appears to have enhanced the progression and mortality due to sea star wasting disease (SSWD; Eisenlord et al., 2016), leading to a major collapse of *P. helianthoides* populations (Harvell et al., 2019), which has in turn had cascading effects on benthic ecosystems in some locations (Schultz et al., 2016). However, several other Blob-related changes have also been documented that are less well understood in terms of broader ecological consequences. In southern California, the Blob appears to have reduced settlement of barnacle larvae, both by decreasing water column stratification and larval abundance (Pineda et al., 2018). Forage fish in Oregon and Washington were found to shift diets, relying more on gelatinous zooplankton during warm years (Brodeur et al., 2018). In deeper epibenthic communities in British Columbia, the Blob may have increased resource availability for large fauna such as tanner crabs, leading to sudden increases in abundance (Chauvet et al., 2018). It may also have facilitated range shifts in numerous benthic taxa (Williams et al., 2018), including northern shifts in several species of nudibranch from southern California to southern Oregon (Goddard et al., 2016). Such range shifts may be ongoing, as the Blob was proceeded by the 2015–2016 El Niño, which is one of the strongest on record by some measures (Goddard et al., 2016; Jacox et al., 2016). Responses of canopy-forming kelps and kelp forest ecosystems to marine heat waves and warming in the region are complex (Bonaviri et al., 2017; Muth et al., 2019), and may differ between *Nereocystis*-dominated beds in the north (Pfister et al., 2018) and *Macrocystis*-dominated beds in the south (Reed et al., 2016b). These responses, as well as other yet unknown consequences of marine heat waves in the north-east Pacific, will likely be further elucidated through ongoing, coordinated data collection programmes.

## 11.7 Importance of Local Environmental Gradients

Assemblage structure at any given time and place within the north-east Pacific is further influenced by local environmental gradients (Blanchard et al., 2017). Localised gradients in abiotic and biotic conditions can lead to considerable differences between seemingly similar sites across small spatial scales. There have been several excellent reviews of abiotic factors and their effect on rocky subtidal ecosystems in the region (Dayton, 1985; North et al., 1986; Schiel and Foster, 2015). The most prominent gradients in subtidal hard-substrate assemblages are those associated with depth and flow (Neushul, 1967; Pirtle et al., 2012; Elahi et al., 2014).

### 11.7.1 Depth Zonation

Assemblage structure varies considerably with depth (Neushul, 1967; Aleem, 1973; Foster and Schiel, 1985; Britton-Simmons et al., 2009). Light is considered to be a primary driver of depth zonation, particularly for macroalgae and invertebrates with algal symbionts (Drew, 1983; Lüning et al., 1990; Gattuso et al., 2006). Macroalgae tend to decrease in abundance and biomass with increasing depth (Aleem, 1973; Britton-Simmons et al., 2009). Light limitation may determine the maximum depth of some kelps (Steneck et al., 2002) and inhibits growth, primary production and competitive interactions of understorey macroalgae in kelp forests (Dayton, 1985; Foster and Schiel, 1985; Clark et al., 2004; Stewart et al., 2009; Miller et al., 2012). Overexposure to ultraviolet light (UV) in shallow water can also be limiting, and affects multiple life stages in

Laminariales (Bischof et al., 2007). Swanson and Druehl (2000) found that UV exposure had a greater negative affect on spore germination in *Macrocystis* and *Pterygophora*, which inhabit moderate depths, than on the spores of shallow-water *Alaria* and *Hedophyllum*. This may be due in part to differences in spore size; meiospores produced by deep-water kelps were smaller and responded more negatively to light stress (Swanson and Druehl, 2000).

Young sporophytes of canopy-forming kelps are increasingly exposed to UV stress as they grow, and expression of stress response and light-harvesting gene transcripts is greater in adult blades near the surface (Konotchick et al., 2013). In *Macrocystis* sporophytes, surface and near-surface blades produce phlorotannins (Swanson and Druehl, 2002) and photoprotective compounds that reduce UV damage and maximise photosynthesis under high irradiance conditions (Colombo-Pallotta et al., 2006). Vertical gradients in irradiance also coincide with gradients in several other environmental parameters experienced by canopy-forming kelps, including hydrostatic pressure, wave energy, fluid transport and particulate deposition (Eckman et al., 1989; Tegner et al., 1991). Hydrostatic pressure below 35 m causes the pneumatocysts of young *Nereocystis* sporophytes to buckle, and may thus limit the maximum depth of *Nereocystis* recruits (Liggan, 2016). Additionally, stratification of water masses can produce considerable differences in temperature, salinity and nutrients between shallow and deep benthic habitats (Tegner et al., 1991). Nitrogen availability may be a particularly important factor for the base of *Macrocystis* thalli; gene transcripts for nitrogen acquisition are upregulated in tissues near the holdfasts of *Macrocystis* in response to seasonal stratification and nutrient limitation (Konotchick et al., 2013). Furthermore, biotic factors such as grazing by urchins can also exhibit vertical gradients (Duggins et al., 2001; Konar et al., 2009), and interactions between abiotic and biotic factors can vary with depth (Arkema et al., 2009). The depth distribution of specific organisms and resulting patterns in assemblage structure that emerge thus is a function of multiple abiotic and biotic factors.

Depth zonation patterns exhibit a high degree of spatial variability, as light attenuation varies throughout north-east Pacific nearshore habitats. The suspended particulate matter that increases light attenuation tends to be greatest near the outlets of major rivers and estuarine systems, such as the San Francisco Bay, but is highly temporally variable (Emmett et al., 2000). In addition, the myriad of other abiotic and biotic factors that affect depth zonation also vary spatially and temporally, leading to complex patterns in assemblage structure with depth across both small and relatively large spatial scales (Neushul, 1967; Konar et al., 2009; Young et al., 2016). In the Gulf of Alaska, Konar et al. (2009) found vertical zonation patterns at a regional scale were difficult to generalise because of high variability between sites. Several studies, each from sites within a relatively small geographical range, have emphasised depth zonation patterns that are important to highlight, however. Some of the earliest work on depth zonation in the north-east Pacific was that conducted by Neushul (1967) in the Salish Sea (Washington). Neushul (1967) suggested three general depth zones based on macroalgal composition on Brown Island, Friday Harbor: a shallow zone dominated by *Laminaria/Saccharina* spp., a middle-depth zone dominated by *Agarum* spp. and a deeper zone dominated by foliose and branching red macroalgae, such as *Callophyllis* and *Rhodoptilum plumosum*. Foster and Schiel (1985) identified a similar pattern in California, with *Egregia menziesii* and *Stephanocystis osmundacea* in shallow areas, *Macrocystis* and/or *Nereocystis* at moderate depths and *Pelagophycus porra* and *Agarum fimbriatum* in deeper areas. Konar et al. (2009) and Chenelot et al. (2006) both found considerable gradients in the distribution of mobile invertebrates with depth, and Dayton (1975) suggests that urchins may control the lower depth limit of kelp forests in Alaska. Depth-related patterns also occur among corallines and green algae inhabiting deeper rocky habitats in California (Spalding et al., 2003). Such patterns may become more evident in other parts of the region as advances in diving technology and robotics make deep subtidal habitats more accessible.

Depth distribution patterns also differ with respect to the orientation of rock surfaces, as vertical surfaces tend to receive less light than sloping or horizontal surfaces and are less susceptible to sedimentation, which can limit settlement and survival (Elahi et al., 2014). In many locations, such as Washington (Salish Sea), the shallowest zone, horizontal and sloped substrate is covered by a dense thicket of kelp blades, several layers thick. Corallines, foliose red algae and invertebrates, such as the holothurian *Cucumaria miniata*, occupy shallow substratum below low-lying kelp. Vertical surfaces in the shallowest zone may have some kelp cover, but are typically dominated by invertebrate assemblages (Elahi et al., 2013). Below about 5–7 m depth, large understorey kelp thin out and a diverse assemblage of red and brown algae form the understorey below large stands of *Nereocytis*. At some sites, the invasive brown alga *Sargassum muticum* now forms a canopy over shallow substrate (Britton-Simmons, 2004, 2006). Vertical rock walls at intermediate depths support a diverse assemblage of invertebrates (Elahi et al., 2014) and crustose algae. Fleshy red crustose algae also occupy vertical surfaces, especially when consumers regularly remove competing invertebrates (Elahi and Sebens, 2012). Deeper rock surfaces are also invertebrate dominated regardless of substrate orientation (Britton-Simmons et al., 2009).

### 11.7.2 Localised Patterns Related to Flow

Water movement is another important driver of localised patterns in subtidal community structure (Eckman et al., 2003; Gasbarro et al., 2018), and serves to link communities separated geographically and along a depth gradient (Britton-Simmons et al., 2009). Water flow affects nutrient uptake (Hurd, 2000), food availability (Sebens, 1984; Witman et al., 1993; Lesser et al., 1994), larval supply and recruitment (Roughgarden et al., 1988; Gaylord and Gaines, 2000; Hurd, 2000; Palardy and Witman, 2011) and survival (Duggins et al., 2001) of sessile flora and fauna, as well as the movement, feeding and behaviour of mobile organisms (Robles et al., 1989; Holsman et al., 2006). Flow speed and turbulence are highly variable over space and time, and may be generated by tidal activity, longshore currents and waves. Peak flow speeds from tidal exchange vary throughout the north-east Pacific, from a few centimetres per second in protected inlets, to well over 1.5 m/s in the Salish Sea (Gaylord et al., 2004). Different types of physiological traits are adaptive for different flow conditions (Duggins et al., 2003; Marchinko and Palmer, 2003) and spatial variation in flow thus leads to gradients in assemblage composition and structure (Elahi et al., 2014).

Differential flow rates are among the factors that influence depth zonation. As a result of vertical gradients in wave energy, fluid transport decreases and particulate deposition is greater with increasing depth, particularly in kelp forests (Eckman et al., 1989). The clear increase in water flow from the bottom to the top of rock walls (Patterson, 1980; Genovese and Witman, 1999) affects the growth rates and size structure of passive suspension feeders on such surfaces (Sebens, 1984). However, strong near-substratum flow also results in increased probability of dislodgement and mortality (Denny, 1987), as well as increased prey encounters (Sebens, 1984; Robinson et al., 2013) and greater larval dispersal (Eckman et al., 1989; Koehl et al., 1993; Gaylord and Gaines, 2000; Gaylord et al., 2007).

Low-flow habitats also present a number of challenges for sessile organisms, including reduced diffusion, reduced particle encounters and accumulation of sediment deposited on their surfaces. Elahi et al. (2014) described a gradient of community structure in the southern Salish Sea which correlated with flow regime; subtidal communities in low-flow habitats (fjords) comprised species assemblages lacking large passive suspension feeders, compared to nearby habitats with strong tidal flows. In general, active suspension feeders, such as sponges, bivalves and barnacles, can cope with low flows by generating their own flows past feeding structures (pumping water or moving appendages). Passive suspension feeders require external flows to effect particle encounters and thus are expected to do poorly in very low flows. Sedimentation in such habitats can cause accumulation on organism surfaces, which can be costly to remove, but organically rich suspended material settling onto surfaces can also be a food source for some suspension feeders. Elahi et al. (2014) documented increased sediment accumulation even on vertical rock wall communities in low-flow habitats, which could limit feeding and/or recruitment to the rock surfaces. Sedimentation rate has a negative effect on kelp spores, gametophytes and young plants (Devinny and Volse, 1978).

In areas with strong tidal or longshore currents, high flow rates can limit particle capture by suspension feeders (Marchinko and Palmer, 2003) and result in reduced larval sediment and recruitment success. Unidirectional currents cause a benthic boundary layer to develop above the seafloor, with flow decreasing towards the substratum (Boudreau and Jørgensen, 2001). Within the benthic boundary layer there is also a diffusional boundary layer closer to the substratum in which diffusion of substances into and away from organism surfaces can become limiting (Nowell and Jumars, 1984; Eckman, 1990). Sessile suspension feeders are thus exposed to a range of flow speeds and feeding conditions depending on their height off the substratum, and this affects the behavioural and physiological traits that are adaptive for low-versus high-flow environments (Marchinko and Palmer, 2003). Several north-east Pacific suspension feeders are known to reorient in response to changing current direction (Young and Braithwaite, 1980), and the physiology of barnacle species in the region correlates with flow (Nishizaki and Carrington, 2014a, 2014b, 2015), with those in high-flow environments having shorter rami with short setae (Marchinko and Palmer, 2003).

High flow also poses risks to sessile organisms with biomechanical properties that increase their risk of breakage or dislodgement. For instance, the biomechanical properties of *Nereocystis* make it well adapted for high wave energy but can cause it to fare poorly in high current, while the understorey kelps *Costaria costata, A. fimbriatum* and *Laminaria complanata* are better adapted to high unidirectional and bidirectional currents, and tend to break under high wave-energy conditions (Duggins et al., 2003). For some macrophytes, there can also be an interaction between the effects of flow and grazing. Duggins et al. (2001) found that *Nereocystis* could withstand forces generated by even the strongest local currents, but its stipe failed in strong currents when grazers had inflicted damage.

Mobile fauna can also be affected directly by high flow, particularly as it influences movement and foraging (Robles et al., 1989; Holsman et al., 2006). The types of predators that can forage effectively in high wave energy are very limited, and prey species that would otherwise be consumed can survive and dominate in high-flow habitats, especially when recruitment is also high (Robles et al., 2001). Water flow can also limit the activity of grazers, such as sea urchins and molluscs (Siddon and Witman, 2003). Berger and Jelinski (2008) found that *S. franciscanus* at low-flow sites in Barkley Sound, British Columbia had higher fecundity than at high-flow sites exposed to the open coast. Urchins may expend more energy seeking protection from wave energy or have fewer opportunities to forage at more exposed sites (Kawamata, 1998; Siddon and Witman, 2003). Food resources may also differ or be more readily available at protected sites, as Berge and Jelinski (2008) detected differences in the stable isotope signatures of urchins collected along a wave exposure gradient. Further manipulative work would be needed to discern the precise mechanism by which flow influences *S. franciscanus* fecundity. Special care, however, is required in the design of empirical studies on the effects of flow on mobile consumers to prevent bias from experimental artefacts, as flow is altered by many standard manipulative field methods (Miller and Gaylord, 2007).

In addition, moving water also transports gametes and larvae (Gaylord and Gaines, 2000). In general, more flow means more dispersal for larvae away from their point of origin, and higher delivery rates of larvae to benthic communities, as long as those larvae can settle without being immediately washed away. For gametes, low flow allows released gametes to accumulate locally and to encounter each other, whereas high flow dilutes gametes rapidly and can thus reduce fertilisation rates. The optimum flow regime for fertilisation is likely to be low oscillatory flow with turbulence, allowing local retention and high rates of encounter. On the larger scale, larvae can be carried long distances by longshore currents, which may move them offshore during periods of upwelling (as discussed earlier).

In northern fjord lands within the north-east Pacific, inshore versus offshore differences in flow are combined with differences in the availability of nutrients. Fjords in Washington, British Columbia and Alaska tend to be rich in iron but nitrogen limited (Stabeno et al., 2004), which can affect phytoplankton, microphytobenthos and benthic macrophytes (Strom et al., 2006). In addition, seasonal influx of glacially driven river input can result in quite low salinity levels for the benthos at intermediate to shallow depths (Royer, 1982). Fjord systems tend to support a greater abundance of high trophic levels (Etherington et al., 2007; Renner et al., 2012), and some demersal consumer species that migrate on the open coast remain relatively stationary throughout the year in inland seas (Andrews and Harvey, 2013).

## 11.8 Anthropogenic Changes

Marine ecosystems are altered considerably by a wide variety of anthropogenic changes. Anthropogenic-induced climate change is a major process influencing marine ecosystems globally (Hoegh-Guldberg and Bruno, 2010; Doney et al., 2012). In the north-east Pacific, warming and climate change are associated with a number of changes in environmental conditions and processes. Glacial melt in the region is among the highest globally, which may have further consequences for nearshore ecosystems (Spurkland and Iken, 2011; O'Neel et al., 2015), and for the organic material that glacial melt provides (Hood et al., 2009). The North Pacific Current is projected to shift further to the north and may have mixed effects on marine ecosystems (Sydeman et al., 2011). Sea-level rise in the north-east Pacific has thus far been relatively limited compared with other part of the world, but this trend may change moving forwards (Bromirski et al., 2011). Most substantial, however, are the direct effects from increased SST and lower pH.

Elevated temperature is known to negatively impact north-east Pacific subtidal organisms, reducing reproductive success, recruitment and survivorship (Miller and Emlet, 1997; Harley and Lopez, 2005; Rogers-Bennett et al., 2010), and may be negatively impacting nearshore populations in combination with exploitation by humans (Rogers-Bennett, 2007). Increases in ocean temperatures periodically occur due to inter-annual and decadal variation, and the combination of El Niño and warm PDO phases can lead to warm surface waters and unusual oceanographic conditions (Bylhouwer et al., 2013). However, comparisons of past and recent El Niños that occurred during warm phases of PDO are remarkably divergent (Smith et al., 2001). Ocean temperature in the north-east Pacific is increasing rapidly (McGowan et al., 1998; Freeland, 2013; Cummins and Masson, 2014), with major potential impacts on meso-scale processes (Bakun et al., 2015), marine ecosystems (Schiel et al., 2004; Harley and Connell, 2009; Kordas et al., 2011; Okey et al., 2014) and economic interests that depend on ecosystem health (Okey et al., 2015).

Temperature is the strongest correlate with instances of SSWD, which decimated populations of at least twenty species of sea star in the north-east Pacific starting in 2013 (Eisenlord et al., 2016; Harvell et al., 2019). While wasting disease in echinoderms had been documented previously in the region during warm periods associated with El Niño (Dungan et al., 1982), the recent die-off of sea stars from SSWD is the largest marine epidemic known to date, resulting in 100 per cent mortality in some locations (Harvell, personal communication). Sea stars infected with wasting disease have greater survival rates under lower temperatures (Kohl et al., 2016), but the epidemic arose during an extended period of warm SST (Eisenlord et al., 2016), and it is currently unclear whether heavily affected populations will persist (Wares and Schiebelhut, 2016). One species that has been particularly affected is the predatory sunflower star, *P. helianthoides* (Figure 11.3; Gudenkauf and Hewson, 2015), which serves as an important predator in subtidal hard-substrate assemblages (Duggins, 1983; Montecino-Latorre et al., 2016). Reductions in *P. helianthoides* abundance have been followed by rapid increases in the density of all three major urchin species (Shultz et al., 2016; Montecino-Latorre et al., 2016; Bonaviri et al.,

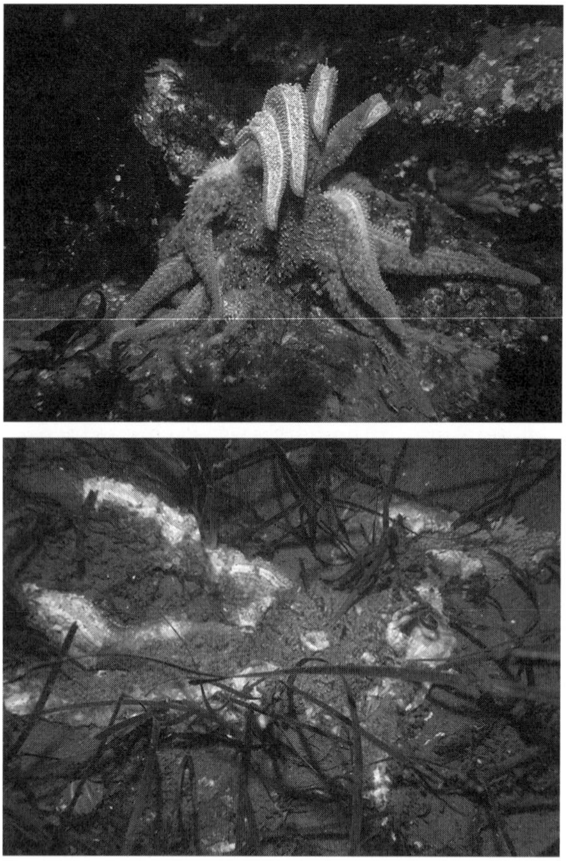

**Fig. 11.3** *P. helianthoides* in the Salish Sea at various stages of SSWD in 2013. Photo by Ed Gullekson. *A black and white version of this figure will appear in some formats. For the colour version, please refer to the plate section.*

2017; Burt et al., 2018). The precise mechanisms driving this shift are not always clear. In the Salish Sea, for instance, newly formed aggregations of *S. droebachiensis* in the years following SSWD comprised individuals larger than would be expected if decreased predation following the epidemic acted primarily on new urchin recruits (Heery and Sebens, unpublished data). Regardless of the underlying mechanism, SSWD-associated increases in urchin density do appear to be impacting kelp stands in at least some parts of the region, particularly in the north (Schultz et al., 2016; Burt et al., 2018).

As ocean temperatures warm as a result of high $CO_2$ concentrations in the atmosphere, the $CO_2$ concentrations in ocean surface layers also increase, reducing the pH of seawater and resulting in ocean acidification. Nearshore environments in the north-east Pacific appear to be particularly vulnerable to ocean acidification as evidenced by their low carbonate saturation state (Gruber et al., 2012). Uptake of $CO_2$ from anthropogenic sources has increased the aerial extent of low pH, undersaturated water during seasonal upwelling (Feely et al., 2008, 2010) and year-round undersaturation is predicted within the next twenty to thirty years (Gruber et al., 2012). Anticipated effects from ocean acidification on benthic flora are mixed, both within and across taxa (Haigh et al., 2015). For instance, higher $pCO_2$ could have potentially positive effects on some non-calcifying macroalgae by increasing photosynthetic rates, reducing the abundance of herbivorous consumers and increasing C:N ratios, which is inversely related to palatability for herbivores (Van Alstyne et al., 2009; Harley et al., 2012), but it could also have negative effects, such as reducing germination and increase mortality for spores (Gaitán-Espitia et al., 2014). Response to acidification among invertebrates is also expected to vary, and will likely involve interactions between the effects of pH, temperature and other factors (Haigh et al., 2015). For instance, acidification appears to alter gene expression in urchins such that they become more susceptible to temperature stress (O'Donnell et al., 2009). The ultimate effects will also depend on the ability of benthic species to adapt to changing pH regimes (Hoshijima and Hofmann, 2019). Low pH appears to cause rapid changes in functional proteins relating to ion transport, biomineralisation, lipid metabolism and cell–cell adhesion in *S. purpuratus* populations (Pespeni et al., 2013). While rapid evolutionary responses may allow some species to survive acidification, the outlook for benthic assemblages as a whole is uncertain (Busch et al., 2013). Brown et al. (2016) found that $CO_2$ enrichment facilitated a shift in the structure of fouling assemblages in British Columbia, reducing cover of the hydroid *Obelia dichotoma*, as well as recruitment by the mussel *Mytilus trossulus*, while increasing cover of the bryozoan *Mebranipora membranacea*. Low pH and higher temperatures also decrease the ability of mussels to maintain their attachment to rock surfaces by changing the mechanical properties of their byssal threads (O'Donnell et al., 2013). Harley et al. (2012) suggest that acidification in the north-east Pacific may cause a shift in assemblage structure from canopy-forming kelp species, calcifying algae and encrusting invertebrates to dominance by low-lying, fleshy macroalgal species. Haigh et al. (2015) provide an excellent review of potential direct and indirect impacts of acidification on marine organisms and the potential implications of these impacts for marine ecosystems more broadly. Although much uncertainty remains, it seems clear that response of benthic assemblages will depend on a variety of interacting factors, localised, historical environmental variation and complex species interactions (Harley et al., 2012; Hoshijima and Hofmann, 2019).

Changes to the marine environment from urbanisation involve major shifts in habitat structure, hydrodynamic conditions, pollutants, species composition and ecosystem functioning (Dafforn et al., 2015). Urbanisation occurs both in a concentrated manner in large metropolitan areas and at a broader spatial scale as coastal zones become increasingly populated, developed and utilised for resource extraction and renewable energy. Both concentrated and broad patterns of urbanisation are particularly evident along the coast of California, which is heavily populated (Small and Nicholls, 2003). California supports two of the largest metropolitan areas in the USA, both of which are coastal (Los Angeles–Riverside–Orange County and San Francisco–Oakland–San Jose), its coast has a greater density of housing units than coastal areas in any other US state, and it contains extensive shoreline armouring, infrastructure for maritime shipping and offshore oil and gas platforms (Crossett et al., 2004; Schroeder and Love, 2004; Pister, 2009). Coastal zones in other parts of the north-east Pacific also support offshore energy projects, fishing and other industries, as well as additional major metropolitan areas, including San Diego–Tijuana, Seattle–Tacoma–Bellevue and the Greater Vancouver Area, which range in population size from 2.3 to 5.1 million (annual estimates of the resident population: 1 April 2010–1 July 2015; Statistics Canada, 2011; Secretaria de Desarrollo Social, 2014; United States Census Bureau, 2016).

Urbanisation alters the chemical composition of nearshore environments, particularly through the introduction of high concentrations of nutrients and contaminants. Nutrient-induced eutrophication can lead to 'green tides' in some north-east Pacific nearshore environments. Green macroalgae that cause green tides include *Ulvaria* spp., which produce dopamine that may have toxic effects on benthic invertebrates (Van Alstyne et al., 2014). Nearshore environments within heavily urbanised areas also contain high concentrations of environmental contaminants, such as heavy metals, polychlorinated biphenyls, polycyclic aromatic hydrocarbons, pharmaceuticals, personal care products, perfluoroalkyl substances and faecal bacteria (Brown et al., 1998; Hornberger et al., 1999; Long et al., 2013). Environmental contaminants are found in the tissues of invertebrates and demersal fish in the north-east Pacific (Hogue and Swig, 2007; Love et al., 2013; Muñoz-Barbosa and Huerta-Diaz, 2013; Courtney-Hogue, 2016). Exposure to contaminants can inhibit growth and reproduction in invertebrates (Kagley et al., 2014), and cause tumours in demersal fish species (Myers et al., 2003). The levels of contaminants delivered by urban run-off are closely tied to storm events, but are also influenced by oceanographic conditions. For instance, the majority of faecal bacteria inputs enter the system during storms (Reeves et al., 2004). Once faecal and pathogenic bacteria have entered coastal environments, their persistence is inversely related to temperature (Walters et al., 2011). Faecal pollution from ocean outfalls, marinas and other sources may be transported relatively long distances via tidal activity, and particularly during upwelling events (Boehm et al., 2002; Jeong et al., 2005), but can also become concentrated in more protected nearshore habitats, leading to stressful and hypoxic conditions that negatively affect benthic organisms (Hendler, 2013). Urban run-off also serves as a novel vector for some marine diseases. For instance, Miller et al. (2010) found that sea otters were contracting and suffering high mortality rates from *Toxoplasma gondii* passed on by cat litter in California.

Most major coastal cities in the north-east Pacific are situated in areas where soft sediment was the dominant habitat type historically, although there are natural rocky subtidal reefs in some large metropolitan areas (Holbrook et al., 1997). Much of the hard substrate that is available for benthic organisms in urban centres is therefore man-made. Artificial structures in the north-east Pacific provide important habitat for rebuilding fish species (Love et al., 2006) but also can support exotic and invasive invertebrates (Page et al., 2006; Simkanin et al., 2012). The north-east Pacific has a long history of invasions by benthic organisms (Carlton, 1987), particularly in the San Francisco Bay (Carlton, 1987, 1996; Cohen and Carlton, 1998). Invasive species are most commonly introduced via maritime shipping, shellfish aquaculture and finfish imports (Levings et al., 2004; Wonham and Carlton, 2005). Populations of introduced species vary in their genetic composition and propensity to establish successfully (Zhan et al., 2012). The relative prevalence of natural habitats and biotic controls they potentially facilitate appear to be an important factor in maintaining resistance to invaders (Thornber et al., 2004; Simkanin et al., 2013).

Although invasions in San Francisco Bay have been extensive, they have also occurred in other parts of the north-east Pacific. Wonham and Carlton (2005) document 123 introduced marine species in coastal water bodies of northern California, Oregon, Washington and British Columbia. Cumulative invasions by molluscs in these areas follow a linear pattern, while cumulative invasions of annelids, bryozoans, cnidarians, ascidians, sponges and flatworms have increased exponentially over time (Wonham and Carlton, 2005). Non-indigenous marine species have had substantial impacts in the Salish Sea. In intertidal and shallow subtidal habitats in the Salish Sea, the Pacific oyster (*Magallana gigas* [previously *Crassostrea gigas*]) and the brown alga *S. muticum*, which is also invasive in California and Mexico (Aguilar-Rosas et al., 2007; Miller et al., 2007; Miller and Engle, 2009), as well as several crabs, snails, clams and mussels, have become established (Byers, 2002; Britton-Simmons, 2004, 2006; Harvey et al., 2009; Grason and Miner, 2012; Grason and Buhle, 2016). *M. gigas* and *S. muticum* have both established denser populations in marine reserves (Klinger et al., 2006) than in nearby comparable non-reserve sites. Several important invaders in southern parts of the north-east Pacific, including the green crab, *C. maenas*, and Asian kelp, *Undaria pinnatifida*, may also be undergoing northward range expansions (Behrens Yamada and Gillespie, 2008; Zabin et al., 2009), particularly in the wake of recent warm ocean conditions (Yamada et al., 2015).

Furthermore, invasive ascidians are of major concern for subtidal rocky assemblages throughout the region (Cordell et al., 2013; McCann et al., 2013; Miller, 2016). The distribution pattern of invasive tunicates may be particularly affected by temperature and salinity (Epelbaum et al., 2009). Some closed waterways, such as Hood Canal in Puget Sound, have experienced substantial ascidian invasions (e.g., *Styela clava, Ciona savignyi, Didemnum vexillum, Molgula manhattensis, Botryllus schlosseri* and *Botrylloides violaceus*; LeClair et al., 2009), while invasive ascidians have remained relatively rare in rocky habitats in other parts of the Salish Sea (Elahi and Sebens, 2012, 2013; Elahi et al., 2014).

## 11.9 Concluding Comments

The structure of and interactions within hard-substrate subtidal ecosystems differ considerably over space and time in the north-east Pacific. Spatially, benthic assemblages exhibit strong biogeographic and phylogeographic differences with latitude. Assemblages in southern parts of the north-east Pacific are heavily influenced by abiotic factors, while there is strong evidence that biotic factors and top-down interactions are particularly important in northern parts of the region, although clearly both factors always have the potential to affect community structure. Localised abiotic factors are also critical drivers of assemblage structure in any given location, although their relative importance may vary spatially as well. Temporally, subtidal assemblage structure undergoes seasonal, annual and decadal fluctuations that are tied to meso-scale processes driven by climate. Anthropogenic-induced warming, acidification, urbanisation and species introductions have had considerable impacts on north-east Pacific subtidal assemblages over the last decade, and are projected to pose increasing risks to marine ecosystems in the region over the coming decades. Knowledge of the patterns and mechanisms by which benthic assemblages are responding is currently limited, but will be necessary to anticipate the effects of anthropogenic change on the ecosystem services that subtidal hard-substrate assemblages provide.

In the last paper co-authored by Dr Robert T. Paine before his death in June 2016, Worm and Paine (2016) identified humans as the ultimate keystone, or 'hyperkeystone' species. They state that 'A new perspective on human influence on ecological interaction chains should help to unify diverse studies on changes in complex biological interactions (e.g., top-down trophic cascades or bottom-up effects…)'. Trophic cascades were identified as a driving force in subtidal assemblages in Alaska in the late 1970s and 1980s, but the extent to which top-down control was generalisable for other parts of the north-east Pacific subsequently became a point of contention. Several decades later, and after the vast extent of anthropogenic impacts has become ever more palpable, it seems worth noting that most of the locations where top-down control of subtidal ecosystems has been clearly documented are along relatively undeveloped coastlines with low human population density. Studies of the relative importance of top-down control are too few and the covariance between coastal development and latitude, flow, temperature, salinity and other factors that influence benthic communities is too great for such an observation to form the basis of substantive hypotheses. However, it seems plausible that the absence of humans in past ecological models is among the reasons for discrepancies observed by researchers working in northern and southern subregions within the north-east Pacific. Humans tend to reduce the abundance of other upper trophic-level species and simultaneously alter bottom-up processes in nearshore environments (Foster and Schiel, 2010). Recent studies have generated an extraordinary amount of information on the interactions between species and between species and the environment throughout the region, and may be sufficient to more systematically evaluate the importance of human activity on species interactions. Whether or not such an endeavour proves fruitful, all indications at present suggest that humans are an essential component of subtidal hard-substrate ecosystems in the north-east Pacific and should be incorporated into conceptual models of assemblage structure and dynamics wherever possible.

Subtidal, hard-substrate assemblages in the north-east Pacific are changing rapidly in response to climate change, ocean acidification and other anthropogenic factors. The methods available to understand these shifts are also evolving rapidly. Next-generation sequencing, novel techniques in chemistry and advances in other laboratory methods will likely provide increasingly valuable insights into the nature of species interactions and species response to abiotic change moving forwards. In addition, advances in diving technology and greater accessibility to deeper hard-substrate habitats should provide greater insights into the importance of these environments, both as refugia for shallow-water species and as connected habitats with functional properties that influence shallow water ecosystems. There is much need for long-term monitoring programmes and before–after–control–impact studies that can more effectively characterise the impacts of anthropogenic change. Support for such studies is difficult to attain presently, as they tend to align poorly with the time horizon of most grants and with fluctuating priorities and interests of major funding agencies. Regardless, documenting long-term trends is essential if we are to understand the broad-reaching effects of anthropogenic change and anticipate impacts on ecosystem services provided by subtidal assemblages.

## REFERENCES

Abbott, I. A. and Hollenberg, G. J. (1976). *Marine Algae of California*. Stanford University Press, Stanford, CA.

Addicott, W. O. (1970). Latitudinal gradients in tertiary molluscan faunas of the Pacific coast. Palaeogeography, Palaeoclimatology, *Palaeoecology*, **8** (4), 287–312, http://dx.doi.org/10.1016/0031-0182(70)90103-3.

Addison, J. A. and Hart, M. W. (2005). Spawning, copulation and inbreeding coefficients in marine invertebrates. *Biology Letters*, **1** (August), 450–3, http://dx.doi.org/10.1098/rsbl.2005.0353.

Aguilar-Rosas, L. E., Aguilar-Rosas, R., Kawai, H. et al. (2007). New record of *Sargassum filicinum* Harvey (Fucales, Phaeophyceae) in the Pacific Coast of Mexico. *Algae*, 22 (1), 17–21.

Alberto, F., Raimondi, P. T., Reed, D. C. et al. (2011). Isolation by oceanographic distance explains genetic structure for *Macrocystis pyrifera* in the Santa Barbara Channel. *Molecular Ecology*, **20** (12), 2543–54, http://dx.doi.org/10.1111/j.1365-294X.2011.05117.x.

Aleem, A. A. (1973). Ecology of a kelp-bed in Southern California. *Botanica Marina*, **16** (2), 83–95, http://dx.doi.org/10.1515/botm.1973.16.2.83.

Van Alstyne, K. L., Harvey, E. L. and Cataldo, M. (2014). Effects of dopamine, a compound released by the green-tide macroalga *Ulvaria obscura* (Chlorophyta), on marine algae and invertebrate larvae and juveniles. *Phycologia*, **53** (2), 195–202, http://dx.doi.org/10.2216/13-237.1.

Van Alstyne, K. L., Pelletreau, K. N. and Kirby, A. (2009). Nutritional preferences override chemical defenses in determining food choice by a generalist herbivore, Littorina sitkana. *Journal of Experimental Marine Biology and Ecology*, **379** (1–2), 85–91, http://dx.doi.org/10.1016/j.jembe.2009.08.002.

De Anda-Montañez, J. A., Martínez-Aguilar, S., Balart, E. F. et al. (2016). Spatio-temporal distribution and abundance patterns of red crab *Pleuroncodes planipes* related to ocean temperature from the Pacific coast of the Baja California Peninsula. *Fisheries Science*, **82** (1), 1–15, http://dx.doi.org/10.1007/s12562-015-0938-8.

Andrews, K. S. and Harvey, C. J. (2013). Ecosystem-level consequences of movement: Seasonal variation in the trophic impact of a top predator. *Marine Ecology Progress Series*, **473**, 247–60, http://dx.doi.org/10.3354/meps10095.

Anthony, R. G., Estes, J. A., Ricca, M. A., Miles, A. K. and Forsman, E. D. (2008). Bald eagles and sea otters in the Aleutian Archipelago: Indirect effects of trophic cascades. *Ecology*, **89** (10), 2725–35, http://dx.doi.org/10.1890/07-1818.1.

Arkema, K. D., Reed, D. D. and Schroeter, S. C. (2009). Direct and indirect effects of giant kelp determine benthic community structure and dynamics. *Ecology*, **90** (11), 3126–37.

Armstrong, R. (1976). Fugitive species: experiments with fungi and some theoretical considerations. *Ecology*, **57** (5), 953–63, http://dx.doi.org/10.2307/1941060.

Arreguín-Sánchez, F., Arcos, E. and Chávez, E. A. (2002). Flows of biomass and structure in an exploited benthic ecosystem in the Gulf of California, Mexico. *Ecological Modelling*, **156** (2–3), 167–83, http://dx.doi.org/10.1016/S0304-3800(02)00159-X.

Azad, A. K., Pearce, C. M. and McKinley, R. S. (2011). Effects of diet and temperature on ingestion, absorption, assimilation, gonad yield, and gonad quality of the purple sea urchin (*Strongylocentrotus purpuratus*). *Aquaculture*, **317** (1–4), 187–96, http://dx.doi.org/10.1016/j.aquaculture.2011.03.019.

Bakun, A., Black, B. A., Bograd, S. J. et al. (2015). Anticipated effects of climate change on coastal upwelling ecosystems. *Current Climate Change Reports*, **1** (2), 85–93, http://dx.doi/10.1007/s40641-015-0008-4.

Barnard, P. L., Short, A. D., Harley, M. D. et al. (2015). Coastal vulnerability across the Pacific dominated by El Niño/Southern Oscillation. *Nature Geoscience*, **8** (10), 801–7, http://dx.doi.org/10.1038/ngeo2539.

Barth, J. A., Menge, B. A., Lubchenco, J. et al. (2007). Delayed upwelling alters nearshore coastal ocean ecosystems in the northern California current. *Proceedings of the National Academy of Sciences of the United States of America*, **104** (10), 3719–24, http://dx.doi.org/10.1073/pnas.0700462104.

Basilio, A., Searcy, S. and Thompson, A. R. (2017). Effects of the Blob on settlement of spotted sand bass, *Paralabrax maculatofasciatus*, to Mission Bay, San Diego, CA. *PLoS ONE*, **12** (11), e0188449.

Behrens Yamada, S. and Gillespie, G. E. (2008). Will the European green crab (*Carcinus maenas*) persist in the Pacific Northwest? *ICES Journal of Marine Science: Journal du Conseil*, **65** (5), 725–9, http://dx.doi.org/10.1093/icesjms/fsm191.

Belanger, C. L., Jablonski, D., Roy, K. et al. (2012). Global environmental predictors of benthic marine biogeographic structure. *Proceedings of the National Academy of Sciences*, **109** (35), 14046–51, http://dx.doi.org/10.1073/pnas.1212381109.

Bell, T. W., Cavanaugh, K. C. and Siegel, D. A. (2015). Remote monitoring of giant kelp biomass and physiological condition: an evaluation of the potential for the Hyperspectral Infrared Imager (HyspIRI) mission. *Remote Sensing of Environment*, **167**, 218–28, http://dx.doi.org/10.1016/j.rse.2015.05.003.

Benes, K. M. and Carpenter, R. C. (2015). Kelp canopy facilitates understory algal assemblage via competitive release during early stages of secondary succession. *Ecology*, **96** (1), 241–51, http://dx.doi.org/10.1890/14-0076.1.

Berger, M. S. and Jelinski, D. E. (2008). Spatial patterns of maternal investment in *Strongylocentrotus franciscanus* along a marine-terrestrial gradient. *Marine Ecology Progress Series*, **364**, 119–27, http://dx.doi.org/10.3354/meps07490.

Bernardi, G. (2000). Barriers to gene flow in *Embiotoca jacksoni*, a marine fish lacking a pelagic larval stage. *Evolution*, **54** (1), 226, http://dx.doi.org/10.1554/0014-3820(2000)054[0226:BTGFIE]2.0.CO;2.

Bernardi, G. (2005). Phylogeography and demography of sympatric sister surfperch species, *Embiotoca jacksoni* and *E. Lateralis* along the California coast: historical versus ecological factors. *Evolution*, **59** (2), 386, http://dx.doi.org/10.1554/04-367.

Bertocci, I., Araújo, R., Oliveira, P. and Sousa-Pinto, I. (2015). Potential effects of kelp species on local fisheries. *Journal of Applied Ecology*, **52** (5), 1216–26, http://dx.doi.org/10.1111/1365-2664.12483.

Bischof, K., Gómez, I., Molis, M. et al. (2007). Ultraviolet radiation shapes seaweed communities. *Life in Extreme Environments*, **2**, 187–212, http://dx.doi.org/10.1007/s11157-006-0002-3.

Blanchard, A. L. (2015). Variability of macrobenthic diversity and distributions in Alaskan sub-Arctic and Arctic marine systems with application to worldwide Arctic Systems. *Marine Biodiversity*, **45** (4), 781–95, http://dx.doi.org/10.1007/s12526-014-0292-6.

Blanchard, A. L., Feder, H. M., Hoberg, M. K. and Knowlton, A. L. (2017). Abiotic/biological interactions in coastal marine communities: insights from an Alaskan fjord. *Estuaries and Coasts*, **40** (5), 1398–417.

Blanchette, C. A., Broitman, B. R. and Gaines, S. D. (2006). Intertidal community structure and oceanographic patterns around Santa Cruz Island, CA, USA. *Marine Biology*, **149** (3), 689–701, http://dx.doi.org/10.1007/s00227-005-0239-3.

Blanchette, C. A., Miner, B. G. and Gaines, S. D. (2002). Geographic variability in form, size and survival of *Egregia menziesii* around Point Conception, California. *Marine Ecology Progress Series*, **239**, 69–82, http://dx.doi.org/10.3354/meps239069.

Blanchette, C. A., Miner, C. M., Raimondi, P. T., Lohse, D., Heady, K. E. K. and Broitman, B. R. (2008). Biogeographical patterns of rocky intertidal communities along the Pacific coast of North America. *Journal of Biogeography*, **35** (9), 1593–607, http://dx.doi.org/10.1111/j.1365-2699.2008.01913.x.

Bodkin, J. L. (1986). Fish assemblages in Macrocystis and Nereocystis kelp forests off central California. *Fishery Bulletin*, **84** (4), 799–808.

Bodkin, J. L. (1988). Effects of kelp forest removal on associated fish assemblages in central California. *Journal of Experimental Marine Biology and Ecology*, **117** (3), 227–38, http://dx.doi.org/10.1016/0022-0981(88)90059-7.

Boehm, A. B., Sanders, B. F. and Winant, C. D. (2002). Cross-shelf transport at Huntington Beach. Implications for the fate of sewage discharged through an offshore ocean outfall. *Environmental Science and Technology*, **36** (9), 1899–906, http://dx.doi.org/10.1021/es0111986.

Bograd, S., Schwing, F., Mendelssohn, R. and Green-Jessen, P. (2002). On the changing seasonality over the North Pacific. *Geophysical Research Letters*, **29** (9), 1333, http://dx.doi.org/10.1029/2001GL013790.

Bograd, S. J., Schroeder, I., Sarkar, N. et al. (2009). Phenology of coastal upwelling in the California Current. *Geophysical Research Letters*, **36** (1), 1–5, http://dx.doi.org/10.1029/2008GL035933.

Bonaviri, C., Graham, M., Gianguzza, P. and Shears, N. T. (2017). Warmer temperatures reduce the influence of an important keystone predator. *Journal of Animal Ecology*, **86** (3), 490–500.

Bond, N. A., Cronin, M. F., Freeland, H. and Mantua, N. (2015). Causes and impacts of teh 2014 warm anomaly in the north-east Pacific. *Geophysical Research Letters*, **42**, 3414–20, http://dx.doi.org/10.1002/2015GL063306.

Botsford, L. (2001). Physical influences on recruitment to California Current invertebrate populations on multiple scales. *ICES Journal of Marine Science*, **58** (5), 1081–91, http://dx.doi.org/10.1006/jmsc.2001.1085.

Botsford, L. W. and Lawrence, C. A. (2002). Patterns of co-variability among California Current chinook salmon, coho salmon, Dungeness crab, and physical oceanographic conditions. *Progress in Oceanography*, **53** (2–4), 283–305, http://dx.doi.org/10.1016/S0079-6611(02)00034-4.

Boudreau, B. P. and Jørgensen, B. B. (2001). *The Benthic Boundary Layer : Transport Processes and Biogeochemistry*. Oxford University Press, Oxford.

Breda, V. A. and Foster, M. S. (1985). Composition, abundance, and phenology of foliose red algae associated with two central California kelp forests. *Journal of Experimental Marine Biology and Ecology*, **94** (1–3), 115–30, http://dx.doi.org/10.1016/0022-0981(85)90053-X.

Briggs, J. C. (1974). *Marine Zoogeography*. McGraw-Hill, New York.

Briggs, J. C. (2003). Guest editorial: marine centres of origin as evolutionary engines. *Journal of Biogeography*, **30** (1), 1–18.

Briggs, J. C. (2007). Marine longitudinal biodiversity: causes and conservation. *Diversity and Distributions*,

**13** (5), 544–55, http://dx.doi.org/10.1111/j.1472-4642.2007.00362.x.

Briggs, J. C. and Bowen, B. W. (2012). A realignment of marine biogeographic provinces with particular reference to fish distributions. *Journal of Biogeography*, **39** (1), 12–30, http://dx.doi.org/10.1111/j.1365-2699.2011.02613.x.

Britton-Simmons, K. H. (2004). Direct and indirect effects of the introduced alga *Sargassum muticum* on benthic, subtidal communities of Washington State, USA. *Marine Ecology Progress Series*, **277**, 61–78, http://dx.doi.org/10.3354/meps277061.

Britton-Simmons, K. H. (2006). Functional group diversity, resource preemption and the genesis of invasion resistance in a community of marine algae. *Oikos*, **113** (3), 395–401.

Britton-Simmons, K. H., Foley, G. and Okamoto, D. (2009). Spatial subsidy in the subtidal zone: Utilization of drift algae by a deep subtidal sea urchin. *Aquatic Biology*, **5** (3), 233–43, http://dx.doi.org/10.3354/ab00154.

Brodeur, R. D., Hunsicker, M. E., Hann, A. and Miller, T. W. (2018). Effects of warming ocean conditions on feeding ecology of small pelagic fishes in a coastal upwelling ecosystem: a shift to gelatinous food sources. *Marine Ecology Progress Series*, https://doi.org/10.3354/meps12497.

Brodeur, R., Ralston, S. and Emmett, R. (2006). Anomalous pelagic nekton abundance, distribution, and apparent recruitment in the northern California Current in 2004 and 2005. *Geophysical Research Letters*, **33** (22), L22S08.

Bromirski, P. D., Cayan, D. R. and Flick, R. E. (2005). Wave spectral energy variability in the northeast Pacific. *Journal of Geophysical Research C: Oceans*, **110** (3), 1–15, http://dx.doi.org/10.1029/2004JC002398.

Bromirski, P. D., Miller, A. J., Flick, R. E. and Auad, G. (2011). Dynamical suppression of sea level rise along the Pacific coast of North America: Indications for imminent acceleration. *Journal of Geophysical Research: Oceans*, **116** (7), 1–13, http://dx.doi.org/10.1029/2010JC006759.

Brown, D. W., McCain B. B., Horness, B. H. et al. (1998). Status, correlations and temporal trends of chemical contaminants in fish and sediment from selected sites on the Pacific Coast of the USA. *Marine Pollution Bulletin*, **37** (1–2), 67–85, http://dx.doi.org/10.1016/S0025-326X(98)00133-7.

Brown, N. E. M., Therriault, T. W. and Harley, C. D. G. (2016). Field-based experimental acidification alters fouling community structure and reduces diversity. *Journal of Animal Ecology*, **85** (5), 1328–39, http://dx.doi.org/10.1111/1365-2656.12557.

Brzezinski, M. A. and Washburn, L. (2011). Phytoplankton primary productivity in the Santa Barbara Channel: Effects of wind-driven upwelling and mesoscale eddies. *Journal of Geophysical Research: Oceans*, **116** (12), http://dx.doi.org/10.1029/2011JC007397.

Burd, B. J., Barnes, P. A. G., Wright, C. A. and Thomson, R. E. (2008). A review of subtidal benthic habitats and invertebrate biota of the Strait of Georgia, British Columbia. *Marine Environmental Research*, **66**, S3–38, http://dx.doi.org/10.1016/j.marenvres.2008.09.004.

Burt, J. M., Tinker, M. T., Okamoto, D. K., Demes, K. W., Holmes, K. and Salomon, A. K. (2018). Sudden collapse of a mesopredator reveals its complementary role in mediating rocky reef regime shifts. *Proceedings of the Royal Society B: Biological Sciences*, **285** (1883), 20180553.

Burton, R. S. (1998). Intraspecific phylogeography across the Point Conception biogeographic boundary. *Evolution*, **52** (3), 734–45.

Busch, D. S., Harvey, C. J. and McElhany, P. (2013). Potential impacts of ocean acidification on the Puget Sound food web. *ICES Journal of Marine Science*, **70** (4), 823–33, http://dx.doi.org/10.1093/icesjms/fst176.

Byers, J. E. (2002). Physical habitat attribute mediates biotic resistance to non-indigenous species invasion. *Oecologia*, **130**, 146–56, http://dx.doi.org/10.1007/s004420100777.

Bylhouwer, B., Ianson, D. and Kohfeld, K. (2013). Changes in the onset and intensity of wind-driven upwelling and downwelling along the North American Pacific coast. *Journal of Geophysical Research: Oceans*, **118** (5), 2565–80, http://dx.doi.org/10.1002/jgrc.20194.

Byrnes, J. E., Reed, D. C., Cardinale, B. C. et al. (2011). Climate-driven increases in storm frequency simplify kelp forest food webs. *Global Change Biology*, **17** (8), 2513–24, http://dx.doi.org/10.1111/j.1365-2486.2011.02409.x.

Calvin, N. I. and Ellis, R. J. (1978). Quantitative and qualitative observations on *Laminaria dentigera* and other subtidal kelps of Southern Kodiak Island, Alaska. *Marine Biology*, **47**, 331–6.

Carey, A. G. (1991). Ecology of North American Arctic continental shelf benthos: a review. *Continental Shelf Research*, **11** (8–10), 865–83, http://dx.doi.org/10.1016/0278-4343(91)90083-I.

Carlton, J. T. (1987). Patterns of transoceanic marine biological invasions in the Pacific-Ocean. *Bulletin of Marine Science*, **41** (2), 452–65.

Carlton, J. T. (1996). Pattern, process, and prediction in marine invasion ecology. *Biological Conservation*, **78** (1–2), 97–106, http://dx.doi.org.10.1016/0006-3207(96)00020-1.

Carrington, E., Moeser, G. M., Dimond, J., Mello, J. J. and Boller, M. L. (2009). Seasonal disturbance to mussel beds: Field test of a mechanistic model predicting wave dislodgment. *Limnology and Oceanography*, **54** (3), 978–86, http://dx.doi.org/10.4319/lo.2009.54.3.0978.

Carter, S. K. and VanBlaricom, G. R. (2002). Effects of experimental harvest on red sea urchins (*Strongylocentrotus franciscanus*) in northern Washington. *Fishery Bulletin*, **100** (4), 662–73.

Carter, S. K., VanBlaricom, G. R. and Allen, B. L. (2007). Testing the generality of the trophic cascade paradigm for sea otters: A case study with kelp forests in northern Washington, USA. *Hydrobiologia*, **579** (1), 233–49, http://dx.doi.org/10.1007/s10750-006-0403-x.

Castilla, J. C. and Camus, P. A. (1992). The Humboldt- El Niño scenario: coastal benthic resources and anthropogenic influences, with particular reference to the 1982/83 ENSO. *South African Journal of Marine Science*, **12** (1), 703–12, http://dx.doi.org/10.2989/02577619209504735.

Castorani, M. C. N., Reed, D. C., Alberto, F. et al. (2015). Connectivity structures local population dynamics: a long-term empirical test in a large metapopulation system. *Ecology*, **96** (12), 3141–52, http://dx.doi.org/10.1890/15-0283.1.

Cavanaugh, K. C., Siegel, D. A., Kinlan, B. P. and Reed, D. C. (2010). Scaling giant kelp field measurements to regional scales using satellite observations. *Marine Ecology Progress Series*, **403**, 13–27, http://dx.doi.org/10.3354/meps08467.

Cavanaugh, K. C., Siegel, D. A., Reed, D. C. and Dennison P. E. (2011). Environmental controls of giant-kelp biomass in the Santa Barbara Channel, California. *Marine Ecology Progress Series*, **429**, 1–17, http://dx.doi.org/10.3354/meps09141.

Cavole, L. M., Demko, A. M., Diner, R. E. et al. (2016). Biological impacts of the 2013–2015 warm-water anomaly in the Northeast Pacific: winners, losers, and the future. *Oceanography*, **29** (2), 273–85.

Chauvet, P., Metaxas, A., Hay, A. E. and Matabos, M. (2018). Annual and seasonal dynamics of deep-sea megafaunal epibenthic communities in Barkley Canyon (British Columbia, Canada): a response to climatology, surface productivity and benthic boundary layer variation. *Progress in Oceanography*, **169**, 89–105.

Chavez, F. P., Strutton, P. G., Friederich, G. E. et al. (1999). Biological and chemical response of the equatorial pacific ocean to the 1997-98 El Nino. *Science*, **286** (5447), 2126–31, http://dx.doi.org/10.1126/science.286.5447.2126.

Checkley, D. M. and Barth, J. A. (2009). Patterns and processes in the California Current System. *Progress in Oceanography*, **83** (1–4), 49–64, http://dx.doi.org/10.1016/j.pocean.2009.07.028.

Chenelot, H., Iken, K., Konar, B. and Edwards, M. (2006). Spatial and Temporal Distribution of Echinoderms in Rocky Nearshore Areas of Alaska. In *The Nagisa World Congress*, 11–28.

Chenillat, F., Rivière, P., Capet, X., Di Lorenzo, E. and Blanke, B. (2012). North Pacific gyre oscillation modulates seasonal timing and ecosystem functioning in the California current upwelling system. *Geophysical Research Letters*, **39** (1), 1–6, http://dx.doi.org.10.1029/2011GL049966.

Clark, R. P., Edwards, M. S. and Foster, M. S. (2004). Effects of shade from multiple kelp canopies on an understory algal assemblage. *Marine Ecology Progress Series*, **267**, 107–19, http://dx.doi.org/10.3354/meps267107.

Clasen, J. L. and Shurin, J. B. (2015). Previously published works. *Ecology*, **96** (3), 862–72, http://dx.doi.org.10.1121/1.4929899.

Cohen, A. N. and Carlton, J. T. (1998). Accelerating invasion rate in a highly invaded estuary. *Science*, **279** (5350), 555–8, http://dx.doi.org/10.1126/science.279.5350.555.

Colombo-Pallotta, M. F., García-Mendoza, E. and Ladah, L. B. (2006). Photosynthetic performance, light absorption, and pigment composition of *Macrocystis pyrifera* (Laminariales, Phaeophyceae) blades from different depths. *Journal of Phycology*, **42** (6), 1225–34, http://dx.doi.org/10.1111/j.1529-8817.2006.00287.x.

Connell, J. H. and Keough, M. J. (1985). Disturbance and Patch Dynamics of Subtidal Marine Animals on Hard Substrata. In S. T. Pickett and P. S. White, eds. *The Ecology of Natural Disturbance and Patch Dynamics*. Academic Press, Orlando, FL, 125–51.

Connell, J. H. and Slatyer, R. (1977). Mechanisms of succession in natural communities and their role in community stability and organization. *The American Naturalist*, **111** (982), 1119–44.

Connolly, S. R., Menge, B. A. and Roughgarden, J. (2001). A latitudinal gradient in recruitment of intertidal invertebrates in the northeast Pacific Ocean. *Ecology*, **82** (7), 1799–1813.

Connolly, S. R. and Roughgarden, J. (1998). A latitudinal gradient in northeast Pacific intertidal community structure: evidence for an oceanographically based synthesis of marine community theory. *The American Naturalist*, **151** (4), 311–26, http://dx.doi.org/10.1086/286121.

Cope, J. M. and Haltuch, M. A. (2012). Temporal and spatial summer groundfish assemblages in trawlable habitat off the west coast of the USA, 1977 to 2009. *Marine Ecology Progress Series*, **451**, 187–200, http://dx.doi.org/10.3354/meps09595.

Cordell, J. R., Levy, C. and Toft, J. D. (2013). Ecological implications of invasive tunicates associated with artificial structures in Puget Sound, Washington, USA. *Biological Invasions*, **15** (6), 1303–18, http://dx.doi.org/10.1007/s10530-012-0366-y.

Courtney-Hogue, C. (2016). Heavy metal accumulation in *Lacistorhynchus dollfusi* (Trypanorhyncha: Lacistorhynchidae) infecting *Citharichthys sordidus* (Pleuronectiformes: Bothidae) from Santa Monica Bay, Southern California. *Parasitology*, **143** (6), 794–9, http://dx.doi.org/10.1017/S003118201600038X.

Cowen, R. K. (1985). Large scale pattern of recruitment by the labrid, *Semicossyphus pulcher*: Causes and implications. *Journal of Marine Research*, 43 (3), 719–42, http://dx.doi.org/10.1357/002224085788440376.

Cowen, R. K., Agegian, C. R. and Foster, M. S. (1982). The maintenance of community structure in a central California giant kelp forest. *Journal of Experimental Marine Biology and Ecology*, **64** (2), 189–201, http://dx.doi.org/10.1016/0022-0981(82)90152-6.

Crossett, K., Culliton, T. J., Wiley, P. C. and Goodspeed, T. R. (2004). *Population trends along the coastal United States: 1980–2008*, Coastal Trends Series Report, 1980–2008. US Department of Commerce, Washington, DC.

Cummins, P. F. and Masson, D. (2014). Climatic variability and trends in the surface waters of coastal British Columbia. *Progress in Oceanography*, **120**, 279–90, http://dx.doi.org/10.1016/j.pocean.2013.10.002.

Dafforn, K. A., Glasby, T. M., Airoldi, L., Rivero, N. K., Mayer-Pinto, M. and Johnston, E. L. (2015). Marine urbanization: an ecological framework for designing multifunctional artificial structures. *Frontiers in Ecology and the Environment*, **13** (2), 82–90, http://dx.doi.org/10.1890/140050.

Datsky, A. V. (2015). Ichthyofauna of the Russian exclusive economic zone of the Bering Sea: 1. Taxonomic diversity. *Journal of Ichthyology*, **55** (6), 792–826, http://dx.doi.org/10.1134/S0032945215060065.

Davenport, A. C. and Anderson, T. W. (2007). Positive indirect effects of reef fishes on kelp performance: the importance of mesograzers. *Ecology*, **88** (6), 1548–61.

Davis, B. J., De Martini, E. E. and McGee, K. (1981). Gene flow among population of a teleost (painted greenling, *Oxylebius pictus*) from Puget Sound to southern California. *Marine Biology*, **65** (1), 17–23, http://dx.doi.org/10.1007/BF00397063.

Dawson, A. G. (1992). *Ice Age Earth: Late Quaternary Geology and Climate*. Routledge, New York.

Dawson, M. (2001). Phylogeography of the tidewater goby, *Eucyclogobius newberryi* (Teleostei, Gobiidae), in Coastal California. *Evolution*, **55** (6), 1167–79.

Dawson, M. N., Waples, R. S. and Bernardi, G. (2006). Phylogeography Chapter. In L. G. Allen, D. J. Pondella and M. H. Horn, eds. *The Ecology of Marine Fishes: California and Adjacent Waters*. University of California Press, Berkeley, pp. 26–54.

Dayton, P. K. (1971). Competition, disturbance, and community organization: the provision and subsequent utilization of space in a rocky intertidal community. *Ecological Monographs*, **41** (4), 351–89.

Dayton, P. K. (1975). Experimental studies of algal canopy interactions in a sea otter dominated kelp community at Amchitka-Island, Alaska. *Fishery Bulletin*, **73** (2), 230–7.

Dayton, P. K. (1985). Ecology of kelp communities. *Annual Review of Ecology and Systematics*, **16**, 215–45.

Dayton, P. K., Currie, V. and Gerrodette, T. I. M. (1984). Patch dynamics and stability of some california kelp communities. *Ecological Monographs*, **54** (3), 253–89.

Dayton, P. K. and Tegner, M. J. (1984). Catastrophic storms, El Niño, and patch stability in a southern California kelp community. *Science*, **224** (4646), 283–5.

Dayton, P. K., Tegner, M. J., Edwards, P. B. and Riser, K. L. (1999). Temporal and spatial scales of kelp demography: the role of oceanographic climate. *Ecological Monographs*, **69** (2), 219–50, http://dx.doi.org/10.1890/0012-9615(1999)069[0219:TASSOK]2.0.CO;2.

Dayton, P. K., Tegner, M. J., Parnell, P. E. and Edwards, P. B. (1992). Temporal and spatial patterns of disturbance and recovery in a kelp forest community. *Ecological Society of America*, **62** (3), 421–45.

Dean, T. A., Haldorson, L., Laur, D. R., Jewett, S. C. and Blanchard, A. (2000). The distribution of nearshore fishes in kelp and eelgrass communities in Prince-William Sound, Alaska: associations with vegetation and physical habitat characteristics. *Environmental Biology of Fishes*, **57**, 271–87.

Debenham, P., Brzezinski, M., Foltz, K. and Gaines, S. (2000). Genetic structure of populations of the red sea urchin, *Strongylocentrotus franciscanus*. *Journal of Experimental Marine Biology and Ecology*, **253** (1), 49–62, http://dx.doi.org/10.1016/S0022-0981(00)00242-2.

Denny, M. W. (1987). Lift as a mechanism of patch initiation in the mussel beds. *Journal of Experimental Marine Biology and Ecology*, **113**, 231–45.

Dethier, M. N., Brown, A. S., Burgess, S. et al. (2014). Degrading detritus: Changes in food quality of aging kelp tissue varies with species. *Journal of Experimental Marine Biology and Ecology*, **460**, 72–9, http://dx.doi.org/10.1016/j.jembe.2014.06.010.

Dever, E. P. (2004). Objective maps of near-surface flow states near Point Conception, California. *Journal of Physical Oceanography*, **34**, 444–61.

Devinny, J. S. and Volse, L. A. (1978). Effects of sediments on the development of *Macrocystis pyrifera* gametophytes. *Marine Biology*, **48** (4), 343–8, http://dx.doi.org/10.1007/BF00391638.

Dominik, C. and Zimmerman, R. (2006). Dynamics of carbon allocation in a deep-water population of the deciduous kelp *Pleurophycus gardneri* (Laminariales). *Marine Ecology Progress Series*, **309**, 143–57, http://dx.doi.org/10.3354/meps309143.

Doney, S. C., Ruckelshaus, M., Duffy, J. E. et al. (2012). Climate change impacts on marine ecosystems. *Annual Review of Marine Science*, **4**, 11–37, http://dx.doi.org/10.1146/annurev-marine-041911-111611.

O'Donnell, M. J., George, M. N. and Carrington, E. (2013). *Mussel byssus* attachment weakened by ocean acidification. *Nature Climate Change*, **3** (4), 1–4, http://dx.doi.org/10.1038/nclimate1846.

O'Donnell, M. J., Hammond, L. M. and Hofmann, G. E. (2009). Predicted impact of ocean acidification on a marine invertebrate: elevated $CO_2$ alters response to thermal stress in sea urchin larvae. *Marine Biology*, **156** (3), 439–46, http://dx.doi.org/10.1007/s00227-008-1097-6.

Drew, E. A. (1983). Physiology of Laminaria. *Marine Ecology*, **4** (3), 227–50, http://dx.doi.org/10.1111/j.1439-0485.1983.tb00298.x.

Druehl, L. D. (1970). The pattern of Laminariales distribution in the northeast Pacific. *Phycologia*, **9** (3–4), 237–47, http://dx.doi.org/10.2216/i0031-8884-9-3-237.1.

Druehl, L. D. (1978). The distribution of *Macrocystis integrifolia* in British Columbia as related to environmental parameters. *Canadian Journal of Botany*, **56** (1), 69–79.

Duggins, D. O. (1980). Kelp beds and sea otters : an experimental approach. *Ecology*, **61** (3), 447–53.

Duggins, D. O. (1983). Starfish predation and the creation of mosaic patterns in a kelp-dominated community. *Ecology*, **64** (6), 1610–19.

Duggins, D. Eckman, J., Siddon, C. and Klinger, T. (2003). Population, morphometric and biomechanical studies of three understory kelps along a hydrodynamic gradient. *Marine Ecology Progress Series*, **265**, 57–76, http://dx.doi.org/10.3354/meps265057.

Duggins, D. O. and Eckman, J. E. (1997). Is kelp detritus a good food for suspension feeders? Effects of kelp species, age and secondary metabolites. *Marine Biology*, **128** (3), 489–95, http://dx.doi.org/10.1007/s002270050115.

Duggins, D. O., Eckman, J. E., Siddon, C. E. and Klinger, T. (2001). Interactive roles of mesograzers and current flow in survival of kelps. *Marine Ecology Progress Series*, **223**, 143–55, http://dx.doi.org/10.3354/meps223143.

Duggins, D. O., Gómez-Buckley, M. C., Buckley, R. M., Lowe, A. T., Galloway, A. W. E. and Dethier, M. N. (2016). Islands in the stream: kelp detritus as faunal magnets. *Marine Biology*, **163** (1), 1–10, http://dx.doi.org/10.1007/s00227-015-2781-y.

Duggins, D. O., Simenstad, C. A. and Estes, J. A. (1989). Magnification of secondary production by kelp detritus in coastal marine ecosystems. *Science*, **245** (4914), 170–3.

Dungan, M. L., Miller, T. E. and Thompson, D. A. (1982). Catastrophic decline of a top carnivore in the Gulf of California rocky intertidal zone. *Science*, **216** (4549), 989–91.

Dunn, R. P. and Hovel, K. A. (2019). Experiments reveal limited top-down control of key herbivores in southern California kelp forests. *Ecology*, e02625.

Ebeling, A. W., Laur, D. R. and Rowley, R. J. (1985). Severe storm disturbances and reversal of community structure in a southern California kelp forest. *Marine Biology*, **84** (3), 287–94, http://dx.doi.org/10.1007/BF00392498.

Eberl, R., Mateos, M., Grosberg, R. K., Santamaria, C. A. and Hurtado, L. A. (2013). Phylogeography of the supralittoral isopod Ligia occidentalis around the Point Conception marine biogeographical boundary. *Journal of Biogeography*, **40** (12), 2361–72, http://dx.doi.org/10.1111/jbi.12168.

Ebert, T. A. and Russell, M. P. (1988). Latitudinal variation in size structure of the west coast purple sea urchin: a correlation with headlands. *Limnology and Oceanography*, **33** (2), 286–94, http://dx.doi.org/10.4319/lo.1988.33.2.0286.

Eckman, J. E. (1990). A model of passive settlement by planktonic larvae onto bottoms of differing roughness. *Limnology and Oceanography*, **35** (4), 887–901, http://dx.doi.org/10.4319/lo.1990.35.4.0887.

Eckman, J. E., Duggins, D. O. and Sewell, A. T. (1989). Ecology of under story kelp environments. I. Effects of kelps on flow and particle transport near the bottom. *Journal of Experimental Marine Biology and Ecology*, **129** (2), 173–87, http://dx.doi.org/10.1016/0022-0981(89)90055-5.

Eckman, J. E., Duggins, D. O. and Siddon, C. E. (2003). Current and wave dynamics in the shallow subtidal: implications to the ecology of understory and surface-canopy kelps. *Marine Ecology Progress Series*, **265**, 45–56, http://dx.doi.org/10.3354/meps 265045.

Edwards, M. S. and Estes, J. A. (2006). Catastrophe, recovery and range limitation in north-east Pacific kelp forests: a large-scale perspective. *Marine Ecology Progress Series*, **320**, 79–87, http://dx.doi.org/10.3354/ meps320079.

Edwards, M. S. and Hernández-Carmona, G. (2005). Delayed recovery of giant kelp near its southern range limit in the North Pacific following El Niño. *Marine Biology*, **147** (1), 273–9, http://dx.doi.org/10 .1007/s00227-004-1548-7.

Eisenlord, M. E., Groner, M. L., Yoshioka, R. M. et al. (2016). Ochre star mortality during the 2014 wasting disease epizootic: role of population size structure and temperature. *Philosophical Transactions of the Royal Society of London B: Biological Sciences*, **371** (1689), http:// dx.doi.org/10.1098/rstb.2015.0212.

Elahi, R., Birkeland, C., Sebens, K. P., Turner, K. R. and Dwyer, T. R. (2013). Limited change in the diversity and structure of subtidal communities over four decades. *Marine Biology*, **160** (12), 3209–19, http://dx .doi.org/10.1007/s00227-013-2308-3.

Elahi, R., Dwyer, T. R. and Sebens, K. P. (2014). Mesoscale variability in oceanographic retention sets the abiotic stage for subtidal benthic diversity. *Marine Ecology Progress Series*, **498**, 117–32, http://dx.doi.org/ 10.3354/meps10642.

Elahi, R. and Sebens, K. P. (2012). Consumers mediate natural variation between prey richness and resource use in a benthic marine community. *Marine Ecology Progress Series*, **452**, 131–43, http://dx.doi.org.10.3354/ meps09603.

Elahi, R. and Sebens, K. P. (2013). Experimental removal and recovery of subtidal grazers highlights the importance of functional redundancy and temporal context. *PLoS ONE*, **8** (11), http://dx.doi.org/10.1371/ journal.pone.0078969.

Emery, W. J. and Hamilton, K. (1985). Atmospheric forcing of interannual variability in the northeast Pacific Ocean: connections with El Nino. *Journal of Geophysical Research*, **90**, 857–68.

Emmett, R., Llansó, R., Newton, J. et al. (2000). Geographic signatures of North American west coast estuaries. *Estuaries*, **23** (6), 765, http://dx.doi.org/10 .2307/1352998.

Epelbaum, A., Herborg, L. M., Therriault, T. W. and Pearce, C. M. (2009). Temperature and salinity effects on growth, survival, reproduction, and potential distribution of two non-indigenous botryllid ascidians in British Columbia. *Journal of Experimental Marine Biology and Ecology*, **369** (1), 43–52, http://dx .doi.org/10.1016/j.jembe.2008.10.028.

Erlandson, J. M., Graham, M. H., Bourque, B. J., Corbett, D., Estes, J. A. and Steneck, R. S. (2007). The Kelp highway hypothesis: marine ecology, the coastal migration theory, and the peopling of the Americas. *The Journal of Island and Coastal Archaeology*, **2** (2), 161–74, http://dx.doi.org/10.1080/ 15564890701628612.

Estes, J. A., Danner, E. M., Doak, D. F. et al. (2004). Complex trophic interactions in kelp forest ecosystems. *Bulletin of Marine Science*, **74** (3), 621–38, www .ingentaconnect.com/content/umrsmas/bullmar/2004/ 00000074/00000003/art00010.

Estes, J. A. and Duggins, D. O. (1995). Sea otters and kelp forests in Alaska: generality and variation in a community ecological paradigm. *Ecological Monographs*, **65** (1), 75–100, http://dx.doi.org/10.2307/2937159.

Estes, J. A., Lindberg, D. R. and Wray, C. (2005). Evolution of large body size in abalone (*Haliotis*): patterns and implications. *Paleobiology*, **31** (4), 591–606, http:// dx.doi.org/10.1666/0094-8373(2005)031[0591:EOLBSI] 2.0.CO;2.

Estes, J. A. and Palmisano, J. F. (1974). Sea otters: their role in structuring nearshore communities. *Science*, **185**, 1058–60, http://dx.doi.org/10.1126/science.185 .4156.1058.

Estes, J. A., Smith, N. S. and Palmisano, J. F. (1978). Sea otter predation and community organization in the Western Aleutian Islands, Alaska. *Ecology*, **59** (4), 822–33.

Estes, J. A., Terborgh J., Brashares J. S. et al. (2011). Trophic downgrading of planet Earth. *Science*, **333** (6040), 301–7.

Estes, J. A., Tinker, M. T., Williams, T. M. and Doak, D. F. (1998). Killer whale predation on sea otters linking oceanic and nearshore ecosystems. *Science*, **282**, 473–6, http://dx.doi.org/10.1126/science.282.5388.473.

Etherington, L. L., Hooge, P. N., Hooge, E. R. and Hill, D. F. (2007). Oceanography of Glacier Bay, Alaska: implications for biological patterns in a glacial fjord estuary. *Estuaries and Coasts*, **30** (6), 927–44, http://dx.doi .org/10.1007/BF02841386.

Fautin, D., Dalton, P., Incze, L. S. et al. (2010). An overview of marine biodiversity in United States waters. *PLoS ONE*, **5** (8), http://dx.doi.org/10.1371/ journal.pone.0011914.

Feehan, C. J., Ludwig, Z., Yu, S. and Adams, D. K. (2018). Synergistic negative effects of thermal stress and altered food resources on echinoid larvae. *Scientific Reports*, **8** (1), 12229.

Feely, R. A., Alin, S. R., Newton, J. et al. (2010). The combined effects of ocean acidification, mixing, and respiration on pH and carbonate saturation in an urbanized estuary. *Estuarine, Coastal and Shelf Science*, **88** (4), 442–9, http://dx.doi.org/10.1016/j.ecss.2010.05.004.

Feely, R. A., Sabine, C. L., Hernandez-Ayon, J. M., Ianson, D. and Hales, B. (2008). Evidence for upwelling of corrosive "acidified" water onto the continental shelf. *Science*, **320** (5882), 1490–2, http://dx.doi.org/10 .1126/science.1155676.

Fiedler, P. C., Methot, R. D. and Hewitt, R. P. (1986). Effects of California El Niño 1982–1984 on the northern anchovy. *Journal of Marine Research*, **44** (2), 317–38, http://dx.doi.org/10.1357/002224086788405365.

Forchhammer, M. C. and Post, E. (2004). Using large-scale climate indices in climate change ecology studies. *Population Ecology*, **46** (1), 1–12, http://dx.doi.org/10.1007/s10144-004-0176-x.

Foreman, M. G. G., Pal, B. and Merryfield, W. J. (2011). Trends in upwelling and downwelling winds along the British Columbia shelf. *Journal of Geophysical Research: Oceans*, **116** (10), 1–11, http://dx.doi.org/10.1029/2011JC006995.

Foreman, R. E. (1977). Benthic community modification and recovery following intensive grazing by *Strongylocentrotus droebachiensis*. *Helgoländer Wissenschaftliche Meeresuntersuchungen*, **30** (1–4), 468–84, http://dx.doi.org/10.1007/BF02207855.

Foreman, R. E. (1984). Studies on Nereocystis growth in British Columbia, Canada. *Hydrobiologia*, **116/117**, 325–32, http://link.springer.com/chapter/10.1007/978-94-009-6560-7_65.

Foster, M. S. (1990). Organization of macroalgal assemblages in the Northeast Pacific: the assumption of homogeneity and the illusion of generality. *Hydrobiologia*, **192**, 21–33.

Foster, M. S. (1992). How Important Is Grazing to Seaweed Evolution and Assemblage Structure in the North-East Pacific. In D. M. John, S. J. Hawkins and J. H. Price, eds. *Plant–Animal Interactions in the Marine Benthos*. Oxford University Press, Oxford, pp. 61–85.

Foster, M. S. and Schiel, D. R. (1985). *The Ecology of Giant Kelp Forests in California: A Community Profile*. US Fish and Wildlife Service, Washington, D.C.

Foster, M. S. and Schiel, D. R. (1988). Kelp Communities and Sea Otters: Keystone Species or Just Another Brick in the Wall?' In *The Community Ecology of Sea Otters*. Springer, Berlin, pp. 92–115, http://dx.doi.org/10.1007/978-3-642-72845-7_5.

Foster, M. S. and Schiel, D. R. (2010). Loss of predators and the collapse of southern California kelp forests (?): alternatives, explanations and generalizations. *Journal of Experimental Marine Biology and Ecology*, **393** (1–2), 59–70, http://dx.doi.org/10.1016/j.jembe.2010.07.002.

Fram, J. P., Stewart, H. L., Brzezinski, M. A. et al. (2008). Physical pathways and utilization of nitrate supply to the giant kelp, *Macrocystis pyrifera*. *Limnology and Oceanography*, **53** (4), 1589–603, http://dx.doi.org/10.4319/lo.2008.53.4.1589.

Francis, R. C., Hare, S. R., Hollowed, A. B. and Wooster, W. S. (1998). Effects of interdecadal climate variability on the oceanic ecosystems of the north-east Pacific. *Fisheries Oceanography*, **7** (1), 1–21, http://dx.doi.org/10.1046/j.1365-2419.1998.00052.x.

Freeland, H. J. (2006). What proportion of the north pacific current finds its way into the Gulf of Alaska?' *Atmosphere-Ocean*, **44** (4), 321–30, http://dx.doi.org/10.3137/ao.440401.

Freeland, H. J. (2013). Evidence of change in the winter mixed layer in the northeast Pacific Ocean: a problem revisited. *Atmosphere-Ocean*, **5900**, 37–41, http://dx.doi.org/10.1080/07055900.2012.754330.

Freeland, H. J. and Denman, K. L. (1982). A topographically controlled upwelling center off southern Vancouver Island. *Journal of Marine Research*, **40** (4), 1069–93.

Fritchman, H. K. (1961). A study of the reproductive cycle in the California Acmaeidae (Gastropoda). *Veliger*, **4**, 41–47, www.sciencedirect.com/science/article/pii/0022098173900579.

Funes-Rodríguez, R., Ruíz-Chavarría, J. A., González-Armas, R., Durazo, R. and Guzmán-del Proó, S. A. (2015). Influence of hydrographic conditions on the distribution of spiny lobster larvae off the west coast of Baja California. *Transactions of the American Fisheries Society*, **144** (6), 1192–205, http://dx.doi.org/10.1080/00028487.2015.1083474.

Gaines, S. and Roughgarden, J. (1985). Larval settlement rate: a leading determinant of structure in an ecological community of the marine intertidal zone. *Proceedings of the National Academy of Sciences of the United States of America*, **82** (11), 3707–11, http://dx.doi.org/10.1073/pnas.82.11.3707.

Gaitán-Espitia, J. D., Hancock, J. R., Padilla-Gamiño, J. L. et al. (2014). Interactive effects of elevated temperature and pCO2 on early-life-history stages of the giant kelp *Macrocystis pyrifera*. *Journal of Experimental Marine Biology and Ecology*, **457**, 51–8, http://dx.doi.org/10.1016/j.jembe.2014.03.018.

Galloway, A. W. E., Lowe, A. T., Sosik, E. A., Yeung, J. S. and Duggins, D. O. (2013). Fatty acid and stable isotope biomarkers suggest microbe-induced differences in benthic food webs between depths. *Limnology and Oceanography*, **58** (4), 1451–62, http://dx.doi.org/10.4319/lo.2013.58.4.1451.

Garbary, D. J., Kim, K. Y., Klinger, T. and Duggins, D. (1999). Preliminary observations on the development of kelp gametophytes endophytic in red algae. *Hydrobiologia*, **398**, 247–52.

Gasbarro, R., Wan, D. and Tunnicliffe, V. (2018). Composition and functional diversity of macrofaunal assemblages on vertical walls of a deep northeast Pacific fjord. *Marine Ecology Progress Series*, **597**, 47–64.

Gattuso, J.-P., Gentili, B., Duarte, C. M., Kleypas, J. A., Middelburg, J. J. and Antoine, D. (2006). Light availability in the coastal ocean: impact on the distribution of benthic photosynthetic organisms and contribution to primary production. *Biogeosciences Discussions*, **3** (4), 895–959, http://dx.doi.org/10.5194/bgd-3-895-2006.

Gaylord, B. and Gaines, S. D. (2000). Temperature or transport? Range limits in marine species mediated solely by flow. *The American Naturalist*, **155** (6), 769–89, http://dx.doi.org/10.1086/303357.

Gaylord, B., Reed, D. C., Washburn, L. and Raimondi, P. T. (2004). Physical-biological coupling in spore dispersal of kelp forest macroalgae. *Journal of Marine Systems*, **49** (1–4), 19–39, http://dx.doi.org/10.1016/j.jmarsys.2003.05.003.

Gaylord, B., Rosman, J. H., Reed, D. C. et al. (2007). Spatial patterns of flow and their modification within and around a giant kelp forest. *Limnology and Oceanography*, **52** (5), 1838–52, http://dx.doi.org/10.4319/lo.2007.52.5.1838.

Gaylord, B. P., Nickols, K. J. and Jurgens, L. (2012). Roles of transport and mixing processes in kelp forest ecology. *Journal of Experimental Biology*, **215** (6), 997–1007, http://dx.doi.org/10.1242/jeb.059824.

Genovese, S. J. and Witman, J. D. (1999). Interactive effects of flow speed and particle concentration on growth rates of an active suspension feeder. *Limnology and Oceanography*, **44** (4), 1120–31.

Gerard, V. A. (1982). Growth and utilization of internal nitrogen reserves by the giant kelp *Macrocystis pyrifera* in a low-nitrogen environment. *Marine Biology*, **66** (1), 27–35, http://dx.doi.org/10.1007/BF00397251.

Gladenkov, A. Y., Oleinik, A. E., Marincovich, Jr, L. and Barinov, K. B. (2002). A refined age for the earliest opening of Bering Strait. *Palaeogeography, Palaeoclimatology, Palaeoecology*, **183** (3–4), 321–8, http://dx.doi.org/10.1016/S0031-0182(02)00249-3.

Glynn, P. W. (1988). El Nino-southern oscillation 1982–1983: nearshore population, community, and ecosystem responses. *Annual Review of Ecology and Systematics*, **19** (1), 309–46, http://dx.doi.org/10.1146/annurev.es.19.110188.001521.

Goddard, J. H. R., Treneman, N., Pence W. E. et al. (2016). Nudibranch range shifts associated with the 2014 warm anomaly in the north-east Pacific. *Bulletin of the Southern California Academy of Sciences*, **115** (1), 15–40.

Gottscho, A. D. (2016). Zoogeography of the San Andreas Fault system: great Pacific fracture zones correspond with spatially concordant phylogeographic boundaries in western North America. *Biological Reviews*, **91** (1), 235–54, http://dx.doi.org/10.1111/brv.12167.

Graham, K. R. and Sebens, K. P. (1996). The distribution of marine invertebrate larvae near vertical surfaces in the rocky subtidal zone. *Ecology*, **77** (3), 933–49.

Graham, M. H. (2004). Effects of local deforestation on the diversity and structure of southern California giant kelp forest food webs. *Ecosystems*, **7** (4), 341–57, http://dx.doi.org/10.1007/s10021-003-0245-6.

Graham, M. H., Kinlan, B. P. and Grosberg, R. K. (2009). Post-glacial redistribution and shifts in productivity of giant kelp forests. *Proceedings of the Royal Society B*, 277 (1680), 399–406, http://dx.doi.org/10.1098/rspb.2009.1664.

Graham, W. M. and Largier, J. L. (1997). Upwelling shadows as nearshore retention sites: the example of northern Monterey Bay. *Continental Shelf Research*, **17** (5), 509–32.

Grantham, B. A., Eckert, G. L. and Shanks, A. L. (2003). Dispersal potential of marine invertebrates in diverse habitats. *Ecological Applications*, **13** (1), 108–16, http://dx.doi.org/10.1890/1051-0761(2003)013[0108:DPOMII]2.0.CO;2.

Grason, E. W. and Buhle, E. R. (2016). Comparing the influence of native and invasive intraguild predators on a rare native oyster. *Journal of Experimental Marine Biology and Ecology*, **479**, 1–8, http://dx.doi.org/10.1016/j.jembe.2016.02.012.

Grason, E. W. and Miner, B. G. (2012). Behavioral plasticity in an invaded system: Non-native whelks recognize risk from native crabs. *Oecologia*, **169** (1), 105–15, http://dx.doi.org/10.1007/s00442-011-2188-5.

Gruber, N., Hauri, C., Lachkar, Z., Loher, D., Frölicher, T. L. and Plattner, G.-K. (2012). Rapid progression of ocean acidification in the California Current System. *Science*, **337** (6091), 220–3, http://dx.doi.org/10.1126/science.1216773.

Gudenkauf, B. M. and Hewson, I. (2015). Metatranscriptomic analysis of *Pycnopodia helianthoides* (Asteroidea) affected by sea star wasting disease. *PLoS ONE*, **10** (5), e0128150, http://dx.doi.org/10.1371/journal.pone.0128150.

Gutt, J. (2001). On the direct impact of ice on marine benthic communities, a review. *Polar Biology*, **24** (8), 553–64, http://dx.doi.org/10.1007/s003000100262.

Haggerty, M. B., Anderson, T. W. and Long, J. D. (2018). Fish predators reduce kelp frond loss via a trait-mediated trophic cascade. *Ecology*, **99** (7), 1574–83.

Haigh, R., Ianson, D., Holt, C. A., Neate, H. E. and Edwards, A. M. (2015). Effects of ocean acidification on temperate coastal marine ecosystems and fisheries in the northeast pacific. *PLoS ONE*, **10** (2), 1–46, http://dx.doi.org/10.1371/journal.pone.0117533.

Hall, C. A. (1964). Shallow-water marine climates and molluscan. *Ecology*, **45** (2), 226–34.

Hallett, T. B., Coulson, T., Pilkington, J. G., Clutton-Brock, T. H., Pemberton, J. M. and Grenfell, B. T. (2004). Why large-scale climate indices seem to predict ecological processes better than local weather. *Nature*, **430** (6995), 71–5, http://dx.doi.org/10.1038/nature02708.

Hamilton, J. and Konar, B. (2007). Implications of substrate complexity and kelp variability for south-central Alaskan nearshore fish communities. *Fishery Bulletin*, **105** (2), 189–96.

Hamilton, S. L., Caselle, J. E., Malone, D. P. and Carr, M. H. (2010). Incorporating biogeography into evaluations of the Channel Islands marine reserve network. *Proceedings of the National Academy of Sciences of the United States of America*, **107** (43), 18272–7, http://dx.doi.org/10.1073/pnas.0908091107.

Hammond, L. M. and Hofmann, G. E. (2010). Thermal tolerance of *Strongylocentrotus purpuratus* early life history stages: Mortality, stress-induced gene expression and biogeographic patterns. *Marine Biology*, **157** (12), 2677–87, http://dx.doi.org/10.1007/s00227-010-1528-z.

Hare, S. R. and Mantua, N. J. (2000). Empirical evidence for North Pacific regime shifts in 1977 and 1989. *Progress in Oceanography*, **47** (2–4), 103–45, http://dx.doi.org/10.1016/S0079-6611(00)00033-1.

Harley, C. D. G., Anderson, K. M., Demes, K. W. et al. (2012). Effects of climate change on global seaweed communities. *Journal of Phycology*, **48** (5), 1064–78, http://dx.doi.org/10.1111/j.1529-8817.2012.01224.x.

Harley, C. D. G. and Connell, S. D. (2009). Shifts in Abiotic Variables and Consequences for Diversity. In M. Wahl, ed. *Marine Hard Bottom Communities: Patterns, Dynamics, Diversity, and Change*. Springer, New York, pp. 257–68, http://dx.doi.org/10.1007/978-3-540-92704-4.

Harley, C. D. G., Hughes, A. R., Hultgren, K. M. et al. (2006). The impacts of climate change in coastal marine systems: Climate change in coastal marine systems. *Ecology Letters*, **9** (2), 228–41, http://dx.doi.org/10.1111/j.1461-0248.2005.00871.x.

Harley, C. D. G. and Lopez, J. P. (2005). The natural history, thermal physiology, and ecological impacts of intertidal mesopredators, *Oedoparena* spp. (Diptera: Dryomyzidae). *Invertebrate Biology*, **122** (1), 61–73, http://dx.doi.org/10.1111/j.1744-7410.2003.tb00073.x.

Harrold, C. and Reed, D. C. (1985). Food availability, sea urchin grazing, and kelp forest community structure. *Ecology*, **66** (4), 1160–9.

Harvell, C. D., Montecino-Latorre, D., Caldwell, J. M. et al. (2019). Disease epidemic and a marine heat wave are associated with the continental-scale collapse of a pivotal predator (*Pycnopodia helianthoides*). *Science Advances*, **5** (1), eaau7042.

Harvey, J. B. J., Hoy, M. S. and Rodriguez, R. J. (2009). Molecular detection of native and invasive marine invertebrate larvae present in ballast and open water environmental samples collected in Puget Sound. *Journal of Experimental Marine Biology and Ecology*, **369** (2), 93–9, http://dx.doi.org/10.1016/j.jembe.2008.10.030.

Hastings, A. (1988). Food web theory and stability. *Ecology*, **69** (6), 1665–8, http://dx.doi.org/10.2307/1941143.

Hawkes, M. W. (1981). *Porphyra nereocystis* and *P. thuretii* (Rhodophyta): gametophyte morphology, distribution, and occurrence. *Syesis*, **14**, 97–108.

Hendler, G. (2013). Recent mass mortality of *Strongylocentrotus purpuratus* (Echinodermata: Echinoidea) at Malibu and a review of purple sea urchin kills elsewhere in California. Bulletin. *Southern California Academy of Sciences*, **112** (1), 19–37, www.bioone.org/doi/abs/10.3160/0038-3872-112.1.19.

Hernández-Carmona, G., Robledod, D. and Serviere-Zaragozab, E. (2001). Effect of nutrient availability on *Macrocystis pyrifera* recruitment and survival near its southern limit off Baja California. *Botanica Marina*, **44**, 221–9.

Hewson, I., Button, J. B., Gudenkauf, B. M. et al. (2014). Densovirus associated with sea-star wasting disease and mass mortality. *Proceedings of the National Academy of Sciences of the United States of America*, **111** (48), 17278–83, http://dx.doi.org/10.1073/pnas.1416625111.

Hickerson, M. J. and Ross, J. R. P. (2001). Post-glacial population history and genetic structure of the northern clingfish (*Gobbiesox maeandricus*), revealed from mtDNA analysis. *Marine Biology*, **138** (2), 407–19, http://dx.doi.org/10.1007/s002270000465.

Hickey, B. M. and Banas, N. S. (2003). Oceanography of the U.S. Pacific northwest coastal ocean and estuaries with application to coastal ecology. *Estuaries*, **26** (4), 1010–31, http://dx.doi.org/10.1007/BF02803360.

Hobday, A. J. (2000). Persistence and transport of fauna on drifting kelp (*Macrocystis pyrifera* (L.) C. Agardh) rafts in the Southern California Bight. *Journal of Experimental Marine Biology and Ecology*, **253** (1), 75–96, http://dx.doi.org/10.1016/S0022-0981(00)00250-1.

Hoegh-Guldberg, O. and Bruno, J. (2010). The impact of climate change on the world's marine ecosystems. *Science*, **328** (5985), 1523–8, http://dx.doi.org/10.1126/science.1189930.

Hogue, C. and Swig, B. (2007). Habitat quality and endoparasitism in the Pacific sanddab *Citharichthys sordidus* from Santa Monica Bay, southern California. *Journal of Fish Biology*, **70** (1), 231–42, http://dx.doi.org/10.1111/j.1095-8649.2006.01298.x.

Holbrook, S. J., Schmitt, R. J., Carr, M. H. and Coyer, J. A. (1990). Effect of giant kelp on local abundance of reef fishes: the importance of ontogenetic resource requirements. *Bulletin of Marine Science*, **47** (1), 104–14.

Holbrook, S. J., Schmitt, R. J. and Stephens, J. S. (1997). Changes in an assemblage of temperate reef fishes associated with a climate shift. *Ecological Applications*, **7** (4), 1299–310, http://dx.doi.org/10.1890/1051-0761(1997)007.

Holsman, K. K., McDonald, P. S. and Armstrong, D. A. (2006). Intertidal migration and habitat use by subadult Dungeness crab *Cancer magister* in a north-east Pacific estuary. *Marine Ecology Progress Series*, **308**, 183–95, http://dx.doi.org/10.3354/meps308183.

Hood, E., Fellman, J., Spencer, R. G. M. et al. (2009). Glaciers as a source of ancient and labile organic matter to the marine environment. *Nature*, **462** (7276), 1044–7, http://dx.doi.org/10.1038/nature08580.

Horn, M. H., Allen, L. G. and Lea, R. N. (2006). Biogeography. In L. G. Allen, D. J. Pondella and M. H. Horn, eds. *The Ecology of Marine Fishes: California and Adjacent Waters*. University of California Press, Berkeley, pp. 3–25, http://dx.doi.org/10.1525/gfc.2008.8.4.24.

Hornberger, M. I., Luoma, S. N., van Geen, A., Fuller, C. and Anima, R. (1999). Historical trends of metals in the sediments of San Francisco Bay, California. *Marine Chemistry*, **64** (1–2), 39–55, http://dx.doi.org/10.1016/S0304-4203(98)80083-2.

Hoshijima, U. and Hofmann, G. (2019). Variability of seawater chemistry in a kelp forest environment is linked to in situ transgenerational effects in the purple sea urchin, *Strongylocentrotus purpuratus*. *Frontiers in Marine Science*, **6**, 62.

Hubbard, C. B., Garbary, D. J., Kim, K. Y. and Chiasson, D. M. (2004). Host specificity and growth of kelp gametophytes symbiotic with filamentous red algae (Ceramiales, Rhodophyta). *Helgoland Marine Research*, **58** (1), 18–25, http://dx.doi.org/10.1007/s10152-003-0162-2.

Hurd, C. L. (2000). Water motion, marine macroalgal physiology, and production. *Journal of Phycology*, **36** (3), 453–72, http://dx.doi.org/10.1046/j.1529-8817.2000 .99139.x.

Huyer, A. (1983). Coastal upwelling in the California Current System. *Progress in Oceanography*, **12** (3), 259–84, http://dx.doi.org/10.1016/0079-6611(83)90010-1.

Huyer, A., Smith, R. L. and Fleischbein, J. (2002). The coastal ocean off Oregon and northern California during the 1997–98 El Niño. *Progress in Oceanography*, **54**, 311–41.

Jackson, G. A. (1977). Nutrients and production of giant kelp, *Macrocystis pyrifera*, off southern California. *Limnology and Oceanography*, **22**, 979–95, http://dx.doi.org/10.4319/lo.1977.22.6.0979.

Jacobs, D. K., Haney, T. A. and Louie, K. D. (2004). Genes, diversity, and geologic process on the Pacific Coast. *Annual Review of Earth and Planetary Sciences*, **32** (1), 601–52, http://dx.doi.org/10.1146/annurev.earth.32.092203.122436.

Jacox, M. G., Hazen, E. L., Zaba, K. D. et al. (2016). Impacts of the 2015-2016 El Niño on the California current system: early assessment and comparison to past events. *Geophysical Research Letters*, **58** (1), 18–25, http://dx.doi.org/10.1002/2016GL069716.

Jensen, M. M. and Denny, M. W. (2015). Experimental determination of the hydrodynamic forces responsible for wave impact events. *Journal of Experimental Marine Biology and Ecology*, **469**, 123–30, http://dx.doi.org/10.1016/j.jembe.2015.04.013.

Jensen, M. M. and Denny, M. W. (2016). Life in an extreme environment: characterizing wave-imposed forces in the rocky intertidal zone using high temporal resolution hydrodynamic measurements. *Limnology and Oceanography*, **61** (5), 1750–61, http://dx.doi.org/10.1002/lno.10327.

Jeong, Y., Grant, S. B., Ritter, S. et al. (2005). Identifying pollutant sources in tidally mixed systems: case study of fecal indicator bacteria from marinas in Newport Bay, Southern California. *Environmental Science and Technology*, **39** (23), 9083–93, http://dx.doi.org/10.1021/es0482684.

Jessee, W. N., Carpenter, A. L. and Carter, J. W. (1985). Distribution patterns and density estimates of fishes on a southern California artificial reef with comparisons to natural kelp-reef habitats. *Bulletin of Marine Science*, **37** (1), 214–26.

Johansson, M. L., Alberto, F., Reed, D. C. et al. (2015). Seascape drivers of *Macrocystis pyrifera* population genetic structure in the northeast Pacific. *Molecular Ecology*, **24** (19), 4866–85, http://dx.doi.org/10.1111/mec.13371.

Johnson, G., Attrill, M. and Sheehan, E. (2007). Recovery of meiofauna communities following mudflat disturbance by trampling associated with crab-tiling. *Marine Environmental Research*, **64** (4), 409–16, www.sciencedirect.com/science/article/pii/S0141113607000438.

Kagley, A. N., Snider, R. G., Kardong, K. E. and Casillas, E. (2014). Effects of chemical contaminants on growth, age-structure, and reproduction of mytilus edulis complex from Puget Sound, Washington. *Bulletin of Environmental Contamination and Toxicology*, **93** (1), 7–12.

Kahn, A. S., Ruhl, H. A. and Smith, K. L. (2012). Temporal changes in deep-sea sponge populations are correlated to changes in surface climate and food supply. *Deep-Sea Research Part I: Oceanographic Research Papers*, **70**, 36–41, http://dx.doi.org/10.1016/j.dsr.2012.08.001.

Kavanaugh, M. T., Nielsen, K. J., Chan, F. T. et al. (2009). Experimental assessment of the effects of shade on an intertidal kelp: do phytoplankton blooms inhibit growth of open coast macroalgae? *Limnology and Oceanography*, **54** (1), 276–88, http://dx.doi.org/10.4319/lo.2009.54.1.0276.

Kawamata, S. (1998). Effect of wave-induced oscillatory flow on grazing by a subtidal sea urchin *Strongylocentrotus nudus* (A. Agassiz). *Journal of Experimental Marine Biology and Ecology*, **224** (1), 31–48, http://dx.doi.org/10.1016/S0022-0981(97)00165-2.

Keever, C. C., Sunday, J., Puritz, J. B. et al. (2009). Discordant distribution of populations and genetic variation in a sea star with high dispersal potential. *Evolution*, **63** (12), 3214–27, http://dx.doi.org/10.1111/j.1558-5646.2009.00801.x.

Kelly, R. P. and Palumbi, S. R. (2010). Genetic structure among 50 species of the northeastern pacific rocky intertidal community. *PLoS ONE*, **5** (1), http://dx.doi.org/10.1371/journal.pone.0008594.

King, J. R., Agostini, V. N., Harvey, C. J. et al. (2011). Climate forcing and the California current ecosystem. *ICES Journal of Marine Science*, **68** (6), 1199–216, http://dx.doi.org/10.1093/icesjms/fsr009.

Klinger, T., Padilla, D. K. and Britton-Simmons, K. (2006). Two invaders achieve higher densities in reserves. *Aquatic Conservation: Marine and Freshwater Ecosystems*, **16** (3), 301–11, http://dx.doi.org/10.1002/aqc.717.

Koehl, M. A. R., Powell, T. M. and Dairiki, G. (1993). Measuring the Fate of Patches in the Water: Larval Dispersal. In S. A. Levin, T. M. Powell and J. H. Steele, eds. *Patch Dynamics*. Springer-Verlag, Berlin, pp. 50–60.

Koenigs, C., Miller, R. and Page, H. (2015). Top predators rely on carbon derived from giant kelp *Macrocystis pyrifera*. *Marine Ecology Progress Series*, **537**, 1–8, http://dx.doi.org/10.3354/meps11467.

Kohl, W. T., McClure, T. I. and Miner, B. G. (2016). Decreased temperature facilitates short-term sea star wasting disease survival in the keystone intertidal sea star *Pisaster ochraceus*. *PLoS ONE*, **11** (4), e0153670, http://dx.doi.org/10.1371/journal.pone.0153670.

Kokkinakis, S. A. and Wheeler, P. A. (1987). Nitrogen uptake and phytoplankton growth in coastal upwelling regions. *Limnology and Oceanography*, **32** (5), 1112–23, http://dx.doi.org/10.4319/lo.1987.32.5.1112.

Konar, B., Iken, K. and Edwards, M. (2009). Depth-stratified community zonation patterns on Gulf of Alaska rocky shores. *Marine Ecology*, **30** (1), 63–73, http://dx.doi.org/10.1111/j.1439-0485.2008.00259.x.

Konotchick, T., Dupont, C. L., Valas, R. E., Badger, J. H. and Allen, A. E. (2013). Transcriptomic analysis of metabolic function in the giant kelp, *Macrocystis pyrifera*, across depth and season. *The New phytologist*, **198** (2), 398–407, http://dx.doi.org/10.1111/nph.12160.

Kordas, R. L., Harley, C. D. G. and O'Connor, M. I. (2011). Community ecology in a warming world: the influence of temperature on interspecific interactions in marine systems. *Journal of Experimental Marine Biology and Ecology*, **400** (1–2), 218–26, http://dx.doi.org/10.1016/j.jembe.2011.02.029.

Kosro, P. M. (2002). A poleward jet and an equatorward undercurrent observed off Oregon and northern California, during the 1997–98 El Niño. *Progress in Oceanography*, **54** (1–4), 343–60, http://dx.doi.org/10.1016/S0079-6611(02)00057-5.

Kusumo, H. T., Pfister, C. A. and Wootton, J. T. (2006). Small-scale genetic structure in the sea palm *Postelsia palmaeformis* Ruprecht (Phaeophyceae). *Marine Biology*, **149** (4), 731–42, http://dx.doi.org/10.1007/s00227-006-0254-z.

Ladah, L. B. and Zertuche-González, J. A. (1999). Giant kelp (*Macrocystis pyrifera*, Phaeophyceae) recruitment near its southern limit in Baja California after mass disappearance during ENSO. *Journal of Phycology*, **35**, 1106–12.

Ladah, L. B. and Zertuche-González, J. A. (2004). Giant kelp (*Macrocystis pyrifera*) survival in deep water

(25-40 m) during El Niño of 1997–1998 in Baja California, Mexico. *Botanica Marina*, **47** (5), 367–72, http://dx.doi .org/10.1515/BOT.2004.054.

Lang, M. A., Marinelli, R. L., Roberts, S. J. and Taylor, P. R., eds. (2013) *Research and Discoveries: The Revolution of Science through Scuba, Smithsonian Contributions to the Marine Sciences*. Smithsonian Institution Press, Washington, DC, www.vliz.be/imisdocs/publications/256301.pdf#page=141.

Lauerman, L. M. L., Kaufmann, R. S. and Smith, K. L. (1996). Distribution and abundance of epibenthic megafauna at a long time-series station in the abyssal northeast Pacific. *Deep-Sea Research Part I: Oceanographic Research Papers*, **43** (7), 1075–103, http://dx.doi .org/10.1016/0967-0637(96)00045-3.

Laur, D., Ebeling, A. and Coon, D. (1988). Effects of Sea Otter Foraging on Subtidal Reef Communities off Central California. In G. R. VanBlaricom and J. A. Estes, eds. *The Community Ecology of Sea Otters*. Springer-Verlag, Berlin, pp. 151–68, http://link .springer.com/chapter/10.1007/978–3-642–72845-7_7.

LeClair, L., Pleus, A. and Schultz, J. (2009) *2007-2009 Biennial Report: Invasive Species Tunicate Response*. Puget Sound Partnership, Olympia, WA.

Lesser, M. P., Witman, J. D. and Sebens, K. P. (1994). Effects of flow and seston availability on scope for growth of benthic suspension-feeding invertebrates from the Gulf of Maine. *Biological Bulletin*, **187** (3), 319–35, http://dx.doi.org/10.2307/1542289.

Lester, S. E., Tobin, E. D. and Behrens, M. D. (2007). Disease dynamics and the potential role of thermal stress in the sea urchin, *Strongylocentrotus purpuratus*. *Canadian Journal of Fisheries and Aquatic Sciences*, **64** (2), 314–23, http://dx.doi.org/10.1139/f07-010.

Levenbach, S. (2008). Community-wide ramifications of an associational refuge on shallow rocky reefs. *Ecology*, **89** (10), 2819–28, http://dx.doi.org/10.1890/07 .0656.1.

Levings, C. D., Cordell, J. R., Ong, S. and Piercey, G. E. (2004). The origin and identity of invertebrate organisms being transported to Canada's Pacific coast by ballast water. *Canadian Journal of Fisheries and Aquatic Sciences*, **61**, 1–11, http://dx.doi.org/10.1139/f03-135.

Levings, C. D., Foreman, R. E. and Tunnicliffe, V. J. (1983). Review of the benthos of the Strait of Georgia and contiguous fjords. *Canadian Journal of Fisheries and Aquatic Sciences*, **40** (7), 1120–41, http://dx.doi.org/10 .1139/f83-131.

Liggan, L. (2016) *Under Pressure: Biomechanics of Buoyancy in Bull Kelp (*Nereocystis leutkeana*)*. University of British Columbia, Vancouver, http://dx.doi.org/10.14288/1.0300227.

Lindstrom, S. C. (2009). The biogeography of seaweeds in Southeast Alaska. *Journal of Biogeography*, **36** (3), 401–9, http://dx.doi.org/10.1111/j.1365-2699.2007 .01855.x.

Lindstrom, S. C. and Foreman, R. E. (1978). Seaweed associations of the Flat Top Islands, British Columbia – comparison of community methods. *Syesis*, **11**, 171–85.

Lindstrom, S. C., Olsen, J. L. and Stam, W. T. (1997). Postglacial recolonization and the biogeography of *Palmaria mollis* (Rhodophyta) along the northeast Pacific coast. *Canadian Journal of Botany*, **75** (11), 1887–96, http://dx.doi.org/10.1139/b97-900.

Litzow, M. A., Mueter, F. J. and Hobday, A. J. (2013). Reassessing regime shifts in the North Pacific: incremental climate change and commercial fishing are necessary for explaining decadal-scale biological variability. *Global Change Biology*, **20** (1), 38–50, http://dx .doi.org/10.1111/gcb.12373.

Long, E. R., Dutch, M., Weakland, S., Chandramouli, B. and Benskin, J. P. (2013). Quantification of pharmaceuticals, personal care products, and perfluoroalkyl substances in the marine sediments of Puget Sound, Washington, USA. *Environmental Toxicology and Chemistry*, 32 (8), 1701–10, http://dx.doi.org/10.1002/etc .2281.

Di Lorenzo, E., Cobb, K. M., Furtado, J. C. et al. (2010). Central Pacific El Niño and decadal climate change in the North Pacific Ocean. *Nature Geoscience*, **3**, 762–5, http://dx.doi.org/10.1038/ngeo984.

Di Lorenzo, E., Schnieder, N., Cobb, K. M. et al. (2008). North Pacific gyre oscillation links ocean climate and ecosystem change. *Geophysical Research Letters*, **35** (8), 2–7, http://dx.doi.org/10.1029/2007GL032838.

Love, M. S. and Larson, R. J. (1978). Geographic variation in the occurrence of tympanic spines and possible genetic differentiation in the kelp rockfish (*Sebastes atrovirens*). *Copeia*, **1978** (1), 53–9.

Love, M. S., Saiki, M. K., May, T. W. and Yee, J. L. (2013). Whole-body concentrations of elements in three fish species from offshore oil platforms and natural areas in the Southern California Bight, USA. *Bulletin of Marine Science*, **89** (3), 717–34, http://dx.doi.org/10 .5343/bms.2012.1078.

Love, M. S., Schroeder, D. M., Lenarz, W., MacCall, A., Bull, A. S. and Thorsteinson, L. (2006). Potential use of offshore marine structures in rebuilding an overfished rockfish species, bocaccio (*Sebastes paucispinis*). *Fishery Bulletin*, **104** (3), 383–90.

Low, N. H. and Micheli, F. (2018). Lethal and functional thresholds of hypoxia in two key benthic grazers. *Marine Ecology Progress Series*, **594**, 165–73.

Lowe, A. T., Galloway, A. W. E., Yeung, J. S., Dethier, M. N. and Duggins, D. O. (2014). Broad sampling and diverse biomarkers allow characterization of nearshore particulate organic matter. *Oikos*, **123** (11), 1341–54, http://dx.doi.org/10.1111/oik.01392.

Lowe, A. T., Whippo, R., Galloway, A. W. E., Britton-Simmons, K. H. and Dethier, M. N. (2015). Sedentary urchins influence benthic community composition below the macroalgal zone. *Marine Ecology*, **36** (2), 129–40, http://dx.doi.org/10.1111/maec.12124.

Lubchenco, J., Navarette, S. A., Tissot, B. N. and Castilla, J. C. (1993). Possible Ecological Responses to Global Climate Change: Nearshore Benthic Biota of Northeastern Pacific Coastal Ecosystems. In H. A. Mooney, E. Fuentes and B. I. Kronberg, eds. *Earth System Responses To Global Change: Contrasts between North and South America*. Academic Press, San Diego, CA, pp. 147–66.

Lucas, A. J., Dupont, C. L., Tai, V., Largier, J. L., Palenik, B. and Franks, P. J. S. (2011). The green ribbon: multiscale physical control of phytoplankton productivity and community structure over a narrow continental shelf. *Limnology and Oceanography*, **56** (2), 611–26, http://dx.doi.org/10.4319/lo.2011.56.2.0611.

Lüning, K. and Freshwater, W. (1988). Temperature tolerance of northeast Pacific marine algae. *Journal of Phycology*, **24** (3), 310–15, http://dx.doi.org/10.1111/j.1529-8817.1988.tb04471.x.

Lüning, K., Yarish, C. and Kirkman, H. (1990) *Seaweeds: Their Environment, Biogeography, and Ecophysiology*. Wiley, New York.

Lyle, M., Koizumi, I., Delaney, M. L. and Barron, J. A. (2000). Sedimentary record of the California current system, middle Miocene to Holocene: a synthesis of leg 167 results. *Proceedings of the Ocean Drilling Program, Scientific results*, **167**, 341–76.

Lynn, R. J. and Simpson, J. J. (1987). The California current system: the seasonal variability of its physical characteristics. *Journal of Geophysical Research: Oceans*, **92** (C12), 12947–66, http://dx.doi.org/10.1029/JC092iC12p12947.

Mace, A. J. and Morgan, S. G. (2006). Biological and physical coupling in the lee of a small headland: contrasting transport mechanisms for crab larvae in an upwelling region. *Marine Ecology Progress Series*, **324**, 185–96, http://dx.doi.org/10.3354/meps324185.

Mackas, D., Peterson, W. and Ohman, M. (2006). Zooplankton anomalies in the California current system before and during the warm ocean conditions of 2005. *Geophysical Research Letters*, **33** (22), L22S07, http://onlinelibrary.wiley.com/doi/10.1029/2006GL027930/full.

Mann, K. H., Kenneth, H. and Lazier, J. R. N. (2006). *Dynamics of Marine Ecosystems Biological–Physical Interactions in the Oceans*, third edn. Blackwell Publishing, Hoboken, NJ.

Mantua, N. J. and Hare, S. R. (2002). The Pacific decadal oscillation. *Journal of Oceanography*, **58**, 35–44.

Mantua, N. J., Hare, S. R., Zhang, Y., Wallace, J. M. and Francis, R. C. (1997). A Pacific interdecadal climate oscillation with impacts on salmon production. *Bulletin of the American Meteorological Society*, **78** (6), 1069–79, http://dx.doi.org/10.1175/1520-0477(1997)078<1069:APICOW>2.0.CO;2.

Marchinko, K. B. and Palmer, A. R. (2003). Feeding in flow extremes: dependence of cirrus form on wave-exposure in four barnacle species. *Zoology*, **106** (2), 127–41, http://dx.doi.org/10.1078/0944-2006-00107.

Markel, R. W. and Shurin, J. B. (2015). Indirect effects of sea otters on rockfish (Sebastes spp.) in giant kelp forests. *Ecology*, **96** (11), 2877–90, http://dx.doi.org/10.1121/1.4929899.

Marko, P. B. (2004). '"What"s larvae got to do with it?' Disparate patterns of post-glacial population structure in two benthic marine gastropods with identical dispersal potential. *Molecular Ecology*, **13** (3), 597–611, http://dx.doi.org/10.1046/j.1365-294X.2004.02096.x.

Matson, P. G. and Edwards, M. S. (2007). Effects of ocean temperature on the southern range limits of two understory kelps, *Pterygophora californica* and *Eisenia arborea*, at multiple life-stages. *Marine Biology*, **151** (5), 1941–9, http://dx.doi.org/10.1007/s00227-007-0630-3.

Mazariegos-Villarreal, A., Casas-Valdez, M., Siqueiros-Beltrones, D. A., Piñon-Gimate, A. and Serviere-Zaragoza, E. (2012). During the 1997 to 1998 El Niño Event in Baja California Sur, Mexico. *Journal of Shellfish Research*, **31** (3), 795–800, http://dx.doi.org/10.2983/035.031.0325.

McCann, L. D., Holzer, K. K., Davidson, I. C. et al. (2013). Promoting invasive species control and eradication in the sea: options for managing the tunicate invader *Didemnum vexillum* in Sitka, Alaska. *Marine Pollution Bulletin*, **77** (1–2), 165–71, http://dx.doi.org/10.1016/j.marpolbul.2013.10.011.

McClain, C. R., Lundsten, L., Ream, M., Barry, J. and DeVogelaere, A. (2009). Endemicity, biogeography, composition, and community structure on a northeast Pacific seamount. *PLoS ONE*, **4** (1), http://dx.doi.org/10.1371/journal.pone.0004141.

McCormick-Ray, J., Warwick, R. M. and Ray, G. C. (2011). Benthic macrofaunal compositional variations in the northern Bering Sea. *Marine Biology*, **158** (6), 1365–76, http://dx.doi.org/10.1007/s00227-011-1655-1.

McGowan, J. A., Cayan, D. R. and Dorman, L. M. (1998). Climate-ocean variability and ecosystem response in the northeast Pacific. *Science*, **281** (5374), 210–17, http://dx.doi.org/10.1126/science.281.5374.210.

Menge, B. A., Chan, F., Nielsen, K. J., Di Lorenzo, E. and Lubchenco, J. (2009). Climatic variation alters supply-side ecology: impact of climate patterns on phytoplankton and mussel recruitment. *Ecological Monographs*, **79** (3), 379–95, http://dx.doi.org/10.1890/08-2086.1.

Menge, B. A., Daley, B. A., Wheeler, P. A., Dahlhoff, E., Sanford, E. and Strub, P. T. (1997). Benthic-pelagic links and rocky intertidal communities: bottom-up effects on top-down control? *Proceedings of National Academy of Sciences*, **94** (26), 14530–5.

Miller-Rushing, A. J., Høye, T. T., Inouye, D. W. and Post, E. (2010). The effects of phenological mismatches on demography. *Philosophical Transactions of the Royal Society of London B*, **365**, 3177–86, http://dx.doi.org/10.1098/rstb.2010.0148.

Miller, B. A. and Emlet, R. B. (1997). Influence of nearshore hydrodynamics on larval abundance and settlement of sea urchins *Strongylocentrotus franciscanus* and *S. purpuratus* in the Oregon upwelling zone. *Marine Ecology Progress Series*, **148** (1–3), 83–94, http://dx.doi.org/10.3354/meps148083.

Miller, E. F. and McGowan, J. A. (2013). Faunal shift in southern California's coastal fishes: a new assemblage and trophic structure takes hold. *Estuarine, Coastal and Shelf Science*, **127**, 29–36, http://dx.doi.org/10.1016/j.ecss.2013.04.014.

Miller, K., Engle, J. M., Uwai, S. et al. (2007). First report of the Asian seaweed *Sargassum filicinum* Harvey (Fucales) in California, USA. *Biological Invasions*, **9** (1), 926–9, http://dx.doi.org/10.1007/s10530-006-9060-2.

Miller, K. and Estes, J. (1989). Western range extension for *Nereocystis luetkeana* in the North Pacific Ocean. *Botanica Marina*, **32** (6), 535–8, www.degruyter.com/view/j/botm.1989.32.issue-6/botm.1989.32.6.535/botm.1989.32.6.535.xml.

Miller, K. A. and Engle, J. M. (2009). The natural history of *Undaria pinnatifida* and *Sargassum filicinum* at the California channel islands: non-native seaweeds with different invasion styles. In *Proceedings of the 7th California Islands Symposium*, 131–40.

Miller, K. B. (2016). Forecasting at the edge of the niche: *Didemnum vexillum* in southeast Alaska. *Marine Biology*, **163** (2), 1–12, http://dx.doi.org/10.1007/s00227-015-2799-1.

Miller, L. P. and Gaylord, B. (2007). Barriers to flow: the effects of experimental cage structures on water velocities in high-energy subtidal and intertidal environments. *Journal of Experimental Marine Biology and Ecology*, **344** (2), 215–28, http://dx.doi.org/10.1016/j.jembe.2007.01.005.

Miller, M. A., Byrne, B. A., Jang, S. S. et al. (2010). Enteric bacterial pathogen detection in southern sea otters (*Enhydra lutris* nereis) is associated with coastal urbanization and freshwater runoff. *Veterinary Research*, **41** (1), http://dx.doi.org/10.1051/vetres/2009049.

Miller, R. J., Harrer, S. and Reed, D. C. (2012). Addition of species abundance and performance predicts community primary production of macroalgae. *Oecologia*, **168** (3), 797–806, http://dx.doi.org/10.1007/s00442-011-2143-5.

Miller, R. J. and Page, H. M. (2012). Kelp as a trophic resource for marine suspension feeders: A review of isotope-based evidence. *Marine Biology*, **159** (7), 1391–402, http://dx.doi.org/10.1007/s00227-012-1929-2.

Miller, R. J., Page, H. M. and Brzezinski, M. A. (2013). δ13C and δ15N of particulate organic matter in the Santa Barbara Channel: Drivers and implications for trophic inference. *Marine Ecology Progress Series*, **474**, 53–66, http://dx.doi.org/10.3354/meps10098.

Miller, R. J., Reed, D. C. and Brzezinski, M. A. (2011). Partitioning of primary production among giant kelp (*Macrocystis pyrifera*), understory macroalgae, and phytoplankton on a temperate reef. *Limnology and Oceanography*, **56** (1), 119–32, http://dx.doi.org/10.4319/lo.2011.56.1.0119.

Miller, S. H. and Morgan, S. G. (2013). Interspecific differences in depth preference: Regulation of larval transport in an upwelling system. *Marine Ecology Progress Series*, **476**, 301–6, http://dx.doi.org/10.3354/meps10150.

Montecino-Latorre, D., Eisenlord, M. E., Turner, M. et al. (2016). Devastating transboundary impacts of sea star wasting disease on subtidal asteroids. *PLoS ONE*, **11** (10), e0163190.

Morales-Zárate, M. V., Arreguín-Sánchez, F., López-Martínez, J. and Lluch-Cota, S. E. (2004). Ecosystem trophic structure and energy flux in the northern Gulf of California, Mexico. *Ecological Modelling*, **174** (4), 331–45, http://dx.doi.org/10.1016/j.ecolmodel.2003.09.028.

Morgan, S. G., Fisher, J. L., Mace, A. J., Akins, L., Slaughter, A. M. and Bollens, S. M. (2009). Cross-shelf distributions and recruitment of crab postlarvae in a region of strong upwelling. *Marine Ecology Progress Series*, **380**, 173–85, http://dx.doi.org/10.3354/meps07913.

Morzaria-Luna, H. N., Ainsworth, C. H., Kaplan, I. C., Levin, P. S. and Fulton, E. A. (2013). Indirect effects of conservation policies on the coupled human-natural ecosystem of the upper Gulf of California. *PLoS ONE*, **8** (5), http://dx.doi.org/10.1371/journal.pone.0064085.

Muñoz-Barbosa, A. and Huerta-Diaz, M. A. (2013). Trace metal enrichments in nearshore sediments and accumulation in mussels (*Modiolus capax*) along the eastern coast of Baja California, Mexico: environmental status in 1995. *Marine Pollution Bulletin*, 77(1–2), 71–81, http://dx.doi.org/10.1016/j.marpolbul.2013.10.030.

Murray, S. N., Littler, M. M. and Abbott, I. A. (1980). Biogeography of the California marine algae with emphasis on the Southern California Islands. In *The California Islands: Proceedings of a Multidiciplinary Symposium*, 35–6.

Muth, A. F., Graham, M. H., Lane, C. E. and Harley, C. D. (2019). Recruitment tolerance to increased temperature present across multiple kelp clades. *Ecology*, e02594.

Myers, M. S., Johnson, L. L. and Collier, T. K. (2003). Establishing the causal relationship between polycyclic aromatic hydrocarbon (PAH) exposure and hepatic neoplasms and neoplasia-related liver lesions in English Sole (*Pleuronectes vetulus*). *Human and Ecological Risk Assessment*, **9** (1), 67–94.

Nakamura, H., Lin, G. and Yamagata, T. (1997). Decadal climate variability in the North Pacific during the recent decades. *Bulletin of the American Meteorological Society*, **78** (10), 2215–25, http://dx.doi.org/10.1175/1520-0477(1997)078<2215:DCVITN>2.0.CO;2.

O'Neel, S., Hood, E., Bidlack, A. L. et al. (2015). Icefield-to-ocean linkages across the northern pacific coastal temperate rainforest ecosystem. *BioScience*, **65** (5), 499–512, http://dx.doi.org/10.1093/biosci/biv027.

Neushul, M. (1967). Studies of subtidal marine vegetation in western Washington. *Ecology*, **48** (1), 83–94.

Neushul, M. and Haxo, F. (1963). Studies on the giant kelp, Macrocystis. I. Growth of young plants. *American Journal of Botany*, **50** (4), 349–53.

Newman, M., Alexander, M. A., Ault, T. R. et al. (2016). The Pacific decadal oscillation, revisited. *Journal of Climate*, 29, 4399–427.

Nidzieko, N. J. and Largier, J. L. (2013). Inner shelf intrusions of offshore water in an upwelling system affect coastal connectivity. *Geophysical Research Letters*, **40** (20), 5423–8, http://dx.doi.org/10.1002/2013GL056756.

Nishizaki, M. T. and Carrington, E. (2014a). Temperature and water flow influence feeding behavior and success in the barnacle *Balanus glandula*. *Marine Ecology Progress Series*, **507**, 207–18, http://dx.doi.org/10.3354/meps10848.

Nishizaki, M. T. and Carrington, E. (2014b). The effect of water temperature and flow on respiration in barnacles: patterns of mass transfer versus kinetic limitation. *Journal of Experimental Biology*, **217** (12), 2101–9, http://dx.doi.org/10.1242/jeb.101030.

Nishizaki, M. T. and Carrington, E. (2015). The effect of water temperature and velocity on barnacle growth: quantifying the impact of multiple environmental stressors. *Journal of Thermal Biology*, **54**, 37–46, http://dx.doi.org/10.1016/j.jtherbio.2015.02.002.

North, W. J. and Hubbs, C. L. (1968) *Utilization of Kelp-Bed Resources in Southern California*. Department of Fish and Game, Sacramento, CA.

North, W. J., Jackson, G. A. and Manley, S. L. (1986). Macrocystis and its environment, knowns and unknowns. *Aquatic Botany*, **26**, 9–26.

North, W. J. and Pearse, J. S. (1970). Sea urchin population explosion in southern California coastal waters. *Science*, **167** (3915), 209.

Nowell, A. R. M. and Jumars, P. A. (1984). Flow environments of aquatic benthos. *Annual Review of Ecology and Systematics*, **15** (1), 303–28, http://dx.doi.org/10.1146/annurev.es.15.110184.001511.

Okamoto, D. K., Reed, D. C. and Schroeter, S. C. (2018). Geographically opposing responses of sea urchin recruitment to changes in ocean climate. *BioRxiv*, 387282.

Okamoto, D. K., Stekoll, M. S. and Eckert, G. L. (2013). Coexistence despite recruitment inhibition of kelps by subtidal algal crusts. *Marine Ecology Progress Series*, **493**, 103–12, http://dx.doi.org/10.3354/meps10505.

Okey, T. A., Alidina, H. M. and Agbayani, S. (2015). Mapping ecological vulnerability to recent climate change in Canada's Pacific marine ecosystems. *Ocean and Coastal Management*, **106**, 35–48, http://dx.doi.org/10.1016/j.ocecoaman.2015.01.009.

Okey, T. A., Alidina, H. M., Lo, V. and Jessen, S. (2014). Effects of climate change on Canada's Pacific marine ecosystems: a summary of scientific knowledge. *Reviews in Fish Biology and Fisheries*, **24** (2), 519–59, http://dx.doi.org/10.1007/s11160-014-9342-1.

Pace, M., Cole, J. J., Carpenter, S. R. and Kitchell, J. F. (1999). Trophic cascades revealed in diverse ecosystems. *Trends in Ecology & Evolution*, **14** (12), 483–8, www.ncbi.nlm.nih.gov/pubmed/10542455.

Page, H. M., Dugan, J. E., Culver, C. S. and Hoesterey, J. C. (2006). Exotic invertebrate species on offshore oil platforms. *Marine Ecology Progress Series*, **325**, 101–7, http://dx.doi.org/10.3354/meps325101.

Page, H. M., Reed, D. C., Brzezinski, M. A., Melack, J. M. and Dugan, J. E. (2008). Assessing the importance of land and marine sources of organic matter to kelp forest food webs. *Marine Ecology Progress Series*, **360**, 47–62, http://dx.doi.org/10.3354/meps07382.

Paine, R. T. and Levin, S. A. (1981). Intertidal landscapes: disturbance and the dynamics of pattern. *Ecological Monographs*, **51** (2), 145–78.

Palardy, J. E. and Witman, J. D. (2011). Water flow drives biodiversity by mediating rarity in marine benthic communities. *Ecology Letters*, **14** (1), 63–8, http://dx.doi.org/10.1111/j.1461-0248.2010.01555.x.

Palumbi, S. R. (2003). Population genetics, demographic connectivity, and the design of marine reserves. *Ecological Applications*, **13** (1), 146–58.

Palumbi, S. R. and Wilson, A. C. (1990). Mitochondrial DNA diversity in the sea urchins *Strongylocentrotus purpuratus* and *S. droebachiensis*. *Evolution*, **44** (2), 403–15.

Parnell, P. E., Miller, E. F., Lennert Cody, C. E., Dayton, P. K., Carter, M. L. and Stebbins, T. D. (2010). The response of giant kelp (*Macrocystis pyrifera*) in southern California to low-frequency climate forcing. *Limnology and Oceanography*, **55** (6), 2686–702, http://dx.doi.org/10.4319/lo.2010.55.6.2686.

Patterson, M. R. (1980). Hydromechanical Adaptations in Alcyonium Sidereum (Octocorallia). In P. J. Schneck, ed. *Biofluid Mechanics 2*. Springer, Boston, 183–201, http://dx.doi.org/10.1007/978-1-4757-4610-5_10.

Pearse, J. S. and Hines, A. H. (1979). Expansion of a central California kelp forest following the mass mortality of sea urchins. *Marine Biology*, **51** (1), 83–91, http://dx.doi.org/10.1007/BF00389034.

Pelc, R. A., Warner, R. R. and Gaines, S. D. (2009). Geographical patterns of genetic structure in marine species with contrasting life histories. *Journal of Biogeography*, **36** (10), 1881–90, http://dx.doi.org/10.1111/j.1365-2699.2009.02138.x.

Pespeni, M. H., Sanford, E., Gaylord, B. et al. (2013). Evolutionary change during experimental ocean acidification. *Proceedings of the National Academy of Sciences*, **110** (17), 6937–42, http://dx.doi.org/10.1073/pnas.1220673110.

Peterson, W., Robert, M. and Bond, N. (2015). The warm blob – conditions in the northeastern Pacific Ocean. *PICES Press*, **23** (1), 36–8, http://search.proquest.com/docview/1665110669?pq-origsite=gscholar.

Pfister, C. A., Berry, H. D. and Mumford, T. (2018). The dynamics of kelp forests in the northeast Pacific Ocean and the relationship with environmental drivers. *Journal of Ecology*, **106** (4), 1520–33.

Piacenza, S. E., Barner, A. K., Benkwitt, C. E. et al. (2015). Patterns and variation in benthic biodiversity in a large marine ecosystem. *PLoS ONE*, **10** (8), 1–23, http://dx.doi.org/10.1371/journal.pone.0135135.

Pineda, J., Reyns, N. and Lentz, S. J. (2018). Reduced barnacle larval abundance and settlement in response to large-scale oceanic disturbances: temporal patterns, nearshore thermal stratification, and potential mechanisms. *Limnology and Oceanography*, **63** (6), 2618–29.

Pirtle, J. L., Ibarra, S. N. and Eckert, G. L. (2012). Nearshore subtidal community structure compared between inner coast and outer coast sites in southeast Alaska. *Polar Biology*, **35** (12), 1889–910, http://dx.doi.org/10.1007/s00300-012-1231-2.

Pister, B. (2009). Urban marine ecology in southern California: the ability of riprap structures to serve as rocky intertidal habitat. *Marine Biology*, http://link.springer.com/article/10.1007/s00227–009-1130-4.

Pondella, D. J., Gintert, B. E., Cobb, J. R. and Allen, L. G. (2005). Biogeography of the nearshore rocky-reef fishes at the southern and Baja California Islands. *Journal of Biogeography*, **32** (2), 187–201.

Rassweiler, A., Schmitt, R. J. and Holbrook, S. J. (2010). Triggers and maintenance of multiple shifts in the state of a natural community. *Oecologia*, **164** (2), 489–98, http://dx.doi.org/10.1007/s00442-010-1666-5.

Reed, D. C., Anderson, T. W., Ebeling, A. W. and Anghera, M. (1997). The role of reproductive synchrony in the colonization potential of kelp. *Ecology*, **78** (8), 2443–57.

Reed, D. C. and Foster, M. S. (1984). The effects of canopy shadings on algal recruitment and growth in a giant kelp forest. *Ecology*, **65** (3), 937–48.

Reed, D. C., Laur, D. R. and Ebeling, A. W. (1988). Variation in algal dispersal and recruitment: the importance of episodic events. *Ecological Monographs*, **58** (4), 321–35.

Reed, D. C., Rassweiler, A., Carr, M. H., Cavanaugh, K. C., Malone, D. P. and Siegel D. A. (2011). Wave disturbance overwhelms top-down and bottom-up control of primary production in California kelp forests. *Ecology*, **92** (11), 2108–16.

Reed, D. C., Rassweiler, A. R. and Arkema, K. D. (2008). Biomass rather than growth rate determines variation in net primary production by giant kelp. *Ecology*, 89(9), 2493–505, http://dx.doi.org/10.1890/07-1106.1.

Reed, D. C., Rassweiler, A. R., Miller, R. J., Page, H. M. and Holbrook, S. J. (2016a). The value of a broad temporal and spatial perspective in understanding dynamics of kelp forest ecosystems. *Marine and Freshwater Research*, **67** (1), 14–24, http://dx.doi.org/10.1071/MF14158.

Reed, D., Washburn, L., Rassweiler, A., Miller, R., Bell, T. and Harrer, S. (2016b). Extreme warming challenges sentinel status of kelp forests as indicators of climate change. *Nature Communications*, **7**, 13757.

Reeves, R. L., Grant, S. B., Mrse, R. D., Copil Oancea, C. M., Sanders, B. F. and Boehm, A. B. (2004). Scaling and management of fecal indicator bacteria in runoff from a coastal urban watershed in southern California. *Environmental Science and Technology*, **38** (9), 2637–48, http://dx.doi.org/10.1021/es034797g.

Renner, M., Arimitsu, M. L. and Piatt, J. F. (2012). Structure of marine predator and prey communities along environmental gradients in a glaciated fjord. *Canadian Journal of Fisheries and Aquatic Sciences*, **69**, 2029–45, http://dx.doi.org/10.1139/f2012-117.

Riosmena-Rodríguez, R., Hinojosa-Arango, G., López-Vivas, J. M. and León-Cisneros, K. (2005). Caracterización espacial y biogeográfica de las asociaciones de macroalgas de Bahía del Rincón, Baja California Sur, México. *Revista de Biologia Tropical*, **53** (1–2), 97–109.

Robinson, H. E., Finelli, C. M. and Koehl, M. A. R. (2013). Interactions between benthic predators and zooplanktonic prey are affected by turbulent waves. *Integrative and Comparative Biology*, **53** (5), 810–20, http://dx.doi.org/10.1093/icb/ict092.

Robles, C., Sweetnam, D. A. and Dittman, D. (1989). Diel variation of intertidal foraging by *Cancer productus* L. in British Columbia. *Journal of Natural History*, **23** (5), 1041–9, http://dx.doi.org/10.1080/00222938900770951.

Robles, C. D., Alvarado, M. A. and Desharnais, R. A. (2001). The shifting balance of littoral predator–prey interaction in regimes of hydrodynamic stress. *Oecologia*, **128** (1), 142–52, http://dx.doi.org/10.1007/s004420100638.

Rocha-Olivares, A. and Vetter, R. D. (1999). Effects of oceanographic circulation on the gene flow, genetic structure, and phylogeography of the rosethorn rockfish (Sebastes helvomaculatus). *Canadian Journal of Fisheries and Aquatic Sciences*, **56** (5), 803–13, http://dx.doi.org/10.1139/cjfas-56-5-803.

Rodriguez, G. E., Rassweiler, A., Reed, D. C. and Holbrook, S. J. (2013). The importance of progressive senescence in the biomass dynam of giant kelp (*Macrocystis pyrifera*). *Ecology*, **94** (8), 1848–58, http://dx.doi.org/10.1890/12-1340.1.

Rogers-Bennett, L. (2007). Is climate change contributing to range reductions and localized extinctions in northern (*Haliotis kamtschatkana*) and flat (*Haliotis walallensis*) abalones ? *Bulletin of Marine Science*, **81** (2), 283–96.

Rogers-Bennett, L., Dondanville, R. F., Moore, J. D. and Ignacio Vilchis, L. (2010). Response of red abalone reproduction to warm water, starvation, and disease stressors: implications of ocean warming. *Journal of Shellfish Research*, **29** (3), 599–611, http://dx.doi.org/10.2983/035.029.0308.

Rosenfeld, L. K., Schwing, F. B., Garfield, N. and Tracy, D. E. (1994). Bifurcated flow from an upwelling center: a cold water source for Monterey Bay. *Continental Shelf Research*, **14** (9), 931–64, http://dx.doi.org/10.1016/0278-4343(94)90058-2.

Rosenthal, R. J., Clarke, W. D. and Dayton, P. K. (1974). Ecology and natural history of a stand of giant kelp, *Macrocystis pyrifera*, off Del Mar, California. *Fishery Bulletin*, **72** (3), 670–84.

Roughgarden, J., Gaines, S. and Possingham, H. (1988). Recruitment dynamics in complex life cycles. *Science*, 1460–6, http://dx.doi.org/10.1126/science.11538249.

Royer, T. C. (1982). Coastal fresh water discharge in the northeast Pacific. *Journal of Geophysical Research*, **87** (1), 2017, http://dx.doi.org/10.1029/JC087iC03p02017.

Royer, T. C. (1983). Observations of the Alaska Coastal Current. In *Coastal Oceanography*. Springer, Boston, pp. 9–30, http://dx.doi.org/10.1007/978-1-4615-6648-9_2.

Ruesink, J. L., Hong, J.-S., Wisehart, L. et al. (2010). Congener comparison of native (*Zostera marina*) and introduced (*Z. japonica*) eelgrass at multiple scales within a Pacific Northwest estuary. *Biological Invasions*, **12** (6), 1773–89, http://dx.doi.org/10.1007/s10530-009-9588-z.

Sackmann, B., Mack, L., Logsdon, M. and Perry, M. J. (2004). Seasonal and inter-annual variability of SeaWiFS-derived chlorophyll *a* concentrations in waters off the Washington and Vancouver Island coasts, 1998–2002. *Deep-Sea Research Part II: Topical Studies in Oceanography*, **51** (10–11), 945–65, http://dx.doi.org/10.1016/j.dsr2.2003.09.004.

Sagarin, R. D., Barry, J. P.,Gilman, S. E. and Baxter, C. H. (1999). Climate-related change in an intertidal community over short and long time scales. *Ecological Monographs*, **69** (4), 465–90.

Sanford, E. and Menge, B. A. (2001). Spatial and temporal variation in barnacle growth in a coastal upwelling system. *Marine Ecology Progress Series*, **209**, 143–57, http://dx.doi.org/10.3354/meps209143.

Saunders, G. W. (2014). Long distance kelp rafting impacts seaweed biogeography in the northeast Pacific: the kelp conveyor hypothesis. *Journal of Phycology*, **50** (6), 968–74, http://dx.doi.org/10.1111/jpy.12237.

Scagel, R. F. (1963). Distribution of Attached Marine Algae in Relation to Oceanographic Conditions in the Northeast Pacific. In M. J. Dunbar, ed. *Marine*

*Distributions*. University of Toronto Press, Toronto, pp. 37–50.

Schiel, D. R. and Foster, M. S. (2015) *The Biology and Ecology of Giant Kelp Forests*. University of California Press, Berkeley.

Schiel, D. R., Steinbeck, J. R. and Foster, M. S. (2004). Ten years of induced ocean warming causes comprehensive changes in marine benthic communities. *Ecology*, **85** (7), 1833–9.

Schmitz, K. and Srivastava, L. M. (1976). The fine structure of sieve elements of *Nereocystis lutkeana*. *American Journal of Botany*, **63** (5), 679–93.

Schroeder, D. M. and Love, M. S. (2004). Ecological and political issues surrounding decommissioning of offshore oil facilities in the Southern California Bight. *Ocean and Coastal Management*, **47** (1–2), 21–48, http://dx.doi.org/10.1016/j.ocecoaman.2004.03.002.

Schroeter, S. C., Reed, D. C. and Raimondi, P. T. (2015). Effects of reef physical structure on development of benthic reef community: a large-scale artificial reef experiment. *Marine Ecology Progress Series*, **540**, 43–55, http://dx.doi.org/10.3354/meps11483.

Schultz, J. A., Cloutier, R. N. and Côté, I. M. (2016). Evidence for a trophic cascade on rocky reefs following sea star mass mortality in British Columbia. *PeerJ*, **4**, e1980, http://dx.doi.org/10.7717/peerj.1980.

Sebens, K. P. (1984). Water flow and coral colony size: Interhabitat comparisons of the octocoral *Alcyonium siderium*. *Proceedings of the National Academy of Sciences of the United States of America*, **81** (17), 5473–7, http://dx.doi.org/10.1073/pnas.81.17.5473.

Sebens, K. P. (1985a). Community Ecology of Vertical Rock Walls in the Gulf of Maine, USA: Small-Scale Processes and Alternative Community States. In P. G. Moore and R. Seed, eds. *The Ecology of Rocky Coasts*. Hodder and Stoughton Educational Press, London, pp. 346–71.

Sebens, K. P. (1985b). The ecology of the rocky subtidal zone. *American Scientist*, **73** (6), 548–57.

Sebens, K. P. (1986). Spatial relationships among encrusting marine organisms in the New England subtidal zone. *Ecological Monographs*, **56** (1), 73–96.

Sebens, K. P. (1991). Habitat Structure and Community Dynamics in Marine Benthic Systems. In S. S. Bell, E. D. McCoy and H. R. Mushinsky, eds. *Habitat Structure: The Physical Arrangement of Objects in Space*, first edn. Chapman and Hall, London, pp. 211–34.

*Secretaria de Desarrollo Social: Catálogo de Localidades* (2014). Mexico City, www.microrregiones.gob.mx/catloc/LocdeMun.aspx?tipo=clave&campo=loc&ent=02&mun=003.

See, K. E. and Feist, B. E. (2010). Reconstructing the range expansion and subsequent invasion of introduced European green crab along the west coast of the United States. *Biological Invasions*, **12** (5), 1305–18, http://dx.doi.org/10.1007/s10530-009-9548-7.

Selkoe, K. A., Gaines, S. D., Caselle, J. E. and Warner, R. R. (2006). Current shifts and kin aggregation explain genetic patchiness in fish recruits. *Ecology*, **87** (12), 3082–94.

Seymour, R. J., Tegner, M. J., Dayton, P. K. and Parnell, P. E. (1989). Storm wave induced mortality of giant kelp, *Macrocystis pyrifera*, in Southerm California. *Estuarine, Coastal and Shelf Science*, **28**, 277–92, http://dx.doi.org/10.1016/0272-7714(89)90018-8.

Shaffer, J. A. (2000). Seasonal variation in understory kelp bed habitats of the Strait of Juan de Fuca. *Journal of Coastal Research*, **16** (3), 768–75.

Shelton, A. O., Harvey, C. J., Samhouri, J. F. et al. (2018). From the predictable to the unexpected: kelp forest and benthic invertebrate community dynamics following decades of sea otter expansion. *Oecologia*, **188** (4), 1105–19.

Shivji, M., Parker, D., Hartwick, B., Smith, M. J. and Sloan, N. A. (1983). Feeding and distribution study of the sunflower sea star *Pycnopodia helianthoides* (Brandt, 1835). *Pacific Science*, **37** (2), 133–40, http://scholarspace.manoa.hawaii.edu/handle/10125/653.

Siddon, C. E. and Witman, J. D. (2003). Influence of chronic, low-level hydrodynamic forces on subtidal community structure. *Marine Ecology Progress Series*, **261**, 99–110, http://dx.doi.org/10.3354/meps261099.

Siddon, E. C., Siddon, C. E. and Stekoll, M. S. (2008). Community level effects of *Nereocystis luetkeana* in southeastern Alaska. *Journal of Experimental Marine Biology and Ecology*, **361** (1), 8–15, http://dx.doi.org/10.1016/j.jembe.2008.03.015.

Simenstad, C. A., Estes, J. A. and Kenyon, K. W. (1978). Aleuts, sea otters, and alternate stable-state communities. *Science*, **200** (4340), 403–11, http://dx.doi.org/10.1126/science.200.4340.403.

Simkanin, C., Davidson, I. C., Dower, J. F., Jamieson, G. and Therriault, T. W. (2012). Anthropogenic structures and the infiltration of natural benthos by invasive ascidians. *Marine Ecology*, **33** (4), 499–511, http://dx.doi.org/10.1111/j.1439-0485.2012.00516.x.

Simkanin, C., Dower, J. F., Filip, N., Jamieson, G. and Therriault, T. W. (2013). Biotic resistance to the infiltration of natural benthic habitats: examining the role of predation in the distribution of the invasive ascidian *Botrylloides violaceus*. *Journal of Experimental Marine Biology and Ecology*, **439**, 76–83, http://dx.doi.org/10.1016/j.jembe.2012.10.004.

Sirenko, B. I. and Gagaev, S. Y. (2007). Unusual abundance of macrobenthos and biological invasions in the Chukchi Sea. *Russian Journal of Marine Biology*, **33** (6), 355–64, http://dx.doi.org/10.1134/S1063074007060016.

Sivasundar, A. and Palumbi, S. R. (2010). Life history, ecology and the biogeography of strong genetic breaks among 15 species of Pacific rockfish, Sebastes. *Marine Biology*, **157** (7), 1433–52, http://dx.doi.org/10.1007/s00227-010-1419-3.

Small, C. and Nicholls, R. (2003). A global analysis of human settlement in coastal zones. *Journal of Coastal Research*, **19** (3), 584–99, www.jstor.org/stable/4299200?seq=1#page_scan_tab_contents.

Smith, K. L., Baldwin, R. J., Ruhl, H. A., Kahru, M., Mitchell, B. G. and Kaufmann, R. S. (2006). Climate effect on food supply to depths greater than 4,000 meters in the northeast Pacific. *Limnology and Oceanography*, **51** (1), 166–76, http://dx.doi.org/10.4319/lo.2006.51.1.0166.

Smith, R. L., Huyer, A. and Fleischbein, J. (2001). The coastal ocean off Oregon from 1961 to 2000: Is there evidence of climate change or only of Los Niños?. *Progress in Oceanography*, 49 (1–4), 63–93, http://dx.doi.org/10.1016/S0079-6611(01)00016-7.

Sousa, W. P. (1984). The role of disturbance in natural communities. *Annual Review of Ecology and Systematics*, **15**, 353–91.

Spalding, H., Foster, M. S. and Heine, J. N. (2003). Composition, distribution, and abundance of deep-water (>30m) macroalgae in Central California. *Journal of Phycology*, **284**, 273–84.

Springer, Y. P., Hays, C. G., Carr, M. H. and Mackey, M. R. (2010). Toward ecosystem-based management of marine macroalgae – the Bull Kelp, *Nereocystis luetkeana*. *Oceanography and Marine Biology: An Annual Review*, **48**, 1–42.

Spurkland, T. and Iken, K. (2011). Kelp bed dynamics in estuarine environments in subarctic Alaska. *Journal of Coastal Research*, **275**, 133–43, http://dx.doi.org/10.2112/JCOASTRES-D-10-00194.1.

Stabeno, P. J., Bond, N. A., Hermann, A. J., Kachel, N. B., Mordy, C. W. and Overland J. E. (2004) Meteorology and oceanography of the Northern Gulf of Alaska. *Continental Shelf Research*, **24** (7–8), 859–97, http://dx.doi.org/10.1016/j.csr.2004.02.007.

Statistics Canada. (2011). *2011 Census. Statistics Canada Catalogue no. 98-316-XWE.* (2012). Ottawa.

Steneck, R., Graham, M. H., Bourque, B. J. et al. (2002). Kelp forest ecosystems: biodiversity, stability, resilience and future. *Environmental Conservation*, **29** (4), 436–59, http://digitalcommons.library.umaine.edu/sms_facpub.

Stenseth, N. C., Mysterud, A., Ottersen, G., Hurrell, J. W., Chan, K. S. and Lima, M. (2002). Ecological effects of climate fluctuations. *Science*, **297**, 1292–6.

Stephens, J. S., Larson, R. J. and Pondella, D. J. (2006). Rocky Reefs and Kelp Beds. In L. G. Allen, D. J. Pondella and M. H. Horn, eds. *Ecology of Marine Fishes: California and Adjacent Waters*. University of California Press, Berkeley, pp. 227–52.

Stewart, H. L., Fram, J. P., Reed, D. C. et al. (2009). Differences in growth, morphology and tissue carbon and nitrogen of *Macrocystis pyrifera* within and at the outer edge of a giant kelp forest in California, USA. *Marine Ecology Progress Series*, **375**, 101–12, http://dx.doi.org/10.3354/meps07752.

Stoker, S. W. (1978) Benthic invertebrate macrofauna of the Eastern Continental Shelf of the Bering and Chukchi Seas. PhD, University of Alaska Fairbanks.

Strom, S. L., Olson, M. B., Macri, E. L. and Mordy, C. W. (2006). Cross-shelf gradients in phytoplankton community structure, nutrient utilization, and growth rate in the coastal Gulf of Alaska. *Marine Ecology Progress Series*, **328**, 75–92, http://dx.doi.org/10.3354/meps328075.

Strub, P. T., Combes, V., Shillington, F. A. and Pizarro, O. (2013) Currents and processes along the eastern boundaries. *International Geophysics*, **103**, 339–84, http://dx.doi.org/10.1016/B978-0-12-391851-2.00014-3.

Sunday, J. M., Popovic, I., Palen, W. J., Foreman, M. G. G. and Hart, M. W. (2014). Ocean circulation model predicts high genetic structure observed in a long-lived pelagic developer. *Molecular Ecology*, **23** (20), 5036–47, http://dx.doi.org/10.1111/mec.12924.

Swanson, A. and Druehl, L. (2002). Induction, exudation and the UV protective role of kelp phlorotannins. *Aquatic Botany*, **73** (3), 241–53, www.sciencedirect.com/science/article/pii/S0304377002000359.

Swanson, A. K. and Druehl, L. D. (2000). Differential meiospore size and tolerance of ultraviolet light stress within and among kelp species along a depth gradient. *Marine Biology*, **136** (4), 657–64, http://dx.doi.org/10.1007/s002270050725.

Sydeman, W., Bradley, R. and Warzybok, P. (2006). Planktivorous auklet *Ptychoramphus aleuticus* responses to ocean climate, 2005: unusual atmospheric blocking? *Geophysical Research Letters*, **33** (22), L22S09, http://onlinelibrary.wiley.com/doi/10.1029/2006GL026736/full.

Sydeman, W. J., Thompson, S. A., Field, J. C. et al. (2011). Does positioning of the North Pacific current affect downstream ecosystem productivity?.

*Geophysical Research Letters*, **38** (12), 1–6, http://dx.doi.org/10.1029/2011GL047212.

Szpak, P., Orchard, T. J., Salomon, A. K. and Gröcke, D. R. (2013). Regional ecological variability and impact of the maritime fur trade on nearshore ecosystems in southern Haida Gwaii (British Columbia, Canada): evidence from stable isotope analysis of rockfish (*Sebastes* spp.) bone collagen. *Archaeological and Anthropological Sciences*, **5** (2), 159–82, http://dx.doi.org/10.1007/s12520-013-0122-y.

Tegner, M., Dayton, P. K., Edwards, P. B. and Riser, K. L. (1997). Large scale, low frequency oceanographic effects on kelp forest succession: a tale of two cohorts. *Marine Ecology Progress Series*, **146**, 117–34, http://dx.doi.org/10.3354/meps146117.

Tegner, M. J. and Dayton, P. K. (1981). Population structure, recruitment and mortality of two sea urchins (*Strongylocentrotus franciscanus* and *S. purpuratus*) in a kelp forest. *Marine Ecology Progress Series*, **5**, 255–68, http://dx.doi.org/10.3354/meps005255.

Tegner, M. J. and Dayton, P. K. (1987). El Nino effects on southern California kelp forest communities. *Advances in Ecological Research*, **17**, 243–79, http://dx.doi.org/10.1016/S0065-2504(08)60247-0.

Tegner, M. J. and Dayton, P. K. (1991). Sea urchins, El Ninos, and the long term stability of Southern California kelp forest communities. *Marine Ecology Progress Series*, **77**, 49–63, http://dx.doi.org/10.3354/meps077049.

Tegner, M. J. and Dayton, P. K. (2000). Ecosystem effects of fishing in kelp forest communities. *ICES Journal of Marine Science*, **57** (3), 579–89, http://dx.doi.org/10.1006/jmsc.2000.0715.

Tegner, M. J. and Levin, L. A. (1983). Spiny lobsters and sea urchins: analysis of a predator–prey interaction. *Journal of Experimental Marine Biology and Ecology*, **73** (2), 125–50, http://dx.doi.org/10.1016/0022-0981(83)90079-5.

Thiel, M. and Gutow, L. (2005). The ecology of rafting in the marine environment. II. The rafting organisms and community. *An Annual Review*, **43**, 279–418.

Thomas, A. and Brickley, P. (2006). Satellite measurements of chlorophyll distribution during spring 2005 in the California Current. *Geophysical Research Letters*, **33** (22), L22S05, http://onlinelibrary.wiley.com/doi/10.1029/2006GL026588/full.

Thomson, D. A., Findley, L. T. and Kerstitch, A. N. (2000) *Reef Fishes of the Sea of Cortez: The Rocky-Shore Fishes of the Gulf of California*. University of Texas Press, Austin.

Thornber, C. (2007). Associational resistance mediates predator-prey interactions in a marine subtidal system. *Marine Ecology*, **28** (4), 480–6, http://dx.doi.org/10.1111/j.1439-0485.2007.00187.x.

Thornber, C. S., Graham, M. H., Kinlan, B. P. and Stachowicz, J. J. (2004). Population ecology of the invasive kelp *Undaria pinnatifida* in California: environmental and biological controls on demography. *Marine Ecology Progress Series*, **268**, 69–80.

Tilman, D. (1994). Competition and biodiversity in spatially structured habitats. *Ecology*, **75** (1), 2–16.

tom Dieck, I. (1993). Temperature tolerance and survival in darkness of kelp gametophytes (Laminariales, Phaeophyta) – ecological and biogeographical implications. *Marine Ecology Progress Series*, **100** (3), 253–64, http://dx.doi.org/10.3354/meps100253.

Torres-Moye, G. and Escofet, A. (2014). Land–sea interactions in Punta China (Baja California, México): addressing anthropic and natural disturbances in a retrospective context. *Journal of Environmental Protection*, **5**, 1520–30.

Trebilco, R., Dulvy, N. K., Stewart, H. and Salomon, A. K. (2015). The role of habitat complexity in shaping the size structure of a temperate reef fish community. *Marine Ecology Progress Series*, **532**, 197–211, http://dx.doi.org/doi:10.3354/meps11330.

Tsurumi, M. and Tunnicliffe, V. (2001). Characteristics of a hydrothermal vent assemblage on a volcanically active segment of Juan de Fuca Ridge, northeast Pacific. *Canadian Journal of Fisheries and Aquatic Sciences*, **58** (3), 530–42, http://dx.doi.org/10.1139/f01-005.

United States Census Bureau. (2016). *Annual Estimates of the Resident Population: April 1, 2010 to July 1, 2015*. United States Census Bureau, Washington, DC.

Vadas, R. L. (1972). Ecological implications of culture studies on *Nereocystis luetkeana*. *Journal of Phycology*, **8** (2), 196–203, http://dx.doi.org/10.1111/j.1529-8817.1972.tb04025.x.

Valentine, J. W. (1966). Numerical analysis of marine molluscan ranges on the extratropical Northeastern Pacific shelf. *Limnology and Oceanography*, **11** (2), 198–211.

Valentine, J. W., Roy, K. and Jablonski, D. (2002). Carnivore/non-carnivore ratios in northeastern Pacific marine gastropods. *Marine Ecology Progress Series*, **228**, 153–63, http://dx.doi.org/10.3354/meps228153.

Vance, R. (1988). Ecological succession and the climax community on a marine subtidal rock wall. *Marine Ecology Progress Series*, **48**, 125–36, http://dx.doi.org/10.3354/meps048125.

Vermeij, G. J. (1991). When biotas meet: understanding biotic interchange. *Science*, **253** (5024), 1099–104, http://dx.doi.org/10.1126/science.253.5024.1099.

Vermeij, G. J. (2012). The evolution of gigantism on temperate seashores. *Biological Journal of the Linnean Society*, **106** (4), 776–93, http://dx.doi.org/10.1111/j.1095-8312.2012.01897.x.

Vetter, E. W. (1998). Population dynamics of a dense assemblage of marine detritivores. *Journal of Experimental Marine Biology and Ecology*, **226** (1), 131–61, http://dx.doi.org/10.1016/S0022-0981(97)00246-3.

Walters, S. P., Thebo, A. L. and Boehm, A. B. (2011). Impact of urbanization and agriculture on the occurrence of bacterial pathogens and stx genes in coastal waterbodies of central California. *Water Research*, **45** (4), 1752–62, http://dx.doi.org/10.1016/j.watres.2010.11.032.

Wares, J. P. and Schiebelhut, L. M. (2016). What doesn't kill them makes them stronger: an association between elongation factor 1-α overdominance in the sea star *Pisaster ochraceus* and "sea star wasting disease". *PeerJ*, **4**, e1876, http://dx.doi.org/10.7717/peerj.1876.

Watanabe, J. M. and Harrold, C. (1991). Destructive grazing by sea urchins *Strongylocentrotus* sin a central California kelp forest: potential roles of recruitment, depth, and predation. *Marine Ecology Progress Series*, **71** (2), 125–41, http://dx.doi.org/10.3354/meps071125.

Watson, J. and Estes, J. A. (2011). Stability, resilience, and phase shifts in rocky subtidal communities along the west coast of Vancouver Island, Canada. *Ecological Monographs*, **81** (2), 215–39, http://dx.doi.org/10.1890/10-0262.1.

Watson, J. R., Mitarai, S., Siegel, D. A., Caselle, J. E., Dong, C. and McWilliams, J. C. (2010). Realized and potential larval connectivity in the Southern California Bight. *Marine Ecology Progress Series*, **401**, 31–48, http://dx.doi.org/10.3354/meps08376.

Weise, M., Costa, D. and Kudela, R. (2006). Movement and diving behavior of male California sea lion (*Zalophus californianus*) during anomalous oceanographic conditions of 2005 compared to those of. *Geophysical Research Letters*, **33** (22), L22S10, http://onlinelibrary.wiley.com/doi/10.1029/2006GL027113/full.

West, J. E., Helser, T. E. and Neill, S. M. O. (2014). Variation in quillback rockfish (*Sebastes maliger*) growth patterns from oceanic to inland waters of the Salish Sea. *Bulletin of Marine Science*, **90** (3), 747–61, http://dx.doi.org/10.5343/bms.2013.1044.

Wheeler, W. N. and Neushul, M. (1981). The Aquatic Environment. In *Physiological Plant Ecology I*. Springer, Berlin, pp. 229–47, http://dx.doi.org/10.1007/978-3-642-68090-8_9.

Williams, J. P., Williams, C. M., Blanchette, C. A., Claisse, J. T., Pondella, D. J. and Caselle, J. E. (2018). Where the weird things are: a collection of species range extensions in the Southern California Bight. *Bulletin, Southern California Academy of Sciences*, **117** (3), 189–203.

Wilson, A. B. (2006). Genetic signature of recent glaciation on populations of a near-shore marine fish species (*Syngnathus leptorhynchus*). *Molecular Ecology*, **15** (7), 1857–71, http://dx.doi.org/10.1111/j.1365-294X.2006.02911.x.

Wing, S. R., Botsford, L. W., Largier, J. L. and Morgan, L. E. (1995). Spatial variability in settlement of benthic invertebrates in a northern California upwelling system. *Marine Ecology Progress Series*, **128**, 199–211.

Wing, S. R., Botsford, L. W., Largier, J. L. and Morgan, L. E. (1996). Spatial structure of relaxation events and crab settlement in the northern California upwelling system. *Oceanographic Literature Review*, **7** (43), 701.

Witman, J. D., Etter, R. J. and Smith, F. (2004). The relationship between regional and local species diversity in marine benthic communities: a global perspective. *Proceedings of National Academy of Sciences*, **101** (44), 15664–9.

Witman, J. D., Leichter, J. J., Genovese, S. J. and Brooks, D. A. (1993). Pulsed phytoplankton supply to the rocky subtidal zone: influence of internal waves. *Proceedings of the National Academy of Sciences of the United States of America*, **90** (5), 1686–90, http://dx.doi.org/10.1073/pnas.90.5.1686.

Wonham, M. and Carlton, J. (2005). Trends in marine biological invasions at local and regional scales: the northeast Pacific Ocean as a model system. *Biological Invasions*, **7** (3), 369–92, http://dx.doi.org/10.1007/s10530-004-2581-7.

Worm, B. and Paine, R. T. (2016). Humans as a hyperkeystone species. *Trends in Ecology & Evolution*, **31** (8), 600–7, http://dx.doi.org/10.1016/J.TREE.2016.05.008.

Yamada, S. B., Peterson, W. T. and Kosro, P. M. (2015). Biological and physical ocean indicators predict the success of an invasive crab, *Carcinus maenas*, in the northern California current. *Marine Ecology Progress Series*, **537**, 175–89, http://dx.doi.org/10.3354/meps11431.

Yang, B., Emerson, S. R. and Peña, M. A. (2018). The effect of the 2013–2016 high temperature anomaly in the subarctic Northeast Pacific (the "Blob") on net community production. *Biogeosciences*, **15** (21), 6747–59.

Yorke, C. E., Miller, R. J., Page, H. M. and Reed, D. C. (2013). Importance of kelp detritus as a component of suspended particulate organic matter in giant kelp *Macrocystis pyrifera* forests. *Marine Ecology Progress Series*, **493**, 113–25, http://dx.doi.org/10.3354/meps10502.

Young, C. M. and Braithwaite, L. E. E. F. (1980). Orientation and current-induced flow in the stalked *Ascidian Styela* montereyensis. *Biological Bulletin*, **159** (2), 428–40.

Young, M. A., Cavanaugh, K., Bell, T. et al. (2016). Environmental controls on spatial patterns in the long-term persistence of giant kelp in central California. *Ecology*, **86** (1), 45–60, http://dx.doi.org/10.1890/15-0267.1.

Zaba, K. D. and Rudnick, D. L. (2016). The 2014–2015 warming anomaly in the southern California current system observed by underwater gliders. *Geophysical Research Letters*, **43** (3), 1241–8, http://dx.doi.org/10.1002/2015GL067550.

Zabin, C. J., Ashton, G. V., Brown, C. W. and Ruiz, G. M. (2009). Northern range expansion of the asian kelp *Undaria pinnatifida* (Harvey) suringar (Laminariales, Phaeophyceae) in Western North America. *Aquatic Invasions*, **4** (3), 429–34, http://dx.doi.org/10.3391/ai.2009.4.3.1.

Zacharias, M. A. and Kushner, D. J. (2006). Sea tempera-tureand wave height as predictors of population size structure and density of *Megastraea (Lithopoma) undosa*: implications for fishery management. *Bulletin of Marine Science*, **79** (1), 71–82, www.ingentaconnect.com/content/umrsmas/bullmar/2006/00000079/00000001/art00004.

Zahn, L. A., Claisse, J. T., Williams, J. P., Williams, C. M. and Pondella II, D. J. (2016). The biogeography and community structure of kelp forest macroinvertebrates. *Marine Ecology*, **37**, 770–85, http://dx.doi.org/10.1111/maec.12346.

Zhan, A., Darling, J. A., Bock, D. G. et al. (2012). Complex genetic patterns in closely related colonizing invasive species. *Ecology and Evolution*, **2** (7), 1331–46, http://dx.doi.org/10.1002/ece3.258.

Zimmerman, R. C. and Kremer, J. N. (1986). In situ growth and chemical composition of the giant kelp, *Macrocystis pyrifera*, response to temporal changes in ambient nutrient availability. *Marine Ecology Progress Series*, **27** (2), 277–85.

Chapter 12

# Consumer–Resource Interactions on an Environmental Mosaic

## *The Role of Top-Down and Bottom-Up Forcing of Ecological Interactions along the Rocky Shores of the Temperate South-Eastern Pacific*

Moisés A. Aguilera, Bernardo R. Broitman, Julio A. Vásquez and Patricio A. Camus

## 12.1 Introduction

Over the past three decades, the biogeography, phylogeography and ecology of the diverse and unique species assemblage that inhabits the subtropical and temperate Pacific shores of South America (5–42°S) has received increasing scientific attention. Different studies have taken place in an interdisciplinary context and are matched with major developments in our understanding of coastal oceanographic processes along the region's shorelines (e.g., Broitman et al., 2001; Navarrete et al., 2005; Lagos et al., 2007; Tapia et al., 2009, 2014). Following seminal work by Camus (2001), the historical biogeographic division of Pacific South America into a warm-temperate Peruvian region (5–42°S) and a cool subantarctic region (42–54°S), has revealed a more complex biogeographic structure than previously thought (e.g., Thiel et al., 2007; Rivadeneira et al., 2012; Valdovinos et al., 2003; see Figure 12.1). The northern warm-temperate Peruvian sector appears separated from the subantarctic province by a diffuse transitional region that spans from 30 to 41°S. Along this region, breaks in species distributions (see Figure 12.2) are clumped around areas where rapid and persistent changes in circulation and hydrographic conditions occur, which are usually associated to large headlands and abrupt changes in the width of the continental shelf (e.g., Hormazábal et al., 2004). These topographically locked oceanographic features determine a persistent spatial structure of the coastal environment along the transitional region, with implications for phylogeography and ecology (e.g., Broitman et al., 2001, 2011; Wieters et al., 2003; Rivadeneira et al., 2012; Haye et al., 2014).

First, and propelled by advances in molecular genetics, studies of the region's biota have revealed the presence of several cryptic species among some of the dominant and habitat-forming algal species of the mid- and low-intertidal zones of the rocky shore, the corticated algae *Mazzaella* and kelps like *Lessonia*, respectively (Tellier et al., 2011; Montecinos et al., 2012). The locations of the breaks between the

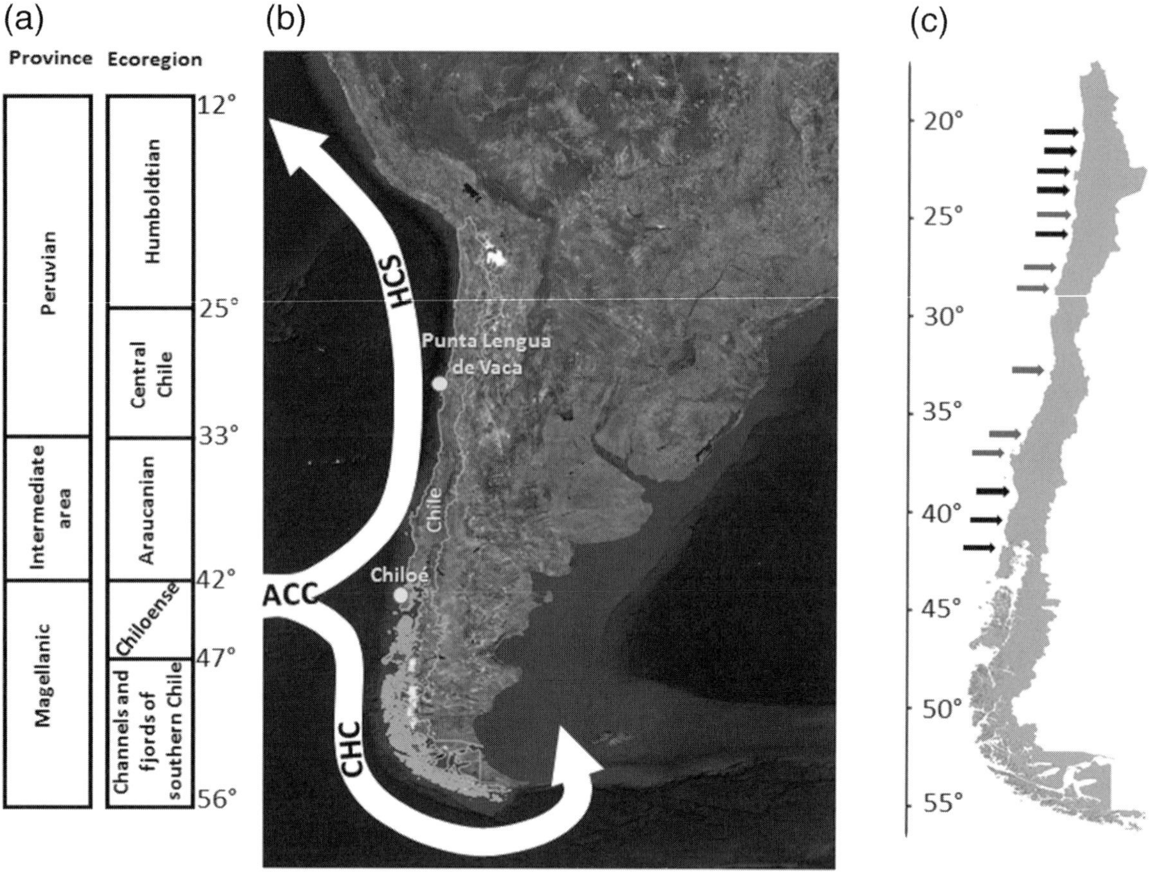


**Fig. 12.1** (a) Biogeographic provinces (after Camus, 2001; Thiel et al., 2007) and ecoregions (after Spalding et al., 2007) along the southeast Pacific coast, indicating their latitudinal boundaries. (b) Simplified view of the major cold flows (white arrows) off western South America, originating from the Antarctic Circumpolar Current (ACC) after reaching the Chilean coast: the poleward Cape Horn Current (CHC), encompassing the southern cone, and the equatorward (HCS), an eastern boundary current formed actually by coastal and oceanic branches. Yellow dots indicate the location of two places often referred to in the text because of their ecological or biogeographical importance. (c) Fine-grained latitudinal breaks in species diversity (after UACH, 2006) along the Chilean coast (excluding the area south of 45°S), detected in analyses combining multiple littoral taxa from rocky and sandy shores. Breaks (associated with the latitude axis on the left) correspond to significant spatial changes in beta diversity (black arrows) and in both alpha and beta diversity (blue arrows). *A black and white version of this figure will appear in some formats. For the colour version, please refer to the plate section.*

cryptic species corresponds well with the geophysical template dictated by coastal oceanography (Hormazábal et al., 2004; Tapia et al., 2009, 2014) and is further supported by the phylogeographic structure of many other coastal species (see Haye et al., 2014). In this context, one of the most important questions is what are the consequences of such distributional changes on species interactions in the rest of the community? For example, different regions separated by the biogeographic break could evidence changes in the functional structure of consumer assemblages (i.e., herbivores and carnivores). That could be related to a replacement from tropical to temperate species occurrences across the HCS (e.g., Lancellotti and Vásquez, 1999; Valdovinos et al., 2003; Hernández et al., 2005; Moreno et al., 2006a). Spatial turnover of species could be influenced by large-scale processes like the El Niño southern oscillation (ENSO), which can generate a gradient in water temperature across this region (Iriarte and González, 2004) affecting

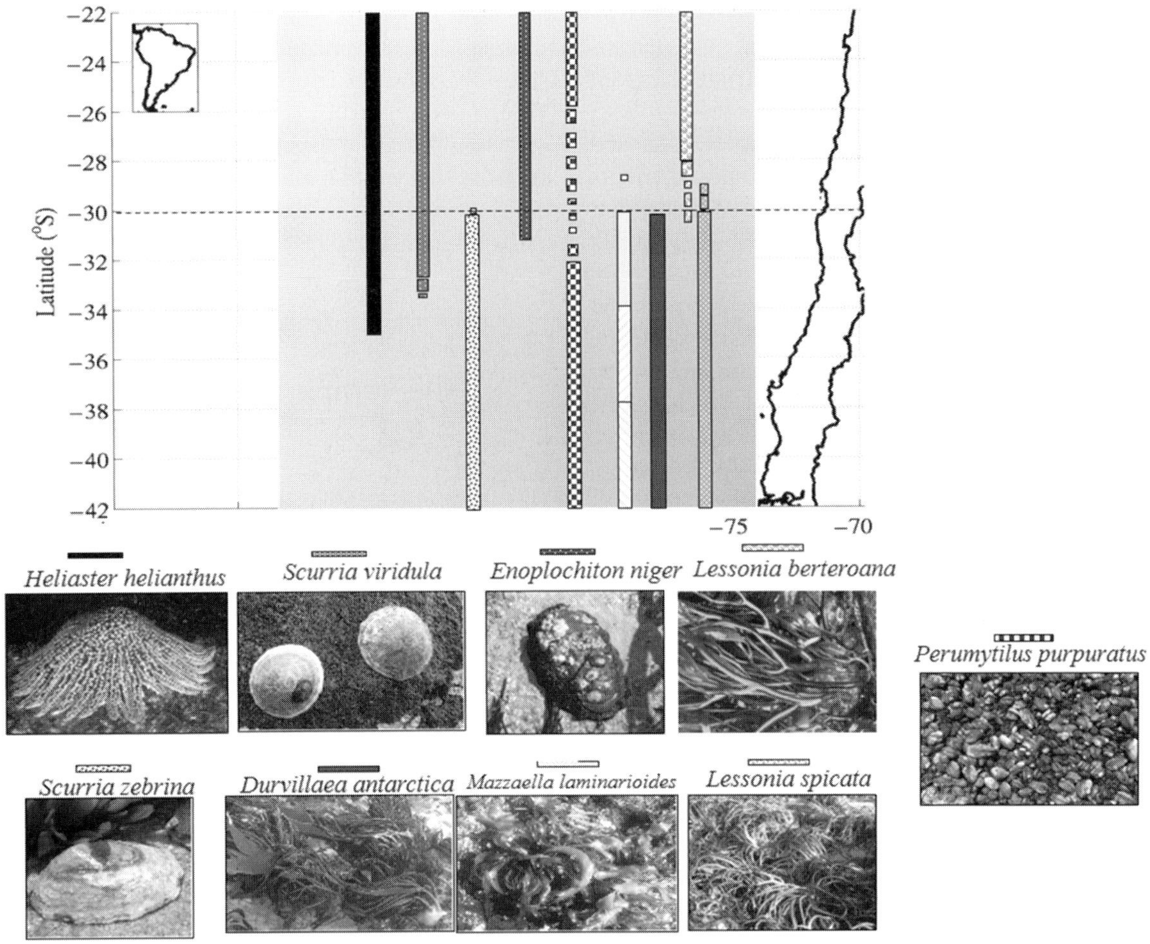


**Fig. 12.2** Scheme of the geographic distributional pattern of different intertidal species, (i.e., distributional species mosaic) present across north and central-south Chile. Most species find their polar or equatorial range edge between the biogeographic transitional zone at Punta Lengua de Vaca (PLV, 30°S), and near 36°S. The sunstar *Heliaster helianthus* and the grazer species; *Scurria viridula*, *Scurria zebrina* and *Enoplochiton niger* are the main consumers in intertidal levels across the range considered. The corticated alga *Mazzaella laminarioides* is dominant in mid- and high-intertidal levels south 30°S, where three clades are reported (see yellow bars, Montecinos et al., 2012). The fucoid *Durvillaea antarctica* and the kelp *Lessonia spicata* are also dominant south PLV in low-intertidal levels. North of this region, these algae are replaced by *Lessonia berteroana* (green bars, e.g., Tellier et al., 2011). Segmented bars correspond to small population densities, or small patches, of invertebrates and algae species present across the biographic transitional zone. For example, the mussel *Perumytilus purpuratus* (brown-grid bars) showed a discontinuous distribution across the transitional zone at 30°S, while a relatively continuous distribution is observed for this species both south and north of this range (Navarrete et al., 2005). *A black and white version of this figure will appear in some formats. For the colour version, please refer to the plate section.*

rocky shore consumer assemblage composition through affecting their early developmental stages in the plankton (e.g., Escribano et al., 2004). Alternatively, there could be a spatially shifting mosaic of consumer functional structure (Aguilera et al., unpublished data) along the two sides of the biogeographic break underlying changes in species composition (e.g., Rivadeneira and Fernández, 2005; Rivadeneira et al., 2012; see Figure 12.2) and matching the patchy spatial structure of oceanographic conditions across the HCS (e.g., Valle-Levinson et al., 2003; Montecinos and Lange, 2009; Thiel et al., 2007; Tapia et al., 2009, 2014; see Figure 12.1).

The deep temporal signature of coastal oceanography on coastal biogeography and phylogeography is underpinned by the spatial structure of bottom-up effects of ecological processes (e.g., Wieters et al., 2003; Wieters, 2005; Nielsen and Navarrete, 2004; Navarrete et al., 2005). Nutrients derived from coastal upwelling determine intertidal algal growth patterns (Wieters et al., 2003; Wieters, 2005), which, in turn, define the role of some herbivores as controllers of algal biomass and important drivers of community composition (Nielsen and Navarrete, 2004; Aguilera and Navarrete, 2012a; Aguilera et al., 2015a). In this way, ecological interactions like the consumer–resource interaction can be strongly modulated by oceanographic context (Nielsen and Navarrete, 2004), providing a mechanistic basis for the seemingly diffuse nature of transitional biogeographic regions.

Across the Pacific coasts, herbivores and carnivores play an important role regulating the structure of rocky shore communities (e.g., Jara and Moreno, 1984; Castilla and Durán, 1985; Paine et al., 1985; Moreno et al., 1986; Vásquez and Buschmann, 1997; Navarrete and Castilla, 2003; Aguilera, 2011). Specifically, given the diversity of impacts that herbivores can have on algal recruitment, abundance and diversity, they can have strong and differentiated effects on rocky shore community succession (Aguilera and Navarrete, 2012a) and overall seascape spatial heterogeneity (Aguilera et al., 2015a). Most species have negative direct effects on their target resource (e.g., algae), but a number of species can have significant positive indirect effects on established algal species (Aguilera et al., 2015b and see later). The tightly linked functional structure results in a low functional compensatory potential in the face of local species loss or extinctions (Aguilera and Navarrete, 2012a), which can be caused by intense human harvesting (Oliva and Castilla, 1986; Castilla, 1999; Moreno, 2001) and/or distributional changes of herbivorous species following climate change (e.g., Rivadeneira and Fernández, 2005).

One of the main challenges for the study of the ecology of rocky shore ecosystems on the temperate Eastern Pacific today, is to examine how coastal oceanographic processes regulate species range expansion and introduction, or contraction, and how biotic interactions and environmental filtering define dynamic biogeographic patterns along spatially structured marine environments (Valdivia et al., 2015). Thus, there is a great interest in exploring the spatial structure of interacting stressors that may shape local and regional ecosystem vulnerability in the face of unprecedented environmental change.

## 12.2 Historical Biogeography of the Temperate Eastern Pacific

Along the south-east Pacific and the Chilean coast (18.5–56°S) in particular, geographic patterns of species diversity are closely linked to meso- and large-scale physical determinants. For instance, mid-low latitudes are dominated by upwelling and land–sea interactions with large oscillations in patterns of physico-biological coupling driven by inter-annual variations and equatorial forcing (Strub et al., 1998; Takesue et al., 2004; Lagos et al., 2007; Thiel et al., 2007). On the other hand, mid-high latitudes are deeply influenced by physical interactions between coastal geomorphology and the ACC (Ahumada et al., 2000; Camus, 2001; Barker and Thomas, 2004; Thiel et al., 2007). On a very large scale, however, the primary importance of the ACC (the world's strongest current) is reflected in the matching between biogeographical patterns and the spatial structure and extent of its two branches, which form after the ACC impinges the coast of South America around 40–45°S (see Figure 12.1). First, the poleward-flowing CHC, that continues eastwards into the Atlantic, forming the equatorwards Malvinas Current, dominates the surface ocean flow around southern South America. Second, the equatorward-flowing Humboldt Current is an eastern boundary current exerting a strong cooling influence up to ~5°S (Sepúlveda et al., 2016). The onset of the ACC, linked to the break-up of Gondwana and the opening of the Drake Passage, involved a substantial reversal of the warm conditions in the late Paleogene (Pfuhl and McCave, 2005), and was the most influential event in the origin of modern biogeographic provinces (see review by Camus, 2001).

Even with some departures reflecting the idiosyncratic histories of some taxa or methodological aspects (e.g., invertebrates: Lancellotti and Vásquez, 1999; Haussermann and Forsterra, 2009; Navarrete et al., 2014; red and brown algae: Meneses and Santelices, 2000; ectoparasites: González and Moreno, 2005), coastal taxa as a whole exhibit a congruent pattern of distributional breaks (Camus, 2001; Thiel et al., 2007) forming three major units (arrows in Figure 12.1), from which the first two show biogeographic anomalies. The first is the warm-temperate Peruvian province (5–30/33°S) that encompasses the coasts of Peru, which are dominated by subtropical taxa, and northern Chile, which is dominated instead by temperate taxa, with a marked paucity of warm-water species (see Santelices, 1980; Brattström and Johanssen, 1983; Camus, 2001). The second is an intermediate area (30/33–42°S) encompassing central-southern Chile, which is dominated by temperate and subantarctic taxa without a common history and showing no gradual pattern of biotic replacement, thus lacking the distinctive biogeographical character of a province or a transition zone. Finally, the cold-temperate Magellanic province (42–56°S, although extended into the Atlantic up to 42.5°S in Argentina), encompasses austral Chile from Chiloé Island to Cape Horn, a complex system of islands, peninsulas, fjords, channels and inner seas shaped by glacial and tectonic action in the Pleistocene. This cold province is characterised by protected hyposaline environments dominated by subantarctic taxa (see Velásquez et al., 2016). The origin of this provincial pattern can be traced back to the establishment of the cold HCS, which is connected with two major events that largely account for the modern irregularities in northern Chile and the intermediate area (Thiel et al., 2007). These events involve the retreat of the former warm-water biota to lower latitudes along with the northward advance of the subantarctic biota (see Brattström and Johanssen, 1983; Camus, 2001) and the onset of the oxygen minimum zone, concurrent with mass extinctions in the current Peruvian province (e.g., about 70 per cent of bivalve species; Rivadeneira and Marquet, 2007). However, extralimital fossil assemblages reveal that strong or long-lasting warming (ENSO-like) events led to repeated invasions of northern Chile by tropical species throughout the Quaternary (Llagostera, 1979; Ortlieb, 1995; Ortlieb et al., 2003), uncovering a complex scenario of extinctions and recolonisations, which may have also been frequent over the last 10,000 years, where ENSO has been continuously active with peaks of high-frequency clusters of events at about 2,000 year intervals (Moy et al., 2002). In fact, phenomena such as isolated or correlated local extinctions and community-wide changes are still common in northern Chile during strong or moderate El Niño events, but they are not observed south of 30°S, where ENSO impacts are comparatively negligible (for reviews and case studies see Castilla and Camus, 1992; Camus, 1994, 2008; Thiel et al., 2007).

The interplay between past and present biogeographic factors in north-central Chile is reflected in the fact that many dominant species show phylogeographic breaks at 30°S, which originated either in the Neogene, the Pleistocene or the Holocene (e.g., Sánchez et al., 2011; Martin and Zuccarello, 2012; Montecinos et al., 2012; Haye et al., 2014). These genetic signatures remain strong in low-dispersal species, but were partially or entirely erased in high-dispersal species following modern gene flow across 30°S (Haye et al., 2014). For instance, major breaks in eddy kinetic activity, wind stress and coastal versus offshore productivity (Hormazábal et al., 2004; Thiel et al., 2007) reveal a physical transition around 30°S that seems to play a crucial biogeographical role (e.g., as a differential barrier to larval transport and dispersal; see Figure 12.2). This putative barrier helps to explain the abrupt distributional breaks at that latitude by numerous taxa with different origins, life cycles, ecologies or habitats (Thiel et al., 2007; Broitman et al., 2011), even including the cephalopod fauna, which is formed entirely by non-endemic, widely ranging and mostly eurithermic species (Ibáñez et al., 2009). In addition, these and other physical factors (analysed in detail in the next section) are among the forcing agents considered by Spalding et al. (2007; based on Sullivan and Bustamante, 1999) to define six coastal ecoregions nested in only two provinces along the temperate south-east Pacific. The larger of these two provinces (warm-temperate south-eastern Pacific) subsumes the Peruvian province and the intermediate area outlined in this chapter (Figure 12.1), although the latter two are supported by the current information. Moreover, fine-grained analyses integrating databases for several rocky- and sandy-shore taxa (UACH, 2006) detected an unusual array of spatial breaks in species richness and beta diversity along the Chilean coast, suggesting that current biogeographical classifications could be masking a more complex patterning of biodiversity.

## 12.3 Species Distribution Patterns and Biotic–Environmental Coupled Processes across Rocky Shores of the Temperate Eastern Pacific

Benthic assemblage composition across the temperate Eastern Pacific rocky shores is dominated by a diverse group of algae and invertebrates endemic to the Peruvian or Magellanic provinces (see Broitman et al., 2001; Moreno et al., 2006a; Rivadeneira et al., 2012). These two biogeographic provinces are abutted by very clear boundaries defined by large-scale oceanographic transitions (Camus, 2001; see Figure 12.1). On the southern end of Pacific South America, the Drake Passage, around 56°S, and energetic ACC flowing through it mark the polar boundary of the Magellanic or subantarctic province. Similarly, the equatorial boundary end of the Peruvian province is marked by the Paita Peninsula (~5°S), where the coast of western South America turns east and the westward-flowing trade winds, running along the equatorial guideline, separate the temperate and tropical realms (Longhurst, 1998). These two biogeographic provinces meet around 40°S, on Chiloé Island, where the eastward-flowing west wind drift impinges on the coast of western South America, determining the ACC (Chaigneau and Pizarro, 2005). The cold, northward-flowing branch of the ACC is commonly known as the Humboldt Current. However, the HCS is made by an ensemble of equatorward- and poleward-flowing currents, some of them superficial or on the subsurface, in proximity to the coast or offshore (Strub et al., 1998). The complex structure of the alongshore flow of the HCS is driven by the spatial and temporal variability of the wind patterns, which is many times locked to coastal geomorphology. Equatorward winds blowing along the coast pushes surface waters along shore which lifts the thermocline (Ekman transport) and following the rotation of the earth (the Coriolis effect) displaces the surface waters to the west. This wind-driven offshore divergence of the equatorward-flowing coastal current brings cold, nutrient-rich waters from the poleward-flowing subsurface waters to the surface, a process known as coastal upwelling (Strub et al., 1998). Importantly, the influx of nutrients to the photic zone in association to coastal upwelling along the HCS maintains some of the most productive fisheries on the planet (Bakun and Weeks, 2008).

### 12.3.1 Productivity Patterns and Invertebrate Larval Supply

The large productivity of the HCS is a defining characteristic of the Peruvian province, where upwelling dynamics also drive energy along and cross-shelf transport process (Hormazábal et al., 2004). The nearshore flow field is an important process influencing the dispersal of planktonic larvae from benthic populations (Poulin et al., 2002). A broad region, limited to the north by a large headland located around 30°S, PLV is recognised as the largest upwelling foci along the region (Aguirre et al., 2012) and marks a division between the Peruvian province to the north and a transitional region that extends to the south until the well-established boundary at Chiloé Island (Thiel et al., 2007 and see earlier). In particular, PLV delimits a region where the rates of larval arrival of rocky shore intertidal invertebrates are up to three orders of magnitude higher that the region northwards (Navarrete et al., 2002, 2005; Valdivia et al., 2015). This spatial pattern seems to be related to coastal geomorphology where bays on the lee of headlands can retain larvae from benthic species, increasing local recruitment (Palma et al., 2006) and the intensity of year-round upwelling around 30°S. The coast north of PLV has also fewer sheltered bays and a narrower continental shelf, thus providing fewer opportunities for nearshore larval retention. In addition, north of PLV, upwelling-favourable winds show a reduced seasonality, implying that, although they are not as strong and intense as in the southern sector, they tend to blow year-round, a process that is related to the latitudinal migration of the South Pacific anticyclone (Rahn et al., 2014). Hence, the narrower continental shelf and temporal distribution of upwelling-favourable winds provide a transport-driven mechanistic underpinning of the ecological and biogeographic patterns discussed later.

### 12.3.2 Environmental Processes across 30°S Transitional Zone

Recent insights from molecular phylogenetics have somehow modified prevailing views of the biogeographic structure of the Peruvian province (Cárdenas et al., 2009; Brante et al., 2012; Haye et al., 2014). A region around the PLV headland is the location where an important number of ecologically important species find their poleward or equatorial boundary (see Figure 12.2 and previous section). In the case of poleward boundaries, some notable examples are the large grazers, the chiton *E. niger* and the limpet *S. viridula* (Broitman et al., 2011, Aguilera et al., 2015a, 2013b). On the other hand, the limpet *S. zebrina* and two algae, the fucoid *D. antarctica* and the gigartinacea *M. laminarioides* have their equatorward boundaries at PLV (see Figure 12.2). As mentioned earlier, molecular phylogenetics have greatly contributed to consolidate the transitional nature of the region (Haye et al., 2014). The modified biogeographic structure has also important ecological implications, as dominant species of the rocky intertidal zone have turned out to be cryptic species with their southern/northern range limit around PLV. For example, *M. laminarioides* is a species complex that actually comprises three species (see yellow bars in Figure 12.2), with the northernmost species having a restricted geographic range around PLV (small yellow dot in Figure 12.2), which seemed to have acted as a refuge during the last glacial maximum about 18,000 years BP (Montecinos et al., 2012). A striking example is the kelp *Lessonia nigrescens*, which dominates the low-intertidal zone along the Peruvian province and is a key component of the assemblage (Thiel et al., 2007 and see later). *Lessonia* has been recently recognised as comprised of two cryptic species, *L. berteroana* to the north and *L. spicata* to the south (green bars in Figure 12.2). The distribution limit between these two cryptic species is a geographic mosaic; i.e., interspersed patches of both species that spans the great Coquimbo bay and extends from PLV up to 29°S (Tellier et al., 2011; González et al., 2012; Figure 12.2). This also corresponds to the area where the northernmost populations of the mid-intertidal species *M. laminarioides* are located (Aguilera et al., 2015b). Several other examples are related to the contrasting phylogeographic structure of other benthic species, not all of them from rocky shores (Haye et al., 2014).

Intertidal zonation patterns across the region are characterised by a low-intertidal zone dominated by a conspicuous belt of kelps of the genus *Lessonia*, which dominate primary space on the low-intertidal shore along with calcareous or fleshy, crustose algae, chiefly the calcareous *Lithothamnion* spp. and the non-calcareous species *Hildenbrandia lecannellieri*, respectively. Fleshy species are more common south of PLV (Broitman et al., 2011), where other corticated algae such as several species of *Gelidium*, and the fucoid *D. antarctica*, among other species, also occupy an important proportion of primary space. Fleshy algae species and *Gelidium* present between *Lessonia* and *Durvillaea* clumps are browsed by large keyhole limpets, mainly *Fissurella limbata* and *Fissurella maxima* while periphyton (microalgae, cyanobacteria and sporelings of macroalgae) are grazed by several chitons, most noticeably the large *Acanthopleura echinata* and *E. niger* (Broitman et al., 2011; Aguilera et al., 2015a). Sea urchin species, *Loxechinus albus* and *Tetrapygus niger*, commonly graze on adult and juvenile *Lessonia* and fleshy algae on rock pools. The small balanoid barnacles, *Nothobalanus flosculus* and *Balanus laevis*, and large clumps of the mussel *Semimytilus algosus* are seasonally abundant in the low shore, but they are rapidly extirpated by carnivores (see later). The mid-intertidal zone to the south of PLV is dominated by the mussel *P. purpuratus*, which are conspicuously absent to the north except for isolated patches inside sheltered bays (Broitman et al., 2001; Navarrete et al., 2005; see brown-mesh bars in Figure 12.2). Corticated algae interspersed or atop mussel beds are abundant, notably the *M. laminarioides* complex and several other species of Gigartinaceae south of 36°S. Small chthamaloid barnacles, mainly *Jehlius cirratus* and *Notochthamalus scabrosus* are common space occupiers (Shinen and Navarrete, 2010), particularly north of PLV where they are the most abundant sessile invertebrates of the mid-intertidal zone. Grazers are abundant, particularly limpets from the genus *Scurria* and several species of chiton, particularly *Chiton granosus* (Aguilera and Navarrete, 2007). The keyhole limpet *Fissurella crassa* is also abundant, especially inside *Perumytilus* beds. The large seastar *H. helianthus* and the muricid gastropod *Concholepas concholepas* are key carnivores across the intertidal zone, exerting strong top-down pressure on invertebrate prey.

The studies highlighted point out to the large role played by environmental filtering, like upwelling intensity and sea water temperature, which can constrain/allow larval settlement and recruitment through large scales (dozens of kilometres; Valdivia et al., 2015). These processes can shape species composition of local assemblages across the large region encompassed by our review. For example, changing coastal wind patterns under climate change in the region are forecasted to increase in intensity and change to poleward direction (Sydeman et al., 2014; Rykaczewski et al., 2015). In turn, this intensification is expected to homogenise coastal upwelling regions (Wang et al., 2015), suppressing the rich spatial structure observed across the HCS. Alternatively, locally intensified upwelling in a warming world may provide for spatially persistent refugia that reinforce the alongshore relationship of oceanographic spatial structure with species composition patterns (Hu and Guillemin, 2016; Lourenço et al., 2016). In this way, a major challenge ahead is to increase our ability to combine environmental models that integrate species' distributions and their physiological limits, and incorporate trophic interactions as determinants of local ecological structure.

## 12.4 Consumer–Resource Interaction: Diversity in Response to Local Environmental Gradients

### 12.4.1 Herbivore–Alga Interaction in Space and Time

Herbivory is related to the consumptive effects that animals impose on populations or communities of plants, and is considered one of the most important ecological processes in marine ecosystems around the world (Lubchenco and Gaines, 1981; Hawkins and Hartnoll, 1983; Poore et al., 2012). Marine herbivores can determine temporal and spatial distribution of algae in both intertidal and subtidal habitats (Lubchenco and Gaines, 1981; Hawkins and Hartnoll, 1983; Jenkins et al., 2005; Poore et al., 2012). Through consumption of mature algae and spores, different herbivores can affect successional pathways, determining algal species composition and abundance (Hawkins and Hartnoll, 1983). As herbivores are also consumed by carnivores, including humans, they also have a central importance in food web dynamics and ecosystem functioning (e.g., Paine, 1980; Ling et al., 2009; Poore et al., 2012).

Seminal studies by Paine (e.g., 1980, 1992) on the functional structure of benthic herbivore assemblages on northern Pacific coasts, showed that very few species have strong impacts on algal community structure. Thus, low compensatory potential was observed for grazers with strong per capita impacts (Paine, 1992; see also Sala and Graham, 2002 for kelp forest ecosystems). Along the coast of Chile, on the southern Pacific, the benthic herbivore guild is characterised by species with diverse functional roles (e.g., Moreno and Jaramillo, 1983; Jara and Moreno, 1984; Nielsen and Navarrete, 2004; Aguilera and Navarrete, 2007, 2012a; and see Vásquez and Buschmann, 1997 and Aguilera, 2011 for review). Keyhole limpets (*Fissurella* spp.), chitons (e.g., *C. granosus*, *E. niger*, *Tonicia* spp.), Scurrinid limpets (e.g., *S. zebrina*, *Scurria ceciliana*, *S. viridula*) and sea urchins (*T. niger*, *L. albus*) are the most abundant herbivores present in rocky intertidal habitats along the Chilean coast (see Vásquez and Buschmann, 1997 and Aguilera, 2011 for review). In subtidal systems, kelp-associated species like the fish *Scartichthys viridis*, the sea urchin *L. albus* and the turban snail *Tegula atra* are among the most important strict herbivore species (Ojeda and Muñoz, 1999; Vásquez and Buschmann, 1997). Omnivorous fish species like *Aplodactylus punctatus* and *Girella laevifrons* are also kelp-associated species (Pérez-Matus et al., 2007), and they can exert important grazing/browsing control of different algae species (Pérez-Matus et al., 2012).

In intertidal habitats, most benthic grazers have strong negative effects on more palatable early successional species (e.g., *Ulva* spp.), that dominate the seascape in the absence of herbivores or under reduced densities of them (see Aguilera, 2011 for review). Thus, equivalence or redundancy in consumptive effects on ephemeral algae is expected for most species of the herbivore assemblage. Experimental studies have shown that most species have diverse functional roles throughout the different stages of community succession (Aguilera and Navarrete, 2012a). Specifically, most herbivores have redundant effects during early succession, but differentiated effects are common during late succession when corticated algae are more dominant (Aguilera and Navarrete, 2012a). Even the magnitude and direction of effects on periphyton (i.e., microalgae and cyanobacteria) community structure can differ among herbivorous species (Aguilera et al., 2013a). Exceptionally, keyhole limpets (*F. crassa*, *Fissurella picta*) have a more differentiated and strong negative effect during late succession consuming the dominant corticated alga *M. laminarioides* (Moreno and Jaramillo, 1983; Aguilera and Navarrete, 2012a). Thus, low compensatory potential in controlling *M. laminarioides* biomass is expected for the other benthic herbivore species if *Fissurella* spp. are absent or if densities of these limpets are reduced by human harvesting, which is intense on Chilean coasts (e.g., Moreno et al., 1984; Oliva and Castilla, 1986; Moreno, 2001; and see later). In the same context, medium-sized grazers (20–30 cm) like chitons and scurrinid limpets tend to have a more positive indirect effect on *M. laminarioides* (Aguilera and Navarrete, 2012a; Aguilera et al., 2015a). It is not clear, however, if positive effects on abundance and recovery of the corticated alga by the 'grazer/scraper' functional group is additive and mediated through removal of the opportunistic, potentially competitive dominant, green algae.

Different studies have shown that, despite phylogenetic inertia i.e., the similarity in phenotypic traits shared by sister species (see Webb et al., 2002), in either diet (e.g., Santelices et al., 1986; Camus et al., 2008) or morphology (e.g. Espoz et al., 2004) among mollusc species, limited redundancy (*sensu* Bellwood et al., 2003; Hoey and Bellwood, 2009) in both temporal and spatial distribution of effects would be expected in this herbivore assemblage. This could be partly related to their broad distribution of body sizes and micro-scale habitat use patterns within the assemblage (e.g., Firth and Crowe, 2008; Aguilera, 2011), which could also mediate herbivore coexistence as observed in other ecosystems (e.g., grasslands; Ritchie and Olff, 1999). Coexistence can also be related to different physiological and/or behavioural responses (Chapman and Underwood, 1992, Klein et al., 2011), which together can account for a functional niche differentiation (see Rosenfeld, 2002 functional model). Both direct and indirect effects, commonly related to differences in behaviour and habitat use at micro-scales (centimetres) observed in some grazers of this guild (e.g., Muñoz et al., 2005; Aguilera and Navarrete, 2011, 2012b), can contribute to the differences in the functional roles different species play on the Chilean rocky intertidal community. Consequently, spatial distribution of algae and bare rock could correlate either negatively or positively with the microspatial (a few centimetres) distribution of herbivores during foraging (Chapman and Underwood, 1992), which creates heterogeneity at the scales of dozens of centimetres to metres in the seascape (Aguilera et al., 2015b). Therefore, within the herbivore guild, some species segregate their functional roles at different successional times, independent of the taxonomic affinities and similarity in feeding capabilities, which were previously considered key traits for functional characterisation (Steneck and Watling, 1982).

Commonly, magnitude of per capita effects of benthic herbivores present on Chilean coasts seem directly related to species body size (Aguilera, 2011; Aguilera and Navarrete, 2012a). Thus, a monotonical increase in effect (per capita) is expected for a given body mass of individual herbivore species present in the guild (Wood et al., 2010). No clear relationship has been observed between body size and the homing range of different species (Aguilera and Navarrete, 2011), modifying the general pattern suggested for the amplitude of different consumer effects. For example, large (i.e., ~7.6 cm shell length) molluscan grazer species like *Fissurella* spp. and/or chitons (e.g., *C. granosus*) exhibit a homing behaviour and have well-constrained foraging ranges (45–120 cm length) (Aguilera and Navarrete, 2011). In addition, factors like predation risk can also constrain the spatial amplitude and frequency of grazing beyond body size per se, determining higher metabolic demands and the consequent (net effect) high grazing pressure in the resource community (e.g., Espoz and Castilla, 2000; Escobar and Navarrete, 2011; Manzur et al., 2014). Thus, non-trophic interactions (e.g., behaviourally mediated effects) are one of the main processes shaping herbivore–algae interaction network and influencing community structure in this system (see Kéfi et al., 2012, 2015).

There is still scant knowledge regarding the role of bottom-up processes influencing herbivore–algae interactions in this system. A study by Nielsen and Navarrete (2004) showed that upwelling intensity can determine variation in grazers' roles on different functional groups of algae over the mesoscale (over 10s to 100s of kilometres). Corticated algae dominate sites with intense upwelling, while ephemeral algae are abundant at sites not exposed to upwelling (Nielsen and Navarrete, 2004). This pattern could define the role that functionally distinct herbivores (i.e., 'scrapers-grazers'; *C. granosus*, *Scurria araucana*, versus 'browsers-grazers'; keyhole limpets, *Siphonaria lessoni*) play at different upwelling regimes. For example, since facultative 'browser' herbivores like *Fissurella* spp. and some fish species like *S. viridis* are capable of controlling dominant corticated algae, like *M. laminarioides* during late successional stages of the community, the role of these herbivores might be greater under intense upwelling. However, human harvesting of these herbivores (see Oliva and Castilla, 1986; Castilla, 1999; Moreno, 2001; Godoy et al., 2010) can reduce their densities, and their demographic pattern (e.g., sex ratio; Borges et al., 2016 and see Fenberg and Roy, 2008 for review), to levels that do not allow them to control dominant late successional algal forms (Nielsen and Navarrete, 2004).

Given the spatial structure of upwelling front across the coast of Chile from north to southern coasts (e.g., Montecinos and Lange, 2009; Tapia et al., 2014 and discussed earlier), herbivore–algae interaction patterns seem to be mostly influenced by nutrient effect rather than latitudinal effects, contrary to previous hypotheses (see Poore et al., 2012 for discussion). This could be also reflected across the northern biogeographic transitional zone (e.g., 30°S), where both quantitative and qualitative characteristics of herbivore–algae interactions can be strongly variable across the biogeographic boundary as can be derived from differences in life history dynamics of populations living at the range edge of their distributions (e.g., Fenberg and Rivadeneira, 2015).

### 12.4.2 Predator Guild Structure in Rocky Shore Habitats; Keystone versus Weak Interactors

Predator–prey interaction strength analyses have provided important insights into the factors that regulate the structure of consumer guilds in marine community structure (e.g., Paine, 1992; Menge et al., 1994; Navarrete and Menge, 1996; Berlow et al., 1999; Sala and Graham, 2002). In Chilean rocky intertidal communities, different quantitative estimates of predation intensity have helped to develop substantial recommendations about conservation and resource management (e.g., Castilla and Durán, 1985; Castilla and Fernández, 1998; Fernández et al., 2000; Gelcich et al., 2010). On the wave-exposed rocky intertidal zone of central Chile, the predator guild preferentially consumes important habitat-forming species such as mussels and barnacles, freeing space for the recruitment of other invertebrate and algal species (Castilla and Paine, 1987; Navarrete and Castilla, 2003). The seastar

*H. helianthus* and the muricid gastropod *C. concholepas* can be considered keystone predators (Navarrete and Castilla, 2003), because they have strong per capita effects on beds of the competitively dominant mussel, *P. purpuratus*, triggering dramatic changes in overall community structure (Castilla and Durán, 1985, Paine et al., 1985; Navarrete and Castilla, 2003). Predatory crabs *Acanthocyclus gayi* and *Acanthocyclus hassleri* also prey on mussels and barnacles at fairly high rates, and can coexist with *Concholepas* and *Heliaster* (e.g., Castilla and Durán, 1985; Navarrette and Castilla, 1988, 1990). Nonetheless, low per capita effects on mussel and barnacle species are observed for these crabs compared with *Concholepas* and *Heliaster* (Navarrete and Castilla, 2003). Also, juveniles and adults of the omnivorous clingfish *Sicyases sanguineus* occasionally feeds on mussels and barnacles at mid- and low-intertidal levels, with low impacts on these species (Cancino and Castilla, 1988). The whelks *Acanthina monodon* and *Crassilabrum crassilabrum* are commonly found on the very-low-intertidal fringe and shallow subtidal areas generating moderate to low per capita impacts on habitat-forming species. Therefore, field and laboratory evidence suggest that few large interactors and multiple weak-interacting species characterises the predator guild in intertidal rocky shore communities present in central Chile (Navarrete and Castilla, 2003). Experimental exclusion of humans from a portion (~500 m long) of the coast in 1982 (actually the marine reserve at Las Cruces: i.e., Estación Costera de Investigaciones Marinas), showed rapid increase of the commercially exploited gastropod *Concholepas* which drastically affected the biomass and productivity of herbivores and algae in the rocky intertidal community (e.g., Castilla and Durán, 1985, Botsford et al., 1997; Fernández et al., 2000; and see also Godoy and Moreno, 1989 for indirect effects of human exclusion). Therefore, studies conducted in the predator guild present in central-northern Chile (see earlier references) are at the core of the most important and challenging ecological research topics dealing with how changes in predator density/diversity can propagate through food webs, influencing ecosystem functioning (Otto et al., 2008). It is not clear, however, if interspecific competition is a strong structuring factor in the predator guild, as observed in molluscan herbivores present in this (Godoy and Moreno, 1989; Aguilera and Navarrete, 2012b) and in other southern coastal systems (e.g., Branch, 1976; Fletcher and Underwood, 1987; Underwood, 1992).

In kelp-dominated subtidal habitats, characterised mainly by *Lessonia trabeculata*, and *Macrocystis intergrifolia*, the predator guild presents from subtropical to temperate habitats is characterised by diverse fish species (e.g., *Graus nigra*; *Semicossyphus pulcher*; *Cheilodactylus variegatus*; *Pinguipes chilensis*; Pérez-Matus et al., 2012). Most species seem to have a strong direct effect on invertebrate assemblages (e.g., shrimps, bivalves, polychaetes, gastropods amphipods, etc.) through feeding on juvenile and adult individuals (Ojeda and Muñoz, 1999; Pérez-Matus et al., 2012). Other important carnivorous species are echinoderms such as *Meyenaster gelatinosus*, *Stichaster striatus*, *H. helianthus* and *Luidia magellanica,* and the omnivorous sea urchins *T. niger* and *L. albus,* all of which share food resources with fish species (Vásquez, 1993a; Pérez-Matus et al., 2012; Vásquez and Donoso, 2013), and therefore can alter the habitat available to fish species in kelp forests (Pérez-Matus et al., 2007). Experimental studies suggest the seastar *M. gelatinosus* has keystone effects on subtidal invertebrate assemblages through intraguild predation on *Heliaster,* one of the stronger interactors of the rocky intertidal (Gaymer and Himmelman, 2008, and see earlier). Large body size (radial amplitude) and higher movement rates seems to confer an advantage to *Meyenaster* over *Heliaster* (Viviani, 1979), with non-lethal effects (behaviourally mediated effect) of the former having considerable impacts on individual performances of the latter. Notwithstanding, poor performances of *Meyenaster* compared with *Heliaster* on wave-exposed habitats suggest the latter could have a predominant role on shallow subtidal communities more exposed to wave action (Viviani, 1979).

Currently, intense harvesting of subtidal kelp species (*Lessonia, Macrocystis*, see later) is reducing habitat availability for predatory species, conforming the guild with additional impacts from fishing eroding the functional structure of this kelp-associated predator guild (Godoy et al., 2010). For example, unregulated spearfishing has shifted the guild structure from large carnivorous species towards smaller-sized omnivores and herbivorous species in temperate reef fish communities (Godoy et al., 2010, 2016). Erosion of the functional structure of this guild correlates with similar human-induced impacts reported in other subtidal ecosystems where depletion of the fish assemblage had entailed a critical regime shift (e.g., coral reefs; Bellwood et al., 2004). Reduction of fish predators can shift the structure of the ecosystem to one where sea urchins (e.g., *Tetrapygus/Loxechinus*) reduce algal productivity and create urchin barrens in subtidal hard-bottom habitats (Vásquez and Donoso, 2013; see Ling et al., 2009 for similar effects for range-shifted sea urchin). Little knowledge is still available on the potential for *Meyenaster* or *Heliaster* to compensate fishing impacts and structuring subtidal invertebrate communities. Redundancy in controlling *Tetrapygus* abundances among the predator species might forestall ecosystem regime shift (from kelp to coralline or bare rock areas; e.g., see Ling et al., 2009 for examples of impacts of fishing spiny lobster *Jasus edwardsii* on sea urchin catastrophic effects on kelps). However, as kelp harvesting is intense in both subtidal and intertidal habitats in northern Chile (e.g., Vásquez et al., 2012; Oróstica et al., 2014; Vega et al., 2014), there would be low potential for overall compensatory effects (Ghedini et al., 2015) in this system. Therefore, management strategies are urgently needed to regulate kelp harvesting and fishing in order to control the reef fish crisis and seemingly inevitable regime shift (e.g., Gelcich et al., 2009, 2010).

### 12.4.3 The Role of Parasite–Host Interactions in Coastal Ecosystems

In general, parasite–host interactions are one of most overlooked interactions in marine communities despite their outstanding ecological and economic effects (Poulin, 1999; Lafferty et al., 2015). In Chile, some studies have focussed on patterns of diversity and richness of parasitic infracommunities (host like molluscs and fish species; e.g; George-Nascimento et al., 2002; Muñoz and George-Nascimento, 2002; González and Moreno, 2005) and their geographic distribution patterns (Oliva and González, 2005). Other studies have considered the primary transmission pathways from primary to final host species in intertidal systems (especially for parasitic trematode species; Loot et al., 2008; Aldana et al., 2009). In this context, more focussed studies have reported on the direct impact of fishing and human activities on the parasitic transmission pathways, infracommunity structure and prevalence (e.g., Loot et al., 2008; Wood et al., 2013). For example, different species like the fish *C. variegatus* and *A. punctatus*, the limpet *Fissurella latimarginata* and the sea urchin *L. albus* are target species for subsidence fisheries in central Chile (33°S), and their densities are commonly low in unprotected areas (Godoy et al., 2016). Different parasites infect these species, and both prevalence and incidence were high in host populations present inside a marine protected area (MPA; Wood et al., 2013). These results showed that fishing drives a decline in parasite abundance at the population level, through reducing the availability of hosts and resource for parasites (Wood et al., 2013).

No studies have explored the alteration of consumer behaviour to parasite load (e.g., Thomas and Poulin, 1998), which can have important indirect consequences on consumer–resource interactions (Poulin, 1999). Similarly, few studies have reported on the direct parasite transmission pathways, which can help to incorporate the parasite–host dimension in these well-studied food webs (Lafferty et al., 2006, 2015). Notwithstanding, the importance of parasite load on interspecific interaction of marine mammal species (i.e., sea lions and fur seals) has been previously suggested (George-Nascimento et al., 1992). Little information is thus available to address the direct and indirect impact of parasites on other consumer interactions, like predator–prey or herbivore–algae dynamics, and also their role as a structuring factor of host communities in this system (see Poulin, 1999 for review). Further studies are thus needed to include parasite–host interactions in the highly diverse food web present in coastal marine ecosystems of Chile (Kéfi et al., 2015).

## 12.5 Drivers for Coastal Biodiversity in Subtidal Kelp-Dominated Ecosystems

Kelps dominate shallow subtidal rocky-bottom areas in temperate and cold seas to a depth of ~40 m (Dayton et al., 1984; Harrold and Pearse, 1987; Vásquez, 1992; Graham et al., 2007). In the south-east Pacific the geographic and local distribution of subtidal Laminarian species are variable: *L. trabeculata* dominates rocky subtidal environments to a depth of ~40 m (Villouta and Santelices, 1984; Vásquez, 1992) and *Macrocystis pyrifera* forms shallow beds south of 42°S. *M. pyrifera* forms coastal belts to intertidal zone to ~15 m in northern-central latitudes (Vega et al., 2005), and to Cape Horn including the fjord zone and the Atlantic coast up to Chubut area in Argentina (Buschmann et al., 2004).

Shallow subtidal environments structured by kelp species (ecosystem engineering; *sensu* Jones et al., 1994) are highly productive and harbour high diversity and abundance of invertebrates and fishes. Kelp holdfasts constitute feeding areas, refuges against predation and bottom currents, spawning, settlement areas and nursery sites (Vásquez and Buschmann, 1997; Vásquez et al., 1998, 2001b; Graham et al., 2007; Teagle et al., 2017; Walls et al., 2017). A wide diversity of algae live under kelp canopy including several Corallinales, *Asparagopsis armata*, *Halopteris paniculata* and *Gelidium* spp; some barnacles and other sessile invertebrates (*Pyura chilensis*, *Phragmatopoma moerchi*, *Aulacomya atra*) are also part of the associated species sheltered by the kelp canopy (Vásquez et al., 1993a, 1993b, 2001a, 2001b; Vásquez and Vega, 2004; Thiel et al., 2007). Invertebrate predators such as the muricid snail *C. concholepas*, seastars (*M. gelatinosus*, *S. striatus*, *H. helianthus* and *L. magellanica*) and intermediate-size coastal fishes (*C. variegatus*, *Semicossyphus darwinii* and *Pinguipes chilensis*) constitute the predator guild in kelp forests along the Chilean coastline. The omnivorous-herbivore guild presents in kelp forest ecosystems include sea urchins (*T. niger* and *L. albus*), gastropods (*Tegula* spp. and *Fissurella* spp.), as well as fishes (*A. punctatus*, *G. laevifrons* and *Kyphosus analogus*; Pérez-Matus et al., 2007, 2012). Marine mammals such as the sea lion *Otaria flavescens* and the sea otter *Lontra felina* distribute widely along the coastal zone of the Humboldt Current, and also use kelp beds as feeding areas (Vega et al., 2005, Vásquez et al., 2006; Thiel et al., 2007).

In absence of kelp assemblages, subtidal rocky reefs commonly form alternative states dominated by crustose coralline algae ('barren-ground'; *sensu* Lawrence, 1975, and see Ling et al., 2009). The most important grazer determining this state are the sea urchins *T. niger* (Molina) and the sympatric but less common species *L. albus* (Molina) (Rodríguez and Ojeda, 1993). The sea urchin *T. niger* is an omnivore, while *L. albus* is an herbivore and feeds on foliose algae and drifting algal rafts (Rodríguez and Ojeda, 1993; Thiel et al., 2007; Vásquez and Donoso, 2013). Although both species can completely destroy kelp beds on a local scale (see Dayton, 1985; Buschmann et al., 2003), *T. niger* is primarily responsible for generation and maintenance of barren areas typically observed in northern Chile (Vásquez and Buschmann, 1997).

The distribution, abundance and diversity patterns of subtidal kelp communities along the south-east Pacific coastline are the result of complex life history strategies and interactions with environmental factors like water movement, nutrient availability, distribution of surface temperature associated with El Niño–La Niña fluctuation, herbivory pressure and a high and intense harvesting by fishermen (Muñoz et al., 2004; Vega et al., 2005; Buschmann et al., 2006; Vásquez et al., 2006; Graham et al., 2007).

Long-term studies of subtidal communities before and after El Niño (Vásquez et al., 2006) suggest that different bottom-up and top-down factors might control ecosystem changes in northern Chile, including the intensity and frequency of upwelling, buffering the positive thermal anomalies of superficial seawater temperature and favoring the persistence of kelps.

Site-dependent oceanographic conditions may generate optimal scenarios for spawning, larval development and recruitment of the echinoderm *T. niger*, which is the main grazer in subtidal rocky coastal areas, and also may decouple the correlation between carnivore abundances and population dynamics of sea urchins. In this context, abundance variability of some predator species as *L. magellanica* or *H. helianthus* may promote population increase of the urchin *T. niger* during El Niño events. These events might thus promote the development of two alternate states: (1) environments dominated by kelp versus and (2) barren ground areas dominated by urchins and coralline crustose algae (Graham et al., 2007; Ling et al., 2009). Nevertheless, other areas of the south-eastern Pacific during El Niño 1997–1998 showed that superficial warming decreased the abundance of kelp on shallow bottoms, inducing migrations of grazers to deeper zones (Fernández et al., 2000; Lleellish et al., 2001). In a wider context, El Niño–La Niña events include large-scale bottom-up and top-down effects which involves various levels of marine subtidal food webs, most of them difficult to predict. In this context, impacts of relative frequency of these large-scale environmental processes on kelps are commonly coupled with changes in human impacts like direct kelp harvesting. For example, human impact involving the landings of brown macroalgae in Chile reaches 4,000,000 wet t year$^{-1}$ (Vásquez et al., 2012), constituting the world's largest landings from natural populations. As was previously mentioned, most of the brown macroalgae are foundation species of marine ecosystems (Graham et al., 2007), forming the basis of coastal food webs (Halpern et al., 2006). In Chile, brown macroalgae contribute significantly to the total biomass of subtidal ecosystems (Santelices et al., 1980; Santelices and Ojeda, 1984). Similarly, brown macroalgae are highly connected with all trophic levels (Vásquez et al., 2006; Thiel et al., 2007), generating a significant impact on the ecosystem in an ecological and evolutionary context (Steinberg et al., 1995; Seeley and Schlesinger, 2012).

Recent comparative studies done in MPAs versus open access areas present in northern Chile show that harvesting has been affecting the population dynamics of kelps and the structure of their associated communities (Vega et al., 2014). Inside MPAs, *Lessonia* recruitment, density and biomass have temporal patterns similar to those described when populations of this brown macroalgae had less extraction (i.e., during 2000). By contrast, in open access areas, the lack of surveillance or assignments of use and property to artisanal fisherman (see Gelcich et al., 2008, 2009) promotes an indiscriminate harvest. Intensification of harvesting negatively affect the structure and population dynamics of subtidal communities associated with kelps compared with populations in areas with some conservation strategies like marine parks or co-management areas with sustainable management plans (Vásquez et al., 2012).

Kelps are useful sentinels of change because they are highly responsive to environmental conditions (Wernberg et al., 2013; Bell et al., 2015; Smale and Vance, 2016) and are directly exposed to many human activities that impact the coastal zone (e.g., harvesting, pollution, sedimentation, invasive species, fishing, recreation). In a global analysis of kelp forest changes over the past fifty years, Krumhansl et al. (2016) identified a high degree of variability in the magnitude and direction of changes across the geographic range of kelps. Although most of the changes worldwide are well correlated with warming or cooling of water surface associated with climate change, the decline of kelps in northern and central Chile is significantly related with the intensity of commercial harvest, despite a regional cooling trend (Krumhansl et al., 2016). This information shows that subtidal kelp ecosystem dynamics in the south-east Pacific, contrary to other coasts, is strongly controlled by processes directly related to disturbances generated by human harvesting, which are directly coupled with economic variation related to kelp exportation rates and market price fluctuations.

## 12.6 Concluding Remarks: Gaps in the Knowledge

Given the advances in understanding the processes influencing variation in distributional patterns and species interactions in rocky shore habitats in this ecosystem, further studies are needed to make explicit predictions regarding the persistence and dynamics of multiple species ranges and changing ecological interactions among species in the face of intensified human harvesting and global change (e.g., Ling et al., 2009; Firth et al., 2009). Hence, the main challenges are thus related, but not constrained to, examining the persistence of biogeographic transitional zones and species range shift under climate-driven impacts (e.g., changes in currents: e.g., Ling et al., 2009; heat waves: Sunday et al., 2012; homogenisation of upwelling patterns: Wang et al., 2015) by examining the stability of species geographic range edges. In this context, information of the geographic variability in the intensity and direction of consumer effects could be of great interest, i.e., under which condition or environmental context herbivore positive effects are more frequent than negative impacts on algal biomass or diversity? Global change is driving rapid variation in assemblage structure in the face of changing environmental conditions (Parmesan and Yohe, 2003; Poloczanska et al., 2008; Sunday et al., 2012; Sydeman et al., 2014). Species ranges are changing in different coasts as a consequence of the synergistic effect of climate warming and human harvesting (e.g., Lima et al., 2006; Parmesan, 2006; Smale and Wernberg, 2013; Fenberg and Rivadeneira, 2015 and see Burrows et al., 2011 and Mieszkowska et al., 2014 for reviews). Thus, one of the most important questions is: how the shift in consumer species (i.e., herbivores, predators) distribution patterns can influence variation in community structure and ecosystem functioning (e.g., Ling et al., 2009)? Recent evidence suggests that the ranges of some herbivore species are shifting across the temperate Eastern Pacific coast (Rivadeneira and Fernández, 2005; Aguilera et al., 2013b). Clear cooling trends (i.e., negative temperature anomalies) are observed across the HCS (e.g., Lima and Wethey, 2012; Rykaczewski et al., 2015; Wang et al., 2015), human harvesting is intensifying (e.g., consumers: Fernández et al., 2000; Godoy et al., 2016; kelps: Krumhansl et al., 2016) and there is potential for the presence of coastal artificial infrastructure could trigger species range shifts as recent studies suggest in other systems (e.g., Airoldi et al., 2015; Dong et al., 2016, and see Firth et al., 2016 for review).

Another important concern is the impact of exotic non-native species on community structure and functioning along the coast of Chile (Camus, 2005; Castilla et al., 2005). At least fifty-one marine non-indigenous invertebrate and algae species have been reported across the country (Castilla and Neill, 2009). However, given the scarce localities studied along Chilean coast the number of exotic species in Chile may be underestimated (see Villaseñor-Parada et al., 2017 for review). Most of these non-native species have been associated with aquaculture operations (e.g., *Codium fragile* var. *tomentosoides, Ciona intestinalis*; Neill et al., 2006 and see Naylor et al., 2001 and Villaseñor-Parada et al., 2017 for review) and transported by ships (e.g., *Pyura praeputialis*) (Castilla and Neill, 2009, see Keller et al., 2011 for discussion). Few studies, however, have been conducted in Chile (but see Dumont et al., 2011) on the role of artificial infrastructures (e.g., pilings) as a habitat for different non-indigenous or exotic species (e.g., Bulleri and Airoldi, 2005; Airoldi et al., 2015 and see Firth et al., 2016 for review). Given the recent proliferation of coastal infrastructure in Chile (e.g., artificial breakwaters; see Aguilera, 2018), an increase of suitable habitat for the spread and establishment of intertidal and subtidal non-indigenous species could be expected (see Firth et al., 2016). Many of the introduced species in Chile also correspond to parasites of both native and exotic (e.g., from aquaculture) host species (Moreno et al., 2006b), which have impacts on the structure of natural communities and on economic activities associated with aquaculture. In addition, there is an important number of introduced bio-engineer species like the ascidian *P. praeputialis* and the sea grass *Heterozostera tasmanica* which provide habitat for a diverse range of intertidal species (Ortiz and Wolff, 2002; Castilla et al., 2005). In contrast to the fact that most exotic species (non-target species) exert negative impacts on economic activities (e.g., like fouling species), the ascidian *P. praeputialis,* which is found exclusively inside the Bay of Antofagasta, ~24°S, is currently under intense exploitation by small-scale fisheries in northern Chile (Castilla et al., 2005; Manríquez et al., 2016). Thus, there is a challenge to deal with the presence and future introduction of non-native species to natural habitats along the coast of Chile (see Soto et al., 2001; Camus, 2005; Castilla and Neill, 2009 for discussion). On the one hand, environmental monitoring plans are required to prevent introduction, and ongoing research based on ballast water risk assessments (Camus et al., unpublished data) will soon provide the first official protocols to be applied in Chilean ports. On the other hand, appropriated strategies are required to restore natural ecosystem through non-native species eradication/control (e.g., Glen et al., 2013), and also to delineate species-specific conservation plans for species which provide important ecosystem services.

A rich ecological knowledge has been accumulated through the study of trophic and non-trophic community interactions along the Chilean rocky shores. Studies have reached a level of resolution that has allowed ecosystem-scale exploration about the structure of complex networks (see Kéfi et al., 2012, 2015). There is still scant knowledge, however, on the consequences of the range contraction or expansion of species on influencing community homogenisation or diversification (e.g., Olden et al., 2004, but see Vergés et al., 2016). For example, through range expansions or invasions, consumer species might displace native species with similar traits or those with a strong overlap in niche space through competitive effects (i.e., as the limiting similarity hypothesis suggests; MacArthur and Wilson, 1967). The functional replacement can drive widespread and unpredictable changes in the functioning of ecosystem through modified interactions at different trophic levels (Stachowicz et al., 2008 and see Ling et al., 2009 for cascade effects of sea urchin range expansion). Conversely, loss of key consumers could diminish resistance of native assemblages to species introductions (Bulleri et al., 2009). Species functional identity, richness and competitive potential seem to be critical traits to predict changes in the community composition and the consumer functional structure (e.g., Arenas et al., 2006; Firth et al., 2009). Thus, field experiments are needed to forecast community responses to changes in community structure driven by human-induced alteration, which operate at different temporal and spatial scales on coastal ecosystems present across the large and heterogeneous south-east temperate Pacific coast.

Strong and consistent changes in the timing, intensity and spatial heterogeneity of coastal upwelling in response to future warming by climate change impacts are predicted for most eastern boundary upwelling systems (Sydeman et al., 2014; Rykaczewski et al., 2015; Wang et al., 2015). For the HCS, an increase in upwelling intensity and duration will result in a substantial reduction of the latitudinal variation in coastal upwelling (Wang et al., 2015). Because there is a strong linkage between upwelling and marine ecosystem productivity, the projected geographic homogenisation in the intensity, timing and spatial structure of coastal upwelling may influence the geographical distribution of marine biodiversity across the south-east temperate Pacific coast owing to biotic homogenisation (e.g., Olden et al., 2004). Hence, synergistic effects of upwelling homogenisation and coastal impacts on local ecosystems through intensive harvesting, urbanisation and associated exotic species proliferation are expected in this system. These large-scale changes pose a strong challenge on the persistence and stability of the large-scale spatial variation in species diversity present across Chile, and on the opportunities to develop general ecosystem management and conservation strategies. This will be a central research topic in the future to contribute to conservation and restoration of local coastal ecosystems tightly connected across the HCS.

## Ackowledgements

MAA was financed by Fondecyt grant # 1160223 and PAI-CONICYT #79150002. BRB was supported by the ICM Center for the study of MUltiple drivers of marine Socio-EcologicaL Systems (MUSELS, NC120086). We thank EcoUrbE-lab for camaraderie and friendship.

### REFERENCES

Aguilera, M. A. (2011). The functional roles of herbivores in the rocky intertidal systems in Chile: a review of food preferences and consumptive effects. *Revista Chilena de Historia Natural*, **84**, 241–61.

Aguilera, M. A. (2018). Artificial defences in coastal marine ecosystems in Chile: opportunities for spatial planning to mitigate habitat loss and alteration of the marine community structure. *Ecological Engineering*, **120**, 601–10, http://doi.org/10.1016/j.ecoleng.2017.04.021.

Aguilera, M. A. and Navarrete, S. A. (2007). Effects of *Chiton granosus* (Frembly, 1827) and other molluscan grazers on algal succession in wave exposed mid-intertidal rocky shores of central Chile. *Journal of Experimental Marine Biology and Ecology*, **349**, 84–98.

Aguilera, M. A. and Navarrete, S. A. (2011). Distribution and activity patterns in an intertidal grazer assemblage: temporal and spatial organization influence inter-specific associations. *Marine Ecology Progress Series*, **431**, 119–36.

Aguilera, M. A. and Navarrete, S. A. (2012a). Functional identity and functional structure change through succession in a rocky intertidal marine herbivore assemblage. *Ecology*, **93**, 75–89.

Aguilera, M. A. and Navarrete, S. A. (2012b). Interspecific competition for shelters in territorial and gregarious intertidal grazers: consequences for individual behaviour. *PLoS ONE*, **7**(9), e46205.

Aguilera, M. A., Navarrete, S. A. and Broitman, B. R. (2013a). Differential effects of grazer species on periphyton of a temperate rocky shore. *Marine Ecology Progress Series*, **484**, 63–78.

Aguilera, M. A., Valdivia, N. and Broitman, B. R. (2013b). Spatial niche differentiation and coexistence at the edge: co-occurrence distribution patterns in *Scurria* limpets. *Marine Ecology Progress Series*, **483**, 185–98.

Aguilera, M. A., Valdivia, N. and Broitman, B. R. (2015a). Herbivore-alga interaction strength influences spatial heterogeneity in a kelp- dominated intertidal community. *PLoS ONE*, **10**(9), e0137287.

Aguilera, M. A., Valdivia, N. and Broitman, B. R. (2015b). Facilitative effect of a generalist herbivore on the recovery of a perennial alga: consequences for persistence at the edge of their geographic range. *PLoS ONE*, **10**, e0146069.

Aguirre, C., Pizarro, O., Strub, P. T., Garreaud, R. and Barth, J. A. (2012). Seasonal dynamics of the near-surface alongshore flow off central Chile. *Journal of Geophysical Research: Oceans*, **117**, 1–17.

Ahumada, R. B., Pinto, L. and Camus, P. A. (2000) The Chilean Coast. In C. R. C. Sheppard , eds. *Seas at the Millennium: An Environmental Analysis*. Pergamon Press, Oxford, pp. 699–717.

Airoldi, L., Turon, X., Perkol-Finkel, S. and Rius, M. (2015). Corridors for aliens but not for natives: effects of marine urban sprawl at a regional scale. *Diversity and Distributions*, **21**, 1–14.

Aldana, M., González, K., Loot, G., Pulgar, J. and Marquet, P. (2009). First intermediate host of the Digenean Trematode *Proctoeces lintoni* (Fellodistomidae) in Chile. *Journal of Parasitology*, **95**, 1408–14.

Arenas, F., Sánchez, I., Hawkins, S. J. and Jenkins, S. R. (2006). The invasibility of marine algal assemblages: role of functional diversity and identity. *Ecology*, **87**, 2851–61.

Bakun, A. and Weeks, S. J. (2008). The marine ecosystem off Peru: what are the secrets of its fishery productivity and what might its future hold? *Progress in Oceanography*, **79**, 290–9.

Barker, P. F. and Thomas, E. (2004). Origin, signature and palaeoclimatic influence of the Antarctic Circumpolar Current. *Earth-Science Reviews*, **66**, 143–62.

Bell, T. W., Cavanaugh, K. C., Reed, D. C. and Siegel, D. A. (2015). Geographic variability in the controls of giant kelp biomass dynamics. *Journal of Biogeography*, **42**, 2010–21.

Bellwood, D. R., Hughes, T., Folke, C. and Nystrom, M. (2004). Confronting the coral reef crisis. *Nature*, 429, 827–33.

Bellwood, D. R., Hoey, A. S. and Choat, J. H. (2003). Limited functional redundancy in high diversity systems: resilience and ecosystem function on coral reefs. *Ecology Letters*, **6**, 281–5.

Berlow, E. L., Navarrete, S. A., Briggs, C. J., Power, M. E. and Menge, B. A. (1999). Quantifying variation in the strengths of species interactions. *Ecology*, **80**, 2206–24.

Borges, C. D., Hawkins, S. J., Crowe, T. P. and Doncaster, C. P. (2016). The influence of simulated exploitation on *Patella vulgata* populations: protandric sex change is size-dependent. *Ecology and Evolution*, **6**, 514–31.

Botsford, L. W., Castilla, J. C. and Peterson, C. H. (1997). The management of fisheries and marine ecosystems. *Science*, **277**, 509–15.

Branch, G. (1976). Interspecific competition experienced by South African *Patella* species. *Journal of Animal Ecology*, **45**, 507–29.

Brante, A., Fernández, M. and Viard, F. (2012). Phylogeography and biogeography concordance in the marine gastropod *Crepipatella dilatata* (calyptraeidae) along the southeastern pacific coast. *Journal of Heredity*, **103**, 630–7.

Brattström, H. and Johanssen, A. (1983). Ecological and regional zoogeography of the marine benthic fauna of Chile. *Sarsia*, **68**, 289–339.

Broitman, B., Navarrete, S. A., Smith, F. and Gaines, S. (2001). Geographic variation of southeastern Pacific intertidal communities. *Marine Ecology Progress Series*, **224**, 21–34.

Broitman, B. R., Véliz, F., Manzur, T. et al. (2011). Geographic variation in diversity of wave exposed rocky intertidal communities along central Chile. *Revista Chilena de Historia Natural*, **84**, 143–54.

Bulleri, F. and Airoldi, L. (2005). Artificial marine structures facilitate the spread of a non-indigenous green alga, *Codium fragile* ssp. *tomentosoides*, in the north Adriatic Sea. *Journal of Applied Ecology*, **42**, 1063–72.

Bulleri, F. Tamburello, L. and Benedetti-Cecchi, L. (2009). Loss of consumers alters the effects of resident assemblages on the local spread of an introduced macroalga. *Oikos*, **118**, 269–79.

Burrows, M. T., Schoeman, D. S., Buckley, L. B. et al. (2011). The pace of shifting climate in marine and terrestrial ecosystems. *Science*, **334**, 652–5.

Buschmann, A. H., García, C., Espinoza, C., Filún, L. and Vásquez, J. A. (2003). Sea Urchin and Kelp (Macrocystis pyrifera) Interaction in Protected Areas in Southern Chile. In J. Lawrence, ed. *Sea Urchins and Fisheries*. CRC, Boca Raton, FL, pp. 120–30.

Buschmann, A. H., Vásquez, J. A., Osorio, P. et al. (2004). The effect of water movement, temperature and salinity on abundance and reproductive patterns of *Macrocystis* spp (Phaeophyta) at different latitudes. *Marine Biology*, **145**, 849–62.

Buschmann, A. H., Moreno, C., Vásquez, J. A. and Hernández-Carmona, M. (2006). Reproduction strategies of *Macrocystis pyrifera* (paheophyta) in southern Chile: the importance of population dynamics. *Journal of Applied Phycology*, **18**, 575–82.

Camus, P. A. (1994). Recruitment of the intertidal kelp *Lessonia nigrescens* Bory in northern Chile: successional constraints and opportunities. *Journal of Experimental Marine Biology and Ecology*, **184**, 171–81.

Camus, P. A. (2001). Marine biogeography of continental Chile. *Revista Chilena de Historia Natural*, **74**, 587–617.

Camus, P. A. (2005). Introducción de especies en ambientes marinos chilenos: no solo exóticas, no siempre evidentes. *Revista Chilena de Historia Natural*, **78**, 155–9.

Camus, P. A. (2008). Understanding biological impacts of ENSO on the eastern Pacific: an evolving scenario. *International Journal of Environment and Health*, **2**, 5–19.

Camus, P. A., Daroch, K. and Opazo, F. L. (2008). Potential for omnivory and apparent intraguild predation in rocky intertidal herbivore assemblages from northern Chile. *Marine Ecology Progress Series*, **361**, 35–45.

Cancino, J. and Castilla, J. C. (1988). Emersion behaviour and foraging ecology of the common clingfish *Sicyases sanguineus* (Pisces: Gobiesocidae). *Journal of Natural History*, **22**, 249–61.

Cárdenas, L., Castilla, J. C. and Viard, F. (2009). A phylogeographical analysis across three biogeographical provinces of the south-eastern Pacific: the case of the marine gastropod *Concholepas concholepas*. *Journal of Biogeography*, **36**, 969–81.

Castilla, J. C. (1999). Coastal marine communities: trends and perspectives from human-exclusion experiments. *Trends in Ecology and Evolution*, **7**, 280–3.

Castilla, J. C. and Camus, P. A. (1992). The Humboldt-El Niño scenario: coastal benthic resources and anthropogenic influences, with particular reference to the 1982/83 ENSO. *South African Journal of Marine Science*, **12**, 111–19.

Castilla, J. C. and Durán, L. R. (1985). Human exclusion from the rocky intertidal zone of Central Chile: the effects on *Concholepas Concholepas* (Gastropoda). *Oikos*, **45**, 391–9.

Castilla, J. C. and Fernández, M. (1998). Small-scale benthic fisheries in Chile: on co-management and sustainable use of benthic invertebrates. *Ecological Applications*, **8**, S124–32.

Castilla, J. C. and Neill, P. (2009). Marine Bioinvasions in the Southeastern Pacific: Status, Ecology Economic Impacts, Conservation and Management. In G. Rilov and J. A. Crooks, eds. *Biological Invasions*. Springer-Verlag, Berlin.

Castilla, J. C. and Paine, R. T. (1987). Predation and community organization in Eastern Pacific, temperate zone, rocky intertidal shores. *Revista Chilena de Historia Natural*, **60**, 131–51.

Castilla, J. C., Uribe, M., Bahamonde, N. et al. (2005). Down under the southeastern Pacific: marine non-indigenous species in Chile. *Biological Invasions*, **7**, 213–32.

Chaigneau, A. and Pizarro, O. (2005). Mean surface circulation and mesoscale turbulent flow characteristics in the eastern South Pacific from satellite tracked drifters. *Journal of Geophysical Research*, **110**, 1–17.

Chapman, M. and Underwood, A. J. (1992). Foraging Behaviour of Marine Benthic Grazers. In D. M. John, S. J. Hawkins and J. H. Price, eds. *Plant–Animal Interactions in the Marine Benthos*. Clarendon Press, Oxford, pp. 289–317.

Dayton, P. K., Currie, V., Gerrodette, T. et al. (1984). Patch dynamic and stability of some Californian kelp communities. *Ecological Monographs*, **54**, 253–89.

Dayton, P. K. (1985). Ecology of kelp communities. *Annual Review of Ecology, Evolution, and Systematics*, **16**, 215–45.

Dong, Y., Huang, X., Wang, W., Li, Y. and Wang, J. (2016). The marine 'great wall' of China: local- and broad-scale ecological impacts of coastal infrastructure on intertidal macrobenthic communities. *Diversity and Distributions*, **22**, 731–44.

Dumont, C. P., Harris, L. G. and Gaymer, C. F. (2011). Anthropogenic structures as a spatial refuge from predation for the invasive bryozoan *Bugula neritina*. *Marine Ecology Progress Series*, **427**, 95–103.

Escobar, J. and Navarrete, S. A. (2011). Risk recognition and variability in escape responses among intertidal molluskan grazers to the sun star *Heliaster helianthus*. *Marine Ecology Progress Series*, **421**, 151–61.

Escribano, R., Daneri, G., Farías, L. et al. (2004). Biological and chemical consequences of the 1997–1998 El Niño in the Chilean coastal upwelling system: a synthesis. *Deep Sea Research Part II: Topical Studies in Oceanography*, **51**, 2389–411.

Espoz, C. and Castilla, J. C. (2000). Escape responses of four Chilean intertidal limpets to seastars. *Marine Biology*, **137**, 887–92.

Espoz, C., Lindberg, D. R., Castilla, J. C. and Simison, B. (2004). Los patelogastrópodos intermareales de Chile y Perú. *Revista Chilena de Historia Natural*, **77**, 257–83.

Fenberg, P. B. and Rivadeneira, M. M. (2015). Range limits and geographic patterns of abundance of the rocky intertidal owl limpet, *Lottia gigantea*. *Journal of Biogeography*, **38**, 2286–98.

Fenberg, P. B. and Roy, K. (2008). Ecological and evolutionary consequences of size-selective harvesting: how much do we know? *Molecular Ecology*, **17**, 209–20.

Fernández, M., Jaramillo, E., Marquet, P. et al. (2000). An overview of the diversity, biogeography and dynamics of nearshore ecosystems in Chile: foundation for marine conservation ecology. *Revista Chilena de Historia Natural*, **73**, 797–830.

Firth, L. B. and Crowe, T. P. (2008). Large-scale coexistence and small-scale segregation of key species on rocky shores. *Hydrobiologia*, **614**, 233–41.

Firth, L. B., Crowe, T. P., Moore, P., Thompson, R. C. and Hawkins, S. J. (2009). Predicting impacts of climate-induced range expansion: an experimental framework and a test involving key grazers on temperate rocky shores. *Global Change Biology*, **15**, 1413–22.

Firth, L. B., Knights, A. M., Bridger, D. et al. (2016). Ocean sprawl: challenges and opportunities for biodiversity management in a changing world. *Oceanography and Marine Biology: An Annual Review*, **54**, 193–269.

Fletcher, W. J. and Underwood, A. J. (1987). Interspecific competition among subtidal limpets: effect of substratum heterogeneity. *Ecology*, **68**, 387–400.

Gaymer, C. F. and Himmelman, J. H. (2008). A keystone predatory sea star in the intertidal zone is controlled by a higher-order predatory sea star in the subtidal zone. *Marine Ecology Progress Series*, **370**, 143–53.

Gelcich, S., Godoy, N. and Castilla, J. C. (2009). Artisanal fisher's perceptions regarding coastal co-management policies in Chile and their potentials to scale-up marine biodiversity conservation. *Ocean and Coastal Management*, **52**, 424–32.

Gelcich, S., Kaiser, M. J., Castilla, J. C. and Edward-Jones, G. (2008). Engagement in co-management of marine benthic resources influences environmental perceptions of artisanal fishers. *Environmental Conservation*, **35**, 36–45.

Gelcich, S., Hughes, T. P., Olsson, P. et al. (2010). Navigating transformations in governance of Chilean marine coastal resources. *Proceedings of the National Academy of Sciences of the United States of America*, **107**, 16794–9.

George-Nascimento, M., Lima, M. and Ortiz, E. (1992). A case of parasite-mediated competition? Phenotypic differentiation among hookworms *Uncinaria* sp. (Nematoda: Ancylostomatidae) in sympatric and allopatric populations of South American sea lions *Otaria byronia*, and fur seals *Arctocephalus australis* (Carnivora: Otariidae). *Marine Biology*, **112**, 527–33.

George-Nascimento, M., Garcías, F. and Muñoz, G. (2002). Parasite body volume and infracommunity patterns in the southern pomfret *Brama australis* (Pisces: Bramidae). *Revista Chilena de Historia Natural*, **75**, 835–9.

Ghedini, G., Russell, B. D. and Connell, S. D. (2015). Trophic compensation reinforces resistance: herbivory absorbs the increasing effects of multiple disturbances. *Ecology Letters*, **18**, 182–7.

Glen, A. S., Atkinson, R., Campbell, K. J. et al. (2013). Eradicating multiple invasive species on inhabited islands: the next big step in island restoration? *Biological Invasions*, **15**, 2589–603.

Godoy, C. and Moreno, C. A. (1989). Indirect effects of human exclusion from the rocky intertidal in Southern Chile: a case of cross-linkage between herbivores. *Oikos*, **54**, 101–6.

Godoy, N., Gelcich, S., Vásquez, J. A. and Castilla, J. C. (2010). Spearfishing to depletion: evidence from temperate reef fishes in Chile. *Ecological Applications*, **20**, 1504–11.

Godoy, N., Gelcich, S., Castilla, J. C., Lima, M. and Smith, A. (2016). Artisanal spearfishery in temperate nearshore ecosystems of Chile: exploring the catch composition, revenue, and management needs. *Marine and Coastal Fisheries Dynamics, Management and Ecosystem Science*, **8**, 436–47.

González, A., Beltrán, J. and Hiriart-Bertrand, L. B. (2012). Identification of cryptic species in the *Lessonia nigrescens* complex (Phaeophyceae, Laminariales). *Journal of Phycology*, **48**, 1153–65.

González, M. T. and Moreno, C. A. (2005). The distribution of the ectoparasite fauna of *Sebastes capensis* from the southern hemisphere does not correspond with zoogeographical provinces of free-living marine animals. *Journal of Biogeography*, **32**, 1539–47.

Graham, M. H., Vásquez, J. A. and Buschmann, A. H. (2007). Global ecology of the giant kelp *Macrocystis*: from ecotypes to ecosystems. *Oceanography and Marine Biology: An Annual Review*, **45**, 39–88.

Halpern, B. S., Cottenie, K. and Broitman, B. R. (2006). Strong top-down control in southern California kelp forest ecosystems. *Science*, **312**, 1230–2.

Harrold, C. and Pearse, J. S. (1987). The Ecological Role of Echinoderms in Kelp Forests. In M. Jangoux and J. M. Lawrence, eds. *Echinoderm Studies*, vol. 2. Balkema, Rotterdam, pp. 137–233.

Haye, P. A., Segovia, N. I., Muñoz-Herrera, N. C. et al. (2014). Phylogeographic structure in benthic marine invertebrates of the southeast pacific coast of Chile with differing dispersal potential. *PLoS ONE*, **9**(2), e88613.

Haussermann, V. and Forsterra, G. (2009). *Marine Benthic Fauna of Chilean Patagonia*. Nature in Focus, Puerto Montt, p. 1000.

Hawkins, S. J. and Hartnoll, R. G. (1983). Grazing of intertidal algae by marine invertebrates. *Oceanography and Marine Biology: An Annual Review*, **21**, 195–282.

Hernández, C. E., Moreno, R. A. and Rozbaczylo, N. (2005). Biogeographical patterns and Rapoport's rule in southeastern Pacific benthic polychaetes of the Chilean coast. *Ecography*, **28**, 363–73.

Hoey, A. and Bellwood, D. (2009). Limited functional redundancy in a high diversity system: single species dominates key ecological process on coral reefs. *Ecosystems*, **12**, 1316–28.

Hormazábal, S., Shaffer, G. and Leth, O. (2004). Coastal transition zone off Chile. *Journal of Geophysical Research*, **109**, C01021.

Hu, Z. M. and Guillemin, M.-L. (2016). Coastal upwelling areas as safe havens during climate warming. *Journal of Biogeography* **43**: 2513–2514.

Ibáñez, C. M., Camus, P. A. and Rocha, F. J. (2009). Diversity and distribution of cephalopod species off the coast of Chile. *Marine Biology Research*, **5**, 374–84.

Iriarte, J. L. and González, H. E. (2004). Phytoplankton size structure during and after the 1997/98 El Niño in a coastal upwelling area of the northern Humboldt current system. *Marine Ecology Progress Series*, **269**, 83–90.

Jara, F. and Moreno, C. (1984). Herbivory and structure in a midlittoral rocky community: a case in southern Chile. *Ecology*, **65**, 28–38.

Jenkins, S., Coleman, R., Santina, P., Hawkins, S., Burrows, M. and Hartnoll, R. (2005). Regional scale differences in the determinism of grazing effects in the rocky intertidal. *Marine Ecology Progress Series*, **287**, 77–86.

Jones, C. G., Lawton, J. H and Shachak, M. (1994). Organisms as ecosystem engineers. *Oikos*, **69**, 73–386.

Kéfi, S., Berlow, E. L., Wieters, E. A. et al. (2012). More than a meal... integrating non-feeding interactions into food webs. *Ecology Letters*, **15**, 291–300.

Kéfi, S., Berlow, E. L., Wieters, E. A. et al. (2015). Network structure beyond food webs: mapping non-trophic and trophic interactions on Chilean rocky shores. *Ecology*, **96**, 291–303.

Keller, R. P., Drake, J. M., Drew, M. B. and Lodge, D. M. (2011). Linking environmental conditions and ship movements to estimate invasive species transport across the global shipping network. *Diversity and Distributions*, **17**, 93–102.

Klein, J. C. Underwood, A. J. and Chapman, M. G. (2011). Urban structures provide new insights into interactions among grazers and habitat. *Ecological Applications*, **21**, 427–38.

Krumhansl, K. A., Okamoto, D. K., Rassweiler, A. et al. (2016). Global patterns of kelp forest change over the past half-century. *Proceedings of Natural Academy of Science of the United States of America*, **113**, 13785–90.

Lafferty, K. D., Dobson, A. P. and Kuris, A. M. (2006). Parasites dominate food web links. *Proceeding of the National Academy of Sciences of the United States of America*, **103**, 11211–16.

Lafferty, K. D., Harvell, C. D., Conrad, J. M. et al. (2015). Infectious disease affect marine fisheries and aquaculture economics. *Annual Review of Marine Science*, **7**, 471–96.

Lagos, N. A., Tapia, F. J., Navarrete, S. A. and Castilla, J. C. (2007). Spatial synchrony in the recruitment of intertidal invertebrates along the coast of central Chile. *Marine Ecology Progress Series*, **350**, 29–39.

Lancellotti, D. A. and Vásquez, J. A. (1999). Biogeographical patterns of benthic macroinvertebrates in the Southeastern Pacific littoral. *Journal of Biogeography*, **26**, 1001–6.

Lawrence, J. M. (1975). On the relationships between marine plants and sea urchins. *Oceanography and Marine Biology: an Annual Review*, **13**, 213–86.

Lima, F. P. and Wethey, D. S. (2012). Three decades of high-resolution coastal sea surface temperatures reveal more than warming. *Nature Communications*, **3**, 1–13.

Lima, F. P., Queiroz, N., Ribeiro, P. A., Hawkins, S. J. and Santos, A. M. (2006). Recent changes in the distribution of a marine gastropod, *Patella rustica* Linnaeus, 1758, and their relationship to unusual climatic events. *Journal of Biogeography*, **33**, 812–22.

Ling, S. D., Johnson, C. R., Frusher, S. D. and Ridgway, K. R. (2009). Overfishing reduces resilience of kelp beds to climate-driven catastrophic phase shift. *Proceedings of the National Academy of Sciences of the United States of America*, **106**, 22341–5.

Llagostera, A. (1979). 9700 years of maritime subsistence on the Pacific: an analysis by means of bioindicators in the north of Chile. *American Antiquity*, **44**, 309–23.

Lleellish, J., Fernández, E. and Hooker, Y. (2001). Disturbancia del bosque submareal de *Macrocystis pyrifera* durante El Niño 1997–1998 en la Bahía de Pucusana. In K. Alveal and T. Antezana, eds. *Sustentabilidad de la biodiversidad. Un problema actual: bases científico técnicas, teorizaciones y proyecciones*. Ediciones Universidad de Concepción, Concepción, pp. 331–50.

Loot, G., Blanchet, S., Aldana, M. and Navarrete, S. A. (2008). Evidence of plasticity in the reproduction of a trematode parasite: the effect of host removal. *Journal of Parasitology*, **94**, 23–7.

Longhurst, A. (1998). *Ecological Geography of the Sea*. Academic Press, San Diego, CA.

Lourenço, C. R. Zardi, G. I., McQuaid, C. D. et al. (2016). Upwelling areas as climate change refugia for the distribution and genetic diversity of a marine macroalga. *Journal of Biogeography*, **43**, 1595–607.

Lubchenco, J. and Gaines, S. D. (1981). A unified approach to marine plant–herbivore interactions. I. Populations and communities. *Annual Review of Ecology and Systematics*, **12**, 405–37.

MacArthur, R. and Wilson, E. (1967). *The Theory of Island Biogeography*. Princeton University Press, Princeton, NJ.

Manríquez, P. H., Castilla, J. C., Ortiz, V. and Jara, M. E. (2016). Empirical evidence for large-scale human impact on intertidal aggregations, larval supply and recruitment of *Pyura praeputialis* around the Bay of Antofagasta, Chile. *Austral Ecology*, **41**, 701–14.

Martin, P. and Zuccarello, G. C. (2012). Molecular phylogeny and timing of radiation in *Lessonia* (Phaeophyceae, Laminariales). *Phycological Research*, **60**, 276–87.

Manzur, T., Vidal, F., Pantoja, J. F., Fernández, M. and Navarrete, S. A. (2014). Behavioural and physiological responses of limpet prey to a seastar predator and their transmission to basal trophic levels. *Journal of Animal Ecology*, **83**, 923–33.

Meneses, I. and Santelices, B. (2000). Patterns and breaking points in the distribution of benthic algae along the temperate Pacific coast of South America. *Revista Chilena de Historia Natural*, **73**, 615–23.

Menge, B. A., Berlow, E. L., Blanchette, C. A., Navarrete, S. A. and Yamada, S. B. (1994). The keystone species concept: variation in interaction strength in a rocky intertidal habitat. *Ecological Monographs*, **64**, 249–86.

Mieszkowska, N., Sugden, H., Firth, L. B. and Hawkins, S. J. (2014). The role of sustained observations in tracking impacts of environmental change on marine biodiversity and ecosystems. *Philosophical Transactions of the Royal Society of London A: Mathematical, Physical and Engineering Sciences*, **372**, 20130339.

Montecinos, V. and Lange, C. B. (2009). The Humboldt current system: ecosystem components and processes, fisheries, and sediment studies. *Progress in Oceanography*, **83**, 65–79.

Montecinos, A., Broitman, B., Faugeron, S., Haye, P. A., Tellier, F. and Guillermin, M. L. (2012). Species replacement along a lineal coastal habitat: phylogeography and speciation in the red alga *Mazzaella laminarioides* along the south east Pacific. *BMC Evolutionary Biology*, **12**, 1–17.

Moreno, C. A. (2001). Community patterns generated by human harvesting on Chilean shores: a review. *Aquatic Conservation: Marine and Freshwater Ecosystems*, **11**, 19–30.

Moreno, C. A. and Jaramillo, E. (1983). The role of grazers in the zonation of intertidal macroalgae of the Chilean coast. *Oikos*, **41**, 73–6.

Moreno, C. A., Sutherland, J. and Jara, F. (1984). Man as predator in the intertidal zone of southern Chile. *Oikos*, **42**, 155–60.

Moreno, C. A., Lunecke, K. M. and Lépez, M. I. (1986). The response of an intertidal *Concholepas concholepas* (Gastropoda) population to protection from man in Southern Chile and the effects on benthic sessile assemblages. *Oikos*, **46**, 359–64.

Moreno, R. A., Hernández, C. E., Rivadeneira, M. M., Vidal, M. A. and Rozbaczylo, N. (2006a). Patterns of endemism in south-eastern Pacific benthic polychaetes of the Chilean coast. *Journal of Biogeography*, **33**, 750–9.

Moreno, R. A., Neill, P. E. and Rozbaczylo, N. (2006b). Native and non-indigenous boring polychaetes in Chile: a threat to native and commercial mollusc species. *Revista Chilena de Historia Natural*, **79**, 263–78.

Moy, C. M., Seltzer, G. O., Rodbell, D. T. Y. and Anderson, D. M. (2002). Variability of El Niño/southern oscillation activity at millennial timescales during the Holocene epoch. *Nature*, **420**, 162–5.

Muñoz, G. and George-Nascimento, M. (2002). *Spiracanthus bovichthys* n. gen. n. sp. Acanthocephala: Arhythmacanthidae), a parasite of littoral fishes of the central-south coast of Chile. *Journal of Parasitology*, **88**, 141–5.

Muñoz, J., Finke, R., Camus, P. and Bozinovic, F. (2005). Thermoregulatory behavior, heat gain and thermal tolerance in intertidal snails: the case of the periwinkle *Echinolittorina peruviana* in central Chile. *Comparative Biochemistry and Physiology A*, **142**, 92–8.

Muñoz, V., Hernandez, M. C., Buschmann, A. H., Graham, M. H. and Vásquez, J. (2004). Variability in *per capita* oogonia and sporophyte production from giant kelp gametophyte (*Macrocystis pyrifera*, Phaeophyceae). *Revista Chilena de Historia Natural*, **77**, 639–47.

Navarrete, A. H., Lagos, N. A. and Ojeda, F. P. (2014). Latitudinal diversity patterns of Chilean coastal fishes: searching for causal processes. *Revista Chilena de Historia Natural*, **87**, 1–11.

Navarrete, S. A. and Castilla, J. C. (1988). Foraging activity of Chilean intertidal crabs *Acanthocyclus gayi* Milne-Edwards et Lucas and *A. hassleri* Rathburn. *Journal of Experimental Marine Biology and Ecology*, **118**, 115–36.

Navarrete, S. A. and Castilla, J. C. (1990). Resource partitioning between intertidal predatory crabs: interference and refuge utilization. *Journal of Experimental Marine Biology and Ecology*, **143**, 101–12.

Navarrete, S. A. and Menge, B. A. (1996). Keystone predation and interaction strength: interactive effects of predators on their main prey. *Ecological Monographs*, **66**, 409–29.

Navarrete, S. A., Broitman, B., Wieters, E. A., Finke, G. R., Venegas, R. M. and Sotomayor, A. (2002). Recruitment of intertidal invertebrates in the southeast Pacific: interannual variability and the 1997–1998 El Niño. *Limnology and Oceanography*, **47**, 791–802.

Navarrete, S. A. and Castilla, J. C. (2003). Experimental determination of predation intensity in an intertidal predator guild: dominant versus subordinate prey. *Oikos*, **100**, 251–62.

Navarrete, S. A., Wieters, E. A., Broitman, B. and Castilla, J. C. (2005). Scales of benthic-pelagic coupling and the intensity of species interactions: from recruitment limitation to top-down control. *Proceedings of the National Academy of Sciences of the United States of America*, **102**, 18046–51.

Naylor, R., Williams, S. and Strong, D. (2001). Aquaculture – a gateway for exotic species. *Science*, **294**, 1655–6.

Neill, P. E., Alcalde, O., Faugeron, S., Navarrete, S. A. and Correa, J. A. (2006). Invasion of *Codium fragile* ssp. *tomentosoides* in northern Chile: a new threat for *Gracilaria* farming. *Aquaculture*, **259**, 202–10.

Nielsen, K. J. and Navarrete, S. A. (2004). Mesoscale regulation comes from the bottom-up: intertidal interactions between consumers and upwelling. *Ecology Letters*, **7**, 31–41.

Ojeda, F. P. and Muñoz, A. (1999). Feeding selectivity of the herbivorous fish *Scartichthys viridis*: Effects on macroalgal community structure in a temperate rocky intertidal coastal zone. *Marine Ecology Progress Series*, **184**, 219–29.

Oliva, D. and Castilla, J. C. (1986). The effects of human exclusion on the population structure of keyhole limpets *Fissurella crassa* and *Fissurella limbata* in the coast of Central Chile. *Marine Ecology*, **7**, 201–17.

Oliva, M. and González, M. T. (2005). The decay of similarity over geographical distance in parasite communities of marine fishes. *Journal of Biogeography*, **32**, 1327–32.

Olden, J .D., Poff, N. L., Douglas, M. R., Douglas, M. E. and Fausch, K. D. (2004). Ecological and evolutionary consequences of biotic homogenization. *Trends in Ecology and Evolution*, **19**, 18–24.

Oróstica, M., Aguilera, M. A., Donoso, G., Vásquez, J. and Broitman, B. R. (2014). Effect of grazing on distribution and recovery of harvested stands of *Lessonia berteroana* kelp in northern Chile. *Marine Ecology Progress Series*, **511**, 71–82.

Ortiz, M. and Wolff, M. (2002). Trophic models of four benthic communities in Tongoy Bay (Chile): comparative analysis and preliminary assessment of management strategies. *Journal of Experimental Marine Biology and Ecology*, **268**, 205–35.

Ortlieb, L. (1995). Paleoclimas cuaternarios en el norte grande de Chile. In J. Argollo and P. Mourguiart, eds. *Cambios Cuaternarios en América del Sur*, ORSTOM-Bolivia, La Paz, pp. 225–46.

Ortlieb, L., Guzmán, N. and Marquardt, C. (2003). A Longer-Lasting and Warmer Interglacial Episode during Isotopic Stage 11: Marine Terrace Evidence in Tropical Western Americas. In A. W. Droxler, R. Z. Poore and L. H. Burckle, eds. *Earth's Climate and Orbital Eccentricity: The Marine Isotope Stage 11 Question*. American Geophysical Union, Washington, DC, Geophysical Monograph 137, pp. 157–80.

Otto, S. B., Berlow, E. L., Rank, N. E., Smiley, J. and Brose, U. (2008). Predator diversity and identity drive interaction strength and trophic cascades in a food web. *Ecology*, **89**, 134–44.

Paine, R. T. (1980). Food webs: linkage, interaction strength and community infrastructure. *Journal of Animal Ecology*, **49**, 666–85.

Paine, R. T. (1992). Food-web analysis through field measurement of per capita interaction strength. *Nature*, **355**, 73–5.

Paine, R. T., Castilla, J. C. and Cancino, J. (1985). Perturbation and recovery patterns of starfish-dominated intertidal assemblages in Chile, New Zealand, and Washington State. *The American Naturalist*, **125**, 679–91.

Palma, A. T., Pardo, L. M., Veas, R. et al. (2006). Coastal brachyuran decapods: Settlement and recruitment under contrasting coastal geometry conditions. *Marine Ecology Progress Series*, **316**, 139–53.

Parmesan, C. (2006). Ecological and evolutionary responses to recent climate change. *Annual Review of Ecology, Evolution and Systematics*, **37**, 637–69.

Parmesan, C. and Yohe, G. (2003). A globally coherent fingerprint of climate change impacts across natural systems. *Nature*, **421**, 37–42.

Pérez-Matus, A., Ferry-Graham, L. A., Cea, A. and Vásquez, J. A. (2007). Community structure of temperate reef fishes in kelp-dominated subtidal habitats of northern Chile. *Marine and Freshwater Research*, **58**, 1069–85.

Pérez-Matus, A., Pledger, S., Díaz, F. J, Ferry, L. A. and Vásquez, J. A. (2012). Plasticity in feeding selectivity and trophic structure of kelp forest associated fishes from northern Chile. *Revista Chilena de Historia Natural*, **85**, 29–48.

Pfuhl, H. A. and McCave, N. I. (2005). Evidence for late Oligocene establishment of the Antarctic Circumpolar Current. *Earth and Planetary Science Letters*, **235**, 715–28.

Poloczanska, E., Hawkins, S. J., Southward, A. J. and Burrows, M. T. (2008). Modeling the response of populations of competing species to climate change. *Ecology*, **89**, 3138–49.

Poore, A. G. B., Campbell, A. H., Coleman, R. A. et al. (2012). Global patterns in the impact of marine herbivores on benthic primary producers. *Ecology Letters*, **15**, 912–22.

Poulin, E., Palma, A. T., Leiva, G. et al. (2002). Avoiding offshore transport of competent larvae during upwelling events: The case of the gastropod *Concholepas concholepas* in central Chile. *Limnology and Oceanography*, **47**, 1248–55.

Poulin, R. (1999). The functional importance of parasites in animal communities: many roles at many levels? *International Journal of Parasitology*, **29**, 903–14.

Rahn, D. A., Rosenbluth, B. and Rutlland, J. A. (2014). Detecting subtle seasonal transitions of upwelling in North-Central Chile. *Journal of Physical Oceanography*, **45**, 854–68.

Rivadeneira, M. and Fernández, M. (2005). Shifts in southern endpoints of distribution in rocky intertidal species along the south-eastern Pacific coast. *Journal of Biogeography*, **32**, 203–9.

Rivadeneira, M. M. and Marquet, P. A. (2007). Selective extinction of late Neogene bivalves on the temperate Pacific coast of South America. *Paleobiology*, **33**, 455–68.

Rivadeneira, M. M., Hernáez, P., Baeza, J. A. et al. (2012). Testing the abundant-centre hypothesis using intertidal porcelain crabs along the Chilean coast: linking abundance and life-history variation. *Journal of Biogeography* **37**, 486–98.

Ritchie, M. E. and Olff, H. (1999). Spatial scaling laws yield a synthetic theory of biodiversity. *Nature*, **400**, 557–60.

Rodríguez, S. R. and Ojeda, F. P. (1993). Distribution patterns of *Tetrapygus niger* Echinodermata: Echinoidea) off the central Chilean coast. *Marine Ecology Progress Series*, **101**, 157–62.

Rosenfeld, R. (2002). Functional redundancy in ecology and conservation. *Oikos*, **98**, 156–62.

Rykaczewski, R. R., Dunne, J. P., Sydeman, W. J., García-Reyes, M., Black, B. A. and Bograd, S. J. (2015). Poleward displacement of coastal upwelling-favorable winds in the ocean's eastern boundary currents through the 21st century. *Geophysical Research Letters*, **42**, 6424–31.

Sala, E. and Graham, M. H. (2002). Community-wide distribution of predator-prey interaction strength in kelp forests. *Proceedings of the National Academy of Sciences of the United States of America*, **99**, 3678–83.

Sánchez, R., Sepúlveda, R. D., Brante, A. and Cárdenas, L. (2011). Spatial pattern of genetic and morphological diversity in the direct developer *Acanthina monodon* (Gastropoda: Mollusca). *Marine Ecology Progress Series*, **434**, 121–31.

Santelices, B. (1980). Phytogeographic characterization of the temperate coast of Pacific South America. *Phycologia*, **19**, 1–12.

Santelices, B., Castilla, J. C., Cancino, J. and Schmiede, P. (1980). Comparative ecology of *Lessonia nigrescens* and *Durvillaea antarctica* (Phaeophyta) in central Chile. *Marine Biology*, **59**, 119–32.

Santelices, B. and Ojeda, P. (1984). Recruitment, growth and durvival of *Lessonia nigrescens* (Phaeophyta) at various tidal levels in exposed habitats of central Chile. *Marine Ecology Progress Series*, **19**, 73–82.

Santelices, B., Vásquez, J. and Meneses, I. (1986). Patrones de distribución ydietas de un gremio de moluscos herbívoros en habitats intermareales expuestos de Chile central. In *Simposio Internacional. Usos y funciones de las algas marinas bentónicas*, 147–71.

Seeley, R. H. and Schlesinger, W. H. (2012). Sustainable seaweed cutting? The rockweed (*Ascophyllum nodosum*) industry of Maine and the Maritime Provinces. *Annals of the New York Academy of Sciences*, **1249**, 84–103.

Sepúlveda, R. D., Camus, P. A. and Moreno, C. A. (2016). Diversity of faunal assemblages associated with ribbed mussel beds along the South American coast: relative roles of biogeography and bioengineering. *Marine Ecology*, **37**, 943–56.

Shinen, J. L. and Navarrete, S. A. (2010). Coexistence and intertidal zonation of chthamalid barnacles along central Chile: Interference competition or a lottery for space? *Journal of Experimental Marine Biology and Ecology*, **392**, 176–87.

Smale, D. A. and Wernberg, T. (2013) Extreme climatic event drives range contraction of a habitat-forming species. *Proceedings of the Royal Society B*, **280**, 2012–29.

Smale, D. A. and Vance, T. (2016). Climate-driven shifts in species' distributions may exacerbate the impacts of storm disturbances on North-east Atlantic kelp forests. *Marine and Freshwater Research*, **67**, 65–74.

Soto, D., Jara, F. and Moreno, C. (2001). Escaped salmon in the inner southern Chile: facing ecological and social conflicts. *Ecological Applications*, **11**, 1750–62.

Spalding, M. D., Fox, H. E., Allen, G. R. et al. (2007). Marine ecoregions of the world: a bioregionalization of coastal and shelf areas. *BioScience*, **57**, 573–83.

Stachowicz, J. J., Graham, M., Bracken, M. and Szoboszlai, A. (2008). Diversity enhances cover and stability of seaweed assemblage: the role of heterogeneity and time. *Ecology*, **89**, 3008–19.

Steneck, R. S. and Watling, L. (1982). Feeding capabilities and limitation of herbivorous molluscs: a functional group approach. *Marine Biology*, **68**, 299–319.

Steinberg, P. D., Estes, J. A. and Winter, F. C. (1995). Evolutionary consequences of food chain length in kelp forest communities. *Proceedings of the National Academy of Sciences of the United States of America*, **92**, 8145–8.

Strub, P., Mesías, J., Montecino, V., Rutlant, J. and Salinas, S. (1998). Coastal Ocean Circulation off Western South America Coastal Segment. In A. Robinson and K. H. Brink, eds. *Global Coastal Ocean*, vol. 11. Harvard University Press, Cambridge, MA.

Sydeman, W. J., García-Reyes, M., Schoeman, M. S. et al. (2014). Climate change and wind intensification in coastal upwelling ecosystems. *Science*, **345**, 77–80.

Sullivan, K. and Bustamante, G. (1999). *Setting Geographic Priorities for Marine Conservation in Latin America and the Caribbean. The Nature Conservancy*, Arlington, VA, p. 141.

Sunday, J. M., Bates, A. E. and Dulvy, N. K. (2012). Thermal tolerance and the global redistribution of animals. *Nature Climate Change*, **2**, 686–90.

Takesue, R. K., van Geen, A., Carriquiry, J. D. et al. (2004). Influence of coastal upwelling and El Niño-southern oscillation on nearshore water along Baja California and Chile: shore-based monitoring during 1997–2000. *Journal of Geophysical Research-Oceans*, **109**, C03009.

Tapia, F. J., Navarrete, S. A., Castillo, M. et al. (2009). Thermal indices of upwelling effects on inner-shelf habitats. *Progress in Oceanography*, **83**, 278–87.

Tapia, F. J., Largier, J., Castillo, M., Wieters, E. A. and Navarrete, S. A. (2014). Latitudinal discontinuity in thermal conditions along the nearshore of central-northern Chile. *PLoS ONE*, **9**, e110841.

Teagle, H., Hawkins, S. J., Moore, P. J. and Smale, D. A. (2017). The role of kelp species as biogenic habitat formers in coastal marine ecosystems. *Journal of Experimental Marine Biology and Ecology*, http://doi.org/10.1016/j.jembe.2017.01.017.

Tellier, F., Tapia, J., Faugeron, S., Destombe, C. and Valero, M. (2011). The *Lessonia nigrescens* species complex (Laminariales, phaeophyceae) shows strict parapatry and complete reproductive isolation in a secondary contact zone. *Journal of Phycology*, **47**, 894–903.

Thiel, M., Macaya, E., Acuña, E. et al. (2007). The Humboldt current system of northern-central Chile: oceanographic processes, ecological interactions and socio-economic feedback. *Oceanography and Marine Biology: An Annual Review*, **45**, 195–345.

Thomas, F. and Poulin, R. (1998). Manipulation of a mollusc by a trophically transmitted parasite: convergent evolution or phylogenetic inheritance? *Parasitology*, **116**, 431–6.

UACH. (2006). Actualización y validación de la clasificación de las zonas biogeográficas litorales. Universidad austral de Chile, Informe Final proyecto FIP 204-28. Fondo de Investigación Pesquera, Santiago, www.fip.cl/Archivos/Hitos/Informes/inffinal%202004-28.pdf.

Underwood, A. J. (1992). Competition and Marine Plant–Animal Interactions. In D. M. John, S. J. Hawkins and J. H. Price, eds. *Plant–Animal Interactions in the Marine Benthos*. Clarendon Press, Oxford, pp. 443–75.

Valle-Levinson, A., Atkinson, L. P., Figueroa, D. and Castro, L. (2003). Flow induced by upwelling winds in an equatorward facing bay: Gulf of Arauco, Chile. *Journal of Geophysical Research*, **108**, 1–14.

Valdivia, N., Aguilera, M. A., Navarrete, S. A. and Broitman, B. R. (2015). Disentangling the effects of propagule supply and environmental filtering on the spatial structure of a rocky shore metacommunity. *Marine Ecology Progress Series*, **538**, 67–79.

Valdovinos, C., Navarrete, S. A. and Marquet, P. (2003). Mollusk species diversity in the Southeastern Pacific: why are there more species towards the pole? *Ecography*, **26**, 139–44.

Vásquez, J. A. (1992). *Lessonia trabeculata*, a subtidal bottom kelp in northern Chile: a case of study for a structural and geographical comparison. In U. Seeliger, ed. *Coastal Plant Communities of Latin America*. Academic Press Inc., San Diego, CA, pp. 77–89.

Vásquez, J. A. (1993a). Patrones de distribución de poblaciones submareales de *Lessonia trabeculata* (Laminariales, Phaeophyta) en el norte de Chile. *Serie Ocasional, Facultad de Ciencias del Mar, Universidad Católica del Norte*, **2**, 187–211.

Vásquez, J. A. (1993b). Abundance, distributional patterns and diets of main herbivorous and carnivorous species associated with *Lessonia trabeculata* kelp beds in northern Chile. Serie Ocasional, Facultad de Ciencias del Mar, Universidad Católica del Norte 2, 213–29.

Vásquez, J. A. and Buschmann, A. (1997). Herbivory-kelp interactions in subtidal Chilean communities: a review. *Revista Chilena de Historia Natural*, **70**, 41–52.

Vásquez, J. A., Camus, P. A. and Ojeda, F. P. (1998). Diversidad, estructura y funcionamiento de ecosistemas costeros rocosos del norte de Chile. *Revista Chilena de Historia Natural*, **71**, 479–99.

Vásquez, J. A., Fonck, E. and Vega, J. A. M. (2001a). Comunidades submareales rocosas dominadas por macroalgas en el norte de Chile: diversidad, abundancia y variabilidad temporal. In K. Alveal and T. Antezana, eds. *Sustentabilidad de la biodiversidad. Un problema actual, bases científico-técnicas, teorizaciones y perspectivas*. Universidad de Concepción, Concepción, pp. 281–92.

Vásquez, J. A., Veliz, D. and Pardo, L. M. (2001b). Biodiversidad bajo las grandes algas. In K. Alveal and T. Antezana, eds. *Sustentabilidad de la biodiversidad. Un problema actual, bases científico-técnicas, teorizaciones y perspectivas*. Universidad de Concepción, Concepción, pp. 293–308.

Vásquez, J. A. and Vega, J. M. A. (2004). El Niño 1997–1998 en el norte de Chile: efectos en la estructura y en la organización de comunidades submareales dominadas por algas pardas. In S. Avaria, J. Carrasco, J. Rutland and E. Yañez, eds. *El Niño-La Niña 1997-2000: su efecto en Chile*. Comité Oceanográfico Nacional, Valparaíso, pp. 115–36.

Vásquez, J. A., Vega, J. M. A. and Buschmann, A. H. (2006). Long term studies on El Niño-La Niña in northern Chile: effects on the structure and organization of subtidal kelp assemblages. *Journal of Applied Phycology* **18**, 505–19.

Vásquez, J. A., Piaget, N. and Vega, J. M. A. (2012). Chilean *Lessonia nigrescens* fishery in northern Chile: how do you harvest is more important than how much do you harvest. *Journal of Applied Phycology*, **24**, 417–26.

Vásquez, J. A. and Donoso, G. (2013). *Loxechinus albus: Biology and Ecology*. Development in Aquaculture and Fisheries Science. Elsevier, Amsterdam.

Vega, J. M. A., Vásquez, J. A. and Buschmann, A. H. (2005). Population biology of the subtidal kelps *Macrocystis integrifolia* and *Lessonia trabeculata* (Laminariales, Phaeophyceae) in an upwelling ecosystem of northern Chile: interannual variability and El Niño 1997–98. *Revista Chilena de Historia Natural*, **78**, 33–50.

Vega, J. M. A., Broitman, B. R. and Vásquez, J. A. (2014). Monitoring the sustainability of *Lessonia nigrescens* complex (Laminariales, Phaeophyta) in northern Chile under string harvest pressure. *Journal Applied Phycology*, **26**, 791–801.

Velásquez, C., Jaramillo, E., Camus, P. A., Manzano, M. and Sánchez, R. (2016). Biota del intermareal rocoso expuesto de la Isla Grande de Chiloé, Archipiélago de Chiloé, Chile: Patrones de diversidad e implicancias ecológicas y biogeográficas. *Revista de Biología Marina y Oceanografía*, **51**, 33–50.

Vergés, A., Doropoulos, C., Malcolm, H. A. et al. (2016). Long-term empirical evidence of ocean warming leading to tropicalization of fish communities, increased herbivory, and loss of kelp. *Proceeding of the National Academy of Science*, **113**, 13791–6.

Villaseñor-Parada, C., Pauchard, A. and Macaya, E. C. (2017). Ecology of marine invasions in continental Chile: what do we know and we need to know? *Revista Chilena de Historia Natural*, **52**, 17.

Villouta, E. and Santelices, B. (1984). Estructura de la comunidad submareal de *Lessonia* (Phaeophyta, Laminariales) en Chile norte y central. *Revista Chilena de Historia Natural*, **57**, 111–22.

Viviani, C. (1979). Ecogeografía del litoral chileno. *Studies on Neotropical Fauna and Environment*, **14**, 65–123.

Walls, A. M., Edwards, M. D., Firth, L. B. and Johnson, M. P. (2017). Successional changes of epibiont fouling communities of the cultivated kelp *Alaria esculenta*: predictability and influences. *Aquaculture Environment Interactions*, **9**, 55–69.

Wang, D., Gouhier, T. C., Menge, B. A. and Ganguly, A. R. (2015). Intensification and spatial homogenization of coastal upwelling under climate change. *Nature*, **518**, 390–4.

Webb, O. C., Ackerly, D. D., McPeek, M. A. and Donoughue, M. J. (2002). Phylogenies and community ecology. *Annual Review of Ecology and Systematics*, **33**, 475–505.

Wernberg, T., Smale, D. A., Tuya, F. et al. (2013). An extreme climatic event alters marine ecosystem structure in a global biodiversity hotspot. *Nature Climate Change*, **3**, 78–82.

Wieters, E. A. (2005). Upwelling control of positive interactions over mesoscales: A new link between bottom-up and top-down processes on rocky shores. *Marine Ecology Progress Series*, **301**, 43–54.

Wieters, E. A., Kaplan, D. M., Navarrete, S. A. et al. (2003). Alongshore and temporal variability in chlorophyll *a* concentration in Chilean nearshore waters. *Marine Ecology Progress Series*, **249**, 93–105.

Wood, S., Lilley, S., Schiel, D. and Shurin, J. (2010). Organismal traits are more important than environment for species interactions in the intertidal zone. *Ecology Letters*, **13**, 1160–71.

Wood, C. L., Micheli, F., Fernández, M., Gelcich, S., Castilla, J. C. and Carvajal, J. (2013). Marine protected areas facilitate parasite populations among four fished host species of central Chile. *Journal of Animal Ecology*, **82**, 1276–87.

# Chapter 13

# Where Three Oceans Meet

## *State of the Art and Developments in Southern African Coastal Marine Biology*[*]

Christopher D. McQuaid and Laura K. Blamey

## 13.1 Introduction

We start by providing an overview of the 'state of the art' of our understanding of coastal ecosystems in southern Africa, then summarise advances that have been made since ~1990. In doing so we emphasise the development of important trends, rather than providing detail on thirty years' worth of findings, and highlight species interactions where these are important. The importance of species interactions differs among ecosystems as some are more strongly shaped by physical drivers than others. Unfortunately, work on these shallow coastal systems is geographically skewed and the important advances within the entire southern African region lie very heavily with developments in South Africa, so that an emphasis on this country is unavoidable. The neighbouring countries with coastlines, Namibia, Angola and Moçambique, have been torn by protracted civil wars and have had priorities other than science, while South Africa remains a major driver of marine research within the continent, contributing to 40 per cent of all papers on coastal ecology emanating from the continent as a whole over the last twenty years. Most of the other publications coming from Africa have involved scientists from North America, Europe and Australia, with other African countries contributing to only 12 per cent (McQuaid, 2010). Even within South Africa, the community of actively publishing marine biologists is small and so directions are disproportionally guided by the interests or the biases of relatively few people.

In addition, different coastal ecosystems predominate in different parts of the country, reflecting contrasting ocean temperatures and productivity. Southern Africa is bound by three oceans: the Southern, west Indian and South Atlantic and the biology of the coastline is deeply shaped by two major currents (Figure 13.1). The cold, eutrophic Benguela Current flows northwards along the west coast, which is characterised by strong summer upwelling driven by south-east winds. The warm, oligotrophic Agulhas Current is a western boundary current that flows from north-east to south-west along the east and south coasts of southern Africa. At its south-west limit it undergoes a massive retroflection along the edge of the Agulhas Bank, a shallow submarine extension of the African land mass that extends south from the Cape of Good Hope. Upwelling along the south and east coasts is less frequent, in places topographically driven,

[*] *Ex Africa semper aliquid novi*: Always something new out of Africa

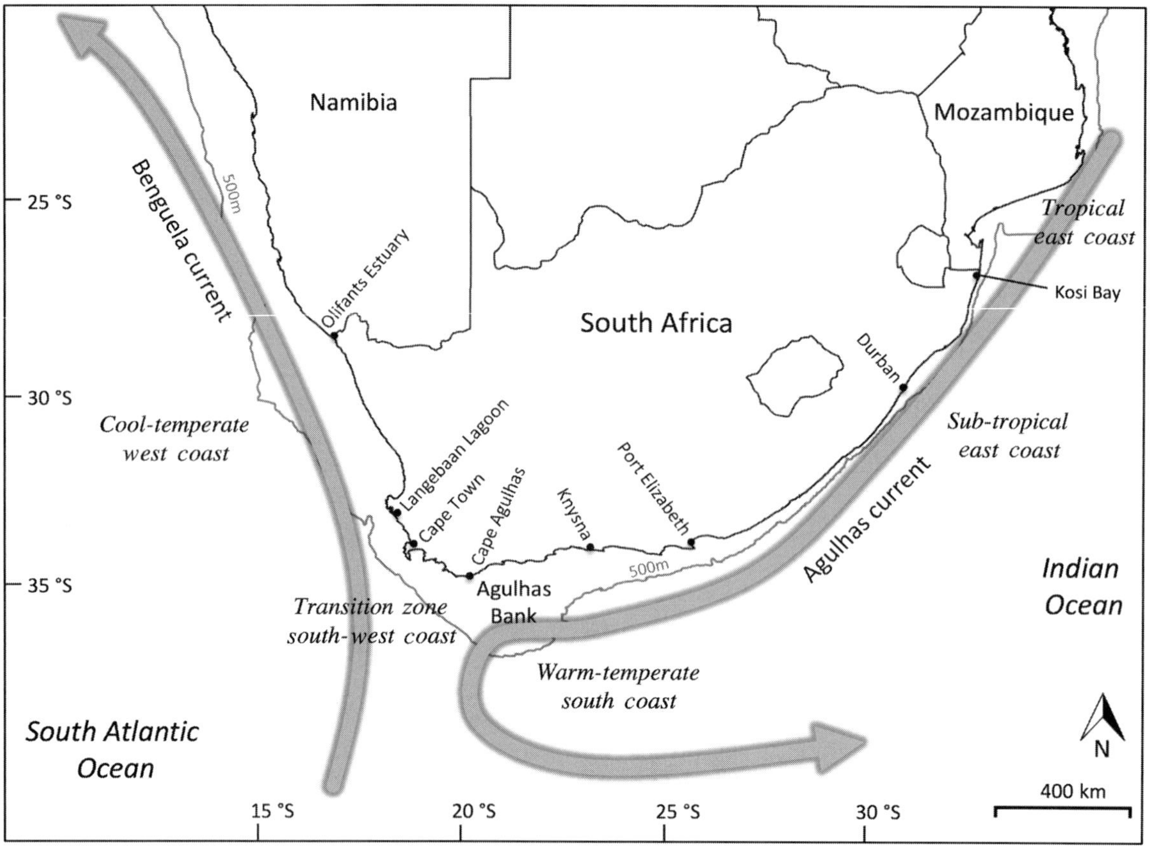


**Fig. 13.1** Map of southern Africa showing the two contrasting currents that shape the major biogeographic regions and biology of the coastline, as well as place names referred to in the text. The 500 m isobath is shown.

more localised and less intense than along the west coast. Together, these two currents shape the biogeography of the region. Setting aside the details, there are three major biogeographic regions: the cool-temperate west coast, the warm-temperate south coast and the subtropical to tropical east coast.

Mangroves and coral reefs are restricted to the warm water east of the region, where the Agulhas Current brings warm oligotrophic waters from the north; while subtidal kelp beds are restricted to the cool-temperate west coast and the transitional south-west coast. This results in geographic biases in research interests. Broadly, intertidal systems between the Namibia/Angola border and the South Africa/Mozambique border comprise around 42 per cent sandy beach, 27 per cent rocky shore and 31 per cent mixed shores where the rock substratum is strongly influenced by periodic sand inundation (Bally et al., 1984). The later have been poorly studied, but show relatively high species richness because they support both psammophobic and psammophilic species as well as species that are restricted to this type of shore (McQuaid and Dower, 1990).

The degree to which these coastal ecosystems are shaped by biotic interactions differs as some are more strongly driven by abiotic conditions than others. In addition, our understanding of the relevant biotic interactions is unbalanced as some systems are more widespread and have received more (or better co-ordinated) attention than others (Griffiths et al., 2010). Generally, marine biologists in South Africa have tended to focus on systems rather than questions. For example, there has been a huge effort on the ecology of estuarine systems, which we do not cover (see recent overview by Allanson and Baird, 2008).

Two research programmes in the period ending in the 1990s were particularly coherent and productive, the Kelp Bed project, based on the west coast, driven by the University of Cape Town and research on sandy beaches largely based on the south coast and driven by the then University of Port Elizabeth. Both made enormous contributions not only to our understanding of local ecosystems, but also to developing a systems-based understanding of community ecology, how ecosystems respond as a whole to their environment and how species interactions guide energy flow and drive those responses (see reviews by Branch, 2008 and Brown and McLachlan, 2010).

The depth of the contribution to global sandy beach ecology is captured by Nel et al. (2014) who note that over the last sixty-three years the literature on these systems has been dominated by a handful of countries, particularly the USA, South Africa, Brazil and Italy. South Africa was second to the USA as the most productive country, contributing 10 per cent of the global total since 1950. This is a remarkable achievement. Additionally, most papers had few authors, which has long typified the normal pattern in South African marine research.

The ongoing history of marine ecology in South Africa can be divided into the what, why and where phases. Early marine ecology involved compiling basic species lists and describing the biogeography of the region, although work defining and describing the biota of the region still continues (e.g., Prochazka and Griffiths, 1992; Bolton and Stegenga, 2002; Bolton et al., 2004; Sink et al., 2005; Maneveldt et al., 2008; Emmerson, 2016). These foundational studies were initially led by European scientists, particularly taxonomists, with important contributions from phycologists and several major expeditions. By the mid-twentieth century there was a clear understanding of the biogeography of the region and how this related to the two major currents, the Benguela and the Agulhas, that dominate the west coast and the east and south coasts, respectively. For intertidal ecosystems, this culminated in the landmark work of Stephenson and Stephenson (1949). This was the 'what' phase, largely concerned with describing what systems existed in the country, and was followed by a period of experimental work aimed at understanding the ecology and trophic interactions of intertidal and shallow subtidal systems using small-scale experiments and later the development of quantitative models to describe energy flow through these systems. 'Why' do these systems look as they do? Here, the community-based approach taken by the Kelp Bed project in the 1970s/1980s and the sandy beach programme in the 1980s/1990s was particularly productive. This phase is slowly merging with the 'where' phase, which is an attempt to predict how coastal systems will change and what they will look like under a regime of increasing anthropogenic pressure and changing environmental conditions. In other words, where are these systems going? This review concerns developments in the 'why' phase and the beginnings of the 'where' phase. Because of their economic importance, there has been an enormous emphasis on the pelagic systems of the Benguela system of the west coast through the fifteen-year Benguela Ecology Programme, spanning the 1980s and 1990s, which led to South African participation in the Benguela Large Marine Ecosystem project (Hempel et al., 2008). We will not attempt to include this research, but will focus on shallow benthic systems.

## 13.2 State of the Art ~1990: Ecosystem Functioning

### 13.2.1 Sandy Beaches

Sandy beaches, including mixed shores, make up ~70 per cent of the coastline in southern Africa, with most of these beaches considered open and fully exposed (McLachlan et al., 1981a). Community composition on sandy beaches is almost totally controlled by physical conditions (McLachlan et al., 1993), the combination of wave action and particle size, which drives morphodynamic state, being the most important factor in structuring these ecosystems (Short and Wright, 1983). The heyday for sandy beach research in southern

Africa was the 1970s, 1980s and early 1990s when the structure and function of the ecosystems they support were extensively studied, particularly by researchers at the (then) University of Port Elizabeth, now the Nelson Mandela Metropolitan University (e.g., McLachlan et al., 1981a, 1981b). There have been relatively few fundamental advances since then, although more recently sandy beach organisms have been used to test general ecological theory (e.g., Baldanzi et al., 2013, 2015).

Basic sandy beach food chains include primary producers (surf zone phytoplankton), consumers (macrofauana), decomposers (interstitial fauna) and transient scavengers/predators (crabs, fish, birds); see Figure 13.2. On exposed beaches, there are two distinct food chains. One is driven by particulate matter, particularly surf zone phytoplankton that is consumed by filter-feeding macrofauna, including fish, which in turn fall prey to fish, birds and crabs, the community structure of the fish being particularly strongly influenced by wave exposure (Romer, 1990). The basis of this web is the diatom *Anaulus australis* (Talbot et al., 1990), and autochthonous productivity within the surf zone is sufficient to support the self-sustained beach/surf zone ecosystem (Du Preez et al., 1990). Temporary wind-driven aggregations of diatoms form foci of energy transfer to pelagic grazers (Odebrecht et al., 2014), while import of diatoms from the surf zone to the intertidal provides food for filter feeders such as the bivalve *Donax* (Donn, 1987), which can reach extremely high densities (e.g., Schoeman et al., 2000). The scavenging food chain is driven by stranded carrion that is consumed by scavenging macrofauna, which are in turn preyed upon by predators (McLachlan et al., 1981a, 1981b). Along open beaches, particulate food is more constant and abundant than carrion and therefore filter feeders dominate the macrofaunal biomass (Brown, 1964; Ansell, 1972; McLachlan et al., 1981b). Deposit feeders are really abundant only on sheltered beaches, where organic matter can accumulate in substantial quantities (McLachlan et al., 1981b) and such shores are rare in South Africa. The interstitial fauna (meiofauana, protozoa and bacteria) feed on decomposing matter as well as dissolved and particulate matter that is flushed into the system through wave energy. Although not eaten by the macrofauna, they excrete nutrients that help fuel the phytoplankton growth and thus indirectly support the macrofauna. Broadly speaking, sandy beaches on the east coast of South Africa, where nearshore nutrient levels are low, are characterised by a macrofauna that is low in biomass and dominated by scavengers/predators, whereas the south and west coasts have moderate to high macrofaunal biomass, dominated by filter feeders and fuelled by surf zone phytoplankton blooms (McLachlan et al., 1981a).

### 13.2.2 Rocky Shores

Intertidal rocky shores are among the best-studied marine ecosystems in southern Africa, with a large body of literature available which was last reviewed at the start of our period by Branch and Griffiths (1988). Figure 13.3 summarises our present understanding of these food webs.

Most of the early work focussed on the composition of the fauna and flora both within rocky shores, and also around the coast. Early work identified the three major biogeographic provinces that are described earlier, each with different biogeographic affinities: Atlantic for the west coast, Indian for the east and a mixture of the two on the south coast, where levels of endemism are highest, although this perception may be partially artefactual (Griffiths et al., 2010). The different biogeographic regions exhibit somewhat different zonation patterns and the same height on the shore can be dominated by different organisms in the cold-temperate, warm-temperate and subtropical bioregions (Stephenson and Stephenson, 1949). All three coasts support four distinct zones with, broadly, animals dominating the upper eulittoral and macroalgae the lower eulittoral and subtidal fringe. These zones include a supralittoral fringe characterised by various littorinid snails and an upper balanoid zone dominated by barnacles and grazers such as limpets and top shells. The lower balanoid zone has dense beds of mussels followed by macroalgae, plus zooanthids on the east coast. Below this on the west and south coasts, there is a dense band of large limpets (*Scutellastra cochlear* and/or *Scutellastra argenvillei*), while the

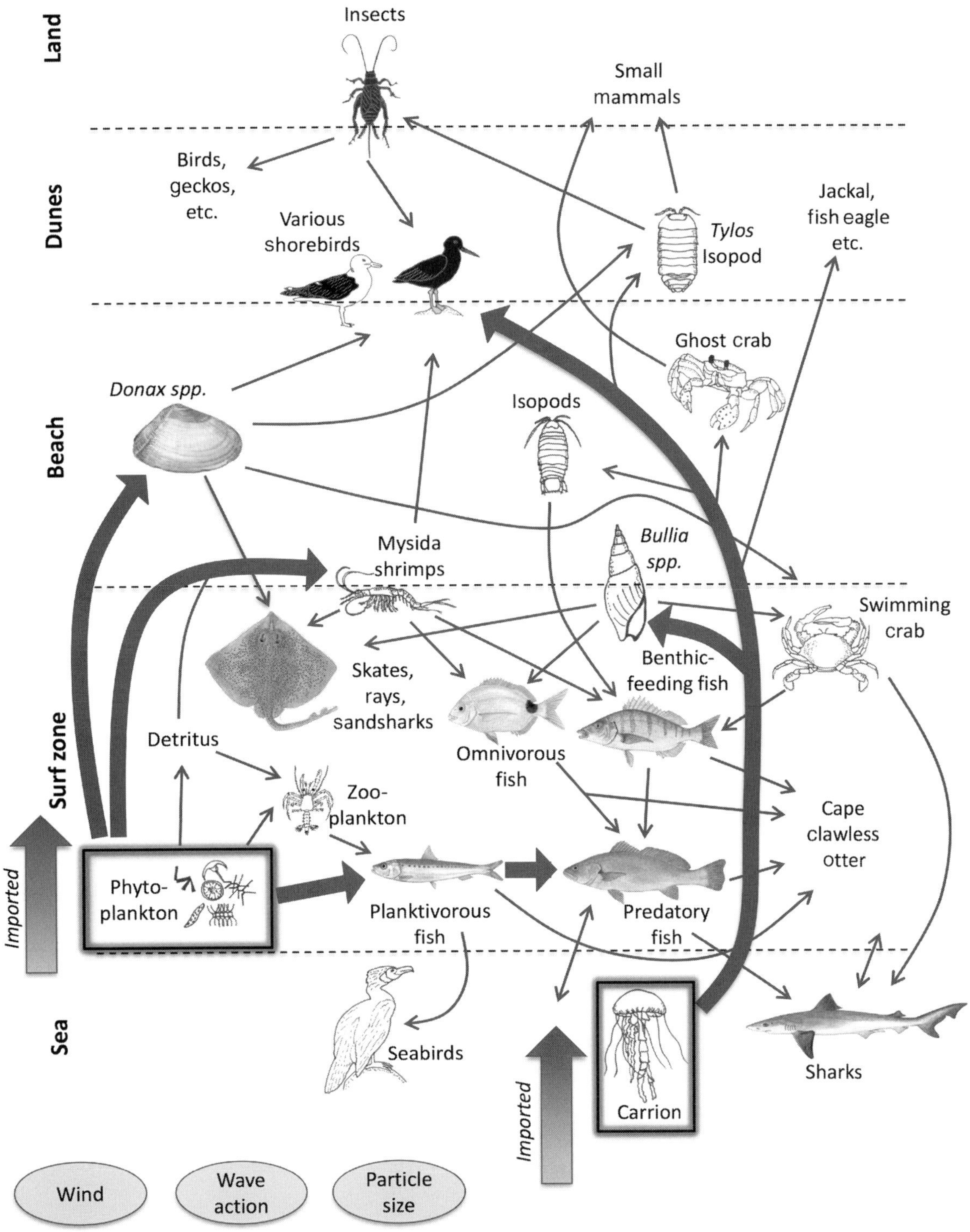


**Fig. 13.2** A typical sandy beach food web. Arrows indicate trophic interactions. Arrow thickness indicates a subjective assessment of the strength of interactions. The main external biotic (boxes) and abiotic (circles) drivers are indicated. Details differ in space and time depending on the morphodynamic state of the shore. Adapted from McLachlan et al. (1981b). Permission to use the copyrighted images was obtained from the South African Institute for Aquatic Biodiversity (SAIAB) and the World Wildlife Fund South Africa (WWF-SA).

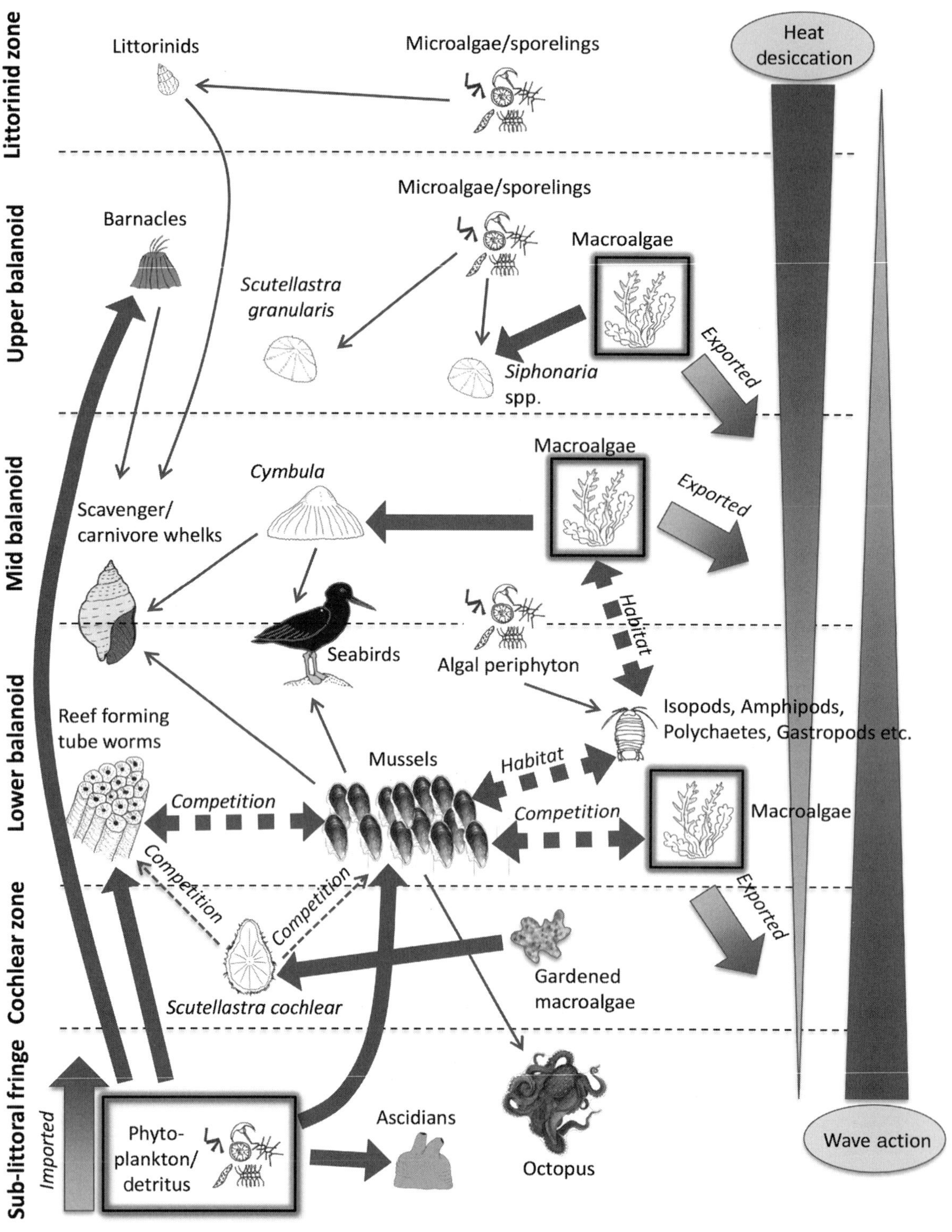


**Fig. 13.3** Rocky shore food web showing trophic (solid arrows) and non-trophic interactions (broken arrows). Arrow thickness indicates a subjective assessment of the strength of interactions. The main external biotic (boxes) and abiotic (circles) drivers are indicated. Note that this is typical for a south coast system. The details of species composition and interaction strength differ for the west and east coasts; they also depend strongly on the intensity of wave action. Permission to use copyrighted images was obtained from the SAIAB and the WWF-SA.

infralittoral fringe is dominated by seaweeds with subtidal kelp beds on the west coast and red algae running into ascidians on the south coast where kelps are limited to a smaller species confined to low shore pools and gullies. The east coast lacks the band of large limpets, but instead, has a dense oyster belt just below the upper balanoid. Physical factors, particularly emersion and wave action, and species tolerances structure these zones, while species interactions help maintain them (Branch and Griffiths, 1988).

Broadscale studies (e.g., Bustamante and Branch, 1996) revealed general patterns around the coast, and also across gradients of wave exposure (e.g., McQuaid and Branch, 1984, 1985). Faunal species richness increases from west to east, while floral species richness remains much the same around the coast. Transition zones between the biogeographic provinces exhibit enhanced species richness. In contrast to species richness, intertidal biomass on the west coast is greater than on the south and east coasts (Bustamante and Branch, 1996) and across all coasts, the trophic structure of these communities varies significantly with wave exposure. On wave-exposed shores, autotrophs, filter feeders and invertebrate predators dominate but diversity is low, while on semi-exposed and sheltered shores, grazers are more abundant and diversity is greater than on exposed shores (Bustamante and Branch, 1996).

Given the variability among shores, no complete energy budget exists for the rocky intertidal (Branch and Griffiths, 1988). Primary producers far outweigh other trophic levels on sheltered shores, which are likely to function as net exporters of material, whereas, on exposed shores, biomass is dominated by filter feeders that depend on a food supply that is imported from the water column (McQuaid and Branch, 1985). Most of the grazer biomass is made up of larger species that don't feed on attached macroalgae, but rather on encrusting, filamentous or microalgae (Branch and Griffiths, 1988). On the west coast, an exceptional biomass of limpets is maintained by the subsidy of subtidal kelps (Bustamante et al., 1995), but this is threatened by invasive mussels (see later).

The most abundant predators on rocky shores include whelks, crabs (more abundant on the east coast), rock lobsters, the spiny starfish *Marthasterias glacialis* (Branch and Griffiths, 1988; Cockcroft and Payne, 1999) and the octopus, *Octopus vulgaris* (Smale and Buchan, 1981). The last three are more prevalent in the subtidal region, but the starfish and the octopus can feed intertidally (e.g., McQuaid, 1994; Oosthuizen and Smale, 2003).

With the exception of harvesting by man, which can drive dramatic changes in community structure (e.g., Lasiak and Field, 1995), predation appears to be a less important factor on South African rocky shores than in parts of the world where invertebrates such as the starfish *Pisaster ochraceus* (Paine, 1969) or the gastropod *Concholepas concholepas* (Moreno et al., 1986) act as keystone predators. For example, Griffiths and Hockey (1987) modelled mortality of intertidal mussels and concluded that intraspecific competition for space is a substantially more important cause of mortality than predation, an interpretation supported by Branch and Steffani (2004). This conclusion mainly concerns predation of adults, and the findings are different for earlier ontogenetic stages. Plass-Johnson et al. (2010) used a variety of cages to show that predation of recent settlers by a combination of benthic and pelagic predators can be intense, resulting in over 50 per cent mortality within two weeks under experimental field conditions, while Porri et al. (2008a) found that cannibalism by filter-feeding adult mussels can remove nearly 80 per cent of conspecific larvae as they attempt to settle.

Rocky shore research in South Africa has followed the global trend in the biological sciences of a growing focus on issues concerned with climate change. There is good evidence that sea temperatures are changing along the South African coast, with cooling on the southern part of the west coast (Rouault et al., 2010) that has been linked to the eastern spread of subtidal kelps (Bolton et al., 2012), while the Agulhas Current is a western boundary current experiencing particularly high rates of warming (Wu et al., 2012). The emphasis on global warming has tended to de-emphasise research on species interactions (although there is a growing recognition that these will alter critically under climate change) and focus attention on species' thermal tolerances. Some of these studies have involved laboratory physiological and modelling approaches such as dynamic energy budget modelling (Tagliarolo et al., 2016), but they have also included the measurement of physiological responses in the field (Tagliarolo and McQuaid, 2015), which is rare, and novel field approaches to conditions actually experienced at the individual level, again in line with international trends (e.g., Helmuth et al., 2010). Some of these have revealed unconventional species interactions. For example, not only does gaping behaviour by the native mussel *Perna perna* result in lower body temperatures for aggregations of individuals (Nicastro et al., 2012), but the non-gaping invasive mussel *Mytilus galloprovincialis* benefits from this effect when the two occur in mixed beds, exhibiting what could almost be called kleptothermy (Lathlean et al., 2016a, 2016b). As another example, shell-eroding endolithic cyanobacteria can be responsible for high rates of mussel mortality by causing shell collapse (Marquet et al., 2013), but by changing the albedo of mussel shells, they can also reduce mussel body temperatures and mortality through heat stress (Zardi et al., 2016).

### 13.2.3 Kelp Forests and Shallow Subtidal Temperate Reefs

The subtidal fringe kelp, *Ecklonia radiata*, occurs on the south coast of South Africa, but extensive kelp forests, comprising mostly *Ecklonia maxima* and *Laminaria pallida*, are restricted to the west and south-west coasts. There they dominate shallow subtidal temperate reefs, extending from southern Namibia in the north, to as far as Cape Agulhas on the south coast (Field et al., 1977; Velimirov et al., 1977). Towards the southern part of the west coast, kelp standing stocks increase, with *E. maxima* typically forming a floating canopy and dominating depths shallower than 9 m, while *L. pallida* forms an understorey bed and penetrates to greater depths (Field et al., 1980a; Branch and Griffiths, 1988). The south-west part of the coast, between Cape Point and the southern tip of the continent at Cape Agulhas, is considered a biogeographic transition zone (Bolton, 1986) and differs from the west coast in terms of kelp forest community composition, although recently the eastward spread of *E. maxima* beyond Cape Agulhas has been linked to a cooling trend in inshore waters (Bolton et al., 2012). The west coast has long been recognised for its large numbers of the west coast rock lobster *Jasus lalandii* (Field et al., 1980a, Branch and Griffiths, 1988), which can exert extreme predation pressure subtidally (Barkai and McQuaid, 1988), and mussels, particularly the black mussel *Choromytilus meridionalis* and the ribbed mussel *Aulacomya atra* (Van Erkom Schurink and Griffiths, 1990; Bustamante and Branch, 1996), as well as dense beds of red algae covering most of the substratum in the shallows (Anderson et al., 1997). In contrast, the south-west coast has a greater abundance of herbivores, predominantly the Cape urchin *Parechinus angulosus*, but also the abalone *Haliotis midae*, the turban snails *Turbo sarmaticus* and *Turbo cidaris* and winkles, *Oxystele* spp. (Field et al., 1980a). The mussels *C. meridionalis* and *A. atra* are scarce (Field et al., 1980b) and foliose algae decrease and are replaced by extensive beds of encrusting corallines (Anderson et al., 1997).

Productivity within these kelp forest ecosystems is enhanced by physical factors such as sun, wind and waves, which play important roles in enhancing photosynthesis and the release of particles through the erosion of kelp plants (Figure 13.4), while the upwelling/downwelling cycle is critical (Field et al., 1977 and see Branch and Griffiths, 1988 and Branch, 2008 for a review). In addition, bacteria play a central role in the mineralisation of kelp debris and faeces, which is necessary to sustain primary production (Field et al., 1977; Newell et al., 1982; Newell and Field, 1983). Indeed, a highlight of research on South African kelp beds was the contribution to the recognition of the central role of bacteria and the microbial loop in marine ecosystems, in the seminal paper by Azam et al. (1983). The most abundant grazer is the sea urchin *P. angulosus*, which feeds predominantly on drift kelp. In contrast to kelp systems in other countries

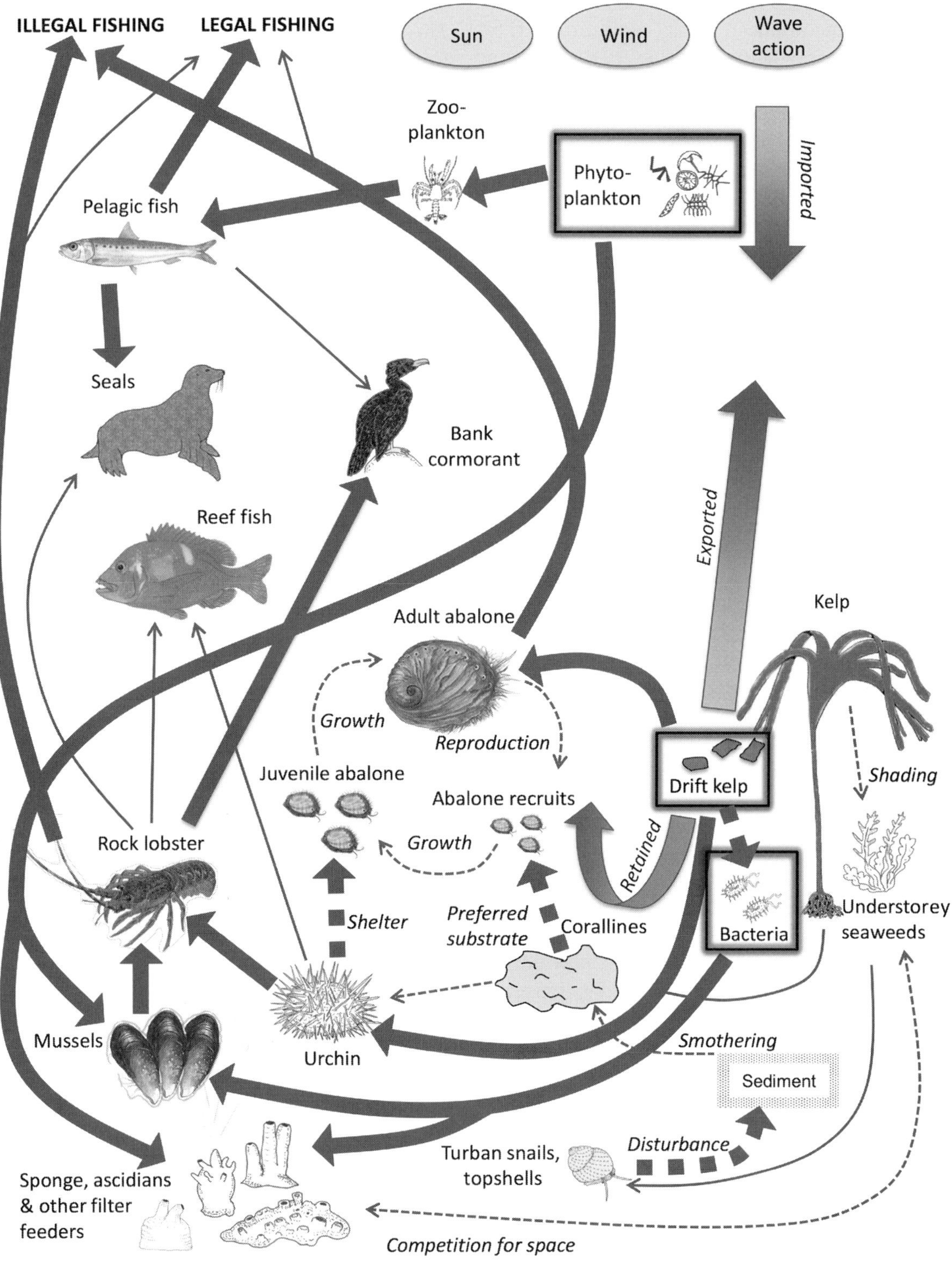


**Fig. 13.4** Kelp forest food web showing trophic (solid arrows), non-trophic interactions (broken arrows) as well as the main external biotic (boxes) and abiotic (circles) drivers. Arrow thickness indicates a subjective assessment of the strength of interactions. Adapted from Day and Branch (2002). Permission to use copyrighted images was obtained from the SAIAB and the WWF-SA.

(e.g., Duggins, 1980), urchins here have little influence on attached algae (Velimirov et al., 1977; Velimirov and Griffiths, 1979; Day and Branch, 2002), but may switch to grazing e.g., kelp sporelings (Fricke, 1979) if drift algae are scarce. While *P. angulosus* doesn't exhibit a critical role in controlling algal cover, it does play an important role in sheltering juvenile abalone, which take refuge from predators beneath its spines (Day and Branch, 2000a, 2000b).

The west coast rock lobster *J. lalandii* is a significant predator in South African temperate reef ecosystems, capable of exerting strong top-down effects on benthic communities. This is largely because they are able to feed on a wide variety of prey, but they are actively selective foragers, specifically seeking out preferred prey such as mussels and urchins (Mayfield et al., 2000a, 2000b, 2001; Haley et al., 2011). When preferred prey are in short supply, *J. lalandii* is capable of surviving on unusual food items such as barnacle recruits, sponges and even mysids (Barkai and Branch, 1988a; Haley et al., 2011). As a consequence, they can radically structure benthic communities (Barkai and Branch, 1988b; Barkai and McQuaid, 1988; Blamey et al., 2010).

Although lobster populations are considerably depleted from pristine conditions, largely due to fishing, but also due to changes in the environment (Blamey et al., 2015), there was an eastward expansion of the species during the early 1990s (Tarr et al., 1996; Cockcroft et al., 2008). This led to a regime shift in kelp ecosystems along the south-west coast. An ecosystem once dominated by urchins, abalone, other grazers and encrusting corallines was transformed into one now dominated by lobsters, sponges and understorey seaweeds (Blamey et al., 2010). Ecosystem models have been used to test hypotheses relating to ecosystem functioning in this region (Blamey et al., 2013) and have complemented empirical research, highlighting complex species interactions and a shift in focus from bottom-up control to top-down control. As in the case of rocky shores, where the effects of harvesting are more localised, largely reflecting human socio-economics, man has a key top-down role as a predator in these subtidal systems (Blamey et al., 2014).

### 13.2.4 Coral Reefs

There are few true coral reefs in South Africa and those that do occur occupy only ~40 km$^2$ along the northern KwaZulu-Natal coast in the north-east (Muthiga et al., 2008). They are considered marginal coral reefs at the extreme southern end of their distribution. They occur from ~8 m to just over 35 m depth and, instead of forming their own substrata (i.e., accretive reefs), they grow on submerged, fossilised dune and sandstone substrata (Riegl et al., 1995; Muthiga et al., 2008). They were first discovered in the 1970s (Heydorn, 1972) but it was only in the 1980s/90s that research on South African coral reefs began in earnest, with a major focus on coral taxonomy and reproduction, initiating long-term monitoring studies and describing community structure and distribution (see reviews by Schleyer 1995, 1999, 2000). While similar research initiatives have continued and extended further south, a more recent focus has been on monitoring coral communities and diversity in the face of climate and anthropogenic-induced changes (e.g., Celliers and Schleyer, 2002, 2008; Schleyer et al., 2008; Floros et al., 2013).

South African coral communities are rich in biodiversity (see review by Schleyer and Celliers, 2003), with species richness greatest among the hard corals (*Scleractinia*), although soft corals (*Alcyonacea*) dominate in terms of coverage (Muthiga et al., 2008). Communities differ along gradients of water motion, sedimentation and light, with alcyonaceans dominating shallow reefs with little sedimentation, while Scleractinians dominate deeper reefs and areas of high sedimentation e.g., deep gullies. The tabular and branching scleractinians are common on deep reefs, but become rare as light diminishes. Instead, the deepest reefs are dominated by sponges, ascidians and sea fans (Riegl et al., 1995). Descriptions of coral community patterns within and between reefs have more recently been refined by Celliers and Schleyer (2008).

A major predator of hard corals is the crown-of-thorns starfish *Acanthaster planci*, known for its ability to aggregate on corals and feed extensively, causing widespread damage to the reef (Moran, 1986). *A. planci* has been recorded on all major reefs in the region (Schleyer, 2000), but only in high densities during an aggregation (spot-outbreak) that occurred in the 1990s, albeit at a much smaller scale than elsewhere in the world (Celliers and Schleyer, 2006). This localised aggregation eventually dispersed and reefs are being monitored for further outbreaks (Schleyer and Cilliers, 2000). Other threats to corals include damage by divers and climate change. A warming of coastal waters appears to be shifting coral communities towards more hard corals, although this change might be transient given that temperatures are approaching the local coral bleaching threshold (Schleyer and Celliers, 2003).

Recent work by South African scientists on coral reefs in East Africa has used stable isotope analysis to examine the diets of both hard corals and coral reef fish (Plass-Johnson et al., 2013, 2015, 2016) and field manipulations to discern the context-dependency of the effects of fish grazing under different reef-management systems (Humphries et al., 2014, 2015).

### 13.2.5 Mangroves

South African mangroves represent some of the most southerly in the world (Giri et al., 2011), are of limited extent and, conforming to global patterns, have shown important reductions in their distribution in the last thirty years, largely through overexploitation as a source of wood (Adams et al., 2004; Rajkaran and Adams, 2007, 2010), although, where industrial developments created suitable conditions, newly established forests have progressed to maturity (Bedin, 2001). Research on these ecosystems has not benefitted from the sort of clear focus or overall ecosystem approach that we have seen in other systems and, instead, has tended to reflect the interests of individual researchers. Apart from interest in the exploitation and management of mangroves (e.g., Taylor et al., 2006; Quisthoudt et al., 2013), scientific attention has mainly focussed on the associated icthyofauna (e.g., Harris and Cyrus, 2000), the physiology of mangroves themselves (e.g., Naidoo et al., 2002; Naidoo, 2006) and mangrove crabs (e.g., Emmerson, 2001, 2016). Encouragingly, some recent studies have used South African mangroves to test general theories of ecology (e.g., Fusi et al., 2015; Kramer et al., 2015).

### 13.2.6 Seagrasses

The eelgrass, *Zostera capensis*, is the dominant seagrass in South Africa, occurring in estuaries and sheltered bays between Kosi Bay on the east coast and the Olifants Estuary on the west coast (see review by Adams, 2016). Although found across the entire South African coastline, seagrasses are most abundant on the west and south coasts, having being lost in many places along the east coast due to disturbance and/or development (Adams, 2016). Like other seagrasses, *Z. capensis* is considered an ecosystem engineer, providing habitat and shelter to other species. It also plays an important role in stabilising sediments (Siebert and Branch, 2006) and in many South African estuaries has been shown to enhance faunal biomass and diversity relative to sediments lacking vegetation (Whitfield et al., 1989; Pillay et al., 2010).

Given the importance of these ecosystems as nursery areas for many fish species, a large amount of research from South African seagrass systems has focussed on this function (e.g., Paterson and Whitfield, 2000a, 2000b). There have also been a number of studies looking at the associated benthos (see Adams, 2016 for a list of the literature), but very few report on species interactions within these ecosystems.

Along the South African south coast, benthic communities associated with *Z. capensis* beds in the Knysna Estuary are fairly well known, where studies have been conducted on their spatial variation relative to wave exposure, shore height and distance upstream (Barnes, 2010; Barnes and Ellwood, 2011, 2012). In terms of species richness and diversity within *Zostera* beds, the Knysna Estuary can be divided into two regions, with the lower, more saline parts of the estuary having a greater faunal diversity and richness than the upper reaches (Barnes and Ellwood, 2012), although this division is less obvious at the functional group level (Barnes and Hendy, 2015). Sheltered areas are dominated by two species of deposit-feeding microgastropods, while semi-exposed areas are dominated by polychaetes and other suspension feeders (Barnes, 2010).

In Langebaan Lagoon, on the west coast of South Africa, *Z. capensis* beds are mainly characterised by the mud prawn *Upogebia africana*, polychaetes, the periwinkle *Assiminea globulus* and crabs. Adjacent unvegetated sandflats are typically dominated by the sand prawn *Callianassa kraussi*, amphipods and polychaetes (Siebert and Branch, 2005b). *Z. capensis* stabilises sediments with its root system, while *C. krausii,* a bioturbator, overturns a large amount of sediment while feeding and burrowing (Siebert and Branch, 2005a; Pillay et al., 2007). Despite their contrasting roles, both play a critical part in structuring benthic communities by altering the composition of sediments, which in turn determines both their own distributions and those of other infauna (Siebert and Branch, 2005a). Experimental research has shown that *Z. capensis* and *C. kraussi* have mutually negative effects on each other and are spatially distinct (Siebert and Branch, 2006), while the mud prawn *U. africana* is frequently found in *Zostera* beds and likely benefits from the exclusion of *C. krausii* as it requires fairly stable sediment for the construction and maintenance of its U-shaped burrows (Siebert and Branch, 2005a, 2006). *Z. capensis* beds are also home to the critically endangered false limpet *Siphonaria compressa*, currently found only in Langebaan Lagoon on the west coast and the Knysna Estuary on the south coast (Allanson and Herbert, 2005; Allanson and Msizi, 2010). Given the importance of seagrass habitats, it is of concern that *Zostera* beds in Langebaan have declined by almost 40 per cent since 1960, with some sites showing up to 98 per cent loss (Pillay et al., 2010).

## 13.3 Synthesis of Recent Trends

Analysing trends in shallow littoral research over a five-year period at the end of the 1980s, McQuaid (1989) identified an emphasis on taxonomy (both algal and faunal) and ecology, while other subjects such as behaviour and evolution were poorly covered and physiology was generally dealt with in a relatively superficial manner, addressing whole organism responses rather than physiological mechanisms. Apart from the more routine inclusion of techniques such as stable isotope analysis or fatty acid analysis that were rare previously, we suggest that three broad trends have emerged in marine ecology since the 1990s. First, the scales of investigation have increased with more emphasis on rigorous experimental design and replicated experiments performed at multiple spatial scales, often across the entire coastline, to allow better generalisation and an understanding of contingency. This has included important developments in modelling approaches and more frequent inclusion of non-trophic species interactions into both experiments and models. Second, the field of marine genetics has been launched, in part as a revolution in taxonomy, with a focus on evolutionary history, phylogeography and cryptic species. Third, the importance of biological invasions has been recognised, including how they have proliferated and the degree to which they affect indigenous systems. We deal with these three areas in turn.

### 13.3.1 Scales of Investigation and Contingency

Marine ecology in South Africa has a strong tradition of field manipulation, particularly on rocky shores, but since the 1990s experimental designs and analyses have become more sophisticated and the scale of experiments has grown enormously from relatively small-scale experiments with little spatial replication to experiments that have been run around the entire coast of the country (e.g., Reaugh-Flower et al., 2011) or even extended to direct comparisons between South Africa and other countries (e.g., Fusi et al., 2015; Liversage et al., 2017). Perhaps the most important point to emerge from this spatial expansion has been the degree of contingency in a wide range of phenomena.

A number of studies have shown that what are perceived as broadscale effects may in fact be highly contingent on the local context. This is unfortunate as ecology seeks generality, but, as with so many aspects of biology, complexity seems to be the rule and even that most obvious global biological pattern of increasing species richness towards the equator still lacks a single convincing, unifying explanation (Rahbek and Graves, 2001). As later described, a simple pattern of habitat segregation between the invasive mussel *M. galloprovincialis* and the indigenous mussel *P. perna* results from complex, interacting effects, including a wide range of species interactions (McQuaid et al., 2015).

The importance of contingency has come out clearly in a set of complementary studies examining the effects of upwelling around the coast of South Africa. Cole and McQuaid (2010) found that upwelling had strong effects on the size structure of mussel beds, but not on the associated fauna found within those beds. Rather, the infauna showed strong effects of biogeography, while local and regional effects overrode those of mussel bed structure (Cole and McQuaid, 2011). Related to this, Lathlean and McQuaid (2017) found that the value of mussels as ecosystem engineers differed among biogeographic regions. Counter-intuitively, the tendency of infaunal species to make use of mussel beds proved greater in cooler regions, highlighting the context-dependent importance of habitat provision. In contrast to mussel size structure, upwelling had no effect on macroalgal morphology or on algal epifauna, which again showed strong biogeographic effects, but also responded to a latitudinal gradient in algal morphology (Mostert, 2011). In a five-year study conducted around the entire coast of South Africa and southern Namibia, Reaugh-Flower et al. (2011) showed that mussel recruitment rates differ profoundly around the coast, mirroring gradients of pelagic productivity, as expected. But they also showed that whether correlations are found between recruit and adult abundances depended on the scale of observation, being more common at very large, biogeographic or very small ($<1$ m) scales and even depended on which species was considered where two mussel species co-occurred.

Working in a different ecosystem, Cramb (2015), like Cole and McQuaid (2010), found that upwelling had a strong influence on the abundances of sandy beach species around the coast, but that the effect was highly dependent on which biogeographic region was considered. In other words, biogeography and region had the strongest effects, while the effects of upwelling were nested within these larger-scale factors. Furthermore, the influence of upwelling became weaker and more varied when species were grouped into higher taxonomic levels. The implication is that predicted changes to upwelling conditions under a regime of climate change will have strong effects on sandy beach fauna, but that the effects will be context dependent.

At a different organisational level, the influence of upwelling on the diets of filter feeders was investigated by Puccinelli et al. (2016a, 2016b) using fatty acid and stable isotope analysis. They found virtually the same result; upwelling had very strong effects on filter-feeder diets, but these effects varied with the nature of the upwelling, specifically its frequency and intensity, and were again nested within the overriding effects of biogeography. Furthermore, when the scope of this investigation was widened to include upwelling systems on other continents, clear upwelling effects were found in three countries, but not a fourth (Puccinelli et al., in press).

Lastly, a different example of contingency comes from the work of Díaz and McQuaid (2011, 2014) who illustrate contingency in our perception of species interactions. They considered the effects of grazers on macroalgal communities at different levels of algal taxonomic resolution from species through functional group to overall biomass (chlorophyll). The detection of grazing effects depended on the taxonomic resolution applied, as did the finding that grazing effects showed a degree of spatial determinism. The same spots on a shore were likely to show grazing effects that were consistently either strong or weak across years, but this was only discernible for foliose species and at the foliose functional level.

Overall, these studies showing context-dependency in a variety of systems and relationships point to the difficulty of obtaining real generality in nature.

### 13.3.2 Marine Genetics

The field of marine genetics only emerged strongly in South Africa after the 1990s and so far has tended to follow three major directions: phylogeography, including its links with biogeography; the relevance of evolutionary history to present-day population genetics and ecology; the relationships between dispersal and genetics.

### Genetic Diversity and Phylogeography

In many ways, South Africa has an enormous geographic advantage when it comes to addressing questions of biogeography and phylogeography. This is because there are three distinct regions around the coast with different biogeographic affinities. There are different views on the degree to which these can be subdivided (e.g., Emanuel et al., 1992) or in defining the transition zones between adjacent regions, but broadly the three regions identified by Stephenson (1939, 1944, 1948) are universally accepted and differences among biogeographic regions in their species composition and overall community structure are clear. As increasing numbers of species have been investigated, many have revealed strong phylogeographic structure that in some, but not all, cases can be related to the transitions between recognised biogeographic provinces.

Teske et al. (2011c) provide an overview of phylogeographic research in southern Africa. Many species that are found in two or more biogeographic regions include morphologically identical populations with strong genetic differences, implying the existence of sister taxa and reduced gene flow across biogeographic breaks (Teske et al., 2007a; von der Heyden et al., 2011). This, of course, has critical implications for understanding how the biota will respond to ongoing climate change because conspecific populations that are allopatric can have very different behavioural and physiological responses to stress (e.g., Baldanzi et al., 2015; Zardi et al., 2015). Importantly, however, phylogeographic structure is not ubiquitous. While some species exhibit phylogeographic breaks that coincide with biogeographic transitions, others lack phylogeographic structure completely or exhibit structure that does not correlate with biogeography. Some species spanning more than one biogeographic region do not exhibit phylogeographic structure (Grant and da Silva-Tatley, 1997), while others with limited potential for dispersal show phylogeographic structure even within a single region, leaving us with genetic patterns that are sometimes, but not always, correlated with dispersal capability, and the recognition that other, generally unidentified, factors can be equally important (Qhaji et al., 2015). Among these factors are issues of evolutionary history.

There is also growing recognition of the importance of intraspecific genetic diversity to the functioning of ecosystems (e.g., Reusch et al., 2005; Hughes et al., 2008; Bolnick et al., 2011) and this comes out very clearly in an example from South Africa that highlights the links among evolutionary history, genetics and the functioning of today's ecosystems. The invasive Mediterranean mussel *M. galloprovincialis* was probably introduced into South Africa in the 1970s and now occurs throughout the west coast of the country and as far east as the transition zone between the warm-temperate south and the subtropical east coasts where its distributional limit has been stable for over a decade (Robinson et al., 2005). The indigenous *P. perna*, with which *M. galloprovincialis* shows partial habitat separation along the south coast (Bownes and McQuaid, 2006), shows a phylogeographic break in exactly this area, with an east and a west lineage overlapping over a stretch of about 200 km where *M. galloprovincialis* reaches its eastern limit (Zardi et al., 2007, 2015). Genetic analyses of specimens of *P. perna* throughout its global range indicate that the species originated in the western Indian Ocean, the eastern South African lineage spreading southwards along the east coast of Africa, while the western lineage expanded through the Tethys Sea and then down the west coast of the continent. Thus, the region where they overlap in South Africa is a point of secondary contact between two relatively distantly related clades (Cunha et al., 2014). Unpublished data indicate that competition with the eastern lineage of *P. perna* contributes to setting the eastern distributional limit of *M. galloprovincialis*, providing an excellent example of the importance of history to understanding present-day ecosystems.

### Evolutionary History

An understanding of evolutionary history can clarify some of the contradictions in patterns of phylogeography described earlier. For example, Teske et al. (2011a) showed that phylogeographic structure in the pulmonate limpet genus *Siphonaria* was not correlated to mode of larval development. One planktonic developer and one direct developer showed clear phylogeographic structure across the subtropical/tropical biogeographic boundary on the east coast, while a sympatric second planktonic developer did not. They interpreted the apparent anomaly of phylogeographic structure in species with different modes of development as coming about through very different evolutionary histories, suggesting that structure in the direct developer is the result of limited gene flow from a small population following a single colonisation event, while in the planktonic developer, genetic structure arose through divergent selection and reduced gene flow between larger initial populations. The importance of historical demography is recognised in several other studies (Teske et al., 2007b; Mmonwa et al., 2015; von der Heyden et al., 2015)

**Gene Flow and Larval Dispersal**

The third major thrust in genetic studies has been the use of these techniques to elucidate patterns of gene flow and larval dispersal. Within South Africa there has been a substantial focus on the control of benthic recruitment, including larval dispersal, in line with a broad interest in 'supply-side ecology' within the discipline (Underwood and Fairweather, 1989), and some of this work occurs at the interface of ecology and genetics, clearly illustrating their interrelatedness.

Considerable ecological work has addressed the dispersal of the larvae of benthic species and how this affects settlement and recruitment rates in South Africa; such work has been done across a wide range of spatial scales (see later). The influence of coastal oceanography on larval dispersal is clearly likely to be important, but nearshore oceanography is inherently complex. Nevertheless, there have been a few successful attempts to understand and model large-scale effects of oceanography on general patterns of larval dispersal and to interpret genetic structure in this light (Zardi et al., 2011; Porri et al., 2014; Assis et al., 2015; Weidberg et al., 2015).

In addition to studies of larval behaviour (e.g., Rius et al., 2010; von der Meden et al., 2010; Porri et al., 2016), this work has included estimates of settlement, post-settlement mortality and recruitment in sessile or sedentary species conducted across a wide range of scales from 100s of metres or less (Porri et al., 2006, 2008b; von der Meden et al., 2015), through kilometre scales (Lawrie and McQuaid, 2001; McQuaid and Lawrie, 2005) to 100s of kilometres (Xavier et al., 2007; Pfaff et al., 2011; von der Meden et al., 2012) to biogeographic scales of 1000s of kilometres (Reaugh-Flower et al., 2011). There were also more ambitious attempts to track larval dispersal or to link events in the water column and on the shore. For example, McQuaid and Phillips (2000) tracked larval clouds in nearshore waters, while Porri et al. (2014) sampled larval abundances between the coast and the inshore edge of the Agulhas Current during large-scale cruises along the entire east and south coasts of the country, matching these with physical characteristics of the water column (Weidberg et al., 2015). Such approaches have seen the combination of ecological and genetic techniques, for example to identify how dispersal and selection interact to shape genetic structure (Zardi et al., 2011).

In addition to these main areas of genetic research, there has, of course, been interest in cryptic species, both indigenous (e.g., Edkins et al., 2007; von der Heyden et al., 2011) and non-indigenous (e.g., Teske et al., 2011b), and this blends into work on invasive species (e.g., Bolton et al., 2011) and on climate change. For example, in a study of invasive ascidians, Rius et al. (2014) found that the studied species were expanding their distributions regardless of either the degree of genetic variability they exhibited or their thermal tolerances, indicating that the spread of non-indigenous species (NIS) is likely to be promoted by climate change.

### 13.3.3 Invasive Species

The history of research into marine invasive species in South Africa began as recently as 1992 (Alexander et al., 2016) and the number of invasive species that have been recorded within the country has increased rapidly since then (Griffiths et al., 2009). The focus has been very much on rocky shore ecosystems, possibly because they are relatively tractable (Alexander et al., 2016), and research on the topic has followed two broad avenues: surveys including adding new species to the list; research on the consequences of invasions, including community-level effects, and species interactions between indigenous and invasive species, with a focus on the mussel *M. galloprovincialis.*

Recognition of new species has been saltatory (Mead et al., 2011) and continues as we write (Micklem et al., 2016), revealing interesting patterns among invasive species. These are most abundant on the west coast, and, on the temperate west and south coasts, invasive species are mostly of northern hemisphere origins. In contrast, the majority of invasive species on the tropical and subtropical east coast have southern hemisphere affinities. Not unexpectedly, on all coasts, the majority are found in harbours or estuaries and only four species occur on the open coast (Mead et al., 2011). Of the nearly ninety invasive species so far identified, the intertidal mussel *M. galloprovincialis* has had the most noticeable impact on indigenous communities (Robinson et al., 2005).

The initial establishment and spread of *M. galloprovincialis* on the west coast, following its arrival there in the mid 1970s is described by Branch and Steffani (2004) and it has come to dominate the mussel zone of the whole of this coast, extending into southern Namibia, with important consequences for the structure and function of the entire system, including top predators. Overall, *M. galloprovincialis* has had strong effects on mid-lower shore intertidal communities, some of which can be interpreted positively. The high fecundity and recruitment rates of the species have resulted in an increase in overall mussel biomass, which has been correlated with increases in populations of one of its predators, the African black oystercatcher, *Haematopus moquini*. Stable isotope signatures indicate that the diet of this bird differs around the coast. It feeds predominantly on a mixture of limpets and mussels along the south coast, where there is a degree of within-pair diet partitioning between the genders, males consuming more limpets and females more mussels (Kohler et al., 2009). On the west coast, the diet of both sexes is dominated by mussels, reflecting the high abundances of *M. galloprovincialis* there. The arrival and spread of the invasive mussel have been linked to concurrent increases in oystercatcher population sizes from about 4,800 in the 1980s to about 6,000 in the early twenty-first century (Hockey, 1983; Hockey et al., 2005) and a concomitant eastward expansion of its breeding range (Vernon, 2004).

Farther down the food chain, high abundances of *M. galloprovincialis* and the fact that it forms multilayered mussel beds that are architecturally more complex than the monolayered beds of the indigenous mussel species ameliorate physical stress and, by increasing habitat complexity, have led to decreased patchiness and marked changes in invertebrate communities (Robinson et al., 2007). Interestingly, the direct effects of *M. galloprovincialis* on indigenous invertebrates can be either positive or negative. In the case of the west coast limpet *Scutellastra granularis*, *M. galloprovincialis* outcompetes adults for space, but facilitates recruitment (Branch et al., 2010). Similarly, on the south coast, *M. galloprovincialis* has a weak facilitating effect on *P. perna* in the mid-mussel zone, while *P. perna* initially facilitates survival of *M. galloprovincialis* by providing protection against wave action, but later excludes it through competition (Rius and McQuaid, 2006, 2009). One striking change has resulted from competition for space between the invasive mussel and the indigenous limpet *S. argenvillei*. Before the arrival of *M. galloprovincialis*, *S. argenvillei* formed dense intertidal populations that could feed collectively by trapping kelp fronds, allowing them to reach unprecedented levels of biomass (Bustamante et al., 1995). Encroachment by the mussel has usurped space, forcing some limpets to live on mussel beds where substratum topography is much more complex and they cannot feed collectively, resulting in small-sized limpets that do not achieve sexual maturity (Steffani and Branch, 2005). The strength of the interaction between the two species depends on the degree of wave exposure, providing the limpet with spatial refuges on shores with moderate wave action (Steffani and Branch 2003a, 2003b). The effects of *M. galloprovincialis* have subsequently been modified by the arrival of later invasive species, particularly the barnacle, *Balanus glandula*, which has largely displaced it from the upper mussel zone (Sadchatheeswaran et al., 2015) and we have reached the point where the only intertidal mussels, over approximately 500 km of the west coast, are the invasive *Semimytilus algosus* and *M. galloprovincialis* (De Greef et al., 2013).

*M. galloprovincialis* appeared later on the south coast (McQuaid and Phillips, 2000). There, both the physical and the biological environment are different, so that it is in competition with a different indigenous mussel species, *P. perna*, under different abiotic conditions. There, it has had much less dramatic effects. On the south coast, *M. galloprovincialis* shows partial habitat segregation with *P. perna*, the two dominating the upper and lower mussel zones, respectively, with co-existence where the two overlap (Bownes and McQuaid, 2006). Considerable effort has been directed to understanding interactions between the two species, including their patterns of recruitment, physiological tolerances and competitive interactions. The overall conclusion is that the very simple pattern of partial habitat segregation they exhibit results from a very complex set of effects (McQuaid et al., 2015), in which trematode parasitism (Calvo-Ugarteburu and McQuaid, 1998a, 1998b), shell-boring by endoliths (Kaehler and McQuaid, 1999; Zardi et al., 2009), zone-dependent competitive hierarchies (Rius and McQuaid, 2006, 2009), different physiological tolerances of wave action (e.g., Zardi et al., 2006a) and differences in their ability to colonise free space (Erlandsson et al., 2006), among other factors, all contribute. This work produced a number of unexpected findings. Predation on adults does not contribute to habitat partitioning (Griffiths and Hockey, 1987; Branch and Steffani, 2004), although levels of both cannibalism of larvae (Porri et al., 2008) and predation of recruits (Plass-Johnson et al., 2010) are high; differences in settlement rates and movement behaviour affect responses to wave-induced freeing of space (Erlandsson et al., 2006; Nicastro et al., 2008); the link between tolerance of anaerobiosis and the need to gape the shell during aerial exposure influences temperature effects (Nicastro et al., 2012; Tagliarollo and McQuaid, 2015) and the ability to withstand sand burial (Zardi et al., 2006b, 2008); the results of direct competitive interactions between *M. galloprovincialis* and *P. perna* differ between the two genetic lineages of *Perna* as the two lineages show differences in behaviour and attachment strength (Zardi et al., 2011, 2015; Hall, 2014).

The consensus is that rates of arrival of NIS are increasing, although we must separate rates of arrival and rates of identification. So far, few have become invasive to a large scale, but many are showing range expansions. Among those that have become widespread, the mussel *M. galloprovincialis* shows complex interactions with local species and has had very dramatic effects on indigenous communities.

## 13.4 Gaps in Knowledge

Needless to say, there are important gaps in our knowledge of marine ecosystems in Southern Africa and how they are likely to respond to large-scale environmental change. Possible alterations to the Benguela and Agulhas currents will be critical, as will changes to the frequency and intensity of upwelling if wind regimes alter. Importantly, while the ecological sciences are relatively strong and marine genetics is strengthening, physiology and behaviour remain relatively weak fields, while taxonomic expertise is disappearing. Apart from differences among disciplines, regionalism and lack of collaboration across bioregions are still problems.

## 13.5 Concluding Comments

South Africa has a long history of research on marine ecosystems and retains a clear leadership role within the continent as a whole. While the research focus has inevitably remained largely within the political boundaries of the country, there has been some outreach to other African countries. Arguably, being an epicentre of research in marine ecology comes with important obligations to the rest of the continent. Overall, the discipline has followed global trends of an increasing focus on the ecosystem-level consequences of global issues associated with climate change and invasive species. Importantly, while southern African marine scientists have made a substantial empirical contribution to the discipline, the contribution at the level of theory has been weak, although this has improved with the wider possibilities for international collaborations following the democratic elections of 1994. Balanced against this are two major threats: the scientific community is very small; the political pressure to make university and government appointments on criteria other than competence and to avoid foreigners make it difficult to attract or retain skilled scientists from both within and outside the country. *Ex Africa quid nunc*?*

## Acknowledgements

This work is based upon research supported by the National Research Foundation of South Africa (Grant number 64801) through the SARChI Chairs Initiative. We are grateful to Stephen Hawkins for the opportunity to make this contribution.

### REFERENCES

Adams, J. B. (2016). Distribution and status of *Zostera capensis* in South African estuaries – a review. *South African Journal of Botany*, **107**, 63–73, http://dx.doi.org/10.1016/j.sajb.2016.07.007.

Adams, J. B., Colloty, B. M. and Bate, G. C. (2004). The distribution and state of mangroves along the coast of Transkei, Eastern Cape Province, South Africa. *Wetlands Ecology and Management*, **12**(5), 531–41.

Alexander, M. E., Simon, C. A., Griffiths, C. L. et al. (2016). Back to the future: reflections and directions of South African marine bioinvasion research. *African Journal of Marine Science*, **38**(1), 141–4.

Allanson B. and Baird, D. (2008). *Estuaries of South Africa*. Cambridge University Press, Cambridge.

Allanson, B. R. and Herbert, D. G. (2005). A newly discovered population of the critically endangered false limpet *Siphonaria compressa* Allanson, 1958 (Pulmonata: Siphonariidae), with observations on its reproductive biology. *South African Journal of Science*, **101**, 95–7.

* What next from Africa?

Allanson, B. R. and Msizi, S. C. (2010). Reproduction and growth of the endangered siphonariid limpet *Siphonaria compressa* (Pulmonata: Basommatophora). *Invertebrate Reproduction and Development*, **54**, 151–61.

Anderson, R. J., Carrick, P., Levitt, G. J. and Share, A. (1997). Holdfasts of adult kelp *Ecklonia maxima* provide refuges from grazing for recruitment of juvenile kelps. *Marine Ecology Progress Series*, **159**, 265–73.

Ansell, A.D. (1972). Distribution, growth and seasonal changes in biochemical composition for the bivalve *Donax vittatus* (da Costa) from Kames Bay, Millport. *Journal of Experimental Marine Biology and Ecology*, **10**, 137–50.

Assis, J., Zupan, M., Nicastro, K. R., Zardi, G. I., McQuaid, C. D. and Serrao, E. A. (2015). Oceanographic conditions limit the spread of a marine invader along southern African shores. *PLoS ONE*, **10** (6), e0128124, http://dx.doi.org/10.1371/journal.pone.0128124.

Azam, F., Fenchel, T., Field, J. G., Gray, J. S., Meyer-Reil, L. A. and Thingstad, F. (1983). The ecological role of water-column microbes in the sea. *Marine Ecology Progress Series*, **10**, 257–63.

Baldanzi, S., McQuaid, C. D., Cannicci, S. and Porri, F. (2013). Environmental domains and range-limiting mechanisms: testing the Abundant Centre Hypothesis using southern African sandhoppers. *PLoS ONE*, **8** (1), e54598, http://dx.doi.org/10.1371/journal.pone.0054598.

Baldanzi, S., Weidberg, N., Fusi, M., Cannicci, S., McQuaid, C. D. and Porri, F. (2015). Contrasting environments shape thermal physiology across the spatial range of the sandhopper *Talorchestia capensis*. *Oecologia*, **179**, 1067–78.

Bally, R., McQuaid, C. D. and Brown, A. C. (1984). Shores of mixed sand and rock: an unexplored marine ecosystem. *South African Journal of Science*, **80**, 500–3.

Barkai, A. and Branch, G. M. (1988a). Energy requirements for a dense population of rock lobsters *Jasus lalandii*: Novel importance of unorthodox food sources. *Marine Ecology Progress Series*, **50**(1), 83–96.

Barkai, A. and Branch, G. M. (1988b). Contrasts between the benthic communities of subtidal hard substrata at Marcus and Malgas islands: a case of alternative stable states? *South African Journal of Marine Science*, **7**(1), 117–37.

Barkai, A. and McQuaid, C. (1988). Predator–prey role reversal in a marine benthic ecosystem. *Science*, **242** (4875), 62–4.

Barnes, R. S. K. (2010). Spatial variation in abundance and diversity of the smaller surface and near-surface eelgrass-associated intertidal macrobenthos within a warm-temperate estuarine bay in the Garden Route National Park, RSA. *Aquatic Conservation*, **20**, 762–72.

Barnes, R. S. K. and Ellwood, M. D. F. (2011). The significance of shore height in intertidal macrobenthic seagrass ecology and conservation. *Aquatic Conservation: Marine and Freshwater Ecosystems*, **21**(7), 614–24.

Barnes, R. S. K. and Ellwood, M. D. F. (2012). Spatial variation in the macrobenthic assemblages of intertidal seagrass along the long axis of an estuary. *Estuarine, Coastal and Shelf Science*, **112**, 173–82.

Barnes, R. S. K. and Hendy, I. W. (2015). Seagrass-associated macrobenthic functional diversity and functional structure along an estuarine gradient. *Estuarine, Coastal and Shelf Science*, **164**, 233–43.

Bedin, T. (2001). The progression of a mangrove forest over a newly formed delta in the Umhlatuze Estuary, South Africa. *South African Journal of Botany*, **67**(3), 433–8.

Blamey, L. K., Branch, G. M. and Reaugh-Flower, K. E. (2010). Temporal changes in kelp forest benthic communities following an invasion by the rock lobster *Jasus lalandii*. *African Journal of Marine Science*, **32**(3), 481–90.

Blamey, L. K., Plagányi, É. E. and Branch, G. M. (2013). Modeling a regime shift in a kelp forest ecosystem caused by a lobster range expansion. *Bulletin of Marine Science*, **89**(1), 347–75.

Blamey, L. K., Plagányi, É. E. and Branch, G. M. (2014). Was overfishing of predatory fish responsible for a lobster-induced regime shift in the Benguela? *Ecological Modelling*, **273**, 140–50.

Blamey, L. K., Shannon, L. J., Bolton, J. J. et al. (2015). Ecosystem change in the southern Benguela and the underlying processes. *Journal of Marine Systems*, **144**, 9–29.

Bolnick, D. I., Amarasekare, P., Araújo, M. S. et al. (2011). Why intraspecific trait variation matters in community ecology. *Trends in Ecology & Evolution*, **26** (4), 183–92.

Bolton, J. J. (1986). Marine phytogeography of the Benguela upwelling region on the west coast of southern Africa: a temperature dependent approach. *Botanica Marina*, **29**(3), 251–6.

Bolton, J. J., Andreakis, N. and Anderson, R. J. (2011). Molecular evidence for three separate cryptic introductions of the red seaweed *Asparagopsis* (Bonnemaisoniales, Rhodophyta) in South Africa. *African Journal of Marine Science*, **33**(2), 263–71.

Bolton, J. J., Anderson, R. J., Smit, A. J. and Rothman, M. D. (2012). South African kelp moving eastwards: the discovery of *Ecklonia maxima* (Osbeck) Papenfuss at De Hoop Nature Reserve on the south coast of South Africa. *African Journal of Marine Science*, **34**(1), 147–51.

Bolton, J. J., Leliaert, F., De Clerck, O. et al. (2004). Where is the western limit of the tropical Indian Ocean seaweed flora? An analysis of intertidal seaweed biogeography on the east coast of South Africa. *Marine Biology*, **144**(1), 51–9.

Bolton, J. J. and Stegenga, H. (2002). Seaweed species diversity in South Africa. *South African Journal of Marine Science*, **24**(1), 9–18.

Bownes, S. J. and McQuaid, C. D. (2006). Will the invasive mussel *Mytilus galloprovincialis* Lamarck replace the indigenous *Perna perna* L. on the south coast of South Africa? *Journal of Experimental Marine Biology and Ecology*, **338**, 140–51.

Branch, G. M. (2008). Trophic interactions in subtidal rocky reefs on the west coast of South Africa. In G. M. Branch and T. R. McClanahan, eds. *Food Webs and the Dynamics of Marine Reefs*. Oxford University Press, New York, pp. 50–78.

Branch, G. M. and Griffiths, C. L. (1988). The Benguela ecosystem. Part V. The coastal zone. *Oceanography and Marine Biology Annual Review*, **26**, 395–486.

Branch, G. M. and Steffani, C. N. (2004). Can we predict the effects of alien species? A case-history of the invasion of South Africa by *Mytilus galloprovincialis* (Lamarck). *Journal of Experimental Marine Biology and Ecology*, **300**(1), 189–215.

Branch, G. M., Odendaal, F. and Robinson, T. B. (2010). Competition and facilitation between the alien mussel *Mytilus galloprovincialis* and indigenous species: moderation by wave action. *Journal of Experimental Marine Biology and Ecology*, **383**(1), 65–78.

Brown, A. C. (1964). Food relationships on the intertidal sandy beaches of the Cape Peninsula. *South African Journal of Science*, **60**(2), 35–41.

Brown, A. C., and McLachlan, A. (2010). *The Ecology of Sandy Shores*. Academic Press, Cambridge, MA.

Bustamante, R. H. and Branch, G. M. (1996). The dependence of intertidal consumers on kelp-derived organic matter on the west coast of South Africa. *Journal of Experimental Marine Biology and Ecology*, **196** (1), 1–28.

Bustamante, R. H., Branch, G. M. and Eekhout, S. (1995). Maintenance of an exceptional intertidal grazer biomass in South Africa: subsidy by subtidal kelps. *Ecology*, **76**(7), 2314–29.

Calvo-Ugarteburu, M. G. and McQuaid, C. D. (1998a). Parasitism and introduced species: epidemiology of trematodes in the intertidal mussels *Perna perna* and *Mytilus galloprovincialis*. *Journal of Experimental Marine Biology Ecology*, **220**, 47–65.

Calvo-Ugarteburu, M. G. and McQuaid, C. D. (1998b). Parasitism and invasive species: effects of digenetic trematodes on mussels. *Marine Ecology Progress Series*, **169**, 149–63.

Celliers, L. and Schleyer, M. H. (2002). Coral bleaching on high-latitude marginal reefs at Sodwana Bay, South Africa. *Marine Pollution Bulletin*, **44**(12), 1380–7.

Celliers, L. and Schleyer, M. H. (2006). Observations on the behaviour and the character of an *Acanthaster planci* (L.) aggregation in a high latitude coral community in South Africa. *Western Indian Ocean Journal of Marine Science*, **5**(1), 105–13.

Celliers, L. and Schleyer, M. H. (2008). Coral community structure and risk assessment of high-latitude reefs at Sodwana Bay, South Africa. *Biodiversity and Conservation*, **17**(13), 3097–117.

Cockcroft, A. C. and Payne, A. I. L. (1999). A cautious fisheries management policy in South Africa: the fisheries for rock lobster. *Marine Policy*, **23**(6), 587–600.

Cockcroft, A. C., Van Zyl, D. and Hutchings, L. (2008). Large-scale changes in the spatial distribution of South African West Coast rock lobsters: an overview. *African Journal of Marine Science*, **30**(1), 149–59.

Cole, V. J. and McQuaid, C. D. (2010). Bioengineers and their associated fauna respond differently to the effects of biogeography and upwelling. *Ecology*, **91**, 3549–62.

Cole, V. J. and McQuaid, C. D. (2011). Broad-scale spatial factors outweigh the influence of habitat structure on the fauna associated with a bioengineer. *Marine Ecology Progress Series*, **442**, 101–9.

Cramb, P. (2015). The influence of coastal upwelling on the biodiversity of sandy beaches in South Africa. PhD thesis, University of St Andrews.

Cunha, R. L., Nicastro, K. R., Costa, J., McQuaid, C. D., Serrão, E. A. and Zardi, G. I. (2014). Wider sampling reveals a non-sister relationship for geographically contiguous lineages of a marine mussel. *Ecology and Evolution*, **4**(11), 2070–81.

Day, E. and Branch, G. M. (2000a). Relationships between recruits of abalone *Haliotis midae*, encrusting corallines and the sea urchin *Parechinus angulosus*. *South African Journal of Marine Science*, **22**(1), 137–44.

Day, E. and Branch, G. M. (2000b). Evidence for a positive relationship between juvenile abalone *Haliotis midae* and the sea urchin *Parechinus angulosus* in the south-western Cape, South Africa. *South African Journal of Marine Science*, **22**(1), 145–56.

Day, E. and Branch, G. M. (2002). Effects of sea urchins (*Parechinus angulosus*) on recruits and juveniles of abalone (*Haliotis midae*). *Ecological Monographs*, **72**(1), 133–49.

De Greef, K., Griffiths, C. L. and Zeeman, Z. (2013). Deja vu? A second mytilid mussel, *Semimytilus algosus*, invades South Africa's west coast. *African Journal of Marine Science*, **35**(3), 307–13.

Díaz, E. R. and McQuaid, C. D. (2011). A spatially explicit approach to trophic interactions and landscape formation: patchiness in small-scale variability of grazing effects along an intertidal stress gradient. *Journal of Ecology*, **99**(2), 416–30.

Díaz, E. R. and McQuaid, C. D. (2014). Short-term spatial stability in trophic interactions. *Journal of Ecology*, **102**(5), 1138–49.

Donn, T. E. (1987). Longshore distribution of *Donax serra* in two log-spiral bays in the eastern Cape, South Africa. *Marine Ecology Progress Series*, **35**(3), 217–22.

Duggins, D. O. (1980). Kelp beds and sea otters: an experimental approach. *Ecology*, **61**(3), 447–53.

Du Preez, H. H., McLachlan, A., Marais, J. F. and Cockcroft, A. C. (1990). Bioenergetics of fishes in a high-energy surf-zone. *Marine Biology*, **106**(1), 1–2.

Edkins, M. T., Teske, P. R., Papadopoulos, I. and Griffiths, C. L. (2007). Morphological and genetic analyses suggest that southern African crown crabs, *Hymenosoma orbiculare*, represent five distinct species. *Crustaceana*, **80**(6), 667–83.

Emanuel, B. P., Bustamante, R. H., Branch, G. M, Eekhout, S. and Odendaal, F. J. (1992). A zoogeographic and functional approach to the selection of marine reserves on the west coast of South Africa. *South African Journal of Marine Science*, **12**(1), 341–54.

Emmerson, W. D. (2001). Aspects of the population dynamics of *Neosarmatium meinerti* at Mgazana, a warm temperate mangrove swamp in the East Cape, South Africa, investigated using an indirect method. In J. P. M. Paula, A. A. V. Flores and C. Fransen, eds. *Advances in Decapod Crustacean Research*. Springer, Amsterdam, pp. 221–9.

Emmerson, W. D. (2016). *A guide to, and checklist for, the Decapoda of Namibia, South Africa and Mozambique*. Cambridge Scholars Publishing, Newcastle upon Tyne.

Erlandsson, J., Pal, P. and McQuaid, C. D. (2006). Recolonisation rate differs between co-existing indigenous and invasive intertidal mussels following major disturbance. *Marine Ecology Progress Series*, **320**, 169–76.

Field, J., Jarman, N., Dieckmann, G., Griffiths, C., Velimirov, B. and Zoutendyk, P. (1977). Sun, waves, seaweed and lobsters: The dynamics of a west-coast kelp bed. *South African Journal of Science*, **73**, 7–10.

Field, J. G., Griffiths, C. L., Griffiths, R. J. et al. (1980a). Variation in structure and biomass of kelp communities along the south-west Cape coast. *Transactions of the Royal Society of South Africa*, **44**(2), 145–203.

Field, J. G., Griffiths, C. L., Linley, E. A., Carter, R. A. and Zoutendyk, P. (1980b). Upwelling in a nearshore marine ecosystem and its biological implications. *Estuarine and Coastal Marine Science*, **11**(2), 133–50.

Floros, C., Schleyer, M. H. and Maggs, J. Q. (2013). Fish as indicators of diving and fishing pressure on high-latitude coral reefs. *Ocean & Coastal Management*, **84**, 130–9.

Fricke, A. H. (1979). Kelp grazing by the common sea urchin *Parechinus angulosus* Leske in False Bay, Cape. *South African Journal of Zoology*, **14**(3), 143–8.

Fusi, M., Giomi, F., Babbini, S. et al. (2015). Thermal specialization across large geographical scales predicts the resilience of mangrove crab populations to global warming. *Oikos*, **124**(6), 784–95.

Giri, C., Ochieng, E., Tieszen, L. L. et al. (2011). Status and distribution of mangrove forests of the world using earth observation satellite data. *Global Ecology and Biogeography*, **20**(1), 154–9.

Grant, W. S. and da Silva-Tatley, F. M. (1997). Lack of genetically-subdivided population structure in *Bullia digitalis*, a southern African marine gastropod with lecithotrophic development. *Marine Biology*, **129**(1), 123–37.

Griffiths, C. L. and Hockey, P. A. R. (1987). A model describing the interactive roles of predation, competition and tidal elevation in structuring mussel populations. *South African Journal of Marine Science*, **5**(1), 547–56.

Griffiths, C. L., Mead, A. and Robinson, T. B. (2009). A brief history of marine bio-invasions in South Africa. *African Zoology*, **44**(2), 241–7.

Griffiths, C. L., Robinson, T. B., Lange, L. and Mead, A. (2010) Marine biodiversity in South Africa: an evaluation of current states of knowledge. *PLoS ONE* **5**(8), e12008, http://dx.doi.org/10.1371/journal.pone.0012008.

Haley, C. N., Blamey, L. K., Atkinson, L. J. and Branch, G. M. (2011). Dietary change of the rock lobster *Jasus lalandii* after an 'invasive' geographic shift: effects of size, density and food availability. *Estuarine, Coastal and Shelf Science*, **93**(2), 160–70.

Hall, M. B. (2014). What limits an invasive? Biotic and abiotic effects on the distribution of the invasive mussel *Mytilus galloprovincialis* on the South African coastline. MSc thesis, Rhodes University.

Harris, S. A. and Cyrus, D. P. (2000). Comparison of larval fish assemblages in three large estuarine systems, KwaZulu-Natal, South Africa. *Marine Biology*, **137**(3), 527–41.

Helmuth, B., Broitman, B. R., Yamane, L. et al. (2010). Organismal climatology: analyzing environmental variability at scales relevant to physiological stress. *Journal of Experimental Biology*, **213**, 995–1003.

Hempel, G., O'Toole, M. and Sweijd, N. (2008). *Benguela: Current of Plenty: A History of International Cooperation in Marine Science and Ecosystem Management*. Benguela Current Commission, Gunther Komnick Studio, Cape Town.

Heydorn, A. (1972). Tongaland's coral reefs-an endangered heritage. *African Wildlife*, **26**, 20–3.

Hockey, P. A. R. (1983). The distribution, population size, movements and conservation of the African Black Oystercatcher *Haematopus moquini*. *Biological Conservation*, **25**, 233–62.

Hockey, P. A. R., Dean, R. W. J., Ryan, P. G. and Maree, S. (2005). *Roberts Birds of South Africa*, seventh edn. John Voelcker, Cape Town, pp. 389–90.

Hughes, A. R., Inouye, B. D., Johnson, M. T., Underwood, N. and Vellend, M. (2008). Ecological consequences of genetic diversity. *Ecology Letters*, **11**(6), 609–23.

Humphries, A. T., McClanahan, T. R. and McQuaid, C. D. (2014). Differential impacts of coral reef herbivores on algal composition and succession in Kenya. *Marine Ecology Progress Series*, **504**, 119–32.

Humphries, A. T., McQuaid, C. D. and McClanahan, T. R. (2015). Context-dependent diversity-effects of seaweed consumption on coral reefs in Kenya. *PLoS ONE*, **10**(12), e0144204, http://dx.doi.org/10.1371/journal.pone.0144204.

Kaehler, S. and McQuaid, C. D. (1999). Lethal and sub--lethal effects of phototrophic endoliths attacking the shell of the intertidal mussel *Perna perna* (L.). *Marine Biology*, **135**, 497–503.

Kohler, S., Bonnevie, B., McQuaid, C. and Jaquemet, S. (2009). Foraging ecology of an endemic shorebird, the African Black Oystercatcher (*Haematopus moquini*) on the south–east coast of South Africa. *Estuarine, Coastal and Shelf Science*, **84**(3), 361–6.

Kramer, R., McQuaid, C. D., Vink, T. J. F., Mostert, B. P. and Wasserman, R. J. (2015). Utilization of mangrove crab-burrow micro-habitats by the goby *Redigobius dewaali*: evidence for dominance hierarchy. *Journal of Experimental Marine Biology and Ecology*, **462**, 1–7.

Lasiak, T. and Field, J. G. (1995). Community-level attributes of exploited and non-exploited rocky infratidal macrofaunal assemblages in Transkei. *Journal of Experimental Marine Biology and Ecology*, **185**, 33–53.

Lathlean, J. and McQuaid, C. D. (2017). Biogeographic variability in the value of mussel beds as ecosystem engineers on South African rocky shores. *Ecosystems*,

**20**, 568–82, http://dx.doi.org/10.1007/s10021-016-0041-8.

Lathlean, J. A., Seuront, L., McQuaid, C. D., Ng, T. P., Zardi, G. I. and Nicastro, K. R. (2016a). Cheating the locals: invasive mussels steal and benefit from the cooling effect of indigenous mussels. *PLoS ONE*, **11**(3), e0152556, http://dx.doi.org/10.1371/journal.pone.0i52556.

Lathlean, J. A., Seuront, L., McQuaid, C. D., Ng, T. P. T., Zardi, G. I. and Nicastro, K. R. (2016b). Size and position (sometimes) matter: small-scale patterns of heat-stress associated with two co-occurring mussels with different thermoregulatory behaviour. *Marine Biology*, **163**, 189, http://dx.doi.org/10.1007/s00227-016-2966-z.

Lawrie, S. M. and McQuaid, C. D. (2001). Scales of mussel bed complexity: structure, associated biota and recruitment. *Journal of Experimental Marine Biology and Ecology*, **257**, 135–61.

Liversage, K., Cole, V., Coleman, R. and McQuaid, C. D. (2017). Availability of microhabitats explains a widespread pattern and informs theory on ecological engineering of boulder reefs. *Journal of Experimental Marine Biology and Ecology*, **489**, 36–42.

Maneveldt, G. W., Chamberlain, Y. M. and Keats, D. W. (2008). A catalogue with keys to the non-geniculate coralline algae (Corallinales, Rhodophyta) of South Africa. *South African Journal of Botany*, **74**(4), 555–66.

Marquet, N., Nicastro, K. R., Gektidis, M. et al. (2013). Comparison of phototrophic shell-degrading endoliths in invasive and native populations of the intertidal mussel *Mytilus galloprovincialis*. *Biological Invasions*, **15**, 1253–72.

Mayfield, S., Atkinson, L. J., Branch, G. M. and Cockcroft, A. C. (2000a). Diet of the West Coast rock lobster *Jasus lalandii*: influence of lobster size, sex, capture depth, latitude and moult stage. *South African Journal of Marine Science*, **22**(1), 57–69.

Mayfield, S., Branch, G. M. and Cockcroft, A. C. (2000b). Relationships among diet, growth rate, and food availability for the South African rock lobster, *Jasus lalandii* (Decapoda, Palinuridea). *Crustaceana*, **73**(7), 815–34.

Mayfield, S., De Beer, E. and Branch, G. M. (2001). Prey preference and the consumption of sea urchins and juvenile abalone by captive rock lobsters (*Jasus lalandii*). *Marine and Freshwater Research*, **52**(5), 773–80.

McLachlan, A., Erasmus, T., Dye, A. H. et al. (1981b). Sand beach energetics: an ecosystem approach towards a high energy interface. *Estuarine, Coastal and Shelf Science*, **13**(1), 11–25.

McLachlan, A., Jaramillo, E., Donn, T. E. and Wessels, F. (1993). Sandy beach macrofauna communities and their control by the physical environment: a geographical comparison. *Journal of Coastal Research, SI*, **15**, 27–38.

McLachlan, A., Wooldridge, T. and Dye, A. H. (1981a). The ecology of sandy beaches in southern Africa. *South African Journal of Zoology*, **16**(4), 219–31.

McQuaid, C. D. (1989). A bibliography and analysis of trends in marine littoral research in South Africa, 1983–1988. *South African Journal of Science*, **85**, 440–6.

McQuaid, C. D. (1994). Feeding behaviour and selection of bivalve prey by *Octopus vulgaris* Cuvier. *Journal of Experimental Marine Biology Ecology* **177**, 187–202.

McQuaid, C. D. (2010). Balancing science and politics in South African marine biology. *South African Journal of Science* **106**(11/12), 1–6.

McQuaid, C. D. and Branch, G. M. (1984). Influence of sea temperature, substratum and wave exposure on rocky intertidal communities: An analysis of faunal and floral biomass. *Marine Ecology Progress Series*, **19** (1), 145–51.

McQuaid, C. D. and Branch, G. M. (1985). Trophic structure of rocky intertidal communities: response to wave action and implications for energy flow. *Marine Ecology Progress Series*, **22**(2), 153–61.

McQuaid, C. D. and Dower, K. M. (1990). Enhancement of habitat heterogeneity and species richness on rocky shores inundated by sand. *Oecologia*, **84**, 142–4.

McQuaid, C. D. and Lawrie, S. M. (2005). Supply-side ecology of the brown mussel, *Perna perna*: an investigation of spatial and temporal variation in, and coupling between, gamete release and larval supply. *Marine Biology*, **147**, 955–63.

McQuaid, C. D. and Phillips, T. E. (2000). Limited wind--driven dispersal of mussel larvae: *in situ* evidence from the plankton and the spread of the invasive species *Mytilus galloprovincialis* in South Africa. *Marine Ecology Progress Series*, **201**, 211–20.

McQuaid, C. D., Porri, F., Nicastro, K. R. and Zardi, G. I. (2015). Simple, scale-dependent patterns emerge from complex effects – an example from the intertidal mussels *Mytilus galloprovincialis* and *Perna perna*. *Oceanography and Marine Biology: An Annual Review*, **53**, 127–56.

Mead, A., Carlton, J. T., Griffiths, C. L. and Rius, M. (2011). Revealing the scale of marine bioinvasions in developing regions: a South African re-assessment. *Biological Invasions*, **13**(9), 1991–2008.

Micklem, J. M., Griffiths, C. L., Ntuli, N. and Mwale, M. (2016). The invasive Asian green mussel *Perna viridis* in South Africa: all that is green is not viridis. *African Journal of Marine Science*, **2**, 1–9.

Mmonwa, K. L, Teske, P. R, McQuaid, C. D and Barker, N. P. (2015). Historical demography of southern African patellid limpets: congruence of population expansions, but not phylogeography. *African Journal of Marine Science*, **37**(1), 11–20.

Moran, P. J. (1986). The acanthaster phenomenon. *Oceanography and Marine Biology*, **24**, 379–480.

Moreno, C. A., Lunecke, K. M. and Lépez, M. I. (1986). The response of an intertidal *Concholepas concholepas* (Gastropoda) population to protection from man in southern Chile and the effects on benthic sessile assemblages. *Oikos*, **1**, 359–64.

Mostert, B. P. (2011). Responses of intertidal macroalgae and associated fauna to interactive processes acting over multiple spatial scales. MSc thesis, Rhodes University.

Muthiga, N., Costa, A., Motta, H., Muhando, C., Mwaipopo, R. and Schleyer, M. (2008). Status of Coral Reefs in East Africa: Kenya, Tanzania, Mozambique and South Africa. In C. Wilkinson, ed. *Status of Coral Reefs of the World*. Global Coral Reef Monitoring Network and Reef and Rainforest Research Centre, Townsville, pp. 91–104.

Naidoo, G. (2006). Factors contributing to dwarfing in the mangrove *Avicennia marina*. *Annals of Botany*, **97** (6), 1095–101.

Naidoo, G., Tuffers, A. V. and von Willert, D. J. (2002). Changes in gas exchange and chlorophyll fluorescence characteristics of two mangroves and a mangrove associate in response to salinity in the natural environment. *Trees*, **16**(2–3), 140–6.

Nel, R., Campbell, E. E., Harris, L. et al. (2014). The status of sandy beach science: Past trends, progress, and possible futures. *Estuarine, Coastal and Shelf Science*, **150**, 1–10.

Newell, R. C. and Field, J. G. (1983). Relative flux of carbon and nitrogen in a kelp-dominated system. *Marine Biology Letters*, **4**(4), 249–57.

Newell, R. C., Field, J. G. and Griffiths, C. L. (1982). Energy balance and significance of microorganisms in a kelp bed community. *Marine Ecology Progress Series*, **8**, 103–13.

Nicastro, K. R., Zardi, G. I., McQuaid, C. D., Pearson, G. A. and Serrao, E. A. (2012) Love thy neighbour: group properties of gaping behaviour in mussel aggregations. *PLoS ONE*, **7**(10), e47382, http://dx.doi.org/10.1371/journal.pone.0047382.

Nicastro, K. R., Zardi, G. I., McQuaid, C. D., Teske, P. R. and Barker, N. P. (2008). Coastal topography drives genetic structure in marine mussels. *Marine Ecology Progress Series*, **368**, 189–95.

Odebrecht, C., Du Preez, D. R., Abreu, P. C. and Campbell, E. E. (2014). Surf zone diatoms: a review of the drivers, patterns and role in sandy beaches food chains. *Estuarine, Coastal and Shelf Science*, **150**, 24–35.

Oosthuizen, A. and Smale, M. J. (2003). Population biology of *Octopus vulgaris* on the temperate south-eastern coast of South Africa. *Journal of the Marine Biological Association of the UK*, **83**(3), 535–41.

Paine, R. T. (1969). A note on trophic complexity and community stability. *The American Naturalist*, **103** (929), 91–3.

Paterson, A. W. and Whitfield, A. K. (2000a). Do shallow-water habitats function as refugia for juvenile fishes?. *Estuarine, Coastal and Shelf Science*, **51**(3), 359–64.

Paterson, A. W. and Whitfield, A. K. (2000b). The ichthyofauna associated with an intertidal creek and adjacent eelgrass beds in the Kariega Estuary, South Africa. *Environmental Biology of Fishes*, **58**(2), 145–56.

Pfaff, M. C., Branch, G. M., Wieters, E. A., Branch, R. A. and Broitman, B. R. (2011). Upwelling intensity and wave exposure determine recruitment of intertidal mussels and barnacles in the southern Benguela upwelling region. *Marine Ecology Progress Series*, **425**, 141–52.

Pillay, D., Branch, G. M. and Forbes, A. T. (2007). The influence of bioturbation by the sandprawn *Callianassa kraussi* on feeding and survival of the bivalve *Eumarcia paupercula* and the gastropod *Nassarius kraussianus*. *Journal of Experimental Marine Biology and Ecology*, **344**(1), 1–9.

Pillay, D., Branch, G. M., Griffiths, C. L., Williams, C. and Prinsloo, A. (2010). Ecosystem change in a South African marine reserve (1960–2009): role of seagrass loss and anthropogenic disturbance. *Marine Ecology Progress Series*, **415**, 35–48.

Plass-Johnson, J. G., McQuaid, C. D. and Hill, J. M. (2013). Stable isotope analysis indicates a lack of inter- and intraspecific dietary redundancy among ecologically important coral reef fishes. *Coral Reefs*, **32**, 429–40.

Plass-Johnson, J. G., McQuaid, C. D. and Hill, J. M. (2015). The effects of tissue type and body size on $\delta^{13}$and $\delta^{15}$N values in parrotfish (Scaridae) from Zanzibar, Tanzania. *Journal of Applied Ichthyology*, **31**, 633–7.

Plass-Johnson, J. G., McQuaid, C. D. and Hill, J. M. (2016). Morphologically similar, coexisting hard corals (*Porites lobata* and *P. solida*) display similar trophic isotopic ratios across reefs and depths. *Marine and Freshwater Research*, **67**, 671–6.

Plass-Johnson, J. G., McQuaid, C. D. and Porri, F. (2010). Top-down effects on intertidal mussel populations: assessing two predator guilds in a South African marine protected area. *Marine Ecology Progress Series*, **411**, 149–59.

Porri, F., Jackson, J. M., von der Meden, C. E. O., Weidberg López, N. and McQuaid, C. D. (2014). The effect of mesoscale oceanographic features on the distribution of mussel larvae along the south coast of South Africa. *Journal of Marine Systems*, **132**, 162–73.

Porri, F., Jordaan, T. and McQuaid, C. D. (2008a). Does cannibalism of larvae by adults affect settlement and connectivity of mussel populations? *Estuarine Coastal and Shelf Science*, **79**, 687–93.

Porri, F., McQuaid, C. D. and Erlandsson, J. (2016). The role of recruitment and behaviour in the formation of mussel dominated assemblages: an ontogenetic and taxonomic perspective. *Marine Biology*, **163**(7), 1–10.

Porri, F., McQuaid, C. D., Lawrie, S. M. and Antrobus, S. J. (2008b). Fine-scale spatial and temporal variation in settlement of the intertidal mussel *Perna perna* indicates differential hydrodynamic delivery of larvae to the shore. *Journal of Experimental Marine Biology and Ecology*, **367**, 213–18.

Porri, F., McQuaid, C. D. and Radloff, S. (2006). Spatio-temporal variability of larval abundance and settlement of *Perna perna*: differential delivery of mussels. *Marine Ecology Progress Series*, **315**, 141–50.

Prochazka, K. and Griffiths, C. L. (1992). The intertidal fish fauna of the west coast of South Africa – species, community and biogeographic patterns. *South African Journal of Zoology*, **27**(3), 115–20.

Puccinelli, E., McQuaid, C. D. and Noyon, M. (2016a). Spatio-temporal variation in effects of upwelling on the fatty acid composition of benthic filter feeders in the Southern Benguela ecosystem: not all upwelling is equal. *PLoS ONE*, **11**(8), e0161919, http://dx.doi.org/10.1371/journal.pone.0161919.

Puccinelli, E., Noyon, M. and McQuaid, C. D. (2016b). Hierarchical effects of biogeography and upwelling shape the dietary signatures of benthic filter feeders. *Marine Ecology Progress Series*, **543**, 37–54.

Puccinelli, E., McQuaid, C. D., Dobretsov, S. and Christofoletti, R. A. Coastal upwelling affects filter-feeder stable isotope composition across three continents. *Marine Environmental Research* (in press).

Qhaji, Y., van Vuuren, B. J., Papadopoulos, I., McQuaid, C. D. and Teske, P. R. (2015). A comparison of genetic structure in two low-dispersal crabs from the Wild Coast, South Africa. *African Journal of Marine Science*, **37**(3), 345–51.

Quisthoudt, K., Adams, J., Rajkaran, A., Dahdouh-Guebas, F., Koedam, N. and Randin, C. F. (2013). Disentangling the effects of global climate and regional land-use change on the current and future distribution of mangroves in South Africa. *Biodiversity and Conservation*, **22**(6–7), 1369–90.

Rahbek, C. and Graves, G. R. (2001). Multiscale assessment of patterns of avian species richness. *Proceedings of the National Academy of Sciences*, **98**(8), 4534–9.

Rajkaran, A. and Adams, J. B. (2007). Mangrove litter production and organic carbon pools in the Mngazana Estuary, South Africa. *African Journal of Aquatic Science*, **32**(1), 17–25.

Rajkaran, A. and Adams, J. B. (2010). The implications of harvesting on the population structure and sediment characteristics of the mangroves at Mngazana Estuary, Eastern Cape, South Africa. *Wetlands Ecology and Management*, **18**(1), 79–89.

Reaugh-Flower, K., Branch, G. M., Harris, J. M. et al. (2011). Scale-dependent patterns and processes of intertidal mussel recruitment around southern Africa. *Marine Ecology Progress Series*, **434**, 101–19.

Reusch, T. B., Ehlers, A., Hämmerli, A. and Worm, B. (2005). Ecosystem recovery after climatic extremes enhanced by genotypic diversity. *Proceedings of the National Academy of Sciences of the United States of America*, **102**(8), 2826–31.

Riegl, B., Schleyer, M. H., Cook, P. J. and Branch, G. M. (1995). Structure of Africa's southernmost coral communities. *Bulletin of Marine Science*, **56**(2), 676–91.

Rius, M., Branch, G. M., Griffiths, C. L. and Turon, X. (2010). Larval settlement behaviour in six gregarious ascidians in relation to adult distribution. *Marine Ecology Progress Series*, **418**, 151–63.

Rius, M., Clusella-Trullas, S., McQuaid, C. D. et al. (2014). Range expansions across ecoregions: interactions of climate change, physiology and genetic diversity. *Global Ecology Biogeography*, **23**, 76–88.

Rius, M. and McQuaid, C. D. (2006). Wave action and competitive interaction between the invasive mussel *Mytilus galloprovincialis* and the indigenous *Perna perna* in South Africa. *Marine Biology*, **150**, 69–78.

Rius, M. and McQuaid, C. D. (2009). Facilitation and competition between invasive and indigenous mussels over a gradient of physical stress. *Basic Applied Ecology*, **10**, 607–13.

Robinson, T. B., Branch, G. M., Griffiths, C. L., Govender, A. and Hockey, P. A. (2007). Changes in South African rocky intertidal invertebrate community structure associated with the invasion of the mussel *Mytilus galloprovincialis*. *Marine Ecology Progress Series*, **340**, 163–71.

Robinson, T. B., Griffiths, C. L., McQuaid, C. D. and Rius, M. (2005). Marine alien species of South Africa – status and impacts. *African Journal of Marine Science*, **27**, 297–306.

Romer, G. S. (1990). Surf zone fish community and species response to a wave energy gradient. *Journal of Fish Biology*, **36**(3), 279–87.

Rouault, M., Pohl, P. and Penven, P. (2010). Coastal oceanic climate change and variability from 1982 to 2009 around South Africa. *African Journal of Marine Science*, **32**(2), 237–46.

Sadchatheeswaran, S., Branch, G. M. and Robinson, T. B. (2015). Changes in habitat complexity resulting from sequential invasions of a rocky shore: implications for community structure. *Biological Invasions*, **17** (6), 1799–816.

Schleyer, M. (1995). *Coral reef communities and their management in southern Africa. Sustainable development of fisheries in Africa*. FISA, Nairobi, pp. 139–40.

Schleyer, M. H. (1999). A synthesis of KwaZulu-Natal coral reef research – South African Association for Marine Biological Research, Durban. *Special Report*, **5**, 1–36.

Schleyer, M. H. (2000). South African Coral Communities. In T. McClanahan, C. Sheppard and D. Obura, eds. *Coral reefs of the Indian Ocean: Their Ecology and Conservation*. Oxford University Press, New York, pp. 83–105.

Schleyer, M. H. and Celliers, L. (2000). The Status of South African Coral Reefs. In D. Souter, D. Obura and O. Lindén, eds. *Coral Reef Degradation in the Indian Ocean*. Stockholm University, Stockholm, pp. 49–50.

Schleyer, M. H. and Celliers, L. (2003). Biodiversity on the marginal coral reefs of South Africa: what does the future hold? *Zoologische Verhandelingen*, **345**, 387–400.

Schleyer, M. H., Kruger, A. and Celliers, L. (2008). Long-term community changes on a high-latitude coral reef in the Greater St Lucia Wetland Park, South Africa. *Marine Pollution Bulletin*, **56**(3), 493–502.

Schoeman, D. S., McLachlan, A. and Dugan, J. E. (2000). Lessons from a disturbance experiment in the intertidal zone of an exposed sandy beach. *Estuarine, Coastal and Shelf Science*, **50**(6), 869–84.

Short, A. D. and Wright, L. D. (1983). Physical Variability of Sandy Beaches. In A. McLachlan and T. Erasmus, eds. *Sandy Beaches as Ecosystems*. Springer, Amsterdam, pp. 133–44.

Siebert, T. and Branch, G. M. (2005a). Interactions between *Zostera capensis*, *Callianassa kraussi* and *Upogebia africana*: deductions from field surveys in Langebaan Lagoon, South Africa. *African Journal of Marine Science*, **27**(2), 345–56.

Siebert, T. and Branch, G. M. (2005b). Interactions between *Zostera capensis* and *Callianassa kraussi*: influences on community composition of eelgrass beds and sandflats. *African Journal of Marine Science*, **27**(2), 357–73.

Siebert, T. and Branch, G. M. (2006). Ecosystem engineers: interactions between eelgrass *Zostera capensis* and the sandprawn *Callianassa kraussi* and their indirect effects on the mudprawn *Upogebia africana*. *Journal of Experimental Marine Biology and Ecology*, **338**(2), 253–70.

Sink, K. J., Branch, G. M. and Harris, J. M. (2005). Biogeographic patterns in rocky intertidal communities in KwaZulu-Natal, South Africa. *African Journal of Marine Science*, **27**(1), 81–96.

Smale, M. J. and Buchan, P. R. (1981). Biology of *Octopus vulgaris* off the east coast of South Africa. *Marine Biology*, **65**(1), 1–12.

Steffani, C. N. and Branch, G. M. (2003a). Spatial comparisons of populations of an indigenous limpet *Scutellastra argenvillei* and an alien mussel *Mytilus galloprovincialis* along a gradient of wave energy. *African Journal of Marine Science*, **25**(1), 195–212.

Steffani, C. N. and Branch, G. M. (2003b). Temporal changes in an interaction between an indigenous limpet *Scutellastra argenvillei* and an alien mussel *Mytilus galloprovincialis*: effects of wave exposure. *African Journal of Marine Science*, **25**(1), 213–29.

Steffani, C. N. and Branch, G. M. (2005). Mechanisms and consequences of competition between an alien mussel, *Mytilus galloprovincialis*, and an indigenous limpet, *Scutellastra argenvillei*. *Journal of Experimental Marine Biology and Ecology*, **317**(2), 127–42.

Stephenson, T. A. (1939). The constitution of the intertidal fauna and flora of South Africa. Part I, *Journal of the Linnean Society*, **40**, 487–536.

Stephenson, T. A. (1944). The constitution of the intertidal fauna and flora of South Africa. Part II, Annals of the Natal Museum, **10**, 261–358.

Stephenson, T. A. (1948). The constitution of the intertidal fauna and flora of South Africa. Part III, Annals of the Natal Museum, **11**, 207–324.

Stephenson, T. A. and Stephenson, A. (1949). The universal features of zonation between tide-marks on rocky coasts. *The Journal of Ecology*, **37**(2), 289–305.

Tagliarolo, M. and McQuaid, C. D. (2015). Sub-lethal and sub-specific temperature effects are better predictors of mussel distribution than thermal tolerance. *Marine Ecology Progress Series*, **535**, 145–59.

Tagliarolo, M., Montalto, V., Sarà, G., Lathlean, L. A. and McQuaid, C. D. (2016). Low temperature trumps high food availability to determine the distribution of intertidal mussels Perna perna in southern Africa. *Marine Ecology Progress Series*, **558**, 51–63.

Talbot, M. M., Bate, G. C. and Campbell, E. E. (1990). A review of the ecology of surf-zone diatoms, with special reference to *Anaulus australis*. *Oceanography and Marine Biology Annual Review*, **28**, 155–75.

Tarr, R. J. Q., Williams, P. V. G. and MacKenzie, A. J. (1996). Abalone, sea urchins and rock lobster: a possible ecological shift that may affect traditional fisheries. *South African Journal of Marine Science*, **17**(1), 319–23.

Taylor, R., Adams, J. B. and Haldorsen, S. (2006). Primary habitats of the St Lucia Estuarine System, South Africa, and their responses to mouth management. *African Journal of Aquatic Science*, **31**(1), 31–41.

Teske, P. R., Froneman, P. W., Barker, N. P. and McQuaid, C. D. (2007a). Phylogeographic structure of the caridean shrimp *Palaemon peringueyi* in South Africa: further evidence for intraspecific genetic units associated with marine biogeographic provinces. *African Journal of Marine Science*, **29**(2), 253–8.

Teske, P. R., Hamilton, H., Matthee, C. A. and Barker, N. P. (2007b). Signatures of seaway closures and founder dispersal in the phylogeny of a circumglobally distributed seahorse lineage. *BMC Evolutionary Biology*, **7** (1), https://doi.org/10.1186/1471-2148-7-138.

Teske, P. R., Papadopoulos, I., Mmonwa, K. L. et al. (2011a). Climate-driven genetic divergence of limpets with different life histories across a southeast African marine biogeographic disjunction: different processes, same outcome. *Molecular Ecology*, **20**(23), 5025–41.

Teske, P. R., Rius, M., McQuaid, C. D. et al. (2011b). "Nested" cryptic diversity in a widespread marine ecosystem engineer: a challenge for detecting biological invasions. *BMC Evolutionary Biology*, **11**, 176, http://dx.doi.org/10.1186/1471-2148-11-176.

Teske, P. R., von der Heyden, S., McQuaid, C. D. and Barker, N. P. (2011c). A review of marine phylogeography in southern Africa. *South African Journal of Science*, **107**, 45–53.

Underwood, A. J. and Fairweather, P. G. (1989). Supply-side ecology and benthic marine assemblages. *Trends in Ecology & Evolution*, **4**(1), 16–20.

Van Erkom Schurink, C. and Griffiths, C. L. (1990). Marine mussels of southern Africa–their distribution patterns, standing stocks, exploitation and culture. *Journal of Shellfish Research* **9**(1), 75–85.

Velimirov, B., Field, J. G., Griffiths, C. L. and Zoutendyk, P. (1977). The ecology of kelp bed communities in the Benguela upwelling system. *Helgoländer wissenschaftliche Meeresuntersuchungen*, **30**(1), 495–518.

Velimirov, B. and Griffiths, C. L. (1979). Wave-induced kelp movement and its importance for community structure. *Botanica Marina*, **22**(3), 169–72.

Vernon, C. J. (2004) Status and abundance of the African black oystercatcher (*Haematopus moquini*) at the eastern limit of its breeding range. *Ostrich*, **75**, 243–9.

von der Heyden, S., Bowie, R. C., Prochazka, K., Bloomer, P., Crane, N. L. and Bernardi, G. (2011). Phylogeographic patterns and cryptic speciation across oceanographic barriers in South African intertidal fishes. *Journal of Evolutionary Biology*, **24**(11), 2505–19.

von der Heyden, S., Toms, J. A., Teske, P. R., Lamberth, S. J. and Holleman, W. (2015). Contrasting signals of genetic diversity and historical demography between two recently diverged marine and estuarine fish species. *Marine Ecology Progress Series*, **526**, 157–67.

von der Meden, C. E. O., Cole, V. J. and McQuaid, C. D. (2015). Do the threats of predation and competition alter larval behaviour and selectivity at settlement under field conditions? *Journal of Experimental Marine Biology and Ecology*, **471**, 240–6.

von der Meden, C. E. O., Porri, F., McQuaid, C. D., Faulkner, K. and Robey, J. (2010). Fine-scale ontogenetic shifts in settlement behaviour of mussels: changing responses to biofilm and conspecific settler presence in *Mytilus galloprovincialis* and *Perna perna*. *Marine Ecology Progress Series*, **411**, 161–71.

von der Meden, C. E. O., Porri, F., Radloff, S. and McQuaid, C. D. (2012). Settlement intensification and coastline topography: understanding the role of habitat availability in the pelagic-benthic transition. *Marine Ecology Progress Series*, **459**, 63–71.

Weidberg, N., Porri, F., von der Meden, C., Jackson, J. M., Goschen, W. and McQuaid, C. D. (2015). Mechanisms of nearshore retention and offshore export of mussel larvae over the Agulhas Bank. *Journal of Marine Systems*, **144**, 70–80.

Whitfield, A. K., Beckley, L. E., Bennett, B. A. et al. (1989). Composition, species richness and similarity of ichthyofaunas in eelgrass *Zostera capensis* beds of southern Africa. *South African Journal of Marine Science*, **8**(1), 251–9.

Wu, L., Cai, W., Zhang, L. et al. (2012). Enhanced warming over the global subtropical western boundary currents. *Nature Climate Change*, **2**(3), 161–6.

Xavier, B. M., Branch, G.M. and Wieters, E. (2007). Abundance, growth and recruitment of *Mytilus galloprovincialis* on the west coast of South Africa in relation to upwelling. *Marine Ecology Progress Series*, **346**, 189–201.

Zardi, G.I., McQuaid, C. D., Teske, P. R. and Barker, N. P. (2007). Unexpected genetic structure of indigenous (*Perna perna*) and invasive (*Mytilus galloprovincialis*) mussel populations in South Africa. *Marine Ecology Progress Series*, **337**, 135–44.

Zardi, G. I., Nicastro, K. R., McQuaid, C. D. et al. (2015). Intraspecific genetic lineages of a marine mussel show behavioural divergence and spatial segregation over a tropical/subtropical biogeographic transition. *BMC Evolutionary Biology*, **15**, 100, http://dx.doi.org/10.1186/s12862-015-0366-5.

Zardi, G. I., Nicastro, K. R., McQuaid, C. D. and Erlandsson, J. (2008). Sand and wave induced mortality in invasive (*Mytilus galloprovincialis*) and indigenous (*Perna perna*) mussels. *Marine Biology*, **153**, 853–8.

Zardi, G. I., Nicastro, K. R., McQuaid, C. D. and Gektidis, M. (2009). Effects of endolithic parasitism on invasive and indigenous mussels in a variable physical environment. *PLoS ONE*, **4**(8), e6560, http://dx.doi.org/10.1371/journal.pone.0006560.

Zardi, G. I., Nicastro, K. R., McQuaid, C. D., Hancke, L. and Helmuth, B. (2011). The combination of selection and dispersal helps explain genetic structure in intertidal mussels. *Oecologia*, **165**, 947–58.

Zardi, G. I., Nicastro, K. R., McQuaid, C. D., Ng, T. P. T., Lathlean, J. and Seuront, L. (2016). Enemies with benefits: parasitic endoliths protect mussels against heat stress. *Scientific Reports*. **6**, 31413, http://dx.doi.org/10.1038/srep31413.

Zardi, G. I., Nicastro, K. R., McQuaid, C. D., Rius, M. and Porri, F. (2006a). Hydrodynamic stress as a determinant factor in habitat segregation between the indigenous mussel *Perna perna* and the invasive *Mytilus galloprovincialis* in South Africa. *Marine Biology*, **150**, 79–88.

Zardi, G. I., Nicastro, K. R., Porri, F. and McQuaid, C. D. (2006b). Sand stress as a non-determinant of habitat segregation of indigenous (*Perna perna*) and invasive (*Mytilus galloprovincialis*) mussels in South Africa. *Marine Biology*, **148**, 1031–8.

Chapter 14

# Rocky Shores of Mainland China, Taiwan and Hong Kong

## *Past, Present and Future*

Gray A. Williams, Benny K. K. Chan and Yun-Wei Dong

## 14.1 Introduction

### 14.1.1 The Coastline of Mainland China, Taiwan and Hong Kong

The coastlines of the north-west Pacific are relatively poorly understood when compared to their American counterparts on the north-east coast which have featured in a variety of pioneering studies and influenced ecological theory, especially in terms of rocky shore ecology (Paine, 1994). Much of the north-west Pacific coastline is dominated by the coast of mainland China, which has a convex coastline running between 3 and 41°N, almost from north to south-west. As such, the coastline of this region encompasses a great diversity of intertidal habitats; ranging from southerly tropical mangroves and coral reefs to temperate rocky shores and mudflats to the north (Figure 14.1). The present review attempts to synthesise the status of our knowledge of the rocky intertidal coasts of this region. The geographical names in this region can be politically sensitive, as such, the aim of this chapter is simply to compare rocky shores within these geographic areas and not to infer any political implications. In this review, the term 'mainland China' includes, geographically, continental China as well as Hainan and Zhoushan Islands; 'Taiwan' refers to the island of Taiwan, and 'Hong Kong' is the Special Administrative Region of China which is located in the southern region of mainland China (Figure 14.1). The review does not cover the shores of North Korea or South Korea, which fall within the same sea system (the Yellow Sea) to the north-east of mainland China, or northern Vietnam, although it is acknowledged that these systems are likely to be similar in their biota compared to the adjacent coasts of mainland China. The overall aim of the chapter is to review the general patterns seen, processes at play and status of rocky shore ecology along the coast of mainland China; however, it does not claim to be an exhaustive review of all the available literature in the region, but draws together key elements from both English- and Chinese-language sources.

The coast of mainland China covers ~18,000 km from Korea in the north to Vietnam in the south and includes ~6,500 islands (reviewed by Liu, 2013). As such, mainland China's coastline extends from the southern, tropical, climatic zone through to the northern, warm-temperate zone and is influenced by a variety of coastal seas and their associated currents (Figure 14.1; Liu, 2013). Within waters in the South China Sea

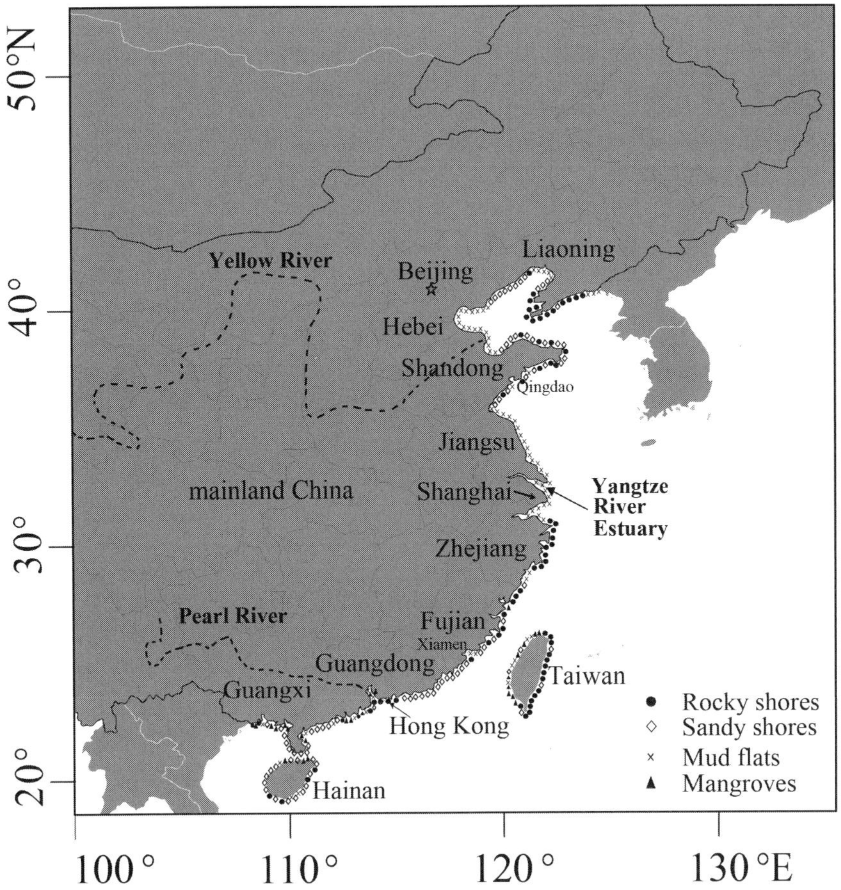


**Fig. 14.1** The coastline, coastal provinces, Pearl, Yangtze and Yellow rivers and intertidal habitats of mainland China, Hong Kong and Taiwan (after Fan, 2002, Wang, 2012).

and the Pacific Ocean, bisected by the Tropic of Cancer at 25°N, lies the island of Taiwan, with a coastline of about 1,139 km, making Taiwan the biggest island in the Indo-West Pacific, with an area of ~36,000 km$^2$. Hong Kong (22°N), which was a British colony but reverted back to the People's Republic of China in 1997, is located further south of Taiwan on the eastern coast of the Pearl River delta (PRD). Although being relatively small (2,755 km$^2$, with a land area of 1,106 km$^2$), Hong Kong has a very complex environment, being composed of a mainland peninsula and over 260 islands, including Hong Kong Island and Lantau.

The intertidal environments and their biota in this region are, as in all parts of the world, strongly influenced by past historic events (e.g., glacial periods), climatic patterns (monsoon systems) and large-scale hydrographic processes (ocean currents) which interact with regional geology to shape coastal environments. This paper summarises the physical conditions that mould the intertidal environment and reviews how this has influenced, and what we know about, rocky shore communities along the coast of mainland China, Taiwan and Hong Kong. To do this, the review examines the influence of historic events on the development of the coastline of mainland

China and adjacent waters; the resulting present-day physical environments and patterns of species distribution and community structure on the rocky shores; what is known of their dynamics and functioning and finally reflects on the future impacts on these communities.

### 14.1.2 A Brief History of Marine Biology in Mainland China, Taiwan and Hong Kong

To appreciate the status of knowledge of the rocky shores of mainland China, Taiwan and Hong Kong it is important to understand the history and evolution of marine biology studies in the region. Scientific studies of marine biology in this region are, compared to Europe and the USA, relatively young (Liu, 2013; Luk, in press). Following the creation of the Republic of China, for example, the first Chinese scientific society (The Science Society of China) was established in 1914 at Cornell University (Wang, 2002) and the Biological Science Society of China was organised in France in 1924 (Tchang, 1946). MacPherson noted that the Science Society of China showed a 'keen interest in marine science' and published a variety of journals, and a number of marine-related programmes or societies were established from which the Marine Biological Association of China was formed at Amoy (Xiamen) University in 1930 (MacPherson, 2003). Amoy Marine Biological Station and the Qingdao Institute of Marine Biology were later established in 1934 and 1936, respectively (Luk, in press). As the foci for marine biological studies, the Marine Research Institute was established in Xiamen 1946 and the Qingdao Marine Biology Laboratory in 1950 – which went on to become the Institute of Oceanology, Chinese Academy of Sciences in 1959 – and which some authors consider marks the beginning of studies on marine biodiversity in China (see Liu, 2013; Luk, in press). The relatively contemporary establishment of these centres/stations in China reflects the more recent development of marine sciences in this region compared with, for example, other institutes such as the European marine stations at Roscoff and Naples founded in the 1870s; and the Marine Biological Laboratory at Woods Hole, USA in 1888.

Most of the research on intertidal shores from mainland China is found in the Chinese literature, some with abstracts in English. As in most parts of the world, the types of studies have tended to evolve from initial descriptions of the taxa present and their biogeographic distribution to more ecological aspects of species diversity and distribution including population patterns (e.g., abundance and biomass) and finally to more community-level studies investigating processes which drive the patterns seen (see Chapter 13). This pattern holds true in mainland China (Liu and Xu, 1963) with initial studies in the 1950s having a more taxonomic approach (reviewed by Liu, 2013), particularly regarding harmful or beneficial species related to marine resources such as fisheries (Tchang, 1959; Tseng and Chang, 1959; Liu, 2013). Notably, C. K. Tseng's work on marine algae and mariculture was groundbreaking (see Neushul and Wang, 2000), detailing the biogeography of marine algae along mainland China's coast and identifying a number of different flora largely based on the temperature tolerances of different species (Tseng and Chang, 1959, 1960). Similar work and approaches were conducted by Tchang (1959) and co-workers (1963) to identify the distribution and geographic relationships of the molluscan fauna, splitting the fauna into three main groups and noting major biogeographic breaks.

Given the historic events after the Second World War in the late 1960s to mid-1970s in mainland China, the study of marine biology was largely halted, and so has only relatively recently made further progress. To date, a literature search for rocky shore ecology publications from the China Knowledge Resource Integrated Database only returns 220 papers for the whole of mainland China (Figure 14.2). The first batch of reports about rocky shore ecology, mainly taxonomic in nature, appeared in the 1930s and publications related to rocky intertidal ecology remained at low levels from 1940 to the 1970s. Later studies (post-1970s) moved onto documenting ecological patterns (population and community-level/diversity investigations), as detailed in later sections. Since the 1950s, there were also several provincial-scale investigations

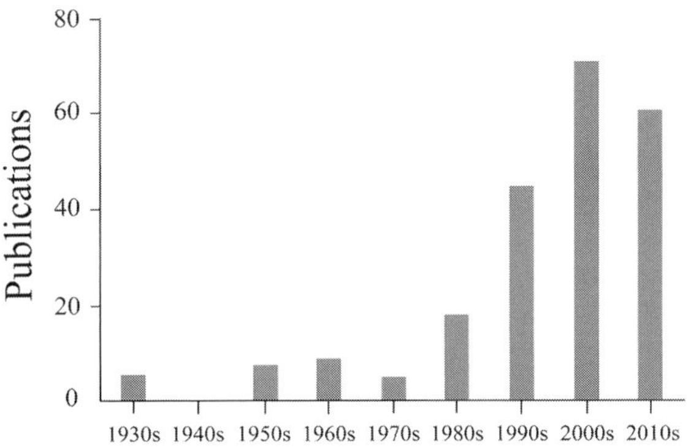


**Fig. 14.2** Publications related to rocky shore ecology in mainland China from the 1930s to 2010s. Dates were extracted from the China National Knowledge Infrastructure and Google Scholar with the key words 'ecology' and 'rocky shore' in Chinese. In total, 220 publications were found.

of the coastal regions of mainland China and, in 1980–1987 and 2005–2011, two national-scale studies were carried out to establish a detailed (presence/absence in different provinces) description of the composition and distribution of marine species along the coastline of mainland China (e.g., Huang, 1994, 2008; Liu, 2008; Huang and Lin, 2012). From the 1980s to 2010s, publications have, therefore, been increasing but most are still focussed on taxonomy and ecological surveys at certain locations of interest, often related to coastal developments or associated with marine institutes, for example Dalian, Qingdao, Zhoushan and Xiamen. As yet, there are relatively few studies dealing with community-level processes, and this is a clear knowledge gap along the coast of mainland China.

A similar evolution is true for the early work on intertidal shores conducted in Taiwan and Hong Kong. In Taiwan, early detailed studies on the diversity of marine fauna started from the nineteenth century, during the Japanese occupation. The Japanese government established the Taiwan Viceroy's Office Museum (today's National Taiwan Museum) to house collections of animals and plants. In 1910, the Natural History Society of Taiwan was established and published *The Transactions of the Natural History Society of Taiwan* in 1911 (Fan, 2012). During this period, many intertidal species including crabs and barnacles were described and reported by Japanese naturalists (e.g., Takahashi, 1934; Sato, 1936a, 1936b, 1936c; Hiro, 1939; see historical review in Chan et al., 2009) as well as extensive studies on the algae (reviewed in Lewis and Norris, 1987). In 1962, the *Bulletin of the Zoological Institute of Academia Sinica* was published (subsequently renamed as *Zoological Studies* in 1994) which also documented the biodiversity of marine species in Taiwan.

In contrast, there is a more developed history of intertidal studies in Hong Kong. Initial papers, as described earlier (from the 1930s onwards), were taxonomic in nature, focussing on crabs, algae and molluscs (detailed in Morton, 1990a and reviewed by MacPherson, 2003; Morton, 2003). Hong Kong's first marine laboratory was established in 1970, at the Chinese University of Hong Kong, and the Marine Biological Association of Hong Kong was founded in 1982. The development of intertidal and shallow-water marine ecology has a great debt to the pioneering workshops organised by B. S. Morton, devoted to understanding the marine biology of Hong Kong and the South China Sea region (summarised in Morton, 2003). These workshops, initiated in 1977, brought together international experts to provide taxonomic identification and keys for the species found in Hong Kong and to undertake research on their distribution and ecology. Much of this work was published in the proceedings emanating from these workshops, which provide fundamental knowledge of the marine biodiversity and ecology of Hong Kong's marine waters (Morton, 2003). This reference base was added to by the creation of a journal, *Asian Marine Biology*, which published regional articles between 1984 and 2001. The subsequent establishment of the Swire Marine Laboratory in 1990 (later renamed the Swire Institute of Marine Science in 1994) helped focus research into shallow-water marine systems and, importantly, stimulated studies into the processes responsible for structuring local communities.

Given this young history, there are relatively few published works, or indeed authoritative collections or identification keys, available for this large area of the north-west Pacific (Morton and Blackmore, 2001; Costello et al., 2010). As a result, species identification in the region is often challenging (see Williams et al., 2016). The 'taxonomic impediment' has also resulted in difficulties in establishing reliable species distribution ranges and hence confounded the monitoring of changes in communities over large spatial scales. Some intertidal taxa are, however, well documented (e.g., economic species such as the algae and fouling organisms). From 1979 to 2014 a total of fifty-four taxonomic volumes were published in the *Fauna Sinica* series, at least thirty of which covered marine fauna as well as seven volumes documenting marine algae (see Ma, 2015), which provided important taxonomic information and keys for identifying marine flora and fauna along the coastline of mainland China. There are also extensive collections based in the marine biological museums at the Chinese Academy of Sciences based in Qingdao and Guangzhou (see review in Liu, 2013). More detailed information is also available from Taiwan, especially an excellent series of species catalogues for the crustacean fauna (McLaughlin et al., 2007; Ahyong et al., 2008; Baba et al., 2009; Chan et al., 2009, 2013, 2014; Chan, 2010; Shih et al., 2015) and a comprehensive check list of the algae (Lewis and Norris, 1987). There are also a number of keys for Hong Kong marine species, and genetic analyses have often been used to complement morphological identification in difficult taxa (e.g., barnacles: Chan et al., 2007a, 2007b, 2007c; oysters: Lam and Morton, 2004). A recent compilation of marine species recorded from Hong Kong reveals a rich biodiversity composed of a mixture of species which occur between the tropical and temperate regions of southern mainland China (Ng et al., 2017a). More regional-based studies are now emerging, utilising large-scale collections and both molecular and morphological approaches which are revealing interesting biogeographic patterns, but also the presence of many cryptic species complexes along the coast of mainland China (e.g., littorinids: Reid, 2007; barnacles: Chan et al., 2007a, 2007b, 2007c; Tsang et al., 2012; gastropods: Dong et al., 2012; Wang et al., 2013, 2015; Li et al., 2015).

## 14.2 The Evolution of the Coastline of Mainland China, Taiwan and Hong Kong: From the Pleistocene to the Present

### 14.2.1 The Coastline during the Last Glacial Maxima

Documenting the impacts of significant historic events, such as the last glacial maximum (LGM, 24,500 BCE), is important in understanding how these events influenced geographic patterns and shaped the evolution and distribution of intertidal species along the mainland China coastline. During the Pleistocene glaciations sea level along, what is today, the coast of mainland China was lower by ~200 m compared to present-day levels (Voris, 2000). As a result, the distribution of landmasses and ocean currents in the Pleistocene were very different from today, and the continental coastlines – including mainland China – were more exposed. In the north-west Pacific, three major marginal seas (Figure 14.3, the Sea of Japan [SJ], the East China Sea [ECS] and South China Sea [SCS]) are believed to have been isolated during the glacial periods (Wang, 1999; Voris, 2000). The Palaeo-Tsushima Current (the branch of the Palaeo-Kuroshio Current flowing into the SJ), for example, was periodically blocked since 3.5 Mya (the Pliocene), indicating that the SJ was regularly isolated (Oba et al., 1991; Kitamura, 2008). Further, the Palaeo-Kuroshio Current did not enter the ECS (Ujiié et al., 2003) or the SJ (Gorbarenko and Southon, 2000) during the LGM. Some of the islands situated in the open ocean (e.g., Okinawa and Ogasawara) also served as refugia for marine species during the glaciation periods (Wu et al., 2014).

After the Pleistocene glaciation (especially the LGM), sea levels rose and the previously exposed coastlines along the continental shelf were flooded, creating new coastline and habitats. The marginal seas in the SJ, SCS and ECS were once again linked, creating new connections among the previously isolated habitats. This uniting of the seas enhanced population expansion and gene flow of marine populations, contributing to the present-day phylogeographic

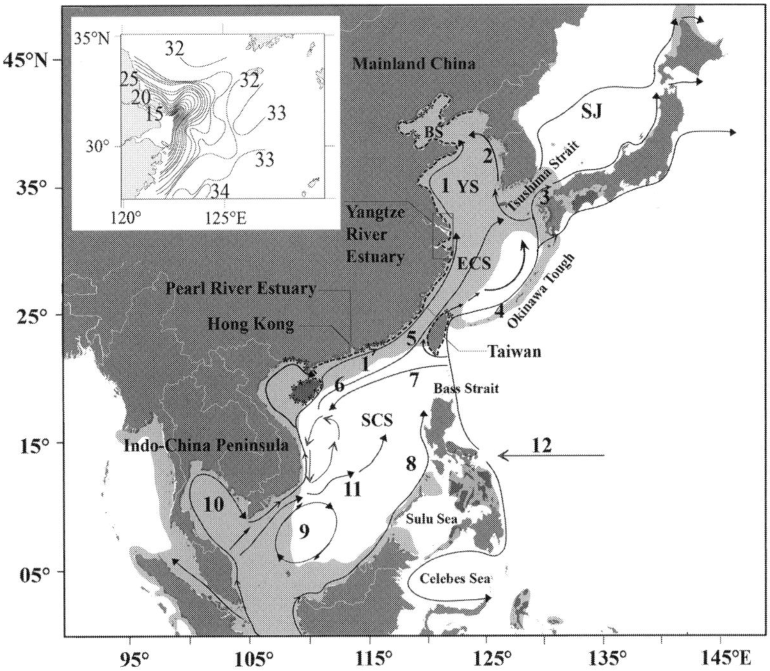


**Fig. 14.3** Marginal seas and major ocean currents affecting mainland China, Taiwan and Hong Kong (adapted after Liu, 2013; Wang et al., 2016). The inset indicates the salinity profile as a result of freshwater discharge from the Yangtze River in spring when this effectively creates a biogeographic barrier to larval transport along the mainland China coast (see Section 14.2.3). The light, shaded sea areas are continental shelves that would have been exposed during periods of low sea level (−130 m) such as in the LGM (see Section 14.2.1). Abbreviations for currents: 1, China Coastal Current; 2, Yellow Sea Warm Current; 3, Tsushima Warm Current; 4, Kuroshio Current; 5, Taiwan Warm Current; 6, SCS Warm Current; 7, SCS branch of Kuroshio; 8, north-west Luzon Coastal Current; 9, SCS Southern Anticyclonic Gyre; 10, circulation in the Gulf of Thailand; 11, south-east Vietnam Offshore Current; 12, North Equatorial Current. Arrows represent directions of major currents. Abbreviations for seas: SJ, Sea of Japan; BS, Bohai Sea; YS, Yellow Sea; ECS, East China Sea; SCS, South China Sea. Dashed lines and asterisks represent semidiurnal and diurnal tides along the coastline of mainland China and Taiwan, respectively (after Qiao, 2012).

pattern of many species seen in this region (see review in Ni et al., 2014).

### 14.2.2 The Present-Day Physical Environment: Ocean Currents, Hydrography, Tidal Cycles and the Influence of Major Rivers

Situated on the western margin of the Pacific Ocean, mainland China is fringed on its eastern and southern coasts by four marginal seas, including the small Bohai Sea (BS) and the Yellow Sea (YS) to the north, the ECS and the SCS (Figure 14.3). Taiwan and Hong Kong fall within the ECS and SCS, respectively, but these areas are all influenced by the prevailing oceanic current systems (Figure 14.3). The ocean current systems in the four marginal seas along China's coast are generally divided into two relatively independent subsystems; the BS, YS and ECS current system, and the SCS current system (Qiao, 2012).

In the YS and ECS, the circulations are dominated by the Kuroshio Current and its branches, and the water circulation is mainly influenced by wind patterns, land topography and the Changjiang River diluted water. The winds over the YS and ECS are monsoonal, being in a north-western direction in the warm and wet season and south-eastward in the cool and dry season (Ichikawai and Beardsley, 2002; Qiao, 2012) (see later discussion of seasons). As a result, the coastal currents flow southward or northward along the Chinese coast in the different seasons. The YS Warm

Current flows northwards, carrying warm and saline water from the Kuroshio to the YS, and is seasonally variable, being stronger in the cool and dry season and relatively weaker in the warm and wet season in summer (Lie and Cho, 2016).

The north and north-east coast of Taiwan are influenced by the ECS (Tseng et al., 2000). The ECS is bound to the east by the Kuroshio Current in the Pacific Ocean (Chen et al., 1995) and to the west by mainland China where it is influenced by run-off from the Yangtze River (Tian et al., 1993). The Kuroshio Current plays a very important role in defining the oceanographic conditions and indeed climate of southern mainland China, Taiwan and Hong Kong (Tseng et al., 2000). The Kuroshio Current is part of the North Equatorial Current and originates from the Philippines, flowing northwards to pass through Taiwan and reach Honshu in Japan (Muromtsev, 1970). The hydrographies of the eastern and south-eastern coastline of Taiwan are, therefore, strongly affected by the warm Kuroshio.

Another branch of the Kuroshio Current (the Kuroshio Branch Current; KBC) separates at the southern tip of Taiwan and enters the Taiwan Strait (Figure 14.3). In the warm and wet season, the KBC flows northwards along the Taiwan Strait and enters the ECS, becoming the Taiwan Warm Current. In the cool and dry season, due to the strong monsoon from northern China, the northwards flow of the KBC in the Taiwan Strait is blocked (Jan et al., 2002) and the Taiwan Strait is influenced by the southward flowing, highly productive, cold China Coastal Current. As a result, during the cool and dry season, seawater temperatures can be as low as 16°C, whereas, in the warm and wet season, seawater temperatures can reach 30°C. Compared to northern Taiwan, the east and south-east coast experiences smaller variation in annual seawater temperatures (mean seawater temperature varies between ~30°C in the warm and wet season and ~20°C in the cool and dry season).

When the Kuroshio passes through the northern tip of Luzon, a branch – the SCS branch – is diverted into the SCS (Figure 14.3; Hu et al., 2000; Hsueh and Zhong, 2004; Yang et al., 2008). This current has a major impact, bringing warm southerly waters to influence Hong Kong and southern mainland China coastlines. During the north-east monsoon, the SCS branch moves towards the eastern coast of Vietnam. Further along the south China coastline, waters are affected by the SCS Warm Current (high temperature and high salinity waters), which has a year-round, northward flow from the eastern coast of Hainan Island towards the northern coast of mainland China (Figure 14.3). In the warm and wet season, the SCS Warm Current is strong and enters the Taiwan Strait, flowing into the ECS, but its influence weakens in the cool and dry season due to the increased southwards flow of the China Coastal Current. Further south, the southern influence of the Kuroshio Current is felt by Hong Kong as it moderates cold waters in the cool and dry season from the Taiwan Current flowing from the north. The cold-water northern Taiwan Current is replaced during the south-east monsoon by the Hainan Current which brings warm waters from the south, resulting in a strong seasonality in average seawater temperatures in Hong Kong (ranging between 18 and 28°C, Morton and Morton, 1983; Morton and Blackmore, 2001).

In most regions of the BS, YS and ECS the tidal cycle is composed of regular semi-diurnal tides, but in some areas of the BS irregular diurnal tides occur, while in some areas of the YS and southern Taiwan Strait tides are irregular semi-diurnal (Qiao, 2012). Tides in Taiwan are mixed with prevalence for semi-diurnal tides. The maximum tidal range varies along the coast of Taiwan, with tidal range around the northern to north-west part of Taiwan reaching a maximum between 1.0 and 1.3 m, whereas along the eastern and south-east coastline ranges vary around 1.5–1.8 m. The west coast facing the narrow Taiwan Strait has a maximum tidal range greater than 3 m. In most regions of the SCS, tides are mainly diurnal or a combination of mixed semi-diurnal (with tides of different amplitude) and diurnal tides. Hong Kong, for example, experiences mixed semi-diurnal tides with a maximum range of ~2.8 m (Morton et al., 1996). Neap tides tend towards being diurnal, with spring tides being mixed semi-diurnal. Importantly for intertidal biota, there is an associated strong seasonal variation in tidal conditions. In the cool and dry season, the lowest tides fall during the early morning; whereas, in the hot and wet season, the lowest tides are on average ~50 cm lower and fall in the afternoon (Kaehler and Williams, 1996). This seasonal interaction of hot air temperatures and the timing of low tides means that the intertidal zone is an extremely harsh environment at this time, when rock surface temperatures can exceed 50°C and organisms are emersed for long periods (Williams, 1994; Ng et al., 2017b).

The influence of freshwater discharge and seasonal storms driven by changes in weather patterns also have a great influence on mainland China's coastal waters. Seasonal discharge from the Yangtze River, the world's third longest river (6,300 km) and fifth in terms of discharge (900 $km^3$/year in the 1970s; Beardsley et al., 1985) results in a huge plume of freshwater entering the ECS in the warm and wet season and causes a great decrease in the salinity of the upper layer of nearshore waters (Figure 14.3). Hong Kong experiences a similar effect from the Pearl River, China's third longest river (2,400 km), which discharges to the west of Hong Kong and creates a gradient in salinity from 12 to 32 per cent from the west to the east coast during the warm and wet season (Morton and Morton, 1983; Chan et al., 2001). The deltas emanating from these rivers at certain times of the year, therefore, create large plumes of fresh, sediment-rich waters, affecting extensive areas of the associated coastlines (Qiao, 2012). The discharge of the Yangtze River has, however, shown a declining trend after the construction of the Three Gorges Dams (TGDs; Dai et al., 2008), and the mean annual post-TGDs water discharge in 2003 was 67 $km^3$/year, 7.4 per cent less than the previous fifty years (1950–2002; Yang et al., 2015), which will strongly affect hydrographic patterns in the region (see Sections 14.2.3 and 14.4.1).

### 14.2.3 The Present-Day Physical Environment: Climate

The extensive coastline of mainland China covers three climatic zones, including tropical, subtropical and temperate zones. The climate in these zones is strongly affected by the prevailing monsoon conditions driven by the East Asian monsoon system (Yancheva et al., 2007; Yi, 2011; reviewed by Liu, 2013). In general, the coast of mainland China, Taiwan and Hong Kong are influenced by the interaction of two dominant air streams: oceanic air from the Pacific which brings southerly warm and moist air; and the Siberian anticyclone which emanates from the north and is characterised by cool and dry air. The seasonal strengthening and waning of the two systems means that the southerly monsoon moves northwards along the coast of mainland China from May to July, after which it is gradually replaced by the northern monsoon. As a result, much of the coastal area receives a warm (hot towards the south) and wet season when the southerly monsoon is strong; followed by a cool and dry season. Depending on prevailing conditions, the periods between monsoons can be of variable duration, and are better seen as transition periods rather than a more temperate system which would have distinct autumn and spring seasons. Associated with the change in the monsoons is the seasonal occurrence of tropical storms and typhoons. Typhoons play an important role in southern China, especially affecting Taiwan, Hong Kong, and the coasts of Guangdong, Fujian and Zhejiang provinces, bringing strong winds and extensive rainfall. As such, typhoons have dramatic impacts on seawater level and tidal currents, and bring destructive waves (Kim et al., 2014).

To the south, the tropical zone, around Hainan Island and the Leizhou peninsula, is strongly affected by the Subtropical Ridge, characterised by warm and windy conditions during the cool and dry season and hot and humid conditions in the hot and wet season. The subtropical zone is situated from the mouth of the Huai River to the Leizhou peninsula, and the temperate monsoon zone is from the mouth of the Huai River to the mouth of the Yalu River (Liu, 2013).

During the cool and dry monsoon there is a dramatic difference in air temperature among the different climate zones, from cold temperate in the north to the more tropical south. Air temperatures are, however, globally high (the average air temperature in July is between 24 and 29°C) along the whole coastline of mainland

China in the hot and wet season months, and the risk of extreme summer heat events has rapidly increased in this region over the past three decades (Qiao, 2012; Sun et al., 2014). Overall, therefore, there is a strong climatic gradient from cold-temperate to subtropical along the north to south gradient during the winter which weakens in the summer as the monsoon winds shift, and these conditions play an important role in shaping intertidal communities along the China coastline.

In Taiwan, the northern coastline also experiences great seasonal variation in temperature. The north of Taiwan experiences a subtropical climate, with a cool season (average air temperature ~16°C) and a hot season (average air temperature ~28°C), whereas the southern part of Taiwan is more tropical, having warmer winters (average air temperature ~20°C) and warmer summers (average air temperature ~28°C). In northern Taiwan, both the cool and hot seasons are wet (average monthly precipitation in January 331 mm, June 301 mm), whereas in southern and eastern Taiwan, the cool season is relatively dry (average precipitation ranges between 62 and 221 mm). Taiwan's physical location means it is often affected by strong typhoons which travel directly across the Pacific, bringing high seas, heavy rainfall and increased freshwater run-off and discharge, including terrestrial vegetation such as logs which further impact intertidal biota.

Hong Kong's location on the coast of southern China, just below the Tropic of Cancer, means that it also experiences a strongly seasonal, monsoon-driven, climate with hot and wet summers and cool, dry winters (Kaehler and Williams, 1996; Morton et al., 1996). The cool and dry season (~November–March) is dictated by the north-east monsoon (mean air temperature ~15°C) and coastal waters are relatively warm (seawater temperature ~17°C). During this period rainfall is slight, but average wind speeds are higher than in the hot and wet season, resulting in, on average, greater wave action in this season. As the south-east monsoon strengthens, however, rainfall intensity increases and Hong Kong experiences a hot and wet season (~May–September) with warm mean air temperatures (28°C) reaching maximums of 36°C. Rainfall between May and September is heavy (with extremes >145 mm/hour, June 2008), and often accompanied by intense winds (up to 160 km/hour) especially during the typhoon season (July–September). Wave action during typhoons is, as a result, extreme and can have strong impacts on the intertidal zone (Hutchinson and Williams, 2003b). The resulting heavy discharge from the PRD at this time creates a strong salinity gradient from the west to east of Hong Kong. Seawater temperatures during this period are high (~27°C) due to the influence of the southerly Hainan Current.

## 14.3 The Intertidal Biota: Patterns of Species Distribution

### 14.3.1 Biogeography, Phylogeographic Patterns and Processes Influencing Large-Scale Patterns of Species Distribution

The interaction between geology, geographic location and hydrography results in a variety of intertidal habitats along the coast of mainland China. The north of mainland China is dominated by flat coastlands of sandy bays or wide tidal flats associated with sedimentation from the major river systems such as the Yellow, Yangtze and, to the south, Pearl rivers (Figure 14.1; Liu, 2013). Sandy shores usually bisect rocky outcrops, making up approximately one third of the mainland China coastline. In contrast, the extent of natural rocky shore is decreasing, from 5,707 km in 1980 to 3,454 km in 2015 due to reclamation and coastal hardening (Xu, 2016). Towards the south, more indented vertical coastlines occur as a result of the more mountainous terrain in this area, and further to the south, mangroves and coral reefs become dominant habitat types (Figure 14.1; Liao and Zhang, 2014). Taiwan and Hong Kong also host a variety of intertidal habitats, although on a smaller geographic scale. Both are influenced by protected waters to the west and more exposed oceanic coastlines on their east coasts. The west coast of Taiwan, protected by the coast of mainland China, is mainly composed of sandy shores and mangroves, whereas the north, east and southeast coast is mostly composed of exposed rocky shores. Hong Kong also has a complex marine environment, due to its geomorphology and the PRD to the west (Morton and Morton, 1983; Chan et al., 2001). As a result, the eastern coast of Hong Kong is oceanic and mostly composed of rocky shores and beaches, while the northern coastal areas are more sheltered and, on the west coast, brackish and estuarine waters characterised by muddy shores and mangroves are abundant (Morton and Morton, 1983; Ng et al., 2017a). In terms of total area, rocky shores comprise 1.26 per cent (~1,424 ha) of the Hong Kong SAR (AFCD, 2017).

As a result of the geographic variation in climate and ocean currents over the latitudinal range of the coast of mainland China, marine diversity increases from the north to the south and can be divided into different biogeographic groups: with temperate species to the north and becoming more influenced by tropical, Indo-West Pacific species to the south (Liu, 2013; Morton and Morton, 1983; Ng et al., 2017a). Four major groups of marine fauna and flora have been proposed, running from the north to south of the waters and adjacent regions of mainland China (described by Liu, 2013). In terms of the coastal areas, Liu (2013) recognises three general characteristics of the marine fauna of the China seas: a tropical fauna found within the SCS which also influences Taiwan; a subtropical fauna in the ECS, including Hong Kong; and temperate fauna in the YS, which is similar to the descriptions of Dong and co-workers for the barnacle fauna (1980), Tchang and co-workers for the molluscan fauna (1963) and Tseng and Chang for macroalgae (1959).

Within these broadscale biogeographic patterns a number of workers have eluded to biogeographic breaks in species distribution patterns associated with large river estuaries or oceanic currents (Figure 14.4; Dong et al., 2012; Liu, 2013). Phylogeographic studies of rocky shore barnacles, for example, have revealed several biogeographic breaks in species distribution patterns (Tsang et al., 2011; Ni et al., 2014). One such break occurs between the continental barnacle populations in mainland China and the oceanic populations along the eastern island chain, including Taiwan. The acorn barnacle *Tetraclita squamosa* was, for example, believed to be widely distributed along the whole Indo-Pacific region (Foster, 1982). Based on molecular analysis, however, *T. squamosa* can be divided into a continental clade in southern China and an oceanic clade in Taiwan and Japan (Figure 14.4; Chan et al., 2007). The separation of the two clades is probably a result of the influences of two major oceanic currents: the SCS Warm Current along the coastline of mainland China and the Kuroshio Current influencing the oceanic islands including Taiwan, Okinawa and Japan (Figures 14.1 and 14.4). The oceanic clade has consistent morphological differences from the continental clades and has, therefore, been described as a new species, *Tetraclita kuroshioensis* (Chan et al., 2007). Similar patterns are also found in the highshore barnacle, *Chthamalus malayensis*, which consists of three genetically distinct populations, namely the SCS clade in mainland China, the Indo-Malay clade in the Indo-Malay Peninsula, and the oceanic Taiwan clade (Figure 14.4; Tsang et al., 2012). In both *Tetraclita* and *Chthamalus*, the continental barnacle populations are younger in geological age than the oceanic clades, as it is believed that their ranges expanded, enabling them to colonise their present habitats after the LGM (Tsang et al., 2012). Using molecular analysis, the midshore barnacle, *Tetraclita japonica japonica,* was also shown to be composed of two populations as a result of oceanographic barriers to gene flow: the northern population being common in mainland Japan and Okinawa and the southern populations present along the South China coast (Tsang et al., 2008).

A number of mobile gastropod species are also widely distributed along the rocky shores of mainland China and their phylogeographic patterns have been relatively well studied (Dong et al., 2012; Ni et al., 2014; Yu et al., 2014; Wang et al., 2015, 2016). Substratum availability, ocean currents, historical events and freshwater discharge are all important factors accounting for the biogeographic patterns seen. The life history traits of individual species largely determine their differential responses to these physical factors, and hence result in contrasting

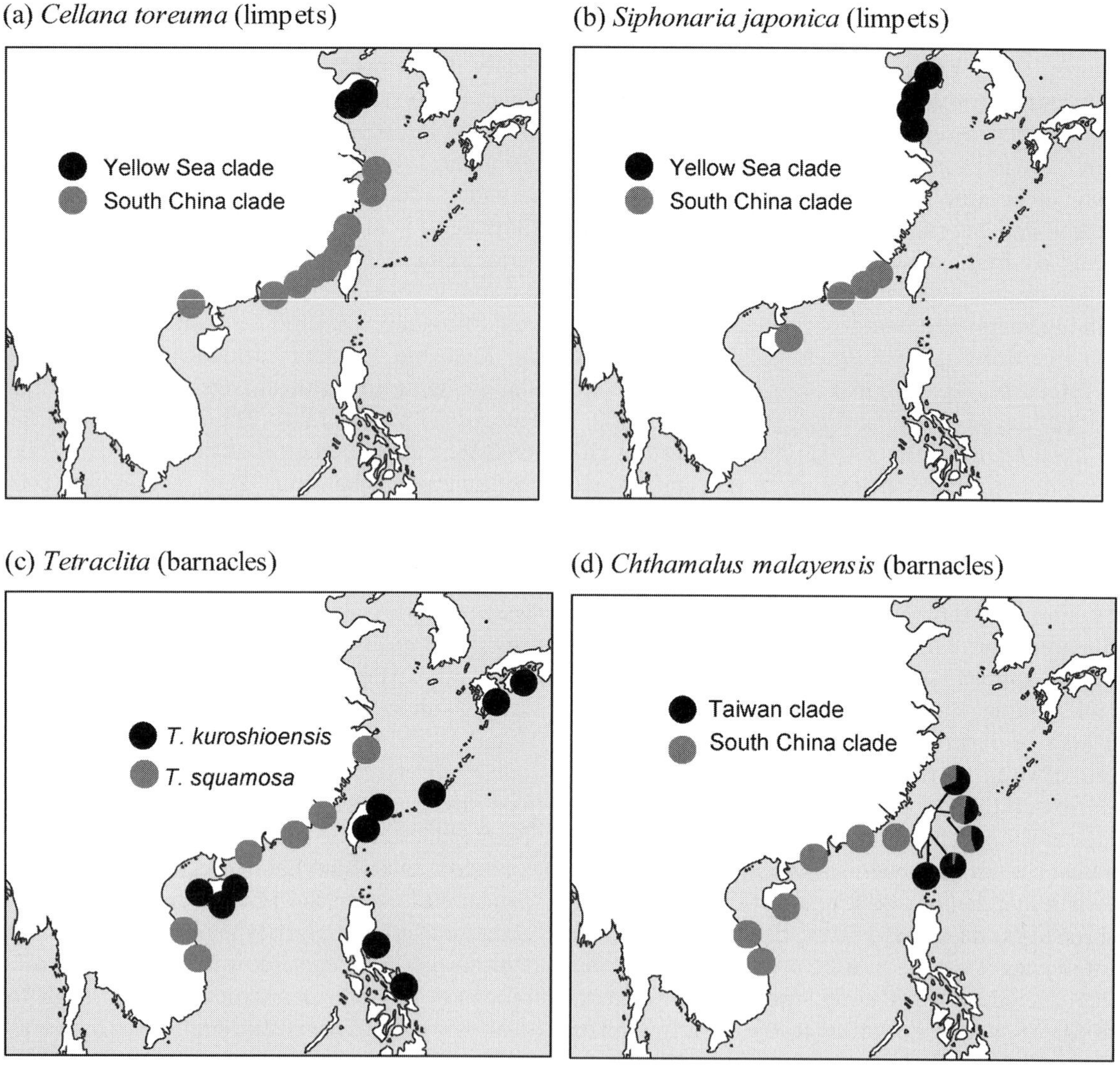


**Fig. 14.4** Representative species distribution patterns to illustrate biogeographic breaks and the influence of oceanographic currents on larval dispersal. (a) the limpet, *Cellana toreuma* (after Dong et al., 2012); (b) the pulmonate limpet, *Siphonaria japonica* (after Wang et al., 2015) and the barnacles (c) *Tetraclita* (after Chan et al., 2007) and (d) *C. malayensis* (after Tsang et al., 2012).

phylogeographic patterns across the same biogeographic barriers (Ni et al., 2014; Wang et al., 2015). There is a clear phylogeographic break for some species, for example *Cellana toreuma*, *Nipponacmea* and *Siphonaria japonica* between the YS and the other marginal seas (ECS and SCS; Figure 14.4; Dong et al., 2012; Ni et al., 2014; Wang et al., 2015, 2016). The phylogeographic break at the Yangtze River Estuary is mainly due to the unsuitable substratum (mudflats) for rocky shore species and freshwater discharge from the river. An extensive mudflat (~600 km of salt marsh) in the Yangtze River Delta, from Lianyungang, Jiangsu province to Shaoxing, Zhejiang province, limits suitable settlement sites for rocky intertidal species and consequently genetic exchange between YS and ECS populations (Figure 14.4; Dong et al., 2012). Run-off from the Yangtze River also creates a low salinity plume in the Yangtze River Estuary, which acts as a dispersal barrier for larvae of rocky shore species (Ni et al., 2017). In a similar case, molecular data from various populations of the widely distributed limpet, *C. toreuma*, reveal that populations from the YS are relatively isolated from those in the ECS and SCS, and a clear phylogeographic barrier is seen at the Yangtze River Estuary (Figure 14.4; Dong et al., 2012). A parallel disjunction has also been recorded for marine algae (e.g., *Sargassum* spp.), associated with historic isolation due to past glacial events (Cheang et al., 2010).

### 14.3.2 Present-Day Patterns and Environmental (Tidal, Wave Exposure and Climatic) Processes Influencing Local-Scale Species' Distribution

**Mainland China**

Many of the descriptive studies of species distribution patterns on shores from mainland China have adopted similar, classical approaches where investigators would survey sites, identify the species present; their abundance and biomass, and calculate diversity indices. Often these studies are of certain localities, usually sites of particular interest such as marine protected areas or offshore islands, and included a variety of intertidal habitats (e.g., north of the Minjiang River, Fujian province: Chen et al., 1989; Nanji Islands, Zhejiang province: Chen et al., 1994; Gao et al., 2007; Zhejiang Province Islands: Shao et al., 1999, 2001; Fujian coast: Zhou et al., 2001; Shengsi Archipelago, Zhejiang province: Liao et al., 2007; Meizhou Bay, Fujian province: Fang et al., 2011; Daya Bay, Guangdong province: Li et al., 1995; Fujian Islands, Fujian province: Zhou et al., 2001; Dachan Island, Zhejiang province: Peng et al., 2012; Zhoushan, Zhejiang province: Cai et al., 1993; Zhu et al., 2006, 2010; Xiao-Yangshan Island, Shanghai: Zhuang et al., 2014). Some studies focussed on certain taxonomic groups (e.g., littorinids on shores of the Jiulong River Estuary, Fujian province: Yi and Li, 1990; marine algae of the Yushan Islands, Zhejiang province: Ruan, 1994; molluscs in Xiamen, Fujian province: Li et al., 1994; mussels, south of Zhejiang province: Zhang et al., 2000; barnacles along the Jiulong River Estuary, Fujian province: Yi and Li, 1990; shellfish on Dongji Island, Zhejiang province: Cai et al., 2013) or specific groups such as fouling organisms (Huang and Tsai, 1962; Zeng et al., 1999a, 1999b, 1999c).

In most cases, however, studies were of all the taxa present on the rocky shores, investigating broadscale tidal levels (e.g., Zhoushan: Cai et al., 1993; Ruan, 1994; Shengsi Archipelago, Zhejiang province: Yang and Chen, 1996, 1998, 1999; Shenshan Island, Zhejiang province: You et al., 1997; Zhu et al., 2006; Xicun Village, Shandong province: Ma et al., 2009; Miaozihu and Huangxing Islands: Liu et al., 2012; Moye Island, Shandong province: Huang et al., 2012; Cai et al., 2013; Dayudao village, Shangdong province: Ma et al., 2013; Zhoushan Archipelago, Zhejiang province: Tang et al., 2016) or wave exposure gradients (Yi and Li, 1990; Chen et al., 1994; Li et al., 1995; Yang and Chen, 1998; Zhou et al., 2001; Cai et al., 1993). Many studies included a temporal component, often surveying shores over longer periods, usually annually, where authors described 'seasonal' changes in community composition (e.g., Yi and Li, 1990; Li et al., 1994, 1995; Ruan, 1994; You et al., 1997; Yang and Chen, 1998; Liao et al., 2007; Ma et al., 2009, 2013; Huang et al., 2012; Cai et al., 2013).

As in earlier studies from Europe and the USA, these studies from mainland China were able to identify different assemblages of species with tidal levels and with different environmental factors (wave exposure, salinity gradients etc.). It is, however, difficult to generalise further from these studies, as often the data were not replicated in space or time and, as a result, rarely were inferential statistics applied. The use of older or inaccurate taxonomic names also hinders comparisons between these studies, although in some cases more recent synonyms can be adopted (these are indicated in brackets later). These studies, therefore, provide useful qualitative information describing patterns of community structure, however, while broadscale patterns can be seen, it is difficult at this time to draw quantitative, comparative patterns between communities along the coast of mainland China.

From the more detailed of these studies, the vertical patterns of species distribution along the north to south axis of the coast of mainland

China can be summarised. In northern China, given the importance of Qingdao as a centre for marine studies, there are a number of accounts of species distributions with tidal height on different shores focussing in Jiaozhou Bay, Shandong province (Fan, 1981; Yang, 1984; Morton, 1990b). These papers describe the species present (all species in Fan, 1981 and Morton, 1990b, but mostly algae in Yang, 1984) and separate them into classical bands or zones according to species composition, but also note seasonal changes associated with the monsoons. Morton (1990b) conducted a qualitative survey of shores around Qingdao in the 1980s and concluded that they supported a rather low species diversity. More sheltered shores were dominated by the rock oyster, *Saccostrea cucullata*, probably as a result of the influence of the relatively low salinity water in Jiaozhou Bay. More exposed shores supported littorinids (*Littorina brevicula*, *Littorina articulata* and *Nodilittorina* [*Echinolittorina*] *radiata* on more wave-exposed shores) and barnacles (*Chthamalus*), below which the mussel *Xenostrobus* was found in crevices in the highshore (but note the records of *Verrucaria* probably refer to the cyanobacterium, *Kyrtuthrix maculans*). The rock oyster, *Saccostrea*, dominated the midshore with *Xenostrobus*, the predatory whelk *Thais* (*Reishia*) *clavigera* and grazers (such as *C. toreuma*, *Patellioda* [*Lottia*] *pygmea*, *Monondonta labio* and *Notoacmea* [*Nipponacmea*] *schrenki* on more sheltered shores), which fed on encrusting algae (*Hildenbrandia* and *Ralfsia*). Lower on the shore more algae were present, including *Sargassum* spp. Of particular note was the great diversity of algal species and seasonal variation in algal assemblages. The seasonal shifts in algal compositions were associated with the change in the monsoons and associated oceanic currents, with many species being temporary and restricted to different seasons in association with their relative environmental tolerances (Yang, 1984; Morton, 1990b).

In Zhoushan, Zhejiang province, central mainland China, Cai and co-workers (1990, 1991) noted a clear increase in species number from the high- to low-intertidal zone on the rocky shores, and associated changes in distribution patterns to be related to salinity and wave exposure. In the high-intertidal zone, *Littoraria scabra* (probably *Littoraria sinensis*) was recorded in wave-protected areas, whereas, in more wave-exposed areas, *Nodilittorina pyramidalis* (*Echinolittorina malaccana*) and *Nodilittorina exigua* (*E. radiata*) were the dominant species. Generally, from wave-protected to exposed shores, *L. scabra*, *Balanus albicostatus*, *S. cucullata* were gradually replaced by *N. pyramidalis* (*E. malaccana*), *N. exigua* (*E. radiata*), *C. toreuma* and *Tetraclita japonica*. In the mid-intertidal zone of more exposed shores the barnacle *Tetraclita* was abundant, along with mobile grazers, such as the limpet *Cellana*, and algae which dominated lower on the shore (Cai et al., 1990, 1991). The mussels, *Septifer virgatus* and barnacles were more important as shores became more wave exposed, while, on more sheltered shores, *Zhe* oysters (*Crassostrea* spp., see Wang et al., 2009) were prevalent. The identification of oyster species is problematic, as these are probably *Saccostrea* species (from the habitat and their location), but in many papers they are variously described as *Zhe* oysters or *Ostrea* species (Fan, 1981; Zhuang et al., 2014).

Further south in Dongshan, Fujian province, the species distribution also showed clear vertical patterns (Zhang et al., 1990). Again, in the high-intertidal zone the common species were the littorinids, *Tectarius granularis* and *Tectarius vilis* (probably *E. malaccana* and *E. radiata*), whereas the mid-intertidal zone supported a variety of species such as the barnacles, *Capitulum mitella* and *T. japonica* and *T. squamosa*, the rock oyster, *Saccostrea echinata*, tubeworms, *Serpulorbis imbricatus*; mobile grazers such as *C. toreuma*, *S. japonica* and the algae *Ishige okamurai* and *Corallina officinalis*. In the low-intertidal zone, the common species were the bivalve *Spondylus* sp., and the alga *Sargassum thunbergii*, and lower down, the coral, *Favia speciosa* and the sponge *Tethya aurantium*.

In many of the studies, temporal replication was also limited, usually surveying shores for a year (see earlier, but see examples >1 year e.g., Ruan, 1994). Researchers in China usually divide the year into a temperate four-season system (winter, spring, summer and autumn), but these are based on the Chinese lunar calendar (and, as such, the seasons are defined every three calendar months and not necessarily related to changes in environmental conditions) and compare attributes between these seasons qualitatively. Given the strong influence of the two dominant monsoon systems along the coast of mainland China (Section 14.2.3), these fixed seasonal classifications may not, however, reflect temporal changes in climate and, as a result, conclusions drawn about the seasonal changes in species assemblages related to climate are difficult to interpret with confidence.

### Taiwan

At present, to our knowledge, there are no quantitative published data on the vertical distribution patterns of intertidal species in Taiwan. Based on our observations (B. K. K. Chan, personal observations) the patterns of distribution of intertidal species from the oceanic north-east coast of Taiwan exposed rocky shores can be described in a qualitative manner. The high-intertidal zone (based on the rocky shore at He Ping Dao, north-east Taiwan) is inhabited by the littorinids *E. malaccana*, *E. radiata* and lower down the shore *Echinolittorina vidua* (Reid, 2007) and the herbivorous crab, *Grapsus albolineatus*. Below the littorinids, the green barnacle, *T. kuroshioensis,* and the pink barnacle, *Tetraclita japonica formosana,* inhabit the mid-intertidal zone (Chan et al., 2007b, 2007c, 2008). Together with the barnacles, the oyster *Saccostrea mordax* and the limpet *Lottia luchuana* are common on the mid-intertidal zone, with *Cellana grata* present in low abundance in the mid–high-intertidal zone. In the low-intertidal zone, the barnacles *Pseudoctomeris sulcata, Megabalanus volcano* and *Megabalanus tintinnabulum* are present (Chan et al., 2009). In the *Megabalanus* zone, the limpet *C. toreuma* and the Siphonariid limpet, *Siphonaria laciniosa*, are also present. Taiwan also supports a high diversity of algae (Lewis and Norris, 1987). In the mid-intertidal zone, the algae *Ulva* and *Gelidium* are common but, during the hot season, these algae die-off, leaving open rock surfaces. In the low-intertidal zone, there are pink encrusting coralline algae on some rock surfaces but most of the space is colonised by a high diversity of red, brown and green algae (see Lewis and Norris, 1987).

### Hong Kong

The descriptive patterns of Hong Kong's rocky intertidal communities are provided in Morton and Morton's definitive book, *The Seashore Ecology of Hong Kong* (1983), and further reviewed by Morton and co-workers (1996). Approximately 80 per cent of Hong Kong's complex coastline is rocky outcrops (Morton et al., 1996) which vary greatly in their exposure to wave action due to their geomorphology and location relative to the east–west estuarine gradient generated by the Pearl River Estuary. Much of the early work concentrated on the qualitative distribution of species over this east–west gradient and between shores of different habitats (Morton and Morton, 1983). Later studies in Hong Kong adopted a more quantitative approach; replicating studies at numerous sites and over longer time frames to incorporate replication of seasonal changes (including Kaehler and Williams, 1996; Hutchinson and Williams, 2001) and from a combination of this work quite a detailed overall picture of the spatial and temporal patterns on Hong Kong shores can be derived.

As with most rocky shores worldwide, the high-intertidal zone is dominated by littorinid snails, in this case *Echinolittorina* species (Ohgaki, 1983; Mak and Williams, 1999) and epilithic biofilms mostly composed of cyanobacteria (Nagarkar and Williams, 1999). A characteristic band of the black, colonial cyanobacterium, *K. maculans* can be found delineating the mid- and high-intertidal zones on sheltered and semi-exposed shores (Williams, 1993a; Kaehler and Williams, 1996). In contrast to patterns seen in temperate areas, where macroalgae dominate the mid- to low-intertidal levels, the mid-intertidal zone on sheltered rocky shores of Hong Kong is dominated by the rock oyster, *S. cucullata*, with numerous mobile gastropods (the chiton *Acanthopleura japonica*, top shell *M. labio* and other gastropods such as *Planaxis sulcatus* and limpets, *Nipponacmea* spp. and *Patelloida* (*Lottia*) spp.) utilising the matrix created by their shells for refuge (Liu, 1993). Sparse cover of the barnacles, *Euraphia withersi* and *T. japonica* are also found with patches of encrusting algae (e.g., *Hildenbrandia*) and, in the cool and dry season, *Ulva*. Lower down on the shore, patches of coralline crusts intersperse with the rock oysters and brown macroalgae, such as *Colpomenia* and *Sargassum*, which grow rapidly during the cool and dry season.

In contrast, semi-exposed shores support a sparse coverage of sessile filter feeders and on casual inspection can appear bare (as noted for tropical shores in Panama, Lubchenco et al., 1984). On closer inspection, however, Hong Kong semi-exposed shores support abundant grazing gastropods and chitons (Morton and Morton, 1983; Williams, 1993a). While the high-intertidal zone is dominated by littorinids, the mid-high zone also supports limpets (*C. grata*), and below the band of the cyanobacterium *Kyrtuthrix* there is a characteristic bare zone of rock with few algae or sessile species (Kaehler and Williams, 1996). Below this, in the mid-intertidal zone, many mobile grazers (molluscs such as limpets, nerites, topshells and chitons, as well as the herbivorous crab, *G. albolineatus*) are abundant, feeding on biofilms (again mostly cyanobacteria, Nagarkar and Williams, 1999) and encrusting algae (such as *Hapalospongidion, Ralfsia* and *Hildenbrandia*; Williams, 1993a, 1993b; Kaehler and Williams, 1996; Kennish et al., 1996) during the hot and wet season. The vertical distribution of many mobile species varies with the monsoons; with individuals migrating to live lower on the shore during the hot and wet season and returning higher on the shore during the cool and dry season (Williams and Morritt, 1995; Harper and Williams, 2001). During the hot and wet season, most mobile species take refuge when emersed in cracks and crevices on the shore, further adding to the impression that these shores are relatively bare of species (Williams, 1993a, 1994; Williams and Morritt, 1995; Ng et al., 2017b). During the cool and dry monsoon, temperate algae such as *Ulva, Enteromorpha, Endarachne, Hincksia* and *Gelidium* can be abundant in the midshore, with *Sargassum* growing rapidly in the low-intertidal–subtidal region (Hodgkiss, 1984; Williams, 1993b; Kaehler and Williams, 1996). At this time of the year, the encrusting algae also proliferate, with coralline algae growing rapidly to colonise the low- to mid-intertidal levels of semi-exposed shores (Williams, 1993b; Kaehler and Williams, 1996). Grazing by sea urchins (*Anthocidaris crassispina*) is important in the shallow subtidal, also affecting the algal assemblage on a seasonal basis (Wai and Williams, 2006).

The splash zone and high-intertidal zone of wave-exposed shores support littorinids, and also the limpets, *Colisella (Lottia) dorsuosa*, and lower down the shore *C. grata*. The herbivorous crab, *G. albolineatus* is common, ranging over the shore to forage on ephemeral macroalgae in the cool and dry season and relying on encrusting algae during the hot and wet season (Kennish et al., 1996; Kennish and Williams, 1997). In common with exposed rocky shores around the globe, sessile filter feeders dominate the rock surfaces of wave-exposed coasts. Highest on the shore, intermingled with the littorinids are the barnacles, *C. malayensis* (Yan et al., 2006), with patches of the stalked barnacle, *C. mitella*, in cracks and crevices in the high- to mid-intertidal levels. The mid-intertidal zone supports dense *Tetraclita* stands, with *T. japonica* higher on the shore than *T. squamosa* (Chan, 2001, 2003; Chan and Williams, 2004), and beds of the mussel, *S. virgatus* (Morton, 1995). As on sheltered shores dominated by oysters, these bivalves provide a matrix and habitat for many mobile species such as chitons and topshells (Ong Che and Morton, 1992; Seed and Brotohadikusomo, 1994). Below the mussels, the large barnacle, *Megabalanus* dominates, with turfs of erect coralline algae (*Corallina* spp.). During the cool and dry season, sessile species and bare rock surfaces provide attachment for many ephemeral algae, such as *Porphyra (Pyropia)* in the splash zone and *Chateomorpha, Endarachne, Ulva* and *Dermonema* in the high- to mid-intertidal zones.

On shores of all exposures, there appear to be no apparent dominant predators, such as starfish or large populations of predatory crabs. On exposed shores, the red-eyed crab (*Eriphia*) is common, but there has been little work to investigate predation impact by crabs (but see Coombes and Seed, 1992). Dog whelks in the genus *Thais (Reishia)* and *Morula (Tenguella)* are common and play an important role on both sheltered and exposed rocky shores, where their behavioural patterns and diets also shift with the seasonal changes in prey species (Tong, 1986; Taylor and Morton, 1996). Compared with shores in other tropical areas, however, there appear to be relatively few subtidal predatory/herbivorous fish and crabs, perhaps as a result of heavy local fishing pressures (see Williams, 1994).

### 14.3.3 What is Known about the Processes Influencing Species Distribution Patterns?

Although there is a growing body of studies on the structure of intertidal communities along the mainland China coast, Taiwan and Hong Kong, there are, as yet, few studies attempting to understand the processes responsible for driving the dynamics of these communities. The use of long-term studies to reveal patterns in community dynamics are also scarce. The field of experimental (manipulative) intertidal ecology, which has been a highlight of research activities in the West since the 1960s (e.g., Paine, 1994; Hawkins et al., 2017), is in its infancy along the coast of mainland China. As a result, there are few studies using a more hypothesis-driven, manipulative approach to investigate the causal factors or processes which drive the patterns seen, with most emanating from Hong Kong. The present summary is, therefore, limited in its generality as it is drawn from few studies, most of which are restricted to a small geographic area.

Summarising the studies that have been conducted, it can be seen that the seasonal shifts associated with the prevailing monsoons and oceanic currents play an important role in the dynamics of rocky shore communities along the coast of China, especially in relation to latitude. Seasonal changes in relation to rainfall, seawater temperature and especially thermal stress during emersion play key roles, again with variation in their timing and importance along the latitudinal gradient of the coast. These seasonal shifts are also associated with variation in the importance of processes such as herbivory and predation, as well as recruitment, although there are only a few localised studies of these processes.

**Seasonality and the Importance of the Monsoons: Variation in Algal Assemblages**

One clear important process is the change in the monsoon systems, with contrasting wind patterns and prevailing climate, and associated changes in oceanic currents that affect coastal waters. As such, this seasonal change affects not only the onshore physical environment, but also the supply and origin of propagules through changes in ocean currents. The warm/hot and wet monsoon sees the dominance of southerly currents, bringing more tropical species; whereas the strengthening of the cool and dry monsoon sees cooler waters from the north affecting China's coastlines.

The strengthening of the south-east monsoon and the warm temperatures and moist airstream that affects the coast from April onwards, moves northwards along the mainland China coast. This change from the cool and dry north-east monsoon can result in abrupt changes in air temperatures on rocky shores, bringing hot and desiccating conditions for organisms, especially in southern China, Hong Kong and central mainland China (Shanghai, Nantong and Xiamen). These increases in temperatures can occur quite quickly if the shift in monsoons is abrupt, resulting in a short transition period. When this rapid switch in weather is associated with calm weather, many species suffer heavy mortalities on the shore (Williams, 1993b, 1994; Liu and Morton, 1994; Williams and Morritt, 1995; Kaehler and Williams, 1996; Kennish et al., 1996; Williams et al., 2005; Chan et al., 2006; Firth and Williams, 2009 and see later).

Seasonal variation in algal assemblages has been noted along the coast of mainland China, Taiwan and especially Hong Kong (Qingdao: Morton, 1990b; Ruan, 1994; Hong Kong: Hodgkiss, 1984; Kaehler and Williams, 1996; Kennish et al., 1996). The change in northern areas is less dramatic, being more of a change in species composition, with the hotter months supporting more tropical and subtropical species (e.g., *Dictyota* and *Padina*), while colder water species (*Porphyra* (*Pyropia*), *Polysiphonia* and *Desmarestia*) dominate in the cooler months (Qingdao: Yang, 1984; Morton, 1990b). Variation in algal patterns (Ruan, 1994; Yang and Chen, 1999), and rocky shore communities in general, have been documented along the coast of mainland China (e.g., You et al., 1997; Yang and Chen, 1998; Shao et al., 2001; Ma et al., 2013); however, it is often difficult to explicitly link the changes recorded to monsoon shifts, as many papers use a simple four season classification and do not link these seasons with environmental changes.

The most extreme changes in rocky shore communities are seen in the south. In Taiwan, although the rocky shores in summer months can still support a high abundance of macroalgae, there is a seasonal die-off of *Ulva* and *Gelidium* in the mid-highshore during the transition from the cool to the hot season. These changes are even more dramatic in Hong Kong, where ephemeral erect macroalgae more typical of temperate waters settle and flourish during the cool and dry season on shores of all exposures (Morton and Morton, 1983; Hodgkiss, 1984; Kaehler and Williams, 1996; Kennish et al., 1996) only to bleach white and die-back to leave the more heat-tolerant, mainly encrusting, species in the hot and wet season (Kaehler and Williams, 1996; Kennish et al., 1996; Nagarkar and Williams, 1999). Many species also colonise higher shore levels during this period compared to during the hot and wet season (e.g., the encrusting corallines: Kaehler and Williams, 1996). In the epilithic biofilm, cyanobacteria occur year-round, but diatoms become more abundant in the cool and dry season (Nagarkar and Williams, 1999). A similar seasonal change is, therefore, also seen in the cyanobacteria assemblages in the epilithic biofilm in terms of species composition; with heat-tolerant species surviving during the hot and wet season and being outcompeted by faster growing and less tolerant species during the cool and dry season (Nagarkar and Williams, 1999). This seasonal variation in algal and cyanobacterial species composition has a great impact on the diets of intertidal herbivores (Kaehler and Kennish, 1996; Kennish and Williams, 1997; Nagarkar et al., 2004).

Overall, therefore, the majority of studies indicate that there are clear seasonal changes in algal assemblages, linked with the environmental conditions and oceanic currents brought about by prevailing monsoon changes.

### Seasonality and the Importance of the Monsoons: Thermal Stress in Mobile and Sessile Invertebrates

Seasonal changes in sessile and mobile invertebrate assemblages have also been recorded, although the visual impact of mortalities due to thermal stress are not as dramatic as that seen for the macroalgae. Changes in abundance of sessile and mobile fauna are less well documented, but also occur with the change to the warm/hot and wet south-east monsoon in Hong Kong and have also been noted further north along the coast of mainland China (Dong et al., 2017).

In terms of sessile species, most species show seasonal settlement patterns following the hot and wet season (Morton, 1995; Chan and Williams, 2004). Individuals which have settled too high on the shore suffer highshore kills, as reported from temperate areas during extreme heat events (Hawkins and Hartnoll, 1985) with the onset of the hot and wet season (e.g., mussels: Liu and Morton, 1994; barnacles: Chan et al., 2006). In the case of barnacles (*Tetraclita* species), settlement and subsequent recruitment can occur in two pulses: after the hot and wet season, and then in the cool and dry season. These different pulses suggest the influence of changes in the oceanic currents supplying propagules to replenish individuals which have died from heat stress in summer (Chan and Williams, 2004). Many species exhibit a similar temporal partitioning of their life histories; suffering heavy mortality during the hot and wet season, but recruiting heavily after this period.

A number of studies have investigated the responses of mobile species to thermal stress, mostly based in Hong Kong, but more recently along the coast of mainland China (Han et al., 2013; Dong et al., 2014). Unlike macroalgae and sessile invertebrates, mobile species can address changes in thermal stress (either long-term seasonal or shorter tidal scales) by adopting behavioural responses to either avoid or reduce the heat they experience (Garrity, 1984 and summarised by Ng et al., 2017b). Seasonally, many mobile gastropods migrate, moving lower on the shore in the hot and wet season, while exploiting higher shore areas during the cool and dry season (Williams and Morritt, 1995; Harper and Williams, 2001). On a shorter temporal scale, the behavioural patterns of species appear strongly adapted to cope with extreme temperatures during emersion. As previously noted, most mobile species partition their activity periods to minimise heat stress, moving while the shore is awash by waves and then seeking refuge in cooler habitats such as cracks and crevices or rock pools when emersed, especially during the hot and wet summer season (Williams, 1994; Williams and Morritt, 1995; Santini et al., 2011; Hutchinson and Williams, 2003a; Ng and Williams, 2006; Stafford et al., 2007; Yeung and Williams, 2012). During emersion, mobile species can also invoke other behaviours to reduce their exposure to thermal stress; including lifting their shells off the rock surface; attaching the shell lip to the substratum with mucus; orientating to minimise exposure to the sun; aggregation; towering and mushrooming behaviour (as described by Garrity, 1984 for mobile species on Panamanian shores; Williams and Morritt, 1995; Williams et al., 2005; Stafford et al., 2008; Seuront and Ng, 2016; for so-called fight responses see review in Ng et al., 2017b). These behaviours are vitally important as rock surface temperatures in Hong Kong can regularly exceed 50°C and low tides fall in the afternoon when conditions are most harsh for intertidal organisms (Williams, 1994; Williams and Morritt, 1995; Kaehler and Williams, 1996; Ng et al., 2017b).

To combat these harsh conditions, many species have extremely high thermal tolerance, indicative of physiological adaptations to the conditions occurring during the hot and wet season. These adaptations include high thermal performance of heart function (Chelazzi et al., 1999, 2001; Dong and Williams, 2011); differential expression of heat shock and other proteins (Williams et al., 2011, Dong and Williams, 2011); metabolic depression (Marshall et al., 2011) and high thermal sensitivity of enzyme functioning (Liao et al., 2017). The cumulative effects of these smaller-scale processes (e.g., body temperature, microclimatic conditions and behavioural thermoregulation) can affect the vulnerability of intertidal species to thermal stress and subsequently their biogeography (Dong et al., 2017). Along the mainland China coastline, eight populations of *E. malaccana*, *L. sinensis* and *Nerita yoldii*, showed variation in their thermal tolerance within and between populations, indicating that not only between, but also within, a population there will be differential individual capacity to survive during extreme hot days (Dong et al., 2017).

The importance of such interactive effects and multiple stressors is being increasingly recognised, especially as the warm and wet monsoon brings changes in rainfall as well as air (and subsequently rock surface) temperatures. Intense rain periods can affect local seawater salinity, and indeed can flood the shore with freshwater when the tide is out (Morritt et al., 2007; Williams et al., 2011). Work in Hong Kong and the mainland of China (Fujian province) has shown a strong interaction between rainfall and hot days – for instance, occurring during typhoons – affecting osmoregulation, heart rate, gene expression and protein expression in limpets (*Cellana* species: Firth and Williams, 2009; Williams et al., 2011; Dong et al., 2014; Dong and Zhang, 2016). Hutchinson and Williams (2003b) demonstrated that the abundance of the epilithic biofilm is reduced and molluscan species can suffer heavy mortality during typhoons in Hong Kong, resulting in the creation of bare rock surfaces. Recovery from such effects is, however, quite rapid and can be completed by the subsequent cool season, as the turnover rates of species on Hong Kong shores is very rapid and many species have settlement pulses after the hot and wet season. Finally, the rapid annual change in season and associated high physical stress seems to have resulted in species in southern China having relatively short life history patterns (see Connell, 1985 for temperate species). The few studies available suggest many species live for 2–3+ years (Liu, 1994a, 1994b; Williams and Morritt, 1995; Chiu, 1998; Chan and Williams, 2003), and some are almost annual (e.g., *S. japonica*: Chan and Williams, unpublished data; *Cellana toruema*: Villarta and Williams, unpublished data).

**Biological Processes: Trophic Interactions, Settlement and Facilitation**

Attempts to integrate the importance of changes in physical environmental conditions with biological processes suggest that certainly the more central and southerly areas of mainland China and Hong Kong are strongly influenced by physical stress mediating biological interactions. Most of this work has been conducted in Hong Kong on shores of moderate exposure, dominated by grazing gastropods and chitons (Williams, 1993a, 1993b). Some of the earliest approaches were simple exclusion experiments, removing grazing molluscs from single locations (Williams 1993a, 1993b; Kaehler and Williams, 1997). While these studies did reveal the seasonal variation in molluscan grazing pressure, their single location confounded the ability to generalise on their impact. One clear response was the interaction of physical and environmental conditions (tidal height and season) with the importance of grazing; with grazing pressure having relatively little impact at higher shore levels or in the hot and wet season, compared to lower shore levels or the cool season when algal supply and productivity were higher (Williams, 1993a, 1993b; 1994; Kaehler and Williams, 1997, 1998). This pattern, with physical constraints limiting the importance of herbivory by slow-moving gastropods and chitons, seems a general phenomenon. Amelioration of physical stress – via experimental roofs to shade the rock or spraying the shore with seawater – can reduce this physical control, and, when this occurs, the impact of grazing becomes more evident (Williams, 1994; Kaehler and Williams, 1997, 1998). Using a combination of fences to remove grazing gastropods and chitons and roofs to reduce physical stress, Chan and Williams (2003) also showed the interactive effect of physical stress and biological control on the settlement and mortality of newly settled barnacles (*Tetraclita* spp.). Both grazing/bulldozing effects from mobile molluscs and thermal stress were responsible for post-settlement mortality of *Tetraclita* on the midshore, while physical stress was the main factor to affect post-settlement mortality in the higher shore.

There is, therefore, a clear link between thermal stress (see Section 14.3.3.2), behaviour of mobile species and their resultant impact on community structure. Physical stress plays an important role in the distribution of grazers, and hence their effectiveness. As a result, mesoscale (crevices) and micro-scale habitats (pits and cracks in the rock) affect grazer effectiveness (via the areas that they can reach to graze on the rock surface without suffering from thermal and desiccation stress, Williams et al., 2000; Hutchinson and Williams, 2003a). Such interactions create spatial and temporal patchiness on shores of moderate wave exposure in Hong Kong (Williams et al., 2000; Hutchinson and Williams, 2001).

Along most of mainland China, Hong Kong and Taiwan, the rock oyster (*Saccostrea*) monopolises the rock surface of more sheltered and estuarine shores, which is in contrast to temperate systems where algae would dominate. On these shores, the rock oyster forms a closely interlocking three-dimensional carpet, which does suggest competition for space is an important process, although there have been very few studies on such interactions and so this remains to be demonstrated. What is clear, however, is that many species utilise the matrix formed by the living and dead oyster shells and that the oysters act as ecological engineers (*sensu* Bruno et al., 2003; McAfee et al., 2016). Similar cases are seen for areas where *Tetraclita* barnacles are abundant on exposed and semi-exposed shores, where they provide shade for mobile molluscs such as littorinids, chitons and limpets (Cartwright and Williams, 2012, 2014), and for beds of the mussel *Septifer*, which is known to provide habitats for a variety of species (Ong Che and Morton, 1992; Seed and Brotohadikusomo, 1994). There is an interesting interaction in Hong Kong between the seasonal use of such biological refuges in the hot and wet season when mobile species move lower on the shore, perhaps to utilise these biogenic structures (Cartwright and Williams, 2012), again linking the close relationship between the strongly seasonal climate and biological interactions.

In summary, based on the limited studies from Hong Kong, shores appear to be largely controlled by physical stress, especially in the hot and wet season. As such, the relative importance of different processes contributing to community structure in this season appears to follow the environmental stress model of Menge and Sutherland (1987). Competition and predation appear less important as structuring forces compared to shores in Europe and the USA where species are longer-lived and do not exhibit such strong intra-annual patterns. With the change in the monsoon system, or by being lower on the shore, physical stress is reduced and species are released from this physical control, and competition and predation will become relatively more important (Williams, 1993a, 1994). Being in the tropics, one would expect predation to be more important low on the shore where physical stress is reduced (Menge et al., 1986). Preliminary observations in Hong Kong, however, suggest this is not the case, and it has been hypothesised that overfishing and local artisanal harvesting may explain the absence of a dominant predatory guild in Hong Kong waters (Williams, 1994; Williams et al., 2000), and other heavily human-impacted areas in south-east Asia (Williams et al., 2016).

Bruno and co-workers further developed the conceptual models of Menge and Sutherland (1987) to include the importance of species which provide habitat amelioration, reducing the impacts of physical stress for associated species (Bruno et al., 2003). In harsh environments, they proposed that species which can ameliorate physical stress will play increasingly important roles in regulating community structure, and evidence from Hong Kong where physical stress is high in the hot and wet season supports this view (Cartwright and Williams, 2012, 2014). This role, as described earlier, is temporal, with many species being released from the controlling effects of physical stress in the cool and dry season, and this is also reflected in the behavioural patterns of mobile species, such as littorinids which do not seek the benefits of such associations during the cool and dry winter season (Cartwright and Williams, 2012).

While experimental evidence for such seasonal switches in the relative importance of different processes responsible for structuring intertidal communities is only available from a limited geographical area, the descriptions of strong seasonal changes in the literature along the coast of mainland China, Taiwan and Hong Kong suggest this is likely to be an important generality, especially towards the central and southern areas where physical stress in the warm/hot and wet season is high. Such strong seasonality also impacts species longevity, and early evidence suggests that many species have relatively short life spans, especially towards the southern ranges (see earlier). Further experimental evidence, combined with more information on the life histories of species, will shed more light on the dynamic changes in processes affecting community structure in such strongly seasonal environments.

## 14.4 The Future of Rocky Shores in Mainland China, Taiwan and Hong Kong: Threats, Knowledge Gaps and Research Imperatives

The coastlines of mainland China, Taiwan and Hong Kong are probably some of the most densely populated in the world and prone to huge anthropogenic pressures. Faced with dense populations needing space to live, food to eat and the requisite infrastructure, the intertidal area is under great threat from development, changes in land use and reclamation (Williams et al., 2016). When these local pressures are compounded by local and regional pollution and the long-term impacts of climate change, it can be seen that the coastline of China is intensely exploited and facing massive challenges. As such, intertidal areas are being lost for development or degraded with consequent impacts to local biota, including shifts in species distribution patterns, changes to connectivity and community functioning (see Firth et al., 2016 for review).

### 14.4.1 Land Reclamation and Coastal Hardening: Rapid Habitat Changes

An excellent example of these rapid threats to the intertidal zone of China's coastline is the reclamation of land for development and the substitution of natural shore lines for sea walls and coastal defences (Ma et al., 2014). While much of the northern shores of mainland China are mudflats and play an important role in aquaculture, a large proportion of the central and eastern coastline is being rapidly transformed by reclamation and coastal hardening. Around 60 per cent of the coastline of mainland China has been modified, with over 10,000 km being transformed to seawalls (Ma et al., 2014), and the same is true for a large proportion of Hong Kong and many of the coastal cities of Taiwan where the shoreline is modified by artificial wave breakers, seawalls and aquaculture facilities. This huge replacement of natural shorelines with artificial (usually non-ecofriendly) habitats usually reduces natural biodiversity and abundance compared to analogous natural rocky reefs (Connell and Glasby, 1999; Chapman, 2003), but also changes the regional physical environment by removing or imposing physical barriers to species colonisation and dispersion (see Bishop et al., 2017 for review). One key example is the artificial hardening of the original softshore areas around the Yangtze River Delta which has created a series of stepping stones for the northwards colonisation of rocky shore species (Huang et al., 2015; Dong et al., 2016). This effect has been compounded by the TGDs, as not only has this coastal development created artificial habitat, but the TGDs have also reduced the freshwater outflow of the Yangtze River, thereby removing a traditional biogeographic break to the larval dispersal of intertidal species (Wang et al., 2015, Dong et al., 2016). The development of large port areas also contributes to providing access points for invasive species and hard substrates for colonisation (e.g., Lambert and Lambert, 2003). The development of the deep-water Yangshan Port in Shanghai, for example, has probably been responsible for the invasion of *Chthamalus challengeri* in Zhoushan since 1991 (Liu et al., 2014, 2015).

### 14.4.2 How Many Species Are There and Where Are They Found?

Predicting or managing the possible change to species distributions, community structure or ecosystem functioning is, however, very difficult in this region. A fundamental prerequisite to achieve this is the ability to identify the distribution of different species along the coastline (Costello et al., 2010; Williams et al., 2016). Such basic information is not only important for understanding natural communities, but also for identifying the influence of invasive species (e.g., *Mytilopsis* and *Xenostrobus*, see Astudillo et al., 2017) and the management of commercial aquaculture species (Williams et al., 2016). As discussed in the previous sections, however, alpha taxonomy and information of species' distributions are often limited, being centred on one or two locations or on a coarse scale (presence/absence based on a province level), a problem which is common in south-east Asia (see Costello et al., 2010). The species names used in much of the literature are often outdated, and issues with misidentifications and synonyms hinders confident identification and construction of species distribution patterns, as well as confounds assessments of biodiversity (see Bouchet, 2006). As a result, we are probably underestimating coastal biodiversity and recent studies confirm this problem, revealing a number of cryptic species complexes (e.g., barnacles: Chu et al., 2010; limpets: Yu et al., 2014 and see later) along the mainland China coastline. A recent survey in Hong Kong, for example, revealed an unsuspected wealth of marine biodiversity, far higher than would have been predicted for the size of Hong Kong's marine environment, prompting calls to promote conservation of habitats in that area (Ng et al., 2017a).

This problem is well recognised and many of these issues are being addressed with more formal taxonomic keys (as noted earlier) as well as the recent release of regional nodes for China and Hong Kong on the World Register of Marine Species (www.marinespecies.org/). Clearly addressing this fundamental problem, using modern molecular techniques combined with traditional morphological approaches and database management will be an important contribution to our understanding of the patterns of intertidal biodiversity along the coast of mainland China. There is a clear need, for example, to link and share the excellent collections already available (e.g., those housed by the Chinese Academy of Sciences), with regional and international taxonomic experts to produce a fine-scale assessment of the distribution of species along the Chinese coastline. Models for this exist in other countries, such as the Marine Biodiversity and Climate Change Project (www.marclim.co.uk/) based in the UK.

### 14.4.3 What Are the Natural Spatial and Temporal Variations in Rocky Shore Communities?

There are very few studies of long-term changes in patterns (mostly studies being based over twelve months) and even less on the processes driving these changes and the functioning of intertidal communities, although basic information is available from Hong Kong. It is clear, however, that the prevailing monsoon changes play a fundamental role in driving assemblage changes along the coast, and this is interlinked with changes in the oceanic currents. These seasonal changes account for the strong temporal patterns seen on most rocky shores, and also the variation in species composition between warm-temperate to tropical species. These changes in monsoons do show latitudinal variation in terms of timing and strength, however, as in other parts of the world, local-scale variation can create mosaic patterns of environmental gradients (see Helmuth et al., 2006). The thermal stresses that different populations face are, therefore, variable over a latitudinal gradient. The thermal sensitivity of the gastropods, *L. brevicula*, *L. sinensis* and *N. yoldii*, for example, show a strong, non-linear relationship with latitude along the Chinese coastline with the most thermally vulnerable populations being found in mid-latitude regions (Dong et al., 2017). The divergent thermal limits and physiological responses among different populations add another layer of complexity to predicting the future dynamics of intertidal species. Such patterns are, again, difficult to document as information on weather conditions and local coastal waters are not easy to access. This is perhaps surprising as there is an extensive literature of near-shore oceanographic processes which detail hydrographic variation (Hu et al., 2000; Ichikawa and Beardsley, 2002). The opportunities to translate this wealth of information into the intertidal habitat have, as yet, to be capitalised upon. Integrating oceanographic data with coastal systems would provide an excellent opportunity to understand links between bottom-up and top-down drivers, as well as disentangle the relative importance of monsoon systems and coastal currents in community regulation. Due to the limited information of environmental data and biodiversity information, potential future distribution shifts of rocky intertidal species in this region in the face of global change remain unclear.

The limited information on processes which structure communities reflects the overall development of intertidal ecology in the region, where we are just emerging from the initial descriptive stage to start to understand the processes that determine the patterns observed (see Underwood, 2000). There are exceedingly few papers documenting the importance of various processes, and most of these have a very small spatial resolution, making generalisation over the coast of China inappropriate (Underwood and Petraitis, 1993). There is an urgent need for more manipulative approaches to test the relative importance of different structuring processes on a regional scale. Such approaches have been adopted in the European Union, where manipulative tests of the importance of settlement, herbivory etc. (Jenkins et al., 2005; Coleman et al., 2006) were conducted on a continental scale to understand the various scales such processes acted on and their relative importance (e.g., the EUROROCK project). Such a programme would need to either follow, or indeed develop with, the establishment of spatial distribution patterns.

## 14.5 Conclusions

Compared to the eastern and western coasts of North America, South Africa, Australia and Europe, we know relatively little about the rocky shore ecology of mainland China, Taiwan and Hong Kong from which scientists can start to predict changes with climate or to inform managers and governments about where to invest or to conserve. While we are able to generalise broadly about similarities in species patterns over vertical tidal gradients and horizontal wave-exposure gradients, it is not possible to contribute to many of the current theories concerning the functioning of these communities, such as the importance of supply-side ecology; bottom-up versus top-down processes, and the interaction between regional- and local-scale processes. Given the size of the coastline of China this is surprising, as the almost north–south axis of the coast makes it ideal to provide insights into latitudinal gradients in temperature, productivity and biogeography. The reasons for this are many-fold and related to the recent history of the region in terms of investment into different activities associated with the intertidal area and a focus largely on aquaculture and offshore processes in China.

What we can see in the case of mainland China, Hong Kong and Taiwan are highly biodiverse intertidal environments which are poorly documented and catalogued. There is an evolving interest in documenting species distribution patterns (in the light of climate change) and understanding the processes that structure and maintain the functioning of these communities. This is in sharp contrast, however, to the rapid and extensive development of coastal areas and degradation of coastal habitats which is occurring due to pollution, overfishing, reclamation and aquaculture. The race is clearly on for scientists to document the intertidal biodiversity and understand more about the processes which maintain the integrity of these systems before they are further degraded or replaced with more simple, opportunist-dominated artificial communities. It is, however, very likely that the systems which we are documenting are already degraded and so it is probably too late to document truly natural systems (see Pauly, 1995).

Finally, given the history of studies of rocky shore ecology along the coast of China, it is clear that a paradigm shift is needed in the way that scientists and managers view the intertidal zone. Historically, intertidal areas have simply been viewed as a resource for exploitation, particularly for aquaculture. Little attention has been given to the other resources these communities supply, or their conservation or sustainable management. This mindset may well explain the lack of studies to understand ecological functioning and processes. There is, therefore, a clear need to move away from the more descriptive approaches and initiate more holistic, process-driven, functional approaches which will enable effective conservation and sustainable management of these resources in the future.

## Acknowledgements

The authors would like to that Xiaoxu Li and Jie Wang (Xiamen University) for chasing down literature; Dr Christine, Yi Lai Luk (The University of Hong Kong) for guiding us through some of the early history, Louise Firth and Anaëlle Lemasson for sympathetic editing and especially Stephen Hawkins for his encouragement and support.

## REFERENCES

AFCD. (2017). www.afcd.gov.hk/english/conservation/hkbiodiversity/habitat/habitat.html.

Ahyong, S. T., Chan, T. Y. and Liao, Y. C. (2008). *A Catalog of Mantis Shrimps (Stomatopoda) of Taiwan*. National Taiwan Ocean University Press, Keelung, p. 190.

Astudillo, J. C., Bonebrake, T. C. and Leung, K. M. Y. (2017). The recently introduced bivalve *Xenostrobus securis* has higher thermal and salinity tolerance than the native *Brachidontes variabilis* and established *Mytilopsis sallei*. *Marine Pollution Bulletin*, **118**, 229–36.

Baba K., Macpherson, E., Lin, C. W. and Chan, T. Y. (2009). *Crustacean Fauna of Taiwan: Squat lobsters (Chirostylidae and Galatheidae)*. National Taiwan Ocean University Press, Keelung, p. 311.

Beardsley, R. C., Limeburner, R., Yu, H. and Cannon, G. A. (1985). Discharge of the Changjiang (Yangtze River) into the East China Sea. *Continental Shelf Research*, **4**, 57–76.

Bishop, M. J., Mayer-Pinto, M., Airoldi, L. et al. (2017). Effects of ocean sprawl on ecological connectivity: impacts and solutions. *Journal of Experimental Marine Biology and Ecology*, **492**, 7–30.

Bouchet, P. (2006). The Magnitude of Marine Biodiversity. In *The Exploration of Marine Biodiversity: Scientific and Technical Challenges*. Fundación BBVA, Bilbao, pp. 33–64.

Bruno, J. F., Stachowicz, J. J. and Bertness, M. D. (2003). Inclusion of facilitation into ecological theory. *Trends in Ecology and Evolution*, **18**, 119–25.

Cai, L. T., Wang, Y. N., Li, X. F. et al. (2013). Preliminary study on the shellfish ecology in intertidal zone of the Dongji Island. *Marine Sciences*, **8**, 47–54.

Cai, R. X., Chen, S. Q., Lu, J. P. and Xue, J. Z. (1993). The distribution and population structures of *Balanus trigonus* in Zhoushan waters. *Donghai Marine Science*, **11**, 22–33.

Cai, R. X., Zheng, F., Chen, Y. S. et al. (1990). Studies on ecology of Zhoushan intertidal zone I: species composition and distribution. *Donghai Marine Science*, **1**, 51–60.

Cai, R. X., Zheng, F., Wang Y. P. et al. (1991). Studies on ecology of Zhoushan intertidal zone. II. Quantities and distributions. *Donghai Marine Science*, **9**, 58–72.

Cartwright, S. R. and Williams, G. A. (2012). Seasonal variation in utilization of biogenic microhabitats by littorinid snails on tropical rocky shores. *Marine Biology*, **159**, 2323–32.

Cartwright, S. R. and Williams, G. A. (2014). How hot for how long? The potential role of heat intensity and duration in moderating the beneficial effects of an ecosystem engineer on rocky shores. *Marine Biology*, **161**, 2097–105.

Chan, B. K. K. (2001). Studies on *Tetraclita squamosa* and *Tetraclita japonica* (Cirripedia: Thoracica) I: adult morphology. *Journal of Crustacean Biology*, **21**, 616–30.

Chan, B. K. K. (2003). Studies on *Tetraclita squamosa* and *Tetraclita japonica* II: larval morphology and development. *Journal of Crustacean Biology*, **23**, 522–47.

Chan, B. K. K., Chen, Y. Y. and Achituv, Y. (2013). *Crustacean Fauna of Taiwan: Barnacles, Vol II – Cirripedia: Thoracica: Pyrgomatidae. Biodiversity Research Center*. Academia Sinica Press, Taipei, p. 364.

Chan, B. K. K., Hsieh, W. P. and Kolbasov, G. A. (2014). *Crustacean Fauna of Taiwan: Barnacles, Vol III – Cirripedia: Acrothoracica. Biodiversity Research Center*. Academia Sinica Press, Taipei, p. 107.

Chan, B. K. K., Morritt, D. and Williams, G. A. (2001). The effect of salinity and recruitment on the distribution of *Tetraclita squamosa* and *Tetraclita japonica* (Cirripedia: Balanomorpha) in Hong Kong. *Marine Biology*, **138**, 999–1009.

Chan, B. K. K., Morritt, D., De Pirro, M., Leung, K. M. Y. and Williams, G. A. (2006). Summer mortality: impacts on the distribution and abundance of the barnacle *Tetraclita japonica* on tropical shores. *Marine Ecology Progress Series*, **328**, 195–204.

Chan, B. K. K., Murata, A. and Lee, P. F. (2008). Latitudinal gradient in the distribution of the intertidal acorn barnacles *Tetraclita* species complex (Crustacea: Cirripedia) in NW Pacific and SE Asian waters. *Marine Ecology Progress Series*, **362**, 201–10.

Chan, B. K. K., Prabowo, R. E. and Lee, K. S. (2009). *Crustacean Fauna of Taiwan: Barnacles, Vol I – Cirripedia: Thoracica excluding the Pyrgomatidae and Acastinae*. National Taiwan Ocean University Press, Keelung, pp. 297–8.

Chan, B. K. K., Tsang, L. M. and Chu, K. H. (2007a). Cryptic diversity of the *Tetraclita squamosa* complex (Crustacea, Cirripedia) in Asia: description of a new species from Singapore. *Zoological Studies*, **46**, 46–56.

Chan, B. K. K., Tsang, L. M. and Chu, K. H. (2007b). Morphological and genetic differentiation of the acorn barnacle *Tetraclita squamosa* (Crustacea, Cirripedia) in East Asia and description of a new species of *Tetraclita*. *Zoologica Scripta*, **36**, 79–91.

Chan, B. K. K., Tsang, L. M., Ma, K. Y., Hsu, C.-H. and Chu, K. H. (2007c). Taxonomic revision of the acorn barnacles *Tetraclita japonica* and *Tetraclita formosana* (Crustacea: Cirripedia) in East Asia based on molecular and morphological analyses. *Bulletin of Marine Science*, **81**, 101–13.

Chan, B. K. K. and Williams, G. A. (2003). The impacts of physical stress and mollusc grazing on settlement, recruitment and vertical distribution of *Tetraclita squamosa* and *Tetraclita japonica* (Cirripedia: Balanomorpha) in Hong Kong. *Journal of Experimental Marine Biology and Ecology*, **284**, 1–23.

Chan, B. K. K. and Williams, G. A. (2004). Population dynamics of the acorn barnacles, *Tetraclita squamosa* and *Tetraclita japonica* (Cirripedia: Balanomorpha), in Hong Kong. *Marine Biology*, **146**, 149–60.

Chan, T. Y. (2010). *Crustacean Fauna of Taiwan: Crab-Like Anomurans (Hippoide, Lithodoidea and Porcellanidae)*. National Taiwan Ocean University Press, Keelung, p. 197.

Chapman, M. G. (2003). Paucity of mobile species on constructed seawalls: effects of urbanization on biodiversity. *Marine Ecology Progress Series* **264**, 21–9.

Cheang, C. C., Chu, K. H. and Ang, P. O. (2010). Phylogeography of the marine macroalga *Sargassum hemiphyllum* (Phaeophyceae, Heterokontophyta) in northwestern Pacific. *Molecular Ecology*, **19**, 2933–48.

Chelazzi, G., De Pirro, M. and Williams, G. A. (2001). Cardiac responses to variation in physical factors in two vertically separated, tropical limpets. *Marine Biology*, **139**, 1079–85.

Chelazzi, G., Williams, G. A. and Gray, D. R. (1999). Field and laboratory measurement of heart rate in a tropical limpet, *Cellana grata*. *Journal of the Marine Biological Association of the United Kingdom*, **79**, 749–51.

Chen, D., Zebiak, S. E., Busalacchi, A. J. and Cane, M. A. (1995). An improved procedure for *El Niño* forecasting: implications for predictability. *Science*, **269**, 1699–702.

Chen, G. T., Yang, X. L. Yang, J. Y. and Gao, A. G. (1994) Ecological and environment qualitative study in the intertidal zone and land area of Nanji Archipelago. *Donghai Marine Science*, **2**, 1–15.

Chen, X. H., Zhou, J. J. and Wang, Y. Y. (1989). The ecology characteristic of Cirripedia the intertidal zone of Fujian. *Journal of Xiamen Fisheries College*, **2**, 26–31.

Chiu, M. C. (1998). Population dynamics of *Saccostrea cucullata* (Bivalvia: Ostreidae) from five shores in Hong Kong. *Asian Marine Biology*, **15**, 73–103.

Chu, K. H., Tsang, L. M., Chan, K. K. and Williams, G. A. (2010). Misinterpreting species distribution patterns:

the impact of cryptic species on ecological studies and climate change predictions. In *International Conference on Environmental Research and Technology, Penang, Malaysia*, 99–104.

Coleman, R. A., Underwood, A. J., Benedetti-Cecchi, L. et al. (2006). A continental scale evaluation of the role of limpet grazing on rocky shores. *Oecologia*, **147**, 556–64.

Connell, J. H. (1985). Variation and Persistence of Rocky Shore Populations. In P. G. Moore and R. Seed, eds. *The Biology of Rocky Coasts*. Hodder and Staughton, London, pp. 57–69.

Connell, S. D. and Glasby, T. M. (1999). Do urban structures influence local abundance and diversity of subtidal epibiota? A case study from Sydney Harbour, Australia. *Marine Environmental Research* **47**, 373–87.

Coombes, M. R. A. and Seed, R. (1992). Predation of the black mussel *Septifer virgatus* by the red-eyed crab *Eriphia laevimana smithii* (Xanthidae). *Asian Marine Biology*, **9**, 245–58.

Costello, M. J., Coll, M., Danovaro, R. et al. (2010). A census of marine biodiversity knowledge, resources, and future challenges. *PLoS ONE*, **5**, e12110.

Dai, Z., Du, J., Li, J., Li, W. and Chen, J. (2008). Runoff characteristics of the Changjiang River during 2006: effect of extreme drought and the impounding of the Three Gorges Dam. *Geophysical Research Letters*, **35**, L07406.

Dong, Y. M., Chen, Y. S. and Cai, R. X. (1980) Preliminary study of the Chinese Cirripedian fauna (Crustacea). *Acta Oceanologica Sinica*, **2**, 124–31.

Dong, Y. W., Han, G. D. and Huang, X. W. (2014). Stress modulation of cellular metabolic sensors: interaction of stress from temperature and rainfall on the intertidal limpet *Cellana toreuma*. *Molecular Ecology*, **23**, 4541–54.

Dong, Y. W., Huang, X. W., Wang, W., Li, Y. and Wang, J. (2016). The marine 'great wall' of China: local and broad-scale ecological impacts of coastal infrastructure on intertidal macrobenthic communities. *Diversity and Distributions*, **22**, 731–44.

Dong, Y. W., Li, X. X., Choi, F. M. P. et al. (2017). Untangling the roles of microclimate, behaviour and physiological polymorphism in governing vulnerability of intertidal snails to heat stress. *Proceedings of the Royal Society B*, **284**, 20162367.

Dong, Y. W., Wang, H. S., Han, G. D. et al. (2012). The impact of Yangtze river discharge, ocean currents and historical events on the biogeographic pattern of *Cellana toreuma* along the China coast. *PLoS ONE*, **7**, e36178.

Dong, Y. W. and Williams, G. A. (2011). Variations in cardiac performance and heat-shock protein expression to thermal stress in two differently zoned limpets on a tropical rocky shore. *Marine Biology*, **158**, 1223–31.

Dong, Y. W. and Zhang, S. (2016). Ecological relevance of energy metabolism: transcriptional responses in energy sensing and expenditure to thermal and osmotic stresses in an intertidal limpet. *Functional Ecology*, **30**, 1539–48.

Fan, T. C. (2012). Taiwan natural history society in the early period of Japanese colonial rule – Japanese naturalists and construction on Taiwan natural history. *Bulletin of Taiwan Historical Research*, **5**, 3–39.

Fan, K. C. (2002). Mangroves in Taiwan: current status and restoration projects. *Bois et forêts des tropiques*, **273**, 43–54.

Fan, Z. G. (1981). Studies on the ecology of intertidal zone of the Jiaozhou Wan I. the intertidal zone of rocky shore. *Acta Ecologica Sinica*, **1**, 117–25.

Fang, S. H., Zhang, Y. P., Luo, Q. Q. and Lv, X. M. (2011). Species diversity and ecological characteristics of polychaetes in Meizhou Bay. *Journal of Oceanography in Taiwan Strait*, **3**, 419–29.

Firth, L. B., Knights, A. M., Thompson, R. C. et al. (2016). Ocean sprawl: challenges and opportunities for biodiversity management in a changing world. *Oceanography and Marine Biology Annual Review*, **54**, 193–269.

Firth, L. B. and Williams, G. A. (2009). The influence of multiple environmental stressors on the limpet *Cellana toreuma* during the summer monsoon season in Hong Kong, *Journal of Experimental Marine Biology and Ecology*, **375**, 70–5.

Foster, B. A. (1982). Shallow Water Barnacles from Hong Kong. In *Proceedings of the Tenth International Marine Biological Workshop: The Marine Flora and Fauna of Hong Kong and Southern China*. Hong Kong University Press, Hong Kong, pp. 207–32.

Gao, A. G., Zeng, J. N. Chen, Q. Z. et al. (2007). Time and space distribution of molluscs of intertidal zone in Nanji Archipalego Marine Nature Reserve. *Acta Oceanologica Sinica*, **2**, 105–11.

Garrity, S. D. (1984). Some adaptations of gastropods to physical stress on a tropical rocky shore. *Ecology*, **65**, 559–74.

Gorbarenko, S. A. and Southon, J. R. (2000). Detailed Japan Sea paleoceanography during the last 25 kyr: constraints from AMS dating and δ 18 O of planktonic foraminifera. *Palaeogeography, Palaeoclimatology, Palaeoecology*, **156**, 177–93.

Han, G. D., Zhang, S., Marshall, D. J., Ke, C. H. and Dong, Y. W. (2013). Metabolic energy sensors (AMPK and SIRT1), protein carboxylation and cardiac failure as biomarkers of thermal stress in an intertidal limpet: linking energetic allocation with environmental temperature during aerial emersion. *Journal of Experimental Biology*, **216**, 3273–82.

Harper, K. D. and Williams, G. A. (2001). Variation in abundance and distribution of the chiton, *Acanthopleura japonica,* and associated molluscs on a seasonal, tropical, rocky shore. *Journal of Zoology*, **253**, 293–300.

Hawkins, S. J., Evans, A. J., Mieszkowska, N. et al. (2017). Distinguishing globally-driven changes from regional-and local-scale impacts: the case for long-term and broad-scale studies of recovery from pollution. *Marine Pollution Bulletin*, **124**, 573–86.

Hawkins, S. J. and Hartnoll, R. G. (1985). Factors determining the upper limits of intertidal canopy-forming algae. *Marine Ecology Progress Series*, **20**, 265–71.

Helmuth, B. S., Broitman, B. R., Blanchette, C. A. et al. (2006). Mosaics in the rocky intertidal zone: implications for climate change. *Ecological Monographs*, **76**, 461–79.

Hiro, F. (1939). Studies on the cirripedian fauna of Japan IV: cirripedes of Formosa (Taiwan) with some geographical and ecological remarks on the littoral forms. *Memoirs of the College of Science, Kyoto University Series B*, **15**, 245–84.

Hodgkiss, I. J. (1984). Seasonal patterns of intertidal algal distribution in Hong Kong. *Asian Marine Biology*, **1**, 49–57.

Hsueh, Y. and Zhong, L. (2004). A pressure-driven South China Sea Warm Current. *Journal of Geophysical Research: Oceans*, **109**, C09014.

Hu, J. Y., Kawamura, H., Hong, H. S. and Qi, Y. Q. (2000). A review on the currents in the South China Sea: seasonal circulation, South China Sea warm current and Kuroshio intrusion. *Journal of Oceanography*, **56**, 607–24.

Huang, H., Li, X. Z. and Wang, H. F. (2012). Characteristics of macrobenthos community from the intertidal zones of Moye Island, Shandong Peninsula. *Marine Sciences*, **36**, 90–7.

Huang, T. K. and Tsai, R. H. (1962). On the fouling organisms of ships and other underwater constructions in Amoy Harbour. *Journal of Xiamen University*, **3**, 163–75.

Huang, X. W., Wang, W. and Dong, Y. W. (2015). Complex ecology of China's seawall. *Science*, **347**, 1079.

Huang, Z. G. (1994). General characteristics of the species in China seas. *Chinese Biodiversity*, **2**, 63–7.

Huang, Z. G. (2008). *Marine Species and Their Distribution in China*, second edn. China Ocean Press, Beijing.

Huang, Z. G. and Lin, M. (2012). *An Illustrated Guide to Species in China's Seas*. Marine Press, Beijing.

Hutchinson, N. and Williams, G. A. (2001). Spatio-temporal variation in recruitment on a seasonal, tropical rocky shore: the importance of local versus non-local processes. *Marine Ecology Progress Series*, **215**, 57–68.

Hutchinson, N. and Williams, G. A. (2003a). An assessment of variation in molluscan grazing pressure on Hong Kong rocky shores. *Marine Biology*, **142**, 495–507.

Hutchinson, N. and Williams, G. A. (2003b). Disturbance and subsequent recovery of mid-shore assemblages on seasonal, tropical, rocky shores. *Marine Ecology Progress Series*, **249**, 25–38.

Ichikawa, H. and Beardsley, R. C. (2002). The current system in the Yellow and East China Seas. *Journal of Oceanography*, **58**, 77–92.

Jan, S., Wang, J., Chern, C. S. and Chao, S. Y. (2002). Seasonal variation of the circulation in the Taiwan Strait. *Journal of Marine Systems*, **35**, 249–68.

Jenkins, S. R., Coleman R. A., Della Santina, P. et al. (2005). Regional scale differences in the determinism of grazing effects in the rocky intertidal. *Marine Ecology Progress Series*, **287**, 77–86.

Kaehler, S. and Kennish, R. (1996). Summer and winter comparisons in the nutritional value of marine macroalgae from Hong Kong. *Botanica Marina*, **39**, 11–18.

Kaehler, S. and Williams, G. A. (1996). Distribution of algae on tropical rocky shores: spatial and temporal patterns of non-coralline encrusting algae in Hong Kong. *Marine Biology*, **125**, 177–87.

Kaehler, S. and Williams, G. A. (1997). Do factors influencing recruitment ultimately determine the distribution and abundance of encrusting algae on seasonal tropical shores? *Marine Ecology Progress Series*, **156**, 87–96.

Kaehler, S. and Williams, G. A. (1998). Algal assemblage development under different regimes of physical and biotic factors on a seasonal tropical rocky shore. *Marine Ecology Progress Series*, **172**, 61–71.

Kennish, R. and Williams, G. A. (1997). Feeding preferences of the herbivorous crab *Grapsus albolineatus:* the differential influence of algal nutrient content and morphology. *Marine Ecology Progress Series*, **147**, 87–95.

Kennish, R., Williams, G. A. and Lee, S. Y. (1996). Algal seasonality on an exposed rocky shore in Hong Kong and the dietary implications for the herbivorous crab *Grapsus albolineatus*. *Marine Biology*, **125**, 55–64

Kim, J. T., Huynh, T. C. and Lee, S. Y. (2014). Wireless structural health monitoring of stay cables under two consecutive typhoons. *Structural Monitoring and Maintenance*, **1**, 47–67.

Kitamura, A. (2008). Paleoceanographic changes of the Sea of Japan during 3.5–0.8 Ma. In *Origin and Evolution of Natural Diversity: Proceedings of the International Symposium, Sapporo*, 187–94.

Lam, K. Y. and Morton, B. S. (2004). The oysters of Hong Kong (Bivalvia: Ostreidae and Gryphaeidae). *Raffles Bulletin of Zoology*, **52**, 11–28.

Lambert, C. C. and Lambert, G. (2003). Persistence and differential distribution of nonindigenous ascidians in harbors of the Southern California Bight. *Marine Ecology Progress Series*, **259**, 145–61.

Lewis, J. E. and Norris, J. N. (1987). A history and annotated account of the benthic marine algae of Taiwan. In *Smithsonian Contributions to the Marine Sciences No. 29*.

Li, F. X., Cai, L. Z. and Dai, P. (1994). Studies on ecology of molluscan on rocky intertidal zone in Xiamen Harbour. *Journal of Oceanography in Taiwan Strait*, **1**, 43–51.

Li, R. G., Jiang, J. X., Wu, Q. Q. et al. (1995). Communities of intertidal benthos near the Daya Bay nuclear power station. *Oceanologia et Limnologia Sinica Supplement*, **5**, 91–101.

Li, S., Li, Q., Yu, H., Kong, L. and Liu, S. (2015). Genetic variation and population structure of the Pacific oyster *Crassostrea gigas* in the northwestern Pacific inferred from mitochondrial COI sequences. *Fisheries Science*, **81**, 1071–82.

Liao, B. W. and Zhang, Q. M. (2014). Area, distribution and species composition of mangroves in China. *Wetland Science*, **12**, 4350440.

Liao, M., Zhang, S., Zhang, G. et al. (2017). Heat-resistant cytosolic malate dehydrogenases (cMDHs) of thermophilic intertidal snails (genus *Echinolittorina*): protein underpinnings of tolerance to body temperatures reaching 55°C. *Journal of Experimental Biology*, **220**, 2066–75.

Liao, Y., Zeng, J., Chen, Q. et al. (2007). Macrobenthos community patterns in the intertidal zone of the Shengsi Archipelago in spring and autumn. *Acta Zoologica Sinica*, **53**, 1000–101.

Lie, H. J. and Cho, C. H. (2016). Seasonal circulation patterns of the Yellow and East China Seas derived from satellite-tracked drifter trajectories and hydrographic observations. *Progress in Oceanography*, **146**, 121–41.

Liu, J. G., Fei, Y. J., Wang, X. L. and Jiang H. (2012). Macrozoobenthos communities in the rocky intertidal zones of Miaozihu Island and Huangxing Island in summer. *Marine Science Bulletin*, **5**, 566–74.

Liu, J. H. (1993). Distribution and Abundance of Rocky Shore Communities at Hoi Ha Wan, Hong Kong. In *The Marine Flora and Fauna of Hong Kong and Southern China (III) Proceedings of the Fourth International Marine Biological Workshop*. Hong Kong University Press, Hong Kong, pp. 823–36.

Liu, J. H. (1994a). The ecology of the Hong Kong limpets *Cellana grata* (Gould, 1859) and *Patelloida pygmaea* (Dunker, 1860): distribution and population dynamics. *Journal of Molluscan Studies*, **60**, 55–67.

Liu, J. H. (1994b) Distribution and population dynamics of three populations of *Siphonaria* on rocky intertidal shores in Hong Kong. *Journal of Molluscan Studies*, **60**, 431–43.

Liu, J. H. and Morton, B. (1994). The temperature tolerances of *Tetraclita squamosa* (Crustacea: Cirripedia) and *Septifer virgatus* (Bivalvia: Mytilidae) on a subtropical rocky shore in Hong Kong. *Journal of Zoology*, **234**, 325–39.

Liu, J. Y. (2013). Status of marine biodiversity of the China Seas. *PLoS ONE*, **8**, e50719.

Liu, R. and Xu, F. (1963). Preliminary studies on the benthic fauna of the Yellow Sea and the East China Sea. *Oceanologia et Limnologia Sinica*, **5**, 306–21.

Liu, R. Y. (2008). *Marine Fauna and Flora of the China Seas*, Science Press, Beijing.

Liu, Y., Wu, H. and Xue, J. (2014). The invasion and its impact for the spread of *Chthamalus challengeri* in Zhoushan sea area. *Journal of Fisheries of China*, **38**, 1047–55.

Liu, Y., Xue, J., Lin, J. and Wu, H. (2015). Invasion and morphological variation of the non-indigenous barnacle *Chthamalus challengeri* (Hoek, 1883) in Yangshan Port and its surrounding areas. *Journal of Ocean University of China*, **14**, 575–83.

Lubchenco, J., Menge, B. A., Garrity, S. D. et al. (1984). Structure, persistence, and role of consumers in a tropical rocky intertidal community (Taboguilla Island, Bay of Panama). *Journal of Experimental Marine Biology and Ecology*, **78**, 23–7.

Luk, C. Y. L. (in press). 'The First Marine Biological Station in Modern China: Amoy University and Amphioxus." In *Why Study Biology by the Sea?* University of Chicago Press, Chicago.

Ma, K. (2015). Species catalogue of China: a remarkable achievement in the field of biodiversity science in China. *Biodiversity Science*, **23**, 137–8.

Ma, L., Li, X. Z., Wang, H. F. et al. (2013). The ecology characterization of macrobenthic community from the intertidal zone on Dayudao Village, Shidao Town, Shandong Peninsula. *Marine Sciences*, **5**, 87–93.

Ma, L., Wang, H. F. and Shuai, L. M. (2009). The ecological report of the macrobenthic community in the intertidal zone of Xicun village, Rushan County, Shandong Peninsula. *Marine Sciences*, **6**, 29–34.

Ma, Z., Melville, D. S., Liu, J. et al. (2014). Rethinking China's new great wall. *Science*, **346**, 912.

MacPherson, K. L. (2003). The History of Marine Science in Hong Kong (1841–1977). Perspectives on Marine Environment Change in Hong Kong and Southern China, 1977–2001. In *Proceedings of an International Workshop Reunion Conference*. Hong Kong University Press, Hong Kong, pp. 7–29.

Mak, Y. M. and Williams, G. A. (1999). Littorinids control high intertidal biofilm abundance on tropical, Hong Kong rocky shores. *Journal of Experimental Marine Biology and Ecology*, **233**, 81–94.

Marshall, D. J., McQuaid, C. D., Williams, G. A. and Dong, Y. W. (2011). Thermal adaptation in the intertidal snail *Echinolittorina malaccana* contradicts current theory by revealing the crucial roles of resting metabolism. *Journal of Experimental Biology*, **214**, 3649–57.

McAfee, D., Cole, V. J. and Bishop, M. J. (2016). Latitudinal gradients in ecosystem engineering by oysters vary across habitats. *Ecology*, **97**, 929–39.

McLaughlin, P. A., Rahayu, D. L., Komai, T. and Chan, T. Y. (2007). *A Catalog of the Hermit Crabs (Paguroidea) of Taiwan*. National Taiwan Ocean University Press, Keelung, p. 365.

Menge, B. A., Lubchenco, J. and Ashkenas, L. R. (1986). Experimental separation of effects of consumers on sessile species of prey in the low zone of a rocky shore in the Bay of Panama: direct and indirect consequences of food web complexity. *Journal of Experimental Marine Biology and Ecology*, **100**, 225–69.

Menge, B. A. and Sutherland, J. P. (1987). Community regulation: variation in disturbance, competition, and predation in relation to environmental stress and recruitment. *The American Naturalist*, **130**, 730–57.

Morritt, D., Leung, K. M. Y., De Pirro, M. et al. (2007). Responses of the limpet, *Cellana grata* (Gould 1859), to hypo-osmotic stress during simulated tropical, monsoon rains. *Journal of Experimental Marine Biology and Ecology*, **352**, 78–88.

Morton, B. (1990a). *A Bibliography of Hong Kong Marine Science 1842–1990*. Hong Kong University Press, Hong Kong, pp. 1–115.

Morton, B. (1990b). The rocky shore ecology of Qingdao, Shandong Province, People's Republic of China. *Asian Marine Biology*, **7**, 167–87.

Morton, B. (1995). The population dynamics and reproductive cycle of *Septifer virgatus* (Bivalvia: Mytilidae) on an exposed rocky shore in Hong Kong. *Journal of Zoology*, **235**, 485–500.

Morton, B. (2003). Hong Kong's International Malacological Wetland and Marine Biological Workshops (1977–1998); Changing Local Attitudes towards Marine Conservation. In *Perspective on marine environmental change in Hong Kong and Southern China (1977-2001) In Proceedings of an International Workshop Reunion Conference*. Hong Kong University Press, Hong Kong, pp. 31–74.

Morton, B. and Blackmore, G. (2001). South China Sea. *Marine Pollution Bulletin*, **42**, 1236–63.

Morton, B. and Morton, J. (1983). *The Sea Shore Ecology of Hong Kong*. Hong Kong University Press, Hong Kong, pp. 3–25.

Morton, B., Williams, G. A. and Lee, S. Y. (1996). The Benthic Marine Ecology of Hong Kong: A Dwindling Heritage? In *Coastal infrastructure development in Hong Kong: A Review*. Hong Kong Government Press, Hong Kong, pp. 233–67.

Muromtsev, A. M. (1970). Some results of investigations of dynamics and thermal structure of the Kuroshio and adjacent regions. In *The Kuroshio, A Symposium on the Japan Current*, 97–106.

Nagarkar, S. and Williams, G. A. (1999). Cyanobacteria dominated epilithic biofilms: spatial and temporal variation on semi-exposed tropical rocky shores. *Phycologia*, **38**, 385–93.

Nagarkar, S., Williams, G. A., Subramanian, G. and Saha, S. K. (2004). Cyanobacteria-dominated biofilms: a high quality food resource for intertidal grazers. *Hydrobiologia*, **512**, 89–95.

Neushul, P. and Wang, Z. (2000). Between the devil and the deep sea: CK Tseng, mariculture, and the politics of science in modern China. *Isis*, **91**, 59–88.

Ng, J. S. S. and Williams, G. A. (2006). Intraspecific variation in foraging behaviour: the influence of shore height on temporal organization of activity in the chiton *Acanthopleura japonica*. *Marine Ecology Progress Series*, **321**, 183–92.

Ng, T. P. T., Cheng, M. C. F., Ho, K. K. Y. et al. (2017a). Hong Kong's rich marine biodiversity: the unseen wealth of South China's megalopolis. *Biodiversity and Conservation*, **26**, 23–36.

Ng, T. P. T., Lau, S. L. Y., Seuront, L. et al. (2017b). Linking behaviour and climate change in intertidal ectotherms: insights from littorinid snails. *Journal of Experimental Marine Biology and Ecology*, **492**, 121–31.

Ni, G., Kern, E., Dong, Y. W., Li, Q. and Park, J. K. (2017). More than meets the eye: the barrier effect of the Yangtze River outflow. *Molecular Ecology*, **26**, 4591–602.

Ni, G., Li, Q., Kong, L. F. and Yu, H. (2014). Comparative phylogeography in marginal seas of the northwestern Pacific. *Molecular Ecology*, **23**, 534–48.

Oba, T., Kato, M., Kitazato, H. et al. (1991). Paleoenvironmental changes in the Japan Sea during the last 85,000 years. *Paleoceanography*, **6**, 499–518.

Ohgaki, S. (1983). Distribution of the Family Littorinidae (Gastropoda) on Hong Kong Rocky Shores. In *Proceedings of the Second International Workshop on the Malacofauna of Hong Kong and Southern China*. Hong Kong University Press, Hong Kong, pp. 457–64.

Ong Che, R. G. and Morton, B. (1992). Structure and seasonal variations in abundance of the macro-invertebrate community associated with *Septifer virgatus* (Bivalvia: Mytilidae) at Cape d'Aguilar, Hong Kong. *Asian Marine Biology*, **9**, 217–33.

Paine, R. T. (1994). *Marine Rocky Shores and Community Ecology: An Experimentalist's Perspective*. Ecology Institute, Oldendorf/Luhe, pp. 1–152.

Pauly, D. (1995). Anecdotes and the shifting baseline syndrome of fisheries. *Trends in Ecology and Evolution*, **10**, 430.

Peng X., Wu H. X., Gao P. C. et al. (2012) Ecology response of intertidal macrobenthos to the construction of stock enhancement and protection area: a case of Dachen Island. *Journal of Marine Sciences*, **1**, 19–26.

Qiao, F. (2012). *Regional Oceanography of China Seas-Physical Oceanography*. Ocean Press, Beijing.

Reid, D. G. (2007). The genus *Echinolittorina* Habe, 1956 (Gastropoda: Littorinidae) in the Indo-West Pacific Ocean. *Zootaxa*, **1420**, 1–103.

Ruan, J. H. (1994) Preliminary study on the intertidal ecology of marine algae on Yushan islands. *Donghai Marine Science*, **4**, 48–57.

Santini, G., Ngan, A. and Williams, G. A. (2011). Plasticity in the temporal organization of foraging in the limpet *Cellana grata*. *Marine Biology*, **158**, 1377–86.

Sato, H. (1936a). The collection of the littoral animals of Formosa (1). *Botany and Zoology*, **4**, 1435–42.

Sato, H. (1936b). The collection of the littoral animals of Formosa (2). *Botany and Zoology*, **4**, 1619–24.

Sato, H. (1936c). The collection of the littoral animals of Formosa (4). *Botany and Zoology*, **4**, 1951–7.

Seed, R. and Brotohadikusumo, N. A. (1994). Spatial Variation in the Molluscan Fauna Associated with *Septifer virgatus* (Bivalvia: Mytilidae) at Cape d'Aguilar, Hong Kong. In *The Malacofauna of Hong Kong and Southern China III: Proceedings of the Third International Workshop on the Malacofauna of Hong Kong and Southern China, Hong Kong*. Hong Kong University Press, Hong Kong, pp. 427–43.

Seuront, L. and Ng, T. P. T. (2016). Standing in the sun: infrared thermography reveals distinct thermal regulatory behaviours in two tropical high-shore littorinid snails. *Journal of Molluscan Studies*, **82**, 336–40.

Shao X. Y., You Z. J., Cai R. X. et al. (1999) Studies on ecology of intertidal zone around islands of Zhejiang Province: I. Species composition and distribution. *Journal of Zhejiang Ocean University (Natural Science)*, **2**, 112–19.

Shao X. Y., You Z. J., Cai R. X. et al. (2001) Study on ecology of intertidal zone around islands of Zhejiang Province II: quantities and distributions. *Journal of Zhejiang Ocean University (Natural Science)*, **4**, 279–86.

Shih, H. T., Chan, B. K. K., Teng, S. J. and Wong, K. J. H. (2015). *Crustacean Fauna of Taiwan: Brachyuran crabs. Vol II-Ocypodoidae*. National Chung Hsing University Press, Taichung, p. 303.

Stafford, R., Davies, M. S. and Williams, G. A. (2007). Computer simulations of high shore littorinids predict small-scale spatial and temporal distribution patterns on rocky shores. *Marine Ecology Progress Series*, **342**, 151–61.

Stafford, R., Davies, M. S. and Williams, G. A. (2008). Self-organization of intertidal snails facilitates evolution of aggregation behavior. *Artificial Life*, **14**, 409–23.

Sun, Y., Zhang, X. B., Zwiers, F. W. et al. (2014). Rapid increase in the risk of extreme summer heat in Eastern China. *Nature Climate Change*, **4**, 1082–5.

Takahashi, S. (1934). An ecological study of the littoral animals near the mouth of Tamsui. *Transactions of the Natural History Society of Formosa*, **24**, 1–14.

Tang, Y. B., Liao, Y. B., Shou, L. et al. (2016). Intertidal zone of the Nanji Islands is a niche for dominant species of the macrobenthic community. *Acta Ecologica Sinica*, **36**, 489–98.

Taylor, J. D. and Morton, B. (1996). The diets of predatory gastropods in Lobster Bay, Cape d'Aguilar, Hong Kong. *Asian Marine Biology*, **13**, 141–65.

Tchang, S. (1946). Progress of investigations of the marine animals in China. *The American Naturalist*, **80** (795), 593–609.

Tchang, S. (1959). Faune des mollusques utiles et nuisibles de la Mer Jaune et al Mer Est de la Chine. *Oceanologia et Limnologia Sinica*, **2**, 27–34.

Tchang, S., Chung Y., Zhang F. S. and Ma S. T. (1963). A preliminary study of the demarcation of marine molluscan faunal regions of china and its adjacent waters. *Oceanologia et Limnologia Sinica*, **2**, 124–38.

Tian, R. C., Hu, F. X. and Martin, J. M. (1993). Summer nutrient fronts in the Changjiang (Yantze River) estuary. *Estuarine, Coastal and Shelf Science*, **37**, 27–41.

Tong, K. Y. (1986). The feeding ecology of *Thais clavigera* (Kuster) and *Morula musiva* (Kiener) (Mollusca: Gastropoda: Muricidae) in Hong Kong. *Asian Marine Biology*, **3**, 163–78.

Tsang, L. M., Chan, B. K. K., Wu, T. H. et al. (2008). Population differentiation of the barnacle, *Chthamalus malayensis*: postglacial colonization and recent connectivity across Pacific and Indian Oceans. *Marine Ecology Progress Series*, **364**, 107–18.

Tsang, L. H., Wu, T. H., Ng, W. C. et al. (2011). Comparative phylogeography of Indo-West Pacific intertidal barnacles. In *Phylogeography and Population Genetics in Crustacea*. CRC Press, Boca Raton, FL, pp. 109–27.

Tsang, L. M., Achituv, Y., Chu, K. H. and Chan, B. K. K. (2012). Zoogeography of intertidal communities in the West Indian Ocean as determined by ocean circulation systems: patterns from the *Tetraclita* barnacles. *PLoS ONE*, **7**, e45120.

Tseng, C., Lin, C., Chen, S. and Shyu, C. (2000). Temporal and spatial variation of sea surface temperature in the East China Sea. *Continental Shelf Research*, **20**, 373–87.

Tseng, C. K. and Chang, C. F. (1959). On the economic marine algal Flora of the Yellow Sea and the East China Sea. *Oceanologia et Limonologia Sinica*, **11**, 43–52.

Tseng, C. K. and Chang, C. F. (1960). An analysis of the nature of marine algal flora. *Oceanologia et Limnologia Sinica*, **3**, 4.

Ujiié, Y., Ujiié, H., Taira, A., Nakamura, T. and Oguri, K. (2003). Spatial and temporal variability of surface water in the Kuroshio source region, Pacific Ocean, over the past 21,000 years: evidence from planktonic foraminifera. *Marine Micropaleontology*, **49**, 335–64.

Underwood, A. J. (2000). Experimental ecology of rocky intertidal habitats: what are we learning? *Journal of Experimental Marine Biology and Ecology*, **250**, 51–76.

Underwood, A. J. and Petraitis, P. S. (1993). Structure of Intertidal Assemblages in Different Locations: How Can Local Processes Be Compared. In *Species Diversity in Ecological Communities*. University of Chicago, Chicago, pp. 38–51.

Voris, H. K. (2000). Maps of Pleistocene sea levels in Southeast Asia: shorelines, river systems and time durations. *Journal of Biogeography*, **27**, 1153–67.

Wai, T. K. and Williams, G. A. (2006). Effect of grazing on coralline algae in seasonal, tropical low-shore rock pools: spatio-temporal variation in primary settlement and persistence. *Marine Ecology Progress Series*, **326**, 99–113.

Wang, H., Qian, L., Wang, A. and Guo, X. (2013). Occurrence and distribution of *Crassostrea sikamea* (Amemiya 1928) in China. *Journal of Shellfish Research*, **32**, 439–46.

Wang, H. Y., Guo, X. M., Liu, X. et al. (2009). Classification of "*Zhe*" oysters from North China. *Marine Sciences*, **33**, 104–6.

Wang, J., Tsang, L. M. and Dong, Y. W. (2015). Causations of phylogeographic barrier of some rocky shore species along the Chinese coastline. *BMC Evolutionary Biology*, **15**, 114.

Wang, P. (1999). Response of Western Pacific marginal seas to glacial cycles: paleoceanographic and sedimentological features. *Marine Geology*, **156**, 5–39.

Wang, Y. (2012). *Regional Oceanography of China Seas-Marine Geomorphology*. Ocean Press, Beijing.

Wang, W., Hui, J. H., Williams, G. A. et al. (2016) Comparative transcriptomics across populations offers new insights into the evolution of thermal resistance in marine snails. *Marine Biology*, **163**, 92, http://dx.doi.org/10.1007/s00227-016-2873-3.

Wang, Z. (2002). Saving China through science: the science society of China, scientific nationalism, and civil society in Republican China. *Osiris*, **17**, 291–322.

Williams, G. A. (1993a). Seasonal variation in algal species richness and abundance in the presence of molluscan herbivores on a tropical rocky shore. *Journal of Experimental Marine Biology and Ecology*, **167**, 261–75.

Williams, G. A. (1993b). The Relationship between Herbivorous Molluscs and Algae on Moderately Exposed Hong Kong Shores. In *The Proceedings of the Second International Conference on the Marine Biology of the South China Sea*. Hong Kong University Press, Hong Kong, pp. 459–70.

Williams, G. A. (1994). The relationship between shade and molluscan grazing in structuring communities on a moderately-exposed tropical rocky shore. *Journal of Experimental Marine Biology and Ecology*, **178**, 79–95.

Williams, G. A., Davies, M. S. and Nagarkar, S. (2000). Primary succession on a seasonal tropical rocky shore: relative roles of spatial heterogeneity and herbivory. *Marine Ecology Progress Series*, **203**, 81–94.

Williams, G. A., De Pirro, M., Cartwright, S. R. et al. (2011). Come rain or shine: the combined effects of physical stresses on physiological and protein-level responses of an intertidal limpet in the monsoonal tropics. *Functional Ecology*, **25**, 101–10.

Williams, G. A., De Pirro, M., Leung, K. M. Y. and Morritt, D. (2005). Physiological responses to heat stress in a tropical limpet, *Cellana grata*: the benefits of mushrooming behaviour. *Marine Ecology Progress Series*, **292**, 213–24.

Williams, G. A., Helmuth, B., Russell, B. D. et al. (2016). Meeting the climate change challenge: pressing issues in southern China and SE Asian coastal ecosystems. *Regional Studies in Marine Science*, **8**, 373–81.

Williams, G. A. and Morritt, D. (1995). Habitat partitioning and thermal tolerance in a tropical limpet, *Cellana grata*. *Marine Ecology Progress Series*, **124**, 89–103.

Wu, T. H., Tsang, L. M., Chan, B. K. K. et al. (2014). Cryptic diversity and phylogeography of the island-associated barnacle *Chthamalus moro* in Asia. *Marine Ecology*, **36**, 368–78.

Xu, N. (2016). Research on spatial and temporal variation of China mainland coastline and coastal engineering. PhD thesis, Yantai Institute of Coastal Zone Research, Chinese Academy of Sciences.

Yan, Y., Chan, B. K. K. and Williams, G. A. (2006). Reproductive development of the barnacle *Chthamalus malayensis* in Hong Kong: implications for the life history patterns of tropical barnacles. *Marine Biology*, **148**, 875–87.

Yancheva, G., Nowaczyk, N. R., Mingram, J. et al. (2007). Influence of the intertropical convergence zone on the East Asia monsoon. *Nature*, **445**, 74–7.

Yang, J., Wu, D. and Lin, X. (2008). On the dynamics of the South China Sea Warm Current. *Journal of Geophysical Research: Oceans*, **113**, C08003.

Yang, S. L., Xu, K. H., Milliman, J. D., Yang, H. F. and Wu, C. S. (2015). Decline of Yangtze River water and sediment discharge: impact from natural and anthropogenic changes. *Scientific Reports*, **5**, 12581

Yang, W. X. and Chen, Y. S. (1996). Community ecology of intertidal zone of Shengsi Archipelago I. Species constitution and seasonal variation of benthic biocommunity in rocky intertidal zone. *Chinese Journal of Applied Ecology*, **7**, 305–9.

Yang, W. X. and Chen, Y. S. (1998). Community ecology of intertidal zone of Shengsi Archipelago II. Community structure of benthic invertebrates in rocky intertidal zone. *Chinese Journal of Applied Ecology*, **9**, 75–8.

Yang, W. X. and Chen, Y. S. (1999). Community ecology of intertidal zone of Shengsi Archipelago III. Species distribution of benthos in rocky intertidal zone. *Donghai Marine Science*, **1**, 60–5.

Yang, Z. D. (1984). A preliminary study on the ecology of marine algae in Jiaozhou Bay. *Studia Marina Sinica*, **22**, 97–114.

Yeung, A. C. Y. and Williams, G. A. (2012). Small scale temporal and spatial variability in foraging behaviour of the mid-shore gastropod, *Nerita yoldii* on seasonal, tropical, rocky shores. *Aquatic Biology*, **16**, 177–88.

Yi, J. S. and Li, F. X. (1990). Ecology characteristics of benthic macro animals on the intertidal hard bed in Jiulong River estuary. *Tropic Oceanology*, **3**, 48–58.

Yi, S. (2011). Holocene Vegetation Responses to East Asian Monsoonal Changes in South Korea. In *Climate Change – Geophysical Foundations and Ecological Effects*. In Tech, Rijeka, chapter 8.

You, Z. J., Hong, J. C., Wong, Y. N. et al. (1997). Distribution pattern of species on the intertidal rocky shore of the Shenshan island, Zhejiang. *Journal of Ningbo University*, **1**, 64–71.

Yu, S. S., Wang, J., Wang, Q. L., Huang, X. W. and Dong, Y. W. (2014). DNA barcoding and phylogeographic analysis of *Nipponacmea* limpets (Gastropoda: Lottiidae) in China. *Journal of Molluscan Studies*, **80**, 420–29.

Zeng, D. G., Cai, R. X., Huang Z. G. et al. (1999a). Studies on marine fouling communities in the East China Sea I. Composition and distribution of species. *Donghai Marine Science*, **1**, 48–55.

Zeng, D. G., Cai, R. X., Huang Z. G. et al. (1999b). Studies on marine fouling communities in the East China Sea II. Biomass distribution. *Donghai Marine Science*, **1**, 56–9.

Zeng, D. G., Cai, R. X., Huang Z. G. et al. (1999c). Studies on marine fouling communities in the East China Sea III. Community structure. *Donghai Marine Science*, **4**, 47–50.

Zhang, Y. P., Zheng, J. and Wang, Y. N. (2000). Ecological characteristics of the intertidal mussels of the islands south of Zhejiang. *Transactions of Oceanology and Limnology*, **3**, 24–8.

Zhang, Y. Z., Chen, C. Z., Zhong, Y. P. and Hu, J. C. (1990). A preliminary study on the ecology of the intertidal zone along the coast of southern Fujian Province, China I. The rocky intertidal zone. *Journal of Xiamen Fisheries College*, **12**, 27–34.

Zhou S. Q., Guo F., Wen P. and Huang J. M. (2001). The studies on composition and distribution of mollusca in intertidal zone, Fujian island. *Journal of Fujian Fisheries*, **2**, 15–19.

Zhu, S. X., Yang, H. L., Wang, P. and Ying, Q. Z. (2006). Studies on inter-tidal zone ecology of Zhoushan archipelago during the summer, 2005. *Journal of Zhejiang Ocean University (Natural Science)*, **4**, 359–72.

Zhu, S. X., Zhou, W. and Zhang, F. J. (2010). Macrobenthic animal community characters in summer in the intertidal zones of Zhoushan archipelago, Zhejiang, China. *Journal of Marine Sciences*, **3**, 23–33.

Zhuang, Y., Jiang, J. F. and Wu, H. X. (2014). Effect of construction of Yangshan port on benthic macrofauna at the intertidal zone of Xiao-Yangshan island. *Transactions of Oceanology and Limnology*, **1**, 155–60.

# Chapter 15

# Biogeographic Comparisons of Pattern and Process on Intertidal Rocky Reefs of New Zealand and South-Eastern Australia

David R. Schiel, A. J. Underwood and M. Gee Chapman

## 15.1 Introduction to Biogeographic Comparisons

The rocky shores of New Zealand (NZ) and Australia provide many interesting comparisons in their intertidal species and structuring processes. Both countries are in the biogeographic realm of temperate Australasia and share many common species and closely related taxa. Here we review similarities and contrasts in communities and structuring processes, especially involving grazing invertebrates and macroalgae. We consider the similarity of the structure of intertidal shores of NZ and south-eastern Australia, a suite of important trophic interactions within and between regions, the utility of local-scale experiments in understanding large-scale processes and how we might better plan for and manage our coasts. The major comparisons are between the warm-temperate areas of northern NZ and New South Wales (NSW), and the cooler areas of southern NZ and south-eastern Australia. In the quest for 'ecosystem'-level understanding, which perforce involves large-scale events, there is an increasing tendency to minimise or ignore the hard-won insights we have gained from well-structured experiments across multiple sites. Because all large-scale effects must be manifested at local sites, it is incumbent on us to determine what scales up or down, and the caveats that make comparisons across biogeographic regions challenging. Here, we discuss these issues using austral shores as models.

Detailed contextual information is a necessary precursor of formulating testable hypotheses about the structuring forces of communities. On a broadscale, such information has often been categorised into biogeographic regions, which have historically given us regional classifications based on taxonomic affinities (e.g., Moore, 1949; Bennett and Pope, 1953). More recent studies have refined classifications with the use of genetic techniques (e.g., Waters et al., 2010) or used large databases to provide global syntheses (Spalding et al., 2007; Blanchette et al., 2009). The impetus for such work is varied, from uses for conservation of biodiversity and marine reserves (Olson et al., 2001; Spalding et al., 2007) to being a basis for understanding biogeographic shifts related to climate change (Thompson et al., 2002; Kortsch et al., 2015; Villarino et al., 2015). At the least, biogeographic studies provide a framework for asking relevant questions about structuring processes, framing them within a spatially relevant framework of similar species or community structure and assisting in integrating finer-scale processes within and across wider regions. In many ways, therefore, biogeographic patterns dictate what we study and the degree of inference appropriate to such studies. In this paper, we consider the 'realm' of 'temperate Australasia' (cf. Spalding et al., 2007) in terms of community structure, underlying processes of organisation and what we know about long-term changes within and across the different ecoregions within this realm.

There is a long list of biogeographic studies of rocky shore regions in eastern Australia and NZ (e.g., Moore, 1949; Bennett and Pope, 1953, 1960; Womersley and Edmonds, 1958; Knox, 1963, 1980; Waters et al., 2010). These show there are strong affinities in the marine flora and fauna between these two countries and many taxa in common. Shared key species include the ubiquitous subtidal kelp *Ecklonia radiata,* which forms a dominant biogenic habitat on subtidal reefs in both countries, the fucoid *Hormosira banksii,* which is common in the mid-intertidal zone and dominates many shores across the realm, and limpets of the genus *Cellana* that are often the most abundant intertidal grazers. In each country, there is a wide range of other grazing invertebrates, filter feeders, echinoderms, barnacles and smaller algal species, many of which are closely related.

Given these affinities, it would be reasonable to expect that similar processes are also shared in terms of their importance to community structure, responses to disturbances and underlying drivers of diversity. Furthermore, there should also be more 'affinities' in processes within the realm than among some other biogeographic zones of the world that do not share taxa to the same extent (e.g., Jenkins et al., 2008). Conversely, the strength and pervasiveness of trophic interactions, population processes and related community dynamics may be uniquely associated with particular conditions, even with similar suites of species. Processes can therefore vary considerably among shores within and across a biogeographic range, masking large-scale biogeographic patterns.

These alternatives have considerable importance if we are to achieve a more accurate and useful global view of how our nearshore systems work, what a changing climate and increasing multiple stressors and cumulative effects might do to them and how and what type of managerial interventions might be effective in preventing them from tipping into impaired states that are difficult or impossible to remediate. It is always the case, however, that our knowledge of the natural history of many species and their ecological interactions in communities can be limited. Consequently, our understanding of communities in different areas will usually depend on which species have actually been investigated, which poses a potential problem for any comparison of bioregions. In an attempt to bypass this issue, it is worth considering the comparability and understanding of processes involving major shared taxa that have been studied with reasonable intensity. We focus on species that we know can be, and often are, important in determining the structure of communities. These are species that have a diversity of types of interactions with other species and a range of responses to different types of disturbances. Throughout this discussion, we use 'community' to mean the assemblage of species found together in particular places or at particular times.

In this paper, our collective research over several decades in NZ and south-eastern Australia offers perspectives that are difficult to glean from a diffuse literature alone. These perceptions are backed by considerable in-the-field observations, sampling and hands-on experiments that lend empirical support to them. With that in mind, we consider the broad questions of:

(1) How similar is the structure of communities on intertidal shores of NZ and south-eastern Australia?
(2) How important are trophic interactions within and between regions?
(3) What might well-structured local-scale experiments tell us about larger-scale climate and stressor forcing?
(4) How can we better plan for and manage extreme events and the unexpected?

## 15.2 Examples of Biogeographic Comparisons from Descriptions

The structure of communities along a latitudinal gradient is obviously affected by a wide range of physical factors, including sea surface temperature (SST), wave-exposure, upwelling and downwelling, coastal morphology and, in some cases, inputs of sediments and nutrients from adjacent terrestrial sources. There have been several large-scale studies attempting to relate intertidal assemblages to latitudinal gradients, particularly with respect to biogeographic boundaries. In a study of the effects of upwelling on intertidal assemblages at a global scale, Blanchette et al. (2009) compared functional diversity among NZ, South Africa, California and Chile, using shores that spanned a latitudinal gradient in each country. At the level of functional groups, assemblages were remarkably similar in all countries, despite the lack of common species. Unfortunately, latitudinal variations in the assemblages within each bioregion were not compared. Wieters et al. (2012) examined changes in abundances of functional groups of biota in twenty locations along 6°S latitude on the Argentinian coast crossing between two biogeographic provinces. They found that the assemblages comprised three to four groups according to total and relative abundances, but these were not related to the latitudinal gradient or associated with the biogeographic boundary.

Considerable shore-to-shore variation at small scales often confounds large-scale patterns of intertidal assemblages. For example, Foster (1990) found that intertidal assemblages along the Californian coast were dominated by unexplained variations from shore to shore, rather than by any broadscale along-coast differences, with more small-scale variability in midshore assemblages than in lowshore assemblages.

### 15.2.1 Large-Scale Patterns of Assemblages in NZ and Australia

NZ is separated from Australia by around 1,500 km and 85 million years. The latitudinal range of NZ is around 13°S, with the mainland extending from ~34.4 to 47.0°S (Figure 15.1a–c). The geographic centres of research discussed here are mainly around Leigh in the north (36.3°S) and Christchurch in the south (43.5°S). The major biogeographic barrier in NZ is Cook Strait, which separates the North and South Islands. In terms of SST, NZ is bathed by subtropically derived waters in the north and cool-temperate waters in the south. The Australian studies discussed here are mainly from NSW (~28–37°S; Figure 15.1b), with many examples centred around Sydney (33.5°S). This region therefore lies in warmer waters than most of NZ.

Complex currents affect both regions. The strongest currents in NZ are the East Auckland Current, derived from the East Australian Current (EAC), which flows around the top of NZ, bringing warm waters to the north-east coast before deflecting offshore at East Cape (Figure 15.1b and 15.1c). The Southland Current brings cool waters to much of southern NZ. It arrives nearshore just north of Fiordland on the south-west coast, wraps around southern NZ, flows up the east coast and is deflected offshore around the north-east coast of the South Island. There are also strong tidal and wind-driven flows through Cook Strait. South-eastern Australia is heavily influenced by the EAC, which brings warm waters from the north, and has a few local gyres and offshoots across the Tasman Sea. The flows derived from this current are largely unidirectional and provide connectivity between Australia and NZ.

There are some similarities in the broadscale patterns of intertidal communities in Australia and NZ. Schiel (2011) presented detailed data on broad biogeographic distributions of key habitat-dominated species along the mid-intertidal zone of shores of NZ. In summary, sites along the east coast from ~34 to 46°S were largely dominated by fucoid algae, primarily a suite of fucoid algae on the lower shore, and the cosmopolitan fucoid *H. banksii* and underlying coralline algae on the midshore (Figure 15.2a). Algae were far less abundant mid-tidally on the west coast, where barnacles and mussels dominated (Figure 15.2b). Domination by mussels along the east coast was restricted to wave-exposed headlands (e.g., 43°S, Banks Peninsula, near Christchurch; Figure 15.2c). Barnacles dominated several midshore sites along the east coast, and were particularly common in more wave-exposed conditions of far southern NZ (Figure 15.2d). Within these trends, however, is considerable shore-to-shore variation. Noteworthy features are the dominance of lowshore fucoids such as *Xiphophora*

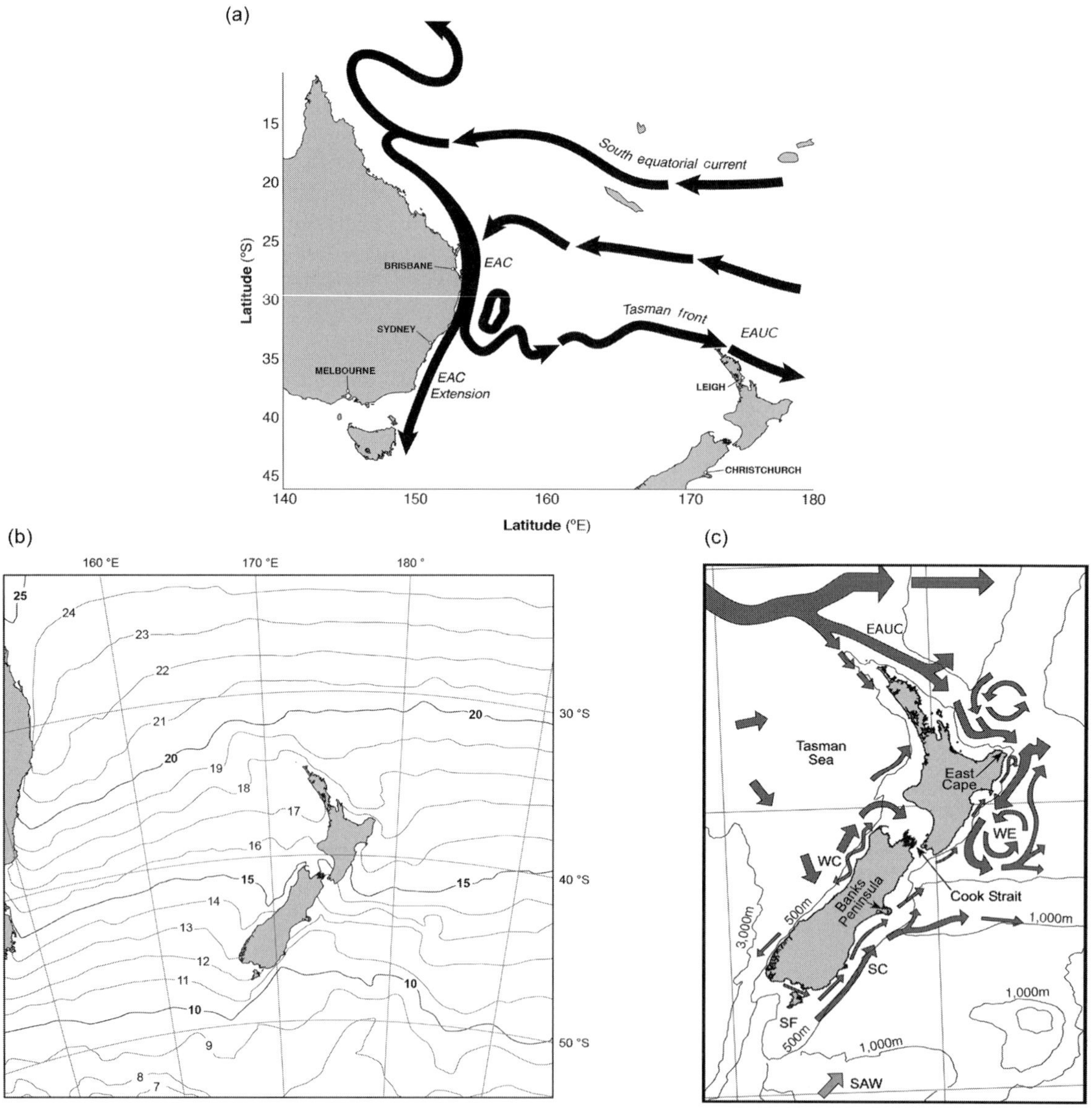


Fig. 15.1 Geographic position of NZ and south-eastern Australia showing (a) the major currents connecting the countries, (b) the SST isoclines showing annual average temperatures and (c) the major currents around NZ. EAC, East Australian Current; EAUC, East Auckland Current; SC, Southland Current.

*chondrophylla*, *Carpophyllum angustifolium* and *Carpophyllum plumosum* in the north and the southern 'bull kelp' (actually a fucoid) *Durvillaea antarctica and Durvillaea poha* in the south, again with much small-scale variation.

On the upper shore, above areas extensively occupied by large macroalgae, the shores throughout NZ are dominated by barnacles, algal crusts, limpets and trochid gastropods. The cosmopolitan turbinid *Lunella smaragdus* (reaching diameters up to ~40 mm) occurs throughout the mid- and low-tidal areas along the east coast and much of the west coast. Prosobranch limpets of the genus *Cellana* are the most common large grazers, but species distributions change south of East Cape (37.7°S, 178.5°E). *Cellana denticulata* is

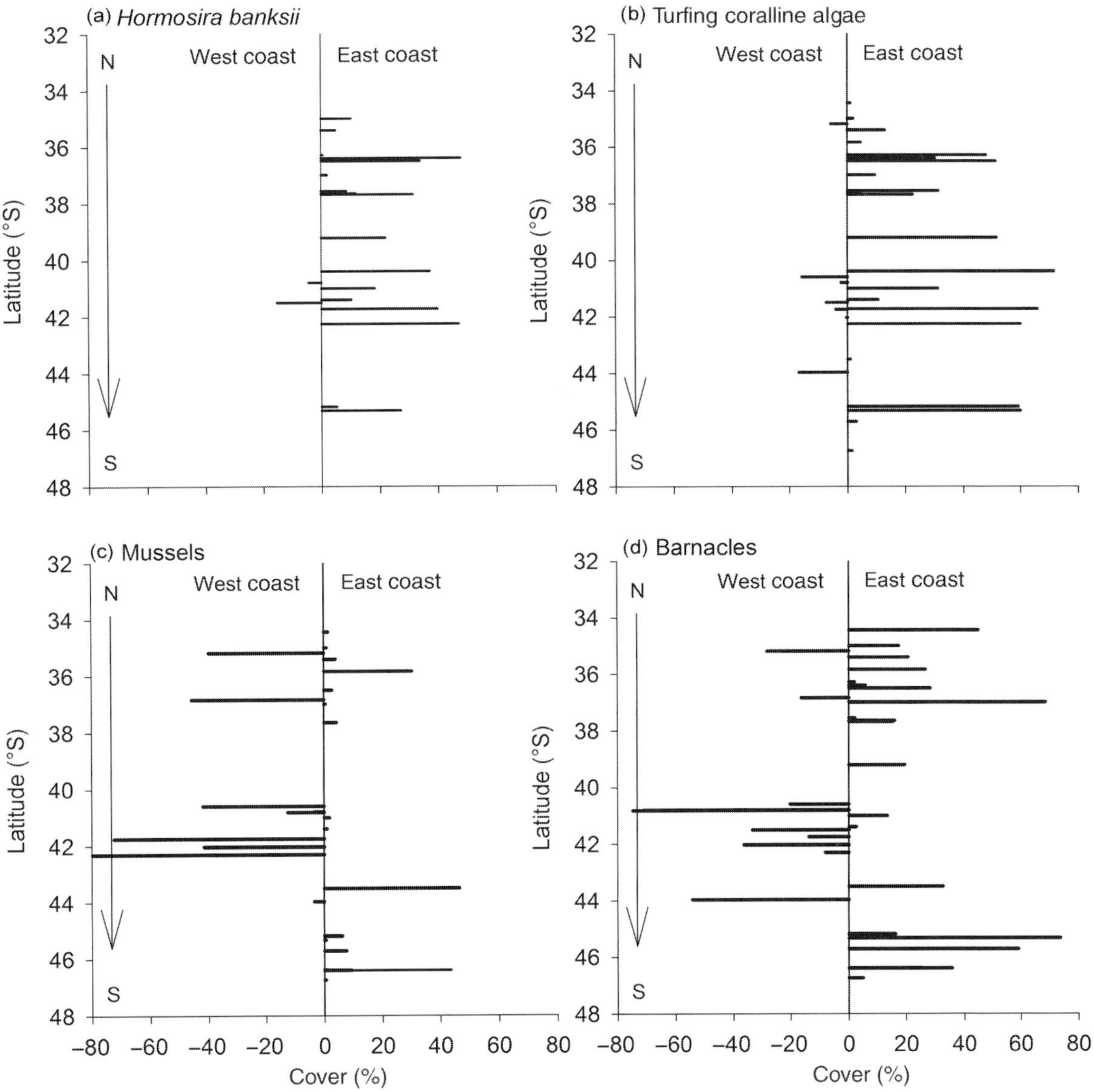


**Fig. 15.2** The mean percentage cover of (a) the fucoid alga *H. banksii*, (b) turfing coralline algae, (c) mussels (*Perna caniliculus* and *Mytilus galloprovincialis*) and (d) barnacles (*Chaemosipho* spp., *Epopella plicata* and *Austrominius modestus*) in the midshore at sites on the east and west coasts around NZ (spanning from 34 to 47°S).

the largest and most abundant of the southern prosobranchs, with large *Cellana flava* abundant on some exposed sites and large *Cellana strigilis* common in the far south. The dominant mussels throughout NZ are *Perna canaliculus* in the low-intertidal zone, *M. galloprovincialis* slightly higher on the shore along with the ribbed mussel *Aulacomya maoriana*. Little black mussels, *Limnoperna pulex*, are often found in dense aggregations in the mid-tidal zone.

Unfortunately, there are few detailed descriptions of intertidal communities along much of the southern parts of south-east Australia, particularly Victoria and South Australia. *Hormosira* can occupy extensive areas of the midshore. Keough and Quinn (1998) found, for example,

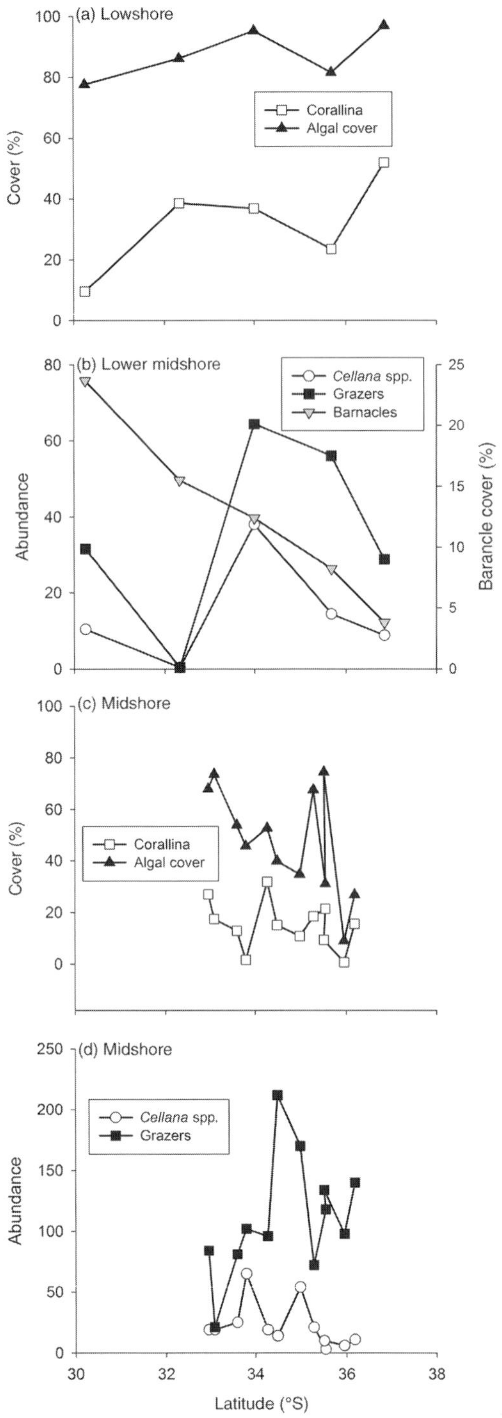


**Fig. 15.3** Some trends in major taxa on (a) lowshore coralline and other algal cover, (b) lower midshore limpets *C. tramoserica*, other grazers and barnacles, (c) midshore coralline and other algal cover and (d) midshore *Cellana* limpets and other grazers along the coast of NSW. Data are means per 50 × 50 cm quadrats. In (b), the right axis is mean percentage cover of barnacles and the left axis is the mean number of all grazers and of the limpet *C. tramoserica*.

that the natural cover of *Hormosira* varied between about 50 and 100 per cent over six years at Mornington Peninsula, Victoria platforms (38.3°S, 144.7°E). Coralline algae are abundant both in the understory and surrounding patches. Superficially, some of these intertidal benches are similar to those along the central coast of the South Island of NZ, with dominant *Hormosira* beds in mid- to low-intertidal areas and a suite of species, such as the fucoids *Phyllospora comosa* and *Cystophora* spp., mentioned later.

Most information on large-scale patterns in intertidal assemblages comes from 700 km of the coast of NSW. Mid-tidal areas of semi-exposed shores are generally dominated by a suite of molluscan grazers, including *Cellana tramoserica* and, on more sheltered shores, *Austrocochlea porcata, Bembicium nanum* and *Nerita atramentosa*. The local turbinid gastropod, *Lunella undulata*, can be abundant on some shores, but is sparse or absent from nearby shores. Barnacles can dominate large areas, with *Tesseropora rosea* and *Catomerus polymerus* most common on more wave-exposed shores and *Chamaesipho columna* on more sheltered shores. Algae include ephemeral species, such as *Ulva* spp., *Gelidium pusillum*, encrusting species, such as *Hildenbrandia rubra* and *Ralfsia verrucosa* and turfing corallines (Dakin, 1953). *H. banksii* is largely confined to tide pools and their margins (O'Gower and Meyer, 1971). These mid-and upper-shores are similar to many along north-eastern NZ.

Lowshore areas along NSW are dominated by a diversity of macroalgae, such as *Sargassum* spp. and coralline turfs (Dakin, 1953). There are generally few mussels, although patches of *Brachidontes hirsutus* can be found on more sheltered shores. The high-intertidal zone is generally occupied only by littorinids (Chapman, 1994).

Sampling over four years along five shores spanning 7°S latitude of NSW showed some latitudinal trends, such as an increase in algal cover (both corallines and other algae), at lowshore levels from north to south (Figure 15.3a), but there were only five pairs of shores over the this latitudinal range. There was a decline in cover of barnacles at the midshore level, (primarily *T. rosea*; Figure 15.3b) and no broadscale changes to numbers of grazers over this spatial scale, including the limpet *C. tramoserica* (Figure 15.3b), but large variations among shores. This suggests that grazers may be less influenced by broadscale environmental differences, or that species that dominate midshore areas may show weaker broadscale patterns, as suggested by Foster (1990).

In other Australian studies, changes in intertidal assemblages and abundances of individual species over ~1,500 km similarly showed variation that was not related to distance among sites, their position on the coast or to environmental variables such as SST or heat stress (Hidas et al., 2010; Lathlean et al., 2015). Settlement and early mortality of the barnacle *T. rosea* also varied idiosyncratically among shores, showing no general latitudinal trends (Lathlean et al., 2010). Nevertheless, sampling the midshore of twelve sites over only 3°S latitude showed there was a decrease in total algal cover from north to south, although this was not apparent for individual species, such as *H. banksii* or *Corallina officinalis* (Chapman and Underwood, unpublished data; Figure 15.3c) and an increase in numbers of grazers, although not of individual species, such as *C. tramoserica* (Figure 15.3d). Again, however, shore-to-shore variation unassociated with latitude dominated the pattern.

There have been few seasonal comparisons of large-scale patterns in intertidal assemblages along coastlines, although it is likely that seasonal conditions may affect the patterns found. Underwood et al. (2008) found a latitudinal trend for increasing numbers of species per site (sampled over two consecutive years) from north to south in winter, which was not present in summer. This seasonal difference disappeared when species density was used (i.e., mean number of species per sample), which gave spurious patterns of changes in species richness (the total number of species sampled per site).

### 15.2.2 Issues About Large-Scale Descriptive Comparisons

There are many problems of knowing how well or how completely assemblages of species have been described when comparing patterns of biogeography (e.g., Nelson et al., 2013). Accumulating species lists by area over periods of a century or more is what comprises most biogeographic databases. Single occurrences of species are, however, common and species lists tend to be longer around easily sampled sites or centres of collection (Gaines and Lubchenco, 1982). When it comes to understanding present distributions of species, it is almost always the case that a complete census of species in any area is impossible. This is not only because of potential problems of finding and identifying small individuals, and the rarity or patchy distribution of some species, but also because there may be large seasonal variation in species' occurrences and they simply may not be present when a census is done. As pointed out by Gaston (1994), biogeography often involves sampling over large spatial scales, which is logistically difficult to do in a relatively short space of time and, as a consequence, studies over a coastline may confound seasonal or other temporal changes in fauna or flora with descriptions of geographic ranges of species. There can also be major problems for interpreting patterns of numbers of species from shore to shore within a bioregion. For example, Connor and Simberloff (1978) found that a major predictor of diversity of species of plants on the various Galapagos Islands was the number of botanists who had visited each island.

The methodology used to sample can also profoundly influence comparisons from one region to another. For example, the 'grain' of sampling is influenced by the size of sampling units used, which can have great effects on what is actually sampled (Gray, 2001). Obviously, the extent of the sample (the overall amount of habitat sampled) will be an important issue (Wiens, 1989). In addition, the focus of a sample (the scale at which the sample units are aggregated or averaged; Scheiner, 2003) will create a specific pattern of numbers of species from place to place along a coastline and therefore a specific interpretation of biogeography. Underwood et al. (2008) demonstrated how differences in frequency of occurrence in sampling units in addition to the variance and skewness of abundances among sampling units can affect the number of species recorded in different places. Unless all these aspects of sampling are comparable, inter-regional descriptions or comparisons are open to different interpretations.

### 15.2.3 Ability to Infer Specific Processes

An insoluble problem of trying to use descriptions of biogeographical patterns to compare the ecology of different places is that, in themselves, they convey little or nothing about the relevant processes involved in creating and maintaining similarities and differences. Any pattern can potentially be explained by quite different conceptual models about processes (see particularly Underwood and Denley, 1984). Trying to attribute processes to patterns requires discrimination among potential models, which is always going to require some agreed process of elimination of models that are irrelevant, inadequate or wrong (Simberloff, 1980; Underwood, 1990). The accepted process of elimination or discrimination is through experiments (well-discussed historically in marine ecology but worth repeating now as more monitoring studies are done worldwide).

## 15.3 Experiments Across Biogeographic Zones

In contrast to broad patterns of distributions of species, ecological processes are difficult to measure over large, biogeographical scales due to the logistics of doing experiments over hundreds of kilometres. Several studies have therefore attempted to test the generality of ecological processes across large spatial scales by comparing the results of small-scale, but well-designed, experiments across multiple sites spanning such scales. These generally come in two forms of comparisons. In a few studies, identical experiments were set-up on replicate shores over many hundreds of kilometres to compare responses directly to experimental manipulations at these large scales. For example, using replicated experimental treatments over large spatial scales, Crowe et al. (2013) examined the effects of disturbance and loss of canopy algae across nine locations in Europe, showing no general patterns. Removal of algal canopies on the underlying community had strong effects at some sites, interacting with physical disturbance on a subset of these, but at other sites there was no effect of canopy removal. Importantly, there were no broadscale trends in effects of canopy removal or physical disturbance, but considerable shore-to-shore variability. Similarly, Martins et al. (2014) showed different responses to removal of canopy algae among sites at the Azores, Canary Islands and Portugal, probably because the different shores were dominated by different algae that would not be expected to respond similarly to the same form of disturbance.

In Australia, an experimental study of recruitment and early development of algal assemblages on intertidal shores across 800 km of NSW similarly showed no biogeographic patterns, but variable patterns among shores and seasons (Chapman and Underwood, 1998). These were due to differences in recruitment among a pool of common species and sporadic arrivals of less common species. Other studies have shown broad latitudinal trends to the same experimental treatments. Coleman et al. (2006) compared the effects of grazing by experimental manipulation of grazers over 17°N latitude in Europe, showing a strong latitudinal effect of grazers on algal cover and spatial variability in algal cover. Similarly, Bulleri et al. (2012) showed latitudinal differences in temporal stability of intertidal assemblages at sites between Italy and the North Atlantic, due to synchronous changes in cover among species further north decreasing temporal stability.

More commonly, however, it is necessary to compare processes at large spatial scales by examining the outcomes of experiments designed in different ways to test similar but not necessarily the same hypotheses. One caveat is therefore that where experiments testing similar processes in different places have used different experimental designs, their results may only be comparable quantitatively and not necessarily quantitatively (Underwood and Petraitis, 1993). Nevertheless, such comparisons can be successful in inferring the relative importance of different processes in structuring intertidal assemblages (e.g., Jenkins et al., 2008). In both Australia and NZ, many experimental studies have investigated ecological processes over different spatial scales using comparable experimental designs. There have, however, been no published experimental studies done simultaneously over several sites using the same experimental designs at the same times (to unconfound any temporal variability) to allow direct quantitative comparisons. Even so, in each country there is good understanding of the role of different ecological processes on intertidal shores, which has been derived from well-designed experimental studies that allow inferences to be drawn of the importance of these processes between Australia and NZ.

## 15.4 Community Processes in NZ and South-Eastern Australia

### 15.4.1 Trophic Dynamics

Here we focus on trophic dynamics involving algae and grazing invertebrates. This type of interaction has historically been the basis of many models of community structure (Lubchenco and Gaines, 1981; Hawkins and Hartnoll, 1983) and continues to be so (Foster and Schiel, 2010). We begin with the shores of NSW, where grazers are often dominant.

Grazing by the limpet *C. tramoserica* has been shown experimentally to control the upper levels of midshore distributions of branching, fleshy algae (Underwood, 1980; Underwood and Jernakoff, 1984). The limpets consume microflora, including blue-green algae, diatoms and the early life history stages of macroalgae, by scraping them from rock surfaces. When limpets are excluded or greatly reduced in density, macroalgal sporelings grow and eventually turfing algae take over the surface and prevent further grazing by limpets (Underwood and Jernakoff, 1981; Figure 15.4a). Lower on the shore, algae grow faster and are less affected by deleterious conditions during low tide, so algae overcome the capacity of grazers to consume them and then dominate the shore (Underwood and Jernakoff, 1981).

Grazing limpets, many species of snails and the starfish *Parvulastra exigua* are in large abundances in many areas. As a result, supplies of

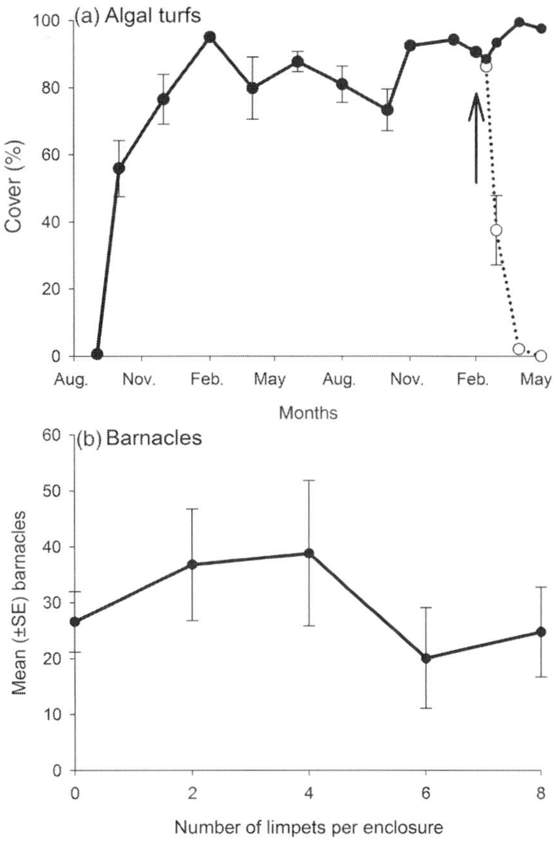


**Fig. 15.4** (a) Mean percentage cover of algal turf (primarily *Ulva* spp.) in areas above the upper limit of turfing algae where limpets were experimentally excluded by fences at midshore levels in NSW (from Underwood, 1980). The arrow indicates removal of the fences from two replicate plots (open circles, dashed line), and the solid black circles represent those plots with fences intact. (b) The density (per 400 $cm^2$) of newly recruited barnacles, *T. rosea*, in experimental areas with different numbers of limpets (from Underwood et al., 1983). SE, standard error.

microalgal food are limiting and there is much competition for food (see later). In contrast, at low levels on the shore, algae grow faster than the limpets can consume them and quickly reach a size that is too large for microalgal grazers to eat them. Even where there are also limpets, such as *Siphonaria denticulata* that eat fleshy macroalgae, they cannot consume entire algae, so there is always a dense macroalgal cover (Underwood and Jernakoff, 1981; Creese and Underwood, 1982).

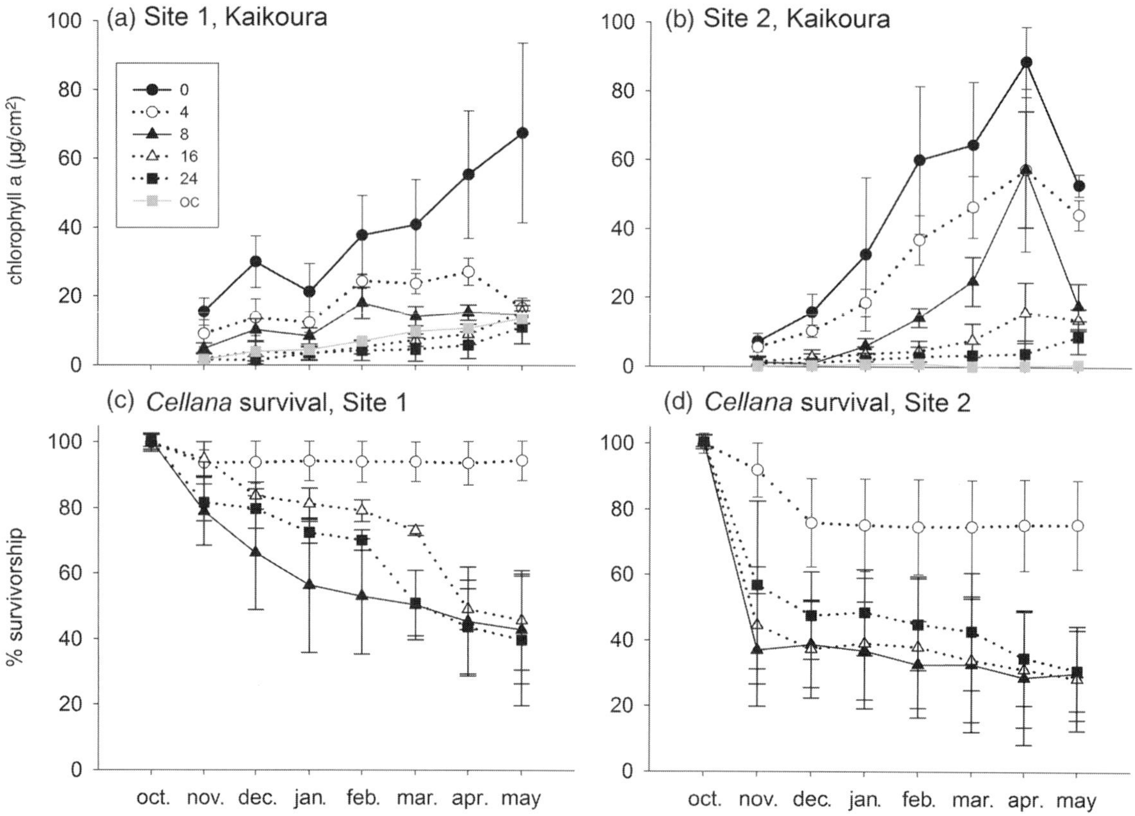


**Fig. 15.5** The mean (±standard error) chlorophyll *a* levels (µg/cm$^2$) in 0.25 m$^2$ plots with 0, 4, 8, 16, 24 limpets (*Cellana ornata*) in southern NZ. OC, open fence control (other controls did not differ significantly from the open control and are not shown for graph clarity) at two Kaikoura sites (a, b) (from Dunmore and Schiel, 2003), and the percentage survival of *C. ornata* through time in the experimental treatments at two Kaikoura sites (c, d).

Also importantly, the grazers, especially limpets, remove newly settled barnacles on sheltered shores (Figure 15.4b). The interactions with barnacles are complex and non-linear because, if grazers are absent or in small numbers, algae grow and make rock surfaces unavailable for barnacle settlement (Underwood et al., 1983). Barnacles only survive well where grazers remove algae, but where they are not so numerous that they also remove the barnacles. On wave-exposed areas, grazers are less numerous and barnacles survive well and are dominant occupiers of space above the levels at which algae thrive.

There has been little work in NZ on interactions between grazers and barnacles, but some research on limpet–algal interactions. In northern NZ, limpets, particularly *Cellana radians*, have much the same effect as *C. tramoserica* in NSW (review by Creese, 1988). In south-eastern NZ, limpets influence the abundance of microalgae and some ephemeral species, particularly on upper-shore levels (Guerry et al., 2009). Dunmore and Schiel (2003) found experimentally at two sites that, as limpet densities increased, there was reduced chlorophyll *a* content on rock surfaces (Figure 15. 5a and 15.5b), which had consequences on limpet survival (Figure 15.5c and 15.5d) and growth. Where limpets were excluded or in low density, low-lying ephemeral species and seasonal blooms of the larger red alga *Porphyra* sp. occupied up to half of the primary substratum. They did not note an effect of limpets on barnacles, but saw indirect effects involving grazing. Where limpet mortality was high in the presence of barnacles, they noted an increase in macroalgae, which smothered barnacles and killed many of them (as in NSW). Despite numerous experiments at sites over 500 km of the South Island, no fucoids ever appeared in patches where limpets were excluded for up to a year (Schiel, unpublished data). This contrasts with the situation in NSW, where grazers slowed down the recovery of *Hormosira* in experimentally cleared areas (Underwood, 1998). Furthermore, in NZ, numerous attempts to transplant limpets into cleared patches within fucoid beds (often only 1–2 m away from 'limpet mounds') were unsuccessful. Limpets (mostly *C. denticulata*) never attached firmly and no algal effects have been detected. Furthermore, transplanted germlings, juvenile and adult *H. banksii*, the dominant and most desiccation-tolerant fucoid, never survived in limpet-dominated areas, even when limpets were excluded for many months.

In NZ, another abundant gastropod, *L. smaragdus,* can variably affect algal assemblages. Extensive experimentation showed few effects of this species, except on ephemeral algae (Schiel, 2006). These algae had greatly reduced abundances in the two higher-density treatments (five and ten *Lunella* per 0.1 $m^2$) in the absence of a fucoid canopy. There was a small effect on *H. banksii* germlings in the absence of a canopy (Walker, 1998). In contrast, Creese (1988) reviewed work done in north-eastern NZ in the 1970s–1980s. Here, the cover of algae in the mid-tidal zone is sparser than in much of southern NZ. *C. radians* removed much of the pre-existing crustose coralline algae and reduced the amount of ephemeral green algae present. Encrusting algae developed at low- and mid-tidal levels in the presence of *Lunella,* but not in the presence of *C. radians,* and foliose red algae developed on the lowshore when *Cellana* was present but not with *Lunella.* Creese concluded that the combined effect of the two grazers was to prevent establishment of macroalgal stands, even at lowshore levels adjacent to the fucoid zone. He also stated that, in northern NZ, as found experimentally in NSW, grazers helped set the upper limit of distribution of many algae, but this was also affected by heat stress. These effects were not seen in southern NZ. The equivalent turbinid in NSW is *L. undulata*, but it seems to have little influence on the distributions or dynamics of intertidal algae in NSW, being confined to pools or very low areas on the shore.

There are therefore clear latitudinal and biogeographic differences in grazing effects, as well as significant differences relating to degrees of wave-exposure and variation among sites. On the upper-shore above fucoid beds in southern NZ, grazing gastropods interact significantly with microalgae, ephemeral algae and barnacles, but they have few if any effects on fucoids. Fucoid beds and large patches dominated by limpets are stable, at least in a broad sense, at some sites. Aerial photographs from the 1970s show pretty much the same configuration of habitats in evidence today. In the hotter conditions of north-eastern NZ and NSW, it appears that grazing gastropods are far more influential on intertidal assemblage structure, particularly in the mid- and upper-shore levels where foliose algae experience harsher conditions.

### 15.4.2 Densities of Gastropods and Competition

In areas where grazers are important, their consumption of resources often reduces available food below their needs, resulting in extensive competition within and between species. In NSW, numerous experiments have demonstrated that intraspecific competition reduces densities of gastropods (particularly *C. tramoserica* and the snail *N. atramentosa*). Interspecific competition is also intense, with *N. atramentosa* outcompeting *C. tramoserica* and both being superior competitors to the snail *B. nanum*. Competition is more intense towards the upper-shore and in summer compared with winter (Underwood and Jernakoff, 1984; Underwood, 1984a, 1984b). The former is because microalgal food is less abundant at higher-shore levels, where there is greater heat stress due to longer periods of emersion. In summer, there is also much more heat stress during low tide than in winter. Competition among the grazers is widespread and complex (Murphy and Underwood, 2006; Underwood and Murphy, 2008).

In northern NZ, *C. radians* has similarly been shown to suffer intraspecific competition at enhanced densities (Creese, 1988). There was, however, no evidence for interspecific competition with a cohabiting snail *L. smaragdus*, although they appear to interact in determining the composition of algal assemblages and their extension towards the upper-shore.

Along the South Island, the limpet *C. ornata* showed strong intraspecific competition in areas where densities were experimentally increased, resulting in reduced growth and increased mortality (Dunmore and Schiel, 2003; Figure 15.5c and 15.5d). It was interesting to note that increased mortality was similar for all densities >4 per 0.25 $m^2$, which are common in many sites. The interactions may be more varied and intense in NSW, possibly due to the diverse suite of abundant grazers there, but competitive interactions for food are a feature of the structure of communities on seashores in NZ, particularly the upper-shore and in more extensive grazer-dominated areas of north-eastern NZ.

### 15.4.3 Disturbances to *H. banksii* Assemblages

Many experiments have been done over many years on the effects of removal of key species of algae. These include *H. banksii* in NZ and Australia (e.g., Underwood, 1998, 1999a; Lilley and Schiel, 2006; Schiel and Lilley, 2007, 2011), other fucoids in southern NZ (Schiel, 2006; Tait and Schiel, 2011) and all algae, particularly *Corallina* spp., in NSW (Chapman and Underwood, 1998). We illustrate effects and trajectories of recovery with examples from each country.

In NSW, a storm with waves of approximately 6 m over two days removed large patches of *H. banksii*. In some cases, only fronds were removed, but, in others, entire plants were removed by a combination of wave-shock, sand-scour and rolling boulders. Where parts of plants remained, the recovery of the canopy took around two years, but where entire plants were removed recovery took at least five years (Figure 15.6a; Underwood, 1998). A follow-up experiment (Figure 15.6b) involving the removal of *H. banksii* demonstrated the potential for consequences of different degrees of removal on the surrounding community. Complete removal of fronds and holdfasts showed almost no recovery compared to controls and removal of fronds only. Because *H. banksii* provides shelter for predatory whelks, removal of the canopy reduces the effectiveness of whelk foraging. Barnacles are then able to survive in areas where, normally, whelks under canopies prevent this (Underwood, 1999a). Comparable experiments have not been published for northern NZ, but the generally sparse cover and often intense grazer effects highlighted in Creese's review (1988) suggest that population recovery of *H. banksii* would be slow as well.

In southern NZ, after experimental removal of *Hormosira*, there were large differences in the trajectories of recovery between sites separated by ~400 km. At Kaikoura (Figure 15.6c), recovery to around 100 per cent cover took about two years after the disturbance, regardless of whether it was a one-off or continual removal of *H. banksii*. Further south at Moeraki (Figure 15.6d), recovery took six to eight years. This was primarily because the substratum became dominated by turfing coralline algae which largely prevented recruitment of *H. banksii* (Lilley and Schiel, 2006, Schiel and Lilley, 2007).

'Recovery', however, has several dimensions other than a return of 100 per cent cover of the original dominant. At least in the algal beds of southern NZ, overall species diversity is highly correlated with the cover of fucoid algae (Schiel and Lilley, 2007). There was around a 17 per cent difference in species diversity between the removal experiments eight years after initiation (Schiel and Lilley, 2011). This was because the *H. banksii* in removal plots had not fully recovered to the lengths and sizes of those in control plots. Furthermore, there was still a significant difference in primary productivity between treatments eight years after initiation (Tait and Schiel, 2011).

### 15.4.4 Disturbance and Mussels

Even greater effects of storms have been shown on sheltered shores in NSW when the large storm described earlier removed the mussel *B. hirsutus* from many areas (Figure 15.7). An initial cover of 40–60 per cent by the mussel dropped to near-zero and stayed there for at least thirty years (Figure 15.7a; Underwood, 1999b). Where mussels were removed, there was increased grazing pressure by limpets (Figure 15.7b) and an expansion of coralline algae (Figure 15.7c). It

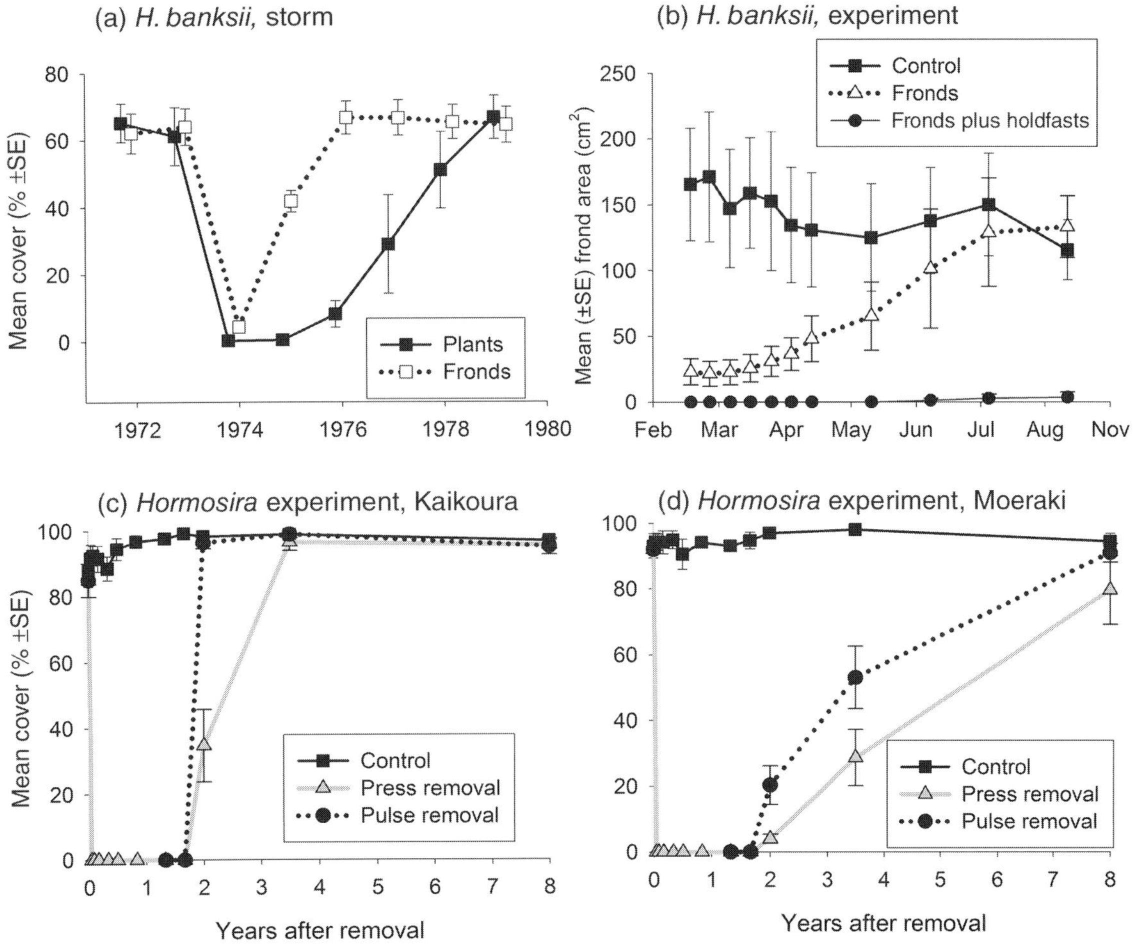


**Fig. 15.6** (a) Recovery of the fucoid alga *H. banksii* in NSW, Australia following a storm in 1974 where some patches had fronds removed, and others had whole plants removed; (b) recovery following an experimental manipulation where fronds and fronds plus holdfasts were removed in NSW. Recovery of *Hormosira* in southern NZ following experimental pulse and press removal of whole plants at (c) Kaikoura and (d) Moeraki. See Underwood et al. (1999) for (a), Underwood et al. (1998) for (b) and Schiel and Lilley (2011) for (c) and (d).

is interesting to note that during several years in the early 1990s when limpets had largely disappeared from one site, there was only a small expansion of corallines but still no resurgence in *Brachidontes*. This long-term failure to reach pre-disturbance assemblage structure was no doubt the consequence of many factors, including grazing, competition with corallines, failure of mussel larvae to arrive in sufficient numbers or predation before they recruited effectively.

In southern NZ, it appears that recruitment dynamics involving larval arrival and predation on recruits are major population controls in mussels. Rilov and Schiel (2006a) experimentally examined survival of small mussel recruits in mussel-dominated sites on Banks Peninsula (near Christchurch). Predation strongly affected survival, but there were considerable site-to-site differences. Unlike what is documented in much of the marine ecological literature, the predators were not starfish, but, instead, a suite of labrid fishes and subtidal crabs that foraged over areas during high tide and were particularly abundant when there were subtidal reefs nearby. This

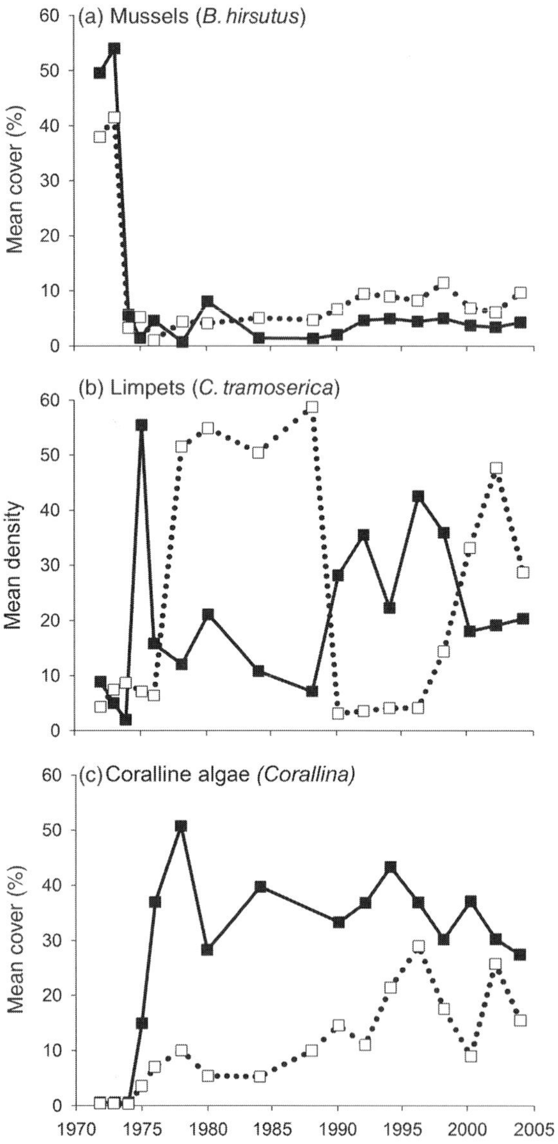


**Fig. 15.7** (a) Percentage cover of mussels, *B. hirsutus*, (b) number of limpets, *C. tramoserica*, and (c) cover of coralline algae, in 50 × 50 cm quadrats in two patches (patch 1 = black symbols, patch 2 = open symbols) (~6–8 $m^2$) where storms in 1974 removed the mussels in NSW (from Underwood, 1999b).

experimental work was extended to the outer coast of north-eastern NZ. In most cases, survival of mussel recruits was greatest at sites distant from subtidal reefs, where there were reduced numbers of mobile predators (Rilov and Schiel, 2006b, 2011). In other studies, recruitment limitation and significant effects of predation by starfish affected mussel-dominated sites along headlands of the east and west coasts of the South Island (Menge et al., 1999).

From these examples, it is difficult to determine the strength of similar structuring processes involving mussel assemblages in the two countries. It is noteworthy, however, that the long-term loss of mussels seen in NSW has not been recorded in NZ. The persistence of coralline algae and grazing limpets is, however, a feature of some algal assemblages in northern NZ, and the expansion of turfing corallines has had long-term effects on fucoid assemblages following disturbances in southern NZ (Lilley and Schiel, 2006; Schiel and Lilley, 2007, 2011).

## 15.4.5 Similarity of Communities and Processes

Given some similarity in the assemblages and component species of south-eastern Australia and NZ and the results of extensive experimental work, it appears that the major processes controlling patterns of composition of the communities are also broadly similar. In more sheltered areas, grazing has a major influence in eastern Australia and northern NZ. There is competition among grazers, and microalgae are generally overgrazed. Predation appears not to be a major process influencing the populations of grazers. The grazers are, however, subject to heat stress during low tide, which can cause them to be associated with crevices or to suffer death when physical conditions are extreme.

In more exposed areas, grazing is not so important, at least in mid- and low-areas, and sessile species dominate. They are partially controlled by predation, but the predators are subject to physical stress due to waves and often confined to crevices or pools. Otherwise, patchiness in space and time of the major users of space is often a result of variations in recruitment. Both regions show great similarity in responses to disturbance by storms. Canopy cover is torn off, with consequent changes in other components of the assemblage. Canopies take several years to recover, so that short-term (one day) disturbances have long-term impacts. In the cooler southern regions of both countries, there is considerable domination by fucoid algae from the midshore downwards. Grazing dynamics by gastropods in large algal beds seem to be relatively unimportant, whereas the demographics of the fucoids – dispersal, settlement, recruitment and growth – in combination with disturbances are dominant processes. The major exception was the almost pervasive effects of an herbivorous fish on the largest, lowshore fucoid *D. antarctica/poha* (Taylor and Schiel, 2010). As well, when disturbances lead to dominance by turfing corallines, there is impaired recovery of *Hormosira* beds.

These similarities in pattern and processes demonstrate important broad similarity in the intertidal ecology of the two areas.

## 15.5 Generalities of Biogeographic Comparisons

### 15.5.1 How Do We Know Enough to Compare the Shores?

This is an important question because our understanding of the make-up of, and processes affecting, intertidal communities comes from well-designed and, sometimes, well-repeated, experimentation. Inevitably, such experimentation is of small spatial and usually short temporal scales. Logistic problems have largely prevented larger-scale, adequately replicated experimentation. Our culture as scientists is not always helpful. For example, a contributing factor to there being few current descriptive studies of some areas of Australian shores (in common with many parts of the world) is that journal editors and reviewers of papers are often of the opinion that descriptive papers rarely can be based on theory, even though they can test current theories (e.g., Foster, 1990). With monitoring programmes being enacted worldwide, probably the most common source of descriptive data is the grey literature of agency reports (which may or may not be peer-reviewed).

Nevertheless, comparisons among biogeographic regions depend on there being an adequate description of the systems being compared. Often, if not usually, these are patchy or of very different scope, intensity or reliability. Here, we also draw the distinction between structure and process. In terms of the available descriptive data, their timescales and spatial extent, comparisons become compromised. Similarly, there is still a great need for small-scale experimentation to provide basic recognition and understanding of the processes that influence the structure of assemblages and how they change. It is also the case that many experiments need to be repeated in time and space to gauge how general the processes and their outcomes actually are. Experimental results often are quite different in different places or at different times (e.g., Chapman and Underwood, 1998), for example, if done during a period of intense El Niño or La Niña conditions. Generalising from single studies in one (or a few) places at one (or a few) times has long been shown to be inadvisable (but it almost inevitably happens anyway).

There is a further caveat in the dependence on small-scale experimentation. Where the experiments are themselves not well designed, the reliability of results is not particularly great. This is so at a very basic level of experimental design, such as Hurlbert's (1984) discovery that more than 25 per cent of papers had no relevant replication. This problem does not go away despite considerable discussion in the ecological literature (e.g., Underwood, 1986) and ever-increasing sophistication in the available help (e.g., Resetarits and Bernardo, 1998; Quinn and Keough, 2002). Again, as with descriptive studies, scientists need to take great care in formulating relevant testable hypotheses, especially as science funding gets increasingly tight, in any attempt to re-do experimental work in new sites, at different times and with better, less confounded designs.

### 15.5.2 What Can Be Scaled Up from Small-Scale Studies?

Small-scale studies are well known to have problems in identifying or dealing with large-scale processes, such as oceanographic processes, large-scale or long-persistent anthropogenically caused changes and, of course, influences of climatic change. Nevertheless, there have been many examples of comparative experimentation along coastlines and across biogeographic scales, using similar experiments in different places. These are often very informative about these issues (e.g., Cole and McQuaid, 2011; Bulleri et al., 2012; Crowe et al., 2013; Menge and Menge, 2013).

One very useful example of scaling-up in biogeographic comparisons is that by Jenkins et al., (2008). They compared the structure, major contributing taxa and analyses of processes between the rocky shores in the eastern (European) and western (American) Atlantic. They had available a wealth of descriptive and experimental studies from which to get the necessary information. Note that the hypotheses being tested and the methods and experimental designs differed among studies. They did not analyse in detail the quality or worth of the many individual studies, but reached conclusions that are undoubtedly robust using the available information. In broad terms, they identified that the assemblages on the two sides of the Atlantic are remarkably similar in structure, yet there are major differences in the processes that are important in regulating the composition of these assemblages. Of particular interest was the absence of patellid limpets from the western Atlantic and the related reduction in importance of grazing in the west and differences in competitive interactions across the Atlantic. This integrated review shows that much can be achieved from data from diverse small-scale studies. Patterns and some major taxa were similar; processes were not (Jenkins et al., 2008).

In contrast here, we infer that several patterns in Australia and NZ were similar and the contributory major taxa were the same or similar, at least within similar coastal latitudinal provinces (cf. Spalding et al., 2007). We also found that, as far as we can tell from the available information, the processes regulating communities are relatively similar between these two biogeographically adjacent areas. Nevertheless, as is often the case in examining ecological concepts, results and outcomes vary from study to study. Comparisons at sites across the Atlantic and one between Australia and NZ do not give the same sorts of interpretations. This sets a limit of how far up we might be able to scale. What happens in comparisons across the Atlantic did not predict what might be happening in the southern hemisphere. This is, of course, not a problem because we should be content with the knowledge that results from one part of the world to another are not the same. It also underpins that more comparisons of biogeographic regions are necessary to clarify these issues.

### 15.5.3 Does Such an Analysis Help with Predictions About Large-Scale Changes?

There have been several reviews of the likely consequences for intertidal communities as a result of climatic change (e.g., Thompson et al., 2002, Helmuth et al., 2006, Johnson et al., 2011). We shall not go over these issues again. What we should do, however, is consider what insights a larger-scale comparative approach to biogeographic regions can offer with respect to climatic change.

The three major predicted changes are to stressors of temperature, storms/wave-force increases and sedimentation from land use and rivers. Air temperatures (during low tide) are predicted to rise by 2°C in eastern Australia over the next thirty years. In NZ, the predicted rises are similar. There have been recent increases in the mean SST at two of three areas examined in the South Island, with average increases of 0.16°C per decade over three decades. The maximum SST did not increase, but the minimum (winter) seawater temperatures did, by up to 0.34°C per decade (Schiel et al., 2011). During the same period, mean annual SST increased in sites of northern NZ (Schiel, 2013). The consequences of increased air temperatures are to some extent predictable. Grazers on sheltered shores where there is no shading by an algal canopy are already prone to heat stress. This will increase.

For example, conditions causing sudden 'heat death' of limpets (*C. tramoserica*) in NSW involve two or more consecutive daytime low tides with air temperatures in excess of about 35°C. These conditions currently occur in only some years, but would occur every year with the predicted rise in temperatures (Underwood, unpublished data). Unless there is rapid adaptation on the part of the limpets, their densities and effectiveness as grazers will be reduced. There are complications with this, in that the limpets that die because of heat stress are those whose physiology is already compromised by shortage of food as a result of intraspecific competition (Underwood, unpublished). Nevertheless, any decrease in effectiveness of grazing will have consequences for the development of macroalgae (Underwood, 1980), unless the algae themselves are also reduced in capacity for recruitment and growth. If this occurs, there will be few refuges for many species (cf. Lilley and Schiel, 2006; Schiel and Lilley, 2007). Such temperatures are likely to cause large and significant changes in the composition of communities and will affect a raft of other species (e.g., Underwood et al., 1983; Lilley and Schiel, 2006).

Increased severity or frequency of wave forces may have profound effects on the communities on shores throughout the two regions. Schiel (2011) and Schiel et al. (2016) have shown that wave forces and large wave events have increased over the past three decades in southern NZ, although this may not be the case in north-eastern NZ (Schiel, 2013). These continuing changes are likely to have numerous consequences, some direct and others indirect. More frequent large storms are likely to remove more of the mussel beds and algal canopies. Each of these biogenic habitats provides shelter for many other species and influences the effectiveness of processes such as recruitment and predation.

Greater wave forces in general would alter the environmental conditions of shores that are currently of intermediate wave stress, presumably shifting their communities to be more like currently wave-exposed places. Some sheltered shores may well become like more wave-exposed sites. This would have consequences for the abundances of populations of the many species that are found predominantly in sheltered and semi-sheltered areas. It may also affect dispersal, settlement and recruitment processes (Taylor et al., 2010).

The issue of sedimentation is pernicious because it can lead to many cumulative changes in coastal ecosystems. There has been relatively little work done on this along the open coast of NSW and NZ. Schiel et al. (2006) showed that even a light dusting of sediment diminishes attachment of fucoid germlings to primary substrata, and Alestra et al. (2014) and Alestra and Schiel (2015) showed that sediments can act synergistically with coralline algae and increased temperature to affect fucoid germling recruitment and survival.

### 15.5.4 Where and How Can We Try to Mitigate Against Future Disturbances?

'Global change' involves more than just climate. Human-induced impacts can be pervasive and these may well interact with climatic change to produce an increased risk of ecological surprises (Lundquist et al., 2016). If we are to manage our coastal resources effectively, we will need to deal with some of the diffuse and large-scale stressors that affect them, which will require some sort of environmental trade-offs (e.g., Foley et al., 2013). We will also have to increase the management 'tools' available to us. Schiel (2013) argued that marine protected areas, for example, will not be an effective management tool if diffuse stressors such as increased sediment and nutrients are affecting a whole coastline. Part of the 'too hard' basket will need also to deal with the fact that many stressors of coastal ecosystems are land derived from a wide variety of intensive uses. Schiel and Howard-Williams (2016) and others have argued that we need to adopt a more holistic mountains-to-sea approach to management if we are to solve these problems.

Despite global stressors, management comes down to local initiatives for particular stretches of coastlines. Attempts to conserve biodiversity depend on being able to identify the species and processes that are at risk and having appropriate mitigating actions. Our current understanding of climate change scenarios is far from complete, so it is likely to change.

There are parallels between our possible responses to large-scale changes in environments and our dealings with other human activities that already damage coastal habitat and species. Human interventions are increasingly altering ecosystems through the elimination of native biota, introduction of non-native biota and major changes to habitats and environmental conditions. Although humans have been altering natural ecosystems for millennia, the current rates of change are much faster than previously, leading to the concept of novel or emerging ecosystems, which have no natural equivalents (Seastedt et al., 2008; Hobbs et al., 2009). These are dominated by a new mix of native and non-native biota and environmental conditions that limit or prevent their recovery to previous conditions. Such habitats need new approaches of management and hard decisions about where to put resources, especially if it is unlikely that highly altered systems will be able to be restored to or maintained at a prior natural state. Management directed at removing undesirable features of novel ecosystems may perpetuate them and perhaps the best that can be done is to attempt to maintain genetic and species diversity and encourage conditions that favour desirable species (Seastedt et al., 2008).

With respect to rocky intertidal shores on open coasts of south-eastern NSW and NZ, there has been very limited direct loss of natural habitats, except near major cities, although these types of losses can be prevalent in estuaries. There is therefore still the potential to manage and conserve rocky shores across a broad biogeographic range in Australia and NZ, especially against direct human exploitation and loss of habitat. Examples might include reduction of tourism effects through controlled pathways across reefs. It has been shown in both NZ and south-eastern Australia that trampling of algal beds can trigger long-term or even more permanent changes to communities through loss of fucoids, dominance by corallines and low-lying turfs and invertebrates (Povey and Keough, 1991; Keough and Quinn, 1998; Schiel and Taylor, 1999). Hundreds of thousands of people may visit iconic rocky reef platforms annually, constituting a 'press' disturbance. We may need to try managing access, such as through designated walkways that are common in sensitive areas of terrestrial parks.

In contrast, control of water-quality effects is far more challenging because of the distance between the sources of many of these contaminants in catchments and the shores that they affect. There has, however, been success in reducing amounts of chemical contamination from point-source discharges at outfalls in coastal and estuarine areas (Romero et al., 2016), but considerably less success with management of diffuse discharges, such as agricultural run-off and sedimentation in the water column (EEA, 2005).

It seems improbable that conservation of intertidal fauna will be possible by reducing the effects of increased air time, low tide temperatures or wave action. This is not actually impossible, but likely to be excessively costly. For example, one could prevent excessive heat during low tide by spraying water over areas of shore. One could prevent effects of increased wave action by the construction of carefully engineered baffles offshore. These solutions are already done in some places to protect coastal infrastructure (Dugan et al., 2011), but require energy and involve very substantial costs. Where ecologists would be involved in this type of conservation is to identify places where it is likely to have maximal effect for minimal cost and to indicate the type and location of the network of locations that should be managed. If people want this sort of conservation and are prepared to pay for it, it is possible.

An alternative is to realise that we cannot actually do much to conserve the status quo in the face of major and rapid large-scale environmental change. These considerations lead to the concept that perhaps we should embrace the facts that habitats are being altered, that environmental conditions are changing and that human activities will continue to impose an expanding set of problems onto natural systems. Although unpalatable to many people, we should perhaps be debating what sort of biodiversity we would like to maintain, rather than wrestling unsuccessfully with trying to conserve everything. If we make better predictions about where communities might finish up along a coastline, we could at least have some advance advice to give about where certain type of shores will be in fifty or a hundred years' time. That should allow planning, for example, for the placement of future marine reserves in an attempt to conserve what will be there, rather than waiting until major shifts have finished and then starting to consider where conservation might best be tried.

At present, many of the pieces of habitat that are visited by people or used for recreation or some sort of gathering of food are already unnatural. Perhaps it would be sensible to try to debate which sorts of species or assemblages would actually be wanted and then to try to plan management to achieve such targets. If other species or interactions were also protected, that would be a bonus.

Whatever becomes required and however we approach the management of large-scale disturbances to complex communities in variable habitats, we shall need increased capacity to understand and predict the effects of interacting disturbances. This will require good data on: (1) distributions across different environmental conditions, (2) understanding of the processes that create and maintain assemblages and (3) how species respond to changing environmental conditions. We shall also need good predictions of what environmental conditions are going to change and how much change is expected over what temporal and spatial scales. As we have tried to demonstrate here, large-scale comparisons across different bioregions will assist in this understanding. Therefore, such comparisons will also assist the management of these habitats and general conservation of intertidal communities in NZ and Australia.

## Acknowledgements

We thank Stacie Lilley for all manner of help with this manuscript. David R. Schiel particularly thanks the Marine Ecology Research Group, especially Spencer Wood and Paul South, for much help with field work over many years, the New Zealand Foundation for Research, Science and Technology and its various replacements for continuous funding to Schiel over decades, the National Science Challenge Sustainable Seas and the New Zealnd National Institute of Water and Atmospheric Research for funding and integration into their programmes. A. J. Underwood and M. Gee Chapman thank the long-term funding (1974–2008) by the Australian Research Council, particularly through its Special Research Centres programme. We thank an army of postdocs, postgraduate students and research assistants for help, advice and hard work.

## REFERENCES

Alestra, T. and Schiel, D. (2015). Impacts of local and global stressors in intertidal habitats: influence of altered nutrient, sediment and temperature levels on the early life history of three habitat-forming macroalgae. *Journal of Experimental Marine Biology and Ecology*, **468**, 29–36.

Alestra, T., Tait, L. and Schiel, D. (2014). Effects of algal turfs and sediment accumulation on replenishment and primary productivity of fucoid assemblages. *Marine Ecology Progress Series*, **511**, 59–70.

Bennett, I. and Pope, E. C. (1953). Intertidal zonation of the exposed rocky shores of Victoria, together with a reassignment of the biogeographical provinces of temperate Australian shores. *Australian Journal of Marine and Freshwater Research*, **4**, 105–59.

Bennett, I. and Pope, E. C. (1960). Intertidal zonation of the exposed rocky shores of Tasmania and its relationship with the rest of Australia. *Australian Journal of Marine and Freshwater Research*, **11**, 182–221.

Blanchette, C. A., Wieters, E. A., Broitman, B. R., Kinlan, B. P. and Schiel, D. R. (2009). Trophic structure and diversity in rocky intertidal upwelling ecosystems: a comparison of community patterns across California, Chile, South Africa and New Zealand. *Progress in Oceanography*, **83**, 107–16.

Bulleri, F., Benedetti-Cecchi, L., Cusson, M. et al. (2012). Temporal stability of European rocky shore assemblages: variation across a latitudinal gradient and the role of habitat-formers. *Oikos*, **121**, 1801–9.

Chapman, M. G. (1994). Small-scale patterns of distribution and size-structure of the intertidal littorinid *Littorina unifasciata* (Gastropoda: Littorinidae) in New South Wales. *Australian Journal of Marine and Freshwater Research*, **45**, 635–52.

Chapman, M. G. and Underwood, A. J. (1998). Inconsistency and variation in the development of rocky intertidal algal assemblages. *Journal of Experimental Marine Biology and Ecology*, **224**, 265–89.

Cole, V. J. and McQuaid, C. D. (2011). Broad-scale spatial factors outweigh the influence of habitat structure on the fauna associated with a bioengineer. *Marine Ecology Progress Series*, **442**, 101–9.

Coleman, R. A., Underwood, A. J., BenedettiCecchi, L. et al. (2006). A continental scale evaluation of the role of limpet grazing on rocky shores. *Oecologia*, **147**, 556–64.

Connor, E. F. and Simberloff, D. (1978). Species number and compositional similarity of the Galapagos flora and avifauna. *Ecological Monographs*, **48**, 219–48.

Creese, R. G. (1988). Ecology of molluscan grazers and their interactions with marine algae in north-eastern New Zealand. *New Zealand Journal of Marine and Freshwater Research*, **22**, 427–44.

Creese, R. and Underwood, A. (1982). Analysis of inter- and intra-specific competition amongst intertidal limpets with different methods of feeding. *Oecologia*, **53**, 337–46.

Crowe, T. P., Cusson, M., Bulleri, F. et al. (2013). Large-scale variation in combined impacts in canopy-loss and disturbance on community structure and ecosystem functioning. *PLoS ONE*, **8**, e66238.

Dakin, W. J. (1953). *Australian Sea Shores*. Angus & Robertson, Sydney.

Dugan, J. E., Airoldi, L., Chapman, M. G., Walker, S. J. and Schlacher, T. (2011). Estuarine and Coastal Structures: Environmental Effects, a Focus on Shore and Nearshore Structures. In E. Wolanski and D. S. MccLusky, eds. *Treatise on Estuarine and Coastal Science*. Academic Press, Waltham, pp. 17–41.

Dunmore, R. and Schiel, D. (2003). Demography, competitive interactions and grazing effects of intertidal limpets in southern New Zealand. *Journal of Experimental Marine Biology and Ecology*, **288**, 17–38.

EEA. (2005). *The European Environment, State and Outlook*. European Environment Agency, Copenhagen.

Foley, M. M., Armsby, M. H., Prahler, E. E. et al. (2013). Improving ocean management through the use of ecological principles and integrated ecosystem assessments. *Bioscience*, **63**, 619–31.

Foster, M. S. (1990). Organization of macroalgal assemblages in the Northeast Pacific: the assumption of homogeneity and the illusion of generality. *Hydrobiology*, **192**, 21–33.

Foster, M. S. and Schiel, D. R. (2010). Loss of predators and the collapse of southern California kelp forests (?): alternatives, explanations and generalizations. *Journal of Experimental Marine Biology and Ecology*, **393**, 59–70.

Gaines, S. D. and Lubchenco, J. (1982). A unified approach to marine plant-herbivore interactions. 2. Biogeography. *Annual Review of Ecology and Systematics*, **13**, 111–38.

Gaston, K. J. (1994). *Rarity*. Chapman & Hall, London.

Gray, J. S. (2001). Marine diversity: the paradigms in patterns of species richness examined. *Scientia Marina*, **65**, 41–56.

Guerry, A.D, Menge, B.A. and Dunmore, R.A. (2009). Effects of consumers and enrichment on abundance and diversity of benthic algae in a rocky intertidal community. *Journal of Experimental Marine Biology and Ecology*, **369**, 155–64.

Hawkins, S. J. and Hartnoll, R. G. (1983). Grazing of intertidal algae by marine invertebrates. *Annual Review of Oceanography and Marine Biology*, **21**, 195–282.

Helmuth, B., Mieszkowska, N., Moore, P. and Hawkins, S. J. (2006). Living on the edge of two changing worlds: forecasting the responses of rocky intertidal ecosystems to climate change. *Annual Review of Ecology, Evolution, and Systematics*, **37**, 373–404.

Hidas, E. Z., Ayre, D. J. and Minchenton, T. E. (2010). Patterns of demography for rocky-shore, intertidal invertebrates approaching their geographica range limits: tests of the abundant-centre hypothesis in south-eastern Australia. *Marine and Freshwater Research*, **61**, 1243–51.

Hobbs, R. J., Higgs, E. and Harris, J. A. (2009). Novel ecosystems: implications for conservation and restoration. *Trends in Ecology and Evolution*, **24**,599–605.

Hurlbert, S. J. (1984). Pseudoreplication and the design of ecological field experiments. *Ecological Monographs*, **54**, 187–211.

Jenkins, S. R., Moore, P., Burrows, M. T. et al. (2008). Comparative ecology of North Atlantic shores: do differences in players matter for process. *Ecology*, **89**, S3–23.

Johnson, C. R., Banks, S. C., Barrett, N. S. et al. (2011). Climate change cascades: shifts in oceanography, species' ranges and subtidal marine community dynamics in eastern Tasmania. *Journal of Experimental Marine Biology and Ecology*, **400**, 17–32.

Keough, M. J. and Quinn, G. (1998). Effects of periodic disturbances from trampling onrocky intertidal algal beds. *Ecological Applications*, **8**, 141–61.

Knox, G. A. (1963). The biogeography and intertidal ecology the Australasian coasts. *Oceanography and Marine Biology*, **1**, 341–404.

Knox, G. A. (1980). Plate tectonics and the evolution of intertidal and shallow-water benthic distribution patterns of the southwest Pacific. *Palaeogeography Palaeoclimatology Palaeoecology*, **31**, 267–97.

Kortsch, S., Primicerio, R., Fossheim, M., Dolgov, A. V. and Aschan, M. (2015). Climate change alters the structure of arctic marine food webs due to poleward shifts of boreal generalists. *Proceedings of the Royal Society B-Biological Sciences*, **282**, 31–9.

Lathlean, J. A., Ayre, D. J. and Minchenton, D. E. (2010). Supply-side biogeography: geographic patterns of settlement and early mortality for a barnacle approaching its range limit. *Marine Ecology Progress Series*, **412**, 141–50.

Lathlean, J. A., McWilliam, R. A., Ayre, D. J. and Minchenton, T. E. (2015). Biogeographical patterns of rocky-shore community structure in south-east Australia: effects of oceanographic conditions and heat stress. *Journal of Biogeography*, **42**, 1538–52.

Lilley, S. A. and Schiel, D. R. (2006). Community effects following the deletion of a habitat-forming alga from rocky marine shores. *Oecologia*, **148**, 672–81.

Lubchenco, J. and Gaines, S. D. (1981). A unified approach to marine plant-herbivore interactions. *I. Population and communities. Annual Review of Ecology and Systematics*, **12**, 405–37.

Lundquist, C. J., Fisher, K. T., Le Heron, R. et al. (2016). Science and societal partnerships to address cumulative impacts. *Frontiers in Marine Science*, **3**, 2.

Martins, G. M., Jenkins, S. R., Ramírez, R., Tuya, F., Neto, A. I. and Arenas, F. (2014). Early patterns of recovery from disturbance in intertidal algal assemblages: consistency across regions within a marine province. *Marine Ecology Progress Series*, **517**, 131–42.

Menge, B. A., Daley, B. A., Lubchenco, J. et al. (1999). Top-down and bottom-up regulation of New Zealand rocky intertidal communities. *Ecological Monographs*, **69**, 297–330.

Menge, B.A. and Menge, D.N. (2013). Dynamics of coastal meta-ecosystems: the intermittent upwelling hypothesis and a test in rocky intertidal regions. *Ecological Monographs*, **83**, 283–310.

Moore, L. B. (1949). The marine algal provinces of New Zealand. *Transactions for the Royal Society of New Zealand*, **77**, 187–9.

Murphy, R. J. and Underwood, A. J. (2006). Novel use of digital colour-infrared imagery to test hypotheses about grazing by intertidal herbivorous gastropods. *Journal of Experimental Marine Biology and Ecology*, **330**, 437–47.

Nelson, W. A., Dalen, J. and Neill, K. F. (2013). Insights from natural history collections: analysing the New Zealand macroalgal flora using herbarium data. *PhytoKeys*, **30**, 1–21.

O'Gower, A. K. and Meyer, G. R. (1971). The ecology of six species of littoral gastropod molluscs. 3. Diurnal and seasonal variations in densities and patterns of distribution in three environments. *Australian Journal of Marine and Freshwater Research*, **22**, 35–40.

Olson, D. M., Dinerstein, E., Wikramanayake, E. D. et al. (2001). Terrestrial ecoregions of the world: a new map of life on Earth a new global map of terrestrial ecoregions provides an innovative tool for conserving biodiversity. *Bioscience*, **51**, 933–8.

Povey, A. and Keough, M. J. (1991). Effects of trampling on plant and animal populations on rocky shores. *Oikos*, **61**, 355–68.

Quinn, G. P. and Keough, M. J. (2002). *Experimental Design and Data Analysis for Biologists*. Cambridge University Press, Cambridge.

Resetarits, W. J. and Bernardo, J. (1998). *Experimental Ecology: Issues and Perspectives*. Oxford University Press, Oxford.

Rilov, G. and Schiel, D. R. (2006a). Seascape-dependent subtidal-intertidal trophic linkages. *Ecology*, **87**, 731–44.

Rilov, G. and Schiel, D. R. (2006b). Trophic linkages across seascapes: subtidal predators limit effective mussel recruitment in rocky intertidal communities. *Marine Ecology Progress Series*, **327**, 83–93.

Rilov, G. and Schiel, D. R. (2011). Community regulation: the relative importance of recruitment and predation intensity of an intertidal community dominant in a seascape context. *PLoS ONE*, **6**, e23958.

Romero, S., Le Gendreb, R., Garniera, J. et al. (2016). Long-term water quality in the lower Seine: lessons learned over 4 decades of monitoring. *Environmental Science & Policy*, **58**, 141–54.

Scheiner, S. M. (2003). Six types of species-area curves. *Global Ecology and Biogeography*, **12**, 441–7.

Schiel, D. R. (2006). Rivets or bolts? When single species count in the function of temperate rocky reef communities. *Journal of Experimental Marine Biology and Ecology*, **338**, 233–52.

Schiel, D. R. (2011). Biogeographic patterns and long-term changes on New Zealand coastal reefs: non-trophic cascades from diffuse and local impacts. *Journal of Experimental Marine Biology and Ecology*, **400**, 33–51.

Schiel, D. R. (2013). The other 93%: trophic cascades, stressors and managing coastlines in non-marine protected areas. *New Zealand Journal of Marine and Freshwater Research*, **47**, 374–91.

Schiel, D. R. and Howard-Williams, C. (2016). Controlling inputs from the land to sea: limit-setting, cumulative impacts and *ki uta ki tai*. *Marine and Freshwater Research*, **67**, 57–64.

Schiel, D. R. and Lilley, S. A. (2007). Gradients of disturbance to an algal canopy and the modification of an intertidal community. *Marine Ecology Progress Series*, **339**, 1–11.

Schiel, D. R. and Lilley, S. A. (2011). Impacts and negative feedbacks in community recovery over eight years following removal of habitat-forming macroalgae. *Journal of Experimental Marine Biology and Ecology*, **407**, 108–15.

Schiel, D. R., Lilley, S. A., South, P. M. and Coggins, J. H. (2016). Decadal changes in sea surface temperature, wave forces and intertidal structure in New Zealand. *Marine Ecology Progress Series*, **548**, 77–95.

Schiel, D. R. and Taylor, D. I. (1999). Effects of trampling on a rocky intertidal algal assemblage in southern New Zealand. *Journal of Experimental Marine Biology and Ecology*, **235**, 213–35.

Schiel, D. R., Wood, S. A., Dunmore, R. A. and Taylor, D. I. (2006). Sediment on rocky intertidal reefs: effects on early post-settlement stages of habitat-forming seaweeds. *Journal of Experimental Marine Biology and Ecology*, **331**(2), 158–72.

Seastedt, T. R., Hobbs, R. J. and Suding, K. N. (2008). Management of novel ecosystems: are novel approaches required? *Frontiers in Ecology and the Environment*, **6**, 547–53.

Simberloff, D. (1980). A Succession of Paradigms in Ecology: Essentialism, Materialism and Probabilism. In E. Saarinen, ed. *Conceptual Issues in Ecology*. Reidel, Dordrecht, pp. 63–99.

Spalding, M. D., Fox, H. E., Halpern, B. S. et al. (2007). Marine ecoregions of the world: a bioregionalization of coastal and shelf areas. *Bioscience*, **57**, 573–83.

Tait, L. W. and Schiel, D. R. (2011). Legacy effects of canopy disturbance on ecosystem functioning in macroalgal assemblages. *PLoS ONE*, **6**, e26986.

Taylor, D., Delaux, S., Stevens, C., Nokes, R. and Schiel, D. (2010). Settlement rates of macroalgal algal propagules: cross-species comparisons in a turbulent environment. *Limnology and Oceanography*, **55**, 66–76.

Taylor, D. I. and Schiel, D. R. (2010). Algal populations controlled by fish herbivory across a wave exposure gradient on southern temperate shores. *Ecology*, **91**, 201–11.

Thompson, R., Crowe, T. and Hawkins, S. (2002). Rocky intertidal communities: past environmental changes, present status and predictions for the next 25 years. *Environmental Conservation*, **29**, 168–91.

Underwood, A. J. (1980). The effects of grazing by gastropods and physical factors on the upper limits of distribution of intertidal macroalgae. *Oecologia*, **46**, 201–13.

Underwood, A. J. (1984a). Vertical and seasonal patterns in competition for microalgae between intertidal gastropods. *Oecologia*, **64**, 211–22.

Underwood, A. J. (1984b). The vertical distribution and seasonal abundance of intertidal microalgae on a rocky shore in New South Wales. *Journal of Experimental Marine Biology and Ecology*, **78**, 199–220.

Underwood, A. J. (1986). The Analysis of Competition by Field Experiments. In J. Kikkawa and D. J. Anderson, eds. *Community Ecology: Pattern and Process*. Blackwells, Melbourne, pp. 240–68.

Underwood, A. J. (1990). Experiments in ecology and management: their logics, functions and interpretations. *Australian Journal of Ecology*, **15**, 365–89.

Underwood, A. J. (1998). Grazing and disturbance: an experimental analysis of patchiness in recovery from a severe storm by the intertidal alga *Hormosira banksii* on rocky shores in New South Wales. *Journal of Experimental Marine Biology and Ecology*, **231**, 291–306.

Underwood, A. J. (1999a). Physical disturbances and their direct effects on an indirect effect: responses of an intertidal assemblage to a severe storm. *Journal of Experimental Marine Biology and Ecology*, **232**, 125–40.

Underwood, A. J. (1999b). History and Recruitment in the Structure of Intertidal Assemblages on Rocky Shores: An Introduction to Problems for Interpretation of Natural Change. In M. Whitfield, J. Matthews and C. Reynolds, eds. *Aquatic Life Cycle Strategies*. Institute of Biology, London, pp. 79–96.

Underwood, A. J., Cole, V. J., Palomo, M. G. and Chapman, M. G. (2008). Numbers and density of species as measures of biodiversity on rocky shores along the coast of New South Wales. *Journal of Experimental Marine Biology and Ecology*, **366**, 175–83.

Underwood, A. J. and Denley, E. J. (1984). Paradigms, Explanations and Generalizations in Models for the Structure of Intertidal Communities on Rocky Shores. In D. R. Strong, D. Simberloff, L. G. Abele and A. Thistle, eds. *Ecological Communities: Conceptual Issues and the Evidence*. Princeton University Press, Princeton, NJ, pp. 151–80.

Underwood, A. J., Denley, E. J. and Moran, M. J. (1983). Experimental analyses of the structure and dynamics of mid-shore rocky intertidal communities in New South Wales. *Oecologia*, **56**, 202–19.

Underwood, A. J. and Jernakoff, P. (1981). Effects of interactions between algae and grazing gastropods in the structure of a low-shore algal community. *Oecologia*, **48**, 221–33.

Underwood, A. J. and Jernakoff, P. (1984). The effects of tidal height, wave-exposure, seasonality and rock-pools on grazing and the distribution of intertidal macroalgae in New South Wales. *Journal of Experimental Marine Biology and Ecology*, **75**, 71–96.

Underwood, A. J. and Murphy, R. J. (2008). Unexpected patterns in facilitatory grazing revealed by quantitative imaging. *Marine Ecology Progress Series*, **358**, 85–94.

Underwood, A. J. and Petraitis, P. S. (1993). Structure of Intertidal Assemblages in Different Locations: How Can Local Processes Be Compared? In R. E. Ricklefs and D. Schluter, eds. *Species Diversity in Ecological Communities: Historical and Geographical Perspectives*. University of Chicago Press, Chicago, pp. 38–51.

Villarino, E., Chust, G., Licandro, P. et al. (2015). Modelling the future biogeography of North Atlantic zooplankton communities in response to climate change. *Marine Ecology Progress Series*, **531**, 121–42.

Walker, N. A. (1998). Grazing in the intertidal zone: effects of the herbivorous *Turbo smaragdus* on macroalgal assemblages. Masters thesis, University of Canterbury, Canterbury.

Waters, J. M., Wernberg, T., Connell, S. D. et al. (2010). Australia's marine biogeography revisited: back to the future? *Austral Ecology*, **35**, 988–92.

Wiens, J. A. (1989). Spatial scaling in ecology. *Functional Ecology*, **3**, 385–97.

Wieters, E. A., McQuaid, C. D., Palomo, M. G., Pappalardo, P. and Navarrete, S. (2012). Biogeographcal boundaries, functional group structure and diversity of rocky shore communities along the Argentinian coast. *PLoS ONE*, **7**, e49725.

Womersley, H. B. S. and Edmonds, S. J. (1958). A general account of the intertidal ecology of South Australian coasts. *Australian Journal of Marine and Freshwater Reserarch*, **9**, 217–60.

# Chapter 16

# The Past and Future Ecologies of Australasian Kelp Forests

Sean D. Connell, Adriana Vergés, Ivan Nagelkerken, Bayden D. Russell, Nick Shears, Thomas Wernberg and Melinda A. Coleman

## 16.1 Introduction

### 16.1.1 Overview

Human history has been shaped by forests of kelp that provide habitat and energy for one of the world's most diverse and productive ecosystems (Erlandson et al., 2007). The ecology of these ecosystems attracts global attention (Krumhansl et al., 2016) because of the wildlife and economic activity they support (Bennett et al., 2016); representing one of humanity's most ancient natural resources (i.e., Indigenous Australians dating back 65,000 years; Thurstan et al., 2018).

The task to understand the past and future ecology of kelp forests has been transformed by technology and global collaboration over the last ten years. Early ecological studies were limited by the challenges of working underwater (Norton, 1999) and fragmentation of research among biogeographic regions (historical review: Connell, 2007a). Today, technological advances and global teamwork allows us to connect finer-scale responses (e.g., molecular and physiological) with broaderscale patterns (e.g., biogeography and oceanography), allowing us to grapple with networks of natural complexity that not only drive community dynamics, but also shape the imprint of human activities across multiple levels of biological organisation.

In this review, we use a biogeographic perspective to synthesise the processes that create variation in the spatial extent and persistence of kelp forests in Australia and New Zealand (NZ). Evolution has created the world's greatest macroalgal diversity in southern Australia (Womersley, 1981; Bolton, 1994) and the divergence of the orders Fucales and Laminariales about 65 Mya (Fraser et al., 2010). We define kelp as canopy-forming algae from both these orders and recognise the contribution of both to the ecology of subtidal ecosystems. This review proceeds by considering fundamental patterns and processes across temperate Australia and NZ (i.e., rocky Australasian coasts) and then considers their future ecology (i.e., response to climate change). These considerations provide sign posts for research into the future ecology of kelp forests, particularly their contraction, stasis or expansion.

### 16.1.2 The Foundations of Austral Subtidal Ecology

Pioneering subtidal work on Australian and NZ rocky coasts focussed on several locations around Australia: Sydney (led by Tony Underwood, Tony Larkum and Peter Steinberg), Adelaide (led by Brian Womersely and Scoresby Shepherd), Tasmania (led by Graham Edgar and Craig Johnson) and Perth (led by Hugh Kirkman and Gary Kendrick) and locations in NZ (led by Howard Choat and David Schiel). Progress accelerated with methodological advances; most notable was the effect of combining scuba technology with manipulative and experimental approaches, computing power and conceptual logic. This combination allowed the ecology of subtidal habitats to be studied with greater intensity and quantitative precision, while renewed attention to logic provided thinking with a lasting imprint through rigorous, repeatable and coherent models and observations (Underwood et al., 2000). The writers and readers of this chapter are beneficiaries of this earlier era of carefully detailed subtidal discovery.

Today, the scale of subtidal research has broadened in spatial and temporal extent to encompass historical ecology and climate change across vast areas of the Austral coast. This broadening has been facilitated by growth: in number of salaried positions for marine ecologists, international interactions and collaborative funding that have been enabled by cheaper travel costs. Regional differences in focus remain among the new generation of laboratories; emphasising local interests in issues such as resource liberation (nitrogen and carbon emissions, South Australia), physical forcing (oceanography, heat waves, warming) and change (tropicalisation, western and eastern Australia) and trophic control (Tasmania, NZ). These differences in focus are, at least partly, founded upon regional differences in vulnerabilities and dominant physical and biological processes of each respective coast (Wernberg et al., 2011a); however, they are also clearly embedded within a modern understanding of the patterns and processes of global seascapes.

Current advances in scientific technology have enabled today's marine ecologists to start thinking about how large-scale ocean processes might drive local ecological patterns, and reconcile information from previously isolated studies and localities into regional-scale understandings of relevant patterns and processes (Connell and Irving, 2008). For example, it is only recently that we have had the power to map kelp forests in meaningful detail at low cost (Griffin et al., 2017); use sophisticated instrumentation to characterise the physical and chemical nature of water masses; and computing and modelling capacity to understand how ocean currents behave. Nevertheless, there is an added perspective that comes from experiencing underwater life that is hard to fathom from experiencing intertidal life after the tide has receded. We champion the value of ecological studies that allows investigators to fully submerse themselves beneath the sea.

## 16.2 Biogeographic Patterns

### 16.2.1 The Oceanographic Canvas of Ecological Patterning

A binding feature of the marine realm is the strong connectivity of ocean life, where oceanography can be viewed as the dynamic 'canvas' upon which much of subtidal and intertidal ecology operates (Menge et al., 2003). Oceanography and historical processes have largely determined the underlying biotic patterns that shape Australia's subtidal ecology. Ocean chemistry and physics drive marine life; not only through their direct effect on organisms and species turnover across gradients (Wernberg et al., 2013b), but also through their interactions (Connell and Russell, 2010).

The vast coastlines of Australia and NZ are diverse (spaning $>15^\circ$ latitude and $40^\circ$ longitude in Australia) and support some of the most speciose marine assemblages globally (1,499 species of algae, 4,100 species of invertebrates and 731 species of reef fishes (Bennett et al., 2016)). The oceanography of this region determines species' persistence within the bounds of its physical and chemical environments, as well as the potential for species to increase (or decrease) in range and, thereby, sets the local context of the ecological interactions we observe. Indeed, oceanography not only helps us place in context the ecological patterns we study, but also anchors the development of our research into scenarios of global change (see Section 16.3).

The continent of Australia is globally unique in that it has two warm, poleward-flowing boundary currents: the East Australian Current (EAC) flowing down eastern Australia and the Leeuwin

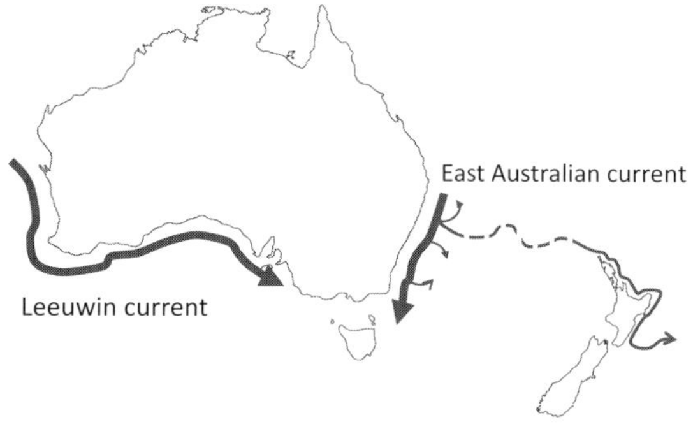


**Fig. 16.1** Map of Australia and NZ showing the flow of the main surface currents. The LC has a relatively uninterrupted and highly connective flow along the west and south coast. The EAC has a more heterogeneous flow and connects Australia with NZ.

Current (LC) flowing down western Australia and extending along the coast of South Australia (Wernberg et al., 2013b; Figure 16.1). These three currents unite over 8,000 km of temperate coast and support many cosmopolitan species (e.g., the common kelp; *Ecklonia radiata*; Coleman et al., 2011b), and provide immense economic value (Bennett et al., 2016). Given these two replicate latitudinal temperature gradients and similar histories of postglacial recolonisation, this geographic situation also allows unique insight into the ecologies of marine flora and fauna relative to other global coastlines that lack such a replicated system.

In contrast, mainland NZ has a ~13° latitudinal gradient and a relatively compressed longitudinal expanse (~12°) with extreme contrasting physical conditions between the highly wave exposed west coast and the more protected east coast, which structures distinct ecologies. The Cook Strait, separating the North and South Islands, adds further complexity to coastal circulation patterns (Chiswell et al., 2015; Figure 16.1).

Eastern Australia and NZ share a contemporary connection via the EAC (Figure 16.1). The EAC departs from the Australian east coast between 28° and 39°S (Cetina-Heredia et al., 2014) and heads east into the Tasman Sea towards NZ. The Tasman Front provides a lose connection (Sutton and Bowen, 2014) between the EAC and East Auckland Current, which heads down the northeastern coast of NZ (Ridgway and Dunn, 2003). Water from the southern Tasman Sea is also entrained around the bottom of NZ in the Southland Current, which is also considered part of the western boundary current complex (Sutton, 2003). Oceanographic connections between Australia and NZ mean there are many shared flora and fauna, most notably, perhaps, the common kelp (*E. radiata*) that structures much of the region's temperate ecologies.

Australia's flora and fauna that dominates the ecology of subtidal reefs can be partitioned into three biogeographic 'provinces' that coincide with both extant ocean currents and the Pleistocene separation by the Bassian isthmus (between Tasmania and the Australian mainland; Waters et al., 2010). Turnover and variation in the composition of algal taxa are thought not only to reflect historic events, but also a heavy imprint of present-day species' temperature tolerances and oceanographic currents (Wernberg et al., 2013b). While historical glaciations did not directly influence the Australian coast, they did cause sea levels to shift as seas rose and fell. This relatively benign history resulted in a stable and diverse algal flora (Phillips, 2001) with genetic signatures that often reflect contemporary (e.g., movement of ocean currents) rather than historic (extinction and recolonisation) processes (Durrant et al., 2015).

Although these boundary current systems physically unite temperate coastlines between western and eastern Australia, and eastern Australia and NZ, different species distributions and ecologies (Connell and Irving, 2009) suggest differences between these currents might influence both historical and contemporary biological patterns (Connell, 2007a, 2007b). The EAC is generally faster and characterised by greater eddy activity (Mata et al., 2006), for example, and is more nutrient rich than the LC (Pearce, 1991; Hanson et al., 2005).

Coastal geomorphology (bays, estuaries), geology (substratum types) and coastal climate (e.g., rainfall) also differ along the east and west Australian coastlines, factors which influence biotic patterns in various ways. These physical differences may alter scales of species dispersal and connectivity (Coleman et al., 2011b), habitat availability and physiological processes. Similarly, in NZ, a complicated coastline and oceanography prevents large-scale generalisations of the main drivers of subtidal ecologies (Shears and Babcock, 2007a; Schiel, 2013).

### 16.2.2 Patterns in Macroalgal Canopies

Canopy-forming algae are a fundamental component of rocky coastal ecosystems in temperate Australasia (Connell and Irving, 2008). Temperate reefs range across more than 8,000 km of exposed coastline, with extensive canopies typically ranging between 3 and 20 m depth in Australia (Connell, 2007a). Canopy-forming algae shape the ecology of other algal species (Irving et al., 2004), invertebrates (Anderson et al., 2005) and fish (Smith et al., 2014), as well as affecting adjacent habitats (Wernberg et al., 2006; Bishop et al., 2010). Despite their ecological value, societal appreciation of the immense value of macroalgal forests is only just beginning to emerge (Bennett et al., 2016).

Macroalgal forests cover most upward-facing, subtidal rock surfaces of western and southern Australia, but not in eastern Australia (~33–37°S) or parts of NZ where their dominance is influenced by the presence of herbivorous sea urchins (Connell and Irving, 2008). In some cases, sea urchins may effectively eliminate macroalgae from large swathes of reef in these areas, forming urchin barrens habitat that is devoid of fleshy macroalgae.

The common kelp, *E. radiata,* (Laminariales) and fucoids (Fucales) are the predominant taxa of these macroalgal forests (Steneck and Johnson, 2014). Fucoids are ubiquitous and dominant, but an understudied component of Australasian subtidal habitats, forming diverse mosaics with common kelp in western and southern Australia and NZ (Fowler-Walker and Connell, 2002; Goodsell et al., 2004; Shears and Babcock, 2007b; Wernberg et al., 2011c; Coleman and Wernberg, 2017) that often support unique ecological assemblages (Marzinelli et al., 2014).

The poleward flow of warm water along Australia's east and west coasts means that there are general latitudinal patterns in macroalgal floras that are replicated on both coasts. Specifically, from low to high latitudes there is a transition from subtropical (coral, *Sargassum* species and turf assemblages) to laminarian dominated (*E. radiata*) to diverse mixed algal stands (laminarians and fucoids). Other macroalgae including giant kelp (*Macrocystis*), *Lessonia* and *Durvillaea* become important at higher latitudes (Tasmania and southern NZ (Shears and Babcock, 2007b).

## 16.3 Biogeographic Dynamics

### 16.3.1 Currents Both Disconnect and Connect Coastal Processes

Ocean currents disperse the propagules of organisms to deliver the basis of local biotic patterns (Coleman et al., 2009; Coleman and Kelaher, 2009). Australia's boundary currents are key to understanding patterns of dispersal and connectivity of macroalgae across regions (Coleman et al., 2009, 2013b; Coleman and Kelaher, 2009) and the continent in general (Coleman et al., 2011b, 2013a). On continental scales, the magnitude of population connectivity follows current strength (Figure 16.1), with greater connectivity within the fast-flowing EAC relative to the slower-flowing LC extension (Coleman et al., 2009; Coleman and Kelaher, 2009). The meso-scale flow of these currents (eddies) as well as habitat availability (Durrant et al., 2018) determines smaller-scale patterns of connectivity and genetic structure of populations (Coleman et al., 2009; Coleman and Kelaher, 2009). Where flow is linear in the LC, for example, populations become increasingly isolated with increasing distance (Coleman et al., 2009; Coleman and Kelaher, 2009), but, where flow is complex, such as in the EAC, populations become more mixed (Coleman et al., 2011a). Ocean currents also structure the flow of nutrients and other resources that have a profound influence on biotic patterning.

### 16.3.2 Resource Enrichment – Propagating Resources from the Bottom-Up

Energy and nutrients underpin life. The availability and transfer of these resources to autotrophs, and their consumption by herbivores, and subsequent consumption by predators is fundamental. The balance between resources and their consumption represents a chain of direct and indirect interactions that are strengthened or weakened by ocean chemistry (Ghedini and Connell, 2017). While the pace and intensity of ecological interactions across some NZ coasts have been characterised by the oceanic delivery of nutrients (e.g., upwelling: Menge et al., 2003), nutrient inputs from rivers and high oceanographic complexity seem to leave stronger ecological imprints. Similarly, such broadscale oceanographic imprints are also difficult to detect in Australia, while smaller-scale variation in nutrient delivery (e.g., urban catchments) leave strong imprints on the ecology of kelp forests (Gorman et al., 2009). Nevertheless, the biophysical nature of currents are detectable; local areas of upwelling within the EAC and LC, albeit weak and infrequent by global standards, explain patchiness in pelagic productivity and kelp cover in the EAC (Jordan, unpublished data). The weak nutrient concentrations across the LC have, for example, been hypothesised to explain more sparse cover of epiphytic growth on kelp relative to the more dense covers on kelp across the EAC (Russell et al., 2005). Nutrient enrichment experiments conducted across several locations of east and south coast Australia show disproportionately larger increases in cover of epiphytes on kelp of the oligotrophic LC compared to the EAC (Connell, unpublished data), suggesting regional differences in response to nutrient elevation. These observations coincide with observations of weak benthic herbivory across most of temperate Australia (i.e., coasts of the LC) and strong herbivory along the EAC (Connell, 2007a; Connell and Irving, 2009).

### 16.3.3 Trophic Dynamics – Consumption from the Top-Down

The striking pattern of urchin barrens dominating coasts of the EAC (Connell and Irving, 2008) and largely lacking across the LC is due to regional differences in urchin herbivory. Much of eastern Australia is characterised by urchins (*Centrostephanus rodgersii*) that create barrens, whereas the Leeuwin coast is characterised by dense forests of canopy-forming algae (Connell and Irving, 2009; Wernberg et al., 2011c). Across the Leeuwin coast, barren-creating urchins like *C. rodgersii* are unable to establish across the biogeographic break that centres on the Bass Strait (Figure 16.1; between the Australian mainland and Tasmania). The urchins that are present along this western coast (e.g., *Heliocidaris erythrogramma*) do not have the capacity to create barrens even at unnaturally elevated densities (Livore and Connell, 2012). Herbivory, nonetheless, can affect the early colonisers of habitat (Russell and Connell, 2005) and mediate the recovery of kelp (Bennett et al., 2015; Ghedini et al., 2015).

Overfishing of vertebrate predators at the top of the food chain (e.g., snapper and spiny lobster) can trigger increases in herbivore populations (primarily sea urchins) in eastern Australia and parts of NZ, leading to widespread kelp deforestation (i.e., a classic 'trophic cascade'; Babcock et al., 2010). Perhaps the world's best-demonstrated example of trophic cascades centres on the east coast of northern NZ (36–37°S) where a ban of fishing in marine protected areas (MPAs) resulted in widespread expansion of kelp forests, reflecting a concomitant decline of sea urchins with increasing densities of predatory fish and lobster (Shears and Babcock, 2002).

However, these effects are not ubiquitous across north-eastern NZ's MPAs. The intensity of trophic cascades and density of urchins varies with factors such as depth, wave exposure and turbidity (Shears et al., 2008). Indeed, the occurrence of urchin barrens is patchy around NZ and there is not a clear understanding of the mechanisms causing variation in the presence of urchin barrens. For example, urchin barrens are lacking from large stretches of coast (e.g., eastern coast of the South Island), but are common in parts of north-eastern NZ, the northern South Island, Fiordland and Stewart Island (Shears and Babcock, 2007b). Schiel (2013) suggests that the trophic effects model is unlikely to apply to much of the coastline of NZ, and that a model involving multiple effects, including bottom-up forces, environmental and climatic influences, species' demographics and catchment-derived sedimentation is more appropriate for kelp communities over most of the country.

Australia and NZ share many of the same habitats (i.e., barrens in eastern Australia and the kelp *E. radiata* across the continent) and have similar predators of urchins (i.e., rock lobster, *Jasus edwardsii* and *Sagmariasus verreauxi*, and snapper, *Pagrus auratus*). Given these broad similarities, similar processes appear be operating across the EAC through Tasmania (Ling et al., 2009). Although MPAs have been established in eastern Australia for up to twenty-five years and higher order predators are returning to protected areas across the east coast (Coleman et al., 2015), preliminary research suggests that such trophic restoration involving kelp has not yet been realised (Coleman et al., 2013a). Such effects can take decades to occur in MPAs (Babcock et al., 2010) and are also dependent on MPAs being designed appropriately to facilitate predator recovery (Edgar et al., 2014) and greatly depend on provision of adequate enforcement (Kelaher et al., 2015).

## 16.4 Future Change: The Diversity of Drivers

Some of the world's most conspicuous and recent loss of kelp forests have centred on Australia's coastline, including the south-, west- and east-facing coasts (Connell et al., 2008; Johnson et al., 2011; Vergés et al., 2016; Wernberg et al., 2016). These losses represent a profound loss of biodiversity and energy. Consequently, a strong theme of Australian subtidal research has been the local and global drivers of kelp loss.

Globally, the loss of kelp forests has been extensive (~38 per cent), although in some places they are expanding (Krumhansl et al., 2016; Laffoley and Baxter, 2016; Filbee-Dexter and Wernberg, 2018). Historically, loss was considered through the narrow prism of the classic top-down model proposed by Steneck and co-authors (2002), but, as this chapter recognises, the drivers of loss are much more diverse. Proposed mechanisms of kelp loss now include deteriorating water quality (nitrogen and sedimentation pollution: Gorman et al., 2009), heatwaves (Wernberg et al., 2016), intensifying herbivory by range-shifting tropical species (Vergés et al., 2016) and changing ocean chemistry (Connell et al., 2013).

### 16.4.1 Changing Oceanography

Changing patterns of dispersal and connectivity set the stage for global alteration of the dynamics of subtidal ecologies. Australia's boundary currents have already undergone change and further alterations are anticipated, with strong implications for dispersal of temperate marine life (Cetina-Heredia et al., 2014). The EAC has been steadily increasing in strength (Ridgway, 2007; Hill et al., 2008) and is predicted to further increase in strength and poleward penetration by 2060 (Sun et al., 2012). Transport in eddies within the EAC is also predicted to increase (Cetina-Heredia et al., 2014). In contrast, the LC and LC extension have weakened in the past fifty years (Feng et al., 2004) and continue to weaken (by a further 15 per cent in 2060; Sun et al., 2012). These changes will have profound implications for organisms inhabiting Australia's coastlines. Already, strengthening of the EAC has induced range expansions of warm-water species, including herbivorous fish and urchins (Ling et al., 2009; Vergés et al., 2014a), and loss of kelp (Johnson et al., 2011; Vergés et al., 2016).

Oceanographic modelling predicts that future changes to ocean currents will alter connectivity and recruitment of many key taxa that interact to structure the benthic ecology of Australian temperate reefs. For example, predicted oceanographic changes will alter recruitment patterns in ecologically and economically valuable species such as lobster (Cetina-Heredia et al., 2015), driving peak recruitment polewards and potentially changing trophic interactions (Provost et al., 2017). Interestingly, however, change in lobster recruitment will be driven more by changes in the movement of water masses and larval transport per se, rather than associated changes in temperature impacts on larval survival (Cetina-Heredia et al., 2015). Similarly, dispersal and connectivity of kelp and urchin populations within the EAC will be altered in subtle ways (Coleman et al., 2017), with important implications for genetic patterns that may mediate vulnerability to climate stress (Wernberg et al., 2018).

For kelp, poleward connectivity will increase in the future, whereas connectivity towards the equator may decrease. These changes may manifest by limiting recolonisation of kelp at warmer latitudes, where they are disappearing through both direct physiological limitations and indirect ecological processes (see Section 16.4.2). Similarly, habitat suitability models for a suite of fifteen prominent habitat-forming algae across southern Australia predict that their current distribution will be reduced on average by 62 per cent (range: 27–100 per cent) even under the most conservative representative concentration pathway 2.6 scenario (Martínez et al., 2018). The distribution and abundance of herbivores and their predators are also predicted to change. Along with concomitant alterations to ocean temperature and chemistry that may directly compromise kelp survival, these future situations will either strengthen or weaken the ecological interactions (Nagelkerken and Connell, 2015) that often underpin persistence (Goldenberg et al., 2018).

The effect of changing oceanography on NZ kelp forests is less well understood. The increasing flow in the EAC extension is expected to result in decreased flows across the Tasman and into the east Auckland Current (Hill et al., 2011). To date, there is little evidence of long-term warming in north-eastern NZ (Bowen et al., 2017) and no reports of increases in tropical species. The continuing intensification of the EAC and weakening of the Tasman Front is, in fact, likely to reduce the connectivity between eastern Australia and northern NZ. This reduced connectivity could lead to declines in populations of tropical and subtropical species that are currently sustained by larval transport across the Tasman Sea, such as the spiny lobster, *S. verreauxi*, which is an ecologically and commercially important species.

Realised and projected changes to ocean currents and connectivity of species also have the potential to alter key genetic parameters that underpin species' abilities to respond to change. For example, given that genetic diversity confers adaptive capacity to respond to change (Sunday et al., 2014), declines in connectivity and a reduction in genetic diversity may render populations and entire ecosystems vulnerable to local extinction (Wernberg et al., 2018). Genetic diversity of kelp and other algae is generally reduced at lower latitudes on Australia's east and west coasts due to a combination of limited connectivity, smaller effective population sizes and interspersion of populations with coral habitat (Coleman et al., 2011a, 2011b). Following a marine heatwave in western Australia (Smale and Wernberg, 2013), for example, populations of kelp with low genetic diversity were completely extirpated, whereas those with higher genetic diversity persisted (Wernberg et al., 2018). The impact of oceanography on dispersal and connectivity may, therefore, further weaken the capacity for kelp persistence at lower latitudes.

### 16.4.2 Ocean Warming: Tropicalisation and Heatwaves

Many species respond to warming by shifting their distributions towards the poles (Wernberg et al., 2011b; Pecl et al., 2017), and alter communities upon arrival. 'Tropicalisation' of temperate coastlines occurs where warm-water species become increasingly dominant, while cool-water species recede towards the poles (Vergés et al., 2014a). Tropicalisation along the warmer latitudes of both the east and the west coasts of Australia has caused the replacement of canopy-forming kelps with either urchin barrens or turfs, and domination by species characteristic of warm-temperate, subtropical and tropical reefs (Johnson et al., 2011; Wernberg et al., 2013a; Bennett et al., 2015; Vergés et al., 2016).

Two main mechanisms appear to be driving climate-mediated declines in kelp forests in Australia. On the one hand, ocean warming can directly impact the physiology of kelp species, which tend to perform best and are most resilient at relatively cool water temperatures (Wernberg et al., 2010). On the other hand, ocean warming and the intensification of poleward-flowing boundary currents such as the EAC are also driving the range expansion of warm-water herbivores, leading to novel interactions between previously separated species and, in some instances, causing the overgrazing of entire kelp forests (Vergés et al., 2014a).

In Tasmania, ocean warming and the increasing influence of nutrient-poor EAC waters has led to important losses of giant kelp forests (*Macrocystis pyrifera*) in the last thirty to forty years. Widespread overgrazing by the recently established sea urchin, *C. rodgersii*, has greatly impacted *E. radiata* beds (Johnson et al., 2011). In turn, the disappearance of kelp has led to a loss of biodiversity and productivity in Tasmanian reefs (Ling, 2008), and is also threatening valuable abalone and rock lobster industries in the region (Johnson et al., 2011).

In western Australia, after decades of ocean warming, a marine heatwave forced a 100 km range contraction of habitat forming algae in 2011 (Wernberg et al. 2016). During this extreme event, warmer seawater anomalies of 2–4°C persisted for more than ten weeks along >2,000 km of coastline (Wernberg et al., 2013a). A community-wide tropicalisation ensued and fundamentally altered key ecological processes, including herbivory. In particular, grazing tropical herbivorous fishes greatly increased in biomass and diversity after the heatwave, displaying grazing rates typical of coral reefs (Bennett et al., 2015; Zarco-Perello et al., 2017). These observations suggest that tropical herbivores contribute to the maintenance of turf-dominated reefs and may be preventing kelp re-establishment (Bennett et al., 2015). As a result, kelps have been replaced by invertebrates, corals and fishes characteristic of subtropical and tropical waters (Wernberg et al., 2013a, 2016).

Kelp forests have also started to disappear from warm latitudes of the EAC through gradual warming. The extent of loss has gradually increased in over a decade, rather than being induced by a single extreme warming event as found on the west coast (Vergés et al., 2016). Herbivory by tropical and subtropical fish species has, however, again been directly implicated in the loss of kelp in eastern Australia, as evidenced by increases in fish bite marks and in the proportion of tropical herbivores in the fish community (Vergés et al., 2016).

A small number of tropical and subtropical herbivorous fishes are capable of consuming adult kelp and can lead to local deforestation of canopy algae. This impact appears to be a global phenomenon, for example, the rabbitfish, *Siganus fuscescens*, has been associated with the large-scale loss of kelp in Japan (Yamaguchi et al., 2010), and eastern (Vergés et al., 2016) and western Australia (Zarco-Perello et al., 2017); while the congeneric species, *Siganus rivulatus* and *Siganus luridus*, appear to have mediated the loss of canopy algae in the eastern Mediterranean (Vergés et al., 2014b). The lack of strong poleward-flowing boundary currents around NZ suggest that tropicalisation is unlikely along these coastlines, or is likely to be driven by other mechanisms. Despite localised warming in some regions, there is so far no evidence of increases in warm-water species along the NZ coast, which highlights how different oceanographic conditions can lead to different ecological responses to climate change.

Marine heatwaves are likely to become more common and intense as oceans warm (Oliver et al., 2018). Heatwaves are defined as discrete periods of anomalously warm water and they represent some of the most pervasive and rapid changes to the ocean environment (Hobday et al., 2016). Such warming events have led to range contraction and a reduction in the abundance of habitat-forming algae (Smale and Wernberg, 2013; Wernberg et al., 2016), although the probability of kelp loss appears to increase at the warmest edges of their range (Wernberg et al., 2013a).

The subsequent shift from a kelp-dominated towards a turf-dominated state with tropicalised fish communities (Vergés et al., 2014b; Bennett et al., 2015) demonstrates that climatic extremes can drive wholesale changes in biodiversity (Oliver et al., 2017). The frequency and intensity of such episodes has implications for predictive models of species' distribution and ecosystem structure, which to date are largely based on gradual warming trends (García Molinos et al., 2015).

### 16.4.3 Ocean Chemistry: Acidification and Resource Enrichment

Carbon and nitrogen are two of the most limiting resources on land and sea and their increasing liberation within both realms is unprecedented (Vitousek, 1994). These resources are fundamental to the extraordinary growth in size of the human population and quality of life. Modern societies are dependent on the intensification of food production (nitrogen fertilisers) removal of human waste (nitrogen effluent) and supply of energy and goods to cities (carbon emissions). Yet, the influence of resource enrichment on marine systems remains a pioneering area of study, particularly carbon enrichment that leads to ocean acidification (Doney et al., 2009).

Cracking the problem of how to understand the ecological effects of ocean acidification (Feely et al., 2004) has been challenging because $CO_2$ also interacts with more easily observed changes (e.g., warming and heatwaves). Ecologists overcame this challenge by first establishing the direct effects of $CO_2$ both as a stressor (i.e., ocean acidification) and as a resource (i.e., carbon enrichment), before trying to solve the broader effects (i.e., indirect effects) that propagate throughout whole ecosystems. Similar to the release of nitrogen from terrestrial sources (eutrophication: Gorman et al., 2009) carbon propagates through food webs (Ghedini and Connell, 2017; Goldenberg et al., 2017), only to be accelerated or buffered by other processes.

In terms of kelp loss, the increase in potential for the competitive displacement of kelps in a high $CO_2$ world (Connell et al., 2018) has been a common interpretation across widely different parts of the globe using contrasting methods (see review by Connell et al., 2013). Enrichment of resources (carbon or nitrogen) supercharges the growth of ephemeral algal species (turfs) while having minor effects on perennial kelps. Kelps do not benefit from resource enrichment (carbon and nitrogen) nearly as much as turfs (Falkenberg et al., 2013a). There is, therefore, a switch in competitive advantage that benefits turfs and leads to their replacement of kelp (Connell et al., 2008; Gorman et al., 2009). Rather than causing the loss of kelp directly, therefore, resource enrichment drives loss by altering this competitive hierarchy (Gorman and Connell, 2009).

Collapse of kelp forests via carbon or nitrogen pollution, therefore, reflects a combination of direct and indirect effects on kelp competitors (Connell et al., 2018). The life history and physiology of turfs not only directly benefit from resource enrichment (Falkenberg et al., 2013a), but they also benefit indirectly by conditions that reduce their consumption by herbivores (Mertens et al., 2015), essentially allowing them to expand unchecked. Hence, the chance of kelp forest collapse is worsened when the increased production of turfs is exacerbated by reduction in its consumption by herbivores (Connell and Ghedini, 2015). Put simply, collapse occurs when resource enrichment reverses the competitive dominance of producers, but consumers then fail to compensate by neutralising the competitor. What this means is that small cumulative increases in enrichment drives a much greater consequence than would be predicted from linear effects measured between competitors (e.g., kelps versus turfs) because these interactions are embedded within a broader network of change that propagates collapse.

The insidious nature of resource enrichment often means that the collapse of an ecosystem can occur as a surprise (Connell et al., 2017). Enrichment drives changes in the strength of species interactions that are not as readily detectable as fluctuations in species densities that can be visually observed. It is, therefore, quite likely that we can mistake the causes of loss, because the more easily observed processes of loss (e.g., heatwaves and tropicalisation) are part of the cumulative increases in loss of resistance and appear to drive a much greater consequence than would be predicted from their isolated effects.

### 16.4.4 Overfishing: Trophic Control and Marine Protection

Historically, overfishing of predators that control herbivore populations was considered to be the leading cause of kelp loss (Steneck et al., 2002). While we now accept there are alternate mechanisms of loss (see earlier), overfishing remains a pervasive impact and large predators are functionally extinct on most coasts. The key issue associated with overfishing of vertebrate predators at the top of the food chain is their effect on herbivore populations, leading to widespread kelp deforestation (i.e., a classic trophic cascade). This loss of kelp normally results from sea urchin grazing, which is controlled by predation where fishing has been minimal, but overgrazing can occur when herbivores are released from predation (Steneck et al., 2002). Despite the large-scale lack of predators on subtidal reefs, it is clear that urchin barrens are not ubiquitous across Australasia (Connell and Irving, 2008), suggesting that there are other, unknown factors beyond overfishing that shape urchin densities. This is also illustrated by differential responses to protection from fishing, with the restoration of kelp forests occurring in some, but not all, MPAs in NZ (Shears et al., 2008, Schiel, 2013).

While the future influence of fishing on kelp forests is difficult to predict, it is already interacting with new stressors associated with climate change. For example, the expansion of the sea urchin, *C. rodgersii,* into Tasmania due to the intensification of the EAC has resulted in substantial loss of kelp forests (Ling et al., 2009). Kelp loss in trophically structured systems may be reversed by MPAs, but this can take decades while predator populations recover. Climate change may in fact counteract the loss of predators and promote recovery of kelp in some areas. For example, blooms of benthic dinoflagellates (*Ostreopsis siamensis*) that occur during warm periods in northern NZ have negative impacts on sea urchins (Shears and Ross, 2010) and can facilitate the recovery of kelp in urchin barrens. Similarly, increased sedimentation in coastal areas is likely with climate change (Seers and Shears, 2015) and has been shown to inhibit sea urchin settlement (Walker, 2007). Conversely, increased sedimentation and associated turbidity will also have negative impacts on kelp, so this may result in shifts to turf-dominated states (e.g., lacking both urchins and kelp). Predicting how existing and new stressors will interact and affect trophic interactions in kelp forests, therefore, poses a significant challenge for the future study of these systems.

## 16.5 Future Persistence: Adjustability and Compensation

Future climate on many coasts will exceed the capacity for kelp survival, yet the same climate may not exceed this capacity on all coasts. Such differences in vulnerability may reflect variation in the capacity for species interactions to adjust and compensate; i.e., ecological compensation (Ghedini et al., 2015). The resulting stability is one of the least understood phenomena in ecology and its study turns out to be surprisingly difficult. For climate science, one challenge centres on the duality of environmental change itself. Environmental change is typically categorised as a 'stressor', but, even at a physiological level, environmental change is biphasic, with abiotic changes exerting negative effects at some intensities, and positive effects at others (Harley et al., 2017). Hence, environmental change acts as either a resource or stressor within a species and its influence on interacting species may buffer ecological change.

We have an opportunity to understand the ecological processes that compensate for environmental change where the difference between collapse and stability is mediated by species interactions, rather than simple breaches of physiological tolerance (Figure 16.2). Consider the positive effect of $CO_2$ enrichment on primary production (turfs) and its consumption (herbivory). While $CO_2$ enrichment boosts turf production, it also intensifies herbivory which adjusts in strength to counter increasing production (i.e., trophic compensation: Ghedini et al., 2015). Appreciating this potential for ecological interactions to buffer environmental change is key to understanding stability (i.e., before tipping points of collapse are reached). Such resistance to change centres on compensatory dynamics that are initiated by disturbance and adjust in proportion to the strength of the disturbance (Ghedini et al., 2015). Understanding such limits to kelp forest decline (e.g., turf driven loss: Filbee-Dexter and Wernberg, 2018) would help identify the processes responsible for kelp forest persistence.

Ecological compensation is often a function of behavioural or population responses to disturbance (Figure 16.2). Herbivores can counter turf expansion through instantaneous increases in per capita consumption (Ghedini et al., 2015) that increases their population size (Connell et al., 2017) through boosted reproduction (Heldt et al., 2016). In cases of resource enrichment, this can increase the nutrient content of algae which causes herbivores to increase their rate of feeding (Falkenberg et al., 2013b) and, as a consequence, enhance their metabolic rates (Ghedini and Connell, 2016). The resulting intensification of

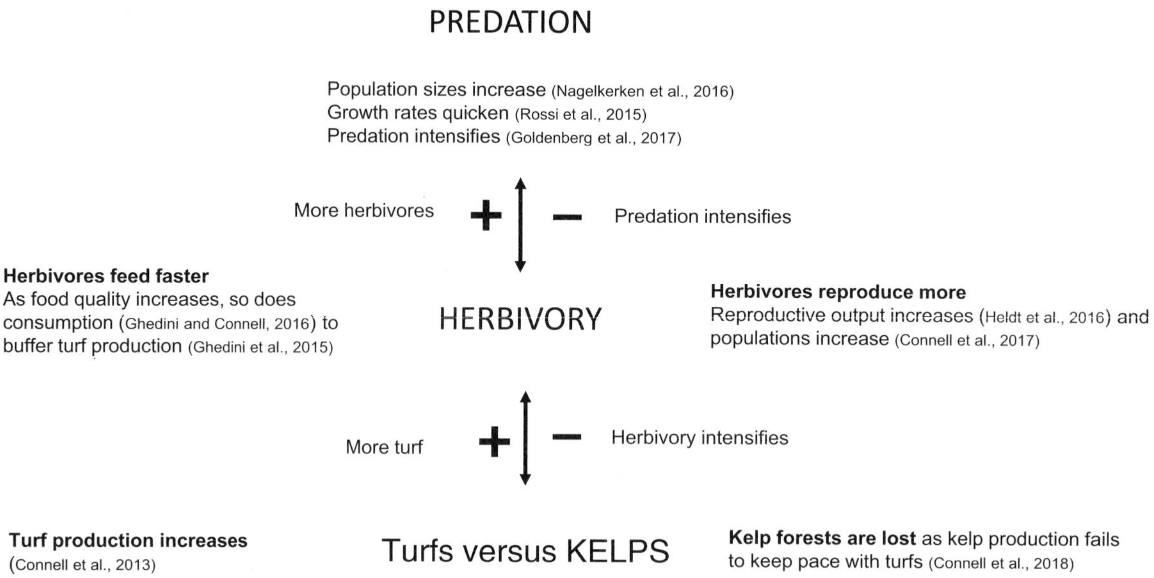
PREDATION
Population sizes increase (Nagelkerken et al., 2016)
Growth rates quicken (Rossi et al., 2015)
Predation intensifies (Goldenberg et al., 2017)
More herbivores
Predation intensifies
Herbivores feed faster
As food quality increases, so does consumption (Ghedini and Connell, 2016) to buffer turf production (Ghedini et al., 2015)
HERBIVORY
Herbivores reproduce more
Reproductive output increases (Heldt et al., 2016) and populations increase (Connell et al., 2017)
More turf
Herbivory intensifies
Turf production increases (Connell et al., 2013)
Turfs versus KELPS
Kelp forests are lost as kelp production fails to keep pace with turfs (Connell et al., 2018)

**Fig. 16.2** As carbon emissions enrich the oceans with $CO_2$, species interactions can stabilise or destabilise kelp communities. Consumption can intensify (predation and herbivory) to compensate for increases in production (herbivore and turf biomass). Yet, competition may tilt in favour of taxa that can better use $CO_2$ as a resource (turfs benefit, but not kelps) and compensation fails (i.e., herbivory fails to negate increased turf growth).

herbivory buffers enhanced productivity of turfs, whether driven by nitrogen enrichment (McSkimming et al., 2015; Russell and Connell, 2007) or carbon enrichment (Ghedini et al., 2015; McSkimming et al., 2016).

Whether herbivore populations expand (Heldt et al., 2016) or are suppressed is not just a function of feeding on enriched or enhanced food sources (Ghedini and Connell, 2016), but also a function of who feeds on them – their predators (Ghedini and Connell, 2017). The size of herbivore populations may vary as a function of food demand by carnivores, which is also under strong control of environmental conditions (Nagelkerken and Connell, 2015). Predators require larger amounts of prey to simply maintain growth rates under warming seas. Consequently, herbivore populations may decline as warming drives predators to intensify their top-down control in the face of elevated metabolic demands (Pistevos et al., 2015). Hence, the trophic compensation by herbivores observed under resource enrichment may be reversed through ocean warming, creating an imbalance between predators and herbivores (Goldenberg et al., 2017).

## 16.6 Conclusions

Constant change characterises the inter-decadal ecology of kelp forests, reflecting natural and human disturbances. If there is a common ecological response to human domination, it is the direct or indirect force we have over the processes that either stabilise (Ghedini et al., 2015; Goldenberg et al., 2018) or destabilise kelp persistence (Vergés et al., 2016; Wernberg et al., 2016). While sudden events, such as heatwaves, drive dramatic change over short timescales (i.e., sudden loss of canopies), over the long-term, adjustments in species interactions seem pivotal in mediating recovery (i.e., survival of recruits) or persistence of canopy loss.

As a global community of researchers, we immerse ourselves under different seas to study a variety of processes and, through good-natured rivalry, we tease each other by oversimplifying the 'fact' that our research area is the more important (e.g., the importance of currents versus warming versus acidification versus fishing). Researchers that work at the warmer limits of kelp distribution emphasise the directly harmful effects of temperature on kelp genetics, physiology and population dynamics, which are measurable and likely to play the primary role in kelp forest loss during sudden climate events. Researchers that work at the cooler range of kelp distribution emphasise indirect effects, particularly how herbivores mediate the effects of climate on kelp. One of the emergent outcomes of this review is the realisation that the diversity in these drivers may combine on some coasts to embed (e.g., persistent turf-dominated states) or buffer change (e.g., persistent kelp-dominated states).

While our alternate explanations for kelp loss appear simple and general, prediction is bafflingly difficult because even the early warning signals of loss are notoriously difficult to interpret. Some systems are remarkably stable in the face of environmental change. Other systems can change slowly, but they can also reveal no outward change before their tipping point is reached and change is sudden (Connell et al., 2017). The inherent complexity of ecological systems tends to buffer environmental change through compensatory dynamics (Goldenberg et al., 2018) such that simple experiments exaggerate change. For example, loss of one type of herbivore can be unexpectedly substituted for by another type to maintain the overall role of herbivores in the system (Hughes, 1994); or herbivory can intensify to compensate for the boosted growth of kelp competitors (Ghedini and Connell, 2016). The erosion of stabilising processes releases the culminated potential of change (Ghedini et al., 2015) such that stability was, actually, long undermined before collapse. Such lagged responses have been considered as 'living dangerously on borrowed time during slow, unrecognized regime shifts' (Hughes et al., 2013). While system collapse and its drivers are in fact readily observable, stability and its drivers remain one of the least studied processes in ecology. Critically, knowledge of such stabilising processes may be useful to managers in circumstances where kelp loss is undesirable and preventable.

Kelp loss is one of the most striking examples of ecosystem collapse because its replacement is a relatively simple habitat. These replacement habitats are either dominated by urchin barrens (Shears et al., 2008) or turfs that form mats (Filbee-Dexter and Wernberg, 2018), both of which represent a profound loss of structural complexity. Whether driven by $CO_2$ enrichment (Connell et al., 2013, 2018) or slow or fast increases in temperature (Wernberg et al., 2016) with concomitant tropicalisation (Vergés et al., 2016; Wernberg et al., 2016), the general prediction is that kelps will be less dominant. Hence, future generations may place greater value on research that seeks to understand the circumstances in which environmental change is resisted.

While constant change characterises subtidal ecology, the discipline seems set to take a deep dive into the biological processes that accelerate and compensate for environmental change. Similar to the technological and collaborative progress that benefitted the present generation of authors, continuing progress may assist future generations of subtidal ecologists to figure out why kelp forests are characterised by global mosaics of long-term loss, gain and stasis (Krumhansl et al., 2016). In what regions and for what reasons might kelp decline or flourish or simply persist future ocean change?

## Acknowledgements

Foremost we thank the pioneers of subtidal ecology in Australia and NZ, many of whom trained or continue collaborations with the authors. We are also indebted to the enthusiastic support of our students, without which the increasingly burdensome regulations on diving would have extinguished scientific diving. Across these generations, we share a passion for the underwater world and hope that scientific diving remains available to future generations.

## REFERENCES

Anderson, M. J., Connell, S. D., Gillanders, B. M. et al. (2005). Relationships between taxonomic resolution and spatial scales of multivariate variation. *Journal of Animal Ecology*, **74**(4), 636–46.

Babcock, R., Shears, N., Alcala, A. et al. (2010). Decadal trends in marine reserves reveal differential rates of change in direct and indirect effects. *Proceedings of the National Academy of Sciences*, **107**(43), 18256–61.

Bennett, S., Wernberg, T., Connell, S. D., Hobday, A. J., Johnson, C. R. and Poloczanska, E. S. (2016). The 'Great Southern Reef': social, ecological and economic value of Australia's neglected kelp forests. *Marine and Freshwater Research*, **67**(1), 47–56.

Bennett, S., Wernberg, T., Harvey, E. S., Santana-Garcon, J. and Saunders, B. J. (2015). Tropical herbivores provide resilience to a climate-mediated phase shift on temperate reefs. *Ecology Letters*, **18**(7), 714–23.

Bishop, M. J., Coleman, M. A. and Kelaher, B. P. (2010). Cross-habitat impacts of species decline: response of estuarine sediment communities to changing detrital resources. *Oecologia*, **163**(2), 517–25.

Bolton, J. J. (1994). Global seaweed diversity: pattern and anomalies. *Botanica Marina*, **37**, 241–5.

Bowen, M., Markham, J., Sutton, P. et al. (2017). Interannual variability of sea surface temperature in the southwest Pacific and the role of ocean dynamics. *Journal of Climate*, **30**(18), 7481–92.

Cetina-Heredia, P., Roughan, M., Van Sebille, E. and Coleman, M. (2014). Temporal variability in the East Australian Current: implications for particle transport along the south-eastern Australian coast. *Journal of Geophysical Research: Oceans*, **119**(7), 4351–66.

Cetina-Heredia, P., Roughan, M., Van Sebille, E., Feng, M. and Coleman, M. A. (2015). Strengthened currents override the effect of warming on lobster larval dispersal and survival. *Global Change Biology*, **21**(12), 4377–86.

Chiswell, S. M., Bostock, H. C., Sutton, P. J. H. and Williams, M. J. M. (2015). Physical oceanography of the deep seas around New Zealand: a review. *New Zealand Journal of Marine and Freshwater Research*, **49** (2), 286–317.

Coleman, M., Bates, A., Stuart-Smith, R. et al. (2015). Functional traits reveal early responses in marine reserves following protection from fishing. *Diversity and Distributions*, **21**(8), 876–87.

Coleman, M., Palmer-Brodie, A. and Kelaher, B. P. (2013a). Conservation benefits of a network of marine reserves and partially protected areas. *Biological Conservation*, **167**, 257–64.

Coleman, M. A., Cetina-Heredia, P., Roughan, M., Feng, M., Sebille, E. and Kelaher, B. P. (2017). Anticipating changes to future connectivity within a network of marine protected areas. *Global Change Biology*, **23**(9), 3533–42.

Coleman, M. A., Chambers, J., Knott, N. A. et al. (2011a). Connectivity within and among a network of temperate marine reserves. *PLoS ONE*, **6**(5), e20168.

Coleman, M. A., Feng, M., Roughan, M., Cetina-Heredia, P. and Connell, S. D. (2013b). Temperate shelf water dispersal by Australian boundary currents: implications for population connectivity. *Limnology and Oceanography: Fluids and Environments*, **3**(1), 295–309.

Coleman, M. A., Gillanders, B. M. and Connell, S. D. (2009). Dispersal and gene flow in the habitat-forming kelp, *Ecklonia radiata*: relative degrees of isolation across an east-west coastline. *Marine and Freshwater Research*, **60**(8), 802–9.

Coleman, M. A. and Kelaher, B. P. (2009). Connectivity among fragmented populations of a habitat-forming alga, *Phyllospora comosa* (Phaeophyceae, Fucales) on an urbanised coast. *Marine Ecology Progress Series*, **381**, 63–70.

Coleman, M. A., Roughan, M., Macdonald, H. S. et al. (2011b). Variation in the strength of continental boundary currents determines continent-wide connectivity in kelp. *Journal of Ecology*, **99**(4), 1026–32.

Coleman, M. A. and Wernberg, T. (2017). Forgotten underwater forests: the key role of fucoids on Australian temperate reefs. *Ecology and Evolution*, **7**(20), 8406–18.

Connell, S. D. (2007a). Subtidal Temperate Rocky Habitats: Habitat Heterogeneity at Local to Continental Scales In S. D. Connell and B. M. Gillanders, eds. *Marine Ecology*. Oxford University Press, Melbourne, pp. 378–401.

Connell, S. D. (2007b). Water Quality and the Loss of Coral Reefs and Kelp Forests: Alternative States and the Influence of Fishing. In S. D. Connell and B. M. Gillanders, eds. *Marine Ecology*. Oxford University Press, Melbourne, pp. 556–68.

Connell, S. D., Doubleday, Z. A., Foster, N. R. et al. (2018). The duality of ocean acidification as a resource and a stressor. *Ecology*, **99**(5), 1005–10.

Connell, S. D., Doubleday, Z. A., Hamlyn, S. B. et al. (2017). How ocean acidification can benefit calcifiers. *Current Biology*, **27**, R83–102.

Connell, S. D. and Ghedini, G. (2015). Resisting regime-shifts: the stabilising effect of compensatory processes. *Trends in Ecology and Evolution*, **30**, 513–15.

Connell, S. D. and Irving, A. D. (2008). Integrating ecology with biogeography using landscape characteristics: a case study of subtidal habitat across continental Australia. *Journal of Biogeography*, **35**, 1608–21.

Connell, S. D. and Irving, A. D. (2009). The Subtidal Ecology of Rocky Coasts: Local-Regional-

Biogeographic Patterns and Their Experimental Analysis. In J. D. Witman and R. Kaustuv, eds. *Marine Macroecology*. University of Chicago Press, Chicago, pp. 392–417.

Connell, S. D., Kroeker, K. J., Fabricius, K. E., Kline, D. I. and Russell, B. D. (2013). The other ocean acidification problem: $CO_2$ as a resource among competitors for ecosystem dominance. *Philosophical Transactions of the Royal Society B-Biological Sciences*, **368**, 20120442.

Connell, S. D. and Russell, B. D. (2010). The direct effects of increasing $CO_2$ and temperature on non-calcifying organisms: increasing the potential for phase shifts in kelp forests. *Proceedings of the Royal Society B-Biological Sciences*, **277**, 1409–15.

Connell, S. D., Russell, B. D., Turner, D. J. et al. (2008). Recovering a lost baseline: missing kelp forests from a metropolitan coast. *Marine Ecology Progress Series*, **360**, 63–72.

Doney, S. C., Fabry, V. J., Feely, R. A. and Kleypas, J. A. (2009). Ocean acidification: the other $CO_2$ problem. *Annual Review of Marine Science*, **1**, 169–92.

Durrant, H. M., Barrett, N. S., Edgar, G. J., Coleman, M. A. and Burridge, C. P. (2015). Shallow phylogeographic histories of key species in a biodiversity hotspot. *Phycologia*, **54**(6), 556–65.

Durrant, H. M. S., Barrett, N. S., Edgar, G. J., Coleman, M. A., and Burridge, C. P. (2018). Seascape habitat patchiness and hydrodynamics explain genetic structuring of kelp populations. *Marine Ecology Progress Series*, **598**, 81–92.

Edgar, G. J., Stuart-Smith, R. D., Willis, T. J. et al. (2014). Global conservation outcomes depend on marine protected areas with five key features. *Nature*, **506** (7487), 216–20.

Erlandson, J. M., Graham, M. H., Bourque, B. J., Corbett, D., Estes, J. A. and Steneck, R. S. (2007). The Kelp Highway Hypothesis: Marine Ecology, the Coastal Migration Theory, and the Peopling of the Americas. *The Journal of Island and Coastal Archaeology*, **2**(2), 161–74.

Falkenberg, L. J., Russell, B. D. and Connell, S. D. (2013a). Contrasting resource limitations of marine primary producers: implications for competitive interactions under enriched CO2 and nutrient regimes. *Oecologia*, **172**, 575–83.

Falkenberg, L. J., Russell, B. D. and Connell, S. D. (2013b). Future herbivory: the indirect effects of enriched $CO_2$ may rival its direct effects *Marine Ecology-Progress Series*, **492**, 85–95.

Feely, R. A., Sabine, C. L., Lee, K. et al. (2004). Impact of anthropogenic $CO_2$ on the $CaCO_3$ system in the oceans. *Science*, **305**(5682), 362–6.

Feng, M., Li, Y. and Meyers, G. (2004). Multidecadal variations of Fremantle sea level: footprint of climate variability in the tropical Pacific. *Geophysical Research Letters*, **31**(16).

Filbee-Dexter, K. and Wernberg, T. (2018). Rise of turfs: a new battlefront for globally declining kelp forests. *Bioscience*, **68**(2), 64–76.

Fowler-Walker, M. J. and Connell, S. D. (2002). Opposing states of subtidal habitat across temperate Australia: consistency and predictability in kelp canopy – understorey associations. *Marine Ecology Progress Series*, **240**, 49–56.

Fraser, C. I., Winter, D. J., Spencer, H. G. and Waters, J. M. (2010). Multigene phylogeny of the southern bull-kelp genus *Durvillaea* (Phaeophyceae: Fucales). *Molecular Phylogenetics and Evolution*, **57**(3), 1301–11.

García Molinos, J., Halpern, B. S., Schoeman, D. S. et al. (2015). Climate velocity and the future global redistribution of marine biodiversity. *Nature Climate Change*, **6**, 83.

Ghedini, G. and Connell, S. D. (2016). Organismal homeostasis buffers the effects of abiotic change on community dynamics. *Ecology*, **97**, 2671–9.

Ghedini, G. and Connell, S. D. (2017). Moving ocean acidification research beyond a simple science: investigating ecological change and their stabilizers. *Food Webs*, **13**(Supplement C), 53–9.

Ghedini, G., Russell, B. D. and Connell, S. D. (2015). Trophic compensation reinforces resistance: herbivory absorbs the increasing effects of multiple disturbances. *Ecology Letters*, **18**, 182–7.

Goldenberg, S. U., Nagelkerken, I., Ferreira, C. M., Ullah, H. and Connell, S. D. (2017). Boosted food web productivity through ocean acidification collapses under warming. *Global Change Biology*, **23**(10), 4177–84.

Goldenberg, S. U., Nagelkerken, I., Marangon, E., Bonnet, A., Ferreira, C. M. and Connell, S. D. (2018). Ecological complexity buffers the impacts of future climate on marine consumers. *Nature Climate Change*, **8**, 229–33.

Goodsell, P. J., Fowler-Walker, M. J., Gillanders, B. M. and Connell, S. D. (2004). Variations in the configuration of algae in subtidal forests: implications for invertebrate assemblages. *Austral Ecology*, **29**, 350–7.

Gorman, D. and Connell, S. D. (2009). Recovering subtidal forests on human-dominated landscapes. *Journal of Applied Ecology*, **46**, 1258–65.

Gorman, D., Russell, B. D. and Connell, S. D. (2009). Land-to-sea connectivity: linking human-derived terrestrial subsidies to subtidal habitat change on open rocky coasts. *Ecological Applications*, **19**(5), 1114–26.

Griffin, K. J., Hedge, L. H., González-Rivero, M., Hoegh-Guldberg, O. I. and Johnston, E. L. (2017). An evaluation of semi-automated methods for collecting ecosystem-level data in temperate marine systems. *Ecology and Evolution*, **7**(13), 4640–50.

Hanson, C. E., Pattiaratchi, C. B. and Waite, A. M. (2005). Seasonal production regimes off south-western Australia: influence of the Capes and Leeuwin Currents on phytoplankton dynamics. *Marine and Freshwater Research*, **56**(7), 1011–26.

Harley, C. D. G., Connell, S. D., Doubleday, Z. A. et al. (2017). Conceptualizing ecosystem tipping points within a physiological framework. *Ecology and Evolution*, **7**(15), 6035–45.

Heldt, K. A., Connell, S. D., Anderson, K., Russell, B. D. and Munguia, P. (2016). Future climate stimulates population out-breaks by relaxing constraints on reproduction. *Scientific Reports*, **6**, 33383.

Hill, K., Rintoul, S., Coleman, R. and Ridgway, K. (2008). Wind forced low frequency variability of the East Australia current. *Geophysical Research Letters*, **35**(8).

Hill, K. L., Rintoul, S. R., Ridgway, K. R. and Oke, P. R. (2011). Decadal changes in the South Pacific western boundary current system revealed in observations and ocean state estimates. *Journal of Geophysical Research: Oceans*, **116**(C1).

Hobday, A. J., Alexander, L. V., Perkins, S. E. et al. (2016). A hierarchical approach to defining marine heatwaves. *Progress in Oceanography*, **141**, 227–38.

Hughes, T. P. (1994). Catastrophes, phase shifts and large-scale degradation of a Caribbean coral reef. *Science*, **265**, 1547–51.

Hughes, T. P., Linares, C., Dakos, V., van de Leemput, I. A. and van Nes, E. H. (2013). Living dangerously on borrowed time during slow, unrecognized regime shifts. *Trends in Ecology & Evolution*, **28**, 149–55.

Irving, A. D., Connell, S. D. and Gillanders, B. M. (2004). Local complexity in patterns of canopy-benthos associations produce regional patterns across temperate Australasia. *Marine Biology*, **144**, 361–8.

Johnson, C. R., Banks, S. C., Barrett, N. S. et al. (2011). Climate change cascades: shifts in oceanography, species' ranges and subtidal marine community dynamics in eastern Tasmania. *Journal of Experimental Marine Biology and Ecology*, **400**(1), 17–32.

Kelaher, B. P., Page, A., Dasey, M. et al. (2015). Strengthened enforcement enhances marine sanctuary performance. *Global Ecology and Conservation*, **3**, 503–10.

Krumhansl, K. A., Okamoto, D. K., Rassweiler, A. et al. (2016). Global patterns of kelp change over the past half-century. *Proceedings of the National Academy of Sciences of the United States of America*, **113**, 13785–90.

Laffoley, D. and Baxter, J. (2016). *Explaining Ocean Warming: Causes, Scale, Effects and Consequences*. IUCN, Gland, 27.

Ling, S. D. (2008). Range expansion of a habitat-modifying species leads to loss of taxonomic diversity: a new and impoverished reef state. *Oecologia*, **156** (4), 883–94.

Ling, S. D., Johnson, C. R., Frusher, S. D. and Ridgway, K. R. (2009). Overfishing reduces resilience of kelp beds to climate-driven catastrophic phase shift. *Proceedings of the National Academy of Sciences of the United States of America*, **106**(52), 22341–5.

Livore, J. P. and Connell, S. D. (2012). Reducing per capita food supply alters urchin condition and habitat. *Marine Biology*, **159**(5), 967–73.

Martínez, B., Radford, B., Thomsen, M. S. et al. (2018). Distribution models predict large contractions in habitat-forming seaweeds in response to ocean warming. *Diversity and Distributions*, **24**(10), 1350–66.

Marzinelli, E., Campbell, A., Vergés, A., Coleman, M., Kelaher, B. P. and Steinberg, P. (2014). Restoring seaweeds: does the declining fucoid *Phyllospora comosa* support different biodiversity than other habitats? *Journal of Applied Phycology*, **26**(2), 1089–96.

Mata, M. M., Wijffels, S. E., Church, J. A. and Tomczak, M. (2006). Eddy shedding and energy conversions in the East Australian Current. *Journal of Geophysical Research: Oceans*, **111**(C9).

McSkimming, C., Russell, B. D., Tanner, J. E. and Connell, S. D. (2016). A test of metabolic and consumptive responses to local and global perturbations: enhanced resources stimulate herbivores to counter expansion of weedy species. *Marine and Freshwater Research*, **67**(1), 96–102.

McSkimming, C., Tanner, J. E., Russell, B. D. and Connell, S. D. (2015). Compensation of nutrient pollution by herbivores in seagrass meadows. *Journal of Experimental Marine Biology and Ecology*, **471**, 112–18.

Menge, B. A., Lubchenco, J., Bracken, M. E. S. et al. (2003). Coastal oceanography sets the pace of rocky intertidal community dynamics. *Proceedings of the National Academy of Sciences of the United States of America*, **100**(21), 12229–34.

Mertens, N. L., Russell, B. D. and Connell, S. D. (2015). Escaping herbivory: ocean warming as a refuge for primary producers where consumer metabolism and consumption cannot pursue. *Oecologia*, **179**(4), 1223–9.

Nagelkerken, I. and Connell, S. D. (2015). Global alteration of ocean ecosystem functioning due to increasing human CO2 emissions. *Proceedings of the National Academy of Sciences of the United States of America*, **112**, 13272–7.

Nagelkerken, I., Russell, B. D., Gillanders, B. M. and Connell, S. D. (2016). Ocean acidification alters fish populations indirectly through habitat modification. *Nature Climate Change*, **6**(1), 89.

Norton, T. A. (1999). *Stars Beneath the Sea: The Pioneers of Diving*. Carroll & Graf Publishers Inc, New York.

Oliver, E. C., Benthuysen, J. A., Bindoff, N. L. et al. (2017). The unprecedented 2015/16 Tasman Sea marine heatwave. *Nature Communications*, **8**, 16101.

Oliver, E. C., Donat, M. G., Burrows, M. T. et al. (2018). Longer and more frequent marine heatwaves over the past century. *Nature Communications*, **9**(1), 1324.

Pearce, A. (1991). Eastern boundary currents of the southern hemisphere. *The journal of the Royal Society of Western Australia*, **74**, 35–45.

Pecl, G. T., Araújo, M. B., Bell, J. D. et al. (2017). Biodiversity redistribution under climate change: impacts on ecosystems and human well-being. *Science*, **355**(6332).

Phillips, J. (2001). Marine macroalgal biodiversity hotspots: why is there high species richness and endemism in southern Australian marine benthic flora? *Biodiversity and Conservation*, **10**(9), 1555–77.

Pistevos, J. C., Nagelkerken, I., Rossi, T., Olmos, M. and Connell, S. D. (2015). Ocean acidification and global warming impair shark hunting behaviour and growth. *Scientific Reports*, **5**, 16293.

Provost, E. J., Kelaher, B. P., Dworjanyn, S. A. et al. (2017). Climate-driven disparities among ecological interactions threaten kelp forest persistence. *Global Change Biology*, **23**(1), 353–61.

Ridgway, K. (2007). Long-term trend and decadal variability of the southward penetration of the East Australian Current. *Geophysical Research Letters*, **34**(13).

Ridgway, K. and Dunn, J. (2003). Mesoscale structure of the mean East Australian Current System and its relationship with topography. *Progress in Oceanography*, **56**(2), 189–222.

Rossi, T., Nagelkerken, I., Simpson, S. D. et al. (2015). Ocean acidification boosts larval fish development but reduces the window of opportunity for successful settlement. *Proceedings of the Royal Society B*, **282**(1821), 20151954.

Russell, B. D. and Connell, S. D. (2005). A novel interaction between nutrients and grazers alters relative dominance of marine habitats. *Marine Ecology-Progress Series*, **289**, 5–11.

Russell, B. D. and Connell, S. D. (2007). Response of grazers to sudden nutrient pulses in oligotrophic v. eutrophic conditions. *Marine Ecology Progress Series*, **349**, 73–80.

Russell, B. D., Elsdon, T. S., Gillanders, B. M. and Connell, S. D. (2005). Nutrients increase epiphyte loads: broad-scale observations and an experimental assessment. *Marine Biology*, **147**(2), 551–8.

Schiel, D. (2013). The other 93%: trophic cascades, stressors and managing coastlines in non-marine protected areas. *New Zealand Journal of Marine and Freshwater Research*, **47**(3), 374–91.

Seers, B. M. and Shears, N. T. (2015). Spatio-temporal patterns in coastal turbidity–Long-term trends and drivers of variation across an estuarine-open coast gradient. *Estuarine, Coastal and Shelf Science*, **154**, 137–51.

Shears, N. T. and Babcock, R. C. (2002). Marine reserves demonstrate top-down control of community structure on temperate reefs. *Oecologia*, **132**, 131–42.

Shears, N. T. and Babcock, R. C. (2007a). Quantitative description of mainland New Zealand's shallow subtidal reef communities. *Science for Conservation*, **280**, 126.

Shears, N. T. and Babcock, R. C. (2007b). *Quantitative Description of Mainland New Zealand's Shallow Subtidal Reef Communities*. Science & Technical Publishing, Wellington.

Shears, N. T., Babcock, R. C. and Salomon, A. K. (2008). Context-dependent effects of fishing: variation in trophic cascades across environmental gradients. *Ecological Applications*, **18**(8), 1860–73.

Shears, N. T. and Ross, P. M. (2010). Toxic cascades: multiple anthropogenic stressors have complex and unanticipated interactive effects on temperate reefs. *Ecology Letters*, **13**(9), 1149–59.

Smale, D. A. and Wernberg, T. (2013). Extreme climatic event drives range contraction of a habitat-forming species, *Proceedings of the Royal Society B*, **280**(1754).

Smith, H. L., Anderson, M. J., Gillanders, B. M. and Connell, S. D. (2014). Longitudinal variation and effects of habitat on biodiversity of Australasian temperate reef fishes. *Journal of Biogeography*, **41**, 2128–39.

Steneck, R. S., Graham, M. H., Bourget, B. J. et al. (2002). Kelp forest ecosystems: biodiversity, stability, resilience and future. *Environmental Conservation*, **29**, 436–59.

Steneck, R. S. and Johnson, C. (2014). Kelp Forests: Dynamic Patterns, Processes, and Feedbacks. In M. D. Bertness, J. F. Bruno, B. R. Sillman and J. J. Stachowicz, eds. *Marine Community Ecology and Conservation*. Sinauer Associates Inc, Sunderland, MA, pp. 315–36.

Sun, C., Feng, M., Matear, R. J. et al. (2012). Marine downscaling of a future climate scenario for Australian boundary currents. *Journal of Climate*, **25**(8), 2947–62.

Sunday, J. M., Calosi, P., Dupont, S., Munday, P. L., Stillman, J. H. and Reusch, T. B. (2014). Evolution in an acidifying ocean. *Trends in Ecology & Evolution*, **29** (2), 117–25.

Sutton, P. J. (2003). The Southland Current: a subantarctic current. *New Zealand Journal of Marine and Freshwater Research*, **37**(3), 645–52.

Sutton, P. J. and Bowen, M. (2014). Flows in the Tasman front south of Norfolk island. *Journal of Geophysical Research: Oceans*, **119**(5), 3041–53.

Thurstan, R. H., Brittain, Z., Jones, D. S., Cameron, E., Dearnaley, J. and Bellgrove, A. (2018). Aboriginal uses of seaweeds in temperate Australia: an archival assessment. *Journal of Applied Phycology*, **30**(3), 1821–32.

Underwood, A. J., Chapman, M. C. and Connell, S. D. (2000). Observations in ecology: you can't make progress on processes without understanding the patterns. *Journal of Experimental Marine Biology and Ecology*, **250**, 97–115.

Vergés, A., Doropoulos, C., Malcolm, H. A. et al. (2016). Long-term empirical evidence of ocean warming leading to tropicalization of fish communities, increased herbivory, and loss of kelp. *Proceedings of the National Academy of Sciences*, **113**(48), 13791–6.

Vergés, A., Steinberg, P. D., Hay, M. E. et al. (2014a). The tropicalization of temperate marine ecosystems: climate-mediated changes in herbivory and community phase shifts. *Proceedings of the Royal Society B*, **281** (1789).

Vergés, A., Tomas, F., Cebrian, E. et al. (2014b). Tropical rabbitfish and the deforestation of a warming temperate sea. *Journal of Ecology*, **102**(6), 1518–27.

Vitousek, P. M. (1994). Beyond global warming: ecology and global change. *Ecology*, **75**(7), 1862–76.

Walker, J. (2007). Effects of fine sediments on settlement and survival of the sea urchin *Evechinus chloroticus* in northeastern New Zealand. *Marine Ecology Progress Series*, **331**, 109–18.

Waters, J. M., Wernberg, T., Connell, S. D. et al. (2010). Australia's marine biogeography revisited: back to the future? *Austral Ecology*, **35**(8), 988–92.

Wernberg, T., Bennett, S., Babcock, R. C. et al. (2016). Climate-driven regime shift of a temperate marine ecosystem. *Science*, **353**(6295), 169–72.

Wernberg, T., Coleman, M. A., Bennett, S., Thomsen, M. S., Tuya, F. and Kelaher, B. P. (2018). Genetic diversity and kelp forest vulnerability to climatic stress. *Scientific Reports*, **8**(1), 1851.

Wernberg, T., Russell, B. D., Moore, P. J. et al. (2011a). Impacts of climate change in a global hotspot for temperate marine biodiversity and ocean warming. *Journal of Experimental Marine Biology and Ecology*, **400** (1–2), 7–16.

Wernberg, T., Russell, B. D., Thomsen, M. S. et al. (2011b). Seaweed Communities in Retreat from Ocean Warming. *Current Biology*, **21**(21), 1828–32.

Wernberg, T., Smale, D. A., Tuya, F. et al. (2013a). An extreme climatic event alters marine ecosystem structure in a global biodiversity hotspot. *Nature Climate Change*, **3**, 78–82.

Wernberg, T., Thomsen, M. S., Connell, S. D. et al. (2013b). The footprint of continental-scale ocean currents on the biogeography of seaweeds. *PLoS ONE*, **8**(11), e80168.

Wernberg, T., Thomsen, M. S., Tuya, F. and Kendrick, G. A. (2011c). Biogenic habitat structure of seaweeds change along a latitudinal gradient in ocean temperature. *Journal of Experimental Marine Biology and Ecology*, **400**(1), 264–71.

Wernberg, T., Thomsen, M. S., Tuya, F., Kendrick, G. A., Staehr, P. A. and Toohey, B. D. (2010). Decreasing resilience of kelp beds along a latitudinal temperature gradient: potential implications for a warmer future. *Ecology Letters*, **13**(6), 685–94.

Wernberg, T., Vanderklift, M. A., How, J. and Lavery, P. S. (2006). Export of detached macroalgae from reefs to adjacent seagrass beds. *Oecologia*, **147**(4), 692–701.

Womersley, H. B. S. (1981). Biogeography of Australasian Marine Macroalgae. In M. N. Clayton and R. J. King, eds. *Marine Botany: An Australian Perspective*. Longman Cheshire Pty Ltd, Melbourne, pp. 292–307.

Yamaguchi, A., Furumitsu, K., Yagishita, N. and Kume, G. (2010). Biology of herbivorous fish in the coastal areas of western Japan. In *Coastal environmental and ecosystem issues of the East China Sea, Nagasaki University, TERRAPUB, Tokyo*, 181–90.

Zarco-Perello, S., Wernberg, T., Langlois, T. J. and Vanderklift, M. A. (2017). Tropicalization strengthens consumer pressure on habitat-forming seaweeds. *Scientific Reports*, **7**(1), 820.

# Chapter 17

# Kropotkin's Garden

## *Facilitation in Mangrove Ecosystems*

Mark Huxham, Uta Berger, Martin W. Skov and Wayne P. Sousa

That is the watchword that comes to us from the bush, the forest, the river, the ocean – therefore combine, practice mutual aid.

*Peter Kropotkin, Mutual Aid 1906*

## 17.1 Introduction

Ecologists have studied 'mutual aid' (or facilitation) since at least the time of Kropotkin (1842–1921). But the 'watchword' which he heard with such force has not always been heeded, remains poorly understood and is rather loosely defined. A seminal early use comes from Connell and Slatyer (1977), where building on the work of Clements (1916), they used 'facilitation' to describe a model of community succession in which pioneer species modify the habitat allowing the colonisation of later ones. While this successional implication remains common in the literature, the word is also applied more broadly to positive interactions between individuals and species. Some authors restrict the term to plant–plant interactions (e.g., Krebs, 2001), while others treat it as synonymous with 'positive species relationships' (He et al., 2013), although Munguia et al. (2009) argue that any relationship that does not cause evolutionary changes to both parties is not a true 'interaction'. Others expand the term to describe any positive relationships between individuals. For example Schöb et al. (2014) state that facilitation describes 'the positive effects of one organism on others', while Singer (2016) defines it as 'interactions between two organisms, or two species, that benefit at least one of them and harm neither'. The latter definition covers interactions at all levels of complexity, from individual behaviour through to ecosystems, and includes obligate symbiosis, commensalism and looser associations.

The term therefore always involves the idea of benefit from association with another organism. This benefit is sometimes mutual, sometimes one-way and sometimes even negative for one partner, particularly over time (the facilitation model of succession allows for the competitive exclusion of the pioneer species by those that they facilitate). Given this broad use, any review of facilitation is in danger of losing focus or becoming overwhelmed. Here, we adopt another influential definition from Bertness and Calloway (1994), which includes interactions within and between species, but restricts the ambit of the term through a focus on stress: 'the benefits to an organism by the minimisation by neighbouring organisms of biotic or physical stress'. Applications of this definition in the literature have emphasised facilitation through the amelioration of physical stresses by habitat modification, biological stresses through associational defences and the existence of scale-dependant 'positive habitat switches' (Wilson and Agnew, 1992), in which sufficiently large or dense associations of organisms permit the presence of others. Here, we consider examples of all of these mechanisms in mangroves. We look at how relationships between individual plants can ameliorate physical and chemical stresses (particularly during establishment and colonisation), and how these effects can manifest at ecosystem scales. We consider how fauna are known to change the physical and chemical environments in favour of tree establishment and growth, and how the biological stress of herbivory may be mitigated by associational defences conferred by spatial aggregation of plants or through indirect interactions with epibionts or predators that defend plants against herbivory. Finally, we discuss the possible function of ecosystem-scale facilitation in the long-term resilience of mangroves to environmental change.

## 17.2 Why Mangroves?

There are theoretical and practical reasons to look for facilitation within mangrove ecosystems. The dominant conceptual model used to explain and predict the occurrence of positive interactions in ecosystems is the stress gradient hypothesis (SGH; Bertness and Callaway, 1994). This suggests that physical and biological stresses will increase the frequency of facilitation, which will help mitigate these stresses through habitat amelioration or associational defences respectively. Later refinements of the theory show that particular outcomes of any interaction depend on the characteristics of the stress factors (resources or non-resources) and on the performance of the involved species (relative stress tolerance and competitiveness), which might change during their life cycles (Maestre et al., 2009).

Intertidal habitats are often used to exemplify stress in the ecological literature; they are places subjected to frequent physical and chemical changes and require adaptations suited for aquatic and terrestrial survival. So it is not surprising that many of the studies demonstrating facilitation and testing the SGH have been conducted in the intertidal, with rocky shores and salt-marsh habitats predominating (see references in He and Bertness, 2014). Mangroves are exposed to the same stresses but are poorly represented in the facilitation literature. Trees and shrubs are the plant forms most likely to show strong facilitative interactions (He et al., 2013) so those growing intertidally – mangroves – should be particularly likely to demonstrate them. Facilitative interactions may also help explain the high productivity of mangrove ecosystems. Many of the adaptations to stress exhibited by mangroves, such as investment in succulent leaves with thick epidermal layers, metabolically costly salt exclusion, high root:shoot ratios and conservative resource capture and growth strategies, are usually associated with relatively low productivity (Chapin et al., 1993). The fact that mangroves show total productivity levels similar to terrestrial tropical forests (Alongi, 2009), despite these adaptive characteristics, suggests we lack a full understanding of how productivity is achieved in mangroves.

A growing literature indicates that fauna help explain this anomaly. The abundant burrows and mounds visible in most mangrove forests show ecosystem engineers – particularly crabs – at work. The activities of large and active bioturbators in areas where the edaphic conditions are stressful to plants creates conditions in which animals are likely to make important changes to growing conditions. While crabs are the focus of most of the relevant literature, there is good evidence of facilitatory importance to forest functioning of other faunal groups, including ants, birds and sponges. Here we review some of that evidence, with emphasis on the facilitatory roles of fauna in the establishment and productivity of mangrove trees.

Large-scale mangrove destruction combined with a growing awareness of the importance of mangroves for people and wildlife have led to global reforestation and restoration efforts. When modelled on silvicultural practices developed for terrestrial forests under benign environmental conditions, these efforts are often unsuccessful (Kodikara et al., 2017). While planting seedlings in straight rows with large, even distances between them might minimize timber loss through density dependent mortality (self-thinning) once the vegetation is closed, growing evidence shows enhanced risk of mortality and reduced growth of single plants without neighbours. Facilitation theory and information on the importance of fauna provides a perspective that may inform restoration and rehabilitation of mangrove habitat (Gedan and Silliman, 2009), particularly in harsh and degraded settings and where positive biotic influences have been overlooked. This chapter considers the evidence for facilitation, as defined by Bertness and Callaway (1994), at the different successional stages of a mangrove forest and life stages of individual mangrove trees. We begin with the initial establishment and growth of seeds and propagules (while colonising new areas and within established forests), consider the growth of seedlings, saplings and young trees, and consider the positive role of biotic interactions in enhancing the long-term persistence of the mangrove ecosystem in the face of environmental change.

## 17.3 Establishment and Colonisation

### 17.3.1 Hydrodynamics at the Seaward Edge

Mangrove seedlings and propagules are particularly vulnerable stages and early growth and establishment is often prevented due to physical and biological stresses. For example, Balke et al. (2011) identify 'windows of opportunity' during which mangroves may establish on tidal flats. These are periods during which hydrodynamic forcing is low enough to allow the initial growth of the plants; without such periods, seedlings are washed away. Because the critical drag force of water required to dislodge a new seedling increases exponentially with the maximum root length (Balke et al., 2011), the very early days during which roots are small are the most vulnerable. Hence, the physical impacts of waves and fast-moving water are critical stresses at the seaward mangrove fringe and on open tidal flats. Hydrological stresses limit establishment of new plants beyond current vegetated areas. The plants themselves modify these stresses by slowing water down, reducing wave heights, stabilising sediments against erosion with their roots and by encouraging sediment accretion and depositing organic material into the sediment. This generates biogeomorphic feedbacks that can lead to alternative states (open mud flat versus dense forest) with very sharp and rapid transitions (both spatially and temporally) between them (Figure 17.1).

Mangroves are effective in attenuating the energy of waves; for example, *Kandelia candel* trees of only five to six years old reduced wave energy by 20 per cent over 100 m (Mazda et al., 1997), while Barbier et al. (2008) show reductions of up to 60 per cent over 100 m of *Sonneratia caseolaris.* Some mangroves, such as *Rhizophora mucronata*, are incapable of establishment without sheltered conditions (Thampanya et al., 2002). In many sites, such species effectively rely on the protection from waves and water movement provided by the seaward forest, consisting of pioneer species such as *Avicennia marina* and *Sonneratia alba* to establish and regenerate. This differential susceptibility to wave impact and biogeomorphic feedbacks between tree density and water movement provides one explanation for species zonation within mangrove forests and is an example of an ecosystem-wide facilitation effect.

Researchers have long discussed whether mangroves can act as active 'land builders' or whether they simply passively follow new accretion (see for example the discussion on p. 106 of Smith (1992) citing examples more than 120 years old). While elements of this debate continue (Alongi, 2008), there is now compelling evidence that mangroves enhance sediment accretion, enable forest surface elevation and reduce erosion (Winterwerp et al., 2005; Mcivor et al., 2012; Krauss et al., 2014; Huxham et al., 2015). Measuring the changes in coastlines over time has shown how mangrove removal can enhance rates of erosion, while forest preservation can reduce erosion and cause land progradation (Thampanya et al., 2006). Hence, mangrove

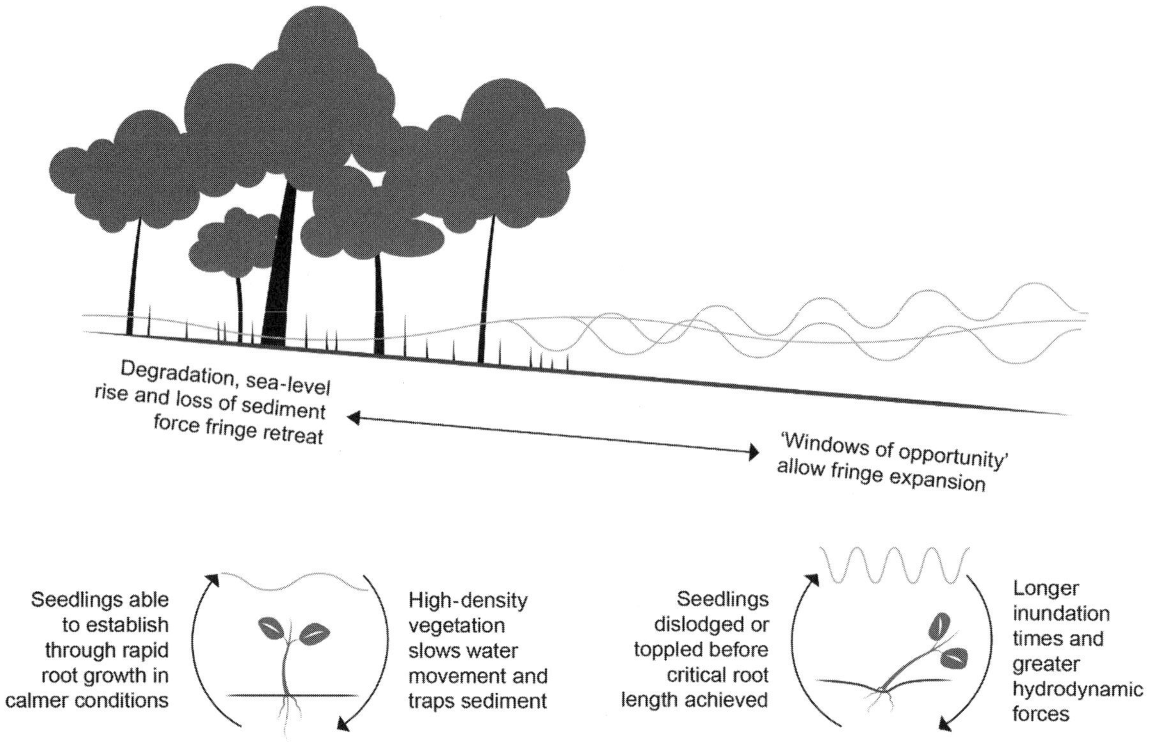


**Fig. 17.1** Biogeomorphic feedbacks between hydrodynamic stresses (including wave height and current speed) and the impacts of mangroves on water movement help to determine the sharp transition zone between the mangrove forest and open shore.

ecosystems help create stable conditions for their persistence and, in some cases, expansion, through the facilitative effects of tree density on water and sediment dynamics.

### 17.3.2 Desiccation at the Landward Edge

At Gazi Bay, southern Kenya, there are many hectares of bare sediment between the landward fringe of the natural mangrove forest and the terrestrial vegetation. These forlorn areas are studded with the stumps of long-dead mangrove trees, testament to an historical transition from forest to bare ground. Local sources suggest this was caused by commercial extraction of wood at least forty years ago (Kirui et al., 2008). Hence, the loss of trees caused an enduring change from forest cover to a new ecosystem, characterised by bare ground and very low biodiversity. This sharp transition between two states is similar to that found at the seaward mangrove fringe and suggests strong feedback processes are operating; in this case, driven not by water energy, but by salinisation and desiccation (Figure 17.2). Canopy removal exposes the sediment to the sun and allows rapid evaporation of pooling saline water. At Gazi this has resulted in salinisation (with sediment pore water at 89 PSU or more), preventing seedling establishment and leaving 'salty barrens' which are hostile to the growth of all higher plants. Under such conditions, forest recovery may follow the opening of natural 'windows of opportunity', such as unusually heavy rainfall events coinciding with periods of seed or propagule production. The lengthy persistence of the Gazi salty barrens suggests that no such windows have opened here over at least the past forty years and active restoration involving the planting-out of seedlings was required to restore mangrove coverage. New seedlings and propagules are very vulnerable to desiccation. However *Avicennia*

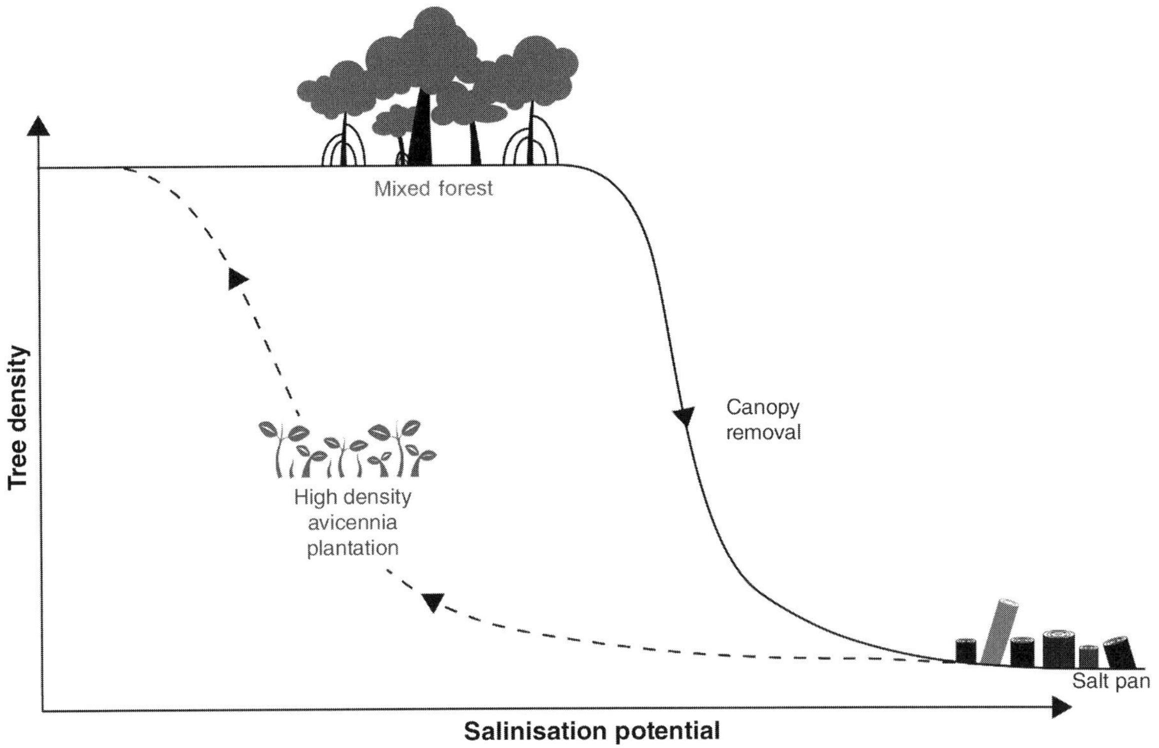


**Fig. 17.2** Canopy removal at the landward fringe can result in salinisation, as salt water evaporates without the protection of shade. This may establish a stable state characterised by hysteresis; rapid transitions and non-linear processes mean that it is difficult to reverse the condition of the ecosystem. Here, planting of high-density *Avicennia marina* seedlings allows mangrove recovery through intra- and inter-specific facilitation, as the seedlings cast shade, enhance soil organic matter and permit colonisation of other species.

*marina* is exceptionally tolerant of salt (Jayatissa et al., 2008), and seedlings older than three to four months were able to survive and grow (Kirui et al., 2008). *Avicennia marina* subsequently acted as a nurse species at this site, allowing the establishment of wild seedlings of other mangrove species under the protection of shade and with enhanced sediment moisture and reduced salinity, an example of inter-specific facilitation (Huxham et al., 2010).

Similar interactions have been reported between salt marsh vegetation and the black mangrove *Avicennia germinans*, which shows enhanced survival and growth when associated with salt-marsh vegetation under high-stress conditions in the USA (Guo et al., 2013). In the Caribbean, forbs and grasses can also act as nurse species for mangroves, both through ameliorating soil conditions and by physically trapping and supporting propagules (McKee et al., 2007b).

Intra-specific facilitation is also common during the early growth stages of intertidal plants, including mangroves, and such interactions can be harnessed to enhance restoration (Gedan and Silliman, 2009). For example, Silliman et al. (2015) showed increased biomass of tidal marsh grass *Spartina* sp. when planted in clumped rather than dispersed configurations at temperate sites. This effect was largely due to the amelioration of soil anoxia, with high densities of plants sharing oxygen leaked from roots. Similar mechanisms, along with mutual shading, may explain why high-density clumps of *Avicennia marina* seedlings showed enhanced survival in Kenya (Huxham et al., 2010).

### 17.3.3 Degradation Due to Changes in Hydrodynamics and the Role of Phenotypic Plasticity in Ecological Recovery

Regular flooding is essential for the normal functioning of most mangrove ecosystems; physical changes that prevent this usually result in the loss or degradation of forests. For example, there was rapid dieback after the construction of a road cutting through the Ajuruteua Peninsula, northern Brazil, in 1974. The road, constructed parallel to the watershed, blocked inundation by tidal channels and led to desiccation and a lethal accumulation of salt of the sediment. The erosion of the uppermost soil layer by wind contributed to a regime shift from a tall mangrove forest (with a maximum canopy height of 30 m; Mehlig et al., 2010) to a bare, hypersaline ground, similar to the salty barrens of Gazi. This lasted almost thirty years before *A. germinans* seedlings were able to recolonize the area (Vogt et al., 2014), presumably during a natural window of opportunity. The colonisers developed as shrubs and thus adapted to the salinity-induced osmotic constraints and water scarcity (Peters et al., 2014). As nurse plants, the shrubs facilitated the survival and growth of con-specific followers (Vogt et al., 2014; Pranchai, 2015), which eventually switched back to tree architecture (Vogt et al., 2014) and an onset of neighbourhood competition (Pranchai, 2015). The change of morphology from shrub to tree architecture was key for this self-rescuing effect and is a feature distinct from other nurse plant systems, e.g., from alpine regions (Schöb et al., 2014), where nurse plants are not conspecifics of the facilitated plants.

### 17.3.4 Faunal Facilitation of Seedling Recruitment

While plant interactions dominate the literature on facilitation during colonisation of new areas, the effects of fauna become increasingly apparent as forests mature. The burrowing and sediment feeding activities of crabs, mud shrimps and mud skipper fish collectively transform what would have been a relatively flat forest floor into a variable-height topography of burrow digging spills, burrow 'hoods' and burrow openings. Minchinton (2001) speculated that such topography provides opportunity for propagules to become trapped on the forest floor during tidal flooding. Burrow mounds were particularly common under canopy cover, accounting for up to 44 per cent of the forest floor. Minchinton (2001) found that 75 per cent more propagules were trapped in areas with mounds than those without, and that propagules were more likely to establish as seedlings when growing on mounds. The establishment of seedlings was also faster on mounds than on flats (Minchinton, 2001), possibly as the window of opportunity for propagule settlement (*sensu* Balke et al., 2011) is extended by the amelioration of hydrological stress by elevation in surface topography. The study showed crab burrowing can have a positive effect on mangrove recruitment in mature forests and in forest gaps and might aid in the recovery of forests after disturbance events. Crabs are also avid propagule predators (Sousa and Dangremond, 2011) and the effective outcome of crabs on seedling establishment is likely to be dependent on contextual variation in offsetting of propagule trapping against propagule predation. Nevertheless, the effect of crab mounds on propagule establishment is similar to that offered by mussels in the establishment of salt marsh vegetation. Angelini et al. (2015) found that mussel patches, which account for a much smaller proportion of marsh area cover than do crab burrows in mangroves, had significant and large-scale facilitating implications for marsh functioning, including the boosting of primary production.

## 17.4 Early Growth – Facilitation in Seedlings and Saplings

### 17.4.1 Sedimentation and Nutrient Supply

Positive density-dependent effects operating over small scales may persist well beyond the initial establishment and growth of mangrove seedlings, although the mechanisms causing this intra- and inter-specific facilitation are likely to change as plants mature. In Sri Lanka, *R. mucronata* seedlings planted at high densities grew faster and showed better survival than those at low densities (Kumara et al., 2010), an effect that persisted for at least six years (Kumara, personal communication). Thampanya et al. (2002) also recorded positive effects of density on the same species in Thailand. Macronutrients, in particular nitrogen and phosphorus, are typically limiting factors of mangrove growth (Alongi, 2009), and biophysical processes that enhance the delivery and absorption of nutrients will usually improve growth. In the Sri Lanka case, denser stands of trees accumulated more allocthonous sediment with high nitrogen content and this was the likely cause of the better performance of high-density plants (Phillips et al., 2017).

### 17.4.2 Associational Resistance to Herbivores

A plant's spatial associations with other plants can alter its detection by and/or vulnerability to herbivores. Interspersion of a host plant with other species can decrease (associational resistance) or increase (associational susceptibility) its herbivore load, compared to what it experiences when growing in monospecific stands (Barbosa et al., 2009). To our knowledge, the beneficial effects of associational resistance to herbivory have not been studied explicitly in mangrove ecosystems. This may be a consequence of the long-standing belief that mangrove vegetation is structured primarily by responses of plants to physical and chemical, rather than biological, stressors. The apparent low rates of folivory in mangroves compared to terrestrial forests has reinforced this view (Komiyama et al., 2008; Alongi, 2009). However, most estimates of folivore damage to mangroves are based on the level of standing leaf damage, which has been shown to greatly underestimate true rates of loss, since leaves that are completely consumed are not included in the estimates. When Burrows (2003) marked individual leaf buds and followed their fates to maturity, he estimated leaf damage of 5–8.3 per cent for *Rhizophora stylosa* and 19.3–29.5 per cent for *Avicennia marina*, substantially higher than earlier estimates based on standing leaf damage. In addition, other, less obvious, herbivorous niches, such as stem wood-borers, are functionally important in mangroves and typically are overlooked when estimating biomass lost to herbivory. For example, stem-boring beetles killed more than 50 per cent of a fringing *Rhizophora mangle* canopy on offshore islands in Belize (Feller and McKee, 1999; Feller, 2002). Hence, it is clear that mangrove stands must cope with both the stressful chemical and physical conditions already discussed and the biological stress imposed by herbivore damage comparable to that observed in other types of forests.

Associational resistance has been well-documented in terrestrial vegetation. Among the first studies to experimentally demonstrate it were Root and colleagues' investigations of herbivore–plant interactions in gardens planted into old field habitat (Tahvanainen and Root, 1972; Root, 1973), which showed an inverse relationship between herbivore load on collard (*Brassica oleracea*) and the diversity of neighbouring vegetation (pure stands of collard versus a diverse mix of meadow grasses and forbs). Alternative (but not mutually exclusive) mechanisms that could account for this relationship include: (1) the enemies hypothesis: diverse vegetation harbours a greater number of natural enemies of herbivores (predators and parasitoids) than single-species stands and (2) the resource concentration hypothesis: a host plant is more likely to be found, particularly by specialist herbivores, if it grows in a dense monoculture of conspecifics than in a mixed-species stand, and such herbivores are more likely to aggregate and reproduce in these areas of high resource availability and are less likely to emigrate from them (Root, 1973; Bach, 1980). Since the publication of this classic garden-meadow study, associational resistance to herbivores has been experimentally demonstrated in a variety of natural communities, including nearshore marine habitats (e.g., Pfister and Hay, 1988), open woodlands with shrub understory (Baraza et al., 2006) and shoreline vegetation (Hambäck et al., 2000).

As described earlier, the resource concentration hypothesis posits that a host plant is at greater risk of herbivory when growing in a monoculture than in a mixed-species stand, particularly if it is primarily fed on by specialists. However, an increase in local density within monocultures could lower the per capita risk of attack if the herbivore does not exhibit a sufficient numerical response (through behavioural aggregation or in situ reproduction) to keep pace with plant growth or compensatory recovery from herbivore damage. In effect, as the local density of host plants increases, the resident herbivore population becomes swamped/satiated and per capita rates of damage and mortality decline (Crawley, 1997; Otway et al., 2005). This is more likely to occur if the specialist herbivore is univoltine (i.e., a single brood of offspring per year), has low fecundity and/or a complex life cycle, with a grazing larval stage that metamorphoses into a mobile, non-grazing adult stage (e.g., Lepidoptera). In the latter instance, the reproductive adult disperses away from the host plant and is unlikely to return to lay eggs in the same patch of plants in which it grazed as a larva (Cromartie, 1975).

As noted earlier, we know of no published investigations that specifically test for associational resistance to herbivory in mangrove habitats. The study that comes closest is Johnstone's (1981), which quantified rates of standing insect grazing damage to mature leaves of twenty-three mangrove species growing in mixed stands in coastal swamps of the Port Moresby region of Papua New Guinea. He found no relationship between rates of herbivory and the species richness of mangroves growing within 25 m of the sampled tree, and no correlation between leaf area eaten and the densities of individual species. However, patterns could have been obscured by large species-specific variation in susceptibility among the twenty-three species he sampled.

New evidence that associational resistance to herbivory occurs in mangroves comes from an unpublished field experiment that one of us (Sousa) conducted to assess the roles of intra- and inter-specific competition on growth and survival of mangroves along the tidal gradient on the central Caribbean coast of Panama. In this experiment, young (~ 1-month-old) mangrove seedlings of the three canopy mangrove species *A. germinans*, *Laguncularia racemosa* and *R. mangle* were planted into single- and mixed-species treatments, both inside and outside of replicate canopy gaps created by past lightning strikes. The complete design and results of this study will be reported elsewhere; here we focus on patterns of survivorship of *A. germinans* seedlings as a function of density and species composition inside three upper-intertidal gaps, within forest stands dominated by conspecific adults. These stands are where populations of the specialist lepidopteran *Junonia genoveva*, or black mangrove buckeye, are concentrated (Sousa, unpublished data). *J. genoveva* larvae feed exclusively on the leaves of *A. germinans* and can strongly impact seedling populations (Elster et al., 1999). We analysed rates of survival over the first six months of the experiment; young seedlings are the most attractive life stage to egg-laying, adult *J. genoveva*, and are the most likely stage to be killed by their larvae (Sousa, unpublished data).

The experiment yielded evidence of both herbivore swamping by high host plant density and interspecific associational resistance (Figure 17.3). Survival rates differed significantly among treatments ($F_{5,12}$ = 12.72, P = 0.0019). Among the *A. germinans* monocultures, the mean rate of survival was higher in the three-fold density treatment than either the two-fold or one-fold seedling densities. Survival was also higher in the 1:1 and 1:2 mixed treatments than the one-fold density monoculture, which contained the same number of *A. germinans* seedlings, a pattern indicative of associational resistance. Survival in the 2:1 mixed treatment appeared to be greater than in the two-fold monoculture, also suggestive of associational resistance, but the difference was not significant (P = 0.095).

This experiment did not directly manipulate *J. genoveva* densities to confirm that grazing by its larvae was responsible for seedling mortality. We did observe numerous larvae feeding on the plants and many adult butterflies flying around the study forests, laying eggs on host plants. In a subsequent experiment of similar design, an herbivore-exclusion treatment was added; it confirmed that *J. genoveva* caterpillars are responsible for most of the mortality of young *A. germinans* seedlings in upper-intertidal basin forests, where this experiment was conducted.

The specific mechanisms accounting for associational resistance in this system are not known. The presence of neighbouring non-host plants may physically interfere with the adult

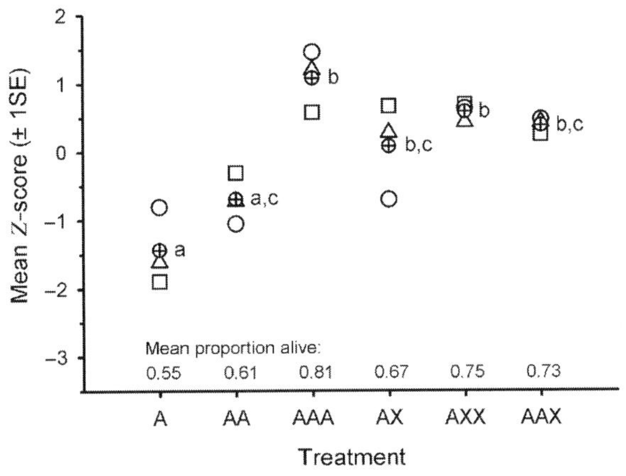


**Figure 17.3** Survival to six months of *A. germinans* (A) seedlings planted in three monospecific treatments of one-, two- and three-fold density (A, AA, AAA), and three mixed-species treatments of 1:1 (AX), 1:2 (AXX) and 2:1 (AAX) density combinations of *A. germinans* with one of two other mangrove species (X = *L. racemosa* or *R. mangle*). Treatments were established in 0.5 $m^2$ plots at densities of twenty-five, forty and eighty-one seedlings per plot; the design was replicated inside three upper-intertidal gaps. The three mixed-species treatments contained twenty-five, twenty-seven and fifty-four *A. germinans* seedlings, respectively. Survival values for the pair of plots (one with *L. racemosa* and the other with *R. mangle*) that contained a 1:1, 1:2 or 2:1 ratio treatment were averaged for the analysis. Because seedlings in one of the gaps suffered nearly four-times higher mortality across all treatments than the other two, we standardised the proportion surviving values from each gap by converting them to z-scores, based on the respective gap means and standard deviations. Open symbols (○, Δ, □) represent values from the three replicate light gaps. The ⊕ symbol marks the mean Z-score of the three replicates. Letters to the right of these symbols summarize results of a Tukey honest significant difference test; those not sharing a letter are significantly different at $P < 0.05$. The mean proportions alive at six months in each of the six treatments are presented just above the x-axis. SE, standard error.

butterfly's ability to locate and oviposit on a host plant. In addition, *A. germinans* leaves contain iridoid glycosides (Fauvel et al., 1995). This class of chemical compounds has been shown to act as oviposition stimulants for *J. genoveva*'s temperate zone congener, *Junonia coenia* (Pereyra and Bowers, 1988); therefore, it is conceivable that there may also be some interference with chemical signalling in patches of mixed-species composition.

## 17.5 Mature Trees and Forests

### 17.5.1 Faunal Enhancement of Mangrove Primary Production

There is good evidence that crabs facilitate mangrove production. Removing crabs from the benthos of *Rhizophora*-dominated stands in Australia led to a reduction in forest primary production and reproductive output (Smith et al., 1991); an outcome that is very similar to the result of crab exclusion experiments in salt marshes (Bertness, 1985). Bioturbation by crabs aerates mangrove soils and reduces soil sulphide content (Smith et al., 1991), a chemical that depresses the primary productivity of coastal wetland plants in general (Bertness, 1985). Burrowing by crabs also regulates other soil characteristics that influence forest production. Smith et al. (2009) found that field exclusion of fiddler crabs reduced the growth of *L. racemosa* seedlings, probably because soil salinity rose in the absence of crab burrowing. Faunal burrows aid the flushing of sediments by tidal water (Stieglitz et al., 2013), which dilutes salinity and reduces sulphate reduction (Kristensen, 2000; Smith et al., 2009). Crab burrowing is also thought to boost nitrification and thereby plant growth (Bertness, 1985), and Skov and Hartnoll (2002) proposed that selective feeding of crabs on mangrove leaf litter facilitates the retention and remineralisation of litter nutrients for the benefit of crabs themselves, as well as forest productivity. So far, the experimental evidence of this happening in coastal wetlands is limited (Daleo et al., 2007; Kristensen et al., 2008).

Crab burrowing might also indirectly enhance forest production by facilitating the activity of other species that stimulate tree growth. In salt marshes, the aeration of soils by crab burrowing facilitated the development of mycorrhizal fungi, which boost the availability of nitrogen for plant production (Daleo et al., 2007). The fungi need soil aeration to grow and the removal of crabs, and their burrows, reduced the growth of salt marsh *Spartina* plants by 35 per cent (Daleo et al., 2007). Mycorrhizal fungi are abundant in the soils of coastal wetlands (Carvalho et al., 2004), and it is likely that crab facilitation of mycorrhiza will also boost mangrove tree production.

Exclusion of crabs does not always result in a reduction of tree growth. A year-long removal of a dominant semi-terrestrial crab from a high-shore *Rhizophora* stand in Brazil did not have any measurable impact on forest growth (Pülmanns et al., 2016). While the lack of effect might be because crab removal was not complete (approximately a third of all crabs were removed), it may equally well be that the facilitatory role of crabs in forest production is context-dependent and influenced by a range of geomorphological, physio-chemical, seasonal or biological factors (Pülmanns et al., 2016). Theoretical and empirical work emphasises the influence of environmental context on the relative importance of facilitation, particularly along gradients of stress (Bruno et al., 2003; Maestre et al., 2009; Schöb et al., 2014). The influence of stress gradients on faunal facilitation has not been addressed in mangroves.

While the evidence for crab facilitation of mangrove production is considerable, other fauna can also positively affect forest production. Ellison and colleagues (Ellison and Farnsworth, 1990; Ellison et al., 1996) described an intriguing facultative mutualism between *Rhizophora* trees and their prop root-colonising sponges. Trees fringing creeks in Belize, with permanently submerged roots, grew adventitious rootlets within sponge tissues to extract nitrogen from sponges. Sponges in return were offered colonisation substrate as well as organic carbon by the trees and both organisms grew substantially faster when coexisting. Facilitation may also be instigated by vertebrates. For example, birds can boost tree production through providing growth-limiting nutrients. Mangroves offer preferential roosting sites to many bird species that feed on adjacent mudflats (Buelow and Sheaves, 2015). Bird roosting sites may be restricted to select areas of the forest, the preference for which may last for years (Pearse, 2010). Bird guano in such habitual roosting areas can offer a locally significant and steady supply of nutrients that ultimately boost tree growth (Onuf et al., 1977; Feller, 1995).

### 17.5.2 Indirect Protection from Herbivores by Epibionts or Predators

A variety of indirect ecological interactions have been shown to reduce the negative impacts of stem-borers on mangrove prop roots and folivores on mangrove leaves. Mats of epibionts, comprised of colonial ascidians and sponges, commonly grow on the surfaces of submerged young *R. mangle* prop roots. As described earlier, these may include organisms that lesson the chemical stress of nutrient limitation through mutualistic associations, and they can also affect biological stresses by deterring attacks by stem-boring isopods (Ellison and Farnsworth, 1990, 1992). The mechanism(s) by which an encrusting layer of ascidians and sponges blocks isopods is unknown; it may act as a physical and/or chemical barrier to their recruitment and boring activity.

Emergent portions of mangrove trees may be indirectly protected from insect herbivory by associated predators, including ants and lizards. In these tri-trophic interactions, the predators reduce the density of insect herbivores, indirectly benefitting the plant. Mangrove protection by ants is well documented, with examples from both comparative and experimental studies. Ozaki et al. (2006) compared rates of predation on populations of the scale insect, *Aulacaspis marina* (Homoptera: Diaspidodae) infesting potted seedlings of *R. mucronata* introduced into mature natural forests and young plantations growing in abandoned shrimp ponds, on Bali Island, Indonesia. In the natural forests, two species of ants (*Monomorium floricola* and *Paratrechina* sp.) preyed heavily on the scale insects, suppressing their density by 88 per cent in seven days. By comparison, only 28 per cent were eaten in the plantation sites, probably because the young trees in these sites afforded few ant nesting sites and the inundated soils prohibited ants from nesting in the ground. A second experiment in which ants were excluded from half the seedlings with sticky barriers applied to their lower stem confirmed that ants accounted for almost all the scale insect mortality.

Offenberg et al. (2004) documented a negative correlation between the density of arboreal weaver ants (*Oecophylla smaragdina*) and levels of herbivory on *R. mucronata* leaves in a Thai mangrove forest. Trees lacking ants had more than three times the herbivore leaf damage of trees with ant nests. On trees with ants, leaves near the nest suffered less damage than those in other areas of the canopy. Most of the leaf damage was caused by chrysomelid beetles and sesarmid crabs; both were deterred by the presence of ants, and ants were observed preying on the beetles. In this case, the relationship might be considered a weak form of ant–plant mutualism, since the ants weave the host mangrove leaves together to build their nest (Offenberg et al., 2004). Weaver ants have long been used for herbivorous insect pest control in a variety of tree crop systems (Van Mele, 2008). An earlier study of insect herbivory rates in a Papua New Guinea mangrove forest (Johnstone, 1981) did not detect a statistically significant relationship between standing leaf damage and the presence of *O. smaragdina* colonies. While the mean percent of leaf area eaten tended to be higher on trees that had no ants, compared to those with ant colonies, there was high variation in herbivory rates among trees with similar densities of ants. The difference in the findings of Offenberg et al. (2004) and Johnstone (1981) likely stems from a fundamental difference in method. The former study focussed on a single host plant species that is frequently occupied by weaver ants, while the latter compared rates for a pooled sample of leaves from twenty-three different species of mangroves, which varied in their attractiveness to the ants. This undoubtedly introduced considerable variation in other leaf attributes that affect rates of herbivory.

Ants also play a defensive role on small Bahamian islands, where the buttonwood mangrove, *Conocarpus erectus,* dominates the shoreline vegetation. This plant has extrafloral nectaries that attract a variety of insects, including nine species of ants. Its leaves come in two morphs, a silver form that has a dense layer of leaf hairs or trichomes on their surface, and a green form that has very few trichomes. Trichomes function as a structural defence against folivorous insects (Schoener, 1987, 1988; Agrawal and Spiller, 2004); silver plants have fewer extrafloral nectaries than green plants, and they produce less nectar and attract fewer ants (Piovia-Scott, 2011). Experimental exclusion of ants from low-trichome plants resulted in higher herbivore damage and lower new leaf production compared to control plants accessible to ants. In contrast, ant exclusion from silver plants had no measurable effect of either leaf damage or leaf production (Piovia-Scott, 2011). *Anolis* lizards also prey on insect folivores of *C. erectus* on these islands; some islands support lizard populations and others lack them. In a survey of seventy-four islands, total leaf damage was 42–59 per cent greater on no-lizard- than lizard-inhabited islands (Schoener, 1988).

### 17.5.3 Facilitation Cascades and the Scaling-Up of Faunal Facilitation of Mangroves

Mangrove trees act as foundation species, creating complex habitats that support multiple other species which together constitute the mangrove ecosystem. When an independent foundation species supports other, dependent foundation species, which themselves facilitate the survival of different organisms, a facilitation cascade may ensue, with facilitatory effects rippling between trophic levels (Angelini et al., 2011). Bishop et al. (2012) describe such a cascade in *Avicennia marina* forest, where beyond a threshold density of pneumatophores capture of algae facilitate fauna that depend on algae to ameliorate environmental stressors, including desiccation risk. Such facilitatory cascades may have profound implications for ecosystem functioning, when the array of facilitated species has distinct functional roles. Angelini et al. (2015) found that colonisation by mussels, which are secondary foundation species in North American salt marshes, facilitated an array of species that together stimulated marsh accretion, carbon sequestration, marsh grass production and faunal functional richness. Crabs could be argued to have a similar secondary foundation role in mangroves, potentially leading to faciliatory cascades of benefit to trees. Their facilitation of other organisms is certainly diverse, from providing habitat to other invertebrates, to facilitating the growth of mycorrhizal fungi and microphytobenthic species (Alongi, 1994; Gillikin et al., 2001; Daleo et al., 2007). Predatory as well as herbivorous crabs may act as facilitators. An observed reduction in salt marsh plant production in the USA has been attributed to 40–80 per cent reductions in predatory blue crab and fish populations and their natural control of herbivorous snails and crabs (Altieri et al., 2012). So far, there is no firm evidence of fauna regulating facilitation cascades in mangroves, but the likelihood of it occurring is high.

## 17.6 Forest Resilience in the Face of Long-Term Change

The intertidal zone is a place of daily, seasonal and decadal flux. Colonising new areas in response to long-term change, driven for example by sea-level rise (SLR) or altering river flows, allows mangroves to adapt. This spatially based resilience relies in part on the facilitative mechanisms discussed earlier; the ability for pioneer plants to exploit windows of opportunity in bare substrate, often supported by interactions with other organisms. But there are many locations where the prospects for forest migration, inshore or offshore, are limited; for example, some mangroves grow on coral islands with no hinterland and forests are increasingly restricted from movement in-land by anthropogenic structures and activities. The absence of room to move does not, however, mean that a mangrove forest lacks resilience and the ability to persist over time. For example, Caribbean mangroves growing in areas without alluvial deposits and with limited opportunities to move spatially are able to adjust to rising sea level through the deposition of organic matter, with some now growing on 10 m of peat representing thousands of years of adjustment (McKee et al., 2007a). This ability of mangroves to elevate their soil surfaces in response to rising sea levels relies on a large range of factors, including the above- and below-ground productivity stimulated by fauna and the sediment trapping assisted by conspecific density (Krauss et al., 2014), and burrowing crabs may also be important agents in helping to store organic carbon below ground (Andreetta et al., 2014).

Hence, facilitatory processes assist in the long-term persistence of mangroves and contribute to the enormous carbon density of these forests (Figure 17.4). Change (over space and time) is a central feature of mangrove systems. Over the past twenty years, the paradigm of 'balance of nature' has largely been replaced among ecologists by 'resilience thinking' (Walker and Salt, 2012), but conservation of the status quo is still a common goal for managers and the broader conservation movement. A focus on resilience mechanisms is essential to identify the relevant spatial and temporal scales at which 'self-similarity' (Jax et al., 1998), 'integrity' (Cumming and Collier, 2005) and 'persistence of relationship' (Holling, 1973) self-guard ecosystem functioning.

## 17.7 Conclusions

Benignity is seldom found in alliance with strength.

*Alexander von Humbolt*

The SGH suggests that searching for facilitation within mangrove ecosystems should be fruitful; these forests are miracles of productivity and persistence in the face of the multiple stresses that they naturally face. Our brief review confirms this prediction. Facilitation can be found at all stages of the lives of individual trees and of the forest ecosystems themselves. While reports of the amelioration of physical and chemical stresses predominate, we also present evidence of how facilitative interactions can help limit the biological stress of herbivory. What remains unclear, however, is whether these diverse observations can inform a coherent theory of facilitation in mangrove ecosystems. While the complex interplay of plant diversity, density and environmental stressors in creating context-specific outcomes is slowly being understood in other, much simpler, plant communities, such as alpine cushion plants (Schöb et al., 2014), a comparable understanding of

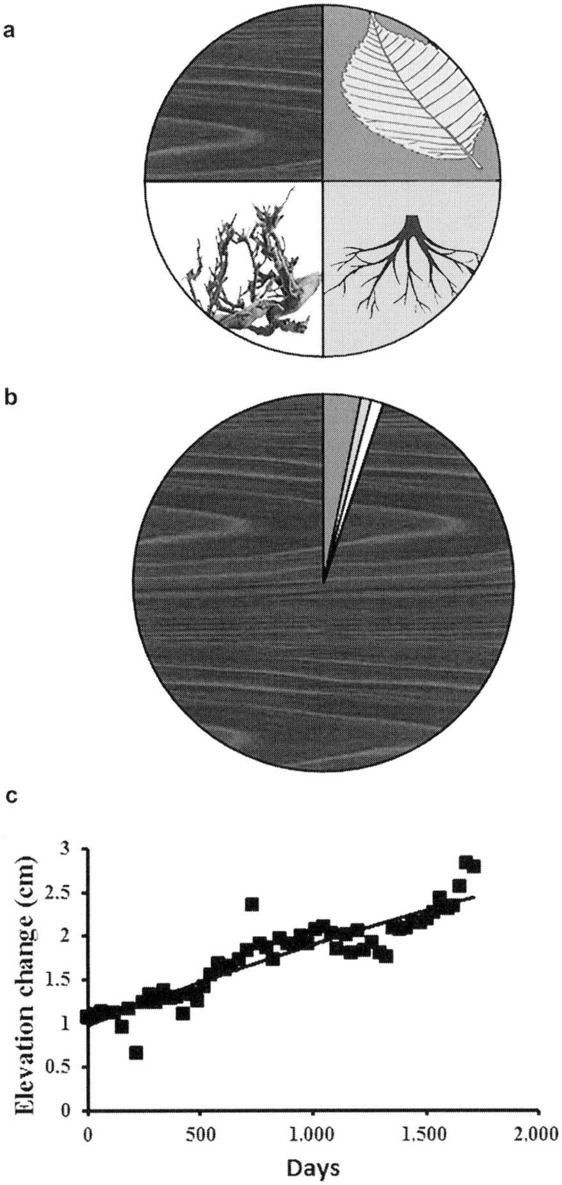


**Figure 17.4** The accumulation of carbon in the *R. mucronata* forest of Gazi Bay, helping with resilience of the forest in the face of SLR and resulting from multiple forms of facilitation. There are four main components of the long-term (years, decades or longer) carbon pool: (a) (clockwise from top-left) soil organic carbon, above-ground biomass, below-ground biomass (living roots) and dead root materials. (b) These four pools, in the same order, but here represented as the percentage contribution to the total pool at Gazi, with soil organic carbon constituting 95 per cent of the total (data from Tamooh et al., 2008; Cohen et al., 2013; Gress et al., 2017; figures converted to t C $ha^{-1}$ equivalents, assuming carbon content of living roots, dead roots and living stems of 0.39, 0.5 and 0.47, respectively). (c) Surface elevation measured in ten plots from Gazi over fifty-eight months, showing 3.1 mm elevation per year, driven by below-ground accumulation of carbon.

contingent ecological outcomes that structure mangrove communities is yet to emerge. In part, this is because we are still in the process of discovering how these unique forests work. Modern methods of ecological research, particularly manipulative field experiments and mathematical as well as computative modelling, have been applied for a much shorter time and by fewer investigators in mangroves than in terrestrial habitats. Not surprisingly, new findings frequently contravene standard explanations derived from studies in other ecosystems. Mangroves contain pronounced stress gradients on relatively compact spatial scales. These gradients consist both of resources (e.g., nitrogen or phosphorous) and of non-resources (e.g., salinity or temperature). Mangrove flora and fauna must cope with multiple, interacting stress factors and all species vary in their stress tolerance and competitiveness. These factors make mangrove systems difficult to understand. But they also mean that mangroves provide an extraordinary laboratory for building a more comprehensive and general understanding of the role that facilitation and other processes play in structuring communities. Achieving this will require a coordinated and hypothesis-driven effort to experimentally disentangle the predictions of the refined stress-gradient hypothesis (Maestre et al., 2009) and its extensions by considering the interactive influences of abiotic and biotic processes across a range of environment settings. Mangrove researchers will need to work together to do this – therefore combine, and practice mutual aid.

## Acknowledgements

We thank Sarah Murray for her help with the figures and Louise Firth for useful editing suggestions. We were inspired by opportunities to meet and share ideas at the Aquatic Biodiversity and Ecosystems Conference in 2015 and the Mangrove and Macrobenthos Meeting 4 in 2016; thank you to all those who organised these meetings and invited us.

## REFERENCES

Agrawal, A. A. and Spiller, D. S. (2004). Polymorphic buttonwood: effects of disturbance on resistance to herbivores in green and silver morphs of a Bahamian shrub. *American Journal of Botany*, **91**, 1990–7.

Alongi, D. M. (1994). Zonation and seasonality of benthic primary production and community respiration in tropical mangrove forests. *Oecologia*, **98**, 320–32.

Alongi, D. M. (2008). Mangrove forests: resilience, protection from tsunamis, and responses to global climate change. *Estuarine, Coastal and Shelf Science*, **76**, 1–13.

Alongi, D. M. (2009). *The Energetics of Mangrove Forests*, Springer, Dordrecht.

Altieri, A. H., Bertness, M. D., Coverdale, T. C., Herrmann, N. C. and Angelini, C. (2012). A trophic cascade triggers collapse of a salt-marsh ecosystem with intensive recreational fishing. *Ecology*, **93**, 1402–10.

Andreetta, A., Fusi, M., Cameldi, I., Cimò, F., Carnicelli, S. and Cannicci, S. (2014). Mangrove carbon sink. Do burrowing crabs contribute to sediment carbon storage? Evidence from a Kenyan mangrove system. *Journal of Sea Research*, **85**, 524–33.

Angelini, C., Altieri, A. H., Silliman, B. R. and Bertness, M. D. (2011). Interactions among foundation species and their consequences for community organization, biodiversity, and conservation. *BioScience*, **61**, 782–9.

Angelini, C., van der Heide, T, Griffin, J. N. et al. (2015). Foundation species' overlap enhances biodiversity and multifunctionality from the patch to landscape scale in southeastern United States salt marshes. *Proceedings of the Royal Society B*, **282**, 20150421.

Bach, C. E. (1980). Effects of plant diversity and time of colonization on an herbivore-plant interaction. *Oecologia*, **44**, 319–26.

Balke, T., Bouma, T. J., Horstman, E. M., Webb, E. L., Erftemeijer, P. L. A. and Herman, P. M. J. (2011). Windows of opportunity: thresholds to mangrove seedling establishment on tidal flats. *Marine Ecology Progress Series*, **440**, 1–9.

Baraza, E., Zamora, R. and Hódar, J. A. (2006). Conditional outcomes in plant-herbivore interactions: neighbours matter. *Oikos*, **113**, 148–56.

Barbier, E. B., Koch, E. W., Silliman, B. R. et al. (2008) Coastal ecosystem-based management with nonlinear ecological functions and values. *Science*, **319**, 321–3.

Barbosa, P., Hines, J., Kaplan, I., Martinson, M., Szczepaniec, A. and Szendrei, Z., (2009). Associational resistance and associational susceptibility: having right or wrong neighbors. *Annual Review of Ecology Evolution and Systematics*, **40**, 1–20.

Bertness, M. D. and Callaway, R. (1994). Positive interactions in communities. *Trends in Ecology & Evolution*, **9**, 191–3.

Bertness, M. D. (1985). Fiddler crab regulation of *Spartina alterniflora* production in a New England salt marsh. *Ecology*, **66**, 1042–55.

Bishop, M. J., Byers, J. E., Marcek, B. J. and Gribben, P. E. (2012). Density-dependent facilitation cascades determine epifaunal community structure in temperate Australian mangroves. *Ecology*, **93**, 1388–401.

Bruno, J. F., Stachowicz, J. J. and Bertness, M.D. (2003). Inclusion of facilitation into ecological theory. *Trends in Ecology and Evolution*, **18**, 119–25.

Buelow, C. and Sheaves, M. (2015). A birds-eye view of biological connectivity in mangrove systems. *Estuarine, Coastal and Shelf Science*, **152**, 33–43.

Burrows, D. W. (2003). The role of insect leaf herbivory on the mangroves Avicennia marina and Rhizophora stylosa. PhD thesis, James Cook University.

Carvalho, L. M., Correia, P. M. and Martins-Loucão, M. A. (2004). Arbuscular mycorrhizal propagules in a salt marsh. *Mycorrhiza*, **14**, 165–70.

Chapin, F. S, Autumn, K. and Pugnaire, F. (1993). Evolution of suites of traits in response to environmental stress. *American Naturalist*, **142**, S78–92.

Clements, F. E. (1916). *Plant Succession: An Analysis of the Development of Vegetation*. Carnegie Institution of Washington, Washington, DC.

Cohen, R., Kaino, J., Okello, J. A. et al. (2013). Propagating uncertainty to estimates of above-ground biomass for Kenyan mangroves: a scaling procedure from tree to landscape level. *Forest Ecology and Management*, **310**, 968–82.

Connell, J. H. and Slatyer, R. O. (1977). Mechanisms of succession in natural communities and their role in community stability and organisation. *The American Naturalist*, **111**, 1119–44.

Crawley, M. J. (1997). Plant–Herbivore Dynamics. In M. J. Crawley, ed. *Plant Ecology*, second edn. Blackwell Scientific Publications, Oxford, pp. 401–73.

Cromartie, W. J., Jr. (1975). The effect of stand size and vegetational background on the colonization of cruciferous plants by herbivorous insects. *Journal of Applied Ecology*, **12**, 517–33.

Cumming, G. S. and Collier, J. (2005). Change and identity in complex systems. *Ecology and Society*, **10**, 29.

Daleo, P., Fanjul, E., Casariego, A. M., Silliman, B. R., Bertness, M. D. and Iribarne, O. (2007). Ecosystem engineers activate mycorrhizal mutualism in salt marshes. *Ecology Letters*, **10**, 902–8.

Ellison, A. M, Farnsworth, E. J and Twilley, R. R. (1996). Facultative mutualism between red mangroves and root-fouling sponges in Belizean mangal. *Ecology*, **77**, 2431–44.

Ellison, A. M. and Farnsworth, E. J. (1990). The ecology of Belizean mangrove-root fouling communities I: epibenthic fauna are barriers to isopod attack of red mangrove roots. *Journal of Experimental Marine Biology and Ecology*, **142**, 91–104.

Ellison, A. M. and Farnsworth, E. J., (1992). The ecology of Belizean mangrove-root fouling communities II: patterns of epibiont distribution and abundance, and effects on root growth. *Hydrobiologia*, **247**, 87–98.

Elster, C., Perdomo, L., Polania, J. and Schnetter, M-L., (1999). Control of *Avicennia germinans* recruitment and survival by Junonia evarete larvae in a disturbed mangrove forest in Columbia. *Journal of Tropical Ecology*, **15**, 791–805.

Fauvel, M-T., Bousquet-Melou, A., Moulis, C., Gleye, J. and Jensen, S. R. (1995). Iridoid glucosides in *Avicennia germinans*. *Phytochemistry*, **38**, 893–4.

Feller, I. C. (1995). Effects of nutrient enrichment on growth and herbivory of dwarf red mangrove (*Rhizophora mangle*). *Ecological Monographs*, **65**, 477–505

Feller, I. C. and McKee, K. L. (1999). Small gap creation in Belizean mangrove forests by a wood-boring insect. *Biotropica*, **31**, 607–17.

Feller, I. C. (2002). The role of herbivory by wood-boring insects in mangrove ecosystems in Belize the role of herbivory by wood-boring insects in mangrove ecosystems in Belize. *Oikos*, **97**, 167–76.

Gedan, K. B. and Silliman, B. R. (2009). Using facilitation theory to enhance mangrove restoration. *Ambio*, **38**, 109.

Gillikin, P.G., de Grave, S. and Tack, J. F. (2001). The occurrence of the semi-terrestrial shrimp *Merguia oligodon* (de Man, 1888) in *Neosaramatium smithi* H. milne Edwards, 1853, burrows in Kenyan mangroves. *Crustaceana*, **74**, 505–7.

Gress, S, Huxham, M., Kairo, J., Mugi, L. and Briers, R. (2017). Evaluating, predicting and mapping belowground carbon stores in Kenyan mangroves. *Global Change Biology*, **23**(1), 224–34, http://dx.doi.org/10.1111/gcb.13438.

Guo, H., Zhang, Y., Lan, Z. and Pennings, S. C. (2013). Biotic interactions mediate the expansion of black mangrove (*Avicennia germinans*) into salt marshes under climate change. *Global Change Biology*, **19**, 2765–74.

Hambäck, P. A., Ågren, J. and Ericson, L. (2000). Associational resistance: insect damage to purple loosestrife reduced in thickets of sweet gale. *Ecology*, **81**, 1784–94.

He, Q. and Bertness, M. D. (2014). Extreme stresses, niches, and positive species interactions along stress gradients. *Ecology*, **95**, 1437–43.

He, Q., Bertness, M. D. and Altieri, A. H. (2013) Global shifts towards positive species interactions with increasing environmental stress. *Ecology Letters*, **16**, 695–706.

Holling, C. S. (1973). Resilience and stability of ecological systems. *Annual Revue of Ecology and Systematics*, **4**, 1–23.

Huxham, M., Emerton, L., Kairo, J. et al. (2015). Applying climate compatible development and economic valuation to coastal management: a case study of Kenya's mangrove forests. *Journal of Environmental Management*, **157**, 168–81.

Huxham, M., Kumara, M. P., Jayatissa, L. P. et al. (2010). Intra- and interspecific facilitation in mangroves may increase resilience to climate change threats. *Philosophical Transactions of the Royal Society B: Biological Sciences*, **365**, 2127–35.

Jayatissa, L. P., Wickramasinghe, W. D. L., Dahdouh-Guebas, F. and Huxham, M. (2008). Interspecific variations in responses of mangrove seedlings to two contrasting salinities. *International Review of Hydrobiology*, **93**, 700–10.

Jax, K., Jones C. and Pickett, S. (1998). The self-identity of ecological units. *Oikos*, **82**, 253–64.

Johnstone, I. M. (1981). Consumption of leaves by herbivores in mixed mangrove stands. *Biotropica*, **13**, 252–9.

Kirui, B. Y. K., Huxham, M., Kairo, J. and Skov, M. (2008). Influence of species richness and environmental context on early survival of replanted mangroves at Gazi bay, Kenya. *Hydrobiologia*, **603**, 171–81.

Kodikara, K. A. S., Mukherjee, N., Jayatissa, L. P., Dahdouh-Guebas, F. and Koedam, N. (2017). Have mangrove restoration projects worked? An in-depth study in Sri Lanka. *Restoration Ecology*, **25**(5), 705–16.

Komiyama, A., Ong, J. E. and Poungparn, S. (2008). Allometry, biomass, and productivity of mangrove forests: a review. *Aquatic Botany*, **89**, 128–37.

Krauss, K. W., Mckee, K. L., Lovelock, C. E. et al. (2014). How mangrove forests adjust to rising sea level. *New Phytologist*, **202**, 19–34.

Krebs, C. (2001). *Ecology: The Experimental Analysis of Distribution and Abundance*, fifth edn. Benjamin Cummings, New York.

Kristensen, E. (2000). Organic Matter Diagenesis at the Oxic/Anoxic Interface in Coastal Marine Sediments, with Emphasis on the Role of Burrowing Animals. In G. Liebezeit, S. Dittmann and I. Kröncke, eds. *Life at Interfaces and Under Extreme Conditions: Developments in Hydrobiology*. Springer, Dordrecht.

Kristensen, E., Buillon, S., Dittmar, T. and Marchand, C. (2008). Organic carbon dynamics in mangrove ecosystems: a review. *Aquatic Botany*, **89**, 201–19.

Kumara, M. P., Jayatissa, L. P., Krauss, K. W., Phillips, D. H. and Huxham, M. (2010). High mangrove density enhances surface accretion, surface elevation change, and tree survival in coastal areas susceptible to sea-level rise. *Oecologia*, **164**, 545–53.

Maestre, F. T., Callaway, R. M., Valladares, F. and Lortie, C. J. (2009). Refining the stress-gradient hypothesis for competition and facilitation in plant communities. *Journal of Ecology*, **97**, 199–205.

Mazda, Y., Michimasa, M., Kogo, M. and Hong, P. N. (1997). Mangroves as coastal protection from waves in the Tong King delta, Vietnam. *Mangroves and Salt Marshes*, **1**, 127–35.

Mcivor, A., Spencer, T. and Möller, I. (2012). *Storm Surge Reduction by Mangroves*, Natural Coastal Protection Series: Report 2. Wetlands International, Nairobi.

McKee, K. L., Cahoon, D. R. and Feller, I. C. (2007a). Caribbean mangroves adjust to rising sea level through biotic controls on change in soil elevation. *Global Ecology and Biogeography*, **16**, 545–56.

McKee, K. L., Rooth, J. E. and Feller, I. C. (2007b). Mangrove recruitment after forest disturbance is facilitated by herbaceous species in the Caribbean. *Ecological Applications*, **17**, 1678–93.

Mehlig, U., Menezes, M. P. M., Reise, A., Schories, D. and Medina, E. (2010). Mangrove vegetation of the Caete estuary. *Mangrove Dynamics and Management in North Brazil*, **211**, 71–107.

Minchinton, T. E. (2001). Canopy and substratum heterogeneity influence recruitment of the mangrove *Avicennia marina*. *Journal of Ecology*, **89**, 888–902.

Munguia, P., Ojanguren, A. F., Evans, A. N. et al. (2009). Is facilitation a true species interaction? *The Open Ecology Journal*, **2**, 83–5.

Offenberg, J., Havanon, S., Aksornkoae, S., Macintosh, D. J. and Nielsen, M. G. (2004). Observations on the ecology of weaver ants (*Oecophylla smaragdina* Fabricius) in a Thai mangrove ecosystem and their effect on herbivory of *Rhizophora mucronata* Lam. *Biotropica*, **36**, 344–51.

Onuf, C. P., Teal, J. M. and Valiela, I. (1977). Interactions of nutrients, plant growth, and herbivory in a mangrove eco- system. *Ecology*, **58**, 514–26.

Otway, S. J., Hector, A. and Lawton, L. H. (2005). Resource dilution effects on specialist insect herbivores in a grassland biodiversity experiment. *Journal of Animal Ecology*, **74**, 234–40.

Ozaki, K., Takashima, S. and Suko, O. (2006). Ant predation suppresses populations of the scale insect *Aulacaspis marina* in natural mangrove forests. *Biotropica*, **32**, 764–8.

Pearse, I. S. (2010). Bird rookeries have different effects on different feeding guilds of herbivores and alter the feeding behavior of a common caterpillar. *Arthropod–Plant Interactions*, **4**, 189–95.

Pereyra, P. C. and Bowers, M.D., (1988). Iridoid glycosides as oviposition stimulants for the buckeye, *Junonia coenia* (Nymphalidae). *Journal of Chemical Ecology*, **14**, 917–28.

Peters, R., Vovides, A. G., Luna, S., Grüters, U. and Berger, U. (2014). Changes in allometric relations of mangrove trees due to resource availability – a new mechanistic modelling approach. *Ecological Modelling*, **283**, 53–61.

Pfister, C. A. and Hay, M. E. (1988). Associational plant refuges: convergent patterns in marine and terrestrial communities result from different mechanisms. *Oecologia*, **77**, 118–29.

Phillips, D. H., Kumara, M. P., Jayatissa, L. P., Krauss, K. W. and Huxham, M. (2017). Impacts of mangrove density on surface sediment accretion, belowground biomass and biogeochemistry in Puttalam Lagoon, Sri Lanka. *Wetlands*, **37**(3), 471–83, http://dx.doi.org/10.1007/s13157-017-0883-7.

Piovia-Scott, J. (2011). Plant phenotype influences the effects of ant mutualists on a polymorphic mangrove. *Journal of Ecology*, **99**, 327–34.

Pranchai, A. (2015). Spatial patterns and processes in a regenerating mangrove forest. Dissertation, TU Dresden.

Pülmanns, N., Mehlig, U., Nordhaus, I., Saint-Paul, U. and Diele, K. (2016). Mangrove crab *Ucides cordatus* removal does not affect sediment parameters and stipule production in a one year experiment in northern Brazil. *PLoS ONE* **11**, e0167375.

Root, R. B. (1973). Organization of a plant-arthropod association in simple and diverse habitats: the fauna of collards (*Brassica oleracea*). *Ecological Monographs*, **43**, 95–124.

Schöb, C., Callaway, R. M., Anthelme, F. et al. (2014). The context dependence of beneficiary feedback effects on benefactors in plant facilitation. *Journal of Physiology*, **204**, 386–96.

Schoener, T. W. (1987). Leaf pubescence in buttonwood: community variation in a putative defense against defoliation. *Proceedings of the National Academy of Sciences*, **84**, 7992–5.

Schoener, T. W., (1988). Leaf damage in island buttonwood, *Conocarpus erectus*: correlations with pubescence, island area, isolation and the distribution of major carnivores. *Oikos*, **53**, 253–66.

Silliman, B. R., Schrack, E., He, Q. et al. (2015). Facilitation shifts paradigms and can amplify coastal restoration efforts. *Proceedings of the National Academy of Sciences*, **112**, 201515297.

Singer, F. (2016). *Ecology in Action*. Cambridge University Press, Cambridge.

Skov, M. W. and Hartnoll, R. G. (2002). Paradoxical selective feeding on a low-nutrient diet: why do mangrove crabs eat leaves? *Oecologia*, **131**, 1–7.

Smith, N. F., Wilcox, C. and Lessmann, J. M. (2009). Fiddler crab burrowing affects growth and production of the white mangrove (*Laguncularia racemosa*) in a restored Florida coastal marsh. *Marine Biology*, **156**, 2255–66.

Smith, T. J. I., Boto, K. G., Frusher, S. D. and Giddins, R. L. (1991). Keystone species and mangrove forest dynamics: the influence of burrowing by crabs on soil nutrient status and forest productivity. *Estuarine Coastal And Shelf Science*, **33**, 419–32.

Smith, T. J. I. (1992). Forest Stucture. In A. I. Robertson and D. M. Alongi, eds. *Tropical Mangrove Ecosystems*. American Geophysical Union, Washington, DC, pp. 101–36.

Sousa, W. P. and Dangremond, E. M. (2011). Trophic interactions in coastal and estuarine mangrove forest ecosystems. *Treatise on Estuarine and Coastal Science*, **6**, 43–93.

Stieglitz, T., Clark, J. F. and Hancock, G. J. (2013). The mangrove pump: the tidal flushing of animal burrows in a tropical mangrove forest determined from radionuclide budgets. *Geochimica et Cosmochimica Acta*, **102**, 12–22.

Tamooh, F., Huxham, M., Karachi, M., Mencuccini, M., Kairo, J. G. and Kirui, B. (2008). Below-ground root yield and distribution in natural and replanted mangrove forests at Gazi bay, Kenya. *Forest Ecology and Management*, **256**, 1290–7.

Tahvanainen, J. O. and Root, R. B. (1972). The influence of vegetational diversity on the population ecology of a specialized herbivore, *Phyllotreta cruciferae* (Coleoptera: Chrysomelidae). *Oecologia*, **10**, 321–46.

Thampanya, U., Vermaat, J. E. and Duarte, C. M. (2002). Colonization success of common Thai mangrove species as a function of shelter from water movement. *Marine Ecology Progress Series*, **237**, 111–20.

Thampanya, U., Vermaat, J. E., Sinsakul, S. and Panapitukkul, N. (2006). Coastal erosion and mangrove progradation of Southern Thailand. *Estuarine, Coastal and Shelf Science*, **68**, 75–85.

Van Mele, P. (2008). A historical review of research on the weaver ant *Oecophylla* in biological control. *Agricultural and Forest Entomology*, **10**, 13–22.

Vogt, J., Lin, Y., Pranchai, A., Frohberg, P., Mehlig, U. and Berger, U. (2014). The importance of conspecific facilitation during recruitment and regeneration: a case study in degraded mangroves. *Basic and Applied Ecology*, **15**, 651–60.

Walker, B. and Salt, D. (2012). *Resilience Thinking: Sustaining Ecosystems and People in a Changing World*. Island Press, Washington, DC.

Wilson, J. B. and Agnew, A.D.Q. (1992). Positive-feedback switches in plant communities. *Advances in Ecological Research*, **23**, 263–336.

Winterwerp, J. C., Borst, W. G. and de Vries, M. B. (2005). Pilot study on the erosion and rehabilitation of a mangrove mud coast. *Journal of Coastal Research*, **212**, 223–30.

# Chapter 18

# Biofilms in Intertidal Habitats

Hanna Schuster, Mark S. Davies, Stephen J. Hawkins, Paula S. Moschella, Richard J. Murphy, Richard C. Thompson and A. J. Underwood

## 18.1 A Short Introduction to Biofilms

### 18.1.1 What Is a Biofilm?

Biofilms are an essential component of aquatic ecosystems (see Lock, 1993 and Cooksey and Wigglesworth-Cooksey, 1995 for reviews), including those on intertidal and subtidal rock (Hawkins et al., 1992). Interactions between microbial films and macrobiota, both bottom-up and top-down, are crucial to the functioning of rocky reef ecosystems (Bustamante et al., 1995; Thompson et al., 2004).

Biofilms consist of a mixed assemblage of microorganisms, including macroalgal propagules, embedded in a matrix of extracellular polysaccharides, which coat every substratum submersed, or periodically submersed, in water (see Wahl, 1989 for marine systems; Lock, 1993 for freshwater). Microbial films play a significant role in aquatic ecosystems in a variety of ways. Primary production (Hawkins et al., 1992; Bustamante et al., 1995; Ahn et al., 1997) and nutrient recycling occur in biofilms (Hamilton and Duthie, 1984; Hillebrand et al., 2000; Frost and Elser, 2002). Biofilms can also lead to bio-erosion, both directly as a consequence of endolithic activity (Donn and Boardman, 1988; Le Campion-Alsumard, 1989; Peyrot-Clausade et al., 1995), and indirectly via their consumption by hard-toothed grazers such as molluscs and echinoderms that scrape the rock and ingest the substratum as well as the surface biofilm (Castenholz, 1961; Dayton, 1971).

Microbial films also influence and interact with the macrobiota in rocky shore assemblages. They represent the initial site for colonisation by many sessile invertebrate larvae and algal propagules, thereby influencing recruitment and settlement of macroorganisms (Crisp, 1974; Keough and Raimondi, 1996; Thompson et al., 1996, 1998; Wieczorek and Todd, 1998; Joint et al., 2000). Biofilms are an important food resource in the diets of many grazers, and thus form the basis of many aquatic food webs (Hawkins et al., 1992; Christian and Luczkovich, 1999), although recent work using both gut contents and stable isotopes has shown that the consumption of living macroalgae (Davies et al., 2007; Lorenzen, 2007; Notman et al., 2016) and detritus by species presumed to be largely microphagous (i.e., Hawkins and Hartnoll, 1983a) are much more important than previously thought.

### 18.1.2 Studying Biofilms

The development of the study of biofilms, particularly their visualisation, identification of constituents and assessment of abundance has been methodologically constrained. Thus, we start by reviewing developments in techniques and approaches over the last forty years. We highlight recent developments in spectroscopy and colour-infrared imaging (CIR), enabling non-destructive studies at wider spatial scale than more traditional approaches using direct sampling of rock chips at the 1–5 $cm^2$ scale. We then discuss the role of biofilms in succession, settlement and recruitment. We consider factors determining their distribution and abundance, both abiotic (tidal elevation, wave action, rock type, light regime) and biotic (primarily grazing, but also shading when this is caused by other organisms), and how these contribute to the spatial and temporal patterns. The respective importance of bottom-up forcing, top-down control and lateral stressors are then considered. The wider roles of biofilms in the functioning of coastal ecosystems are discussed, including a consideration of the ecosystem services they provide. We then identify gaps in knowledge and the opportunities offered by modern techniques to fill these.

The focus of our review is on intertidal biofilms forming on natural rocky shores and artificial structures built of rock (for a recent review of biofilms on other surfaces see Salta et al., 2013). We have also concentrated primarily on the photosynthetic element of the film (cyanobacteria, diatoms, early stages of macroalgae), reflecting our own interests and their importance in the interactions structuring intertidal assemblages.

### 18.1.3 What Is in a Biofilm?

Biofilms occur on all surfaces in the sea. On rocky shores they coat what appears as bare rock to the naked eye, as well as macroalgae, the shells of barnacles, mussels and other sessile invertebrates and mobile species (Thompson et al., 1996). Areas devoid of macroalgae are usually maintained primarily by grazing (Underwood 1980, 1984a, 1984b, 1984c; Maclulich 1986, 1987), but also by sweeping by canopy-forming macroalgae (Hill and Hawkins, 1991; Jenkins et al., 1999a, 1999b). Nonetheless biofilm covers these apparently bare surfaces. Harsh physical conditions on the highshore that are too severe for macroalgae still support biofilms. On virgin rock surfaces that develop as a result of erosional processes or disturbance, such as sand scouring or mobile boulders, biofilms form during primary succession (Wahl, 1989).

The primary settlers of a biofilms generally are rod-shaped bacteria, which arrive first, followed by coccoids and finally by stalked and filamentous forms (Costerton et al., 1995). The biofilm composition varies with time during succession until a steady state is reached. Steady-state biofilms consist of a combination of prokaryotes (bacteria, archaea) including photosynthetic cyanobacteria, fungi, diatoms and the early stages of macroalgae. Protists also occur in biofilms. Cyanobacteria and diatoms often dominate biofilms in intertidal habitats; Proteobacteria dominate in subtidal habitats. Additionally, Actinobacteria, Bacteroidetes and Planctomycetes are major groups present in both the subtidal and intertidal zone (Lee et al., 2014).

In addition to the organisms themselves, there is a matrix of exo-polymeric substances secreted by the microorganisms and macroalgal propagules forming an essential part of the film. This matrix is further enhanced by foraging molluscs that lay down mucus trails (Figure 18.1; see Section 18.5.8; Connor, 1986; Davies et al., 1990, 1992a, b). This matrix also binds the cells of the film together and to the rock surfaces, provides mechanical stability and interconnects the individual cells with each other. It also acts as an external digestive system because it keeps extracellular enzymes close to the corresponding excreting cell (Flemming and Wingender, 2010). In addition to extracellular enzymes adhering to the matrix, suspended phytoplankton cells are able to stick more easily to the biofilm. This

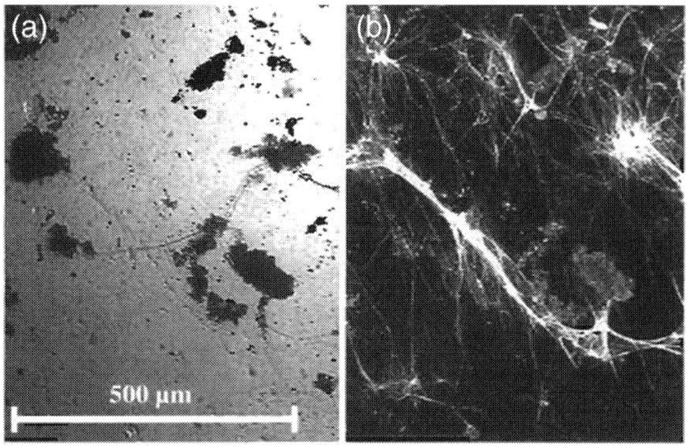


**Fig. 18.1** Limpet mucus on a glass microscope slide viewed under (a) transmitted light microscopy and (b) confocal laser scanning microscopy (CLSM) after staining with fluorescein isothiocyanate-labelled concanavalin-A. From Norton et al. 1998.

increases its food value, prompting suggestions that some molluscs may use the mucus they secrete as a means of gardening (Connor, 1986; Davies et al., 1992a) or at least recycling some of the energetic costs of mucus (Davies et al., 1992a, b; Davies and Beckwith, 1999) by reingestion (see Davies and Hawkins, 1998; Ng et al., 2013).

## 18.2 Methods for Identifying Components of Biofilms

A wide variety of techniques, methods and protocols have been developed to quantify the biomass, abundance and species composition of the microbenthic community.

### 18.2.1 Microscopy

Different types of microscopy have been widely used to identify biofilms at all stages (Figures 18.1–18.2)

**Epifluorescent Microscopy**

Epifluorescence light microscopy has been widely used to enumerate microalgal cells (Nicotri, 1977; MacLulich, 1987), but is less useful for identification of other microorganisms. Nagarkar and Williams (1997) were able to observe filamentous and unicellular cyanobacteria, but other groups were not clear. This technique is also not suitable for thick biofilms due to lack of light penetration. This technique, as described by Jones (1974), is based on the property of chlorophyll *a* to fluoresce if excited by blue light of a certain wavelength (435 µm). Using different filters, fresh samples of biofilm can be observed directly under a microscope while still intact on their substratum (rock chip). Additional use of acridine orange can be useful for differentiating green algae from cyanobacteria, as this stain links the DNA in the prokaryotic forms only. While most eukaryotic cells will fluoresce in red light (chlorophyll *a*), other clades that use additional pigments (e.g., phycoerythrin and phycocyanin of red algae) give yellow fluorescence without staining (French and Young,

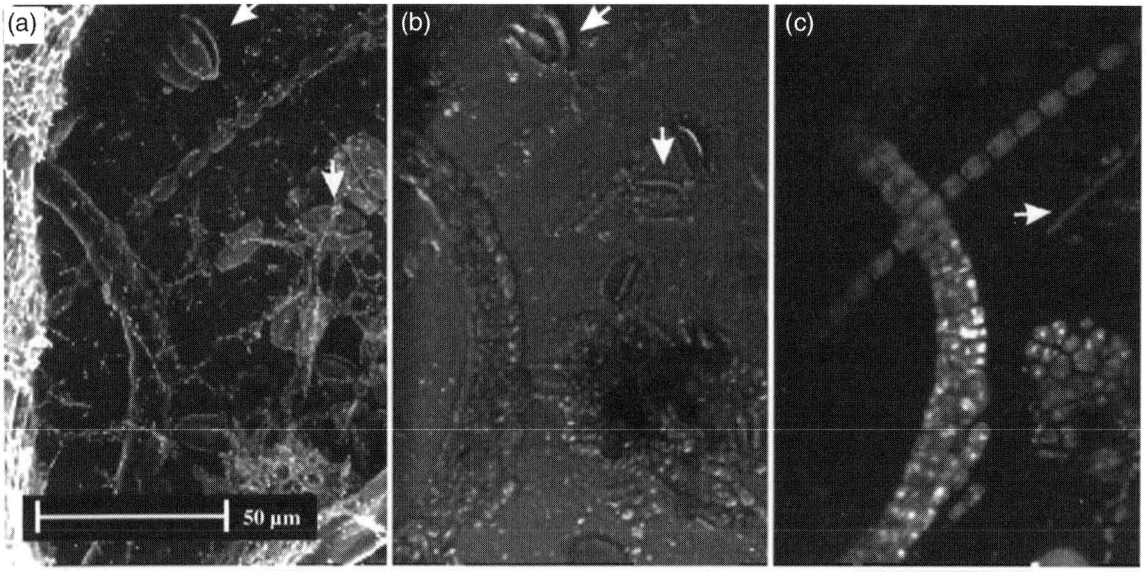


**Fig. 18.2** Biofilm on a glass microscope slide after forty-five-day immersion in the Princes Dock, Liverpool using different techniques. The same field was imaged by (a) Scanning electron microscopy (SEM), (b) phase contrast microscopy and (c) CLSM. The arrows in (a) and (b) show empty diatom frustules, as seen under phase contrast and SEM, but not visible using CLSM. The arrow in (c) indicates a cyanobacterial strand clearly visible only under CLSM. From Norton et al., 1998.

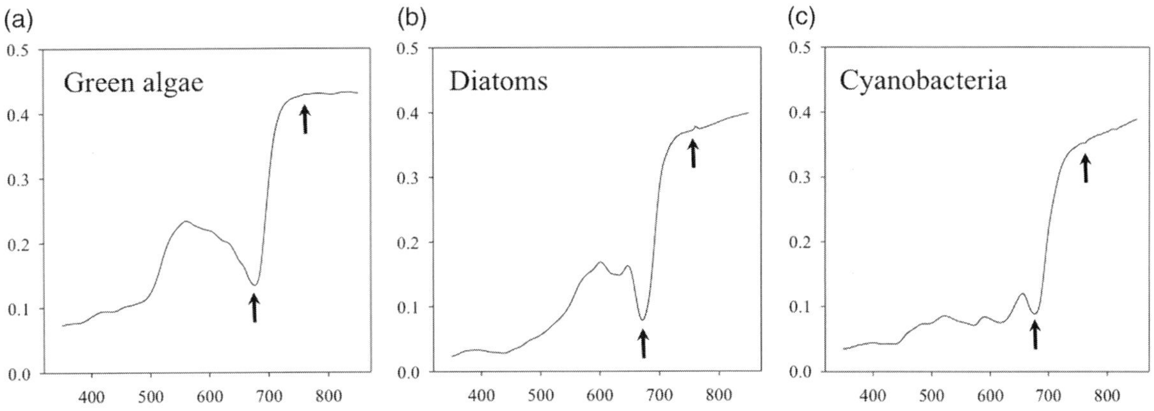


**Fig. 18.3** Reflectance spectra (350–850 nm) of different microalgal groups: (a) green algae, (b) diatoms and (c) cyanobacteria. The red (670 nm) and NIR (750 nm) wavelengths used to construct the ratio to quantify chlorophyll *a* are indicated in each graph. Note the different shapes of the spectral curves caused by absorption by different pigments. From Murphy et al., 2005.

1952). Samples can also be fixed in glutaraldehyde and then stained (Marszalek et al., 1979). Other components of biofilms can be visualised by using different stains. The mucopolysaccharine matrix can also be labelled and visualised by lectin concanavalin-A (Figure 18.1, Norton et al., 1998). Another dye used for bacterial counts is fluorescciamine, which reacts specifically with protein and aminoacids of prokaryotic cells (Poglazova et al., 1996).

Wolfaardt et al. (1991) tested the use of 4'-6-diamidino-2-phenylidole, a sensitive fluorescein DNA stain which allows the quantification of sessile bacteria. The stained bacteria fluoresced with a bright blue colour, while all the other organisms or materials appeared pale yellow. One advantage of this method is the short time needed for sample preparation and the longevity of fixed samples (which can be stored for up to twenty-four weeks). Unfortunately, the method only visualises bacteria, and it is not possible to discriminate between living and non-living cells.

Becker et al. (1997) measured the fluorescence intensities of epilithic microalgae grown on clay tiles by analysing the surface directly using an epifluorescence microscope photometer. This technique has the advantage of being non-destructive, enabling biofilms to be studied intact and then returned to field or laboratory experiments. This method is very suitable for flat rock surfaces, but high magnification cannot be achieved with substrata of greater roughness because of the short distance between the lens and the surface (Jones, 1974). Epifluorescence microscopy is not suitable for species identification within the microbial community, as very often it is not possible to identify microorganisms even to generic taxonomic level. In addition, MacLulich (1986) observed that this technique only works satisfactorily with fresh living microalgal samples, as the intensity of fluorescence decreases rapidly one to two hours after collection. Furthermore, the intensity rapidly declines to zero after short exposure to ultraviolet (UV) light. Thus, he did not advocate this technique for routine sampling.

### Confocal Microscopy

CLSM allows observation of the biofilm and its interaction with the substratum (Surman et al., 1996). Biofilms can be observed in detail by using a variety of fluorescent antibodies, lectins and nucleic acid stains which label the different microbial components such as bacteria, microalgae and exopolymers (Lawrence et al., 1998; Norton et al., 1998). CLSM seems to have several advantages compared to traditional microscopy (Figure 18.2). It gives clear images of microorganisms embedded in the mucopolysaccharine matrix and can also discriminate between living and dead cells. Furthermore, the three-dimensional structure of thick biofilms can also be observed (Norton et al., 1998).

**Scanning Electron Microscopy**

SEM is an extremely useful technique for examination of intact microscopic organisms on the surface of small fragments of rock or other hard substrata removed from the field (MacLulich, 1986; Hill and Hawkins, 1990) and for determining their three-dimensional structure (Veltkamp et al., 1994). SEM provides an excellent resolution of biofilm components, although sample preparation (fixation, dehydration) can cause artefacts (such as condensation of glycocalyx) which obscure cellular structures (Surman et al., 1996). Norton et al. (1998) pointed out that SEM analysis might exaggerate the diatom component in microbial film, while fungi and bacteria are difficult to visualise. This observation may also help to explain the difference in the composition of biofilms found in the guts of limpets (analysed using phase-contrast microscopy after acid cleaning and centrifugation) when compared to the rock surface (analysed using SEM) on which they grazed (Hill and Hawkins, 1990). SEM gives a good estimate of microalgal abundance (i.e., numbers of cells) and has been shown to give a good correlation with chlorophyll *a* content and in vivo fluorescence (MacLulich, 1986; Hill and Hawkins, 1990; Thompson et al., 1996), although there was more variability among individual measurements in the latter (Becker et al., 1997).

Fixation and drying living material prior to SEM examination is an important step in the analysis. Thus, different methods have been compared. Rock chips can be fixed in 2.5 per cent glutaraldehyde solution in filtered seawater for approximately three hours and then air dried overnight, before coating with gold palladium (MacLulich, 1986; Hill and Hawkins, 1990). This method showed a greater species number and abundance of biofilms than the use of critical-point drying that damaged morphology and removed loosely attached species (Nagarkar and Williams, 1997). It can be used for routine quantitative analysis because it requires minimal preparation time compared with the other methods. However, glutaraldehyde is very toxic and requires special facilities for handling. Another advantage is that components of the biofilm are not damaged or detached from the substratum. One possible drawback is that some species, such as cyanobacteria (Norton et al., 1998) and other fragile microorganisms may not be adequately visualised (MacLulich, 1986; Figure 18.2). Hill and Hawkins (1990) observed a great diversity of epilithic diatoms and filamentous algae (including cyanobacteria misidentified at the time as juvenile macroalgae) using this method.

Critical-point drying is more time-consuming, as dehydration in an alcohol series is required prior to it. Veltkamp et al. (1994) suggested a freeze dehydration technique for fixing freshwater epilithic specimens. This technique consists of repeated cycles where the samples are frozen and then cooled in the fridge. The advantages of this technique are that it is less time-consuming and less expensive than the traditional method for dehydration through an ethanol series. Critical-point drying can, however, cause more damage to the specimens than the air-drying method, as Nagarkar and Williams (1997) observed on the alga *Hapalospongidion*. It can also underestimate the diversity in a biofilm by removing species from the surfaces (Hill and Hawkins, 1990; Norton et al., 1998). Nevertheless, some specimens are more clearly visible and easier to identify using this technique. With this method, Hill and Hawkins (1990) observed a greater abundance of algal sporelings and germlings, but fewer diatoms, compared with air-dried samples.

Cryo-stage electron microscopy allows samples to be viewed in a hydrated state, revealing the three-dimensional structure of the biofilm (Hill and Hawkins, 1990). Samples are freeze dried and then observed under low-temperature cryo-stage SEM. As for critical-point drying, this method can cause damage to micro-organisms. For instance, Nagarkar and Williams (1997) noticed that the surface morphology of the filamentous cyanobacteria was not well preserved. This technique allows better observation of the bacterial component of biofilms compared with air drying, but reveals fewer species due to the organic matrix (Hill and Hawkins, 1990).

**Environmental SEM**

Another technique for studying biofilms is environmental SEM (ESEM), which enables structural analysis of hydrated organisms (Walker et al., 1998; Priester et al., 2007). ESEM is a modified version of SEM, where the specimen chamber is differentially pumped, allowing it to operate with up to 10 torr of water vapour, thus normal sample preparation such as fixation, dehydration and staining are not necessary (Surman et al., 1996). Additionally, the shrinkage and artefacts seen in SEM do not occur, although the electron beam may cause damage after a few minutes of analysis.

### 18.2.2 Reflectance Spectrometry

A major advance in the last decade or so has been the use of non-destructive spectrometry. Different algal groups (blue-green, green, red and brown algae) contain different photosynthetic pigments. The ability to detect the relative abundances of these pigments provides information about the types of microflora comprising an algal assemblage. Different photosynthetic pigments absorb light at specific wavelengths or regions of wavelength. Reflectance spectrometry, which measures the amount of light reflected from surfaces across a wide range of visible near infrared (NIR) wavelengths (typically 400–900 nm), detects and quantifies absorption by different photosynthetic pigments (Murphy et al., 2005, 2008; Figure 18.3). Although pigments can be identified from their specific absorptions in reflectance spectra, different pigments can have broad and/or overlapping absorptions, making it difficult to quantify them separately from these spectra. Derivative analysis (e.g., by numerical differentiation; Savitzky and Golay, 1964) enables weak or overlapping features to be enhanced, separated and quantified from reflectance spectra themselves or from transforms of reflectance such as pseudo-absorbance (log (1/reflectance); Murphy et al., 2011). The height of the peaks above the zero baseline at a particular wavelength represents the intensity of absorption of the pigment absorbing at this wavelength, and together these peaks can be used to provide information on the composition of the algal assemblage (e.g., Chapman et al., 2010; Iveša et al., 2010; Jackson et al., 2010). It should be noted that, if absolute measures of pigments are required, the height of the derivative peak representing the relative amount of absorption by each pigment would need to be calibrated against a known pigment standard.

Reflectance spectrometry enables rapid and numerous measurements to be acquired from a substratum without the need for destructive sampling or the use of toxic chemicals such as glutaraldehyde or time-consuming laboratory assays. Reflectance measurements can easily be acquired using a field spectrometer using either natural (sunlight) or artificial light (Murphy et al., 2011). Both these types of measurements require the spectrum to be normalised to the amount of incident light by acquiring a spectrum from a calibration standard of known brightness. The use of natural light enables spectra to be acquired from different-sized patches of rock platform (ranging up to tens of centimetres or more, depending upon the distance of the sensor from the rock platform). Artificial light enables spectra to be acquired under relatively stable illumination conditions using a reflectance probe with an integrated halogen light source. In this case, the size of the rock platform to be sampled is limited by the diameter of the measuring window of the probe (~2 cm). The decreasing cost of field spectrometers means that the use of reflectance spectrometry for intertidal ecology is becoming increasingly accessible.

### 18.2.3 Genomics

In recent years, the rapid development of high throughput ‘omics’ techniques (i.e., genomics, transcriptomics, proteomics, metabolomics) has provided a better understanding of the biodiversity and functioning of biofilms, particularly in medical sciences and antifouling research. Studies using these techniques are also increasingly applied to characterise biofilms on natural marine rocky reefs (Russell et al., 2013; Lee et al., 2014; Taylor et al., 2014). Next-generation sequencing has mainly been used to identify which taxa are present in biofilms. The advantage of the use of these techniques is the high taxonomic resolution that can be achieved, particularly of prokaryotes. Thus, changes in species composition that could not be detected before can now be detected. For example, an astonishing estimate of over 3,000 operational taxonomic units and 8,000 species were recovered in some samples from subtropical habitats using pyrosequencing (Lee et al., 2014).

Analyses of the transcriptome and proteome of biofilms are still rare, but could reveal the functional differences between different biofilm species and further highlight how species interact when close to each other (Mosier et al., 2014). The combination of genetic and microscopic investigation on monospecific biofilms in the laboratory have laid a foundation, and may also enhance the understanding of quorum sensing and behavioural responses of biofilms to changes in temperature and $CO_2$, as well as grazing pressure and settlement cues.

## 18.3 Methods for Quantifying Biofilms

### 18.3.1 Removal of Biofilm for Sampling

A variety of different methods have been used to collect biofilms from the rock surface. Nicotri (1977) sampled microalgae from the shore by scraping the surface with a toothbrush and washing it periodically in glass vials filled with seawater and a few drops of formalin as fixative. This method provides a sample containing mostly intact cells, which can be used to make direct counts of microbial abundance. Using this method, estimates of microalgal biomass on the shore are approximate because the endolithic component is not included in the sample (MacLulich, 1986). A further disadvantage in brushing the rock surface is the efficiency of removal of biofilm, which varies with the rock hardness, surface complexity and the capacity of the microorganisms to adhere to the surface (Hill and Hawkins, 1990; Nagarkar and Williams, 1997). Alternatively, the rock surface may be scraped with a sharp knife or blade, which allows the microalgae to be collected with minimal quantities of detritus (Underwood, 1984b). One drawback with this methodology is that samples must be processed quickly before the pigments degrade (Underwood, 1984b; MacLulich, 1986). Jones (1974) found that the scraping method may seriously underestimate the microalgal population in terms of the number of cells, when compared with direct counts on the rock surface, mainly caused by breakage of cells into uncountable fragments. Both methods – toothbrushing and scraping – need to be used carefully and in good weather conditions, because wind, rain or rough seas create the possibility of losing some of the material collected (Underwood, 1984b).

Chiselling the rock to remove a thin layer is a very good method for preserving intact microbial film, including the endolithic component (Hill and Hawkins, 1990). It is particularly suitable for the identification of microorganisms, but is destructive, especially when many replicates must be collected (MacLulich, 1986).

Sonication is a method that is used for removing bacteria from artificial substrata such as steel or glass surfaces, although attention must be paid to the power and duration applied to the sample to avoid possible destruction of the cells and release of chlorophyll (Lindsay and Von Holy, 1997; Claret, 1998). Hirsch et al. (1995) adopted a series of physical and chemical methods to detach epilithic and endolithic microorganisms from rock samples. The samples were treated with a solution of NaCl and detergent followed by vortexing and low-power sonication. The NaCl solution is intended to minimise charge interactions between cells and substratum and disrupt the protein binding of the cells. The authors also suggested that, for organisms that produce extracellular glycoproteins, such as microalgae, the concentration of the reagent and the strength of physical treatments must be increased, while trying not to cause damage to the cells.

### 18.3.2 Artifical Substrata

The use of artificial substrata for studying microbenthic communities reduces many problems in sampling and processing. They can be easily transported to the laboratory without losing any component of the film and are more suitable for analysis by light microscopy and optical density (MacLulich, 1986). They need, however, a considerable amount of preparation and installation to secure them on the shore and, more importantly, the microbial community which develops may be very different from that on the natural substrata (Snoeijs, 1991).

### 18.3.3 Using Pigments to Quantify Biofilm

Chlorophyll *a* concentration has been long used to estimate microalgal standing stock in phytoplankton (Antia et al., 1963; Castenholz, 1963), and on soft shores and subtidal sediments (Wright et al., 2005). It is a reliable, repeatable measure of biomass for epilithic microphytobenthos (Underwood, 1984b). This technique consists of submersing samples in solvents, which extract chlorophyll from the algal cells. A variety of different solvents and protocols have been used (e.g., Wright et al., 2005). Additional steps during processing such as sonication and centrifuging have also been used in some studies, but without any significant increases in extracted chlorophyll (Thompson et al., 1999).

The concentration of chlorophyll released in the solvent is then determined by spectrophotometric analysis. For chlorophyll *a*, absorbance is measured at 665 nm, and background absorbance at 750 nm is then subtracted (Thompson et al., 1999). This analysis can give a good estimate of the amount of chlorophyll in the biofilm, although other chlorophyll pigments (*b* and *c*) and degradation products may affect the results if present. For that reason, it is usually better to use a range of wavelengths for mixed assemblages in a biofilm (Underwood, 1984b; Wright et al., 2005).

Additional sources of variation may result from the variety of protocols adopted for the analysis. The state of hydration of epilithic biofilms before extraction has been shown to influence the amount of chlorophyll extracted from samples (Thompson et al., 1999), although this factor has not been considered in most studies (e.g., Underwood, 1984a; Hill and Hawkins, 1990; Dye and White, 1991). Thompson et al. (1999) observed that hydrated samples released more than twice the amount of chlorophyll *a* than did dry biofilms, and they suggested that very small chlorophyll concentrations during summer reported by some authors (Underwood, 1984c; Hill and Hawkins, 1990) could, in part, be a consequence of the extremely dry conditions experienced by the microbial community at this time of the year. This is, however, not particularly likely because, once chlorophyll has been extracted, further application of solvents fails to extract more pigments (Underwood, unpublished data). In addition, MacLulich (1987) showed the same general patterns of seasonal change in microalgae by direct counts of cells, as had been found using measures of cholorophyll. Colour Infrared (CIR) is not sensitive to wetting: in an experiment that included wetting of the substratum, the amount of wetness made only a very small (and not statistically significant) difference to any of the results (Murphy et al., 2006). This is clearly not a problem for this technique.

### 18.3.4 Reflectance Spectrometry

Reflectance spectrometry, in addition to providing information on the composition of algal assemblages (see earlier section), presents a relatively straightforward way of acquiring measurements of algal (chlorophyll *a*) biomass from rocky shores and intertidal sediments. Murphy et al. (2005) showed that a ratio of reflectances at 750 nm and ~670 nm was strongly and positively correlated with amounts of chlorophyll *a*. These findings were in agreement with (Méléder et al., 2003) using monospecific cultures of diatoms. The use of these wavelengths quantifies the sharp rise in reflectance between ~670 nm, where chlorophyll *a* is maximally absorptive, and 750 nm, where there is no absorption by photosynthetically active pigments, but where scattering processes caused by the cell membrane and organelles are dominant. Thus, a ratio of the reflectance at these wavelengths quantifies the amount of absorption by chlorophyll *a* at ~670 nm. Ratios using reflectance at other wavelengths can be used, but the use of wavelengths around 670 nm, which are physically linked to chlorophyll a absorption, make this ratio specific to chlorophyll a.

Different substrata (e.g., concrete, sandstone, granite) have different spectral responses in the visible and NIR, which may affect the measurements of acquired biofilms, particularly when the biofilm is thin and the concentration of chlorophyll a is small. In cases where biofilm needs to be measured on substrata of different composition, the ratio would need to be converted to absolute amounts of chlorophyll *a* using the

**Fig. 18.4** (a) NIR reflectance – chlorophyll not absorptive; (b) red reflectance – chlorophyll highly absorptive and (c) a ratio image derived from reflectance at NIR and red bands (NIR/red). In this case, the image has been calibrated to absolute amounts of chlorophyll with increasing pixel brightness representing increasing chlorophyll abundance. Replicate plots are where algae had been experimentally removed and allowed to recolonise the rock surface for different amounts of time: (A) three days; (B) sixteen days; (C) thirty days; (D) forty-five days and (U) unscraped areas (after Murphy et al., 2006)

relationship between the ratio and amounts of chlorophyll *a* for each type of substratum. This approach was used by Iveša et al. (2010) in their analysis of biofilms growing on concrete and sandstone tiles.

### 18.3.5 Colour Infrared Photography

Although reflectance spectrometry provides excellent estimates of chlorophyll *a*, it can only provide measurements from discrete areas of the rock surface, each a few centimetres in size. Methods to quantify chlorophyll *a* from CIR were therefore developed (Murphy et al., 2004, 2006). The technique uses the NIR and red bands acquired by a digital CIR. To compensate for variation in incident illumination data acquired from reflectance, standards were used to calibrate the image to reflectance. A ratio (NIR/red) was then constructed from these data to provide an index of the amount of absorption by chlorophyll *a* on the substratum (Figure 18.4). The advantage of this method is that each contiguous pixel in the image represents a single independent measurement of chlorophyll *a*. Thus, the entire 'foodscape' within the area bounded by the image can be quantified. The method quantifies all chlorophyll a on the substratum, contained within the microalgal biofilm and any encrusting algae that are present. Consistent with conventional methods of measuring chlorophyll by direct sampling, if measurements of chlorophyll *a* are required only from microalgae, then the substratum would first need to be cleared of all algae and a fresh microalgal biofilm allowed to develop before data were acquired.

The development of imaging techniques to quantify chlorophyll *a* has allowed new kinds of experiments to be devised which were not possible before (e.g., Murphy and Underwood, 2006; Murphy et al., 2008; Jackson et al., 2009). This is because, conventionally, the data could not be acquired at the appropriate spatial resolution, data were not contiguous or because data could only be obtained by destructive sampling.

### 18.3.6 Productivity

A good system to quantify the relative abundance of the autotrophs (microalgae and cyanobacteria) and heterotrophs (bacteria) in a microbial film is the analysis of the electron transport system activity (ETS) as an indirect index of respiration. ETS is the activity of the enzymatic system, which controls the respiration in mitochondria and microsomes. ETS can be measured quite easily using spectrophotometric analysis (Christensen and Packard, 1979. Arístegui and Montero (1995) demonstrated for oceanic microbial communities a significant positive correlation between respiration and ETS activity. Bhosle et al. (1994) analysed the ETS activity of marine biofilms grown on steel panels and found a significant correlation with relative measurements of chlorophyll *a*. Relexans (1996) found some possible problems in relating ETS activity to respiration in marine system and suggested the method could be better used as an index of biomass.

## 18.4 Succession of Biofilms

The first phase of a biofilm forming consists of the adsorption of organic compounds (mainly glycoproteins, proteoglucans and polysaccharides) to the substratum just a few seconds after the substratum has been submerged (Wahl, 1989). This organic film seems to be adsorbed to any kind of substratum, which then acquires similar physical and chemical properties at the surface/liquid interface. Within one hour of submersion in seawater, bacteria start to colonise the substratum.

Exposed surfaces in nature are inoculated with microorganisms by means of physical processes (i.e., advection, diffusion), followed by adsorption and attachment of cells and finally by growth (Wahl, 1989; Characklis et al., 1990). Bacterial adsorption and adhesion are a consequence of interacting processes involving electrostatic forces, hydrodynamics, chemotaxis, surface rugosity, exopolymer production and substratum wettability (Marshall et al., 1971; Marszalek et al., 1979; Lock et al., 1984; Pringle and Fletcher, 1986; Zheng et al., 1994; reviewed in Cooksey and Wigglesworth-Cooksey, 1995; Baty et al., 1996; Taylor et al., 1997)

After this process, the physical and chemical properties of the substratum again change significantly depending on bacterial succession (Marszalek et al., 1979). In general, rod-shaped bacteria arrive first, followed by coccoids and finally by stalked and filamentous forms (Costerton et al., 1995). Observations on isolated cultures of marine cyanobacteria showed that the production of variable quantities of extracellular polymeric substances (EPSs) plays an important role in adhesion of these early colonisers (Scott et al., 1996). Colonisation of the substratum continues during the next few days with the arrival and settlement of unicellular eukaryotes such as yeasts, fungi, protozoans and diatoms. This settlement is enhanced by EPSs secreted by the first settlers. Diatoms are usually more abundant than other taxa and can form a dense mat (Patil and Anil, 2005), with benthic motile diatoms generally preceding stalked forms. Diatoms, similarly to cyanobacteria, produce exopolymeric substances, which also enhance movements of the motile forms (Smith and Underwood, 1998). At this early stage of colonisation, biotic interactions such as competition for space and food resources may occur within biofilms (reviewed by Wimpenny, 1996).

This classic successional sequence of conditioning, organic film, bacteria and diatoms, which generally characterises the first phases of microbial colonisation, does not always occur. Tuchman and Blinn (1979) observed diatoms as primary colonisers of artificial substrata, followed by cyanobacteria after the first week of submersion. In marine sediments, Underwood (1980) showed that colonisation was characterised by a sequence of diatoms and then green algal assemblages, while the occurrence of cyanobacteria was independent of this successional sequence. The type of microorganisms that first colonise the substrata can also be influenced by seasonal variability in the abundance of colonising organisms as well as the prevailing environmental conditions (Anderson, 1995).

In general, it takes more than three weeks for a substratum to be colonised by a mature microbial community. In freshwater systems, the microalgal community appears to reach a mature state after twenty to twenty-five days, with no further changes in species composition and cell density (Tuchman and Blinn, 1979; Blinn et al., 1980; Hamilton and Duthie, 1984). In more disturbed and unstable systems, however, such as an epipsammic microalgal community, colonisation can be kept in a pioneer state (Miller et al., 1987). On intertidal rocky shores, Wimpenny (1996) suggested that the time required for a recolonisation of the rock surface is three to four weeks, although a longer period of approximately six weeks has been demonstrated to be necessary for a recovery of microbenthic assemblages (Underwood, 1984b; Murphy et al., 2008).

Colonisation seems also to be characterised by a diverse and apparently unstructured community during the first week, and the early stages of recolonisation of biofilm can have a markedly different spatial distribution from that in a mature one (Murphy et al., 2008). After this period, the assemblages become dominated by one or few species (Niell and Varela, 1984).

However, this does, not seem to be the case for all biofilms, with some being characterised by great biodiversity even in their mature state, for example, in subtropical habitats. There, bacterial communities were shown to vary with season and tidal level on experimentally deployed polystyrene blocks, and contained a large number of unknown species and many taxa in small abundances both in intertidal and subtidal habitats (Lee et al., 2014). Furthermore, differences between substrata apparent in the initial phases of adhesion and settlement of periphyton gradually disappeared by later stages (Hamilton and Duthie, 1984). After several days to weeks, depending on environmental conditions, the microbial community, evolving continuously in response to abiotic factors and biotic interactions, is enriched by the settlement of larvae and algal propagules. This leads to the establishment of macroscopic assemblages (Characklis, 1981). The early stages of colonisation of macroalgae also show greater spatial variability than is the case when they are more fully developed (Chapman and Underwood, 1998).

The succession as described earlier (also reviewed by Wahl, 1989) assumes that the starting point is a virgin surface. Scour by sand and ice can produce such surfaces, as can human activity, but the majority of the marine benthic substrata are rarely virgin. Although grazing (Dayton, 1971; Underwood, 1980; Hawkins and Hartnoll, 1983a; Hawkins et al., 1992) and possibly algal sweeping (Hill and Hawkins, 1991) can prevent the establishment of macroflora and fauna, these processes result in a microbial 'lawn' on rock surfaces, thereby arresting succession well before mature macroalgae grow.

As mentioned in earlier sections and further detailed later (Section 18.5.8), mobile grazers and predators, particularly molluscs, secrete mucus as part of their locomotory activity (Davies and Hawkins, 1998; Ng et al., 2013; Figure 18.1). Given typical densities and patterns of mobility of grazers and the lengthy persistence of mucus (Davies et al., 1992a), much of the biofilm is likely to have a significant molluscan-derived mucus component, although this might be reduced in the tropics (Davies and Williams, 1995). Hence, grazers both crop biofilms and enhance their development through production of mucus. This provides nutrition and a 'sticky' surface, and may influence succession in biofilms.

## 18.5 Ecology of Biofilms

### 18.5.1 Vertical Elevation on the Shore

Desiccation, insolation and thermal stresses appear to be the most important physical factors controlling the growth and abundance of benthic microflora (Underwood, 1984b; MacLulich, 1987; Blanchard et al., 1997; Mak and Williams, 1999; Thompson et al., 2004, 2005). In upper areas of a shore, these factors have a stronger impact on microbiota and are the main cause of the decline of microalgae during the summer (Underwood, 1984a; Williams, 1993a; Nagarkar and Williams, 1999). In mid- and lower-areas of a shore, physical factors are still very important, but grazing contributes to a greater extent to the control of microbial abundance and species composition (Underwood, 1984b; Williams, 1994; Thompson et al., 1996).

The vertical distribution of microbial films across a shore does not follow a common pattern. Generally, microphytobenthos seems to be more abundant, and have a greater biomass and diversity towards the lower shore (Aleem, 1950; Underwood, 1984a; MacLulich, 1987). In contrast, during winter, Thompson et al. (2004, 2005) observed maximal standing stock and abundance of diatoms on the upper shore, and minimal values on the lower shore. However, as the authors pointed out, these differences are likely to be more related to seasonal variation, as the inverse occurred during summer.

Considerable variability in biofilm distribution can be also found at smaller spatial scales. Many investigators have concluded that these microbial communities are characterised by a high degree of patchiness from scales of a few metres squared down to micrometres squared (Nicotri, 1977; MacLulich 1987; Hill and Hawkins, 1990, 1991; Thompson et al., 1996; Jackson et al., 2009). Differences in biomass and species composition of biofilms occur between microhabitats such as barnacles plates, gastropods shells, macroalgae and bare rock, but not in terms of species diversity (Hill and Hawkins, 1991; Thompson et al., 1996).

### 18.5.2 Wave Action

Another factor, which should be considered, especially when comparing microalgal communities between different sites, is the wave exposure of the shore. Aleem (1950) suggested that the degree of exposure might directly affect the distribution of diatoms; this was confirmed by Davies et al. (1992b). Microalgal standing stock and primary productivity can also differ between sheltered and exposed shores (Bustamante et al., 1995; Jenkins and Hartnoll, 2001; Thompson et al., 2005), although Davies et al. (1992b) found this only to be true when surfaces were covered with molluscan mucus. There are also differences between areas that are or are not dominated by an algal canopy (Thompson et al., 2005), which, in turn, is influenced by wave exposure.

### 18.5.3 Type of Rock

Microalgal standing stock and abundance have been shown to be much greater on rocks characterised by greater porosity and microrugosity, such as sandstone, than on smoother limestone, basalt or slate (Blinn et al., 1980; Edyvean et al., 1985a). Thompson et al. (2000) observed the same pattern on limestone rocks with different rugosities. The remarkable spatial differences which develop early in succession (shown by Edyvean et al., 1985a) disappeared completely in the following weeks. This suggests that the typical microtopography and chemical composition of the substrata, which possibly affected the initial colonisation, were soon modified and minimised by a uniform film of organic matter.

Other properties of the substratum, such as surface tension, are likely to affect colonisation and growth of a biofilm. High surface tension, typical of hard substrata (e.g., granite), can interfere with the adsorption of preconditioning film and the bacterial adhesion, resulting in low colonisation rates (Hamilton and Duthie, 1984). Topography also appears to influence the species composition and diversity of microalgae. Colonisation generally starts in small pits and microcrevices and then propagates onto smoother surfaces (Hamilton and Duthie, 1984; Miller et al., 1987). The general texture of surfaces also appears to influence species composition within microalgal communities. Adnate diatoms, which are tightly attached to the substratum, have been found on rougher surfaces, while stalked forms, filamentous colonies and long chain-forming diatoms dominated smooth substrata (Hamilton and Duthie, 1984; Miller et al., 1987; Snoeijs, 1991). The opposite result was obtained by Sabater et al. (1998), who observed that adnate forms were more abundant on flat surfaces. These contrasting observations may be explained by large differences in the type of substratum considered for the study and by the prevailing environmental conditions. Microtopography can also be modified by biological features such as barnacles (Thompson et al., 1996). More microalgae have been found to colonise concrete than sandstone surfaces, but these differences disappeared in the presence of grazing limpets (Iveša et al., 2010). Further investigations are clearly needed for a better understanding of the factors interacting between the substratum and the microbial film (Holmes et al., 2002).

### 18.5.4 Light and Shade

Light is a strong limiting factor for intertidal biofilms, which consist mainly of autotrophic organisms (Castenholz, 1963; MacLulich, 1987; Hill and Hawkins, 1991). The effect of light on microalgae has long been known. Aleem (1950) and Castenholz (1961) suggested that the amount of exposure to direct solar radiation during low tide was probably the most important factor determining the vertical distribution of marine littoral diatoms and the relative species composition. UV – particularly UV-B radiation – may also contribute to the significant reduction of microalgae during the summer, but contrasting results have also been obtained. Richardson (1983) showed that UV radiation caused serious damage to the microalgal photosynthetic system, and Bothwell et al. (1993) demonstrated that diatoms were directly inhibited by UV-B.

Conversely, Peletier et al. (1996) and Hill et al. (1997) did not find any significant effects of UV radiation on microalgal biomass and diversity. Furthermore, it seems that this factor does not represent a limiting factor during summer time and has much less effect than does the amount of insolation. Experimental reduction of UV-B radiation did not cause an increase in microalgal biomass. The abundance of diatoms, however, increased significantly when insolation was experimentally reduced by shading (to 50 or 85 per cent of ambient). Insolation seemed to have a strong effect on diatoms, but not on cyanobacteria, which appeared to be regulated to some extent by light limitation in the winter (Thompson et al., 2004). This may also help explain why biofilms on shores in lower latitudes are dominated by cyanobacteria, as in New South Wales (NSW) and Queensland (Jackson et al., 2010), and Hong Kong (Martinez, Coleman and Murphy, unpublished data).

### 18.5.5 Seasons

Microalgal assemblages change significantly with seasons. Many studies have demonstrated that on tropical and temperate shores microalgae are abundant during winter and spring and then decrease in abundance during the warmer summer months (Castenholz, 1961; Nicotri, 1977; Cubit, 1984; Underwood, 1984a; MacLulich, 1987; Dye and White, 1991; Hill and Hawkins, 1991; Williams, 1993a). A similar seasonal pattern has been observed for microflora living in sandy intertidal marine sediments (Sundbäck et al., 1997; Murphy et al., 2009).

Diatoms appear to reach maximal biomass and abundance in spring, followed by a second peak of lower magnitude in the autumn (Aleem, 1950; MacLulich, 1987; Hill and Hawkins, 1991). In fact, the photosynthetic activity, which determines microalgal primary production, shows a peak in spring when light intensity, temperature and nutrients are near to the optimal values required by photosynthesis (Blanchard et al., 1997). Cyanobacteria are less affected by seasonal variation, possibly because of their greater tolerance and resistance to physical stresses such as temperature, desiccation and insolation (Mak and Williams, 1999). Many authors (Underwood, 1984b; MacLulich, 1987; Thompson et al., 2004) have shown that, although the abundance of blue-green algae decreased in summer, a reduced cover still persisted on the shore; in some areas, however, diatoms disappeared completely.

### 18.5.6 Gradients of Carbon Dioxide

A meta-genomic study of epilithic bacteria along a $pCO_2$ gradient in south Italy from a natural volcanic vent showed that Cyanobacteria, Proteobacteria and Bacteroidetes dominated the biofilms in all sites. The composition of taxa within these groups was, however, significantly different between sites and dependent on $pCO_2$ levels (Taylor et al., 2014).

### 18.5.7 Biological Interactions

Grazers play an important role in limiting algal abundance and distribution on rocky shores (for reviews see Lubchenco and Gaines, 1981; Hawkins and Hartnoll, 1983a; Sousa, 1992; Poore et al., 2012). Although many studies have demonstrated that grazing only affects macroalgae while they are still at microscopic stages – as sporelings or germlings (Southward, 1964; Underwood, 1980, 1984a; Hawkins, 1981a, b) – several investigations have focussed on the effects of grazing on microalgae and, more generally, on microbial films (Nicotri, 1977; Cubit, 1984; Underwood, 1984b; Hill and Hawkins, 1990, 1991; Williams, 1993b; Anderson, 1995; Mak and Williams, 1999; Sommer, 1999). A large variety of grazers have been shown to feed on microflora. On tropical shores, herbivorous fish regulate microalgal growth on coral reefs. In temperate regions, caddisfly and mayfly larvae in freshwaters (Fuller et al., 1998), marine isopods (Salemaa, 1987; Schaffelke et al., 1995; Sommer, 1997, 1999), crustaceans (Fleeger et al., 1999) and amphipods (Jernakoff and Nielsen, 1997) also have a microphagous diet.

On rocky shores worldwide, molluscan gastropods are the predominant grazers (see for reviews Underwood, 1979; Hawkins and Hartnoll, 1983a; Norton et al., 1990). A wide variety of gastropods live in the eulittoral zone. These include prosobranchs and pulmonate limpets, trochids and littorinids (Underwood, 1979; Hawkins and Hartnoll, 1983a; Hawkins et al., 1989; Dye and White, 1991; Anderson and Underwood, 1997). The differing radula morphology, feeding mechanisms and grazing behaviour of molluscs result in a diverse exploitation of microbial films as a food resource (Steneck and Watling, 1982; Norton et al., 1990). As a consequence, the spatial and seasonal distribution and abundance of intertidal microflora can be significantly affected by grazers.

Nicotri (1977) demonstrated that grazing by different species of acmaeid limpets in California had different impacts on the microflora. In particular, chain-forming diatoms appeared to be selected, but most probably because once the radula encountered any part of a chain, the whole chain tended to be ingested. The prosobranch snail *Bembicium nanum* on rocky shores in NSW has been demonstrated to be particularly ingesting early stages of life history of macroalgae when grazing over a biofilm (Murphy and Underwood, 2006), in contrast to *Nerita atramentosa*, which had no effect on this component of the microflora. Cubit (1984) demonstrated that the exclusion of acmaeid limpets from high on the shore during summer produced a thin film consisting of blue-green algae, diatoms and algal sporelings. Grazing may, however, only exaggerate the seasonal and spatial variation of microalgal primary production (Dye and White, 1991). At highshore levels, physical stresses can minimise the impact of these grazers on microflora by limiting their feeding activity (Williams, 1994).

The impact of grazers on a microalgal community can be more or less effective depending also on competition (Underwood, 1984c), densities of the grazers (Mak and Williams, 1999) and their feeding capabilities (Haven, 1973; Underwood and Jernakoff, 1981). As a striking example of the different influences of grazers on biofilms, the small intertidal starfish *Parvulastrea exigua* feeds by direct contact of its everted stomach onto the substratum, rather than by scraping over the surface. As a result, the starfish can remove pretty much all of the underlying biofilm, in small patches the size and shape of the everted stomach. This creates quite distinct haloes of relatively bare space scattered wherever the animals have been feeding (Jackson et al., 2009; Figure 18.5).

The most complete analysis so far of competition for microalgal food resources has been on shores in NSW. Food is always in short supply on most shores, particularly in summer and at higher shore levels (Underwood, 1984c). As a result, competition among grazing gastropods is particularly intense at higher shore levels and in warmer parts of the year (Underwood, 1984c). There is a hierarchy of competitive abilities (Underwood, 1978), but there can also be apparent facilitation among species; in the absence of the snail *N. atramentosa*, which facilitates biofilm growth, there was less food available for the limpet *Cellana tramoserica* (Underwood and Murphy, 2008).

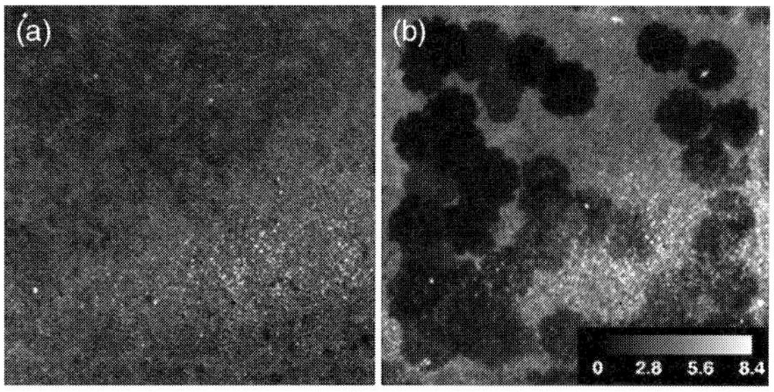


**Fig. 18.5** Amounts of chlorophyll on an experimental sandstone tile: (a) on day 0 before grazing and (b) after five days of grazing by two *P. exigua* starfish. Scale bar in (b) indicates the amount of chlorophyll (μg $cm^{-2}$). Note large areas the size and shape of the everted stomach of *P. exigua*, where nearly all chlorophyll has been removed from the substratum by grazing (after Jackson et al., 2009).

Mak and Williams (1999) suggested that the impact of grazers can be strong enough to affect the abundance of epilithic biofilm, even on the upper shore in tropical areas. Where littorinids were excluded from upper-intertidal areas, the standing crop of biofilm increased significantly in all the experimental sites. This suggested that littorinids can influence biofilm assemblages at this tidal height, despite the relatively harsh environmental conditions. The influence of grazing becomes much stronger lower on the shore, where physical factors are less important, and even during the hot season grazers prevented settlement and establishment of algal propagules and reduced algal abundance (Williams, 1993b, 1994).

Nicotri (1977) showed a great decrease in microalgal biomass in low-intertidal rocky shores in the presence of grazing. As a general consideration, it appears that grazers may play an important role in regulating the distribution and abundance of microalgae, but the impact is always mediated by physical factors which determine the seasonal and spatial patterns of the microbial community (Haven, 1973; Underwood, 1984b; Anderson, 1995).

Highshore areas in cold-temperate regions such as the British Isles tend to be covered by both a microbial film (mainly cyanobacteria) in summer, and a lawn of small ephemeral macroalgae in winter (Hawkins and Hartnoll, 1983b). Grazers are often restricted to topographic features such as pits, cracks and crevices, which provide refuge from the harsh physical environment, and emerge to feed on surrounding algal lawns (Skov et al., 2010). In winter, growth of a biofilm and then ephemeral green (*Blidingia, Ulva, Ulothrix, Prasiola*) and red (*Porphyra* spp.) can then be so great that it swamps the grazers. In summer, haloes around these pits expand again.

### 18.5.8 Mucus Trails

Many studies have also examined the longer-term effects of grazing on biofilms (e.g., Skov et al., 2010) and there is often a tendency to ascribe any effects to the action of the radula or bulldozing by the shell, removing some of the biofilm, whereas effects may also arise from the deposition of pedal mucus immediately after the action of the radula or shell. This mucus, secreted by many grazing molluscs (Figure 18.1), is mostly water; the functioning mucin is a glycoconjugate, specifically a form of glycosaminoglycan (for details see reviews by Davies and Hawkins, 1998 and Smith and Morin, 2002). Pedal mucus is an energy-rich product (Davies et al., 1990; Davies and Blackwell, 2007) likely to aid nutrition of some biofilm components through the provision of both carbon and nitrogen.

Once ‘trapped’ in a layer of mucus, any settling organism, including autotrophs, will experience a potentially utilisable organically rich microenvironment (Connor, 1986). The mucus may be further enriched by layering of trails where multiple gastropods follow each other’s trails, for example, to mate or seek refuge (Tankersley, 1990; Chapman, 1998; Ng et al., 2013). Although the effects of layering within biofilms could have a large effect on the structure and functioning of biofilms, they remain understudied.

Furthermore, owing to its sticky properties, mucus laid down in trails plays an important role in biofilm development; settlement of propagules and adhesion of normally planktonic cells and other organic and inorganic materials is greater on mucus-covered surfaces (Connor, 1986; Davies et al., 1992a), although, unsurprisingly, the first colonisers are likely to be bacteria (Herndl et al., 1989; Peduzzi and Herndl, 1991; Guo et al., 2009). Although mucus can also be produced by other components of the biofilm, especially diatoms, and again may be an important energy source for grazers (see Huang and Boney, 1983), this type of mucus may not be as effective as molluscan mucus in developing the biofilm further through larval settlement (Gallardo and Buen, 2003).

The effects of all grazers on biofilms are, however, not equal. Santelices and Bobadilla (1996) found great variations in the species of macroalgal propagules trapped in pedal mucus and their subsequent growth rates, depending on the species of trail-laying intertidal gastropod. In field experiments, the mucus of the limpet *Patella vulgata* trapped much more organic material than did that of *Littorina littorea*. Additionally, this efficacy of limpet mucus increased on wave-swept shores (Davies et al., 1992a). In similar field experiments, unspecified microalgae grew better in the mucus of homing limpets (*Lottia gigantea* and *Macklintockia scabra*) than in the mucus of a non-homing limpet (*Lottia digitalis*) or that of a carnivorous gastropod (*Nucella emarginata*) (Connor and Quinn, 1984; Connor, 1986), leading these authors to suggest that a fertiliser is added to mucus (i.e., mucus contains substances that enhance biofilm growth by fertilising the substratum). Certainly, some substance that manipulates biofilms, thereby probably enhancing their value as nutrition for grazers, is present in mucous trails of some species. One candidate is gamma-aminobutyric acid, found in pedal mucus of abalone (*Haliotis* spp.) and other gastropods, which can promote settlement of gastropod larvae (Hadfield, 1978; Hadfield and Scheuer, 1985; Laimek et al., 2008).

The function of a mucus-dominated biofilm in settlement depends on both the trail-laying species and the settling species, and can be beneficial or detrimental. Mucus can promote settlement and survival of the settler – the presence of *P. vulgata* mucus increases the settling and survival of *Semibalanus balanoides* (Holmes et al., 2002) – or it can deter the settler from settling – for instance, *Balanus glandula* specimen would have a high chance of being eaten when settling on a *Nucella lamellose* trail (Johnson and Strathmann, 1989). Avoidance responses by larvae may be an alternative explanation to biological disturbance in generating different localised patterns of recruitment. Whether the cue is positive or negative, settling larvae can clearly detect some component of mucus biofilms and use it to stimulate or inhibit settlement, the function of which may not be immediately apparent. This shows adaptation to the widespread presence and often persistent nature of mucus biofilms on hard surfaces in the sea.

The abrasive action of seawater on mucus-heavy biofilms can, however, significantly affect mucus-biofilm composition. Mucus of *P. vulgata* loses weight at a rate approaching 2%/h when submerged in seawater (Davies et al., 1992b), irrespective of whether the mucus was dehydrated or not before being submerged. While biofilms higher on the shore will be less prone to abrasive loss, they will dehydrate to an extent that they may not fully rehydrate during the next tide (Davies et al., 1992b).

Longer-term effects on the community composition of mucus-dominated biofilms are largely unknown, but Williams et al. (2000) reported an increase in macroalgal cover over a period of months in field treatments coated with herbivore mucus in comparison to those not so coated. Production of mucus trails by molluscs can contribute significantly to the nature of biofilms, affecting both their composition and capacity for development.

### 18.5.9 Bottom-Up versus Top-Down Control of Biofilms

Concepts of top-down and bottom-up effects were historically viewed as providing mutually exclusive explanations of community structuring and ecosystem functioning. Over the last twenty years, it has become more widely acknowledged that these processes operate simultaneously and interactively (Menge, 1992, 2000). Research in intertidal habitats (e.g., Nielsen, 2001; Harley, 2002) and in particular on marine biofilms on rocky shores has illustrated the need for a refinement of the bottom-up–top-down dichotomy to incorporate the role of physical stresses (Figure 18.6).

Based on manipulative experiments with biofilms on moderately exposed shores and earlier work in intertidal habitats (e.g., Newell, 1970; Cubit, 1984; Lüning, 1990; Somero, 2002), Thompson et al. (2004) suggested that photosynthetic microbial assemblages on temperate rocky shores in the north-east Atlantic were ultimately regulated by environmental stresses, particularly insolation stress, but were proximately regulated by grazing intensity. This was in agreement with earlier pioneering studies (Castenholz, 1961; Lamontagne et al., 1989), demonstrating that physiological responses to excessive insolation could depress growth of biofilms, leading to a decline in productivity during the summer. Thompson et al. (2004) also indicated that the activity of grazers that feed on these films was positively regulated by temperature, leading to an increased demand for microbiota during the summer. The outcomes of interactions between these factors vary according to weather conditions. Cool, overcast summers favour growth of photosynthetic microbiota and macroalgal germlings, leading to the establishment of a macroalgal canopy. Conversely, warm, sunny summers are likely to reduce the growth of photosynthetic microbiota and macroalgae (Thompson et al., 2000). This balance between macroalgae and grazers is fundamental to the structure of assemblages on rocky shores in the north-east Atlantic (Southward, 1964; Hartnoll and Hawkins, 1985; Hawkins et al., 1992). Similarly, thermal stress can regulate the feeding behaviour of predators and filter feeders (Dahlhoff et al., 2001; Sanford and Menge, 2001), and desiccation can regulate macroalgal productivity (Johnson et al., 1998). Hence, physical stresses can influence marine biofilms and macrobiota, thus modifying assemblages 'laterally' rather than simply from the top or bottom of trophic cascades. The extent to which physical factors

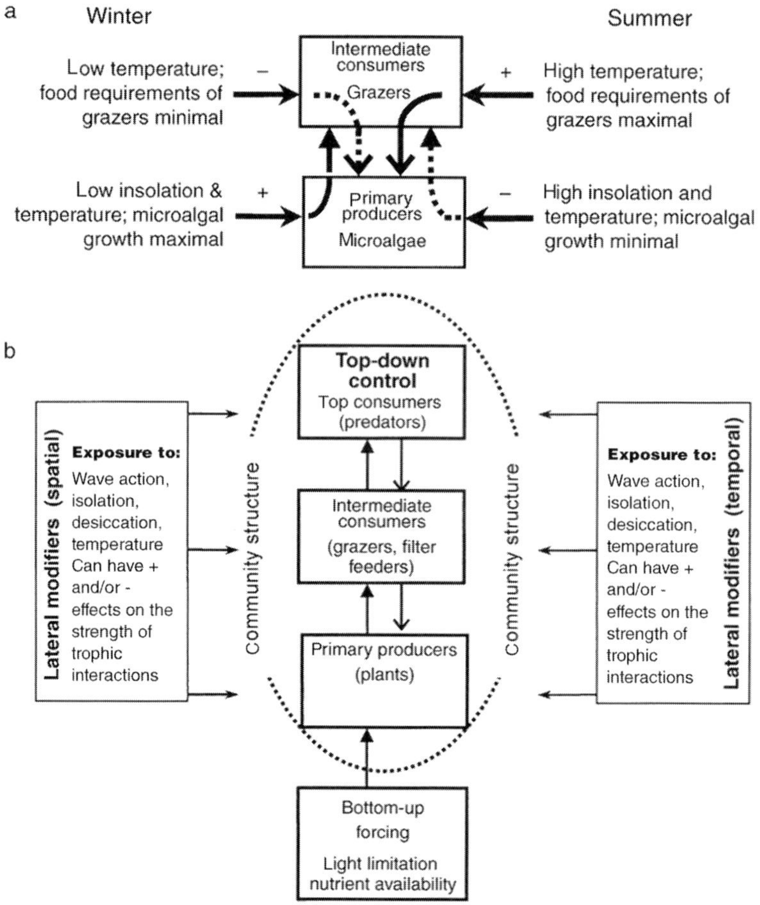


**Fig. 18.6** (a) Regulation of the balance between producers (photosynthetic microbiota) and consumers (grazers) in the rocky intertidal during summer and winter. The positive (+) and negative (-) lateral effect of contrasting physical factors on bottom-up and top-down control of trophic interactions are illustrated. Strong effects are shown as solid lines, weak effects as dashed lines. (b) Conceptual model to incorporate the role of physical stresses operating laterally at all trophic levels (sideways-facing arrows) together with bottom-up forcing (physical and chemical limiting factors; upward arrows) and top-down control (predation and grazing; downward arrows) in regulating community structure on temperate rocky shores (Thompson et al., 2004).

interact with top-down and bottom-up processes varies in time and space across a range of scales. For example, there is temporal variability associated with seasonal or climatic change and spatial variability associated with wave exposure and emersion gradients, or with orientation and surface topography.

Consequently, quite subtle changes in physical conditions can lead to complex and potentially unpredictable interactions. Thompson et al. (2004) presented a conceptual model incorporating possible direct and indirect effects of physiological stresses together with bottom-up forcing and top-down control in regulating intertidal biofilms and associated macrobiotic assemblages (Figure 18.6)

### 18.5.10 Ecosystem Functioning and Services

Biofilms are the major primary producers on shores devoid of macroalgae, yet surprisingly few studies have been made to measure production and to compare it to macroalgal production or that due to the arrival of phytoplankton and detritus.

Biofilms are, however, an essential part of the provisioning services that rocky shores impart to coastal ecosystems. Given their relatively small biomass in comparison to macrophytes, it is tempting to conclude that biofilms contribute relatively little to overall production in intertidal habitats. In the absence of appropriate measures of productivity, this conclusion may be completely incorrect. As is well-known in planktonic assemblages, the standing stock of phytoplankton is small, but productivity is generally large (i.e., the so-called paradox of the plankton). The same may turn out to be the case for biofilms. Biofilms do, however, contribute much to the diet of many herbivorous snails and limpets. Much macroalgal production ends up in the detrital food chain; in contrast, biofilm production passes up the food web to herbivores.

Little is known about the roles of biofilms in coastal ecosystems. They do, however, provide a resource for early stages of macrobiota, which then provide diverse goods and services when they grow up. Other ecosystem services are less obvious: biofilms are possibly involved in regulating services such as removal of nutrients, but probably to a much lesser extent than are macroalgae and macrophytes such as seagrasses.

In terms of disservices, biofilms contribute to fouling and corrosion of artificial fixed and floating surfaces (Edyvean et al., 1985b; Safriel and Erez, 1987).

## 18.6 Knowledge Gaps and Future Directions

The study of biofilms has always been limited by the available techniques to visualise and assess biomass. Microscopy is challenging and non-transparent, solid, uneven rock surfaces cannot easily be examined. Chlorophyll extraction from biofilms is also more challenging than from water samples used in planktonic work and from sediment surfaces. These constraints mean that sampling in the past was destructive and often at the scale of 1–5-cm rock chips or surrogate unnatural substrata such as microscope slides, plastic strips and tiles. Despite these difficulties, work since the 1980s has shown how important biofilms are in rocky shore ecosystems.

More recently, new spectroscopic techniques have revolutionised the ability to do non-destructive work across larger and smaller spatial scales than was previously possible. Modern genomic methods have also enabled biodiversity to be estimated using bar-coding methods, supplemented by proteomics and transcriptomics to understand functioning. The interest in biofilms from a medical perspective, in terms of infection and disease, has led to much development of techniques that could be deployed in an ecological context. Thus, the study of biofilms is likely to accelerate in the coming years.

Most of the work on natural substrata has been on the photosynthetic component of biofilms, cyanobacteria and diatoms, because of their ease of visualisation and, in the case of diatoms, identification. Many components of biofilms can also be quantified by spectroscopy and CIR (Murphy et al., 2005, 2008, 2009; Murphy and Underwood, 2006). While there has been much attention to the non-photosynthetic bacterial and archaea on artificial fouled surfaces, there has been less attention from rocky shore ecologists because of the need for specialised skills for identification and culturing. Recent advances in metagenomic approaches have enhanced our ability to identify these taxa (e.g., bacteroidetes), and next-generation sequencing has made such assays both affordable and more precise (Shokralla et al., 2012). This enables not only descriptive surveys of prokaryotic assemblages on natural rock surfaces, but also the ability to use such data as a community-level response variable to experimental manipulations of abiotic and biotic factors. If such genomic approaches can be coupled with proteomics, the functional attribution of the biofilm could be probed. Transcriptomic approaches would further enable gene expression to be both qualified and quantified, giving deeper insights into the processes occuring within biofilms. These data could be used as a first indication of the response of biofilms to any changes of biotic and abiotic factors.

Insights from biofouling and biomedically orientated studies of biofilms suggest that consortia are formed aided by cell-to-cell signalling (Schuster and Markx, 2013), leading to the biofilm operating as a 'super-organism', rather than individual cells acting individually, and therefore showing early steps to multicellularity (Miller and Bassler, 2001; Dang and Lovell, 2016). If further investigated in naturally occurring biofilms on rocky shores, this could provide further insights into both positive facilitatory and negative inhibitory interactions during succession and in steady-state biofilms.

Another major gap in our knowledge of the role of biofilms in rocky shore ecosystems is the relative contribution of microbial and macrophytic primary production. In Hawkins et al. (1992), some very preliminary estimates were given of the relative importance of microbial and macroalgal primary production. More needs to be done using both direct measurements (e.g., oxygen electrodes/optodes) and fluorimetry-based approaches to quantify rates of production. These could then be linked with spectroscopic estimates of biomass to estimate production at realistic spatial scales. Indirect methods such as isotope geochemistry, combined with more traditional gut content analysis (e.g., Notman et al., 2016) could be used to understand trophic connections on rocky shores. Notman et al. (2016) showed that the biofilm contributed much less to the diet of grazing limpets (*P. vulgata*), but that living macroalgae and algal detritus made a major contribution. Direct observations have also shown that *P. vulgata* is more than capable of eating large macroalgae such as *Ascophyllum* when available (Davies et al., 2007), confirming earlier work by Jones (1948) that has been overlooked in recent years.

## REFERENCES


Ahn, I.-Y., Chung, H., Kang, J.-S. and Kang, S.-H. (1997). Diatom composition and biomass variability in nearshore waters of Maxwell Bay, Antarctica, during the 1992/1993 austral summer. *Polar Biology*, **17**, 123–30.

Aleem, A. A. (1950). Distribution and ecology of British marine littoral diatoms. *Journal of Ecology*, **38**, 75–106.

Anderson, M. J. (1995). Variations in biofilms colonizing artificial surfaces: seasonal effects and effects of grazers. *Journal of the Marine Biological Association of the United Kingdom*, **75**, 705–14.

Anderson, M. J. and Underwood, A. J. (1997). Effects of gastropod grazers on recruitment and succession of an estuarine assemblage: a multivariate and univariate approach. *Oecologia*, **109**, 442–53.

Antia, N. J., McAllister, C. D., Parsons, T. R., Stephens, K. and Strickland, J. D. H. (1963). Further measurement of primary production using a large-volume plastic sphere. *Limnology and Oceanography*, **7**, 166–83.

Arístegui, J. and Montero, M. F. (1995). The relationship between community respiration and ETS activity in the ocean. *Journal of Plankton Research*, **17**, 1563–71.

Baty, A. M., Frølund, B., Geesey, G. G., Langille, S., Quintero, E. J. and Suci, P. A.,, (1996). Adhesion of biofilms to inert surfaces: a molecular level approach directed at the marine environment. *Biofouling*, 10, 111–21.

Becker, G., Holfeld, H., Hasselrot, A. T., Fiebig, D. M. and Menzler, D. A. (1997). Use of a microscope photometer to analyze in vivo fluorescence intensity of epilithic microalgae grown on artificial substrata. *Applied and Environmental Microbiology*, **63**, 1318–25.

Bhosle, N. B., Tulaskar, A. S. and Wagh, A. B. (1994). Electron transport system activity of microfouling material: relationships with biomass parameters. *Biofouling*, **8**, 1–11.

Blanchard, G. F. Guarini, J.-M., Gros, P. and Richard, P. (1997). Seasonal effect on the relationship between the photosynthetic capacity of intertidal microphytobenthos and temperature. *Journal of Phycology*, **33**, 723–8.

Blinn, D. W., Fredericksen, A. and Korte, V. (1980). Colonization rates and community structure of diatoms on three different rock substrata in a lotic system. *British Phycological Journal*, **15**, 303–10.

Bothwell, M. L., Sherbot, D., Roberge, A. C. and Daley, R. J. (1993). Influence of natural ultraviolet radiation on lotic periphytic diatom community growth, biomass accrual, and species composition: short-term versus long-term effects. *Journal of Phycology*, **29**, 24–35.

Bustamante R. H., Branch G. M. and Eekhout S. (1995). Maintenance of an exceptional intertidal grazer biomass in South Africa: subsidy by subtidal kelps. *Ecology*, **76**, 2314–29.

Castenholz, R. W. (1961). The effect of grazing on marine littoral diatom populations. *Ecology*, **42**, 783–94.

Castenholz, R. W. (1963). An experimental study of the vertical distribution of littoral marine diatoms. *Limnology and Oceanography*, **8**, 450–62.

Chapman, M. G. (1998). Variability in trail-following and aggregation in *Nodilittorina unifasciata* Gray. *Journal of Experimental Marine Biology and Ecology*, **224**, 48–71.

Chapman, M. G., Tolhurst, T. J., Murphy, R. J. and Underwood, A. J. (2010). Complex and inconsistent patterns of variation in benthos, micro-algae and sediment over multiple spatial scales. *Marine Ecology Progress Series*, **398**, 33–47.

Chapman, M. G. and Underwood, A. J. (1998). Inconsistency and variation in the development of rocky intertidal algal assemblages. *Journal of Experimental Marine Biology and Ecology*, **224**, 265–89.

Characklis, W. G. (1981). Bioengineering report: fouling biofilm development: a process analysis. *Biotechnology and Bioengineering*, **23**, 1923–60.

Characklis, W. G., McFeters, G. A. and Marshall, K. C. (1990). Physiological ecology in biofilm systems. *Biofilms*, **37**, 67–72.

Christian, R. R. and Luczkovich, J. J. (1999). Organizing and understanding a winter's seagrass foodweb network through effective trophic levels. *Ecological Modelling*, **117**, 99–124.

Christensen, J. P. and Packard, T. T. (1979). Respiratory electron transport activities in phytoplankton and bacteria: comparison of methods. *Limnology and Oceanography*, **24**(3), 576–83.

Claret, C. (1998). A method based on artificial substrates to monitor hyporheic biofilm development. *International Review of Hydrobiology*, **83**, 135–43.

Connor, V. M. and Quinn, J. F. (1984). Stimulation of food species growth by limpet mucus. *Science*, **225**, 843–5.

Connor, V. M. (1986). The use of mucous trails by intertidal limpets to enhance food resources. *Biological Bulletin*, **171**, 548–64.

Cooksey, K. E. and Wigglesworth-Cooksey, B. (1995). Adhesion of bacteria and diatoms to surfaces in the sea: a review. *Aquatic Microbial Ecology*, **9**, 87–96.

Costerton, J. W., Lewandowski, D. R., Caldwell, D. E., Korber, D. R. and Lappin-Scott, H. M. (1995). Microbial biofilms. *Annual Review of Microbiology*, **49**, 711–45.

Crisp, D. J. (1974). Factors Influencing the Settlement of Marine Invertebrate Larvae. In P. T. Grant and A. M. Mackie, eds. *Chemoreception in Marine Organisms*. Academic Press, London, pp. 177–265.

Cubit, J. D. (1984). Herbivory and the seasonal abundance of algae on a high intertidal rocky shore. *Ecology*, **65**, 1904–17.

Dahlhoff, E., Buckley, B. A. and Menge, B. A. (2001). Physiology of the rocky intertidal predator *Nucella ostrina* along an environmental stress gradient. *Ecology*, **82**, 2816–29.

Dang, H. and Lovell, C. R. (2016). Microbial surface colonization and biofilm development in marine environments. *Microbiology and Molecular Biology Reviews*, **80**(1), 91–138.

Davies, A. J., Johnson, M. P. and Maggs, C. A. (2007). Limpet grazing and loss of *Ascophyllum nodosum* canopies on decadal time scales. *Marine Ecology Progress Series*, **339**, 131–41.

Davies, M. S. and Beckwith, P. (1999). Role of mucus trails and trail-following in the behaviour and nutrition of the periwinkle *Littorina littorea*. *Marine Ecology Progress Series*, **179**, 247–57.

Davies, M. S. and Blackwell, J. (2007). Energy saving through trail following in a marine snail. *Proceedings of the Royal Society of London B*, **274**, 1233–6.

Davies, M. S. and Hawkins, S. J. (1998). Mucus from marine molluscs. *Advances in Marine Biology*, **34**, 1–71.

Davies, M. S., Hawkins, S. J. and Jones, H. D. (1990). Mucus production and physiological energetics in *Patella vulgata* L. *Journal of Molluscan Studies*, **56**, 499–503.

Davies, M. S., Hawkins, S. J. and Jones, H. D. (1992a). Pedal mucus and its influence on the microbial food supply of two intertidal gastropods, *Patella vulgata* L. and *Littorina littorea* (L.). *Journal of Experimental Marine Biology and Ecology*, **161**, 57–77.

Davies, M. S., Jones, H. D. and Hawkins, S. J. (1992b). Physical factors affecting the fate of pedal mucus produced by the common limpet *Patella vulgata*. *Journal of the Marine Biological Association of the United Kingdom*, **72**, 633–43.

Davies, M. S. and Williams, G. A. (1995). Pedal mucus of a tropical limpet, *Cellana grata* (Gould): energetics, production and fate. *Journal of Experimental Marine Biology and Ecology*, **186**, 77–87.

Dayton, P. K. (1971). Competition, disturbance and community organization: the provision and subsequent utilization of space in a rocky intertidal community. *Ecological Monographs*, **41**, 351–89.

Donn, T. F. and Boardman, M. R. (1988). Bioerosion of rocky carbonate coastlines on Andros Island, Bahamas. *Journal of Coastal Research*, **4**(3), 381–94.

Dye, A. H. and White, D. R. A. (1991). Intertidal microalgal production and molluscan herbivory in relation to season and elevation on two rocky shores on the east coast of Southern Africa. *South African Journal of Marine Science*, **11**, 483–9.

Edyvean, R. G. J., Rands, G. A. and Moss, B. L. (1985a). A comparison of diatom colonization on natural and artificial substrata in seawater. *Estuarine, Coastal and Shelf Science*, **20**, 233–8.

Edyvean, R. G. J., Terry, L. A. and Picken, G. B. (1985b). Marine fouling and its effects on offshore structures in the North Sea: a review. *International biodeterioration*, **21**(4), 277–84.

Fleeger, J. W., Carman, K. R., Webb, S., Hilbun, N. and Pace, M. C. (1999). Consumption of microalgae by the grass shrimp *Palaemonetes pugio*. *Journal of Crustacean Biology*, **19**, 324–36.

Flemming, H.-C. and Wingender, J. (2010). The biofilm matrix. *Nature Reviews Microbiology*, **8**, 623–33.

French, C. S. and Young, V. K. (1952). The fluorescence spectra of red algae and the transfer of energy from phycoerythrin to phycocyanin and chlorophyll. *The Journal of general physiology*, **35**(6), 873–90.

Frost, P. C. and Elser, J. J. (2002). Effects of light and nutrients on the net accumulation and elemental composition of epilithon in boreal lakes. *Freshwater Biology*, **47**, 173–83.

Fuller, R. L., Ribble, C., Kelley, A. and Gaenzle, E. (1998). Impact of stream grazers on periphyton communities: a laboratory and field manipulation. *Journal of Freshwater Ecology*, **13**, 105–14.

Gallardo, W. G. and Buen, S. M. A. (2003). Evaluation of mucus, *Navicula*, and mixed diatoms as larval settlement inducers for the tropical abalone *Haliotis asinina*. *Aquaculture*, **221**, 357–64.

Guo, F., Huang, Z.-B., Huang, M.-G., Zhao, J. and Ke, C.-H. (2009). Effects of small abalone, *Haliotis diversicolor*, pedal mucus on bacterial growth, attachment, biofilm formation and community structure. *Aquaculture*, **293**, 35–41.

Hadfield, M. G. (1978). Metamorphosis in Marine Molluscan Larvae: An Analysis of Stimulus and Response. In F.-S. Chia and M. E. Rice, eds. *Settlement and Metamorphosis of Marine Invertebrate Larvae*. Elsevier North Holland Biomedical Press, New York.

Hadfield, M. G. and Scheuer, D. (1985). Evidence for a soluble metamorphic inducer in *Phestilla*: ecological, chemical and biological data. *Bulletin of Marine Science*, **37**, 556–66.

Hamilton, P. B. and Duthie, H. C. (1984). Periphyton colonization of rock surfaces in a boreal forest stream studied by scanning electron microscopy and track autoradiography. *Journal of Phycology*, **20**, 525–32.

Harley, C. D. G. (2002). Light availability indirectly limits herbivore growth and abundance in a high rocky intertidal community during the winter. *Limnology and Oceanography*, **47**, 1217–22.

Hartnoll, R. G. and Hawkins, S. J. (1985). Patchiness and fluctuations on moderately exposed rocky shores. *Ophelia*, **24**, 53–63.

Haven, S. B. (1973). Competition for food between the intertidal gastropods *Acmaea scabra* and *Acmaea digitalis*. *Ecology*, **54**, 143–51.

Hawkins, S. J., Watson, D. C., Hill, A. S. et al. (1989). A comparison of feeding mechanisms in microphagous, herbivorous, intertidal, prosobranchs in relation to resource partitioning. *Journal of Molluscan Studies*, **55**, 151–65.

Hawkins, S. J. (1981a). The influence of season and barnacles on the algal colonization of *Patella vulgata* exclusion areas. *Journal of the Marine Biological Association of the United Kingdom*, **61**, 1–15.

Hawkins, S. J. (1981b). The influence of *Patella* grazing on the fucoid/barnacle mosaic on moderately exposed rocky shores. *Kieler Meeresforschungen*, **5**, 537–43.

Hawkins, S., and Hartnoll, R. (1983b). Changes in a rocky shore community: an evaluation of monitoring. *Marine Environmental Research*, **9**, 131–81.

Hawkins, S. J., and Hartnoll, R. (1983b). Grazing of intertidal algae by marine invertebrates. *Oceanography and Marine Biology: An Annual Review*, **21**, 195–282.

Hawkins, S. J., John, D. M. and Price, J. H. (1992). *Plant–Animal Interactions in the Marine Benthos*. Oxford University Press, Oxford.

Herndl, G. J., Peduzzi, P. and Fanuko, N. (1989). Benthic community metabolism and microbial dynamics in the Gulf of Trieste (Northern Adriatic Sea). *Marine Ecology Progress Series*, **53**, 169–78.

Hill, A. S. and Hawkins, S. J. (1990). An investigation of methods for sampling microbial films on rocky shores. *Journal of the Marine Biological Association of the United Kingdom*, **70**, 77–88.

Hill, A. S. and Hawkins, S. J. (1991). Seasonal and spatial variation of epilithic micro algal distribution and abundance and its ingestion by *Patella vulgata* on a moderately exposed rocky shore. *Journal of the Marine Biological Association of the United Kingdom*, **71**, 403–23.

Hill, W. R., Dimick, S. M., McNamara, A. E. and Branson, C. A. (1997). No effects of ambient UV radiation detected in periphyton and grazers. *Limnology and Oceanography*, **42**(4), 769–74.

Hillebrand, H., Worm, B. and Lotze, H. K. (2000). Marine microbenthic community structure regulated by nitrogen loading and grazing pressure. *Marine Ecology Progress Series*, **204**, 27–38.

Hirsch, P., Eckhardt, F. E. W. and Palmer, R. J. (1995). Methods for the study of rock-inhabiting microorganisms – a mini review. *Journal of Microbiological Methods*, **23**, 143–67.

Holmes, S. P., Cherrill, A. and Davies, M. S. (2002). The surface characteristics of pedal mucus: a potential aid to the settlement of marine organisms? *Journal of the Marine Biological Association of the United Kingdom*, **82**(1), 131–9.

Huang, R. and Boney, A. D. (1983). Effects of diatom mucilage on the growth and morphology of marine algae. *Journal of Experimental Marine Biology and Ecology*, **67**, 79–89.

Iveša, L., Chapman, M. G., Underwood, A. J. and Murphy, R. J. (2010). Differential patterns of distribution of limpets on intertidal seawalls: experimental investigation of the roles of recruitment, survival and competition. *Marine Ecology Progress Series*, **407**, 55–69.

Jackson, A. C., Murphy, R. J. and Underwood, A. J. (2009). *Patiriella exigua*: grazing by a starfish in an overgrazed intertidal system. *Marine Ecology Progress Series*, **376**, 153–63.

Jackson, A. C., Underwood, A. J., Murphy, R. J. and Skilleter, G. A. (2010). Latitudinal and environmental patterns in abundance and composition of epilithic microphytobenthos. *Marine Ecology Progress Series*, **417**, 27–38.

Jenkins, S. R., Hawkins, S. J. and Norton, T. A. (1999a). Direct and indirect effects of a macroalgal canopy and limpet grazing in structuring a sheltered intertidal community. *Marine Ecology Progress Series*, **188**, 81–92.

Jenkins, S. R., Norton, T. A. and Hawkins, S. J. (1999b). Interactions between canopy forming algae in the eulittoral zone of sheltered rocky shores on the Isle of Man. *Journal of the Marine Biological Association of the United Kingdom*, **79**(2), 341–9.

Jenkins, S. R. and Hartnoll, R. G. (2001). Food supply, grazing activity and growth rate in the limpet *Patella vulgata* L.: a comparison between exposed and sheltered shores. *Journal of Experimental Marine Biology and Ecology*, **258**, 123–39.

Jernakoff, P. and Nielsen, J. (1997). The relative importance of amphipod and gastropod grazers in *Posidonia sinuosa* meadows. *Aquatic Botany*, **56**, 183–202.

Johnson, L. E. and Strathmann, R. R. (1989). Settling barnacle larvae avoid substrata previously occupied by a mobile predator. *Journal of Experimental Marine Biology and Ecology*, **128**, 87–103.

Johnson, M. P., Hawkins, S. J., Hartnoll, R. G. and Norton, T. A. (1998). The establishment of fucoid zonation on algal-dominated rocky shores: hypotheses derived from a simulation model. *Functional Ecology*, **12**, 259–69.

Joint, I., Callow, M. E., Callow, J. A. and Clarke, K. R. (2000). The attachment of Enteromorpha zoospores to a bacterial biofilm assemblage. *Biofouling*, **16**, 151–8.

Jones, J. G. (1974). A method for observation and enumeration of epilithic algae directly on the surface of stones. *Oecologia*, **16**, 1–8.

Jones, N. S. (1948). Observations and experiments on the biology of *Patella vulgata* at Port St. Mary, Isle of Man. In *Proceedings and Transactions of the Liverpool Biological Society*, 60–77.

Keough, M. J. and Raimondi, P. T. (1996). Responses of settling invertebrate larvae to bioorganic films: effects of large-scale variation in films. *Journal of Experimental Marine Biology and Ecology*, **207**, 59–78.

Laimek, P., Clark, S., Stewart, M. et al. (2008). The presence of GABA in gastropod mucus and its role in inducing larval settlement. *Journal of Experimental Marine Biology and Ecology*, **354**, 182–91.

Lamontagne, I., Cardinal, A. and Fortier, L. (1989). Environmental forcing versus endogenous control of photosynthesis in intertidal epilithic microalgae. *Marine Ecology Progress Series*, **51**, 177–87.

Lawrence, J. R., Wolfaardt, G. M. and Neu, T. R. (1998). *The Study of Biofilms Using Confocal Laser Scanning Microscopy: Digital Image Analysis of Microbes*. Wiley, Chichester, pp. 431–65.

Le Campion-Alsumard, T. (1989). Les cyanobactéries marines endolithes. Bulletin de la Société Botanique de France. *Actualités Botaniques*, **136**, 99–112.

Lee, O. O., Chung, H. C., Yang, J. et al. (2014). Molecular techniques revealed highly diverse microbial communities in natural marine biofilms on polystyrene dishes for invertebrate larval settlement. *Microbial Ecology*, **68**, 81–93.

Lindsay, D. and Von Holy, A. (1997). Evaluation of dislodging methods for laboratory-grown bacterial biofilms. *Food Microbiology*, **14**, 383–90.

Lock, M. (1993). Attached Microbial Communities in Rivers. In T. Ford, ed. *Aquatic Microbiology: An Ecological Approach*. Blackwell Scientific Publications, Boston, 113–38.

Lock, M. A., Wallace, R. R., Costerton, J. W., Ventullo, R. M. and Charlton, S. E. (1984). River epilithon: toward a structural-functional model. *Oikos*, **42**, 10–22.

Lorenzen, S. (2007). The limpet *Patella vulgata* L. at night in air: effective feeding on *Ascophyllum nodosum* monocultures and stranded seaweeds. *Journal of Molluscan Studies*, **73**, 267–74.

Lubchenco, J. and Gaines, S. D. (1981). A unified approach to marine plant-herbivore interactions. I. Populations and communities. *Annual Review of Ecology and Systematics*, **12**, 405–37.

Lüning, K. (1990). *Seaweeds: Their Environment, Biogeography, and Ecophysiology*, John Wiley and Sons, New York.

MacLulich, J. H. (1986). Experimental evaluation of methods for sampling and assaying intertidal epilithic microalgae. *Marine Ecology Progress Series*, **34**, 275–80.

MacLulich, J. H. (1987). Variations in the density and variety of intertidal epilithic microflora. *Marine Ecology Progress Series*, **40**, 285–93.

Mak, Y. M. and Williams, G. A. (1999). Littorinids control high intertidal biofilm abundance on tropical, Hong Kong rocky shores. *Journal of Experimental Marine Biology and Ecology*, **233**, 81–94.

Marshall, K. C., Stout, R. and Mitchell, R. (1971). Mechanism of the Initial Events in the sorption of marine bacteria to surfaces. *Journal of General Microbiology*, **68**, 337–48.

Marszalek, D. S., Gerchakov, S. M. and Udey, L. R. (1979). Influence of substrate composition on marine microfouling influence of substrate composition on marine microfouling. *Applied and Environmental Microbiology*, **38**, 987–95.

Méléder, V., Barillé, L., Launeau, P., Carrere, V. and Rincé, Y. (2003). Spectrometric constraint in analysis of benthic diatom biomass using monospecific cultures. *Remote Sensing of Environment*, **88**, 386–400.

Menge, B. A. (1992). Community regulation – under what conditions are bottom-up factors important on rocky shores. *Ecology*, **73**, 755–65.

Menge, B. A. (2000). Bottom-Up: Top-Down Determination of Rocky Intertidal Shorescape Dynamics. In G. A. Polis, M. E. Power and G. Huxel, eds. *Foodwebs at the Landscape Level*. University of Chicago Press, Chicago.

Miller, A. R., Lowe, R. L. and Rotenberry, J. T. (1987). Succession of diatom communities on sand grains. *Journal of Ecology*, **75**, 693–709.

Miller, M. B. and Bassler, B. L. (2001). Quorum sensing in bacteria. *Annual Reviews in Microbiology*, **55**(1), 165–99.

Mosier, A. C., Li, Z., Thomas, B. C., Hettich, R. L., Pan, C. and Banfield, J. F. (2014). Elevated temperature alters proteomic responses of individual organisms within a biofilm community. *ISME Journal*, **9**, 180–94.

Murphy, R. J., Klein, J. C. and Underwood, A. J. (2011). Chlorophyll *a* and intertidal epilithic biofilms analysed in situ using a reflectance probe. *Aquatic Biology*, **12**, 165–76.

Murphy, R. J., Tolhurst, T. J., Chapman, M. G. and Underwood, A. J. (2004). Estimation of surface chlorophyll on an exposed mudflat using digital colour-infrared (CIR) photography. *Estuarine, Coastal And Shelf Science*, **59**, 625–38.

Murphy, R. J., Tolhurst, T. J., Chapman, M. G. and Underwood, A. J. (2009). Seasonal distribution of chlorophyll on mudflats in New South Wales, Australia measured by field spectrometry and PAM fluorometry. *Estuarine, Coastal and Shelf Science*, **84**, 108–18.

Murphy, R. J. and Underwood, A. J. (2006). Novel use of digital colour-infrared imagery to test hypotheses about grazing by intertidal herbivorous gastropods. *Journal of Experimental Marine Biology and Ecology*, **330**, 437–47.

Murphy, R. J., Underwood, A. J. and Pinkerton, M. H. (2006). Quantitative imaging to measure photosynthetic biomass on an intertidal rock-platform. *Marine Ecology Progress Series*, **312**, 45–55.

Murphy, R. J., Underwood, A. J., Pinkerton, M. H. and Range, P. (2005). Field spectrometry: new methods to investigate epilithic micro-algae on rocky shores. *Journal of Experimental Marine Biology and Ecology*, **325**, 111–24.

Murphy, R. J., Underwood, A. J., Tolhurst, T. J. and Chapman, M. G. (2008). Field-based remote-sensing for experimental intertidal ecology: case studies using hyperspatial and hyperspectral data for New South Wales (Australia). *Remote Sensing of Environment*, **112**, 3353–65.

Nagarkar, S. and Williams, G. A. (1997). Comparative techniques to quantify cyanobacteria dominated biofilms on tropical rocky shores. *Marine Ecology Progress Series*, **154**, 281–91.

Nagarkar, S. and Williams, G. A. (1999). Cyanobacteria dominated epilithic biofilms: spatial and temporal variation on semi-exposed tropical rocky shores. *Phycologia*, **38**, 385–93.

Newell, R. C. (1970). *Biology of Intertidal Animals*, Marine Ecological Surveys. Logos Press Limited, London.

Ng, T. P. T., Saltin, S. H., Davies, M. S., Johannesson, K. and Williams, G. A. (2013). Snails and their trails : the multiple functions of trail-following in gastropods. *Biological Reviews of the Cambridge Philosophical Society*, **88**, 683–700.

Nicotri, M. E. (1977). Grazing effects of four marine intertidal herbivores on the microflora. *Ecology*, **58**, 1020–32.

Niell, F. X. and Varela, M. (1984). Initial colonization stages on rocky coastal substrates. *Marine Ecology*, **5** (1), 45–56.

Nielsen, K. J. (2001). Bottom-up and top-down forces in tide pools: test of a food chain model in an intertidal community. *Ecological Monographs*, **71**, 187–217.

Norton, T. A., Hawkins, S. J., Manley, N. L., Williams, G. A. and Watson, D. C. (1990). Scraping a living: a review of littorinid grazing. *Hydrobiologia*, **193**, 117–38.

Norton, T. A., Thompson, R. C., Pope, J. et al. (1998). Using confocal laser scanning microscopy, scanning electron microscopy and phase contrast light microscopy to examine marine biofilms. *Aquatic Microbial Ecology*, **16**, 199–204.

Notman, G. M., McGill, R. A. R., Hawkins, S. J. and Burrows, M. T. (2016). Macroalgae contribute to the diet of *Patella vulgata* from contrasting conditions of latitude and wave exposure in the UK. *Marine Ecology Progress Series*, **549**, 113–23.

Patil, J. S. and Anil, A. C. (2005). Biofilm diatom community structure: influence of temporal and substratum variability. *Biofouling*, **21**(3–4), 189–206.

Peduzzi, P. and Herndl, G. J. (1991). Mucus trails in the rocky intertidal: a highly active microenvironment. *Marine Ecology Progress Series*, **179**, 267–74.

Peletier, H., Gieskes, W. W. C. and Buma, A. G. J. (1996). Ultraviolet-B radiation resistance of benthic diatoms isolated from tidal flats in the Dutch Wadden Sea. *Marine Ecology Progress Series*, **135**, 163–8.

Peyrot-Clausade, M., Le Campion-Alsumard, T., Harmelin-Vivien, M. et al. (1995). La bioerosion dans le cycle des carbonates; essais de quantification des processes en Polynesie francaise. *Bulletin de la Société Géologique de France*, **166**, 85–94.

Poglazova, M. N., Mitskevich, I. N. and Kuzhinovsky, V. A. (1996). A spectrofluorimetric method for the determination of total bacterial counts in environmental samples. *Journal of Microbiological Methods*, **24**, 211–18.

Poore, A. G., Campbell, A. H., Coleman, R. A. et al. (2012). Global patterns in the impact of marine herbivores on benthic primary producers. *Ecology Letters*, **15**(8), 912–22.

Priester, J. H., Horst, A. M., Van De Werfhorst, L. C., Saleta, J. L., Mertes, L. A. K. and Holden, P. A. (2007). Enhanced visualization of microbial biofilms by staining and environmental scanning electron microscopy. *Journal of Microbiological Methods*, **68**, 577–87.

Pringle, J. H. and Fletcher, M. (1986). Influence of substratum hydration and adsorbed macromolecules on bacterial attachment to surfaces. *Applied and Environmental Microbiology*, **51**, 1321–5.

Relexans, J. C. (1996). Measurement of the respiratory electron transport system (ETS) activity in marine sediments: state-of-the-art and interpretation. 1. Methodology and review of literature data. *Marine Ecology Progress Series*, **136**, 277–87.

Richardson, M. D. (1983). Standing stock and vertical distribution of benthic fauna from the Venezuela Basin. *EOS*, **64**, 1086.

Russell, B. D., Connell, S. D., Findlay, H. S., Tait, K., Widdicombe, S. and Mieszkowska, N. (2013). Ocean acidification and rising temperatures may increase biofilm primary productivity but decrease grazer consumption. *Philosophical Transactions of the Royal Society of London. Series B*, **368**, 20120438.

Sabater, S., Gregory, S. V. and Sedell, J. R. (1998). Community dynamics and metabolism of benthic algae colonizing wood and rock substrata in a forest stream. *Journal of Phycology*, **34**, 561–7.

Safriel, U. N. and Erez, N. (1987). Effect of limpets on the fouling of ships in the Mediterranean. *Marine Biology*, **95**(4), 531–7.

Salemaa, H. (1987). Herbivory and microhabitat preferences of *Idotea* spp.(Isopoda) in the northern Baltic Sea. *Ophelia*, **27**, 1–15.

Salta, M., Wharton, J. A., Blache, Y., Stokes, K. R. and Briand, J.-F. (2013). Marine biofilms on artificial surfaces: structure and dynamics. *Environmental Microbiology*, **15**, 2879–93.

Sanford, E. and Menge, B. A. (2001). Spatial and temporal variation in barnacle growth in a coastal upwelling system. *Marine Ecology Progress Series*, **209**, 143–57.

Santelices, B. and Bobadilla, M. (1996). Gastropod pedal mucus retains seaweed propagules. *Journal of Experimental Marine Biology and Ecology*, **197**, 251–61.

Savitzky, A. and Golay, M. J. E. (1964). Smoothing and differentiation of data by simplified least squares procedures. *Analytical Chemistry*, **36**, 1627–39.

Schaffelke, B., Evers, D. and Walhorn, A. (1995). Selective grazing of the isopod *Idotea baltica* between *Fucus evanescens* and *F. vesiculosus* from Kiel Fjord (western Baltic). *Marine Biology*, **124**, 215–18.

Schuster, J. J. and Markx, G. H. (2013). Biofilm architecture. *Advances in Biochemical Engineering*, **146**, 77–96.

Scott, C., Fletcher, R. L. and Bremer, G. B. (1996). Observations on the mechanisms of attachment of some marine fouling blue-green algae. *Biofouling*, **10**, 161–73.

Shokralla, S., Spall, J. L., Gibson, J. F. and Hajibabaei, M. (2012). Next-generation sequencing technologies for environmental DNA research. *Molecular Ecology*, **21**(8), 1794–805.

Skov, M. W., Volkelt-Igoe, M., Hawkins, S. J., Jesus, B., Thompson, R. C. and Doncaster, C. P. (2010). Past and present grazing boosts the photo-autotrophic biomass of biofilms. *Marine Ecology Progress Series*, **401**, 101–11.

Smith, A. M. and Morin, M. C. (2002). Biochemical differences between trail mucus and adhesive mucus from marsh periwinkle snails. *Biological Bulletin*, **203** (3), 338–46.

Smith, D. J. and Underwood, G. J. C. (1998). Exopolymer production by intertidal epipelic diatoms. *Limnology and Oceanography*, **43**, 1578–91.

Snoeijs, P. J. M. (1991). Monitoring pollution effects by diatom community composition. *A Comparison of Sampling Methods*, **121**, 497–510.

Somero, G. N. (2002). Thermal physiology and vertical zonation of intertidal animals: optima, limits and costs of living. *Integrative and Comparative Biology*, **42**, 780–9.

Sommer, U. (1997). Selectivity of *Idothea chelipes* (Crustacea: Isopoda) grazing on benthic microalgae. *Limnology and Oceanography*, **42**, 1622–8.

Sommer, U. (1999). The impact of herbivore type and grazing pressure on benthic microalgal diversity. *Ecology Letters*, **2**, 65–9.

Sousa, W. P. (1992). Grazing and Succession in Marine Algae. In *Plant–Animal Interactions in the Marine Benthos*. Oxford University Press, Oxford, pp. 425–41.

Southward, A. J. (1964). *Limpet Grazing and the Control of Vegetation on Rocky Shores*. Blackwell, Oxford, pp. 265–73.

Steneck, R. S. and Watling, L. (1982). Feeding capabilities and limitation of herbivorous mollusks – a functional-group approach. *Marine Biology*, **68**, 299–319

Sundbäck, K., Odmark, S., Wulff, A., Nilsson, C. and Wängberg, S. Å. (1997). Effects of enhanced UVB radiation on a marine benthic diatom mat. *Marine Biology*, **128**(1), 171–9.

Surman, S. B., Walker, J. T., Goddard, D. T. et al. (1996). Comparison of microscope techniques for the examination of biofilms. *Journal of Microbiological Methods*, **25**, 57–70.

Tankersley, R. A. (1990). Trail following in *Littorina irrorata*: the influence of visual stimuli and the possible role of tracking in orientation. *Veliger*, **33**, 116–23.

Taylor, G. T., Zheng, D., Lee, M., Troy, P. J., Gyananath, G. and Sharma, S. K. (1997). Influence of surface properties on accumulation of conditioning films and marine bacteria on substrata exposed to oligotrophic waters. *Biofouling*, **11**, 31–57.

Taylor, J. D., Ellis, R., Milazzo, M., Hall-Spencer, J. M. and Cunliffe, M. (2014). Intertidal epilithic bacteria diversity changes along a naturally occurring carbon dioxide and pH gradient. *FEMS Microbiology Ecology*, **89**, 670–8.

Thompson, R. C., Wilson, B. J. Tobin, M. L. Hill, A. S. and Hawkins, S. J. (1996). Biologically generated habitat provision and diversity of rocky shore organisms at a hierarchy of spatial scales. *Journal of Experimental Marine Biology and Ecology*, **202**, 73–84.

Thompson, R. C., Roberts, M. F., Norton, T. A. and Hawkins, S. J. (2000). Feast or famine for intertidal grazing molluscs: a mis-match between seasonal variations in grazing intensity and the abundance of microbial resources. *Hydrobiologia*, **440**, 357–67.

Thompson, R. C., Tobin, M. L., Hawkins, S. J. and Norton, T. A. (1999). Problems in extraction and spectrophotometric determination of chlorophyll from epilithic microbial biofilms: towards a standard method. *Journal of the Marine Biological Association of the UK*, **79**, 551–8.

Thompson, R. C., Norton, T. A. and Hawkins, S. J. (2004). Physical stress and biological control regulate the producer–consumer balance in intertidal biofilms. *Ecology*, **85**, 1372–82.

Thompson, R. C., Moschella, P. S., Jenkins, S. R., Norton and Hawkins, S. J. (2005). Differences in photosynthetic marine biofilms between sheltered and moderately exposed rocky shores. *Marine Ecology Progress Series*, **296**, 53–63.

Thompson, R. C., Norton, T. A. and Hawkins, S. J. (1998). The influence of epilithic microbial films on the settlement of *Semibalanus balanoides* cyprids–a comparison between laboratory and field experiments. *Hydrobiologia*, **375**, 203–16.

Tuchman, M. and Blinn, D. W. (1979). Comparison of attached algal communities on natural and artificial substrata along a thermal gradient. *British Phycological Journal*, **14**, 243–54.

Underwood, A. J. (1978). An experimental evaluation of competition between three species of intertidal prosobranch gastropods. *Oecologia*, **33**, 185–202.

Underwood, A. J. (1979). The ecology of intertidal gastropods. *Advances in Marine Biology*, **16**, 111–210.

Underwood, A. J. (1980). The effects of grazing by gastropods and physical factors on the upper limits of distribution of intertidal macroalgae. *Oecologia*, **46**, 201–13.

Underwood, A. J. (1984a). Microalgal food and the growth of the intertidal gastropods *Nerita atramentosa* Reeve and *Bembicium nanum* (Lamarck) at four heights on a shore. *Journal of Experimental Marine Biology and Ecology*, **79**, 277–91.

Underwood, A. J. (1984b). The vertical distribution and seasonal abundance of intertidal microalgae on a rocky shore in New South Wales. *Journal of Experimental Marine Biology and Ecology*, **78**, 199–220.

Underwood, A. J. (1984c). Vertical and seasonal patterns in competition for microalgae between intertidal gastropods. *Oecologia*, **64**, 211–22.

Underwood, A. J. and Jernakoff, P. (1981). Effects of interactions between algae and grazing gastropods on the structure of a low-shore intertidal algal community. *Oecologia*, **48**, 221–33.

Underwood, A. J. and Murphy, R. J. (2008). Unexpected patterns of facilitatory grazing revealed by quantitative imaging. *Marine Ecology Progress Series*, **358**, 85–94.

Veltkamp, C. J., Chubb, J. C., Birch, S. P. and Eaton, J. W. (1994). A simple freeze dehydration method for studying epiphytic and epizoic communities using the scanning electron microscope. *Hydrobiologia*, **288**, 33–8.

Wahl, M. (1989). Marine epibiosis. I. Fouling and antifoulinng: some basic aspects. *Marine Ecology Progress Series*, **58**, 175–89.

Walker, J. T., Hanson, K., Caldwell, D. and Keevil, C. W. (1998). Scanning confocal laser microscopy study of biofilm induced corrosion on copper plumbing tubes. *Biofouling*, **12**, 333–44.

Wieczorek, S. K. and Todd, C. D. (1998). Inhibition and facilitation of settlement of epifaunal marine invertebrate larvae by microbial biofilm cues. *Biofouling*, **12**, 81–118.

Williams, G. A. (1993a). Seasonal variation in algal species richness and abundance in the presence of molluscan herbivores on a tropical rocky shore. *Journal of Experimental Marine Biology and Ecology*, **167**, 261–75.

Williams, G. A. (1993b). The relationship between herbivorous molluscs and algae on moderately exposed Hong Kong shores. In *Proceedings of the 1st International Conference on the Marine Biology of the South China Sea*. Hong Kong University Press, Hong Kong. pp. 459–70.

Williams, G. A. (1994). The relationship between shade and molluscan grazing in structuring communities on a moderately-exposed tropical rocky shore. *Journal of Experimental Marine Biology and Ecology*, **178**, 79–95.

Williams, G. A., Davies, M. S. and Nagarkar, S. (2000). Primary succession on a seasonal tropical rocky shore: the relative roles of spatial heterogeneity and herbivory. *Marine Ecology Progress Series*, **203**, 81–94.

Wimpenny, J. (1996). Ecological determinants of biofilm formation. *Biofouling*, **10**, 43–63.

Wolfaardt, G. M., Archibald, R. E. M. and Cloete, T. E. (1991). The use of DAPI in the quantification of sessile bacteria on submerged surfaces. *Biofouling*, **4**, 265–74.

Wright, S. W., Jeffrey, S. W. and Mantoura, R. F. C. (2005). *Phytoplankton Pigments in Oceanography: Guidelines to Modern Methods*. UNESCO Publishing, Paris.

Zheng, D., Taylor, G. T. and Gyananath, G. (1994). Influence of laminar flow velocity and nutrient concentration on attachment of marine bacterioplankton. *Biofouling*, **8**, 107–20.

# Chapter 19

# Interactions in the Deep Sea

A. Louise Allcock and Mark P. Johnson

## 19.1 Introduction

The deep-ocean floor extends over two thirds of the world's surface and is thus the largest benthic habitat on the planet. The myth of depauperate deep-sea communities was debunked in the 1960s by the pioneering work of Hessler and Sanders (1967; Sanders and Hessler, 1969), with their newly developed epibenthic sled. They showed deep-sea diversity to be equivalent to that found in shallow tropical marine habitats, and greater than in boreal tropical and temperate estuaries and boreal shallow marine habitats. They also identified depth as the most important correlate of faunal abundance and as a factor driving community composition. Technological developments over the last fifty years have continued to drive advances in our knowledge of this diverse and heterogeneous biome. Efforts to enumerate and catalogue the diversity have led to claims of high levels of endemism (e.g., Wolff, 1970; Belyaev, 1989; Stocks and Hart, 2007; Ebbe et al., 2010), but poor knowledge of the global species pool and uneven regional sampling probably artifactually inflate these estimates (e.g., Rowden et al., 2010; Clark et al., 2012).

The term 'deep sea' encompasses many different habitats, shaped by their physical characteristics – geographic location, slope, depth – which determine their biodiversity and dominant fauna, and their connectivity. Here, we first explore biogeography and phylogeography of these habitats, and consider some of the molecular work which is testing various biogeographic schemes. We then look briefly at some of the abiotic parameters that characterise various deep-sea habitats. Interactions in the deep sea are many, but they are often not well investigated. Few studies on competition and predation match the detail of those conducted in shallow waters. In contrast, symbioses have been better researched, being the basis of much productivity at hydrothermal vents and cold seeps, and highly prevalent in filter-feeder-dominated habitats on the steep slopes of submarine canyons and seamounts. Finally, we explore where sufficient bodies of work exist to allow us to infer processes from patterns, and conclude that very much more work on ecological interactions in the deep sea is needed.

## 19.2 Biogeography and Phylogeography

The sparseness of biological samples on a global scale has made delineation of biogeographic boundaries based on faunal differences extremely challenging, although depth certainly plays an important role. Carney (2005) discussed the potential impacts of high pressure on biochemical reactions and the need for specialised membranes and proteins to counteract this. Reviewing studies of zonation on the continental slope, Carney (2005) recognised several transition zones at bathyal depths, but found faunal boundaries in the bathyal zone to be 'fuzzy' and taxon-specific, although sharp local boundaries could be associated with local hydrographic regimes.

Attempting a global biogeography of the deep-ocean floor, Watling et al. (2013) treated the lower bathyal (800–3,500 m) as a single depth zone, excluding the shallower upper bathyal (300–800 m), which mostly falls within national exclusive economic zones, from their classification. They defined fourteen biogeographic provinces within the lower bathyal (Arctic, subantarctic, Antarctic, Indian Ocean, seven Pacific Ocean provinces and three Atlantic provinces) and a further fourteen in abyssal (3,500–6,000 m) depths (Arctic, east and west Antarctic provinces, Indian Ocean, six Pacific provinces and four Atlantic provinces). The areas within lower bathyal regions are discontinuous, since they are formed by seamounts, mid-ocean ridges and the lower slopes of continental margins and oceanic islands. In contrast, the abyssal provinces tend to extend across ocean basins, where they may be incised by mid-ocean ridges (for example, the abyssal Indian Ocean province, the abyssal North Atlantic province). Ten hadal provinces (deeper than 6,500 m) were also described and follow the scheme of Belyaev (1989), originally published in the Russian literature. The ten hadal provinces comprise either individual trenches (e.g., the Java hadal province for the Java Trench; the Peru–Chile hadal province for the Atacama trench) or collections of two or more geographically close trenches (e.g., the Mariana hadal province for the Volcano, Mariana, Yap and Palau trenches). Despite an increase in hadal research in recent years (e.g., Fujii et al., 2013, Blazewicz-Paszkowycz et al., 2015; Lacey et al., 2016; Linley et al., 2016, 2017), there remain insufficient data to test whether the hypothesised hadal provinces reflect community diversity.

Both the bathyal and abyssal provinces of Watling et al. (2013) were proposed mostly based on hydrographic data (temperature and salinity) and fluxes of particulate organic carbon (POC), and are strongly influenced by water masses; for example, the North Atlantic abyssal province is the area north of the equator influenced by North Atlantic deep water, and the South Atlantic bathyal province encompasses the area south of the equator to the Antarctic convergence, where Antarctic intermediate water dominates. Although faunal data are limited, there is at least some support for these divisions. For example, Watling et al. (2013) report that the composition of cumaceans (Watling, 2009) and protobranch bivalves (Allen and Sanders, 1996) in the North Atlantic appear to reflect the proposed abyssal biogeographic provinces, although not all taxa show such differences.

There appear to be fewer extensive bathyal data sets with which to test the bathyal provinces (Watling et al., 2013), but hydrothermal vent studies provide a subset of relevant data. Vent habitats occur at mid-oceanic ridges, convergent plate boundaries and subduction zones. They are well known for clear differences in dominant fauna among locations, with, for example, caridean shrimps, such as *Rimicaris excoculata*, and bathymodiolid mussels dominating Atlantic sites, *Riftia* tubeworms dominating sites on the East Pacific Rise and the smaller *Ridgea* tubeworms dominating in the north-east Pacific (Ramirez-Llodra et al., 2007). An analysis of the community composition at sixty-three vent fields (Bachraty et al., 2009) yielded a biogeographic model with six provinces, which mapped reasonably well (Table 19.1) to the bathyal provinces of Watling et al. (2013). A previous analysis of vent provinces also yielded a six-province solution but with different boundaries (Van Dover et al., 2002). Tyler and Young (2003) proposed a nine-province solution that included less well-explored areas at higher latitudes. As vent studies progress and extend (e.g., to the Reykjanes Ridge in the Northern Atlantic Boreal province), correlations between proposed models can be better tested.

Cold seeps have been grouped into five geographic provinces (Gulf of Mexico, Atlantic, Mediterranean, East Pacific and West Pacific; Tyler et al., 2003) and these map less well to the provinces of Watling et al. (2013). However, numerous seep sites have been discovered since biogeographic provinces based on seep communities were originally proposed, including new sites in the Black Sea (Klaucke et al., 2006) and Antarctica (Domack et al., 2005). Furthermore, in-depth

Table 19.1 | Comparison of proposed biogeography schemes.

| Bathyal provinces (Watling et al., 2013) | Vent provinces (Bachraty et al., 2009) | Cold Seep provinces (Tyler et al., 2003) |
|---|---|---|
| BY1. Arctic | | Atlantic |
| BY2. Northern Atlantic Boreal | | Atlantic |
| BY3. Northern Pacific Boreal | North-West Pacific and North-East Pacific | Some deep 'West Pacific' seeps are in the North Pacific abyssal province |
| BY4. North Atlantic | Northern Mid-Atlantic Ridge | Gulf of Mexico, Mediterranean, Atlantic |
| BY5. South-East Pacific ridges | Southern East Pacific Rise | |
| BY6. New Zealand–Kermadec | South-West Pacific | |
| BY7. Cocos Plate | Northern East Pacific Rise | East Pacific |
| BY8. Nazca plate | | East Pacific |
| BY9. Antarctic | | |
| BY10. Subantarctic | | |
| BY11. Indian | South-West Pacific | |
| BY12. West Pacific | South-West Pacific and slight overlap with North-West Pacific | West Pacific |
| BY13. South Atlantic | | Atlantic |
| BY14. North Pacific | | |

analyses across the Atlantic equatorial belt suggest seep faunal communities are influenced far more by depth than by geographic location (Olu et al., 2010).

Genetic data can be used to test whether barriers to gene flow are congruent with biogeographic hypotheses. For example, even where species transcend biogeographic boundaries, genetic subdivisions may be found that correspond to those boundaries. Populations of several hydrothermal vent species on the northern East Pacific Rise (Cocos Plate province) are genetically different from those on the southern East Pacific Rise (south-east Pacific ridges province), including those of the extremophile *Alvinella pompejana* (Jang et al., 2016), and the giant tube worms *Tevnia jerichonana* (Zhang et al., 2015) and *Riftia pachyptila* (Coykendall et al., 2011). Further genetic subdivision is seen in the southern region of the southern East Pacific Rise at the Easter microplate in *A. pompejana* and *T. jerichonana*. Furthermore, two species of bythograeid crabs meet at the Easter microplate (Mateos et al., 2012), as do two species of *Bathymodiolus*, the latter hybridising in this region (Johnson et al., 2013b). Although areas south of the Easter microplate are distinguished as the northern Pacific Antarctic Ridge, they are still part of the southern East Pacific Rise biogeographic province (*sensu* Bachraty et al., 2009). The ephemeral nature of vents, and therefore the historical local extinction of and decolonisation by specialised vent fauna, may not render these animals best suited to test biogeographic provinces, because their biogeographic history is affected by forces to which other animals are not subjected. For example, populations of *R. pachyptila* that were less than half a kilometre apart on the northern East Pacific Rise were genetically distinguishable using AFLPs, and hydrodynamic forces, which entrain larvae in hydrothermal plumes, might lead to dispersal and delivery of larvae at the cohort level (Shank and Halanych, 2007).

Genetic studies of deep-water corals, often the dominant fauna in other bathyal habitats, are in their infancy, but are increasingly being used to delimit species and better define the distribution of species that often have variable morphologies. For example, in the cosmopolitan genus of bamboo coral *Acanella*, recent DNA sequencing has confirmed the distinctiveness of several Pacific and Atlantic species, and led the authors to resurrect a Mediterranean species, synonymise two widely used names (*Acanella eburnea* junior to *Acanella arbuscula*), describe new species, and provide expanded range limits for species (Saucier et al., 2017). Similar reviews of genera such as *Anthothela* and *Primnoisis* (Moore et al., 2016, 2017), *Narella* (Baco and Cairnes, 2012) and the Pacific genera *Pacifigorgia*, *Leptogorgia* and *Eugorgia* (Soler-Hurtado et al., 2017) are helping to clarify the taxonomic landscape and provide new information on species ranges and the distances over which populations can interact. Such basic information is crucial to understanding biogeography. Few studies on deep-water coral population genetics have been conducted to date; however, the scleractinian *Stephanocyathus spiniger* and the antipatharian *Stichopathes filiformis* seem to maintain panmixia across ocean expanses (Miller et al., 2010), while other species such as the scleractinians *Lophelia pertusa* (Morrison et al., 2011; Dahl et al., 2012) and *Desmophyllum dianthus* (Miller et al., 2011), the octocorals *Corallium rubrum* (Constantini et al., 2011), *Paragorgia arborea* (Herrera et al., 2012), *Callogorgia delta* (Quattrini et al., 2015a) and species of *Narella* (Baco and Cairnes, 2012), and the antipatharians *Antipathes robillardi* and *Stichopathes variabilis* (Miller et al., 2010) show distinct genetic structuring over varying spatial and/or bathymetric scales.

Genetic differentiation by depth, hypothesised to occur due to physiological gradients (see Rex and Etter, 2010) has been identified in other benthic deep-sea invertebrates apart from corals, including molluscs (Jennings et al., 2013, Strugnell et al., 2017) and polychaetes (Schüller, 2011; Brasier et al., 2017) among others (reviewed in Morrison et al., 2017 and Taylor and Roterman, 2017). Noting the relatively wide depth band of their lower bathyal, Watling et al. (2013) acknowledged a potential need for further subdivision by depth in some oceans and this appears to be supported by available molecular data.

## 19.3 Abiotic Interactions

With the exception of chemosynthetic habitats, deep-sea communities are highly dependent on the amount of POC reaching the sea floor. A predictive model of POC flux based on sediment trap data, net primary production data from remote sensing, sea-surface temperature and global bathymetry data (Lutz et al., 2007) shows wide discrepancies in POC supply to deep benthic communities. POC flux to the sea floor is strongly affected by depth and is consequently substantially higher around continental margins, ridges and seamount chains. Other processes also act to enhance POC supply in these regions. For example, modifications to local currents caused by submarine canyon topography can cause local upwelling, enhancing net primary production, while downwelling may develop over the canyons themselves, trapping POC within the canyon system (Fernandez-Arcaya et al., 2017). Nepheloid layers of suspended particulate material often form in canyons due to local hydrodynamics (e.g., Wilson et al., 2015), as can internal waves, and these facilitate the development of high biomass benthic communities in these areas (e.g., Johnson et al., 2013a). Particle modelling has demonstrated increased supply at the sea floor near shelf breaks (Thiem et al., 2006), a suggested factor in the prevalence of corals and other filter feeders in these locations (e.g., Quattrini et al., 2015b).

Local hydrodynamics also strongly affect food availability to benthic filter feeders at seamounts (Clarke et al., 2010), although the exact mechanism is disputed. There is little observational evidence for the classical 'Taylor's cone' theory, whereby flow across the top of the seamount creates anticyclonic flow around it and associated upward movement of nutrient-rich waters (Genin and Dower, 2007), but tidally forced circulation may facilitate retention of POC (White et al., 2007). Additionally, trapping of vertically migrating zooplankton around seamounts likely plays a role in enhanced food availability (Genin and Dower, 2007). Seasonal effects on reproductive development of macrofauna (e.g., Sumida et al., 2000) illustrate the importance of benthopelagic coupling in bathyal regions. At abyssal sites in geographic areas with marked climate seasonality, e.g., the Porcupine Abyssal Plain, an annual pulse of phytodetritus still reaches the benthos despite the extreme depth, and is an important food resource (Iken et al., 2001).

## 19.4 Ecological Interactions in the Deep Sea

The difficulties of gathering data in the deep sea limit evaluations of ecological interactions. There are few observations of phenomena like detailed time series of overgrowth of sessile species (Elliott et al., 2016) or responses of predator populations to prey outbreaks (Witman et al., 2003). The full range of ecological interactions undoubtedly occurs. Time series studies have identified dynamic benthic assemblages, with variation in feeding group dominance (Taylor et al., 2017), but the importance of ecological interactions generally remains to be characterised.

### 19.4.1 Competition and Predation

The arrival of a new resource like a whale carcass or a baited camera provides examples of interactions between predatory/scavenging species. For example, Yeh and Drazen (2009) saw events they interpreted as *Histiobranchus* sp. preying on aristeid shrimp around baited cameras. Abundances of other species were negatively associated, which could be interpreted as a competitive interaction (Yeh and Drazen, 2009), although negative associations are only suggestive of an interaction given other potentially confounding processes. A further example of a potential predatory interaction is seen in the avoidance behaviour of squat lobsters, *Munidopsis crassa*, leaving a porpoise carcass when *Benthoctopus* sp. was present (Kemp et al., 2006). A whale fall, which provides a sudden pulse of energy to the deep-sea floor, becomes the platform for a unique ecosystem (see Smith et al., 2015) and is the site of numerous ecological interactions. Nematode abundance around an artificially placed whale carcass was shown to decrease close to the carcass, presumably due to either predation by, or competition with, other organisms attracted to the carcass (Debenham et al., 2004). Conversely, nematode abundance 30 m from the carcass increased over a year and a half, potentially due to organic enrichment of the surrounding sediment as the carcass decomposed. Enhanced prey availability at whale falls must lead to enhanced predator interactions, although these are not well documented. Observations of pycnogonids at whale carcasses, sunken wood and cold seeps provide evidence that these enriched habitats do attract predators (Braby et al., 2009). In Monterey Bay, at depths of 3,000 m, two species of pycnogonids were found to be in higher-than-normal abundance at these habitats, where they were seen feeding on the anemones *Anthosactis pearseae*, which attaches directly to whale bone, and *Liponema brevicornis*, which favours soft sediment. The anemones were presumed to accumulate in these habitats due to local enrichment, but also due to the topography that disrupted currents in a manner favourable to filter feeders.

Vent communities can be highly structured, with zonation of species distributions thought to reflect environmental gradients (Cuvelier et al., 2011; Kim and Hammerstrom, 2012; Marsh et al., 2012). There is potential for species to compete for space on environmentally optimal surfaces. For example, Lenihan et al. (2008) found fewer invertebrate recruits on blocks with high densities of *Bathymodiolus*, even when transplanted to locations with otherwise high recruitment. Relatively low recruitment was interpreted to reflect potential interactions like predation (either by filter feeding on larvae by mussels or by organisms associated with mussels) and/or competition involving space pre–emption mediated by larvae choosing not to settle in mussel beds or by other means.

A rare caging experiment indicated the role of predatory fish at vents (Micheli et al., 2002). The dominant zoarcid fish (*Thermarces cerberus*) feeds on gastropods and amphipods. When predatory fish were excluded with cages, gastropods and amphipods increased in number and sessile invertebrates became less common (Micheli et al., 2002). This effect on the community was stronger at treatments placed nearest to hydrothermal vents.

### 19.4.2 Symbioses, Facilitation and Parasitism

The roles of chemoautotrophic microbial symbionts in sustaining deep-sea food chains are well known. Vestimentiferan worms, bathymodiolid mussels, vesicomyid clams and *Rimicaris* shrimps, among others, harbour chemoautotrophic symbiotic bacteria. Several squat lobster species of the genera *Kiwa* and *Shinkaia* even garden chemoautotrophic bacteria on dense specialised setae, which they consume potentially as their primary food source (Baeza, 2011).

More broadly, commensal invertebrates seem fairly common (Figure 19.1), with species like scale worms recognised from mussel hosts (e.g., Pettibone, 1986). Where organisms like sea pens, corals, sponges and xenophyophores provide structure or attachment points, a diverse assemblage of facultative and potentially obligate symbionts can be found (Beaulieu, 2001; Buhl-Mortensen et al., 2010; Figure 19.1a and 19.1d–19.1f). Some species associations may occur as hard substrate is not otherwise available or they may reflect fitness improvements from better feeding opportunities or avoidance of predation (Barry et al., 2017). The importance of structural supports to biodiversity may increase with depth, reflecting associated declines in habitat complexity and productivity (Buhl-Mortensen et al., 2010).

Among corals, which are well known for enhancing habitat substrate, eleven families of deep-water Alcyonacea are regularly found with symbionts (see review by Watling et al. (2011) which includes a complete list of octocoral hosts and their symbionts). Common octocoral symbionts include scale worms (family Polynoidae), copepods, ascothoracids (parasitic crustaceans previously considered to be cirripedes), brittle stars and various amphipods. Some families seem to have evolved more numerous symbioses than others. Watling et al. (2011) list *Anthomastus grandiflorus* as the only deep-water species of Alcyoniidae (true soft corals) to host a symbiont (the commensal polynoid *Harmothoe acanellae*) and *Victogorgia josephinae* as the only anthothelid to host associates. In contrast, they report *P. arborea* (family Paragorgiidae) to host a commensal amphipod (pleustid), sphaerodorid worm and an anemone, as well as a parasitic zoanthid and copepod, while other *Paragorgia* species are also reported to host associates. Among the Primnoidae, Watling et al. (2011) reported seventeen symbionts from eleven species. Many of these relationships are unique to the species involved. For instance, each of five species of *Chrysogorgia* (Octocorallia, Chrysogorgiidae) hosts a different species of *Thalassomembracis* (Crustacea, Ascothoracidae). Conversely, there are symbiont species reported from more than one host and these hosts may be quite different; for example, *Gorgoniapolynoe muzikae* has been reported from species of both *Candidella* (family Primnoidae) and *Acanthogorgia* (family Acanthogorgiidae). Few of these relationships have been studied in detail, although the brittle star *Ophiocreas oedipus* is thought to colonise its *Metallogorgia* host when both are young, with the single symbiont growing with the coral until the latter dies (Mosher and Watling, 2009). In contrast, multiple *Gorgoniapolynoe caeciliae* must colonise the primnoid *Candidella imbricata* because the longevity of the symbiont species is much less than that of the coral, leading to considerable turnover of symbionts during the coral's life (Watling et al., 2011). Interestingly, the presence of the worm affects the morphology of the coral providing a refuge for the symbiont species (Eckelbarger et al., 2005). Similar host manipulation is seen in the symbiosis between species of eunicid worms and deep-water hermatypic scleractinians (Oppelt et al., 2017), and in the relationship between eunicids and black corals, where the hosts' skeletal spines are modified (Molodtsova and Budaeva, 2007).

The benefit for the symbiont is often obvious, even when the coral is not induced to produce a specialist refugium. The increased height above the substrate is beneficial to filter feeders, and it has further been suggested that octocorals, with their presumed secondary-metabolite-laden tissues (to protect themselves from predation) also offer protection from predation to the symbiont. Most of the known relationships between octocorals and their symbionts, if not clearly parasitic, have been classed as commensal, i.e., with no obvious benefit to the host. Parasites include the zoanthid *Isozoanthus primnoidus*, associated with the cold-water coral *Callogorgia*

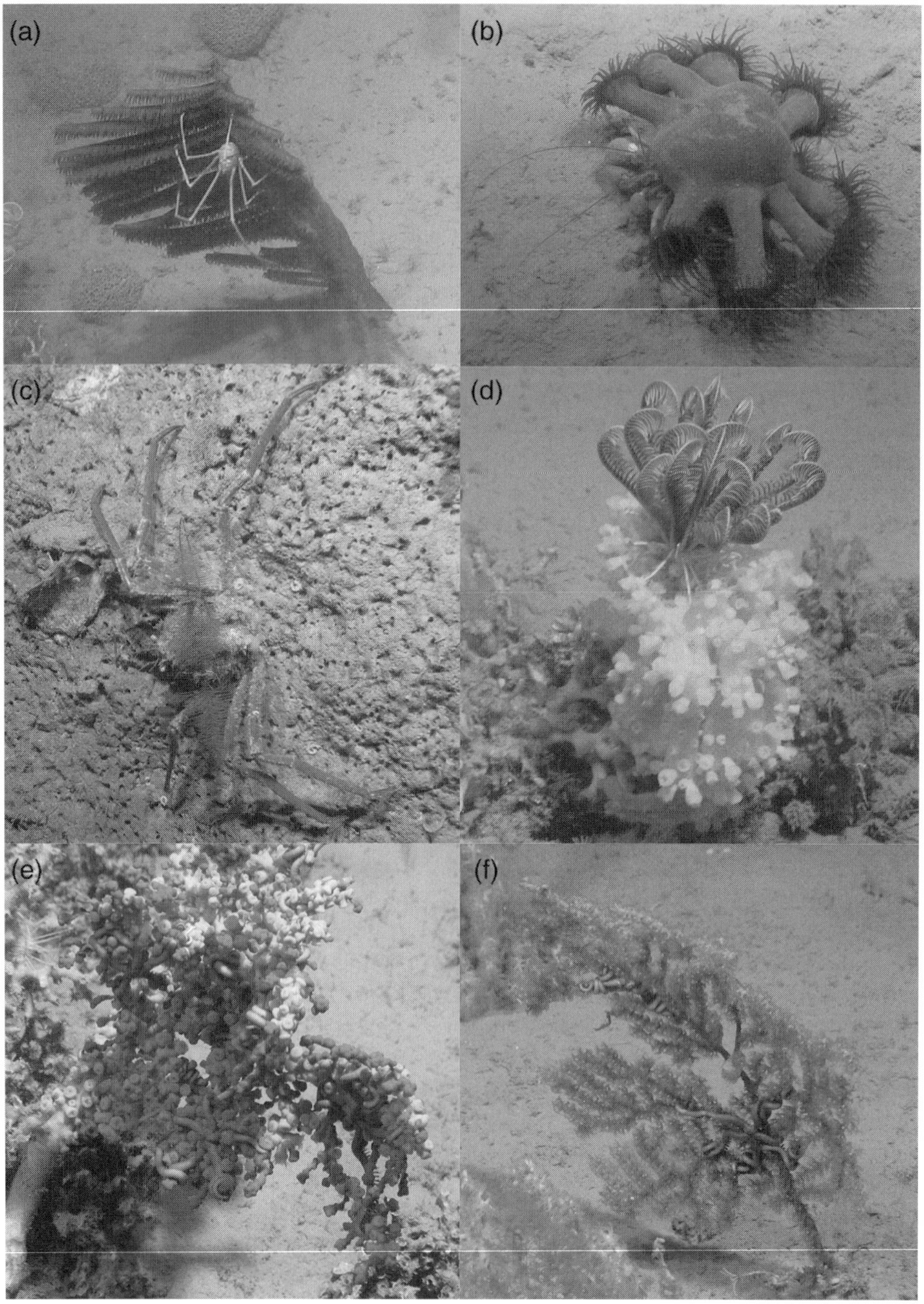


**Fig. 19.1** Animal–animal associations in the deep sea: (a) chirostylid on black coral *Bathypathes;* (b) the zoanthid *Epizoanthus paguriphilus* providing a shell for the hermit crab *Parapagurus pilosimanus*; (c) the carrier crab *Paramola cuvieri* carrying a black coral, probably *Parantipathes*; (d) yellow zoanthid on the sponge *Aphrocallistes beatrix*; (e) yellow zoanthid and brittle star *Asteroschema* on the bubblegum coral *Paragorgia*; (f) brittle star *Asteroschema* on the plexaurid coral *Paramuricea*. All photos taken by NUI Galway during cruise CE14009 aboard RV Celtic Explorer using ROV Holland I, copyright Marine Institute. *A black and white version of this figure will appear in some formats. For the colour version, please refer to the plate section.*

*verticillata*, which appears to progressively harm its host, overgrowing the coral coenchyma and polyps, and incorporating coral sclerites into its own tissue for protection (Carreiro-Silva et al., 2011). A similarly parasitic relationship has also been found for a species of *Epizoanthus* which overgrows *P. arborea* (Buhl-Mortensen and Buhl-Mortensen, 2004). In this intensive study, forty-seven species were found associated with *P. arborea* and ninety-seven with *Primnoa resedaeformis*. While some of these species were feeding on the host corals, a large number were identified as using the corals as substrate or for refuge, indicating the enormous role octocoral gardens can play in biodiversity through commensal relationships.

A recent study, however, has suggested that there may be tangible host benefits. Girard et al. (2016) studied the relationship between the ophiuroid symbiont *Asteroschema clavigerum* on the plexaurid octocoral *Paramuricea biscaya*. They monitored damage and recovery of *Paramuricea* following the Deepwater Horizon oil spill in the Gulf of Mexico. They found that recovery of the colony was negatively correlated with how far away that part of the colony was from a symbiotic ophiuroid, and suggested that ophiuroids both removed material that was deposited on polyps and inhibited settlement of hydroids on the colony. Although these effects were associated with a catastrophic impact event, the authors suggested that some benefit from sedimentation in natural conditions might be conferred on the coral host.

The presence of potential hosting benefits might explain the prevalence of host–symbiont associations in black corals. Black corals (Hexacorallia, Antipatharia), unlike other corals, are not able to retract their polyps and are thus particularly vulnerable to sedimentation events. In a comprehensive review of black coral ecology, Wagner et al. (2012) listed known symbionts. Among deep-water black corals, chirostylids are particularly prevalent, and have been recorded on species of *Bathypathes, Leiopathes, Parantipathes* and *Antipathes* (Wienberg et al., 2008; Le Guilloux et al., 2010). Chirostylids appear to feed on particles trapped in the mucus secreted by the black corals, and hence obtain their food while potentially cleaning the host (Le Guilloux et al., 2010). In shallow-water associations between decapods and antipatharians, it has further been suggested that the decapod might deter other predators (e.g., Glynn, 1980), although some decapods have also been shown to feed on antipatharian tissue (Wirtz and d'Udekem d'Acoz, 2001).

Aside from emergent structures, burrowers and other agents of disturbance may facilitate other species by creating small-scale heterogeneity. Fish associate with sponge and coral reefs, potentially giving the habitats value in sustaining both fish populations and fisheries. Unfortunately, testing these sorts of hypotheses is difficult: statistical models of fish–habitat associations are subject to issues of scale, temporal variability and potential confounding variables (Kutti et al., 2014; Pham et al., 2015).

Just over a decade ago, de Buron and Morand (2004) asked the question why there were not more deep-sea vent parasites. They identified over fifteen vent examples and 126 non-vent parasite species. In answer to the question 'why not more?' their answer was that the level of collection and identification was limiting the apparent prevalence of parasitic relationships.

## 19.5 Inference of Process from Pattern

It has been tempting to make broad statements about the importance of trophic interactions in shaping deep-sea ecology and evolution. Examples include Sanders' (1968) stability-time hypothesis and Rex's explanations for bathyal diversity peaks (Rex, 1981). The stability-time hypothesis addresses a similar issue to the paradox of the plankton (Hutchinson, 1961): richness in a habitat expected for other reasons to be species poor. In Sanders' hypothesis, the richness is a result of a long evolutionary niche diversification in a stable environment. Rex (1981) suggested that interactions between competition, predation and productivity shape the change in species richness with depth, with a humped diversity–productivity relationship occurring along the depth gradient. Such broad statements about the causes of patterns in species diversity are not widely accepted as sufficient explanations. Stability-time ideas have been contradicted by observations of disturbance and population turnover, even at abyssal depths (Levin et al., 2001, although see McClain and Schlacher, 2015, contrasting the relevant timescales of disturbance). A general explanation invoking species interactions for the peaking of diversity in the mid-bathyal zone needs consistent explanations for competitive exclusion mechanisms, and a framework for explaining why the pattern does not seem to be universal (Stuart and Rex, 2009; Rex and Etter, 2010). Terrestrial ecology has grappled with whether a humped diversity–productivity relationship is common (Adler et al., 2011), what models of competition are actually predicted under high productivity (Abrams, 1995), as well as the low predictive value of single variables (Grace et al., 2016). Deep-sea ecology will need a similar approach, an understanding of proxy variables and inclusion of additional variables (Etter and Grassle, 1992), particularly as the $r^2$ is often low in diversity depth regressions (Rex and Etter, 2010). With respect to the absence of a single accepted theory that successfully integrates an understanding of species interactions into an explanation of pattern, deep-sea ecology is no different from many other areas of ecology. Brown (1981) describes a widespread belief during the 1960s and 1970s that interspecific competition was the main factor limiting diversity. This belief was followed by disappointment as many of the models for competition were often shown to rely on unrealistic assumptions and to produce ambiguous predictions that were not clearly testable.

The difficulties in inferring process have contributed to a situation where ‘every old hypothesis is also a new one’ (McClain and Schlacher, 2015). Older hypotheses have not been refuted or built on as the demonstration of processes like competition or niche partitioning is difficult and some of the predictions cannot be distinguished from the predictions of alternative hypotheses. Wider applications of tools like stable isotopes and sequence analysis are likely to identify more detailed examples of trophic specialisation (e.g., Levesque et al., 2003; Govenar et al., 2015).

## 19.6 Concluding Remarks

We are, however, still some distance from an evaluation of the effects of ecological interactions in shaping deep-sea communities (McClain and Rex, 2015). Levin et al. (2001) offer a synthesis of the roles of ecological interactions and other processes in a series of conceptual models spanning gradients of food input, flow strength, oxygenation, sediment heterogeneity and disturbance. While an important step, such syntheses do not generate testable predictions that would allow ecological interactions to be evaluated. Levin and Dayton (2009) identified links between ecological theory and management of continental margins. This is potentially the area where the clearest picture of the role of ecological interactions will be developed in the near future: not in the description of general patterns, but in the management of environmental change and human impacts.

## Acknowledgements

The authors’ deep-sea research is supported by Science Foundation Ireland (SFI) and the Marine Institute under Investigators Programme Grant Number SFI/15/1A/3100, co-funded under the European Regional Development Fund 2014-2020.

### REFERENCES

Abrams, P. A. (1995). Monotonic or unimodal diversity–productivity gradients: what does competition theory predict? *Ecology*, **76**, 2019–27.

Adler, P. B., Seabloom, E. W., Borer, E. T. et al. (2011). Productivity is a poor predictor of plant species richness. *Science*, **333**, 1750–3.

Allen, J. A. and Sanders, H. L. (1996). The zoogeography, diversity and origin of the deep–sea protobranch bivalves of the Atlantic: the epilogue. *Progress in Oceanography*, **38**, 95–153.

Bachraty, C., Legendre, P. and Desbruyeres, D. (2009). Biogeographical relationships among deep–sea hydrothermal vent faunas at global scale. *Deep-Sea Research Part I – Oceanographic Research Papers*, **56**, 1371–8.

Baco, A. R. and Cairnes, S. D. (2012). Comparing molecular variation to morphological species designations in the deep–sea coral *Narella* reveals new insights into seamount coral ranges. *PLoS ONE*, **7**, e45555.

Baeza, J. A. (2011). Squat Lobsters as Symbionts and in Chemo-Autotrophic Environments. In *The Biology Of Squat Lobsters*. CRC Press, Boca Raton, FL, pp. 249–70.

Barry, J. P., Taylor, J. R., Kuhnz, L. A. and De Vogelaere, A. P. (2017). Symbiosis between the holothurian *Scotoplanes* sp. A and the lithodid crab *Neolithodes diomedeae* on a featureless bathyal sediment plain. *Marine Ecology*, **38**, e12396.

Beaulieu, S. E. (2001). Life on glass houses: sponge stalk communities in the deep sea. *Marine Biology*, **138**, 803–17.

Belyaev, G. M. (1989). *Deep Sea Ocean Trenches and Their Fauna*. Nauka, Moscow.

Blazewicz-Paszkowycz, M., Pabis, K. and Jozwiak, P. (2015). Tanaidacean fauna of the Kuril–Kamchatka Trench and adjacent abyssal plain – abundance, diversity and rare species. *Deep-Sea Research Part II – Topical Studies in Oceanography*, **111** (SI), 325–32.

Braby, C. E., Pearse, V. B., Baine, B. A. and Vrijenhoek, R. C. (2009). Pycnogonid–cnidarian trophic interactions in the deep Monterey Submarine Canyon. *Invertebrate Biology*, **128**, 359–63.

Brasier, M. J., Wiklund, H., Neal, L. et al. (2017). DNA barcoding uncovers cryptic diversity in 50% of deep-sea Antarctic polychaetes. *Royal Society Open Science*, **3**, 160432.

Brown, J. H. (1981). Two decades of homage to Santa Rosalia: toward a general theory of diversity. *American Zoologist*, **21**, 877–88.

Buhl-Mortensen, L. and Buhl-Mortensen, P. (2004). The Distribution and Diversity of Species Associated with Deep Sea Gorgonian Corals off the Atlantic Canada. In A. Freiwald and R. J. Murray, eds. *Cold-Water Corals and Ecosystems*. Springer, Berlin, pp. 849–79.

Buhl-Mortensen, L., Vanreusel, A., Gooday, A. J. et al. (2010). Biological structures as a source of habitat heterogeneity and biodiversity on the deep ocean margins. *Marine Ecology*, **31**, 21–50.

Carney, R. S. (2005). Zonation of deep biota on continental margins. *Oceanography and Marine Biology: An Annual Review*, **43**, 211–78.

Carreiro-Silva, M., Braga-Henriques, A., Sampaio, I., de Matos, V., Porteiro, F. M. and Ocaña, O. (2011). *Isozoanthus primnoidus*, a new species of zoanthid (Cnidaria: Zoantharia) associated with the gorgonian *Callogorgia verticillata* (Cnidaria: Alcyonacea). *ICES Journal of Marine Science*, **68**, 408–15.

Clark, M. R., Rowden, A. A., Schlacher, T. et al. (2010). The ecology of seamounts: structure, function, and human impacts. *Annual Review of Marine Science*, **2**, 253–78.

Clark, M. R., Schlacher, T. A., Rowden, A. A., Stocks, K. I. and Consalvey, M. (2012). Science priorities for seamounts: research links to conservation and management. *PLoS ONE*, **7**, e29232.

Constantini, F., Rossi, S., Pintus, E., Cerrano, C., Gili, J.-M. and Abbiati, M. (2011). Low connectivity and declining genetic variability along a depth gradient in *Corallium rubrum* populations. *Coral Reefs*, **30**, 991–1003.

Coykendall, D. K., Johnson, S. B., Karl, S. A., Lutz, R. A. and Vrijenhoek, R. C. (2011). Genetic diversity and demographic instability in *Riftia pachyptila* tubeworms from eastern Pacific hydrothermal vents. *BMC Evolutionary Biology*, **11**, 96.

Cuvelier, D., Sarrazin, J., Colaco, A. et al. (2011). Community dynamics over 14 years at the Eiffel Tower hydrothermal edifice on the mid-Atlantic ridge. *Limnology and Oceanography*, **56**, 1624–40.

Dahl, M. P., Pereyra, R. T., Lundälv, T. and André, C. (2012). Fine-scale spatial genetic structure and clonal distribution of the cold-water coral *Lophelia pertusa*. *Coral Reefs*, **31**, 1135–48.

Debenham, N. J., Lambshead, P. J. D., Ferrero, T. J. and Smith, C. R. (2004). The impact of whale falls on nematode abundance in the deep sea. *Deep–Sea Research Part I–Oceanographic Research Papers*, **51**, 701–6.

de Buron, I. and Morand, S. (2004). Deep-sea hydrothermal vent parasites: why do we not find more? *Parasitology*, **128**, 1–6.

Domack, E., Ishman, S., Leventer, A., Sylva, S., Willmott, V. and Huber, B. (2005). A chemotrophic ecosystem found beneath Antarctic Ice Shelf. *EOS Transactions of the American Geophysical Union*, **86**, 269–76.

Ebbe, B., Billett, D., Brandt, A. et al. (2010). Diversity of Abyssal Marine Life. In *Life in the World's Oceans: Diversity, Distribution and Abudnance*. Blackwell Publishing, Oxford, pp. 139–60.

Eckelbarger, K. J., Watling, L. and Fournier, H. (2005). Reproductive biology of the deep–sea polychaete *Gorgoniapolynoe caeciliae* (Polynoidae), a commensal species associated with octocorals. *Journal of the Marine Biological Association of the UK*, **85**, 1425–33.

Elliott, J., Patterson, M., Summers, N., Miternique, C., Montocchio, E. and Vitry, E. (2016). How does the proliferation of the coral–killing sponge *Terpios hoshinota* affect benthic community structure on coral reefs? *Coral Reefs* **35**, 1083–95.

Etter, R. J. and Grassle, J. F. (1992). Patterns of species diversity in the deep sea as a function of sediment particle size diversity. *Nature*, **360**, 576–8.

Fernandez-Arcaya, U., Ramirez–Llodra, E., Aguzzi, J. et al. (2017). Ecological role of submarine canyons and need for canyon conservation: a review. *Frontiers in Marine Science*, **4**, 5.

Fujii, T., Kilgallen, N. M., Rowden, A. A. and Jamieson, A. J. (2013). Deep-sea amphipod community structure across abyssal to hadal depths in the Peru–Chile and Kermadec trenches. *Marine Ecology Progress Series*, **492**, 125–38.

Genin, A. and Dower, J. F. (2007). Seamount Plankton Dynamics. In *Seamounts: Ecology, Fisheries, and Conservation*, Blackwell Fisheries and Aquatic Resources Series, vol. 12. Blackwell Publishing, Oxford, pp. 86–100.

Girard, F., Fu, B. and Fisher, C. R. (2016). Mutualistic symbiosis with ophiuroids limited the impact of the Deepwater Horizon oil spill on deep–sea octocorals. *Marine Ecology Progress Series*, **549**, 89–98.

Glynn, P. W. (1980). Defense by symbiotic Crustacea of host corals elicited by chemical cues from predator. *Oecologia*, **47**, 287–90.

Govenar, B., Fisher, C. R. and Shank, T. M. (2015). Variation in the diets of hydrothermal vent gastropods. *Deep-Sea Research Part II–Topical Studies in Oceanography*, **121**, 193–201.

Grace, J. B., Anderson, T. M., Seabloom, E. W. et al. (2016). Integrative modelling reveals mechanisms linking productivity and plant species richness. *Nature*, **529**, 390–93.

Herrera, S., Shank, T. M. and Sanchez, J. (2012). Spatial and temporal patterns of genetic variation in the widespread antitropical deep-sea coral *Paragorgia arborea*. *Molecular Ecology*, **21**, 6053–67.

Hessler, R. R. and Sanders, H. (1967). Faunal diversity in the deep-sea. *Deep-Sea Research*, **14**, 65–78.

Hutchinson, G. E. (1961). The paradox of the plankton. *The American Naturalist*, **95**, 137–45.

Iken, K., Brey, T., Wand, U., Voigt, J. and Junghans, P. (2001). Food web structure of the benthic community at the Porcupine Abyssal Plain (NE Atlantic): a stable isotope analysis. *Progress in Oceanography*, **50**, 383–405.

Jang, S–K., Park, E., Lee, W.–K., Johnson, S. B., Vrijenhoek, R. C. and Won, Y.–J. (2016). Population subdivision of hydrothermal vent polychaete *Alvinella pompejana* across equatorial and Easter Microplate boundaries. *BMC Evolutionary Biology*, **16**, 235.

Jennings, R. M., Etter, R. J. and Ficarra, L. (2013). Population differentiation and species formation in the deep sea: the potential role of environmental gradients and depth. *PLoS ONE*, **8**, e77594.

Johnson, M. P., White, M., Wilson, A. et al. (2013a). A vertical wall dominated by *Acesta excavata* and *Neopycnodonte zibrowii*, part of an undersampled group of deep–sea habitats. *PLoS ONE*, **8**, e79917.

Johnson, S. B., Won, Y.–J., Harvey, J. B. and Vrijenhoek, R. C. (2013b). A hybrid zone between *Bathymodiolus* mussel lineages from eastern Pacific hydrothermal vents. *BMC Evolutionary Biology*, **13**, 21.

Kemp, K. M., Jamieson, A. J., Bagley, P. M. et al. (2006). Consumption of large bathyal food fall, a six month study in the NE Atlantic. *Marine Ecology Progress Series*, **310**, 65–76.

Kim, S. and Hammerstrom, K. (2012). Hydrothermal vent community zonation along environmental gradients at the Lau back–arc spreading center. *Deep–Sea Research Part I–Oceanographic Research Papers*, **62**, 10–19.

Klaucke, I., Shaling, H., Weinrebe, W. et al. (2006). Acoustic investigation of cold seeps offshore Georgia, eastern Black Sea. *Marine Geology*, **231**, 51–67.

Kutti, T., Bergstad, O. A., Fossa, J. H. and Helle, K. (2014). Cold–water coral mounds and sponge–beds as habitats for demersal fish on the Norwegian shelf. *Deep–Sea Research Part II–Topical Studies in Oceanography*, **99**, 122–33.

Lacey, N. C., Rowden, A. A., Clark, M. R. et al. (2016). Community structure and diversity of scavenging amphipods from bathyal to hadal depths in three South Pacific Trenches. *Deep-Sea Research Part I – Oceanographic Research Papers*, **111**, 121–37.

Le Guilloux, E., Hall–Spencer, J. M., Söffker, M. K. and Olu, K. (2010). Association between the squat lobster *Gastroptychus formosus* and the cold–water corals in the North Atlantic. *Journal of the Marine Biological Association of the UK*, **96**, 1363–9.

Lenihan, H. S., Mills, S. W., Mullineaux, L. S., Peterson, C. H., Fisher, C. R. and Micheli, F. (2008). Biotic interactions at hydrothermal vents: recruitment inhibition by the mussel *Bathymodiolus thermophiles*. *Deep-Sea Research Part I – Oceanographic Research Papers*, **55**, 1707–17.

Levesque, C., Juniper, S. K. and Marcus, J. (2003). Food resource partitioning and competition among alvinellid polychaetes of Juan de Fuca Ridge hydrothermal vents. *Marine Ecology Progress Series*, **246**, 173–82.

Levin, L. A. and Dayton, P. K. (2009). Ecological theory and continental margins: where shallow meets deep. *Trends in Ecology and Evolution*, **24**, 606–17.

Levin, L. A., Etter, R. J., Rex, M. A. et al. (2001). Environmental influences on regional deep-sea species diversity. *Annual Review of Ecology and Systematics*, **32**, 51–93.

Linley, T. D., Gerringer, M. E., Yancey, P. H., Drazen, J. C., Weinstock, C. L. and Jamieson, A. J. (2016). Fishes of the hadal zone including new species, in situ observations and depth records of Liparidae. *Deep-Sea Research Part I – Oceanographic Research Papers*, **114**, 99–110.

Linley, T. D., Stewart, A. L., McMillan, P. J. et al. (2017). Bait attending fishes of the abyssal zone and hadal boundary: community structure, functional groups and species distribution in the Kermadec, New Hebrides and Mariana trenches. *Deep-Sea Research Part I – Oceanographic Research Papers*, **121**, 38–53.

Lutz, M. J., Caldeira, K., Dunbar, R. B. and Behrenfeld, M. J. (2007). Seasonal rhythms of net primary production and particulate organic carbon flux to depth describe the efficiency of biological pump in the global ocean. *Journal of Geophysical Research*, **112**, C10011.

Marsh, L., Copley, J. T., Huvenne, V. A. I. et al. (2012). Microdistribution of faunal assemblages at deep–Sea hydrothermal vents in the Southern Ocean. *PLoS ONE*, **7**, e48348.

Mateos, M., Hurtado, L. A., Santamaria, C. A., Leignel, V. and Guinot, D. (2012). Molecular systematics of the deep–sea hydrothermal vent endemic Brachyuran family Bythograeidae: a comparison of three Bayesian species tree methods. *PLoS ONE*, **7**, e32066.

McClain, C. R. and Rex, M. A. (2015). Toward a conceptual understanding of biodiversity in the deep–sea benthos. *Annual Review of Ecology, Evolution, and Systematics*, **46**, 623–42.

McClain, C. R. and Schlacher, T. A. (2015). On some hypotheses of diversity of animal life at great depths on the sea floor *Marine Ecology*, **36**, 849–72.

Micheli, F., Peterson, C. H., Mullineaux, L. S. et al. (2002). Predation structures communities at deep–sea hydrothermal vents. *Ecological Monographs*, **72**, 365–82.

Miller, K., Williams, A., Rowden, A. A., Knowles, C. and Dunshea, G. (2010). Conflicting estimates of connectivity among deep-sea coral populations. *Marine Ecology*, **31**(Suppl. 1), 144–57.

Molodstova, T. N. and Budaeva, N. (2007). Modifications of the corallum morphology in black corals as an effect of associated fauna. *Bulletin of Marine Science*, **81**, 469–79.

Moore, K. M., Alderslade, P. and Miller, K. J. (2016). A taxonomic revision of the genus *Primnoisis* Studer [&Wright], 1887 (Coelenterata: Octocorallia: Isididae) using morphological and molecular data. *Zootaxa*, **4075**, 1–141.

Moore, K. M., Alderslade, P. and Miller, K. J. (2017). A taxonomic revision of *Anthothela* (Octocorallia: Scleraxonia: Anthothelidae) and related genera, with the addition of new taxa, using morphological and molecular data. *Zootaxa*, **4304**, 1–212.

Morisson, C. L., Baco, A. R., Nizinski, M. S. et al. (2017). Population Connectivity of Deep-Sea Corals. In *The State of Deep-Sea Coral and Sponge Ecosystems of the United States*, NOAA Technical Memorandum. NOOA, Silver Spring, MD, pp. 379–407.

Morrison, C. L., Ross, S. W., Nizinski, M. S. et al. (2011). Genetic discontinuity among regional populations of *Lophelia pertusa* in the North Atlantic Ocean. *Conservation Genetics*, **12**, 713–29.

Mosher, C. V. and Watling, L. (2009). Partners for life: a brittle star and its octocoral host. *Marine Ecology Progress Series*, **397**, 81–8.

Olu, K., Cordes, E. E., Fisher, C. R., Brooks, J. M., Sibuet, M. and Desbruyères, D. (2010). Biogeography and potential exchanges among the Atlantic equatorial belt cold–seep faunas. *PLoS ONE*, **5**, e11967.

Oppelt, A., Matthias Lopez, C. and Rocha, C. (2017). Biogeochemical analysis of the calcification patterns of cold-water corals *Madrepora oculata* and *Lophelia pertusa* along contact surfaces with calcified tubes of the symbiotic polychaete *Eunice norvegica*: evaluation of a 'mucus' calcification hypothesis. *Deep-Sea Research Part I – Oceanographic Research Papers*, **127**, 90–104.

Pettibone, M. H. (1986). A new scale–worm commensal with deep–sea mussels in the seep–sites at the Florida escarpment in the eastern Gulf of Mexico (Polychaeta, Polynoidae, Branchipolynoinae). *Proceedings of the Biological Society of Washington*, **99**, 444–51.

Pham, C. K., Vandeperre, F., Menezes, G., Porteiro, F., Isidro, E. and Morato, T. (2015). The importance of deep–sea vulnerable marine ecosystems for demersal fish in the Azores. *Deep-Sea Research Part I – Oceanographic Research Papers*, **96**, 80–8.

Quattrini, A. M., Baums, I. B., Shank, T. M., Morrison, C. L. and Cordes, E. E. (2015a). Testing the depth-differentiation hypothesis in a deepwater octocoral. *Proceedings of the Royal Society B*, **282**, 20150008.

Quattrini, A. M., Nizinski, M. S., Chaytor, J. D. et al. (2015b). Exploration of the canyon–incised continental margin of the northeastern United States reveals dynamic habitats and diverse communities. *PLoS ONE*, **10**, e0139904.

Ramirez-Llodra, E., Shank, T. M. and German, C. R. (2007). Biodiversity and Biogeography of hydrothermal vent species. *Oceanography*, **20**, 30–41.

Rex, M. A. (1981). Community structure in the deep–sea benthos. *Annual Review of Ecology, Evolution, and Systematics*, **12**, 331–53.

Rex, M. A. and Etter, R. J. (2010). *Deep Sea Biodiversity: Pattern and Scale*. Harvard University Press, Cambridge, MA.

Rowden, A. A., Dower, J. F., Schlacher, T. A., Consalvey, M. and Clark, M. R. (2010). Paradigms in seamount ecology: fact, fiction, and future. *Marine Ecology*, **31** (Suppl. 1), 226–39.

Sanders, H. L. (1968). Marine benthic diversity: a comparative study. *The American Naturalist*, **102**, 243–82.

Sanders, H. L. and Hessler, R. R. (1969). Ecology of the deep-sea benthos. *Science*, **163**, 1419–24.

Saucier, E. H., Sajjadi, A. and France, S. C. (2017). A taxonomic review of the genus *Acanella* (Cnidaria: Octocorallia: Isididae) in the North Atlantic Ocean, with descriptions of two new species. *Zootaxa*, **4323**, 359–90.

Schüller, M. (2011). Evidence for a role of bathymetry and emergence in speciation in the genus *Glycera* (Glyceridae, Polychaeta) from the deep Eastern Weddell Sea. *Polar Biology*, **34**, 549–64.

Shank, T. M. and Halanych, K. M. (2007). Toward a mechanistic understanding of larval dispersal: insights from genomic fingerprinting of the deep–sea hydrothermal vent tubeworm *Riftia pachyptila*. *Marine Ecology*, **28**, 25–35.

Smith, C. R., Adrian, G. G., Treude, T., Higgs, N. D. and Amon, D. J. (2015). Whale–fall Ecosystems: recent insights into ecology, paleoecology, and evolution. *Annual Reviews in Marine Science*, **7**, 571–96.

Soler-Hurtado, M. M., Lopez–Gonzalez, P. J. and Machordom, A. (2017). Molecular phylogenetic relationships reveal contrasting evolutionary patterns in Gorgoniidae (Octocorallia) in the Eastern Pacific. *Molecular Phylogenetics and Evolution*, **111**, 219–30.

Stocks, K. I. and Hart, P. J. B. (2007). Biogeography and Biodiversity of Seamounts. In *Seamounts: Ecology, Fisheries, and Conservation*, Blackwell Fisheries and Aquatic Resources Series, vol. 12. Blackwell Publishing, Oxford, pp. 255–81.

Strugnell, J. M., Allcock, A. L. and Watts, P. C. (2017). Closely related octopus species show different spatial genetic structures in response to the Antarctic seascape. *Ecology and Evolution*, **7**, 8087–99.

Stuart, C. T., and Rex, M. A. (2009). Bathymetric patterns of deep–sea gastropod species diversity in 10 basins of the Atlantic Ocean and Norwegian Sea. *Marine Ecology*, **30**, 164–80.

Sumida, P. Y. G., Tyler, P. A., Lampitt, R. S. and Gage, J. D. (2000). Reproduction, dispersal and settlement of the bathyal ophiuroid *Ophiocten gracilis* in the NE Atlantic Ocean. *Marine Biology*, **137**, 623–30.

Taylor, J., Krumpen, T., Soltwedel, T., Gutt, J. and Bergmann, M. (2017). Dynamic benthic megafaunal communities: assessing temporal variations in structure, composition and diversity at the Arctic deep-sea observatory HAUSGARTEN between 2004 and 2015. *Deep-Sea Research Part I – Oceanographic Research Papers*, **122**, 81–94.

Taylor, M. L. and Roterman, C. N. (2017). Invertebrate population genetics across Earth's largest habitat: the deep-sea floor. *Molecular Ecology*, **26**, 4872–96.

Thiem, Ø., Ravagnan, E., Fosså, J. H. and Berntsen, J. (2006). Food supply mechanisms for cold–water corals along a continental shelf edge. *Journal of Marine Systems*, **60**, 207–19.

Tyler, P. A., German, C. R., Ramirez–Llodra, E. and Van Dover, C. (2003). Understanding the biogeography of chemosynthetic ecosystems. *Oceanologica Acta*, **25**, 227–41.

Tyler, P. A. and Young, C. M. Y. (2003). Dispersal at hydrothermal vents: a summary of recent progress. *Hydrobiologia*, **503**, 9–19.

Van Dover, C. L., German, C. R., Speer, K. G., Parson, L. M. and Vrijenhoek, R. C. (2002). Evolution and biogeography of deep-sea vent and seep invertebrates. *Science*, **295**, 1253–7.

Wagner, D., Luck, D. G. and Toonen, R. J. (2012). The biology and ecology of black corals (Cnidaria: Anthozoa: Hexacorallia: Antipatharia). *Advances in Marine Biology*, **63**, 67–132.

Watling, L. (2009). Biogeographic provinces in the Atlantic deep sea determined from cumacean distribution patterns. *Deep-Sea Research Part II – Topical Studies in Oceanography*, **56**, 1747–53.

Watling, L., France, S. C., Pante, E. and Simpson, A. (2011). Biology of deep–water octocorals. *Advances in Marine Biology*, **60**, 41–122.

Watling, L., Guinotte, J., Clark, M. R. and Smith, C. R. (2013). A proposed biogeography of the deep ocean floor. *Progress in Oceanography*, **111**, 91–112.

White, M., Bashmachnikov, I., Arístegui, J. and Martins, A. (2007). Physical Processes and Seamount Productivity. In *Seamounts: Ecology, Fisheries, and Conservation*, Blackwell Fisheries and Aquatic Resources Series, vol. 12. Blackwell Publishing, Oxford, pp. 65–84.

Wienberg, C., Beuck, L., Heidkamp, S. et al. (2008). Franken Mound: Facies and biocoenoses on a newly-discovered "carbonate mound" on the western Rockall Bank, NE Atlantic. *Facies*, **54**, 1–24.

Wilson, A. M., Raine, R., Mohn, C. and White, M. (2015). Nepheloid layer distribution in the

Whittard Canyon, NE Atlantic Margin. *Marine Geology*, **367**, 130–42.

Wirtz, P. and d'Udekem d'Acoz, C. (2001). Decapoda from Antipatharia, Gorgonia and Bivalvia at the Cape Verde Islands. *Helgoland Marine Research*, **55**, 112–5.

Witman, J. D., Genovese, S. J., Bruno, J. F., McLaughlin, J. W. and Pavlin, B. I. (2003). Massive prey recruitment and the control of rocky subtidal communities on large spatial scales. *Ecological Monographs*, **73**, 441–62.

Wolff, T. (1970). The concept of the hadal or ultra–abyssal fauna. *Deep-Sea Research*, **17**, 983–1003.

Yeh, J. and Drazen, J. C. (2009). Depth zonation and bathymetric trends of deep-sea megafaunal scavengers of the Hawaiian Islands. *Deep-Sea Research I – Oceanographic Research Papers*, **56**, 251–66.

Zhang, H., Johnson, S. B., Flores, V. R. and Vrijenhoek, R. C. (2015). Intergradation between discrete lineages of *Tevnia jerichonana*, a deep-sea hydrothermal vent tubeworm. *Deep-Sea Research Part II – Topical Studies in Oceanography*, **121**, 53–61.

# Chapter 20

# Overview and Synthesis

Stephen J. Hawkins, Katrin Bohn, Anaëlle J. Lemasson, Gray A. Williams, David R. Schiel, Stuart R. Jenkins and Louise B. Firth

## 20.1 Introduction

This volume has achieved a large coverage of the experimentally well-studied areas of the temperate and subtropical coasts of the world (see Figure 1.1) – venturing into the tropics in some regions (Chapter 14, South-East Asia) and including mangroves (Chapter 17). Coral reef systems have not been considered. Much of the emphasis has been on rocky habitats as this is where the majority of experimental work on interactions has been done (but see Chapter 6). As well as reviewing regions where there has been a long history of experimental research (e.g., Chapters 2–4, 6, 10, 11, 13, 15, 16), areas of emerging experimental research in the last twenty-five years (e.g., Chapter 8, western Mediterranean; Chapter 12, south-east Pacific) and understudied regions (e.g., Chapter 7, Argentina; Chapter 14, South-East Asia) have also been included, allowing more comprehensive insights into the processes important for shaping these communities. In this short synthesis chapter, we first consider the main processes determining patterns covered by the previous chapters. We then consider major human impacts in these regions. Finally, we identify gaps in knowledge and make some suggestions for the way forward. We make the case for combining phylogeographic studies with macro-ecology and biogeography, coupled with well-designed hypothesis testing experiments, to better understand processes generating patterns on micro-evolutionary (hundreds to thousands of years) and ecological (up to hundreds of years) time scales.

## 20.2 Relative Importance of Different Processes

Underwood and Fairweather (1986) asked the thought provoking question 'Are there different ecologies or different ecologists?' This question was a prompt for the 1990 Liverpool conference (see John et al., 1992) and is still valid today. At the follow-up conference in 2015, and among the lead authors of the chapters in this volume, there was surprising consensus on the major processes shaping patterns in their respective systems (see Table 20.1, derived from a questionnaire of the lead authors of the chapters in this volume). Over the last three decades, broader-scale studies have emphasised the importance of regional-scale oceanographic processes interacting with biogeographic differences in assemblage composition – and hence the possible range of species interactions important in determining patterns of community structure. At smaller scales, these in turn interact with the sharp local environmental gradients of bathymetry and tidal elevation, exposure to wave action, sediment particle size and salinity (see Raffaelli and Hawkins, 1996).

**Table 20.1** Dominant processes shaping community structure in the various study systems covered in this volume, as identified by the leading chapter authors. The system's characteristics that are shaped by these processes is listed in brackets. LW, low water

| Region [chapter] | Dominant processes (impacted characteristic of the system) | Other |
|---|---|---|
| North-east Atlantic – rocky intertidal [2] | Cold-water refuge (distribution); upwelling and coastline topography (distribution, recruitment and productivity); grazing by major grazers – limpets above LW and urchins below LW | Tidal elevation, wave exposure |
| North-east Atlantic – rocky subtidal [3] | Regional oceanography (distribution and recruitment); temperature (distribution); kelp forest (productivity); coastal topography (distribution and assemblages); grazing, important but regional variations – sea urchins in the infralittoral; starfish (predation on mussels) | Fjords and caves, wave action, tidal currents |
| North-west Atlantic – rocky intertidal [4] | Coastal topography (distribution); regional oceanography (distribution and recruitment); temperature (recruitment); nutrients and mixing (productivity); grazing, important but variable – gastropods above LW | Tidal elevation, wave exposure |
| North-west Atlantic – rocky subtidal [5] | Hurricanes and regional oceanography (distribution); unknown (recruitment); upwelling (productivity); ice-cover (seasonality and algal cover above LW); grazing by major grazers, but variable – urchins below LW, starfish (predation on blue mussels) | Overfishing |
| North-west Atlantic – soft sediment [6] | Coastal topography (distribution); regional oceanography and temperature (distribution and recruitment); megalopolis and eutrophication (productivity); disturbances (predation and refugia) | |
| Argentina – all habitats [7] | Mesoscale processes (recruitment); competition for space (distribution and recruitment) | Tidal elevation, wave exposure |
| Western Mediterranean – all habitats [8] | Cold-water refuge and regional oceanography (distribution and connectivity); coastal topography and nutrients (productivity); grazing by major grazers – mosaics created by limpets and urchins | Competition for space, extreme events – heatwaves and storms |
| Eastern Mediterranean – all habitats [9] | Coastal topography, regional oceanography and winds (distribution and recruitment); seasonality (productivity – generally low); grazing, important – limpets in midshore and rabbitfish below vermetid platform | Extreme events – heatwaves and storms, invasive species, climate change |
| North-east Pacific – rocky intertidal [10] | Upwelling (distribution, recruitment and productivity); grazing (major – limpets in midshore); starfish (predation); competition for space (distribution and recruitment) | Predation (including human), tidal elevation, wave exposure |

(cont)

**Table 20.1** (continued)

| Region [chapter] | Dominant processes (impacted characteristic of the system) | Other |
|---|---|---|
| North-east Pacific – rocky subtidal [11] | Upwelling and regional oceanography (distribution and recruitment); El Niño/La Niña and decadal oscillations (recruitment and productivity); grazing by major grazers, especially subtidally; starfish (predation) | Top-down control (urchins) |
| South-east Pacific – all habitats [12] | Upwelling and coastal topography (distribution); upwelling (recruitment); latitude (productivity); grazing by major grazers but modulated by upwelling – urchins below LW; starfish (predation – keystone) | Top-down control (urchins) |
| South Africa – all habitats [13] | Regional oceanography and wave action (distribution); coastal topography, west–east gradient, regional oceanography and small-scale hydrodynamics (recruitment); upwelling and nutrients (productivity); location dependent; grazing, important – limpets at high- and mid-shores and urchins subtidally | Competition for space, heat and desiccation (highshore) |
| South-East Asia – rocky intertidal [14] | Biogeographic breaks and seasonality in current patterns (distribution); rivers (recruitment); productivity and grazing poorly understood; grazing (important lowshore in the hot season) | Monsoon system (hot summers/cool winters, south and mid-China), typhoons |
| Australasia – rocky intertidal [15] | Largely unknown/upwelling (distribution); upwelling/ downwelling and coastal topography (recruitment); upwelling and rivers (productivity); regional oceanography (distribution, recruitment and productivity); grazing, variable importance – limpets in high- and mid-shore – urchins below LW, starfish (predation – patchy on mussels, but not common) | Disturbances (storms, human trampling, earthquakes) |
| Australasia – rocky subtidal [16] | Regional oceanography (distribution, recruitment and productivity); nutrients (productivity); grazing (major below 5 m – latitudinal gradient with herbivores tropicalisation | Mesoscale processes, physical framework |
| Mangroves [17] | Latitude, east–west gradient and upwelling (distribution); upwelling (recruitment); tides, trees and freshwater inputs (productivity); grazing (ubiquitous but not intense – snails and crabs) | Crabs (key fauna) |
| Deep sea [19] | Mesoscale processes, topography and hydrodynamism (distribution and productivity); recruitment poorly understood | Habitat variation (canyons, seamounts, abyss) |

### 20.2.1 Physical Setting

Mesoscale processes, operating through the interaction of coastal hydrography and geomorphology, are important in shaping temperature, nutrient and recruitment regimes. This is perhaps best epitomised in regions that are strongly influenced by upwelling: the Pacific coasts of North America (Chapters 10 and 11, north-east Pacific) and South America (Chapter 12, south-east Pacific), the Atlantic coasts of the Iberian Peninsula (Chapter 2, north-east Atlantic) and southern Africa (Chapter 13). In these regions, deep water abuts the coast and the combination of winds and Coriolis effects drive the upwelling of cold, nutrient-rich water. Equally important is when upwelling is relaxed as winds change, on both short time scales and during El Niño Southern Oscillation episodes that can last a year or more (Okumura and Deser, 2010).

Upwelling does not occur in regions with an extensive shallow coastal shelf. In those regions, the topography of the coastline becomes more important, with larvae and nutrient concentrations often being higher in the water column of embayed areas or gulfs (Burrows et al., 2010) than around isolated islands (Johannesson, 1988), headlands or towards the tips of large peninsulas (Crisp and Southward, 1958). Inputs from estuaries can be important, both in terms of nutrients and, when large, as a potential biogeographic barrier due to low salinity plumes (Chapter 14, South-East Asia). Headlands are also known to act as important hydrographic barriers that can influence the geographic distributional limits of species (e.g., in Chapter 11, north-east Pacific: Gaylord and Gaines, 2000; Chapter 2, north-east Atlantic: Crisp and Southward, 1958; Keith et al., 2011). These mesoscale features can modulate the relative importance of bottom-up forcing by physico-chemical factors and top-down control by consumers. For example, dense recruitment resulting from these mesoscale features can intensify both intra- and inter-specific competition (e.g., around Anglesey, Jenkins et al., 2005; the Firth of Clyde, Connell, 1961; off the coast of Plymouth, Gordon and Knights, 2018). In subtropical and warm-temperate upwelling areas, tide-out conditions can often be harsh, especially at higher levels on the shore. Thus, there is a potentially sharp gradient with subtidal and low-shore areas bathed in colder water and mid- and upper-shore areas stressed by high temperatures and low relative humidity.

Differences in recruitment regimes are generated by hydrography (upwelling, coastal currents, wind-driven processes) interacting with coastal morphology. Interactions are more intense in high-recruitment areas due to greater saturation of space, and hence greater intra- and inter-specific interactions (see Chapter 10, north-east Pacific; Chapter 11, north-west Pacific; Chapter 12, south-east Pacific; Chapter 13, South Africa). In lower-recruitment areas, space can be undersaturated, leading to much less competition. Variable recruitment can often drive fluctuations in community structure and ecosystem functioning. Some systems seem particularly prone to fluctuations, driven by a combination of external extreme events, such as storms, and/or major recruitment episodes of important space occupiers or consumers (e.g., Chapter 15, Australasia; Underwood, 1998; moderately exposed shores in north-west Europe, Chapter 2, north-east Atlantic). High recruitment is often linked to high productivity (i.e., in upwelling areas), leading to bottom-up forcing in highly productive communities of primary-producing plants and secondary-producing filter feeders and grazers using biofilms, living macroalgae and/or detritus. Fast growth and high recruitment can intensify competitive interactions.

Complex coastlines can also lead to sharp local gradients in wave action around islands, into bays and up rias and fjords. These are largely absent in areas with linear coasts: much of central and northern Chile (Chapter 7, Argentina), southern Africa (Chapter 13) and the Atlantic coast of Portugal (Chapter 2, north-east Atlantic) are exposed to oceanic swell regimes with very little local shelter, except within estuaries. These gradients on complex coastlines will influence the local pool of species and the balance of assemblages over relatively small spatial scales.

The physical setting of a shore is, therefore, an ultimate factor in determining the potential species pool in a locality and their distributions. The exact physico-chemical environment is due to a variable set of influences: broader-scale climatic differences often associated with latitude, mesoscale processes such as upwelling or coastal topography influencing recruitment/productivity, coupled within the well-established local template of vertical depth/tidal elevation and horizontal wave action gradients. Physical context affects the likely sign of biological interactions, their intensity and hence probability of outcomes of biotic interactions that have a direct proximate effect. At a microscale, spatial heterogeneity can influence survival and the outcome of interactions by providing refuges from both physical factors and biological interactions, in some cases leading to lower-density haloes of prey around refuges occupied by consumers.

### 20.2.2 Biological Interactions

In some regions, particular taxa that have a strong role in community structuring are present due to the outcomes of past biogeographic processes. A classic example is large predatory seastars (starfish) in the temperate rocky intertidal zone of the Pacific (North and South American coasts, Chapter 5, north-west Atlantic; Chapter 7, Argentina; Chapter 10, north-east Pacific) and in limited parts of New Zealand (mainly west coast, Chapter 15, Australasia), which in some places have a true keystone role (Paine, 1969). Seastars (starfish) are, however, much less common on most Atlantic coasts, with those that are present smaller and restricted primarily to the subtidal zone. In the north-east Atlantic, patellid limpets have an important role in controlling algae in the midshore region. In contrast, these limpets are absent from the north-west Atlantic, having not recolonised from the eastern Atlantic after the last ice age, unlike other taxa (Jenkins et al., 2008). In the north-west Atlantic (Chapter 4) the most important grazer is *Littorina littorea*, probably an introduced species (Chapman et al., 2007). Therefore, the types and intensity of interactions depend on the presence of the species and their functional roles.

The views and philosophies of ecologists are often shaped by the model systems on which they cut their ecological teeth, rather than by overarching theories or laws. Fifty to sixty years ago, the power of the small-scale experimental approach led to much interest in the role of biological interactions, replacing the more classical descriptive approach that largely correlated patterns to physical factors (Stephenson and Stephenson, 1949, 1972; Lewis, 1964). In more recent years, physical contexts have been more appreciated at larger scales through the emergence of macro-ecological approaches aided by spatially explicit databases and better appreciation of coastal oceanography, often visualised by remote sensing and interpreted through modelling. At local scales, major advances in instrumentation and data logging have enabled in situ measurement of wave forces (Bell and Denny, 1994; Figurski et al., 2011) and temperatures (e.g., Fitzhenry et al., 2004; Lima and Wethey, 2009; Denny et al., 2011; Seabra et al., 2011; Zhang et al., 2014; Chan et al., 2016; Helmuth et al., 2016; Lima et al., 2016) actually experienced by organisms, especially the extremes. The physical framework often determines the probability of the outcomes of an interaction. In the northern north-east Atlantic, canopy-forming fucoid seaweeds always swamp grazing limpets on sheltered shores; conversely, in exposed areas grazers virtually always prevent the establishment of seaweeds (Chapter 2, north-east Atlantic). On moderately exposed shores, however, sometimes fucoids win and sometimes limpets and barnacles win, leading to fluctuating mosaics of patches. Similarly, the probability of fucoid seaweeds surviving on warmer shores in southern Europe is very much less than further north because of a combination of physical stress and grazing pressure (Ferreira et al., 2015).

Before 1990, forty years of experimental ecology on biological interactions emphasised negative interactions such as competition, grazing, predation and biologically mediated disturbance. Since 1990, a greater appreciation of the role of positive, facilitative interactions has grown (Bruno et al., 2003), especially at the harsher end of environmental gradients (Bertness and Callaway, 1994), whether in saltmarshes (Bertness and Shumway, 1993) or rocky shores (Bertness, 1989). The importance of habitat provision and amelioration by ecosystem engineers (Jones et al., 1994) has been recognised. This has, perhaps, culminated in the concept of facilitation cascades (Altieri et al., 2010; Bishop et al., 2013;

Thomsen et al., 2016; Gribben et al., 2017). Interestingly, some of the first experimental work on rocky shores by Hatton (1938) had appreciated the importance of positive interactions (Hawkins et al., 2016). Mangroves are an area where positive interactions are the major structuring force above and below high water (Chapter 17): reducing water flow and wave action, thereby increasing sedimentation as well as providing a variety of habitats and refuges for a variety of species.

In soft-sediment systems (see Chapter 6, north-west Atlantic soft sediments), widespread use of mesocosms has enabled interactions between species to be explored, with physical factors such as temperature and salinity being manipulated. These approaches have been particularly powerful in understanding biodiversity–ecosystem functioning relationships (Emmerson et al., 2001). They have also been put to good use in understanding the interactions of global warming with ocean acidification (e.g., Kroeker et al., 2013) and other stressors (e.g., Fitch and Crowe, 2011). Mesocosms have been less used in rocky shore systems. Where used, they have yielded insights into invasion biology (Arenas et al., 2006) and biodiversity and ecosystem functioning (Griffin et al., 2008). Rock pools have also been used as field analogues of mesocosms to understand relationships between biodiversity and ecosystem functioning (O'Connor and Crowe, 2005; Griffin et al., 2010).

### 20.2.3 The Intertidal–Subtidal Vertical Gradient on Rocky Coasts

Hawkins and Hartnoll (1983a) and Hawkins et al. (1992) suggested that the classical Stephensonian three-zone system (Stephenson and Stephenson, 1949, 1972) could be mechanistically explained in areas with greater than moderate wave exposure. The analysis was also extended into the subtidal zone. The region either side of lowest astronomical tides is dominated by large algae (usually kelps or kelp-like species, such as *Durvillaea*), algal turfs and occasionally space-occupying animals (e.g., large tunicates such as *Pyura*). These are free from grazing or consumer control by dint of their ability to outpace grazers and other consumers, which are often constrained by strong wave action either side of low water. A balance occurs between the ability of algae to grow and the ability of grazers to control them, mediated by the emersion gradient. This often leads to a sharp transition between areas where the seaweeds win and those where grazed areas are dominated by barnacles and mussels (Underwood and Jernakoff, 1984, work in Australia; Boaventura et al., 2002, work in the North Atlantic). In many parts of the world, algae are controlled in the midshore by grazing, particularly by limpets in less sheltered areas (Pacific coasts of North and South America, Chapters 10 and 12, repectively), the warmer areas of Australasia (Chapter 15) and the north-east Atlantic (Chapter 2). There are strong interactions between grazing intensity and physical drivers, such as seasonal patterns of temperature (Thompson et al., 2004). In tropical, monsoon-influenced systems, for example, heat stress limits both grazer effectiveness but also survival of the most erect macroalgae in the hot season; while the cool season sees the proliferation of ephemeral algae which often grow rapidly to escape the top-down control by grazers (Chapter 14, South-East Asia).

In the subtidal zone, grazing by sea urchins often leads to reduced algal cover – so-called barrens. These have been reported in many regions (Chapter 3, north-east Atlantic; Chapter 16, Australasia; Chapter 8, western Mediterranean; and the Pacific coast of California, Schiel and Foster, 2015). In very sheltered conditions urchins can graze right up to the low watermark. There are also species such as *Paracentrotus lividus* in the Atlantic and Mediterranean (Trudgill et al., 1987; Chapter 2, north-east Atlantic) that can excavate refuges in softer rocks and persist either side of low water. Such species also can occur in low- and mid-shore pools. In wave-exposed areas, a refuge from urchin grazing can occur (e.g., north-east and north-west Atlantic subtidal, Chapters 3 and 5, respectively). Deeper in the subtidal zone, as light fades, algae give way to animals, especially on vertical surfaces and overhangs. Grazing can set the lower depth limit of some kelps, rather than light limitation (e.g., in parts of the north-east Atlantic).

In the high-intertidal zone, physical stresses reign supreme. In temperate areas, the splash zone at the top of the shore becomes covered by ephemeral algae in the winter, which then decrease in abundance in spring as desiccation stress predominates (Hawkins and Hartnoll, 1983b). In colder-temperate areas at higher latitudes, especially those with high rainfall, ephemeral algae can persist all summer. Grazers are restricted to refuges such as crevices and pits around which seasonally dependent grazing haloes can develop – with the grazers winning in the summer but being swamped by algae in winter (Skov et al., 2011). In more tropical areas, the highshore appears barren, but supports thin films of heat-resistant cyanobacteria, which are grazed by littorinid snails. These snails are present in highshore areas all around the world, with the exception of Antarctica, and have evolved effective physiological and behavioural adaptations to exploit this harsh environment (Ng et al., 2017). With the shift in monsoon systems, however, even shores in the subtropics can see proliferation of ephemeral algae during the cool season, especially on more exposed shores (Chapter 14, South-East Asia).

## 20.3 Global Change and Major Impacts at Regional and Local Scales

Anthropogenically forced change occurs globally by the interaction of impacts at a variety of scales. At a planetary scale, climate change due to greenhouse gas emissions is coupled with alterations of major biogeochemical processes as excess atmospheric $CO_2$ dissolved in the ocean leads to changes in the carbonate chemistry and reductions in pH of the ocean (referred to as 'ocean acidification'). These processes are operating on decadal and centennial timescales across the oceans of the whole planet (Table 20.2). Climate change is not just warming, but also rising sea levels (Rahmstorf, 2007) and stronger winds, with mounting evidence that the oceans are getting more stormy (e.g., Chapter 15, Australasia; Grevemeyer et al., 2000). Many scientists have highlighted the importance of variability and extremes in weather patterns (Bates et al., 2018), which are often not factored into many models and subsequent predictions of impacts to natural communities (Helmuth et al., 2014).

These global changes interact with regional- and local-scale impacts (Hawkins et al., 2017) to modify patterns of marine biodiversity and functioning of ecosystems. Global change probably also influences ocean-basin patterns over a span of a few years to decades, such as the North Atlantic oscillation, the Pacific decadal oscillation and the El Niño southern oscillation, with consequences at regional scales such as reduction in or temporary switching-off of upwellings (e.g., Folland et al., 2002). There is growing evidence that teleconnections exist between the oceans (e.g., Salinger et al., 2001), with events in the Pacific influencing the Atlantic (Müller et al., 2008). Paradoxically, in a warmer and windier world, upwelling may intensify in some systems as equatorwards winds strengthen (Bakun, 1990). This could lead to much greater tide-out stress in the mid- and high-intertidal zones, but lowshore and subtidal conditions might be ameliorated by stronger upwelling of colder waters.

In temperate areas, warming is likely to favour ectothermic grazers by increasing grazing rates; increased wave action is also likely to lead to greater dislodgement of large macroalgae (Jonsson et al., 2006; Hawkins et al., 2008, 2009). Together, these processes are likely to reduce the establishment of macroalgae and their subsequent persistence, especially in places like the north-east Atlantic where grazing is very important. Extreme temperatures and rainfall may also lead to deaths of large macroalgae in shallow waters and the intertidal in response to marine 'heatwaves' (Mills et al., 2013; Pearce and Feng, 2013; Wernberg et al., 2013). Mass mortalities of species are regularly observed in subtropical areas (Firth and Williams, 2009; Chapter 14, South-East Asia) and are becoming of more concern in other temperate regions as temperatures rise (Chapter 16, Australasia; Chapter 8, western Mediterranean; Chapter 9, Eastern Mediterranean).

Our survey of the lead authors of chapters in this volume (Table 20.2) indicated that responses to warming are being seen in most coastlines across the world. Species are exhibiting range shifts, in turn influencing assemblage composition and

**Table 20.2** The most important local- and regional-scale impacts affecting the various study systems covered in this volume, as identified and ranked by the leading chapter authors. EAC, East Australian Current; LC, Leeuwin Current; NNS, non-native species; OA, ocean acidification; SLR, sea-level rise

| Region [chapter] | Severity of impacts | Other |
|---|---|---|
| North-east Atlantic – rocky intertidal [2] | North Europe: NNS > coastal modification > recreation > seafood harvest > pollution and litter<br>South Europe: seafood harvest > NNS > tourism and recreation > pollution > coastal modification | Climate change, wave energy (potential) |
| North-east Atlantic – rocky subtidal [3] | Fishing (mobile gear) > overfishing > eutrophication = pollution > fishing (static gear) > seafood harvest | Construction, dredge spoil disposal, anchoring and mooring (localised), NNS (current and potential), climate change |
| North-west Atlantic – rocky intertidal [4] | North of Cape Cod: seafood harvest > climate change > NNS > OA > eutrophication<br>South of Cape Cod: coastal modification = pollution > seafood harvest > NNS > climate change > eutrophication | OA (current) |
| North-west Atlantic – rocky subtidal [5] | Fishing > climate change > disease > NNS > aquaculture | |
| North-west Atlantic – soft sediment [6] | Disturbance > climate change > eutrophication | OA (current) |
| Argentina – all habitats [7] | Pollution and litter > coastal modification > tourism and recreation > seafood harvest > NNS | |
| Western Mediterranean – all habitats [8] | Sediment input/anoxia > NNS > climate change > coastal modification > eutrophication | SLR (potential), OA (potential), interaction of climate change and NNS |
| Eastern Mediterranean – all habitats [9] | NNS > climate change > overfishing > pollution and litter > sedimentation | Desalination (potential), OA (potential), interaction climate change and NNS, climate change |
| North-east Pacific – rocky intertidal [10] | Disease > overfishing = seafood harvest > climate change = OA > pollution | Wave energy (potential), interaction of OA and disease and hypoxia (potential) |
| North-east Pacific – rocky subtidal [11] | El Niño = climate change = disease = overfishing = coastal modification = urbanisation | OA (potential) |
| South-east Pacific – all habitats [12] | Kelp harvest > overfishing = seafood harvest > coastal modification > urbanisation > tourism and recreation | Wave energy (potential), climate change, OA (potential) |
| South Africa – all habitats [13] | NNS (west coast) and seafood harvest (south and east coasts) > large-scale oceanographic effects on recruitment > coastal modification (west coast) > oil spills > pollution | El Niño (suspected), climate change |

*(cont)*

Table 20.2 (continued)

| Region [chapter] | Severity of impacts | Other |
|---|---|---|
| South-East Asia – rocky intertidal [14] | Human pressures > habitat loss = coastal modification > overfishing > pollution | Lack of baseline knowledge, climate change (current), OA (potential) |
| Australasia – rocky intertidal [15] | Human pressures and seafood harvest = tourism and recreation > coastal modification > NNS > sediment input > pollution = eutrophication > climate change (Australia) | Climate change (north New Zealand and Australia), wave energy/height (south New Zealand), SLR |
| Australasia – rocky subtidal [16] | EAC (at warm latitudes) and LC (at warm and cool latitudes) > extension of EAC to Tasmania > OA and ocean warming (cool water coasts of EAC and LC) | |
| Mangroves [17] | Aquaculture (south Asia)/removal of timber and wood (Africa) > agriculture (Asia) > coastal modification = urbanisation > damming and removal of freshwater > climate change | Wave energy (potential) |
| Deep sea [19] | Fishing > mining > climate change > pollution and litter | OA (potential) |

community structure with consequences for ecosystem functioning (Hawkins et al., 2008, 2009). The exceptions may include New Zealand where the complex current systems (Chiswell et al., 2015; Bowen et al., 2017) mean that latitudinal temperature gradients are not unidirectional. Mangroves, however, appear to be on the march southwards in New Zealand (Saintilan et al., 2014). Although not a range shift, there has been a recent large localised die-off of the large low-intertidal fucoid *Durvillaea* due to a combination of exceptionally warm sea temperatures, very low tides and high air temperatures (Schiel, personal observations).

Laboratory and mesocosm experimental studies have shown the likely consequences of ocean acidification at future predicted levels (Kuffner et al., 2008; Wood et al., 2008). Interactions with rising temperature have also been shown. Although likely to occur globally, observed effects to date have been seen only in the north-east Pacific (Cooley et al., 2017), with suspected impacts on larval stages in the north-west Atlantic and south-west Pacific, especially those with calcified larvae.

The spread of non-native invasive species from one biogeographical realm or ocean to another can also be seen as a global issue, often with dire ecological and economic consequences locally. Non-native species are proliferating worldwide, leading to biotic homogenisation of assemblages (McKinney and Lockwood, 1999). Opportunistic, generalist and resilient species tend to make good invaders. Not surprisingly, these tend to thrive in the more extreme conditions associated with climate change, but also in already degraded environments. Many successful invaders come from eastern seaboards (e.g., Asia, North America), perhaps reflecting an ability to deal with continentally influenced weather (both colder winters and hotter summers), as well as reflecting the direction of maritime trade. There are exceptions: *Carcinus maenas* from the north-east Atlantic seems to thrive worldwide once it gets established. In some regions, whole novel functional groups that were previously absent consist entirely of non-native species (e.g., Chapter 7, Argentina: canopy-forming and foliose fleshy algae, barnacles (Chapter 2); north-east Atlantic: reef-forming oysters on mid- and low-shore levels of rocky shores and intertidal sedimentary systems). Non-native species feature in the top five of perceived impacts in most regions globally. Lessepsian migration from the species-rich Red Sea and wider Indian Ocean into the eastern basin of the Mediterranean and beyond is perhaps the most pernicious example, with vast numbers of diverse non-native species invading via the Suez Canal (Chapter 9, eastern Mediterranean). These species, facilitated by warmer conditions, are ousting native species, changing trophic interactions and having a major impact on the ecology and functioning of the eastern Mediterranean basin. They are now advancing westwards throughout the Mediterranean Sea.

Climate change can interact with regional-scale impacts, such as overfishing in continental shelf ecosystems (Genner et al., 2010; Williams et al., 2016). Eutrophication acting at the regional scale of semi-enclosed seas (Baltic, Adriatic) can interact with climate change in various ways, including increasing the probability of harmful algal blooms (Hallegraeff, 2010). These kinds of impacts can spill inshore, sometimes killing near-shore and intertidal invertebrates (Southgate et al., 1984; Smayda, 2000). Overfishing will influence mobile predators, such as fish and crabs that feed in the shallow subtidal zone and migrate into the intertidal zone when the tide comes in. The role of such species is poorly studied (but see work by Rilov and Schiel 2006a, 2006b; Silva et al., 2008, 2010), so the impacts of overfishing them may be underappreciated.

## 20.4 Knowledge Gaps and Future Directions

There remains a bias in the amount of research and indeed understanding of the processes important for shaping shallow-water coastal systems across the globe. Although this volume has been able to partially address this bias with reviews from China (Chapter 14, South-East Asia) and Argentina (Chapter 7), it is clear that even basic taxonomic information and descriptions of patterns are scarce or not published in the international literature in some regions such as South-East Asia, Russia, India and West Africa, especially at lower latitudes. There are many historical reasons for this disparity in focus, but it is ironic that many hotspots for marine biodiversity are areas for which we have little information on assemblage composition. We know almost nothing about the key structuring processes in non-coral communities (but see Chapter 17, mangroves). As a result, most textbooks promulgate a very temperate picture of the patterns and processes that are important for shallow-water coastal systems, especially in terms of the relationship between abiotic factors and biotic interactions. The importance of monsoon systems, for example, is rarely mentioned and yet these systems dominate much of the coasts of tropical Africa and Asia.

The impacts of human influence are found everywhere, as evidenced by references in nearly all chapters in this volume. Almost all of the chapters include the impacts of climate change, which is in stark contrast to the 1992 volume, where the huge implications of increasing $CO_2$ and ocean warming were yet to be truly realised. Both of these issues are globally recognised, yet perhaps both scientists and society have been less sensitive to more direct regional and local human impacts on coastal systems. Insidious effects such as increasing nutrient, pollutant and sedimentation levels are mentioned, but probably the most direct impact is habitat loss through urbanisation and land-use intensification, with direct flows into nearshore ecosystems (e.g., Firth et al., 2016; Schiel and Howard-Williams, 2016, Williams et al., 2016). The replacement of natural systems such as mangroves, wetlands and rocky shores is prevalent throughout the world, but in some areas such as China is accelerating at a phenomenal rate (Chapter 14, South-East Asia). Habitat and subsequent biodiversity loss is the obvious impact, but the homogenisation of coastlines creates corridors for invasive species and removes the natural biogeographic barriers that have maintained different communities over centuries (Dong et al., 2016).

Overfishing is another double-edged impact on coastal systems. The direct loss of higher-order consumers can be seen on coastal economies, but the insidious effects on trophic dynamics has resulted in regime shifts in some areas (Connell, 2007; Ling et al., 2015). In many parts of the world where overfishing is intense and largely unregulated (e.g., South-East Asia), we are already studying very different ecosystems and observing nature from a greatly shifted baseline, which affects our interpretation of the important processes structuring local communities.

One aspect that has been neglected because of operational difficulties is the behaviour and influence of mobile predators such as fish, crabs and octopuses in the shallow subtidal zone and the intertidal zone when the shore is immersed (Silva et al., 2014). New technologies such as cheap underwater telemetry, digital photography and video capabilities are making their study much easier. Similarly, the role of birds as predators in the intertidal and shallow subtidal zones (e.g., diving eider ducks) has been under-investigated (but see Lindberg et al., 1987; Bosman et al., 1989; Coleman et al., 1999).

One area of research which has moved from heroic exploitation to technology-aided hypothesis testing research in the last thirty years is the deep ocean. New remotely operated vehicles and landers have enabled experiments to be conducted in deep waters (Chapter 19).

Technological advances have progressed knowledge of biofilms (Chapter 18) that are an ubiquitous feature of all marine habitats. On rocky shores, the use of spectroscopic methods to determine chlorophyll as a proxy of biomass (Murphy et al., 2005a, 2005b, 2009) has enabled greater spatial coverage and non-destructive experiments (Murphy and Underwood, 2006; Sanz-Lázaro et al., 2015). Next-generation sequencing has enabled meta-genomic studies of the diversity of microbial communities, especially the prokaryotic component. Confocal microscopy has enabled three-dimensional visualisation of living biofilms (Norton et al., 1998). Thus, the contribution of biofilms to the functioning of both hard- and soft-sediment communities has been explored, including better understanding of their role in biological interactions (Murphy and Underwood, 2006).

Reading through the 1992 volume and this book it is interesting to reflect on the similarities, differences and advances made during the intervening period. It is clear that in all regions the approach to shallow-water coastal systems goes through stages: an early discovery phase based on identification of species and patterns; a more processes-driven experimental phase, seeking to understand the small-scale (within similar shores in a region) functioning of these communities; and then adopting a larger-scale, regional approach exploring linkages to try to tackle the global issues facing these systems. There is clear evidence of movement towards developing more predictive power to contribute towards better ways to conserve and mitigate against the huge array of anthropogenic pressures coastal systems face (climate change, invasive species, loss of habitat and artificial hardening, cumulative stressors, see Table 20.2).

In this volume (and comparing with the 1992 volume) we can see that different areas are at different stages of this trajectory. Some are very well advanced and offer mechanistic and predictive understanding of how to manage these systems better, whereas others are still unravelling the interactions which drive these patterns. Perhaps most striking is that some regions are still at the very early stages of documenting species distributions and the basic ecology of patterns at individual species and assemblage level. Such work is a fundamentally important but we may be at risk of skipping over gaining basic ecological understanding and moving straight to what may be perceived as more high-impact research projects – especially in this day and age of research assessment exercises and output-driven research programmes.

The role of certain schools and research groups has also driven the development of these ideas in different regions. The dominance of the northern temperate schools in the early days of the field-based research has led to these regions contributing greatly to our knowledge of important ecological interactions and later the role of wider processes such as upwelling and ocean currents in driving coastal processes. This work has often attained paradigm-like status, dominating the ecological textbooks. The contribution of

Australia and, to a lesser extent, New Zealand to our knowledge of processes through carefully designed experimental approaches is also notable (Chapters 15 and 16), and this has been emulated in the Mediterranean (Chapter 8, western Mediterranean) and South Africa (Chapter 13), building on the tradition of classic descriptive studies in both regions. Elsewhere, there are hotspots of knowledge development, for example in Chile (Chapter 11, south-east Pacific); but again these are largely focussed around a limited number of marine laboratories or research groups. As a consequence, where such research hubs do not exist, or where funding is devoted to more immediate priorities such as aquaculture and fisheries or other economically important aspects of marine science, research is still at the early stages of development (e.g., Chapter 7, south-west Atlantic). The shallow-water coastal ecosystems of these regions remain poorly understood and little studied – but perhaps these areas may yet yield new insights (e.g., poleward increases in diversity in Argentina, Chapter 7). These areas are also the ones where anthropogenic development and impacts are accelerating, and, therefore, are perhaps the regions where application of research to inform conservation and aid prediction would have the most positive impact but are impeded by lack of baseline data (e.g., Chee et al., 2017).

## 20.5 The Way Forward

Seeking generality or appreciating idiosyncratic, context-dependent pattern and underlying process is fraught with challenges when comparing biogeographic realms. Schiel et al. (Chapter 15, Australasia) make some excellent observations on the challenges: they note that Jenkins et al. (2008) were able to make fairly robust comparisons across the Atlantic where many of the canopy- and space-occupying (foundation) species are similar, but differences from the north-east Atlantic (Chapter 2) occur due to the absence of the major grazers (patellid limpets) in the north-west Atlantic (Chapter 4) due to differential postglacial biogeographic assembly processes (Chapter 2). Patterns were superficially similar, but processes differed. Their own comparisons between south-east Australia and New Zealand were made easier by very similar clades being present in both places that are not widely separated.

As Schiel et al. (Chapter 15, Australasia) point out, synthesising from a series of local small-scale processes that address specific locally relevant questions and hypotheses can be a challenge – and can be compromised by poor experimental design. One attempt to undertake testing hypotheses across broad scales using the same experimental designs was the EUROROCK project funded by the European Commission (see Chapter 2, north-east Atlantic), spanning the British Isles and Ireland, Sweden down through Spain, Portugal and into the Mediterranean Sea. This led to an understanding of key processes, such as recruitment (e.g., Jenkins et al., 2000) and grazing (e.g., Coleman et al., 2006), over thousands of kilometres and several degrees of latitude. Similar experimental approaches have been used in North America (Chapter 4, north-west Atlantic; Chapter 10, north-east Pacific) and New Zealand (Chapters 15 and 16) by Menge and colleagues (e.g., 1999; Broitman et al., 2008). Global networks such as the Zostera Experimental Network (http://zenscience.org) and the World Harbour Project (www.worldharbourproject.org) are emerging, enabling global-scale comparisons of patterns and processes.

Ideally, a combination of approaches needs to be employed. Phylogeography using modern next-generation sequencing techniques on large groups of species can show their biogeographic origins and give insights into the direction of spread following postglacial colonisation processes, now accelerated by global climate change and interwoven with the spread of non-native species. Broadscale surveys can measure rates of change in response to climate (reviewed in Chapters 2–6 and 9–12), charting range extensions and retractions (Wethey et al., 2011), as well as fluctuations of abundances within range leading to local extinctions and fragmentation of populations. Deeper analysis of patterns can be made using macro-ecological approaches where possible (e.g., Burrows et al., 2009, 2010), making use of publically available data. The causes of such patterns can then be investigated by field experiments at a range of locations (Firth et al., 2009). Only with planned hypotheses-testing field experiments will underlying processes be determined. These insights can then be used to build models enabling simulation and forecast (Burrows and Hawkins, 1998; Johnson et al., 1998) of both local- and broader-scale patterns, including responses to climate change (Poloczanska et al., 2008). A promising new approach enabling scaling-up from individual to populations is dynamic energy budget modelling (Sarà et al., 2011).

Funding mechanisms have precluded experimental comparisons across transoceanic, biogeographic realm comparisons to date. These are possible, but are best done by manipulating analogous functional groups (e.g., canopy- or turf-forming algae, grazing gastropods in the intertidal zone or sea urchins in the subtidal zone, guilds of predators). Ideally, a new generation of scientists will seek collaborative funding to make such comparisons of processes across the globe. Meta-analyses (Poore et al., 2012) make important contributions and are a good start; but understanding can only truly occur by coordinated empirical work delineating spatial and temporal patterns and processes using similar methods, and then testing hypotheses derived from these observations using in situ field experimentations – hopefully integrated by modelling. Our bets are that there will be some generalities, but context-specific differences in assemblage composition – and hence identity of key species – combined with the physical setting will be of key importance and lead to a much greater understanding of differences and similarities between regions and localities.

We fervently hope that major scientific journals continue to recognise and value high-quality empirical research contributions from around the world, especially hypothesis-testing field experiments. It is also beholden on marine scientists to make every effort to find and scrutinise the burgeoning relevant literature, and not just the parts that fit their paradigm. Context dependency can be viewed as 'just-so' stories from other parts of the world, or as underpinning valuable knowledge about our increasingly connected marine ecosystems.

It has been more than twenty-five years between these two volumes (John et al., 1992). Our knowledge of the patterns we have been investigating has improved, including charting range shifts of species due to climate change and the consequences for communities. This volume has charted much progress in our understanding of processes, at ever increasing scales, from different parts of the world. In another twenty-five years it will be interesting to reflect where our field stands. Given the anthropogenic impacts discussed in the present volume, it is likely that coastal communities and habitats will look very different in twenty-five years. Significant changes in communities in a new, human-impacted world are increasingly being documented. It is, therefore, crucial that coastal ecologists commit to improve our understanding of the linkages between smaller-scale, local-, regional- and global-scale processes and drivers to contribute towards efforts to manage and conserve these coastal systems. We are, however, encouraged by the increasing efforts of marine ecologists worldwide to work with governments and their agencies, non-governmental organisations, the public, schools and new generations of students to help increase appreciation and understanding of the oceans and our coasts in the context of a rapidly changing world. Only by working together can better stewardship of marine ecosystems, firmly based in sound science, be promoted.

## REFERENCES

Altieri, A. H., van Wesenbeeck, B. K., Bertness, M. D, and Silliman, B. R. (2010). Facilitation cascade drives positive relationship between native biodiversity and invasion success. *Ecology*, **91** (5), 1269–75.

Arenas, F., Sanchez, I., Hawkins, S. J. and Jenkins, S. R. (2006). The invasibility of marine algal assemblages: role of functional diversity and identity. *Ecology*, **87**, 2851–61.

Bakun, A. (1990). Global climate change and intensification of coastal ocean upwelling. *Science*, **247** (4939), 198–201.

Bates, A. E., Helmuth, B., Burrows, M. T. et al. (2018). Biologists ignore ocean weather at their peril. *Nature*, **560**, 299–301.

Bell, E. C. and Denny, M. W. (1994). Quantifying "wave exposure": a simple device for recording maximum velocity and results of its use at several field sites. *Journal of Experimental Marine Biology and Ecology*, **181** (1), 9–29.

Bertness, M. D. (1989). Intraspecific competition and facilitation in a northern acorn barnacle population. *Ecology*, **70** (1), 257–68.

Bertness, M. D. and Callaway, R. (1994). Positive interactions in communities. *Trends in Ecology & Evolution*, **9** (5), 191–3.

Bertness, M. D. and Shumway, S. W. (1993). Competition and facilitation in marsh plants. *The American Naturalist*, **142** (4), 718–24.

Bishop, M. J., Fraser, J. and Gribben, P. E. (2013). Morphological traits and density of foundation species modulate a facilitation cascade in Australian mangroves. *Ecology*, **94** (9), 1927–36.

Boaventura, D., Alexander, M., Della Santina, P. et al. (2002). The effects of grazing on the distribution and composition of low-shore algal communities on the central coast of Portugal and on the southern coast of Britain. *Journal of Experimental Marine Biology and Ecology*, **267**, 185–206.

Bosman, A. L., Hockey, P. A. and Underhill, L. G. (1989). Oystercatcher predation and limpet mortality: the importance of refuges in enhancing the reproductive output of prey populations. *Veliger*, **32**, 120–9.

Bowen, M., Markham, J., Sutton, P. et al. (2017). Interannual variability of sea surface temperature in the southwest Pacific and the role of ocean dynamics. *Journal of Climate*, **30** (18), 7481–92.

Broitman, B. R., Blanchette, C. A., Menge, B. A. and Lubchenco, J. (2008). Spatial and temporal patterns of invertebrate recruitment along the west coast of the United States. *Ecological Monographs*, **78** (3), 403–21.

Bruno, J. F., Stachowicz, J. J. and Bertness, M. D. (2003). Inclusion of facilitation into ecological theory. *Trends in Ecology & Evolution*, **18** (3), 119–25.

Burrows, M. T., Harvey, R., Robb, L. et al. (2009). Spatial scales of variance in distributions of intertidal species on complex coastlines: effects of region, dispersal mode and trophic level. *Ecology*, **90** (5), 1242–54.

Burrows, M. T. and Hawkins, S. J. (1998). Modelling patch dynamics on rocky shores using deterministic cellular automata. *Marine Ecology Progress Series*, **167**, 1–13.

Burrows M. T., Jenkins, S. R., Robb, L. and Harvey, R. (2010). Spatial variation in size and density of adult and post-settlement *Semibalanus balanoides*: disentangling effects of oceanography and local conditions. *Marine Ecology Progress Series*, **398**, 207–19.

Chan, B. K., Lima, F. P., Williams, G. A., Seabra, R. and Wang, H. Y. (2016). A simplified biomimetic temperature logger for recording intertidal barnacle body temperatures. *Limnology and Oceanography: Methods*, **14** (7), 448–55.

Chapman, J. W., Carlton, J. T., Bellinger, M. R. and Blakeslee, A. M. (2007). Premature refutation of a human-mediated marine species introduction: the case history of the marine snail *Littorina littorea* in the Northwestern Atlantic. *Biological Invasions*, **9** (8), 995–1008.

Chee, S. Y., Othman, A. G., Sim, Y. K., Adam, A. N. M. and Firth, L. B. (2017). Land reclamation and artificial islands: walking the tightrope between development and conservation. *Global Ecology and Conservation*, **12**, 80–95.

Chiswell, S. M., Bostock, H. C., Sutton, P. J. and Williams, M. J. (2015). Physical oceanography of the deep seas around New Zealand: a review. *New Zealand Journal of Marine and Freshwater Research*, **49** (2), 286–317.

Coleman, R. A., Goss-Custard, J. D., dit Durell, S. E. L. V. and Hawkins, S. J. (1999). Limpet *Patella* spp. consumption by oystercatchers *Haematopus ostralegus*: a preference for solitary prey items. *Marine Ecology Progress Series*, **183**, 253–61.

Coleman, R. A., Underwood, A. J., Benedetti-Cecchi, L. et al. (2006). A continental scale evaluation of the role of limpet grazing on rocky shores. *Oecologia*, **147**, 556–64.

Connell, J. H. (1961). The influence of interspecific competition and other factors on the distribution of the barnacle *Chthamalus stellatus*. *Ecology*, **42**, 710–23.

Connell, S. D. (2007). *Water Quality and the Loss of Coral Reefs and Kelp Forests: Alternative States and the Influence of Fishing*. Oxford University Press, Melbourne, pp. 556–68.

Cooley, S. R., Cheney, J. E., Kelly, R. P. and Allison, E. H. (2017). Ocean acidification and Pacific oyster larval failures in the Pacific Northwest United States. In *Global Change in Marine Systems: Societal and Governing Responses*. Routledge, Abingdon.

Crisp, D. J. and Southward, A. J. (1958). The distribution of intertidal organisms along the coasts of the English Channel. *Journal of the Marine Biological Association of the United Kingdom*, **37** (1), 157–203.

Denny, M. W., Dowd, W. W., Bilir, L. and Mach, K. J. (2011). Spreading the risk: small-scale body temperature variation among intertidal organisms and its implications for species persistence. *Journal of Experimental Marine Biology and Ecology*, **400** (1–2), 175–90.

Dong, Y.-W., Huang, X.-W., Wang, W., Li, Y. and Wang, J. (2016). The marine 'great wall' of China: local and broad-scale ecological impacts of coastal infrastructure on intertidal macrobenthic communities. *Diversity and Distribution*, **22**, 731–44.

Emmerson, M. C., Solan, M., Emes, C., Paterson, D. M. and Raffaelli, D. (2001). Consistent patterns and the idiosyncratic effects of biodiversity in marine ecosystems. *Nature*, **411** (6833), 73.

Ferreira, J. G., Hawkins, S. J. and Jenkins, S. R. (2015). Physical and biological control of fucoid recruitment in range edge and range centre populations. *Marine Ecology Progress Series*, **518**, 85–94.

Figurski, J. D., Malone, D., Lacy, J. R. and Denny, M. (2011). An inexpensive instrument for measuring wave exposure and water velocity. *Limnology and Oceanography: Methods*, **9** (5), 204–14.

Firth, L. B., Crowe, T. P., Moore, P., Thompson, R. C. and Hawkins, S. J. (2009). Predicting impacts of climate-induced range expansion: an experimental framework and a test involving key grazers on temperate rocky shores. *Global Change Biology*, **15** (6), 1413–22.

Firth, L. B., Knights, A. M., Bridger, D. et al. (2016). Ocean Sprawl: Challenges and Opportunities for Biodiversity Management in a Changing World. In *Oceanography and Marine Biology*. CRC Press, Boca Raton, FL, pp. 201–78.

Firth, L. B. and Williams, G. A. (2009). The influence of multiple environmental stressors on the limpet *Cellana toreuma* during the summer monsoon season in Hong Kong. *Journal of Experimental Marine Biology and Ecology*, **375** (1–2), 70–5.

Fitch, J. E. and Crowe, T. P. (2011). Combined effects of temperature, inorganic nutrients and organic matter on ecosystem processes in intertidal sediments. *Journal of Experimental Marine Biology and Ecology*, **400** (1–2), 257–63.

Fitzhenry, T., Halpin, P. M. and Helmuth, B. (2004). Testing the effects of wave exposure, site, and behavior on intertidal mussel body temperatures: applications and limits of temperature logger design. *Marine Biology*, **145** (2), 339–49.

Folland, C. K., Renwick, J. A., Salinger, M. J. and Mullan, A. B. (2002). Relative influences of the interdecadal Pacific oscillation and ENSO on the South Pacific convergence zone. *Geophysical Research Letters*, **29** (13), 21-1.

Gaylord, B. and Gaines, S. D. (2000). Temperature or transport? Range limits in marine species mediated solely by flow. *The American Naturalist*, **155** (6), 769–89.

Genner, M. J., Sims, D. W., Southward, A. J. et al. (2010). Body size-dependent responses of a marine fish assemblage to climate change and fishing over a century-long scale. *Global Change Biology*, **16** (2), 517–27.

Gordon, J. M. and Knights, A. M. (2018). Revisiting Connell: competition but not as we know it. *Journal of the Marine Biological Association of the United Kingdom*, **98** (6), 1253–61.

Grevemeyer, I., Herber, R. and Essen, H. H. (2000). Microseismological evidence for a changing wave climate in the northeast Atlantic Ocean. *Nature*, **408** (6810), 349.

Gribben, P. E., Kimbro, D. L., Vergés, A. et al. (2017). Positive and negative interactions control a facilitation cascade. *Ecosphere*, **8** (12), e02065.

Griffin, J. N., De La Haye, K. L., Hawkins, S. J., Thompson, R. C. and Jenkins, S. R. (2008). Predator diversity and ecosystem functioning: density modifies the effect of resource partitioning. *Ecology*, **89** (2), 298–305.

Griffin, J. N., Laure, M. L. N., Crowe, T. P. et al. (2010). Consumer effects on ecosystem functioning in rock pools: roles of species richness and composition. *Marine Ecology Progress Series*, **420**, 45–56.

Hallegraeff, G. M. (2010). Ocean climate change, phytoplankton community responses, and harmful algal blooms: a formidable predictive challenge. *Journal of Phycology*, **46** (2), 220–35.

Hatton, H. (1938). Essais de bionomie explicative sur quelques especes intercotidales d'algues et d'animaux. *Annls Inst Oceanogr Monaco*, **17**, 241–348.

Hawkins, S. J., Evans, A. J., Mieszkowska, N. et al. (2017). Distinguishing globally-driven changes from regional-and local-scale impacts: the case for long-term and broad-scale studies of recovery from pollution. *Marine Pollution Bulletin*, **124** (2), 573–86.

Hawkins, S. J. and Hartnoll, R. G. (1983a). Grazing of intertidal algae by marine invertebrates. *Oceanography and Marine Biology: An Annual Review*, **21**, 195–282.

Hawkins, S. J. and Hartnoll, R. G. (1983b). Changes in a rocky shore community: an evaluation of monitoring. *Marine Environmental Research*, **9**, 131–81.

Hawkins, S. J., Hartnoll, R. G., Kain, J. M. and Norton, T. A. (1992). Plant–Animal Interactions on Hard Substrata in the Northeast Atlantic. In D. M. John, S. J. Hawkins and J. H. Price, eds. *Plant-Animal Interactions in the Marine Benthos*. Oxford University Press, Oxford, pp. 1–32.

Hawkins, S. J., Mieszkowska, N., Firth, L. B. et al. (2016). Looking backwards to look forwards: the role of natural history in temperate reef ecology. *Marine and Freshwater Research*, **67** (1), 1–13.

Hawkins, S. J., Moore, P., Burrows, M. T. et al. (2008). Complex interactions in a rapidly changing world: responses of rocky shore communities to recent climate change. *Climate Research*, **37**, 123–33.

Hawkins, S. J., Sugden, H. E., Mieszkowska, N. et al. (2009). Consequences of climate driven biodiversity changes for ecosystem functioning of North European Rocky Shores. *Marine Ecology Progress Series*, **396**, 245–59.

Helmuth, B., Choi, F., Matzelle, A. et al. (2016). Long-term, high frequency *in-situ* measurements of intertidal mussel bed temperatures using biomimetic sensors. *Scientific Data*, **3**, 160087.

Helmuth, B., Russell, B. D., Connell, S. D. et al. (2014). Beyond long-term averages: making biological sense of a rapidly changing world. *Climate Change Responses*, **1** (1), 6.

Jenkins, S. R., Åberg, P., Cervin, G. et al. (2000). Spatial and temporal variation in settlement and recruitment of the intertidal barnacle *Semibalanus balanoides* (L.)(Crustacea: Cirripedia) over a European scale. *Journal of Experimental Marine Biology and Ecology*, **243** (2), 209–25.

Jenkins, S. R., Coleman, R. A., Hawkins, S. J., Burrows, M. T. and Hartnoll, R. G. (2005). Regional scale differences in determinism of grazing effects in the rocky intertidal. *Marine Ecology Progress Series*, **287**, 77–86.

Jenkins, S. R., Moore, P., Burrows, M. T. et al. (2008). Comparative ecology of North Atlantic shores: do differences in players matter for process? *Ecology*, **89** (sp11), S3–23.

Johannesson, K. (1988) The paradox of Rockall: why is a brooding gastropod (*Littorina saxatilis*) more widespread than one having a planktonic larval dispersal stage (*L. littorea*)? *Marine Biology*, **99**, 507–13.

John, D. M., Hawkins, S. J. and Price, J. H., eds. (1992). *Plant–Animal Interactions in the Marine Benthos*, The Systematics Association Special volume no. 46, Clarendon Press, Oxford.

Johnson, M. P., Burrows, M. T. and Hawkins, S. J. (1998). Individual based simulations of the direct and indirect effects of limpets on a rocky shore *Fucus* mosaic. *Marine Ecology Progress Series*, **169**, 179–88.

Jones, C. G., Lawton, J. H. and Shachak, M. (1994). Organisms as Ecosystem Engineers. In *Ecosystem Management*. Springer, New York, pp. 130–47.

Jonsson, P. R., Granhag, L., Moschella, P. S., Åberg, P., Hawkins, S. J. and Thompson, R. C. (2006). Interactions between wave action and grazing control the distribution of intertidal macroalgae. *Ecology*, **87** (5), 1169–78.

Keith, S. A., Roger, J. H., Norton, P. A., Hawkins, S. J. and Newton, A. C. (2011). Individualistic species limitations of climate-induced expansions generated by meso-scale dispersal barriers. *Diversity and Distributions*, **17**, 275–86.

Kroeker, K. J., Kordas, R. L., Crim, R. et al. (2013). Impacts of ocean acidification on marine organisms: quantifying sensitivities and interaction with warming. *Global Change Biology*, **19** (6), 1884–96.

Kuffner, I. B., Andersson, A. J., Jokiel, P. L., Ku'ulei, S. R. and Mackenzie, F. T. (2008). Decreased abundance of crustose coralline algae due to ocean acidification. *Nature Geoscience*, **1** (2), 114.

Lewis, J. R. (1964). *The Ecology of Rocky Shores*. English Universities Press, London.

Lima, F. P., Gomes, F., Seabra, R. et al. (2016). Loss of thermal refugia near equatorial range limits. *Global Change Biology*, **22** (1), 254–63.

Lima, F. P. and Wethey, D. S. (2009). Robolimpets: measuring intertidal body temperatures using biomimetic loggers. *Limnology and Oceanography: Methods*, **7** (5), 347–53.

Lindberg, D. R., Warheit, K. I. and Estes, J. A. (1987). Prey preference and seasonal predation by oystercatchers on limpets at San Nicolas Island, California, USA. *Marine Ecology Progress Series*, **39**, 105–13.

Ling, S. D., Scheibling, R. E., Rassweiler, A. et al. (2015). Global regime shift dynamics of catastrophic sea urchin overgrazing. *Philosophical Transactions of the Royal Society B*, **370** (1659), 20130269.

McKinney, M. L. and Lockwood, J. L. (1999). Biotic homogenization: a few winners replacing many losers in the next mass extinction. *Trends in Ecology & Evolution*, **14** (11), 450–3.

Menge, B. A., Daley, B. A., Lubchenco, J. et al. (1999). Top-down and bottom-up regulation of New Zealand rocky intertidal communities. *Ecological Monographs*, **69** (3), 297–330.

Mills, K. E., Pershing, A. J., Brown, C. J. et al. (2013). Fisheries management in a changing climate: lessons from the 2012 ocean heat wave in the Northwest Atlantic. *Oceanography*, **26** (2), 191–5.

Müller, W. A., Frankignoul, C. and Chouaib, N. (2008). Observed decadal tropical Pacific–North Atlantic teleconnections. *Geophysical Research Letters*, **35** (24).

Murphy, R. J., Tolhurst, T. J., Chapman, M. G. and Underwood, A. J. (2005a). Estimation of surface chlorophyll-*a* on an emersed mudflat using field spectrometry: accuracy of ratios and derivative-based approaches. *International Journal of Remote Sensing*, **26** (9), 1835–59.

Murphy, R. J. and Underwood, A. J. (2006). Novel use of digital colour-infrared imagery to test hypotheses about grazing by intertidal herbivorous gastropods.

*Journal of Experimental Marine Biology and Ecology*, **330** (2), 437–47.

Murphy, R. J., Underwood, A. J. and Jackson, A. C. (2009). Field-based remote sensing of intertidal epilithic chlorophyll: techniques using specialized and conventional digital cameras. *Journal of Experimental Marine Biology and Ecology*, **380** (1–2), 68–76.

Murphy, R. J., Underwood, A. J., Pinkerton, M. H. and Range, P. (2005b). Field spectrometry: new methods to investigate epilithic micro-algae on rocky shores. *Journal of Experimental Marine Biology and Ecology*, **325** (1), 111–24.

Ng, T. P., Lau, S. L., Seuront, L. et al. (2017). Linking behaviour and climate change in intertidal ectotherms: insights from littorinid snails. *Journal of Experimental Marine Biology and Ecology*, **492**, 121–31.

Norton, T. A., Thompson, R. C., Pope, J. et al. (1998). Using confocal laser scanning microscopy, scanning electron microscopy and phase contrast light microscopy to examine marine biofilms. *Aquatic Microbial Ecology*, **16** (2), 199–204.

O'Connor, N. E. and Crowe, T. P. (2005). Biodiversity loss and ecosystem functioning: distinguishing between number and identity of species. *Ecology*, **86** (7), 1783–96.

Okumura, Y. M. and Deser, C. (2010). Asymmetry in the duration of El Niño and La Niña. *Journal of Climate*, **23** (21), 5826–43.

Paine, R. T. (1969). A note on trophic complexity and community stability. *The American Naturalist*, **103** (929), 91–3.

Pearce, A. F. and Feng, M. (2013). The rise and fall of the "marine heat wave" off Western Australia during the summer of 2010/2011. *Journal of Marine Systems*, **111**, 139–56.

Poloczanska, E. S., Hawkins, S. J., Southward, A. J. and Burrows, M. T. (2008). Modelling the response of populations of competing species to climate change. *Ecology*, **89** (11), 3138–49.

Poore, A. G., Campbell, A. H., Coleman, R. A. et al. (2012). Global patterns in the impact of marine herbivores on benthic primary producers. *Ecology Letters*, **15** (8), 912–22.

Raffaelli, D. and Hawkins, S. J. (1996). *Intertidal Ecology*. Springer Science & Business Media, Amsterdam.

Rahmstorf, S. (2007). A semi-empirical approach to projecting future sea-level rise. *Science*, **315** (5810), 368–70.

Rilov, G. and Schiel, D. R. (2006a). Seascape-dependent subtidal-intertidal trophic linkages. *Ecology*, **87** (3), 731–44.

Rilov, G. and Schiel, D. R. (2006b). Trophic linkages across seascapes: subtidal predators limit effective mussel recruitment in rocky intertidal communities. *Marine Ecology Progress Series*, **327**, 83–93.

Saintilan, N., Wilson, N. C., Rogers, K., Rajkaran, A. and Krauss, K. W. (2014). Mangrove expansion and salt marsh decline at mangrove poleward limits. *Global Change Biology*, **20** (1), 147–57.

Salinger, M. J., Renwick, J. A. and Mullan, A. B. (2001). Interdecadal Pacific oscillation and south Pacific climate. *International Journal of Climatology*, **21** (14), 1705–21.

Sanz-Lázaro, C., Rindi, L., Maggi, E., Dal Bello, M. and Benedetti-Cecchi, L. (2015). Effects of grazer diversity on marine microphytobenthic biofilm: a 'tug of war' between complementarity and competition. *Marine Ecology Progress Series*, **540**, 145–55.

Sarà, G., Kearney, M. and Helmuth, B. (2011). Combining heat-transfer and energy budget models to predict thermal stress in Mediterranean intertidal mussels. *Chemistry and Ecology*, **27** (2), 135–45.

Schiel, D. R. and Foster, M. S. (2015). *The Biology and Ecology of Giant Kelp Forests*. University of California Press, Berkeley.

Schiel, D. R. and Howard-Williams, C. (2016). Controlling inputs from the land to sea: limit-setting, cumulative impacts and ki uta ki tai. *Marine and Freshwater Research*, **67** (1), 57–64.

Seabra, R., Wethey, D. S., Santos, A. M. and Lima, F. P. (2011). Side matters: microhabitat influence on intertidal heat stress over a large geographical scale. *Journal of Experimental Marine Biology and Ecology*, **400** (1–2), 200–8.

Silva, A. C. F., Boaventura, D. M., Thompson, R. C. and Hawkins, S. J. (2014). Spatial and temporal patterns of subtidal and intertidal crabs excursions. *Journal of Sea Research*, **85**, 343–8.

Silva, A. C. F., Hawkins, S. J., Boaventura, D. M., Brewster, E. and Thompson, R. C. (2010). Use of the intertidal zone by mobile predators: influence of wave exposure, tidal phase and elevation on abundance and diet. *Marine Ecology Progress Series*, **406**, 197–210.

Silva, A. C. F., Hawkins, S. J., Boaventura, D. M. and Thompson, R. C. (2008). Predation by small mobile aquatic predators regulates populations of the intertidal limpet *Patella vulgata* (L.). *Journal of Experimental Marine Biology and Ecology*, **367** (2), 259–65.

Skov, M. W., Hawkins, S. J., Volkelt-Igoe, M., Pike, J., Thompson, R. C. and Doncaster, C. P. (2011). Patchiness in resource distribution mitigates habitat loss: insights from high-shore grazers. *Ecosphere*, **2** (5), 1–17.

Smayda, T. J. (2000). Ecological features of harmful algal blooms in coastal upwelling ecosystems. *African Journal of Marine Science*, **22**, 219–53.

Southgate, T., Wilson, K., Cross, T. F. and Myers, A. A. (1984). Recolonization of a rocky shore in SW Ireland following a toxic bloom of the dinoflagellate, *Gyrodinium aureolum*. *Journal of the Marine Biological Association of the United Kingdom*, **64** (2), 485–92.

Stephenson, T. A. and Stephenson, A. (1949). The universal features of zonation between tide-marks on rocky coasts. *The Journal of Ecology*, **37**, 289–305.

Stephenson, T. A. and Stephenson, A. (1972). *Life between Tidemarks on Rocky Shores*. W. H. Freeman, San Francisco, CA.

Thompson, R. C., Norton, T. A. and Hawkins, S. J. (2004). Physical stress and biological control regulate the producer–consumer balance in intertidal biofilms. *Ecology*, **85** (5), 1372–82.

Thomsen, M. S., Metcalfe, I., South, P. and Schiel, D. R. (2016). A host-specific habitat former controls biodiversity across ecological transitions in a rocky intertidal facilitation cascade. *Marine and Freshwater Research*, **67** (1), 144–52.

Trudgill, S. T., Smart, P. L., Friederich, H. and Crabtree, R. W. (1987). Bioerosion of intertidal limestone, Co. Clare, Eire – 1: *Paracentrotus lividus*. *Marine Geology*, **74** (1–2), 85–98.

Underwood, A. J. (1998). Grazing and disturbance: an experimental analysis of patchiness in recovery from a severe storm by the intertidal alga *Hormosira banksii* on rocky shores in New South Wales. *Journal of Experimental Marine Biology and Ecology*, **231** (2), 291–306.

Underwood, A. J. and Fairweather, P. G. (1986). Intertidal communities: do they have different ecologies or different ecologists. *Proceedings of the Ecological Society of Australia*, **14**, 7–16.

Underwood, A. J. and Jernakoff, P. (1984). The effects of tidal height, wave-exposure, seasonality and rockpools on grazing and the distribution of intertidal macroalgae in New South Wales. *Journal of Experimental Marine Biology and Ecology*, **75** (1), 71–96.

Wernberg, T., Smale, D. A., Tuya, F. et al. (2013). An extreme climatic event alters marine ecosystem structure in a global biodiversity hotspot. *Nature Climate Change*, **3** (1), 78.

Wethey D. S., Woodin, S. A., Hilbish, T. J., Jones, S. J., Lima, F. P. and Brannock, P. M. (2011). Response of intertidal populations to climate: effects of extreme events versus long term change. *Journal of Experimental Marine Biology and Ecology*, **400** (1–2), 132–44.

Williams, G. A., Helmuth, B., Russell, B. D., Dong, Y. W., Thiyagarajan, V. and Seuront, L. (2016). Meeting the climate change challenge: pressing issues in southern China and SE Asian coastal ecosystems. *Regional Studies in Marine Science*, **8**, 373–81.

Wood, H. L., Spicer, J. I. and Widdicombe, S. (2008). Ocean acidification may increase calcification rates, but at a cost. *Proceedings of the Royal Society of London B*, **275** (1644), 1767–73.

Zhang, S., Han, G. D. and Dong, Y. W. (2014). Temporal patterns of cardiac performance and genes encoding heat shock proteins and metabolic sensors of an intertidal limpet *Cellana toreuma* during sublethal heat stress. *Journal of Thermal Biology*, **41**, 31–7.

# Index

# Systematics Association Special Volumes

1. The New Systematics (1940)[a]
   *Edited by J. S. Huxley* (reprinted 1971)
2. Chemotaxonomy and Serotaxonomy (1968)*
   *Edited by J. C. Hawkes*
3. Data Processing in Biology and Geology (1971)*
   *Edited by J. L. Cutbill*
4. Scanning Electron Microscopy (1971)*
   *Edited by V. H. Heywood*
5. Taxonomy and Ecology (1973)*
   *Edited by V. H. Heywood*
6. The Changing Flora and Fauna of Britain (1974)*
   *Edited by D. L. Hawksworth*
7. Biological Identification with Computers (1975)*
   *Edited by R. J. Pankhurst*
8. Lichenology: Progress and Problems (1976)*
   *Edited by D. H. Brown, D. L. Hawksworth and R. H. Bailey*
9. Key Works to the Fauna and Flora of the British Isles and Northwestern Europe, fourth edition (1978)*
   *Edited by G. J. Kerrich, D. L. Hawksworth and R. W. Sims*
10. Modern Approaches to the Taxonomy of Red and Brown Algae (1978)*
    *Edited by D. E. G. Irvine and J. H. Price*
11. Biology and Systematics of Colonial Organisms (1979)*
    *Edited by C. Larwood and B. R. Rosen*
12. The Origin of Major Invertebrate Groups (1979)*
    *Edited by M. R. House*
13. Advances in Bryozoology (1979)*
    *Edited by G. P. Larwood and M. B. Abbott*
14. Bryophyte Systematics (1979)*
    *Edited by G. C. S. Clarke and J. G. Duckett*
15. The Terrestrial Environment and the Origin of Land Vertebrates (1980)*
    *Edited by A. L. Panchen*
16. Chemosystematics: Principles and Practice (1980)*
    *Edited by F. A. Bisby, J. G. Vaughan and C. A. Wright*
17. The Shore Environment: Methods and Ecosystems (two volumes) (1980)*
    *Edited by J. H. Price, D. E. C. Irvine and W. F. Farnham*
18. The Ammonoidea (1981)*
    *Edited by M. R. House and J. R. Senior*

19. Biosystematics of Social Insects (1981)*
*Edited by P. E. House and J.-L. Clement*

20. Genome Evolution (1982)*
*Edited by G. A. Dover and R. B. Flavell*

21. Problems of Phylogenetic Reconstruction (1982)*
*Edited by K. A. Joysey and A. E. Friday*

22. Concepts in Nematode Systematics (1983)*
*Edited by A. R. Stone, H. M. Platt and L. F. Khalil*

23. Evolution, Time and Space: The Emergence of the Biosphere (1983)*
*Edited by R. W. Sims, J. H. Price and P. E. S. Whalley*

24. Protein Polymorphism: Adaptive and Taxonomic Significance (1983)*
*Edited by G. S. Oxford and D. Rollinson*

25. Current Concepts in Plant Taxonomy (1983)*
*Edited by V. H. Heywood and D. M. Moore*

26. Databases in Systematics (1984)*
*Edited by R. Allkin and F. A. Bisby*

27. Systematics of the Green Algae (1984)*
*Edited by D. E. G. Irvine and D. M. John*

28. The Origins and Relationships of Lower Invertebrates (1985)‡
*Edited by S. Conway Morris, J. D. George, R. Gibson and H. M. Platt*

29. Infraspecific Classification of Wild and Cultivated Plants (1986)‡
*Edited by B. T. Styles*

30. Biomineralization in Lower Plants and Animals (1986)‡
*Edited by B. S. C. Leadbeater and R. Riding*

31. Systematic and Taxonomic Approaches in Palaeobotany (1986)‡
*Edited by R. A. Spicer and B. A. Thomas*

32. Coevolution and Systematics (1986)‡
*Edited by A. R. Stone and D. L. Hawksworth*

33. Key Works to the Fauna and Flora of the British Isles and Northwestern Europe, fifth edition (1988)‡
*Edited by R. W. Sims, P. Freeman and D. L. Hawksworth*

34. Extinction and Survival in the Fossil Record (1988)‡
*Edited by G. P. Larwood*

35. The Phylogeny and Classification of the Tetrapods (two volumes) (1988)‡
*Edited by M. J. Benton*

36. Prospects in Systematics (1988)‡
*Edited by J. L. Hawksworth*

37. Biosystematics of Haematophagous Insects (1988)‡
*Edited by M. W. Service*

38. The Chromophyte Algae: Problems and Perspective (1989)‡
*Edited by J. C. Green, B. S.C. Leadbeater and W. L. Diver*

39. Electrophoretic Studies on Agricultural Pests (1989)‡
*Edited by H. D. Loxdale and J. den Hollander*

40. Evolution, Systematics and Fossil History of the Hamamelidae (two volumes) (1989)‡
*Edited by P. R. Crane and S. Blackmore*

41. Scanning Electron Microscopy in Taxonomy and Functional Morphology (1990)[‡]
*Edited by D. Claugher*

42. Major Evolutionary Radiations (1990)[‡]
*Edited by P. D. Taylor and G. P. Larwood*

43. Tropical Lichens: Their Systematics, Conservation and Ecology (1991)[‡]
*Edited by G. J. Galloway*

44. Pollen and Spores: Patterns and Diversification (1991)[‡]
*Edited by S. Blackmore and S. H. Barnes*

45. The Biology of Free-Living Heterotrophic Flagellates (1991)[‡]
*Edited by D. J. Patterson and J. Larsen*

46. Plant–Animal Interactions in the Marine Benthos (1992)[‡]
*Edited by D. M. John, S. J. Hawkins and J. H. Price*

47. The Ammonoidea: Environment, Ecology and Evolutionary Change (1993)[‡]
*Edited by M. R. House*

48. Designs for a Global Plant Species Information System (1993)[‡]
*Edited by F. A. Bisby, G. F. Russell and R. J. Pankhurst*

49. Plant Galls: Organisms, Interactions, Populations (1994)[‡]
*Edited by M. A. J. Williams*

50. Systematics and Conservation Evaluation (1994)[‡]
*Edited by P. L. Forey, C. J. Humphries and R. I. Vane-Wright*

51. The Haptophyte Algae (1994)[‡]
*Edited by J. C. Green and B. S. C. Leadbeater*

52. Models in Phylogeny Reconstruction (1994)[‡]
*Edited by R. Scotland, D. I. Siebert and D. M. Williams*

53. The Ecology of Agricultural Pests: Biochemical Approaches (1996)[**]
*Edited by W. O. C. Symondson and J. E. Liddell*

54. Species: The Units of Diversity (1997)[**]
*Edited by M. F. Claridge, H. A. Dawah and M. R. Wilson*

55. Arthropod Relationships (1998)[**]
*Edited by R. A. Fortey and R. H. Thomas*

56. Evolutionary Relationships among Protozoa (1998)[**]
*Edited by G. H. Coombs, K. Vickerman, M. A. Sleigh and A. Warren*

57. Molecular Systematics and Plant Evolution (1999)[‡‡]
*Edited by P. M. Hollingsworth, R. M. Bateman and R. J. Gornall*

58. Homology and Systematics (2000)[‡‡]
*Edited by R. Scotland and R. T. Pennington*

59. The Flagellates: Unity, Diversity and Evolution (2000)[‡‡]
*Edited by B. S. C. Leadbeater and J. C. Green*

60. Interrelationships of the Platyhelminthes (2001)[‡‡]
*Edited by D. T. J. Littlewood and R. A. Bray*

61. Major Events in Early Vertebrate Evolution (2001)[‡‡]
*Edited by P. E. Ahlberg*

62. The Changing Wildlife of Great Britain and Ireland (2001)[‡‡]
*Edited by D. L. Hawksworth*

63. Brachiopods Past and Present (2001)[‡‡]
*Edited by H. Brunton, L. R. M. Cocks and S. L. Long*

64. Morphology, Shape and Phylogeny (2002)[‡‡]
*Edited by N. MacLeod and P. L. Forey*

65. Developmental Genetics and Plant Evolution (2002)[‡‡]
*Edited by Q. C. B. Cronk, R. M. Bateman and J. A. Hawkins*

66. Telling the Evolutionary Time: Molecular Clocks and the Fossil Record (2003)[‡‡]
*Edited by P. C. J. Donoghue and M. P. Smith*

67. Milestones in Systematics (2004)[‡‡]
*Edited by D. M. Williams and P. L. Forey*

68. Organelles, Genomes and Eukaryote Phylogeny (2004)[‡‡]
*Edited by R. P. Hirt and D. S. Horner*

69. Neotropical Savannas and Seasonally Dry Forests: Plant Diversity, Biogeography and Conservation (2006)[‡‡]
*Edited by R. T. Pennington, G. P. Lewis and J. A. Rattan*

70. Biogeography in a Changing World (2006)[‡‡]
*Edited by M. C. Ebach and R. S. Tangney*

71. Pleurocarpous Mosses: Systematics and Evolution (2006)[‡‡]
*Edited by A. E. Newton and R. S. Tangney*

72. Reconstructing the Tree of Life: Taxonomy and Systematics of Species Rich Taxa (2006)[‡‡]
*Edited by T. R. Hodkinson and J. A. N. Parnell*

73. Biodiversity Databases: Techniques, Politics, and Applications (2007)[‡‡]
*Edited by G. B. Curry and C. J. Humphries*

74. Automated Taxon Identification in Systematics: Theory, Approaches and Applications (2007)[‡‡]
*Edited by N. MacLeod*

75. Unravelling the Algae: The Past, Present, and Future of Algal Systematics (2008)[‡‡]
*Edited by J. Brodie and J. Lewis*

76. The New Taxonomy (2008)[‡‡]
*Edited by Q. D. Wheeler*

77. Palaeogeography and Palaeobiogeography: Biodiversity in Space and Time (2011)[‡‡]
*Edited by P. Upchurch, A. McGowan and C. Slater*

---

[a] Published by Clarendon Press for the Systematics Association
[*] Published by Academic Press for the Systematics Association
[‡] Published by Oxford University Press for the Systematics Association
[**] Published by Chapman & Hall for the Systematics Association
[‡‡] Published by CRC Press for the Systematics Association